This is an authorized facsimile, made from the master copy of the original book.

Out-of-Print Books on Demand is a publishing service of UMI. The program offers xerographic reprints of more than 100,000 books that are no longer in print.

The primary focus is academic and professional resource materials originally published by university presses, academic societies and trade book publishers worldwide.

U·M·I Out-of-Print Books on Demand

University Microfilms International
A Bell & Howell Information Company
300 N. Zeeb Road, Ann Arbor, Michigan 48106
800-521-0600 OR 313/761-4700

Printed in 1992 by xerographic process on acid-free paper

36002

BACTERIOPHAGE T4

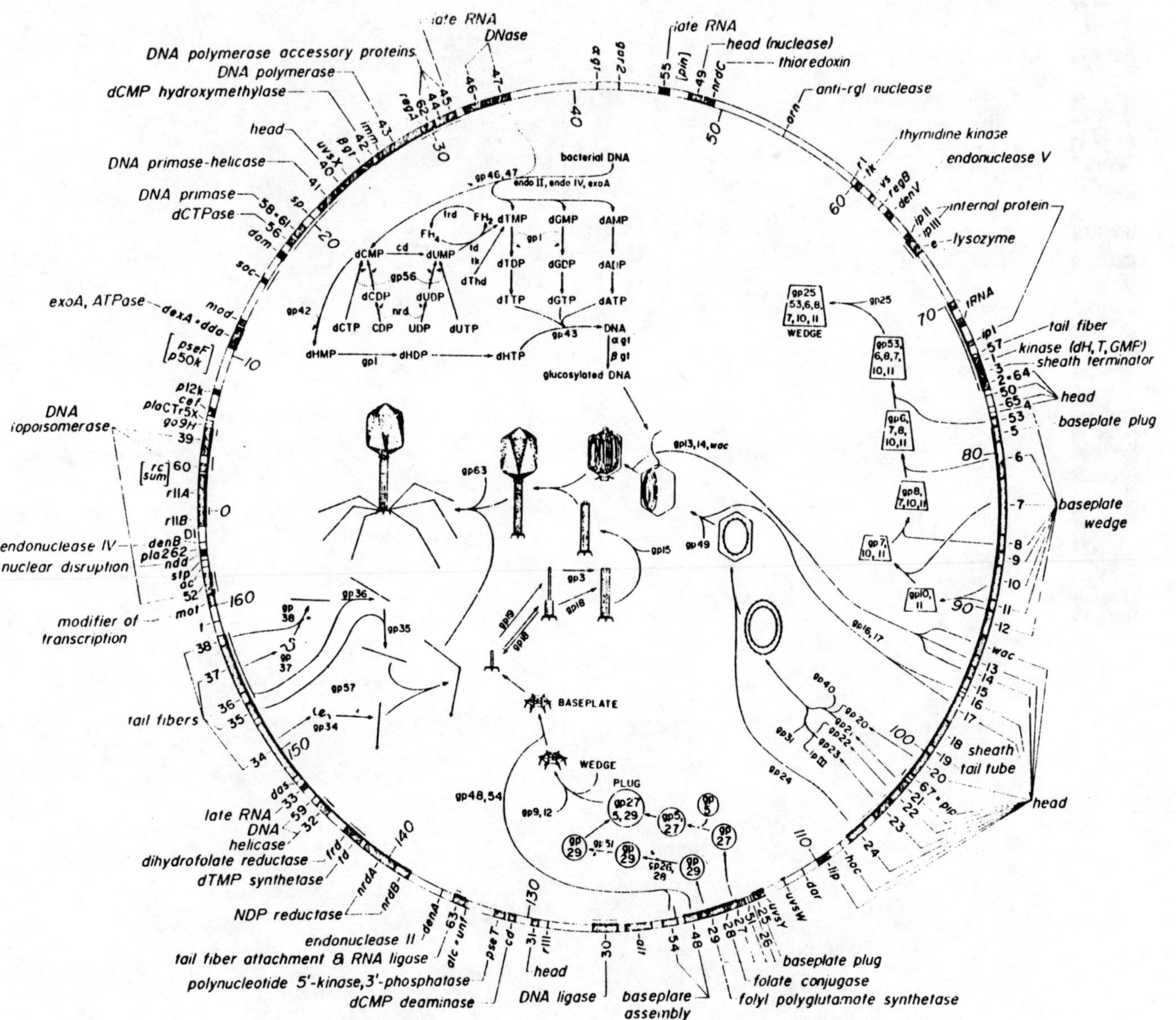

Genomic map of bacteriophage T4.

BACTERIOPHAGE T4

Christopher K. Mathews
Department of Biochemistry and Biophysics
Oregon State University
Corvallis, Oregon 97331

Elizabeth M. Kutter
The Evergreen State College
Olympia, Washington 98505

Gisela Mosig
Department of Molecular Biology
Vanderbilt University
Nashville, Tennessee 37235

Peter B. Berget
Department of Biochemistry and Molecular Biology
University of Texas Medical School
and Graduate School of Biomedical Science
Houston, Texas 77225

1983 • American Society for Microbiology, Washington, D.C.

1913 I Street, N.W.
Washington, DC 20006

Library of Congress Cataloging in Publication Data

Main entry under title:

Bacteriophage T4.

Bibliography: p. 375
Includes index.
1. Bacteriophage T4—Congresses. I. Mathews, Christopher K., 1937- . II. American Society for Microbiology. [DNLM: 1. T-phages—Congresses.

QW 161.5.C6 B131 1981]
QR342.B33 1983 589.9'0234 83-11945

ISBN 0-914826-56-5

Table of Contents

I. T4 Structure and Initiation of Infection

II. DNA Metabolism

A. Enzymes and Proteins of DNA Metabolism

B. DNA Metabolism In Vivo

III. Regulation of Gene Expression

A. Transcription

B. Processing and Translation

IV. Morphogenesis

V. Structure, Organization, and Manipulation of the Genome

VI. Some Complexities of T4 Genes, Gene Products, and Gene Product Interactions

Top row: Jim Watson; Dave Krieg and Martha Chase; John Wiberg. Middle: Karin Karlson and Claes Linder; Max and Manny Delbrück (center); Sydney Brenner (center); Salvador Luria. Bottom: Mark Adams; Nick Visconti, Jill and Al Hershey.

Top: (seated) Al Hershey, Mark Adams, Dick Epstein, Alan Garen; (standing) Gus Doermann, Dave Krieg, Giuseppe Bertani, Cy Levinthal, Frank Stahl, (?), Norton Zinder, Martha Chase, (?), Dorothy Fraser (at a Cold Spring Harbor Phage Meeting sometime between 1953 and 1956). Middle: Gunther Stent; Matt Meselson and George Streisinger. Bottom left (at the dedication of the Hershey Laboratory at Cold Spring Harbor, August, 1979): Frank Stahl, Jim Ebert, Al Hershey, Max Delbrück, Salvador Luria, Jim Watson. Bottom right (at Evergreen Phage Meeting, 1980): Fred Eiserling, Burt Guttman, Dick Epstein.

Introduction to the Early Years of Bacteriophage T4

A. H. DOERMANN

Department of Genetics, University of Washington, Seattle, Washington 98195

The bacterial virus T4
Can exult in its honors galore,
But this book will reveal
That many still feel
Its Golden Age lies to the fore.

—Alan Christensen

This book is the story of a remarkable organism that played a vital role in the history of molecular biology, namely, the bacterial virus T4. Although bacteriophages had been discovered as early as 1915 by Twort and 1917 by d'Herelle, their role in the emergence of molecular biology was not felt until Max Delbrück set out, in 1939, to understand their multiplication. Most of his early experiments, many in collaboration with S. E. Luria, were done with the phages α (later renamed T1), γ (later T2), and δ (later T7). T4 was born into that phage family in 1945 when Demerec and Fano reported its isolation from a phage mixture received from T. Rakieten, and they included it in a set of seven phages, T1 through T7, all of which are active on the *Escherichia coli* strain B. In the same year Delbrück decreed that the membership of the "Delbrück School" would restrict their researches to this set which he affectionately referred to as "Snow White and the Seven Dwarfs." Within the T set, T2, T4, and T6 form a subset which is part of a larger family generally called the T-even group. The T-even phages share many properties that attest to their very close relatedness. For example, their DNAs normally contain hydroxymethylcytosine instead of cytosine (Wyatt and Cohen, 1953) and show sequence homology of greater than 85% (Kim and Davidson, 1974), genetic recombination occurs in all pairwise combinations of T2, T4, and T6 (Delbrück and Bailey, 1946), and they are serologically related (Delbrück, 1946; Lanni and Lanni, 1953). Because of the closeness of the relationship among the T-even phages, the members of this group have often been discussed interchangeably, implying that whatever is learned about one will also apply to the others. While T4 will remain the focal point of this chapter, that practice will also be followed here to a limited extent.

The extensive research that has been carried out in many laboratories has given rise to the existence of distinguishable strains of both T2 and T4. All T2 strains originated from Bronfenbrenner's strain, which he called PC. Delbrück's isolate, first known as phage γ, became T2L used by Luria. The line isolated and used by Hershey in most of his experiments is T2H, and T2K is a variety isolated from strain PC by Kalmanson. All of these can be distinguished by their differing host ranges on a variety of indicator bacteria (Hershey, 1946a). Two lines of T4 have also been widely used, namely, T4B, which was obtained by Benzer from Luria's laboratory, and T4D, which originated from a single plaque which Doermann isolated from Delbrück's stock. The plaque morphology segregants from crosses of T4B by T4D indicate that they differ by at least three mutations (Edgar, personal communication).

Presumably because I have been acquainted with T4 nearly from its birth, and because it has occupied by far the greatest part of my scientific efforts since that time (probably a claim no one else can make), I have been asked to write the introductory chapter to this volume. With such an extended association, that might be considered an easy task. To me it is just the opposite. In that period too much progress has been made to permit me to give a general account that does even modest justice to the remarkable history of T4. This chapter will be limited to the early history of T4, highlighting only three contributions that T4 has made to the science of molecular biology as we know it today and with more limited comments on several others. An account of many other contributions may be obtained from the following chapters of this volume and from the many papers referenced in this chapter as well as in those following. Another rich source that should not be overlooked is the book *Phage and the Origins of Molecular Biology*, published by the Cold Spring Harbor Laboratory in 1966, honoring Max Delbrück on his 60th birthday and recognizing his preeminent role in the origin and early development of the new discipline.

T4 AND THE VIRAL LIFE CYCLE

For an appreciation of the impact that research with the T-even phages had on the emergence and evolution of molecular biology, it is important to realize that this science was still in an early prenatal stage when Snow White and her Seven Dwarfs appeared on the scene. The stage had been set by the classic experimental work of Beadle, Tatum, and their co-workers that gave rise to the "one gene-one enzyme hypothesis" (Beadle, 1945). On a related front, however, even though suggestive evidence was available, it was by no means considered certain that genetic information is encoded in nucleotide sequence. Avery and his co-workers (1944) had, of course, shown that the transformation of pneumococci could be attributed to the DNA fraction of their bacterial extracts, but the problem of accounting for the enormous diversity of genetic information with only four nucleotides was, nonetheless, a substantial barrier to identifying genes with nucleic acid. For example, as late as 1948, Beadle wrote, "Indirect and circumstantial evidence indicates that they (genes) are nucleo*proteins*" (italics mine). When, in 1952, Hotchkiss noted that reducing the protein contamination from a higher level to 0.02% did not quantitatively affect the activity of a pneumococcal transforming preparation, DNA, apart from protein, finally began to be accepted as the material that encodes genetic information. Nevertheless, as discussed below, the establishment of the general principle awaited the support of the classic injection experiments done by Hershey and Chase (1952).

In 1945, the viral life cycle was equally poorly understood. After studying the status of the viral particle once it is attached to the host cell, Wollman and Wollman (1937, 1938) proposed that a phage assumes a different state once associated with a host bacterium, because they failed to recover infectious phage from infected *Bacillus megaterium* cells prematurely lysed with lysozyme. In 1946, citing examples from both plant and animal viruses, Pirie similarly hypothesized that part, at least, of the intracellular virus is complexed with host cell material. The classic one-step growth experiment (Ellis and Delbrück, 1939) had succeeded in identifying highly reproducible latent periods for the phages α (T1), γ (T2), and δ (T7) (Delbrück, 1945), but it was not yet realized that more than one infecting viral particle can multiply within a single bacterium. Delbrück and Luria (1942) had shown that when two unrelated phages infected a bacterium, one of them was prevented from multiplying (mutual exclusion), and when Luria (1945) attempted to test whether both T2 and the host range mutant T2*h* can grow within individual bacteria, the conditions of his experiment were such that T2*h* was excluded from participation in the mixedly infected cells with a high enough frequency that he was forced to conclude: "The results confirm the conclusion that only one of the infecting particles succeeds in growing in each bacterial cell." Several major breakthroughs were made in 1946. Hershey (1946b) reported that he had obtained mottled plaques when bacteria were mixedly infected with a mixture of wild-type and *r*-mutant T2 and inferred that both types of virus can, in fact, grow in a single cell. At the Cold Spring Harbor Symposium, Delbrück and Bailey (1946) reported that not only can two related viruses (any pair of the T2, T4, T6 group) multiply within the same host cell, but genetic information can be exchanged among them. Luria then discovered that T-even phage particles which had been rendered inviable by small doses of UV radiation could still collaborate in multiple infections and produce viable phage progeny (multiplicity reactivation). He proposed a recombination-type mechanism to explain his results (Luria, 1947; Luria and Dulbecco, 1949). Immediately after the appearance of Luria's 1947 paper, two papers appeared from Hershey's laboratory (Hershey and Rotman, 1948; Hershey and Rotman, 1949) in which linkage and crossing over were clearly demonstrated and which provided the foundation for all of the subsequent investigations of genetic structure and recombination in viruses.

The discovery that two compatible viruses can multiply in a single bacterial cell, and that they can even exchange genetic information, was soon supplemented with additional reports that helped to clarify the nature of the intracellular state of the viral components. In 1948 I devised an effective procedure for lysing infected bacteria before their normal bursting time and was able to show that no infectious phage are found during the first half of the latent period (the interval between infection and normal cell lysis) of T2, T3, and T4 (Doermann, 1948). About halfway through the latent period, new infective particles begin to accumulate. Of particular importance, the earliest detectable infectious phage found, using both T2 and T4, included genetic recombinants that were randomly (not clonally) distributed among individual cells and in only slightly lower frequency than found at the end of the latent period. The latter observation showed unmistakably that the basic processes of multiplication and recombination occur in a noninfectious condition that has been referred to as the vegetative state (Doermann, 1953; Doermann and Dissosway, 1949).

At about the same time, Anderson (1950) discovered that the T-even phage shell acts as an osmotic membrane and that sudden dilution from high salt concentrations causes rupture of the membrane and inactivation of the virus (osmotic shock), leaving empty-headed phage ghosts. Herriott (1951) then showed that such inactivated particles consist of nearly pure protein and noted that these DNA-free ghosts are still able to attach to, and kill, bacteria that are sensitive to T-even phage. In the same period, Anderson (1951) developed the critical-point method of preparing specimens for electron microscopic study and with this method was able to eliminate the surface tension artifacts that ordinarily attend the drying of samples being prepared for electron microscopic study. The stereoscopic photographs which resulted demonstrated convincingly for the first time that T-even phage particles attach to cells tail first (Anderson, 1952).

But it was the insight and direct experimental approach of Hershey that provided the climax to these observations. The "blendor experiment" of Hershey and Chase (1952) showed that the function of the phage particle's protein shell was completed once it had transferred its DNA cargo to the host bacterium. The great majority of the shell could thereafter be stripped off the cell without interfering with phage production. The DNA still remaining inside the cell is evidently not expendable, because the combined observations of Anderson (1950) and Herriott (1951) show that DNA-free ghosts of phage, although they can kill bacteria, cannot bring about production of new phage. Thus it was indicated that the DNA of the phage particle evidently carries its genetic endowment, and for the first time it became clear that, to understand the viral life cycle, it would be essential to learn how the phage particle's DNA invades the cellular machinery and brings about the production of progeny phage. Much of that work has been chronicled by Cohen (1968). It may also be noted that recognition of the genetic role of DNA in the T-even phages had a strong influence in diminishing the skepticism of those who still doubted that genes are composed of DNA.

The 1953 Cold Spring Harbor Symposium was entitled "Viruses." Organized by Max Delbrück, its proceedings serve as an excellent account of the state of knowledge about bacterial viruses in particular and especially about the viral life cycle. That symposium also included a paper which at the time seemed only distantly related to the main topic: Jim Watson made one of the first announcements of a proposal by Francis Crick and himself that the structure of DNA embodies a double helix composed of two complementary strands. As all of us now realize, that conception of DNA has probably had the greatest influence of all on the development of molecular biology. It provided a way to think about mechanisms for replication and recombination that logically included the precision required for the substance that encodes genetic information. The framework for structural analysis of mutation was also intrinsic in the concept. And, finally, it brought into focus the problem of translating the genotypic information in the DNA of the organism into the molecules that affect its phenotype. In fact, it would probably be no exaggeration to assert that molecular biology was born in 1953.

T4 AND THE NATURE OF THE GENETIC CODE

The idea that in some way the sequence of nucleotides in DNA directs the specificity of the amino acid sequence in proteins (sequence hypothesis) had gained general acceptance by the late 1950s. At that stage understanding the code that made this translation possible became a problem of paramount importance. A doublet sequence, providing only 16 combinations of 4 nucleotides taken 2 at a time, was clearly inadequate to account for 20 amino acids, and Brenner (1957) had already shown that a universal code with overlapping triplets provided far too few codons to account for the amino acid sequences already known at that time. Through an ingenious and elegant set of experiments with T4, Francis Crick and his collaborators were able to show that the code is read from a fixed starting point in triplets and that it is degenerate (i.e., an amino acid can be coded by one of several triplets).

Crick et al. (1961) used a strictly genetic approach to the problem, taking advantage of the T4 *r*II-K(λ) system developed by Benzer (1959, 1961). They concentrated their efforts on a short segment of the *r*IIB gene which appeared to be expendable for the gene's function (Champe and Benzer, 1962a). The experiments of the Cambridge group were based on the conviction that the mutagenic property of acridines such as proflavine is due to their causing the insertion or deletion of one or more nucleotides in the DNA sequence (Brenner et al., 1961). In a sequentially translated linear system, such mutants would suffer a shift in reading frame resulting in an altered sequence of amino acids distal to the point of mutation. It also followed from their hypothesis that a base insertion could be suppressed with the nearby deletion of a base that would restore the normal reading frame with only a few abnormal amino acids inserted in the translation product. Their first experiment was to isolate 20 spontaneous revertants of a proflavine-induced *r*IIB mutant strain, FC-0. Of the 20 revertants, 18 proved to be due to second *r* mutations closely linked to FC-0, but which could be separated from it by crossing over. Six of the FC-0-suppressing mutations were tested and, like FC-0, found to be revertable by a second spontaneous *r* mutation closely linked to the original. The hypothesis predicted that FC-0, although reverted when combined with any one of its first-round suppressors, would not be reverted by being combined with any of the suppressors that suppressed its own suppressors. Thus, if FC-0 is arbitrarily assigned the symbol (+) and its suppressors are assigned the symbol (−), the suppressors of the (−) mutations should again be (+), and recombining a (+) with a (+) should not restore the correct reading frame. A second prediction was that combining any (+) with any (−) should restore the proper reading frame after the distal mutation. The experiments bore out both predictions, except for those cases in which the interval between the (+) and (−) mutations was shifted in such a way that an unacceptable code word (barrier) resulted.

After this strong confirmation of their frameshift hypothesis, Crick et al. were able to show that the basic unit of the code is the nucleotide triplet by constructing six triple-frameshift mutants, five (+)(+)(+) and one (−)(−)(−), being careful to avoid unacceptable shifts across barrier locations. Each of the six triples displayed the ability to grow on K(λ), thereby identifying the wild phenotype. With these and other well-reasoned experiments, the Cambridge group was able to demonstrate convincingly what has subsequently proved to be entirely accurate, namely, that: (i) a group of three consecutive nucleotides codes for one amino acid; (ii) the code is not overlapping; (iii) the sequence of bases is read from a fixed starting point; and (iv) the code is degenerate. This classic paper by Crick et al. (1961) is an exceptional example of the powerful tool that a genetic attack can provide.

The next urgent problem became one of matching the nucleotide triplets to the amino acids they specify. Although the three bases in a triplet could be determined for some amino acids, their order in the triplet could not be unambiguously established until Leder and Nirenberg introduced their triplet binding method in 1964, a method that was soon followed by Khorana's procedure which involved use of synthetic repeating deoxyribonucleotides. By the time of the 1966 Cold Spring Harbor Symposium, the code was completely worked out except for one codon (UGA) which subsequently turned out to be a nonsense or chain termination codon.

A striking proof of the in vivo correctness of many of these codon assignments as well as of the conclusions of the Cambridge group regarding frameshift mutations and the nature of the genetic code was provided by Streisinger and his collaborators, who also presented a paper at the 1966 Cold Spring Harbor Symposium. Because they wished to compare the amino acid sequence of a wild-type protein with the sequence of its pseudo-wild-type homolog produced by strains carrying mutually correcting frameshift mutations, that group turned to the T4 lysozyme gene, whose product could readily be isolated and for which they could isolate and identify frameshift mutants. They (Streisinger et al., 1966) constructed a set of pseudo-wild-type recombinants and assembled the following combinations of frameshift mutations: (+)(−), (−)(+), (−)(−)(+)(+), and (+)(+)(−). All of those strains produced a lysozyme that could be isolated and sequenced. With that material the Streisinger team made three crucial observations. First, each pseudo-wild-type sequence differed from the amino acid sequence of the wild-type protein in a set that began at its leftmost frameshift mutation and ended at the mutation located farthest to the right, after which the wild-type amino acid sequence was again observed. Second, both of their (+)(+)(+) combinations had an additional amino acid included in the mutated sequence. Finally, with 63 of the 64 triplets already recognized from the work of Nirenberg, Khorana, and others, they could test a prediction of the prevailing conception of the genetic code. They could ask whether it is possible to assign a sequence of triplets to the wild-type lysozyme gene which, with additions and deletions of bases indicated for the pseudo-wild-type combinations, would yield base sequences that encode the amino acid sequences found for the pseudo-wild-type enzymes. The test was affirmative in every case. The genetic analysis of the *r*IIB frameshift mutations, from which the Cambridge group deduced the nature of the genetic code as well as the in vivo correctness of many codon assignments, was thereby provided a most convincing confirmation.

THE GENOME OF T4

Much of the success that has attended T4 investigations has been at least partly, and often completely, dependent on the availability of suitable genetic material. Early examples already mentioned include the role of genetic recombinants in proving the existence and elaborating the nature of the vegetative state in the phage life cycle, and the elegant use of frameshift mutations in the *r*II and lysozyme genes for elucidating the nature of the genetic code. A multitude of additional examples will be found in the chapters that follow. In the vast majority of cases they depend on the detailed information available about the T4 genome. In 1976 Wood and Revel, in a comprehensive review of the literature, listed 135 genes that had been identified in this DNA molecule which is composed of only 166,000 nucleotide pairs. On the assumption that an average T4 gene is composed of about 1,000 nucleotide pairs, about 80% of the T4 genes were known at that time, and more have since been added to the list (Mosig, appendix to this volume). This extent of genetic saturation is matched only by a few other smaller phages such as lambda, T7, and ϕX174, each of which must depend on a significantly larger number of host proteins for the completion of its life cycle than does T4.

The history of the identification of T-even genes began with Luria's (1945) recognition of host range mutations and Hershey's (1946a) study of the *r* (rapid lysis) mutations, both in T2. The usefulness of host range variants and plaque morphology variants was immediately demonstrated by Delbrück and Bailey (1946) in their previously mentioned experiments showing multiple infections and recombination among the T-even phages, and by Hershey and Rotman (1948, 1949), who demonstrated linkage and crossing over in T2. Other types of hereditary variant were the biochemical (adsorption cofactor) mutants of Delbrück (1948) and the turbid halo (*tu*) mutations studied by Doermann and Hill (1953). With these materials Streisinger and Bruce (1960) carried out linkage experiments sensitive enough to show that all of the known mutations are included in a single linkage group.

In spite of the striking successes of Hershey, Benzer, Crick, Streisinger, and others with the mutant material available to them, Robert Edgar and Richard Epstein realized that the variety of mutant types available severely limited the prospect of gaining a comprehensive picture of the molecular genetics of T4. In response to that problem, they resorted to exploiting two general categories of mutations that are conditionally lethal. The temperature-sensitive mutations they isolated were completely viable at 25°C but abortive at 42°C. The amber mutants were able to multiply in the *E. coli* strain CR63, but their growth proved defective on strain B and its derivatives such as S and S/6. The difference was subsequently shown to be dependent on an amber suppressor which is present in CR63 but absent from B strains. Edgar (1966) gives an interesting account of how amber mutations acquired their somewhat incongruous name. In the original search for mutants that have lost their ability to multiply on strain B but are still able to grow in strain CR63, Dick Epstein and Charley Steinberg enlisted the volunteer assistance of Harris Bernstein with the bribe that if mutants were indeed found they would be named after his mother. That promise was kept, "amber" being the English equivalent of the German "Bernstein." Mutations of both types were already recognized. Allan Campbell (1961) had isolated, mapped, and made complementation studies of both temperature-sensitive and amber-type mutants of phage lambda, and Benzer and Champe (1961, 1962) had investigated the mechanism of suppression with *r*II mutants of T4. Garen and Siddiqi (1962) had made a similar analysis of suppressor-sensitive mutations in the alkaline phosphatase gene of *E. coli*. The work of Edgar, Epstein, and their many collaborators was unique, however, in that it not only made a systematic effort to saturate the T4 genetic map with conditional-lethal mutations, but it also identified the mutations with specific phenotypic effects. By the time of the 1963 Cold Spring Harbor Symposium, they had distinguished and mapped no less than 50 previously unknown genes by complementation tests and had classified the 50 genes into groups with related phenotypes. These genes were numbered 1 through 50 in their map sequence, and the precise functions of most of them were gradually identified. The Cold Spring Harbor Symposium paper of Epstein et al. (1963) must rank as one of the classic accomplishments of T4 molecular genetics. Although significantly more T4 genes have been identified and characterized since that time, the paper has served, and still serves, as the foundation from which the subsequent successes with T4 have emerged. It is the reason T4 has become what Kornberg (1980) referred to as ". . . the favored member of the T-even family."

T4 IN A HANDFUL OF OTHER APPLICATIONS

What has been described to this point must, in the evolution of molecular biology, be considered ancient history. Obviously, even that has been restricted to a very few of the major events from a much longer series in which the T-even phages have played a significant role. I would like here to broaden the picture somewhat by simply enumerating, with limited details, just a few of the other early events that emphasize the impact that T4 has had on molecular biology.

Fine Structure of the Gene

In a series of papers that began in 1955, Seymour Benzer focussed his efforts on the *r*IIA and *r*IIB genes of T4. His genetic fine-structure analysis will not soon, if ever, be surpassed in the extent to which he achieved map saturation of two genes. The methodology he developed included the first conditional-lethal system used in T4, namely, the lethality of *r*II mutants on lambda lysogens in contrast to their excellent viability on strains not harboring lambda. Without this well-characterized system, the investigation of Crick and his colleagues into the general nature of the genetic code would, at best, have been long delayed. Another far-reaching effect of Benzer's work was its influence on the general conception of the nature of the gene. The difficulties and uncertainties that, as late as 1954, hindered the development of a clear concept of what the gene is have been eloquently set forth in a paper entitled "The Gene," written by one of the most eminent of geneticists, L. J. Stadler (1954). Benzer, by using the powerful *r*II–K12(λ) system he developed and by adopting precise operational definitions (Benzer, 1957) for the muton, the recon, and the cistron, provided the tools, the data, and the insight which, only a few years after Stadler's article, moved the mystery of the gene well on its way to solution.

Circularity, Terminal Redundancy, and the Headful of DNA

From the viewpoint of a geneticist, especially one interested in replication and recombination of the phage chromosome, there is probably no more satisfying insight in phage genetics than one which must be credited to George Streisinger. He was concerned about two apparently contradictory observations. After the work of Berns and Thomas (1961), it was becoming more and more certain that the DNA of T2 and T4 particles is a duplex composed of two single strands running from end to end without interruptions. As mentioned above, Streisinger and Bruce (1960) had also shown that all of the T4 genes recognized at that time are located on a single linkage group. On the other hand, Doermann and Boehner (1963), studying the progeny phage produced from individual multimarked heterozygotes, had found that the segregation products showed a distinct polarity which, they reasoned, required discontinuities of some sort at the margins of the heterozygous segments. They had proposed an overlapping structure composed of a short region of diploidy (tetraplex). Another related conclusion that had been reached by Nomura and Benzer (1961) was that two kinds of heterozygotes emerge from T4 crosses. In one type deletions as well as point mutations can participate in the heterozygosity, whereas in the other type deletions appear to be forbidden. With a single model (first suggested by Meselson) Streisinger unified the apparently discordant ideas. He proposed the hypothesis that the DNA molecule has a short segment of redundant genetic information at its two ends and that, among the phage particles of a population, the redundancy is permuted rather than located at one place in the genetic map. That implied, of course, that two genetic markers, while quite closely linked in most phage particles of a population, might be at opposite ends of the DNA molecule in some others. Such a hypothesis predicts not only that all markers should be on a single linkage group, but that the genetic map must be circular. In one paper by Streisinger and his collaborators (Streisinger et al., 1964) it was shown that the genetic map of T4 is indeed circular. In a second paper (Sechaud et al., 1965) this group provided evidence that the two kinds of heterozygotes distinguished by Nomura and Benzer do exist, one occurring when the terminal redundancy has different alleles in the right and left ends of the DNA molecule (deletions permitted), and the other being internal in the DNA molecule where the two DNA strands carry different genetic information (deletions not permitted). Finally, Streisinger realized that a permuted terminal redundancy model makes it difficult to conceive of a way in which the T4 chromosome could determine its own length. In a third paper (Streisinger et al., 1967) a model was proposed, and evidence was presented, that it is the capacity of the capsid rather than the DNA molecule that determines the length of the DNA molecule encapsidated, a model generally referred to as the headful model. The correctness of this model has since been verified by other workers showing that the T4 particles with short capsids (petites) contain only a fractional genome (Eiserling et al., 1970; Mosig, 1963) and that giant-headed T4 phage contain multiple-genome concatemers (Doermann et al., 1973b).

Finally, it should be mentioned that the permuted terminal redundancy model was proved not only by genetic experiments, but also by another approach that was just being developed in the mid-1960s. I refer to what Charles Thomas has called "molecular cytogenetics." The possibility of visualizing DNA in the electron microscope and the technical capability of distinguishing single-stranded segments in a double-stranded molecule enabled Thomas and a group of associates to demonstrate directly the permuted nature of T2 chromosomes, to prove the existence of terminal redundancies, and, in fact, to measure the lengths of the redundancies (Thomas, 1967). Thus, with Streisinger's insights proved by both genetic and cytological methods, the foundation was laid for all subsequent thinking about the mechanical aspects involved in the replication, recombination, and packaging of T4 chromosomes.

Demonstrating the Existence of mRNA

On a somewhat different tack from the two preceding examples of achievements with T4 is the fact that T4-infected *E. coli* was the first system used to prove experimentally the existence of mRNA. In 1960 it was widely believed that the transfer of genetic information from DNA to the protein-synthesizing mechanism was mediated through RNA, and it was generally thought that the stable RNA of the ribosome itself carried that information. In the following year Jacob and Monod (1961), in a review article on genetic regulatory mechanisms in protein synthesis, detailed the inconsistencies between the existing data and that hypothesis. They concluded that the informational intermediate must have properties significantly different from those of rRNA. In particular, they pointed out that the intermediate should have a nucleotide composition similar to that of the DNA from which the RNA originates, that it should have a temporary association with ribosomes, and, finally, that it would be expected to have a high turnover rate. The first compelling evidence that eliminated ribosomes themselves as the intermediate was obtained by Brenner et al. (1961) with T4-infected bacteria. By using density labeling and radioactive-isotope labeling they were able to show that no new rRNA was synthesized after infection. The experiments also showed that RNA synthesized after infection did become temporarily associated with ribosomes that had been constructed before infection and that nascent protein was associated with the ribosomes before it appeared in the intracellular pool. At this stage it was also realized that a likely candidate for the messenger molecules had already been isolated in 1956 by Volkin and Astrachan from T2-infected bacteria and later by Volkin et al. (1958) from T7-infected cells. In each case the nucleotide composition of the RNA fraction was very similar to the nucleotide composition of the infecting phage.

Colinearity of the Gene with the Polypeptide Chain

The sequence hypothesis (Crick, 1958), proposing that the amino acid sequence of a protein is specified by the nucleotide sequence of the gene that encodes it, was widely accepted in the early 1960s. Even though the paper by Crick et al. in 1961 pointed strongly to the triplet nature of the genetic code and Nirenberg and Matthaei (1961) had discovered that polyuridylic acid encodes polyphenylalanine, it had not yet been demonstrated experimentally that the gene and its protein product are, in fact, colinear. The experimental verification of colinearity is another first for T4, although it should be pointed out that another proof of that hypothesis was more or less simultaneously being provided by Yanfosky et al. (1964) with the A protein of the *E. coli* tryptophan synthetase. The T4 demonstration was made by Sarabhai et al. (1964), who had assembled and mapped by conventional crosses a set of 10 amber mutations, all located at different sites in the gene encoding the major protein in the T4 capsid. They found that the amber mutations lead to the production of fragments of the normal capsid protein. The fragments were digested by trypsin and by chymotrypsin, and the peptide products were compared with similar digests of wild-type gene 23 protein. Each peptide represented a certain segment of the polypeptide chain, and it was shown that they could be arranged in a unique sequence by the following criterion: for each amber fragment any peptide present was accompanied by every fragment located on its left, and, conversely, when a peptide was missing from a fragment, all of the fragments to its right were also missing. By that ordering technique, 7 of the 10 amber mutations could be placed with respect to each other and to a contiguous set of the 3 others that were themselves not distinguished by this method. The order obtained in that way proved to be identical to the order of the mutations found by conventional three-point genetic mapping. Thus the gene and its protein product were shown to be colinear, and the data gave strong support for the idea that amber mutations in a nonsuppressing environment cause premature termination of the polypeptide chain.

In Vitro Assembly

A final example of an area in which T4 has had an early and significant impact on molecular biology is in the study of the assembly of complex virus particles. One of the most powerful approaches for investigating the general principles of macromolecular assembly as well as the specific roles of particular genes in such constructions depends on the development of a method for carrying out, in vitro, specific steps in the assembly process. That kind of approach for T4 was begun in 1966 with a paper by Edgar and Wood. They showed that incomplete, noninfectious phage can be completed and made infectious by appropriate extracts added in vitro (Edgar and Wood, 1966). Making use of the extensive collection of T4 amber mutants (Epstein et al., 1963) mentioned above, they carried out experiments of the following type. T4 particles lacking tail fibers, and therefore noninfectious, were obtained by chloroform-induced lysis of bacteria infected with phage carrying amber mutations in four genes that encode tail fiber components. Tail fiber extracts were prepared from cells infected with T4 that had an amber mutation in gene 23, which is responsible for the major protein of the capsid. When the tail extract was added to the fiberless particles, infectious phage was produced in direct proportion to the number of fiberless particles in the mixture. Clearly the attachment of tail fibers to phage had been accomplished in an in vitro system. They also carried out many additional experiments directed to the question of whether extracts from cells infected by other amber mutants could complement headless extracts (gene 23 amber) or tail-defective extracts (gene 6 amber). On the basis of the infectious phage produced, they were able to distinguish genes involved with steps in head assembly from those responsible for the assembly of tails. Since that time, this technique has been extensively modified and refined and has provided elegant analyses of some of the assembly complexities that are involved in the construction of T4 particles. No more need be said here, however, because those developments are discussed in later papers of this volume, in particular those by Wood and Crowther and by Berget and King.

CONCLUSION

Having discussed the ancient history of T4, I must end this introduction before I become more entangled in questions that really belong to the modern and future history of that phage. I repeat my earlier warning that this is not meant to tell the complete story of even the early days of T4. Many important topics have been left untouched, but they have been confidently left to the authors of subsequent chapters of this book.

To conclude, it is clear that the T-even phages have played a critical role in the birth and growth of molecular genetics. Surely T4, with the majority of its essential genes identified and its minimal dependence on functions carried out by its host, promises great rewards for continued investigation of problems of virology, genetics, and molecular biology. There is, however, the danger that this remarkable system may, in the not too distant future, be relegated to history. In the last few years financial support for research with bacterial viruses has eroded. Among the large number of molecular geneticists being trained today, few would dare to hope that a new grant application for support of a T4 project would be warmly received by the agencies responsible for distributing research funds. It is hoped that this trend will be reversed. If it is, there can be no doubt that, for many years to come, T4 will remain in the vanguard of progress in molecular genetics. Its Golden Age is not yet over. That contention is what this book is all about. Read on!

Overview

BURTON GUTTMAN AND ELIZABETH KUTTER

The Evergreen State College, Olympia, Washington 98505

Bacteriophage T4 is large and complex, with close to 200 genes in its genome and an elaborate, tightly organized developmental process. Much of this process is now understood in considerable detail, as described in the various chapters of this book, while significant new questions continue to arise. Since it is easy for those who are new to the field to get lost in the complexities, in this introduction we will give a brief overview of the events that occur during the course of T4 infection and of some of the analytical methods that have led to this understanding.

Like all viruses, phage T4 consists of minute particles, or virions, each made of a protein coat, or capsid, that surrounds and protects the viral genome. T4 is a member of the large class of urophages, which are characterized by a large head filled with double-stranded DNA and a tail through which the DNA is extruded during infection. This structure is discussed in detail in the chapter by Eiserling in this volume.

T4 has many advantages for experimental work. High-titer stocks are readily prepared, and infection can be initiated by a single virion. Because of its elaborate tail assembly, DNA injection is so efficient that virtually every virion is capable of initiating a productive infection. Because adsorption is also rapid, timed events of infection are highly synchronized throughout the cells of a population.

Most work with T4 is based on the single-step growth experiment first devised by Ellis and Delbrück (1939) (see Doermann's prefatory chapter). A concentrated sample of phage is mixed with exponential-phase bacteria to give a multiplicity of infection (the ratio of phage to cells) of about 5 to 10. The infected cells are incubated at 37°C with good aeration. (Many workers routinely carry out their experiments at 30°C, where all events are slowed by a factor of approximately 1.5 to 1.7.) When samples are taken periodically and plated, they yield a constant number of plaques up to about 23 min; then the number of plaques suddenly begins to increase and levels off at about 200 to 400 times the original value. This result shows that the phage multiply inside the intact bacteria during a latent period until each cell contains a few hundred particles, at which time the cell bursts open, or lyses, liberating its contents. The burst size is the average number of phage produced per cell.

When the cells are artificially lysed at various times (Doermann, 1952)—for instance, by adding chloroform—one finds that the number of phage per cell increases smoothly during the latent period, except that no phage at all are detectable during an eclipse period that lasts until about 11 min. This is because only the DNA is injected into the cell, along with a few auxiliary proteins (Hershey and Chase, 1952). Thus, during the first few minutes of infection, the cell contains no infectious particles.

Many refinements of these experiments in recent years have permitted a detailed analysis of phage multiplication. During the latent period, the synthesis of major polymers is measured by labeling with appropriate radioactive tracers, either continuously throughout infection or in pulses. For example, DNA synthesis can be followed by means of labeled thymidine, and the synthesis of individual proteins is determined routinely by labeling with short pulses of radioactive amino acids. Proteins are then separated on polyacrylamide gels, either one-dimensional (Hosoda and Levinthal, 1968) or two-dimensional (Burke et al., this volume), and visualized on X-ray film by autoradiography.

The course of phage infection is fairly typical of most viruses (Fig. 1). The virion is merely a vehicle for conveying the viral genome to a new cell. Once inside the cell, the genome takes over the host by stopping cellular DNA synthesis and actually destroying the host genome (Snustad et al., this volume). It turns the cell into a minute factory for making new virions by specifying both enzymes for phage DNA replication and all of the proteins of which new phage capsids are made. But the whole program of multiplication is intricately choreographed by means of complex regulatory mechanisms that still elude our understanding.

In the first minutes after infection, two major events occur. First, all synthesis of host proteins and mRNA ceases very quickly; inducible enzymes of the host cannot be induced after infection. Second, the host RNA polymerase begins to transcribe a series of genes for early enzymes. Some of these are nucleases and, apparently, other kinds of control proteins that stop host transcription and reduce the host DNA to nucleotides. Others are enzymes for synthesis of T4 DNA. T4 DNA replication is accomplished by a special new polymerase embedded in a complex of at least six different proteins (Nossal and Alberts, this volume). Furthermore, T4 DNA contains the unusual base 5-hydroxymethylcytosine instead of the usual cytosine, and special enzymes are required for its synthesis; these and the other enzymes of deoxyribonucleotide biosynthesis also function as a complex, probably tied in to the replication complex (Mathews and Allen, this volume). After synthesis, the DNA is glucosylated (Revel, this volume). These various modifications protect the DNA against host restriction and also allow discrimination between viral and host DNA in various control processes (Snustad et al., this volume).

Synthesis of these early enzymes ceases at about 12 min after infection at 37°C. Meanwhile, beginning at about 5 min, DNA replication has begun, along with the transcription of another class of genes, those which code for the so-called late proteins. These are largely the components of new viral capsids; about 40 genes have been identified that have this function. However, even though the capsid is so complex, its components assemble themselves in a remarkably

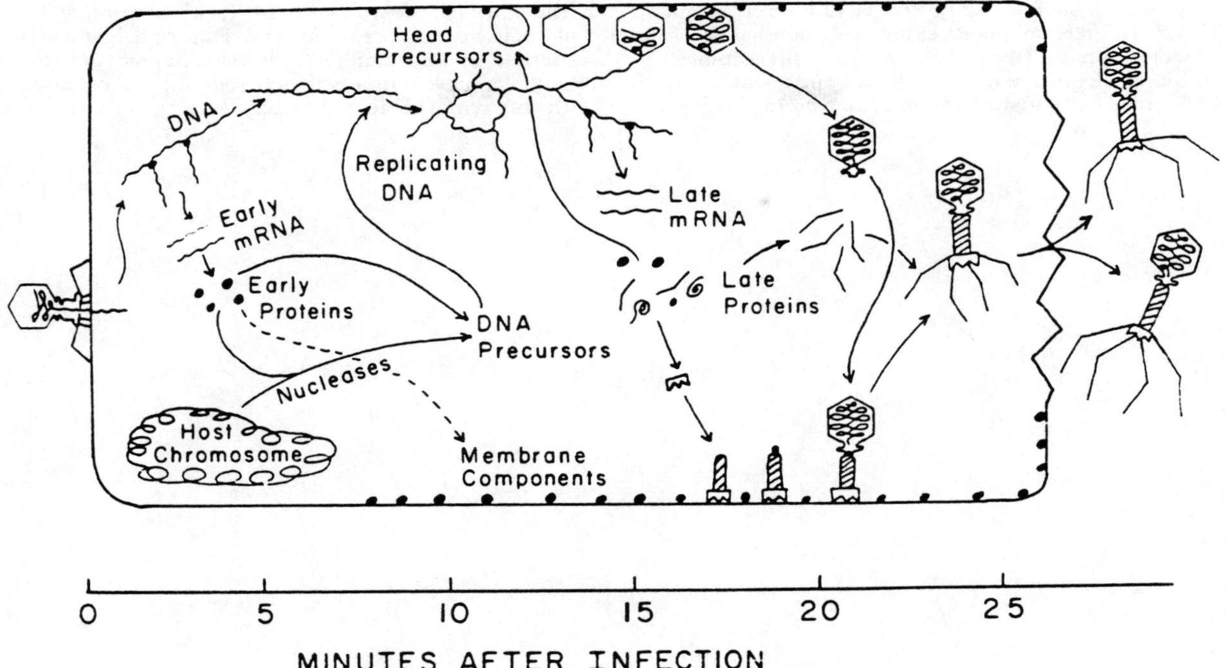

FIG. 1. An overview of the T4 reproductive cycle. From Mathews (1977).

efficient manner. The phage heads are made in one pathway, the tail fibers in a second, and the baseplate of the tail in two others which then converge into a common pathway for tail assembly (see the chapters in this volume by Black and Showe, Berget and King, and Wood and Crowther). The heads are then filled with DNA and joined to the tails, and the tail fibers are added. The process occurs rapidly enough to make at least 200 particles by 25 to 30 min after infection at 37°C in rich medium.

At the end of the multiplication cycle, the cell lyses. Although the details are not clear, lysis appears to be accompanied by a sudden cessation of respiration (Mukai et al., 1967). Furthermore, lysis requires the still-unknown product of the phage *t* gene (Josslin, 1970) plus a lysozyme coded by the phage *e* gene which attacks the murein layer of the cell wall, although respiration stops at the normal time even in lysozymeless mutants. T4 exhibits the still-mysterious phenomenon of lysis inhibition, whereby normal lysis is inhibited by secondary infection of cells at least 3 min after the primary infection. The superinfecting phage are effective even if their DNA does not replicate and, indeed, even if they have been killed by X-radiation. Mutants in the *r*I, *r*II, and *r*III genes and others in a gene known as *spackle* (Emrich, 1968) are not subject to lysis inhibition.

The details of phage structure and its multiplication cycle have been elucidated primarily through the use of mutants defective in various functions. The most useful strains for this purpose are the two types of conditional-lethal mutants first found by Edgar and by Epstein (Epstein et al., 1963): temperature-sensitive (*ts*) mutants that will grow at 30 but not at 42°C, and amber mutants that will grow only in bacterial strains carrying amber suppressors. As first shown by Hershey and Rotman (1948), recombination occurs between two or more mutants when they infect cells simultaneously, and this phenomenon can be used to create a genetic map in which the frequency of recombination between two markers is used as a measure of the distance between their sites (Doermann, this volume). There is a certain amount of negative interference in phage crosses; that is, there is a higher frequency of double crossovers in neighboring regions than is expected on the basis of recombination frequencies within each region by itself. Also, an unusually high negative interference is observed when two markers are very close together. However, when these factors are taken into account, a very good mapping function can be derived which relates frequency of recombination to map distance (Stahl et al., 1964).

An unusual feature of phage T4 is that its genetic map is circular even though the DNA of each virion is linear (Streisinger et al., 1964). In seeking to resolve various puzzles about T4 genetics and structure, Streisinger hypothesized that each phage genome contains a short terminal redundancy, so that genes found on one end of the DNA are repeated on the other end. Recent experiments, discussed in this volume by Black and Showe, confirm his suggestion that after DNA replication has begun, replica molecules can "mate" with one another through their terminal redundancies and, by recombination, create very long molecules, equivalent to many genomes in length. Then each virion is made by packaging a headful of DNA cut from such a molecule, where one headful is on the order of 105% of a genome. This process generates a population of phage carrying all possible circular permutations of the genome, and genetic crosses with such a population yield a circular map. This model has now been extensively confirmed.

Bacteriophage T4 has been called a gift to molecular biology from a benevolent (and omniscient?) deity. And, indeed, the history of molecular biology is largely the story of research with T4, its close relative T2, and their host *Escherichia coli*. This rich history is presented in Doermann's prefatory chapter, and the rest of the book explores the current state of research with this remarkable creature.

T4 STRUCTURE AND INITIATION OF INFECTION

Structure of the T4 Virion†

FREDERICK A. EISERLING

Department of Microbiology and Molecular Biology Institute, University of California, Los Angeles, Los Angeles, California 90024

INTRODUCTION

Bacteriophage Structural Studies

Bacteriophage T4 structure studies differ from the usual static descriptions of virus structure because the results, even at low resolution, are directly related to the essential biological functions of infection, reproduction, and morphogenesis. Structural studies have provided immediate insight into viral functions such as adsorption to the cell surface, DNA uncoiling and escape, and interactions of structural components with the host cell membrane during maturation. The T4 virion is considerably more than a passive container for nucleic acid. It undergoes a number of major structural transformations during infection and maturation which are as essential to replication as is any part of the program of viral gene expression. Bacteriophage T4 devotes more than 40% of its total genetic information to the design, production, and assembly of additional structural components beyond a simple container needed to hold its DNA, which in principle could be made from a few hundred copies of a single protein. A few examples of essential structural interactions in T4 assembly which require the intervention of additional components are the assembly of the capsid, in which both length and width are determined, the positioning and function of accessory capsid protein components gp*hoc* and gp*soc*, the structural switches during capsid assembly, including covalent cleavages of gp22, gp23, and gp24, lattice expansion, and the assembly role of the core. Major structural rearrangements are required for DNA entry into proheads, exit of core proteins, DNA condensation, and preparation for ejection.

The structural participation of the head-tail connector in head morphogenesis and DNA packaging poses the question of symmetry matching between the fivefold icosahedral capsid vertex and the sixfold tail structure. The tail itself provides a remarkable wealth of structural interactions related to viral function. The sequential assembly of baseplate proteins and joining of six wedge subassemblies establishes the sixfold symmetry of the baseplate. The complexity of the baseplate is almost overwhelming. As it is assembled into a "strained" conformation as a hexagonal plate, short tail fibers are added in a "stored" position ready to be rearranged when the entire structure is switched from hexagon to star conformation upon receipt of signals conveyed by the long tail fibers. When the entire machinery is completed, binding sites created for tail tube and sheath proteins promote the assembly of a precisely regulated, energetically strained tail, terminated by a component that fits into the head structure and permits "docking" in a configuration that allows the DNA to align itself with the tail tube in preparation for ejection. The attachment of environmental sensors, the whiskers and tail fibers, requires subassemblies to join, attach to the baseplate and head-tail connector, and prepare the structure for the reversible switching steps used to detect host cell surfaces and to initiate the entire infectious cycle.

Historical Survey

The first electron micrographs of phages were made by Ruska (1940) in Germany, and the concept of a head-and-tail structure was developed in the 1942 publication of Luria and Anderson. Estimates of sizes had been made by other biophysical techniques, but direct measurements were possible only after the development of the electron microscope. By 1943, two major concepts resulting from structural studies were generated: phages adsorb to the cell surface, and development is intracellular. Herriott's (1951) biochemical determinations showed that these empty-headed "ghosts" lacked DNA. In 1953, Anderson localized DNA in the head of T4 by osmotic shock experiments and visualization of empty heads. At about the same time, Anderson developed the critical-point method of specimen preservation, which gave direct evidence that phages attach by their tails. Through structural analysis, the details of the infective process were then defined.

The work of Herriott and Barlow (1952), Williams and Fraser (1956), Kellenberger and Arber (1955), and Kozloff and Lute (1959) defined the functional components of phage T2. The tail components were revealed by partial disruption, using chemical treatments or freeze-thawing. High-quality metal-shadowed electron micrographs by Williams and Fraser (1956) and Kellenberger and Arber (1955) showed the sheath, tail tube, and tail fibers.

The fibers were only visible by electron microscopy using air-drying techniques. Freeze-drying methods show a blob at the tail tip, and since freeze-drying was thought to give more reliable preservation, the fibers were interpreted as an artificial disruption of tail structure (Williams and Fraser, 1953). That these fibers were real phage structures was strongly sug-

† With contributions by Edouard Kellenberger, Microbiology Department, Biozentrum, Basel University, Basel 4056, Switzerland.

gested by the later work of Williams and Fraser (1956), but final proof required further studies which demonstrated their biological role in adsorption and extension-retraction (Kellenberger et al., 1965).

Since there was no way of determining whether the tail tube was hollow, speculation at that time (1955) centered on whether there was partial or complete entry of the tube into the cell. Sheath shortening upon infection was described, but demonstration of contractility required more detailed electron microscopy, even though it was proposed by Williams and Fraser in 1956. The application of negative-contrast methods to phage by Brenner et al. (1959) extended structural studies to molecular dimensions and, in addition to defining sheath contraction and a central channel in tail tubes through which DNA might pass, revealed hexagonal baseplates with long tail fibers attached near each vertex. Head structure was not well defined because of lack of detail caused by the smooth surface, image superposition, and the lack of a theory to guide structural interpretation. With the development of the theory of Caspar and Klug (1962) of virus structure and the application of optical diffraction technology by DeRosier and Klug (1968), some progress was made during the 1960s, primarily by Moody, who inferred icosahedral head symmetry (Moody, 1965) and also defined in molecular detail the subunit rearrangements during sheath contraction (Moody, 1973). During this same period, biochemical studies began to define the sizes and number of proteins composing the phage subassemblies at about the same time that genetics and extracellular assembly studies defined the number and kinds of functional parts of the phage. This resulted in a detailed model for phage function, including baseplate hexagon-star transformation (Simon and Anderson, 1967a; Simon and Anderson, 1967b), tail fiber movement, and short tail fiber attachment. During the 1970s, these fundamental discoveries were refined, the detailed structure of the head was described, and X-ray diffraction yielded results on head size, DNA arrangement, and tail tube structure (Aebi, ten Heggeler, Onorato, et al., 1977; Aebi, van Driel, Bijlenga, et al., 1977; Earnshaw, King, and Eiserling, 1978; Earnshaw, King, Harrison, and Eiserling, 1978; Moody and Makowski, 1981). These results, while satisfying in providing a superficial description of phage structure and function, are now on the verge of providing molecular explanations for major switching events such as DNA ejection, fiber-cell surface interactions, and shape and size determination.

STRUCTURE OF T4 HEAD

General Considerations

All bacteriophage head shells analyzed in sufficient detail are based on an icosahedral design. The original models of Caspar and Klug (1962) established hierarchical levels of structural organization. The basic structure unit or protomer, equivalent to the crystallographic asymmetric unit, can consist of a single polypeptide chain or of several polypeptide chains that associate to form the fundamental structural repeat unit. To date, such protomers in T4 and other large, complex bacteriophage capsids have been found to be single polypeptide chains, but there is no requirement that this be so. The clustering of protomers about local and strict axes of symmetry gives rise to morphological units (capsomers) visible at low resolution in the electron microscope. With improvements of electron microscopy technology it is now possible, using image processing techniques on suitable specimens, to resolve clearly the capsid structure units themselves. In some cases it has been shown that the capsomer is a definite subassembly of the capsid, although for phage T4 this is still uncertain. Structural studies of the T4 head (Ishii and Yanagida, 1977) have established that additional polypeptide chains (gp*hoc*, gp*soc*, and gp24) are placed on or into the basic lattice design at symmetry-related locations (Aebi, van Driel, Bijlenga, et al., 1977; Yanagida, 1977) (Fig. 1). These adjunct proteins are known to be added after the assembly of the basic lattice formed by gp23, and the subunits are selectively removable by detergent treatment. Their function may be to provide additional stability to the capsid or to regulate ion passage across it (Yanagida, 1977). An additional feature of T4 and most large bacteriophage capsids is the provision of a specialized subunit structure at one of the icosahedral vertices, which serves as the attachment point for an adsorption structure such as the baseplate or tail. It is likely that in T4 this structure is also the remnant of the initiation point for capsid assembly at a site on the cell membrane.

The identification of the exact number of structural components of bacteriophage heads, or of any virus, is not straightforward. Heads are usually prepared by use of mutants defective in tail attachment, followed by chemical and physical purification. These manipulations usually cause the loss of DNA and the associated internal proteins. The molecular weights of capsid components are sometimes very similar, and major capsid protein bands on sodium dodecyl sulfate gels can obscure minor components. Viral nonstructural proteins may adhere nonspecifically. A number of protein bands identified as head or capsid components might actually be part of the connecting structure between head and tail. Depending upon growth and isolation conditions, these connector proteins might form aberrant associations with either head or tail, or be lost (Coombs and Eiserling, 1977). Thus, the absolute number of copies of viral capsid proteins is not known to better than ±10% in most cases. Estimates for the number of copies of the gp23 coat protein of T4 range from 800 (Onorato et al., 1978) to over 1,000 (Aebi et al., 1974).

The general architectural plan of isometric phage heads is a major structural framework constructed from a fairly large (35,000- to 40,000-dalton) protein according to the icosahedral surface lattice properties predicted by Caspar and Klug (1962). Assembly takes place via a prohead precursor in which the structural proteins are tightly packed into a thick-walled structure (Black and Showe, this volume). Maturation is accompanied by a displacive expansion of the previously determined surface lattice, which may then incorporate a small (10,000- to 13,000-dalton) protein in numbers nearly equivalent to those of the major protein.

T4 Head

The structure of anisometric viral capsids requires the regulation of both diameter and length during

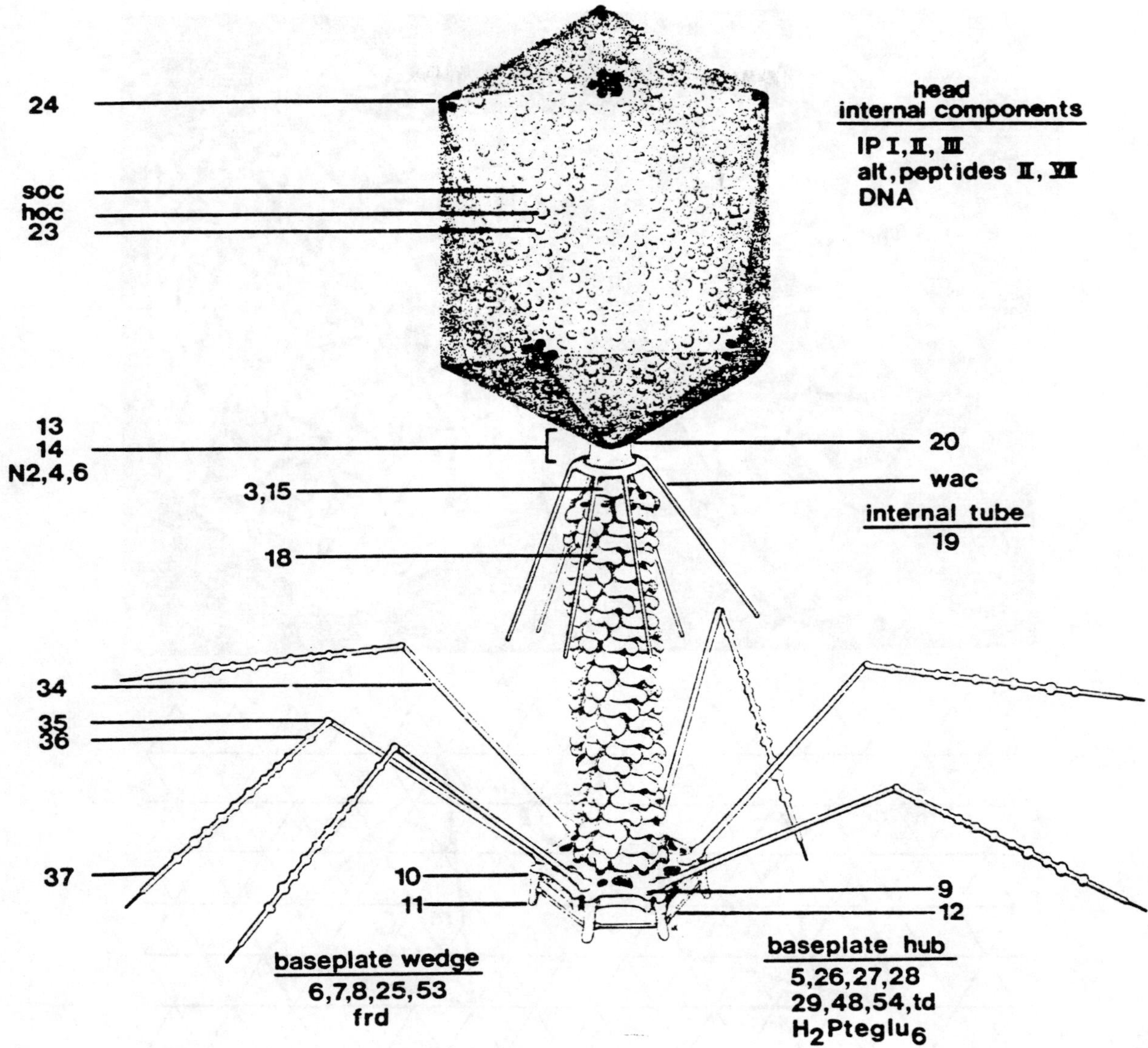

FIG. 1. Structure of bacteriophage T4, based on electron microscopic structure analysis to a resolution of about 2 to 3 nm. Near the head and tail are shown the locations of the known major and minor proteins. The icosahedral vertices are made of cleaved gp24. The gene 20 protein is located at the connector vertex, bound to the upper collar of the neck structure. The six whiskers and the collar structure appear to be made of a single protein species, gp*wac*. The gp18 sheath subunits fit into holes in the baseplate, and the gp12 short tail fibers are shown in a stored position. The baseplate is assembled from a central plug and six wedges, and although the locations of several proteins are unknown, they are included here with the plug components.

assembly. Proheads are built with the anisometric form already determined (Eiserling et al., 1970; Vinuela et al., 1976). The initiation of a properly designed cap with the appropriate triangulation number and built-in curvature must be followed by elongation via addition of capsid subunits and additional regulatory protein components. In the case of T4, an assembly core plays a role in both diameter selection and length determination, but the detailed structural arrangement of either the core proteins or the initiation complex has not yet been determined. The core is, however, highly organized (Black and Showe, this volume).

Phage T4 represents the extreme in structural complexity among bacterial viruses. Twenty genes are involved in head morphogenesis, and the head itself (depending on the presence of connector proteins and on whether some protein bands seen on gels are truly head proteins) has from 9 to 19 different polypeptide components. Four of these are internal proteins (Black and Ahmed-Zadeh, 1971; Horvitz, 1974b), and two are internal peptides (Kurtz and Champe, 1977) packaged

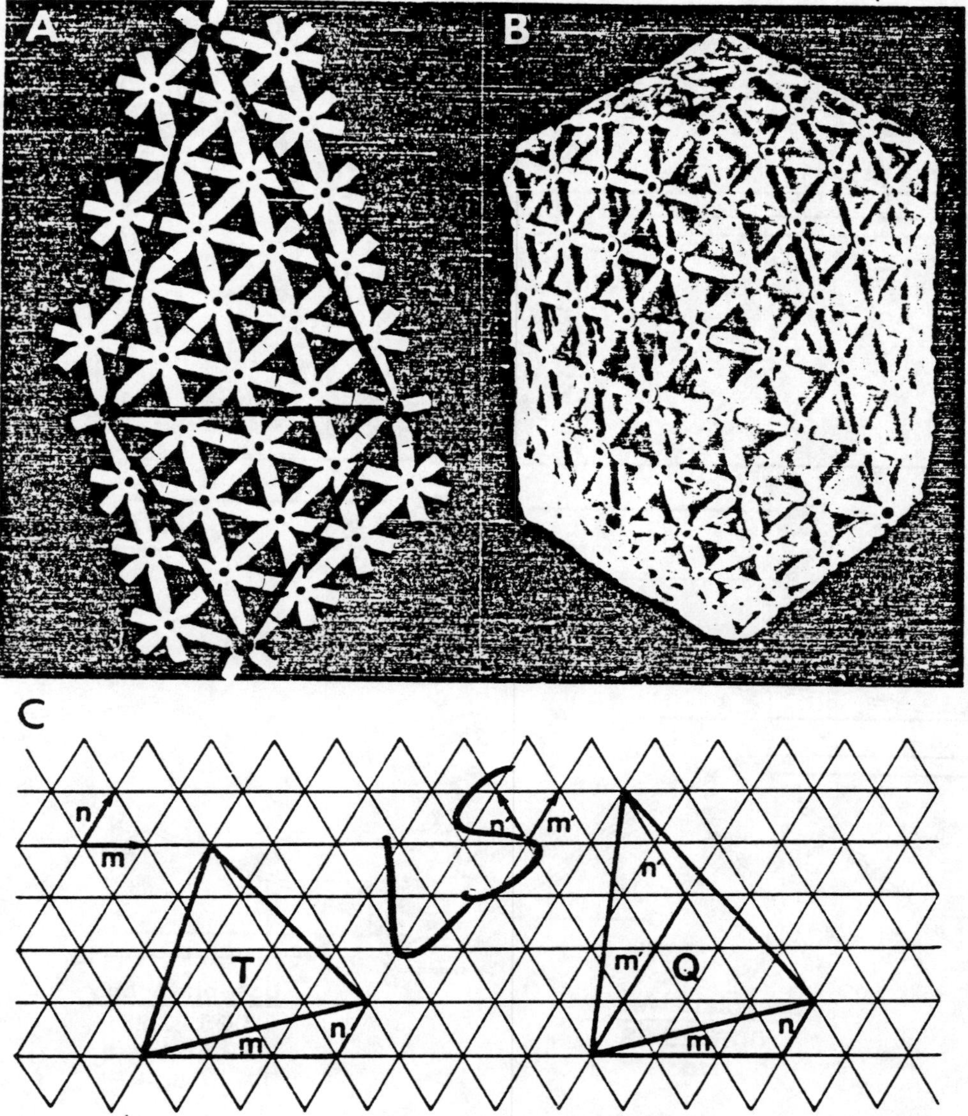

FIG. 2. Surface lattice parameters of bacteriophage T4. The model (B) shows the T4 head elongated along a fivefold icosahedral axis, and a flattened portion of the model (A) indicates the pathway between fivefold vertices ($T = 13$) and along the extended axis ($Q = 21$). (C) T is $T(m,n) = m^2 + mn + n^2$; $Q(m',n')$ is the Q number and is any integer greater than T (see Table 1).

inside the head along with the DNA and released from it upon infection. The T4 head has an icosahedral surface lattice of $T = 13$, elongated along a fivefold axis (Aebi et al., 1974; Branton and Klug, 1975; Moody, 1965). The length of the T4 head is defined by a number, Q, related to the icosahedral T number. The derivation of the Q number is shown in Fig. 2 and is illustrated on a model of the T4 head. The evidence presented by Branton and Klug (1975) for $Q = 17$ depends on the ability to count capsomers between fivefold vertices along the elongated head axis and to some extent on the length/width ratio. This was defined in their freeze-fracture studies as length/width = 1.35, based on those favorable cleavage planes which exposed the maximum length of the phage. In our view, this method could lead to an underestimate of the length/width ratio. Table 1 shows the relation between Q number and head dimensions for a model of an elongated icosahedron based on $T = 13$.

Studies of head length variants of T4 demonstrated that functional viruses can be assembled with shorter and longer DNA (Eiserling et al., 1970; Mosig, Carnigan, Bibring, et al., 1972; Cummings et al., 1973; Doermann et al., 1973b) (Fig. 3). The petite T4 variant containing 0.70 normal DNA length appears isometric ($T = 13$, $Q = 13$). Intermediate-length particles have also been described (Mosig, Carnigan, Bibring, et al., 1972; Doermann et al., 1973b). Under certain growth

TABLE 1. Variation in T4 head size, Q number, and volume[a]

$Q(m',n')$	Head length (nm)	Relative head vol (nm^3 [$\times 10^{-5}$])
13(3,1)	95	3.42
14(2,2)	98	3.64
15(5,0)	101	3.85
16(4,1)	104	4.01
17(3,2)	107	4.20
18(2,3)	110	4.41
19(5,1)	113	4.60
20(4,2)	116	4.77
21(3,3)	119	4.96
22(2,4)	121	5.18
23(5,2)	124	5.36
24(4,3)	127	5.53

[a] Modified from theoretical calculations by U. Aebi (unpublished data), assuming an icosahedral lattice constant of 13.9 nm and a shell thickness of 5 nm. The T number for phage T4 is $13l$ ($T = m^2 + mn + n^2$, $m = 3$, $n = 1$). An extended icosahedron is defined by a Q number that indicates the extent of elongation; Q is any integer greater than T and can be visualized in Fig. 2 by using the values of m' and n' given above. A change in Q number by ± 1 corresponds to an axial length difference of ± 3.0 nm or ± 30 gp23 copies.

conditions with *ptg* mutants (Doermann et al., 1973), normal, intermediate, and petite phages are produced, and when prepared for electron microscopy they show a narrow width distribution and lengths which fall into three main classes (Fig. 4). Fractional genome length measurements by biological methods (Mosig, Bowden, and Bock, 1972) give lengths of 0.71, 0.86, and 1.03 for the DNA (Doermann and Boehner, unpublished data). These data fit with Q numbers of 13 for petite phages, 17 for intermediates, and 21 for normal-length phages.

The absolute size of T4 heads can probably be measured accurately only on particles in solution. This has been done by using X-ray diffraction (Earnshaw et al., 1976; Earnshaw, King, and Eiserling, 1978), which gives the width of T4 as 85 nm, measured on both giant and petite phages. Volume measurements by M. Moncany, F. Borle, and E. Kellenberger (unpublished data) of T-even phages in CsCl, Percoll, metrizamide, and metrizoate agree well with a diameter of 85 nm, a length of 110 to 115 nm, and a Q value of 21 or 22. The agreement of the solution-determined values at around 85 by 110 nm is satisfying in terms of volumes and capsid proteins. The remaining difference could arise from isotropic contraction of the structure due to the recently described "wrapping" phenomenon, in which samples for electron microscopy settle into the deformable carbon film substrate and can shrink uniformly as they dry (Kellenberger and Wurtz, 1982). This probably is the case for the particles measured in Fig. 4. For a curved structure such as the T4 head, the inner radius of the protein shell is compressed with respect to the outer radius. Assuming that the lattice constant at the inner radius is 12.5 nm, then calculations by M. Wurtz and E. Kellenberger show that the lattice constant at the outer radius would be 14.7 nm, a value consistent with a diameter for the head of 85 nm. Also, measurements on thin sections of T4 heads give outer dimension values of 82 by 105 nm (Wunderli et al., 1977). We conclude that the best value for the T4 head size is 85 by 115 nm, based on $T = 13$ and $Q = 21$. The inner volume of the head is about 4×10^{-16} ml, and the 111 megadaltons of DNA is packaged with 75 to 85% (vol/vol) water (Earnshaw, King, and Eiserling, 1978; Moncany, Kellenberger, et al., unpublished data).

The T-even phage surface lattice design was determined directly by freeze-fracture techniques on phage T2 (Branton and Klug, 1975), which lacks the T4 minor proteins gp*hoc* and gp*soc*. In the case of T4, these additional proteins made the structural determination very difficult until the discovery of giant phages, elongated much more than normal along the fivefold axis (Cummings et al., 1973; Doermann et al., 1973b) (Fig. 5). These elongated heads permitted detailed optical diffraction analysis of the electron micrographs because of the larger number of repeats and, thus, a higher signal/noise ratio. The location of the proteins gp*soc* (small outer capsid) and gp*hoc* (highly antigenic outer capsid) was determined by Ishii and Yanagida (1977) and by Aebi, van Driel, Bijlenga, et al. (1977) using giant T4 phage. gp*hoc* lies at the center of a ring (capsomer) of cleaved gp23, whereas there are six copies of gp*soc* around each six-membered capsomer of cleaved gp23. The location of the other capsid components has been accomplished by Müller-Salamin et al. (1977). Using antibody directed against purified gp24, they showed the most likely location of cleaved gp24 to be the icosahedral vertices, excluding the tail-joining vertex. In principle, the fivefold vertices could also be made of cleaved gp23, but they are not. Differential extraction of T4 capsid (in 7 M urea, pH 11) removes all of the cleaved gp24 as seen by sodium dodecyl sulfate gel electrophoresis, and gaps appear at the vertices, confirming this location.

Although gp*hoc* and gp*soc* may confer some additional capsid stability, osmotic shock resistance is controlled by the gp24 vertex proteins (Leibo et al., 1979). The gp24 vertex protein can be replaced, presumably by gp23, in certain mutants of gene 24 (McNicol et al., 1977).

The neck of T4 is a complex structure which includes gp13, gp*wac* (whiskers), gp14, gp20, and two to three other proteins (Coombs and Eiserling, 1977). The upper knob of the neck is the site of gp20 binding to the neck and to the capsid (Driedonks et al., 1981). This structure, which has sixfold symmetry, consists of a disk of 12 copies of gp20. The six whiskers made up from 18 gp*wac* molecules are attached to the lower of the two knobs protruding from the connector structure (Fig. 6). These serve a role in the attachment, extension, and retraction of long tail fibers (Conley and Wood, 1975).

The T4 prohead expands during maturation. The early prohead lacks minor proteins, and, as seen also for the phage lambda D protein, gp*hoc* and gp*soc* are added later, after expansion of the lattice of gp23 from 11.2 nm (Aebi et al., 1974) to 12.9 nm (Aebi et al., 1976) followed by the cleavages of gp23 and gp24. Most of the T4 head proteins have now been localized in the head structure; however, some protein components have not yet been correlated with genetic lesions. Since the products of five genes known to be involved with head morphogenesis (genes 2, 4, 50, 64, and 65)

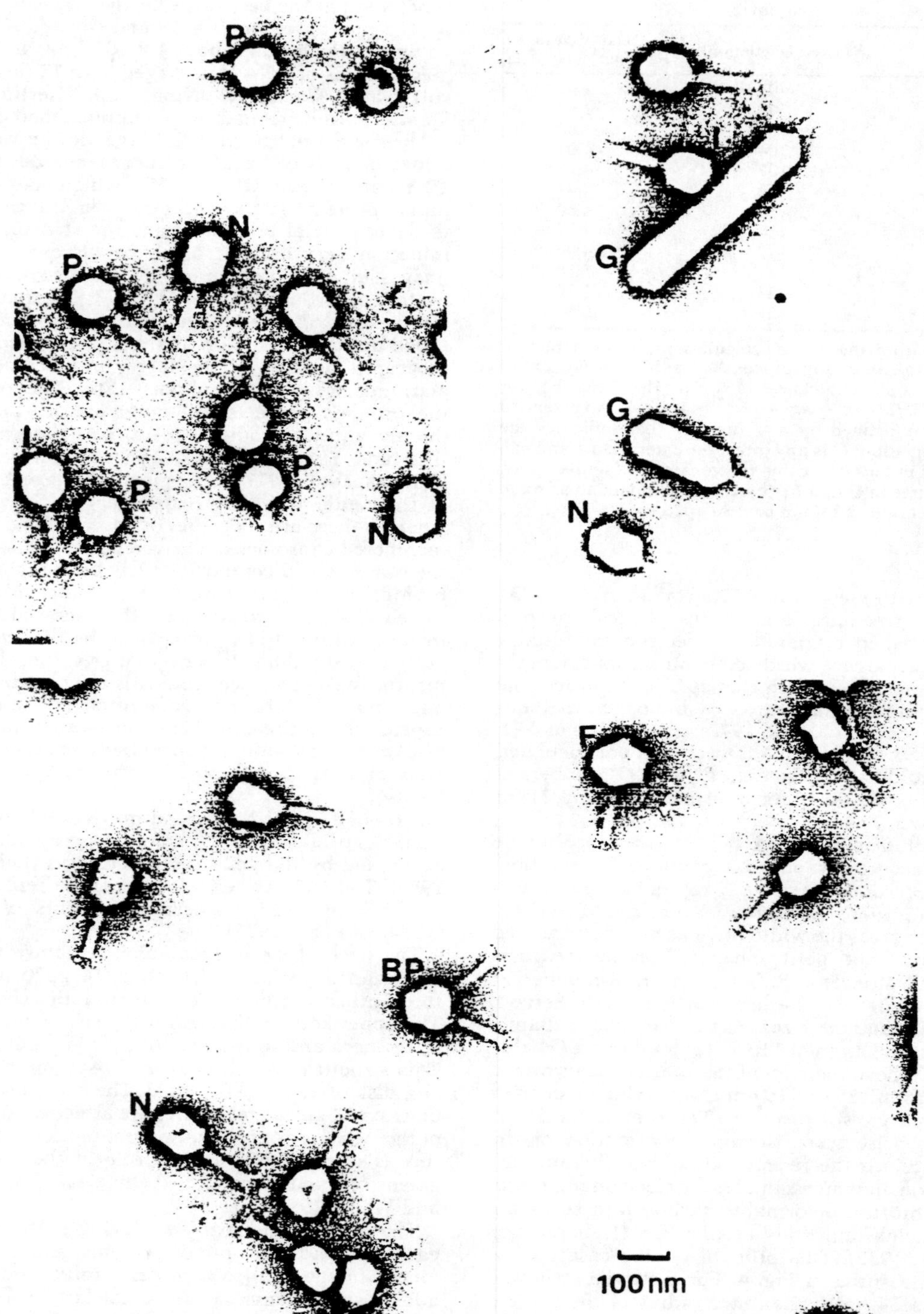

FIG. 3. Head size variants of T4. Normal (N), intermediate (I), petite (P), giant (G), biprolate (BP), and "fat" (F) phages are shown. The normal, intermediate, petite, and biprolate phages have been discussed elsewhere (Moody, 1965; Doermann et al., 1973b). Extension along a single fivefold icosahedral axis explains petite, intermediate, normal, and giant phages. Biprolate phages arise spontaneously and by missense mutations in gene 22. The fat phages appear to be larger in all dimensions and have only been seen in gene 22 mutants (P. Niederberger and A. H. Doermann, personal communication). The lower micrograph was taken by P. Niederberger.

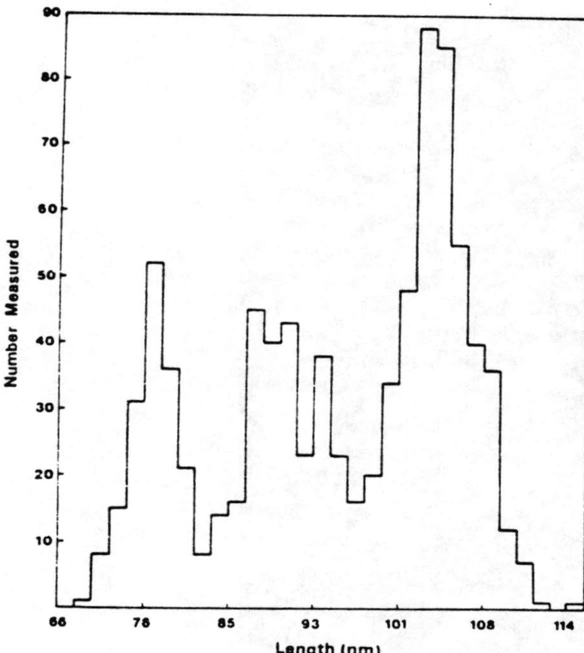

FIG. 4. Distribution of head lengths of *ptg* mutants (Doermann et al., 1973b), including petite isometric, intermediate, and normal phage. The average length of phage in the first, intermediate, and last peaks is 78, 92, and 105 nm, respectively. A total of 841 phage were measured on micrographs of uranyl acetate-soaked phage, using a mask device to measure sizes. The head width measured under these conditions was 71 ± 2.1 nm (previously unpublished data of U. Aebi, A. H. Doermann, and F. A. Eiserling).

have not yet been identified as protein bands on gels. It is likely that several of these gene products will turn out to be head internal proteins, neck proteins, or proteins injected with DNA (Goldberg, 1980).

The initiation event in head assembly must include several shape-determining steps. The logical structural site for head initiation is the part of the head where the tail attaches, referred to as the neck or connector. For all phages described, some such structure exists; these include the core and neck proteins of T7 (Stevens et al., 1983), phage P22 (King, 1980), the four to five proteins of the λ connector (Hendrix, 1978), the gene 10 and 11 products of ϕ29 (Hagen et al., 1976), and the six neck proteins of T4 (Coombs and Eiserling, 1977) plus gp20 (Müller-Salamin et al., 1977). In all cases where data are available, the connector components have 6- or 12-fold symmetry, and the structure is inserted into the 5-fold icosahedral vertex. The symmetry mismatch was pointed out by Moody (1965) and may have some significance or may simply reflect the junction of two structures of different symmetry which are mechanically joined or trapped by the constriction at the head vertex. This seems possible, since relatively mild osmotic shock of T4 and lambda produces tails with the connector structure still attached (Coombs and Eiserling, 1977; Katsura and Kuhl, 1974) (Fig. 6). Müller-Salamin et al. (1977) have localized gp20 in the head at the vertex where the tail joins; thus, gp20 is also part of the head vertex which provides a binding site to the connector structure. The role of T4 gp20 is a key one in head assembly. In its absence, head proteins are assembled into tubular structures without icosahedral caps, and for this reason it has been postulated that gp20 is the icosahedral cap initiator. The gene 20 protein has an estimated molecular weight of 64,000 and is present in 12 copies per phage (Driedonks et al., 1981). Hsiao and Black (1977) have shown that a cold-sensitive mutant in gene 20 shows an unusual phenotype: it makes empty heads with cleaved proteins which appear normal and contain gp20. Since DNA is not packaged, this vertex protein appears to have a second role in structure determination. Using the method of Jarvik and Botstein (1975) for ordering the steps in assembly with *ts* and *cs* mutants, Hsiao and Black (1977) showed that the gp20 vertex protein functions both early in assembly, at the icosahedral cap stage, and later, at the time of DNA packaging.

The initiation of head assembly appears to take place on the cell membrane in intact cells (Simon, 1972; Black and Showe, this volume). Mutations in gene 40 result in irreversible binding of gp20 to the membrane, as do certain *ts* mutations of gene 20 (Brown and Eiserling, 1979a; Brown and Eiserling, 1979b). Regulation of T4 head length must be exerted between the stages of icosahedral cap initiation, interaction with the assembly core, and elongation (Black and Showe, this volume).

One aspect of head structure determination which has received little attention is the choice of direction for the skew classes of icosahedral surface lattices such as $T = 7l$ for lambda (Hohn and Katsura, 1977) and T7 (Stevens et al., 1983) and $T = 13l$ for T2 (Branton and Klug, 1975). Mixtures of *d* and *l* lattice types are not observed, so some positive control mechanism must exist. One suggestion, made by Leonard et al. (1972) and by Lake and Leonard (1974), uses the notion that the capsid penton vertex protein (such as gp24) is different from the hexagonal lattice protein (gp23). Depending on the triangulation number, the axis of the vertex protein cluster will point in different directions because the fivefold axis is inclined at an angle to the icosahedral face, and Lake and Leonard suggest that this tilt could establish the handedness. Another, possibly more likely, explanation is that the interaction between core protein and capsid protein would create a specific location on the vertex protein. Assembly could be thought to begin by interaction of the membrane-bound gp20 disk with core proteins and successive rings of gp23 hexamers. The first ring would contain 5 capsomers, the second would contain 10, and the third would contain 15. At the addition of the next ring of 20 hexamers, five five-coordinated units must be added to complete the $T = 13$ icosahedral cap. The combined curvatures of the core and capsid proteins could then create binding sites for the gp24 vertex protein, and the *l* hand of the lattice would be established.

STRUCTURE OF T4 TAIL AND TAIL FIBERS

T4 Tail

Phage tails are attached to the specialized vertex used to initiate capsid assembly and are designed to

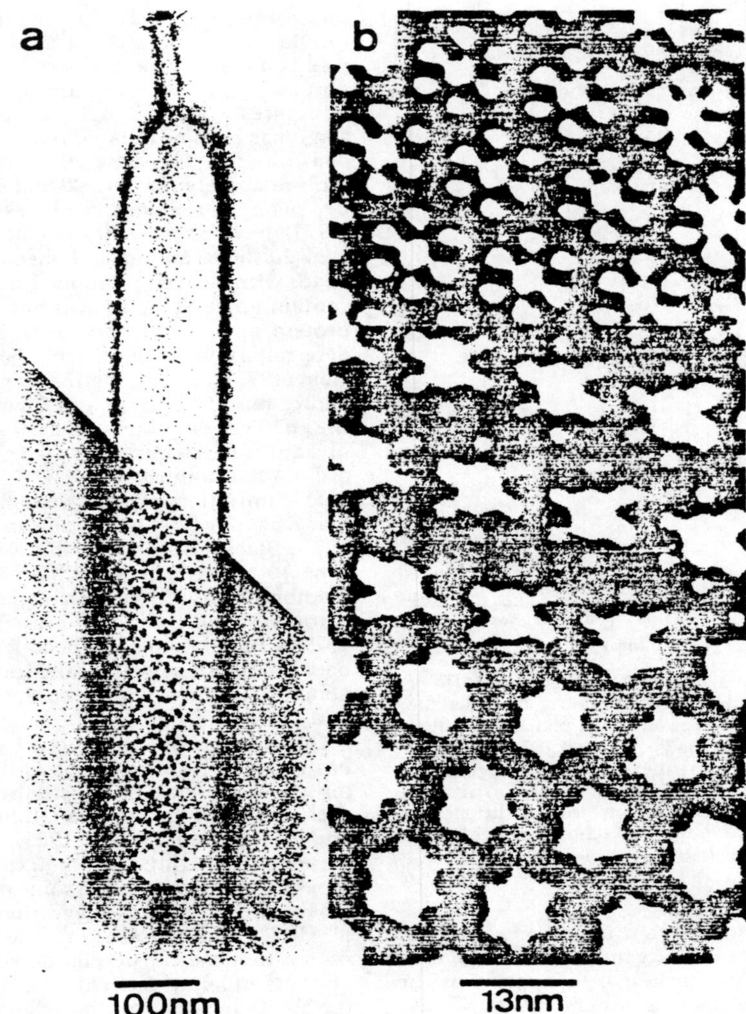

FIG. 5. Capsomer arrangement on T4 giant phage capsids. (a) Electron micrograph of (6 + 1)-type capsids (top) which have been labeled with an excess of anti-gp*hoc* Fab fragments (bottom). (b) Optical filtration of areas of the micrographs to the left. The central panel is a filtration of a capsid labeled with an undersaturating amount of Fab fragments. (c) Electron micrograph of a giant phage capsid containing all components (6 + 6 + 1) (top), labeled with an excess of anti-gp*soc* Fab fragments (bottom). (d) Optical filtration of areas of the micrographs to the left. The central panel shows a schematic superposition of the unlabeled pattern (solid lines) and the Fab-labeled pattern (dotted lines). Magnifications are the same for panels a and c and for panels b and d. Data are from U. Aebi.

attach to cell surfaces and initiate nucleic acid transfer to the host cell. Phages with contractile tails, such as T4, have the most complex arrangement of viral proteins, including an intricate baseplate and tail fibers composed of many parts which change conformation during infection, resulting in sheath contraction and DNA injection. Detailed studies of the T4 sheath (Amos and Klug, 1975; Krimm and Anderson, 1967; Moody, 1973; Smith et al., 1976) and the baseplate (Crowther et al., 1977) have given perhaps the most complete structure analysis for any complex virus tail. Because of the large number of repeating units in the extended sheath and the fact that they are arranged with helical symmetry, a single electron micrograph presents many different views of the same subunit, which has permitted a three-dimensional analysis of the structure of the T4 tail sheath. It is composed of a single protein, gp18 (Dickson, 1974; King and Mykolajewycz, 1973). On sodium dodecyl sulfate gels the protein appears to have an M_r of nearly 80,000 (Dickson, 1974), but other chemical and physical studies established a mass of 67,000 daltons (Tschopp et al., 1979). Hexosamine is present at about 3% by weight, or 10 residues per subunit (Tschopp et al., 1979).

The 144 copies of gp18 are arranged in 24 rings of six subunits each. The rings are spaced 4.1 nm apart, and the sheath is 22 nm across in the extended

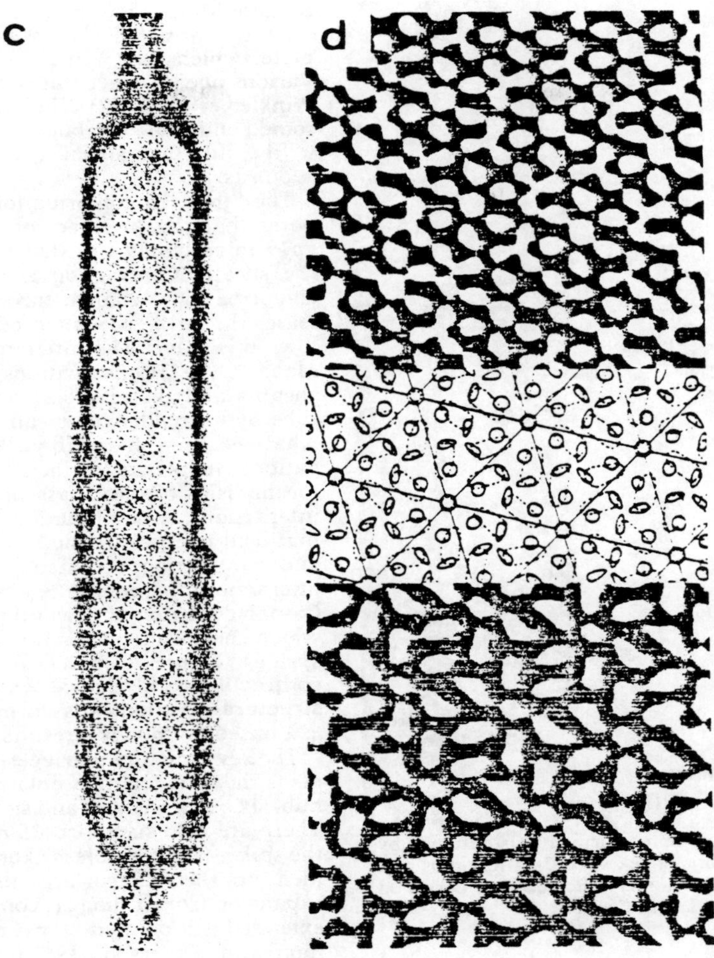

FIG. 5—*Continued*

configuration. The distal row of subunits is in contact with the baseplate proteins (Fig. 7), and the row of gp18 nearest the head binds to the terminator made of gp3 and gp15. All subunits are in close contact with those of the central tube, made of gp19. The subunits in each ring are rotated by about 17° to the right with respect to the ones below, giving rise to the helical lines visible in electron micrographs (Fig. 7). The stacked disk structure is reoriented in the original viewing direction every seven stacks of disks, giving 42 subunits in each repeat. This represents 21 different views of the subunit in a single electron micrograph, since half of the views are the same. The main features of the structure are six helical "tunnels" and two sets of grooves on the surface. The deepest grooves give rise to the prominent right-handed helix seen on the surface (Fig. 1). The gp18 subunit has protruding knobs at the tips, which are likely to fit into corresponding holes in the baseplate at the terminal annulus. Amos and Klug (1975) portray each gp18 in the extended sheath as sloping downward from inner to outer radius. We have used this feature in constructing Fig. 1, although it should be noted that there is little downward slope in the model presented by Smith et al. (1976).

Sheath contraction has been shown by Moody (1973) to be a displacive transition from the extended to the contracted state. Upon contraction, the axial repeat decreases from 4.1 to 1.5 nm, and the twist angle changes from 17 to 32° (Fig. 7). As shown by studies of polysheath (an aberrant assembly form of gp18 in the contracted state) (Kellenberger and Boy de la Tour, 1964) and three-dimensional reconstructions of extended sheath and polysheath (Amos and Klug, 1975), the dramatic change in the shape of the sheath upon contraction is related to relatively small changes in the overall conformational form of gp18. Both Moody (1973) and Donelli et al. (1972) have shown micrographs of partially contracted sheaths, the latter from a *Bacillus subtilis* phage, phage G. Moody was able to trap contracting tails at an intermediate stage by using alkaline formaldehyde, and he could demonstrate directly the displacive nature of the sheath subunit rearrangement and show that for

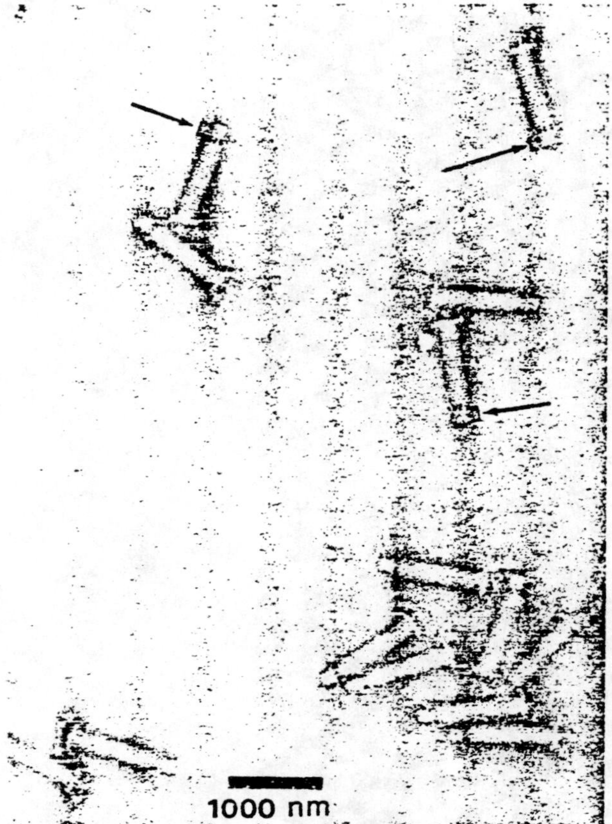

FIG. 6. Electron microgram of purified tails obtained by disjoining heads by osmotic shock with CsCl (Coombs and Eiserling, 1977). The neck region is indicated by brackets on three tails. Micrograph by David Coombs.

T4, the gp18 conformational change occurs before sheath contraction itself is completed. The contraction process begins when signals, transmitted to the baseplate by the tail fibers, trigger a rearrangement of the baseplate. The connection of the annulus of sheath subunits at the surface of the baseplate to the central tube weakens, and the gp18 subunits move apart. As the ring expands, subunits above move down into the spaces created by the baseplate expansion, setting off the contraction process and driving the tail tube downward into the bacterial cell. The next step in infection, the ejection of DNA, requires additional signals and responses (Goldberg, 1980).

Several models have been discussed for the nature of the protein interactions during sheath contraction (Caspar, 1980; Kellenberger and Boy de la Tour, 1964; Moody, 1973). In one, which we call the "loaded spring," the sheath is held in the extended form by bonds to gp18 subunits above and below, with no significant contact to the tail tube. Contraction can be triggered by baseplate expansion and release of the bonds holding the tail tube to the baseplate or, artificially, by release of the sheath-baseplate attachment by urea or detergents. Sheath contraction normally moves the tail tube through the baseplate, which remains attached to the sheath. In some cases, however, the contracted sheath separates from the baseplate, which then remains at the end of the tube. Various agents effect this separation (To et al., 1969; Winkler et al., 1962). These results show that under some conditions the bonds connecting the baseplate to the tube are stronger than those to the sheath subunits.

The "induced conformational change" model proposes that sheath extension is maintained by gp18-gp19 interactions and that there is no tension along the sheath. Contraction is triggered by a wave of conformational change through the tube, which releases the gp18-gp19 interaction, and contraction follows this wave. The latter model is consistent with Moody's (1973) observations of partially contracted sheaths and the artificial cross-linking of sheath to tube by formaldehyde treatment. We prefer it somewhat because it fits with various experimental observations involving artificial contraction and with mechanisms of sheath assembly involving gp18-gp19 interactions. A combination of models, with conformational energy supplied by gp18-gp18 interactions and contraction initiated by release of gp18-gp19 interactions, is represented in Caspar's (1980) model. Caspar constructed a working mechanical device in which energy for contraction is provided by stretched springs. Figure 7 shows in detail the steps in the contraction of this device, which illustrates clearly the structural relationships during the contraction process, based on Moody's results.

The key structure in triggering the contraction process is the hexagonal baseplate. It consists of a central hub, six outer wedges, and six tail spikes (King, 1980). There are also six short (35-nm) tail fibers bound to the spikes (Kells and Haselkorn, 1974). During attachment to the cell surface, the baseplate undergoes expansion from a thicker, compact hexagon to a thin, extended, six-pointed star (Crowther et al., 1977; Simon and Anderson, 1967a; Simon and Anderson, 1967b). Using rotational filtering to improve the signal-to-noise ratio of electron micrographs of isolated baseplates, Crowther et al. (1977) determined the architecture of the hexagonal and star-shaped baseplates to about a 3.0-nm resolution. Having described the complete structures (Fig. 8), they examined mutants lacking various protein components, with the goal of localizing these components in the structure. Fortunately for that analytical approach, three of the proteins, gp9, gp11, and gp12, account for a considerable fraction (40%) of the total mass of the baseplate, and gp9 and gp11 are added after the hexagonal structure is completed. This permits a relatively unambiguous determination of their locations (Fig. 1 and 8). Since gp9 is needed for tail fiber attachment and is located near the site where tail fibers join to the baseplate, and since antiserum directed at gp9 determinants blocks fiber attachment (M. Urig, S. Brown, and W. B. Wood, personal communication), it is reasonable to assume that gp9 is the site of tail fiber binding.

The gene 11 product has been shown to be the distal portion of the tail pin. If gp11 is missing, gp12 cannot bind; thus, gp11 must supply the gp12 binding site. Kells and Haselkorn (1974) showed that gp12 forms the six 35-nm short fibers. In some wild-type phage

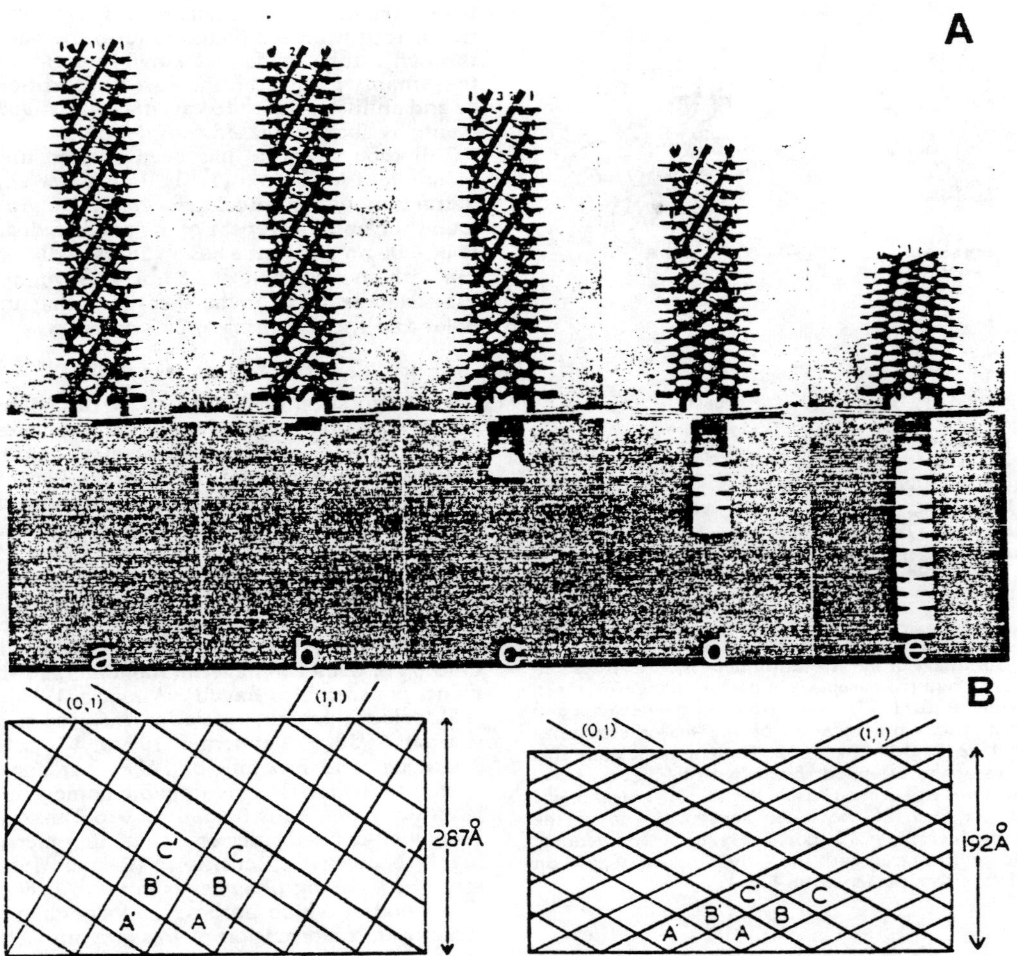

FIG. 7. (A) Steps in the contraction of the model constructed by D. Caspar (1980). (a) Fully extended sheath. (b–d) Successively more connector units are unplugged from their bonding to the central tube, showing intermediate stages of contraction. The black lines following the helix grooves are metal springs in the model. (e) Sheath is fully contracted. The numbers at the top, spaced 60° apart on the top of the tube, show its rotation during contraction. (B) Schematic drawing, modified from Amos and Klug (1975), of the extended and contracted tail sheath lattice of phage T4, using the terminology of Moody (1973). Left: the "long-pitch helices" (1,1) give rise to the strong, right-handed helical grooves seen in Fig. 1. The lattice repeats after seven rings of six subunits. The shallower, left-handed helices are indicated by the (0,1) notation. Right: upon contraction, each protein per unit cell changes conformation, but without major changes in the bonding relations to neighboring subunits. The subunits are numbered to show this relationship.

and in isolated tails (Fig. 6), these 35-nm fibers are not always visible extended from the baseplate, suggesting that they occupy a stored or folded position (Fig. 8).

Using the same analytical approach, Crowther et al. (1977) determined the structural changes which take place during the hexagon-star transformation. The major events in this process are the outward extension of the tail pins, in which gp11 folds out to form the points of the star form, the switch of gp9 from the outside of the hexagon toward the inside of the star, and the switch in the position of gp12 from a folded position in hexagons to an extended 35-nm fiber in stars. The hexagon is stabilized by the presence of gp12; in its absence, many isolated baseplates spontaneously convert from hexagon to star form. The sheath contraction process is probably controlled by the long tail fibers, since Yamamoto and Uchida (1975) and Arscott and Goldberg (1976) have shown that fiberless phage are more resistant to induced sheath contraction than are phage with the normal complement of long tail fibers. While moving over the outer-membrane lipopolysaccharides, the long fibers are supposed to transmit receptor-modulated binding and conformational-change information to the baseplate, possibly via the gene 9 product, which may act as an inhibitor of activation (Crowther, 1980). The fiber-binding step is thought to activate the baseplate, which is then triggered into the hexagon-star transformation by the contact of the gp12 short tail fibers with

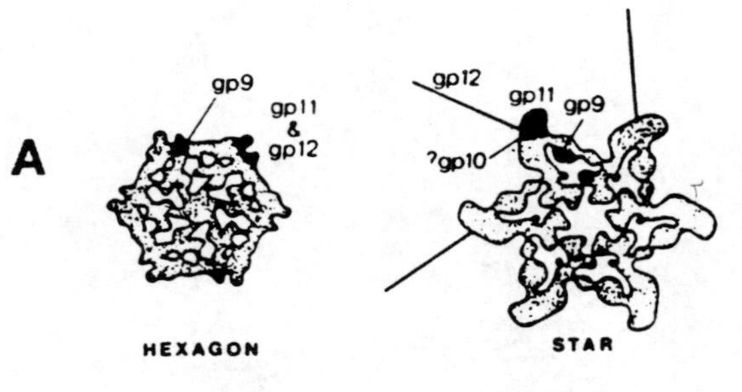

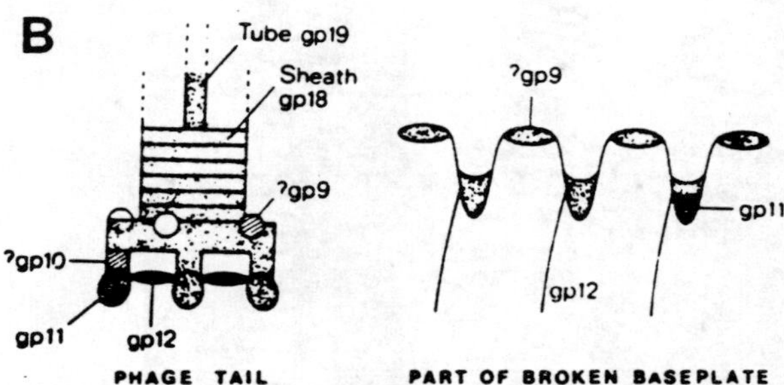

FIG. 8. Summary of the structural details of the hexagon-to-star transition in the baseplate of bacteriophage T4, taken from Crowther et al. (1977). (A) Anatomy of the hexagon and star conformations and location of the major structural proteins gp9, gp10, gp11, and gp12. (B) Schematic diagrams of side views of the extended sheath with hexagonal baseplate and of a piece of broken baseplate. The latter is thought to represent a form of the outer hexagonal rim of the baseplate, in which the gp12 short fibers are fully extended. This should be compared with the "stored" form of gp12 on the lower left, also represented in Fig. 1.

the cell surface receptor proteins. This switch to the star form is the key event in inducing sheath contraction. T4 and other phages are known to adsorb to adhesions between inner and outer membranes (Bayer, 1968), but the significance of this localization is still a mystery. The baseplate also contains structural proteins which are known to have enzymatic activities (reviewed by Kozloff, this volume). These include dihydrofolate reductase, thymidylate synthetase, gp28 (γ-glutamyl hydrolase), gp29 (folyl polyglutamate synthetase), gp5 (cell wall breakdown enzyme), and possibly gp12 (unknown catalysis related to zinc metalloproteins). Of these, the role of gp5 (Kao and McClain, 1980a) seems to be clear, in that infection is aided by a baseplate murolytic enzyme. The functions of the other enzyme components in the infectious cycle are less clear.

Contractile phage tails, as pointed out by Bradley (1967), have a remarkably constant structural design, i.e., an inner tubular part composed of a ca. 20,000-dalton protein arranged in 4.0-nm stacked disks, surrounded by a sheath with the same periodicity (1) which is made of a larger, ca. 60,000-dalton protein. Clearly, such structures represent an evolutionary increase in complexity over the phages such as lambda, in which conformational changes in a simpler basal structure are sufficient to permit DNA to be transferred from the phage head to the bacterial cell through a tube made of a single protein species. T4 presumably gains both increased adsorption efficiency and ability to adapt to varying cell surface environments by such increased complexity.

Tail tube structure has been studied in detail by Moody and Makowski (1981). Using low-angle X-ray diffraction on oriented gels of T2 polytubes, they found the annular repeat of the stacked disks of gp19 to be 4.06 nm. The tube has an average diameter of 9.0 nm with an inner hole of 3.5 nm. The symmetry of the tube matches that of the sheath, in that the tube is polar and the annuli have C6 symmetry.

Tail Length Regulation

The regulation of the length of the bacteriophage tail tube is a model problem in the search for biological length-determining mechanisms. Tail tube assembly is similar for all tailed phages: a basal structure initiates the polymerization of tail subunits which are present in large numbers but which are designed to remain unassembled within the cell (King, 1980). The basal structure presumably imposes a conformational change upon the first bound tail subunits, which in turn exposes new binding sites. This permits the addition of more subunits in a manner similar to the binding of bacterial flagellin monomers to short fragments of assembled flagella (Asakura, 1970).

Models for length regulation have been discussed (Caspar, 1980; Kellenberger, 1972b; King, 1980; Wagenknecht and Bloomfield, 1977), but none has yet been established. The remarkable homogeneity in the length of phage tails (usually ±1 to 2 disks) requires some well-defined ruler or a switching mechanism of highly cooperative sensitivity (Caspar, 1980; Harrison, 1980). Many phage tails contain over 50 disks; thus, models which propose uniform changes in the structural conformation of the growing structure at the addition of each disk seem unlikely to give such a narrow length distribution (Caspar, 1980).

The most completely studied system for measuring phage tail length at present is that of phage lambda. Studies by Katsura and collaborators have established the pathway for tail assembly and demonstrated the requirement for an initiator structure and for a termination protein. The normal initiator, like the phage T4 baseplate, requires many (at least seven) gene products for its formation. Once properly formed, the initiator facilitates the assembly of about 200 molecules (32 disks) of the major tail protein (gene *V* product) (Katsura and Kuhl, 1975b). In the presence of the gene *U* termination protein, tail assembly stops at the proper length, and the stable structure is ready for joining to heads. The mechanism of action of the gene *U* product is unknown, but it appears to bind to the tail protein by adding a single ring at the head attachment end (Katsura and Kuhl, 1974). Even in the absence of gp*U*, however, the growing tail apparently stops at the normal length, although this structure is unstable and later is converted into an aberrant tail of indeterminant length, incapable of joining to heads (Katsura 1976; Katsura and Kuhl, 1975a). One component of the initiator structure is gp*H*, a 79-kilodalton protein that has been proposed as a possible "ruler" molecule for tail length regulation (Katsura, 1981a;

King, 1980). Although internal deletions of gp*H* give viable phage with normal tail length (Katsura, 1981a; Katsura, 1981b), gp*H* is still the only candidate for such a ruler.

The phage T4 tail tube has an assembly sequence similar to that of phage lambda. The initiator is the baseplate, itself composed of 19 gene products. After the addition of gp48 and gp54, the baseplate can bind the unassembled tube protein. In vitro studies on assembly by Wagenknecht and Bloomfield (1975, 1977) show the same result as for phage lambda: purified tube protein gp19 adds to baseplates to form structures of near the normal length, but these too are unstable and eventually grow past the normal length. Treatment of baseplates with proteases gives tails of shorter length in vitro (King, 1980), and extra-long tails produced in vitro on T4 (Tschopp and Smith, 1978) or by mutations in phage SPO1 (Parker and Eiserling, 1983) show partial stain penetration down to the normal tail length, suggesting that there is an internal component. Purified T4 tube-baseplates treated with guanidine hydrochloride dissociate (Duda and Eiserling, 1982). The released tail tubes are of normal length but show partial stain penetration at the head end and a short fiber protruding from the baseplate end (Fig. 9). These structures contain near-normal amounts of gp48 but not of other tail proteins, suggesting that gp48 may be involved in tail length determination. A drawing summarizing the structural features of the phage tails is shown in Fig. 10.

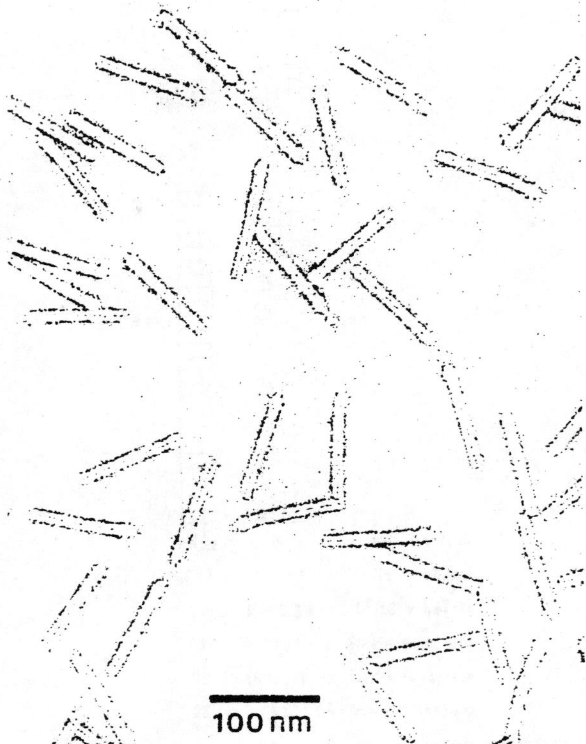

FIG. 9. Electron micrograph of T4D tail tubes after treatment with 3 M guanidine hydrochloride (Duda and Eiserling, 1982). Many of the tubes have tapered tips from which a short fiber protrudes. The most likely candidate for the fiber is gp48. Micrograph by Robert Duda.

Tail Fibers

The six long tail fibers of T4 function as sensors of the environment, attach the virus to the cell surface, and signal conformational changes to the baseplate. Initial contact of the tapered tip of the distal half of the tail fiber with diglucosyl residues in the outer membrane lipopolysaccharide results in reversible binding of the phage to the cell (Goldberg, 1980). Fibers are about 3.0 to 4.0 nm thick and 150 nm long and are made by the joining of two half-fibers (King, 1980). The half-fiber bound to the baseplate is constructed from two molecules of gp34 which have a terminal thickening near the attachment site. The half-fiber that binds to the cell surface is more complex and is made of one copy of gp35 and two copies each of gp36 and gp37. The tip of gp37 contacts the cell surface. The two gp37 molecules are oriented in the same direction, colinear with the fibers and with the C terminus near the distal tip (Beckendorf, 1973). It has been proposed from combined X-ray diffraction and electron microscopic studies that the distal half-fiber is composed of a set of globular domains at specific regions along the half-fiber and that much of the polypeptide chains are in a cross-conformation, with face-to-face packing of both gp36 and gp37 (Earnshaw et al., 1979; Oliver and Crowther, 1981). The globular domains are indicated in Fig. 1 and in the article by Wood and Crowther in this volume.

DYNAMIC ASPECTS OF T4 STRUCTURE

Luria (1950) wrote that one of the first tasks of the then-new field of quantitative phage biology was "to uncover the relation of the virus particle as we know it in the extracellular state to what is replicated inside the infected host." He articulated the view of the phage school at that time that there was a schism between the dynamic, intracellular replicating form and the static extracellular virion. This focus led to essential and fundamental discoveries in virology and molecular biology, although studies on virus structure and assembly received less attention. Indeed, the view then was that it is "unlikely that even the most careful and painstaking work on the physical properties of extracellular virus particles . . . can throw much light on the fundamental problem of virology: virus reproduction." In the longer view, this has turned out not to be the case; studies of virus structure have now entered the dynamic stage. In that same symposium in 1950, Luria emphasized the search for intermediates in phage reproduction and wondered what they might be like. Ten years later, with the discovery and study of phage conditional-lethal mutants, a major change began in phage research, and many intermediates, including assembly intermediates, became available. Detailed studies of the mature T4 virion have shown that the structure is not static and that its dynamic functioning is an essential part of the process of viral reproduction.

From this growth of structural information on bacteriophage T4 assembly, it has become clear that an impressive number of structural transformations of the virion take place during assembly and upon infection. These include at least eight major structural

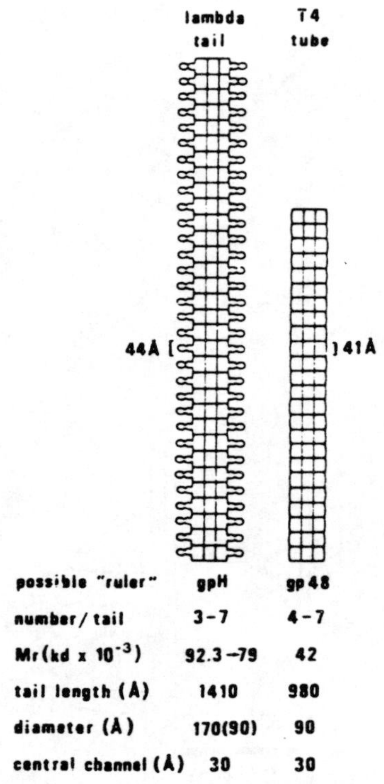

FIG. 10. Comparison of the bacteriophage lambda tail structure and the tail tube of T4, showing similarities and properties of the potential "ruler" proteins.

rearrangements of T4 components: during head assembly (prohead lattice expansion and addition of accessory proteins, entry and exit of core proteins, entry and packaging of DNA, and possible movement of neck proteins); during tail fiber assembly (positioning of long-fiber subassemblies on whiskers); during adsorption and injection (whisker-fiber-sheath interactions during long tail fiber extension and retraction, short tail fiber extension, baseplate hexagon-star rearrangement, tail sheath contraction, tail core movement, and DNA ejection). More of these structural transitions remain to be discovered, and many are known in comparable outline for other phage systems. The elucidation of viral structure at the atomic level will provide even more fundamental insights. D. L. D. Caspar's idea that the purposeful switching among different conformational states exerts self-control of the formation and action of protein assemblies is clearly demonstrated with virus structures. Tobacco mosaic virus disk-to-helix switching initiates the assembly with RNA, and movable polypeptide "jaws" bind the RNA sequence. Icosahedral plant virus structural details at high resolution reveal how a single flexible protein subunit can provide substructural domains to build its own scaffolding function and how that viral shell responds to environmental changes (Harrison, 1980). The elucidation of the molecular mechanisms of these transformations in virus structure over the next few years will continue to provide insights into the problems of viral assembly, adsorption, penetration, environmental sensing, and the more general problems of the genesis and function of subcellular structures.

The T4 research in the laboratory of F.A.E. is supported by a Public Health Service grant from the Institute of Allergy and Infectious Diseases.

L. Anderson, R. Duda, B. J. Mueller, M. C. Rayner, and R. Sweet contributed substantially to this chapter. Special thanks to M. Wurtz for calculations on head dimensions.

In many ways this paper is a tribute to the pioneering and supportive efforts of Edouard Kellenberger over the past 30 years. Many students of T4 structure and morphogenesis are indebted to him for his hospitality, first in his laboratory in Geneva and more recently in Basel. We have greatly benefited from his boundless enthusiasm and his constant efforts to promote molecular biology in Switzerland and Europe, and, in particular, from his insistence on the highest experimental standards for quantitative structural microbiology, which he established with his own research results. We also acknowledge the supportive environment provided by his colleagues in Geneva by Richard H. Epstein in particular and also by Grete Kellenberger-Gujer, Janine Séchaud, and Edouard Boy de la Tour.

The T4 Particle: Low-Molecular-Weight Compounds and Associated Enzymes

LLOYD M. KOZLOFF

Department of Microbiology, University of California, San Francisco, San Francisco, California 94143

It is assumed that strong selection pressure during viral reproduction results in viral particles in which every gene and gene product are involved in some phase of forming the new particles. With this in mind, one can examine the T4 particles for those structures necessary for host cell recognition and invasion, those responsible for control of host metabolism, and those involved in replication and assembly. Many of these fundamental processes are understood reasonably well in outline, but most others are only perceived sketchily. Surprisingly, some gene products, including some normal structural head proteins and tail proteins, are thought to be nonessential and do not necessarily appear in the progeny. Yet the classical approach of determining the chemical nature of the virion has offered unexpected insights into the processes of invasion and assembly. In this chapter, the discovery and identification of some of the low-molecular-weight compounds found in phage particles are presented. In addition, the role of these compounds in phage reproduction is assessed, and, finally, information on the phage genes responsible for their production is also considered.

FOLATES IN T4

Discovery and Chemical Nature

Studies of the interaction of cofactors for T4B adsorption, such as tryptophan, or inhibitors of T2H adsorption, such as indole, indicated that these small molecules probably formed a charge transfer complex (Kanner and Kozloff, 1964) with some component of the phage tail. In non-cofactor-requiring phage, such as T4D, there apparently is an endogenous tryptophan which plays a role similar to that of the added tryptophan (Gamow and Kozloff, 1968). In this complex, the indole ring was presumed to be donating or sharing its electrons with some compound which had the property of serving as an electron acceptor. A search for such compounds led to the discovery (Kozloff and Lute, 1965) that all of the T-even phages possess a polyglutamyl form of folate. In these experiments, highly purified phage particles were first osmotically shocked to rupture the head and remove the phage DNA. The DNA was digested with DNase, and the phage ghost particles were then boiled in the presence of cysteine or ascorbate to release the folate. The initial assays were carried out microbiologically, using both *Streptococcus faecalis* and *Lactobacillus casei* as test organisms. Conventional microbiological assays for folates have included a digestion step with an extract from hog kidneys to hydrolyze polyglutamyl forms of folic acid to smaller molecules which can be taken up by these bacterial cells. The *Streptococcus* cells can use the mono- and diglutamyl forms, whereas the *Lactobacillus* cells can use molecules containing as many as three glutamate residues. No growth activity for either organism was found in phage extracts without a prior enzymatic treatment with the hog kidney preparation. After digestion, two to five molecules of folate per phage particle were found in T-even phage preparations. The number of glutamyl residues in the undigested phage folate compounds was initially determined by using a paper electrophoresis method, and mobilities were observed of biological activity corresponding to folates possessing five to six glutamyl residues.

These observations led to experiments to examine the effect of the partially purified hog kidney enzyme on viable phage particles. It was found that T4 and the other phages were irreversibly inactivated upon incubation with this enzyme. Further, the conditions required for the inactivation reaction to proceed optimally were those required for the enzyme to act on known folate substrates. Finally, it was shown that the addition of a known folyl polyglutamate compound as a competitive substrate to the reaction mixture protected the phage particles from inactivation. These observations demonstrated that a folyl polyglutamate was an essential constituent of the T-even phage particle.

The precise chemical nature of the phage folate compound was examined with improved methods somewhat later (Kozloff, Lute, Crosby, et al., 1970). The oxidation state and the presence of possible substituents on the pteridine ring were of interest. The compound was extracted from a large amount of T4 phage, and its fluorescent excitation and emission spectra were measured. These spectra corresponded to those of dihydrofolate and differed from those of the other known folate compounds. The determination of the number of glutamyl residues on the phage compounds was carried out on phage folate which had been degraded to the *p*-aminobenzoyl glutamyl compound. This portion of the phage folate behaves on DEAE columns in a manner that is determined by the number of glutamyl residues. These analyses showed that the phage compound was a dihydropteroyl hexaglutamate (Nakamura and Kozloff, 1978).

Location on the Phage Particle

The initial supposition from the early cofactor experiment that the folate compound was involved in phage tail function was confirmed when it was observed that in the absence of tryptophan, T4B phage particles were not susceptible to attack by the hog kidney enzyme. Measurements of the folate content of isolated phage substructures showed that the substructure consisting of the baseplate plus the tail tube was highly enriched in folate per milligram of protein as compared with whole particles. Rather surprisingly, isolated and purified baseplates (isolated from T4D *54⁻ am*-infected *Escherichia coli* B cultures) contained little or no folate. In view of the evidence that folate is a baseplate component, it must be concluded that the absence of the gene 54 product (as well as the gene 48 and gene 19 products) permits the folate to be lost from this baseplate. Supporting this view is the finding that these isolated baseplates were not active in complementation assays. The precise location of the folate was studied further by using complementation reactions in which incomplete phage particles could be examined for the exposures of different elements of the phage folate compound. For example, antiserum against the pteridine moiety inhibited the addition of the gene 11 product to the phage tail baseplate (Kozloff, Lute, and Crosby, 1975). This antiserum had no effect on intact phage or on phage particles lacking either the gene 12 product or the long tail fibers. It can be concluded that the pteridine portion was on the distal surface of the baseplate partially covered by the gene 11 product. Studies on phage inactivation by conjugase led Dawes and Goldberg (1973b) to similar conclusions regarding dihydropteroyl hexaglutamate as part of the adsorption organelle. [*Editor's note*: Studies from the laboratories of Kozloff and Goldberg disagree regarding the reversibility of the effect of conjugase upon phage infectivity.]

The location of the polyglutamate portion of the folate compound was also studied (Kozloff et al., 1979). Antiserum prepared against a polyglutamate antigen had a small inhibitory effect on the attachment to the phage of the long tail fibers, but did not affect other assembly reactions. It was also found that oligopeptides of L-glutamate linked via the gamma-carboxyl groups, especially the gamma-L-triglutamate compound, were potent inhibitors in complementation reactions of long tail fiber attachment. Other non-folate-like glutamyl peptides had little or no effect. These observations showed clearly that the folate compound is a tail baseplate compound and that it is tightly buried in the baseplate structure.

Calculations of the number of different chemical bonds that the phage folyl polyglutamate can form and of the total distance that the extended molecule can cover have been presented (Kozloff, 1980; Kozloff, 1981) (Fig. 1). My group has proposed that the pteridine portion of this molecule is bound to the active site of the baseplate dihydrofolate reductase, a component of the baseplate outer wedge structure (Mosher and Mathews, 1979). The polyglutamate portion is known to have a high affinity for the thymidylate synthase which is found in the hub of the baseplate (Kozloff and Zorzopulos, 1981), and it is suggested that, given the total bonding energy and the distance the molecule could span, the dihydropteryl hexaglutamate could act as a cross-link between the baseplate wedges and the baseplate center hub. Additionally, since folate compounds with only two to three glutamate residues have been shown by Kisliuk et al. (1979) to bind maximally to thymidylate synthase, the other three to four glutamate residues of the phage compound are free to participate in binding to other baseplate components, including two additional

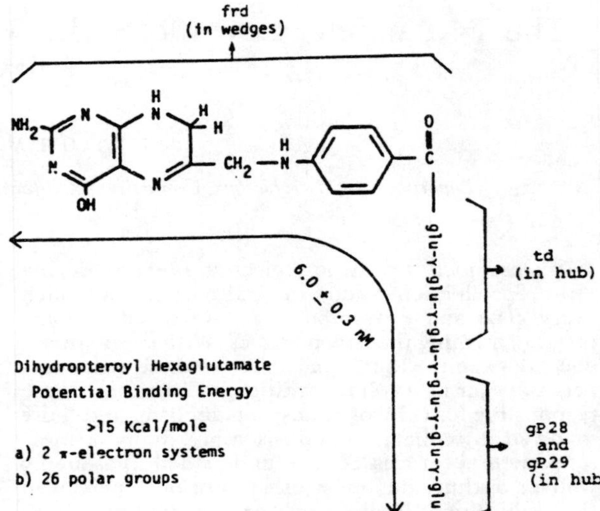

FIG. 1. Structure of dihydropteroyl hexaglutamate and possible interactions with baseplate wedge and hub components. frd, Dihydrofolate reductase; td, thymidylate synthase.

folate enzymes in the baseplate hub (see below), as well as to the long tail fibers.

Biosynthesis After Infection

Previous understanding of the polyglutamyl folate pools in bacteria was based on the results from the limited analytical methods then available, but these analyses implied that the major folate compound in bacterial cells was the triglutamyl form. Recent analyses have confirmed that the major component is the triglutamyl form but have indicated that bacterial cells contain a population of the polyglutamyl forms. A recent report indicated that *E. coli* contains folates with up to nine glutamyl residues (Nakamura and Kozloff, 1978). The total folate pool in *E. coli* contains about 500,000 molecules per bacterial cell (Kozloff and Lute, 1965). If, in an uninfected cell, the folyl hexaglutamate compounds were 2% of the total folate pool, that would be 10,000 molecules. If 300 phage particles were formed per cell, that would require 1,800 folate molecules, each with the six glutamate residues. While the cell has that many molecules with six glutamate residues, some of them are in an oxidation state other than the dihydro form or may be substituted by methyl, formyl, or other groups. Since phage infection normally increases the pool of all precursors severalfold, it seemed likely that phage infection would result in significant changes in the distribution of different folate compounds in the cell. Analysis of folate polyglutamate pools showed that infection resulted in a marked increase in the relative amount of the pteroyl hexaglutamate compound (Kozloff and Lute, 1973; Nakamura and Kozloff, 1978).

Experiments were carried out to identify the phage gene product(s) responsible for the increase in the amount of hexaglutamate compound. It was initially assumed that one viral gene product might be responsible for the formation of the folic acid. Extracts were made of nonpermissive cells infected with single T4 amber mutants. These were then supplemented with chemically synthesized pteroyl hexaglutamate, incubated, and assayed for newly formed phage (Kozloff et al., 1973; Kozloff, Lute, and Crosby, 1970). Only extracts made after infection with T4D *28⁻ am* and, to a considerably lesser extent, T4D *29⁻ am* were stimulated to form small amounts of new phage particles. Analysis of the folyl polyglutamates in *E. coli* B infected with T4D *28⁻ am* revealed a considerable increase in higher-molecular-weight folyl polyglutamates and the accumulation of a folate compound containing possibly 12 to 14 glutamate residues. This unusual compound amounted to 8% of the total folate pool in the infected cell (Nakamura and Kozloff, 1978). This observation suggested that the T4D gene 28 product might be acting as a carboxypeptidase to cleave high-molecular-weight folyl polyglutamates to the size needed for viral assembly. However, it also seemed clear that phage infection stimulated glutamate addition to smaller folate molecules. Recent evidence has implicated the T4D gene 29 product as the likely folyl polyglutamate synthetase, and those data are presented below.

Folates in Other Phage Particles

Only a preliminary survey of other bacterial viruses for the presence of a folyl polyglutamate has been carried out. The direct determination of small amounts of folate in phage preparations is difficult and open to criticism if the phage is not highly purified and then freed of its DNA. However, the sensitivity of T4, T2, and T6 to the folyl polyglutamate carboxypeptidase (called conjugase in earlier work) from hog kidney does offer a simple procedure for examining other phage particles. It was reported that *E. coli* phages without a hexagonal baseplate, such as T1, T3, T5, T7, and lambda, were all unaffected by this enzyme. Recently, two phages attacking *Pseudomonas syringae* were isolated whose morphology was similar to that of the T-even particles (Kozloff, 1981). Both types of particles were slowly inactivated by the hog kidney enzyme. When a competing folyl polyglutamate substrate was added to the reaction mixture of phage plus enzyme, the *Pseudomonas* phages were protected from inactivation. It should be noted that *Pseudomonas* spp. are quite unrelated to enteric bacteria such as *E. coli* and that T4 does not attack *P. syringae*, nor do these phages attack *E. coli*. It has been known for some time that many phages attacking highly diverse bacteria have similar morphological features, including long tail fibers, hexagonal baseplates, and a contractile tail protein. It seems likely or at least possible that folyl polyglutamates will be found as a constituent of hexagonal-type baseplates on many different phages. The evolutionary origin of this unusual use of a particular folate is not clear, especially since its formation requires enough genetic information to make at least two enzymes.

FOLATE ENZYMES

Dihydrofolate Reductase

With the identification of the phage folate compound as a dihydropteroyl hexaglutamate, phage particles were examined for the presence of virus-induced proteins which could be binding the folate compound in the tail structure. The binding constants of dihydrofolate compounds for the enzyme dihydrofolate reductase (the phage gene is called *frd*) are known to be extremely high, and this enzyme was known to be induced by phage infection (Mathews, 1967). Assays for dihydrofolate reductase enzymatic activity of purified intact particles were negative, but relatively mild denaturing treatment with urea or formamide uncovered a weak but definite reductase activity (Kozloff, Verses, Lute, and Crosby, 1970). The presence of this enzyme in phage particles and the fact that it is essential to this particle were confirmed in a variety of experiments which bear directly on its function during virus assembly and phage infection.

Briefly, these results include: (i) the demonstration of phage inactivation by antiserum prepared against homogeneous phage-induced dihydrofolate reductase (Mathews et al., 1973); (ii) experiments showing that mutations in the *frd* gene change the heat stability of T4 particles (particles made after infection with a mutant which is temperature sensitive for dihydrofolate reductase are heat labile, whereas revertants of *ts frd* mutants produce heat-stable particles [Kozloff, Crosby, Lute, and Hall, 1975]); and (iii) the correlation of phage viability with the enzymological properties of dihydrofolate reductase (Kozloff, Verses, Lute, and Crosby, 1970). The dihydrofolate reductase enzymological properties reported include the following. (i) T4 infectivity was inactivated by NADPH (but not by NADH) and partially reactivated upon subsequent incubation with NADP. Remarkably, NADPH inactivation kinetics were logarithmic, and no NADPH-resistant mutants could be detected in a preparation containing 10^{12} T4 particles. [*Editor's note*: There are probably additional virion targets for NADPH, since urea-treated phage, which lack virion dihydrofolate reductase are partially inactivated by NADPH (Dawes and Goldberg, 1973b).] (ii) Analogs of NADPH such as adenosine diphosphoribosylphosphate, which itself is a good inhibitor of dihydrofolate reductase, reacted with phage particles and both changed their heat lability and decreased their adsorption rate to host cells (Male and Kozloff, 1973). It should be emphasized that enzymatic activity of the particle-bound dihydrofolate reductase is not required for successful phage infection to occur. This was evident from the viability of *frd⁻* phage, once it had been established that the *frd* gene codes for virion dihydrofolate reductase (Mathews, 1971b). Phage containing amber *frd* mutations are viable and can also be inactivated by T4 dihydrofolate reductase antiserum, indicating that a truncated portion of the dihydrofolate reductase molecule suffices for the structural role (Dawes and Goldberg, 1973b; Kozloff, Crosby, Lute, and Hall 1975).

All of the evidence described above clearly pointed to the baseplate as the site of dihydrofolate reductase. Further, isolated baseplates, when treated with urea, did show dihydrofolate reductase activity, which was 40-fold increased, on a weight basis, over that found in whole phage particles. It is not surprising that dihydrofolate reductase activity would be poorly expressed when the enzyme was incorporated into a complex structure such as a baseplate. Further, in whole phage particles, the dihydropteroyl group would be tightly bound to the active site, preventing catalytic activity. In fact, one could expect only those dihydrofolate reductase molecules not binding dihydrofolate to be even potentially enzymatically active. From this viewpoint, then, the number of dihydrofolate reductase molecules per phage would not be expected to more than the number of folate molecules. Up to five to six folate molecules per particle have been found in different phage preparations (Kozloff and Lute, 1965), and, therefore, possibly six potentially active dihydrofolate reductase catalytic sites could be available in each phage particle. The catalytic activities measured have always been considerably below this number.

The exposure of the presumed binding site for NADPH on the dihydrofolate reductase molecule is in accord with all of the other data on the location of the enzyme on the baseplate. It has been proposed that the NADPH-binding site on the baseplate may interact with the various NAD compounds leaking from host cells during the infectious process. This could provide a trigger mechanism to open the central hole in the baseplate.

Two recent reports from the laboratory of C. K. Mathews on the chemical nature of the phage dihydrofolate reductase and its location have been important in considering the role of this molecule in phage assembly. Mosher and Mathews (1979) have found that dihydrofolate reductase is a component of the outer wedges of the baseplate. This enzyme had not previously been detected in wedges, and the use of labeled antibody was necessary to detect its presence in gels. Additionally, Purohit et al. (1981) have reported that phage dihydrofolate reductase exists as a dimer of two 20,000-dalton monomers. This observation, together with the finding of a maximum of only six folate molecules per particle and of only three thymidylate synthase dimers (see below), supports the view that there are only three dihydrofolate reductase dimers per phage particle. An attractive possibility is that each wedge contains one dihydrofolate reductase monomer and that lateral binding of the wedges to each other is enhanced by the formation of the normal dihydrofolate reductase dimers.

An interesting feature of the structure of the phage-induced dihydrofolate reductase molecule (and the thymidylate synthase molecule [see below]) is that the carboxyl end of the molecule is not required for the role it plays in the baseplate. A T4 amber mutant in *frd*, called in older terminology wh11, forms viable phage although the phage contain no enzymatically active dihydrofolate reductase. According to D. H. Hall (personal communication), the product of *frd* wh11 possibly lacks about 10% of the carboxyl end of its polypeptide chain. This suggests that the carboxyl end of the dihydrofolate reductase molecule is only loosely attached to the baseplate structure. My group has found that phage-induced dihydrofolate reductase is rapidly inactivated by incubation with carboxypeptidase B, an exo-carboxypeptidase specific for hydrolyzing carboxyl-terminal arginine or lysine residues. The host enzyme is not affected by carboxypeptidase B. These properties of dihydrofolate reductase offer an explanation for the earlier observations (Shapiro and Kozloff, 1970) that the tail structure of the T-even phages was disrupted upon incubation with arginine or arginine analog and that T4 tail assembly in extracts was specifically inhibited by the presence of arginine or by treatment with carboxypeptidase B. The main step in assembly that was sensitive was the addition of the initial sheath monomers to the baseplate and tail tube structure. Based on this observation, we propose that the carboxyl end of the phage-induced dihydrofolate reductase molecule, an arginine residue, participates in binding tail sheath units largely by polar interactions involving the carboxyl and guanidyl ends of the arginine molecule. This view of the location of the molecule is in accord with the finding that dihydrofolate reductase is not in the baseplate hub, which might be covered by the tail tube, but is located in the peripheral wedge elements of the baseplate.

Thymidylate Synthase

Thymidylate synthase, induced by a virus gene called *td*, was shown by Capco and Mathews in 1973 to be a phage component. Phage infectivity was neutralized by antiserum against the purified homogenous enzyme, and transfer of the *td* gene from phage T6 into phage T4 resulted in T4 particles with altered heat sensitivity. This last observation implied that thymidylate synthase was a component of the baseplate, since the phage baseplate is the most heat-labile component of the phage particle. This last observation was amplified in studies of *td* mutants (Kozloff, Crosby, and Lute, 1975). It was observed that single mutations in the *td* gene produced temperature-sensitive enzymes which altered the heat sensitivity of phage particles. Phenotypic reversions of this mutation again resulted in changes in the phage heat sensitivity. These results confirmed that the *td* gene product was a phage structural component and pointed to the baseplate as the most likely substructure for its location. However, like with dihydrofolate reductase, there were doubts about this location, since analytical results obtained by standard techniques did not reveal a component of the molecular size of the thymidylate synthase molecule in baseplates. Further, two T4 mutants containing amber mutations in the *td* gene were able to produce viable phage particles (Krauss et al., 1973) under restrictive conditions.

Careful measurements of thymidylate synthase activity in wild-type phage preparations showed small levels of activity, and no activity was found in preparations of particles containing the *td* amber mutation. Treatment of particles with denaturing agents analogous to those used to expose the buried dihydrofolate reductase activity destroyed all thymidylate synthase activity, as might have been expected from the lability of thymidylate synthase molecule. However, incomplete T4 particles lacking the gene 11 and gene 12 products on the distal surface of the baseplate had more thymidylate synthase activity than did intact particles (Kozloff, Crosby, and Lute, 1975) and were more sensitive to thymidylate synthase antiserum than were whole T4 particles. These last results supported the earlier indications that thymidylate synthase is a baseplate component.

The two known T4 *td* amber mutants had been isolated by using T4B rather than T4D as the starting wild-type phage. While particles produced in restrictive cells were viable, it was not immediately apparent whether or how fragments of the thymidylate synthase molecule which lacked catalytic activity could be incorporated into the baseplate structure. To look at less obvious properties of the mutant phage baseplates, advantage was taken of a well-known property of the tryptophan cofactor requirements of the T4B particle. T4B particles are known to plate on minimal medium even in the absence of L-tryptophan if the temperatures is raised to 42 to 44°C. However, the two T4B *td* amber mutants, which generated quite different polypeptide lengths, had responses markedly different from those of wild-type phage and each other when plated at elevated temperatures (Kozloff, 1981). The properties of these two *td* mutants support the view that, like in dihydrofolate reductase, only the N-terminal end of the thymidylate synthase molecule is required as a structural component of the phage particle and that possibly 10 to 30% of the carboxyl end of the thymidylate synthase molecule is nonessential in forming new particles.

While the results described above indicated that the thymidylate synthase molecule was in the baseplate, they did not give a precise indication of the number of molecules or of whether the thymidylate synthase was a wedge or a hub component. Enzymatic measurements of thymidylate synthase incorporated into the baseplate did not yield a value for number of molecules. Additionally, the location of the thymidylate synthase under the gene 11 product or even as one component of the baseplate which controlled cofactor requirements and, thus, long tail fiber orientation or heat sensitivity was not sufficiently precise to locate these molecules.

A highly sensitive chase-labeling procedure in complementation experiments devised to detect even minor components of the baseplate hub finally answered these questions (Kozloff, 1981; Kozloff and Zorzopulos, 1981). These experiments showed that hubs contained a small amount of a previously undetected component with a molecular weight of 29,000. Since thymidylate synthase monomers have this size and since no 29,000-dalton component has been found in baseplate wedges, it seems likely that thymidylate synthase is a hub component. All of the properties ascribed to structurally bound thymidylate synthase would be in accord with this location of the molecule.

The number of thymidylate synthase monomers in the hub can be calculated relative to the number of the other hub components. Given that the hexagonal hubs contained six molecules each of gp29, gp27, and gp5, there appeared to be about one-half as much thymidylate synthase, or three monomers. However, thymidylate synthase normally exists as a dimer, and, therefore, one might expect that there would be at least two complete molecules, or four monomers. This value agrees only reasonably well with the number of folate molecules found per particle and with the proposed three dihydrofolate reductase dimers (six monomers) in the wedges. These results, even with the difficulties in the measurements, do imply that there are six dihydropteroyl hexaglutamates, three dihydrofolate reductase dimers in baseplate wedges, and about two to three thymidylate synthase dimers in the hub baseplate of each T4 particle.

Gene Product 28, a Gamma-Glutamyl Carboxypeptidase

In complementation experiments using the sensitive ^{14}C chase-labeling technique, T4D gp28 was shown to be a component of the central hub of the baseplate and to have a molecular size of 24,000 daltons (Kozloff and Zorzopulos, 1981). This component had not been detected previously (Kikuchi and King, 1975c) because of its relatively small size and because of the small amount present in the phage baseplate.

In addition to serving as a component of the baseplate, the gene 28 product was also shown to be a cleavage enzyme or carboxypeptidase attacking the gamma-glutamyl bonds in folyl polyglutamates. In *E. coli* B infected with T4D *28⁻ am*, i.e., in the absence of the gene 28 product, as described above, large folyl polyglutamate compounds accumulated, amounting to as much as 8% of the total folates. When an assay was devised to measure this activity, it was found that the gene 28 product was a gamma-glutamyl carboxypeptidase (Kozloff and Lute, 1981) which could be readily distinguished from an endogenous host carboxypeptidase. The virus-induced enzyme had a pH optimum of 6.0 to 6.5, whereas the host enzyme had a pH optimum of about 8.0. The characterization of the gene 28 product is preliminary, and the enzyme appears to be quite labile (Kozloff and Lute, 1981).

The determination of the amount of gene 28 product in phage particles is based on the radioautography after polyacrylamide gel electrophoresis of labeled T4 baseplate hubs. Compared with the amounts of the major hub components, gp29, gp27, and gp5, there appear to be only three molecules of gp28 per particle. It should be noted that phage particles themselves have a weak carboxypeptidase activity, suggesting that the active site of the gene 28 product is exposed sufficiently to be able to bind a folyl polyglutamate substrate. This suggestion for the location of the catalytic site of the gene 28 product is supported by the findings that (i) antiserum to baseplates inactivates the catalytic activity of phage particles, and (ii) pyrimethamine, a folate analog, inhibits the carboxypeptidase activity of phage particles.

Gene Product 29, a Folyl Polyglutamyl Synthetase

The increase in the relative amount of folates containing six glutamate residues which occurs in cells infected with wild-type T4D suggested that phage infection was inducing a new folyl polyglutamate synthetase. This view was supported by the observation described above that in cells infected with T4D *28⁻ am*, folates containing 12 to 14 glutamate residues accumulated. Efforts to detect phage induction of a new synthetase were initially unsuccessful because of the presence of both the host- and the phage-induced carboxypeptidases. However, a modification of the assay procedure of Shane (1980) so that extracts of uninfected and infected cells were incubated for only 30 s did show that T4D infection induced a threefold increase in the rate of addition of labeled glutamates to a folate substrate. The T4D gene 29 product was found to be the ATP-requiring synthetase, since infection of cells with gene 29 mutants did not result in any increase in synthetase activity whereas infection with all other phage mutants tested, including those with mutations in genes 27 and 28, did result in increased levels of synthetase activity. The enzymatic characterization of this synthetase has just been started.

The amount of gene 29 product in the baseplate has been estimated by Crowther et al. (1977) as six molecules per baseplate. This value would be in accord with later estimates of the stoichiometry of the baseplate constituent (Kozloff, 1981). There is convincing evidence that the gene 29 product is a component of the central hub. The proposal of my group is that the gene 29 product and possibly the gene 28 product bind to the distal three glutamates of the pteroyl hexaglutamates (Fig. 1). A synthetase and a specific carboxypeptidase would be expected to have some affinity for glutamate residues linked via their gamma-carboxyl groups.

METAL IONS—ROLE OF ZINC IN T4 STRUCTURE

A Zinc Metalloprotein, gp12

In 1957, Kozloff et al. reported that complexes of the zinc group metals such as $Zn(CN)_3^-$ or $Cd(CN)_3^-$ rapidly and irreversibly inactivated T2, T4, and T6 bacteriophages but not the other *E. coli* bacteriophages. The site of attack was shown to be the tail structure by several experiments. T4B was only attacked at a significant rate in the presence of the adsorption cofactor L-tryptophan. Electron micrographs showed that inactivated particles had contracted tail sheaths, exposed tail tubes, and an open channel through which the DNA could be readily released. From current information, it is apparent that incubation with $Cd(CN)_3^-$ converted the terminal baseplate from its intact hexagonal form to the star form. The action of these metal complexes on phage particles resembled changes produced by a variety of other reagents which were known to attack buried sulfhydryl groups or even labile bonds such as thiol esters. Probably the most likely possibility was that the zinc complexes were acting to displace some naturally occurring baseplate zinc metalloprotein from its attachment or binding site. This view was supported by the observation that zinc-chelating agents reacted with phage particles and interfered with DNA injection but not with attachment or host cell killing.

Direct analysis of highly purified T2 and T4D phage gave values of three to six zinc atoms per particle (Kozloff and Lute, 1977). A number of additional observations clearly supported the presence of protein-bound zinc in the baseplate. These included the increase in heat lability of T4 in the presence of *o*-phenanthroline and the inhibition of tail fiber addition to fiberless particles by *o*-phenanthroline. Perhaps most convincing were the experiments showing that phage inactivated at 60°C in the presence of zinc-chelating agents could be reconstituted by incubation with Zn^{2+} and less efficiently with Co^{2+} or Ni^{2+}. Later, it was shown that T4 grown in the presence of 2×10^{-6} M cobaltous chloride contained about four atoms of cobalt per host particle. Rather surprisingly, it was found that phage particles grown in the presence of Co^{2+} or Ni^{2+} were resistant to inactivation by either $Cd(CN)_3^-$ or hog kidney carboxypeptidase. The results support the conclusion that the zinc metalloprotein interacts directly or indirectly with the tail baseplate folyl polyglutamates.

In 1978, Zorzopulos and Kozloff reported that gp12, which forms the short tail fibers extending from the distal surface of the baseplate, contained zinc as an essential component. Purified gp12 was found to have one atom of zinc per 55,000-dalton monomer. Removal of the zinc by gentle heating in the presence of chelating agents inactivated the gp12 as judged by activity in complementation experiments. This activity could be restored by the addition of zinc, cobalt, or nickel ions. Since there are six short tail fibers, each containing three monomers, one might expect to find at least 18 zinc atoms per phage particle. An analysis of phage particles after removal of the DNA by osmotic shock (i.e., phage ghosts) gave values of only three to six zinc atoms per particle, indicating that osmotic shock may cause a loss of either the short tail fibers or their zinc atoms. There is support for the possibility that osmotic shock causes a loss of short tail fiber activity; ghost particles have only one-third of the killing power of whole particles. The loss of gp12 zinc atoms would adequately account for this decrease in ghost killing power since gp12 has been shown by Kells and Haselkorn (1974) to be the phage component responsible for host cell killing.

Zinc Metabolism in T4-Infected Bacteria

There have been very few studies of metal ion requirements or metabolism in phage-infected cells. In view of the demonstration that the formation of each phage particle requires at least 18 zinc atoms, there is the possibility that the availability of zinc may limit phage production. Further, it should be noted that at least two important enzymes, the T4D-induced, DNA-dependent DNA polymerase and the T4D-modified, DNA-dependent RNA polymerase, are zinc-containing enzymes. A calculation of the minimal amount of zinc required indicates that 1 liter of infected bacteria containing 4×10^8 cells per ml and yielding 300 T4 particles per cell would require about 8×10^{15} zinc atoms per liter for the synthesis of only gp12. If each cell also required, minimally, another 10^4 zinc atoms per cell for polymerases and other enzymes, then the total requirement would be at least 1×10^{16} atoms per liter or about 2×10^{-8} mol. Since minimal media, such as M-9, contain at most 5×10^{-8} mol, it is apparent that zinc may be an unsuspected limiting requirement. This suggestion was supported by the finding that the T4 yield could be stimulated by the addition to M-9 medium of 1×10^{-6} to 2×10^{-6} M zinc (Kozloff, 1978). Cobalt also stimulated phage production, supporting the finding that it can substitute for zinc in forming active gp12. Although nickel ions, as well as zinc and cobalt ions, can reactivate zinc-depleted gp12, nickel ions were found to be toxic to infected as well as uninfected cells. Studies were also carried out on zinc transport or uptake before and after T4 infection (Kozloff and Zorzopulos, 1978). Not surprisingly, the rate of zinc uptake was stimulated two- to threefold after infection with wild-type T4.

LYTIC ACTIVITY OF PHAGE PARTICLES

Although a variety of lytic enzymes has been found associated with many types of bacteriophage particles, the nature of the lytic activity associated with T4 particles is still unsettled. The description of the phenomenon of phage-initiated lysis-from-without over 40 years ago suggested that the T-even particles might contain a lytic enzyme as a structural component. In 1954, Barrington and Kozloff reported that material was released from isolated cell walls by the action of T2 particles. It was suspected initially that the phage lysozyme, the product of the *e* gene, was part of the virion and was responsible for cell wall breakdown. Phage lysozyme has been detected in phage preparations, but its presence is probably adventitious, and Loeb (1974) has shown that T4 phage particles produced after infection with e^- particles still can cause lysis-from-without and can release cell wall materials. Kao and McClain (1980a, 1980b) have presented indirect evidence indicating that the product of T4 gene 5 has lytic activity on cell walls. Their results show that in cells infected with T4D e^-, gp5 is responsible for lysis-from-within and that gp5 in phage particles is probably also responsible for lysis-from-without. While gp5 has not been isolated and demonstrated to have lytic activity, the genetic and biological evidence is reasonably convincing. It should be noted that gp5 is a component of the central hub of the baseplate and therefore is in a reasonable location to attack the cell wall. Further, this preliminary identification of gp5 as an enzyme means that of the six structural components of the central hub, four have enzymatic activity, namely, gp5, gp*td*, gp28, and gp29 (see above). Only gp26 and gp27, the remaining two hub components, have not been shown to play a catalytic role.

OTHER SMALL MOLECULES—NUCLEOTIDES

In the course of describing the contractile protein in the T2 and T4 tail, Kozloff and Lute (1959) reported that each phage particle contained about 144 molecules of ATP, one-third of which was deoxy ATP. The ATP was measured by using a sensitive luciferin assay method. While there was no doubt that the ATP was present in highly purified phage preparations and that there was about one ATP molecule per protein subunit in the contractile tail sheath protein, its significance was difficult to establish. Phage ghost particles did not contain ATP, suggesting at the very least that the binding of ATP molecules to the sheath was very weak or that the ATP found in the preparations of phage particles may have been packed with the DNA inside the head of the particle. Three other observations are of interest in evaluating the significance of the presence of ATP. (i) T4 phage particles do have a very weak ATPase activity, although currently this activity could be ascribed to the polyglutamyl synthetase, gp29, or even to an enzyme suspected to be at the neck of the phage particle which plays a role in pumping the DNA into the head, rather than to the action of the contractile tail protein. (ii) Upon binding to the cell wall, the phage-associated ATP is largely broken down to ADP and P_i. The location of the ATPase responsible for this breakdown is uncertain. (iii) Finally, the original suggestion that there might be one ATP molecule per contractile tail sheath subunit has been supported by studies showing that ATP does bind to isolated contractile sheaths and does influence its configuration (Sarkar et al., 1964). It is clear that the interpretation of the presence of ATP and ATPase on the phage particle is more complicated than originally thought.

The preparation of this manuscript was aided by Public Health Service Research Grant AI 18370 from the National Institute of Allergy and Infectious Diseases.

Recognition, Attachment, and Injection

EDWARD GOLDBERG

Department of Molecular Biology and Microbiology, Tufts University School of Medicine, Boston, Massachusetts 02111

The sophistication of the T4 virion permits it to recognize and infect appropriate bacteria efficiently. The efficiency of plating (EOP) of T4 can approach 1, whereas the maximum EOP of phages such as λ, T5, or T7 does not exceed 0.2 to 0.3, and phages such as φX174 are even less efficient. The sophistication of T4 is apparent in the morphology of its tail, whose sheath contracts about a central tube which conducts the DNA from the phage head into the host cell cytoplasm. In addition, there is a broad baseplate composed of at least 16 different proteins with a complicated, symmetrical, hexapartite construction. The baseplate changes configuration during the irreversible attachment process. This leads to sheath contraction and extension of the tube through the periplasm to the outer surface of the cytoplasmic membrane. There are six stiff, double-jointed tail fibers arranged regularly about the baseplate. These serve not only as phage feelers for the host surface, but as reversible attachment organelles, enabling the phage to remain attached while wandering over the surface probing for further recognition sites with its baseplate. In addition, the arrangement of the partially flexible elbow in the middle of each of the six tail fibers helps to keep the baseplate parallel to the cell surface and to monitor the distance between the baseplate and the cell surface. Cooperative action of the tail fibers is needed at two places: at the distal end to permit lateral diffusion on the cell surface while still remaining attached, and at the proximal end to trigger both the contraction of the sheath and the irreversible attachment of the baseplate to the cell surface when it is aligned at the proper site.

In this chapter I review some current ideas about the various stages of recognition of the host cell by T4 phage as well as some unsolved problems relating to the signals and mechanisms that direct the passage of phage DNA from the phage head into the host. For a more general discussion of these problems with respect to different types of phages, see my recent review (Goldberg, 1980).

It has been recognized for over 30 years (Stent and Wollman, 1952) that there is a reversible stage of T4 attachment to *Escherichia coli* that precedes the irreversible stage. This implies, of course, two types of recognition. Later morphological and physiological evidence suggests yet a third stage of tail tube attachment to the cytoplasmic membrane (Benz and Goldberg, 1973; Simon and Anderson, 1967). These stages might, of course, be subdivided and further defined by new experimental evidence or hypotheses. However, since that sort of approach creates a continuum of definitions and a semantic morass, I will try to minimize the refinement.

REVERSIBLE ATTACHMENT

Reversible attachment involves interaction between the distal part of the phage tail fibers and the lipopolysaccharide (LPS) of the *E. coli* surface. Anderson (1953) first showed that the phage attach tail first. In their classic study, Simon and Anderson (1967a) showed clearly that T4 phage tail fibers attach to the cell surface by their distal ends. This work has been corroborated by the physical and morphological evidence of Wilson et al. (1970) showing that T4 tail fibers bind specifically to purified *E. coli* LPS. Jesaitis and Goebel (1953) identified LPS as the specific receptor for T4 adsorption by using phenol extracts of *E. coli*. Dawes (1975) was able to interfere with phage T4 adsorption by use of specific mono- and disaccharides. Finally, by chemical analysis of a set of T4-resistant *E. coli* LPS mutants, Prehm et al. (1976) identified the glucosyl α-1,3-glucose terminus of the rough LPS core of *E. coli* B as the essential moiety for T4 phage recognition.

Cell Surface Receptors

The T-even phages were among the first to be characterized with respect to adsorption. Since they cross-react immunologically, look alike physically, and complement and recombine genetically with one another, their divergence in host receptor specificity is generally accepted as being a later evolutionary variation. Hence, the differences in receptor specificity (Schwartz, 1980; Wright et al., 1980) among T6 (specific for *tsx* protein [26,000 daltons]), T2 (specific for outer membrane protein OmpF [27,000 daltons]), and T4 (specific for the terminal diglucosyl moiety of rough LPS [see above]) are considered to be minor variations in recognition on a structural and functional theme to accommodate recognition of different host cell surface components. Thus, the major difference between the completed virions of these phages lies in the types of tail fiber they possess.

Recently Yu and Mizushima (1982) have shown that although the LPS of *E. coli* B is sufficient as a receptor for T4 phage, in *E. coli* K the situation is somewhat different. Phage workers have long known that it is easy to find T4 host range (*h*) mutants to *E. coli* K/4 but not to B/4 hosts. (This was first transmitted to me by Hershey in 1961 when I was trying to isolate a T4*h* mutant on *E. coli* B.) Since most B/4 strains have no glucose at the termini of their LPS molecules, there can be no reversible adsorption, which is a necessary prelude to infection. Many K/4 strains, on the other hand, still permit reversible adsorption. It now appears that there are two separate types of receptors for reversible T4 tail fiber adsorption to *E. coli* K: LPS and the cell surface protein OmpC (36,000 daltons).

TABLE 1. Role of LPS and OmpC receptors in determining EOP of T4 on *E. coli* K

LPS[a]	OmpC	EOP
Wild type	+	1.0
Mutant I	+	1.2
Mutant II	+	10^{-3}
Wild type	−	10^{-3}
Mutant I	−	0.7
Mutant II	−	$<10^{-7}$

[a] LPS (from *E. coli* K-12) structures were as follows:

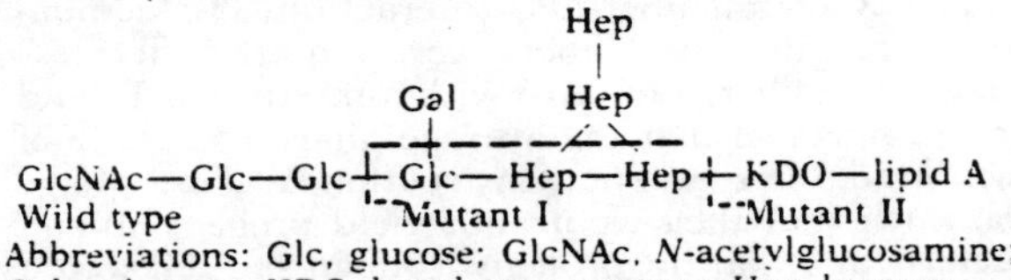

```
                                        Hep
                                         |
                      Gal               Hep
                  ----+-------------/--\---
GlcNAc—Glc—Glc-+ Glc— Hep—Hep+-KDO—lipid A
Wild type       '-Mutant I       '-Mutant II
```

Abbreviations: Glc, glucose; GlcNAc, *N*-acetylglucosamine; Gal, galactose; KDO, ketodeoxyoctanoate; Hep, heptose.

Table 1 (data from Yu and Mizushima, 1982) shows that in the absence of either OmpC (wild type, $OmpC^-$) or the LPS core distal to ketodeoxyoctanoate (mutant II, $OmpC^+$), the EOP is reduced to 10^{-3}. In the absence of both receptors (mutant II, $OmpC^-$), the EOP is reduced to $<10^{-7}$. Thus, both surface components must be independently effective as tail fiber receptors. Partial truncation of the LPS core so as to leave a terminal glucose leads to a normal EOP even in the absence of OmpC (mutant I, $OmpC^-$). This implies that the presence of sugars distal to glucose on the LPS reduces its effectiveness as a T4 receptor.

Interaction of Tail Fibers with Their Cell Surface Receptors

How does the phage recognize the cell surface receptors? There are six tail fibers symmetrically situated about the hexagonal baseplate. If one of the tail fibers interacts with an LPS molecule, it is more likely that others will follow suit, since the effective concentration of tail fibers of that phage in the region of receptor sites will be high. Moreover, if a bound tail fiber becomes unbound, it is likely to bind again nearby, since the LPS concentration on the cell surface is high. (There are about 10^6 LPS monomers [Smit et al., 1975] on a cell surface of about 2×10^6 nm^2, based on a cylindrical bacterium of 1-μm length and 0.5-μm diameter. This is equivalent to an average distance of approximately 1.4 nm between adjacent LPS molecules.) Any of the six tail fibers may bind the phage to the cell, and yet this phage binding is reversible if the next stage of baseplate interaction (i.e., the irreversible expansion of the baseplate and attachment of the short baseplate fibrils) does not occur. This is the case for $gp11^-$ phage. Reversibility of this stage of phage binding suggests that the individual tail fiber-LPS interaction is weak. For example, if at equilibrium of the individual tail fiber-LPS interaction, half of the tail fibers were bound and half were unbound, the phage would be bound by at least one of its six fibers 98% [$1 - (1/2)^6$] of the time. (The equilibrium binding of single isolated tail fibers would be even lower due to the concentration effect described at the beginning of this section.) The fact that tail fiber LPS binding is weak was borne out by the work of Wilson et al. (1970) who demonstrated that only about 30% (see their Table 5) of isolated tail fibers bind specifically to excess isolated LPS. Weak binding of six individual tail fibers to a dense LPS lawn on the cell surface would permit phage to wander over the two-dimensional surface but would restrict them most of the time from rising from the cell surface into the medium.

IRREVERSIBLE ATTACHMENT

The transition from reversible to irreversible attachment involves a number of phage and host organelles and many proteins. There are also many changes in structure between the unattached and the irreversibly attached phage (i.e., in the relation of the phage proteins to one another and in the structure of some of the proteins themselves). It was the morphological evidence of Simon and Anderson that first implicated the baseplate fibrils as the organelles irreversibly attaching phage to the cell surface. The work of Simon et al. (1970) and Kells and Haselkorn (1974) showed that these fibrils contain the product of gene 12 (gp12), and thus the organelle has been called P12. Due to the beautiful in vitro complementation system of Kikuchi and King (1975a, 1975b, 1975c) and the brilliant morphological analysis of mutant T4 baseplates by Crowther et al. (1977), we now know that P12 is attached to P11 on the baseplate during morphogenesis. Therefore, a gene 11 mutant, which lacks gp12 as well, cannot attach irreversibly to the host cell surface. The actual P12 organelle in the attached phage is not derived directly from the bumps (or spikes) sticking out perpendicular to the baseplate (which is P11) but rather is derived from the material which connects the bumps, parallel to the baseplate.

Transition Between Reversible and Irreversible Attachment

How does reversible binding lead to irreversible binding? The tail fiber must signal the baseplate that the cell surface is appropriate for infection; that is, it must recognize an appropriate area of the host surface and signal this recognition to the phage so that the baseplate binds irreversibly, thereby leading it to inject its DNA. The tail fiber has two working ends, the distal tip, which has a recognition site for a terminal glucosyl moiety, and the proximal end, which signals the phage via its interaction with the phage baseplate, probably with gp9. I think that expansion of the baseplate is triggered by the simultaneous occurrence of a certain position of a few of the tail fibers plus the proximity of the baseplate to an appropriate receptor region on the cell surface. This implies that a phage cannot attach irreversibly at any site on the cell surface but must wander over the surface, probing intermittently with its baseplate (by a vibrational mode normal to the cell surface) until it finds and aligns itself properly with such a site. The symmetrical array of tail fibers tends to keep the baseplate parallel to the cell surface. Bayer (1968) has found adhesion zones between inner and outer membranes at sites of T4 phage attachment. Whether these zones are specific to sites of potential phage attachment or result from the attachment process, e.g., by stabilizing potential or transient adhesions, is not yet clear. Bayer estimates their number at a couple of hundred, which is in agreement with the number of T4 injection sites and the number of membrane subunits conserved during cell division (Green and Schaechter, 1972). If, in fact, the membrane is composed of a few hundred subunits, there should be a corresponding minimum for the number of copies of some membrane components. Nevertheless, some functional subunits (e.g., pili) may be less frequent due to membrane inheritance, poor synthesis of a component required for function but not for growth, or an active function precluding an alternative function.

Relation Between Host Range and Ease of Baseplate Triggering

Phage T4 host range mutants are of two general types, having mutations either in the distal part of the tail fiber (gene 37) or in one of the baseplate genes. Gene 37 host range mutations are found in a region about 50 to 100 amino acids from the C terminus (unpublished data), in the distal tip. It is not clear how these mutants extend the host range. Do they interact with normal LPS receptors with an altered affinity, or do they recognize an altered LPS receptor more readily? In this context, does the distal tip (the thin piece, at the C-terminal end) also bind to OmpC? With this in mind it might be useful to investigate the occurrence and specific location of host range mutations in phage which can infect LPS^- $OmpC^-$ and LPS^- $OmpC^+$ (i.e., mutant II) bacteria (see Table 1).

Host range mutations that affect the baseplate are understood somewhat better. These mutant phage are generally sensitive to incubation at high temperatures (e.g., 60°C), whereas wild-type phage are not (Beckendorf, unpublished data: Dawes and Goldberg, 1973b). This property is called thermolability to distinguish it from temperature sensitivity, which is the inability of vegetative phage to develop or mature inside the host cell at elevated temperature. Inactivation of these thermolabile baseplate *h* mutant virions at high temperature is accompanied by normal expansion of the baseplate and contraction of the tail in the absence of cells (Arscott and Goldberg, 1976). (The reason that wild-type T4 phage have not evolved such an extended host range might be that baseplate *h* mutants are less stable; furthermore, during attachment they may trigger abortively with a less-than-perfect positioning of the baseplate, similar to exclusion of superinfecting phage DNA, thus lowering the EOP.) This inactivation requires tail fibers to trigger it (Arscott and Goldberg, 1976; Yamamoto and Uchida, 1975) since fiberless phage, whether wild type or mutant in the baseplate, are not inactivated at high temperature and can, in fact, be activated in vitro by adding tail fibers after heating. The requirement of tail fibers for baseplate expansion and the localization of *h* mutations in the genes for a number of baseplate proteins, in both the central core and the outer wedges, imply a cooperative interaction of several tail fibers to trigger a series of baseplate protein interactions leading to baseplate expansion and sheath contraction. The requirement for several tail fibers in phage infection was first shown by Wood and Henninger (1969), who also calculated that the number needed for wild-type infection is about three. Crawford and Goldberg (1977) showed that when the likelihood of simultaneous tail fiber-LPS interactions per phage is reduced, infectivity decreases much more rapidly than does the attachment of reversibly bound phage to the cell surface. Two independent techniques were used to prove this: (i) a bacterial mutant that is temperature sensitive in LPS synthesis and whose LPS surface density depends on the growth temperature, and (ii) addition of the competitive inhibitor sucrose, an analog of the LPS terminus. In both cases, wild-type phage were more sensitive than baseplate *h* mutants to a reduction in frequency of tail fiber-LPS interactions. To demonstrate that three tail fibers were required, fiberless phage were fibered in vitro with a mixture of T4 and T6 tail fibers so that the average phage had three of each kind. (The genetic construction of these phage was such that they would not yield progeny in su^- bacteria and that the progeny from an su^+ cell would have the host range of T2.)

Table 2 (data from Crawford and Goldberg, 1980) shows that these phage infect T4-resistant ($Bsu^-/4$) and T6-resistant ($Bsu^-/6$) *E. coli* about equally well, but more poorly than they infect totally sensitive *E. coli* Bsu^+. This is because most phage, having a majority of one kind of fiber and therefore two or fewer of the other type, cannot infect the type of cell requiring the minority fiber for attachment. Both types of fibers can obviously function in a complementary fashion to infect cells such as *E. coli* Bsu^+, which have both types of receptors. If the phage are first preadsorbed to *E. coli* B (and then shaken with $CHCl_3$ to kill infected cells), irreversible adsorption is complete and no viable phage are left to infect either strain of resistant bacteria (Table 2). Phage preadsorption to either resistant type (B/4 or B/6) permits infection by all phage of the appropriate type (i.e., those with sufficient T6 or T4 tail fibers, respectively) but leaves all of the other type reversibly adsorbed or unadsorbed. Therefore, the phage which can infect -/4 or -/6 bacteria are of two different, mutually exclusive classes. Since they add up to only 74% of the viable phage particles, there is also a class (25%) which infects neither type. This class is demonstrated explicitly by preadsorption with *E. coli* B/4 plus B/6 (Table 2), after which a class (19%) is left which can infect normal bacteria. This class must be composed mostly of phage with equal numbers of T4 and T6 tail fibers. As one example of a model that would explain these results, we can postulate a random distribution of T4 or T6 tail fibers on all six vertices of each phage. The respective fractions of phage with zero, one, two, or three fibers of a specific type would be (according to a binomial distribution) 0.016, 0.094, 0.234, and 0.312. If we assume that at least three fibers of one type are

TABLE 2. Infectivity of T4 phage having both T4 and T6 tail fibers

Preadsorption on *E. coli*:	Infectivity (%) for *E. coli*:		
	Bsu^-	$Bsu^-/4$	$Bsu^-/6$
None	100	34	40
B	2	<1	<1
B/4	60	1	35
B/6	55	35	2
B/4 + B/6	19	<1	<1

required and the arrangements:

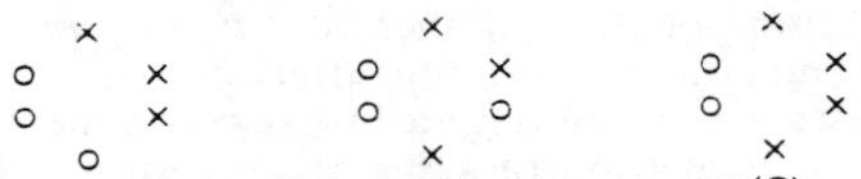

```
   ×         ×         ×
o     ×   o     ×   o     ×
o     ×   o     o   o     ×
   o         ×         ×
  (A)       (B)       (C)
```

are not infective, then 26% would infect neither -/4 nor -/6 bacteria. Thermolabile baseplate *h* mutants show no significant noninfection class when titers on -/4 and -/6 are summed. Half will infect -/4 and half -/6, and the sums range from 85 to 105%. Thus, the more easily triggered baseplates require fewer simultaneously cooperating tail fibers to trigger infection than do wild-type baseplates. If, however, arrangement A (see above) was still uninfective, there should still be a 10% uninfective class, which the data are probably not accurate enough to detect. An experiment with T4 *h* phage similar to the one shown in Table 2 (B/4 plus B/6) might have settled this question.

Roles of Baseplate Gene Products in Phage T4 Infection

After the morphological analysis of baseplate structure by image enhancement of electron microscope photomicrographs, Crowther (1980) attempted to reexamine the role of gene 9 in the mechanism of baseplate triggering. In the absence of gp9, tail fibers are not attached to the T4 baseplate. Thus, it has been generally supposed that tail fibers attach to the baseplate via gp9. This supposition was reinforced by the finding that gp9 stabilizes the baseplate both to triggering on contact with its host and to abortive triggering of isolated phage particles in low-ionic-strength media. If gp9 prevents abortive triggering, then its interaction with tail fibers might well relieve this stabilization at the appropriate moment. Crowther was able to show that particles without gp9 (and without tail fibers) can infect *E. coli* Bb as well as some -/4 strains, although adsorption is probably slow. He also isolated phage mutants (*pfp* [permit fiberless plating]) which, when combined with gene 9 mutations, form turbid plaques on *E. coli* Bb with almost normal efficiency. *pfp* mutations were found in different baseplate genes (e.g., genes 6, 7, and 27) and made the $gp9^-$ phage even more unstable in low-Mg^{2+} medium. The T4 9(am) *pfp* particles, though of lower stability than normal fibered particles, are infective and may resemble a more primitive form of T-even phage. As suggested by Crowther (1980) these mutants could be quite useful in identifying the host receptor for gp12, that is, for the short baseplate fibrils which attach the baseplate to the cell surface irreversibly. To my knowledge this has yet to be done.

Various laboratories have investigated the role of specific baseplate proteins in the infection process. We can divide the baseplate proteins into two categories: those required for baseplate morphogenesis and those that can be added to an otherwise complete particle (i.e., gp9, gp11, and gp12). I have already discussed the role of gp9 (see above). The product of gene 12 forms a fibrous trimer which comprises the short baseplate fibril (Kells and Haselkorn, 1974; Simon and Anderson, 1967b). This is seen on thin sections of phage-infected bacteria, connecting the baseplate to the cell surface. The interaction between gp12 and its receptor may well represent the main killing function of T-even phage particle adsorption, since $gp12^-$ phage do not kill cells (King, 1968), and isolated gp12 can kill cells under certain conditions (Kells and Haselkorn, 1974). The difference between ghost and phage killing may depend to some extent on the ability of the injected phage DNA to repair a colicin-like action of the gp12 organelle. The composition, structure, and function of gp12 is discussed further in Kozloff's chapter in this volume.

It was thought that the protrusions from the bottom of the baseplate of phage particles were gp12, which elongated on attachment. Crowther et al. (1977) showed that, in fact, gp12 was positioned as a hoop girding the middle of these protrusions, which themselves were composed in part of gp11 and possibly of gp10 as well. This arrangement may explain the greater lability of $gp12^-$ particles (Benz and Goldberg, 1973; Crowther et al., 1977; King, 1968; Simon et al., 1970). The baseplates of $gp12^-$ particles are triggered after adsorption to the cell surface, but the particles do not bind irreversibly, and usually the heads of the released contracted particles are still full of DNA. Since such particles can infect spheroplasts, the contracted form may well resemble a normal intermediate of phage infection (Benz and Goldberg, 1973). Particles that lack gp11 also lack gp12 (since gp12 normally binds to gp11 and, in addition, gene 12 expression may be polar to gene 11). It is also likely that the putative baseplate recognition site on the cell surface may be recognized by gp11, since particles possessing gp11 and lacking gp12 are triggered but cannot attach irreversibly, whereas those lacking gp11 attach reversibly to the host cell (King, 1968; Simon et al., 1970) but do not contract or irreversibly attach.

During tail morphogenesis, six wedges array themselves about a central hub (Berget and King, this volume). The inner tail tube (also known as the needle or core) polymerizes onto the baseplate only after gp54 and gp48 have been added to the otherwise completed baseplate. When the baseplate expands during infection from the hexagon to the star form, an opening is made in the center of the baseplate through which the tail tube extrudes. Although it is not certain whether the hub rearranges to form the opening, it seems likely to me that at least part of the hub must be at the distal tip of the tail tube even after the contraction stage. This would mean that the hub portion pushes through the murein layer and makes contact with the cytoplasmic membrane. For many years it was thought that the product of gene *e* (the major phage lysozyme gene) attached to the tip of the tail tube and was required to penetrate the murein layer. Emrich and Streisinger (1968) showed that although lysozyme activity (from gp*e*) is bound to phage from normal lysates, phage from lysates of gene *e* mutants, which have no such associated activity, infect their host just as well. They still found about 1 to 3% of the usual lysozyme activity associated with these phage particles, but it was not inhibited by antibody to gp*e*. In this regard it is interesting to note that a hub protein, gp5, was recently shown by Kao and McClain (1980a, 1980b) to have lysozyme activity. For further discussion on the activities of hub proteins, see the chapter by Kozloff in this volume.

Dawes and Goldberg (1973a) attempted to establish the stages at which various baseplate proteins are required during infection. They compared infection by phage with that by urea-treated particles (where the interaction is between contracted particles and bacterial spheroplasts whose cytoplasmic membrane is already exposed). They showed that when thermolabile baseplate mutants with mutations in genes 5, 6, 7, and 8 were used, high temperature inactivated infection by urea-treated phage to the same extent as it inactivated normal phage infection. On the other hand, a gene 10 mutant treated with urea was as active after heat treatment as before. Since urea-treated particles are contracted, they presumably initiate infection by attachment of the tail fiber tip to the cytoplasmic membrane. Thus, it was concluded that gp10 is involved only in the early stages of T4 adsorption, whereas the other genes may be involved at later stages (as well as at earlier stages). The complex interrelationship among all of the proteins in the baseplate that leads to baseplate expansion should make us wary of drawing firm conclusions about the particular protein affected in such experiments. Nevertheless, since gp10 is in the outer wedge, it seems reasonable that it is not involved at the level of interaction between the tail tube tip and the membrane. This also seems to be the case for T4 dihydrofolate reductase (FRD), an enzyme shown to be a baseplate component and, more specifically, a wedge component (Mosher and Mathews, 1979). T4 phage are inactivated by antiserum against T4 FRD. (For a more complete discussion on the presence and role of FRD, see the chapter by Kozloff in this volume.) Dawes and Goldberg (1973a) showed that after contraction (by either urea treatment or reversible adsorption of particles lacking gp12), infection of contracted particles was no longer sensitive to anti-T4 FRD. The anti-T4 FRD antibody bound to the $gp12^-$ particles but not to urea-treated phage. (Urea treatment seems to remove FRD from the particle.) A contracted particle lacking FRD, or with FRD complexed to anti-FRD on its elevated wedge component, infects normally. Therefore, Dawes and Goldberg concluded that FRD is not required in the later stages of infection.

Dihydropteroyl hexaglutamate is another phage baseplate component. Kozloff (see his chapter) discovered that T4 phage can be inactivated by conjugase, a gamma-glutamyl hydrolase. Dawes and Goldberg (1973a) showed that infection by contracted phage particles was also blocked by conjugase and concluded that dihydropteroyl hexaglutamate might be required in the later stages of the infection process. However, the action of conjugase on T4 and urea-treated T4 particles is reversible by urea and, in some cases (T4 ac_{41} phage particles), by trypsin. Thus, the inhibition of infection by conjugase (and by anti-T4 FRD) might be due to steric hindrance of phage attachment and not to specific inactivation or impediment of the specific molecule to which the conjugase binds.

DNA INJECTION

Classically, adsorbed phage were thought to "contract" and thereby "inject" their DNA into the cell cytoplasm. The analogy was made to a hypodermic syringe. Subsequent work over the last 30 years has refined our view so that we now understand that the "contraction" of the tail sheath is not a muscle-like phenomenon (i.e., a reversible alteration in tail sheath length by actin-like and myosin-like components), but rather the relaxation of a metastable state. The tail sheath goes from a long, narrow cylinder, covering the inner tail tube, to a wider and shorter cylinder only partially covering the tail tube. This contraction of the sheath is triggered by expansion of the baseplate to which it is attached at the distal end (Moody, 1973). Injection of DNA is probably not directly into the cytoplasm, since the tail tube tip seems only to reach the outer surface of the cytoplasmic membrane after tail sheath contraction (Simon and Anderson, 1967a). In fact, injection, in the hypodermic syringe analogy, implies that an external force reduces the head volume, thereby forcing out the DNA. There is no evidence to support this hypothesis. Thus, the terms DNA "ejection" and "uptake" have become more common in describing the exit of DNA from the phage particle and the entry of DNA into the cell cytoplasm, respectively. If these two processes are separate, we can ask how the end of the DNA starts to enter the membrane. Is there a pore? If so, how specific is it? Does the phage or cell or both provide the components for this specificity? Is energy required by the cell for uptake of DNA? If so, can this requirement sometimes be filled by the phage, which seems to have enough stored energy or entropy (Zarybnicky, 1969) to permit DNA ejection in vitro? What signals the phage to release its DNA, and how are these signals modulated by the cell's physiological state? These questions have been addressed more generally in my recent article (Goldberg, 1980), so I will limit myself here to phage T4, with the caveat that other phages may solve some of these problems in a different manner.

Specificity of the DNA Pore

If contact of the tail tube tip with a specific site on the cytoplasmic membrane surface is the trigger for phage DNA release, we can postulate that the cell surface site becomes an extension of the tail tube, which then becomes an elongated pore, or a pipeline into the cell cytoplasm. Thus, during entry, DNA would not be subject to periplasmic nucleases. Other phages, such as T5, are not as efficient as T4; a fraction of the phage eject abortively, and the released DNA is degraded by periplasmic enzymes (Labedan and Legault-Demare, 1974). The efficiency of infection may well depend upon the quality of interaction between the point where DNA exits from the phage and the point where it enters the cytoplasm. The sophisticated adsorption system of T4 may have developed, in part, to ensure the quality of this interaction, to position DNA for direct entry and prevent untoward effects of periplasmic components.

The extensive host range of the tail tube tip-membrane interaction is illustrated by the ability of urea-treated T4 phage to infect spheroplasts of *Aerobacter*, *Salmonella*, *Proteus*, and *Serratia* species, in addition to *E. coli* (Wais and Goldberg, 1969). Normal T4 does not infect or even attach to these other bacterial species. Therefore, it would seem that we are seeing a more primitive interaction. Neither *Bacillus* species nor the gram-negative *Pseudomonas* species could be infected by urea-treated T4 phage, suggesting that membranes of enteric gram-negative bacteria have a common receptor site with which the T4 tail tube tip interacts. This commonality implies either that the site is ubiquitous and nonspecific (e.g., a phospholipid layer specific to enteric gram-negative bacteria) or that it is some common important protein which has not diverged much since the time that the outer surfaces of enteric bacteria speciated. However, since it is not known why urea-treated T4 will not grow on *Bacillus* or *Pseudomonas* species, we must still question whether events after penetration may simply be inadequate to produce T4 progeny in these bacteria.

Pilot Protein

Kornberg (1974) postulated a pilot protein for phage infection whose function is to provide specificity and possibly structural help in transferring the phage DNA from the virion into the host cell cytoplasm and to aid in initiation of DNA replication. This concept was defined most fully in phage ϕX174 (Jazwinsky et al., 1975). In T4 the pilot protein would be a head protein which is attached to one or both DNA termini and resides at the vertex of the head to which the tail joined. It would pass through this junction as well as through the tail tube and the part of the baseplate hub which may form the distal tip as an extension of the tail tube. It is this distal tip which probably interacts with a cytoplasmic membrane protein or complex (the DNA pore) to permit transport of DNA with the pilot protein into the cell cytoplasm. I have suggested that gp2, alone or in a protein complex, may serve as the T4 pilot protein.

In restrictive (su^-) hosts, gene 2 *am* mutants can produce normal-appearing progeny which in turn attach normally to *E. coli* and kill the bacteria (Granboulan et al., 1971; Silverstein and Goldberg, 1976). DNA from such particles (designated $2^-.su^-$) enters the cytoplasm and is degraded to acid-soluble fragments. This same phenotype is found for $64^-.su^-$ particles. Both $2^-.su^-$ and $64^-.su^-$ particles can infect and grow in *E. coli su*$^-$ *recBC* mutants. The lack of exonuclease V (the product of the *recBC* locus) in the host suppresses the gene 2 and gene 64 mutations because the entering DNA is no longer degraded. Moreover, Oliver and Goldberg (1977) showed that giant $2^-.su^-$ phage, with a chromosome about three genomes long, can grow even in su^- bacteria. In this case, although exonuclease V starts to degrade the entering T4 chromosome exonucleolytically, a complete T4 genome still remains by the time exonuclease V action is stopped (probably by an anti-exonuclease V activity coded for by the phage [Behme et al., 1976]). Thus, it seems that gp2 and gp64 protect the termini of parental DNA from exonucleolytic degradation.

Granboulan et al. (1971) showed that $2^-.su^-$ particles are more unstable and lose their DNA more readily than do wild-type or $64^-.su^-$ particles. On the other hand, $64^-.su^-$ particles are more likely to display tails attached at an incorrect angle to the head and a more fragile head-tail junction than are $2^-.su^-$ or wild-type particles. Furthermore, Granboulan et al. showed that empty heads cannot attach to tails. These facts fit the idea that a protein or complex attached to the tip of the DNA at an apical vertex helps to define the site of, and participates in, tail attachment. More recently in my laboratory, B. Lipinska (unpublished data) has sequenced the DNA corresponding to the gene 2-64 region and has found only one open reading frame, for a single 25,000-dalton protein which contains 20% lysine and arginine and no cysteine. Such a protein should have good ability to bind DNA. Furthermore, contrary to expectation this open reading frame is in the counterclockwise direction (as is, we find, the only open reading frame in the neighboring gene 3 region). It has yet to be established whether it is expressed early or late. The published and unpublished data for genes 2 and 64 do not indicate that they are in different complementation groups, and, in fact, they now seem to be segments of a single peptide. Further work is required to separate and identify this elusive and still putative protein and to determine whether there is any processing during phage development. D. Hamilton (unpublished data), in my laboratory, showed that $gp64^-$ phage heads can be complemented in vitro with a $gp64^+$ extract and subsequent addition of tails.

While much work suggests a role for gp2/64 in phage assembly, DNA ejection, and transport, there is as yet no direct demonstration that gp2/64 is a head protein. It is still possible that this protein or complex alters DNA or another protein to give the required phenotype. Furthermore, if gp2/64 were required for assembly and infection, how would a $2^-.su^-$ or $64^-.su^-$ particle succeed in infecting a bacterium? Though it is possible to postulate truncated *am* peptide fragments which serve a pilot function but no longer prevent exonuclease V digestion, the actual explanations are still unclear.

DNA Ejection

In vitro complementation experiments suggest that the total T4 genome is packaged in the phage head before addition of the tail. The release of DNA from the phage particle must therefore require an "unplugging" at the head-tail junction. Since this junction is about 100 nm away from the baseplate from which the signal must come to release DNA for injection, it is possible that contact with the membrane receptor (or pore) induces a change in the tail tube tip which is propagated up the tube and culminates in unplugging. An alternative hypothesis is that a long protein such as gp48 (see the chapter by Berget and King in this volume) reaches from the tail tube tip to the head-tail junction. I prefer yet a third possibility: that the DNA enters the tail tube at the time of tail attachment to the head. With this scheme, no long-distance signal transfer is required at the time of adsorption. This implies that the tail of the completed virion has DNA inside and that the pilot protein complex becomes part of the baseplate. This would also mean that unplugging occurs twice, once at the head-tail junction during morphogenesis and once at the tail tube tip at the time of infection. There is some support for these ideas from other systems. During morphogenesis of phages T5 and λ, the DNA does enter the tail partially (Saigo, 1975), and in phage T5, not only does the DNA unplugging occur at the level of both the head-tail junction and the tail-membrane junction, but DNA entry is stopped after 8% of the DNA has been taken up. This segment of DNA must be expressed before the DNA penetration apparatus is unplugged once more to permit the rest of the DNA to enter the cell.

When phages are mixed with LPS preparations, often the phage is contracted and DNA is released from the head. When the LPS is extracted repeatedly with alcohol, DNA release is prevented and may be restored by the addition of specific fatty acids (Jesaitis and Goebel, 1953). Isolated LPS, however, is very inefficient in binding to phage, and high concentrations of LPS (or cell debris) are needed to inactivate phage (Wilson et al., 1970). Most likely this is why T4 phage are not inactivated to any great extent by cell debris and free LPS in a bacterial culture. Contracted particles can also be produced by treatment with urea (van Arkel et al., 1961) or by adsorption of 12^-.*su*$^-$ particles to *E. coli* (Simon et al., 1970). The 12^-.*su*$^-$ particles adsorb, contract, and are then released from the cell surface (King, 1968; Simon et al., 1970). Contracted particles (either 12^-.*su*$^-$ or by urea) cannot infect normal bacteria, but about 0.2% of them can infect spheroplasts of *E. coli* (Wais and Goldberg, 1969). I do not understand why such a small fraction of the contracted particles is infective on spheroplasts. Are most of them potentially infective but only a small fraction of the particle-spheroplast interactions lead to a productive infection? Or is but a small fraction potentially infective in the first place, and if so, might they be not contracted at all but modified in some other way, not readily apparent in the electron microscope?

Ejection of DNA from contracted particles can be induced by addition of phosphatidylglycerol but not phosphatidylethanolamine (Baumann et al., 1970; Benz and Goldberg, 1973). Thus, it is possible to cause ejection of DNA in vitro with a specific phospholipid found in the membrane. This may have some physiological significance since contracted particles are normally found attached to the host cell surface. Nevertheless, it is unlikely that phosphatidylglycerol is the exclusive receptor component. If that were the case, there would be little or no specificity for a membrane interaction site, and all contraction which normally leads to contact between the tail tube tip and the cytoplasmic membrane surface should lead to DNA ejection. The presence of many contracted phage particles with full heads on superinfection shows that this is not the case. When T4 phage superinfect bacteria which have been previously infected with T4, they attach and contract normally, but only about half inject their DNA (Anderson and Eigner, 1971). (The photograph in Fig. 6 of the article by Simon and Anderson [1967a], showing few empty heads among many contracted particles after primary infection at a high multiplicity of infection, may really represent superinfection due to slow adsorption of most of the phage.) We can infer from this in vitro and in vivo evidence that contraction per se does not necessarily lead to DNA ejection.

The superinfecting phage which eject their DNA do so abortively into the periplasm (superinfection exclusion), where it is degraded (superinfection breakdown) by endonuclease I (Anderson and Eigner, 1971). Superinfection exclusion in T4 depends on the expression of a phage-coded gene called immunity (*imm*), which produces a protein (gp*imm*) within 1 to 2 min after infection (Vallée and Cornett, 1972). Thus, even release of DNA from the phage after a seemingly normal adsorption and contraction does not inevitably lead to its uptake into the cell cytoplasm. Alteration of the cytoplasmic membrane by a specific protein from within can abort this process. Not much has been done to investigate this phenomenon since the untimely death of M. Vallée, but it would certainly be of great interest to clone the *imm* gene and see whether its product is lethal and with what membrane components it interacts.

It should be clear by now that although contraction, DNA ejection, and DNA uptake normally follow hard upon one another in the infection process, each of these stages can be separated from the next.

Energy Requirement for Initiation of DNA Transport

Recently it was found that the primary infecting T4 DNA transport across the cytoplasmic membrane can be regulated by the environment (or cellular physiology) (for recent reviews, see Grinius, 1982, and Labedan and Goldberg, 1982). Some energy poisons, such as cyanide (a respiratory inhibitor), azide, and 2,4-dinitrophenol and carbonyl cyanide-*m*-chlorophenyl hydrazone (protein ionophore uncouplers) reduce the chemiosmotic gradient and prevent phage DNA uptake (Kalasauskaite and Grinius, 1979; Labedan and Goldberg, 1979). Reduction of phosphate potential, on the other hand, does not inhibit T4 DNA penetration (Kalasauskaite et al., 1980). More specifically, the membrane potential and not the pH gradient is required for DNA uptake (Labedan and Goldberg, 1979). DNA uptake (as measured by the acid-soluble fraction of infecting 2^-.*su*$^-$ phage DNA) is prevented by colicin K or by a high potassium ion concentration in the presence of its ionophore valinomycin. Both of these treatments collapse the membrane potential but leave the pH gradient intact. At pH 8 or in the presence of nigericin, which collapses the pH gradient while leaving the membrane potential intact, DNA uptake is normal. When the membrane potential is adjusted by using varying concentrations of K^+ in the presence of valinomycin, there is a threshold value (ca. 60 mV) below which DNA no longer enters the cytoplasm and above which DNA enters normally (Labedan et al., 1980). This suggests a "gating" phenomenon which would imply that initiation of DNA penetration through the membrane pore or channel is regulated by the potential across the membrane. The conformation of the pore proteins, and therefore their interaction with the tip of the DNA (and its pilot protein) or with the tail tube tip, would be voltage dependent. Thus, reducing the voltage across the membrane would prevent DNA exit from the phage or permit abortive ejection into the periplasm. This phenomenon is quite reminiscent of the immunity function discussed above, which might also act by altering the DNA pore.

N. Suresh showed in our lab (unpublished data) that addition of CCCP to infected cells does not prevent the usual exclusion and breakdown of superinfecting T4 phage DNA. If both gp*imm* and CCCP act to "close" the DNA pore, why is the excluded DNA degraded during superinfection but not during primary infection in the presence of CCCP? Maybe the alteration by CCCP is different and prevents not only DNA transport (into the cytoplasm) but also prevents DNA ejection (from the head into the periplasm). Alternatively, gp*imm* may not only prevent DNA transport but also may activate or decompartmentalize the periplasmic endonuclease I so that periplasmic DNA can be degraded. We have not yet been able to determine definitively whether the DNA of primary infecting phage, in the absence of $\Delta\psi$, is ejected into the periplasm or left in the phage head.

It is of interest that both Labedan and Letellier (1981) and Kalasauskaite et al. (1983) have shown that T4 infection causes a transient depolarization of the host cell membrane. This effect is also seen in the presence of chloramphenicol and when T4 ghosts are used. It is not yet clear what causes the depolarization, nor what signals the recovery. It will be of interest to see what physiological role this phenomenon plays or what process it reflects during T4 phage infection.

Another, still unsolved, problem is whether the nucleotide-pair-by-nucleotide-pair transport of DNA across the membrane also requires energy, and if so, in what form. Though there is enough energy or entropy in the phage head to eject in vitro, cells can also take up DNA in the absence of this force during transfection. It seems to me that the phage-host interaction at this level, as at others, will prove to have a series of transport control reactions which can regulate its ability to transport DNA. It is also likely that different phages may regulate this process in different manners, and to some extent, the mode of regulation of ejection and initiation of transport (i.e., finding and entering the DNA pore) may determine to a great extent the efficiency of plating of different phages.

This work was supported in part by grants from the National Institute of General Medical Sciences and the National Science Foundation.

Effects on Host Genome Structure and Expression

D. PETER SNUSTAD,[1] LARRY SNYDER,[2] AND ELIZABETH KUTTER[3]

Department of Genetics and Cell Biology, University of Minnesota, St. Paul, Minnesota 55108[1]; Department of Microbiology and Public Health, Michigan State University, East Lansing, Michigan 48824[2]; and The Evergreen State College, Olympia, Washington 98505[3]

At the time of infection of an *Escherichia coli* cell with a T4 bacteriophage, that cell is presented with two sets of genetic instructions dictating two completely different pathways of metabolic activity. In the commonly studied strains of *E. coli* grown under standard laboratory conditions, such T4-infected cells respond invariably to the dictations of the viral genophore. Within 2 to 5 min after infection, host DNA synthesis (excluding some repair synthesis), RNA synthesis, and protein synthesis virtually terminate. Subsequent macromolecular synthesis in the infected cell is almost entirely directed by the phage.

Unfortunately, we still do not know all of the mechanisms by which host macromolecular synthesis is so rapidly and substantially arrested. We do know that the heterocatalytic expression of the host chromosome is specifically inhibited at both the transcription and the translation levels.

In this chapter, we focus on the T4-induced modifications of the host chromosome and arrest of transcription from cytosine-containing DNA. Modifications of the translational machinery of the host cell and their possible roles in the shutoff of host translation are discussed by Wiberg and Karam (this volume).

HISTORICAL BACKGROUND

The first studies of the effects of T-even phage infection on host cell metabolism were carried out in the mid- to late 1940s by S. Cohen and colleagues (see Cohen [1968] for a detailed account of these pioneering studies). Some 35 years later, we are still trying to elucidate these complex interactions between the T-even phages and the metabolic machinery of infected *E. coli* cells.

Shutoff of Host Gene Expression

Cohen and Anderson (1946) demonstrated that the rate of respiration (and, thus, the overall metabolic rate) remains constant in cells infected with T2 phage. In a classic paper on macromolecular synthesis in T2-infected cells, Cohen (1947) showed that (i) there is a brief pause in DNA synthesis shortly after infection, but thereafter DNA synthesis resumes and proceeds at 5 to 10 times the preinfection rate; (ii) net accumulation of RNA ceases shortly after infection; and (iii) protein synthesis continues at the preinfection rate.

Host protein synthesis. Monod and Wollman (1947) showed that the *E. coli* enzyme β-galactosidase cannot be induced in T-even-phage-infected cells. Subsequent studies have demonstrated that T-even phage infection shuts off the synthesis of all host enzymes that have been so examined (Cohen, 1968). These results all suggested that host protein synthesis stops shortly after infection and that subsequent synthesis is phage specific. Hershey et al. (1954) showed that most of the proteins synthesized later than 10 min postinfection were phage structural proteins. The nature of certain of the proteins synthesized at earlier times after infection was first established by the discovery of new enzyme activities in T2 phage-infected cells (Flaks and Cohen, 1957). One of these enzymes, deoxycytidylate hydroxymethylase (HMase), is not present in uninfected *E. coli* cells. The other, thymidylate synthetase, is present in uninfected *E. coli*, but its level increases sharply after T-even phage infection due to expression of the T4 *td* gene.

The use of polyacrylamide gel electrophoresis to resolve pulse-labeled proteins synthesized in uninfected *E. coli* cells and in cells at various times after infection with phage T4 confirmed that the vast majority of the host proteins cease to be synthesized within 2 to 3 min after infection (Levinthal et al., 1967).

The shutoff of the synthesis of certain outer-membrane host proteins is apparently slower than the shutoff of the synthesis of other host proteins (Beckey et al., 1974; Huang, 1975; Pollock and Duckworth, 1973). This may be explained in part by these proteins being synthesized from mRNAs with longer-than-average half-lives (Hirashima et al., 1973; Lee and Inouye, 1974). However, this does not explain why their synthesis is not shut off at the translation level.

Host transcription. Whereas Cohen (1947) showed that there was no net accumulation of RNA in T-even-phage-infected cells, ^{32}P-labeling experiments indicated that at least some RNA synthesis occurred in infected cells (Cohen, 1948). The key to understanding the nature of the RNA synthesized in T-even-phage-infected cells was provided by the results of Volkin and Astrachan (1956), in studies which played a major role in the discovery of mRNA. They showed that the RNA synthesized in T2 phage-infected cells has (i) a high turnover rate and (ii) the nucleotide composition characteristic of T2 phage DNA, rather than that of *E. coli* DNA.

Nomura et al. (1960) and Hall and Spiegelman (1961) subsequently demonstrated that the labeled RNA synthesized in T-even-phage-infected cells is distinct from the unlabeled RNA present in cells before infection and is homologous to the phage DNA. They showed that the labeled RNA synthesized after infection hybridizes with denatured phage DNA, but not with denatured *E. coli* DNA. DNA-RNA hybridization experiments have shown that host RNA synthesis continues at decreasing rates for about the first 2 to 5 min after infection at 30 to 37°C (Fig. 1) (Adesnik and Levinthal, 1970; Kennell, 1968; Kennell, 1970; Landy

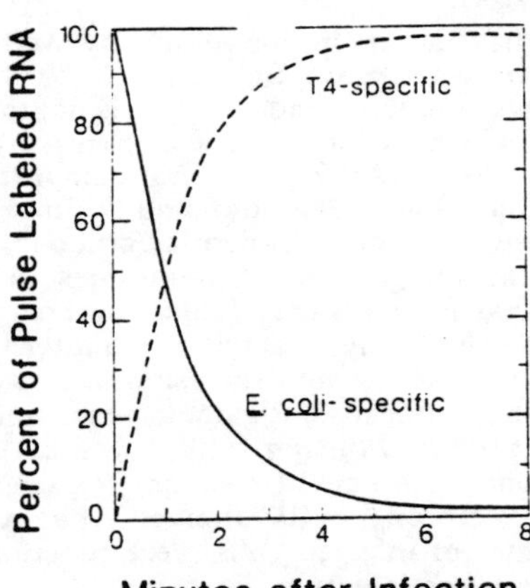

FIG. 1. Kinetics of switchover from *E. coli*-specific RNA synthesis to phage T4-specific RNA synthesis after infection of *E. coli* B/5 by T4D⁻ at 30°C. Based on the data of Snustad and Bursch (1977) and Tigges et al. (1977).

and Spiegelman, 1968; Nomura et al., 1966). However, it should be noted that Sirotkin et al. (1977) found that substantial amounts of RNA pulse-labeled from 14 to 16 min after infection of *E. coli* B834 with wild-type T4 hybridized with *E. coli* DNA. Thus, there may be a resurgence of *E. coli* transcription late in infection under some conditions.

Most of the attempts to determine the mechanism of the T4-induced shutoff of host transcription have focused on the phage-induced modifications of the host RNA polymerase. These studies are discussed below and in the chapter by Rabussay in this volume.

Effects of ghosts. The inhibition of bacterial metabolism by T-even phage ghosts (phage coats from which the DNA has been released by osmotic shock), reviewed in detail by Duckworth (1970a), has been a confounding factor in attempts to elucidate the mechanism by which T-even phage shut off host macromolecular synthesis. Because little information has been added since Duckworth's review, the specifics of these experiments with ghosts will not be discussed here. There appear to be three major differences between the effects of T-even ghosts and intact T-even phage on host metabolism. (i) Ghosts appear to inhibit host macromolecular synthesis almost immediately after attachment (Duckworth, 1970b; French and Siminovitch, 1955; Herriott and Barlow, 1957; Lehman and Herriott, 1958), whereas host macromolecules continue to be synthesized, albeit at declining rates, for 2 to 5 min after infection with intact phage (Hayward and Green, 1965; Hosoda and Levinthal, 1968; Kennell, 1968; Kennell, 1970; Landy and Spiegelman, 1968; Scofield et al., 1974). (ii) The complete arrest of host macromolecular synthesis requires protein synthesis after phage infection, but not after ghost attachment (Cohen and Ennis, 1965; Duckworth, 1971; Goldman and Lodish, 1973; Nomura, Okamoto, and Asanor, 1962; Nomura et al., 1966). (iii) The arrest of host macromolecular synthesis is reversible in a portion of the ghost-infected cells, particularly when incubated in minimal medium for extended periods of time (Duckworth and Bessman, 1965; French and Siminovitch, 1955).

Host translation. In addition to the shutoff of the heterocatalytic expression of the host chromosome at the transcription level, phage T4 is known to rapidly arrest the translation of host and phage lambda mRNAs (Hattman and Hofschneider, 1967; Kennell, 1968; Kennell, 1970), even when the cessation of host transcription is delayed (Pearson and Snyder, 1980; Sirotkin et al., 1977; Tigges et al., 1977). The T4-induced alterations of the translational machinery of the host cell and their possible roles in host translation shutoff are discussed by Wiberg and Karam (this volume).

Replication of host DNA. A recurrent problem in measuring host DNA synthesis is the difficulty in distinguishing between host and phage DNA replication. This differentiation was aided by the discovery by Wyatt and Cohen (1952) that T-even phage DNA contained 5-hydroxymethylcytosine (HMC) instead of the cytosine present in *E. coli* DNA. Using this biochemical distinction, Hershey et al. (1953) concluded that most of the DNA synthesized in T2-infected cells is phage DNA. Subsequent studies suggested a more complex picture, however, at least for the DNA synthesized during the first 10 min after infection.

The DNA synthesis occurring after T-even phage infection clearly includes three distinct components: (i) host DNA replication continuing at a declining rate during the first 5 min after infection (Duckworth, 1971; Scofield et al., 1974; Snustad, Bursch, Parson, and Hefeneider, 1976); (ii) phage DNA replication starting at 3 to 4 min after infection (Carlson, 1973; Miller et al., 1970); and (iii) host and phage DNA repair synthesis continuing for at least the first 10 min after infection (Scofield et al., 1974).

The kinetics of the shutoff of replication of the host chromosome after T-even phage infection are still not very clearly established. There are two major confounding factors. First, T-even phages code for their own thymidylate synthetase. This decreases the rate of incorporation of exogenous thymidine after infection, a factor not considered in the early experiments. Second, in most of the studies, the net accumulation of acid-insoluble radioactivity has been measured by using labeled thymidine as a precursor. Because of the subsequent degradation of nascent host DNA, these experiments underestimate the amount of host DNA synthesis.

Scofield et al. (1974) measured DNA synthesis in a thymidylate synthetase-deficient host infected with a phage T4 double mutant deficient in thymidylate synthetase and in host DNA degradation. They reported that host DNA synthesis occurred at 40 to 80% of the preinfection rate as late as 10 min after infection. Their results indicated that initially this was primarily replicative synthesis, but that at later times repair synthesis predominated. On the other hand, Miller and Kozinski (1970) detected no host DNA in the DNA pulse-labeled with [^{3}H]thymidine from 3 to 3.5 min after infection. Snustad, Bursch, Parson, and Hefeneider (1976) used host DNA degradation-deficient, phage DNA synthesis-deficient T4 multiple mutants in conjunction with 5-fluorodeoxyuridine to inhibit thymidylate synthetase activity to examine the shutoff of host DNA synthesis. Their results indicated that under these conditions, host DNA synthesis continues at decreasing rates for the first 3 or 4 min after infection. Thereafter, little host DNA synthesis was observed, as long as the infecting phage carried a functional *ndd* gene (see below).

For additional details of studies on the effects of T-even phage infection on host metabolism, see the earlier reviews by Cohen (1968), Duckworth (1970a), Koerner and Snustad (1979), Mathews (1971a, 1977), and Stent (1963).

Effects on Host Nucleoid Structure

Cytological studies by Luria and Human (1950) and Murray et al. (1950) showed that the nucleoids of cells infected with T-even phage undergo a gross rearrangement within 2 to 3 min after infection. During this process, called nuclear disruption, the host DNA moves from its largely central location into juxtaposition with the cell membrane. Nuclear disruption was shown to be induced by UV light-inactivated phage (Luria and Human, 1950), but not by phage ghosts (Bonifas and Kellenberger, 1955) or in the presence of chloramphenicol (Kellenberger et al., 1959).

Nuclear disruption was initially believed to be the cytological counterpart of an early step in host DNA degradation (Stent, 1963). However, Snustad et al. (1972) have shown that nuclear disruption occurs normally in cells infected with degradation-deficient T4 mutants. Furthermore, nuclear disruption-defective (*ndd*) T4 mutants have now been isolated and have been shown to induce degradation of host DNA with normal kinetics (Snustad et al., 1974) (see below).

The chromosomes of bacteria such as *E. coli* exist as compact structures consisting of numerous negatively supercoiled domains (Stonington and Pettijohn, 1971; Worcel and Burgi, 1972). Phage T4 has been shown to induce the "unfolding" of these so-called "folded genomes" within 5 min after infection (Tutas et al., 1974), a process under the control of the *unf* ("unfoldase")/*alc* gene, which is also involved in blocking transcription of cytosine-containing DNA (see below). Unfolding occurs independently of host DNA degradation and nuclear disruption (Sirotkin et al., 1977; Snustad, Tigges, Parson, et al., 1976) (see below).

Infection by phage T4 thus induces three major modifications of the host nucleoid: degradation of the host DNA, nuclear disruption, and unfolding of the nucleoid structure. Each of these modifications is effected by a different T4 gene or genes, and the three seem to act largely independently of one another.

DEGRADATION OF HOST DNA

The demonstration during the early 1950s that the host DNA is degraded after T-even phage infection raised two very interesting questions. How does a virus induce the degradation of the DNA of its host without committing suicide by degrading its own DNA? And, is the degradation of the host DNA responsible for the shutoff of host gene expression?

For T4, the answer to the first question is simple. The virus codes for (i) three enzymes that modify the DNA precursors so that the viral DNA contains HMC rather than the cytosine present in host DNA, and (ii) two nucleases that are specific for cytosine-containing DNA (see the chapters in this volume by Mathews and Allen and by Warner and Snustad).

The answer to the second question is no; the shutoff of host gene expression precedes extensive degradation of the host DNA. This was first demonstrated by Nomura, Matsubara, Okamoto, and Fujimura (1962), who used sucrose and CsCl density gradient analyses to study the integrity of chromosomes in infected cells. By that method, they found the host chromosomes to be physically intact at 5 min after infection, by which time host gene expression had stopped. In addition, the synthesis of β-galactosidase in recombinant cells produced during conjugation between T4-infected donor cells and T4-resistant, β-galactosidase-deficient recipient cells showed that the host chromosomes of infected cells were functionally intact at 5 min postinfection.

The first indication that components of host DNA were utilized in the synthesis of T-even progeny viruses was Cohen's (1948) demonstration that about one-third of the phosphorus in progeny phage was derived from the infected host cells. His results were soon verified and extended by others (Koch et al., 1952; Kozloff, 1953; Kozloff and Putnam, 1950; Weed and Cohen, 1951), clearly showing that the host DNA is degraded after infection and that the breakdown products (mononucleotides) are reutilized in the synthesis of progeny viral DNA.

Kinetics of Host DNA Degradation

Hershey et al. (1953) first examined the kinetics of host DNA degradation after T-even phage infection by using colorimetric assays to measure the amounts of HMC- and cytosine-containing DNAs. They found that the amount of cytosine-containing DNA began to decrease at about 10 min after infection. By 20 min after infection, two-thirds of the cytosine-containing DNA had been degraded. Their results have been verified and extended in several studies using radioactive labels to follow the fate of the host DNA.

To measure the conversion of labeled host DNA to acid-soluble degradation products directly, one must use a DNA synthesis-deficient mutant as the infecting virus. Otherwise, the host DNA degradation products are incorporated into progeny viral DNA so rapidly that they cannot be detected in the intracellular nucleotide pool (Fig. 2A).

The most complete study of the kinetics of host DNA degradation after phage T4 infection was that of Kutter and Wiberg (1968). They demonstrated that the degradation of host DNA to acid-soluble products occurred very rapidly between 7 and 20 min after infection and then continued at a much slower rate thereafter (Fig. 2A). Note, however, that about 10 to 15% of the host DNA remains undegraded (as fragments of about 10^6 daltons) at the time of cell lysis.

Kutter and Wiberg (1968) also examined the accumulation of single- and double-strand breaks in host DNA after T4 infection, by alkaline and neutral sucrose density gradient analyses. Neutral sucrose gradients of labeled host DNA from cells harvested at 6 min after infection at 37°C showed that the DNA occurred in a single peak containing fragments with an estimated average size at least 2×10^8 daltons (Fig. 2B). This size corresponds well with other estimates of the molecular weights of the largest DNA molecules

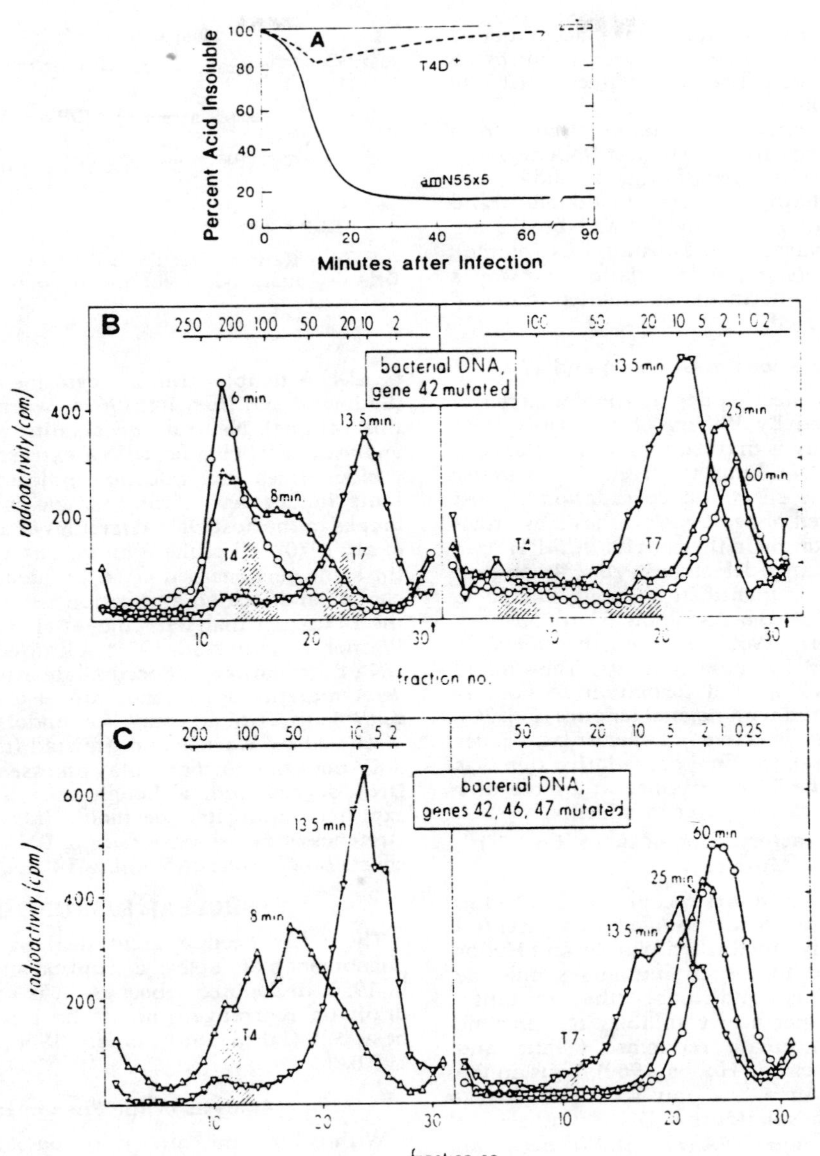

FIG. 2. Kinetics of host DNA degradation after T4 infection. (A) Conversion of prelabeled host DNA to acid-soluble form after infection of *E. coli* B with T4D⁺ or the DNA synthesis-deficient mutant *am*N55x5 (defective in gene 42 [dCMP HMase]) at 37°C. Based on the data of Kutter and Wiberg (1968). (B) Neutral sucrose density gradient profile of [^{3}H]deoxythymidylate-labeled *E. coli* B DNA at various stages during normal degradation (using an HMase amber mutation to block T4 DNA synthesis). The left panel is with a lysate from 5 × 10^6 cells per gradient; the right panel is with 10^7 cells at 13.5 min and 2.25 × 10^7 cells at 25 and 60 min. (C) Same as B, but infection was with an amber mutant defective in genes 42, 46, and 47; 5 × 10^6 cells per gradient were used throughout. From Kutter and Wiberg (1968).

obtained from bacterial chromosomes by these procedures with high-speed centrifugation. No further studies have been done since the development of the low-speed centrifugation procedures for measuring sizes of DNA molecules as large as intact *E. coli* chromosomes, i.e., about 2.5 × 10^9 daltons (Kavenoff, 1972; Levin and Hutchinson, 1973; Rubenstein and Leighton, 1974), so no information is available on the time of occurrence of the first nicks and breaks in the host DNA.

By 8 min after infection, a significant number of double-strand breaks were detected, with many fragments of about 5 × 10^7 daltons. The average size of the fragments decreased quite rapidly thereafter, to about 2 × 10^6 and 1 × 10^6 daltons at 25 and 60 min, respectively, after infection. Alkaline gradient analysis showed that single-strand breaks accumulated slightly earlier than double-strand breaks, but by 25 min postinfection these had all been converted to double-strand breaks.

Two important conclusions can be drawn from these results. First, an early step in host DNA degradation involves the introduction of single-strand breaks. These are subsequently converted to double-strand breaks, probably after the single-strand breaks are enlarged to gaps (Warner and Snustad, this volume). Second, the last step(s) in the degradation pathway is very fast, since no intermediates smaller than 10^6 daltons can be detected.

Effects of Mutations in Genes 46 and 47

The first mutants affecting degradation of bacterial DNA were identified by Wiberg (1966). Host DNA degradation normally is marked by the simultaneous synthesis of phage DNA. Thus, to screen known amber mutants for possible effects on degradation of host DNA, Wiberg blocked phage DNA synthesis by crossing in an amber mutation in the gene for dCMP HMase (gene 42). Only mutants defective in gene 46 or gene 47 or both were thus identified as being unable to degrade the host DNA to acid-soluble form.

Kutter and Wiberg (1968) analyzed the step(s) in degradation affected by these mutants. They found that the initial nicking and formation of double-strand breaks occurred with normal kinetics (Fig. 2C). However, in mutants defective in either or both genes (and in dCMP HMase), the final degradative step was blocked and all of the DNA remained at a molecular weight of about 10^6 daltons later in infection.

Isolation and Characterization of *denA* (EndoII) Mutants

The isolation of T4 mutants specifically blocked in host DNA degradation, with no effects on essential functions, was greatly aided when Warner and Hobbs (1969) demonstrated that such mutants should be hydroxyurea-sensitive conditional-lethal mutants, since hydroxyurea specifically inhibits the enzyme ribonucleoside diphosphate reductase (Sinha and Snustad, 1972; Warner and Hobbs, 1969). Thus, in the presence of hydroxyurea, the only source of deoxyribonucleoside triphosphates for DNA synthesis is from degradation of host DNA (Fig. 3). Warner et al. (1970) used this screening procedure in conjunction with the hydroxylamine mutagenesis procedure of Tessman (1968) to isolate the first phage T4 mutant with a defect in an early step in host DNA degradation.

Independently, Hercules et al. (1971) used a mutant enrichment scheme to isolate additional mutants with the same phenotype. They labeled host DNA with [^{3}H]thymidine at high specific activity, infected with phage T4 at a low multiplicity of infection, and then stored the progeny phage to allow inactivation of those containing [^{3}H]thymidine by radioactive decays. Progeny phage from cells infected with mutants deficient in host DNA degradation should be nonradioactive and thus not inactivated during storage.

Both sets of T4 mutants blocked in host DNA degradation were found to be deficient in the production of T4 endonuclease II (endoII) (Sadowski et al., 1971), an enzyme that catalyzes the formation of single-strand breaks in double-stranded cytosine-containing DNA (Sadowski and Hurwitz, 1969a; Warner and Snustad, this volume). Neutral and alkaline sucrose gradient analyses of labeled host DNA extracted from cells at various times after infection indicate that these mutants induce few, if any, single- or double-strand breaks in the host DNA (Hercules et al., 1971; Warner et al., 1970). All of these mutations appear to map in the same gene, named *denA* for "deficient in endonuclease" and located at position 136 kilobases (kb) on the T4 linkage map (Hercules et al., 1971; Ray, Sinha, Warner, and Snustad, 1972). A limited amount of host DNA degradation still occurs late after infection with *denA* mutants unless they are also defective in the *denB* gene, which codes for endoIV. However, as discussed by Warner and Snustad (this volume), endoIV does not normally play an essential role in host DNA degradation, although such a role might be expected from its specificity (for single-stranded stretches of cytosine-containing DNA) and its in vivo effects on cytosine-containing T4 DNA (see below).

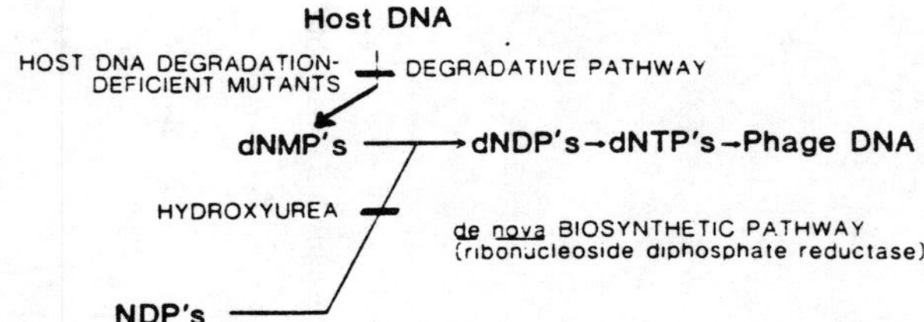

FIG. 3. Rationale for the hydroxyurea sensitivity of host DNA degradation-deficient mutants of phage T4.

NUCLEAR DISRUPTION

There has been a great deal of interest in the phenomenon of nuclear disruption since its discovery in 1950, undoubtedly because it is such a rapid and dramatic rearrangement of the morphology of the host cell (Luria and Human, 1950; Murray et al., 1950).

Analysis of the Phenomenon

Within 2 to 3 min after infection of *E. coli* with a T-even bacteriophage, the cytoplasm of the host cell essentially turns inside out, the DNA moving from the central region of the cell to the periphery and the ribosomes moving to the center of the cell (Fig. 4). Several procedures have been developed for observing nuclear disruption by phase-contrast and thin-section electron microscopy (Bonifas and Kellenberger, 1955; Epstein et al., 1963; Kellenberger et al., 1959; Snustad et al., 1972; Snustad and Conroy, 1974). The simplest of these merely involves spreading the cells on a thin layer of 17.5% gelatin and observing them by phase-contrast microscopy (Fig. 4) (Snustad and Conroy, 1974).

Nuclear Disruption-Defective (*ndd*) Mutants

All of the classical conditional-lethal mutants isolated by Epstein et al. (1963) were shown to induce nuclear disruption normally. We now know that conditional-lethal mutations are not found in the gene

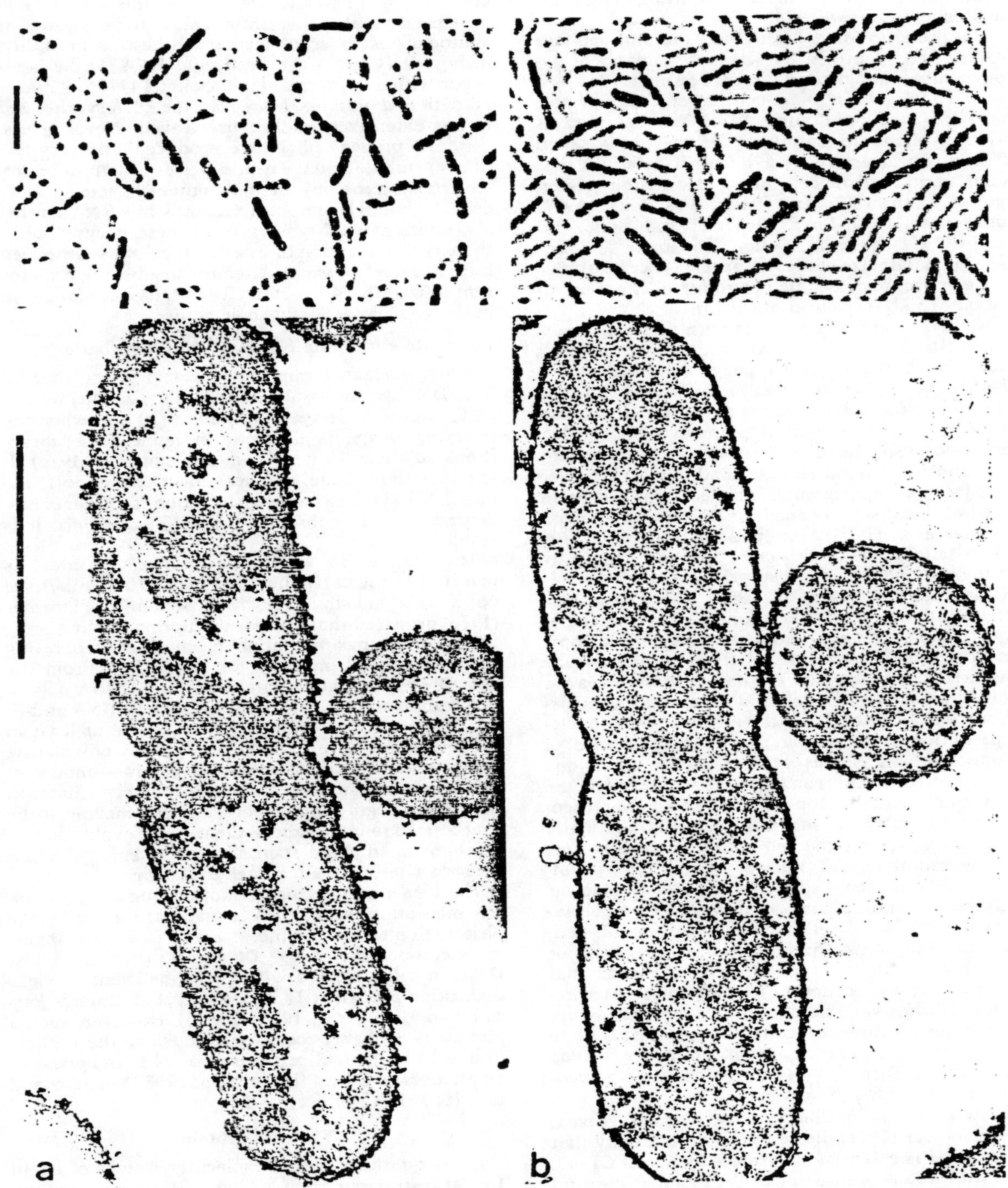

FIG. 4. Nuclear disruption in T4-infected *E. coli* B/5 cells observed by phase-contrast microscopy (top) or thin-section electron microscopy (bottom). (a) Uninfected B/5 cells. (b) B/5 cells fixed at 10 min after infection with the nuclear disruption-proficient T4 double mutant *am*N82 (defective in gene 44, DNA synthesis deficient)-*nd*28x6 (*denA* [endoII], host DNA degradation deficient) at 30°C. The use of this double mutant to observe nuclear disruption avoids the confounding effects of host DNA degradation and phage DNA synthesis. Bars on left, 1 μm.

responsible for nuclear disruption because the gene is unessential in standard laboratory strains of *E. coli* (Snustad and Conroy, 1974).

Nuclear disruption-deficient (*ndd*) mutants of phage T4 were first identified by screening isolates of hydroxylamine-mutagenized phage (Tessman, 1968), using the 17.5% gelatin phase-contrast microscopy procedure (Snustad and Conroy, 1974). The *ndd* mutants have been shown to map in a gene previously called *D2b* and characterized by a set of *r*II deletions that extend various distances into a dispensible region adjacent to the *r*IIB gene (Bautz and Bautz, 1967; Bruner et al., 1972; Dove, 1968; Sederoff, Bolle, and Epstein, 1971; Sederoff, Boller, Goodman, and Epstein, 1971). The size of the *ndd* gene has been estimated by electron microscopic heteroduplex mapping to be about 0.5 kb (Depew et al., 1975).

Host DNA degradation occurs with the same kinetics in cells infected with *ndd* mutants as in cells infected with wild-type T4 phage (Snustad et al., 1974). However, in the absence of T4 endoII (*denA* mutant infections), the absence of nuclear disruption results in an alternate, slow pathway of host DNA degradation (Snustad et al., 1974) that is dependent on T4 endoIV (Parson and Snustad, 1975). Whether this is just a compartmentation effect or is due to an altered physical state of the host DNA is not known.

Snustad et al. (1974) examined the state of the host DNA in cells infected with ndd^+ and ndd^- phage by alkaline and neutral sucrose gradient analyses, M-band analysis (Earhart et al., 1968), "folded genome" analysis (Stonington and Pettijohn, 1971; Worcel and Burgi, 1972), and thin-section electron microscopy. The most important results of these studies can be summarized as follows. (i) The host chromosomes are unfolded with normal kinetics in the absence of nuclear disruption. This excludes the possibility that nuclear disruption is merely a passive consequence of disruption of the forces that maintain the folded genome structure. This conclusion is reinforced by the observation, made by thin-section electron microscopy, that the DNA of the host nucleoid gradually disperses throughout the cell in the absence of host DNA degradation and nuclear disruption ([*denA*, *denB*, *ndd*] infections). (ii) The process of nuclear disruption does not require extensive nicking or cleavage of host DNA. In cells infected with endoII-deficient (*denA*) mutants, there are few, if any, nicks or breaks present in the host DNA at the time that nuclear disruption occurs. (iii) The simplest interpretation of all the available data is that nuclear disruption involves multiple attachment of the host DNA to the cell membrane, presumably mediated by the *ndd* gene product. Direct evidence for multiple attachment sites is lacking, however.

Depew et al. (1975) and Snustad, Bursch, Parson, and Hefeneider (1976) discovered independently that the *ndd* gene is essential for growth in *E. coli* CT447, one of the California Institute of Technology "hospital strains" (Wilson, 1973). Genetic and biochemical studies of strain CT447 have suggested that this restriction is imposed by a large plasmid present in this strain (S. Hefeneider, M.S. thesis, University of Minnesota, St. Paul, 1975; Casey, Herman, and Snustad, unpublished data).

Koerner and Snustad (1979 and unpublished data) have examined several parameters of T4 growth in strain CT447 in an attempt to determine why the *ndd* gene is essential in this strain. These studies show that shutoff of host macromolecular synthesis, degradation of host DNA, DNA synthesis, and RNA synthesis all occur with near-normal kinetics in CT447 cells infected with *ndd* mutants. Moreover, the DNA synthesized is not extensively nicked or cleaved. Nevertheless, very few progeny phage are produced, and very few filled or unfilled phage heads are seen by thin-section electron microscopy of *ndd* mutant-infected CT447 cells. It should be emphasized that while RNA synthesis occurs at wild-type rates in these infected cells, there is no indication of whether the RNA synthesized is qualitatively normal. Attempts to label the proteins synthesized in T4-infected CT447 cells have failed, so nothing is known about their nature.

Role in Shutoff of Host DNA Synthesis

While nuclear disruption-proficient phage shut off host DNA synthesis at about 4 min postinfection at 30°C, shutoff is delayed until about 10 min when the infecting phage is deficient in nuclear disruption (Snustad, Bursch, Parson, and Hefeneider, 1976). If the infecting phage produces an active endoII, the host DNA synthesized after infection is subsequently degraded. In the absence of functional endoII, it is stable.

The manner in which the *ndd* gene product is involved in the early shutoff of host DNA synthesis is not known. Snustad, Bursch, Parson, and Hefeneider (1976) proposed that nuclear disruption might merely act as a compartmentation mechanism, separating the biosynthetic pathway for phage DNA from the host DNA degradative pathway. The T4 DNA polymerase binds to nicks and ends of both T4 DNA and *E. coli* DNA and also replicates both DNAs, at least in vitro (Goulian et al., 1968). If T4 DNA polymerase exhibits these properties in vivo, a large amount of this enzyme might become bound to intermediates in host DNA degradation, requiring the enzyme to be synthesized in large excess. Clearly, the evolution of a mechanism to avoid such a waste of energy should provide a selective advantage to the virus.

The idea that nuclear disruption might be a compartmentation mechanism is made attractive by the observation of two distinct pools of DNA, one apparently composed of phage DNA and the other of host DNA, in cells infected with endoII-deficient, nuclear disruption-proficient T4 phage (Snustad, Bursch, Parson, and Hefeneider, 1976) (Fig. 5). However, such a picture is probably too simplistic, since the replication forks in the replicative pool of T4 DNA appear to be membrane bound (Huberman, 1968; Kozinski and Lin, 1965; R. C. Miller, 1972).

The *ndd* Protein

A basic protein with a molecular weight of about 15,000 (estimated from its mobility during sodium dodecyl sulfate-polyacrylamide gel electrophoresis) has tentatively been identified as the product of the *ndd* gene (Koerner et al., 1979; Snustad, unpublished data). This protein is synthesized very early after infection, with the maximum rate of synthesis occurring between 3 and 6 min (Koerner et al., 1979). By 12 min after infection, no synthesis of the *ndd* protein can

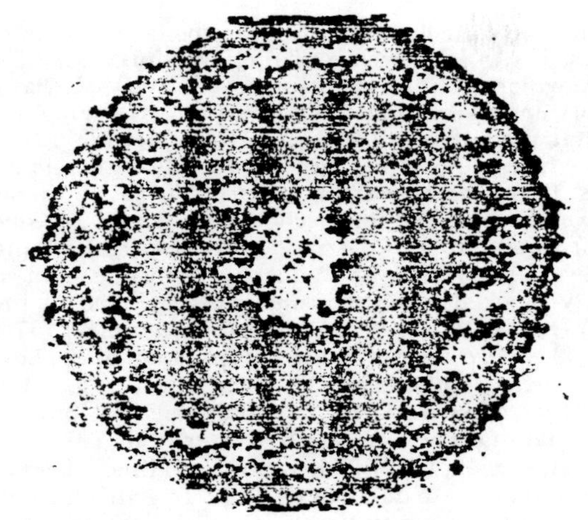

FIG. 5. Two apparent pools of DNA in a T4-infected *E. coli* B/5 cell: host DNA at the periphery of the cell as a result of nuclear disruption and phage DNA in the central region of the cell. Fixation was at 20 min after infection with mutant *nd*28x6 (*denA*) at 30°C.

be detected. It is synthesized in large quantities (>4,000 molecules per infected cell), and gene dosage experiments indicate that the protein is required in stoichiometric amounts in restrictive (CT447) host cells (Snustad, Bursch, Parson, and Hefeneider, 1976).

The *ndd* protein is associated with the particulate fractions of cell extracts (Koerner et al., 1979), suggesting that it may be bound to the cell envelope. There are several interesting questions to be asked regarding the *ndd* protein. Does the protein contain a hydrophobic domain for association with the cell membrane? Does it have DNA binding properties? If so, is the binding specific for cytosine-containing DNA, or for specific sites?

UNFOLDING OF THE HOST NUCLEOID AND TRANSCRIPTIONAL ALTERATIONS

The goals of studies on the T4-induced unfolding of the host nucleoid (Tutas et al., 1974) are threefold: (i) to gain additional insight into the structure and function of the *E. coli* nucleoid; (ii) to determine how phage-induced modification of this structure alters its function; and (iii) to examine the possibility that the T4 gene(s) responsible for unfolding might be responsible for the shutoff of host transcription, since *E. coli* nucleoids are also unfolded after treatment with rifampin (Dworsky and Schaechter, 1973; Pettijohn and Hecht, 1973), which also results in the arrest of transcription.

Kinetics of Unfolding

A simple viscosity test ("pour test") was developed and used by Snustad, Tigges, Parson, et al. (1976) to demonstrate that the chromosomes of T4-infected cells remain folded until 3 min after infection at 30°C. Thereafter, they rapidly unfold, becoming completely unfolded by 5 min after infection. The kinetics of unfolding are the same in the presence or absence of nuclear disruption and host DNA degradation. Experiments in which protein synthesis was inhibited by the addition of chloramphenicol at various times after infection showed that the T4-induced unfolding of the host nucleoid is dependent on protein synthesis occurring between 2 and 3 min postinfection. Unfolding does not occur after infection with T4 phage ghosts.

Isolation and Characterization of "Unfoldase"-Defective (*unf*) Mutants

Snustad, Tigges, Parson, et al. (1976) used the pour test to identify T4 mutants deficient in unfolding of the host nucleoid, which were named *unf* (for "unfoldase"-defective) mutants. One of them, *unf*39, was characterized in some detail, and the *unf* mutation was mapped at a position near 131 kb on the T4 linkage map. It was shown to have little or no effect on most parameters of the T4 life cycle (Snustad, Tigges, Parson, et al., 1976; Tigges et al., 1977). Host DNA degradation, nuclear disruption, early enzyme synthesis, phage DNA synthesis, and the shutoff of host DNA and protein synthesis all occurred with normal or near-normal kinetics. Only the switchover from host RNA synthesis to phage RNA synthesis was altered significantly (see below). There may also be an interaction with the nuclear disruption function, since the presence of an *ndd* mutation markedly affects the results of the pour test. If the phage are *ndd*$^-$, the viscosity of *unf*$^-$ phage lysates is increased (Sirotkin and Snyder, unpublished data). It is not clear whether this is an in vitro artefact of the folded genome assay or whether it indicates a relationship between the *unf* and *ndd* activities.

Subsequently, mutants that allow late transcription of cytosine-containing phage DNA (*alc* mutants; see below) were isolated by Snyder et al. (1976). Many *alc* mutants are also defective in the T4-induced unfolding of the host nucleoid, and there is other evidence that *alc* and *unf* are the same gene (see below).

Isolation of *alc* Mutants, Which Allow Late Transcription of Cytosine-Containing T4 DNA

The isolation of mutants defective in the shutoff of host transcription has been facilitated by the search for viable T4 mutants that can incorporate cytosine rather than HMC into their own DNA. It early became clear that T4 might well exploit its use of glucosylated HMC in place of cytosine to distinguish between its own and foreign DNA during the shutoff of host function. Any such mechanism should also discriminate against cytosine-containing T4 DNA. The now-successful efforts to isolate T4 mutants that are able to produce fully cytosine-containing T4 phage have indeed revealed new aspects of the shutoff of host transcription, as well as interesting nuances of the physiology of T4-infected cells. The study of these mutants has also led to additional puzzles and questions.

The mutations required to allow T4 to incorporate cytosine rather than HMC into its DNA are apparent from the pathways shown in Fig. 6. A T4-directed dCTPase degrades dCTP (and dCDP) to dCMP, thereby excluding cytosine from T4 DNA. The inactivation of this enzyme is sufficient to allow T4 to replicate DNA to contain cytosine (Wiberg, 1966); the T4 DNA po-

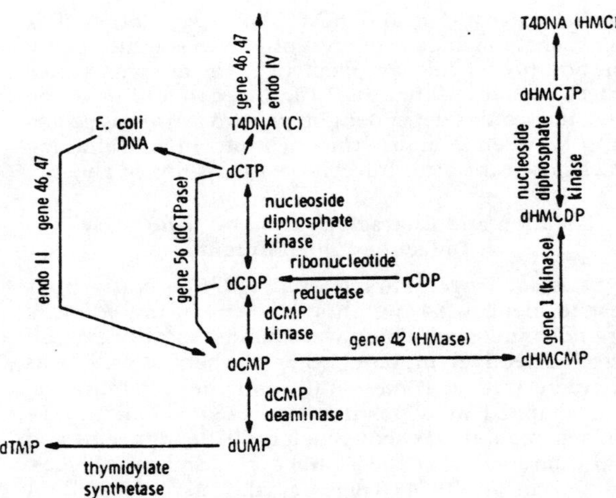

FIG. 6. Pathways affecting the availability of deoxycytosine and dHMC triphosphates in T4-infected cells.

lymerase makes no distinction between cytosine and HMC nucleotides (see also Goulian et al., 1968). But T4 DNA with cytosine is susceptible to the nucleases described above and by Warner and Snustad (this volume) which are involved in the breakdown of host DNA. Thus, mutants lacking the T4 dCTPase make a small amount of DNA, but this DNA is rapidly degraded (Kutter and Wiberg, 1968; Wiberg, 1966). Such T4 dC-DNA is particularly sensitive to endoIV; however, it is significantly less vulnerable to endoII, which nicks double-stranded DNA, or at least the damage done by the latter enzyme can be largely repaired. Genes 46 and 47 are also involved in the later stages of degradation of T4 dC-DNA; mutants altered in genes 46 and 47 and in dCTPase ([46, 47, dCTPase] mutants) make virtually as much DNA as do [46, 47] mutants, but the DNA of the former has an average molecular weight of only 10^7 (Kutter and Wiberg, 1968; Warner and Snustad, this volume).

It is intriguing that T4 dC-DNA is so much more sensitive to endoIV than to endoII, since the reverse is true of host DNA; as discussed above and by Warner and Snustad (this volume), endoIV plays only a minor role in host DNA degradation, whereas endoII is a key component. There may well be more single-stranded regions, susceptible to endoIV, in T4 DNA, which is actively undergoing replication, transcription, and recombination, than in host DNA, where such functions are inhibited after infection. This explanation is supported by the work of Elliott et al. (1973). They treated cells with toluene 12 min after infection at 37°C, making them permeable to added nucleoside triphosphates. They found that in the presence of ATP and the four normal deoxyribonucleoside triphosphates, host DNA synthesis could resume although it had stopped at least 10 min earlier in vivo; this resumption occurred only in the absence of functional endoIV but was not inhibited by the presence of endoII. Thus, host DNA is sensitive to endoIV when it contains single-stranded regions produced during replication.

EndoII and endoIV appear to be the only enzymes significantly involved in initiating the degradation of T4 dC-DNA; [dCTPase, endoIV] mutants make T4 DNA which is only slightly nicked, and [dCTPase, endoII, endoIV] mutant DNA is virtually the same size as that made by wild-type T4 (Kutter et al., 1975). Nevertheless, no phage are produced by these mutants. This is because the synthesis of T4 late proteins is severely impaired under conditions where cytosine is incorporated into T4 progeny DNA (Kutter and Wiberg, 1968; Kutter and Wiberg, 1969) and is almost totally blocked when cytosine replaces most of the HMC (Kutter et al., 1975; Wu and Geiduschek, 1975) (Fig. 7). There are some interesting anomalies. Less late-protein synthesis occurs after infection by mutants that make dC-DNA than after infection by mutants that make no DNA at all (Kutter et al., 1975; Wu and Geiduschek, 1975). This is not an indirect effect of dCTP, but is due to dC-DNA itself; adding an additional gene 43 mutation to the dCTPase mutant phage to block DNA replication actually enhances the rate of late-gene expression (Wu, Geiduschek, and Cascino, 1975). Also, Wu and Geiduschek (1975) found significantly more r-strand (late) transcription after infection by dCTPase mutants than would be expected from the almost total absence of late proteins; this RNA was not further characterized. Finally, the defect

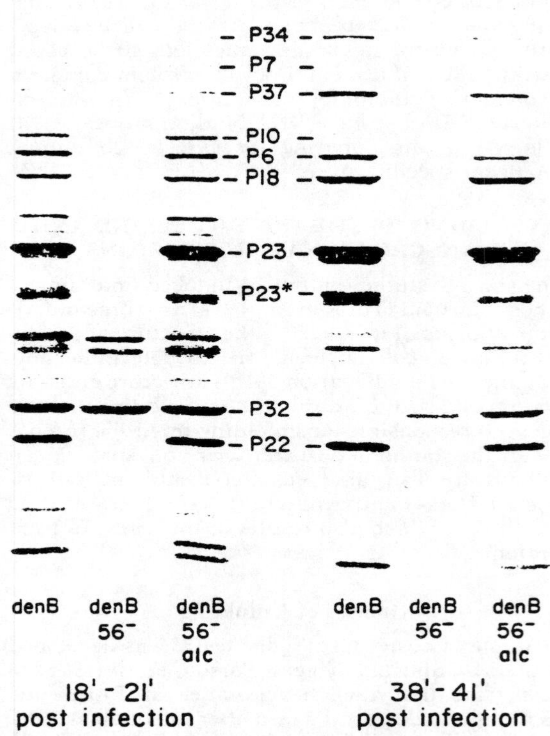

FIG. 7. Patterns of late-protein synthesis after infection of *E. coli* B at 28°C with *denB*, [*denB*, 56], and [*denB*, 56, *alc*] mutants. From Snyder et al. (1976).

in late-gene expression is somewhat temperature dependent, for at temperatures over 40°C, a small amount of some late proteins is made late in infection (unpublished data of Kutter and of Snyder).

In summary, with the right combination of mutations, T4 will replicate its DNA to contain cytosine. However, no phage are produced because the T4 dC-DNA does not direct the expression of the late genes. It is perhaps not surprising that the T4-altered transcription apparatus, which normally recognizes DNA with HMC and does not transcribe cytosine-containing *E. coli* DNA, is unable to transcribe T4 DNA with cytosine. A priori, a number of gene products could have been involved. In fact, as we discuss next, the phage makes only one gene product, that of the *alc* gene, that actively prevents the transcription of T4 dC-DNA late in infection.

Extensive efforts to isolate pseudorevertants of [endoII, endoIV, dCTPase] mutants that would allow late-gene expression from dC-DNA (*alc* mutants) were unsuccessful at first, due, it turned out, to the effects of host restriction. However, when [dCTPase, endoIV] mutants were plated on nonsuppressing bacteria, plaques arose at a frequency of about 10^{-4}, due to suppressing (*alc*) mutations which overcame the block in late-gene expression (Snyder et al., 1976). It was found that these phage have only about 60% of their HMC replaced by cytosine; as seen in Fig. 6, being endoII⁺, they break down host DNA, which then supplies dCMP for HMase, and thus can incorporate enough HMC to let them largely escape the host restriction system. To obtain *alc* mutants with higher levels of cytosine substitution, phage also lacking endoII, HMase, or dHMP kinase must be plated on bacteria that lack the normal restriction-modification systems. By using such hosts, cytosine-containing T4 strains have now been isolated in which cytosine almost totally replaces the HMC (see appendix to this chapter).

The sensitivity of cytosine-containing T4 to the host restriction-modification system was somewhat unexpected. The incoming T4 DNA is effectively modified because it contains HMC. The progeny DNA should then be modified to escape restriction in subsequent generations, but it is not. Either T4 inhibits the host modification system, or the rate of T4 DNA replication outstrips the rate of modification.

Evidence that the Same Gene Directs *unf* and *alc* Functions

All of the mutations which allow late-gene expression from cytosine-containing DNA (*alc* mutations) that have been tested fail to complement each other and so are presumably in the same gene. Mapping of the gene was complicated by the fact that the only known phenotypes occurred when the mutations were combined with two or more other T4 mutations. However, the mapping was aided by the observation that a particular *alc* strain had a conditional-lethal *ts* phenotype which was at least closely linked to the *alc* mutation and so could be used to map it (Sirotkin et al., 1977). In retrospect, the temperature sensitivity was probably not due to the *alc* mutation itself, but for the purposes of initial mapping it was sufficient that they were close.

Once *alc* mutations were mapped to the 131-kb region, the next step was to determine whether they were in any of the known genes in this region. Mutations in genes 63, *pseT*, and 31 did not affect—and were complemented by—*alc* mutations. However, a relationship with the *unf* gene was clearly apparent, for many *alc* mutations were also measurably unfoldase deficient (Sirotkin et al., 1977).

Mutants are initially characterized as *alc* or *unf* on the basis of different phenotypic criteria. The Unf⁺ or Unf⁻ phenotype is determined primarily on the basis of the pour test assay (see above), which measures viscosity of lysates, whereas the Alc⁻ phenotype is determined by the growth of the mutant, with an appropriate genetic background (lacking dCTPase and endoIV), on a nonsuppressing, nonrestrictive host such as *E. coli* B834. However, a number of considerations made it seem likely that *unf* and *alc* are the same gene. First, it makes some theoretical sense. Transcription appears to be required to hold the nucleoid together (Pettijohn and Hecht, 1973), and the *alc* function blocks at least some types of transcription (see below). Second, the genetic evidence supports this interpretation. The original *unf* mutation, *unf*39, characterized by Snustad, Tigges, Parson, et al. (1976), was shown to produce the *alc* phenotype when crossed into a [dCTPase, endoIV] genetic background, and nine independent spontaneous *unf* mutations identified by the *alc* selection procedure all failed to complement *unf*39 as indicated by pour test assays in pairwise complementation tests (Tigges et al., 1977). In addition, when two *alc unf* mutants were crossed with each other, the Alc⁻ recombinants were also Unf⁺ (Runnels, unpublished data). Thus, the same mutation was causing both phenotypes, and these *alc* and *unf* mutations, at least, could not be deletions encompassing two neighboring genes. Finally, preliminary mapping evidence indicates that mutants which show only the Alc⁻ phenotype and those which are both Unf⁻ and Alc⁻ have intermingled mutations (Runnels and Snyder, unpublished data).

If *alc* and *unf* are the same gene, then why are only some *alc* mutants measurably Unf⁻? The simplest interpretation is that only the most completely inactivating *alc* mutations are also measurably deficient in the *unf* function. It is a matter of general experience that those *alc* mutations selected in mutants with higher levels of cytosine substitution—or which better allow multiplication with dC-DNA—are more commonly also Unf⁻. However, the possibilities remain that the Alc and Unf functions are due to overlapping genes or are different activities of the same gene product. Hereafter, we shall nevertheless treat these as properties of a single gene, *alc/unf*.

Recently, Herman and Snustad (1982) have demonstrated that the presence of the F1 incompatibility group plasmid pR386 Tc renders laboratory strains of *E. coli* relatively restrictive to the growth of *alc/unf* mutants; this discovery should help determine the relationship between *alc* and *unf*. These restrictive host strains are also being used to construct a fine-structure map of the *alc/unf* gene (Payzant and Kutter, unpublished data) and to isolate conditional-lethal *alc/unf* mutants (Manjula and Snustad, unpublished data). Work with these host strains is complicated, however, by the effects of other T4 mutations on the plating efficiency of *alc* mutants, and some *alc* mutants are less well restricted than others on many of the plasmid-bearing strains.

Precise localization of the *alc/unf* gene has depended on the analysis of deletions *pseT*Δ1 and *pseT*Δ3, which were first isolated as *alc* mutations (*alc*1 and *alc*3) but were shown to contain deletions extending through all known *pseT* point mutations (Sirotkin et al., 1978) and on through *cd* (dCMP deaminase) (Hall, personal communication). Using heteroduplex mapping, Kutter (unpublished data) showed that the two strains both contain appropriate deletions of about 3.5 kb. These deletions have been mapped more precisely by analyzing their effects on the fragments observed after digestion with the restriction enzyme *Eco*RI (Fig. 8). The *pseT*Δ1 deletion removes most or all of fragments 33 and 21 but leaves fragment 31 virtually intact (Mileham et al., 1980), whereas *pseT*Δ3 also removes fragments 33 and 21 and extends about 200 base pairs into fragment 31 (Gram and Rüger, personal communication). Payzant and Kutter (manuscript in preparation) find that both the *alc*1 and *alc*3 mutations map at the distal end of the *alc* gene and show no recombination with each other, but recombine with all but one other mutation which has been mapped.

Taken together, this evidence seems to indicate that the *alc* gene is largely or totally contained within fragment *Eco*RI-31. It appears likely that the *alc* gene is located just distal to the early promoter on this fragment (Fig. 8), as might be expected from the early functioning of this gene. Sequencing of the DNA in this region from various mutants should answer many questions about *alc* and *unf* functions. Fragment *Eco*RI-31 also contains most or all of the RNA ligase gene (gene 63) and is the one *Eco*RI fragment in the region which has not been clonable, even from *alc* mutants (Mileham et al., 1980), but subfragments covering the known gene 63 markers have been cloned and partly sequenced (S. Brown, personal communication) *Eco*RI fragment 21 has been cloned in λ (Mileham et al., 1980) and in pBR325 (Rice and Snyder, unpublished data).

Assuming that gp*alc* is a protein, Kutter and Drivdahl have made extensive efforts to identify it, using two-dimensional acrylamide gel electrophoresis (Fig. 9). Unambiguous identification has been complicated by such factors as the lack of amber *alc* mutants, the tendency of *alc* strains to pick up extra mutations (see below), and the need to develop methods to fully resolve the very large number of very small T4 proteins. It appears very likely, however, that the approximately 18,000-dalton protein marked "ALC?" in Fig. 9 is gp*alc*; it is shifted in charge or size or both in at least four *alc* mutants (*unf*39, *alcGT*7, *alcD*22, and *alcE*2) and is missing in deletion mutants *alc*Δ1, *alc*Δ3, and *alcD*32. In *alcD*32, one new, smaller spot appears, which might be the residual N-terminal portion of gp*alc*.

Herman, Haas, and Snustad (unpublished data) have independently shown that the 18,000-dalton protein marked "ALC?" in Fig. 9 is altered or missing in cells infected with *unf*39×5, *alc*Δ1, and two other *unf/alc* mutants. Moreover, an 18,000-dalton protein indistinguishable (by two-dimensional polyacrylamide gel electrophoresis) from this protein is synthesized in cells infected with each of three independent Unf$^+$ revertants of *unf*39×5. This indicates that the 18,000-dalton protein is either the product of the *unf/alc* gene or a protein whose synthesis requires a functional *unf/alc* gene.

Gram (thesis, University of Bochum, West Germany 1982) has used coupled transcription-translation of T4 dC-DNA fragment *Bgl*II-6 (isolated from gels) to analyze the proteins produced from the two immediate early promoters on this fragment (Fig. 8). Since he started with the intracellular DNA from an Alc$^+$ [dCTPase, HMase, endoIV] mutant, gp*alc* should be among the proteins produced. In addition to the kinase phosphatase (gp*pseT*, 30,000 daltons), at least five small proteins are produced, of about 5,000 to 18,000 daltons; these appear likely to be the same as those identified by Kutter and Drivdahl as missing in *alc*Δ1 and *alcD*32 mutants (Fig. 9), and some are presumably the products of as-yet-unidentified genes in this region.

EFFECTS OF gp*alc/unf* ON TRANSCRIPTION OF T4 DNA

As discussed above, T4 late genes are not transcribed when the progeny DNA contains cytosine unless gp*alc/unf* is mutationally inactivated; this is the basis on which *alc* mutations were initially isolated and identified. More detailed analyses of the effects of gp*alc/unf* on transcription of T4 dC-DNA should provide clues as to the methods of action of this versatile gene product.

Transcriptional Patterns when the Infecting DNA Contains HMC and the Progeny T4 DNA Contains Cytosine

Late proteins are made at essentially the normal rate and in the normal relative ratios by most of the

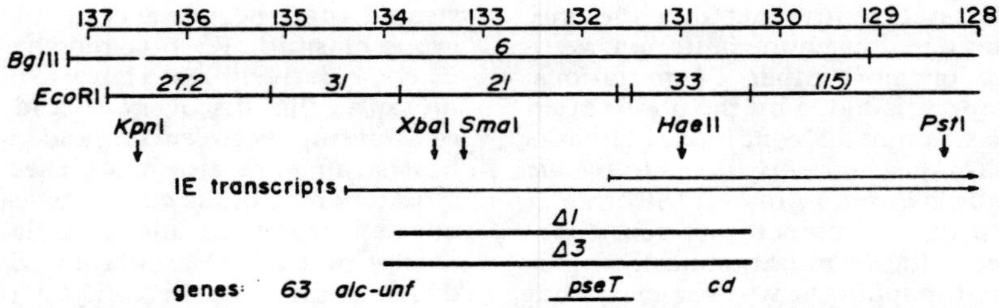

FIG. 8. Map of the gene 63-*alc/unf-pset* region of T4 (Mileham et al. 1980; position of promoters and of gene 43 from Gram and Rüger [unpublished data]). IE, Immediate early.

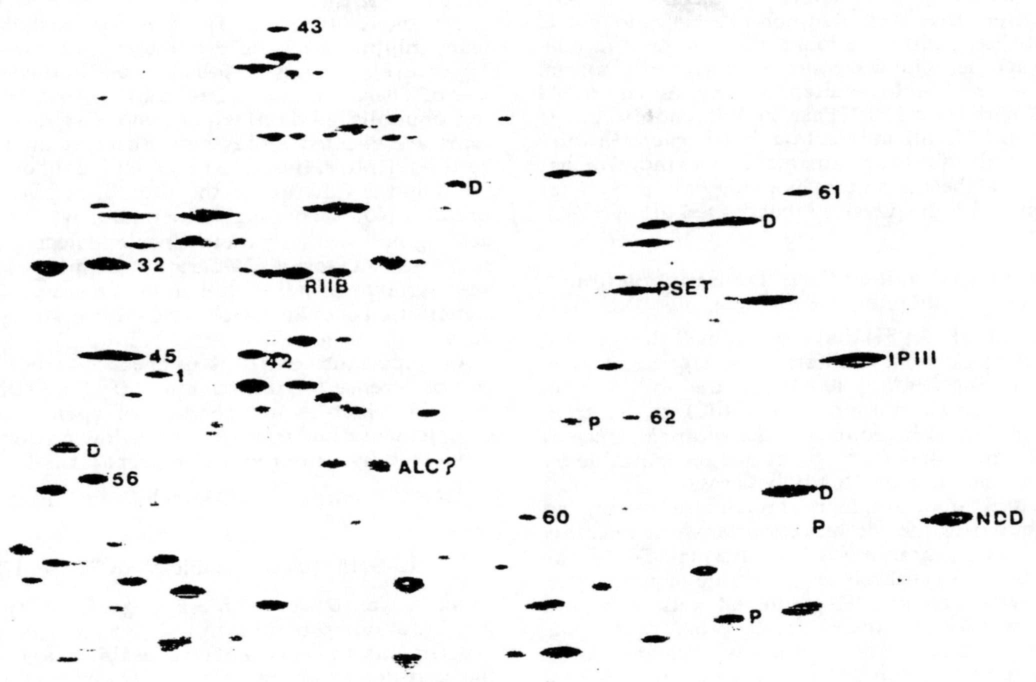

FIG. 9. Two-dimensional gel electrophoresis of T4D⁻ proteins labeled 3 to 8 min after infection at 37°C. For this gel, the method of sample preparation and running of the nonequilibrium pH gradient electrophoresis first-dimension gel was virtually identical to that of Hosoda (method 3, Burke et al., this volume), with the gel run for 1,600 V-h; the second dimension is a simple 10 to 17% sodium dodecyl sulfate-polyacrylamide gradient gel. The labeled spots were identified in this laboratory, except for gp60, gp62, and gp61, which are identified on the basis of the studies reported by Burke et al. (this volume) and labeled here to facilitate cross comparisons. The spots labeled P are missing in *pseT* deletions Δ1 and Δ3, along with PSET and ALC? and one additional very small, acidic protein lost at the bottom of this gel. Those labeled D are among the proteins missing in deletion $(39\text{-}56)_{12}$; the topmost one is identified as *dda* by Burke et al. (this volume).

[*alc*, dCTPase, endoIV] and [*alc*, dCTPase, endoII, endoIV] mutants analyzed to date (Fig. 7) (Morton et al., 1978; Snyder et al., 1976). This is interpreted as implying that gp*alc* simply interferes with the interaction between RNA polymerase and cytosine-containing DNA and plays no more direct role in promoter selection or transcriptional specificity. However, maturation-related processing of structural proteins is somewhat slow when the progeny DNA contains cytosine (Fig. 7), and phage production is often only 20 to 80% of that seen with T4D⁻ (Kutter et al., manuscript in preparation); the reasons are not yet clear, but may reflect packaging problems.

The patterns of early-enzyme synthesis are also essentially normal for these mutants, although often the shutoff of synthesis of some proteins is delayed. When the phage are endoII⁻, the overproduction of DNA polymerase is much more marked and is accompanied by overproduction of gp30 (DNA ligase), gp32, and gp46 (Kutter et al., unpublished data). This might be a response to nicking of the cytosine-containing phage DNA by endoII; if so, it suggests that production of gp30 and gp46 may be functionally regulated in some way, as has been demonstrated for gp32, gp43, and gp*regA*, discussed elsewhere in this book. However, one must be careful in interpreting such patterns. It appears, for example, that the extensive overproduction of gp52 seen in many of the *alc denB* mutant phage is a consequence of the *denB* deletion linking gene 52 to the *r*IIB promoter, rather than a reflection of some metabolic alteration (T. Mattson, personal communication).

Gp*alc* appears to function directly and reversibly in blocking transcription of progeny T4 dC-DNA, as has been shown with the aid of a temperature-sensitive *alc* mutant (*alcts*3 mutant) isolated by Slocum and Snyder. Kutter, Drivdahl, Redmond-Payne, and d'Acci (unpublished data) have found that the level of late-protein synthesis jumps from zero to virtually normal within a few minutes after shifting *E. coli* B834 infected with [*alcts*3, dCTPase, endoII, endoIV] phage from 27 to 39°C late in infection. Further temperature shift and pulse shift experiments seem to indicate that the gp*alc* of the *ts*3 mutant is made only early after infection and is irreversibly inactivated at high temperature.

Effects of gp*alc*/*unf* on Early Transcription from Infecting T4 dC-DNA

Kutter et al. (1981) have examined the general ability of gp*alc*/*unf* to inhibit transcription of cytosine-containing DNA by looking at the ability of infecting cytosine-containing T4 (T4dC) to be transcribed in the presence or absence of gp*alc*, since at least the immediate-early genes are recognizable by the unmodified *E. coli* RNA polymerase.

In the *alc*$^-$ condition, there is no difference that can be attributed to the *alc* mutation between infecting with cytosine-containing T4 or normal T4 in the kinetics of DNA synthesis or in the rate of production of any early protein detectable on gels. It would appear, therefore, that any factor other than gp*alc* which affects *E. coli* transcription is not significantly blocking the transcription of cytosine-containing early T4 genes.

It is more difficult to compare infecting T4 dC- versus HMdC-DNA as templates under *alc*$^+$ conditions, since only *alc* mutants can make cytosine-containing T4 phage, and no satisfactory conditional-lethal *alc* mutants are available. Kutter et al. (1981) therefore performed a series of complementation tests between cytosine- and HMdC-containing T4 phage, in which the *alc*$^+$ function was provided by any of a series of amber or deletion mutants with HMdC-DNA and an early function(s), X, missing in each mutant (such as DNA polymerase, gp45, gp52, or gp55) must be read off the parental dC-DNA to obtain phage growth or enzyme synthesis or both:

HMdC-DNA: [endoIV, X]

+

dC-DNA: [*alc*, endoIV, dCTPase, HMase]

Complementation was measured by densitometry of the protein bands on polyacrylamide slab gels or in terms of phage, DNA, and late-protein production, as appropriate. It was found that production of all early enzymes is drastically reduced when their production must be directed by infecting DNA containing mostly cytosine. Virtually no synthesis was observed of those proteins which could be carefully quantitated, such as gp45, gp42, and DNA polymerase, as long as the extent of cytosine substitution was 95 to 100%. However, when the DNA contains only 75% cytosine, there is very extensive complementation for all early enzymes examined, including those classified as delayed early, and almost complete complementation for some, such as *r*IIB.

These experiments indicate that gp*alc* acts extensively quite early in infection and also that it affects the recognition of all early promoters. However, as little as 25% HMC in the DNA is enough to significantly circumvent the effect. This contrasts with the significant inhibition of late protein synthesis with much lower levels of cytosine substitution, studied by using a *ts* dCTPase mutant (Kutter and Wiberg, 1969; Kutter, unpublished data). Furthermore, some early proteins are affected more than others at intermediate cytosine/HMC ratios, and the observed differences are not solely a function of the time of synthesis of the protein. Not surprisingly, therefore, we seem to be dealing here with promoter-specific effects in the *alc* response to cytosine/HMC ratios. Similar effects are seen for late proteins, but at much lower levels of substitution of cytosine for HMC (Kutter, unpublished data).

An important control is to check whether there is any preference for transcription of HMdC-DNA over dC-DNA which is independent of gp*alc*. Therefore, complementation tests of the following form were conducted by Kutter et al. (in preparation):

HMdC-DNA: [*alc*, *unf*39, endoIV, dCTPase, X]

+

dC-DNA: [*alcGT*7, endoIV, dCTPase, HMase]

Here, X was DNA polymerase, gp32, or gp55. The pattern of late-protein synthesis was normal in these experiments, indicating extensive ability to transcribe the cytosine-containing DNA. However, the initial rate of production of both DNA polymerase and gp32 was substantially lower than expected, even though it did reach the expected levels later in infection for polymerase, and DNA and phage production was reduced. The effects on DNA and phage were not observed when the phage contained only 95% cytosine, and the extent of the effect was much more sensitive to the relative multiplicity of infection and fraction of cytosine than in the parallel *alc*$^+$ experiments. Largely complementary results were obtained after infection with the one known usable *ts alc* mutant.

These experiments indicate that there is some preference for transcribing glucosylated HMdC-DNA over dC-DNA even in the absence of gp*alc*, but gp*alc* also plays a major role in inhibiting transcription of T4 early genes from dC-DNA; both effects are likely to contribute to the shutoff of host transcription after normal T4 infection. In view of the in vitro results with RNA polymerase (see below), it appears most likely that gp*alt* is involved in this additional level of cytosine-HMC distinction.

EFFECT OF gp*alc*/*unf* ON THE SHUTOFF OF TRANSCRIPTION FROM OTHER CYTOSINE-CONTAINING DNA

The initial expectation in isolating *alc* mutants was that they would also affect the transcription of other cytosine-containing DNAs and thus provide clues to the mechanism(s) of T4 inhibition of host transcription. Various direct attempts to show a general effect of gp*alc*/*unf* on *E. coli* transcription have led to some surprising difficulties and contradictions. Also, it is clear that gp*alc*/*unf* can directly block phage λ late transcription, but it seems to have little or no role in the shutoff of λ early transcription after T4 infection.

As discussed above, synthesis of T4 early proteins, on the other hand, is affected by gp*alc*/*unf* when the infecting phage contain dC-DNA, but other factors also contribute to this restriction.

Effect of gp*alc*/*unf* on Shutoff of Host Transcription

While there is general agreement that gp*alc*/*unf* affects the shutoff of *E. coli* transcription, the details of its action remain obscure. Sirotkin et al. (1977) found that substantially more of the RNA made after infection hybridized to *E. coli* DNA when the cells were infected by an *alc*/*unf* mutant. Surprisingly, they found that considerable amounts of *E. coli* RNA were made in any case as late as 15 min after infection under their conditions. Tigges et al. (1977) demonstrated that the switchover from host transcription to T4 transcription was 2 to 3 min slower in cells infected with *unf*$^-$ phage than in cells infected with *unf*$^+$ phage; the differences in the amounts of *E. coli* and T4 RNAs synthesized in *unf*$^+$ and *unf*$^-$ phage-infected cells were shown to be statistically significant between 2 and 6 min after infection. However, in their experiments shutoff of host transcription was essentially complete by 10 min after infection with *unf*$^+$ or *unf*$^-$ phage. These results indicate that the *alc*/*unf* gene is involved in the switchover from host RNA synthesis to phage RNA synthesis, probably acting in concert with other T4 genes.

Most other work supports this general model. P. Dennis (personal communication) has examined the shutoff of transcription of several specific *E. coli* genes, using the DNA of various λ transducing phages as probes and measuring RNA pulse-labeled 9 to 10 min after infection with T4D$^+$ or with *alc*Δ1 or its *alc*$^+$ parent. He observed no differences in transcription levels of the *E. coli lac*, *ilv*, or RNA polymerase β and β′ genes. A very low but detectable transcription of the ribosomal protein genes was observed only after infection with the *alc*Δ1 mutant, but the amount seemed to be proportional to the number of surviving uninfected cells. He therefore concluded that although the *alc*Δ1 mutant may be somewhat slower in turning off host transcription, virtually no synthesis of these products occurs after 9 min postinfection. R. Buckland and A. A. Travers (personal communication) find that cells infected with *unf*39 mutants synthesize several times as much ribosomal RNA between 5 and 10 min after infection as do cells infected with T4D$^+$. However, the amount still seems to be small; Guttman (personal communication) sees no continued synthesis of 16S, 23S, or 30S RNA detectable on sucrose gradients after either T4D$^+$ or *alc* mutant infection.

Effect of gp*alc*/*unf* on Expression of T4 Late Genes from Plasmids

Considering the devastating effect that gp*alc*/*unf* has on the expression of T4 late genes from dC-DNA during infection, it seems almost axiomatic that it would also block the expression of T4 late genes cloned in plasmids. Such experiments were performed as part of an analysis of the ability of cloned T4 genes to complement T4 amber mutations in late genes 11, 23, and 24 (Jacobs et al., 1981). It was found that complementation did not depend upon the presence of an *alc*/*unf* mutation, although there was some enhancing effect. This may be explained by the fact that their phage strains with the *alc*/*unf* mutation replicate with cytosine-containing DNA, whereas those without it replicate their DNA to contain HMC. In the latter case, if there is plasmid replication after T4 infection, as indicated by the results of Takahashi and Saito (1982a, 1982c), the plasmids would contain HMC and would be insensitive to *alc*/*unf*. When these experiments were performed with phage which replicated their DNA to contain cytosine but did not have an *alc*/*unf* mutation, there was no expression of T4 late genes from either the plasmid or the infecting T4 genome (P. Geiduschek, personal communication). The interpretation that plasmid replication to contain HMC allows the cloned T4 late genes to escape gp*alc*/*unf* inhibition is supported by the observation that gp43, the T4 DNA polymerase, is required for the expression of a late gene from a plasmid only if the phage replicate by using HMC and do not have an *alc*/*unf* mutation. Thus, there is no evidence at this point to indicate a difference between effects of gp*alc*/*unf* on plasmid-cloned late genes and on late genes carried on the T4 genome.

Effect of gp*alc*/*unf* on λ Transcription in Mixed Infections

It has been known for some time that T4 shuts off λ transcription (Hayward and Green, 1965; Kennell, 1970). There are advantages to studying the effect of gp*alc*/*unf* on λ transcription rather than on T4 or host transcription. First, there should be no recombination between λ and T4 genomes to complicate the analysis. Second, by studying the effect on λ transcription at various stages of λ development, one can selectively determine the effects on various types of transcription. Third, the λ genome has been extensively characterized and the regulation of its transcription is fairly well understood. Fourth, the λ genome itself is fairly simple compared with either the T4 or the *E. coli* genome, making an analysis of the effect of the gp*alc*/*unf* on λ DNA structures easier.

A role for gp*alc*/*unf* in the shutoff of λ transcription was clearly evident (Pearson and Snyder, 1980). The basic experiment was to infect cells with λ or to induce λ lysogens, wait various lengths of time, and then superinfect the cells with T4 phage, using either an [*alc*/*unf*, endoII, endoIV, dCTPase] mutant or its *alc*$^+$ parent. The RNAs were labeled after T4 superinfection and hybridized to total λ DNA.

The effect of gp*alc*/*unf* on λ transcription was found to depend on the time of T4 superinfection. If the λ-infected cells were late into λ development and thus were making λ late RNA, the effect of gp*alc*/*unf* was dramatic. There was little or no shutoff of the λ transcription after superinfection by an *alc*/*unf* mutant, whereas T4 without an *alc*/*unf* mutation halted λ transcription within about 4 min after superinfection. The effect on λ early transcription was less dramatic; shutoff proceeded more slowly and was essentially independent of gp*alc*/*unf*.

Subsequent work suggests that the shutoff of λ late transcription depends on the structure of the λ DNA template as well as on the promoter from which the λ late genes are transcribed (Green, Rice, and Snyder, manuscript in preparation). Normally, λ late transcription occurs from templates which have been concatemerized by "rolling circle" replication or by recombination (Enquist and Skalka, 1973). If the λ is γ^- and the host is *recA*$^-$, concatemers never appear and the λ DNA remains as supercoiled covalently closed circles or relaxed circles. However, λ late transcription occurs at almost normal rates (Enquist and Skalka, 1973; Green and Snyder, unpublished data). Surprisingly, the λ late transcription which occurs from circles is not as sensitive to gp*alc*/*unf*, as though the structure of the DNA determines whether gp*alc*/*unf* can block transcription.

The promoter which is used also appears to be important. Most λ late transcription is thought to occur from a promoter PR[1] to the right of *Q* and is dependent on *Q* function. However, in the absence of *Q*, some transcription of the λ late region occurs, presumably from the early promoter PR (Herskowitz and Signer, 1970). It is of interest to determine whether the λ late transcription in the absence of *Q* is sensitive to gp*alc*/*unf*, since λ early transcription is relatively insensitive. Preliminary evidence indicates that late transcription is not as sensitive (Green, Rice, and Snyder, in preparation). This leads to the somewhat paradoxical conclusion that whether gp*alc*/*unf* can block transcription depends on the structure of the DNA from which it is transcribed, as well as the promoter from which it is initiated.

One possible explanation for the ability of gp*alc*/*unf* to unfold the bacterial nucleoid is that it is a topoisomerase. However, work with λ revealed that gp*alc*/*unf* could not remove the supercoils from covalently closed λ DNA in vivo. Thus, if gp*alc*/*unf* has a topoisomerase activity, it is probably not active against circular λ DNA (Pearson and Snyder, 1980).

RNA POLYMERASE CHANGES AND THE INHIBITION OF HOST TRANSCRIPTION

Most of the attempts to elucidate the mechanism of the T4-induced shutoff of host transcription have focused on the phage-induced modifications of the host RNA polymerase, discussed in detail by Rabussay (1982a, 1982c, and this volume). To summarize briefly, T4 utilizes the *E. coli* DNA-dependent RNA polymerase throughout infection, modifying it in two ways. The first is ADP ribosylation, in which one of the α subunits is immediately ADP-ribosylated ("altered") by gp*alt*, which is injected with the DNA; gp*mod*, produced a few minutes later, ADP-ribosylates ("modifies") both α subunits (Goff and Weber, 1970; Rabussay et al., 1972). The second modification is the addition of at least four new subunits: first the proteins p10k and p15k, whose genes are unidentified, and then gp33 and gp55, which are required for T4 late transcription (Stevens, 1972). In addition, gp45 binds to the phage-modified polymerase in vitro, although these proteins normally do not copurify (Ratner, 1974a).

The degree to which each of these changes in T4-directed polymerase is involved in the shutoff of host transcription remains uncertain. It is difficult to interpret in vitro results, since unknown amounts of the added subunits are lost during purification and the final preparations are clearly heterogeneous; that is, most of the added subunits are found in less than stoichiometric amounts (Rabussay and Geiduschek, this volume; Stevens, 1972). In vitro, there is little decrease in overall UTP incorporation directed by *E. coli* DNA using polymerase isolated from T4-infected rather than uninfected cells, but functional mRNA is synthesized at a much lower rate (Mailhammer et al., 1975). When T7 DNA is used as the template, T4 infection markedly inhibits polymerase activity (Chamberlin, personal communication).

On the basis of in vitro polymerase subunit reassembly experiments, Mailhammer et al. (1975) concluded that "modification of the α subunit of RNA polymerase is sufficient for inhibition of host transcription." However, in vivo, T4 *alt* and *mod* mutants have been found to shut off host transcription normally (Goff and Setzer, 1980; Horvitz, 1974b). Thus, ADP-ribosylation of RNA polymerase cannot be solely responsible for shutting off host transcription, although it may well be a contributing factor.

As discussed above, the only T4 gene clearly shown to be involved in shutoff of host transcription is *alc*/*unf*. Because some *alc*/*unf* mutations affected the binding of p15k to the RNA polymerase (Sirotkin et al., 1977), it was proposed that this protein might actually be gp*alc*/*unf*. However, this observation has not been confirmed, and the situation remains puzzling. The only mutation known to affect p15k produces a split in the protein band on gels (Poteete, Ratner, and Horvitz, unpublished data); this mutation maps in or near gene 60 (Goff, unpublished data).

R. Drivdahl is examining the effects of the *alc* mutation on in vitro RNA polymerase activity and contrasting them with the effects of *alt* and *mod* mutations (Drivdahl et al., 1981; Drivdahl and Kutter, 1983 and unpublished data), using cell extracts which are fractionated on heparin agarose. Within 45 s at 37°C, Drivdahl observes approximately a 50% drop in activity on both T7 and T4 dC-DNA templates after infection with either T4D$^+$ or various *alc* mutants (Fig. 10). This initial drop occurs even when the infection is carried out in the presence of chloram-

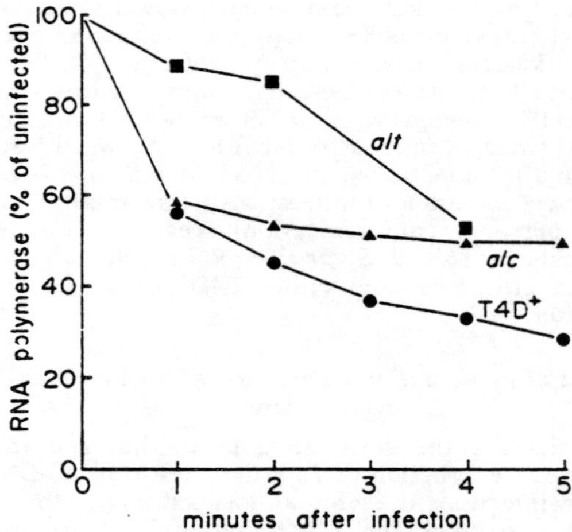

FIG. 10. RNA polymerase activity observed at various times after infection of *E. coli* B834 at 37°C, using T7 DNA as the template. Here, the soluble portion of the lysate has been fractionated on a heparin agarose column, but similar results are obtained with crude extracts. The *alc* mutant used here is *unf*39; similar results have been observed with several other *alc* mutants as well.

phenicol and is largely dependent on gp*alt* (Fig. 10). Mailhammer et al. (1975) demonstrated that ADP-ribosylation of both α subunits severely interferes with the ability of the polymerase to productively transcribe at least some bacterial genes in vitro; Drivdahl's work indicates that a significant contribution is made by ribosylation of even one α subunit per molecule.

There is little further drop in activity over the next 5 min after infection with *alc* mutants, but with T4D⁺ infections the ability to transcribe T7 DNA continues to drop to 20 to 30% of the uninfected level by 5 min (Fig. 10). This second, *alc*-dependent portion of the drop is blocked by adding chloramphenicol before infection. When glucosylated HMdC-DNA is used as template, much less of a drop is seen in activity, and there is far less difference between T4D⁺-, *alt*-, and *alc/unf*-infected cells. Thus, gp*alt* and gp*alc/unf* each produce a relatively specific impairment of the ability to read dC-DNA, and there is not just a general loss of polymerase activity after infection.

Later in infection, the activity on both dC- and glucosylated HMdC-DNA drops further in all strains examined, eventually reaching only about 5 to 10% of the uninfected level; this presumably reflects the additions of gp55 and gp33 and the shift to the form of the polymerase that transcribes T4 late genes.

Unfortunately, the differences in activity between the polymerases from *alc* mutant- and T4D⁺-infected cells are largely lost during further purification by any of several different methods; the easiest interpretation of this observation is that gp*alc* is only loosely bound to the polymerase.

It is known that σ is less tightly bound in the polymerase from T4-infected cells than in that from uninfected cells. Stevens and Rhoton (1975) and Stevens (1976) have presented evidence that the small polymerase-associated T4 protein p10k alters the binding of this subunit in vitro, even though the sigma factor remains active in the cell; it is, in fact, involved in transcription throughout T4 infection (Zograff, 1982). p10k may thus play a key role in the shutoff of host transcription, but since the relevant gene has not yet been identified, no mutants are available to test this possibility in vivo.

The possibility has not been excluded that *alc* affects the binding of p10k. Preliminary analyses of salt effects in vitro (Stevens, 1976) would be consistent with this possibility (Drivdahl, unpublished data). It is unlikely, however, that lack of σ alone is responsible for the observed differences in activity between dC- and HMdC-DNA templates, since σ acts catalytically, is recycled rapidly, and is required under both conditions.

CONCLUSIONS AND OUTSTANDING ISSUES

This chapter has discussed the many advances which have increased our understanding of how T4 disrupts host metabolism, in particular, how it destroys the structure of the host nucleoid and prevents host gene expression and replication. It is becoming more and more apparent that these studies are leading into one of the major unsolved problems in procaryotic molecular biology: how is the normal folded structure of the bacterial chromosome maintained, and how does this structure affect replication and gene expression? It is perhaps not surprising that progress has been rather slow.

Useful and interesting advances have been made in the course of these studies. We have learned how to make cytosine-containing T4 DNA (see appendix to this chapter), which has made possible the cloning of T4 genes and the construction of restriction maps, among other applications. The T4 genes responsible for host DNA degradation, nuclear disruption, and unfolding of the nucleoid have been identified and their roles have been substantially characterized with the aid of appropriate mutants, along with the first gene which is clearly involved in the shutoff of transcription from cytosine-containing DNA. Further characterization of these genes and gene products is certain to yield important insights into the structure of procaryotic nucleoids.

A number of important specific questions have also arisen from these studies. These include the following. (i) What are the mechanisms of action of gp*ndd* and gp*alc/unf*? (ii) What is the relationship between the DNA unfolding and inhibition of dC-DNA transcription functions of gp*alc/unf*; i.e., are the *alc* and *unf* phenotypes really manifestations of the same function? (iii) What are the mechanisms of restriction of *ndd* and *alc/unf* mutants in certain plasmid-carrying strains? (iv) Are there any functional relationships among *alc/unf*, *pseT*, and RNA ligase, as suggested by the map positions of these genes and the variety of interesting observations discussed by Snyder (this volume)? (v) What can we learn about the structure of the folded bacterial genome and the way(s) it is maintained by looking at the T4 unfolding process? (vi) What are the factors in addition to gp*alc* which shut off host transcription? Is gp*alt* involved, as would appear from in vitro results? Are there additional mechanisms? (vii) What are the mechanisms by which T4-coded polypeptides and ADP-ribosylation alter the transcriptional specificity of RNA polymerase? (viii) Why is gp*alc/unf* the only one of the transcription shutoff mechanisms that affects late transcription from T4 dC-DNA? (ix) What is the translational shutoff mechanism(s) directed against host mRNA? (x) Why is there such a discrepancy between the breakdown of host DNA and T4 dC-DNA in mutants which are endoII⁻ but endoIV⁻, where host DNA is degraded more rapidly, if anything, than in endoII⁺ endoIV⁻ infections, but the new T4 dC-DNA appears to only be somewhat nicked? Does repair or recombination (or both) play a significant role in the difference? (xi) What is the mechanism(s) of ghost killing of cells? What portions of that mechanism are also operative in productive infections, and how are the other aspects reversed after infection?

Whatever the answers to these questions, we anticipate that more surprises and new questions are in store as we pursue the elusive mechanisms of the prompt, virtually total shutoff of host functions after T4 infection.

We express our appreciation to our students and colleagues whose unpublished observations we have cited; to Rolf Drivdahl for the section on effects on RNA polymerase in vitro; to Burt Guttman for extensive support, discussion, and critical reading of the manuscript; and to our respective secretaries for their cheerful assistance.

The original work reported here was supported by National Science Foundation grants PCM 7905626 to E.K. and PCM 8003877 to L.S. and by Public Health Service grant GM 25417 from the National Institutes of Health to P.S.

APPENDIX

Preparation of Cytosine-Containing T4 Phage

ELIZABETH KUTTER AND LARRY SNYDER

Cytosine-containing T4 phage (T4dC strains) are useful for a variety of purposes, including a whole range of experiments discussed by Snustad et al. (this volume). They have also been used for restriction mapping and cloning of T4 DNA, which was previously difficult because the presence of 5′-hydroxymethyldeoxycytosine (HMdC) makes normal T4 DNA refractory to most restriction endonucleases (Kutter and Rüger, this volume). These phage also can be used for generalized transduction (Takahashi and Saito, 1982a; Takahashi and Saito, 1982c; Wilson et al., 1979), packaging very long pieces (perhaps entire headfuls) of *Escherichia coli* DNA. In addition, Casna and Shub (1982) have constructed derivatives which appear to be potentially useful as cloning vectors.

All T4dC strains must at least be defective in genes 56 (dCTPase) and *denB* (endonuclease IV) (see Fig. 6 of Snustad et al., this volume). Since *denB* mutations confer no detectable phenotype, most of the strains used have carried deletions extending into *denB* from the neighboring *r*II or *ac* regions. (The use of these deletions, incidentally, has led to a gap in most restriction maps of this region.) To make viable phage, the strains must also have a mutation in the *alc/unf* gene, which allows the transcription of genes for capsid proteins from dC-DNA. *alc/unf* mutants deficient in dCTPase and endonuclease IV can make viable phage in which cytosine replaces 40 to 70% of the HMC; such phage (termed T4dC* strains, as suggested by Takahashi et al. [1978]) are useful for generating a random population of partial digests for cloning work, but not for isolating specific bands or for constructing restriction maps. The fraction of dC in the DNA can be raised to about 95% by incorporating a *denA* (endonuclease II) mutation in the strain, thus preventing degradation of host DNA and eliminating one source of dCMP for the hydroxymethylase (HMase) (Hercules et al., 1971; Warner et al., 1970). Alternatively, it can be increased to 100% by including a gene 42 mutation that eliminates the HMase (Morton et al., 1980; Takahashi et al., 1978; Wilson et al., 1977) or a gene 1 mutation, eliminating dHMP kinase (Cowan, Kutter, and Karlson, manuscript in preparation). Unfortunately, the 1^- and especially the 42^- strains multiply rather poorly (Fig. 1), for reasons that are not completely understood but may be related to a possible direct role for the HMase and dHMP kinase in T4 DNA replication (Chiu and Greenberg, 1968; Mathews and Allen, this volume; Morton et al., 1980). Phage production, however, is much more drastically affected than DNA production (Fig. 2), particularly in infections with 42^- strains; this is related to aberrations in the rate of production of many early and late proteins (Morton et al., 1980). Phage production is even more inhibited at a low multiplicity of infection. The recently isolated 1^- T4dC strains grow significantly better than the 42^- strains, despite making similar amounts of DNA (Fig. 1 and 2): such mutants may therefore become the strains of choice for making T4 dC-DNA.

It should be noted that all T4dC strains will have HMdC if their DNA is propagated on amber-suppressing *E. coli* such as strain K803; they are then called T4dC^H stocks, to avoid confusion. T4dC strains will be restricted on most *E. coli* strains when their DNA contains cytosine, due to sensitivity to the host restriction system (Snustad et al., this volume). Thus, to prepare phage with dC-DNA, they must be propagated on nonsuppressing, nonrestricting *E. coli*.

In general, T4 with dC-DNA multiply more slowly than their HMdC-containing counterparts. Revertants of the *denA*, gene 1, or gene 42 mutations will help the phage multiply and are therefore favored, as are any other mutations affecting the pathways shown in Fig. 6 of Snustad et al. (this volume) which increase the amount of HMC. In addition, T4 DNA containing mostly dC or unglucosylated HMdC seems to have a high probability of undergoing rearrangements; Chow and Broker (personal communication) have used homoduplex mapping to show that 5% of such phage have major insertions, deletions, or inversions. These are recurring problems if one is isolating *alc* mutants or preparing T4 dC-DNA.

All in all, to get a useful stock of these phage containing 100% dC-DNA, one should propagate them routinely on strain K803 and then infect an $r^- m^- su^-$ host such as strain B834 at a multiplicity of infection of ≥3, waiting at least 90 min before lysing the cells. Two cycles of growth on strain B834 are usually needed to get the HMdC content down to acceptable levels, as determined by plating on strain 5KR1 (see below).

Some experimentation with T4dC strains requires that another mutation(s) be crossed into them. If a phage containing the mutation of interest is crossed with one containing the four or five mutations required for dC-DNA, a plethora of recombinant types must be screened to find the desired one. To help alleviate these technical problems and reverse the selection for pseudorevertants incorporating extra HMdC, Runnels and Snyder (1978) constructed an *E. coli* strain, B834 *galU56*, that is selective for T4 with dC-DNA. The selection is based on the fact that HMdC-containing DNA is subject to restriction by the *rgl* system unless it is glucosylated (Revel, 1967). Glucosylation cannot occur in a host with the *galU* mutation, which prevents the synthesis of uracil diphosphoglucose, the donor of glucose to T4 DNA (Fukasawa and Saito, 1964), but if the phage contain dC, they will escape restriction. B834 *galU56* is useful

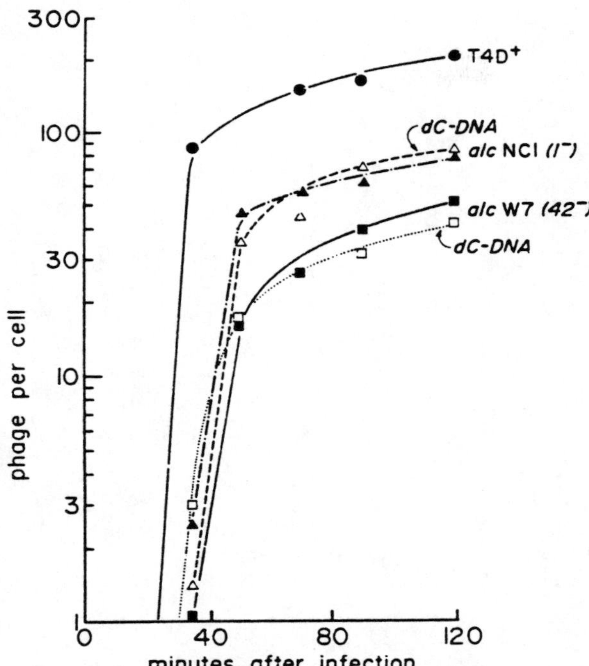

FIG. 1. Phage production after infection of *E. coli* B834 at 37°C with T4D$^+$ or various T4dC^H or T4dC strains. Symbols: ●, T4D$^+$; ▲ and △, alcNC1 (*r*IIB endonuclease IV$^-$ dCTPase$^-$ dNMP kinase$^-$), with stock grown on strain K803 (i.e., containing glucosylated HMdC) or on strain B834 (i.e., containing dC), respectively; ■ and □, alcGT7 (*r*IIB endonuclease IV$^-$ dCTPase$^-$ HMase$^-$), with stock grown on strain K803 or on strain B834, respectively. From Cowan, Kutter, and Karlson, in preparation.

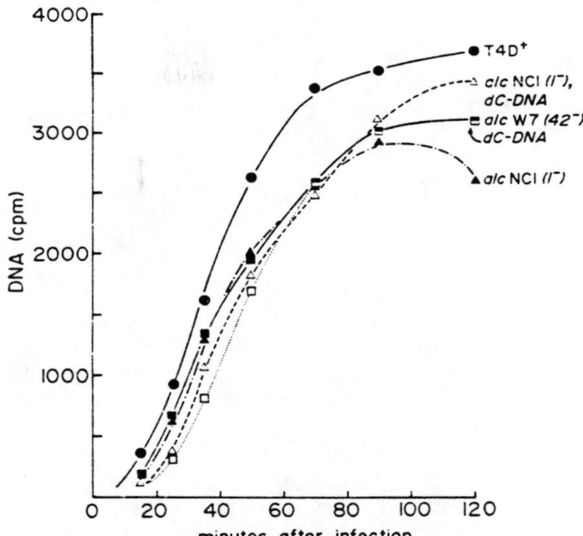

FIG. 2. DNA production from the experiment of Fig. 1, measured in terms of incorporation of [^{3}H]deoxythymidylate (0.5 μCi/ml and 10 μg/ml) in the presence of dA (100 μg/ml). Symbols used are the same as for Fig. 1.

for preparing T4dC stocks; it is somewhat less useful for detecting recombinants with dC-DNA or for isolating mutants, because HMdC-containing phage are not totally restricted. Doermann (this volume) provides additional suggestions for constructing multiple mutants involving T4dC strains, using UV rescue techniques. The *alc*-restrictive strains described by Snustad et al. (this volume) are also very helpful.

Once a stock of a T4dC strain has been prepared, there are a number of methods for estimating its cytosine composition, short of determining the base composition of the DNA. Since T4dC strains are restricted by the normal restriction-modification system, an indication of cytosine content can be obtained by determining the relative efficiency on r^+ and r^- *E. coli*. It is our experience that the ratio of plating efficiencies is about 10^{-1} for phage with 55% cytosine, about 10^{-3} to 10^{-4} for those with 95% cytosine, and about 10^{-5} for those with 100% cytosine. The sensitivity to restriction is highest if the host 5KR1 is used, carrying the gene for *Eco*RI on a small plasmid.

In addition, reversion of the amber mutation in gene 56 can be detected by using the fact that the dCTPase is also a dUTPase responsible for excluding uracil from T4 DNA. Thus, if the amber mutation is reverted or suppressed, the phage will be able to multiply on hosts positive for uracil-DNA-*N*-glycosylase after one cycle of growth on mutants of *E. coli* deficient in uracil-DNA-*N*-glycosylase and dUTPase (*ung*$^-$ *dut*$^-$ mutants).

Many of the T4dC strains are also, surprisingly, quite efficient transducing phage (Takahashi and Saito, 1982a; Takahashi and Saito, 1982c; Wilson et al., 1979), particularly those which are also 42$^-$. It is not clear how much, if any, of the T4 genome is carried by these transducing phage. Thus, for some such strains, such as alcGT7, a significant portion of the phage may contain bacterial genes, although the fraction is not large enough to have been detected to date by various physical methods.

All in all, preparation of cytosine-containing T4 phage takes some special care and precautions. However, as seen throughout this book, the T4dC strains have proved themselves well worth the special effort, making possible the generation of a detailed restriction map of T4, the cloning and sequencing of many parts of the genome, the mapping of several new genes, and the localization and characterization of a variety of control sites.

DNA METABOLISM

Enzymes and Proteins of DNA Metabolism

DNA Precursor Biosynthesis

CHRISTOPHER K. MATHEWS AND JAMES R. ALLEN

Department of Biochemistry and Biophysics, Oregon State University, Corvallis, Oregon 97331

The enzymology of DNA precursor biosynthesis in T-even-phage-infected bacteria represents one of the classical areas of T4 biochemistry. Beginning with the discovery of deoxycytidylate hydroxymethylase by Flaks and Cohen (1957), all of the enzymes involved had been described by 1970. Most of this information has been reviewed (Cohen, 1968; Mathews, 1971a; Mathews, 1977) and is covered here only briefly. Of greater current interest are the complex biological roles of some of the early enzymes, particularly their roles as virion proteins (reviewed in the chapter by Kozloff in this volume); the functioning of most of the enzymes of dNTP biosynthesis in an organized multienzyme complex; the action of this complex vis à vis the rate and fidelity of DNA replication; and the use of T4 as a model for understanding relationships between DNA replication and precursor biosynthesis in other organisms.

(The following abbreviations are used in this chapter: dNMP, dNDP, and dNTP, deoxyribonucleoside mono-, di-, and triphosphate, respectively; rNDP, ribonucleoside diphosphate; hm-dCMP, hm-dCDP, and hm-dCTP, 5-hydroxymethyldeoxycytidine mono-, di-, and triphosphate, respectively; HMC, hydroxymethylcytosine.)

PATHWAYS OF DNA PRECURSOR BIOSYNTHESIS IN T4 INFECTION

T4 infection drastically alters the flow of precursors into nucleic acids. Concomitant with a shutoff of all host RNA synthesis is a flow of ribonucleotides into pools of DNA precursors. However, although the rate of DNA synthesis increases up to 10-fold relative to that seen in uninfected *Escherichia coli*, the pools of dNTPs do not increase substantially (Mathews, 1972). One qualitative change seen is the disappearance of the dCTP pool and its replacement by hm-dCTP; this accounts for the substitution in T-even phage DNA of HMC for cytosine. A further modification, the glucosylation of the hydroxymethyl groups in HMC, occurs after deoxyribonucleotides have been polymerized into nascent DNA (Erikson and Szybalski, 1964).

To sustain the higher rates of dNTP synthesis seen in infected bacteria requires augmentation of several enzyme activities present in uninfected bacteria. The virus accomplishes this by synthesizing enzymes, such as thymidylate synthetase or dUTPase, which duplicate and enhance preexisting activities but which can readily be shown to be separate proteins, encoded by the T4 genome. Other phage-induced enzymes, such as those of hm-dCTP biosynthesis, catalyze reactions which are unique to the infected cell. A final group of enzymes includes two nucleotide kinases of the host cell which carry out the same roles in phage dNTP biosynthesis: nucleoside diphospho kinase and (d)AMP kinase. As summarized in Fig. 1, these enzymes do not have phage-specific counterparts. A third enzyme in this category is dCTP deaminase, which is an important enzyme of thymine nucleotide biosynthesis in *E. coli* (O'Donovan et al., 1971). However, T4 induces a dCMP deaminase which renders the dCTP deaminase nonessential; the disappearance of dCTP from infected cells probably makes the latter enzyme nonfunctional.

Aside from nucleotide synthesis de novo, a substantial source of DNA precursors is the bacterial chromosome, whose breakdown can yield up to 20 phage-equivalent units of each of the deoxyribonucleotides. Properties of the nucleases and other proteins involved in degradation of the host chromosome are discussed by Warner and Snustad in this volume.

Structural genes for all of the known T4 proteins involved have been identified and mapped. Some of these fall in a cluster between genes 32 and 63 (Fig. 2). Most of the rest (with the notable exception of gene 1) lie near other genes controlling DNA replication, between 18 and 30 kilobases on the T4 genome.

PROPERTIES, REGULATION, AND FUNCTIONS OF THE ENZYMES INVOLVED

Physical and Catalytic Properties

Table 1 summarizes information on each of the enzymes of T4 nucleotide metabolism, with respect to reaction(s) catalyzed, molecular weight, and availability of mutants altered in the production of each respective enzyme. We include here information on the glucosyltransferases which alter HMC residues in newly replicated DNA. Even though these are not, strictly speaking, enzymes of DNA precursor metabolism, they do lead to production of modified nucleo-

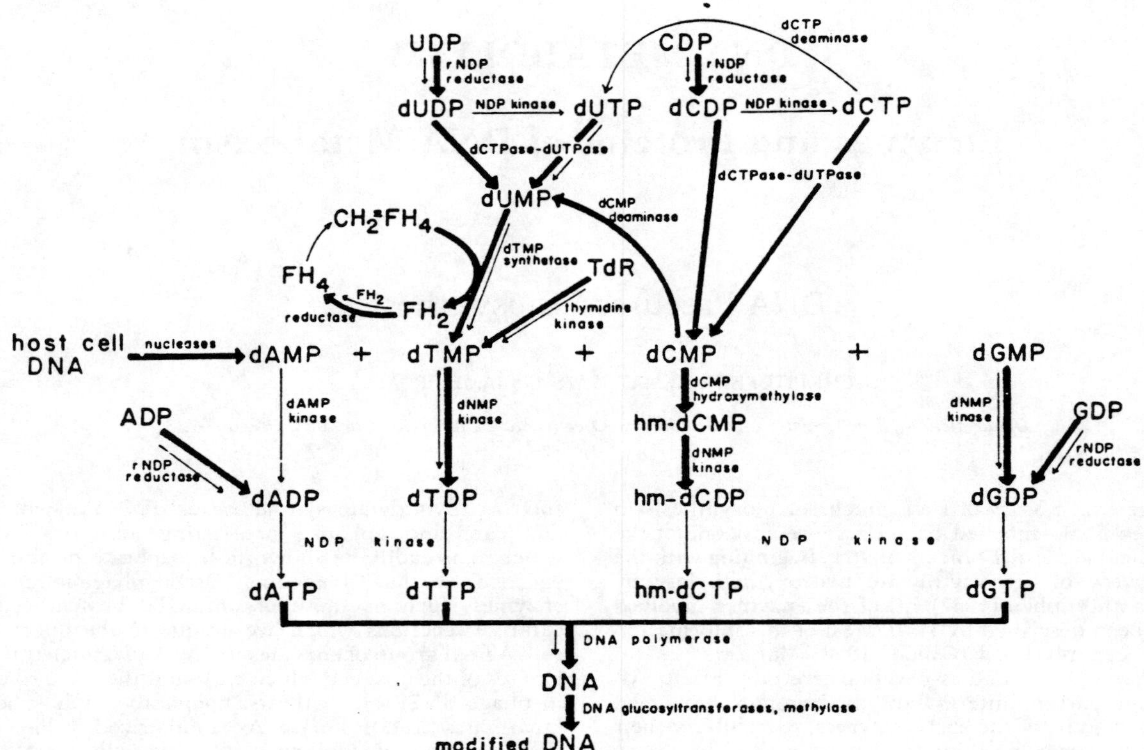

FIG. 1. Reactions of DNA precursor biosynthesis in T4 phage-infected *E. coli*. Reactions catalyzed by virus-coded enzymes are identified with heavy arrows, and those catalyzed by preexisting host cell enzymes are denoted with light arrows.

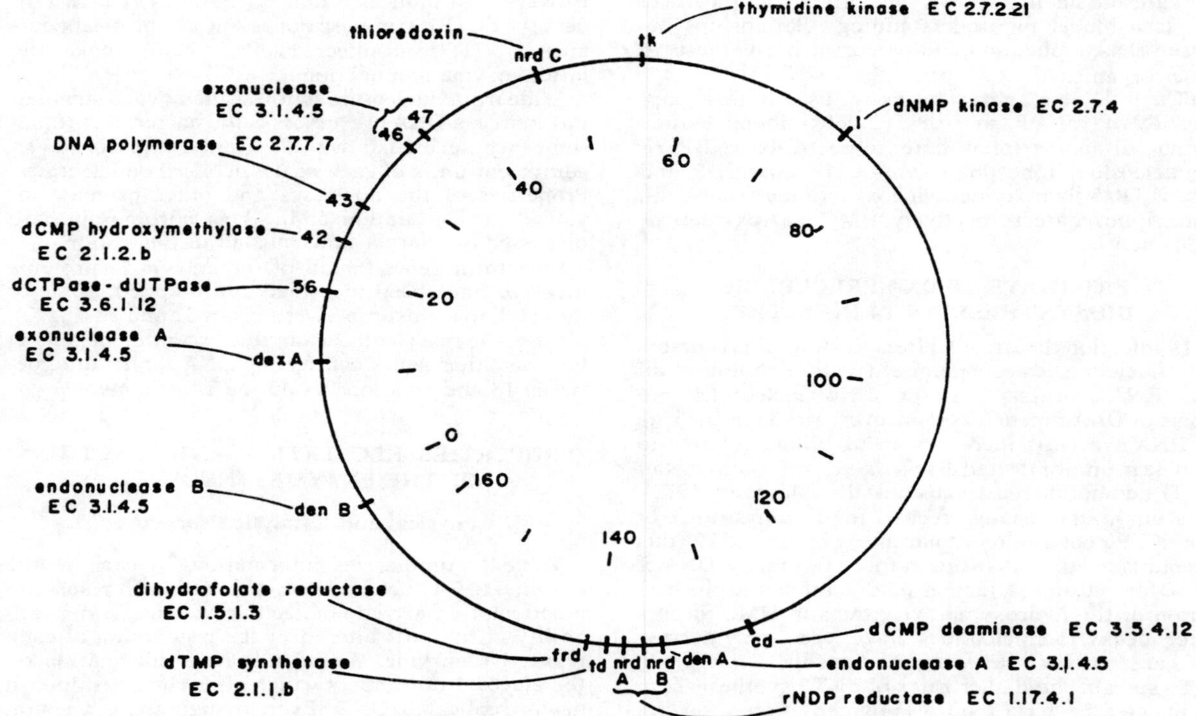

FIG. 2. Map positions of T4 genes coding for enzymes of DNA precursor biosynthesis. The numbers in the interior represent distance in kilobases from the *r*IIA/*r*IIB cistron divide.

TABLE I. Enzymes in T4 DNA precursor synthesis[a]

Enzyme	Reaction(s) catalyzed	Structural gene	Mutants available	Mol wt (× 103)	
				Native	Subunit
dCMP hydroxymethylase	dCMP + CH_2=FH_4 → hm-dCMP + FH_4	42	Amber, *ts*	60	27, 25[b]
dTMP synthetase	dUMP + CH_2=FH_4 → dTMP + FH_2	*td*	Nonlethal amber, missense, deletions	58	29
Dihydrofolate reductase	FH_2 + NADPH + H^+ → FH_4 + $NADP^+$	*frd*	Nonlethal amber, deletions, missense	44.5[c]	23[c]
dNMP kinase	dTMP, dGMP, hm-dCMP + ATP → dTDP, dGDP, hm-dCDP + ADP	1	Amber, *ts*	Unknown	22
(d)AMP kinase	AMP, dAMP + ATP → ADP, dADP + ADP	Host gene		Unknown	Unknown
NDP kinase	NDP, dNDP + NTP → NTP, dNTP + NDP	Host gene	One *cs* in *Salmonella*	110[d]	15.5[c]
dCTPase-dUTPase	dCTP, dUTP + H_2O → dCMP, dUMP + PP_i; dCDP, dUDP + H_2O → dCMP, dUMP + P_i	56	Amber, *ts*	59	15
dCMP deaminase	dCMP + H_2O → dUMP + NH_3	*cd*	Nonlethal missense	124, 129[f]	20.2
Thymidine kinase	TdR + ATP → dTMP + ADP	*tk*	Nonlethal missense; BUdR resistant	86	28
rNDP reductase	CDP, UDP, GDP, ADP + reduced thioredoxin → dCDP, dUDP, dGDP, dADP + oxidized thioredoxin	*nrdA*, *nrdB*	Nonlethal missense, amber, deletions, folate analog resistant[e]	225	85 (*nrdA*), 35 (*nrdB*)
Thioredoxin	see above	*nrdC*	Nonlethal missense	10.4	10.4
DNA α-glucosyltransferase	DNA-HMC + UDP-glucose → α-glucosyl DNA-HMC + UDP	*α-gt*	Amber, missense (lethal in combination with *β-gt*)	Unknown	Unknown
DNA β-glucosyltransferase	DNA-HMC + UDP-glucose → β-glucosyl DNA-HMC + UDP	*β-gt*	Amber, missense (lethal in combination with *α-gt*)	Unknown	46[h]

[a] Except as otherwise indicated, data are from sources cited in Mathews (1977).
[b] North and Mathews (1977); O'Farrell, et al. (1973).
[c] Purohit et al. (1981).
[d] Roisin and Kepes (1978).
[e] G. W. Lasser, unpublished data.
[f] Maley et al. (1972); Scocca et al. (1969).
[g] Johnson et al. (1976).
[h] Huang and Buchanan (1974).

tide residues which can be recovered from phage DNA (Lehman and Pratt, 1960; Lichtenstein and Cohen, 1960). Surprisingly, although these enzymes were described quite early (Josse and Kornberg, 1962) and the biology of DNA glucosylation has been extensively studied (Revel, this volume), little information is available regarding the molecular properties of these enzymes. All of the other enzymes have been obtained in highly purified or homogeneous form. Relatively little work has been done on the structures of these proteins, although T4 thioredoxin has been crystallized, and a three-dimensional model has been generated from X-ray diffraction data (Söderberg et al., 1978). The availability of cloned segments of the T4 genome is permitting primary structural analysis of T4 enzymes from DNA sequence determination and also large-scale enzyme isolations from T4 genes cloned into high-expression vectors (Belfort et al., 1983). The results of such analyses should yield interesting evolutionary comparisons between a bacterial virus-induced enzyme and its counterparts in higher organisms. For example, sequence data on the T4 *frd* gene, which codes for dihydrofolate reductase, shows considerable homology between the phage enzyme, on one hand, and bacterial and eucaryotic dihydrofolate reductases, on the other (S. Purohit and C. Mathews, Fed. Proc., in press). However, the enzyme is also similar in some respects to plasmid-encoded dihydrofolate reductases specified by certain trimethoprim resistance transfer factors, and the possibility of a common evolutionary origin is quite intriguing.

Regulation

Genetic regulation is discussed elsewhere in this volume. Suffice it to say here that all of the genes coding for enzymes of dNTP biosynthesis are expressed in the prereplicative phase. Some, like genes 42 and *frd*, are immediate-early genes, transcribed by an unmodified *E. coli* RNA polymerase, whereas at least one, dNMP kinase, shows a distinctive quasi-late pattern of expression (see Brody et al., this volume). Others, such as *nrd* and *cd*, have not been carefully analyzed. Ribonucleotide reductase activity, as measured in extracts, develops late relative to the other activities (Berglund et al., 1969; Yeh et al., 1969), but this may be a function not of transcription of the *nrd* genes, but of assembly of a multienzyme complex in which the reductase is a key constituent (see below).

Three of the T4 enzymes of dNTP biosynthesis have counterparts in higher cells which are subject to complex feedback regulation by dNTPs themselves: thymidine kinase, dCMP deaminase, and ribonucleotide reductase. Since T4 must synthesize a great deal of DNA in a short time during its life cycle, it is interesting to compare these phage enzymes with their cellular counterparts, to see whether similar control mechanisms have either developed or been conserved in the viral proteins. Most thymidine kinases are activated by dCTP and inhibited by dTTP. Ritchie et al. (1974) reported that T4 thymidine kinase is not subject to feedback regulation, but Iwatsuki (1977) reported specific inhibition by dTTP at physiological concentrations (85% inhibition at 0.29 mM). Activation by cytosine or HMC nucleotides was either undetectable or quite small. On the other hand, phage-coded dCMP deaminase displays both inhibition by dTTP and a nearly absolute dependence upon dCTP for activity, just as do eucaryotic dCMP deaminases (Maley and Maley, 1982; Maley et al., 1967; Scocca et al., 1969). Although the allosteric activator in vivo is probably hm-dCTP, it is noteworthy that dCTP is considerably more effective as an activator in vitro.

dCMP deaminase has been reported to play a special role in regulating the activity of a multienzyme complex of T4 dNTP-synthesizing enzymes, via mechanisms more complex than those just described (Chiu et al., 1977). This is discussed further below.

In all organisms, ribonucleotide reductase represents the branch point between pathways of RNA and DNA precursor biosynthesis. Since the enzyme reduces all four common rNDPs and since the resultant dNTPs are used for little else than the synthesis of DNA, the enzyme is closely regulated so as to produce a balanced supply of DNA precursors. In both bacterial and eucaryotic reductases, complex patterns of activation and inhibition are seen with different nucleoside triphosphates as effectors. Thelander and Reichard (1979) have presented a model for the *E. coli* enzyme in which the B1 subunit (analogous to the *nrdA* product of T4) contains two classes of sites for binding effectors: (i) an activity site, which binds either ATP or dATP and tends to increase or decrease all activities, respectively; and (ii) a specificity site, which binds ATP, dATP, dTTP, or dGTP and affects the activity of the enzyme for reduction of particular ribonucleotides. Berglund (1972) showed that activities of the T4 enzyme are subject to allosteric activation of a type shown by other rNDP reductases, but that the only negative effect is a weak inhibition of GDP reduction by dATP; dATP inhibits all activities of this enzyme from other organisms. Thelander and Reichard (1979) explain the T4 data in terms of the existence in this enzyme of specificity sites but no activity sites. However, the specificity sites of the T4 enzyme must be somewhat distinctive, because Berglund (1975) showed hm-dCTP to be a potent activator of the UDP and CDP reductase activities of this enzyme. The cellular counterpart, dCTP, plays little or no role in regulating rNDP reductase from other sources.

Although the effect of dATP on GDP reduction by the T4 enzyme is not pronounced, there is evidence that this regulatory mechanism is operative in vivo. Mathews (1972) found that when T4 DNA synthesis was blocked by mutation, intracellular dNTP pools expanded by up to 50-fold, with the exception of dGTP, which barely doubled. These pool data suggest also that the *E. coli* enzyme is not active during normal infection, for if it were, one would expect dATP accumulation to reduce the production of other dNTPs. The same conclusion is supported more directly by studies on the redox properties of *E. coli* and T4 thioredoxins (Berglund and Holmgren, 1975; Holmgren, 1978). The two proteins show no amino acid sequence homology, and the oxidation-reduction potentials differ enough that thioredoxin reductase can catalyze the reduction of T4 thioredoxin by *E. coli* thioredoxin. Since each ribonucleotide reductase can react only with its own thioredoxin, the *E. coli* reductase is presumed to become inactive once all of its own thioredoxin has become oxidized at the expense of T4 thioredoxin. On the other hand, Chiu et al. (1980) have shown that T4 *nrd* mutants can synthesize deoxyribonucleotides in infected cells (although at a much reduced rate), suggesting that the host enzyme is active under these conditions. This does not imply that the host reductase is active in normal infection, because in nrd^- infection, T4 thioredoxin presumably becomes fully reduced and is not able thereafter to oxidize *E. coli* thioredoxin.

T4 thioredoxin is unlike other thioredoxins in that it can be reduced directly by glutathione (*E. coli* thioredoxin reductase is the catalyst). In this respect it resembles glutaredoxin, an *E. coli* protein which can transfer electrons directly from glutathione to ribonucleotide reductase (Holmgren, 1976). Since glutathione itself is maintained in the reduced state by a very active NADPH-dependent glutathione reductase, this puts virtually all of the reducing equivalents in the cell at the disposal of the ribonucleotide reductase system via the following reductive electron transport chain: NADPH → glutathione → T4 thioredoxin → ribonucleotides. This is probably a key event in sustaining the extremely high rates of DNA synthesis characteristic of T4-infected bacteria.

Host Enzymes in T4 DNA Precursor Biosynthesis

Precursors to T4 DNA are derived largely via ribonucleotide reduction but also, as mentioned above, via the degradation of bacterial DNA. All of the steps involved in converting these distal precursors to dNTPs and thence to DNA are catalyzed by virus-coded enzymes, with two exceptions, as mentioned above: the phosphorylation of dAMP to dADP and the phosphorylation of all four dNDPs to their respective triphosphates. These observations were made primarily in Bessman's laboratory (Bessman, 1959; Bello and Bessman, 1963a; Duckworth and Bessman, 1967) as a result of investigations of the phage-induced dNMP kinase. This enzyme phosphorylates hm-dCMP, an activity undetectable in host cells and hence essential for phage DNA synthesis. The kinase also acts upon dGMP and dTMP, with all three substrates bound at the same site on the enzyme. The fourth monophosphate, dAMP, is the substrate for an extremely active bacterial AMP-dAMP kinase (although, we should note, dAMP is not an intermediate in the major pathway for dNTP biosynthesis, which goes through rNDP reductase). Also, whereas hm-dCMP kinase activity is absent in uninfected cells, Bello and Bessman (1963b) found high activity for phosphorylation of hm-dCDP to hm-dCTP. This was shown to be associated with nucleoside diphosphokinase, an enzyme of extremely high activity and low specificity for nucleotide substrates. In crude extracts the activity of this enzyme is some 20-fold higher than that of the fully induced T4 dNMP kinase, thus obviating the need for the virus to produce its own diphosphokinase. However, as discussed below, only a small fraction of this host enzyme is actually available to function in phage DNA precursor biosynthesis.

Phenotypes of Mutants Defective in DNA Precursor Biosynthesis

Genes 1, 42, and 56. Since valuable clues to the biological role(s) played by an enzyme can be obtained by analysis of enzyme-deficient mutants, it is worthwhile to consider the phenotypes of T4 mutants defective in the synthesis of enzymes of DNA precursor biosynthesis. This information is summarized in Table 1. From this we see that conditional-lethal amber and *ts* mutations exist only in those genes which play indispensable roles in the synthesis of hm-dCTP, namely, genes 1, 42, and 56. So far as has been studied, amber mutants with mutations in genes 1 or 42 display an absolutely DNA-negative phenotype (Mathews, 1968; Warner and Hobbs, 1967), whereas gene 56 mutants show a small accumulation of labeled precursors into DNA (Kutter and Wiberg, 1968; Warner and Hobbs, 1967; Wiberg, 1967). Actually, gene 56 mutants synthesize ample amounts of cytosine-containing DNA because the dCTP pool is not depleted; however, this DNA is rapidly degraded by phage-coded nucleases, which readily cleave DNAs in which cytosine has not been completely replaced by HMC. As discussed in the chapter by Snustad et al. in this volume, one can force T4 to accept cytosine in its DNA, but one must silence dCTPase and at least one cytosine-specific phage nuclease. An additional mutation in the *alc* gene is necessary for this DNA to express late gene functions to the extent that viable progeny phage can be formed.

α-*gt* and β-*gt*. In phages T2 and T6, only conditional-lethal mutations are available in genes coding for DNA glucosyltransferases. This is because all HMC residues in these phage DNAs are α-glucosylated, and inactivation of the gene abolishes all glucosylation. Phages with DNA containing non-glucosylated HMC are subject to a host-controlled restriction system which is still not fully characterized (but see Dharmalingam et al., 1982; Fleischmann et al., 1976; Hewlett and Mathews, 1975; Oliver and Goldberg, 1977; and Revel, this volume). This causes breakdown of phage DNA, at the cell surface, upon its entry into bacterial cells. In T4 70% of the HMC residues are α-glucosylated, and 30% are modified in the β configuration. The absence of either glucosyltransferase causes a sufficient number of HMC residues to be modified by the other glucosyltransferase that the phage escapes restriction. However, α-*gt* β-*gt* double mutants do show a conditional-lethal phenotype (Georgopoulos and Revel, 1971), just as do α-glucosyltransferase mutants of T2 or T6.

Obviously glucosylation plays an indispensable role in protecting phage DNA from host restriction. However, there is evidence for additional roles played by glucosylation. Montgomery and Snyder (1973) described a T4 mutant selected for its ability to grow on a rifampicin-resistant *E. coli* mutant with an altered RNA polymerase, which supports wild-type T4 growth poorly. The fast-growing mutant was found to lack β-glucosyltransferase, suggesting a role for glucosylation in transcriptional regulation (Snyder and Montgomery, 1974). More recently Levy and Goldberg (1980a) described a genetic recombination system in T4, which is specific for glucosylated DNA and acts only in certain regions of the genome. So far the biological significance of this system has not been elucidated.

***td, frd, cd, nrd,* and *tk*.** For the most part, phage genes controlling steps in the metabolism of a "standard" DNA precursor—dGTP, dTTP, or dATP—do not display conditional-lethal phenotypes. Specific selection techniques have been devised for the isolation of mutants altered in these genes. Perhaps the most ingenious is that of Hall et al. (1967), which allows selection of *td* or *frd* mutants. A host strain defective both in de novo pyrimidine biosynthesis and in cytidine-deoxycytidine deaminase is grown with cytidine as the sole pyrimidine source. This generates a dUMP deficiency, which can be remedied if the cells are infected with a mutant defective either in dTMP synthetase or in dihydrofolate reductase. Both mutants cause dUMP to accumulate (Fig. 1), which can leak from infected cells and feed neighboring cells. This causes mutant plaques to be surrounded by a white halo, which represents an area of more abundant cell growth than elsewhere on the plate. For reasons still unexplained, *frd* mutants display a thicker halo than do *td* mutants.

The white-halo phenotype provides a basis for the isolation of other strains with mutations affecting DNA precursor synthesis. Referring to Fig. 1, one can see that inactivation of either dCMP deaminase or ribonucleotide reductase should alleviate the dUMP overproduction seen in *td* or *frd* mutants. As expected, missense or nonlethal nonsense mutations in either *cd* or one of the three *nrd* genes suppress the white-halo phenotype of *td* or *frd* mutants when both primary and secondary mutations lie on the same genome (Hall, 1967; Yeh et al., 1969). These and related selections, plus brute-force mutant isolation, have allowed mapping of these genes and functional analysis of the gene products.

In general, where quantitative studies of phage growth and DNA synthesis have been performed, it has been found that these genes play useful but not indispensable roles in the life of these phage on common laboratory hosts. Two- to threefold reductions in the rate of DNA synthesis are seen when one of these gene products is inactive. Indeed, deletions are available which span large portions of the genome between genes 63 and 32 (Homyk and Weil, 1974). However, although viable, these deletion mutants grow very poorly on most hosts and not at all on about one-third of a set of wild-type *E. coli* strains called the Cal Tech collection (Mosher et al., 1977; Wilson, 1973). Although the bases for these defective phenotypes have not been investigated, it seems likely that the gene products missing in the deletion-bearing strains are essential for growth in some hosts. Consistent with this, we have isolated T4 point mutants which are restricted on one of the Cal Tech strains, CT526. These mutants are defective in induction of thymidylate synthetase (C. Mathews, unpublished data).

Multiple Roles of Some Early Enzymes

The functions of several T4 early enzymes are clearly more complex than simply the catalysis of reactions in intermediary metabolism. Here we mention two examples of this complexity, one of which is developed further below and the other of which is treated in detail in the chapter by Kozloff in this volume. The first example pertains to the near certainty that the T4 enzymes of dNTP biosynthesis function as part of an organized multienzyme complex, which also contains at least two host-coded enzymes of nucleotide metabolism. This concept was derived originally from the observation of Chiu and Greenberg (1968) that in gently lysed T4-infected *E. coli*, most of the dCMP hydroxymethylase activity was recovered in an aggregate which sedimented through sucrose gradients much more rapidly than did the free enzyme. This suggested at first that the enzyme protein plays a second role, as part of a DNA replication complex. However, evidence accumulating since then suggests that the aggregate of which dCMP hydroxymethylase is part is actually a dNTP-synthesizing complex which is juxtaposed in the cell with the DNA polymerization apparatus. As such it maintains local concentration gradients of DNA precursors at replication sites and hence contributes toward maintaining high rates of replication. This is discussed further below.

The second area of complexity relates to the fact that at least two of the early enzymes—thymidylate synthetase and dihydrofolate reductase—play both catalytic roles, in metabolism, and structural roles, as elements of the baseplate of the tail. This seemingly bizarre finding derived from the original observation of Kozloff and Lute (1965) that T-even phage tails contain a folic acid-like compound, localized in the baseplate. Once the folate had been identified as dihydropteroylhexaglutamate, a search for proteins which might bind this compound revealed the presence of a small amount of dihydrofolate reductase activity, also localized to baseplates (Kozloff, Lute, and Crosby, 1970). It seems clear that the protein is an integral part of the virion structure and not an adventitious contaminant, for several reasons. First, the *frd* gene, which codes for dihydrofolate reductase, has been shown also to be a determinant of the heat lability of virion infectivity (Kozloff, Crosby, Lute, and Hall, 1975; Mathews, 1971b). Second, antibodies raised against purified T4 dihydrofolate reductase neutralize the infectivity of T4 particles (Mathews et al., 1973). Evidence suggests that the target for this antiserum is in fact the virion-bound dihydrofolate reductase (Mosher et al., 1977) and that it is located in the radial arms of the baseplate (Mosher and Mathews, 1979). Comparable evidence has been presented for T4 thymidylate synthetase, and it is virtually certain that this protein is also a minor virion baseplate constituent (Capco and Mathews, 1973; Kozloff, 1981; Kozloff, Crosby, and Lute, 1975).

A seemingly anomalous aspect of these findings is the existence of viable mutants bearing long deletions which extend into or through the *td* or *frd* gene (or both genes) (Homyk and Weil, 1974). If the abovementioned enzyme proteins are essential structural elements, then these mutants should be nonviable. To be sure, they grow very slowly in laboratory strains of *E. coli*. However, the puzzling observation is that these mutants synthesize proteins which react immunologically with antibodies either to T4 dihydrofolate reductase or to T4 thymidylate synthetase (Capco and Mathews, 1973; Mosher et al., 1977). Such proteins can be detected in extracts or by the observation that antiserum to either enzyme neutralizes infectivity of deletion mutant particles actually more efficiently than it neutralizes that of wild-type phage. A possible answer to this dilemma may be found in the complex pedigree of the deletion mutants, raising the possibility that they contain at least part of the *frd* gene (Homyk and Weil, 1974; G. Mosig, personal communication). We hope that nucleotide sequencing studies, now underway, will help to resolve this problem (Purohit and Mathews, in press).

In other bacterial species dihydrofolate reductase regulates expression of its own gene, apparently by acting as a genetic regulatory protein (Sirotnak and McCuen, 1973). Gronenborn and Davies (1981) have shown that the enzyme in *Lactobacillus casei* binds to *L. casei* DNA, evidently at a specific site which may represent a regulatory region. No such findings have been made for the T4 enzyme, but intriguing genetic evidence depicts interactions between the *frd* gene product and two specific gene products involved in the DNA replication process per se, namely, the helicase-primase complex encoded by genes 41 and 61 (Macdonald and Hall, 1981). The nature of these interactions has not yet been analyzed biochemically but remains a fascinating question for further investigation.

MULTIENZYME COMPLEXES AND DNA PRECURSOR COMPARTMENTATION

DNA replication is more closely coordinated with the synthesis of its precursors than are other macromolecular biosynthetic processes. This relates at least partly to the facts that in most organisms DNA replication occurs during only part of the life cycle and that replication uses specialized precursors, the deoxyribonucleotides, which are not required for other metabolic processes. Thus, for the most part dNTPs need be present only when required for DNA replication. Two additional aspects of procaryotic DNA replication deserve comment: first, DNA chains grow very rapidly (500 to 1,000 nucleotides per s at 37°C), and second, the affinity of the replication apparatus for dNTPs is low, as judged from the properties of in vitro replication systems (Mathews and Sinha, 1982). All of these factors combine to support a model of replication in T4 which involves functional compartmentation of DNA precursors. This is maintained by a multienzyme complex which efficiently synthesizes dNTPs at replication forks, maintains high local concentrations despite rapid turnover of dNTPs, and also balances the synthesis of each of the four dNTPs at rates corresponding to the nucleotide composition of DNA. Such a complex could serve as a substrate channel and functionally compartmentalize DNA precursors, even in the absence of physically distinct compartments (Reddy and Mathews, 1978; Wovcha et al., 1976). Although data supporting this type of model have accumulated from studies of bacterial and mammalian cells (reviewed in Mathews et al., 1979), our most detailed evidence has come from studies of the T4 phage system.

Early Evidence for a T4 dNTP-Synthesizing Complex

As stated above, Chiu and Greenberg (1968) found that gently lysed T4-infected *E. coli* contained a rapidly sedimenting form of dCMP hydroxymethylase. This initially pointed toward a direct role for the gene 42 protein in the replication process, an idea which gained support when cells infected with gene 42 mutants displayed defective DNA synthesis in vitro, even when the metabolic block was bypassed by provision of hm-dCTP (Collinsworth and Mathews, 1974; North et al., 1976; Wovcha et al., 1973). On the other hand, Morton et al. (1978) reported that the requirement for active dCMP hydroxymethylase could be largely bypassed in vivo, at least under special conditions allowing dCTP incorporation and accumulation of cytosine-containing DNA. Two genetic observations pointed also toward interactions between the gene 42 protein and the replication apparatus. First, *ts* mutations in gene 42 altered the fidelity of replication, as seen by increased spontaneous mutation rates at other loci (Chiu and Greenberg, 1973; Drake, 1973; Williams and Drake, 1977). Second, Chao et al. (1977) described a nonsense mutation in gene 42 which was suppressed by a mutation in gene 43, suggesting direct or indirect interactions between dCMP hydroxymethylase and DNA polymerase.

It now seems likely that dCMP hydroxymethylase functions not directly in replication but as a part of a dNTP-synthesizing multienzyme complex. Two observations pointed in this direction. First, Tomich et al. (1974) devised an ingenious assay for the activity of dCMP hydroxymethylase and dTMP synthetase in intact infected cells. [5-^{3}H]uridine is fed to cells and is converted intracellularly to labeled substrates for both enzymes, each of which displaces the tritium from the 5 position. Thus, the in vivo activity of both enzymes is monitored by the production of tritiated water. Using this assay, Tomich and colleagues found that in vivo action of these enzymes starts at the same time as DNA replication, at least 5 min after enzyme activities can be detected in cell-free extracts. They postulated that the enzymes must be assembled into a complex to become activated in vivo.

The other pertinent observation was made by Mathews (1976), who followed thymine nucleotide pool expansion after a reversible blockade of DNA synthesis. Surprisingly, dTDP accumulated nearly as rapidly as dTTP, even though the activity of NDP kinase in extracts is some 20-fold higher than that of dNMP kinase, such that dTDP should have been immediately converted to dTTP. One way to explain this observation is to postulate that only a small fraction of the NDP kinase in a cell is available to participate in dNTP synthesis, perhaps by its incorporation into a specific complex.

Isolation and Properties of a T4 dNTP-Synthesizing Complex

When Reddy et al. (1977) sedimented gently prepared lysates of T4-infected *E. coli* through sucrose gradients, they confirmed the observation of Chiu and Greenberg (1968) that dCMP hydroxymethylase sediments as part of a large aggregate. Several other enzymes behaved similarly, including dTMP synthetase, dNMP kinase, and 3 to 5% of the host cell NDP kinase. Since the original observations, several other enzymes have been found in this aggregate: dCTPase-dUTPase, dCMP deaminase, rNDP reductase (but apparently not thioredoxin [J. W. Booth, unpublished data]), thymidine kinase, and host cell (d)AMP kinase (Allen et al., 1980). With modifications in the lysis procedure, we have found that dihydrofolate reductase, originally not found in the aggregate, is indeed associated with the other enzymes (Allen et al., 1983).

Sucrose gradient centrifugation yields a sedimentation coefficient for this enzyme aggregate of 15 to 20S, corresponding to an estimated molecular weight of 7 $\times$ 10^5 to 10 $\times$ 10^5. Gel filtration chromatography yields an estimated molecular weight of about 10^6 (Allen et al., 1983). This is close to the sum of the molecular weights of the individual enzymes known to be present in the aggregate. Since the aggregate to date has been purified to only about 20% of homogeneity, we have not yet been able to analyze for the presence of additional proteins or other components; however, these considerations suggest that the complex as isolated does not contain much additional material. If the aggregate does represent an organized multienzyme complex, then different approaches must be used to determine whether that complex is associated with the replication apparatus in intact cells.

What basis is there for believing that the aggregate does represent a specific multienzyme complex? First, control experiments rule out the possibility of artefactual aggregation after lysis of the cells (Allen et al., 1983). More importantly, the enzymes are kinetically coupled to one another (Allen et al., 1980; Reddy et al., 1977). As discussed by Gaertner (1978), kinetic coupling reduces the transit time for catalysis of a multistep reaction pathway; i.e., it reduces the time interval between addition of a substrate and attainment of the maximal rate of final-product formation. This is because intermediates released by one enzyme need not diffuse to reach the catalytic site of the next enzyme in the sequence, since the two enzymes are juxtaposed. This greatly reduces the time needed for accumulation of saturating levels of substrate for the second enzyme, and the same is true for the third and subsequent enzymes in a pathway. At the same time, intermediates accumulate to much lower concentrations than those seen in uncoupled systems, because saturating substrate levels need be reached only in the immediate vicinity of the enzyme complex.

We originally applied this reasoning to analysis of the three-step sequence dUMP → dTMP → dTDP → dTTP (Reddy et al., 1977). The T4 enzyme aggregate shows virtually no lag in dTTP production from dUMP, and the one intermediate which could be assayed, dTDP, accumulates to a much lower level than predicted for a corresponding mixture of unaggregated enzymes (Fig. 3). Computer simulations allowed an estimate that the local concentrations of intermediates in the vicinity of the complex were maintained at levels some one to two orders of magnitude higher than the average concentration in the reaction mixture.

This system was used to test an explanation for the requirement for a functional gene 42 product for T4 DNA synthesis in vitro, as described above (Reddy and Mathews, 1978). We proposed that in minimally disrupted cell preparations, and in vivo, the replication apparatus preferentially uses dNTPs which are generated in situ, by the action of an associated dNTP-synthesizing complex. When dCMP hydroxymethylase in the complex is inactivated, other activities of the complex are functionally inhibited. Consistent with this, we found that complexes prepared from cells infected with a *ts* gene 42 mutant are unable to synthesize dTTP from dUMP at 42°C, even though dCMP hydroxymethylase is not directly involved in this sequence (Reddy and Mathews, 1978).

Kinetic coupling has now been demonstrated in several other reaction sequences catalyzed by enzymes in the aggregate. In the dUMP-to-dTTP assay, we find that substitution of dihydrofolate and NADPH for tetrahydrofolate increases the activity of the system, indicating that dihydrofolate reductase is kinetically linked to the other three enzymes (Allen et al., 1983). Another readily demonstrated pathway is dCTP → dCMP → dUMP → dTMP. This confirms the existence of dCTPase and dCMP deaminase activities as part of the complex and also provides a convenient spectrophotometric assay for a coupled system, based upon the standard spectrophotometric assay for dTMP synthetase. The longest sequence yet demonstrated is UDP → dUDP → dUTP → dUMP → dTMP → dTDP → dTTP. Since the gene 56 nucleotidase acts on both dUDP and dUTP, dUTP may not be an obligatory intermediate in this pathway. Control experiments indicate dUTP accumulation only to low levels in this system, but that is expected even if it is an intermediate (Allen et al., 1980). At any rate, this experiment demonstrates coupling among five activities, including the key enzyme ribonucleotide reductase.

During the past several years Greenberg's laboratory has conducted parallel investigations of the isolated T4 dNTP-synthesizing complex. Their results, recently published in detail (Chiu et al., 1982), are in good agreement with ours. The functional integrity of their preparations was monitored by a simple but ingenious assay which follows the coupled sequence CDP → dCDP → dCMP → hm-dCMP. Two-dimensional electrophoretic analysis of their preparations revealed the presence of several DNA-associated proteins, including the products of genes 32, 46, *r*IIA, and *r*IIB, subunits of the T4 topoisomerase, and DNA polymerase. Further fractionation of the complex will probably be necessary to confirm that these proteins are physically associated with the several dNTP-synthesizing activities in the complex. The preparations of Chiu et al., like ours, seem to be free of membrane components or DNA, even though all may well be associated in vivo. Preparations from both laboratories are readily dissociated by dilution or by moderate salt concentrations, accentuating the difficulty of purifying a dNTP-synthesizing complex to homogeneity.

Functional Compartmentation of DNA Precursors

A major question at this stage is whether the dNTP-synthesizing complex as isolated from infected bacteria is actually associated with the replication apparatus in the cell. Evidence on this point, largely indirect but persuasive in the aggregate, comes from both our laboratory and Greenberg's laboratory. Our evidence is derived from studies with infected cells permeabilized by sucrose plasmolysis. These cells incorporate distal DNA precursors—rNDPs or dNMPs—into DNA severalfold more rapidly than they incorporate proximal precursors, namely, the dNTPs (Reddy and Mathews, 1978). This suggests that dNTPs generated through action of the complex are preferred substrates for DNA synthesis. Indeed, we found that dNTPs added to sucrose-plasmolyzed cells must break

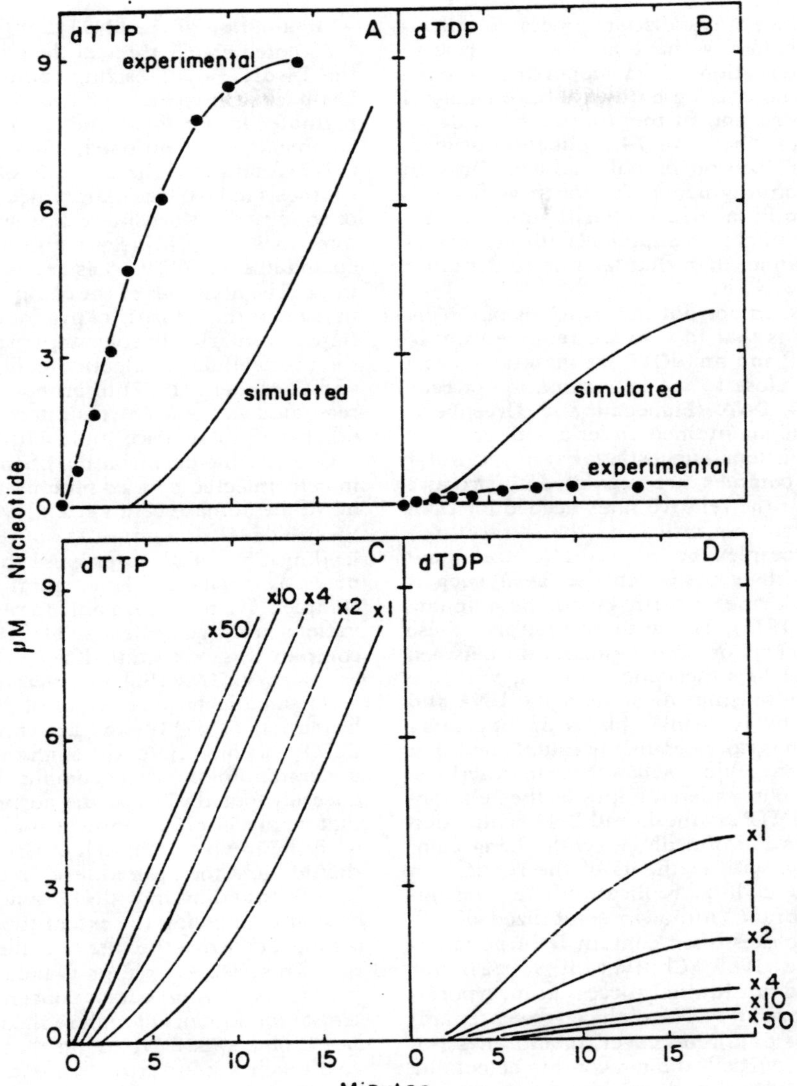

FIG. 3. Production of dTTP from dUMP by aggregated enzymes. (A and B) Experimental data, along with simulated results which would have been expected if the enzymes involved were not kinetically linked. (C and D) Simulations based upon increases in concentrations of intermediates by arbitrary factors of 2, 4, 10, and 50.

down before being incorporated into DNA. To the extent that the plasmolyzed cell can provide a model for an intact infected cell, we can visualize two functionally distinct compartments of dNTPs: a small, rapidly turning over, highly concentrated pool at or near the replication fork, which feeds the replication apparatus, and a large, slowly replenished, and more dilute pool which may be distributed throughout the cell and can be used in DNA repair or recombination (or both). The replication pool is seen as being very small in comparison with the latter pool, because plasmolyzed cells maintain vigorous nucleotide incorporation into DNA even when levels of dNTPs in the cells are so low as to be undetectable (Reddy and Mathews, 1978).

Using sucrose-plasmolyzed cells, Greenberg's laboratory has described two distinct DNA synthetic processes in these cells (Wovcha et al., 1976). One utilizes deoxyribonucleotide substrates, either mono- or triphosphates. The other can use ribonucleotides, but does not require a complement of four added precursors. Wovcha et al. suggested that exogenous precursors in the latter system contribute to, or monitor, an ongoing process which uses endogenously generated precursors. Both processes represent semiconservative replication, and both require functional DNA replication genes. Both are seen as substrate channels of the type we are discussing, with the deoxyribonucleotide-dependent system perhaps representing utilization of host cell DNA breakdown products and the other system representing the utilization of nucleotides synthesized de novo. None of our data argues against the existence of two distinct precursor utilization systems, although we have not directly tried to confirm these observations. Data supporting the existence of distinct, nonmixing pathways have emerged in a study by Morris and Bittner (unpublished data). These workers described five T4 replication origins, based upon hybridization of pulse-labeled DNA to mapped restriction fragments. Two of these five origins are inefficiently labeled by [^{3}H]thymidine, suggesting that they might be supplied with precursors by a different channel than that leading to the other origins.

One of the most important observations of Greenberg's laboratory is that in vivo the relative rates of synthesis of dTMP and hm-dCTP are maintained at a ratio of 2:1, very close to the representation of each nucleotide in T4 DNA (Flanegan and Greenberg, 1977). This ratio is maintained under a wide range of physiological conditions, suggesting that intracellular activities of the complex are regulated to produce DNA precursors at the relative rates needed for DNA synthesis. However, this ratio, and the intracellular activities of the complex, are altered after treatment of infected cells with agents which alter DNA, such as mitomycin C, proflavine, or methyl methane sulfonate (Flanegan et al., 1977). The authors interpret these observations in terms of direct interaction between DNA and the dNTP-synthesizing complex, with the complex somehow sensing the state of the DNA and being controlled by it. While this is an appealing interpretation, it has not been firmly established that DNA is in fact the target for each of these interactions.

A giant step in our understanding of the relationship between T4 dNTP synthesis and DNA replication would be the isolation of a dNTP-synthesizing complex in association with elements of the replication apparatus and the cellular replication site, presumably at the membrane. Although solubilized dNTP-synthesizing complexes often contain DNA polymerase (Allen et al., 1980; Chiu et al., 1982), our laboratory has had but limited success at incorporating distal precursors into DNA with cell-free preparations (unpublished data). Moreover, membranes isolated from T4-infected cells did not contain detectable amounts of dNMP kinase or dCMP hydroxymethylase when the proteins were displayed on gels (North and Mathews, 1977). Nevertheless, there is good reason to believe that dNTP synthesis is membrane associated in vivo. The best evidence is that of Wirak and Greenberg (1980). Using the tritium release assay described above, they showed that mutations in gene 39, and to a lesser extent in the other DNA delay genes, affect the rate of deoxyribonucleotide synthesis in vivo. The DNA delay genes 39, 52, and 60 encode T4 topoisomerase (Kreuzer and Huang, this volume). Two of these proteins (gp39 and gp52) have been found among the DNA-binding proteins of T4 (Huang and Buchanan, 1974), and the same two proteins have been detected in membranes of infected cells (Takacs and Rosenbusch, 1975). These observations led Wirak and Greenberg (1980) to propose that the topoisomerase plays a critical role in assembling, at the membrane, complexes for dNTP biosynthesis and DNA replication. The presence of topoisomerase subunits in the solubilized preparations of Chiu et al. (1982) lends support to this model.

Regulation of the dNTP-Synthesizing Complex

As noted above, three of the constituent enzymes of the T4 dNTP-synthesizing complex are known to be feedback controlled in vitro. Do such mechanisms regulate activities of these enzymes in the intact complex? As mentioned, Flanegan and Greenberg (1977) found that the 2:1 ratio of dTMP to hm-dCMP synthesis in vivo was maintained under a wide variety of conditions. Most interesting was what was seen in infections by DNA-negative mutants, where, as discussed above, dNTP pools expand by some 50-fold or more. The activities of the complexed enzymes, and in particular the 2:1 ratio of pyrimidine dNMP synthesis, remain remarkably constant despite huge changes in the intracellular nucleotide pools. This led Flanegan and Greenberg (1977) to propose that the complex is regulated in some intrinsic fashion, by mechanisms other than those discernible with individual enzymes.

One specific perturbation of the 2:1 ratio was seen in cells infected by a *cd* mutant, unable to synthesize dCMP deaminase (Chiu et al., 1977). These cells synthesize dTMP and hm-dCMP at relative rates of 0.6:1, leading Chiu et al. to propose that dCMP deaminase plays a special role in regulating activities of the complex. We have been able to reproduce these observations in vitro (Allen et al., 1983). When isolated complex was exposed to dCMP, it formed hm-dCMP in excess over dTMP (Fig. 4). However, in the presence of dCTP, an allosteric activator of dCMP deaminase, the formation of dTMP was activated and that of hm-dCMP was inhibited. Although this leaves us far from understanding how the complex is controlled to produce all four dNTPs at the correct rates, it suggests that analysis of our preparations in vitro, coupled with the in vivo approach of Greenberg's laboratory, should yield the desired level of comprehension.

Understanding how dNTP synthesis is regulated is crucial to knowing the extent to which operation of a complex controls the rate and fidelity of DNA replication. There is good reason to believe that the complex maintains DNA precursor concentration gradients essential for support of high replication rates (Mathews and Sinha, 1982). We found that DNA replication in vivo reaches its maximal rate at a dTTP pool which corresponds to an average intracellular concentration of 65 μM. However, DNA replication in the T4 purified protein system, an excellent model for chain elongation (Sinha et al., 1980), requires nearly 250 μM of each dNTP to achieve saturation. To the extent that one can compare in vivo and in vitro conditions, this suggests at least a fourfold dNTP gradient near replication forks.

Understanding the relationship to fidelity of DNA replication will be more difficult. It is clear, from studies on in vitro replication systems, that dNTP levels represent a determinant of fidelity and that one can force substitution errors by imposing large pool biases (Fersht, 1979; Hibner and Alberts, 1980; Sinha and Goodman, this volume; Sinha and Haimes, 1981). However, it seems unlikely that such large asymmetries are present in vivo, although the difficulty of measuring effective dNTP concentrations at forks seems to rule out testing this. What seems more intriguing are the mutator phenotypes displayed by some gene 42 *ts* mutants, antimutator phenotypes of some other gene 42 mutations (Williams and Drake,

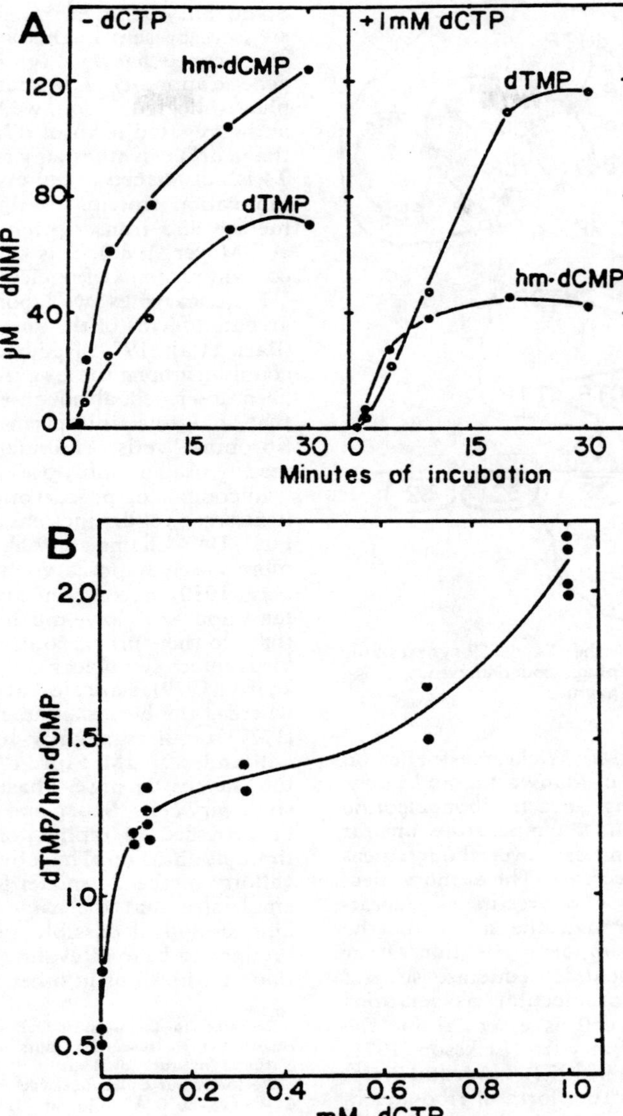

FIG. 4. Regulation of pyrimidine dNMP synthesis by dCTP in vitro. (A) Partially purified T4 complex was incubated with [^{3}H]dCMP in the presence or absence of 1 mM dCTP. Synthesis of dTMP and hm-dCMP was monitored with time. (B) Ratios of dTMP to hm-dCMP synthesized in 20-min incubations are plotted against dCTP concentrations. Data are combined from several different experiments.

1977), and, more recently, an antimutator phenotype of a *ts* gene 1 mutation (C. Mathews, unpublished data). It seems unlikely, although it has not yet been tested, that these mutations affect dNTP steady-state pools in some gross way, particularly since they are tested at temperatures permissive for the *ts* strains. However, through alteration of one component of the complex, its overall activities might change so as to affect concentrations in the microenvironment of the replication fork. Although this cannot be tested in vivo, the in vitro system with partially purified complex seems amenable to this type of analysis.

Applicability of the T4 Model

A simplified, and still speculative, model of the T4 dNTP-synthesizing complex is shown in Fig. 5. To what extent is this a suitable model for understanding relationships between DNA precursor biosynthesis and DNA replication in other organisms? The idea of an organized complex which generates substrates at their sites of utilization has some appeal a priori and has been used to explain various observed complexities of DNA metabolism, in both procaryotes and eucaryotes (Manwaring and Fuchs, 1979; Rode et al.,

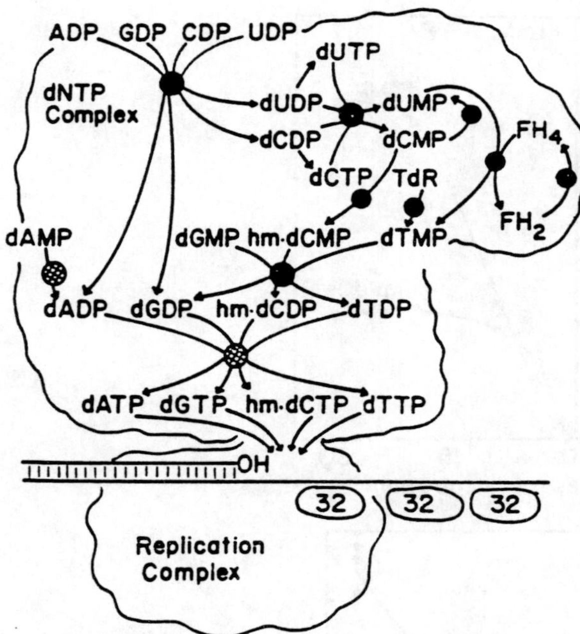

FIG. 5. Speculative view of the T4 dNTP-synthesizing complex. Closed circles denote phage-coded enzymes: cross-hatched circles denote *E. coli* enzymes.

1980; Scott and Forsdyke, 1980; Wickremasinghe and Hoffbrand, 1980). The data of Manwaring and Fuchs were particularly interesting. In a *ts* ribonucleotide reductase mutant of *E. coli*, a temperature upshift immediately halted DNA synthesis, even though measured dNTP pools did not decrease. The authors speculated that only the small pool of precursors generated at replication sites through the action of the reductase was available to support replication. Other studies on *E. coli* ribonucleotide reductase suggest that maintenance of intramolecular associations which exist in the intact cell is essential for the enzyme to be highly active in vitro (Eriksson, 1975; Lunn and Pigiet, 1979; Warner, 1973). Lunn and Pigiet described a rapidly sedimenting form of ribonucleotide reductase in gently lysed spheroplasts of *E. coli*. Some, but not all, of the enzymes of dNTP biosynthesis were present in this aggregate, whose biological significance has not yet been tested by kinetic or genetic approaches. In another procaryotic system, T5 phage-infected *E. coli*, we have found no evidence for an aggregated form of dTMP synthetase, suggesting that a dNTP-synthesizing complex of the type found in T4 is not formed (Allen et al., 1983). In T5 the major replication proteins are tightly coupled to RNA polymerase and transcription-regulatory proteins (Ficht and Moyer, 1980). This complex as purified does not contain proteins of nucleotide biosynthesis.

For eucaryotes two laboratories have described aggregated forms of the enzymes of dNTP biosynthesis (Baril et al., 1974; Reddy and Pardee, 1980). Kinetic coupling among these aggregated enzymes has not yet been described, although Reddy and Pardee reported that in lysolecithin-permeabilized Chinese hamster fibroblast cells, ribonucleoside diphosphates are readily used as substrates for DNA synthesis.

In comparing procaryotic and eucaryotic DNA replication, two distinctions must be borne in mind. First, DNA chains grow about an order of magnitude more slowly in procaryotes than in eucaryotes (Kornberg, 1980). Second, eucaryotic DNA replication systems appear to have much higher affinities for dNTPs than do their procaryotic counterparts. The polyoma virus-infected nuclear system of Eliasson and Reichard (1979) saturates at as low as 5 μM dNTPs, whereas the herpes simplex virus system of Francke (1977) requires slightly higher concentrations, but still under 25 μM. Since dNTP pools are localized to the nucleus during S phase, at least in Chinese hamster ovary cells (Skoog and Bjursell, 1974), dNTPs can be provided to replication sites at adequate rates through diffusion. This obviously does not negate the validity of the T4 model for eucaryotes, but it does emphasize that the same combination of physical, kinetic, and, if possible, genetic evidence should be brought to bear in evaluating any precursor-replication relationships in other systems.

Research in the authors' laboratory was supported by Public Health Service research grants AI-15145 and CA-24323 from the National Institutes of Health.

We thank our colleagues who contributed to this work, notably, G. W. Lasser, D. A. Goldman, J. W. Booth, S. Purohit, and R. G. Sargent.

Mechanism of DNA Replication Catalyzed by Purified T4 Replication Proteins

NANCY G. NOSSAL[1] AND BRUCE M. ALBERTS[2]

Laboratory of Biochemical Pharmacology, National Institute of Arthritis, Diabetes, and Digestive and Kidney Diseases, Bethesda, Maryland 20205,[1] and Department of Biochemistry and Biophysics, University of California, San Francisco, San Francisco, California 94143[2]

Our long-range goal has been to obtain a complete set of highly purified T4 bacteriophage DNA replication proteins that can be mixed back together to reconstruct a functional replication apparatus in vitro. Ideally, this apparatus should synthesize DNA on double-helical DNA templates with the same fork geometry, fork rate, and base-pairing fidelity that is observed for T4 replication forks within the cell. In addition, it should initiate replication forks at replication origins by the correct biological mechanism.

The marriage of genetics and biochemistry has been a fruitful one in these studies of T4 DNA replication. Studies of the 47 complementation groups, or genes, in the original set of conditionally lethal T4 mutations (Epstein et al., 1964) revealed 10 genes that cause a noticeable defect in DNA replication. Subsequent work has enlarged the list of mutations, and now about 30 genes that severely affect T4 DNA synthesis are known (Mathews, 1977; Wood and Revel, 1976). Fortunately, not all of these genes directly affect replication fork mechanisms; for example, a sizable number are known to encode enzymes needed for nucleotide precursor biosynthesis (Cohen, 1968; Mathews, 1977; Mathews and Allen, this volume).

Genetically identified components of the T4 DNA replication apparatus have been purified by using in vitro complementation assays (Barry and Alberts, 1972; Morris, Hama-Inaba, Mace, et al., 1979; Morris, Moran, and Alberts, 1979; Nossal, 1979). In deciding which gene products to purify, we initially chose those whose alteration by mutation causes the most severe defects in DNA replication, without depleting the normal deoxyribonucleoside triphosphate precursor pools (a subclass of the "DO" [no DNA synthesis] mutants of Epstein et al. [1964], as defined by the experiments of Mathews [1972]). In this way, the proteins corresponding to genes 44/62, 45, and 41 were purified. Other components, such as the DNA polymerase (gene 43 product), the gene 32 DNA helix-destabilizing protein, the gene 61 priming protein, the *dda* DNA helicase, and the DNA topoisomerase, have been purified by more standard procedures, in which a direct property or activity of the protein was monitored. But in all cases, the phenotype of a bacteriophage carrying a mutant version of the protein in question has been crucial in establishing that the protein has an important role in DNA replication.

To date, T4 replication proteins corresponding to the products of 11 T4 genes have been purified to near homogeneity and extensively characterized (Table 1). Our studies have established definite roles for 8 of these 11 gene products in the process of replication fork movement, as is discussed in detail in this chapter (reviewed in Table 1). When these proteins are mixed together, replication forks form in vitro that in all respects closely resemble the replication forks found in vivo. However, the forks formed in the in vitro system are initiated by a non-biological mechanism, since they start by strand displacement synthesis from a random nick on a double-helical template, rather than from a replication bubble that forms at a specific origin DNA sequence.

Our present knowledge of the mechanism of replication fork initiation at a T4 replication origin has been derived from studies carried out on whole cells. Such in vivo studies suggest that both the T4 DNA topoisomerase and the host *Escherichia coli* RNA polymerase are likely to be involved in the true fork initiation process (see the chapters by Mosig, Kozinski, and Kreuzer and Huang in this volume). Attempts to reconstitute this biological fork initiation process in vitro have thus far been unsuccessful (unpublished data). We therefore confine our attention in this review to the details of the mechanism of replication fork movement.

T4 DNA POLYMERASE

The first of the T4 replication enzymes to be isolated was the DNA polymerase (Table 1). The analogous T2 enzyme was identified in extracts from infected cells as a polymerase differing from *E. coli* DNA polymerase I in its chromatographic properties and in its strong preference for single-stranded DNA templates (Aposhian and Kornberg, 1962). T4 DNA polymerase was subsequently identified and shown to be the product of gene 43 (de Waard et al., 1965; Warner and Barnes, 1966).

The purified T4 enzyme, a monomer of about 110 kilodaltons, by itself catalyzes three reactions: (i) 5'-to-3' polymerization on a primed single-stranded DNA template; (ii) 3'-to-5' exonucleolytic hydrolysis of single-stranded DNA and, at a slower rate, of a 3'-OH end on duplex DNA; and (iii) primer-template-dependent turnover of deoxyribonucleoside triphosphates to the corresponding monophosphates. (Englund, 1971; Goulian et al., 1968; Hershfield and Nossal, 1972). This turnover appears to result from hydrolysis of newly incorporated nucleotide at the 3' terminus; thus, in addition to requiring a primer-template, turnover is enhanced when further polymerization is delayed by pause sites in the template (regions of special secondary structure or sequence) or when the polymerase is blocked at the end of a template or by the lack of the next required complementary nucleotide (Englund, 1971; Nossal and Hershfield, 1973; Roth et al., 1982).

TABLE 1. T4 DNA replication proteins[a]

Designation	T4 gene	Activities of protein alone[b]	Role in DNA replication		
			RNA primer synthesis	Synthesis on single-stranded templates	Strand displacement synthesis
DNA polymerase	43	5′-to-3′ polymerase on ssDNA templates only; 3′-to-5′ exonuclease	—[c]	Required	Required
Helix-destabilizing protein (gene 32 protein)	32	Cooperative binding to ssDNA; helix destabilization	—	Stimulates	Required
Polymerase accessory proteins					
Gene 44/62 protein	44/62	ssDNA termini-dependent ATPase (dATPase)	—	Stimulates	Required
Gene 45 protein	45	Stimulates gene 44/62 protein ATPase	—	Stimulates	Required
Priming proteins					
Gene 61 protein	61	Binds ssDNA	Required	—	—
Gene 41 protein	41	Long ssDNA-dependent GTPase and ATPase; DNA unwinding (5′-to-3′ helicase)[d]	Required	—	Stimulates
T4 DNA helicase	*dda*	ssDNA-dependent ATPase (dATPase); DNA unwinding (5′-to-3′ helicase)[d]	—	—	Stimulates
T4 DNA topoisomerase	39/52/60	dsDNA-dependent ATPase; type II topoisomerase (dsDNA strand passage); dsDNA cleavage when inhibitors added	—	—	—

[a] References and further details are given in the text.
[b] Abbreviations: ss, single stranded; ds, double stranded.
[c] —, No role has been demonstrated.
[d] Helicase direction is indicated as the direction of movement on the strand that is not displaced.

The conclusion that turnover is a two-step process of polymerization followed by excision is supported by recent studies with α-thio analogs of deoxyribonucleoside triphosphates; these show inversion of the configuration at the α-phosphorus in the nucleotidyl transfer reaction during polymerization (Romaniuk and Eckstein, 1982) and retention of this configuration in the deoxyribonucleoside monophosphate produced by the net turnover reaction (Gupta et al., 1982). The same studies argue against the formation of a covalent deoxyribonucleotide-polymerase complex as an intermediate in the polymerization process; in this case, a retention of configuration during polymerization would be expected due to the two-step nature of the polymerization event.

The 3′-to-5′ exonuclease of T4 polymerase is more active than that of other procaryotic DNA polymerases. This makes it the enzyme of choice for labeling DNA fragments by hydrolysis and repair at the 3′ terminus (O'Farrell et al., 1980), but results in a considerable wasting of deoxyribonucleoside triphosphates due to turnover to the monophosphates even under optimal in vitro conditions for DNA synthesis (Hershfield and Nossal, 1972). The preference of the nuclease for non-base-paired nucleotides at the 3′ terminus of duplex DNA led to the proposal that it acts as a proofreader to remove most incorrect nucleotides inserted during polymerization (Brutlag and Kornberg, 1972; Muzyczka et al., 1972). This proposal is supported by the finding that the ratio of incorrect nucleotides turned over to those stably incorporated is decreased for the polymerase produced by some gene 43 mutants with a mutator phenotype and increased for the polymerase produced by some gene 43 mutants with an antimutator phenotype (Bessman et al., 1974; Gillin and Nossal, 1976a; Muzyczka et al., 1972). Replication fidelity can also be decreased by a reduction in the ability of mutant T4 DNA polymerases to select the correct nucleotide for insertion (Gillin and Nossal, 1976b; Hershfield, 1973; Reha-Krantz and Bessman, 1981; Sinha and Goodman, this volume.)

Purified T4 DNA polymerase by itself cannot carry out several of the reactions needed to copy duplex DNA: (i) the enzyme repairs single-stranded gaps in duplex DNA and copies primed single-stranded templates (Goulian et al., 1968; Masamune et al., 1971), but the processivity (nucleotides added per enzyme-binding event) is less than 10 at salt concentrations that approach physiological values (Das and Fujimura, 1979; Newport et al., 1980); (ii) unlike *E. coli* DNA polymerase I and bacteriophage T5 DNA polymerase, the T4 enzyme does not catalyze strand displacement synthesis on a duplex DNA, where polymerization is initiated by covalent addition to the 3′-OH end at a nick and is accompanied by unwinding of the template helix (Masamune and Richardson, 1971; Nossal, 1974); and (iii) like other DNA polymerases, T4 polymerase is unable to initiate the synthesis of a new DNA chain, as required to start each Okazaki fragment on the lagging strand. These characteristics of the purified T4 DNA polymerase, as well as the genetic evidence for the existence of other replication genes, made it clear that other T4 enzymes would be needed to reconstruct the reactions at a DNA replication fork.

GENE 32 HELIX-DESTABILIZING PROTEIN AND GENE 44/62 AND GENE 45 POLYMERASE ACCESSORY PROTEINS

The T4 gene 32 helix-destabilizing protein (single-stranded-DNA-binding protein) was isolated on the basis of its strong affinity for single-stranded DNA cellulose and identified by the fact that the protein is missing when *E. coli* is infected by any one of several phage mutants with mutations in gene 32 (Alberts et al., 1969). The gene 32 protein binds cooperatively to single-stranded DNA and destabilizes but does not melt natural duplex DNA (Alberts and Frey, 1970). These binding reactions have subsequently been studied in detail, as described in the chapters by von Hippel et al. and Williams and Konigsberg in this volume.

The products of T4 genes 44, 62, and 45 were purified by in vitro complementation assays that measure DNA synthesis with endogenous DNA templates in crude extracts made from *E. coli* infected with the corresponding mutant phage (Barry and Alberts, 1972; Morris, Hama-Inaba, Mace, et al., 1979; Nossal, 1979). The products of genes 44 and 62 purify as a complex that complements either 44^- or 62^- extracts (Table 1). This complex has a single-stranded-DNA-dependent ATPase (or dATPase) activity that is stimulated by the purified gene 45 protein (D. Mace, Ph.D. thesis, Princeton University, Princeton, N.J., 1975; Piperno et al., 1978).

Stimulation of DNA Synthesis on a Single-Stranded DNA Template

The gene 32 protein, on the one hand, and the combination of the three polymerase accessory proteins, on the other, act independently to stimulate the copying of primed single-stranded DNA templates by T4 DNA polymerase, while not stimulating other *E. coli* DNA polymerases (Huberman et al., 1971; Mace, thesis; Piperno and Alberts, 1978). The gene 32 protein and the accessory proteins act by different mechanisms. The gene 44/62 and gene 45 proteins are jointly required to increase the rate and processivity of DNA synthesis by a mechanism that requires ATP hydrolysis. Studies with oligonucleotide-primed fd DNA and restriction fragment-primed φX174 DNA templates have shown that regions of secondary structure such as palindromic hairpins act as strong kinetic barriers that cause the polymerase to pause when copying single-stranded templates (Challberg and Englund, 1979; Huang et al., 1981; Roth et al., 1982). The gene 44/62 and gene 45 proteins do not appear to change the structure of these template regions, but nevertheless facilitate synthesis up to and through the pause sites. These accessory proteins appear to act by decreasing the rate of dissociation of the polymerase from the primer-template end. Thus, their effect is greatest at low polymerase concentrations where the diffusion-mediated process of polymerase reassociation with primer-template ends is rate limiting in their absence (Huang et al., 1981). The rate of nucleotide turnover by a polymerase molecule detained at a pause site in the presence of the accessory proteins and ATP is much higher than that observed with the polymerase alone on the same template-primer, presumably reflecting repeated attempts at polymerization through a pause site by a polymerase molecule that is tied down on the template in a complex with the accessory proteins (Roth et al., 1982). The addition of gene 32 protein to the mixture of the polymerase and accessory proteins tends to remove the regions of secondary structure in the template, thereby reducing the pauses, increasing the rate of synthesis, and reducing nucleotide turnover.

ATP hydrolysis, not just ATP binding, is required for the stimulation by the accessory proteins, since the stimulation is prevented by the addition of excess riboadenosine 5′-*O*-(3-thiotriphosphate) (ATP-γ-S). However, less than 1 ATP molecule is hydrolyzed for every 10 deoxyribonucleotides incorporated on single-stranded templates (Piperno and Alberts, 1978). It has been proposed that ATP hydrolysis is required for the assembly of a complex of the accessory proteins and the polymerase, but not for the movement of this complex along a single-stranded template (Alberts et al., 1980; Newport et al., 1980). This proposal is based on two findings. First, at short times (less than 45 s), the number of oligo(dT) primers extended on poly(dA) in the presence of the products of genes 43, 44/62, 45, and 32 increases with increasing ATP concentration, with no change in the average product length (Newport et al., 1980). Second, the effect of the accessory proteins on oligonucleotide primer extension on a single-stranded fd DNA template is partially resistant to the addition of excess ATP-γ-S once the reaction has been started, with the effect being reported to persist for at least 10 min (Huang et al., 1981). However, the degree of resistance of the accessory protein-stimulated synthesis to the addition of ATP-γ-S varies with the template-primer. Thus, after the addition of excess ATP-γ-S, the amount of elongation of a restriction fragment annealed to single-stranded ϕX174 DNA is equivalent to less than 30 s of synthesis (Nossal, unpublished data), and synthesis on a nicked duplex template that requires the accessory proteins stops in less than 10 s (Alberts et al., 1980).

From these results, we conclude that a periodic event, dependent on ATP hydrolysis, is required to maintain the stimulation by the accessory proteins. Since synthesis on both single-stranded and nicked duplex templates is extremely processive in the presence of the accessory proteins, it is clear that the complex cannot completely dissociate from the primer terminus between each ATP hydrolysis event. As a working hypothesis, it has therefore been proposed that the accessory proteins assemble into a "sliding clamp" composed of two "half-clamps" around the polymerase. Each half-clamp can keep the polymerase on the template during periodic ATP-dependent resetting of the other half-clamp (Alberts et al., 1980). In this view, the accessory protein-polymerase complex is especially labilized on certain templates and under certain conditions. In these cases relatively frequent ATP hydrolysis will be required to maintain the accessory protein effect on the polymerase (Bedinger and Alberts, J. Biol. Chem., in press).

Stimulation of Polymerase Exonuclease Hydrolysis of Duplex DNA

In addition to stimulating DNA polymerase movement in the forward direction, the accessory proteins and the gene 32 protein also stimulate the reverse reaction: hydrolysis of duplex DNA by the 3′-to-5′ exonuclease of the polymerase (Alberts et al., 1980; Bedinger and Alberts, in press, Venkatesan and Nossal, 1982). The stimulation requires a single molecule of the gene 44/62 complex per DNA end (Bedinger and Alberts, in press). Maximum stimulation by the accessory proteins requires that there is a single-stranded extension of the complementary strand that protrudes more than three nucleotides beyond the 3′-OH end of the duplex region to be hydrolyzed. This single-stranded DNA extension may serve as an important part of the binding site for one or more of the accessory proteins (Venkatesan and Nossal, 1982). Nucleotides behind the 3′-OH terminus of the strand being degraded are also required for the accessory protein action, since the accessory proteins do not stimulate hydrolysis by the polymerase exonuclease once the strand being degraded has been reduced to a chain length of 7 to 11. In contrast, the polymerase exonuclease alone yields a limit product of mono-, di-, and trinucleotides (Venkatesan and Nossal, 1982).

All of the stimulatory effects of the gene 44/62, gene 45, and gene 32 proteins seem to require specific protein-protein interactions. Neither the gene 32 protein nor the accessory proteins increase the 3′-to-5′ exonuclease activity (Bedinger and Alberts, in press) or the polymerase activity (Kolodner and Nossal, unpublished data) of bacteriophage T7 DNA polymerase. The accessory proteins also fail to stimulate the exonuclease activity of the amber fragment produced by the *am*B22 mutant in T4 gene 43. This suggests that the carboxyl terminal 20% of the wild-type polymerase, which is the region missing in the *am*B22 amber fragment (Nossal and Hershfield, 1971), may be involved in the polymerase interaction with the accessory proteins (Venkatesan and Nossal, 1982).

RNA PRIMER SYNTHESIS ON THE LAGGING STRAND

Because DNA polymerases only synthesize DNA in the 5′-to-3′ chain direction, all of the DNA synthesized on the lagging strand of a replication fork must be made discontinuously, as a series of short Okazaki fragments that are subsequently joined together to create an intact daughter strand (Fig. 1) (reviewed in Kornberg, 1980). In T4 bacteriophage-infected cells, an average Okazaki fragment is about 2,000 nucleotides long, and, at the observed polymerization rate of roughly 500 nucleotides per s, it should take about 4 s to complete. Furthermore, since the DNA polymerases

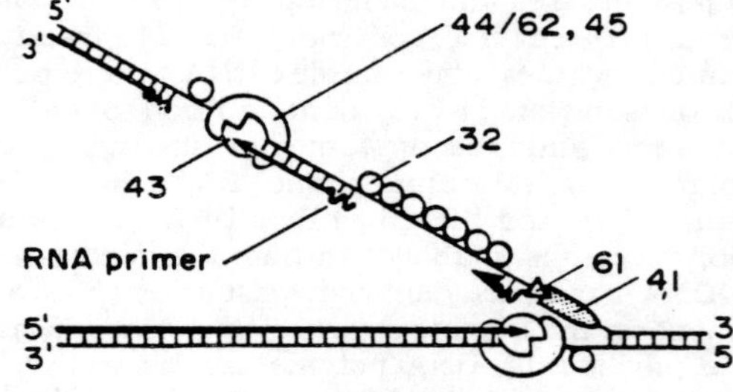

FIG. 1. Model of a replication fork showing the roles of bacteriophage T4 proteins in the continuous synthesis on the leading strand and in the discontinuous synthesis on the lagging strand. See text for details.

studied so far are unable to start a new polynucleotide chain, special mechanisms involving either terminal proteins or RNA primers to which deoxyribonucleotides can be added are required to begin new DNA chains (reviewed in Kornberg, 1980, and Kornberg, 1982). In the T4 DNA replication system, each Okazaki fragment begins with a short base-paired RNA primer that is synthesized from ribonucleoside triphosphates (rNTPs) by the T4 gene 41 and gene 61 proteins.

The process of T4 RNA primer synthesis has been characterized by studying rNTP-dependent DNA synthesis on single-stranded DNA templates. Any natural single-stranded DNA, as well as the synthetic polymer poly(dI,dT), can be used as a substitute for the lagging-strand template. This single-stranded DNA, the gene 41 and gene 61 proteins, ATP, and CTP are the minimum requirements for RNA primer synthesis (Liu and Alberts, 1980; Liu and Alberts, 1981b; Liu et al., 1979; Nossal, 1980; Silver and Nossal, 1979). In the presence of the other replication proteins, more than 90% of the RNA oligonucleotides synthesized by the gene 41 and gene 61 proteins become covalently attached to the 5′ end of newly initiated DNA chains, from which they can be recovered by nuclease digestion (Liu and Alberts, 1980; Liu and Alberts, 1981b; Nossal, 1980).

Virtually all of the primers made in vitro in the presence of four rNTPs are pentaribonucleotides. Sequence analysis of 5′-terminally labeled primers reveals that the major products have the sequence pppApCpNpNpN, where (at least in positions 3 and 4) N can be any one of the four nucleotides (Liu and Alberts, 1980; Liu and Alberts, 1981b; Nossal, 1980). Whereas only pppA starts primers on single-stranded hydroxymethyldeoxycytidine-containing T4 DNA, additional primers that begin with pppG are also found when single-stranded deoxycytosine-containing T4 DNA or bacteriophage fd DNA (Liu and Alberts, 1981) or bacteriophage φX174 DNA (Nossal, 1980) is used as the template. The size and sequence of the RNA primers made in vitro on hydroxymethyldeoxycytidine-containing T4 DNA agree completely with those isolated from T4-infected cells (Kurosawa and Okazaki, 1979).

With only ATP and CTP, di-, tri-, and tetranucleotides beginning with pppApC are found in addition to pentanucleotides. The shorter products apparently result when synthesis is prematurely terminated due to the absence of the next nucleotide required by the template. Primers containing only A and C are efficiently elongated by T4 DNA polymerase and the accessory proteins (Liu et al., 1979; Silver and Nossal, 1979). Although there are more than 40 sites on fd DNA which could serve as a template for pentanucleotides pppApCpNpNpN, where N is A or C, only about 10 start sites for new DNA chains were observed on this template with ATP and CTP (Liu and Alberts, 1980). This finding suggests that there may be some requirement for a start site beyond the sequence complementary to the primer.

The catalytic roles of the gene 41 and gene 61 proteins are not yet clear. There is no primer synthesis in the absence of either protein (Alberts et al., 1980; Burke and Alberts, manuscript in preparation; Liu et al., 1979; Nossal, 1980; Silver and Nossal, 1982). The gene 61 protein was first identified by Liu et al. (1979) as a protein present in some preparations of gene 32 protein that was required for rNTP-dependent DNA synthesis with the T4 gene 43, 44/62, 45, and 41 proteins. It was subsequently shown to be a very basic protein with a molecular weight of about 44,000 that is missing in extracts made from *E. coli* infected with T4 gene 61 mutants (Alberts et al., 1980; Liu et al., 1979; Silver and Nossal, 1979; Silver and Nossal, 1982). The phenotype of gene 61 mutants is a delayed synthesis of DNA (Yegian et al., 1971). In contrast, gene 41 mutants were originally classified as DO (no DNA synthesis) (Epstein et al., 1964), although there is some accumulation of single-stranded DNA (Oishi, 1968). The more severe defect in gene 41 mutants may be due to the additional role of gene 41 protein in stimulating leading-strand DNA synthesis (see below).

The gene 41 protein is a single-stranded-DNA-dependent nucleotidase that hydrolyzes rATP, rGTP, dATP, and dGTP (Liu and Alberts, 1981a; Morris, Moran, and Alberts, 1979; Nossal, 1979; Venkatesan et al., 1982). Liu and Alberts (1981a) found that long DNA chains or single-stranded DNA circles are the best cofactors for the gene 41 protein nucleotidase activity and that there is only limited RNA primer synthesis when excess GTP-γ-S is added to inhibit this nucleotidase. On the basis of these results, they suggested that the gene 41 protein moves along a DNA single strand, driven by conformational changes induced by GTP hydrolysis. At the replication fork, the gene 41 protein was proposed to act as a "mobile promoter" which marks the locations where RNA primers are to be synthesized on the lagging strand. Liu and Alberts (1981a) also showed that incubation of a high concentration of gene 41 protein with GTP or GTP-γ-S at 30°C increases its ability to catalyze rNTP-dependent DNA synthesis and that incubation with GTP-γ-S increases the sedimentation coefficient of the protein from 4.9 to 6.1S, consistent with the formation of an active dimer (or higher oligomer). RNA primer synthesis and rNTP-dependent DNA synthesis normally have a linear dependence on gene 61 protein concentration, but a sigmoidal dependence on gene 41 protein concentration. Silver and Nossal (1982) showed that the response to gene 41 protein concentration becomes linear when gene 41 protein is first activated by incubation with GTP and that there is then a stoichiometric relationship between the gene 41 and gene 61 proteins. They have therefore proposed that a large oligomer of gene 41 protein interacts with a monomer of gene 61 protein to form a complex that is active in primer synthesis.

A model consistent with these observations is shown in Fig. 1. A complex of the gene 41 and gene 61 proteins is envisioned to move along the lagging strand in the 5′-to-3′ direction, in a reaction dependent on nucleotide hydrolysis. At intervals of about 2,000 nucleotides, the complex stops at appropriate sites to make pentaribonucleotide primers. These primers probably remain attached to the priming proteins until elongated by the DNA polymerase, since added pentaribonucleotides by themselves are not efficiently used as primers. Further evidence for the proposed movement of the priming proteins, and the effect of this movement on leading strand DNA synthesis, is described below.

To produce a continuous DNA chain from the many

Okazaki fragments made on the lagging strand, a special DNA repair system is needed behind the fork. The minimal requirements are for an RNase H enzyme to degrade the RNA primer, a DNA polymerase molecule to replace the ribonucleotides removed, and a DNA ligase to form a phosphodiester linkage between the 3′ end of each fragment and the 5′ end of the preceding one. The T4 DNA polymerase (gene 43 protein), T4 DNA ligase (gene 30 protein), and a T4-induced RNase H of 44,000 daltons whose gene has yet to be identified are all thought to be involved (V. Chandler and B. M. Alberts, and R. Crouch and N. G. Nossal, unpublished data). However, this Okazaki fragment repair process has not yet been studied in detail in the T4 system.

LEADING-STRAND SYNTHESIS

The DNA synthesis on the leading strand of a replication fork requires rapid unwinding of the DNA duplex ahead of the growing chain (Fig. 1). Strand displacement synthesis that starts from a preexisting nick in duplex DNA has been used as a model for this type of synthesis. Five T4 proteins are the minimum required for this type of synthesis: DNA polymerase, gene 32 protein, and the three gene products that act as polymerase accessory proteins (Table 1) (Liu et al., 1979; Nossal and Peterlin, 1979). The products of this five-protein reaction are duplex DNA molecules from which long single-stranded branches protrude. Distinct pause sites are revealed by analysis of the products that begin at a single nick at a defined location on duplex circular ϕX174 DNA or fd DNA (Alberts et al., 1983; Venkatesan et al., 1982).

Gene 32 protein appears to play an active role in destabilizing the duplex, since the rate of fork movement increases steadily with increasing free gene 32 protein concentration, reaching a rate of about 180 nucleotides per s at a gene 32 protein concentration of 200 μg/ml (Alberts et al., 1980). The proteolytic product called 32* I protein, which is missing 8,000 daltons from the carboxy-terminal end of gene 32 protein, can replace intact gene 32 protein in this reaction (Burke et al., 1980). Although T4 DNA polymerase cannot use a duplex template by itself, it also must contribute to destabilizing the helix. Thus, the polymerase made by an antimutator phage, T4 tsCB120, copies single-stranded homopolymers at rates comparable to those of the wild-type enzyme, but is more inhibited by hairpin pause sites in single-stranded natural DNA and poly(dA-dT) templates (Gillin and Nossal, 1976a). This tsCB120 mutant polymerase is stimulated by gene 32 protein and the accessory proteins on single-stranded DNA templates, but it is unable to catalyze strand displacement synthesis on double-helical DNA templates with the gene 32, gene 44/62, and gene 45 proteins present. Further addition of the gene 41 protein, which stimulates strand displacement synthesis with the wild-type DNA polymerase present (see below), is still insufficient for strand displacement synthesis with the mutant polymerase (Nossal, unpublished data).

During synthesis on a double-helical template, the gene 44/62 and gene 45 proteins presumably act to keep the polymerase continuously on the growing 3′-OH chain end where it can take advantage of gene 32 protein-catalyzed breathing in the template helix ahead. Polymerase dilution experiments reveal that the DNA synthesis obtained is highly processive, with the same DNA polymerase molecule synthesizing stretches of DNA that are tens of thousands of nucleotides long (Alberts et al., 1980; Barry, unpublished data). As noted above, very frequent ATP hydrolysis by the accessory proteins is required for synthesis on duplex templates, although, with duplex templates as well as single-stranded templates, less than 1 molecule of ATP is hydrolyzed for every 10 deoxyribonucleotides polymerized (Liu et al., 1979).

Even at high gene 32 protein concentrations, the rate of fork movement in the five-protein reaction described above is significantly lower than the rate at which the fork moves in vivo (Alberts et al., 1980; Liu et al., 1979). The in vitro rate is enhanced by the addition of either of two T4-induced DNA helicases: the gene 41 protein or the product of the *dda* gene (Table 1). As discussed above, the gene 41 protein is a single-stranded-DNA-dependent nucleotidase that is required together with the gene 61 protein for RNA primer synthesis. The addition of gene 41 protein greatly increases the rate of fork movement above that obtained in the five-protein reaction, with the largest stimulation being observed when limiting concentrations of gene 32 protein are present (Alberts et al., 1980; Liu et al., 1979; Venkatesan et al., 1982). However, gene 41 protein does not remove the requirement for each of the five other replication proteins. The gene 41 protein is also a DNA helicase that can unwind short DNA fragments (50 to 400 base pairs) that are base paired to a complementary DNA strand; as expected, this reaction is dependent on GTP or ATP hydrolysis (Venkatesan et al., 1982). Helicase activity requires a single-stranded DNA extension adjacent to the duplex region to be unwound, and the unwinding begins at the 3′ end of the fragment. This direction of helix unwinding is consistent with the proposal (Alberts et al., 1980) that the gene 41 protein destabilizes the helix ahead of the replication fork as it moves 5′ to 3′ on the lagging strand of the fork toward the next site for primer synthesis (Fig. 1). The helicase activity of gene 41 protein is stimulated by both gene 61 protein and the combination of the three polymerase accessory proteins, suggesting that all of these proteins may be closely associated with each other at the replication fork (Venkatesan et al., 1982).

The T4 *dda* protein was initially identified as a single-stranded-DNA-dependent ATPase (Purkey and Ebisuzaki, 1977) and later shown to be an ATP-dependent DNA helicase (Krell et al., 1979). The helicase activity of the *dda* protein, like that of the gene 41 protein, requires a short single-stranded extension beyond the 3′ OH of the duplex region to be unwound, supporting the conclusion that the *dda* protein unwinds DNA in the same direction as does the gene 41 protein (Krell et al., 1979). Unlike the gene 41 protein, the *dda* helicase can unwind very long DNA helices. However, in this reaction it acts stoichiometrically with approximately one protein molecule required for each two base pairs unwound. At a low concentration, the *dda* helicase increases the rate of fork movement of the five-protein replication reaction on a nicked DNA template, mimicking the action of the gene 41 protein in this regard. The *dda* protein neither further stimulates the fork rate when gene 41 protein is present nor substitutes for the gene 41 protein in RNA priming reactions.

A clue to the biological role of the *dda* protein was obtained in experiments that examined the effect of template-bound RNA polymerase molecules on the movement of T4 replication forks in vitro. Replication forks formed in the absence of the *dda* protein, whether or not they contain the gene 41 protein, are stopped by template-bound RNA polymerase molecules. The addition of small amounts of the *dda* helicase completely removes this inhibition, allowing the replication fork to move rapidly past RNA polymerase molecules that are bound to the DNA template (Alberts et al., 1983; Bedinger et al., submitted for publication). RNA polymerase molecules that are not transcribing are displaced by the DNA replication enzymes, but the fate of transcribing RNA polymerase molecules has not been determined (Bedinger et al., submitted for publication). It seems likely that the removal of barriers to DNA replication caused by RNA polymerase and other DNA-bound proteins is an important function of the type of helicase encoded by the T4 *dda* gene. This might explain why more than one 5'-to-3' DNA helicase seems to be required at a T4 replication fork.

There is genetic evidence to support a role for the *dda* helicase in T4 DNA replication. Although T4 bacteriophage that have the *dda* gene deleted grow normally in most *E. coli* hosts (Behme and Ebisuzaki, 1975), they are unable to grow in an *E. coli optA* mutant (Gauss et al., 1983). The *optA*1 mutant was originally selected as a nonpermissive host for the growth of bacteriophage T7 mutants in gene 1.2, a gene of unknown function that is located close to the T7 replication origin (Saito and Richardson, 1981; Saito et al., 1980). The *optA*1 mutant also fails to grow certain bacteriophage T4 temperature-sensitive mutations in the polymerase gene, including the antimutator tsCB120, even at low temperatures (Gauss et al., 1983). As noted above, the tsCB120 polymerase is defective in strand displacement synthesis in vitro. On the basis of these studies, Gauss and his colleagues have proposed that the *E. coli optA* gene encodes a host DNA helicase that interacts with the T4 DNA replication proteins and is required unless both the T4 *dda* helicase and the full helix-destabilizing activity of the wild-type T4 DNA polymerase are present.

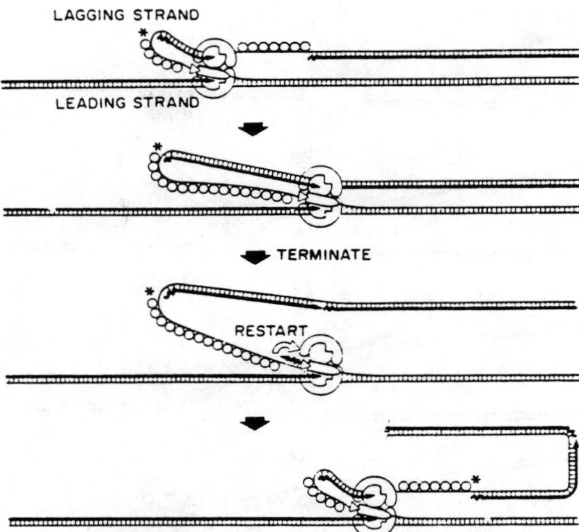

FIG. 2. Model of T4 proteins at a replication fork that accounts for the recycling of the polymerase on the lagging strand and the stimulation of leading-strand synthesis by the gene 41 protein primase-helicase. T4 DNA replication proteins are represented by the symbols used in Fig. 1. In this model the polymerase-accessory protein complex on the lagging strand is bound to the corresponding complex on the leading strand, and it is assumed that the two strands are replicated at the same rate (adapted from Alberts et al., 1983).

COUPLING OF LEADING-STRAND AND LAGGING-STRAND DNA SYNTHESIS

The simple replication fork model illustrated in Fig. 1 implies that a completely separate complexe of the polymerase and the accessory proteins elongates the leading and lagging strands. In this model, the complex of the polymerase and accessory proteins on the leading strand is assumed to remain permanently bound to the template as the fork moves. Thus, synthesis on the leading strand is continuous and highly processive, accounting for the finding of a minor fraction of template DNA molecules on which there has been extensive synthesis (>50,000 nucleotides added) under conditions where there has been little or no synthesis on the majority of other template molecules (Nossal and Peterlin, 1979; Sinha et al., 1980). In contrast, the model in Fig. 1 predicts that the polymerase complex on the lagging strand dissociates from its template each time that it finishes the 3' end of an Okazaki fragment and encounters the 5' end of the previous fragment. If this model is correct, decreasing the polymerase concentration to very low levels should delay the start of each Okazaki fragment by making the collision of a free DNA polymerase molecule with an RNA primer unlikely, while having little effect on further synthesis on the leading strand. In fact, this model is contradicted by recent studies in which the concentration of T4 DNA polymerase was varied over a wide range, with each of the other six replication proteins present in excess. In these reactions, DNA synthesis on the lagging strand was unaffected by extreme polymerase dilution, which is only explicable if the DNA polymerase molecule on the lagging strand is recycled for multiple rounds of Okazaki fragment synthesis and thus remains bound to the fork at all times (Alberts et al., 1983; Barry and Alberts, manuscript in preparation). Similar experiments suggest that the gene 41 protein and gene 61 protein also remain bound to the fork for prolonged periods (Barry, unpublished data).

Two closely related models of the replication fork which account for the recycling of the polymerase and priming proteins on the lagging strand, as well as the coupling of leading- and lagging-strand synthesis by the gene 41 protein helicase, are illustrated in Fig. 2 and 3. In the "trombone" model (Fig. 2) proposed by Alberts et al. (1983), which is based in part on protein-protein interactions determined by protein affinity column chromatography (Alberts et al., 1983; Formosa et al., Proc. Natl. Acad. Sci. U.S.A., in press), the

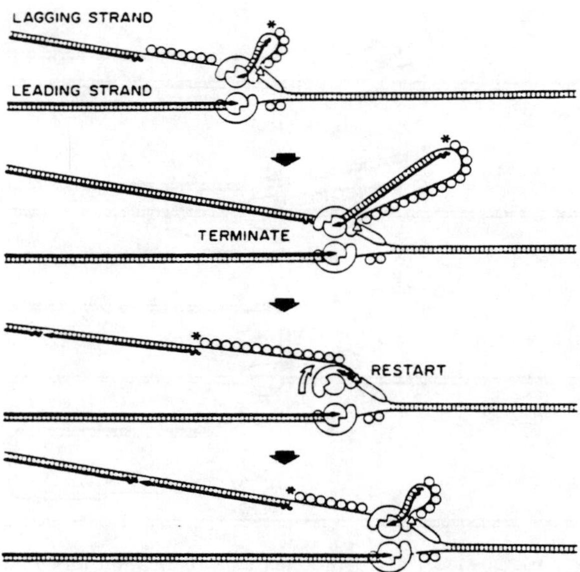

FIG. 3. A second possible model of T4 proteins at a replication fork that, like the model in Fig. 2, accounts for recycling of the polymerase-accessory protein complex on the lagging strand and the stimulation of leading-strand synthesis by the gene 41 protein primase-helicase. T4 proteins are represented by the symbols used in Fig. 1. In this model the polymerase-accessory protein complex is bound by the gene 41-gene 61 protein primase-helicase complex that remains on the lagging strand.

polymerase-accessory protein complex on the lagging strand is tied to the corresponding polymerase complex on the leading strand. The synthesis of each Okazaki fragment involves the enlargement of a large loop of DNA, half of which is single stranded and half of which is double stranded. The crucial points in the proposed cycle (Fig. 2) are the termination and restart steps, which involve movements of the DNA on the lagging strand of the fork around a fixed lagging-strand DNA polymerase molecule, as indicated. Primer synthesis will take place at an appropriate sequence reached after termination of the preceding fragment.

In the trombone model, it was assumed that the two polymerase molecules joined in the complex must elongate the two growing DNA strands at the same rate so that there is always a constant amount of single-stranded DNA on the lagging strand. This was proposed to make it possible for the gene 32 protein molecules bound to the loop to be continuously recycled on the lagging strand. In this case, the length of the single-stranded DNA that accumulates in the loop (Fig. 2)—which will determine the size of the next Okazaki fragment—will always be equal to the length of the previously synthesized fragment. Therefore, each Okazaki fragment on an individual lagging strand should be about the same size as the preceding fragment (Alberts et al., 1983). The proposition that the Okazaki fragments are of constant size and the idea that the gene 32 protein molecules recycle remain to be tested.

In the model in Fig. 3, the polymerase-accessory protein complex on the lagging strand is tied to the fork by being bound to the gene 41 protein-gene 61 protein primase-helicase complex that moves 5′ to 3′ on the lagging strand of the fork. Synthesis on the two strands is coupled because the gene 41 protein helicase on the lagging strand helps to destabilize the helix ahead of the new leading strand. The rate of elongation of the polymerase-accessory protein complexes on the leading and lagging strands would not in general be precisely the same; the relative rates would be expected to vary due to the different distribution of pause sites on the two template strands. Since T4 Okazaki fragments are about 2,000 nucleotides long, the gene 41 protein-gene 61 protein complex must pass by many potential priming sites. Nossal has proposed (unpublished) that the release of the polymerase-accessory protein complex from the 3′-OH end of a completed Okazaki fragment alters the conformation of the attached primase-helicase complex and that this conformational change acts as a signal to the priming proteins to make a new primer at the next appropriate sequence. This proposition, which also remains to be tested, predicts that the length of each Okazaki fragment will be determined by the particular distance covered by the gene 41 protein helicase during the time necessary for the polymerase to complete the previous fragment.

It is important to realize that the models in Fig. 2 and 3 are closely related. The proposed conformational change in the priming proteins when the polymerase completes an Okazaki fragment could also occur if the polymerase complexes on the leading and lagging strands were physically linked to each other. Furthermore, the polymerase molecules could be linked as shown in Fig. 2 and still copy the two strands at different rates. In that case, the remaining difference between the two models is that in Fig. 3 the proteins are arranged so that the 3′-OH termini of the two new strands enter the polymerase complex from opposite sides, whereas in Fig. 2 it is proposed that the two growing chains enter the tightly coupled protein complex from the same side.

Regardless of the details, an important feature of the type of models shown in Fig. 2 and 3 is that the replication apparatus lays down unsealed fragments that must then be ligated. Assuming that the polymerase is released before the RNA primer is removed from the preceding fragment, a second polymerase molecule will be needed to repair the gap.

The different possibilities discussed above will need to be resolved by further experiments. However, the major focus of both models is the coupling of leading- and lagging-strand synthesis and the recycling of the replication proteins that are characteristic of the T4 DNA replication system. The picture that emerges is one of a large "replication machine" whose component parts move relative to each other during different stages of DNA replication without disassembling from the complex. The enzymes that synthesize nucleotide precursors also seem to be part of the replication complex (Chiu et al., 1982; Mathews and Allen, this volume). The precise assembly of these large replication protein aggregates may be a rate-limiting step in the process of replication fork initiation at T4 replication origins.

SIMILARITIES BETWEEN T4 DNA REPLICATION AND OTHER DNA REPLICATION SYSTEMS

The goal of the enzymatic studies that have been carried out on the T4 replication proteins has been to understand in detail the types of reactions that are required to replicate duplex DNA. It is therefore satisfying that mechanisms similar to those described for the T4 proteins have also been demonstrated in the bacteriophage T7 and *E. coli* DNA replication systems (reviewed in Nossal, Annu. Rev. Biochem., in press). Analogous reactions are beginning to be found in eucaryotic systems as well.

DNA Polymerases and Accessory Proteins

Like the T4 replication system, the T7 bacteriophage and *E. coli* DNA replication systems utilize a single type of DNA polymerase to elongate both the leading and lagging strands (Table 2). The T7 DNA polymerase is relatively simple and does not require accessory proteins analogous to the T4 gene 44/62 and gene 45 proteins. In contrast, *E. coli* DNA polymerase III (PolIII) is quite complex. PolIII core, composed of three subunits (α, ε, and θ), can be purified as a complex with at least four other subunits (β, γ, δ, and τ). This whole complex has been designated "PolIII holoenzyme." This holoenzyme can be converted to smaller complexes by standard protein purification procedures (reviewed in McHenry and Kornberg, 1981). Alternatively, an activity equivalent to the PolIII holoenzyme can be reconstructed from PolIII core by the addition of three separate factors (EFI, EFIII, and *dnaZ* protein) that have been purified on the basis of their ability to allow PolIII core to use primed single-stranded DNA templates (reviewed in Wickner, 1978). The factors or subunits have functions analogous to the T4 polymerase accessory proteins: factors γ and δ (*dnaZ* and EFIII) increase the processivity of PolIII core and are required, together with β (EFI), for the binding of the polymerase core to primed single-stranded templates.

The accessory proteins that function in concert with the T4 and *E. coli* DNA polymerases seem to be important for keeping the polymerase tightly bound to the template, and they presumably reduce the frequency with which major "geometrical errors" occur in replication. Such errors include both "template slippage" errors that can cause groups of nucleotides to be either added or deleted in the new DNA chain and "template switching" errors. The latter arise in a reaction first discovered with *E. coli* DNA polymerase I, in which the DNA polymerase turns around to copy a previously replicated region, thereby creating a region of self-complementary sequence detected as "reversibly denaturable" DNA (reviewed in Kornberg, 1980). Reversibly denaturable DNA is a major product of the limited synthesis on nicked duplex templates catalyzed by T4 DNA polymerase in the presence of only the gene 32 protein (Nossal, 1974), but it can not be detected in the products made in the T4 five-protein reaction that includes the polymerase accessory proteins (Nossal and Peterlin, 1979; Sinha et al., 1980).

Why should bacteriophages T4 and T7, with progressively smaller genomes and larger copy numbers, require simpler DNA polymerase complexes than *E. coli* for their replication? It is tempting to speculate that the replication of longer single-copy genomes requires an increasingly complex array of accessory polypeptide chains to reduce the frequency of polymerase errors, since it is crucial that the essential regions of the single copy be replicated without a geometrical mistake. In this view, the genome of bacteriophage T7 is so small (40,000 base pairs) and present at such a high copy number that it can tolerate the errors made by a relatively simple form of DNA polymerase.

Strand Displacement Synthesis from a Nick on a Double-Helical Template

One component that is apparently quite similar in all three of the DNA replication systems described above is that designated as a helix-destabilizing protein (HD protein) by some workers (Alberts and Sternglanz, 1977) and as a single-stranded-DNA-binding protein (SSB protein) by others (Kornberg, 1980). The *E. coli* SSB protein closely resembles the T7 gene 2.5 protein and the T4 gene 32 protein in its properties, and, like the T4 protein, it increases the rate and processivity of DNA synthesis on single-stranded templates catalyzed by its homologous DNA polymerase (the PolIII holoenzyme).

The demonstrated requirements in the *E. coli* system for strand displacement synthesis on a nicked duplex template are considerably more complex than in the T4 system; they include the PolIII holoenzyme, the *E. coli* SSB protein, and a 3′-to-5′ helicase (*rep* protein); in addition, at least one other protein is involved at the start of synthesis (for example, the bacteriophage fd gene 2 protein [Geider et al., 1982]) (Table 2). Moreover, at least three other *E. coli* DNA helicases (helicases I, II, and III) exist which unwind DNA in the same direction as the T4 gene 41 and *dda* proteins; however, the role of these helicases in replication is still not clear (reviewed in Geider and Hoffmann-Berling, 1981). The 3′-to-5′ DNA helicases are thought to move on the leading strand of the fork, 5′-to-3′ DNA helicases, like the T4 gene 41 protein, move on the lagging strand (Fig. 4).

In contrast to the complex *E. coli* system, the T7 DNA polymerase catalyzes efficient strand displacement synthesis on nicked duplex templates with only the T7 gene 4 protein present. This gene 4 protein is a multifunctional protein with primase, DNA-dependent nucleotidase, and helicase activities (reviewed in Richardson, *in* Y. Becker, ed., *Replication of Viral and Cellular Genomes*, in press). While a helix-destabilizing protein is not required, DNA synthesis is stimulated by either the *E. coli* SSB protein or the T7 gene 2.5 protein, and it would appear that one or another of these two proteins is required for T7 DNA replication in vivo.

RNA Primer Synthesis

The T7 bacteriophage also has the simplest priming system. The T7 gene 4 primase-helicase recognizes the template sequence N_2N_1GTC and synthesizes tetranucleotide primers beginning with pppApC on all natural single-stranded DNAs tested (Romano et al., 1979; Scherzinger et al., 1977; Tabor and Richardson, 1981). By comparing the relative frequency with which dif-

TABLE 2. Comparison of T4 DNA replication proteins with those of other procaryotic DNA replication systems

Source of proteins	Proteins					
	T4	T7[a]	*E. coli* and bacteriophage G4, ϕX174, and fd systems[b]			
			General priming	G4	ϕX174	fd, M13
Primer synthesis (ssDNA)[c]						
Primase	T4 gene 61[d] T4 gene 41[d]	T7 gene 4	*dnaG*	*dnaG*	*dnaG*	RNA polymerase
Other	None	None	*dnaB*	SSB	SSB; n′ (Y); n,n″ (Z); i (X), *dnaB*; *dnaC*	SSB
DNA chain elongation						
Polymerase	T4 DNA polymerase (gene 43)	T7 DNA polymerase (T7 gene 5 + *E. coli* thioredoxin)	PolIII core[e]	PolIII core	PolIII core	PolIII core
Polymerase accessory	T4 gene 44/62 T4 gene 45	None	PolIII holoenzyme subunits[f]	PolIII holoenzyme subunits	PolIII holoenzyme subunits	PolIII holoenzyme subunits
SSB (helix destabilizing)	T4 gene 32	T7 gene 2.5 or *E. coli* SSB	*E. coli* SSB	*E. coli* SSB	*E. coli* SSB	*E. coli* SSB
DNA helicases[g]	T4 gene 41 (5′ to 3′) T4 *dda* (5′ to 3′)	T7 gene 4 (5′ to 3′)		*rep* (3′ to 5′)	*rep* (3′ to 5′)	*rep* (3′ to 5′)
Site-specific nicking	None	None		G4 gene *A*	ϕX gene *A*	fd gene 2

[a] Reviewed in Richardson (1982).
[b] Reviewed in Kornberg (1982), Meyer and Geider (1982), Reinberg et al. (1981), and Wickner (1978). All are *E. coli* proteins, except the phage-encoded site-specific nicking enzymes.
[c] ssDNA, Single-stranded DNA.
[d] Both T4 gene 41 protein and T4 gene 61 protein are required to observe any primer synthesis. Gene 41 protein also has a single-stranded-DNA-dependent nucleotidase activity analogous to that of the T7 gene 4 and *E. coli* n′ (Y) and *dnaB* priming proteins.
[e] *E. coli* DNA PolIII core subunits are α (*dnaE*), ε (*mutD*, *dnaQ*), and θ.
[f] Subunits present in PolIII holoenzyme but not in PolIII core include γ (*dnaZ*), δ (EFIII), β (EFI, *dnaN*), and τ. These function like "accessory proteins" in increasing the processivity of the core polymerase.
[g] Helicase direction is indicated as the direction of movement on the strand that is not displaced.

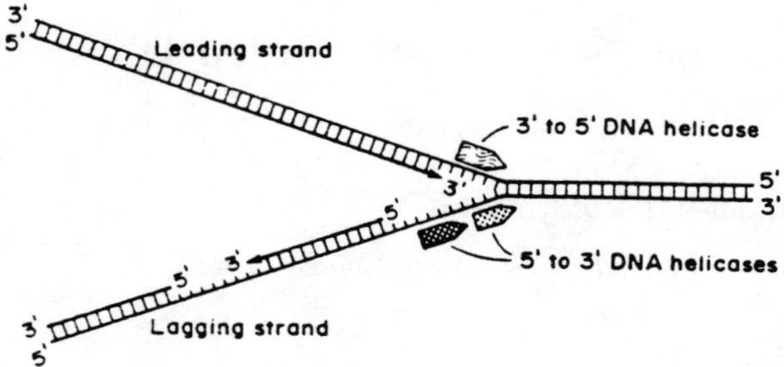

FIG. 4. General model of a replication fork, showing the proposed location of 5'-to-3' DNA helicases on the lagging strand and 3'-to-5' DNA helicases on the leading strand. By convention, DNA helicases are designated by their direction of movement on the strand that is not displaced.

ferent priming sites are used on ϕX174 DNA, it has been inferred that the gene 4 protein binds at random and travels 5' to 3' on the single-stranded DNA template strand (Tabor and Richardson, 1981).

In *E. coli*, primer synthesis has been studied by using the replication of a variety of single-stranded circular bacteriophage DNAs as model systems. Either *E. coli* RNA polymerase or the *dnaG* protein serves as the primase, depending on which bacteriophage DNA is the template (Table 2) (reviewed in Kornberg, 1982; Meyer and Geider, 1980; Reinberg et al., 1981; Wickner, 1978). In the general priming reaction, the *dnaG* primase requires the *dnaB* protein in order to make short ribooligonucleotide (or mixed ribo- and deoxyribooligonucleotide) primers on natural single-stranded DNA and on poly(dT). This priming reaction most resembles those found in the T4 and T7 bacteriophage systems.

Upon addition of *E. coli* SSB protein, the *dnaG* primase reaction becomes dependent on specific recognition structures or sequences on the DNA. On the DNA of G4 and related phages, *dnaG* protein (in the presence of only the SSB protein) makes a 26- to 29-nucleotide primer at one specific site. In contrast, on ϕX174 DNA, a complex of at least six additional proteins is required for primer synthesis. Two of these proteins, the n' (Y) and the *dnaB* proteins, are DNA-dependent nucleotidases. The priming complex, which has been termed a "primosome," is initially assembled on the DNA at a specific site recognized by the n' (Y) protein but then moves in the 5' to 3' direction on the template, as judged by the frequency of primers made at different locations (Arai and Kornberg, 1981; Arai et al., 1981).

The primosome complex moves in the same direction along a DNA strand as the T4 gene 41 and T7 gene 4 priming proteins and thus may also serve as a 5'-to-3' helicase at the replication fork. While helicase activity has not been shown directly, the inhibition of leading-strand DNA synthesis on the ColE1 plasmid by antisera to the *dnaB*-primosome protein has led to the proposal that this protein helps to unwind the DNA duplex ahead of the leading strand (Staudenbauer et al., 1979).

CONCLUSION

Studies of DNA replication in the T4 and T7 bacteriophages and the *E. coli* systems have indicated that efficient and accurate replication requires an impressive array of interacting proteins. Helicases, helix-destabilizing (SSB) proteins, and polymerase complexes all contribute to helix destabilization. A dual role of the priming proteins as helicases helps to coordinate leading- and lagging-strand synthesis. A dimeric form of DNA polymerase has been proposed to copy the leading and lagging strands simultaneously in both the T4 (Alberts et al., 1983) and *E. coli* (Kornberg, Burgers, and Stayton, cited in Kornberg, 1982) DNA replication systems.

The manner in which the bacteriophage T4 DNA replication proteins are physically connected and the mechanisms by which the activities of the T4 replication proteins are coordinated with each other (and with the activity of the enzymes synthesizing nucleotide precursors) remain to be determined. Current progress in locating the bacteriophage T4 DNA replication origins (see the chapters by Mosig, Kozinski, and Kreuzer and Huang in this volume) should greatly facilitate studies on the mechanism of fork initiation by the T4 DNA replication proteins.

We thank the colleagues who have made important contributions to these studies: Jack Barry, Pat Bedinger, Michael Bittner, Rae Lyn Burke, Monique Davies, Tim Formosa, Hiroko Hama-Inaba, John Hearst, Ula Hibner, Chiao-Chain Huang, Victor Jongeneel, Ken Kreuzer, Chung-Chen Liu, Leroy Liu, David Mace, Larry Moran, C. Fred Morris, Jeanette Piperno, and Navin Sinha; and Michael Hershfield, Frances Gillin, Matija Peterlin, Annette Roth, Paul Englund, Lynn Silver, Malabi Venkatesan, and Deborah Hinton. N.N. expresses her appreciation to Deborah Hinton for constructive criticism of the manuscript and to Helen Jenerick for carefully typing the manuscript.

Structure-Function Relationships in the T4 Single-Stranded DNA Binding Protein

KENNETH R. WILLIAMS AND WILLIAM H. KONIGSBERG

Department of Molecular Biophysics and Biochemistry, Yale University, New Haven, Connecticut 06510

The protein encoded by gene 32 of bacteriophage T4 (gp32) has served as a prototype for a class of proteins which bind relatively non-specifically to single-stranded DNA (ssDNA) and have no other enzymatic activity (Kowalczykowski, Bear, and von Hippel, 1981). Similar proteins have been purified from bacteriophage T7, *Escherichia coli*, fungus, meiotic cells of lily plants, mouse tissues grown in culture, rat liver, and calf thymus (for recent reviews, see Coleman and Oakley, 1980; Kowalczykowski, Bear, and von Hippel, 1981; and Williams and Konigsberg, 1981). The requirement for gp32 has been amply demonstrated for T4 DNA replication (Epstein et al., 1963), repair (Wu and Yeh, 1973), and recombination (Tomizawa et al., 1966). In addition, gp32 controls its own rate of synthesis (Gold et al., 1976; Krisch et al., 1974). Most, if not all, of the various functions assigned to gp32 in vivo have been explained on the basis of its ability to bind tightly and cooperatively to single-stranded nucleic acids (for a review, see Doherty et al., *in* J. Kane, ed., *Multifunctional Proteins: Regulatory and Catalytic/Structural*, in press). Binding of gp32 to DNA results in the removal of the hairpin-like, base-paired loops usually found in ssDNA and the imposition of a highly extended structure in which the nucleotide bases are exposed. Besides being an ideal substrate for T4 DNA polymerase, the ssDNA in a gp32-ssDNA complex is resistant to attack by nucleases, yet it can readily base pair with homologous ssDNA. Depending upon the ionic strength, temperature, and the presence of other ssDNA or double-stranded DNA (dsDNA) binding proteins, gp32 therefore has the potential to either denature or renature dsDNA. Because gp32 binds ssDNA more tightly than it binds ssRNA (Newport et al., 1981), it can effectively regulate its own rate of synthesis (von Hippel et al., this volume). As gp32 is synthesized in vivo, it binds first to all of the ssDNA that is present as a result of normal T4 DNA replication, repair, and recombination. At this point the free gp32 concentration rises until it reaches a certain threshold level necessary for gp32 to bind specifically to its own mRNA and to prevent further gp32 synthesis. The 10^1- to 10^4-times-higher affinity (depending upon the particular homopolymer tested) of gp32 for ssDNA over ssRNA assures that all of the intracellular ssDNA will be saturated with gp32 before a significant amount of binding to mRNA takes place (Newport et al., 1981). The specificity for this translational repression may arise from some unique features that are inherent in the gp32 mRNA molecule itself (Doherty et al., *Multifunctional Proteins*, in press; Krisch et al., 1980; Russel et al., 1976).

Although the binding of gp32 to ssDNA and its cooperative gp32-gp32 protein interactions would appear to be sufficient to account for all of the above-described activities ascribed to gp32, there is some evidence that gp32 also interacts with several other proteins involved in T4 DNA metabolism. In vitro experiments indicate that gp32 can bind to at least four of the other six proteins required for T4 DNA synthesis. These include T4 DNA polymerase, gp43 (Burke et al., 1980; Huberman et al., 1971), the polymerase accessory protein complex gp42-gp62 (Alberts et al., 1977), and the RNA priming protein gp61 (Burke et al., 1980). While it is tempting to speculate that these various gp32-protein interactions are important for maintaining the structural integrity of an intact T4 DNA replication complex, more definitive experiments are required to substantiate this idea. G. Mosig and her collaborators have conducted an impressive number of genetic experiments which suggest that gp32 may, in addition, interact with several other proteins, including T4 DNA ligase, T4 and *E. coli* nucleases involved in recombination, T4 topoisomerase, membrane proteins including *r*IIA and *r*IIB, and the products of T4 gene 41 (involved in RNA primer synthesis) and gene 17 (involved in DNA packaging) (see Karam et al., this volume). Most of these gp32-protein interactions were inferred from an analysis of mutations in gene 32 that inactivate some, but not all, functions of gp32 and from the ability of mutations in other genes to either suppress or enhance the original defect in the gene 32 mutant being studied (Mosig, Dannenberg, Ghosal, et al., 1979; Mosig, Luder, Garcia, et al., 1979). Once the relative positions of the various gene 32 mutants were located and each correlated with a particular functional defect, Mosig was able to map several discrete domains in gp32 that appear to be specific for each of these various gp32-protein interactions. Doherty et al. (*Multifunctional Proteins*, in press), however, have argued that many of these in vivo genetic experiments can also be explained on the basis of more indirect "metabolic" effects which do not require direct contact between gp32 and the other proteins implicated by Mosig. In this chapter we summarize existing in vitro data which correlate gp32 functions with discrete regions in the gp32 primary structure, and we present our preliminary results on gp32 isolated from a T4 strain carrying the *tsP*7 mutation in gene 32. These latter results provide an interesting opportunity to compare conclusions derived from the in vivo and in vitro study of a mutant gene 32 protein.

STRUCTURE OF THE GENE 32 PROTEIN

Both the amino acid sequence of the gp32 (Williams et al., 1980) and the nucleotide sequence of gene 32 (Krisch and Allet, 1982) have now been determined. The amino acid sequence predicted by translation of

10 20
Met-Phe-Lys-Arg-Lys-Ser-Thr-Ala-Glu-Leu-Ala-Ala-Gln-Met-Ala-Lys-Leu-Asn-Gly-Asn-

30 40
Lys-Gly-Phe-Ser-Ser-Glu-Asp-Lys-Gly-Glu-Trp-Lys-Leu-Lys-Leu-Asp-Asn-Ala-Gly-Asn-

50 60
Gly-Gln-Ala-Val-Ile-Arg-Phe-Leu-Pro-Ser-Lys-Asn-Asp-Glu-Gln-Ala-Pro-Phe-Ala-Ile-

70 80
Leu-Val-Asn-His-Gly-Phe-Lys-Lys-Asn-Gly-Lys-Trp-Tyr-Ile-Glu-Thr-Cys-Ser-Ser-Thr-

90 100
His-Gly-Asp-Tyr-Asp-Ser-Cys-Pro-Val-Cys-Gln-Tyr-Ile-Ser-Lys-Asn-Asp-Leu-Tyr-Asn-

110 120
Thr-Asp-Asn-Lys-Glu-Tyr-Ser-Leu-Val-Lys-Arg-Lys-Thr-Ser-Tyr-Trp-Ala-Asn-Ile-Leu-

130 140
Val-Val-Lys-Asp-Pro-Ala-Ala-Pro-Glu-Asn-Glu-Gly-Lys-Val-Phe-Lys-Tyr-Arg-Phe-Gly-

150 160
Lys-Lys-Ile-Trp-Asp-Lys-Ile-Asn-Ala-Met-Ile-Ala-Val-Asp-Val-Glu-Met-Gly-Glu-Thr-

170 180
Pro-Val-Asp-Val-Thr-Cys-Pro-Trp-Glu-Gly-Ala-Asn-Phe-Val-Leu-Lys-Val-Lys-Gln-Val-

190 200
Ser-Gly-Phe-Ser-Asn-Tyr-Asp-Glu-Ser-Lys-Phe-Leu-Asn-Gln-Ser-Ala-Ile-Pro-Asn-Ile-

210 220
Asp-Asp-Glu-Ser-Phe-Gln-Lys-Glu-Leu-Phe-Glu-Gln-Met-Val-Asp-Leu-Ser-Glu-Met-Thr-

230 240
Ser-Lys-Asp-Lys-Phe-Lys-Ser-Phe-Glu-Glu-Leu-Asn-Thr-Lys-Phe-Gly-Gln-Val-Met-Gly-

250 260
Thr-Ala-Val-Met-Gly-Gly-Ala-Ala-Ala-Thr-Ala-Ala-Lys-*Lys-Ala-Asp-Lys-Val-Ala-Asp-*

270 280
Asp-Leu-Asp-Ala-Phe-Asn-Val-Asp-Asp-Phe-Asn-Thr-Lys-Thr-Glu-Asp-Asp-Phe-Met-Ser-

290 300
Ser-Ser-Ser-Gly-Ser-Ser-Ser-Ser-Ala-Asp-Asp-Thr-Asp-Leu-Asp-Asp-Leu-Leu-Asn-Asp-Leu

FIG. 1. Amino acid sequence of bacteriophage T4 gp32. The B region (residues 1 to 21) and the A region (residues 254 to 301) are italicized. The eight tyrosines in gp32 are boxed. (From Krisch and Allet, 1982, and Williams et al., 1980.)

the structural gene is in good agreement with that previously determined by protein chemistry. gp32 contains 301 amino acids, giving a molecular weight of 33,488 and a net charge of approximately −10 at a pH of 7 (Fig. 1) (Krisch and Allet, 1982; Williams et al., 1981). These values are in good agreement with the previously reported molecular weight of 35,000, as determined by sodium dodecyl sulfate-polyacrylamide gel electrophoresis (Alberts and Frey, 1970), and with the isoelectric point of 5.0, as obtained from isoelectric focusing in urea-containing polyacrylamide gels (Hosoda and Moise, 1978). The charge distribution within gp32 is asymmetric, with the NH_2-terminal half having a net charge of +10 and the COOH-terminal half having a net charge of −20. Even though intact gp32 is acidic and carries a net negative charge, the electrostatic nature of the gp32-ssDNA interaction (Alberts and Frey, 1970; Kowalczykowski, Lonberg, Newport, and von Hippel, 1981) requires that the region of gp32 in direct contact with DNA contain at least three positively charged residues spaced in such a way to enable it to interact with the phosphodiester backbone of ssDNA.

A secondary structure has been proposed for gp32, based on an empirical analysis (Chou and Fasman, 1978b) of its amino acid sequence. gp32 is predicted to contain 36% α-helix, 18% β-sheet, and 46% random coil (Fig. 2) (Williams et al., 1981). These figures are in reasonable accord with circular dichroism measurements which indicated that gp32 contains 22% α-helix, 26% β-sheet, and 52% random coil (Greve et al., 1978a). Based on its predicted secondary structure, gp32 can be divided into three domains. The NH_2-terminal (residues 1 to 35) and COOH-terminal (residues 187 to 301) regions are primarily α-helical, whereas the central region (residues 36 to 186) con-

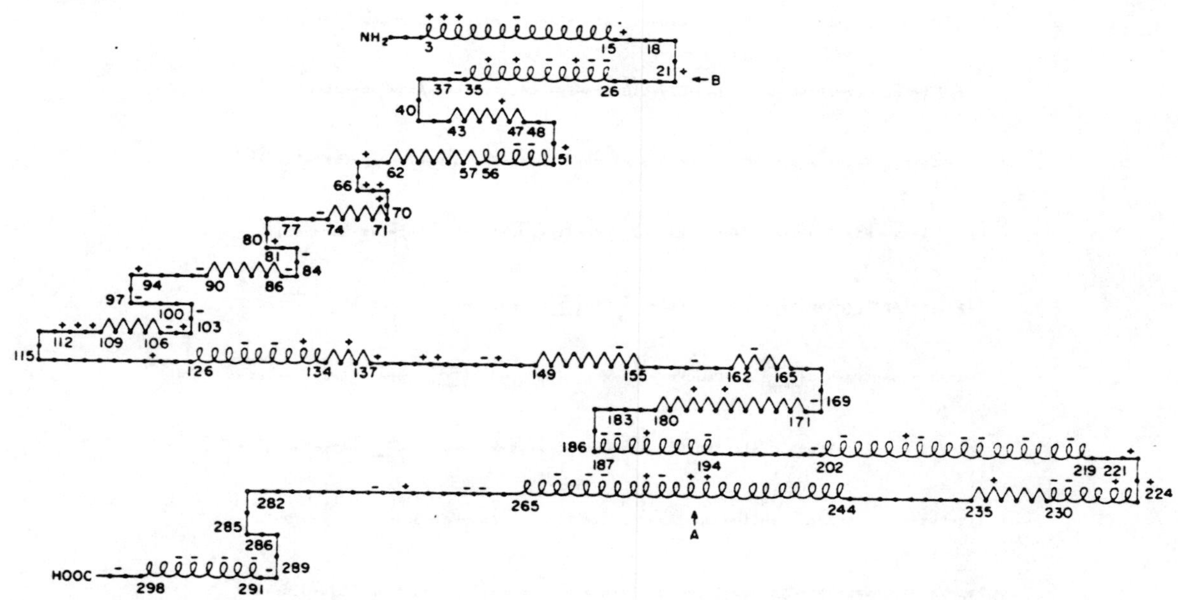

FIG. 2. Schematic diagram of the predicted secondary structure of gp32. Residues are presented in helical (ℓ), β-sheet (Λ), and coil (—) conformations. β-turns are denoted by chain reversals. The position of charged residues are indicated, and conformational-boundary residues are numbered. The arrows pointing from A and B indicate the trypsin-sensitive bonds in the native protein that when both are cleaved give rise to gp32*-(A+B). (From Williams et al., 1981.)

tains most of the β-sheet and 11 of the 15 predicted β-turns (Fig. 2). This large number of β-turns suggests a compact structure which is consistent with the resistance of this central "core" region to proteolytic cleavage (see below).

In solution gp32 appears to be quite elongated, behaving as though it were a prolate ellipsoid with an axial ratio of 4:1 and an overall length approaching 12 nm (Alberts and Frey, 1970). Since in its complex with ssDNA each gp32 molecule may only span a distance of about 2.8 nm (assuming a binding-site size of 6 to 7 nucleotides [Jensen et al., 1976; Kelly et al., 1976; Spicer et al., 1979] and an internucleotide spacing of about 0.47 nm [Delius et al., 1972]), adjacent molecules of gp32 probably overlap one another. This overlap presumably contributes to the cooperativity that is observed when gp32 binds ssDNA (Alberts and Frey, 1970; Jensen et al., 1976). Similar gp32-gp32 protein interactions can apparently occur even in the absence of ssDNA. Although gp32 exists as a monomer in dilute solution (<0.025 mg/ml), the protein has been shown to undergo indefinite self-association at higher concentrations (Carroll et al., 1975). The apparent size of these aggregates increases with protein concentration so that at a concentration of 3.4 mg/ml, which is the estimated total gp32 concentration in a T4-infected bacterium, gp32 exists as a linear aggregate containing about 10 monomers (Carroll et al., 1975).

The observation (Anderson and Coleman, 1975; Hosoda et al., 1974; Moise and Hosoda, 1976) that the NH_2 and COOH termini of gp32 can be removed by limited proteolysis has greatly facilitated efforts to identify regions in its primary structure that are responsible for its various functions. The regions spanning residues 9 to 21 and 253 to 275 are particularly susceptible to cleavage by a wide variety of proteinases, including *Staphylococcus aureus* protease, trypsin, and chymotrypsin (Hosoda and Moise, 1978; Moise and Hosoda, 1976; Tsugita and Hosoda, 1978; Williams and Konigsberg, 1978; Williams et al., 1980). Cleavage at any point between residues 9 and 21 removes the basic, NH_2-terminal "B" region and produces gp32*-B. Cleavage at any point between residues 253 and 275 removes the acidic, COOH-terminal "A" region and produces gp32*-A. Cleavage at both sites results in a 26,000-dalton fragment called gp32*-(A+B). Since the in vitro properties of these individual cleavage products appear to be identical irrespective of the particular enzyme used in their preparation (Hosoda et al., 1980), it would appear that the regions between residues 9 to 21 and 253 to 275 have no function other than to serve as hinges for the A and B regions. Limited proteolysis studies (Fig. 3) indicate that both of these hinges can apparently move in response to a conformational change that is brought about by cooperative binding to ssDNA. Figure 3 shows the effect of increasing trypsin concentrations on gp32 in the absence (lanes 2 to 4) and presence (lanes 5 to 7) of ssDNA. When the native gp32 is incubated with trypsin, cleavage occurs first after lysine 253, producing gp32*-A (Fig. 3, lane 2), and then after lysine 21, which converts all of the gp32*-A to gp32*-(A+B) (Fig. 3, lane 4) (Williams and Konigsberg, 1978). Binding of gp32 to ssDNA increases the rate of removal of the COOH-terminal A region (compare lanes 2 and 5 in Fig. 3) and decreases the rate of removal of the NH_2-terminal B region (compare lanes 4 and 7 in Fig. 3) (Hosoda and Moise, 1978; Williams and Konigsberg, 1978).

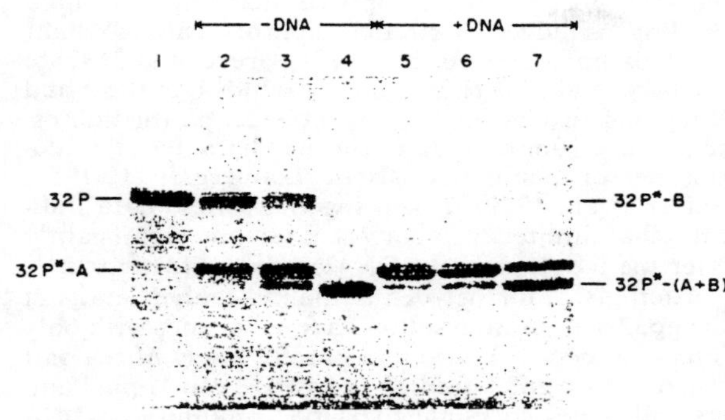

FIG. 3. Partial trypsin digestion of gp32 (32P) in the absence (lanes 2 to 4) and presence (lanes 5 to 7) of fd ssDNA (ssDNA/gp32 base ratio = 9.4), as previously described (Williams and Konigsberg, 1978). The digests were continued for 60 min at 21°C with 0.27 (lanes 2 and 5), 0.54 (lanes 3 and 6), and 8.8 (lanes 4 and 7) μg of trypsin per ml. Lane 1 contains 10.5 μg of gp32.

THE NH_2-TERMINAL B REGION IS ESSENTIAL FOR COOPERATIVE BINDING TO ssDNA

The function of the NH_2-terminal B region, defined here as the first nine amino acids in gp32 (Met-Phe-Lys-Arg-Lys-Ser-Thr-Ala-Glu-), has been approached by comparing the properties of gp32*-A with those of gp32*-(A+B) and, more recently, by characterizing gp32*-B directly. Regardless of which approach is taken, the conclusion seems to be the same: the NH_2-terminal B region is required for gp32-gp32 protein interactions whether they occur between gp32 molecules free in solution or between those bound to a ssDNA lattice. Thus, both gp32*-B and gp32*-(A+B) elute from a Sephadex G-100 column at a position corresponding to their monomeric molecular weight, whereas native gp32, because of its ability to undergo indefinite aggregation in solution, elutes in the void volume even on a Bio-Gel A-5M column (Bittner et al., 1979; Hosoda et al., 1980; Williams and Konigsberg, 1981).

Both gp32*-B and gp32*-(A+B) can be eluted from ssDNA-cellulose with less than 0.5 M NaCl (Hosoda et al., 1980; Williams and Konigsberg, 1981), compared with 2.0 M NaCl for intact gp32, which suggests that the B region makes a significant contribution to the affinity of gp32 for ssDNA. Differential scanning microcalorimetry suggests that the decreased affinity of gp32*-B for ssDNA is due to a loss of cooperative protein-protein interactions rather than to any impairment in the actual protein-ssDNA interaction. The native gp32 denatures at 56.3°C, compared with 54.1°C for gp32*-B and 51.0°C for gp32*-A (Fig. 4). While ssDNA increases the denaturation temperature of all three proteins (Fig. 4) (Williams et al., 1979; Williams and Konigsberg, 1981), only intact gp32 and gp32*-A undergo narrow thermal transitions indicative of cooperative protein-protein interactions. Quantitative binding constants for the interaction of these proteins with ssDNA can be readily determined by taking advantage of the intrinsic fluorescence of gp32. Binding to nucleotide ligands quenches the intrinsic fluorescence of gp32 from approximately 10% for dinucleotides to 20 to 57% for oligonucleotides and polynucleotides longer than seven residues (Kowalczykowski, Lonberg, Newport, and von Hippel, 1981). Maximal fluorescence quenching occurs at a nucleotide-to-protein ratio of approximately 5 to 6 (Kelly et al., 1976; Spicer et al., 1979), which is close to the binding-site size of about 7 as estimated by other methods (Jensen et al., 1976). On the basis of fluorescence quenching measurements, gp32, gp32*-A, and gp32*-(A+B) all bind oligonucleotides such as $d(pT)_8$ with an affinity of about 10^6 M^{-1} (Spicer et al., 1979). While limited proteolysis has no significant effect on the affinity of gp32 for $d(pT)_8$, loss of the B region decreases the affinity of gp32 for poly(dT) by at least 200-fold. Thus, the binding constant for the gp32-poly(dT) complex is at least 7×10^8 M^{-1}, whereas that for the gp32*-(A+B)-poly(dT) complex is only 3×10^6 M^{-1} (Spicer et al., 1979). More recent fluorescence studies have further confirmed that this decrease is

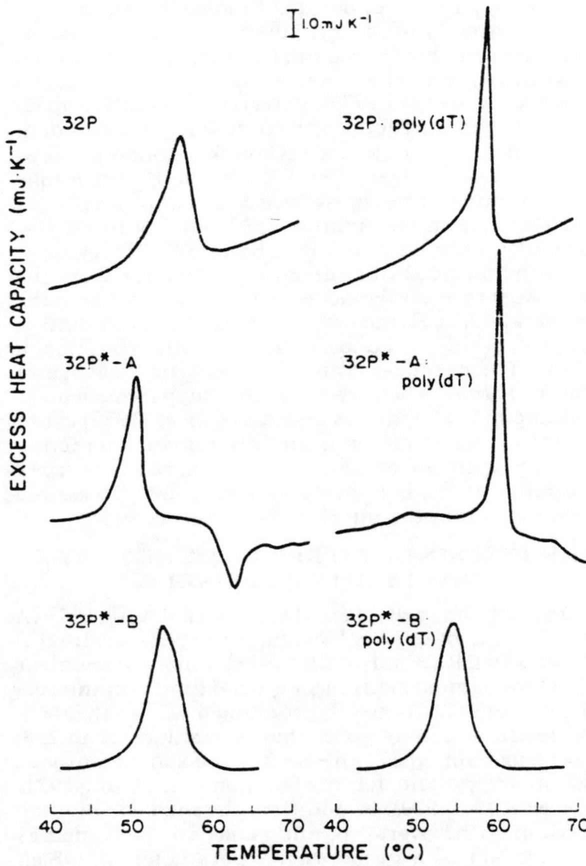

FIG. 4. Thermal denaturation of gp32 (32P) and its partial proteolysis products in the absence and presence of poly(dT). gp32 (29 μM), gp32*-A (22 μM), and gp32*-B (31 μM) were subjected to differential scanning microcalorimetry as previously described (Williams et al., 1979; Williams and Konigsberg, 1981).

entirely due to the loss of cooperative gp32-gp32 protein interactions. While the cooperativity parameter (defined as the equilibrium constant for moving a protein molecule from an isolated to a contiguous binding position on an ssDNA lattice) for the gp32-ssDNA interaction is about 10^3, that for the gp32*-(A+B)–ssDNA interaction is only 1 (Lonberg et al., 1981).

Although there is no doubt that the first nine amino acids in gp32 are required for both cooperative binding to ssDNA and indefinite self-aggregation, it has not yet been demonstrated that the NH_2-terminal B region is directly involved in either of these processes. Moreover, differences in salt sensitivity and the estimated free energy of gp32-gp32 interaction in each case suggest that the actual gp32-gp32 contact sites involved in cooperative DNA binding may be different from those responsible for self-association (Kowalczykowski, Bear, and von Hippel, 1981; Williams and Konigsberg, 1981). If the B region is directly involved in gp32-gp32 protein contacts, then it would seem likely that these interactions arise from electrostatic contacts between the cluster of positively charged amino acids in the B region (the Lys-Arg-Lys sequence at residues 3 to 5) of one gp32 molecule and a negatively charged site on a neighboring gp32 molecule. As predicted by this simple model, gp32 self-association has been shown to be inhibited by high salt concentrations (Carroll et al., 1975); however, cooperative gp32-gp32 protein interactions are completely independent of ionic strength (Kowalczykowski, Lonberg, Newport, and von Hippel, 1981). Thus, the B region may play a direct role in self-association but only an indirect role in cooperative gp32 binding to ssDNA. One such indirect role might be that the B region is required for gp32 to undergo the transition from the non-cooperative, oligonucleotide binding mode to the cooperative, polynucleotide binding conformation (Kowalczykowski, Lonberg, Newport, and von Hippel, 1981). The decreased rate of proteolytic cleavage of the B region when gp32 is in the polynucleotide binding mode (Williams and Konigsberg, 1978) could therefore result either from this region interacting directly with an anionic site on an adjacent gp32 molecule or from it being put in a less accessible location within the interior of the molecule.

THE TYROSINE-RICH REGION OF gp32 MAY BE INVOLVED IN DNA BINDING

Although an earlier report suggested that the ssDNA binding site of gp32 is located in the NH_2-terminal B region (Tsugita and Hosoda, 1978), it now seems clear that those amino acids most crucial for direct interactions with ssDNA must be contained within the trypsin-resistant core of gp32, that is, residues 22 to 253. This fragment, gp32*-(A+B), has the same affinity as the intact protein for $d(pT)_8$ (Lonberg et al., 1981; Spicer et al., 1979). In addition, the intrinsic binding constants (the overall binding constant is defined as the product of a cooperativity parameter, ω, which reflects gp32-gp32 interactions, and the intrinsic binding constant, K, which reflects gp32-ssDNA interactions) for gp32*-(A+B) binding to various polynucleotides closely resemble those of the native protein (Lonberg et al., 1981; Spicer et al., 1979). Similarly, gp32, gp32*-A, and gp32*-(A+B) all bring about comparable polynucleotide lattice deformation upon binding, as judged by circular dichroism and UV light spectroscopy (Greve et al., 1978a; Greve et al., 1978b; Lonberg et al., 1981). Proteolytic removal of the A and B regions also has no significant effect on the fluorescence spectrum of gp32 or on the characteristic fluorescence quenching by ssDNA (Lonberg et al., 1981; Spicer et al., 1979). Taken together, these data indicate that limited proteolysis does not significantly alter the topology of the DNA binding site of gp32.

Attempts to further define the molecular details of the gp32-ssDNA interaction have so far met with only limited success. Based on a careful study of the salt sensitivity of gp32 binding to ssDNA, von Hippel and his colleagues conclude that gp32 binding to ssDNA probably involves three ionic interactions between nucleotide phosphates and basic amino acids (Kowalczykowski, Lonberg, Newport, and von Hippel, 1981). Chemical modification experiments suggest that tyrosine residues might also be directly involved in the binding interaction. Treatment of the native gp32 with tetranitromethane nitrates four or five of the eight tyrosines in this protein and abolishes the tight binding to ssDNA. In contrast, no tyrosines are modified when the reaction is done in the presence of ssDNA (Anderson and Coleman, 1975). Additional evidence implicating tyrosine residues comes from preliminary nuclear magnetic resonance studies on gp32 containing [^{19}F]fluorotyrosine. gp32 binding to ssDNA shifts several of the fluorotyrosine resonances, suggesting that tyrosine intercalation may be involved in polynucleotide binding. In contrast, none of the five tryptophan resonances is altered by ssDNA binding (Kowalczykowski, Bear, and von Hippel, 1981). If four or five tyrosines are involved in gp32 binding to ssDNA, then the tyrosine-rich region of gp32 (residues 72 to 116) must contain at least part of the DNA binding site. Six of the eight tyrosines in gp32 are nearly evenly spaced in this region, occurring at an interval of 7 to 11 amino acids apart (Williams et al., 1981). In addition, there is some primary and secondary structure homology between the tyrosine-rich region of gp32 and the DNA binding site of the bacteriophage fd ssDNA binding protein encoded by gene 5 (Williams and Konigsberg, 1981). This homology is interesting because tyrosine intercalation is known to be involved in the binding of gene 5 protein to ssDNA (Coleman and Oakley, 1980). The tyrosine-rich region of gp32 is also unusual in that it contains a large number of comparatively rare amino acids, including three of the four cysteines, two of the five tryptophans, and one of the two histidines in gp32 (Williams et al., 1981). Preliminary chemical modification experiments and sequence homologies with RNA binding proteins from type C viruses (Henderson et al., 1981) suggest that these cysteines are near the DNA binding site of gp32 (Williams and Konigsberg, 1981).

These general conclusions concerning the location of the binding interface between gp32 and DNA are also supported by genetic studies on mutations mapping within gene 32. Gold and his collaborators have recently isolated over 100 new missense gene 32 mutants and identified 22 mutational sites (Doherty et al., 1982). Fourteen of these sites were shown to occur between residues 36 and 125. In addition, all four of the "classical" temperature-sensitive mutations in gene 32 (*P7*, *P401*, *L171*, and *G26*) also map within this same region (Doherty et al., 1982; Mosig et al., 1977). This clustering of temperature-sensitive mutations, along with the available in vitro data, also prompted these investigators to conclude that the region including residues 72 to 116 might be involved directly in ssDNA binding. Based on earlier in vivo studies, Mosig suggested that even a ~10,000-dalton amber peptide (A453, which is now known to contain residues 1 to 116 [Krisch and Allet, 1982]) can still bind to ssDNA and retains all of the functions necessary for one round of T4 DNA replication (Breschkin and Mosig, 1977 [however, see Doherty et al., *Multifunctional Proteins*, in press, for an alternative interpretation of these earlier studies which does not require that the A453 peptide retain the ability to bind ssDNA]). While all of these in vivo and in vitro data suggest that the DNA binding site of gp32 is within the first one-third of the molecule, perhaps between residues 72 and 116, more definitive results may soon be forthcoming now that the gp32*-(A+B) fragment has been crystallized in a form that appears suitable for high-resolution X-ray crystallographic studies (McKay and Williams, 1982).

THE COOH-TERMINAL A REGION IS ESSENTIAL TO LIMIT THE HELIX-DESTABILIZING ACTIVITY OF gp32

The lysine-lysine bond at position 253 to 254 is the most trypsin-sensitive bond in gp32 (Hosoda and Moise, 1978; Williams and Konigsberg, 1978; Williams et al., 1980). Cleavage at this bond produces a 29,000-dalton fragment, gp32*-A, which lacks residues 254 to 301 but retains the ability to bind tightly and cooperatively to ssDNA (Spicer et al., 1979). Differential scanning microcalorimetry (Fig. 4) demonstrates that loss of the A region (residues 254 to 301) decreases the denaturation temperature of gp32 from 56.3 to 51.0°C (Williams et al., 1979), suggesting that the A region must interact with some other site on the gp32 molecule in such a way as to increase the thermostability of the native conformation of gp32. Since residues 253 to 275 are extremely sensitive to proteolysis (and therefore probably exposed in native gp32), the stabilizing influence of the A region probably results from an interaction between the very acidic COOH terminus (approximately residues 275 to 301) and a basic region located elsewhere on the molecule. These ideas are consistent with several models that have been proposed for gp32 binding to ssDNA (Kowalczykowski, Lonberg, Newport, and von Hippel, 1981; Moise and Hosoda, 1976; Williams and Konigsberg, 1978). Although cooperative gp32 binding to ssDNA results in a conformational change that increases the exposure of the protease-sensitive bonds (between residues 253 and 275) to cleavage (Hosoda and Moise, 1978; Williams and Konigsberg, 1978), it has not yet been demonstrated that this conformational change extends into the acidic region at the COOH terminus (residues 276 to 301).

With the exception of the gp32*-A–ssDNA complex being more sensitive to increasing ionic strength than the gp32-ssDNA complex (Lonberg et al., 1981), the ssDNA binding properties of gp32 and gp32*-A are virtually identical. This fragment binds ssDNA cooperatively (Fig. 4) and with an affinity that is two- or threefold greater than that of gp32 (Lonberg et al., 1981). This small difference in binding affinity is most easily demonstrated by poly[d(A-T)]–poly[d(A-T)] melting experiments. gp32 lowers the denaturation temperature of this double-stranded polynucleotide from about 63 or 64°C to less than 20°C (Greve et al., 1978b). gp32*-A, because of its tighter binding to ssDNA, lowers the denaturation temperature of this polynucleotide by approximately an additional 10°C, so that it is completely melted even at 11°C (Greve et al., 1978a). A similar effect is also seen when gp32*-A is mixed with native dsDNA. In 0.01 M NaCl, gp32*-A lowers the T_m of T4 dsDNA from about 70°C to less than 1°C (Greve et al., 1978a). By analogy with the poly[d(A-T)]–poly[d(A-T)] melting experiments we would expect gp32 to lower the T_m of T4 dsDNA to about 10°C in 0.01 M NaCl. However, at least up to about 55°C, at which point gp32 is itself denatured (Williams et al., 1979), gp32 has no effect on the T_m of naturally occurring dsDNAs (Jensen et al., 1976). Apparently, there is a "kinetic block" which usually prevents gp32 from denaturing dsDNA. Removal of the acidic COOH terminus of gp32, although it has a negligible effect on the affinity of this protein for ssDNA (Spicer et al., 1979), completely eliminates this kinetic block. While more detailed studies are required, it has been suggested that gp32 cannot denature dsDNA because the ssDNA loops that normally occur in native DNA as a result of "breathing" are too transient to permit efficient nucleation of cooperative gp32 binding (Jensen et al., 1976). While gp32 may bind noncooperatively to these short ssDNA stretches in native DNA, dissociation apparently occurs before gp32 can undergo the relatively slow conformational change (Lohman, 1980) that is required to initiate the tight cooperative binding that is capable of denaturing native dsDNA. Removal of the A region either facilitates this conformational change (Williams and Konigsberg, 1981) or allows gp32 to denature dsDNA via an alternative kinetic pathway (Lonberg et al., 1981; Spicer et al., 1979).

While the exact mechanism by which the A region prevents gp32 from denaturing dsDNA is not yet clear, the possible in vivo consequences of a gp32 lacking the A region are readily apparent in the in vitro T4 DNA replication system. Studies with this system have demonstrated that, although gp32*-A can substitute for gp32 in leading-strand synthesis, there is complete absence of lagging-strand synthesis in the presence of gp32*-A (Burke et al., 1980). This complete absence of RNA primer extension (lagging-strand synthesis) appears to be due to destabilization of the 3′-hydroxy chain terminus by gp32*-A. Although these in vitro studies demonstrate that the helix-destabilizing activity of gp32 is greatly increased by removal of its acidic, COOH-terminal domain, it is unclear whether the primary in vivo function of this region is to provide a means for controlling or merely limiting the ability of gp32 to denature dsDNA. Since there is no evidence that any significant partial proteolysis of gp32 occurs in vivo, it has been suggested that heterologous protein-protein interactions involving the COOH-terminal A region of gp32 and other T4 replication proteins may result in an effect similar to that of the actual removal of this domain (Burke et al., 1980).

If this notion is correct, then the A region would enable other T4 replication proteins to specifically enhance the helix-invasion potential of gp32 only when gp32 is situated in front of an advancing replication fork (Burke et al., 1980). In support of this idea, in vitro studies demonstrate that gp32, but not gp32*-A, interacts with both gp61 (the suspected T4 RNA primase) and gp43 (the T4 DNA polymerase); thus, either one, or both, of these proteins might be involved in controlling the helix-destabilizing activity of gp32 (Burke et al., 1980). Even if this control is not fully realized in vivo, it seems clear that the A region is essential to limit the helix-invasion potential of gp32, thus allowing lagging-strand DNA synthesis to occur and preventing random denaturation of T4 dsDNA.

This mechanism for limiting the helix-destabilizing ability of an ssDNA binding protein may be a general one, not limited to gp32. The bacteriophage T7 ssDNA binding protein, although it has no significant amino acid sequence homology with gp32, also contains a very acidic region at its COOH terminus. In this instance, there are 15 negatively charged amino acids within the last 21 residues at the COOH terminus (Dunn and Studier, 1981). Once again, preliminary in vitro studies on a mutant T7 DNA binding protein lacking the last 17 amino acids at the COOH terminus indicate that the removal of these residues has an effect very similar to that seen with gp32; that is, unlike the wild-type T7 protein, this mutant protein (Araki and Ogawa, 1981) can denature dsDNA and fails to stimulate RNA primer-dependent DNA synthesis by T7 DNA polymerase (H. Araki, personal communication). Similarly, the ssDNA binding protein from *E. coli*, which is functionally homologous but structurally unrelated to gp32 (Williams and Konigsberg, 1981), has an acidic, COOH-terminal region that can be removed by partial proteolysis, resulting in a fragment with a stronger helix-destabilizing activity (Williams et al., 1983). Like gp32, cooperative binding of the *E. coli* ssDNA binding protein to ssDNA results in a conformational change that increases the exposure of the COOH terminus to partial proteolysis (Williams et al., 1983). Evidence has previously been presented that eucaryotic ssDNA binding proteins may also have similar functional domains essential for controlling their helix-destabilizing activity (Burke et al., 1980). Indeed, recent results on high-molecular-weight members of the (eucaryotic) high-mobility group nonhistone chromatin proteins reveal that these proteins have extremely acidic COOH-terminal regions (Walker et al., 1980) that can be removed by partial proteolysis, resulting in increased affinity for DNA (Isackson and Reeck, 1982 and personal communication). It appears that in gp32, bacteriophage T4 has once again provided an experimentally accessible prototype for a class of functionally homologous proteins that may be ubiquitously associated with biological systems undergoing DNA replication.

THE *P7* MUTATION IN gp32 RESULTS IN A THERMOSENSITIVE PROTEIN THAT UNFOLDS AT TEMPERATURES ABOVE 37°C

The crucial function that gp32 plays in T4 DNA metabolism in vivo is clearly shown by studies on bacteriophage T4 containing the temperature-sensitive *P7* mutation in gene 32. When cells infected with the *P7* mutant are shifted from 25 to 42°C, T4 DNA synthesis (Curtis and Alberts, 1976; Riva et al., 1970a), and repair (Wu and Yeh, 1973) ceases, and the T4 DNA is rapidly degraded (Curtis and Alberts, 1976). In vivo genetic studies suggest that whereas the *P7* gp32 is unable to bind ssDNA at temperatures above 37°C, it nonetheless retains its ability to interact directly with the T4 gene 46/47-controlled nuclease and the *E. coli recBC* nuclease (Mosig and Bock, 1976). Thus, it was proposed that the *P7* mutation results in an unfavorable amino acid substitution at the DNA binding site of gp32 such that the gp32-ssDNA interaction becomes temperature sensitive but the gp32-nuclease interactions remain normal even at restrictive temperatures (Breschkin and Mosig, 1977a) (the wild-type gp32 can bind ssDNA even at temperatures exceeding 50°C [Williams et al., 1979]). Similar in vivo studies on other gp32 mutants have resulted in the proposal that gp32 may contain six or more discrete functional domains, several of which can be selectively inactivated by exposing a suitable temperature-sensitive or amber gene 32 mutant to restrictive conditions (Mosig, Dannenberg, Ghosal, et al., 1979; Mosig, Luder, Garcia, et al., 1979). In an attempt to localize the DNA binding domain in gp32, we identified the mutation that has occurred in the *P7* mutant and initiated a number of in vitro studies to determine the effect of this mutation on the stability and DNA binding properties of gp32.

During the isolation of the *P7* gp32 we verified earlier in vitro results (Curtis and Alberts, 1976) which indicated that this temperature-sensitive gp32 binds normally to ssDNA-cellulose at permissive temperatures, requiring 2 M NaCl for elution, but is rapidly eluted from this chromatographic support by raising the temperature above 37°C even when the NaCl concentration is only 0.05 M (K. R. Williams, L. Sillerud, E. Spicer, and M. Lo Presti, unpublished data). Amino acid analysis of the purified *P7* gp32 suggested that it might contain a cysteine in place of one of the four arginine residues normally present in gp32. Comparative peptide mapping by high-pressure liquid chromatography confirmed this conclusion and localized the mutation at arginine 46 (Williams et al., 1982). Differential scanning microcalorimetry (Fig. 5) indicates that gp32 from the *P7* strain undergoes a single thermal transition which is centered at 37.5°C and is unaffected by the presence of ssDNA. The inability of ssDNA to sharpen the *P7* gp32 transition (Fig. 5) suggests that at permissive temperatures this mutant protein may bind less cooperatively than the wild-type protein does to ssDNA. That the single transition at 37.5°C corresponds to complete unfolding of the *P7* gp32 was verified by limited proteolysis studies. These latter experiments demonstrated that whereas at 25°C, gp32*-(A+B) can be easily obtained by limited trypsin digestion of gp32 from either the wild type of *P7* temperature-sensitive strains, at 43°C, this proteolytic fragment can only be obtained from the wild-type protein. At this restrictive temperature, trypsin immediately cleaves the *P7* gp32 into a large number of small peptides, none of which is larger than 5,000 daltons. Differences in the enthalpy of denaturation and relative rates of chemical modification of

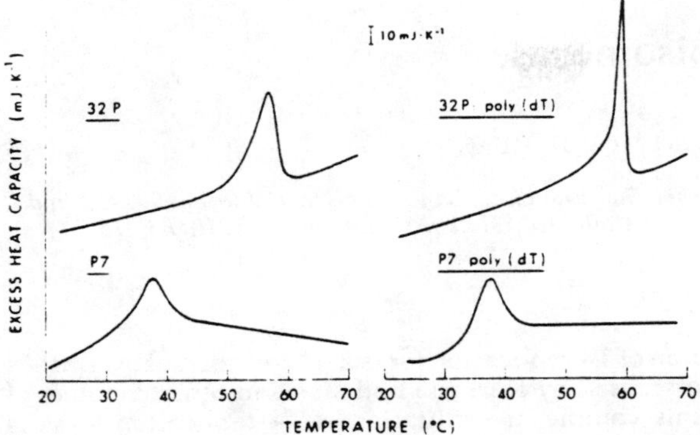

FIG. 5. Thermal denaturation of the wild-type and *P7* temperature-sensitive gp32 (32P) in the absence and presence of poly(dT). The wild-type (29 μM) and *P7* (59 μM) proteins were subjected to differential scanning microcalorimetry as previously described (Williams et al., 1979; Williams and Konigsberg, 1981).

the *P7* and wild-type gp32 suggest that there may be important differences in the conformations of these two proteins even at permissive temperatures. As calculated from the area under the curves in Fig. 5, the molar enthalpy of denaturation of the *P7* protein is less than 20% of that for the wild-type protein. These differences, coupled with the apparent decreased cooperativity of binding of ssDNA by the *P7* gp32, may account for some unusual properties of this temperature-sensitive protein that are observed even at temperatures below 37°C. These include the inability of the *P7* gp32 to catalyze renaturation of dsDNA in vitro (Huberman et al., 1971) as well as the overproduction of the *P7* gp32 (Gold et al., 1976) and the recombination deficiencies (Mosig et al., 1977) that have been observed in vivo in T4 strains carrying the *P7* mutation.

Based on our preliminary in vitro studies, we conclude that the primary effect of the *P7* mutation is to decrease the thermostability of gp32. The inability of the *P7* gp32 to bind ssDNA at temperatures above 37°C appears to be due to unfolding of the entire protein rather than to any specific effect at the DNA binding site. We have not found any in vitro evidence to support the conclusion derived from in vivo studies (Breschkin and Mosig, 1977a) that the *P7* gp32 retains enough of its native conformation at temperatures above 37°C to interact in a functionally meaningful way with nucleases involved in recombination.

CONCLUSION

The ease of purification, wealth of genetic data (see especially Doherty et al., *Multifunctional Proteins*, in press; Doherty et al., 1982; Mosig, Dannenberg, Ghosal, et al., 1979; Mosig, Luder, Garcia, et al., 1979) and autoregulatory properties (von Hippel et al., this volume) of gp32 make this protein an extremely valuable prototype for the study of ssDNA binding proteins and their interaction with nucleic acids and other proteins involved in DNA replication, repair, and recombination. The completion of the amino acid sequence of gp32 (Williams et al., 1980) and the successful crystallization of a gp32 fragment (McKay and Williams, 1982) should ultimately allow a molecular understanding of the diverse interactions in which this protein participates. By utilizing data derived from in vitro and in vivo analyses, some regions of gp32 have been found that are essential for cooperative binding and the control of the helix-destabilizing activity of this protein. However, as previously noted by Doherty et al. (*Multifunctional Enzymes*, in press), the A and B regions of gp32 may not be structural domains in a strict sense, in that neither region has been demonstrated to independently possess any functional activity once it has been cleaved from the native protein. Much remains to be learned about the exact mechanisms involved in gp32 binding to single-stranded nucleic acids, its autoregulatory properties, and its contribution to the structural integrity of the protein complexes involved in T4 DNA replication, repair, and recombination.

T4 DNA Topoisomerase

KENNETH N. KREUZER[1] AND WAI MUN HUANG[2]

Department of Biochemistry and Biophysics, University of California, San Francisco, San Francisco, California 94143,[1] and Department of Cellular, Viral and Molecular Biology, University of Utah Medical Center, Salt Lake City, Utah 84132[2]

Topoisomerases are a class of enzymes that catalyze a remarkable set of reactions resulting in dramatic alterations in DNA structure. These alterations are, for the most part, simple topological changes in the state of circular DNA, involving, for example, the introduction or removal of supercoils, the formation or resolution of interlocked circles (catenanes), and the knotting or unknotting of circular DNA.

Type I topoisomerases, such as *Escherichia coli* topoisomerase I (ω protein), act by a mechanism involving a transient break in one strand of the double helix (Brown and Cozzarelli, 1981; Tse and Wang, 1980). The type II topoisomerases, on the other hand, make a transient double-strand break in the helix backbone; the topological alterations in DNA structure are executed by the passage of another region of duplex DNA through this transient break (for reviews, see Cozzarelli, 1980; Gellert, 1981). The first type II topoisomerase discovered was *E. coli* DNA gyrase (Gellert et al., 1976), which can perform all of the above-mentioned topological interconversions with duplex DNA. A second type II topoisomerase was found in extracts of *E. coli* after bacteriophage T4 infection, and this enzyme has been shown to be encoded by the phage genes 39, 52, and 60 (L. F. Liu et al., 1979; Stetler et al., 1979) (see below). The T4 enzyme is similar to the *E. coli* DNA gyrase in many of its properties (Kreuzer and Alberts, manuscript in preparation; L. F. Liu et al., 1979), but the phage enzyme is not able to introduce negative supercoils into covalently closed circular DNA substrates in vitro (see below).

Although precise details concerning the physiological role of the T4-induced topoisomerase are still obscure, it is clear that the enzyme plays an important role in T4 DNA synthesis. Conditional mutations in any of the three subunit structural genes result in the "DNA-delay" phenotype, with a marked delay in the acceleration of DNA synthesis as infection proceeds (Mufti and Bernstein, 1974). While the topoisomerase-deficient mutants do synthesize some DNA in wild-type *E. coli*, the additional inactivation of the host DNA gyrase results in little or no net T4 DNA synthesis (McCarthy, 1979). This implies that host gyrase can partially compensate for a deficiency in phage topoisomerase and that there is an absolute requirement for type II topoisomerase activity in T4 DNA synthesis.

In the absence of the T4 topoisomerase (but the presence of host gyrase), the rate of replication fork movement in vivo is identical to that in wild-type infections, but the number of functioning replication forks is reduced (McCarthy et al., 1976). Therefore, the topoisomerase is thought to be involved in the initiation of T4 replication forks at chromosomal origins. As discussed by Kozinski and Mosig in other chapters of this volume, the initiation of T4 replication forks is thought to include two distinct modes: a primary mode, which requires host RNA polymerase, and a secondary mode, dependent on T4 recombination proteins. In this chapter, we summarize primarily the in vitro properties of the T4 topoisomerase and relate these properties to the possible functions of the enzyme.

(The following abbreviations are used in this chapter: T4 dC-DNA, unmodified cytosine-containing T4 DNA prepared from 56⁻ 42⁻ *denB*⁻ *alc*⁻ phage; dsDNA, double-stranded DNA; glu T4 HMdC-DNA, modified [glucosylated and hydroxymethylated] T4 DNA prepared from wild-type phage; ssDNA, single-stranded DNA.)

ENZYME STRUCTURE

Purification of T4 DNA Topoisomerase

T4 topoisomerase can readily be purified from T4 phage-infected *E. coli* cells, using as an assay its ability to relax superhelical DNA in an ATP-dependent reaction (L. F. Liu et al., 1979; Stetler et al., 1979). Alternatively, the enzyme may be assayed by its ability to stimulate T4 DNA replication in vitro in a T4 DNA-delay mutant-infected lysate. Mutants with mutations in any one of the three DNA-delay genes (gene 39, 52, or 60) can be used as the recipient in a complementation assay using semipermeable cellophane disks as the supporting matrix (Stetler et al., 1979; see Huang, this volume, for details).

T4 phages defective in DNA polymerase (43⁻) or in the phage maturation genes (33⁻ 55⁻) are used for the production of the phage topoisomerase, since they do not cause cell lysis and infected cells accumulate early T4 proteins. Two related methods of purification have been developed; one of these will not be discussed further since it is published elsewhere (Kreuzer and Jongeneel, 1983). Alternatively, an *E. coli* strain that carries the T4 gene 39 on a plasmid is used as the starting bacterium. This facilitates the recovery of the topoisomerase, since the gene 39 protein is the least stable subunit and is prone to protease degradation. The subunits of T4 DNA topoisomerase have been shown to bind DNA tightly and are membrane associated (Huang 1975; Huang and Buchanan, 1974), and these properties have also been incorporated in the purification scheme. First, a gently lysed fast-sedimenting membrane fraction of the infected culture is prepared. All of the topoisomerase activity remains associated with this fraction, but approximately 90%

of the contaminating proteins are removed. The topoisomerase can then be extracted from the membrane fraction with 2 M NaCl and further purified by hydroxylapatite and DNA-cellulose chromatography. Further details concerning this method of purification have been reported (Stetler et al., 1979). Using this procedure, one can obtain approximately 1.8 mg of the T4 DNA topoisomerase from 65 liters of the infected cells.

Subunit Structure

The purified T4 DNA topoisomerase, when examined by sodium dodecyl sulfate-polyacrylamide gel analysis, shows three protein bands of approximately equal molar concentration. The monomer molecular weights of these three subunits have been estimated to be 64,000, 51,000, and 12,000 by Huang's laboratory, in reasonable agreement with values (56,000, 46,000, and 17,500, respectively), reported by Kreuzer and Jongeneel (1983). Earlier reports indicated the presence of a minor 110,000-dalton band (Stetler et al., 1979) as well as a 23,000-dalton band (L. F. Liu et al., 1979); these were apparently contaminants as they are absent in more recent preparations. The 64,000- and 51,000-dalton bands correspond to the products of T4 genes 39 and 52, respectively (O'Farrell et al., 1973). More recently, rabbit antibody prepared against the purified T4 DNA topoisomerase was used to show immunologically that amber mutants defective in T4 gene 60 do not synthesize the 12,000-dalton subunit, thereby providing evidence that this protein is the product of T4 gene 60 (Huang, unpublished data). Furthermore, this is consistent with the observation that *E. coli* extracts prepared from cells infected with T4 39^-, 52^-, or 60^- mutants lack the ATP-dependent topoisomerase activity (L. F. Liu et al., 1979).

By direct examination in the electron microscope, the native T4 DNA topoisomerase appears as spherical globular particles with a circumference of 47.4 ± 3.6 nm after fixation with formaldehyde and glutaraldehyde and then rotary shadowing with tungsten (Huang, unpublished data). This value corresponds to a particle diameter of 15.1 ± 1 nm and is consistent with globular particles of approximately 500,000 daltons in mass. When the enzyme was analyzed by gel filtration in Sephacryl S-300, it eluted in a broad peak, equivalent to a Stokes radius of 6.1 nm and a native molecular weight of approximately 380,000 to 400,000. If the protomer of the enzyme, which has the three different components, is taken to have a molecular weight of 127,000, then the gel filtration analysis suggests that the native protein consists of three protomers (three of each of the three subunits). Since the native enzyme should have a twofold symmetry when its interactions with DNA are considered (see below), and because of the broadness of the elution profile, the gel filtration data are also consistent with the model that both dimeric and tetrameric forms of the protomers are functional, thus yielding a broad elution profile in gel filtration analysis, centering around an average trimer value; however, the size determination from electron microscopy is most consistent with the tetrameric structure. It should be noted that in sucrose gradient analysis, the native enzyme sediments at 6.5S.

ENZYMATIC REACTIONS

Strand Passage Reactions

The T4 topoisomerase enzymatically relaxes both positively and negatively supercoiled DNA and also performs knotting, unknotting, catenation, and decatenation of closed circular duplex DNA substrates. As depicted schematically in Fig. 1, all of these reactions are explained by the double-strand passage mechanism of the type II topoisomerases (for reviews, see Cozzarelli, 1980; Gellert, 1981).

The T4 topoisomerase was initially identified by its ability to catalyze the ATP-dependent relaxation of negatively supercoiled DNA (L. F. Liu et al., 1979; Stetler et al., 1979), and, as expected from the strand passage model, two supercoils are removed for each reaction cycle (Liu et al., 1980). With roughly equal efficiencies, the enzyme relaxes various circular plasmid and phage DNAs. While this result indicates a lack of substrate specificity for the topoisomerase, the true in vivo substrate for the enzyme, namely, modified glu HMdC-T4 DNA, has not been properly tested for relaxation or supercoiling. A model proposed by L. F. Liu et al. (1979) is that the enzyme is capable of site-specific supercoiling of DNA at the origin(s) of T4 DNA replication. This could be accomplished, for example, by the recognition of two oriented DNA sequences bracketing the origin, allowing unidirectional strand passage leading to active negative super-

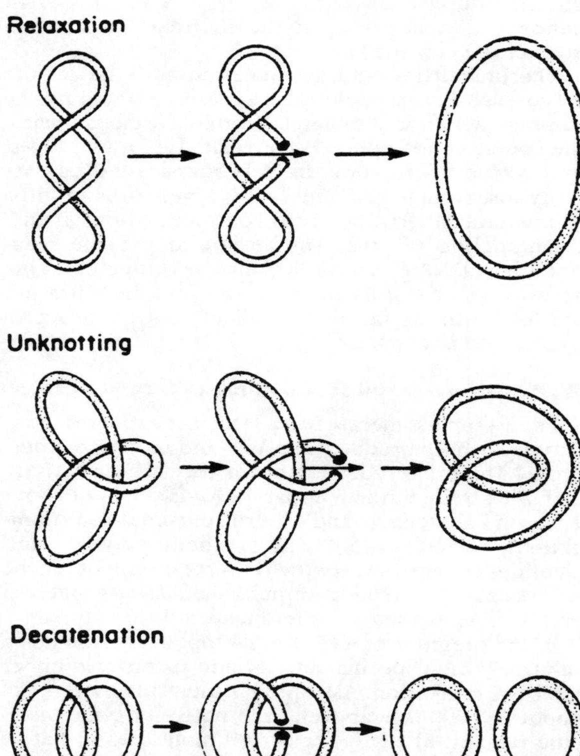

FIG. 1. Reactions catalyzed by the type II topoisomerases.

coiling. The finding that T4 DNA is grossly altered in topoisomerase cleavage specificity because of the cytosine modifications (see below) strengthens the possibility that some unique topological interconversions occur on native T4 DNA. Tests are currently in progress to search for such an activity, using closed circular DNA substrates constructed from fragments of native T4 DNA.

The interconversions of knotted and catenated DNA forms by the T4 topoisomerase are strongly influenced by the effect of the reaction conditions on DNA structure. Catenation or knotting (or both) of circular DNA occurs under conditions where the DNA is in a condensed state, for example, when complexed with histone H1, spermidine, or polymin P (Liu et al., 1980). Under the standard relaxation reaction conditions, unknotting and decatenation are catalyzed by the enzyme, since catenanes and knotted DNA forms are energetically unfavorable compared with simple monomeric circles in the absence of DNA-condensing agents. In general, then, the role of the topoisomerase is limited to the strand passage reaction, and whether catenanes or knots are produced or destroyed simply reflects the state of DNA aggregation dictated by the reaction conditions (see Krasnow and Cozzarelli, 1982, for a detailed analysis of the effect of aggregation on catenation by DNA gyrase). There is one exception; namely, at high enzyme concentrations, the T4 topoisomerase itself apparently promotes some form of DNA condensation as well as catalyzing the strand passage reaction, resulting in the knotting of circular duplex DNA (Liu et al., 1980). This DNA condensation may relate to the multiple DNA binding sites of the enzyme (see below).

The unknotting and decatenation activities of the T4 topoisomerase could play a role in untangling the complex network of phage DNA that develops during the course of infection (Huberman, 1969). As judged by electron microscopy, the topoisomerase does partially disentangle isolated T4 DNA networks in vitro (Burke and Alberts, personal communication). If this disentangling involves simple strand passage reactions, host DNA gyrase should also be competent. This activity would not be expected to effect the initiation of DNA synthesis, but it could effect subsequent segregation into phage heads.

ATP Hydrolysis and Its Coupling to Strand Passage

The T4 topoisomerase has a DNA-dependent ATPase activity, whose products are ADP and inorganic phosphate. The preferred cofactor for the ATPase activity is duplex DNA, with an apparent K_m for DNA of about 1.5 μg/ml (Jongeneel and Alberts, personal communication). Various ssDNAs and synthetic polydeoxynucleotides can also satisfy the DNA requirement for the ATPase activity (Huang, unpublished data; Jongeneel and Alberts, personal communication).

In the presence of ATP, the T4 topoisomerase has a relatively high specific activity and turnover number for DNA relaxation. Assuming a molecular weight of about 260,000 (see above), each native enzyme molecule relaxes on the order of 200 molecules of native pBR322 (corresponding to over 2,000 strand passage events) in 30 min at 30°C. As a cofactor for the strand passage reactions, ATP can be replaced by dATP, but by none of the other ribo- or deoxyribonucleoside triphosphates (Huang, unpublished data; Zetina-Rosales and Alberts, personal communication).

Rough calculations indicate that one or two ATP molecules are hydrolyzed per strand passage event, and it has been suggested that ATP hydrolysis is involved in the turnover of the enzyme for repeated cycles of relaxation (L. F. Liu et al., 1979). In the case of DNA gyrase, ATP hydrolysis is required for turnover of the enzyme, and the binding of ATP (or a nonhydrolyzable analog) is sufficient for one reaction cycle (Sugino et al., 1978). However, the ATP coupling may be more complex with the T4 enzyme, since stoichiometric amounts of the highly purified enzyme induce a low level of DNA relaxation even in the absence of any nucleoside triphosphate (unpublished data; also see below).

The ATP requirement for enzymatic relaxation by T4 topoisomerase poses a paradox, in that the relaxation of superhelical DNA is already a thermodynamically favorable reaction. Relaxation is carried out by other topoisomerases, such as the type I enzymes and DNA gyrase, in the absence of ATP. Therefore, it is possible that the role of ATP in the T4 topoisomerase reaction is not directly related to energy coupling, but rather involves a change in the catalytic state of the enzyme, in a manner similar to the role of ATP in the type I restriction enzymes (Yuan, 1981). It is also possible that ATP hydrolysis is coupled to some energy-requiring reaction in vivo, such as in the site-specific supercoiling model discussed above, and that some other unidentified cellular protein or cofactor is required for the coupling of ATP hydrolysis to such an energy-requiring reaction.

Interactions with DNA

Two of the subunits of the T4 topoisomerase, namely, the products of genes 39 and 52, have previously been shown to bind DNA tightly, as they are retained on DNA-cellulose columns (Huang and Buchanan, 1974). The interactions of the native enzyme with DNA have also been examined by direct visualization in the electron microscope (Fig. 2A). In the absence of ATP, T4 topoisomerase forms a stable complex with superhelical pBR322 DNA. The enzyme is seen as globular particles located exclusively at the intersection of the DNA duplex strands, and the circular DNA molecules appear as multiply looped structures with the enzyme at the base of the loops. In some instances, two or three individual pBR322 molecules are bound together to form a complex with one large globular protein. The size distribution of the proteins in the observed protein-DNA complexes is more heterogeneous than that of a corresponding preparation without DNA. The circumference of the protein particles in the complexes ranges from 60 to 94 nm. These values correspond to a particle diameter of 21 to 30 nm, which is larger than the values obtained for the proteins alone (about 15 nm [see above]). The heterogeneity in the size of the proteins bound to DNA is observed regardless of the initial protein concentration.

To further examine the stoichiometry, the enzyme-DNA complex, formed by using ^{3}H-labeled pBR322 DNA and subsequently fixed by formaldehyde and glutaraldehyde, was also analyzed in a shallow CsCl density gradient. The result is consistent with the

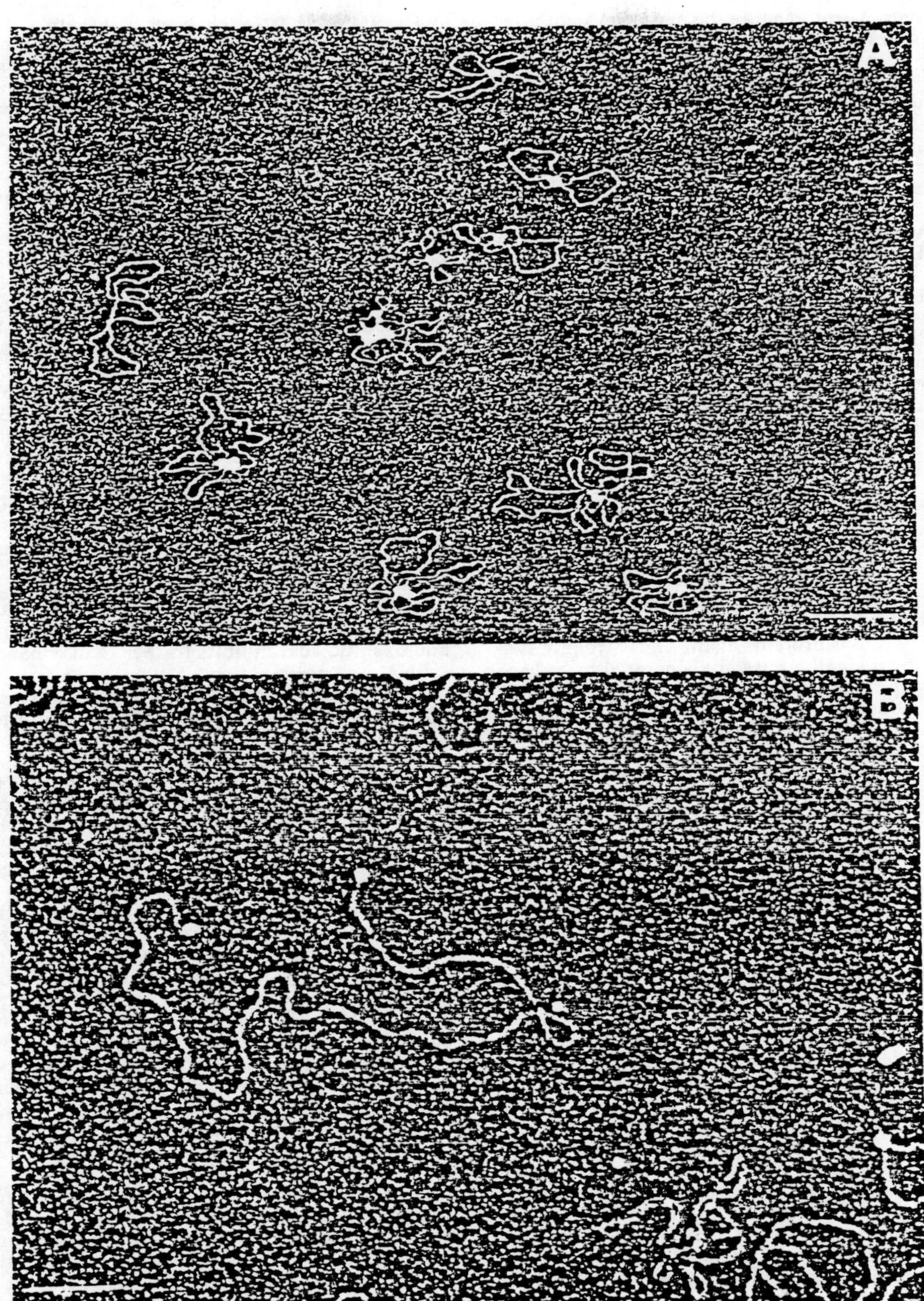

FIG. 2. Visualization of T4 topoisomerase-DNA complexes by electron microscopy. (A) Superhelical pBR322 interaction with T4 topoisomerase in the absence of ATP. Bar = 200 nm. (B) Linearized pBR322 DNA with protein molecules attached to both ends as a result of the induced cleavage reaction. Bar = 100 nm.

notion that both dimeric and tetrameric forms of the 127,000-dalton protomer are the most prevalent interacting species of the T4 topoisomerase (Huang et al., manuscript in preparation).

The DNA-protein complex with a multiply looped DNA structure is also observed when relaxed closed circular DNA is used. Under the same conditions, linear duplex DNA does not readily form an enzyme-DNA complex. These results are consistent with the notion that the preferred binding sites of the enzyme are located at the intersections or crossover points of DNA, and the structural constraint of the closed circular DNA makes them better substrates for stable complex formation.

Since T4 topoisomerase can be induced to cleave its DNA substrate by the addition of a detergent such as sodium dodecyl sulfate (also see below), the products of such a reaction were also examined by electron microscopy. As expected for a type II enzyme, superhelical pBR322 DNA is cleaved to form linear molecules with a protein attached to each end (Fig. 2B). Furthermore, smaller relaxed circles, each having an associated protein particle, are also observed. These small circles are of varying length, shorter than unit length pBR322. It is apparent that these small circular structures are held together by the associated proteins, presumably one at each end of the DNA molecule, since proteinase treatment linearizes these molecules. The appearance of the small circles with the attached proteins and the absence of linear DNA molecules without proteins at the ends strongly suggest that these small circles are generated by at least two cleavage events occurring in the helical DNA substrate. Since one subunit of the enzyme is transferred to each of the two newly broken ends of a DNA molecule for each cleavage event, this observation is also consistent with the notion that tetrameric forms of the protomer are also a functioning unit of the enzyme.

Isolated DNA Cleavage Reactions

Type II topoisomerases, such as DNA gyrase, can be induced to cleave a duplex DNA substrate, as described above, yielding a product with protein covalently attached to the DNA. This reaction complex of the T4 topoisomerase can be induced by terminating the standard reaction with detergent, and, as with DNA gyrase, the reaction product is broken duplex DNA with free 3'-hydroxyl groups and 5'-phosphoryl groups blocked by the covalent attachment of a topoisomerase subunit.

Two different conditions have been found to increase the yield of this complex. First, higher yields are obtained from reactions containing superhelical DNA in the absence of ATP, indicating that a superhelical structure of the substrate somehow enhances the DNA cleavage activity of the enzyme (Stetler and Huang, unpublished data). Second, the addition of the DNA gyrase inhibitor oxolinic acid greatly enhances the cleavage activity of the enzyme on all duplex substrates tested (Kreuzer and Alberts, manuscript in preparation). Oxolinic acid also inhibits relaxation by the T4 topoisomerase, with half-maximal inhibition occurring at a drug concentration of about 250 μg/ml. Except for the fact that higher drug levels are required for the T4 enzyme, oxolinic acid seems to act identically on the T4 topoisomerase and *E. coli* DNA gyrase.

The induced DNA cleavage sites are convenient markers for determining the enzyme binding sites. Several very strong cleavage sites have been mapped on native T4 DNA by using this cleavage reaction, and their locations suggest possible physiological relevance (see below). An analysis of oxolinic acid-enhanced cleavage of the more simple φX174 and pBR322 DNAs indicated that the T4 topoisomerase recognizes a hierarchy of sites, with very weak cleavage sites occurring frequently (at least on the order of one every 50 base pairs) and very strong cleavage sites occurring quite infrequently (one every few thousand base pairs) (Kreuzer and Alberts, unpublished data). There were no obvious rules of sequence recognition obtained by comparing various cleavage sites. Similar conclusions were reached in an analysis of host DNA gyrase cleavage specificity (Morrison and Cozzarelli, 1979), and, in fact, the cleavage sites on duplex φX174 DNA for host DNA gyrase and the T4 topoisomerase overlap considerably, implying a functional or evolutionary relatedness between the two enzymes.

A second class of DNA cleavage events induced by the T4 topoisomerase distinguishes the enzyme from host DNA gyrase. The T4 enzyme cleaves ssDNA templates, and, as in the cleavage of dsDNA, leaves protein covalently attached to the newly formed breaks (Kreuzer, manuscript in preparation). Unlike dsDNA cleavage, the cleavage of ssDNA does not require detergent treatment of the reaction products, and oxolinic acid inhibits (rather than enhances) the ssDNA cleavage reaction.

The sites of cleavage in φX174 ssDNA correlate to regions of probable secondary structure in the DNA, and these sites were shifted by changes in temperature (Kreuzer, manuscript in preparation). Therefore, the actual structure recognized by the enzyme may be a dsDNA-ssDNA junction, rather than just a particular stretch of ssDNA. Gene 32 protein, which coats ssDNA and eliminates any secondary structure, blocks the cleavage of ssDNA in vitro. It seems likely, then, that the topoisomerase does not generally attack ssDNA in vivo, since such DNA is coated with the gene 32 protein. Perhaps the enzyme does recognize some form of secondary structure, such as cruciforms, in duplex DNA. The cleavage activity itself could be involved in secondary initiation of T4 replication if one postulates that the topoisomerase cleaves either particular recombination intermediates (containing short regions of ssDNA) or regions of secondary structure in the duplex DNA (possibly the recombination hot spots of the T4 genome [see below]). The free 3'-hydroxyl ends generated by such cleavage could be suitable primers for chain extension by the T4 replication complex.

The two classes of cleavage events can be understood as follows. Cleavage of dsDNA requires denaturation of the enzyme (as by detergent treatment) to reveal the DNA breaks. This implies either that the denaturation of the enzyme causes the cleavage event (perhaps by disassociation of the enzyme subunits, for example) or that the enzyme denaturation simply traps a reaction intermediate that existed just before the denaturant was added. In the latter case, the enzyme must have been in a form in which it was holding the broken DNA ends in a position where resealing could have occurred if the enzyme was not denatured. This form of the enzyme would be a normal reaction intermediate of transiently broken DNA, either before, during, or after the passage of the intact segment of dsDNA through the break. The ssDNA cleavage reaction does not require denaturation of the enzyme, and so the two broken DNA ends have presumably dissociated from each other during the course of the reaction. This implies that the enzyme creates an unstable reaction intermediate with ssDNA and thereby commits suicide by attempting such a reaction. As expected from these considerations, ssDNA inhibits the relaxation of dsDNA.

Site Specificity on T4 DNA

The dsDNA cleavage activity of the T4 topoisomerase has been used to study the binding sites on T4 DNA templates, in an attempt to determine whether the topoisomerase is indeed a replication origin-specific DNA binding protein. To analyze cleavage sites on the large and circularly permuted T4 chromosome, a filter-binding assay was developed to select out, by virtue of their covalently attached protein, only those T4 DNA restriction fragments which had been cleaved by topoisomerase (Kreuzer and Alberts manuscript in preparation).

It seemed imperative to compare cleavage of native modified glu T4 HMdC-DNA with that of the multiply mutant unmodified T4 dC-DNA, since there are many physiological disturbances in infections by the T4 dC-DNA strains (see Snustad et al., this volume). This became possible when it was discovered that the restriction enzyme *Taq*I cleaves fully modified glu T4 HMdC-DNA (Coit and Alberts, personal communication). By using the filter-binding assay with radioactive *Taq*I restriction fragments of T4 dC and glu T4 HMdC-DNAs, it was found that the cytosine modifications dramatically alter topoisomerase cleavage specificity (Kreuzer and Alberts, manuscript in preparation). The native glu T4 HMdC-DNA is recognized with greater specificity, showing only on the order of eight strong cleavage sites, and is the only DNA substrate known for which the addition of ATP has a major effect on the cleavage products recovered. The alteration in cleavage specificity was traced specifically to the glucosyl groups on the modified cytosine residues, since DNA containing the hydroxymethylcytosine alone was recognized essentially identically to unmodified T4 dC-DNA.

Since specific topoisomerase recognition requires native glu T4 HMdC-DNA, but a restriction map has been generated only for T4 dC-DNA (Kutter and Rüger, this volume), the topoisomerase cleavage sites could be mapped only by hybridization to known restriction digests of T4 dC-DNA. This was done by the "Southern cross-blot" procedure (Sato et al., 1977), which allowed the mapping of several of the cleavage sites simultaneously. Figure 3 summarizes the locations of four strong topoisomerase cleavage sites; several additional cleavage fragments have not yet been mapped for technical reasons (Alberts et al., in press; also see Addendum in Proof).

One of the primary origins of T4 replication is thought to be near gene 56 (Morris and Bittner, personal communication; Mosig et al., 1981). A strong topoisomerase cleavage site maps in this vicinity (Fig. 3), and recent experiments suggest that this site is about 300 base pairs downstream from a strong early promoter at about 19 kilobases on the T4 map (Kreuzer and Alberts, unpublished data; Kutter and Rüger, this volume). Primary initiation of T4 replication has been reported to require the recognition of an early promoter by host RNA polymerase (Luder and Mosig, 1982), so it seems possible that primary initiation involves some interaction of the phage-induced topoisomerase and the host RNA polymerase at these adjacent binding sites. However, recent results suggest that the site where DNA synthesis starts is actually at about 16 kilobases (Kozinski, this volume; Mosig, this volume), or about 3 kilobases downstream from these two binding sites. Further analysis will be necessary to confirm the precise location where DNA synthesis starts and to determine which protein binding sites in this region of the DNA are required for replication origin function.

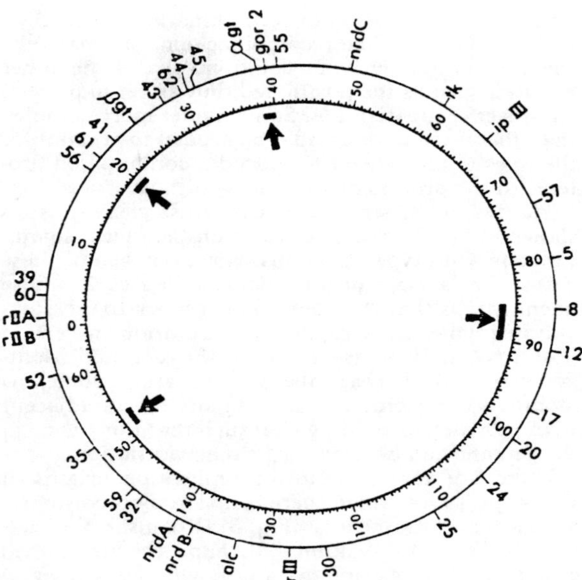

FIG. 3. Map locations of strong T4 topoisomerase cleavage sites on native T4 DNA The bars on the inside represent the locations of four strong T4 topoisomerase cleavage sites on native T4 DNA. The sites map somewhere within the regions delineated by the bars and were determined by hybridization to restriction digests of T4 dC-DNA (Alberts et al., in press; also see Addendum in Proof).

A second strong cleavage site maps near the recombination hot spot in the region of gene 35 (Fig. 3). The finding that the enhanced recombination from this hot spot requires glucosylation of the participating DNAs (Levy and Goldberg, 1980) strengthens the possibility that this topoisomerase binding site is functionally involved in the enhanced recombination, since topoisomerase cleavage specificity is also strongly affected by glucosylation (see above). Secondary initiation of T4 replication depends on recombination, and secondary origins of replication seem to correlate to recombination hot spots, including the one near gene 35 (see Mosig, this volume). As discussed above, DNA cleavage by the topoisomerase could be directly involved in initiation of replication and could also stimulate recombination, since DNA ends are known to be recombinogenic. Alternatively, the role of the enzyme could involve some other activity, such as the postulated (but as yet unproven) site-specific supercoiling. Some model more complex than these will undoubtedly be needed to account for the roles of the other T4-induced recombination proteins in the process of secondary initiation.

The two other strong topoisomerase cleavage sites shown in Fig. 3 have no obvious physiological significance in wild-type infections. However, one of these maps near a class of mutations, called *gor2*, whose phenotype is that they overcome a block to (presumably primary) DNA replication initiation in certain host RNA polymerase mutants (Snyder and Montgomery, 1974). Perhaps the *gor2* mutations somehow modify a topoisomerase binding site (or an adjacent RNA polymerase binding site) such that this region of the genome can be used for primary initiation.

There are several putative replication origins of wild-type T4 for which there is no strong topoisomerase cleavage site nearby (Fig. 3) (Kozinski, this volume; Mosig, this volume). Perhaps the unmapped cleavage sites referred to above will be shown to represent these remaining origins. However, it is important to emphasize that correlation of topoisomerase cleavage sites with replication origins and recombination hot spots does not prove that these cleavage sites actually have physiological significance. Further experiments will be necessary to confirm or deny the importance of these correlations.

CONCLUDING REMARKS

Type II DNA topoisomerases, with their multiple capacities to alter DNA topology, have been shown to be involved in DNA replication. However, their role(s) in the replicative process remains to be clearly defined. Supportive evidence has been accumulated to show that DNA gyrases are required for the replication of some bacterial, plasmid, and phage chromosomes. In T4, mutants that are defective in the phage-induced topoisomerase have a phenotype characterized by a delay in the onset of DNA replication, but a normal rate of fork movement, and they are cold sensitive for DNA replication. The replication defects in these mutants can be partially restored in vitro by the addition of the purified T4 topoisomerase, suggesting that it is required for T4 replication. Since T4 replication has been suggested to occur on the bacterial membrane, the findings that the phage topoisomerase subunits are membrane associated and that topoisomerase mutants are more sensitive to the actions of inhibitors of membrane biosynthesis (Huang, 1979a) make it tempting to hypothesize that the enzyme may play a role in the membrane association of replicating DNA.

It has been established that the product of T4 gene 39 is specific for the replication of T4 DNA, and it cannot be used to replicate even the closely related T6 DNA (Huang, 1979b). The specificity in gene 39 protein utilization is formally analogous to that of O protein in phage lambda DNA replication. The lambda O protein has been clearly documented to be involved in the initiation process, although purified O protein does not seem to have topoisomerase activity (Furth et al., 1978; Tsurimoto and Matsubara, 1981). These considerations suggest that the T4 DNA topoisomerase has two different roles in DNA replication. One is the more obvious role of providing non-sequence-specific relief of topological constraints generated during DNA chain elongation and fork movement. This is illustrated by the ability of the purified enzyme to relax non-specifically any superhelical DNA in vitro. In addition, the enzyme may have a sequence-specific function in recognizing the origin of replication. Recent experiments designed to explore the function of the T4 enzyme have centered around the second possibility. Unfortunately, the locations of the origins of T4 DNA replication are still not clearly defined (Mosig, this volume; Kozinski, this volume). Nonetheless, two approaches are being taken: (i) examination of the interactions of the enzyme with the cloned DNA fragments containing the putative origins of replication, and (ii) study of the induced cleavage reaction of the enzyme, using T4 DNA, on the assumption that the cleavage sites are related to the primary binding sites. It is hoped that analysis of these target sites will help reveal the general features of the function of T4 topoisomerase in the initiation process.

T4 DNA replication and recombination are thought to be coupled at late times of phage infection (Mosig, this volume). It is intriguing to find that when the T4 gene 39 sequences are inserted into a plasmid and cloned in *E. coli* cells under conditions in which high levels of the protein are synthesized, the growth of these cells is retarded. Furthermore, the host *recA* protein, which is responsible for the major recombination pathway, is required if the cells are to tolerate the presence of high levels of gene 39 protein. The plasmid-containing cells are temperature sensitive if the host contains a temperature-sensitive *recA* mutation, and survivors isolated under nonpermissive conditions carry insertion elements in the plasmid, causing inactivation of the synthesis of gene 39 protein (Huang, unpublished data). It is not clear what specific effects a high concentration of gene 39 protein has on cellular metabolism in the absence of the other two topoisomerase subunits. However, it seems possible that one of the many functions or partially activities of the topoisomerase, presumably carried out by the gene 39 protein alone or in conjunction with other host cellular components, causes DNA damage and thereby induces SOS function to repair the damage in a *recA*$^+$ host. The elucidation of the relationship between gene 39 protein and *recA* protein might provide an important insight concerning the role of T4 topoisomerase in DNA metabolism.

ADDENDUM IN PROOF

The strong topoisomerase cleavage sites on native glu T4 HMdC-DNA have now been mapped, using DNA fragments generated with *Aha*III, a new restriction enzyme that cleaves glu T4 HMdC-DNA (Kreuzer and Alberts, unpublished data). This analysis confirmed the locations of the four strong cleavage sites described above and also revealed three additional sites, mapping within the intervals 25 to 30 kilobases, 108 to 111 kilobases, and 133 to 137 kilobases on the T4 genome. The first two of these correlate to regions of the genome which have been implicated by others to contain both recombination hot spots and origins of replication.

APPENDIX

T4 DNA Replication on Cellophane Disks

WAI MUN HUANG

Department of Cellular, Viral and Molecular Biology, University of Utah Medical Center, Salt Lake City, Utah 84132

To decipher the complex process of DNA replication, which involves the interactions of multienzyme complexes and possibly other cellular components, the development of an in vitro replication system capable of mimicking the in vivo process is usually a necessary approach. Historically, treatments which render the *Escherichia coli* cells permeable to deoxyribonucleoside triphosphates have been used to establish in vitro cell-free systems, since these triphosphates are the immediate precursors of DNA synthesis. Dicou and Cozzarelli (1973) described a T4-directed DNA synthesis in toluene-treated cells. Although the system has provided a relatively efficient and long-lived in vitro DNA synthesis, it is not permeable to high-molecular-weight substances. Therefore, its usefulness was limited. A more gentle method using sucrose-plasmolyzed phage-infected *E. coli* cells has also been used to examine T4 DNA replication and metabolism (Chiu et al., 1977; Stafford et al., 1977). Although much has been learned regarding some of the T4 gene products that are required for the replicative process and especially in the biosynthesis of DNA precursors (Mathews and Allen, this volume), this system still has the disadvantage of being impermeable to large molecules. Since it has been suggested that T4 DNA replication might involve the bacterial membrane, my co-workers and I have developed an in vitro system based on the method introduced by Schaller et al. (1972) for *E. coli* replication. The system involves a gentle and complete lysis of infected cells on the surface of semipermeable cellophane disks. In this fashion, very concentrated extracts can be obtained and manipulated. Most importantly, the system will allow the introduction of macromolecules into the reaction, so in vitro complementation experiments become feasible. This T4 DNA replication system does retain all of the known features characteristic of the in vivo process, including the requirement for the DNA-delay products, which the system was designed to investigate (Huang et al., 1976). A similar system, using different host and phage strains, has been reported by Imae and Okazaki (1976). Currently it is the most efficient in vitro T4 DNA replication system available. (Here I consider only in situ systems, where intracellular associations are maintained. Purified-protein systems are discussed elsewhere by Nossal and Alberts [this volume].) Under optimal conditions, 15 phage equivalents of T4 DNA are synthesized per cell per h at 25°C, which is approximately one-third of the in vivo yield at this temperature. Furthermore, it is the only system in which the regulation and specificity of DNA synthesis have been demonstrated and examined. T4-infected lysates prepared on cellophane disks have also been used successfully to examine the pattern of late transcription (Rabussay and Geiduschek, 1979a; Rabussay and Geiduschek, 1979b).

DESCRIPTION OF THE SYSTEM

The preparation of the extract is quite straightforward. Controlled lysis of concentrated cells is done by the combined action of the nonionic detergent Brij 58 and lysozyme in the presence of sucrose at 4°C. Since highly concentrated lysates are viscous and difficult to manipulate by pipetting, the lysis is carried out on a semipermeable membrane which is placed on the surface of an agar plate. Thereafter all of the manipulation can be done by simply picking up the membrane with forceps. The first agar plate contains sucrose to maintain osmotic pressure. After the initial lysis, the membrane is transferred to a second agar plate which does not contain sucrose in order to complete the lysis and effectively dialyze the lysate. It is in this step that small molecules, including nucleoside triphosphates, are removed from the extract. The lysate is further concentrated with a continuous air flow blowing on the surface of the disk. In this manner we can easily put 2×10^8 cells in 2 µl on a 1.5-cm-diameter disk (Kalle AG, Wiesbaden-Biebrich, West Germany). Subdivision of samples into aliquots after they have been lysed on the cellophane disks is not practical. Thus, for experiments involving time points, a separate disk is used for each defined incubation condition. The compositions of the two kinds of agar plates used are the same as those originally described by Schaller et al. (1972). They work well for the T4 system, provided that the surface of the agar plates remains moist; they should be stored in the cold and used within 3 days after they are prepared, and the agar solution should be melted by boiling instead of by autoclaving.

We use a *polA E. coli* B strain (JT1200) as the host for the in vitro DNA replication study to reduce the incorporation due to host repair processes. Since T4 DNA replication requires phage-induced proteins, the infection is best done at 25°C (to slow down the infectious cycle) for 15 min before the cells are harvested. Under these conditions, all of the T4 prereplicative proteins are already synthesized (O'Farrell et al., 1973), and T4 DNA replication is still minimal. The infected culture is collected by centrifugation and washed once with 0.1 M Tris-hydrochloride (pH 7.4) containing 0.1 M KCl, 1 mM $MgCl_2$ and 4 mM $CaCl_2$. The resulting cell pellet can be quickly frozen in liquid nitrogen and stored at −70°C for future use without significant loss of activity. The infected cells are resuspended in the same buffer to 10^{11} cells per ml. A 2-μl portion of the concentrated cells is delivered onto the surface of a cellophane disk on a sucrose-containing agar plate, and 2 μl of a freshly prepared solution of lysozyme (1 mg/ml) containing 0.5% Brij 58 is added to the same disk. The two drops are mixed together gently with a glass rod and allowed to incubate in the cold for 20 min. To complete the lysis, the disk is then transferred to a second, sucroseless agar plate, for another 15 to 20 min with air blowing. At this point, the lysate appears as a shiny surface on top of the cellophane disk, and it is ready for in vitro DNA synthesis by being placed on top of a drop of reaction mixture.

The optimal reaction mixture for T4 DNA replication on cellophane disks is given in Table 1. The high concentration of KCl (0.1 M) is needed to facilitate the transport of the highly charged triphosphate molecules across the membrane. ATP is absolutely required by the system at a moderate concentration and is not replaceable by an ATP-generating system. Although the other three ribonucleoside triphosphates stimulate incorporation, they are dispensable. Since normal T4 DNA contains hydroxymethylated and glucosylated cytosine residues, the use of the appropriate precursors in the reaction mixture, such as the addition of hydroxymethyl dCTP (HMdCTP) and UDP-glucose, does influence the extent and the length of time of the synthesis. Figure 1 shows the time course of DNA synthesis in the presence of either dCTP or HMdCTP as the fourth deoxyribonucleoside triphosphate. When UDP-glucose is added along with the HMdCTP, T4 DNA replication proceeds for at least 60 min at 20°C, accumulating more than 730 pmol of incorporated dTMP as acid-insoluble material per disk. This level corresponds to approximately 15 phage equivalents of DNA synthesized per cell. It is important to note that during the layering of the cellophane disk on the drop of reaction mixture, care should be exercised to ensure that no reaction mixture is spilled on top of the disk, thereby causing a dilution of the lysate. When this happens, a lower level of DNA incorporation is always obtained. For this reason it is not advisable to use a larger drop of the reaction mixture than is necessary to cover the area of the disk. For a 1.5-cm disk, we use 60 μl of reaction mixture placed on a flat surface such as a petri plate. At the end of the incubation period, we find it convenient to stop the synthesis by inverting the cellophane disk onto a GF/C (Whatman) filter which has been soaked with 0.5 M NaOH–10 mM EDTA–0.1 M sodium pyrophosphate. A few more drops of the same solution can

TABLE 1. Reaction mixture for T4 DNA replication on cellophane disks

Component	Concn (mM)
Morpholinopropane sulfonic acid (pH 7.5)	20
KCl	100
$(NH_4)_2SO_4$	17
$MgCl_2$	5
EDTA	0.1
ATP	1
UTP, CTP, GTP	0.05 (each)
dATP, HMdCTP, dGTP, [^{3}H]dTTP	0.17 (each)
UDP-glucose	0.33

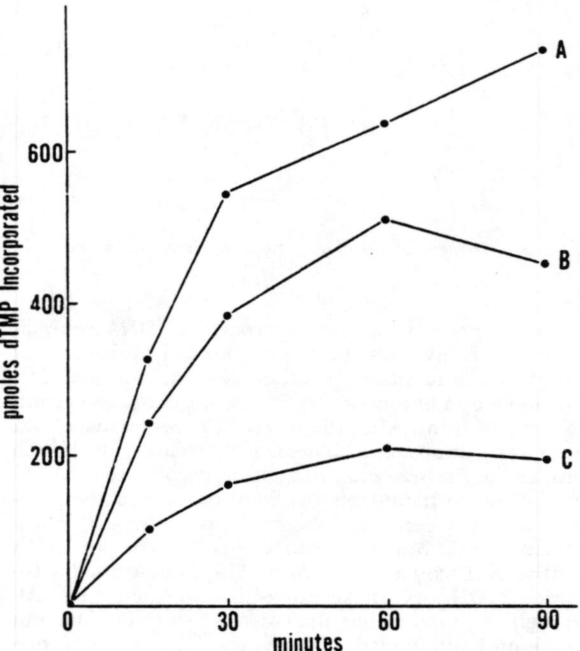

FIG. 1. Time course of DNA synthesis at 20°C. T4 DNA was used to infect *E. coli* JT1200 (*polA*) at a multiplicity of 10 at 25°C for 15 min, and 2 × 10^8 cell equivalents were incubated with a complete reaction mixture as given in Table 1 (A). Similar reactions where done with UDP-glucose omitted from the complete mixture (B) and with dCTP substituted for HMdCTP (C).

be added on top of the disk-filter "sandwich" about 5 min later to complete the transfer. After air drying for another 5 min, the cellophane disk can be peeled off, and the remaining GF/C filter can be washed batchwise in trichloroacetic acid (10%) and alcohol to determine the level of nucleotide incorporation.

Some of the requirements and effects of perturbation on the synthetic reaction are given in Table 2. Deoxyribonucleoside monophosphates can be used instead of the corresponding triphosphates as the precursors, although with less efficiency; however, the system cannot incorporate thymidine. This is different from the sucrose-plasmolyzed system, in which both thymidine and deoxyribonucleoside monophosphates are better than the corresponding triphosphates as substrates for the in vitro synthesis. The system uses endogenous T4 DNA as template for synthesis; T4 DNA added on top of the disk has little effect on the level of incorporation (Fig. 1). Similarly, purified T4 DNA polymerase does not seem to further stimulate the reaction; it may actually cause a decrease in the observed synthetic activity, possibly due to excessive handling of the materials during mixing. This reaction is sensitive to the addition of DNase but not RNase. Although no exogenously added protective reagent for sulfhydryl groups is needed for optimal reaction, the presence of sulfhydryl inhibitors such as *N*-ethylmaleimide aborts the synthetic reaction. Inhibitors of *E. coli* DNA replication, specifically, those

TABLE 2. Requirements of T4 DNA replication on cellophane disks

Conditions	Activity (%)
Complete mixture[a]	100
− KCl	10
− ATP	6
− HMdCTP	4
− 4dNTP; + dAMP, dCMP, dGMP, [^{3}H]dTMP (0.17 mM each)	25
− 4dNTP; + [^{3}H]thymidine (0.16 mM)	2
+ Dithiothreitol (1 mM)	110
+ *N*-Ethylmaleimide (1.6 mM)	8
+ T4 DNA (3 μg/disk)[b]	84
+ T4 DNA polymerase (3 μg/disk)[b]	50
+ DNase (2 μg/disk)[b]	6
+ DNase (10 μg/ml)[c]	60
+ RNase (10 μg/ml)[c]	80
+ Nalidixic acid (80 μg/ml)[c]	88
+ Novobiocin (50 μg/ml)[c]	95
+ Rifampin (500 μg/ml)[c]	106

[a] The reaction mixture (60 μl) is given in Table 1 for 2 × 10^8 cells per disk. Incubation was for 60 min at 20°C. An activity of 100% corresponds to 400 pmol of dTMP incorporated.

[b] Added on top of the disk at the time the lysate was prepared.

[c] Added in the reaction mixture.

that inhibit the host gyrase function such as nalidixic acid and novobiocin, have little effect on the replication of wild-type T4 DNA, as predicted from the in vivo replication (McCarthy, 1979). Rifampin (500 μg/ml), when added to the reaction mixture, also appears to have little effect on the incorporation. The optimal temperature of the reaction is 20 to 30°C. At higher temperatures, the synthetic machinery appears to be less stable. Although a more rapid initial rate of synthesis is achieved under these conditions, both the final level of synthesis and the length of time that synthesis can proceed are less. At 37 and 42°C the levels of synthesis are 80 and 55%, respectively, of that at 2°C.

PHAGE PROTEINS REQUIRED FOR THE SYSTEM

As predicted from the in vivo replicative process, the cellophane disk system has an absolute requirement for the T4 proteins coded by the DNA-negative genes (Warner and Hobbs, 1967). Extracts prepared from cultures infected with T4 32⁻, 41⁻, 42⁻, 43⁻, 44⁻, 45⁻, and 62⁻ mutants give no significant synthesis. Furthermore, the system also requires the products of the T4 DNA-delay genes 39, 52, 58, and 60 (gene 58 is now known to be identical to gene 61 [Kutter and Rüger, this volume]). Mufti and Bernstein (1974) have shown that T4 DNA-delay mutants are cold sensitive in that they are defective in DNA synthesis only at low temperature; at 37°C, near normal levels of DNA replication are observed due to the compensation effect exerted by the host gyrase (McCarthy, 1979). This is also true in the cellophane disk system. The levels of DNA synthesis in extracts prepared from some of these mutant-infected cultures are given in Fig. 2. DNA synthesis in all of the DNA-negative and DNA-delay mutants can be partially restored by the addition on the cellophane disk of a second extract which contains the missing protein in a complementation experiment (see below).

PRODUCTS OF THE SYNTHETIC REACTION

In the presence of HMdCTP and UDP-glucose, the product of the reaction is fully modified T4 DNA, in terms of its density in Cs_2SO_4 and the ratio of the incorporated glucose to total DNA synthesized. The fidelity of the in vitro reaction can be confirmed by examining the products of similar reactions with T2 and T6 phages used for the infections. In each of these cases, the reaction product is characteristic of the infecting phage in its level of glucosylation. The size of the DNA product at the end of a 30- to 60-min incubation with a wild-type T4-infected extract is about 160 kilobases long, as judged by alkaline sucrose gradient sedimentation analysis (Fig. 3). Furthermore, no material larger than unit-sized T4 DNA is synthesized, suggesting that recombination, which is coupled with replication in vivo, is probably minimal in vitro. To determine whether the product is covalently attached to the parental DNA, T4-infected extract was prepared from ^{32}P-labeled phage, and the reaction was carried out with the density label dBUTP instead of dTTP, with [^{3}H]dATP as the labeled nucleotide. The resulting product is of intermediate density (Fig. 4A). If the product is sheared before it is subjected to the Cs_2SO_4 equilibrium centrifugation analysis, both heavy and hybrid materials are found (Fig. 4B).

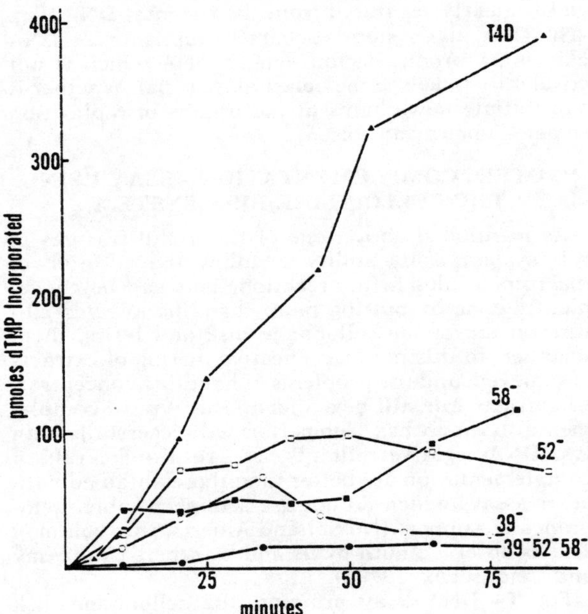

FIG. 2. Requirements of T4 gene products for in vitro replication on the cellophane disks. T4 mutants were used to infect *E. coli* JT1200 (*polA*) at a multiplicity of infection of 10 for 15 min at 25°C, and 2 × 10^8 cell equivalents were lysed on the cellophane disk and incubated with a complete reaction mixture. The mutants used were 39⁻ (amN116), 52⁻ (amH17), and 58⁻ (amHL627).

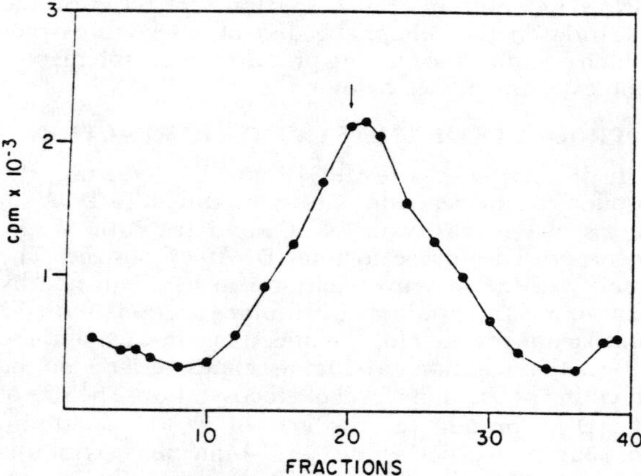

FIG. 3. Alkaline sucrose gradient analysis of the synthetic product. T4 DNA synthesis was performed for 45 min at 20°C in a complete mixture. The reaction product was dissolved in 0.5 M NaOH and layered on a 5 to 25% sucrose gradient containing 0.3 M NaOH, 0.8 M NaCl, and 2 mM EDTA, with 0.2 ml of 70% sucrose placed in the bottom of the tube to act as a cushion. Sedimentation was at 38,000 rpm for 2 h and 30 min at 20°C in an SW41 rotor. The position of mature T4 DNA is marked by an arrow.

This suggests that both strands of the T4 DNA are being replicated at least partially. Furthermore, if the reaction product is analyzed under alkaline conditions (without shearing), the newly synthesized material is clearly separated from the parental DNA (Fig. 4C). Thus, the system appears to replicate T4 DNA efficiently, producing unit-length DNA which is not covalently linked to the parental material. Whether it can initiate new chains at the origins of replication remains unclear at present.

PROTEIN COMPLEMENTATION ASSAY USING THE CELLOPHANE DISK SYSTEM

As mentioned above, one of the useful features of this system is its ability to allow the addition of macromolecules in the reaction. This can be conveniently done by putting more than one infected culture on top of the cellophane disk and lysing them together. In this manner, effective mixing of extracts is achieved, and the problems of handling concentrated extracts can still be avoided. This type of complementation assay has been performed successfully with two DNA-negative mutants. The resulting levels of complementation are better than those obtained with other assays which do not use semipermeable membranes as support (Nossal and Alberts, this volume). This is observed both by us and by others (Rabussay and Geiduschek, 1979b).

For T4 DNA-delay proteins, the cellophane disk system is the only method with which DNA complementation assays are feasible. When the product of one of the DNA-delay genes (gene 39, 52, 58, or 60) is absent, there is a characteristic delay in the onset of DNA synthesis. At later times in the infectious cycle, depending on the temperature and the host, significant amounts of DNA replication and phage are ob-

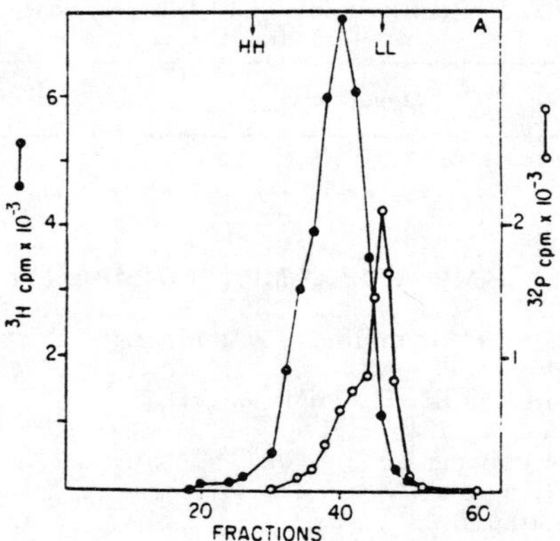

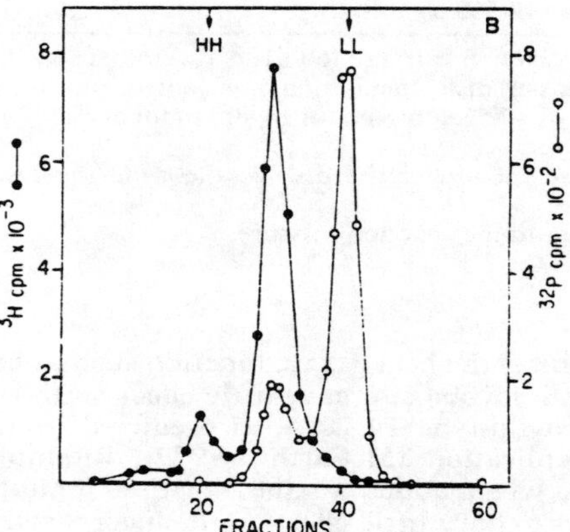

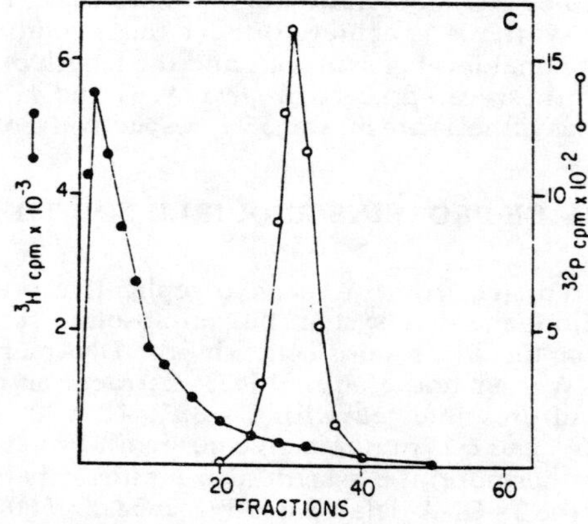

FIG. 4. Equilibrium density gradient analysis of synthetic products. ^{32}P-labeled T4D was used to infect *E. coli* JT1200 (*polA*). The extract was incubated with a complete mixture containing [^{3}H]dATP and dBUTP instead of dTTP. Incubation was for 60 min at 20°C. The reaction was terminated by the addition of Sarkosyl (0.5%). (A) The product was analyzed in a Cs_2SO_4 density gradient containing 50 mM Tris (pH 7.5). (B) The product was sheared by passing through a hypodermic needle (25 gauge) before the analysis as in panel A. (C) The product (without shearing) was analyzed in a CsCl gradient containing 0.3 M NaOH. Centrifugation was for 72 h at 20°C in a 50 Ti rotor.

tained. This is interpreted to mean that the DNA-delay proteins are not absolutely required for replication, and their roles may be in the regulation of the replicative process. Since mutants defective in these genes synthesize normal levels of all the other prereplicative proteins, including the DNA-negative proteins, a cell-free extract in solution does yield high levels of DNA synthesis, making a simple mixing experiment inadequate as an assay. However, DNA-delay mutants are cold sensitive, and the DNA-delay proteins have been shown to associate with the bacterial membrane (Huang, 1979). We chose to take advantage of these properties and to prepare the DNA-delay mutant-infected cultures at 25°C and make the lysates on cellophane disks. The complementation assay of T4 gene 39 protein is shown in Fig. 5. In this case, a 43^- phage-infected lysate, which contains the normal level of the gene 39 protein, is used as the donor, and the recipient is the 39^- mutant-infected culture. The level of DNA synthesis beyond those given by the individual extracts is taken as an index of the contribution of the gene 39 protein in the complementation assay. By use of this assay, T4 gene 39 protein has been purified from a T4 43^- mutant-infected culture (Stetler et al., 1979). Similarly, T6 gene 39 protein has also been purified from a T6 43^- phage-infected culture. Since the recipient 39^- mutant-infected cells can be prepared in advance and stored at −70°C, this assay is equivalent to a simple biochemical reaction. Although the absolute levels of stimulation due to the gene 39 protein may vary up to 50% from day to day, mainly because of the variability in the way the infected cultures are lysed on the cellophane disks, the relative amounts of stimulation within one set of experiments are reliable and are proportional to the amounts of gene 39 protein added.

This type of in vitro complementation assay was also used to address the question of whether there is any specificity in the functioning of the DNA-delay proteins among the T-even phages. Specifically, we asked whether these proteins may be used interchangeably to stimulate DNA synthesis in a heterologous mixture of two different kinds of phage-infected cultures, without knowing the exact role these proteins have in the replicative process. We have learned that T4 gene 39 protein is specific for T4 DNA replication and that it does not stimulate DNA replication in a T6 39^- mutant-infected culture and vice versa. On the other hand, gene 52 proteins of the three T-even phages are identical; they can be used interchangeably in any combination of pairwise mixtures in the complementation assay (Huang, 1979a). Thus, it appears that gene 39 protein is a determinant which positively regulates the specific replication of T-even phages and is formally analogous to the O protein of lambda phage (Furth et al., 1978; Furth et al., 1979).

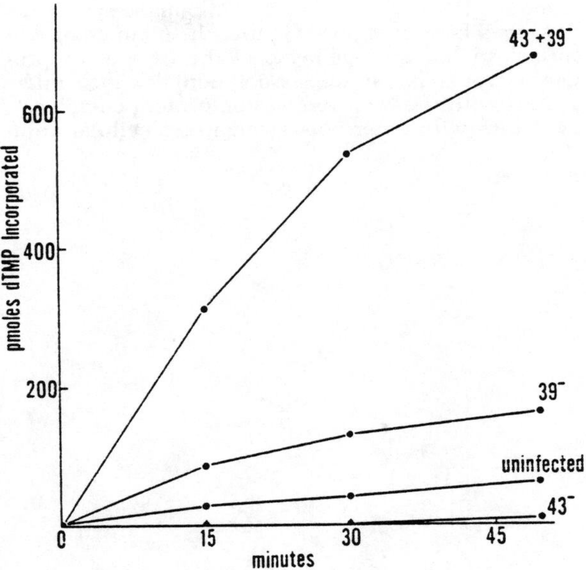

FIG. 5. Protein complementation assay for DNA-delay proteins. The T4 mutants used were 43^- (am4315) and 39^- (amN116). The preparation of extracts and the reaction conditions are described in the text.

CONCLUDING REMARKS

DNA replication on cellophane disks is a gentle but relatively manipulatable system which retains all of the known features of the in vivo replicative process. The system does offer many advantages, mainly in the provision of a very concentrated extract presently not easily achieved by other systems and the ability to add exogenous macromolecules to the reaction. Because it is possible to use frozen cells as the starting material, the system can be less time consuming and is comparable to a more conventional biochemical reaction. For processes that involve the participation of the bacterial membrane, this system appears to represent a promising approach, possibly because of the way cell lysis is controlled and the fact that all the intracellular macromolecular components are retained. Thus far it is the only system in which initiation of chromosomal replication in *E. coli* involving the *dnaC* protein can be demonstrated and examined (Nusslein-Crystalla et al., 1982). In T4, the role of the DNA-delay proteins in the replicative process is still not clearly established. Three of the DNA-delay proteins, gene products 39, 52, and 60, when purified, form a complex which gives a nonspecific topoisomerase activity which relaxes superhelical DNA (Kreuzer and Huang, this volume; L. F. Liu et al., 1979; Stetler et al., 1979). Various lines of indirect evidence suggest that these proteins may be involved in the initiation process, since the gene 39 protein appears to regulate specific DNA replication (Huang, 1979a). Furthermore, these proteins are found to bind DNA tightly and are membrane associated (Huang, 1975; Huang and Buchanan, 1974), and it has been postulated that T4 DNA replication occurs on the bacterial membrane (Huang, 1979b; Siegel and Schaechter, 1973). However, when the purified topoisomerase is added in solution to a reaction mixture containing all of the T4 proteins needed for the process of chain elongation and fork movement (Nossal and Alberts, this volume), DNA synthesis is not stimulated as predicted by the in vivo reaction. On the other hand, these proteins do stimulate DNA synthesis in the cellophane system, as described here. It is possible that the maintenance of special spatial arrangements of the DNA-delay proteins (through membrane association) and their interactions with the DNA replication protein complexes, as well as with other not-yet-identified cellular components, are all necessary to fully reconstitute the complete replicative machinery in vitro. The specificity and regulation of DNA replication remain difficult to examine by direct biochemical means. Further exploration of the cellophane disk system appears to be a valuable and needed approach.

T4 DNA Nucleases

HUBER R. WARNER[1] AND D. PETER SNUSTAD[2]

Department of Biochemistry[1] and Department of Genetics and Cell Biology,[2] University of Minnesota, St. Paul, Minnesota 55108

All viruses are dependent on the metabolic machinery of their host cells for growth. However, the degree to which viruses supplement the metabolic potential of host cells is highly variable. The T-even bacteriophages represent one extreme in this respect; they direct the synthesis of many enzymes with activities that are the same as or very similar to those of enzymes present in uninfected host cells. This is very evident in comparing the activities of the nucleases coded for by the phage T4 genome (Table 1) with those of the nucleases present in uninfected *Escherichia coli* cells (Table 2).

Nucleases specified by the phage T4 genome have roles in (i) the degradation of cytosine-containing DNA (normally host DNA), (ii) DNA synthesis, (iii) DNA repair, (iv) recombination, and (v) packaging of DNA during phage maturation. Only two of these nucleases are essential for efficient phage reproduction (gene 46 and 47 exonuclease and gene 49 endonuclease); the others are unessential for growth, but provide an apparent selective advantage. However, no example is known of an *E. coli* nuclease providing an essential function for replication of either wild-type or mutant T4 phage.

NUCLEASES INVOLVED IN THE DEGRADATION OF CYTOSINE-CONTAINING DNA

One mechanism by which a virus can maximize its reproductive capacity under conditions where the supply of nutrients is limiting is to degrade components of the host cell and recycle the breakdown products into viral components. Phage T4 has evolved a set of genes that direct the degradation of the host DNA to mononucleotides, which are subsequently reutilized. This degradation of host DNA also required the virus to evolve a mechanism by which it can distinguish its own genetic material from that of the host, so as not to commit suicide in the process of degrading the host's genetic material.

The genetic system that has evolved in phage T4 provides an elegant mechanism to assure that the virus can distinguish its own DNA from that of its host, *E. coli*, through the use of hydroxymethylcytosine (HMC) rather than cytosine in the T4 DNA and the subsequent glucosylation of the HMC. It is now quite clear that the glucosylation protects T4 DNA from degradation by at least one host restriction endonuclease that acts on 5-HMC-containing DNA (Revel, this volume), whereas the hydroxymethylation of T4 DNA prevents it from being degraded by the T4 nucleases responsible for the degradation of the host DNA and also by most bacterial restriction enzymes.

Host DNA Degradation

Phage T4 codes for two endonucleases that are specific for cytosine-containing DNA. One, T4 endonuclease II, produces single-strand breaks in double-stranded DNA (Sadowski and Hurwitz, 1969a). In vitro, it yields 3′-hydroxyl and 5′-phosphate termini and shows some nucleotide specificity because 70% of the 5′ termini are dGMP and dCMP; it also appears to cleave DNA in a sequence-specific manner in vivo (see below). The second enzyme, T4 endonuclease IV, preferentially cleaves single-stranded DNA, yielding 3′-hydroxyl and 5′-phosphate termini in vitro (Sadowski and Hurwitz, 1969b). It cuts adjacent to cytosine residues, leaving 5′ termini containing exclusively dCMP. The limit digestion product obtained by treating single-stranded cytosine-containing fd DNA with T4 endonuclease IV in vitro is 25% acid soluble and consists of oligonucleotides with an average length of about 50 nucleotides (Sadowski and Bakyta, 1972). However, a shorter period of treatment of fd DNA yields eight sharp bands, indicating significant sequence preference (Kutter, personal communication).

The role of T4 endonuclease II in catalyzing the first step in host DNA degradation is firmly established (Snustad et al., this volume). Host DNA degradation-deficient mutants of phage T4 have been isolated (Hercules et al., 1971; Warner et al., 1970) and have been shown to be unable to induce T4 endonuclease II (Sadowski et al., 1971). The affected gene has been named *denA* (deficiency in endonuclease) and maps near 136 kilobases on the T4 linkage map (Hercules et al., 1971; Ray, Sinha, Warner, and Snustad, 1972).

The role of T4 endonuclease IV in host DNA degradation is less clear. Its specificity suggests that it might act after T4 endonuclease II has nicked the DNA, and an exonuclease has enlarged this nick to a gap. Endonuclease IV could then cleave the second strand at the exposed single-stranded regions in the double-stranded DNA. However, T4 mutants which fail to induce the synthesis of endonuclease IV have been identified (Bruner et al., 1973; Sadowski and Vetter, 1973; Vetter and Sadowski, 1974), and the lack of T4 endonuclease IV has been shown to have little or no effect on the degradation of host DNA in the presence of T4 endonuclease II and nuclear disruption (Souther et al., 1972). The small amount of host DNA degradation that occurs in *E. coli* cells infected with *denA* mutants does not occur in cells infected with *denA denB* double mutants (Souther et al., 1972). Thus, T4 endonuclease IV, possibly in conjunction with other host or phage nucleases, can catalyze limited host DNA degradation in the absence of T4 endonuclease II. In addition, T4 endonuclease IV is required for the alternate pathway of host DNA degradation that occurs in the absence of both T4 endonuclease II and nuclear disruption (Parson and Snustad, 1975; Snustad et al., 1974). Thus, the precise role of T4 endonuclease IV in host DNA degradation appears to depend on what other nucleases are present and on

TABLE 1. T4-induced nucleases active on DNA

Enzyme	Time of expression	Specificity[a]	Role during infection	Gene
Exonucleases				
Exonuclease associated with DNA polymerase	Early	3′ → 5′ on ssDNA	Proofreading during DNA replication	43
Exonuclease A	Early	Oligonucleotides	Unknown	*dexA*
Exonuclease B	Early	5′ → 3′ on nicked, irradiated dsDNA; produces oligonucleotides	Excise pyrimidine dimers from nicked parental DNA	Unknown
Exonuclease C	Early	5′ → 3′ on nicked, irradiated dsDNA; also 3′ → 5′?	Excise pyrimidine dimers from nicked parental DNA	Unknown
Endonucleases				
Endonuclease V	Immediate early	Nick dsDNA containing pyrimidine dimers	Repair parental DNA before replication begins; repair progeny DNA to permit packaging	*denV*
Endonuclease II	Early	Nick cytosine-containing dsDNA	Host DNA degradation	*denA*
Endonuclease III	Early?	Break HMC- or cytosine-containing ssDNA	Unknown	Unknown
Endonuclease IV	Early	Break cytosine-containing ssDNA	Host DNA degradation	*denB*
Endonuclease (I)	Late	Nick HMC- or cytosine-containing dsDNA	Unknown	Unknown
Endonuclease VI	Late	Nick HMC- or cytosine-containing ssDNA or dsDNA; mixture of endonucleases (I) and III?	Unknown	Unknown
Endonuclease VII	Late	Break dsDNA in gapped regions	Remove recombination branches during packaging	49
Endonuclease associated with ghosts	?	Nick and break cytosine- or HMC-containing dsDNA	DNA packaging	17?

[a] ssDNA, Single-stranded DNA; dsDNA, double-stranded DNA.

the organization or intracellular location of the host DNA.

Genes 46 and 47 were the first T4 genes shown to specify products required for normal host DNA degradation (Kutter and Wiberg, 1968; Snustad et al., this volume; Wiberg, 1966). A mutation in either of these genes results in the accumulation of host DNA fragments of size 10^6 to 10^7 daltons. Recently, Mickelson and Wiberg (1981) have presented evidence indicating that genes 46 and 47 control the synthesis of a membrane-bound exonuclease. A mutation in either of these genes totally eliminates this activity. The gene 46 and 47 products were identified in the membrane fraction of infected cells and were shown to bind to both T4 DNA and *E. coli* DNA. Because genes 46 and 47 are also required for T4 DNA synthesis and recombination in vivo (Nossal and Alberts, this volume; Sinha and Goodman, this volume), the "46,47 exonuclease" must act on both T4 DNA and *E. coli* DNA. Because these two gene products have not yet been extensively purified, their exact roles remain uncertain. However, the data of Prashad and Hosoda (1972) suggest they may constitute a nuclease which enlarges nicks to gaps.

Other phage T4-induced nucleases that might be involved in host DNA degradation include endonuclease I (Altman and Meselson, 1970), endonuclease III (Sadowski and Bakyta, 1972), endonuclease VI (Kemper and Hurwitz, 1973), and exonuclease A (Warner et al., 1972). However, there is as yet no evidence for any such role for any of them.

Phage DNA Degradation

Whereas T4 endonuclease II clearly plays the major endonucleolytic role in host DNA degradation, with T4 endonuclease IV playing a minor role, just the reverse occurs in the case of the degradation of the cytosine-containing T4 DNA synthesized in cells infected with dCTPase-deficient mutants (Bruner et al., 1972; Kutter et al., 1975). That is, cytosine-containing T4 DNA undergoes extensive degradation in the absence of endonuclease II, but remains largely intact in the absence of endonuclease IV. Kutter et al. (1975) have proposed that the difference in the susceptibility of cytosine-containing phage and host DNAs to endonuclease IV may be that the phage DNA has extensive single-stranded regions near the multiple replication forks and in regions being transcribed. These single-stranded regions should be susceptible to endonuclease IV cleavage. Alternatively, the difference could be due to different intracellular locations of the enzymes and the DNAs.

In any case, the "46,47 exonuclease" appears to be involved at the same stage in the degradation of cytosine-containing DNA of both T4 and *E. coli*. Kut-

TABLE 2. *E. coli* nucleases active on DNA

Enzyme	Specificity[a]	T4 enzyme with comparable activity	Reference
Exonucleases			
Exonuclease I	3′ → 5′ on ssDNA	DNA polymerase	Lehman, 1960
Exonuclease II (associated with DNA polymerase I)	3′ → 5′; ss > ds	DNA polymerase	Lehman and Richardson, 1964
Exonuclease III	3′ → 5′ on dsDNA; AP endonuclease	None	Richardson et al., 1964
Exonuclease IV	Oligonucleotides?	Exonuclease A	Jorgensen and Koerner, 1966
Exonuclease V	3′ → 5′ and 5′ → 3′ on dsDNA; produces mono- and oligonucleotides	None	Muskavitch and Linn, 1981
Exonuclease VI (associated with DNA polymerase I)	5′ → 3′ on dsDNA	None	Klett et al., 1968
Exonuclease VII	3′ → 5′ and 5′ → 3′ on ssDNA; produces oligonucleotides	Exonucleases B and C	Chase and Richardson, 1974
Endonucleases			
Endonuclease I	Makes breaks in dsDNA	None	Lehman et al., 1962
Endonuclease II (associated with exonuclease III)	AP endonuclease	None	Hadi and Goldthwait, 1971
Endonuclease III	Thymine glycol-DNA glycosylase; AP endonuclease	None	Gates and Linn, 1977b
Endonuclease IV	AP endonuclease	None	Ljungquist, 1977
Endonuclease V	DNA containing various distorted regions	None	Gates and Linn, 1977a
Endonuclease VI (same as endonuclease II)	AP endonuclease	None	Verly and Pacquette, 1972
uvrABC complex	Pyrimidine dimers and other distorted regions in DNA	None	Seeberg, 1981
Restriction endonucleases	Recognize specific sequences and make breaks in dsDNA	None	

[a] ssDNA, Single-stranded DNA; dsDNA, double-stranded DNA; AP, apyrimidinic.

ter and Wiberg (1968) have shown that cytosine-containing T4 DNA is cleaved to fragments with an average size of about 10^7 daltons in restrictive host cells infected with gene 56 (dCTPase deficient), 46, 47 triple mutants. Carlson and Wiberg (personal communication) have recently obtained evidence for the presence, in vivo, of discrete, genetically specific fragments of cytosine-containing T4 DNA. These "restriction fragments" appear in *E. coli* B infected with T4 phage containing amber mutations in genes 56 (dCTPase), 42 (dCMP hydroxymethylase), and 46 (DNA exonuclease). Formation of the fragments is essentially blocked by introduction of a mutation in the T4 *denA* gene (endonuclease II), but is only minimally affected by mutations in the T4 *denB* gene (endonuclease IV); it is not clear whether the role of endonuclease II is direct or indirect. The gene 46 mutation prevents degradation of the fragments to acid-soluble products. At least 16 fragments are detectable by gel electrophoretic analysis, ranging in size from less than 2 to more than 20 kilobase pairs. That at least some of the fragments are gene specific was shown by hybridization against cloned T4 DNA fragments or T4 DNA fragments generated in vitro by standard restriction enzymes. These results appear to represent one of the first demonstrations of discrete-size restriction fragments in vivo; Lee and Sadowski (1982) recently made a related observation with phage T7.

NUCLEASES INVOLVED IN METABOLISM OF T4 DNA

DNA Synthesis

Procaryotic DNA polymerases uniformly have been shown to contain associated exonuclease activity. The best-characterized bacterial enzyme is *E. coli* DNA polymerase I, which contains both 5′-to-3′ and 3′-to-5′ exonuclease activities (Klett et al., 1968; Lehman and Richardson, 1964). In contrast, the T4-induced DNA polymerase has only a 3′-to-5′ exonuclease activity (Goulian et al., 1968; Huang and Lehman, 1972), but the turnover number of the 3′-to-5′ exonuclease of the T4 polymerase is about 250 times that of *E. coli* DNA polymerase I. It produces 5′-deoxyribonucleotides, has a marked preference for single-stranded DNA, and is equally active on T4 and *E. coli* DNA. The exonuclease activity is not inhibited by deoxyribonucleoside triphosphates when single-stranded DNA is the substrate, but is inhibited when double-stranded DNA is the substrate.

T4 DNA polymerase is the product of gene 43, and some gene 43 amber mutants have been isolated which retain normal or increased levels of 3′-to-5′ exonuclease but are devoid of DNA polymerase activity (Nossal and Hershfield, 1971). Experiments with other gene 43 mutants, which retain both exonuclease and polymerase activities, suggest that the major role of the 3′-to-5′ exonuclease is to monitor whether the correct deoxyribonucleotide has been incorporated during replication and repair synthesis. As discussed by Goodman and Sinha (this volume), Bessman and his co-workers have studied the relationship between the level of 3′-to-5′ exonuclease activity of T4 DNA polymerase and mutation frequency (Muzyczka et al., 1972). They showed that mutant DNA polymerases with a high ratio of exonuclease-to-polymerase activity correlated with an antimutator phenotype, whereas mutant DNA polymerases with a low ratio correlated with a mutator phenotype. In enzymatic studies, the ability to remove either a matched or a mismatched nucleotide from the 3′ primer terminus of double-stranded DNA was directly proportional to the amount of 3′-to-5′ exonuclease activity retained in the mutant DNA polymerase. Finally, the misincorporation of base analogs is inversely proportional to the amount of 3′-to-5′ exonuclease activity retained. Although they have not ruled out a role for the DNA polymerizing activity in preventing misincorporation of bases, Goodman et al. (1974) concluded that the magnitude of the 3′-to-5′ exonuclease activity exerts the greatest influence on the error correction capacity of T4 DNA polymerase. These results support the initial suggestion by Brutlag and Kornberg (1972) that the function of the 3′-to-5′ exonuclease of T4 DNA polymerase (and other DNA polymerases) is to remove mispaired nucleotides which have been incorrectly incorporated. However, the exonuclease level is unchanged in at least one of the strongest mutator polymerases, *ts*L88 (Hershfield, 1973). Furthermore, its activity is stimulated by the polymerase accessory proteins (Nossal, personal communication).

The products of genes 46 and 47 control a membrane-associated exonuclease activity involved in host DNA degradation, as discussed above. These gene products are also essential for normal T4 DNA synthesis. When *E. coli* is infected by gene 46 and 47 mutants, DNA synthesis commences normally but arrests after about 10 min (Kutter and Wiberg, 1968). The arrest of DNA synthesis in the absence of the products of genes 46 and 47 is correlated with the inability to synthesize late proteins (Hosoda et al., 1971). Luder and Mosig (1982) have provided a possible explanation for these observations (Mosig, this volume). It appears that primary initiation of replication requires *E. coli* RNA polymerase, and the gp55 modification of the polymerase necessary for late protein synthesis abolishes this activity. Thus, recombinational intermediates would contain the only remaining initiation sites for replication. Production of these intermediates in turn would be blocked in the absence of the gp46-47 exonuclease (see below).

DNA Repair

Bacteriophage T4 may be unique among bacterial viruses for its ability to induce an endonuclease specific for DNA containing pyrimidine dimers (Bernstein and Wallace, this volume). The presence of such an enzyme was implied by the isolation of UV-sensitive T4 mutants by Harm (1963). The enzyme was partially purified by Friedberg and King (1971) and then purified to homogeneity by Nakabeppu and Sekiguchi (1981).

The mechanism by which this enzyme incises DNA has only recently been elucidated. The highly purified enzyme preparation has both pyrimidine dimer-DNA glycosylase and apurinic/apyrimidinic endonuclease activities in the same protein (McMillan et al., 1981; Nakabeppu and Sekiguchi, 1981; Warner et al., 1981). The incision occurs in two steps; hydrolysis of the *N*-glycosylic bond of the 5′ nucleotide of the dimer to produce an apyrimidinic site is followed by phosphodiester cleavage of the DNA on the 3′ side of this site (Fig. 1). Thus, incision produces a nick with deoxyribose 5-phosphate at the 3′ terminus and a nucleotide containing a pyrimidine dimer at the 5′ terminus. This mechanism is apparently different from that used by the *uvr* system of the *E. coli* host to incise irradiated DNA, although the details of the latter have not yet been elucidated (Seeberg, 1981). The major role of the T4 UV endonuclease enzyme may be to repair parental DNA before replication occurs, because the enzyme is induced immediately after infection (Warner et al., 1981) and appears to be active on parental DNA only until about 5 to 7 min after infection (Pawl et al., 1976), which is before any T4 DNA replication has occurred.

Final repair of UV-irradiated DNA incised by the T4 UV endonuclease requires the excision of the deoxyribose 5-phosphate from the 3′ terminus and of the pyrimidine dimer-containing nucleotide from the 5′ terminus. It is not clear how this is actually accomplished in vivo, but T4-infected *E. coli* contain several enzymes which may be involved. Shimizu and Sekiguchi (1976) and Ohshima and Sekiguchi (1972) have described two enzyme activities in T4-infected cells which are not present in uninfected *E. coli*. These have been designated exonucleases B and C. Exonuclease B is capable of excising pyrimidine dimers from irradiated DNA previously treated with the T4-induced UV endonuclease. It exhibits preferential attack at the 5′ terminus or at internal nicks created by the UV endonuclease. Hydrolysis is limited in vivo and has been estimated to be as low as four nucleotides per pyrimidine dimer excised (Radany and Friedberg, 1982; Yarosh et al., 1981). The major product of the reaction is short oligonucleotides. Because the T4 DNA polymerase contains no 5′-to-3′ exonuclease activity associated with it as does *E. coli* polymerase I, exonuclease B may act in concert with the T4 DNA polymerase to excise the pyrimidine dimer and replace the damaged nucleotides. Exonuclease C has not been characterized as well as exonuclease B but appears to have similar properties.

The replacement of damaged nucleotides cannot occur until the deoxyribose 5-phosphate has been excised from the 3′ terminus. The 3′-to-5′ exonuclease activity of T4 DNA polymerase has not been tested on DNA containing deoxyribose 5-phosphate at the 3′ terminus of nicks, but such DNA is a poor substrate for the 3′-to-5′ exonuclease activity of *E. coli* DNA polymerase I (Mosbaugh and Linn, 1982). However, *E. coli* contains at least two activities which can excise deoxyribose 5-phosphate from a 3′ terminus; these include *E. coli* endonucleases IV and VI (Warner et al., 1980), both of which have apurinic/apyrimidinic endonuclease activity. At present, there is no evidence that T4 induces any comparable activity.

The possible roles of T4-induced and *E. coli* nu-

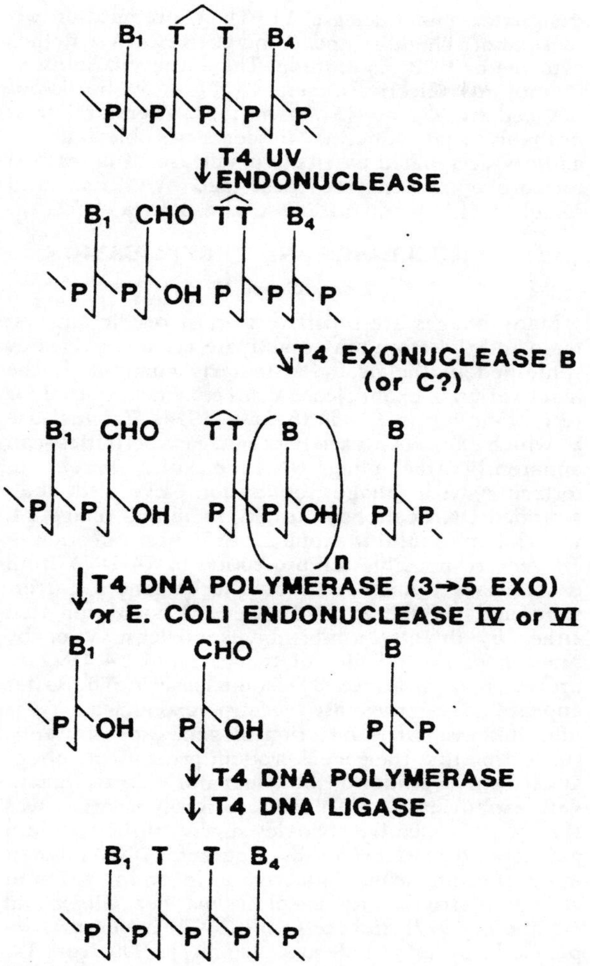

FIG. 1. Hypothetical pathway for repair of irradiated DNA in T4-infected *E. coli*.

cleases for repair of T4 parental DNA containing pyrimidine dimers are shown in Fig. 1.

DNA Packaging

Head maturation involves the sequential formation of proheads I, II, and III as precursors of mature heads (Black and Showe, this volume). Proheads I and II contain little, if any, DNA, whereas prohead III is partially filled with DNA. Prohead II must interact in some way with concatemeric DNA to begin the packaging reaction, which is prevented by mutations in gene 16, 17, or 20 (King, 1968). The conversion of prohead III to the mature head is prevented by mutations in gene 49 (Laemmli and Favre, 1973; Luftig and Ganz, 1972). The exact role of any of these four gene products in DNA packaging is unknown, but one or more of them apparently code for endonuclease activities.

Minagawa (1977) has reported that phage particles disrupted by osmotic shock or guanidine contain an endonuclease which makes both nicks and double-strand breaks in DNA. The enzyme is active on native T4 DNA and makes 10 to 15 times more nicks than breaks. This activity is present in capsids prepared under restrictive conditions with 16^- or 49^- phage, but is missing in capsids prepared with 17^- phage. However, some mutations in gene 17 also overcome the packaging defect caused by mutations in gene 20 (Hsiao and Black, 1977), suggesting that the products of genes 17 and 20 interact and play some common role in initiating DNA packaging. Thus, it is not clear what the role of the gene 17-associated nuclease activity is in packaging, or even whether it is coded by gene 17. Instead, it might be coded by gene 20, or some other gene, and indirectly controlled by the product of gene 17.

When gene 49 mutants are grown under restrictive conditions, unpackaged progeny DNA accumulates in the cell in a very high-molecular-weight form (Frankel et al., 1971). Cells infected with wild-type phage contain an activity which cleaves this DNA (1,400 to 2,000S) into smaller pieces of about 200S, and this activity is missing in cells infected with 49^- mutants (Minagawa and Ryo, 1978). A partially purified preparation of this enzyme was not active on single-stranded DNA or on intact or nicked double-stranded DNA but was active on gapped DNA, provided that the 5′ termini were phosphorylated (Minagawa and Ryo, 1978; Nishimoto et al., 1979). However, using a more purified enzyme preparation, Kemper and co-workers (Kemper and Garabett, 1981; Kemper et al., 1981) found that the enzyme is active on single-stranded DNA, but only in the absence of helix-destabilizing proteins (e.g., T4 gp32).

The gene 49 nuclease, known as endonuclease VII, could play either of two roles in DNA packaging. If gaps in the DNA inhibit completion of packaging, then the activity of the gene 49 nuclease could cleave in the gapped region at or near the 5′ terminus of the gap. Nicks apparently interfere with packaging (Zachary and Black, 1981), so gaps presumably would also interfere. However, this gap-directed cleavage would only produce viable phage particles if it occurred at genome length intervals in the DNA. Alternatively, endonuclease VII may cleave the DNA at branches arising from recombinational events (Kemper and Brown, 1976; Minagawa, 1977). Such branches would certainly interfere with packaging and would account for the phenotype of gene 49 mutants. Mizuuchi et al. (1982) have recently shown that this enzyme will cleave the recombination intermediates known as Holliday structures to produce two linear molecules, each containing a nick in one strand. If endonuclease VII does resolve recombination intermediates in vivo, then which nuclease is involved in producing genome length DNA molecules remains obscure.

Recombination

Although it is obvious that endonucleolytic events must occur during recombination of T4 DNA, no T4-induced nucleases have been shown to be required for this process. However, as mentioned above, the products of genes 46 and 47 are involved in recombination as well as in host DNA degradation. Bernstein and Wallace (this volume) discuss the genetic evidence for their role in recombination. Whether these gene products act as nucleases during recombination or play some other role is not currently known. As discussed above, the endonuclease controlled by gene 49 appears to play a terminal role in recombination by removing recombinational branches so packaging can proceed to completion.

NUCLEASES OF UNKNOWN FUNCTION

Exonucleases

Oleson and Koerner (1964) demonstrated that infection of *E. coli* with bacteriophage T2 resulted in a large increase in exonuclease activity in the cells. This activity is induced with kinetics typical of early enzymes and has been designated exonuclease A. A similar increase occurs in cells infected with phage T4 (Warner et al., 1972) and presumably also in cells infected with phage T6. The exonuclease activity is specific for DNA, but does not distinguish between native and denatured DNA, neither of which is a good substrate for the enzyme. The best substrate was obtained by pretreating native DNA with pancreatic DNase until about 15% of the DNA had become acid soluble. The products of the action of exonuclease A on oligonucleotides are deoxyribonucleoside 5′-monophosphates. Dinucleotides are not substrates for the enzyme.

A mutant (*dexA*) unable to induce exonuclease A has been isolated (Warner et al., 1972). The gene is also absent from some deletions located between genes 39 and 56. These mutants replicate normally and show no deficiency for host DNA degradation or phage DNA recombination. Therefore, even though T4 exonuclease A accounts for 75 to 80% of the total exonuclease activity in extracts of infected cells, an essential role for this enzyme has not been demonstrated. Recent work by Gauss et al. (1983) suggests that *dexA* and *dda* (DNA-dependent ATPase) may be two functions of the same gene.

Endonucleases

Whereas T4 endonucleases II and IV have been implicated in host DNA breakdown by virtue of their substrate specificity and the phenotypes of mutants unable to induce these enzymes, three other nucleases without known function have been detected and partially described. None of these activities has been defined genetically.

The endonuclease described by Altman and Meselson (1970) produces nicks in native DNA, whether cytosine or HMC-containing, and is not synthesized in cells infected with phage with *am* mutations in early genes. The enzyme first appears about 5 min after infection and reaches its maximum level 20 to 30 min after infection. It is synthesized in cells infected with phage with amber mutations in genes 16, 17, 49, *denV*, and *x*. This enzyme has been designated endonuclease I by Wood and Revel (1976).

Sadowski and Bakyta (1972) described an enzyme active on single-stranded DNA which is similar to T4 endonuclease IV, but which was active on HMC-containing DNA as well as cytosine-containing DNA. The time of induction of this enzyme has not been clearly established. They designated this activity endonuclease III.

Kemper and Hurwitz (1973) partially purified (about 36×) an endonuclease activity which they designated endonuclease VI. Their preparation attacked both single-stranded and native DNA, whether cytosine or HMC containing. The activity is induced by mutants defective in gene 16, 17, or 49, but is not induced by a gene 43 *am* mutant. Therefore, it is probably a late function. It seems possible that the activity designated as T4 endonuclease VI is really a mixture of the enzymes described by Altman and Meselson (1970) and Sadowski and Bakyta (1972).

HOST NUCLEASES AND T4 REPLICATION

Exonuclease V

Many phages are sensitive to host restriction systems and therefore may inactivate various nucleases contained in the host. Particularly common is the inactivation of exonuclease V, an enzyme required for recombination in *E. coli* (Sakaki, 1974). Exonuclease V, which exhibits a variety of nuclease activities, can apparently attack phage DNA molecules, thereby interfering with phage replication. Every double-stranded DNA coliphage tested, including phage T4, was shown to inhibit exonuclease V after infection. A T4 gene responsible for protection of T4 DNA from exonuclease V has been shown to be gene 2 (Silverstein and Goldberg, 1976). Gene 2 could function either by directly inhibiting exonuclease V or by binding to the termini of the parental T4 DNA to protect it from attack by exonuclease V. The latter appears to be the case because exonuclease V is inhibited even after infection by gene 2 *am* mutants, suggesting that the gene 2 product protects the phage DNA from degradation but does not directly inactivate exonuclease V. An interesting observation was that gene 2-defective particles survive if they contain at least two genome lengths of parental DNA, indicating that exonuclease V inactivates incoming parental DNA in a strictly exonucleolytic manner (Oliver and Goldberg, 1977). Behme et al. (1976) have partially purified a protein (molecular weight, 12,000) from T4-infected cells which inhibits exonuclease V. The properties of this inhibitor suggest that it is probably not the product of gene 2, but this has not been firmly established.

Endonuclease I

E. coli endonuclease I also presents a potential barrier to T4 DNA injection. This enzyme is localized in the periplasmic space, and the infecting DNA must pass through this space to enter the cell. The DNA from a primary infecting phage can apparently traverse this space without incident, whereas superinfecting phage DNA does not and becomes degraded to acid-soluble fragments by *E. coli* endonuclease I (Anderson and Eigner, 1971). Endonuclease I-deficient mutants are not able to degrade superinfecting phage DNA to acid-soluble fragments, although limited degradation does occur.

Restriction endonucleases

Unglucosylated HMC-containing T4 DNA is attacked by at least two *E. coli* restriction systems (Dharmalingam and Goldberg, 1976a; Dharmalingam and Goldberg, 1976b; Revel, 1967; Revel, this volume). One has been designated r_6, is localized in the membrane, and restricts only incoming DNA (Hewlett and Mathews, 1975). It is inhibited by phage infection, even in the absence of expression of the phage genome. The other nuclease has been designated $r_{2,4}$ and is located inside the cell. This enzyme is inhibited after infection, but only if phage expression is permitted; the kinetics of inhibition are typical of the synthesis of an immediate-early protein. It is possible that the nuclease acted on by this inhibitor is a class I restriction endonuclease, such as *Eco*B.

SUMMARY AND CONCLUSIONS

The array of nucleases present in T4-infected cells is impressive. Between 20 and 25 different activities have been identified and are presumably present in active form at some time during the replication of phage T4. The protection of replicating DNA from deleterious attack by these nucleases has received some attention in the past, and some general strategies have become clear. T4 phage may use at least four mechanisms to prevent unwanted degradation. These are: (i) base modification (HMC versus cytosine, and glucosylation); (ii) cellular location (e.g., *E. coli* endonuclease I is located in the periplasm); (iii) phage-induced inhibitors (inhibition of *E. coli* exonuclease V); and (iv) protection by DNA binding proteins (gp2). It seems possible that other mechanisms remain to be discovered.

In spite of the large amount of research already expended in studying nucleases in T4-infected cells, the details of many processes involving nucleases remain to be established. These include the following. (i) What enzymes are involved in gap formation during host DNA degradation and recombination? (ii) What enzyme(s) helps degrade host but not cytosine-containing T4 DNA in cells infected with mutants lacking endonuclease IV? (iii) What enzyme(s) is involved in final degradation of host DNA to 5′-deoxyribonucleotides? (iv) What enzymes (if any) initiate recombination by nicking the DNA? (v) What enzyme cuts DNA concatemers into genome lengths?

These uncertainties indicate there are interesting questions remaining to be resolved in the area of DNA metabolism in T4-infected cells.

DNA Metabolism In Vivo

Origins of T4 DNA Replication

ANDRZEJ W. KOZINSKI

Department of Human Genetics, University of Pennsylvania School of Medicine, Philadelphia, Pennsylvania 19104

Student: We want understanding and you give us doubt. Professor: Natural science has recognized limits, beyond which lie freedom of will and the ethical force of European man. Student: You teach meaninglessness. Professor: I recognize incomprehensibility.

Russell McCormmach, *Night Thoughts of a Classical Physicist*

Several lines of evidence indicate that T4 DNA replication can be initiated from multiple origins. Electron microscopy of partially replicated parental T4 DNA molecules revealed multiple replicative loops (Delius et al., 1971). The existence of multiple locations of initiation was also demonstrated by gradual shearing of partially replicated molecules (Howe et al., 1973). It was demonstrated that initiations must be specific, as short progeny fragments did not hybridize randomly to mature phage DNA (Howe et al., 1973), but displayed no bias in hybridization to separated T4 DNA strands (Kozinski, 1969). With the development of cloning techniques, it has become possible to explore where, more specifically, those origins are located. This chapter will deal with site specificity of the initiation of T4 phage DNA replication. Specific sections include: (i) the genetic specificity of initiation of parental DNA; (ii) gene amplification—an accumulation of partial replicas specific for the areas of initiation; and (iii) specificity of reinitiations at late times after infection.

SITE SPECIFICITY OF EARLY PROGENY DNA ASSOCIATED WITH PARENTAL MOLECULE

Multiple Sites of Initiation and Bidirectionality of T4 DNA Replication: Early Research

In the past there has been some controversy concerning the number and location of origins used in the replication of T4 DNA. In previous papers from this laboratory, physicochemical (Howe et al., 1973) and electron microscopic (Delius et al., 1971) evidence was presented showing that the initiation of T4 DNA replication can occur at multiple but specific sites. The electron microscopic study showed that replicative loops expand bidirectionally, with the 3′ end leading the 5′ end by approximately 0.25 μm (Delius et al., 1971) (see Fig. 1). In addition, loops displaying secondary and tertiary reinitiations were demonstrated. Since induction of mutations at later times after infection led to clonal distribution of mutants, the basic mode of DNA replication must remain exponential throughout both early and late stages of DNA replication (Kosturko and Kozinski, 1976). These results argue against the possibility of a shift to a rolling-circle mode of replication at late times. It was demonstrated that DNA concatemers used in the production of phages are assembled by recombination: upon coinfection of bacteria with two different mutants, giant progeny phage package concatenated DNA in which consecutive genomes differ genetically (Kozinski and Kosturko, 1976).

In contrast, early genetic evidence presented by others (Marsh et al., 1971; Mosig, 1970a) suggested that replication initiates from a single origin located between genes 42 and 43 and proceeds unidirectionally clockwise along the circular DNA template. This conclusion was drawn as a result of the study of replication of petite phages. However, in this laboratory, we have shown that petite phages with terminal deletions replicate completely and repeatedly either in the presence or absence of chloramphenicol (CM) (Kozinski and Doermann, 1975), indicating that, given the presence of essential proteins, a phage need not circularize to replicate successfully.

Localization of Origins of Replication by Hybridization to Specific Segments of T4 DNA

Here I discuss the genetic site specificity of the early progeny label which remains hydrogen bonded to the partially replicated parental molecule. The genetic composition of such progeny was determined by hybridization to cloned genetic segments.

To isolate such early replicative progeny DNA, it is necessary to be able to distinguish, and to isolate, parental molecules which have replicated only slightly. If cells are infected with parental phage whose DNA contains a heavy-density label, and then the light, newly synthesized progeny DNA are labeled with [^{3}H]thymidine, it is possible to observe and purify these partially replicated molecules by their position in CsCl density gradients (see outline of the procedure in Fig. 2). To do this, cells were infected with 5-bromodeoxyuridine-labeled (heavy) parental T4 phage. Shortly after infection, [^{3}H]thymidine was added. Samples were withdrawn at intervals, and DNA was extracted. The DNA samples were then fractionated in CsCl density gradients. In the 5-min sample, most of the progeny ^{3}H label associated with the parental DNA banded at the heavy location. Thus, progeny contribution was very small, presumably restricted to the areas of origins. This DNA was then used for hybridization to cloned genetic segments immobilized separately on small nitrocellulose filters. As an internal reference ^{32}P-labeled mature phage DNA was cohybridized. This allows one to represent data as a ratio of experimental DNA (^{3}H-labeled progeny) to ^{32}P-labeled reference DNA (Halpern et al., 1979).

(The procedure of hybridization to cloned segments immobilized on individual small filters has an advan-

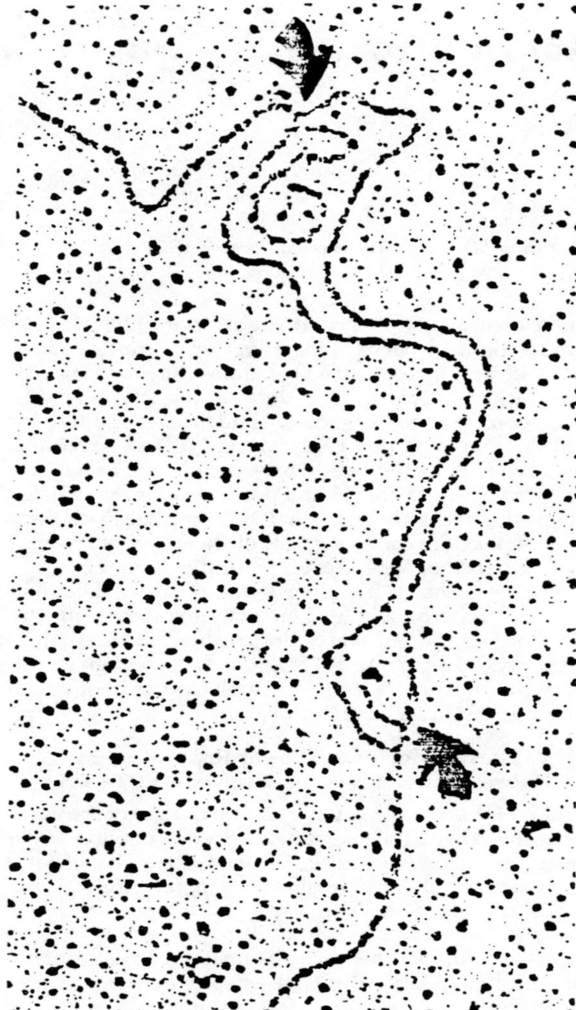

FIG. 1. Early replicative loop. Note two single-stranded whiskers in *trans* position. Those whiskers are susceptible to exonuclease I and thus are 3′-hydroxyl terminated (from Delius et al., 1971).

tage over the hybridization to a Southern blot of the complete restriction digest. There is no ambiguity where one segment ends and the other begins, and there is no partial overlap of genetically nonproximal areas. Importantly, one can ensure that adequate amounts of cloned segments are adsorbed to each filter and can thereby achieve a comparable range of counts of hybridized DNA. In the Southern blot method, of course, there is a sizable overlap of many segments with most restriction enzymes. Longer segments hybridize larger amounts of radioactivity than do short fragments. This makes it difficult to obtain reliable and comparable errors for all areas tested. In fact, there was no attempt to present possible errors in those papers for which Southern blots were used for probing specificity of initiation.

(Ling et al. [1981] provide a detailed description of the hybridization conditions used in our laboratory, with the range of errors, controls, rationale, and mathematical derivations. One important point should be made here which likely applies for all [^{3}H]thymidine uptake experiments. The possibility that the uptake patterns reflect a high versus low frequency of adenine-thymine base pairs in the different cloned fragments is eliminated by the observation that in the control cohybridization experiments with [^{3}H]thymidine and ^{32}P-labeled DNA [random labeling], ^{3}H to ^{32}P ratios were identical for all tested genetic areas [Kozinski et al., 1980]. In other words, all of the tested areas have, on the average, the same proportional amount of thymidine to phosphorus. Second, sonication of [^{3}H]adenine and ^{32}P-labeled heavy DNA revealed no significant skew of the specific activities of the two labels after CsCl gradient analysis [Howe et al., 1973].)

Results obtained in the filter hybridization experiments (Fig. 3) allowed us to conclude that, among the genes tested, the region of genes 50 through 5 shows a strong initiation site for DNA replication. The region of genes *w* through 29 shows a weaker site, whereas the area of genes 40 through 43 shows no initiation; on the contrary, we observed the lowest ^{3}H-to-^{32}P ratios in this area (Halpern et al., 1979). The maximum/minimum ratio of ^{3}H/^{32}P annealed to different cloned segments at early times was 40/1. Similar results were obtained in experiments using a low multiplicity of infection (MOI) (Kozinski et al., 1980), indicating that the observed specificity was not an artefact of interparental recombination. (It should be noted that at a low MOI things are happening more slowly, likely due to the gene dosage effect [Hutchin-

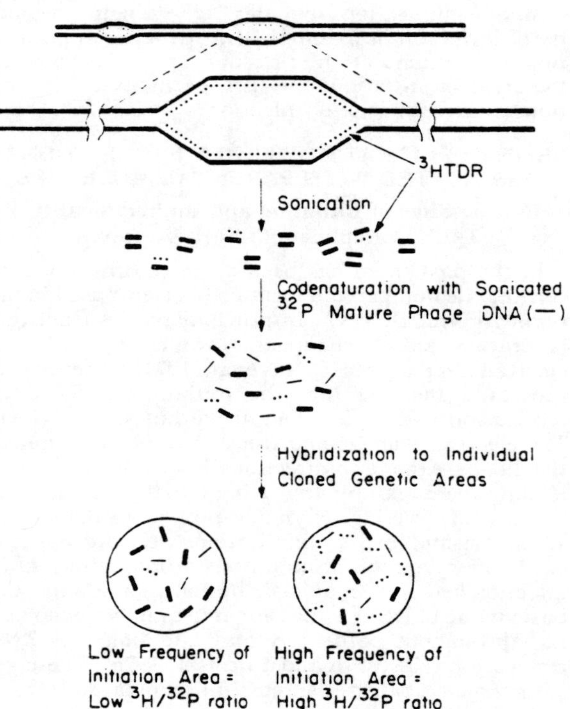

FIG. 2. Schematized procedure of isolation and identification of areas of origin.

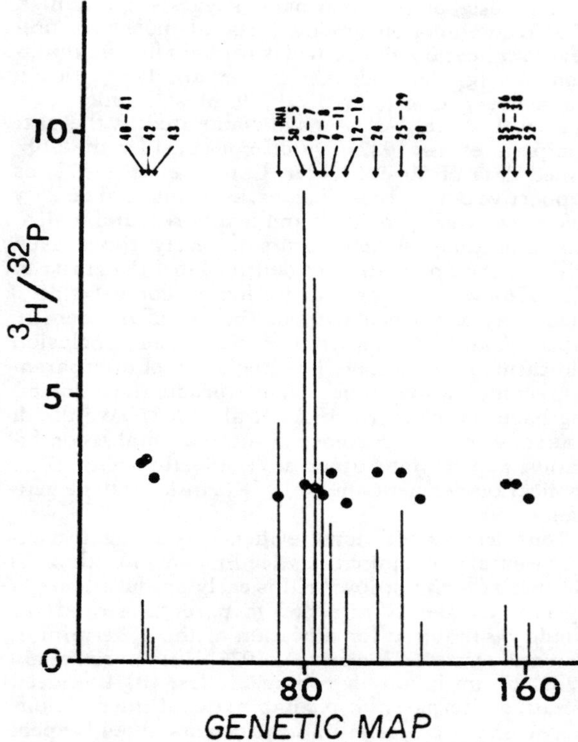

FIG. 3. Typical pattern of hybridization of the early progeny DNA to cloned genetic segments (vertical bars). Black dots represent hybridization of progeny DNA at the advanced stages of replication when CM was added at 10 min postinfection. Higher ratios of $^3H/^{32}P$ correspond to preferentially replicated areas (from Halpern et al., 1979).

son et al., 1978]. Therefore, for a given time after infection, replication is more advanced at a high MOI than at a low MOI.) The results are not affected by the use of 5-bromodeoxyuridine density labeling; similar results were seen when the densities were reversed, i.e., the host bacteria and media were heavy and the parental phage were light (M. E. Halpern, Ph.D. thesis, University of Pennsylvania, Philadelphia, 1983), or when no density label was used (Halpern et al., 1982; Kozinski and Ling, 1982) (Fig. 4B). It should be emphasized that in actuality every filter tested did hybridize some [^{3}H]thymidine, with the lowest counts found in the area of genes 40 through 43. There are two possible explanations for this phenomenon. First, all of the DNA molecules probably do not initiate replication at exactly the same moment. Therefore, at the time of sampling, some replicative loops will have elongated farther on the genome than others. Second, which we consider likely, there is a high probability of initiation at certain very active origins from which most replication initiates, as well as a very low probability of initiation at more numerous points throughout the genome. In this case, the low ^{3}H-to-^{32}P ratios could be due to these infrequent initiations, whereas the high ratios could be due to frequent initiations at the most active origins. In drawing these conclusions, we emphasize that we may not have detected all of the T4 origins, because we have not tested a full genomic complement. Indeed, Bittner and Morris (personal communication) suggest additional locations (see below; also see Mosig, this volume). We also stress that this analysis cannot discriminate between a single potent origin and several less potent origins located close together. The predominance of initiation in the area of genes 50 through 5 (Fig. 3) could therefore be due to frequent initiation at a unique single origin or to initiations at a cluster of origins in this area, each with less probability of being initiated than the single origin in the first case. Similarly, the smaller peak observed at genes *w* through 29 could be due to a less "attractive" single origin, i.e., an origin whose probability of initiation is lower than that of genes 50 through 5, or to a decreased number of origins in this area. The discrimination between the single-versus-clustered origin possibilities requires further experimentation. From the patterns observed (see Fig. 3), one can draw more subtle inferences. For example, the group of genes 40–41, 42, and 43 represents the pattern of high, lower, and lowest ratios. Such pattern occurred in 15 of 18 hybridizations performed in independent experiments. Likewise, the rightmost group of genes 35–36, 37–38, and 52 represents the pattern of high, low, and high ratios which occurred in 13 of 18 hybridizations. One should realize the improbability that these repeatable patterns resulted from random errors. Therefore, the tendency of increasing values on the left side of gene 40 and on the left side of gene 35 has been interpreted (Halpern et al., 1979) as indicative of the presence of other origins located to the left of those two areas. The other important aspect of observed patterns of hybridization is that areas on both sides of origin display, as a function of time, gradual gain in the amount of progeny label. This confirms the bidirectionality of T4 phage DNA replication (Halpern et al., 1979).

Since the validity of interpretation of experimental results for localization of initiation sites depends on the general fate of parental and, at later stages, of progeny DNA, it is important to review some physicochemical properties of both parental and progeny DNA during various stages of the vegetative DNA pool.

First of all, one must critically review evidence for or against interparental recombination which, if occurring before initiation of DNA replication, would provide for initiation at recombinational intersections, as frequently stated. This is unlikely. In the past, the concept of initiation at recombinational intersections was presented both from our laboratory (Hutchinson et al., 1978) and from Mosig's group (Mosig, Dannenberg, Ghosal, et al., 1979), even though neither laboratory proposed specific molecular mechanisms or elaborated on the relevance to an understanding of high negative interference. One should not, however, overlook the difference between those two groups in assigning the time of the occurrence of such initiations. Our laboratory has shown that such initiations occur relatively late in infection, when enzymes involved in recombination have been expressed, some 9 min past infection at 37°C, and when a large amount of progeny DNA has already accumulated in the cell (Halpern et al., 1982; Hutchinson et al.,

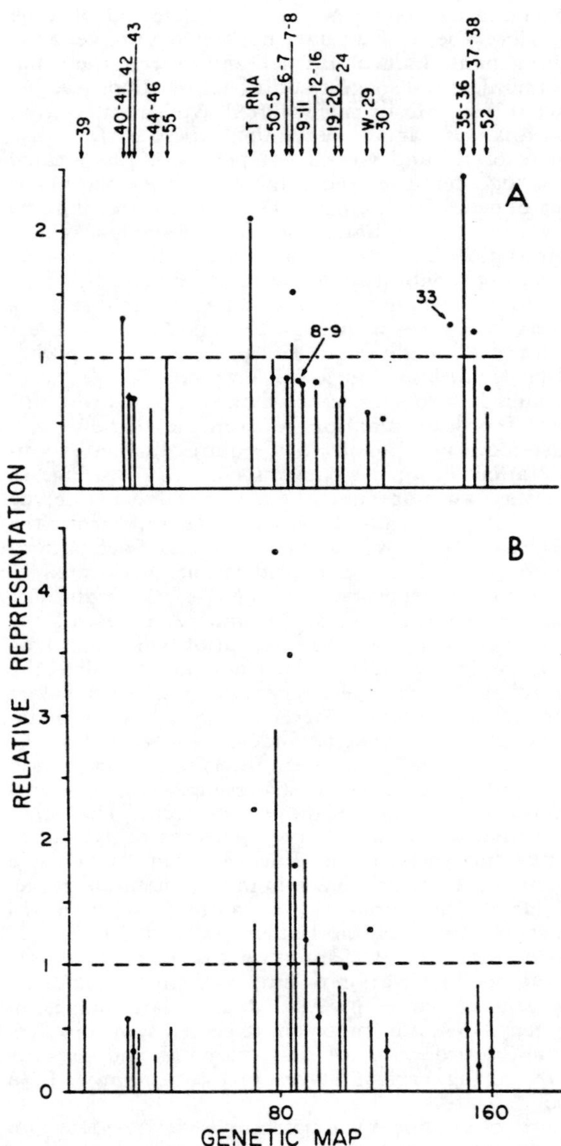

FIG. 4. Comparison of typical initiation at early times after infection occurring in the absence (A) or in the presence (B) of gene 44. No density label was used in this experiment. Dots and bars represent results obtained in two separate experiments. Dots in (B) were sampled at 6 min; bars in (B) were sampled at 7 min. Cells infected with T4*am*N82 (Am44) were sampled at 30 min in both experiments (A). Note the "typical" pattern of initiation in cells infected with wild-type phage (B), and the drastically different pattern in cells infected with the gene 44-deficient mutant (A). Note also the coincidence of hybridization values obtained in independent experiments. The results of hybridization are expressed as relative representations (RR) of genetic segment x in the progeny DNA (Kozinski et al., 1980). The $^3H/^{32}P$ ratio observed for a filter charged with a given genomic area, x, was divided by the ratio of the sums of ^3H and ^{32}P hybridized to all of the nitrocellulose filters in a set: $RR_x = (^3H_x/^{32}P_x)/(\Sigma^3H/\Sigma^{32}P)$. (From Kozinski and Ling, 1982.)

1978). Mosig, on the other hand, suggests that at high MOI recombination among parental molecules precedes replication and provides primers for the initiation (Mosig, Benedict, Ghosal, et al., 1980; Mosig, Dannenberg, Ghosal, et al., 1979; Mosig, Luder, Carcia, et al., 1978). As we have elaborated in the past (Halpern, et al., 1982), we do not consider the published data of Mosig, Luder, Garcia, et al. (1978) as supportive of those conclusions, as no internal density references were provided, and firmly repeatable density alteration, which occurs at early times even without the application of density label (Kozinski et al., 1976), was ignored. In addition, no control experiment was performed without the use of the density label. We therefore adhere to our previous conclusion indicating an extremely low frequency of interparental recombination among phages productively infecting bacterial cells (Kozinski et al., 1963). (We do, of course, support occurrence of interparental recombination at very late times after infection with DNA replication-deficient mutants [Kozinski and Felgenhauer, 1967]).

Considerable additional evidence exists against interparental recombinations leading to initiation. (i) Neither at high nor low MOI is early produced progeny ever covalently attached to parental strands as would be required for initiation of the 3′ termini of parental strands (Howe et al., 1973; Hutchinson et al., 1978) or initiation in rolling circles. (ii) If indeed putative interparental recombinational intersections were responsible for initiation, then one should expect differences in the genetic specificity of the early progeny depending on whether the experiment was performed at high or low MOI. This is not what has been observed (Halpern et al., 1982; Kozinski et al., 1980). (iii) Loops resulting from interparental recombination should have 3′ ending strands in *cis* configuration (see Fig. 5 in Halpern et al., 1982). In contrast, early replicative loops had 3′ termini in *trans* configuration (Delius et al., 1971) (see Fig. 1). (iv) Finally, normal initiative events as revealed by hybridization to cloned genes do not occur at locations which would correspond to those of high frequency of genetic recombination (Halpern et al., 1979; Halpern et al., 1982).

Taking together these arguments, one might conclude that locations of origins as described cannot be attributed to interparental recombinations.

At this point, I would like to comment upon and discuss data of others. Originally, it was suggested by Mosig (1970a: 1971) that initiation occurs at gene 43 and proceeds unidirectionally. However, recently Mosig has modified this concept; Mosig et al. (1980) describe evidence for a single major replicative origin located within the area of *Sal*I restriction fragment 2, spanning genes *r*II and 42; it has now been localized further to *Xba* fragment 17 (Mosig, this volume). Genes 39 and 42 (which reside within the *Sal*I fragment) were both found by us to be negative in initiation (see Fig. 4B). There may, however, be an origin located counterclockwise (to the left) of gene 42, as we suggested while commenting on the repetitively observed "slope" in the hybridization pattern of genes 40-43.

More recently, Morris and Bittner (personal communication) hybridized early progeny DNA to complete sets of restriction fragments. With $^{32}P_i$ as a precursor, they observed five apparent replication origins: site 1, genes 56 through 41 (at 17 to 22 kilobases [kb]; see Kutter and Rüger, this volume); site 2, genes *nrd*C through *tk* (at 47 to 58 kb); site 3, genes 1 through 5 (at 73 to 78 kb); site 4, genes *w* through 29 (at 110 to 120 kb); and site 5, genes *nrd*A through 34 (at 136 to 150 kb). This study confirmed our finding that the site near gene 5 (their site 3) is dominant in initiation at 5 to 7 min postinfection, with the genes *w* through 29 (site 4) being next most important. Morris and Bittner observed that if cells are concentrated before labeling, "this step drastically retards the normal T4 replication process and results in an alteration of normal phage replication" (personal communication). Under these latter conditions they observed the gene 56–41 region (site 1) to be dominant in uptake of label for short intervals immediately after infection. As part of their DNA isolation procedure, Morris and Bittner used alkaline digestion; consequently, one cannot determine whether the labeled DNA was associated with parental molecules. The possibility cannot be ruled out that the label hybridizing to the gene 56–41 region is derived from the DNA portion of the RNA:DNA copolymer described previously (Buckley et al., 1972).

Thus, the unpublished data of Morris and Bittner, though based upon less pronounced biases in labeling ratios than our data, seem to confirm two of our reported locations and likely add new sites, one of them in the same general region as defined by Mosig's work. These five initiation sites might well correspond to those originally suggested by gradual DNA shearing (Howe et al., 1973) and by electron microscopic analysis (Delius et al., 1971).

Still more recently, King and Huang (1982) performed hybridization of the label (^{32}P) associated with early-synthesized, fast-sedimenting complexes to *Eco*RI restriction fragments. Radioautography of the material bound to these fragments revealed two peaks of similar intensity, which were mapped by analysis with other restriction enzymes. One of these was a 5.6-kb fragment corresponding to genes *reg*A through 46, and the second was a 1.9-kb fragment at the map position 62 to 64 kb from the *r*II cistron divide. Neither of these corresponds to peaks observed by other authors. We did not test the second area in our laboratory, but (see Fig. 3) we tested the cloned genes 44 through 45 and found those negative.

How can we reconcile our consistent failure to observe initiation in these areas with the results of King and Huang? Let me present a hypothesis which might be valid. We assume that label incorporated at 2.5 min postinfection does not correspond to DNA residing in the initiative loop of parental DNA but rather represents a DNA component of the RNA:DNA copolymer. Indeed, results of Cs_2SO_4 isopycnic analyses in the study of King and Huang (1982) might support this explanation. The incorporated label in their study banded at an equilibrium density much higher than that of reference DNA (we should note that non-glucosylated DNA or DNA complexed to "membranes" will either band at lighter locations or float, respectively). The best explanation of this unusual, but unremarked, result, is that the dense, labeled material from the Cs_2SO_4 gradient is an RNA:DNA copolymer. This, of course, could be so only if there were not complete digestion of RNA by alkali. It is also possible that the observed peak represents polyphosphates. If the hybridizing material represents DNA:RNA copolymer, it remains to elucidate further whether the early appearance of the copolymer is in any way related to initiation of DNA replication or rather represents an mRNA containing DNA inclusions (which, as was shown by transformation [McNicol, 1973], contains gene 43 and the gene for lysozyme; both these genes are somewhat close to those areas which were shown by King and Huang [1982] to hybridize ^{32}P). A more trivial alternative interpretation of the results of King and Huang might be that the hybridized label is an early mRNA. The extraction procedure used called for isolation in a sucrose gradient of fast-sedimenting complexes. Those were digested with RNase. Upon digestion, sodium dodecyl sulfate (an RNase inhibitor) was added, followed by deproteinization, and the resulting extract was used for performing hybridization. (No purification in isopycnic gradients was performed.) Since digestion with RNase should not destroy the RNA:DNA hybrid which conceivably might be preserved in the complexes, and since King and Huang (1982) observed that hybridized label corresponded to early genes, the possibility that the hybridized label represented RNA cannot be excluded. Importantly, there was no mention of a control showing whether the ^{32}P-labeled moiety which hybridized was DNase sensitive or alkali resistant. Clearly, additional chemical characterization of the hybridizing polymer is in order.

GENE AMPLIFICATION IN T4

Appearance of Bistranded Progeny Segments Before Completion of Replication of the Parental Molecule

A simple model for DNA replication in T4 could begin with the following sequence of events: first, initiation at the origins would produce parental molecules containing small replicative loops; second, extension of the loops along the full length of the parental molecule would produce hybrid molecules consisting of half parental DNA and half progeny DNA; and third, an additional round of replication would produce molecules consisting of two strands of progeny DNA in addition to hybrid molecules. Consequently, if the parental DNA were labeled with 5-bromodeoxyuridine (heavy, nonradioactive) and the newly synthesized progeny DNA were labeled with [^{3}H]thymidine (light, radioactive), the model predicts that immediately after initiation of replication, the incorporated [^{3}H]thymidine would band very close to the heavy (HH) density in CsCl gradients. Then, as replication proceeds through the first round, the [^{3}H]thymidine would move toward the hybrid (HL) density, while the amount of label at the heavy density would decline. Finally, light (LL)-density molecules would be observed only after a second round of replication is completed.

Experimental data did not, however, conform to these simplified predictions (Kozinski et al., 1980); at very early times after infection, the progeny label was found to band not only at the expected heavy density, but also at the light density. At this time, there was still no detectable peak of the progeny label at the hybrid density, which indicates that less than one complete round of replication at the parental molecule had occurred.

Site Specificity of Early Progeny Subunits

When this light moiety of progeny DNA, which was not attached to the parental molecule, was isolated and assayed for its genetic make-up, it was shown to be biased in its genetic representation. This was especially observable at low MOI (Kozinski et al., 1980). It was shown that, by timely addition of chloramphenicol (CM), one can freeze replication in this apparent amplified mode while there is sizable accumulation of partial replicas of areas corresponding to initiation sites. Up to 50 copies of partial replicas of genes 5 through 50, for instance, can be accumulated by 40 min postinfection when CM is added at 5 min (Kozinski et al., 1980). Addition of CM at 10 min, on the other hand, results in the production of progeny DNA representing all tested areas equally (see Fig. 3). Since the addition of CM at 5 min prevents molecular recombination, whereas addition of CM at 10 min allows full expression of molecular recombination (Kozinski et al., 1967), we postulate (see Specificity of Reinitiations, below) that without recombination, progeny DNA cannot "extend all the way," and completion of the replication of the parental molecule is prohibited.

We have shown that amplified DNA can replicate autonomously. The strategy of the experiment was similar to that used for "petite" phages (Kozinski and Doermann, 1975). We allowed accumulation of ^{3}H-labeled light amplified DNA and then removed the radioisotope and added unlabeled 5-bromodeoxyuridine. Upon further incubation, most of the originally light ^{3}H-labeled DNA assumed HL density. Therefore, amplification cannot be due to consecutive "firing" of origins on the parental molecule. If this were the case, amplified DNA should not be able to replicate autonomously, being produced exclusively from the parental master template (Hutchinson, thesis, 1983). Whether amplified DNA is detached from the parental molecule in vivo or remains associated by short terminal complementarity, thus resulting in a loop resembling a longitudinal cross-section of an onion, is of no consequence in respect to this conclusion. It is possible that such a multilayered structure is fragile and the weak terminal complementarity is broken during extraction. This is suggested by an experiment in which we followed the transfer (maturation) of both ^{32}P-labeled parental and ^{3}H-labeled amplified DNA to the progeny phage. We found that the relative specific activity of ^{3}H/^{32}P remained the same in the progeny phage as that observed in the vegetative pool. Moreover, ^{3}H-labeled subunits separated by shearing from the DNA molecules of progeny phage retained their genetic bias for initiative areas. (This excludes the possibility of breakdown and reutilization of amplified DNA.) The best explanation of this result is that in the amplifying mode, partial replicas remain in a multilayered structure. Branches of such a structure replicate autonomously but do not extend their length. Upon removal of CM, layers of partial replicas could presumably expand along the parental template. Such multilayered loops have been observed, albeit infrequently, in electron microscopic studies of partially replicated molecules (Delius et al., 1971) (see Fig. 5).

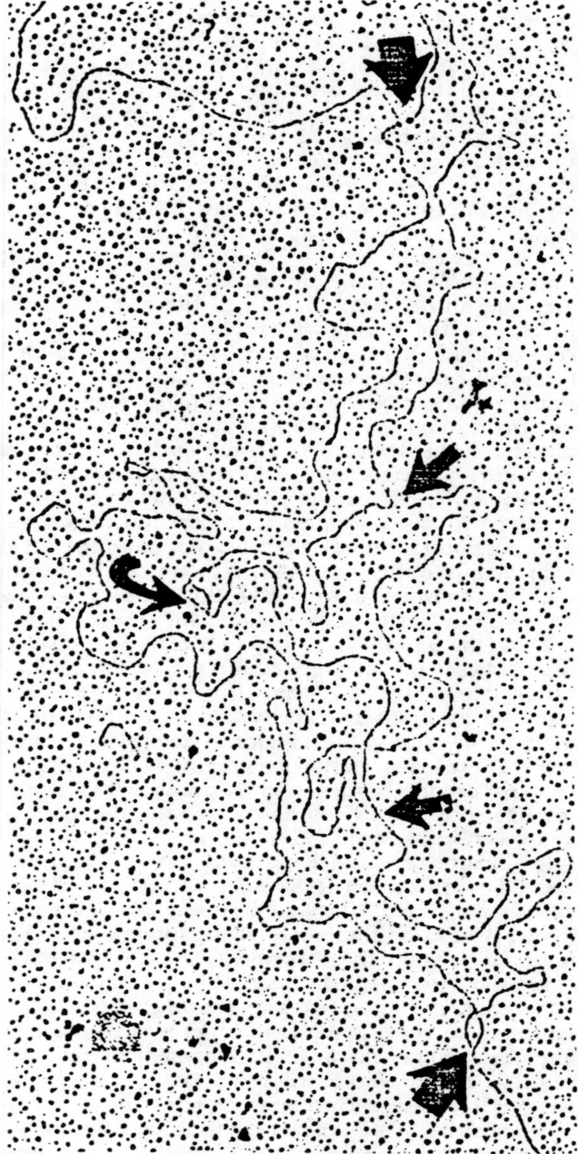

FIG. 5. Reinitiation of T4 DNA at early times after infection as visualized by electron microscopy (from Delius et al., 1971).

What determines the observed genetic bias of highly represented and poorly represented initiative areas in the pool of amplified DNA? There are two possibilities. (i) The origins could contain qualitatively different signals, thus being "better" and "less attractive" origins. If this were the case, observed bias should increase as a function of time after infection. (ii) Observed bias could be due to a primary event in the initiation of the parental molecule (for instance, sequential initiation caused by preferential site association of the parental molecule with the replicative complex). This would provide for delay in initiation of certain areas. Once replicated, however, such areas could reinitiate with an efficiency equal to areas which were first to replicate from the parental template. If this were correct, bias in the representation of different genetic areas in the amplified DNA should remain constant for an extended period of time. This latter is what was found when we followed the intensity of genetic bias over extended periods of the amplified DNA synthesis (Hutchinson, thesis, 1983).

It would be premature to propose any detailed mechanism for such partial replication. Very likely, replicative loops which are initiated at the origins of intact parental molecules might be prevented from extending very far down the molecule by, for instance, the existence of encoded terminator signals, DNA cross-links, terminator proteins, or supercoiling of the parental molecule. With any of these terminating events, since arms of loops can rotate about the axis, it would be possible to initiate another round of replication within the original, hindered loop. Branch migration could release (either in vivo or upon extraction) double-stranded progeny and "close" the parental duplex into apparently nonreplicated molecules. (It should be mentioned that light amplified DNA is resistant to single-strand-specific endonucleases [S1]; thus, putative single-stranded termini must be short [Hutchinson, thesis, 1982]).

Another form of gene amplification is observed upon infection of bacteria with UV-inactivated heavy phages, where a considerable amount of phage DNA synthesis occurs in cells not productively infected. This DNA is biased in its genetic representation, and the accumulated regions correspond to the areas shown above known to contain major origins of T4 DNA replication (Ling et al., 1981). These data are consistent with the assumption that UV irradiation, acting upon heavy DNA, imposes termination (most likely by cuts occurring within the host cell). This termination in turn leads to the replication and amplification of those segments which are endowed with origins and therefore results in abundant accumulation of preferred initiative areas. The sequestration by UV of autonomously replicating subunits complements data on specificity of origins. It also offers a molecular explanation of observed patterns in marker rescue experiments (Womack, 1965) in which certain genetic areas are rescued better than others.

The process of amplification could have regulatory functions. For instance, it might be involved in late transcription. Most of the amplified areas contain genes that code for late-function proteins, and it is known that DNA replication is required for the expression of late function genes (Riva et al., 1970a); therefore, it is possible that amplification is needed to provide proper transcriptional templates for these genes. Preferential and partial replication of specific areas of the genome and putative role of amplified DNA in transcription could then resemble chromosomal events in the salivary gland of *Drosophila*.

SPECIFICITY OF REINITIATIONS OF DNA REPLICATION AT LATE TIMES AFTER INFECTION

Role of Recombination in Late Initiation

In the previous two sections, we have stated that initiation of replication of the parental molecule of T4 phage occurs at several specific locations as demonstrated by electron microscopic studies (Delius et al., 1971) and by hybridization of early progeny DNA to cloned genetic fragments (Halpern et al., 1979). It was also observed that early after infection, there is amplification of specific areas of the genome (Kozinski et al., 1980). The amplifying mode of replication can be "fixed" by the early addition of CM, which allows a large number of copies of partial replicas to be accumulated over a long period of incubation (Kozinski et al., 1980). Yet the addition of CM, at times when molecular recombination is fully expressed, results in the accumulation of progeny DNA which does not display a genetic bias. The genetic representation of the progeny DNA accumulated at later times after infection without CM is also unbiased.

What is the mechanism of this equalization? There have been several experiments designed to follow site specificity of reinitiation at late times postinfection. Here, it is important to note that if bacteria which have accumulated a large pool of HH progeny DNA are superinfected with radioactive phage of light (LL) density, the superinfecting DNA recombines with the pool of progeny DNA very soon after infection. Such recombinants band, in CsCl gradients, at a "progeny-like" (heavy) density. Sonication of the recombinant reveals, however, that the parental subunit remains nonreplicated, i.e., bistranded LL. (Since most of the parental contribution resides in the LL density, the major mechanism of observed recombination calls for terminal joining of bistranded parental subunits to progeny DNA, in contrast to the "invasion" of progeny DNA by a single parental strand. Of course, such a contribution should, after sonication, assume the form of a hybrid [HL].) Denaturation of the recombinant releases light strands of superinfecting phage DNA quantitatively; thus, such recombinants are not repaired. At later times after superinfection, the superinfecting genome becomes replicated (i.e., after sonication, parental label assumes an HL location). At this time, invariably, an equivalent volume of parental label becomes covalently joined in the progeny molecules (Hutchinson et al., 1978). Taken together, these observations lead to the hypothesis that recombination might provide for initiation at the 3′ end of a recombinational intersection (Hutchinson et al., 1978). Such initiation should be random or drastically different from the initiative origins observed early after infection. As demonstrated by Kozinski et al. (1967), timely addition of CM can serve as a fine dissecting tool, separating various replicative events. When CM is added at 5 to 6 min postinfection, there is no parent-to-progeny recombination, and the progeny DNA displays a genetic bias, being fixed in the amplifying mode (Kozinski et al., 1980). Addition of CM at 9 to 13 min allows for recombination and covalent joining of the parental segments to the progeny DNA. Therefore, if the working hypothesis of random initiation at recombinant intersections is correct (as contrasted to specific initiation observed before the onset of recombination repair), the following could be predicted.

(i) In the absence of recombination, a short pulse of the label applied at late times will be divided between label which was involved in the process of reinitiation of origins occurring during the time span of the pulse (which should be of pure density and likely site

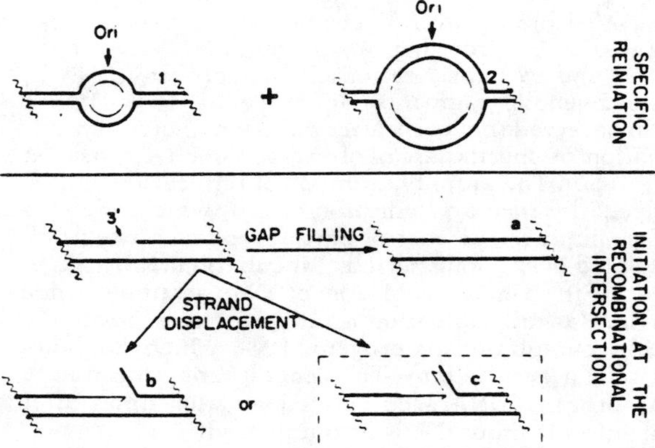

FIG. 6. Two modes of reinitiation (from Halpern et al., 1982) at late times postinfection. Schematic diagram of incorporation of short-pulse light (and ^3H) label (thin line) when a large pool of heavy, unlabeled progeny (thick line) has accumulated. The upper drawings show incorporation when the process of reinitiation occurs at genetically specific sites and proceeds bidirectionally. Note that the radioactive progeny is made up of two distinct classes: (1) radioactive progeny that exists in a replicative loop that was initiated during the pulse (this progeny is of light density upon denaturation), and (2) radioactive progeny that is attached to elongating heavy, unlabeled progeny strands (this progeny assumes intermediate density upon denaturation). The lower drawings represent events when initiation occurs at the 3′ end of a recombinational intersection. The possibility framed by dashed lines is favored. Note that in this case the progeny produced on the displaced strand will be light upon denaturation but is not site specific.

specific) and label added to the ends of already initiated, elongating progeny strands, which would band in CsCl at an intermediate density (Fig. 6A).

(ii) In contrast, when recombination is expressed and if recombinational 3′ ends act as randomly located primers, then the short pulse label should find its way either to elongating structures of intermediate density or to progeny strands of pure density in those events when strand displacement and its (presumably) discontinuous replication occur during the span of the labeling period. In this case, the light moiety should not display site specificity (see Fig. 6B). The isolation of the light (L) strands and the hybridization of these to cloned genes should permit discrimination between reinitiation of specific sites and random initiation.

The experiments of Halpern et al. (1982) point to the second of these alternatives. Light single-stranded moiety was isolated upon short pulse labeling and hybridized to cloned segments. In the branch of experiments where recombination was inhibited (by CM added at 6 min), reinitiation occurred at areas reasonably similar to those involved in the initiation of the parental molecule, even though there was approximately 50 PEU (phage equivalent units) of DNA accumulated in the cell. When CM was not added until 13 min postinfection and abundant molecular recombination was observed, the uptake of [^{3}H]thymidine to light strands revealed no apparent site specificity. This is what one would expect, assuming near randomness of recombinational events.

On the basis of these observations, mechanisms were proposed for the establishment of covalently joined recombinants. It was postulated that repair involves mostly replication and displacement of the strand, which is followed in turn by its replication (presumably the discontinuous process of the lagging strand), rather than by the filling in of gaps, ligation, the excision of single-stranded branches (Halpern et al., 1982), or all of these. Indeed, Kozinski and Kozinski (1969) observed that there is an intriguingly efficient covalent repair of recombinant molecules during infection with ligase-deficient mutant phages. Furthermore, although both the replicative parental and the progeny DNA are of a very small size, they replicate autonomously as if such small fragments contained origins of replication (Kozinski and Kozinski, 1968). At first glance, one would be at a loss to explain the covalent repair and replication of recombinant molecules in the absence of ligase. However, according to the model presented, the repair and replication of the recombinants could proceed quite efficiently in the absence of ligase simply by the proposed mechanism involving initiation at the 3′ termini of recombination intersections. Many topological features and genetic consequences of such mechanisms were discussed by Halpern et al. (1982). The most important is the fitness of the model to achieve high negative interference.

In experiments of completely different design, in lambda phage, Stahl's group arrived at the conclusion that recombination is achieved by "break and copy" (Stahl et al., 1973; Stahl and Stahl, 1974; Stahl, 1979b). There is a similarity between Stahl's model and proposed events in T4. Most recently Luder and Mosig (1982) restated a model of initiation at molecular intersections; in their present schematic drawings, initiation at specific origins is suggested before the recombinational events. They observed that continued synthesis of DNA was sensitive to rifampin (an RNA polymerase inhibitor) if the drug was added at later times after infection with recombination-defective mutants. Thus, they postulate that specific initiation at origins is primed by RNA polymerase, whereas at late times primers could be provided by recombinational intersections. There is a similarity between models presented in their paper and those presented by Halpern et al. (1982). Although interruption of the continued synthesis of DNA by rifampin was interpreted by Luder and Mosig (1982) as indicative of the role of rifampin-sensitive RNA polymerase (of host origin) in repriming specific origins, an alternative explanation should be considered. One can assume that in the absence of recombination, reinitiation demands continued transcription (which is sensitive to rifampin) as a factor "conditioning" DNA strands for reinitiation. For instance, transcription (rifampin-sensitive process) from one of the DNA strands leaves the other, complementary strand in a single-stranded form. Such single-stranded DNA can now self-anneal, in areas of inverted repeats, to form an intrastranded duplex, a hairpin-like structure. Such putative secondary structures might act now as a functional origin. Initiation of the replication at such-activated origins need not be performed by host RNA polymerase. Such a sequence of events could explain how palindromes are being recognized by the priming enzymes. The proposed hypothesis would, however, demand further explanation for the role of gene 33 and 53 products.

Evidence for covalent addition of precursor to the recombinational intersections is strengthened by the discovery by Kozinski and Ling (1982) that, in the absence of gene 44 protein, a late and aborted initiation occurs most likely at recombinant intersections. Such progeny DNA, in contrast to the progeny DNA observed in normal initiation, is covalently attached to the parental DNA strands and displays dramatically different site specificity from that observed in the presence of gene 44 protein (see upper panel of Fig. 4). One preferentially initiated area corresponds to the area of high frequency of genetic recombination in the gene 35–36 region described by Mosig (1968). At the other extreme, the lowest values of hybridizations are observed in areas of low frequency in genetic recombination. Presumably, initiation at recombinational intersections is allowed in the absence of gene 44, but such progeny does not considerably elongate.

In conclusion, we postulate (Halpern et al., 1982) that there are two distinct stages of T4 DNA replication. The first stage occurs early, when replication initiates at specific origins but the progeny strands are not allowed to extend "all the way." Rather, replicative extension is terminated, yet reinitiation is possible. This leads to gene amplification, where the areas of origin are significantly overrepresented. The first stage might be prolonged, or for that matter fixed, by timely addition of CM. The second stage of replication commences at later times when there is recombination and, moreover, a putative protein allowing the covalent addition of progeny to recombinants is expressed. This would be a critical stage in replication, providing for the equalization of all the genes, a fact observed in the progeny DNA synthesized late after infection, coincidentally with the onset of molecular recombination and covalent joining of recombinants. The interdependence of those two stages determines successful T4 reproduction; that may be why each T4 bacteriophage has a history of a number of molecular recombinations, resulting in discrete subunits which are invariably joined covalently (Kozinski, 1961; Kozinski and Kozinski, 1963), and there are no true recombination-negative T4 mutants.

Relationship of T4 DNA Replication and Recombination

GISELA MOSIG

Department of Molecular Biology, Vanderbilt University, Nashville, Tennessee 37235

DNA replication ultimately determines proliferation and growth of living things and thus must be coordinated with other metabolic processes. Perhaps for this reason evolution has designed some of the most sophisticated control circuits to regulate DNA replication. The replicon model (Jacob et al., 1963; Pritchard, 1978) has provided a successful framework to analyze this control: DNA replication is regulated at the level of initiation by interactions of positive and negative control elements with specific DNA initiation sequences, i.e., origins of replication. This regulation ultimately achieves tight coordination of DNA replication with cell division.

Viruses are thought to provide a window into their host's control mechanisms, because they depend on some or all of their host's control circuits at least immediately after infection. Eventually, however, they have to escape that control. It is perhaps for this reason that many, if not all, DNA viruses have evolved two different modes of DNA replication: a so-called "early" mode which utilizes some or all of the host's control elements, and a so-called "late" mode which is less dependent on them. This review will consider the sequential use of different initiation modes as an important aspect of the intricate transition from host-type to virus-type controls with the hope that the example of T4 helps us to better understand this aspect of virus-host interactions.

The well-known genetics and biochemistry of T4 replication proteins (Nossal and Alberts, this volume) offer great advantages for analyzing elements of different replication modes in detail. We believe that the apparent discrepancies on the number and locations of T4 origins (Halpern et al., 1979; Halpern et al., 1982; King and Huang, 1982; Kozinski et al., 1980; Macdonald et al., 1983; Marsh et al., 1971; Morris and Bittner, personal communication; Mosig, 1970a; Mosig, Luder, Rowen, et al., 1981) result largely from different effects of certain experimental procedures on one or the other mode of initiation (see below). It now appears that "firing" of different T4 origins depends on the activity of certain genes as well as on growth and labeling conditions (Macdonald et al., 1983; Morris and Bittner, personal communication). Thus, there is a strong hope that the original difficulties in characterizing initiation of T4 DNA replication will ultimately help to understand better how different origins and different initiation modes are controlled. I shall discuss these controls with the premise that transcription and recombination are intimately related to DNA replication.

There is considerable evidence that the first mode of T4 DNA replication depends on de novo initiation from origin sequences (Kozinski, this volume) and that it requires functioning *Escherichia coli* RNA polymerase (Macdonald et al., 1983; Snyder and Montgomery, 1974). Since modification of RNA polymerase during T4 development eventually interferes with recognition of early promoters, and thus of early origin promoters (Rabussay, 1982b; Rabussay, 1982c), subsequent DNA replication should depend on alternative initiation modes. In principle, these could use different promoters or entirely different mechanisms.

Several reasons, most importantly the different genetic requirements for the first and subsequent rounds of replication (Breschkin and Mosig, 1977a; Breschkin and Mosig, 1977b; Dannenberg and Mosig, 1981; Marsh et al., 1971; Mosig et al., 1980; Mosig, Dannenberg, Ghosal, et al., 1979; Mosig, Luder, Garcia, et al., 1979; Mosig, Luder, Rowen, et al., 1981; Mosig and Werner, 1969; Mosig, Bowden, and Bock, 1972), the enhanced recombination potential of chromosomal tips (Doermann and Boehner, 1963; Doermann and Parma, 1967; Mosig, 1963; Mosig et al., 1971; Womack, 1963), and the distinct segregation pattern of terminal markers (Mosig et al., 1971), as well as the resistance to rifampin of late T4 replication (Karam and Speyer, 1970; Melamede and Wallace, 1977; Rabussay, 1982b; Rabussay, 1982c), led us to propose that in this alternative mode, T4 DNA replication forks are initiated from recombinational intermediates (Luder and Mosig, 1982; Mosig, Dannenberg, Ghosal, et al., 1979; Mosig, Luder, Garcia, et al., 1979; Mosig et al., 1980; Mosig, Luder, Rowen, et al., 1981), a process which we call "recombinational initiation." Although this proposal appeared controversial at first, there is now general agreement and ample experimental evidence that replication forks can be initiated from recombinational intermediates (Dannenberg and Mosig, 1983; Halpern et al., 1982; Luder and Mosig, 1982; Mosig, Luder, Rowen, et al., 1981). I believe that the remaining controversies about timing of recombination (Broker and Doermann, 1975; Dannenberg and Mosig, 1983; Halpern et al., 1982; Kozinski et al., 1976; Kozinski et al., 1967) are mainly due to different experimental conditions, e.g., use of metabolic inhibitors (see below). Although the information required for origin initiation is found in specific DNA sequences, recombinational initiation depends on the specific structure of DNA intermediates in recombination. We have recently demonstrated that this type of initiation is essential for T4 growth because of the T4 transcriptional program (Luder and Mosig, 1982). For reasons of clarity, I shall first describe our general model (Fig. 1). Because this model is based on the premise that competing interactions between components of transcription, recombination, repair, and packaging (see Fig. 1 and 4) regulate T4 development, I will then review certain aspects of these processes to present a comprehensive and integrated view of the relationship between replication and recombination and its importance in T4 develop-

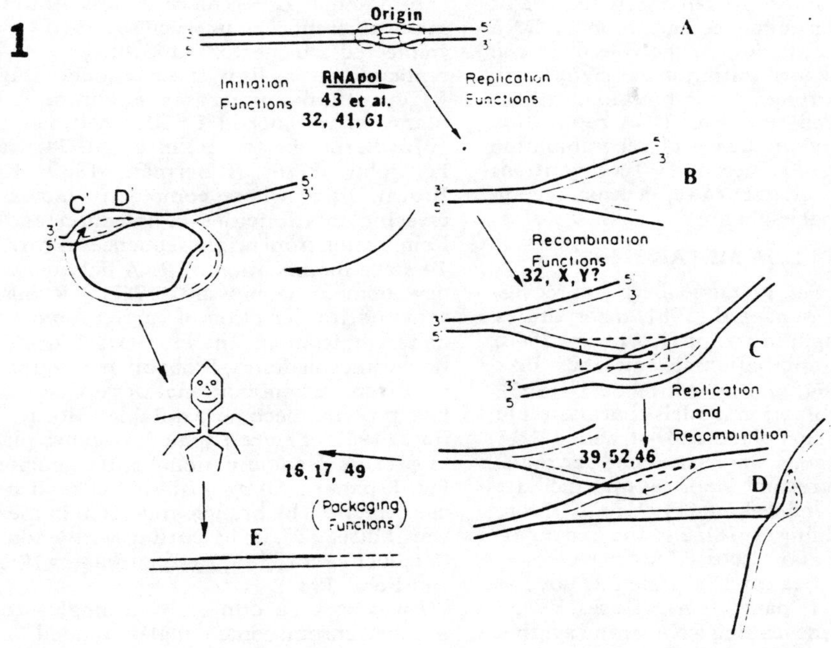

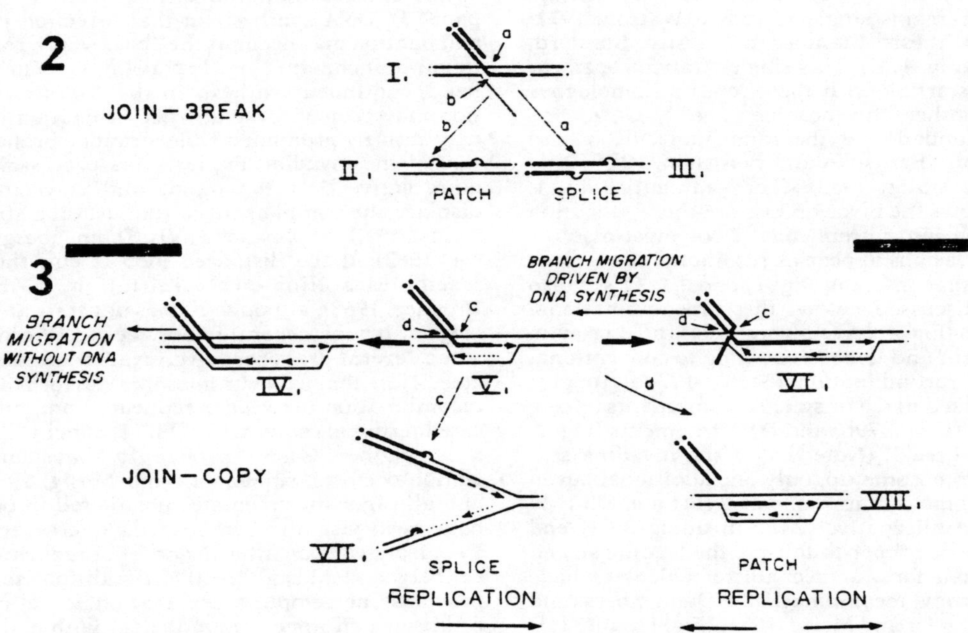

FIG. 1. Model of T4 DNA metabolism. Panel 1 illustrates the pathway from origin initiation through recombinational initiation and packaging described in the text. Heavy solid lines indicate parental DNA; thin solid lines, DNA synthesized during the first round of replication; broken lines, DNA synthesized during subsequent rounds of replication. Numbers refer to some of the genes which function in specific steps shown. (Participation of other genes and additional functions of the gene products shown in other steps are anticipated.) Panels 2 and 3 show the resolution of recombinational intermediates (boxed area in panel 1, C) by conventional join-break mechanisms (panel 2) or by join-copy mechanisms (panel 3). Symbols: dashed lines, elongation from 3′ ends (leading strands); dotted lines, lagging strands; small arrows (a through d), endonuclease cuts; semicircles: ligation; dots at the ends, DNA strands continue. Further details are given in the text. For simplicity, reinitiation from the same origin is not shown. It does occur, but it is rare (Dannenberg and Mosig, 1983; Delius et al., 1971).

ment. Specifically I will discuss separately five major factors which together influence initiation of DNA replication by different modes: (i) the role of *E. coli* RNA polymerase in de novo initiation at origin(s); (ii) early promoters in a strong T4 origin region; (iii) the mutual dependence and timing of DNA replication and recombination; (iv) influence of recombination patterns on replication of alleles; and (v) competitions between proteins involved in DNA replication, recombination, repair, and packaging.

MODEL FOR T4 DNA METABOLISM

The model shown in Fig. 1 (Dannenberg and Mosig, 1983; Mosig, Luder, Rowen, et al., 1981) incorporates current ideas about origin initiation (Kornberg, 1980, 1982); the general recombination schemes first proposed by Holliday (1964) and by Whitehouse (1963); those subsets of recombination models that postulate as a first event the invasion of a double-stranded DNA molecule by a homologous single-stranded terminus to generate a branched recombinational intermediate (Broker, 1973; Broker and Lehman, 1971; Fox, 1966; Mosig et al., 1971; Radding, 1978); and the five major factors just mentioned and discussed further below.

The first round of replication is initiated de novo at origin sequences (Fig. 1, panel 1, A); it uses *E. coli* RNA polymerase to prime leading DNA strand synthesis (Luder and Mosig, 1982), and T4 primase (see Nossal and Alberts, this volume) to prime Okazaki pieces. When a growing point reaches one end of the chromosome, the 3′ end of the template for the lagging strand remains single stranded (Watson, 1972; Broker, Ph.D. thesis, Stanford University, Stanford, Calif., 1972) (Fig. 1, B). This single-stranded segment can invade (starting with the 3′ end) a homologous region of another chromosome (Fig. 1, C), or the terminal redundancy of the same molecule (C′) to generate a ring (Bernstein and Bernstein, 1974; Dannenberg and Mosig, 1983). (For simplicity, Fig. 1, panel 1, C shows the invasion into another replication loop. Invasion into unreplicated DNA must also occur.) Such invasions appear as recombinational forks (e.g., rectangular insert in Fig. 1, panel 1, C; see also Fig. 4). As discussed below, these recombinational intermediates (Fig. 4) have pivotal roles in T4 growth. They can be cut and joined according to conventional pathways of recombination (Stahl, 1979b) to give covalently linked patch or splice recombinants (Mosig et al., 1971; Stahl, 1979b) and DNA fragments (Fig. 1, panel 2, join-break). (Note that if the invading segment is a chromosome tip, only one additional cut in that chromosome is required to generate a patch-type recombinant.) Alternatively, the invading 3′OH end can be used as a primer to initiate the leading strand of a replication fork, a mechanism which we have called "join-copy recombination" (Dannenberg and Mosig, 1983; Luder and Mosig, 1982) (Fig. 1, panel 1, D and D′, and panel 3). Okazaki pieces on the lagging strand are then initiated by T4 primase. Alternatively, in the absence of primase, synthesis of the other strand can be primed from an appropriate nick in the recombinational intermediate (Mosig et al., 1980; A. Luder, Ph.D. thesis, Vanderbilt University, Nashville, Tenn., 1981). Since the ends of T4 chromosomes are circularly permuted, similar invasions of single-stranded termini into homologous regions of other chromosomes occur more or less simultaneously at many different regions when several T4 particles have coinfected a bacterium, and subsequent initiation of replication forks from these branches (Fig. 1, panel 1, D) eventually generates a complex network of branched and looped T4 DNA (Altman and Lerman, 1970; Bernstein and Bernstein, 1974; Hamilton and Pettijohn, 1976; Huberman, 1969; Kemper and Brown, 1976) whose complexity increases with increasing multiplicities of infection and with time. Reinitiation from origin sequences is rare in wild-type T4 since modification of RNA polymerase during T4 development (Rabussay, 1982b; Rabussay, 1982c) shuts off transcription of early promoters. Most secondary initiations therefore start from recombinational intermediates. Probably packaging is also initiated from recombinational or replicative forks when late proteins become available (Mosig, Ghosal, and Bock, 1981). Figure 1 (panel 3) shows plausible ways to process recombinational intermediates (insert in Fig. 1, panel 1, C) by initiation of replication forks in one direction, by branch migration in the other (backward) direction, or by cutting with endonuclease VII (Mizuuchi et al., 1982) and packaging (Mosig, Ghosal, and Bock, 1981).

Invasion of a duplex by a single-stranded tip of another chromosome (single-stranded branch migration) may or may not require concomitant DNA synthesis (Fig. 1, panel 2, I and panel 3, V). We assume that Okazaki pieces can be initiated on the displaced strand of a recombinational intermediate (see Fig. 1, panel 3). DNA synthesis in that direction may stop, and ligation may occur at the "backward" recombinational branchpoint (Fig. 1, panel 3, IV). On the other hand, continued synthesis in the "backward" direction may occur (in the absence of ligation) and thus may actively promote double-stranded branch migration of the invading duplex. This process would require active DNA unwinding and may temporarily displace the complement of the invading strand as a "whisker" (Fig. 1, panel 3, VI) (Dannenberg and Mosig, 1983). If the displaced loop is cut, the process resembles rolling-circle initiation (Gilbert and Dressler, 1968), although it does not start at bona fide origins. It may generate "firewheels" (Werner, 1969) when several branches have invaded the same molecule. (Note that a T4 chromosome can form a circle by recombination between a replicated and an unreplicated terminal redundancy [Fig. 1, panel 1, C′ and D′], although there is a low probability that such circularization occurs [Dannenberg and Mosig, 1983]). This and all other intermediates postulated in our model have been visualized in electron micrographs of T4 DNA isolated soon after the onset of replication (Dannenberg and Mosig, 1983). In addition, this model provides the simplest explanation for a variety of results which appear paradoxical within the framework of other models (Hershey, 1958) and for the puzzling early observation that in T4 the number of recombinational pairings during one infectious cycle ("rounds of mating") appears to equal the number of rounds of replication (Levinthal and Visconti, 1953; Visconti and Delbrück, 1953). Additional evidence supporting different aspects of our model will now be discussed in more detail, following the five points mentioned above.

TRANSCRIPTION, RECOMBINATION, AND PACKAGING INFLUENCE DNA REPLICATION

De Novo Initiation Requires *E. coli* RNA Polymerase

There has been controversial evidence for (Snyder and Montgomery, 1974) and against (Karam and Speyer, 1970; Melamede and Wallace, 1977; Rosenthal and Reid, 1973) a direct role of RNA polymerase in DNA replication. Since RNA polymerase is certainly required to transcribe the T4 replication genes, it was difficult to prove a direct effect on DNA replication. More recent results from our laboratory (Luder and Mosig, 1982) have now clearly shown that functional RNA polymerase is required for origin initiation. Since the leading strand is initiated normally, even when both the T4 and *E. coli* primases are inactivated by mutations (Mosig et al., 1980; Mosig, Luder, Rowen, et al., 1981b; Luder, thesis), we believe that RNA polymerase synthesizes an obligatory primer for leading-strand DNA synthesis and is not only needed for "transcriptional activation" of an origin. We showed that continued DNA replication of recombination-deficient mutants, which occurs when genes 33 and 55 are also mutated (Hosoda et al., 1971; Lembach et al., 1969; Shah and Berger, 1971; Shalitin and Naot, 1971), requires continued de novo initiation from origins by RNA polymerase and remains sensitive to inhibitors of RNA polymerase at late times, when all required replication proteins have been synthesized (Luder and Mosig, 1982). On the other hand, consistent with our model, DNA synthesis accelerates in spite of the presence of rifampin if recombination is restored after a shift to permissive conditions for a recombination-deficient (46^-) mutant.

These results imply that preventing the modification of RNA polymerase by accessory proteins (gp33 and gp55) allows continuing transcription from early origin primer promoters. We postulate that the use of chloramphenicol at critical times can prevent synthesis of gene 33 and 55 products and thus allows continuing origin initiation. This effect of chloramphenicol is responsible for the reversal of DNA arrest in recombination-deficient mutants (Broker and Doermann, 1975; Bernstein and Wallace, this volume). The probability of reinitiation from the same origin, and thus the residual DNA replication under recombination-deficient conditions, depends on the temperature, on the multiplicity of infection (Dannenberg and Mosig, 1983), on the genetic background (Dannenberg and Mosig, 1981), and probably on other growth conditions as well. Thus, it should not be surprising that incomplete T4 chromosomes, lacking terminal redundancies and thus unable to recombine after single infection, replicate to different extents, depending on their relative length. Mosig and Werner (1969) and Marsh et al. (1971) found that most incomplete T4 chromosomes measuring 2/3 of the normal length, after single infection, arrest DNA replication after one round. Kozinski and Doermann (1975), in contrast, investigated replication of incomplete T4 chromosomes of 86% genome length and found that most of them replicated at least twice. We had considered two possibilities to explain the failure of the 2/3 chromosomes to replicate repeatedly (Marsh et al., 1971): these chromosomes cannot recombine and thus cannot circularize or they lack certain genes that are required for reinitiation. Experiments originally designed to test these ideas (Breschkin and Mosig, 1977a; Breschkin and Mosig, 1977b; Mosig, Bowden, and Bock, 1972) clearly showed that many incomplete 2/3 chromosomes use substituting host functions for missing T4 genes. It now appears that these factors influence origin selection and directionality of replication as well as the probability of reinitiation. Obviously, the longer incomplete chromosomes used by Kozinski and Doermann (1975) were missing fewer T4 replication genes than the shorter ones we used. Thus, both sets of results are readily compatible with the model presented in Fig. 1.

Promoters in a Strong T4 Origin Region Resemble *E. coli* Promoters

We have mapped a strong origin of replication between genes *dda* and *dam* (Fig. 2) and isolated a short nascent initiator DNA species that is synthesized from this region (Macdonald et al., 1983). The region upstream from the initiator DNA sequence contains two sequences which resemble *E. coli* promoters (see Fig. 3). One or both of these promoters are prime candidates for primer promoters. They should not be recognized by the T4-modified RNA polymerase, since late T4 promoters are different (Christensen and Young, this volume).

To map origin regions we have isolated early replicated T4 DNA labeled in vivo and have used it as probes for Southern hybridization (Southern, 1975) to T4 restriction fragments cut from T4 dC-DNA, i.e., T4 DNA that contains deoxycytosine instead of hydroxymethyldeoxycytosine (HMdC) in wild-type T4 (Snyder et al., 1976; Wilson et al., 1977). Hybridization of replicated ^{32}P-labeled HMdC-DNA (isolated early after infection) to these immobilized restriction fragments reveals a strong origin region spanning the junction of *Xba*I restriction fragments 15 and 17 (shaded 1° area in Fig. 2) and the 680-base pair *Eco*RI fragment overlapping that *Xba*I site (see Fig. 3) (Macdonald et al., 1983). At 25°C, replication of wild-type T4 or of recombination-deficient gene 46 mutants begins at first from this region undirectionally in the counterclockwise direction (the major direction of early T4 transcription). Soon thereafter, replication becomes bidirectional. In a T4 gene 39 (topoisomerase-deficient) mutant, replication is initiated in the same region but, in contrast to wild-type T4, remains unidirectional until late after infection. This shows that unidirectional initiation in the major direction of early T4 transcription does not depend on T4 topoisomerase, but that bidirectional replication in the opposite direction does require T4 topoisomerase. Gratuitously, this observation allowed more precise mapping of the origin region at the transition between the most and the least frequently labeled restriction fragments in a 39^- infection (Fig. 2).

We now discuss the search for primer promoters in this region. As discussed above, *E. coli* RNA polymerase is required to initiate leading-strand synthesis. We therefore reasoned that leading-strand primers should be longer than the pentaribonucleotides synthesized by T4 primase (Nossal and Alberts, this volume) and, therefore, DNA attached to such primers should have an apparent density that is intermediate between

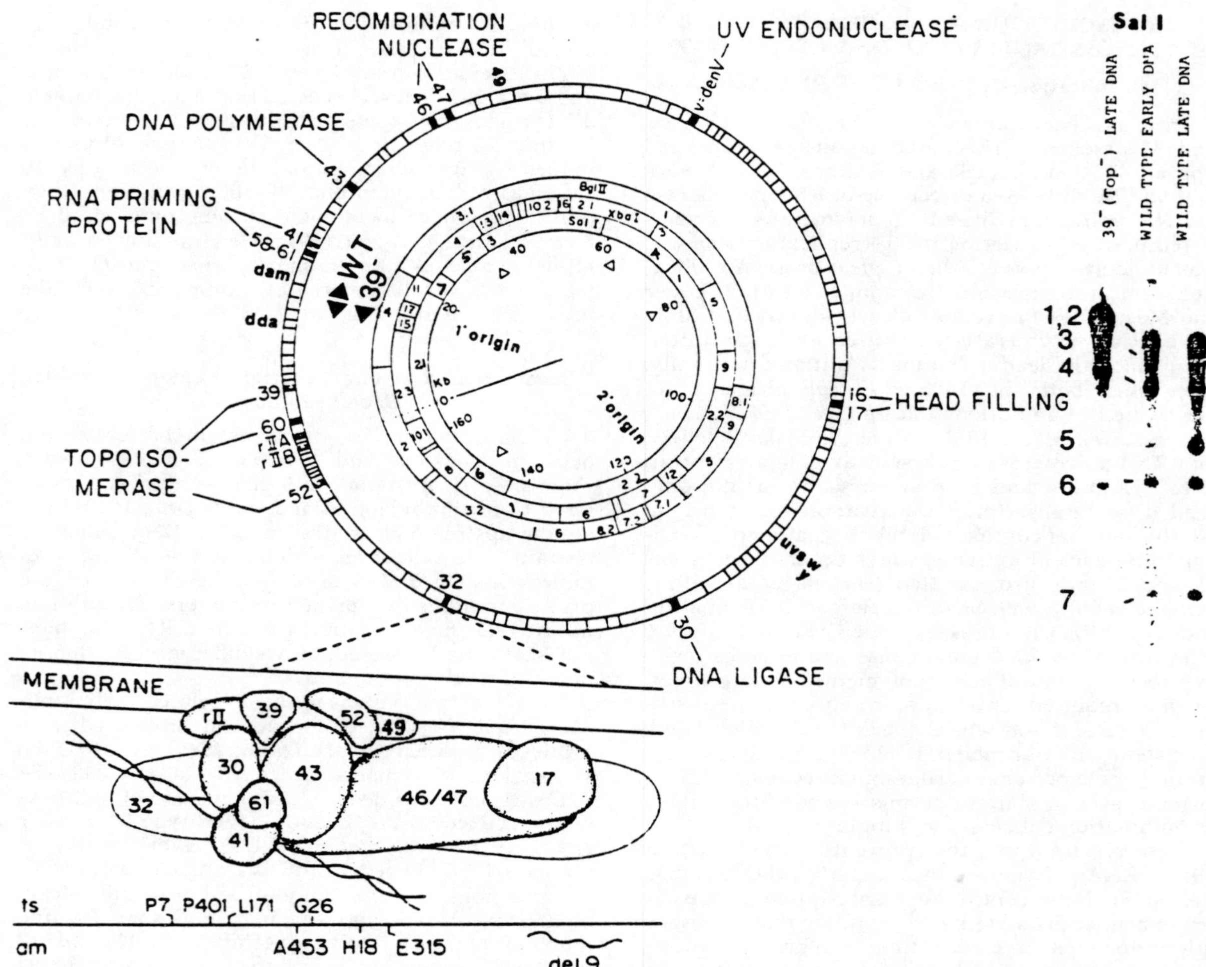

FIG. 2. T4 map showing relative positions of DNA fragments cut by restriction enzymes *Sal*I, *Xba*I, and *Bgl*I (Kutter and Rüger, this volume), the positions of the two origins that we found (stippled areas), the origins described by Halpern et al. (1979), King and Huang (1982), and Morris and Bittner (personal communication) (open triangles in the inner circle), and a cartoon of the T4 proteins which interact with the single-stranded DNA-binding protein, gp32 (deduced from genetic studies, summarized by Mosig, Luder, Garcia, et al. [1979] and Mosig, Ghosal, and Bock [1981]; interaction of gp49 with *r*II, Mosig and McAndless, unpublished data). Positions of the gene 32 *ts* and *am* mutations are from Mosig et al. (1977). The representative Southern hybridizations of early vegetative T4, right panel, show that the *Sal*I fragments 7, 5, and 3 are underreplicated in early wild-type DNA (12 min at 25°C) as well as in later DNA (24 min at 25°C) from the gene 39 mutant N116. Hybridization to digests by other restriction fragments confirmed these assignments. The numbering system of the *Sal*I fragments shown here differs slightly from the numbering system of Yee and Marsh (1981) which we have used previously (Macdonald et al., 1983). Two large *Sal*I fragments which comigrate in our agarose gels were named no. 2 and no. 3 by Yee and Marsh (1981) and are renamed here no. 2.1 and no. 2.2, respectively. Consequently, the *Sal*I fragments 6 and 8 of Yee and Marsh are renamed here 5 and 7, respectively.

those of DNA and RNA. Following this reasoning, we have isolated a putative initiator DNA-RNA copolymer from infected *E. coli* by repeated Cs_2SO_4 gradient centrifugation. Fractions of intermediate density (between RNA and single-stranded DNA) were subjected to mild alkaline hydrolysis to remove the RNA, and the DNA moiety was fractionated in high-resolution denaturing gels (Macdonald et al., 1983). Surprisingly, there were several DNA species of precise lengths. The major species was about 80 nucleotides long. The precise length of the nascent initiator suggests that there is a preferred pause or arrest site for DNA synthesis, or that the origin DNA is cleaved by a site-specific endogenous nuclease.

Southern hybridization (Southern, 1975) of this short nascent DNA (after it was eluted from the gel) to both cloned and uncloned T4 DNA restriction fragments (Fig. 2) revealed that it is synthesized from a narrow region spanning the junction between *Xba*I fragments 15 and 17 and the overlapping 680-base pair *Eco*RI fragment (Macdonald et al., 1983), i.e., from the strong origin region mentioned (Fig. 2). We conclude, therefore, that the DNA species is an arrested (or cut) initiator DNA. We suspect that at that site certain accessory proteins may become essential for continuation of the replication fork and that interactions with certain proteins determine the precise length of this nascent DNA.

E. coli RNA polymerase initiates transcription in vitro from linear restriction fragments spanning this region (Fig. 3) near P1 (−10 region at 284–279 and −35 at 307–302) in the same direction as overall early T4 transcription, as well as from other promoters in that region (Macdonald et al., 1983; P. M. Macdonald, unpublished data). This is the direction of the earliest wild-type DNA replication at 25°C and the major direction of replication in the T4 topoisomerase mutants (see above). Since P1 lies upstream from the initiator sequence (boxed area in Fig. 3), it is reasonable to assume that it could serve as a primer promoter for leading-strand DNA synthesis in that direction. Upstream from P1 there is a palindrome followed by four AT pairs. This sequence might terminate transcripts originating upstream, e.g., within *Bgl*II fragment 4 (Niggemann et al., 1981), allowing for separate control of transcription in the origin region. The short initiator DNA was originally isolated from Cs_2SO_4 gradients at a density that is consistent with its association with RNA initiated from promoter P1 (in Fig. 3), but we do not yet know the termination site of the P1 transcript nor the exact position of the putative RNA-DNA junction. The hybridization pattern of the arrested initiator DNA (Fig. 3) suggests that the RNA-DNA transition occurs at or near the position of a second promoter-like sequence, P2. Surprisingly, this sequence, though it also resembles *E. coli* promoter sequences, points in the direction opposite to the direction of early transcription. The function of the P2 sequence is unknown. We consider, however, four possibilities (which are not mutually exclusive), as follows.

(i) Binding of RNA polymerase to P2 may terminate opposing transcription from another promoter (e.g., P1) and thus facilitate transition from RNA to DNA synthesis in its vicinity.

(ii) P2 may be used as a primer promoter to initiate leading-strand synthesis in the clockwise direction. This probability is especially intriguing because the topoisomerase mutants fail to initiate in this direction. If in vivo transcription from P2 required T4 topoisomerase, the unidirectional initiation of the topoisomerase mutants could be explained.

(iii) The primers may be synthesized as primer precursors (Itoh and Tomizawa, 1980) and terminate somewhere downstream from the putative RNA-DNA transition site. Binding of RNA polymerase to P2 may aid in correct processing of the preprimer transcript, in keeping this transcript inside the DNA template, and in binding of replication proteins, thereby allowing transition from RNA to DNA synthesis.

(iv) P2 could promote synthesis of a short RNA species that regulates origin primer processing, analogous to RNA I of ColE1 (Conrad and Campbell, 1979; Itoh and Tomizawa, 1980).

The strong origin between *dda* and *dam* is used predominantly, if not exclusively, when [^{3}H]thymidine or a low level of ^{32}P is used for labeling in vivo, or when the DNA initiated de novo is isolated in a density-shift experiment and then labeled in vitro by nick translation (Mosig et al., 1980; Mosig, Luder, Rowen, et al., 1981). Obviously, this origin could not have been detected by Halpern et al. (1979), since their collection of T4 clones did not include this region. In addition to this strong origin, we found another origin (shaded area near *uvs*Y in Fig. 2) when a high level of ^{32}P was present before and during infection. It is in the same region as the minor origin described by Halpern et al. (1979) and by Morris and Bittner (personal communication). Since we have detected activity of this origin in wild-type T4 after heavy radiation damage inflicted upon the host cells and after the induction of filamentous growth before infection, it is possible that this origin is under SOS control. The origin near *uvs*Y is used almost exclusively in certain viable deletion mutants (Homyk and Weil, 1974) that lack sequences downstream from the primary origin but contain the template sequences for the short initiator DNA described above (Fig. 3). We suspect that in these deletions, one or more genes are missing that are essential for late steps in initiation at the origin near *dam*. It is noteworthy that the origin region near *uvs*Y, like the strong origin described above, is transcribed early, although it is sandwiched between late genes (Takahashi and Saito, 1982b). Interestingly, this region codes for largely nonessential DNA repair and recombination functions (see Bernstein and Wallace, this volume; Drake and Ripley, this volume). Perhaps its genes and its origin were originally transposed from other replicons.

Other origins which we have not detected under our conditions of infection (25°C) have been found by other investigators (Halpern et al., 1979; King and Huang, 1982; Morris and Bittner, personal communication). Most of these origins are located in early regions, but one, near gene 5, appears to be in a late region. It remains to be seen, however, whether this origin region really contains only late promoters and, in fact, whether all described origins initiate replication de novo. Morris and Bittner (personal communication) have shown that different origins are activated under different growth conditions. Preferred sites for recombinational initiation may appear as "partial origins" (Mosig, Luder, Rowen, et al., 1981), in particular if they contain preferred binding sites for T4 topoisomerase (McCarthy et al., 1976; Kreuzer and Huang, this volume) and perhaps also for accessory replication proteins (Macdonald et al., 1983; Nossal and Alberts, this volume). In this context it is noteworthy that the unusually recombinogenic region spanning genes 34 and 35 (Mosig, 1966; Mosig, 1968), is preferentially labeled in gene 44 mutants, i.e., when true replication does not occur (Kozinski and Ling, 1982).

When recombination could not occur, most, if not all, replicating T4 DNA molecules showed only a single replication loop in the electron microscope early after infection (Dannenberg and Mosig, 1981; Dannenberg and Mosig, 1983). This argues strongly that in T4, as in other systems with multiple origins (e.g., phage T7 [Tabor et al., 1981] or R factors [Stalker et al., 1981]), most individual chromosomes use only one origin to initiate DNA replication de novo.

Interrelationship and Timing of DNA Replication and Recombination

The pioneering work of Hershey and Rotman (1949), who analyzed exchanges between differentially

1

```
-70         -60        -50        -40        -30        -20        -10
TCGAATAAAAC TTTTTCATTA CTTCAGATTC CCGAGCTTTT CTTTCTGCCC TATTTTGCCC TATACGTTGT
AGCTTATTTTG AAAAAGTAAT GAAGTCTAAG GGCTCGAAAA GAAAGACGGG ATAAAACGGG ATATGCAACA
TaqI
```

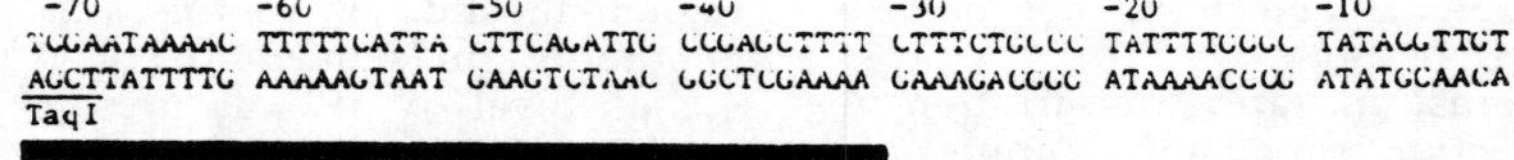

del 12 endpoint (+/-10bp)

```
        10         20         30         40         50         60         70
GAATTCTTCA TCCGAGTTTT TGTTTCAGGA TTGTTTAATC TAAGTTTATG TTCTAATCGT TGTTGCTCAG
CTTAAGAAGT AGGCTCAAAA ACAAAGTCCT AACAAATTAG ATTCAAATAC AAGATTAGCA ACAACGAGTC
EcoRI

        80         90        100        110        120        130        140
TCCATTTTCT TCCTTCTCCA CCCTGACCAC CAGGAGAAAT ATTAATGCAA GTATCTGGGT GTTTACGTTT
AGGTAAAAGA AGGAAGAGGT GGGACTGGTG GTCCTCTTTA TAATTACGTT CATAGACCCA CAAATGCAAA

       150        160        170        180        190        200        210
TAATGCAGAT ATTAGCTCAC GTTCAACTTC ATATGATTTT TCTCTAGAAC CATGCCATTT TGACCATCGT
ATTACGTCTA TAATCGAGTG CAAGTTGAAG TATACTAAAA AGAGATCTTG GTACGGTAAA ACTGGTAGCA
                                                XbaI     -35(P2)

       220        230        240        250        260        270        280
ATTTTATAAT TAAATCCATA CTTACGATAT ATGTTCCATA GTATTTTACC TGAACCCGGA TATTTATCAT
TAAAATATTA ATTTAGGTAT GAATGCTATA TACAAGGTAT CATAAAATGG ACTTGGGCCT ATAAATAGTA
  -10(P2)

       290        300        310        320        330        340        350
TATATGGATT TACAATAAAT GATTCATGTT TCCCAGCGTA CCAAAAGACG CCTTTCGGCG TCCTAACTTT
ATATACCTAA ATGTTATTTA CTAAGTACAA AGGGTCGCAT GGTTTTCTGC GGAAAGCCGC AGGATTGAAA
-10(P1)                 -35(P1)

       360        370        380        390        400        410        420
TACTATATAT GTTCTATAAA ATTTTTGTTT AATATCCATT ATCTTGACGT TCAAAATTAT GTTTGTTCTT
ATGATATATA CAAGATATTT TAAAAACAAA TTATAGGTAA TAGAACTGCA AGTTTTAATA CAAACAAGAA

       430        440        450        460        470        480        490
CAGATAATAA AGTTTAAAGA TTTCTTCCGC ATTCATTCCA AGCCCAACAA ACATATTTAA TACGAAATGA
GTCTATTATT TCAAATTTCT AAAGAAGGCG TAAGTAAGGT TCGGGTTGTT TGTATAAATT ATGCTTTACT

       500        510        520
AATATATCCA CAAGCTCAAA TTTAATTTCG A
TTATATAGGT GTTCGAGTTT AAATTAAAGC T
                             TaqI
```

2

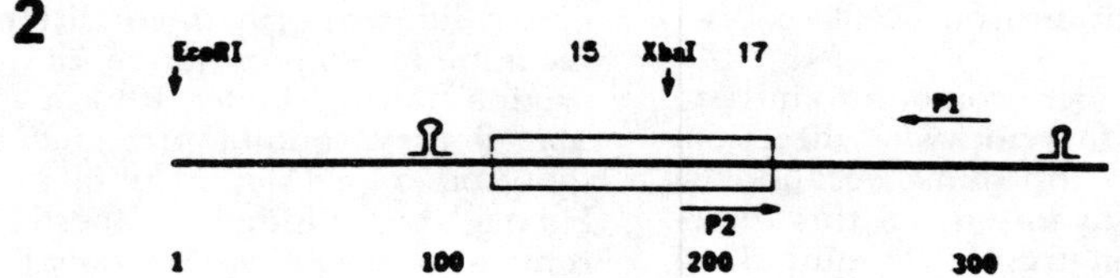

3

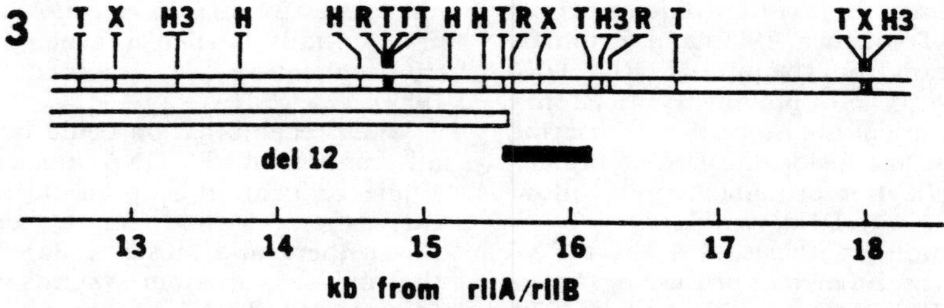

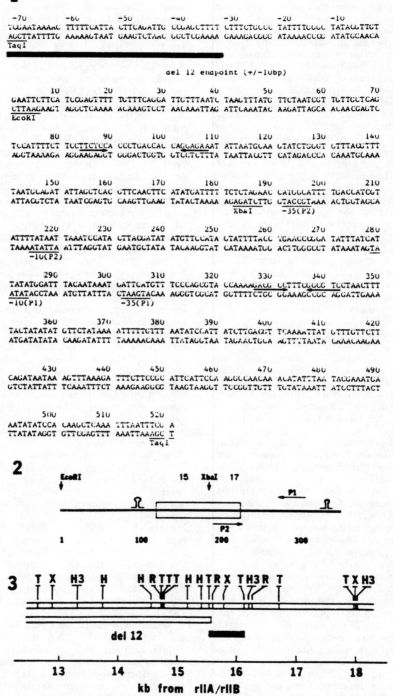

FIG. 3. Fine structure of a T4 origin region. (Panel 1) DNA sequence of the 587-base pair *Taq*I restriction fragment spanning the *Xba*I 15–17 junction (see panel 3). The G of the *Eco*RI restriction site to the right of the *Taq*I site is taken to be nucleotide 1. (Panel 2) Landmarks in a portion of that sequence; the open boxed area is the region to which the arrested initiator DNA hybridizes. Promoter-like sequences and pronounced palindromes are also indicated. The endpoint of the deletion del(39-56)-12 (Homyk and Weil, 1974) was determined by hybridizing the *Taq*I fragment shown in panel 1, from a nondeletion DNA, with a *Taq*I fragment from DNA of the deletion mutant and subsequent S1 mapping. (Panel 3) Map of restriction sites in the region surrounding this origin. The solid bar corresponds to the sequence shown in panel 1. T, *Taq*I; X, *Xba*I; H3, *Hin*dIII; H, *Hae*II; R, *Eco*RI. Data are from Macdonald et al. (1983) and Macdonald and Mosig (manuscript in preparation). The position of the rightmost *Taq*I site in panel 3 is only approximate; there probably is an additional *Taq*I site in that region.

marked T-even genomes, established that T-even chromosomes recombine with high frequency and that limited regions of "hybrid" or "heteroduplex" overlaps are intermediates in this process (Hershey, 1958). This work paved the way for understanding the mechanisms of recombination in terms of interactions between homologous DNA molecules and the proteins that mediate these interactions.

Since the work on T-even recombination has been well summarized (Broker and Doermann, 1975; Hershey, 1958; Miller, 1975a; Mosig, 1970b; Stahl, 1979b; Symonds, 1976), experiments and conclusions discussed in the literature will not be repeated here except for those results which were not fully compatible with the models discussed then and which led to the formulation of the model shown in Fig. 1.

It is well established that during a lytic growth cycle the genomes of the infecting T4 chromosomes recombine by breakage and joining with homologous segments of different parentage and that their DNA is thereby distributed over several progeny particles (see Kozinski, this volume). Electron micrographs in the pioneering studies of Broker and Lehman (1971) and Broker (1973) clearly demonstrated that recombinational intermediates can be formed without prior or concomitant DNA replication and that recombination generates branched intermediates which can be initiated by pairing of complementary single-stranded regions or by invasions of single-stranded segments into duplex DNA. Once they are initiated, heteroduplex regions are expanded by branch migration. Since it seemed impossible to distinguish replicating from recombining DNA, Broker (Broker, 1973; Broker and Lehman, 1971) chose to investigate recombining T4 DNA in mutants which were defective in T4 DNA polymerase (gene 43) and thus were blocked in replication. Under these conditions, pairing between differentially labeled parental chromosomes occurs only late after infection. By this time a large proportion of the infecting molecules have been partially degraded and fragmented by the combined action of endo- and exonucleases (Anraku and Tomizawa, 1965). Recombinational intermediates accumulate even more when the T4 gene 43 mutations are combined with additional mutations in other genes, all of which primarily enhance formation of single-stranded nicks and gaps. The conversion of these base-paired recombinational intermediates (so-called "joint molecules") to covalently linked recombinants requires DNA polymerase(s), i.e., T4 DNA polymerase (Miller, 1975b) or *E. coli* DNA polymerase I (Anraku and Tomizawa, 1965; Mosig, 1974). Thus, it appears that under replication-deficient conditions, there is a large contribution of fragmentation and branch migration to recombination.

On the other hand, certain aspects of the early recombination analyses, particularly non-reciprocity of exchanges (Hershey and Rotman, 1949) and the segregation of alleles in terminal regions of T4 chromosomes (Doermann and Boehner, 1963; Mosig, 1963; Womack, 1963), indicated that the outcome of recombination was intimately related to DNA replication. For this reason, initially some kind of copy-choice mechanism had been suggested. Of course, the idea of simple copy-choice recombination became rather unpopular since it seemed to violate the rules of semiconservative DNA replication. However, attempts to explain all aspects of T-even genetic recombination by a simple breakage-reunion mechanism led to serious paradoxes (Hershey, 1958). The model in Fig. 1 resolves these paradoxes.

When replication is permitted, recombinational intermediates appear much earlier after infection, as soon as the first growing points (initiated de novo at origin sequences) have reached an end (Dannenberg and Mosig, 1983; Mosig et al., 1980; Mosig, Dannenberg, Ghosal, et al., 1979; Mosig, Luder, Garcia, et al., 1979). Since T4 chromosomes are circularly permuted (Streisinger et al., 1967) and origins are located at variable distances from the ends, recombinational initiation from more or less randomly located recombinational intermediates occurs within a minute or less after the onset of replication (Dannenberg and Mosig, 1983). Thus, the preferential replication of the strong origin region near *dam* can be detected only within a short time, unless recombinational initiation is inhibited or delayed by certain mutations (Dannenberg and Mosig, 1981; Dannenberg and Mosig, 1983; Macdonald et al., 1983; Mosig, Luder, Rowen, et al., 1981). While some of the recombinational forks formed under replication-proficient conditions resemble recombinational forks in the unreplicated intermediates of Broker (1973) and Broker and Lehman (1971), several differences are striking: (i) at early times after infection no recombinational forks are seen at the density position of unreplicated DNA; and (ii) at this time, few, if any, of the branched molecules can unambiguously be identified as "H-structures"; instead, most of the forks show displaced single-stranded "whiskers" but no gaps, like a minority class of unreplicated recombinational intermediates shown by Broker and Lehman (1971). Most striking is the observation that the vast majority of single-stranded chromosome tips are actually caught in the process of invading homologous duplexes (Dannenberg and Mosig, 1983). It is clear from these and other results that replication greatly enhances the competence of DNA for recombination, primarily because replication generates single-stranded termini (see Fig. 1), but that a limited extent of replication is sufficient and no single-stranded gaps or nicks are required in the recipients that are being invaded. The immediate recombination of superinfecting T4 chromosomes (Hutchinson et al., 1978) is readily explained, if they are invaded by tips of the resident replicating T4 chromosomes.

The particular role of single-stranded termini is most obvious in experiments with gene 46 mutants: when analyzed under replication-deficient (43^-) conditions, gene 46 (recombination nuclease-deficient) mutants do not form branched recombinational intermediates (Broker, 1973). In contrast, replication proficient gene 46 mutants do form branched recombinational intermediates, whose displaced single-stranded whiskers are longer than the whiskers found in vegetative 46^- DNA (A. Luder, M. Blankenship, and G. Mosig, manuscript in preparation). This is predicted by the model in Fig. 1 and shows that nucleolytic degradation of one strand is not required to form recombinogenic single-stranded regions, when replication can generate single-stranded tips. In addition, these results indicate that the gene 46^- mutants are blocked in recombination in a step subsequent to the formation of branched recombinational intermediates. This suggests that their continuous replication can be rescued not only by permitting continuous origin initiation (e.g., by 33^- or 55^- mutations, as discussed above), but also by mutations which bypass or eliminate the block to initiation of DNA replication from recombinational intermediates. Mutations in genes *das*, *w*, and *dar*, all of which restore DNA synthesis and some progeny production in recombination-deficient infections (Cunningham and Berger, 1978; Hamlett and Berger, 1975; Mickelson and Wiberg, 1981; Wu and Yeh, 1975; Ebisuzaki, personal communication), probably suppress the 46^- mutations by the second mechanism.

Recombination Pattern Influences Replication of Alleles

In both the join-break and the join-copy pathways (Fig. 1), patch or splice heteroduplexes can be formed from the same type of recombinational intermediate, depending on the position of the ultimate resolving cut. Thus, the high recombination potential of chromosomal ends and the occurrence of both patch and splice recombination in the terminal regions seem to be equally compatible with two hypotheses, as follows. (i) Repeated replication of linear infecting chromosomes is followed by independent recombination between the replicated copies (Broker and Doermann, 1975; Doermann, 1973; Halpern et al., 1982; Stahl, 1979b). (ii) According to the join-copy pathway of Fig. 1, recombinogenic chromosomal tips end up in the interior of the concatemers formed by recombination (and thus are no longer recombinogenic). However, since these first recombinants are preferentially replicated, they contribute disproportionately to total recombination frequencies.

I now want to emphasize that certain aspects of recombination in chromosomal tips (Mosig, 1963; Mosig et al., 1971) are inconsistent with the idea of extensive repeated replication and subsequent independent recombination. In fact, these results first suggested to us alternative (ii) above. Note that this model involves recombination between partially replicated, and not between unreplicated chromosomes (as stated, e.g., by Halpern et al., 1982). Nonetheless it involves parental DNA strands. The model shown in Fig. 1 predicts that, if the first (unreplicated) chromosome tip invades another molecule and join-copy recombination resolves the recombinational intermediate into a splice (Fig. 1, panel 3, VII), subsequent replication produces predominantly single exchange recombinants and, therefore, positive interference. In contrast, if the recombinational intermediate is resolved to a patch heteroduplex (Fig. 1, panel 3, VIII), subsequent replication produces predominantly double exchange recombinants which appear with higher frequency than expected from the frequency of single exchanges (negative interference).

In 19-factor crosses between single complete and incomplete T4 chromosomes (analyzed in single bursts) (Mosig et al., 1971), the ends of individual infecting chromosomes could be identified genetically, and thus recombination of terminal and subterminal markers could be analyzed for patch or splice recombination (which we had called at that time insertion-type or crossover-type, respectively) as well as for the frequency with which a certain allele was replicated. This analysis revealed a striking correlation: terminal markers that participate in patch recombination (negative interference) consistently (18 of 20 ends) appear among the progeny with higher (13) or equal (5) frequency than adjacent subterminal markers from the same parental chromosome. In contrast, almost all terminal markers (44 of 45 ends) that participate in splice recombination (positive interference) appear among the progeny with lower frequency than adjacent subterminal markers from the same parental chromosome. The pattern among the patch recombinants (which was not discussed by Broker and Doermann [1975] or Stahl [1979b]) is inconsistent with the postulate that extensive repeated replication of linear chromosomes is responsible for the high recombination potential of their ends: both the underreplication of the terminal segments (Watson, 1972) and the preferential elimination of terminal markers by recombination (Doermann and Boehner, 1963) should reduce the overall frequency of terminal markers. In contrast, the observed pattern is predicted by the join-copy pathway in Fig. 1, if patch recombination occurs and if initiation at origins is rare as compared with initiation from recombinational intermediates. Even if the terminal marker is underreplicated during the first round, if inserted as a patch it becomes the first marker to initiate a new replication fork and therefore is replicated equally or more frequently than its original neighbors. If resolved into a splice, the initial underreplication of the terminal marker is not compensated by subsequent replications. (Fig. 1).

Neither the repeated replication and subsequent recombination postulated by Broker and Doermann (1975) nor the extensive repeated replication from origins and subsequent late initiation from recombinational intermediates postulated by Halpern et al. (1982) can adequately account for these results, since independent recombination among the many copies should obliterate the pronounced difference between patch and splice recombination and the striking correlation with the segregation patterns mentioned above. Womack (1963) had observed similar segregation patterns in multiply marked terminal redundancy heterozygotes. While those patterns are readily compatible with the model in Fig. 1, they have been explained by the (unsubstantiated) assumption that recombination between the terminal redundancies occurs before replication. In addition, interpretations of her results and those of Doermann and Parma (1967) probably are complicated by heteroduplex repair (Berger and Pardoll, 1976; Shcherbakov et al., 1982).

Interactions and Competitions Between Replication, Recombination, and DNA Packaging Proteins

The central structures in recombinational initiation are branched recombinational intermediates (Fig. 4) whose displaced single-stranded segments are mostly covered with gene 32 protein. Other processes in T4's life cycle must compete with this initiation (Fig. 1). Since gene 32 protein interacts both with DNA and with other proteins which act in different processes,

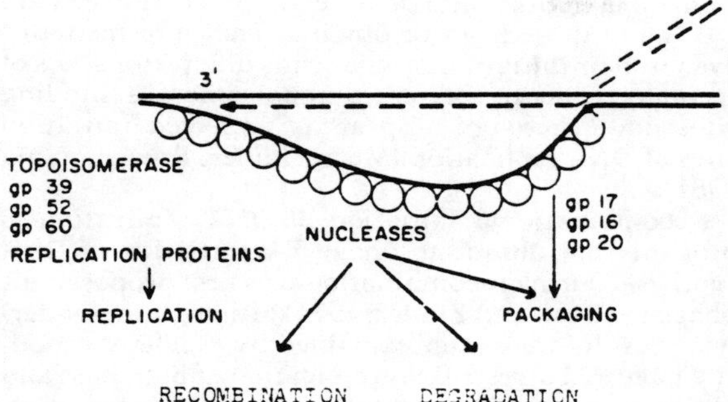

FIG. 4. Recombinational intermediate whose displaced single-stranded segment is covered with gene 32 proteins. We propose that interactions of this structure with different proteins have pivotal regulatory roles in the decision whether to initiate new replication forks, to resolve by join-break recombination, to package the DNA, or to degrade it.

we propose that these interactions have pivotal roles in the decision whether an intermediate, once it is formed, is converted to a replication fork (Fig. 1, panel 3), becomes a "true" (but still heteroduplex) recombinant (Fig. 1, panel 2), is used to initiate packaging, or is degraded (Mosig, Ghosal, and Bock, 1981; Mosig, Luder, Rowen, et al., 1981).

To detect proteins which might interact with or substitute for gp32 in vivo, we and others have isolated spontaneous "secondary" mutations which alter the phenotype of "primary" gene 32 mutations (under semipermissive conditions). In addition we have analyzed various aspects of DNA metabolism, progeny production, and heterozygosity in the single and double mutants under permissive, semipermissive, and restrictive conditions. These results show that different mutations differentially affect different processes, by affecting different interactions (for the rationale and a summary, see Karam et al., this volume). Most of the single "secondary" mutations, when separated from the "primary" gene 32 mutations, cause a *ts* phenotype and, as expected, map in known DNA genes. Most of them affect simultaneously recombination and initiation of replication forks, suggesting that they affect recombinational initiation (Breschkin and Mosig, 1977a; Breschkin and Mosig, 1977b; Mosig et al., 1980; Mosig, Dannenberg, Ghosal, et al., 1979; Mosig, Luder, Garcia, et al., 1979; Mosig, Luder, Rowen, et al., 1981) probably by altering protein interactions involved in recombinational initiation. In contrast, several other secondary suppressor mutations do not affect DNA synthesis. Several of these mutations map in the DNA packaging gene 17 (Mosig et al., 1980; Mosig, Ghosal, and Bock, 1981).

We believe that specific gene 17 mutations are selected in response to certain gene 32 mutations under conditions when DNA replication is normal but DNA maturation is partially defective. The fact that all of the compensating mutations have been found in gene 17 and none in gene 16 or 20 (whose products interact with gp17 [Hsiao and Black, 1977]) suggests that gp17 specifically interacts with gp32 (Mosig et al., 1980; Mosig, Ghosal, and Bock, 1981). The gene 17 mutations enhance replication of topoisomerase mutants. This suggests that packaging and formation of replication forks compete for the same substrate: displaced single-stranded regions covered with gene 32 protein (Fig. 4). The allele-specific interactions that we and others have so far detected are summarized by Karam et al. (this volume) and in Fig. 2 (see also Williams and Konigsberg, this volume). Recently, many of these interactions have been confirmed by biochemical studies (Nossal and Alberts, this volume; Hosoda, personal communication). These results show that the two major domains of gp32 are involved in different interactions. The N-terminal domain is mainly involved in binding to DNA and to proteins involved in recombinational initiation, whereas the C-terminal domain has other functions, e.g., interactions with nucleases. In addition, the N-terminal segment that is accessible to proteases affects cooperativity, and a C-terminal tail affects binding to single-versus double-stranded DNA.

It is remarkable that most known gene 32 *ts* mutations, though they affect interactions with different proteins, map in the promoter-proximal third of the gene (Fig. 2 of Mosig et al. [1977]). In fact, they alter the region which is also involved in binding to DNA (Krisch et al., 1980; Williams et al., 1981). A subsequent extensive mutant hunt yielded additional mutants mainly in this region (Doherty et al., 1982). This crowding of essential mutational sites in the gene and interaction sites on the protein implies that different interactions in the N-terminal domain are not mediated by different interaction sites, but that many of the potential interactions involve subtle allosteric modifications mediated by binding to DNA and to other proteins and that they are competitive. Competition between proteins of different pathways for the same substrate (e.g., between topoisomerases and gp17, see Fig. 4) provides an easy way to channel the same intermediate into different pathways of DNA metabolism during different times in the life cycle of T4. Such subtle allosteric changes in response to different interactions are exquisitely suited to coordinate the precise functioning, in time and space, of the many enzymes that have to act sequentially during initiation of replication forks even though their genes are not clustered (Fig. 2).

It is noteworthy that most of the DNA-protein-protein interactions that are detected by these kinds of studies (Fig. 2) are of primary importance in recombinational initiation and not in origin initiation or DNA elongation. It is tempting to speculate that this is related to the two modes of control mentioned in the Introduction: the specificity of recognition between different proteins is expected to be less stringent when T4-coded proteins have to interact with host proteins, e.g., RNA polymerase, during origin initiation, than when they interact mainly with other T4 proteins during recombinational initiation. In this context it is important to note that several host replication proteins (*dna*G, *dna*C, *pol*A [Breschkin and Mosig, 1977b; Mosig, Bowden, and Bock, 1972], and *ssb* protein [Mosig, unpublished data]) can substitute for certain defective or missing T4 replication proteins in primary replication (see also Murray and Mathews, 1969), but they cannot sustain T4 replication when recombinational initiation becomes essential.

The roles of most of these proteins in recombination

and in replication have been extensively discussed (Broker and Doermann, 1975; Miller, 1975a; Mosig, Dannenberg, Ghosal, et al., 1979). The role of T4 topoisomerase requires some additional comments. T4 topoisomerase mutants (Liu et al., 1980; Stetler et al., 1979), which had been classified as so-called DNA delay mutants (Yegian et al., 1971), are defective in initiation of replication forks (McCarthy et al., 1976; Kreuzer and Huang, this volume) for two apparent reasons: (i) at origin(s) they initiate only unidirectional replication (Macdonald et al., 1983; see Fig. 2 above), and (ii) they are delayed in recombinational initiation (Mosig et al., 1980; Mosig, Luder, Garcia, et al., 1979; Mosig, Luder, Rowen, et al., 1981). These mutants are not recombination defective (Cunningham and Berger, 1978; Hamlett and Berger, 1975; Leung et al., 1975; Mufti and Bernstein, 1974), which suggests either that the recombinational intermediates are formed but that they are mainly resolved by join-break recombination, or that those molecules that recombined were preferentially packaged. As discussed by Bernstein and Wallace (this volume), the T4 topoisomerase mutants are also defective in multiplicity reactivation, a process which most likely depends on recombinational initiation involving the partial replicas generated from the damaged chromosomes (Barricelli, 1960; Barricelli and Doermann, 1961; Rayssiguier et al., 1980). We consider it likely that recombinational initiation of replication is part of the process of multiplicity reactivation and that the presence of topoisomerase would facilitate the conversion of recombinational intermediates to replication forks (Dannenberg and Mosig, 1983), a step which is defective in the mutants.

CONCLUSIONS

It has been postulated by others that T4 DNA replication occurs in two different modes: an early mode, in which there is no recombination and many unit-length daughter molecules are generated, and a late mode, which coincides with recombination and with the formation of long concatemers (Broker and Doermann, 1975; Halpern et al., 1982; Kozinski et al., 1976; Kozinski et al., 1967; Stahl, 1979b). Our earlier results (Mosig et al., 1971) argue against a defined switch from an early to a late mode and against the arguments that repeated replication of linear T4 chromosomes is required before recombination. Instead, all kinds of recombinational intermediates, postulated in our model are seen within 2 min after the onset of replication (Dannenberg and Mosig, 1983). Since recombinational forks can be formed as soon as the first growing points have reached an end and can then be converted to replication forks, no clear-cut distinction can be made between replication forks initiated from bona fide origins (Delius et al., 1971; Howe et al., 1973) and those from recombinational intermediates even if their initiation mechanisms differ. Hot spots of recombination or preferred topoisomerase binding sites may, therefore, appear as preferred initiation sites of DNA replication (Mosig, Luder, Rowen, et al., 1981).

Recombinational initiation of DNA replication is probably not unique to phage T4: initiation of DNA synthesis during recombination was first proposed for phage f1 (Boon and Zinder, 1971) although the underlying results were later explained by a different model. Phage P22 arrests DNA replication, if both host and phage recombination systems are inactivated (Botstein and Matz, 1970). Phages T7 (Powling and Knippers, 1976) and lambda (Skalka, 1974; Stahl, 1979b) initiate some DNA synthesis from recombinational intermediates, although their growth does not depend on it. Introducing recombinational hot spots ("Chi sites"; Stahl, 1979a) into lambda results in marked stimulation of DNA synthesis (Henderson and Weil, 1975). Transposition of transposable elements, including phage Mu, is thought to involve DNA synthesis initiated from an intermediate in this illegitimate recombination (Galas and Chandler, 1981; Grindley and Sherratt, 1979; Harshey et al., 1982; Shapiro, 1979) and is, in turn, stimulated by replication of the chromosome (Teifel and Schmieger, 1981). The recently described DNA intermediates in retrovirus replication (Junghans et al., 1982) resemble T4 DNA intermediates described by Dannenberg and Mosig (1983). The *recA*-dependent SOS replication of *E. coli* (Kogoma et al., 1981; Lark et al., 1981) does not depend on *ori* C (Torrey and Kogoma, 1982) and may depend on initiation from recombinational intermediates (Kogoma et al., 1981). Similarly, integration of donor DNA during transformation in *Bacillus subtilis* has been proposed to generate new replication forks (Laird, Ph.D. thesis, Stanford University, Stanford, Calif., 1967).

Recombinational initiation of replication forks is ideally suited to rapidly accelerate overall DNA synthesis in virus-infected cells (or in damaged cells) and to escape the cellular control mechanisms designed to coordinate cell division and DNA replication. Phage T4 depends on this initiation mode because the RNA polymerase-dependent priming from origin promoters is turned off as a consequence of the general developmental program (Luder and Mosig, 1982; Rabussay, 1982b; Rabussay, 1982c).

I thank Paul Macdonald, Andreas Luder, and Richard Dannenberg for stimulating discussions, and Paul Macdonald, Chris Mathews, Helen Revel, and Betty McFall for critical readings of this manuscript.

Research in our laboratory was supported by Public Health Service grant GM-13221 from the National Institutes of Health.

Fidelity of DNA Replication

NAVIN K. SINHA[1] AND MYRON F. GOODMAN[2]

Waksman Institute of Microbiology, Rutgers University, Piscataway, New Jersey 08854,[1] *and Department of Biological Sciences, Molecular Biology Section, University of Southern California, University Park, MC-1481, Los Angeles, California 90089*[2]

The formal study of biological organisms through the observation of genetic change as well as the evolution of the organisms themselves is dependent on the fidelity of replicating and translating genetic information. The biochemical means by which a cell is able to duplicate itself and all its functions with exceptionally high fidelity has only recently begun to be explored on a molecular level. It is well to remember that when one carries on mutation studies in bacteria, viruses, or animal cells, the normal mutation frequencies encountered are on the order of 10^{-7} to 10^{-9} per base pair per round of replication. When mutagens or strong mutator alleles are employed to achieve increased mutagenesis, the highest mutation frequency compatible with viability in haploid organisms is about 10^{-4} to 10^{-5}. Hence even under highly mutagenic conditions, biochemical events involved in replicating DNA are accurate to from 1 part in 10^4 to 1 part in 10^5. The subject of this chapter is the fidelity of DNA replication, and we will attempt to include in our discussion many of the current questions related to the biochemical basis of mutagenesis.

A molecular basis to explain spontaneous mutations was first proposed by Watson and Crick (1953a, 1953b) in their classic papers on the structure of DNA. Their proposal allowed that the four common nucleotides, if present in solution in unfavored tautomeric forms, could form stable hydrogen bonds with incorrect partners. If, for example, the imino form of adenine were present at the replication site on the DNA polymerase, then it could form two hydrogen bonds with a template cytosine, resulting in an A·C mispair. Failure to remove the misinserted adenine deoxynucleotide would, upon an additional round of DNA replication, result in an A·T base pair being located at the original G·C site. A·T → G·C and G·C → A·T mutational transitions involving all possible incorrect purine-pyrimidine partners could, in principle, be accounted for by increasing the base-pairing possibilities via keto-enol (for Gua and Thy) and amino-imino (for Ade and Cyt) tautomeric shifts. As proposed by Topal and Fresco (1976), transversion mutations in which purine-purine but not pyrimidine-pyrimidine base pairs occur could also be accommodated in Watson-Crick base-pairing conformations, provided that the deoxyribose rings exist in the double helix in a *syn* conformation.

As attractive in their logical simplicity as the tautomer models for spontaneous mutagenesis are, several problems have interfered with their verification. An observation of rare tautomeric forms of the natural bases in solution has not been made because their expected equilibrium frequencies, ca. 10^{-5} (Katritzky and Waring, 1962; Wolfenden, 1969), are too small to be measured directly. However, a second problem also arises on inferential grounds because these hypothetical rare tautomer magnitudes are themselves three to five orders of magnitude too large to account directly for spontaneous mutation frequencies.

There is currently a substantial and growing body of evidence in procaryotic systems that error correction processes play a prominent role in reducing mutations in DNA far below the frequencies at which heteroduplex base mispairs are formed. These editing processes include DNA polymerase-associated 3′-exonuclease proofreading (Brutlag and Kornberg, 1972; Muzyczka et al., 1972) and DNA methylation-directed postreplication mismatch repair (Glickman and Radman, 1980; Nevers and Spatz, 1975; Wagner and Meselson, 1976). There is as yet no evidence for the presence of either of the these two editing mechanisms in animal systems, and we regard the question of fidelity in higher eucaryotes as a wide-open area. In procaryotes, however, and especially in the T4 system and its host *Escherichia coli*, progress has been made to the point where in vitro model systems have been developed to test individual protein components and complex protein aggregates to determine their role in affecting deoxynucleotide misinsertion and excision frequencies at specific loci on defined DNA templates.

The T4 bacteriophage system has been particularly valuable in the study of fidelity by permitting a direct link to be made between genetic studies on mutations in vivo and biochemical studies using purified protein components in vitro. Thus, the early "mutator-antimutator" genetic studies (Drake and Allen, 1968; Drake et al., 1969; Freese, 1959a; Freese and Freese, 1967; Speyer, 1965; Speyer et al., 1966), which clearly implied that gene 43 (which codes for T4 DNA polymerase) has a prominent role in controlling mutation rates, were followed by biochemical studies (Bessman et al., 1974; Gillin and Nossal, 1976a; Gillin and Nossal, 1976b; Hershfield, 1973; Lo and Bessman, 1976; Muzyczka et al., 1972; Nossal and Hershfield, 1973; Reha-Krantz and Bessman, 1977) in which the excision and insertion activities of the mutant and wild-type DNA polymerases were related logically to the phenotypic behavior of the T4 alleles in vivo. Subsequent work has shown that other proteins in the T4 replication complex also affect mutagenesis to an important, yet significantly lower extent than the polymerase (Bernstein et al., 1972; Mufti, 1979; Watanabe and Goodman, 1978). The stage is now set to relate DNA replication fidelity to measured mutation frequencies at specific sites in the genome as a function of hydrogen bonding, base selection by the polymerase alone and in conjunction with other proteins, controlled deoxyribonucleoside triphosphate (dNTP) pool size perturbations, nearest-neighbor nucleotides and more distant nucleotide sequences, and DNA secondary structure.

METHODS FOR STUDYING FIDELITY OF DNA REPLICATION IN VITRO

Use of Homopolymer Templates

Misincorporation assay. With natural DNA templates, it is difficult to tell when an error in copying has occurred because every nucleotide can be a correct one depending upon what is present in the template strand. This problem can be circumvented by using homopolymer template-primer systems containing only two of the four nucleotides. One can use either homopolymers such as poly(dA)·oligo(dT) or alternating copolymers such as poly(dA·dT). With such templates, errors are detected as the incorporation of the noncomplementary nucleotides (such as C or G in the case of the templates given above). Many workers have made use of such templates (see Loeb and Kunkel [1982] for a recent review).

Using poly(dC)·oligo(dG) as the template primer and measuring the incorporation of dTMP as the incorrect base (relative to dGMP, the correct base), Hall and Lehman (1968) showed that the error rate for wild-type T4 DNA polymerase was about 1/50,000 (at equal concentration of the competing substrates). More recently, an error rate of 1/42,300 for the incorporation of dGMP into a poly(dA·dT) copolymer template by T4 DNA polymerase was obtained by Kunkel et al. (1979). These error rates are near the lower limits of detection in these assays.

There are several problems with this assay. The template and the nucleotide substrate must be scrupulously clean of contaminating nucleotides to avoid artificially high error rates. Since the error rate for T4 DNA polymerase is so low, one has to use large quantities of labeled incorrect nucleotides to detect any misincorporation at all. This is quite expensive, and also it is difficult to prepare high-specific-activity radioactive nucleotides with sufficiently low levels of acid-insoluble matter that rare errors can be detected above the background noise. Furthermore, since these homopolymer templates have a rather simple repeating structure, the possibility exists that the error rates might be higher on these templates because the mispaired residue can easily loop out.

Turnover assay. The sensitivity of the homopolymer assay can be increased by doing a turnover, rather than an incorporation, assay. Because of the presence of a proofreading 3′,5′-exonuclease activity in most procaryotic DNA polymerases, the great majority of the mistakes at the nucleotide insertion step do not survive as stable misincorporations. However, these errors can still be detected as a conversion of an incorrect dNTP substrate to a dNMP.

Using such an assay, Nossal and co-workers have compared the accuracy of the *ts*L88 mutator and CB120 antimutator DNA polymerases to that of the wild-type enzyme (Gillin and Nossal, 1976a; Gillin and Nossal, 1976b; Hershfield, 1973; Hershfield and Nossal, 1972; Nossal and Hershfield, 1973). It was shown that the *ts*L88 enzyme has a normal level of proofreading exonuclease activity but a reduced discrimination between the correct and the incorrect substrates at the nucleotide insertion step. In contrast, the CB120 antimutator polymerase possessed an increased accuracy at the nucleotide selection step as well as an enhanced proofreading exonuclease activity.

The nature of mispairs during these replication errors was not carefully examined in these studies. More recently, Topal et al. (1980) have reported such studies. Each homopolymer template such as poly(dA)·poly(dT) has two primer ends and thus two potential places where an incoming wrong dNTP can be inserted. By using competition with a correct dNTP substrate or blocking one of the two primer ends with a dideoxynucleotide, they were able to show that base-stacking interactions are more important than base-pairing interactions for the insertion of a mismatched deoxynucleotide when T4 DNA polymerase is acting alone. However, the turnover of matched substrate nucleotides appeared to be independent of stacking partner.

The interpretation of these turnover experiments is somewhat complicated because there is little net synthesis during these experiments and, therefore, the possibility exists that the DNA polymerase might not behave the same way as during active DNA synthesis. Other uncertainties concerning the behavior of the replication enzymes on these templates arise because the homopolymers clearly have a structure different from natural DNA. For example, it is not clear what effect the extensive strand slippage known to occur with some homopolymer templates has on the accuracy of the DNA polymerase.

A complementary observation, made by using DNA-synthesizing conditions, implied that hydrogen-bonding forces appeared to play a dominant role over base-stacking forces in controlling nucleotide insertion rates when H-bonding was possible (Watanabe and Goodman, 1981). As the stability of H-bonding increased for various base pairs, the effects of different base-stacking partners on nucleotide insertion frequencies were eliminated. Since, in general, base-stacking forces tend to be stronger than H-bonding forces, a possible explanation of the data may be that the DNA polymerase imposes a set of steric constraints on substrate-primer-template interactions to maximize the information-bearing substrate-template interaction at the expense of the information-lacking substrate-primer interaction. In the absence of stable H-bonding, e.g., base mispairs, the only remaining forces are base stacking and the weak, nonspecific polymerase-triphosphate site-binding force. In this case, the base-stacking forces appear dominant.

Studies with the Base Analog 2-Aminopurine

2-Aminopurine (AP) has been employed as a potent mutagen in *E. coli* and especially in T4-infected *E. coli* (see, e.g., Ronen [1979]). AP is able to drive transition mutations bidirectionally, A·T ⇆ AP·T ⇆ AP·C ⇆ G·C. The currently accepted explanation for the mutagenic action of AP invokes an ambiguity in the analog's base-pairing properties. As proposed by Freese (1959), in its common amino form AP can form two hydrogen bonds with thymine, whereas in its rare imino form, AP can form two hydrogen bonds with cytosine.

The mispaired intermediates AP·T and AP·C in the two transition mutation pathways have been investigated in vitro (Bessman et al., 1974; Clayton et al., 1979; Goodman et al., 1983; Watanabe and Goodman, 1981; Watanabe and Goodman, 1982) and in vivo (Goodman, Hopkins, and Gore, 1977; Hopkins and Goodman, 1979; Ripley, 1981a). The formation of each heteroduplex base pair is asymmetric in the following sense. An AP·T base pair in the A·T → G·C pathway is formed by the substrate deoxyaminopurine triphosphate (dAPTP) competing with dATP for insertion opposite a template thymine site; in the G·C → A·T direction, the AP·T base pair is formed with AP present on the template and dTTP competing with dCTP for insertion opposite AP. In the A·T → G·C direction, the AP·C base pair is formed with dCTP competing with dTTP for incorporation opposite a template AP, whereas in the G·C → A·T direction, dAPTP is competing with dGTP for incorporation at a template C site. As discussed below, each of these intermediates can be investigated in vitro and in vivo by using the T4 system.

Looking first at the AP·T base pair intermediate, we note that DNA polymerases purified from T4 mutators L56 and L98, wild-type 43^+, antimutators L141 and L42, *E. coli*, and α polymerase from calf thymus all misinsert dAPTP in competition with equimolar amounts of dATP opposite template thymines at frequencies between 12 and 15% (Bessman et al., 1974; Clayton et al., 1979). This frequency represents a free energy difference of about 1.1 kcal/mol at the polymerase insertion site, favoring the formation of A·T over AP·T base pairs (Clayton et al., 1979; Galas and Branscomb, 1978).

For the T4 DNA polymerase mutator and antimutator alleles containing varying amounts of proofreading 3′-exonuclease activities, the net AP misincorporation frequency (the difference between misinsertion and excision) is consistent with the organism's phenotype. Hence, the strong L56 mutator polymerase incorporates about threefold more AP into DNA than the strong antimutator L141 polymerase. Similar differences in AP misincorporation, comparing L56, 43^+, and L141 alleles, have also been measured in vivo (Goodman, Hopkins, and Gore, 1977). The absolute levels of AP substituted in the phage DNA also reflect the metabolism of AP conversion to dAPTP in T4-infected *E. coli*. On the basis of the studies using purified T4 polymerase to measure AP incorporation opposite T when dAPTP and dATP are present at equimolar concentrations, one can directly deduce that if similar polymerase fidelity mechanisms are operative in vivo and in vitro, then the dAPTP/dATP pool size ratios measured in vivo should be in the range of about 1 to 5% (Goodman, Hopkins, and Gore, 1977). Recent measurements place this ratio at about 1% in 43^+ infected cells (R. Hopkins and M. F. Goodman, manuscript in preparation).

The second intermediate in the AP-induced A·T → G·C pathway, the AP·C mispair, has been measured by two different genetic approaches in vivo (Hopkins and Goodman, 1979; Ripley, 1981a) and also by using DNA polymerases in a cell-free assay to insert dCTP in competition with dTTP opposite a template AP site (Watanabe and Goodman, 1981). The in vivo and in vitro measurements are in excellent agreement, suggesting that AP·C heteroduplex heterozygotes are formed at a frequency of about 5%. That is, the insertion of dTTP is favored by about 20:1 over the insertion of dCTP opposite a template AP.

The in vitro data also provide a quantitative estimate of the mutagenic potential of AP when situated on the DNA template. It was observed that when AP replaces A to the 5′ side of a nearest-neighbor C on a synthetic DNA template, the misinsertion of dCTP at the AP site is increased 35-fold; when AP replaces A next to a nearest-neighbor A, the misinsertion of dCTP is increased by at least 250-fold (Watanabe and Goodman, 1981). These data appear consistent with observed order-of-magnitude effects of AP on mutagenesis frequencies in *E. coli* and T4, where roughly 10- to 1,000-fold enhancements over spontaneous mutation frequencies are commonly observed when using different marker alleles.

It is worth noting that unlike bromodeoxyuridine-induced mutagenesis, in which G·C → A·T transitions are strongly favored over A·T → G·C transitions (see, e.g., Champe and Benzer, 1962b), apart from certain mutagenic hot spots, AP mutagenesis probably does not exhibit a strong pathway preference. While it is true that due to the asymmetry in the formation of the mispaired intermediates discussed above, generation of AP·C base pairs in the G·C → A·T direction is about 1,000-fold less likely than in the A·T → G·C direction, a roughly equal net transition rate is expected since the formation of AP·T base pairs is also a low-probability step in the A·T → G·C pathway (about 10^{-3} for 43^+) because of the low pool size of dAPTP (Goodman, Hopkins, and Gore, 1977). Thus, whereas the very infrequent formation of AP·C base pairs is the only slow step in the G·C → A·T pathway, there appear to be two slow steps occurring in the A·T → G·C direction: the formation of AP·T and AP·C base pairs. It may possibly be coincidental that the product of the AP·T and AP·C generation rates in the A·T → G·C direction is roughly equal to the rate of generating AP·C base pairs in the G·C → A·T direction.

Studies Measuring Reversion of Mutant Codons

The traditional method for studying mutation rates has been to measure the reversion of an easily selectable marker. In recent years several workers have applied this method to the study of fidelity of DNA replication in vitro (Fersht, 1979; Fersht and Knill-Jones, 1981; Hibner and Alberts, 1980; Kunkel and Loeb, 1980; Kunkel et al., 1979; Liu, Burke, Hibner, et al., 1979; Sinha and Haimes, 1980; Sinha and Haimes, 1981; Weymouth and Loeb, 1978). In all such studies so far, the DNA of an amber mutant form of the phage φX174 was duplicated in vitro, and the error rate during this replication was quantitated as reversion at the site of the amber mutation to a viable revertant.

The assay used by Liu, Burke, Hibner, et al. (1979) and by Sinha and Haimes (1980) measures the accuracy of DNA replication in vitro by using a seven-enzyme complex (the products of phage T4 genes 32, 41, 43, 44, 45, 61, and 62). By use of this assay and DNA from several mutants of φX174, it has been shown that the accuracy of this complex is quite comparable to that observed during φX174 replication in vivo. The in vitro replication process is so accurate that no increase in the frequency of revertants over that originally present in the DNA template was seen at any of the amber mutant sites examined. The error rates could, however, be increased by using imbalances in the concentration of the dNTP substrates. It was found that in all cases the frequency of revertants increased linearly with increasing pool bias. Moreover, by the use of appropriate pool imbalance combinations the error rates for several different mispairs could be determined at a single amber site

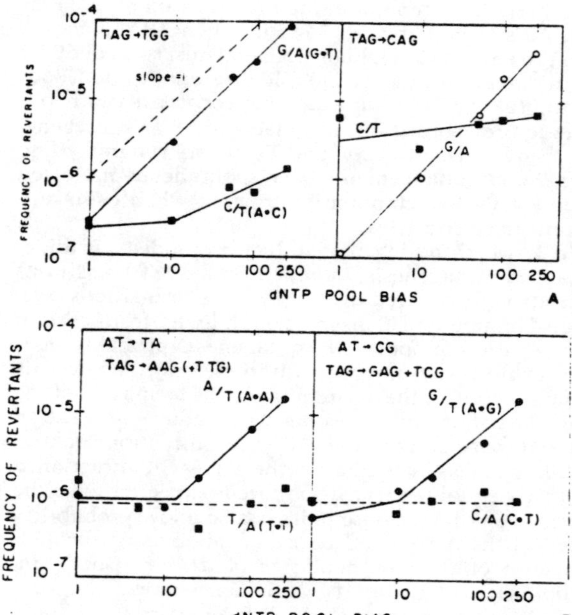

FIG. 1. Induction of mutations at the φX174 *am*16 site by dNTP pool imbalances during DNA replication in vitro. DNA was synthesized on *am*16 double-stranded DNA templates, using purified proteins coded by T4 genes 32, 41, 43, 44, 45, 61, and 62 for 30 min at 37°C. The dNTP substrates not involved in the pool bias were kept at 200 μM. The nucleotide being biased against (such as dATP in a G/A bias) was kept at 20 μM. Increasing pool bias was achieved by raising the concentration of the competing nucleotide (such as dGTP) from 20 μM to 5 mM. After DNA synthesis, the DNA was prepared for infectivity assays as described by Sinha and Haimes (1981). (A) Induction of AT → GC transitions. (B) Induction of AT → TA or AT → CG transversions.

(Fig. 1). Since this seven-enzyme complex synthesizes both the strands of DNA simultaneously, error rates for complementary mispairs leading to the same base pair change at a given nucleotide could be determined.

The results for a number of φX174 amber mutants are summarized in Table 1. Several conclusions concerning the accuracy of phage T4 replication complex can be drawn. The relative order of nucleotide mispairing events is seen to be G·T > (A·C ≅ A·G ≅ A·A) > (C·T ≅ T·T). In other words, AT → GC transitions primarily occur through G·T rather than A·C mispairs. AT → CG and AT → TA transversions occur primarily through purine-purine (A·A or A·G) mispairs. In general, transition mutations are more frequent than transversions at any given site. Error rates for a given mispair are quite different at different sites. In one case (*am*16), it was possible to measure the error rates for the same mispairs (G·T and A·C) leading to AT → GC transitions at adjacent nucleotides (TAG → CAG and TAG → TGG). These are observed to be approximately fivefold different. Clearly, neighboring DNA sequence is an important determinant of error rate in DNA replication.

Using this assay and appropriate mutant forms of φX174 DNA as template, one can readily measure the effect of a variety of perturbations in the conditions of DNA synthesis on accuracy of replication, through both transition and transversion pathways. Variations in temperature (20 to 45°C), pH (from 7 to 9.5), the concentrations of potassium (0 to 150 mM), Mg^{2+} (5 to 30 mM), and all four dNTPs together (from 20 μM to 2 mM each), and the addition of dNMPs (up to 3 mM each), sodium pyrophosphate (up to 3 mM), or spermidine (up to 6 mM) had no effect on the accuracy of the T4 replication apparatus, even though these perturbations altered the rate of DNA synthesis greatly (T. Razzaki and N. Sinha, unpublished data). Only replacement of dTTP by 5-bromodeoxyuridine triphosphate or of Mg^{2+} by Mn^{2+} was found to be mutagenic. No antimutagenic conditions were observed.

The effect of changes in the concentrations of the next correct nucleotide was also examined at a number of φX mutant sites (*am*3, *am*86, *am*18, *am*33, and *am*16). Due to experimental constraints, this could only be varied from 20 μM to 5 mM, a range where the rate of DNA synthesis was altered more than 10-fold. In all cases, the error rate was totally unaffected (N. Sinha, unpublished data).

The main drawback of this assay is that it requires the simultaneous presence of all seven T4 phage replication proteins. Therefore, the relative contribution of each of the components of the replication complex cannot be determined by this method. Loeb and co-workers have described an alternative assay which can measure the accuracy of the DNA polymerase alone and which, in principle at least, can be used to determine the relative contribution of each of the

TABLE 1. Frequencies of mispairs at different sites on φX174 DNA

Reversion pathway	Mispair	*am*33	*am*18	*am*3	*am*86	*am*9	*am*16
AT→ GC transition	G·T	9.8×10^{-6}	2×10^{-6}	5×10^{-6}	7×10^{-7}		3×10^{-7} (TAG→TGG) 6.4×10^{-8} (TAG→CAG)
	A·C	$<1 \times 10^{-6}$	$<2 \times 10^{-7}$	9×10^{-8}	$<7.4 \times 10^{-8}$		4.3×10^{-8} (TAG→TGG) $\leq 1 \times 10^{-8}$ (TAG→CAG)
Transversions							
AT→CG	A·G	8×10^{-6}	4.5×10^{-7}				6×10^{-8}
	C·T	$<1 \times 10^{-6}$	$<1.5 \times 10^{-7}$				$<4 \times 10^{-9}$
AT→TA	A·A	5×10^{-6}	5×10^{-7}				6×10^{-8}
	T·T	$<1 \times 10^{-6}$	$<2 \times 10^{-7}$				$<4 \times 10^{-9}$
AT→CG + AT→TA	A·A+A·G			5.6×10^{-7}	2.9×10^{-7}	4×10^{-8}	
	C·T+ T·T			$<1.2 \times 10^{-7}$	$<1.2 \times 10^{-7}$	$<2.5 \times 10^{-9}$	

components of the replication complex (Weymouth and Loeb, 1978). The assay measures the formation of A·C (or A·G or A·A) mispairs at the *am*3 site of φX174. However, due to a number of limitations (incomplete copying of the template DNA, poor expression of the in vitro product, and attempts to measure relatively rare mispairs), so far it has not been possible to accurately measure even the accuracy of the T4 DNA polymerase alone. Only a value of $<1 \times 10^{-7}$ for A·C mispairs has been reported (Kunkel et al., 1981).

Polymerase Accessory Proteins

A large number of genetic studies have examined the effect of mutations in the phage T4 replication genes on overall mutation rates (Bernstein, 1971; Bernstein et al., 1972; Drake and Allen, 1968; Koch et al., 1976; Mufti, 1979; Speyer et al., 1966; Watanabe and Goodman, 1978; Williams and Drake, 1977). Strong mutator and antimutator effects were observed for some of the mutants in the gene for DNA polymerase. Much smaller mutator effects were seen for many of the other replication genes. Also, some of the genes involved in the synthesis of nucleotide precursors (genes for dCMP hydroxymethylase and thymidylate synthetase) showed mutator effects, presumably because the intracellular nucleotide pools were abnormal.

Role of helix-destabilizing (gene 32) protein. In addition to the genetic evidence mentioned above, biochemical evidence implicating the helix-destabilizing (gene 32) protein in the determination of accuracy of DNA replication in vitro was reported by Gillin and Nossal (1976a, 1976b). However, the exact mechanism of this enhancement in fidelity was not determined. Topal et al. (1980) had observed that stacking interactions tended to dominate over base-pairing interactions in the selection of mismatched deoxynucleotides for insertion at a primer terminus when T4 DNA polymerase was working alone. This led them to propose that gene 32 protein might enhance the accuracy of the DNA polymerase by stretching the backbone in the single-stranded template strand, thus reducing stacking interactions between the incoming nucleotide substrate and the primer terminus. Addition of gene 32 protein of T4 DNA polymerase was found to greatly reduce the error rate. Titration studies suggested that gene 32 protein increased the accuracy of interaction with the DNA template and not through interaction with the DNA polymerase (Topal and Sinha, submitted for publication).

Other polymerase accessory proteins. Genetic evidence in vivo suggests that all the proteins in the replication complex can perturb the accuracy of DNA replication, at least to some degree (Mufti, 1979; Watanabe and Goodman, 1978). When the accuracy of T4 DNA polymerase was measured by the nucleotide turnover assay using homopolymer primer templates, only gene 32 and 45 proteins had any effect on the fidelity of replication. Gene 41, 44, 61, and 62 proteins had no effect. Titration studies suggested that the gene 45 protein increased fidelity by interacting with the DNA polymerase rather than by changing the template conformation (Topal and Sinha, unpublished data). The effects of the addition of T4-coded topoisomerase (the complex of gene 39, 52, and 60 proteins; Liu, Liu, and Alberts, 1979) or the precursor-synthesizing enzyme complex (Mathews et al., 1979) have not yet been determined.

HOW IS HIGH FIDELITY GENERATED?

The means by which high fidelity is generated during DNA replication have been the subject of active experimental and theoretical investigation since 1953. One point seems clear: perturbations in replication complex proteins and particularly DNA polymerase significantly alter base substitution and frameshift mutation frequencies. The classic example is found in the T4 system, where amino acid substitutions in both gene 43 and the replication complex proteins can result in strongly mutagenic and antimutagenic phenotype reversion frequencies (Bernstein et al., 1972; Drake and Allen, 1968; Freese and Freese, 1967; Mufti, 1979; Reha-Krantz and Bessman, 1977; Ripley, 1975; Speyer, 1965; Watanabe and Goodman, 1978). A question which has generated some lively discussion concerns the mechanism(s) by which the polymerases, both in the presence and absence of accessory proteins, achieve their high standard of replication fidelity.

Discrimination at the Insertion Step

Enzyme-catalyzed reactions generally exhibit a high degree of substrate specificity. In the case of DNA synthesis, however, a high degree of specificity is already available from base-pairing interactions. A question arises as to whether DNA polymerase can reduce misinsertion frequencies below what can be attributed to free energy differences between matched and mismatched base pairs. Is there a source of free energy available in addition to base-pairing free energy differences which might allow the polymerase to adjust its active site conformation to reduce the rate of phosphodiester bond formation for a potentially mispaired nucleotide?

A measurement of an effective free energy difference between matched and mismatched base pairs during DNA synthesis in vitro has been carried out using AP. When dAPTP is allowed to compete with dATP for insertion at a template T site, a free energy difference of 1.1 kcal/mol is computed for the ratio of AP·T to A·T base pairs formed (Clayton et al., 1979; Galas and Branscomb, 1978):

$$\frac{I(\mathrm{AP})}{I(\mathrm{A})} = \frac{[\mathrm{dAPTP}]}{[\mathrm{dATP}]} \exp(-\Delta G/RT)$$

where R is the universal gas constant and T is temperature. Here $I(\mathrm{AP})/I(\mathrm{A})$ is the ratio of AP to A insertions measured at a given [dAPTP] to [dATP] concentration pool bias; ΔG is the computed free energy difference for the insertion of AP·T versus A·T base pairs. The AP·T measurement was made with 43^-, L56 mutator, and L141 antimutator T4 DNA polymerases, as well as *E. coli* polymerase I and mammalian DNA polymerase α. By a similar protocol, the free energy difference between AP·T and AP·C base pairs was measured and found to be about 1.8 kcal/mol (Watanabe and Goodman, 1981).

These measurements may be relevant to the mechanism of polymerase fidelity at the insertion step. A·T is formed in preference to AP·T base pairs by a ratio of 7:1, and AP·T is formed in preference to AP·C base pairs by a ratio of 20:1. A kinetic measurement of the K_m values of forming A·T, AP·T and AP·C base pairs showed that the K_m ratios were similar to the measured misinsertion ratios, i.e., $K_m{}^{AP}/K_m{}^{A} \approx 7$ (Clayton et al., 1979) and $K_m{}^{C}/K_m{}^{AP} \approx 22$ (Watanabe and Goodman, 1982), whereas the maximum velocities of forming A·T, AP·T, and AP·C base pairs were similar to one another. One can conclude that at least for the case of making AP·T and AP·C base mispairs, the insertion specificity of the polymerase is governed by a K_m rather than V_{max} discrimination mechanism (Clayton et al., 1979; Goodman et al., 1980).

One can offer an additional conjecture concerning the mechanisms governing insertion fidelity, using the data on AP mentioned above in conjunction with a theoretical model on fidelity set up by Galas and Branscomb (1978). The ratio of right to wrong insertions, as mentioned earlier, was found to be equal, within experimental error, to the ratio of wrong to right K_m values. The Occam's razor interpretation of a K_m (as opposed to V_{max}) mechanism for insertion fidelity based on the model is that the polymerase does not control fidelity by selecting one base in preference to another in a literate response to each template base. Instead, the ratio of wrong to right insertions is governed by the relative residence times of the two competing nucleotides, which depends, in turn, primarily on the free energy difference (in the presence of the polymerase) between the matched and mismatched base pairs. Thus, according to this model, differences in base-pairing free energies (hydrogen bonding plus base stacking), and not literate nucleotide selection by the polymerase, play a dominant role in determining the fidelity of nucleotide insertion.

Proofreading

Brutlag and Kornberg (1972) showed for *E. coli* polymerase I, and Muzyczka et al. (1972) showed for T4 polymerase, that the 3′-exonuclease activity associated with these procaryotic DNA polymerases plays the role of a proofreader. When presented with a mismatched 3′-OH primer terminus, the exonuclease activity is able to excise an incorrect nucleotide, and the polymerase activity is then subsequently able to insert a correct nucleotide in its place. It was observed in the T4 study that the 3′-exonuclease activity under conditions where DNA synthesis either is or is not occurring is much larger for antimutator polymerases than for 43^+, whereas several mutator polymerases contained significantly less exonuclease activity than the wild-type enzyme.

A logical relationship exists between the biochemical data and the mutagenic behavior of the organism. Bessman and co-workers (1974) proposed that the presence of increased editing activity in polymerases exhibiting high nuclease-to-polymerase (N/P) ratios compared to 43^+ should result in antimutator phenotypes; enzymes with low N/P ratios compared to 43^+ are expected to confer mutator phenotypes on their respective phage. This prediction was borne out in an elegant study by Reha-Krantz and Bessman (1977) in which a group of *am*43 mutants were each grown in three suppressor strains of *E. coli*, thereby generating a sizable class of polymerases having single amino acid substitutions. Mutator and antimutator phenotypes measured for the different alleles were found to correspond as predicted to low and high N/P ratios measured for the purified polymerases.

In the experiments discussed earlier in which AP was used to study discrimination at the insertion step (Bessman et al., 1974; Clayton et al., 1979), discrimination at the exonucleolytic proofreading step was also evaluated. The basic conclusion was that the proofreading specificities of the mutator, 43^+, and antimutator polymerases used in the study were all similar to one another. From the viewpoint of the K_m discrimination model, one can again speculate that base-pairing free energy differences govern proofreading specificity rather than an intrinsic recognition of wrong versus right base pair by the polymerase-associated 3′-exonuclease.

After insertion, a mispaired nucleotide is more likely to be melted out than a properly paired nucleotide. The T4 polymerase-associated 3′-exonuclease is known to be much more active on single-stranded than on double-stranded DNA. Hence, the very active proofreading exonuclease of the L141 antimutator polymerase, for example, should remove a mispaired nucleotide more efficiently under rapid DNA-synthesizing conditions than the low-N/P-ratio L56 mutator polymerase despite the fact that the proofreading specificities are similar for the two enzymes. Since even properly paired nucleotides can be expected to exhibit some degree of transient single-stranded character at the growing fork, a corollary relationship which should also occur is the removal of a relatively excessive amount of properly paired nucleotides by the high-N/P-ratio L141 polymerase. This point has been verified both for A·T (Bessman et al., 1974; Clayton et al., 1979) and for highly stable G·C (D. Mhaskar and M. F. Goodman, unpublished data) base pairs.

In addition to the class of mutant polymerases which appear to differ mainly in their relative abilities to proofread DNA, another class of T4 mutator polymerases has been identified, first (*ts 43* L88) by Hershfield and Nossal (Hershfield, 1973; Nossal and Hershfield, 1973) and later (*ts 43* M19) by Reha-Krantz and Bessman (1981). These mutant polymerases appear to have N/P ratios similar to 43^+ yet insert incorrect nucleotides into DNA at elevated frequencies. In the two examples reported, the misinsertion mutators were purified from phage which exhibit strong mutator phenotypes.

Cost of High Fidelity

Rare tautomeric forms of bases capable of mispairing are estimated to exist naturally at a frequency of 10^{-4} to 10^{-5}. Therefore, an error rate in DNA replication lower than this might not be possible even with complete discrimination between a correct and an incorrect precursor unless some kind of a proofreading step is employed. If the discrimination at the insertion step and the proofreading step is absolute, then the highest accuracy theoretically attainable (10^{-8} to 10^{-10}) can be reached. In reality, however, discrimination between the right and the wrong base is less than 100% because of the cost involved in achieving perfect discrimination. Galas and Branscomb (1978) and Fersht et al. (1982) have described mathematical treatments for the cost of discrimination using an exonuclease for proofreading. The cost of proofreading can simply be expressed as the fraction of correctly polymerized nucleotides that are mistakenly hydrolyzed to achieve a certain level of discrimination between the right and the wrong nucleotide precursors. The cost for DNA polymerases of *E. coli* at maximal rates of DNA synthesis was found to be about 6% for dCTP and dGTP and about 12 to 15% for dATP and dTTP (Fersht et al., 1982). With the T4 polymerase alone (Clayton et al., 1979; Muzyczka et al., 1972) and with the seven-enzyme T4 replication apparatus during the copying of double-stranded ϕX174 DNA template (N. Sinha, unpublished data), with each nucleotide present at saturating concentrations, the cost was about 35% for A and T nucleotides and about 15% for C and G nucleotides. The cost goes up dramatically as the rate of DNA synthesis is reduced by lowering the concentration of the nucleotide precursors.

Hydrolysis of such a high percentage of correct nucleotide precursors, even at maximal rates of DNA synthesis, suggests a substantial contribution by the exonuclease proofreading step to the ultimate accuracy of the T4 replication complex. Since its replication complex has such a high proofreading exonuclease activity and is, therefore, so wasteful, it becomes intelligible why bacteriophage T4 has gone to the trouble of assembling a precursor-synthesizing complex that will generate a high concentration of dNTP precursors at the replication site (Mathews et al., 1979; Mathews and Sinha, 1982).

Postreplication Mismatch Repair

As discussed above, the maximum accuracy theoretically possible during DNA replication using a two-step discrimination system is about 10^{-10}. Since the average mutation rate in phage T4 is estimated to be about 10^{-8}, this would appear to be sufficient and no further accuracy-enhancing steps need be present. The absence of general mutators (except in replication genes) and the existence of a high level of heteroduplex-type heterozygotes (Sechaud et al., 1965) in phage T4 indicates the absence of postreplication mismatch correction mechanisms. In *E. coli*, where the average mutation rate is about 10^{-10} to 10^{-11}, clearly there is need for further corrections beyond what can be achieved at the DNA replication stage. A large body of data suggests that in *E. coli* there is postreplication mismatch correction. At least some of this involves the use of editing systems that use methylation in the parental strand to decide the direction of correction (Glickman and Radman, 1980; Wagner and Meselson, 1976).

Research in our laboratories is supported by Public Health Service grants GM-24391 (to N.K.S.) and GM-21422 and CA-17358 (to M.F.G.) from the National Institutes of Health.

DNA Repair

CAROL BERNSTEIN[1] AND SUSAN S. WALLACE[2]

Department of Molecular and Medical Microbiology, University of Arizona College of Medicine, Tucson, Arizona 85724,[1] and Department of Microbiology, New York Medical College, Valhalla, New York 10595[2]

HISTORICAL STUDIES

The first systematic study of the repair of genetic material was by Luria (1947) with the T-even phages. His work on multiplicity reactiviation was the start of studies on recombinational repair. Soon after, Kelner demonstrated photoreactivation of UV-treated conidia of *Streptomyces griseus* and communicated his results to Dulbecco. In a few weeks Dulbecco found photoreactivation in UV-treated T phage, including phage T4. Kelner (1949) and Dulbecco (1949) published at nearly the same time, thus initiating studies in this field of photobiology. Luria (1947) and Luria and Dulbecco (1949) found that phage T4 was about twice as resistant to UV irradiation as phages T2 and T6. In an abstract, Luria (1949) suggested that this was due to a single genetic unit. Streisinger (1956) confirmed this explanation, calling the locus involved the "u" gene, and postulated that this gene was involved in repair. This turned out to be the start of studies in excision repair in phage T4 which were then extended by Harm (1958; 1961). Harm found the "u" gene to be a wild-type complementing allele of T4 which conferred resistance to T2 in mixed infection. He renamed the "u" gene *v*. Thus, the initial studies on recombinational repair and photoreactivation, and some of the earliest work on excision repair, were with phage T4. For other recent reviews of T4 DNA repair see Friedberg (1975) and Bernstein (1981).

PATHWAYS OF REPAIR

Photoreactivation

Photoreactivation of DNA consists of the enzyme-mediated, light-dependent monomerization of pyrimidine dimers, resulting in repair of the DNA and restoration of its biological integrity (Sutherland, 1978). Dulbecco (1950) showed that a host enzyme was likely to be responsible in T-even photoreactivation, since this process occurs in phage T2 as early as 10 s after adsorption. The action spectrum for photoreactivation had a maximum at about 365 nm. The fraction of lethal lesions that can be photoreactivated (photoreactivable sector) varied from 0.20 for phages T4 and T5 to 0.68 for phage T1. Harm (1963) showed that the size of the photoreactivable sector depends on the overlap of photoreactivable damages with damages reactivable by other pathways. The magnitude of the photoreactivable sector indicates that photoreactivation is a major repair mode for UV-induced damage in T4.

Photoreactivation of phage T2r DNA within infected cells was shown to consist of a first dark reaction followed by an irreversible light reaction, each of these reactions following first-order kinetics (Bowen, 1953a; Bowen, 1953b).

Helene et al. (1976) also found that the phage T4 gene 32 protein (gp32) can promote the photosensitized splitting of thymine dimers in vitro. It is not known whether this reaction, or possibly other similar reactions carried out by tryptophan-containing proteins which bind to DNA (Helene, 1978), are of significance in repairing pyrimidine dimer-containing DNA in vivo. There is some indication (Childs et al., 1978) that glucosylated hydroxymethyl cytosine-containing pyrimidine dimers are formed in phage T4 and that such glucosylated dimers are not photoreactivable.

Excision Repair

In the phage T4 system, excision repair of UV-induced pyrimidine dimers, defined by the *denV* pathway and controlled by the *v* gene, has been studied in great detail. For a recent extensive review of the enzymology of excision repair of pyrimidine dimers, see Friedberg et al. (1981).

Recognition and incision. In early studies, Streisinger (1956) and Harm (1961) showed that for UV-induced lesions there is an overlap between *denV* gene-mediated reactivation and photoreactivation, thus demonstrating that the substrate for the *denV* gene product was probably the pyrimidine dimer. Setlow and Carrier (1968) analyzed both the acid-insoluble and acid-soluble fractions of UV-irradiated T4$^+$ and T4 *denV* DNA after infection of both *Escherichia coli* B or *E. coli* B_{S-1} and found a rapid excision of dimers from the phage DNA in the presence of the wild-type allele of *denV* but not in its absence. Further, in mixed infection with T4 *denV*$^+$ and T4 *denV*, excision also occurred. Similar results were also obtained by Takagi et al. (1968) and Sekiguchi et al. (1970), who used cell-free extracts of phage-infected and uninfected cells and showed a preferential release of dimers into the acid-soluble fraction when UV-irradiated DNA was incubated with extracts of T4$^+$-infected cells. An endonucleolytic action of the T4 *denV* gene product was established when Friedberg and King (1969) demonstrated that extracts of wild-type T4-infected *E. coli* were capable of incising UV-irradiated *E. coli* DNA as measured by the production of strand breaks with alkaline sucrose gradient sedimentation as an assay. Extracts from T4 *denV* mutant-infected cells showed fewer strand breaks than extracts from wild-type-infected cells. DNA degradation was also measured and was shown to be greater in wild-type T4-infected cells than in T4 *denV* mutant-infected cells. These results were confirmed in similar studies done by Yasuda and Sekiguchi (1970a). These latter workers, using temperature-sensitive mutants of T4 *denV*, also showed unambiguously that the *denV* gene product is the structural gene for a UV endonuclease (Sekiguchi et al., 1975).

The T4 *denV* gene product is an early enzyme, being expressed within 2 min after phage infection (Friedberg and King, 1971). Furthermore, the enzyme appears to act only at early times after infection, the excision of dimers reaching a maximum within 5 to 7 min after infection and then maintaining a plateau value (Pawl et al., 1976). Sato and Sekiguchi (1976) also showed that the *denV* gene product was required at early but not late times after infection.

T4 *denV* gene product, DNA endonuclease V, was partially purified and characterized and shown to have a molecular weight of about 18,000 as determined by sodium dodecyl sulfate-gel electrophoresis and Sephadex filtration. The enzyme exhibited a broad pH optimum, was not inhibited by sulfhydryl inhibitors, and was completely active in the presence of 10 mM EDTA (Friedberg and King, 1971; Yasuda and Sekiguchi, 1970b). Ganesan (1974) showed that all UV-light-induced endonuclease-sensitive sites detectable in alkaline sucrose gradients were photoreversible, thus implicating the pyrimidine dimer as a substrate for the enzyme. Friedberg (1975) obtained similar results in vitro by using purified photoreactivating enzymes. In addition, using supercoiled simian virus 40 DNA, Friedberg and Clayton (1972) demonstrated that, at an average of one dimer per molecule, the endonuclease V-induced nicking of the DNA was quantitative and occurred on only one strand. Furthermore, DNA treated in vitro with chemical agents such as mitomycin C (MMC) and 4-nitroquinoline-*N*-oxide is not a substrate for endonuclease V, although DNA alkylated with nitrogen mustard does undergo degradation when incubated with this enzyme (Friedberg, 1972). However, T4 *denV* mutants are not more sensitive than wild-type T4 to nitrogen mustard or other alkylating agents, indicating that the *denV* gene plays no in vivo role in the repair of damages produced by these agents (Friedberg, 1972). The nicking by endonuclease V of alkylated DNA, which contains apurinic sites, is compatible with the fact that this enzyme has apurinic endonuclease activity, an observation to be described below.

The *uvr*A/B/C endonuclease of the host (Seeberg, 1978) does not appear to act on UV-irradiated T4 DNA since no difference in UV survival of either wild-type or T4 *denV* mutant phage was observed when the phage were plated on an *E. coli uvrA* strain (Wallace and Melamede, 1972). Furthermore, Strike (1978) showed that infection of *E. coli* with phage T4 results in inactivation of the host UV-specific endonuclease. However, T4 *denV*$^-$ phages can complement the activity of the host *uvr*A/B/C gene by rescuing a fraction of UV-irradiated excision-defective host cells (Friedberg, 1972; Harm, 1968).

Recent studies have indicated that the T4 *denV* gene product, endonuclease V, has two enzymatic activities. These studies were stimulated by the observation of Grossman et al. (1979) and Haseltine et al. (1980) that the analogous UV endonuclease of *Micrococcus luteus* possesses a glycosylase activity which cleaves the glycosylic bond between the 5′ pyrimidine of a dimer and the corresponding deoxyribose. Work by Demple and Linn (1980), Gordon and Haseltine (1980), Radany and Friedberg (1980), Seawell et al. (1980), and Warner et al. (1980) indicates that the T4 endonuclease, like the *M. luteus* UV endonuclease, contains a pyrimidine dimer glycosylase activity as well as an apurinic/apyrimidinic endonuclease activity. It was shown that endonuclease V first cleaves the glycosylic bond at the 5′ half of a pyrimidine dimer and subsequently catalyzes cleavage of the phosphodiester bond originally linking the two nucleotides of the dimer, leaving an apyrimidinic 3′ terminus in the newly nicked DNA.

Nakabeppu and Sekiguchi (1981) have purified the T4 *denV* product to apparent physical homogeneity. Utilizing a UV-irradiated poly(dA)-poly(dT) template and differentiating the production of alkali-labile apyrimidinic sites (the glycosylase activity) from the formation of nicks (the endonuclease activity), these investigators showed that both activities copurify and are found in a single 16,000-dalton polypeptide. Further, enzyme fractions from cells infected with T4 *denV* mutants exhibit a simultaneous loss of both activities, and suppression of the mutation renders both activities partially active. These investigations, as well as those of Seawell et al. (1980), showed that both activities exhibit differential thermal sensitivity, suggesting two active sites. In independent studies, McMillan et al. (1981) and Warner et al. (1981) also showed that endonuclease V contains both pyrimidine dimer DNA glycosylase and apyrimidinic endonuclease activities. The former study showed that DNA containing apyrimidinic sites competed for the pyrimidine dimer glycosylase activity against UV-irradiated DNA. The latter study demonstrated that both activities copurify and that the kinetics of induction of both activities are the same in infected cells. In contrast to Seawell et al. (1980) and Nakabeppu and Sekiguchi (1981), Warner et al. (1981) showed that both the glycosylase and apyrimidinic activities have the same thermal lability. Recently, however, Nakabeppu et al. (1982) confirmed that the *denV* gene product contains two distinct active sites on one polypeptide chain, one for the apyrimidinic endonuclease and one for the glycosylase. These workers demonstrated that when a particular *denV* amber mutant, *uvs*-13, was suppressed, only the glycosylase activity of the *denV* gene product was restored. The action of T4 endonuclease V on UV-irradiated DNA appears to be processive, since incisions appeared at most dimer sites in some molecules of closed circular ColE1 DNA (25 dimers per molecule) before any scissions were found in other molecules (Lloyd et al. 1980).

Recently, Radany and Friedberg (1982) have shown that the *denV* gene product also acts as a dimer glycosylase in vivo. The approach used here was to infect *uvr E. coli* with wild-type T4 and isolate the acid-soluble material containing the pyrimidine dimers. This acid-soluble fraction was then directly photoreversed. Free thymine was produced. This could only have occurred if the initial enzymatic step in vivo had been a glycosylase action which freed one of the thymines of the dimerized pair from the DNA backbone.

Because of its action at the site of pyrimidine dimers, T4 endonuclease V has often been used as a sensitive probe for UV-induced damage (Ganesan, 1973; Van Zeeland, 1978). Further, the *denV* gene product has been introduced into permeabilized cells of *E. coli*, increasing the UV light resistance of *uvr* mutants (Shimizu and Sekiguchi, 1979). It has also been introduced into permeabilized isolated nuclei from human fibroblasts derived from patients with xeroderma pigmentosum (Ciarrocchi and Linn, 1978; Smith and Hanawalt, 1978; Tanaka et al., 1977). Addition of the *denV* product complemented the defect in an early step of excision repair. The T4 *denV* gene has been recently cloned into plasmid pBR322 and found to complement *uvr E. coli*, enhancing bacterial survival after UV treatment. The cloned fragment also enhanced the survival of UV-irradiated bacteriophage lambda or UV-irradiated T4 *denV* mutants (Lloyd and Hanawalt, 1981).

Excision, repolymerization, and ligation. The excision of pyrimidine dimers in vivo must be catalyzed by an enzyme that excises in the 5′ → 3′ direction. In vitro, such excision can be accomplished on endonuclease V-treated, UV-irradiated DNA by the 5′ → 3′ exonuclease activity associated with *E. coli* DNA polymerase I (Friedberg and Lehman, 1974) as well as *E. coli* exonuclease V (Tanaka and Sekiguchi, 1975). Also, the 5′ → 3′ exonuclease activity of *E. coli* DNA polymerase III (Livingston and Richardson, 1975) and *E. coli* exonuclease VII (Chase and Richardson, 1974) can excise UV-irradiated DNA incised by the *M. luteus* UV endonuclease. Furthermore, a phage-coded 5′ → 3′ exonuclease activity can be isolated which also effects dimer excision from incised DNA in vitro (Friedberg et al., 1974; Ohshima and Sekiguchi, 1972; Shimizu and Sekiguchi, 1976).

Both Maynard-Smith et al. (1970) and Wallace and Melamede (1972) have shown in vivo that UV-irradiated T4 has reduced survival in hosts defective in DNA polymerase I activity. This reduced survival is not seen after T4 *denV* mutant infection, thus indicating that DNA polymerase I acts in the same pathway as the *denV* enzyme. Pawl et al. (1976) examined the kinetics of the loss of thymine-containing dimers from the acid-insoluble fraction after infection of *E. coli* with tritium-labeled, UV-irradiated phage T4. Their results, which indicated very little dimer excision in DNA polymerase I-defective *E. coli* mutants, suggest that the major enzyme responsible for the excision of dimers in T4 DNA in vivo is the 5′ → 3′ exonuclease activity of DNA polymerase I. Host exonuclease V and VII and the phage-coded 5′ → 3′ exonuclease were found not to be involved in dimer excision in vivo. This is in keeping with the observations that host exonuclease V is completely inhibited in a noncatalytic reaction with a protein synthesized after phage T4 infection (Behme et al., 1976) and that the phage 5′ → 3′ exonuclease was only expressed at late times after infection, when excision repair may have been completed (Pawl et al., 1976).

The repolymerization step in excision repair is also probably catalyzed by the host *E. coli* DNA polymerase I since the 5′ → 3′ exonuclease activity and the 5′ → 3′ polymerizing activity of DNA polymerase I have been shown to act in concert during the excision of pyrimidine dimers (Friedberg and Lehman, 1974). Bromodeoxyuridine photolysis studies of the repaired regions in T4 DNA (Yarosh et al., 1981) and calculations based on acid-soluble radioactivity associated with excision (T. Bonura, E. H. Radany, S. McMillan, J. D. Love, R. A. Schultz, H. J. Edenberg, and E. C. Frieberg, Biochimie, in press) indicate that the average patch size is between 4 and 7 nucleotides, smaller than had been observed for the host cell, which ranges between 10 and 30 nucleotides.

The final sealing step in the excision repair process probably depends on the product of gene 30, DNA ligase. Baldy (1968; 1970) has shown that certain gene 30 temperature-sensitive mutants exhibit increased UV sensitivity when assayed under semipermissive conditions, and Maynard-Smith and Symonds (1973) showed that the gene 30 ligase is in the *denV* excision pathway. Wallace and Melamede (1972) found no differences in the survival of UV-irradiated wild-type T4 when assayed either on wild-type or ligase-defective hosts. Thus, host ligase does not appear to function in the in vivo repair of UV-irradiated wild-type T4.

A question that remains unanswered regarding the excision repair of a pyrimidine dimer in T4-infected cells is how the apyrimidinic lesion, present at the 3′-hydroxyl terminus of the endonuclease V incision site, is removed. Although this apyrimidinic site would be a substrate for *E. coli* endonuclease VI, coded for by the *E. coli xth* gene, McMillan et al. (1981) have shown that T4 is no more UV sensitive in *xthA* strains than it is in wild type. An apyrimidinic site on the 3′-hydroxyl side of the endonuclease V-induced nick does not efficiently bind *E. coli* DNA polymerase I (Warner et al., 1980). While it is possible that in vivo the site binds polymerase I efficiently enough to allow for repolymerization, this would result in the fixing of an apyrimidinic site in the repaired DNA, a potentially lethal lesion (Kudrna et al., 1979; Schaaper and Loeb, 1981). This lesion itself would have to be acted upon by a repair system. Alternatively, the apyrimidinic endonuclease activity of endonuclease V may not function in vivo since temperature-sensitive mutants of T4 *denV* with reduced levels of the apyrimidinic endonuclease activity are not UV sensitive (Bonura et al., in press). Thus, it remains to be seen whether endonuclease V nicks the DNA backbone in vivo, requiring subsequent removal of an apyrimidinic site by an as yet undetermined phage or host enzyme, or whether the DNA backbone at the apyrimidinic site is initially nicked in a different manner by an unknown phage or host enzyme. A diagrammatic representation of the excision repair process in T4-infected cells, including a hypothetical removal of the apyrimidinic site, is presented in Fig. 1.

Involvement of other T4 genes in excision repair. Genes can be allocated to the same repair pathway by a survival-of-phenotype test (Ebisuzaki, 1966; Maynard-Smith and Symonds, 1973). For the *denV* excision repair pathway, this test can be carried out by measuring the ability of UV-irradiated *am*$^+$ *denV*$^+$ phage to complement coinfecting *am denV*$^+$ phage and comparing this with the ability of UV-irradiated *am*$^+$ *denV* phage to complement unirradiated coinfecting *am denV* phage. The survival of the ability of *am*$^+$ to complement *am*$^-$ measures whether the function of the *am*$^+$ gene is being carried out. If the *am*$^+$ acts in the same pathway of UV repair as the *denV* function, then it will not matter if a *denV*$^+$ or *denV* allele is present. Once the *am*$^+$ function is inactivated by UV damage, the *denV* pathway could not be carried out whether gp*denV* is functional or not. However, if the function of the *am*$^+$ gene product is in a different pathway from gp*denV*, then the ability of the *am*$^+$

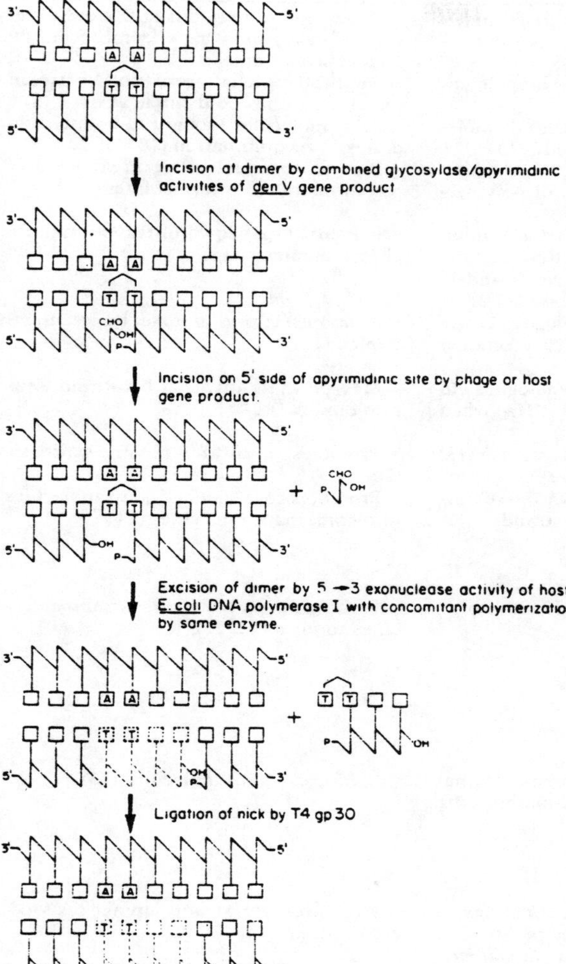
Incision at dimer by combined glycosylase/apyrimidinic activities of denV gene product
Incision on 5' side of apyrimidinic site by phage or host gene product.
Excision of dimer by 5 → 3 exonuclease activity of host E. coli DNA polymerase I with concomitant polymerization by same enzyme.
Ligation of nick by T4 gp30

FIG. 1. Excision repair in T4-infected cells.

function to complement would survive at a higher level in UV-irradiated *denV*⁻ phage (where the *am*⁺ gene, damaged by UV, could be repaired by the *denV*⁺ function) than in *denV*⁻ phage. By this criterion, the products of genes 1, 30, 42, 45, and 56 were found to be involved in excision repair whereas the products of genes 32, 41, 43, and 49 were not (Maynard-Smith and Symonds, 1973). Genes 1, 42, and 56 code for enzymes necessary for the formation of hydroxymethyl-dCTP (see Mathews and Allen, this volume) and thus might be expected to be required for appropriate gap-filling after excision of the dimer. The gene 30 product is DNA ligase (Fareed and Richardson, 1967). The involvement of gp45 is not easily explained since its main role seems to be as an accessory protein in DNA replication (Alberts et al., 1980; Nossal and Alberts, this volume), and the T4-encoded DNA polymerase (gp43) and its other accessory proteins do not appear to be involved in excision repair.

Recombinational Repair

Postreplication recombinational repair. In his initial studies of the UV-sensitive *v* and *x* mutants, Harm (1963) found that these mutants affected different repair processes, since the double mutant *v*, *x* was more UV sensitive than either the *v* or *x* mutants. (This is the double-mutant test for placing gene functions in separate pathways.) In 1964 he analyzed the *x* mutant further (Harm, 1964), showing it has reduced ability to carry out marker rescue (a recombinational process) and genetic recombination. He proposed that "to explain the increased UV-sensitivity of single-infecting T4*x* phage, it is tentatively assumed that the *x* allele, in contrast to the *x*⁺ allele, does not permit a repair process which has some mechanism in common with genetic recombination." As a possible model for such repair he suggested: "In a double-stranded DNA, base sequence and polarity in a given part of strand A are identical with those in a homologous part of a newly synthesized complement of strand B. Therefore, a UV-hit in strand A can theoretically be circumvented by any recombination with the complement of strand B . . ." The concept that recombinational processes have a role in overcoming DNA damages was apparently conceived independently by Harm, as indicated by the above quotation, and by Howard-Flanders (see personal communication in Clark and Margulies [1965]). The designation "postreplication recombinational repair" (PRRR) is applied to the process briefly outlined by Harm in the above quotation. PRRR has been characterized both biochemically and genetically in *E. coli* (for reviews see Hanawalt et al., 1979; Howard-Flanders, 1975; and Smith, 1978).

In phage T4, if a mutant is found (i) to be more sensitive to a DNA-damaging agent than wild type, and (ii) to have reduced genetic recombination, this is usually taken as suggestive evidence for a defect in PRRR. By these criteria, the products of genes 32, 46, 47, 59, *uvsW*, *uvsX*, and *uvsY* (see Table 1) are considered to be required for PRRR. By the double-mutant test, evidence was obtained that the protein phage products of genes 58–61, 59, *uvsW*, *uvsX*, and *uvsY* and the ligase of *E. coli* act in the same pathway (Table 1). Maynard-Smith and Symonds (1973), using the survival-of-phenotype test, found nine genes involved in the same UV repair pathway as the product of gene *uvsY*. As indicated in Table 1, these are genes 1, 30, 32, 41, 42, 43, 44, 45, and 56.

Combining the results of the double-mutant test and the survival-of-phenotype test, one can conclude that 15 of the 17 genes listed in Table 1 act in the same pathway. The two others, genes 46 and 47, could not be tested by these methods since amber mutants defective in these genes are too leaky to be used in survival-of-phenotype tests (Maynard-Smith and Symonds, 1973), and the double mutants, constructed from *am* alleles of genes 46 or 47 combined with *am* alleles of *uvsX* or 59, are unable to survive (Wakem and Ebisuzaki, 1981). However, genes 46 and 47 are likely to act in the same PRRR pathway as genes *uvsW*, *uvsX*, *uvsY*, 32, and 59 since mutants defective in these genes share the phenotypes of reduced spontaneous recombination and sensitivity to many agents (Table 1). In addition, mutations in genes 46, 47, 59,

TABLE 1. Gene functions involved in PRRR

Gene	Basis for inclusion[a,b]	Known function(s) that may be involved in repair	Hypothetical step in repair[c]
32	rsr (1)[d] sens [1.4]UV (2) [2.6]MNNG (3) [1.1]MMC (4) [1.3]HNO_2 (1) [rd]EMS (25) spt (6)	Binds cooperatively to single-stranded DNA (7) and stabilizes single-strand regions (8).	Opens up local regions of DNA (8). Prevents single-strand gaps from becoming double-strand breaks (9).
		Promotes renaturation of complementary DNA (8).	Facilitates helix formation between matching, complexed single strands (8).
		Required for formation of joint molecules of DNA in vivo (10, 11, 12, 13).	Joint molecules thought to be intermediates in recombination (10, 11, 12, 13).
		Binds to single-strand termini, blocking 3' to 5' exonuclease activity of polymerase (14).	Prevents exonucleolytic breakdown of recombinational intermediates.
		The C-terminal domain of the protein moderates the nucleolytic activities of the exonuclease controlled by genes 46 and 47 and of the host RecBC nuclease (15).	Prevents breakdown of recombinational intermediates (15).
		Enhances uptake of homologous single strands by duplex DNA when promoted by *E. coli recA* protein (16).	Promotes strand uptake by recipient duplex (16).
		Enhances uptake of homologous fd DNA fragments by fd RFI DNA when promoted by T4 gp*uvsX* (17).	May promote invasion by strand ends into duplex DNA.
		Binds to gene 43 DNA polymerase (18) and to gene 58–61 primase (19).	Promotes accurate repair synthesis (20).
		Interacts with gene 30 DNA ligase during ligation of recombining strands (21).	Promotes conversion of joint molecules to recombinant molecules (21).
46/47	rsr (22, 23) sens [1.4/1.2]UV (2) [1.5/1.5]PUVA (24, 27) [3.1/3.4]MNNG (3) [1.4/1.7]MMC (4) [1.2/1.3]HNO_2 (1) [rm/rm]EMS (5)	Controls exonuclease (25).	
		Expands nicks into gaps (13, 26).	Produces gaps which allow joint molecules to form (13, 26).
59	rsr (28, 29) sens [1.9]UV (30, 32) [2.8]MMS (30) [1.8]X rays (31)	May be necessary in concatemer formation (30) or in repairing and maintaining concatemers (33).	"Matures" joint molecules (33).
uvsW	rsr (34, 35) sens [1.5]UV (33, 34, 35, 36) [1.4]PUVA (37, 38) [1.8]MMS (35) dmt (35, 36)	May supply some of same functions as *E. coli recA* protein since the presence of *uvsW* and *y* on a plasmid in an *E. coli recA* strain provides protection against UV damage (39).	May promote strand uptake as does *recA* protein (16, 40).
		$uvsW^-$ mutation can suppress deficiencies in recombination and UV resistance caused by mutation in gene 59 (33).	
		$uvsW^-$ mutation causes overproduction of gene 32 protein, implying excess single-stranded DNA (41).	
uvsX	rsr (42) sens [1.7]UV (42) [1.6]PUVA (38) [1.9]MNNG (3) [1.8]MMS (35, 43, 44) [1.5]MMC (4) [1.2]HNO_2 (45) [rd]EMS (5, 46) [1.7][e] X rays (44, 47, 48) dmt (36, 48)	gp*uvsX* binds to double-stranded DNA in presence of ATP, binds to double-stranded or single-stranded DNA in absence of ATP. The complex is rodlike (17).	Prepares DNA for pairing.
		Has single-strand DNA-dependent ATPase activity (17).	
		Promotes pairing of sonicated fd DNA with RFI DNA in vitro in presence of ATP and Mg^{2+} (17).	Promotes strand uptake by recipient duplex.
		$uvsX^-$ mutation reduces formation of joint molecules in vivo (17).	
		$uvsX^-$ mutation carried on infecting phage reduces marker rescue of gene 18 cloned on a plasmid (17).	
uvsY	rsr (50) sens [1.7]UV (50) [1.6]PUVA (38) [1.6]MNNG (3)	May supply some of same functions as *E. coli recA* protein since the presence of *uvsW* and *uvsY* on a plasmid in an *E. coli recA* strain provides protection against UV damage (39).	May promote strand uptake as in function of *recA* (16, 40).

TABLE 1—*Continued*

Gene	Basis for inclusion[a,b]	Known function(s) that may be involved in repair	Hypothetical step in repair[c]
	[1.9]MMS (43) [1.5]MMC (4) [1.2]HNO_2 (45) [rd]EMS (5, 46) [1.4][e] X rays (47, 50) dmt (36, 49) spt (6)		
30	sens [1.2]UV (2) [2.9]MMS (51) [rd]EMS (5, 46) spt (6)	DNA ligase (52).	Final sealing step in recombination (53).
41	sens [1.2]UV (2, 54) [2.1]MMS (55) spt (6)	DNA-helicase; is needed for primase activity in conjunction with gp58–61 for Okasaki fragments on lagging strand (56, 57).	Gp41 is thought to function in two repair processes. In PRRR its function may be during DNA synthesis (58), presumably in producing gaps opposite a lesion.
58–61	sens [1.8]UV (35) dmt (35)	gp58–61 and gp41 act in conjunction as primase for Okasaki fragments on lagging strand (56, 57).	May act in DNA synthesis in producing gaps opposite a lesion.
43	sens [1.2]PUVA (27) [rd]EMS (46) spt (6)	DNA polymerase (59).	May act in DNA synthesis in producing gaps opposite a lesion. May repair gaps in recombinants (60).
1, 42, 44, 45, 56	spt (6)	Synthesis of nucleotides and polymerization of DNA; see reference 61 for references. Gp1, gp42, and gp56 may be in a nucleotide synthetase complex (62, 63) acting at the replication fork with the replication complex including gene proteins 32, 41, 43, 44, 45, and 58–61 (56).	DNA synthesis (6) as a prelude to producing gaps.
lig of *E. coli*	sens [1.1][e] X rays (47) dmt (47)	DNA ligase.	May contribute to final sealing step.

[a] Code: rsr, mutations in this gene cause reduced spontaneous recombination; sens, mutations in this gene cause increased sensitivity to the agents listed, compared to wild-type sensitivity; dmt, this gene acts in the same UV-repair pathway as *uvsX* or *uvsY* by the double-mutant test; spt, this gene acts in the same UV-repair pathway as *uvsY* by the survival-of-phenotype test.

[b] When applied to free phage or to phage-host complexes, all agents caused inactivation of the ability to produce infective centers. When the log of the surviving fraction was plotted versus dose of the agent, the survival curve was generally a straight line, although in the case of some UV and X-ray irradiation curves there was a very small initial shoulder. Except for these few small shoulders, survival curves represented killing with single-hit kinetics, following the equation $N/N_0 = e^{-kd}$, where N_0 is the number of phage present in an initially chosen population, N is the number of surviving phage after treatment, d is the dose of the inactivating agent, and k is the number of lethal lesions delivered per unit dose. The numbers listed represent $k_{mutant}/k_{wild\ type}$, which is the ratio of lethal hits delivered to the mutant compared to wild type at any dose. The values listed within brackets were the largest of the values of $k_{mutant}/k_{wild\ type}$ tabulated in Bernstein (1981) for alleles of the gene, except where a superscript e is used. Then the values are calculated from curves presented in the reference indicated. In the case of EMS, there was a pronounced shoulder followed by a straight-line inactivation curve; rm, removal of the shoulder; rd, reduction of the shoulder. For the straight-line portion of the inactivation curve, $k_{mutant}/k_{wild\ type}$ was usually about 1.0.

[c] Where the hypothetical step, or an analogous step, has been proposed in the literature, this is indicated by inclusion of a reference. Each of the steps listed in this column is correlated with the adjacent function listed in the column to the left.

[d] Numbers in parentheses refer to these references: (1) Nonn and Bernstein, 1977; (2) Baldy, 1970; (3) Schneider et al., 1978; (4) Holmes et al., 1980; (5) Johns et al., 1978; (6) Maynard-Smith and Symonds, 1973; (7) Alberts et al., 1969; (8) Alberts and Frey, 1970; (9) Wu and Yeh, 1973; (10) Tomizawa et al. 1966; (11) Kozinski and Felgenhauer, 1967; (12) Broker and Lehman, 1971; (13) Broker, 1973; (14) Huang and Lehman, 1972; (15) Mosig and Bock, 1976; (16) Shibata et al., 1980; (17) results obtained by Yonosaki and Minagawa (personal communication, Teiichi Minagawa, Kyoto University); (18) Huberman et al., 1971; (19) Hosoda et al., 1980; (20) Mufti, 1980; (21) Mosig and Breschkin, 1975; (22) Bernstein, 1968; (23) Berger et al., 1969; (24) Johns, personal communication; (25) Mickelson and Wiberg, 1981; (26) Prashad and Hosoda, 1972; (27) Yarosh et al., 1980; (28) Shah, 1976; (29) Davis and Symonds, 1974; (30) Wu, Wu, and Yeh, 1975; (31) Wu and Yeh, 1981; (32) Wakem and Ebisuzaki, 1981; (33) Cunningham and Berger, 1977; (34) van den Ende and Symonds, 1972; (35) Hamlett and Berger, 1975; (36) Symonds et al., 1973; (37) Green and Drake, 1974; (38) Zerler and Wallace, 1979; (39) DeVries and Wallace, submitted for publication; (40) McEntee et al., 1979; (41) Krisch and van Houwe, 1976; (42) Harm, 1964; (43) Ebisuzaki et al., 1975; (44) Mortelmans and Friedberg, 1972; (45) Harm, 1974; (46) Ray et al., 1972; (47) Wallace and Melamede, 1972; (48) Childs, 1980; (49) Boyle, 1969; (50) Boyle and Symonds, 1969; (51) Baldy et al., 1971; (52) Fareed and Richardson, 1967; (53) Anraku and Lehman, 1969; (54) van Minderhout et al., 1978; (55) van Minderhout and Grimbergen, 1976; (56) Alberts et al., 1980; (57) Silver et al., 1980; (58) Cupido et al., 1980; (59) Goulian et al., 1968; (60) Miller, 1975b; (61) Wood and Revel, 1976; (62) Reddy et al., 1977; (63) Wovcha et al., 1973.

[e] These values were calculated from figures in the references indicated.

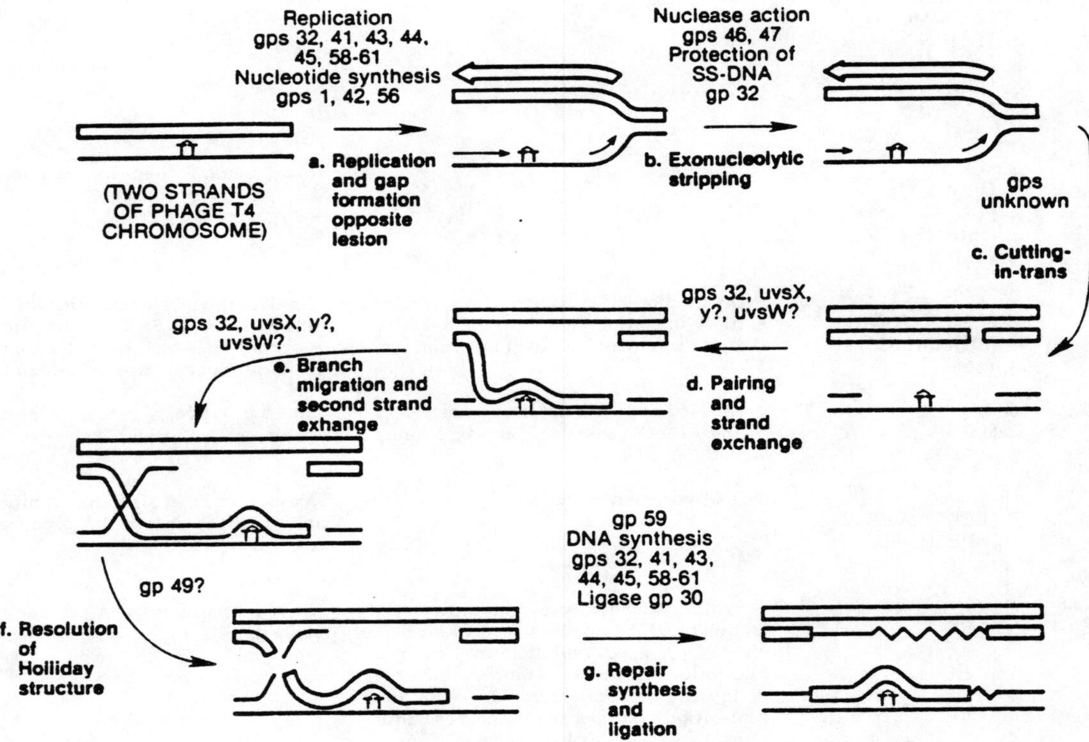

FIG. 2. Model of PRRR in T4-infected cells.

uvsX, and *uvsY* can suppress gene 49 mutations and have been proposed to be in the same pathway by this criterion (Wakem and Ebisuzaki, 1981).

The known functions of the 17 gene products and the molecular steps which have been proposed to be promoted by these gene functions are summarized in Table 1. A PRRR model for T4, very similar to those presented by West et al. (1981), Howard-Flanders (1982), and Livneh and Lehman (1982) for *E. coli*, is shown in Fig. 2. This figure indicates the steps in the pathway and where each of the gene functions listed in Table 1 may act. The resolution of the single-strand cross-over structure (Holliday structure) shown in step f may require the gene 49 protein, which is a nuclease that can resolve such structures in vitro (Mizuuchi et al., 1982; Warner and Snustad, this volume). However, gp49 has not yet been shown to have a role in recombinational repair in vivo. The thymine dimers which remain in the chromosome may be transmitted into progeny phage (Sauerbier and Hirsch-Kauffmann, 1968). The functions of gp*uvsX* have recently been shown by Yonesaki and Minagawa (Minagawa, personal communication) to be very similar to those of the *E. coli recA* protein (Radding, 1981; Shibata et al., 1980), and these functions are listed in Table 1. In addition, Miyazaki and Minagawa showed that gp*uvsX* had a molecular weight of about 40,000, similar to that of *recA* protein (Minagawa, personal communication). The *recA* protein has been shown to be essential for PRRR in *E. coli* (Smith and Meun, 1970).

Further substantiation of the similarity between *uvsX*-, *uvsY*-, and *uvsW*-mediated PRRR and *recA*-mediated PRRR has been provided by J. K. DeVries and S. S. Wallace (unpublished data). These workers have shown that *uvsY* and *uvsW* genes, cloned in plasmid pBR322, are expressed in *E. coli*, as measured by their ability to complement T4 *uvsY* and *uvsW* mutants in terms of survival after UV irradiation, and that plasmids containing both the *uvsY* and *uvsW* genes can enhance the survival of *E. coli recA* mutants treated with UV, methyl methane sulfonate (MMS), and ethyl methane sulfonate (EMS). Recently, Junko Hosoda (personal communication) has found that the gene products of *uvsX* and *uvsY* bind to a gp32 column and give molecular weights of 42,000 and 17,000, respectively, as determined by sodium dodecyl sulfate-gel electrophoresis.

A direct link between *uvsX*, *uvsY*, and *uvsW* functions, DNA replication, and structural DNA replicative intermediates has been established in a number of laboratories. Melamede and Wallace (1977, 1978, 1980a) have shown that the incorporation of thymidine into DNA in cells infected with *uvsX*, *uvsY*, and *uvsW* mutants depends on the concentration of thymidine used to measure synthesis. When high concentrations of thymidine were used and synthetic capacity was measured over the course of infection, *uvsX*, *uvsY*, and *uvsW* mutant-infected cells exhibited a small but reproducible reduction in DNA synthesis. When low concentrations of thymidine were used to pulse-label infected cells at various times after infection, no differences were observed in mutant-infected cells early in infection. However, when this same procedure was used late in infection, T4 *uvsX* and *uvsY* mutant-infected cells exhibited greatly reduced DNA synthetic capacity.

Melamede and Wallace (1977) have also demonstrated that *uvsX* and *uvsY* are early genes, that is, their transcripts are synthesized before 2 min and translated before 8 min after infection at 37°C. Early transcripts have also been observed by use of cloned gene segments containing *uvsY* and *uvsW* (Young et al., 1980).

Late in infection, T4 *uvsX*, *uvsY*, and *uvsW* mutant-infected cells exhibit reduced formation of concatemeric DNA (Cunningham and Berger, 1977; Hamlett and Berger, 1975; Wakem and Ebisuzaki, 1976). Also, *uvsW* mutants exhibit fast-sedimenting DNA replicative intermediates, whereas *uvsX* and *uvsY* mutants have slow-sedimenting DNA intermediates as compared to wild type (Melamede and Wallace, 1980a; Melamede and Wallace, 1980b). Furthermore, Yonesaki and Minagawa (Minagawa, personal communication) have shown greatly reduced formation of joint molecules in *uvsX*- and *uvsY*-infected cells compared to wild type, whereas wild-type levels of joint molecules were found in cells infected with *uvsW* mutants.

The presence of a *uvsW* mutation (but not *uvsX* or *uvsY*) suppresses the DNA arrest phenotype of gene 46–47 and 59 mutants, but does not affect the gene 46-47 mutant recombination defects and only partially suppresses the gene 59 mutant recombination defect (Cunningham and Berger, 1977). These data imply that the replication- and recombination-deficient phenotypes of gene 46, 47, and 59 mutants are separable. Wu and Yeh (1975; 1978) have isolated a mutation called *dar*, which also suppresses the DNA arrest phenotype of gene 59, is sensitive to hydroxyurea, maps between genes 24 and 25, and has properties similar to *uvsW* with respect to DNA intermediates. T4 *dar*, however, is not UV sensitive or recombination deficient and might represent an interesting allele of *uvsW*.

Two mutations, *fdsA* and *fdsB*, isolated as suppressors of a gene 49 mutation (Dewey and Frankel, 1975), turned out to be alleles of the *uvsX* and *uvsY* genes, respectively (Cunningham and Berger, 1977; Shah and DeLorenzo, 1977). Mutants bearing these genes also exhibit aberrant DNA replicative intermediates and show patterns of thymidine incorporation similar but not identical to those of the *uvsX* and *uvsY* mutants originally studied (Melamede and Wallace, 1980b). DNA intermediates formed in *uvsX* 49 double mutants contain single-stranded regions that can be protected by gp32 (Shah and DeLorenzo, 1977). It is likely that the ability of *uvsX* and *uvsY* mutants to suppress the gene 49 defect is related to their abnormally small DNA replicative intermediates, which enables them fortuitously to supply phage-size DNA lengths.

Recently, a number of amber (Wakem and Ebisuzaki, 1981) and temperature-sensitive (Conkling and Drake, personal communication) alleles of *uvsX* and *uvsY* have been identified which exhibit some or all of the phenotypes elaborated above; that is, some of the phenotypes can be uncoupled. This observation agrees with the earlier studies in which epistasis of *uvsX*, *uvsY*, and *uvsW* was observed with repair but not recombination (Hamlett and Berger, 1975).

In recent studies using *uvsW* and *uvsX* mutant-infected plasmolysed cells, Melamede and Wallace (two articles in Mol. Gen. Genet., in press) have demonstrated that aberrant DNA incorporation patterns are also observed in situ and are identical to the patterns observed in vivo. These workers have interpreted their data as reflecting unique and separate precursor feeds to the replicative fork where thymidine is a preferred exogenous precursor. The mutant incorporation patterns are presumed to result from perturbations in DNA replication induced by the recombinational defects of *uvsX*, *uvsY*, and *uvsW* mutants.

Taken together, these data indicate that the *uvsX*, *uvsY*, and *uvsW* functions are involved in repair, recombination, DNA replication, and processing events which occur late in T4 infection.

As indicated in Table 1, defects in most of the 17 genes listed have been shown to cause significantly increased levels of sensitivity to various agents. These increased levels of sensitivity indicate that the *uvsX* system, originally discovered by Harm (1964), is an important repair pathway in phage T4. Since seven gene functions in this pathway are required for recombination and a further eight are required for replication, and since gp*uvsX* is similar in function to *E. coli recA*, we have inferred that this pathway is similar to the well-established pathway of PRRR in *E. coli*.

Multiplicity reactivation. When phage are treated with an agent causing reduced survival, it is often seen that allowing multiple phage to infect each cell gives a level of survival of infective centers that is greater than would be expected from the presence of multiple independent targets. This phenomenon, called multiplicity reactivation (MR), implies that the presence of more than one phage chromosome within a cell facilitates a DNA repair process. There is a substantial amount of genetic evidence indicating that MR is a recombination repair process. Recombination is implicated, first, because MR depends on the presence of two or more genomes (Luria, 1947). Second, for the DNA-damaging agents UV, HNO_2, and MMS, MR depends on several gene functions required for normal levels of spontaneous recombination (Table 2 lists the gene products on which MR depends, and Table 1 lists many of their functions). Third, under conditions where MR occurs, the frequency of genetic recombination increases. This has been shown for lesions caused by UV (Epstein, 1958), nitrous acid (HNO_2; Fry, 1979), *N*-methyl-*N*′-nitro-*N*-nitrosoguanidine (MNNG; Schneider et al., 1978), EMS (Johns et al., 1978), MMC (Holmes et al., 1980), psoralen plus near-UV light (PUVA; et al., 1982), Miskimins ^{32}P (Symonds and Ritchie, 1961), and X rays (Harm, 1958). Fourth, a mutation in gene 46 or 47 reduces or eliminates both MR of HNO_2-damaged phage (Nonn and Bernstein, 1977) and the increased genetic recombination caused by HNO_2 damage (Fry, 1979).

For most lesions MR is probably similar to PRRR carried out during single infections since there is a large overlap of gene functions involved in the two repair processes. Sixteen gene functions have been shown to be required for normal levels of MR, and 10 of them also seem to be required for PRRR (Table 2). The remaining seven functions implicated in PRRR have not been tested for MR involvement.

Six functions required for MR, however, were tested for a role in PRRR and not found to be required (Table 2). Thus, there appear to be some nonoverlapping reactions as well. There may be more than one MR pathway, since gp46 and gp47 are required for MR of HNO_2-, MMC-, and UV-damaged phages, but not for the MR of EMS- and MNNG-damaged phages (Table 2). In addition, for some kinds of MR there seem to be two components to the process. For three agents, namely, UV, X rays, and EMS, when survival is plotted against dose, there is a resistant shoulder followed by single-hit kinetics (see Table 3). For MR of UV-damaged phage, the resistant shoulder can be removed or reduced by mutations in genes *uvsX*, *uvsY*, *uvsW* (strain 1206), 32, 44, 46, 47 (*ts*L86 at 31°C), or 59 without substantially affecting the exponential portion of the survival curve (Boyle and Symonds, 1969; Davis and Symonds, 1974; Nonn and Bernstein, 1977; Symonds, Heindl and White, 1973; Van den Ende and Symonds, 1972). Thus it was proposed that there may be a saturable but efficient kind of MR generating the shoulder that depends on gene proteins 32, 44, 46, 47, 59, *uvsX*, *uvsY*, and *uvsW*, whereas a second form of MR may be responsible for the exponential portion of the survival curve (Nonn and Bernstein, 1977; Symonds et al., 1973). Mutations in genes *denV*, 47 (*ts*L86 at 34.5°C), and host *polA* reduce the slope of the exponential portion (rather than the shoulder) of the MR curve of phage treated with UV (Maynard-Smith et al., 1970; Nonn and Bernstein, 1977; Symonds et al., 1973).

Krisch and Van Houwe (1976) showed that the gene 46- and 47-controlled exonuclease, the gene 43 DNA polymerase, and, to some extent, the gp*uvsY* function before gp32 during MR.

Luria (1947) and Luria and Dulbecco (1949) showed that the presence of UV-damaged genomes of a T-even phage within a cell does not affect the ability of a single undamaged T-even genome in the same cell to carry out a successful infection. This implies that in multiple infections lethal damages are not usually recombined into otherwise functional DNA. Many thymine dimers induced by UV light, however, may be transmitted into progeny phage during MR (Sauerbier and Hirsch-Kauffmann, 1968). The MR pathway which acts on UV-induced lesions in phage T4 is also apparently error free, since it overcomes lethal lesions without introducing new mutations (Yarosh, 1978).

One of the first models proposed for MR was by A. H. Sturtevant, the distinguished *Drosophila* geneticist, in a footnote to a paper by Luria and Dulbecco (1949). He proposed "a process of zipperwise replication of the various units of a phage particle. When in this process an inactive unit was reached, replication could continue only if the partial replica came in contact with another phage particle in which that unit was active. The process would then continue by addition of replicas of the active units of the second phage particle. If repeated several times, such a mechanism would provide for selective recombination of all ac-

TABLE 2. Comparison of gene products involved in MR and PRRR

Product of gene:	Survival of phage damaged by the agent shown does not (0) or does (+) depend on the gene product indicated: MR	PRRR[a]
30	0 HNO_2 (1)[b]	0 HNO_2 (2), +UV, +MMS, +EMS
32	+HNO_2 (1), +MMC (3), +UV (4)	+HNO_2, +MMC, +UV, +EMS, +MNNG
41	+UV (5, 6)	+UV, +MMS
44	+UV (4)	+UV
46	+HNO_2 (1), +MMC (3), +UV (4), 0 EMS (7), 0 MNNG (8)	+HNO_2, +MMC, +UV, +EMS, +MNNG, +PUVA
47	+HNO_2 (1), +MMC (3), +UV (1, 4), 0 EMS (7), 0 MNNG (8)	+HNO_2, +MMC, +UV, +EMS, +MNNG, +PUVA
58–61	Not tested	+UV
59	+UV (4)	+UV, +MMS
uvsW	+UV (6a, 9)	+UV, +PUVA, +MMS
uvsX	+UV (6, 9, 10, 11), +HNO_2 (1), +MMC (12), 0 MMC (3)	+UV, +HNO_2, +MMC, +MMS, +MNNG, +PUVA, +EMS
uvsY	+UV (9, 10), +HNO_2 (1), +MMC (3)	+UV, +HNO_2, +MMC, +MMS, +MNNG, +PUVA, +EMS
45	+MMC (13)	+UV
1, 42, 43, 56	Not tested	+UV
E. coli lig	Not tested	0 UV (14), +X ray
denV	+UV (9, 10, 11, 15)	0 UV (5, 11, 16, 17, 18)
E. coli polA	+UV (19)	0 UV (16, 20), 0 MMS (20)
E. coli recA	+UV (21)	0 UV (20), 0 MMS (20)
39, 52, 60[c]	+PUVA (22)	0 PUVA (22)

[a] This information is from Table 1 unless a reference is given.

[b] Numbers in parentheses indicate these references: (1) Nonn and Bernstein, 1977; (2) Bernstein et al., 1976; (3) Holmes et al., 1980; (4) Davis and Symonds, 1974; (5) Cupido et al., 1980; (6) van Minderhout and Grimbergen, 1976; (6a) van den Ende and Symonds, 1972; (7) Johns et al., 1978; (8) Schneider et al., 1978; (9) Symonds et al., 1973; (10) Boyle and Symonds, 1969; (11) Harm, 1963; (12) Sekiguchi and Takagi, 1960; (13) Schneider, personal communication; (14) Wallace and Melamede, 1972; (15) van Minderhout and Grimbergen, 1975; (16) Ebisuzaki et al., 1975; (17) van Minderhout et al., 1978; (18) Harm, 1964; (19) Maynard-Smith et al., 1970; (20) Mortelmans and Friedberg, 1972; (21) Priemer and Chan, 1978; (22) Miskimins et al., 1982.

[c] The products of these three genes jointly form a DNA toposiomerase (Liu, Liu, and Alberts, 1979; Stetler et al., 1979).

tive units." This kind of model can accommodate most of the presently known features of MR. It involves the interaction of two homologous genomes and it predicts a dependence on gene functions required for recombination. It also implies that lesions should stimulate recombination.

With some modification Sturtevant's model would also fit with the data of Rayssiguier and Vigier (1972; 1977), who analyzed the progeny produced as a result of MR from individual single bursts when a cross was performed between differentially marked phage that had been UV irradiated with an average of about seven phage lethal hits. In their experiments the parental phages differed from each other by markers at either 26 or 35 well-distributed sites. In 30% to 70% of the single bursts examined in three experiments, a clone of progeny phage with either one or the other of the parental combinations of markers was found. When one parental genotype was present, the other parental genotype was generally absent. The average size of these parental marker clones was 9 to 22 phage in the three experiments. The single bursts also contained different recombinant types. About 95% of the observed recombinants occurred only once in a burst. These results suggest that during MR of UV-damaged phage there is an early appearance in a fraction of the bacteria of an undamaged parental genome which is able to replicate, and there is also the subsequent formation of many recombinants. To accommodate these data the Sturtevant model could be modified in the following way. During replication, when a lesion is reached, replication would stop and then could continue only at a point somewhat past the lesion, leaving an unreplicated single-stranded gap. This gap could be repaired by addition of a single-stranded region from a second chromosome. The process of replication and single-strand-gap filling at lesions would continue until an entire active chromosome was completed. This active chromosome would then replicate a number of times, accounting for the clonal occurrence of phage with the parental configuration.

Another approach to interpreting MR data was the partial replica model (Luria, 1947). Luria assumed "that a phage particle contains a certain number of different self-reproducing 'units' (loci) . . ." This number of units was postulated to be independent of whether damage was present or not, and for phage T4 it was about 32 units. When UV irradiation had occurred and multiple phage infected a cell, then "each of the active units [units undamaged by UV] can reproduce copies of itself in excess of the number needed for multiplication of the phage particle as a whole. The copies of each unit might then either become incorporated into other phage particles that missed them, or come together to reconstitute active phage particles, independently of the origin of the individual units from one or another of the infecting particles."

This partial replica model has been modified over the years. To accommodate the inactivation kinetics at high UV doses, Barricelli (1956) proposed that the subunits of the phage are part of a continuous genetic structure. This is especially pointed out in Epstein's study (1958), a paper in which Barricelli is also thanked for his advice. To accommodate the fact that lesions stimulate recombination, Barricelli (1960) further modified the partial replica model to allow recombination acts to be initiated at the site of lesions. By 1980, Rayssiguier et al. (1980) added to the partial replica model the change "that DNA replication can be initiated only at a small number of specific sites (replication origins) . . ." The partial replica model is now similar to the Sturtevant model in that both assume sequential replication along a parental genome and stimulated recombination at the site of a lesion. Rayssiguier et al. (1980) further showed that during MR: (i) there is replication of 76% of UV-damaged genomes when each received an average of 10 phage lethal hits; (ii) for most of these genomes this replication is repetitive; and (iii) newly synthesized progeny DNA originating from UV-irradiated templates appears as shorter single-strand genome segments than progeny DNA from non-UV-irradiated templates.

A likely major pathway of MR is presented in Fig. 3. In this figure only a segment of the chromosome to be repaired is shown. Other lesions may be present in the parental strands outside the diagrammed segment. The pathway is assumed to be similar to the pathway for PRRR (Fig. 2) because of the large overlap of gene functions involved in these two pathways. Although similar, MR and PRRR are also distinct in that MR substantially enhances survival of damaged phage even when the PRRR pathway is present. PRRR may be more constrained than MR because PRRR is necessarily between two daughter chromosomes resulting from replication of a parental chromosome, whereas MR may be between a replicating chromosome and an unrelated, nonreplicating chromosome (as in Fig. 3). A damaged chromosome which is being replicated may have a number of gaps and stripped-back single-stranded regions in both newly replicated daughter arms, as well as a single-stranded region at the replicative fork due to lagging strand synthesis. These regions of single-stranded DNA could not participate in recombinational repair. If PRRR were taking place, the potential for recombinational repair between the daughter chromosomes may be limited by these single-stranded regions. With MR, however, the single-strand segments which are needed to patch over a lesion (Fig. 3, step d) can come from a chromosome which has not yet replicated and therefore does not have single-stranded gaps or stripped-back regions.

All the gene products known to act in MR of MMC lesions, HNO_2 lesions, and UV lesions repaired during "shoulder repair" (Table 2) have been placed in the pathway shown in Fig. 3. The gene products involved in these types of MR overlap to a large extent (Table 2), making it reasonable to suppose that a common MR process is used. The kind of recombination act shown, transfer of a single-stranded patch, may be the most frequent type of recombination occurring during repair, since the data of Rayssiguier and Vigier (1972; 1977) imply that even after seven recombinational bypasses of lesions a parental configuration of markers can often be maintained.

Error-Prone Repair

The fidelity of DNA replication has been more actively studied in phage T4 than in any other single system and is discussed in detail by Drake and Ripley and by Goodman and Sinha (this volume). The avail-

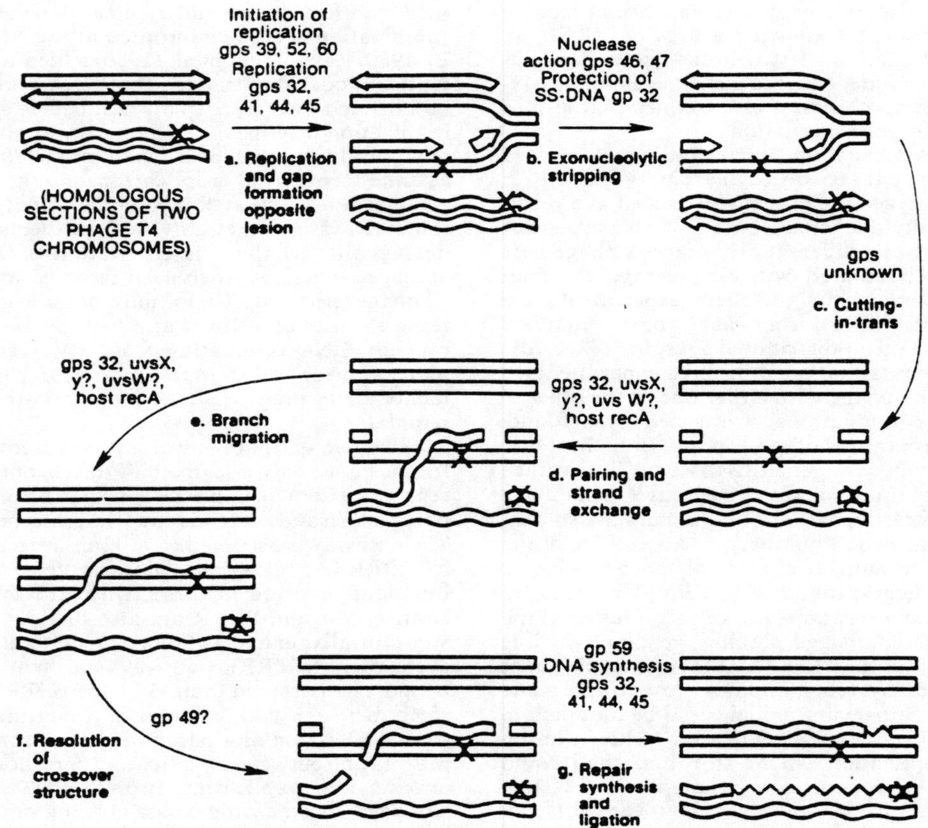

FIG. 3. Model of MR in T4-infected cells.

ability of mutator and antimutator mutants defective in the gene 43 polymerase has led to the hypothesis that the fidelity of DNA replication depends on the specificity of the polymerizing function of the enzyme and on the efficiency of its 3′-exonuclease editing function (Gillin and Nossal, 1976a; Lo and Bessman, 1976). Yarosh et al. (1980) presented evidence that the DNA polymerase is involved in error-prone repair of UV and PUVA lesions. Furthermore, replication fidelity during error-prone repair is promoted by an intact functional replication complex in which gp43 appropriately interacts with gp44, gp45, and gp62 (accessory proteins), gp32 (helix-destabilizing protein), and, probably, gp41 and gp61 (RNA priming proteins) (Mufti, 1980).

Meistrich and Drake (1972) showed that in T4-infected cells thymine dimers are mutagenic. However, they also showed that almost all transition mutations caused by thymine dimers involve GC to AT transitions. Thus, the error takes place at a G or C site that is not directly opposite the thymine dimer. This implies error-prone synthesis rather than mispairing opposite a lesion. Error-prone repair in phage T4 has been found to depend on the products of *uvsX*, *uvsY*, and *uvsW*, since UV-induced mutation is reduced in mutants defective in these genes (Green and Drake, 1974). A number of other lesions are also subject to the *uvsX*, *uvsY*, *uvsW* error-prone repair system, including those induced by MMS and 8-methoxypsoralen plus white light (Green and Drake, 1974) and by gamma rays (Bleichrodt and Roos-Verheij, 1979; Conkling et al., 1976).

Bernstein et al. (1976), working with HNO_2 mutagenesis, and Yarosh (1978), working with UV mutagenesis, showed that a temperature-sensitive allele of the T4 DNA ligase eliminated induced mutagenesis under conditions where the burst size was still 35% that of normal. Thus, HNO_2- or UV-induced mutation apparently depends on a ligase-mediated repair process rather than on overall DNA synthesis. Whereas mutants which have an antimutator DNA polymerase phenotype or a ligase defect can eliminate mutagenesis by UV, PUVA, or HNO_2 (Bernstein et al., 1976; Drake and Greening, 1970; Yarosh et al., 1980), these mutants have almost or completely wild-type levels of resistance to inactivation by these agents (Bernstein et al., 1976; Yarosh et al., 1980). Thus the error-prone repair pathway blocked by these mutations must be utilized for only a small proportion of UV, PUVA, or HNO_2 lesions, or else it can be replaced, when blocked, with a more accurate pathway.

Other Repair Pathways

A mutant, *mms*1, has increased sensitivity to MMS when combined with a mutation in either gene *uvsX* or *uvsY*, as well as an increased sensitivity to UV when combined with T4 *denV* (Ebisuzaki et al., 1975). Thus, this mutant appears to function on both MMS-induced and UV-induced damages in a repair pathway distinct from the already defined pathways.

Three other UV-sensitive mutants, *uvs*58 (gene 41), *uvs*79 (gene 41), and *uvs*57 (*uvsZ*), were characterized by van Minderhout and Grimbergen (1976), van Minderhout et al. (1978), Cupido et al. (1980), and Cupido et al. (1982). Similarly to *mms*1, these mutants also showed increased UV sensitivity in combination with either an amber *denV* mutant or a *uvsX* or *uvsY* mutant. Gene 41 codes for an enzyme which functions both as a helicase and a primase and is required for T4 synthesis in vitro and in vivo (Alberts et al., 1980; Epstein et al., 1963; Silver et al., 1980).

In addition to its role in excision repair of pyrimidine dimers, the *E. coli* DNA polymerase I (*polA*) gene has been implicated in the repair of lesions induced by MMS (Ebisuzaki et al., 1975; Nishida et al., 1976), EMS (Ray et al., 1972), and X rays (Wallace and Melamede, 1972). This repair appears to be distinct from the *uvsX*, *uvsY*, *uvsW* PRRR repair pathway since mutants defective in these genes exhibit reduced survival after MMS or X-ray treatment when plated on a *polA* mutant host.

TYPES OF LESIONS REPAIRED AND EFFICIENCIES OF REPAIR

We have summarized (Table 1) the rates of lethal hits delivered to mutants compared to wild type at any dose. The largest relative inactivation rate ($k_{mutant}/k_{wild\ type}$) in Table 1 was 3.4 for a gene 47 mutant treated with MNNG. By using this value, it can be calculated that when the gene 47 product is functional, (3.4 − 1.0)/3.4 or 71% of the lethal MNNG lesions which are present during a mutant infection are repaired by a pathway using the gene 47 function in a wild-type infection. Calculations such as this may underestimate the fraction of lethal lesions which are reparable, since another repair pathway, not involving the tested gene function, might remove additional lesions caused by this agent. In cases of UV and MMS damages there are three pathways of repair that operate during a single infection (see above). In these cases we can use $k_{mutant}/k_{wild\ type}$ values of mutants defective in all three pathways. These values are 5.5 for UV (Cupido et al., 1982) and 2.8 for MMS (Ebisuzaki et al., 1975). The percentages of lethal lesions known to be repaired have been calculated and are given for each agent in Table 3. They indicate that phage T4 is able to repair a substantial percentage of many different types of DNA lesions during single infections.

TABLE 3. Measured levels of repair during single and multiple infections

Agent	Single infections: % lesions repaired[a]	Multiple infections	
		MR factor	Shoulder repair
UV	82	4.2 (1)[b]	18 (1)
MNNG	71	2.9 (2)	None
MMS	64	—[c]	—
MMC	41	36.0 (3)	None
X rays	41	4.0 (4)	27 (4)
PUVA[d] trioxalen	38	5.5 (5)	None
4′-Aminomethyl-4,5′, 8-trimethyl-psoralen	—	11.0 (6)	None
HNO_2	23	5.0 (7)	None
^{32}P	—	4.6 (8)	None
EMS	3 lethal hits[e]	1.0 (9)	6 (9)

[a] These values are calculated using $k_{mutant}/k_{wild\ type}$ values from Table 1 except for UV and MMS. For UV and MMS the calculations are described in the text. Other values were calculated in an analogous fashion.

[b] Numbers in parentheses refer to the indicated references: (1) Average of 12 separate determinations which are as follows for MR factor and shoulder repair: 2.3 and 12, Cramer and Uretz (1966); 2.3 and 9, Nonn and Bernstein (1977); 2.3 and 12, van Minderhout and Grimbergen (1975); 2.4 and 20, Harm (1963); 2.5 and 20, Harm (1956); 3.3 and 30, Symonds et al. (1973); 3.4 and 6, Epstein (1958); 3.6 and 16, Boyle and Symonds (1969); 4.0 and 25, Dulbecco (1952); 5.0 and 34, van den Ende and Symonds (1972); 7.0 and 7–9, Yarosh (1978); 12.0 and 19, Davis and Symonds (1974). (2) Schneider et al. (1978). (3) This is the average of two determinations: 10, Holmes et al. (1980); 62, Miskimins et al. (1982). (4) Harm (1958). (5) Miskimins et al. (1982); similar results, using methoxalen, but with biphasic curves, were obtained by Strike et al. (1981). (6) Schneider, personal communication. (7) Nonn and Bernstein (1977). (8) Symonds and Ritchie (1961). (9) Johns et al. (1978).

[c] —, Not tested.

[d] There are a number of kinds of psoralen (Isaacs et al., 1977) that can be used with near-UV light (UVA) to create various ratios of four different monoaddition products and a cross-linking diadduct (Kanne et al., 1982).

[e] The EMS survival curve, under single-infection conditions, has a shoulder equal to three phage lethal hits (of the exponential portion of the curve), and this shoulder is removed by a mutation in gene 46 or 47 (Johns et al., 1978).

In general, data from MR experiments are presented by plotting the log of surviving infective center-forming ability of multiply infected cells (multicomplexes) versus dose of an inactivating agent. These multicomplex inactivation curves are usually straight lines, or straight lines with an initial shoulder, depending on the agent used and the mutations present in the phage. Over the straight-line regions of these curves, the inactivation kinetics can be represented, in a form similar to the form of the inactivation kinetics of singly infected cells, as $N/N_0 = e^{-kd}$. To compare the survival of infective center-forming abilities of singly infected cells (monocomplexes) and multiply infected cells (multicomplexes), a term called the MR factor can be calculated. The MR factor for any strain is given by k_{mono}/k_{multi}, where the k values are taken from the straight-line portions of the curves. Table 3 shows the MR factors of wild-type phage after treatment with each of nine different agents. For three of the nine inactivating agents listed in Table 3 (UV, X rays, and EMS), the multicomplex inactivation curves had substantial shoulders. The length of the shoulder is defined by the dose of inactivating agent needed to reach the transition to the straight-line portion of the curve. This dose can be represented in terms of lethal hits delivered to monocomplexes (observed in experiments performed in parallel with the multicomplex experiment). Table 3 gives the amounts of shoulder repair measured in these terms.

The MR factor and the shoulder repair value together indicate the additional repair available in multicomplexes compared with monocomplexes. For MMC lesions, which had one of the highest levels of MR, with an MR factor of 36 and no shoulder repair, 36-fold fewer lethal lesions remain in an average multicomplex compared with the number present in an average monocomplex for any given dose of treatment. For UV irradiation, there is both a large amount of shoulder repair and a significant MR factor. In the experiments of Dulbecco (1952), inactivation of multicomplexes was barely measurable at a UV dose which reduced survival of monocomplexes to 10^{-5}. Thus, when MR occurs it is often a very dramatic phenomenon.

Lesions caused by lethal agents. (i) Thymine dimers. Work in phage T4 was important in defining the relative significance of thymine dimers in UV inactivation. UV light produces a number of photoproducts in DNA. The most frequently produced lesions include pyrimidine dimers (see Patrick and Rahn [1976] and Rahn [1979] for review of these and other lesions unless a reference is given), an alkali-labile PydC lesion (Lippke et al., 1981), dihydrothymine, pyrimidine adducts, and DNA cross-links. Other lesions include cytosine hydrate, single-strand breaks, and DNA-protein cross-links. Meistrich (1972) has shown that after UV irradiation of phage T4, thymine dimers (as distinct from all pyrimidine dimers) account for 56% of lethal hits if the major repair pathways dependent on the products of *denV* and *uvsX* are absent. Meistrich (1972) also showed that a double mutant defective in both the *denV* repair pathway and the *uvsX* repair pathway had a ratio of 2.5 thymine dimers created per lethal hit to the phage. This implies that 22% of all thymine dimers (56% divided by 2.5) induce lethality in the double mutant used. Meistrich was also able to show that the *denV* pathway preferentially repairs thymine dimers, whereas the *uvsX* pathway acts on all UV-induced lethal lesions.

Pawl et al. (1976) showed that during a wild-type infection about 50% of the thymine-containing dimers in a phage chromosome are removed by repair during the first 5 min after infection. The remaining dimers are not removed with an added 10 min of incubation whether DNA replication occurred or was inhibited.

In cells infected by UV-treated phage T4, the UV photoproducts left after photoreactivation have a higher probability of undergoing MR than the combination of lesions present before photoreactivation (Dulbecco, 1952). As discussed above, MR of UV-treated phage T4 probably occurs by a recombinational mechanism. Thus non-photoreactivable UV-induced lesions may especially stimulate recombinational repair.

(ii) Alkylation products. Alkylating agents produce a wide variety of changes in DNA (Strauss et al., 1975). When wild-type phage T4 is treated with EMS, inactivation curves (plotted on semilogarithmic coordinates) have a shoulder, followed by a straight-line exponential decline (Brooks and Lawley, 1963; Johns et al., 1978). Approximately 500 ethylations of DNA must occur before any lethality is detected (Brooks and Lawley, 1963). This corresponds to the shoulder portion of the curve. In the exponential portion, there are about 400 additional ethylations for each lethal hit. When mutations in gene 32, 46, or 47 are present, the shoulder of the inactivation curve is removed (Johns et al., 1978). The shoulder is reduced by a mutation in phage gene *uvsX* (Johns et al., 1978; Ray et al., 1972), *uvsY* (Johns et al., 1978), or 30 (Johns et al., 1978; Ray et al., 1972) or in a *polA* host (Ray et al., 1972). The remainder of the inactivation curve generally has a slope which is about the same as the slope in the exponential portion of the wild-type inactivation curve. These results suggest that the initial lesions introduced by EMS are subject to efficient repair by one or more pathways involving the products of phage genes 30, 32, 46, 47, *uvsX*, and *uvsY* and the host gene *polA*, but that this repair system can be saturated by a certain number of lesions in the DNA. MNNG, another alkylating agent, causes inactivation of T4-host complexes with single-hit kinetics (Zampieri et al., 1968). When a mutant defective in gene 32, 46, 47, *uvsX*, or *uvsY* is used, single-hit inactivation is again observed, but the rate of inactivation in all cases is greater than the rate for the wild-type (Schneider et al., 1978). There was no initial shoulder on these MNNG inactivation curves, indicating that the repair pathway(s) coded for by these genes is(are) not saturable, in contrast to the EMS repair pathway(s). Loveless (1966) summarized and interpreted the previously reported effects of alkylating agents on phages, including phage T2.

(iii) DNA cross-links. Strike et al. (1981) used PUVA treatment to induce both monoadducts and cross-links in phage T4 DNA. A sample of PUVA-treated phage was also dialyzed to remove unbound psoralen and reirradiated with near-UV light to convert some of the monoadducts into cross-links. Strike et al. then measured both the frequency of phage that were free of cross-links and the surviving fraction of phage. From this they inferred that monoadducts cause a substantial amount of lethality. However, on converting monoadducts to diadducts there was increased lethality, indicating that cross-links are more lethal than monoadducts. Krasin et al. (1976) also presented data indicating that cross-links formed upon decay of [2-^{3}H]adenine were lethal in phage T4.

Nitrous acid induces cross-links at a frequency which is close to the frequency of induction of lethal hits in phage T2 (Becker et al., 1964). However, nitrous acid also produces three oxidatively deaminated bases in DNA with sufficient frequency to account for the lethality (Vielmetter and Schuster, 1960).

Many alkylating agents are bifunctional and induce DNA cross-links. Loveless (1966) comprehensively reviewed the quantitative inactivating effects of bifunctional cross-linking agents on phage.

Other lesions. Other lesions which are subject to repair after phage T4 infection are mismatched bases (Berger and Pardoll, 1976) and heteroduplex loops (Berger and Benz, 1975). Repair of the latter depends on the *denV* product. X rays produce a variety of lesions in DNA including strand breaks, base damages, and cross-links (for a review, see Ward [1975]). In the case of T4, most X-ray inactivation is due to double-strand breaks (Freifelder, 1968). Target theory calculations, which indicate that T4 is very X ray resistant for its size, led Freifelder to conclude that X-ray-induced base damages are efficiently repaired in T4 as compared to other coliphages. ^{32}P decays can produce single- and double-stranded breaks in phage T4 DNA (Miller, 1970). The double-strand breaks seem to cause most lethality (Levy, 1975; Ley and Krisch, 1974). Some of this lethality can be reversed by MR (Symonds and Ritchie, 1961).

CONCLUSIONS

Repair in Phage T4 as a Useful Model System

As we have seen in Table 3, phage T4 can repair large fractions of the lesions induced by many agents. The repair pathways in phage T4, including excision repair, recombinational repair, photoreactivation, and error-prone repair, are probably similar to the repair systems used by most organisms. Thus we consider that an understanding of DNA repair in the experimentally accessible phage T4 system should clarify the essential features of repair in higher organisms, including humans.

Sexual Reproduction

Sexual reproduction has evolved as the principal means of reproduction in most extant multicellular organisms. It is usually assumed that the key element of the sexual cycle, which explains the selective value of this means of reproduction, is genetic recombination. It was proposed by Fisher (1930) and Muller (1932) that the selective advantage of genetic recombination, and hence of sexual reproduction, arises from the production of genetic variation among the progeny produced by mating. Only recently has it been shown that there are profound difficulties in explaining the selective advantage of sexual reproduction solely in terms of the variation hypothesis (see, e.g., the authoritative books by Williams [1975] and Maynard-Smith [1978]).

An alternative hypothesis, proposed recently, is that sexual production may be advantageous, in large part, because it promotes recombinational repair of germ line DNA (Bernstein, 1977; Bernstein, 1979; Bernstein et al., 1981; Martin, 1977). Some of the best examples of efficient recombinational repair are in the phage T4 system, as reviewed above.

Although the sexual cycle varies greatly among diverse organisms, from phages to humans, the main steps of the cycle are probably of general occurrence. These are: (i) two genomes or parts of genomes come together within a shared cytoplasm; (ii) the genomes pair, so that homologous sequences are adjacent; (iii) accurate exchange of genetic material occurs between the two genomes; and (iv) the exchange is followed by separation of the products of the interaction. The steps of the sexual cycle are the same as those presumed to be required for recombinational repair, including MR, a major form of repair in phage T4. If spontaneous or induced lesions similar to those produced by the nine agents listed in Table 3 are a significant problem to survival of various organisms in their natural condition, they could have provided the selective basis for the evolution of recombinational repair and hence sexual reproduction.

Support was provided by Public Health Service grants GM27219 (to C.B.) and CA33657 (to S.S.W.) from The National Institutes of Health.

DNA Modification: Methylation

STANLEY HATTMAN

Department of Biology, University of Rochester, Rochester, New York 14627

Bacteriophages T2 and T4 contain minor amounts of the methylated base N^6-methyladenine (m^{6}A) (Dunn and Smith, 1958; Gefter et al., 1966); however, the closely related phage T6 lacks any detectable level of m^{6}A. In accord with these findings is the fact that T2 and T4, but not T6, each control the production of a DNA-adenine methylase activity (Fujimoto et al., 1965; Gold et al., 1964; Hausmann and Gold, 1966). These enzymes are distinct from the major host DNA-adenine methylase (J. E. Brooks, Ph.D. thesis, University of Rochester, Rochester, N.Y., 1977; van Ormondt et al., 1975) specified by the *Escherichia coli dam* gene (Marinus and Morris, 1973). The host enzyme methylates the palindromic sequence G-A-T-C to produce G-m^{6}A-T-C (Hattman, Brooks, and Masurekar, 1978; Lacks and Greenberg, 1977); the phage enzymes are also capable of methylating this sequence (Brooks, thesis; Hattman, van Ormondt, and de Waard, 1978). Mutants of both T2 and T4 have been isolated which alter or abolish the enzyme activity (C. P. Georgopoulos, Ph.D. thesis, Massachusetts Institute of Technology, Cambridge, 1969; Hattman, 1970). However, methylation is a nonessential function because unmethylated T2, T4, and T6 are viable (Georgopoulos, thesis; Hattman, 1970). In this connection, one is able to obtain virtually unmethylated virion DNA, suggesting that the host *dam*$^-$ methylase poorly utilizes T-even phage DNA (whether glucosylated or not) as a substrate, because presence of 5-hydroxymethylcytosine in the phage DNA interferes with host enzyme recognition. Alternatively, the host enzyme may be present at a relatively low concentration compared to the induced phage enzyme. This is supported by the fact that *dam*$^-$ mutant phage DNA is methylated after growth in a host that overproduces the *E. coli dam*$^-$ methylase (our unpublished data). Here I will describe several aspects of phage DNA methylase synthesis, a special case in which DNA methylation is essential for phage growth, and recent developments in cloning and mapping of the T4 *dam* gene. Because of the high degree of sequence homology (Cowie et al., 1971; Schildkraut et al., 1962) and the similarity in genetic maps (R. L. Russell, Ph.D. thesis, California Institute of Technology, Pasadena, 1967) I will assume that T2 and T4 produce similar enzymes, although this is one of the few cases in which a gene may map in different positions in T2 and T4 (see below).

DNA METHYLASE

Specificity

The first evidence for a phage-induced DNA methylase was reported in 1964 (Gold et al., 1964); extracts made from *E. coli* B cells infected with T1, T2, or T4 (but not T6) exhibited substantial increases in DNA methylase activity. It is interesting that T2-infected cells had a 10-fold higher specific activity than T4-infected cells (this will be discussed further below). Kinetic analysis showed that DNA methylase activity begins to increase within a few minutes after infection (Hausmann and Gold, 1966; Sellin et al., 1966). Addition of protein synthesis inhibitors blocked any increase in enzyme activity, indicating that de novo protein synthesis is required for methylase induction.

The T2 and T4 DNA methylases modify only adenine residues (to m^{6}A) (Fujimoto et al., 1965), and recognition is nucleotide sequence specific (Vanyushin et al., 1971). As for all other DNA methylases, the methyl donor is *S*-adenosyl-L-methionine. The T2 DNA methylase appears to be a small protein, with an estimated molecular weight of about 15,000 (M. Masurekar and S. Hattman, unpublished data; J. E. Brooks and S. Hattman, unpublished data); it is not precluded that the active enzyme is an oligomeric protein, comprised of identical subunits.

The virion DNA of wild-type phage contains only about 1 m^{6}A per 200 A residues (Hattman, 1970). This level is affected by the degree of 5-hydroxymethylcytosine glucosylation; e.g., *gt*$^-$ mutants (defective in glucosyl transferase activity and lacking glucose) exhibit 30 to 100% higher m^{6}A contents than the parental (*gt*$^+$) wild-type phage or *gt*$^+$ revertants (Hattman, 1970). Thus, it appears that glucose residues sterically hinder methylation at many recognition sites.

The m^{6}A content is lower for wild-type T4 than for wild-type T2 and is lower for T4 α*gt*$^-$ β*gt*$^-$ than for T2 α*gt*$^-$ (Hattman, 1970); it is not clear whether this is attributable to the lower level of DNA methylase activity induced by T4 (Gold et al., 1964; Hausmann and Gold, 1966), to different sequence specificity, or to some other property. These observations indicate, at least, that the T2 and T4 methylases (or their regulation) might not be identical.

As will be discussed in more detail below, my colleagues and I have been able to isolate T2 and T4 mutants, designated *dam*^{h}, which produce an altered DNA methylase (Hattman, 1970; Hehlmann and Hattman, 1972; Revel and Hattman, 1971). The *dam*^{h} forms methylate their DNA to a several-fold higher m^{6}A content than the wild-type (*dam*$^+$) parent; in addition, with some exceptions, heterologous DNAs are methylated to higher extents in vitro by *dam*^{h} than by *dam*$^+$. One consequence of this hypermethylation is that the DNA becomes modified to a form resistant to P1 restriction (Brooks and Hattman, 1978; Hattman, 1970).

We attempted to compare the sequence specificity of each of the two methylase forms; however, fingerprint and sequence analysis of labeled oligonucleotides obtained from *Micrococcus luteus* or calf thymus DNA extensively methylated in vitro did not reveal any clear qualitative differences between *dam*$^+$ and *dam*^{h} methylation (Hattman, van Ormondt, and de

Waard, 1978). For example, each enzyme had only G as the 5′-nearest neighbor to m^{6}A, and both T and C as 3′-nearest neighbors (with T preferred). Furthermore, the m^{6}A-containing tri- and tetranucleotides were similar for both enzymes. However, later studies comparing T2 *dam*$^+$ with T2 *dam*^{h} methylation of phage λ DNA provided new insight (Brooks and Hattman, 1978). First, *dam*^{h} methylation was shown to convert λ DNA to a form resistant to P1 restriction (demonstrated by transfection of *E. coli* K-12 [P1] spheroplasts and by resistance to cleavage in vitro by purified *Eco*P1 restriction enzyme). Under the same conditions, the T2 *dam*$^+$ enzyme methylated the λ DNA to a much lower extent and did not confer P1 resistance. On the other hand, both enzymes modified unmethylated λ DNA to complete resistance against cleavage by the *Mbo*I restriction nuclease. Thus, both methylases readily modify *Mbo*I sites (G-A-T-C → G-m^{6}A-T-C), but only *dam*^{h} can modify the P1 site (A-G-A-C-C → A-G-m^{6}A-C-C) (Bächi et al., 1979; Hattman, Brooks, and Masurekar, 1978). Furthermore, analysis of the nearest neighbor of m^{6}A produced during the early stages of in vitro methylation revealed a further difference between *dam*$^+$ and *dam*^{h} (Brooks and Hattman, 1978). Both enzymes appear to methylate G-A-T sites equally well, but the *dam*$^+$ form was relatively inefficient at methylating G-A-C sites. This result is consistent with the differential ability in modifying the P1 sequence.

In conclusion, although our understanding of the *dam*$^+$ and *dam*^{h} sequence specificity has increased through these studies, factors controlling in vivo methylation remain to be elucidated. For example, the 2 to 3% m^{6}A content produced by in vivo methylation of T2 *gt*$^-$ *dam*^{h} DNA is lower than the predicted 8% m^{6}A expected for methylation of G-A-Py, although we have been able to achieve this level of methylation in vitro (Masurekar and Hattman, manuscript in preparation). Likewise, we would expect about 5% m^{6}A (methylation of G-A-T) in T2 *gt*$^-$ *dam*$^+$ DNA, but we have observed 0.6 to 0.9% m^{6}A (Hattman, 1970).

We have looked for other differences in properties of the T2 enzymes resulting from the *dam*$^+$ → *dam*^{h} mutation. However, with the exception of a slightly elevated thermal lability for *dam*^{h} (Hehlmann and Hattman, 1972), the two enzymes are indistinguishable with respect to chromatographic behavior, pH and temperature optima (7.0 to 8.5, and 37°C, respectively), stimulation by mono- and divalent cations, K_m values for DNA and methyl donor, isoelectric point (9.6 to 9.9), and the nature and degree of product inhibition (Brooks, thesis; Masurekar and Hattman, in preparation).

Regulation

Under normal infection conditions, the rapid rise in enzyme activity terminates at about 10 to 15 min postinfection. In this regard, T2 DNA methylase synthesis appears similar to the other known "early enzymes."

Regulation of early enzyme production is known to be strongly influenced by conditions which block DNA synthesis; e.g., UV radiation (Delihas, 1961; Dirksen et al., 1960) and DO mutations (mutations in early genes essential for DNA synthesis) lead to a loss in the normal shutoff mechanism and a several-fold overproduction in enzyme activity (Wiberg et al., 1962). However, the pattern with T2 DNA methylase is not so clear cut. At a UV dose sufficient to reduce phage survival to 10^{-3}, both the rate of increase and the final level of methylase activity were reduced (compared to the unirradiated control); however, the shutoff was delayed beyond the normal time (Hausmann and Gold, 1966). At a higher UV dose (4×10^{-7} survival), where dTMP kinase synthesis was not reduced, T2 DNA methylase synthesis was almost abolished. The extent of this inhibition decreased with increasing multiplicity of infection; in contrast, with non-irradiated phage the level of enzyme is not dependent on multiplicity of infection. Since the response to UV irradiation was considered atypical compared to other early enzymes, it was suggested that regulation of DNA methylase synthesis might occur by a different mechanism (Hausmann and Gold, 1966). However, this question remains open because synthetic capacity, as a function of UV dose, is not the same for all early enzymes (Dirksen et al., 1960).

The effect of DO mutations on DNA methylase production was also investigated (Hausmann and Gold, 1966). With three different amber DO mutants, three different responses in methylase synthesis were observed: T2 *am*8 (gene 32) produced a normal enzyme level, T2 *am*146 (gene 43) overproduced by about 50%, and T2 *am*3 (gene 42) induced a threefold overproduction. DNA methylase overproduction was also noted with T4 amber DO mutants in genes 43 and 44. In my laboratory, however, we observed normal kinetics and level in methylase synthesis with T2 *am*3 (gene 42) (Brooks, thesis). These disparate results might be related to different degrees of "leakiness" among the different mutant strains or to different hosts. Thus, it would be worthwhile to reassess the question of DNA methylase regulation.

BIOLOGICAL ROLE OF DNA METHYLATION

As mentioned earlier, the T2 and T4 DNA methylases have no essential function in viral development under standard laboratory conditions. This follows from the fact that the *dam*$^-$ phage mutants grow normally (Georgopoulos, thesis; Hattman, 1970) and produce unmethylated viral DNA. Under appropriate conditions, however, a biological effect of DNA methylation has been observed.

Protection against P1 restriction

Non-glucosylated T-even phages (Hattman and Fukasawa, 1963; Revel et al., 1965; Shedlovsky and Brenner, 1963; Symonds et al., 1963) are subject to certain restriction systems that do not act on the glucosylated forms (Luria and Human, 1952; Revel and Luria, 1970). One such system is that controlled by phage P1 (Klein, 1965; B. Molholt, Ph.D. thesis, University of Indiana, Bloomington, 1967; Revel and Georgopoulos, 1969). For example, T2 α*gt*$^-$ phage is strongly restricted by P1-lysogenic cells, and the designation *rP1* was applied to denote "restricted by P1" (Revel and Georgopoulos, 1969). (The P1 restriction and modification enzymes recognize the same pentanucleotide sequence; the modification enzyme methylates the sequence, A-G-A-C-C, to A-G-m^{6}A-C-C [Bächi et al., 1979; Hattman, Brooks, and Masurekar, 1978]. Non-glucosylated T-even phage DNA is restrict-

TABLE 1. Phenotype/genotype of various T-even phage forms with respect to P1 restriction[a]

$P1^r$	→	$P1^s$	→	$P1^r$	→	$P1^s$
T2 αgt^+	→	T2 αgt^- *rP1*	→	T2 αgt^- *uP1*	→	T2 αgt^- u^RP1
T4 αgt^+ βgt^+	→	T4 αgt^- βgt^- *rP1*	→	T4 αgt^- βgt^- *uP1*	→	T4 αgt^- βgt^- u^RP1
T6 αgt^+	→	T6 αgt^- *rP1*				
T2 αgt^+ dam^+	→	T2 αgt^- dam^+	→	T2 αgt^- dam^h	→	T2 αgt^- dam^h *dam*-1
T4 αgt^+ βgt^+ dam^+	→	T4 αgt^- βgt^- dam^+	→	T4 αgt^- βgt^- dam^h	→	T4 αgt^- βgt^- dam^h *dam*-1
T6 αgt^+ dam^-	→	T6 αgt^- dam^-				

[a] The old nomenclature (*rP1*, *uP1*, u^RP1) has been replaced with a notation describing the *dam* genotype. It should be noted that certain gt^- mutants are conditionally sensitive to P1 restriction (Revel and Georgopoulos, 1969), i.e., they become sensitive to P1 only after they are propagated in a host which cannot synthesize UDP-glucose. Although these mutant phage, designated r_cP1, have barely detectable levels of DNA glucose, it appears to be sufficient to inhibit P1 restriction unless they are grown on strains lacking UDP-glucose (Revel and Georgopoulos, 1969). $P1^r$, Resistant to P1; $P1^s$, sensitive to P1.

ed but not modified by the P1 methylation; apparently, 5-hydroxymethylcytosine blocks methylation but not cleavage.) However, spontaneous mutants could be isolated which were "unrestricted by P1" (designated *uP1*) (Revel and Georgopoulos, 1969); these mutants remained sensitive to the restriction systems that degrade non-glucosylated T-even phage. Both T2 and T4, but not T6, gt^- phages yield *uP1* variants. Finally, *uP1* mutants readily revert to a form, u^RP1, that is once again sensitive to P1 (Revel and Hattman, 1971). These findings are summarized in Table 1.

The mutation from *rP1* to *uP1* has been shown to be correlated with an increase in viral DNA methylation, both in the virion DNA (Hattman, 1970) and in intracellular replicating DNA (Hattman, 1972); mutation from *uP1* to u^RP1 results in almost complete loss of m^6A in virion DNA (Revel and Hattman, 1971). These and other findings have led to a better understanding of the role of DNA methylation in these events. In summary, presence of a wild-type (dam^+) phage methylase is a prerequisite for the mutation to P1 resistance (Table 1). This mutation, dam^+ to dam^h, produces an altered enzyme (Brooks, thesis; Brooks and Hattman, 1978; Hattman, 1970; Hehlmann and Hattman, 1972; Revel and Hattman, 1971) which, by virtue of its hypermethylation capability, confers resistance to P1. Second-site mutations within the dam^h gene lead to the complete loss of methylase activity and render the DNA once again susceptible to P1 restriction.

Effect on mutation

Attempts in this laboratory to assess the effect of m^6A on mutation suggested that high m^6A content increases the frequency of spontaneous reversion (unpublished data cited in Hattman, 1981). However, more recent studies revealed that we were observing differential growth advantage of revertants in dam^h compared to dam^h *dam*-1 (unpublished data). Thus we still do not know whether m^6A content influences mutation in phage T2 gt^-.

MAPPING OF THE *dam* GENE

Because dam^+ methylation is generally a nonessential function, it was not possible to carry out genetic

rIIB rIIA dda mod [dam, 56] αgt 49

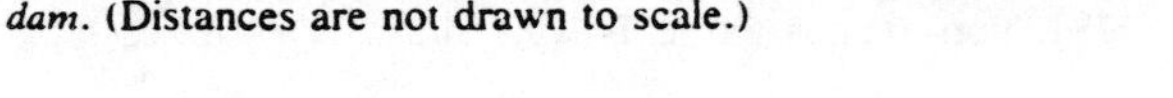

FIG. 1. Schematic map of the relative genetic position of *dam*. (Distances are not drawn to scale.)

mapping until mutants and a conditional lethal system were available. Thus, the discovery of P1 restriction against gt^- phage and the isolation of P1-resistant variants set the stage for genetic analysis. As discussed above, the *rP1* → *uP1* mutation is, in reality, $dam^+ \rightarrow dam^h$. Thus, the original genetic study localizing the *uP1* locus (Revel and Georgopoulos, 1969) was a mapping of the *dam* gene. Crosses between appropriate T4 strains suggested a map sequence of βgt^- αgt^- *uP1* (dam^h); the *uP1* locus was 26 map units clockwise from αgt and 40 map units from βgt. Moreover, in a cross between T2 wild type and T2 αgt^- *uP1*, it was observed that *uP1* was 26 map units clockwise from αgt (Revel and Georgopoulos, 1969). In another study with T2, the *dam* locus was mapped between genes 49 and *r*I (Brooks and Hattman, 1973). The first doubts arose later when we analyzed several T4 deletion mutants for m^6A content. One mutant, *far*P13 (Chace and Hall, 1975), has an extensive deletion spanning through genes *r*I and *nrdC* toward gene 49; yet, *far*P13 contained the normal level of m^6A (S. Hattman, D. Hall, and K. Chace, unpublished data).

Molecular cloning of the T4 dam^+ gene (Schlagman and Hattman, 1983) has allowed us to physically map the T4 dam^+ locus (by Southern blot hybridization of ^{32}P-labeled, nick-translated *dam* probe to restriction nuclease digests of T4 DNA). These studies reveal that *dam* is located at a position different from that deduced by genetic mapping; the relative position of *dam*, illustrated by the schematic map in Fig. 1, is on a 1.8-kilobase *Hin*dIII fragment which is located 16.2 to 18.0 kilobases from the *r*IIA-*r*IIB junction; the position of *dam* relative to gene 56 is not known, and this is indicated by the brackets.

Because this result is inconsistent with the genetic mapping data, we have reinvestigated this question. These studies, however, showed that the T4 dam^h allele is, in fact, closely linked to gene 56 (M. Myers, S. Schlagman, and S. Hattman, unpublished data).

OUTLOOK

Cloning the DNA methylase genes of T2 and T4 is an important objective. With dam^+ and dam^h clones we can pursue several lines of investigation: (i) purification of the methylases from plasmid-containing cells will allow us to characterize and compare the T4 enzymes (and their relationship to the T2 methylases); (ii) after subcloning, we can establish the nucleotide sequences of the promoter and structural genes for *dam*$^{+}$ versus *dam*^{h}; (iii) we can probe T2 and T6 for sequence homology to T4 *dam*; (iv) we can study regulation of *dam* expression at the level of transcription; and (v) we can transfer the *dam*$^{+}$ and *dam*^{h} plasmids into organisms lacking m^6A and assess the effect on mutability. These and other studies should contribute further to our understanding of the specificity, control, and biological effect of DNA methylation.

This work was supported by Public Health Service grant no. GM-29227 from the National Institutes of Health.

DNA Modification: Glucosylation

HELEN R. REVEL

Department of Biophysics and Theoretical Biology, University of Chicago, Chicago, Illinois 60637

T-even bacteriophage DNA bears several modifications. Not only does 5-hydroxymethylcytosine (HMC) replace cytosine (Wyatt and Cohen, 1952; Wyatt and Cohen, 1953), but the phage DNAs are glucosylated and methylated as well. Glucose is covalently linked through carbon 1 with the hydroxymethyl group of HMC (Jesaitis, 1956; Sinsheimer, 1954; Volkin, 1954). Glucosylation patterns, both abundance of glucose and stereochemistry of glycosidic bonds, are phage specific (Lehman and Pratt, 1960), a reflection of the specificity and regulation of the phage-controlled HMC-glucosyltransferases (Kornberg et al., 1959; Kornberg et al., 1961). The DNA of T2 and T4, but not that of T6, is further modified by methylation of a small fraction (0.5 to 1.5%) of adenines to give N^6-methyladenine (Hattman, 1970; see Hattman, this volume). In contrast to HMC substitution, which occurs at the level of DNA precursor biosynthesis, both glucosylation and methylation are modifications of the completed polynucleotide chain catalyzed by phage-controlled enzymes utilizing glucosyl and methyl donors provided by the host.

A major role of these DNA structural alterations appears to be a protective one, facilitating the T-even phage lytic life style. Not only can these phages with HMC DNA scavenge deoxyribonucleotides from the host DNA by induction of DNA endonucleases specific for cytosine-containing DNA, but also they can shuttle, with impunity, between bacterial strains harboring cytosine-specific restriction-modification endonucleases. Glucose modification of HMC extends this autonomy to include hosts that have HMC-specific endonucleases (Rgl restriction) or carry plasmids that encode some restricting enzymes. The role of methylation is unknown, except for the special case in which an altered methylase of T2 protects non-glucosylated HMC (Glu$^-$) from P1 restriction (Hattman, 1970); its usefulness may have been superseded by the more encompassing protection afforded by glucosylation.

The evolution of glucose modification may have been accompanied by subtle alterations in the transcription apparatus. Although the yield of T-even phage growing on a nonmodifying *galU* host (T* progeny) is similar to the phage yield on modifying *galU*$^+$ hosts (wild-type progeny) (Dharmalingam et al., 1982; Fukasawa and Saito, 1964; Georgopoulos and Revel, 1971), phages mutant in glucosyltransferases (*gt*) have reduced bursts on permissive hosts (Georgopoulos and Revel, 1971; Revel, 1967). In addition, in an RNA polymerase mutant host, T4 β-glucosylation is detrimental to T4 growth (Montgomery and Snyder, 1973). A third effect of glucosylated DNA is the stimulation of a region-specific recombination (Levy and Goldberg, 1980a, 1980b, 1980c). Whereas the latter could be attributed to effects of altered DNA structure, the first two are perhaps more easily interpreted as alterations in genetic specificity.

The observation that wild-type T-even phage grow normally in *galU* hosts has prompted the notion that Rgl restriction activity against non-glucosylated T-even DNA may reside in the membrane, where it monitors entering phage DNA (Fukasawa, 1964). However, as discussed below, there are two Rgl functions which appear to be differentially located and regulated within the host cell, and only one of them is membrane associated.

The mechanism of DNA glucosylation has been thoroughly reviewed (Cohen, 1968), as have its genetic determination and a descriptive analysis of its major role in phage-host interactions, i.e., the protection against HMC-specific host restriction mechanisms (Revel and Luria, 1970). This chapter will summarize the earlier work, review the properties of permissive mutants of *Escherichia coli* B and K-12, and concentrate on more recent developments which utilize these mutants in the analysis of the mechanism, localization, and control of restriction of Glu$^-$ DNA in vivo and in vitro, as well as the expression of the Glu$^-$ phage genome under various experimental conditions.

GLUCOSYLATION OF T-EVEN DNA

HMC Glucosyltransferases

The glucosylation reaction involves the transfer of glucose from uridinediphosphoglucose (UDPG) to the HMC residues in double-stranded polydeoxyribonucleotide chains (Kornberg et al., 1959; Kornberg et al., 1961). Neither denatured HMC DNA, the free nucleotides, nor cytosine DNA acts as an acceptor. The glucosyltransferases that catalyze the reaction are induced early in phage infection, and at least four have been defined as phage enzymes by the isolation of *gt* mutations (see below). The properties of the purified enzymes from T2-, T4-, and T6-infected cells can account for most of the phage-specific patterns of glucosylation of virion DNA. In T4 DNA all of the HMC is modified, 70% with α-glucosyl and 30% with β-glucosyl residues, whereas in T2 and T6 25% of the HMCs are not glucosylated. α-Glucosyl and β-1,6-glucosyl-α-glucosyl (gentiobiose) groups are present to the extent of 70% and 5% in T2 DNA and 3% and 72% in T6 DNA, respectively. All three phages induce an HMC α-glucosyltransferase. The T2 α-glucosyltransferase adds a single glucose in α-linkage to a fixed fraction of synthetic HMC DNA, is unreactive with the free HMC bases in T2 DNA, and adds normal T2 levels of α-glucose to the glucose-deficient DNA of T2 α*gt*$^-$ and T*2 made in vivo (C. P. Georgopoulos, Ph.D. thesis, Massachusetts Institute of Technology,

Cambridge, 1969; Kornberg et al., 1961; Revel et al., 1965). The T4 and T6 α-glucosyltransferases act similarly with both synthetic and in vivo-generated HMC DNA but can also add a small amount of α-glucose (≤6%) to normally glucosylated T2 DNA (Georgopoulos and Revel, 1971; Josse and Kornberg, 1962). A second enzyme induced by T4, HMC β-glucosyltransferase, genetically and physically separable from the T4 α enzyme and differing in many biochemical parameters, adds monoglucosyl units in β linkage to the majority of sites in synthetic HMC DNA, to the unmodified HMC residues in T2, T4 α*gt*$^+$ β*gt*$^-$, and T6 DNA, as well as to some HMC residues normally α-glucosylated in these phage DNAs (Georgopoulos and Revel, 1971; Zimmerman et al., 1962). The T4 α*gt*$^-$ β*gt*$^+$ HMC DNA is virtually 100% modified in vivo and is not a substrate in vitro (Georgopoulos, 1967; Georgopoulos and Revel, 1971; Revel and Georgopoulos, 1969). A fifth enzyme, not yet genetically defined, is a β-glucosyl-HMC-α-glucosyltransferase induced by T6 which adds a second glucose in β linkage to the carbon 6 of preexisting α-glucosyl HMC DNA (Kornberg et al., 1961). Thus the two T6 enzymes, acting in succession, account for the predominance of gentiobiose in T6 DNA. The enzymatic basis for 5% diglucosyl residues in T2 DNA may be similar but the enzyme has not been demonstrated, possibly due to its instability or low abundance.

Genetic Control of Glucosylation

In T2 and T6, where glucosylation is totally dependent on the α-glucosyltransferases, glucose-deficient mutants (*gt*) have been isolated by using a mixture of permissive and restrictive bacteria as indicator cells (Revel et al., 1965). Isolation of *gt* mutants of T4, where either the α- or β-glucosyltransferase alone protects against restricting hosts, depended on the fortuitous discovery of a T4 α*gt* mutation in a gene 30 amber mutant stock (Hosoda, 1967). T4 α*gt1* allowed the generation of a variety of T4 mutants with defects in the β*gt* enzyme as determined by the mixed indicator. Reversion of double mutants to α*gt*$^+$ β*gt*$^-$ and a second round of mutagenesis and selection gave additional α*gt*$^-$ mutants as well (Georgopoulos, 1967). A temperature-sensitive T2 α*gt* (Molholt and de Groot, 1969) and the production of temperature-sensitive glucosyltransferases by suppression of nonsense *gt* mutations in all three T-even phages provided evidence that mutations were in the structural genes for the enzymes (Georgopoulos and Revel, 1971). The T4 β*gt* gene maps between genes 41 and 42 (Georgopoulos, 1968) and codes for a polypeptide of about 46,000 daltons (Huang and Buchanan, 1974). The α*gt* genes of T2 and T4 map between genes 47 and 55 (Georgopoulos, 1968; Russell, 1974); the gene product sizes are unknown. The patterns of glucosylation of the *gt* mutant phage DNA (Table 1) conform with the specificities of the glucosyltransferases described above. Thus, T2 α*gt* and T6 α*gt* mutant DNAs are devoid of glucose; T4 α*gt*$^-$ β*gt*$^-$ is 100% glucosylated, and T4 α*gt*$^+$ β*gt*$^-$ has about 80% glucose.

The glucosyl donor for the glucosylation reaction in vivo, UDPG, is supplied by the host bacterium. *galU* host mutants, defective in the enzyme UDPG pyrophosphorylase which converts uridine triphosphate (UTP) to UDPG in the presence of glucose 1-phosphate, eliminate UDPG, and so do other host mutants affecting the metabolic flow of glucose in the UDPG synthetic pathway. T-even phages grown in these hosts contain phenotypically unglucosylated DNA and have been given the designation T* as discussed below. Phage mutations in genes responsible for the substitution of HMC for cytosine would, of course, eliminate the HMC acceptor DNA.

Specificity and Regulation of Glucosylation

Specificity for certain base sequences in the glucosylation reaction is suggested by the failure of the α-glucosyltransferases to modify all available HMC residues: about 25% of these bases remain unglucosylated when only these enzymes are present. The basis for this differential glucosylation lies in the nucleotide sequences adjacent to HMC residues (de Waard et al., 1967; Lunt and Newton, 1965). The α-glucosyltransferases cannot add glucose to HMC linked 5′ to another HMC. The β enzyme is not so limited and will glucosylate HMC-rich regions of the DNA as well as HMC residues normally α-glucosylated (Georgopoulos and Revel, 1971; Kornberg et al., 1961). In the absence of the T4 α-glucosyltransferase in vivo (T4 α*gt*$^-$ β*gt*$^-$), the DNA is virtually 100% modified.

The normal ratio of α/β glucosyl residues in T4 DNA is 70:30. Some factor other than enzyme specificity must be involved, since the activities of both enzymes (Kornberg et al., 1961) and their respective mRNAs (Black and Gold, 1971; Young and Van Houwe, 1970) appear in vivo with the kinetics of early enzyme functions. De Waard et al. (1967) have shown in vitro that the ratio of α to β glucose varies widely with the magnesium concentration in the reaction mixture, but they were unable to demonstrate this in vivo, presumably because the internal Mg^{2+} level was controlled by the cell. McNicol and Goldberg (1973), using antibodies specific for α- or β-glucosylated DNA, have subsequently examined the kinetics of glucosylation of entering phenotypically Glu$^-$ parental wild-type phage DNA in vivo. Their findings support the idea of a physiological partitioning of the two enzyme activities. The α-glucosylation reaction is completed by 8 to 10 min postinfection, whereas β-glucosylation is only 20% complete at that time. Thus it appears that the α enzyme can modify a specific subset of HMC residues at or just after polymerization. The β enzyme would subsequently modify the remaining residues. The mechanism remains as mysterious as ever.

MODIFICATION AND RESTRICTION

Host-Controlled Modification and the Rgl Restriction Phenotype

A major role of glucosylation is protection against host restricting activities. Host-controlled modification of T-even phage, first described by Luria (Luria, 1953; Luria and Human, 1952), can now be explained by the mechanism of glucosylation described above. T2 normally grows on *E. coli* strains B and K-12 and on *Shigella dysenteriae*. Luria observed that growth of T2 on a specific *E. coli* B strain, B/4_0, produced altered progeny, T*2, able to grow on *S. dysenteriae* but not on *E. coli* B. A single cycle of growth in the permissive host *S. dysenteriae* restored T*2 to normal T2 capable of growth on both *E. coli* B and *S. dysenteriae*. Strain

TABLE 1. Correlation of DNA glucosylation with restriction by various bacterial strains

Phage				Bacterial strain (phenotypes)[a]						
Strain	*gt* genotype	DNA glucosylation[b]		*S. dysenteriae*	*E. coli* B		*E. coli* K-12			
		α[c]	β	RglA⁻ RglB⁻	RglA⁺ RglB⁺	RglA⁻ RglB⁺	RglA⁺ RglB⁺	RglA⁻ RglB⁺	RglA⁺ RglB⁻	RglA⁻ RglB⁻
T2	α^+	75	0	1.0	1.0	1.0	0.1	0.1	1.0	1.0
$T^*2 \cdot B/4_0$	α^+	<3	0	1.0	10^{-4}	0.5	10^{-4}	10^{-3}	10^{-2}	1.0
T2 αgt	α^-	<3	0	1.0	10^{-6}	0.5	10^{-7}	10^{-7}	10^{-6}	1.0
T4	$\alpha^+ \beta^+$	70	30	1.0	1.0	1.0	1.0	1.0	1.0	1.0
$T^*4 \cdot W4597$	$\alpha^+ \beta^+$	<3	<3	1.0	10^{-3}	0.1	10^{-3}	10^{-2}	10^{-2}	1.0
T4 αgt	$\alpha^- \beta^+$	0	100	1.0	0.3	0.3	0.1	0.1	1.0	1.0
T4 βgt	$\alpha^+ \beta^-$	80	0	1.0	1.0	1.0	1.0	1.0	1.0	1.0
T4 αgt βgt	$\alpha^- \beta^-$	<3	<3	1.0	10^{-5}	0.05	10^{-6}	10^{-6}	10^{-5}	1.0
T6	α^+	75	0	1.0	1.0	1.0	1.0	1.0	1.0	1.0
$T^*6 \cdot B/4_0$	α^-	<3	0	1.0	10^{-3}	1.0	10^{-2}	1.0	10^{-2}	1.0
T6 αgt	α^-	<3	0	1.0	10^{-6}	1.0	10^{-5}	1.0	10^{-5}	1.0

[a] Values are plaque titers normalized to plaque titer on *S. dysenteriae*. Assays were performed at 37°C. Data have been compiled from Georgopoulos, 1967; Georgopoulos, thesis; Georgopoulos and Revel, 1971; Lehman and Pratt, 1960; Revel, 1967; and Revel and Georgopoulos, 1969.

[b] Values are percent of HMC glucosylated.

[c] Includes both α-glucosyl and β-1,6 glucosyl-α-glucosyl HMC DNA.

$B/4_0$ and strains W4597, U95, and B/3,4,7, which also produce T* progeny when infected with T-even phage, are *galU* mutants deficient in UDPG pyrophosphorylase and are thus unable to supply UDPG for the glucosylation reaction (Hattman and Fukasawa, 1963; Shedlovsky and Brenner, 1963; Symonds et al., 1963). The progeny DNA is not glucosylated, and the T* phage are subject to restriction in *E. coli* strains B and K-12 but not in *S. dysenteriae*. Since the permissive host *S. dysenteriae* can make UDPG, growth of T* phage in *Shigella* returns the phage DNA to its normal glucosylated state, no longer susceptible to restriction.

The restriction of Glu⁻ phage DNA in *E. coli* strains B and K-12 was shown by early investigators (Fukasawa, 1964; Hattman, 1964; S. Hattman, Ph.D. thesis, Massachusetts Institute of Technology, Cambridge, 1965; Hattman et al., 1966; Luria, 1953; Luria and Human, 1952). It is now known to be due to a specific restricting system called the Rgl system (Revel, 1967). Glu⁻ phage absorb normally and inject their genomes, but neither DNA synthesis nor late proteins are detected. In fact, characteristically, the Glu⁻ DNA is rapidly but incompletely (50%) broken down to acid-soluble fragments. This breakdown was presumed to be the primary mechanism of restriction. The restricted Glu⁻ phage function ineffectively in host killing, in complementation of early genes, and in early enzyme synthesis. The genetic control of Rgl restriction of Glu⁻ phage, its mechanism, localization, and control, and expression of the restricted genome will be discussed more fully in the following sections.

Genetic Control of Rgl Restriction

The restriction of Glu⁻ T-even phages by *E. coli* strains B and K-12 is controlled by at least two genes, *rglA* and *rglB* (formerly *r6* and *r24*). These genes are not essential to the host since mutants defective in either or both Rgl functions have been isolated in various K-12 strains (Georgopoulos, 1968; Georgopoulos and Revel, 1971; Revel, 1967; Revel and Georgopoulos, 1969); only RglA⁻ mutants have been described for *E. coli* B (Eigner and Block, 1968; Molholt and Fraser, 1968; Revel, 1967). All these mutants are stable. Table 1 correlates the extent of glucose modification of the T-even phages and their Glu⁻ derivatives with the restricting activity of various bacterial strains. Growth of phenotypically Glu⁻ T* phage measures the probability of successful infection of a restricting host; *gt* phage growth represents *gt*⁺ revertant frequencies, except on *E. coli* B *rglA*⁻ *rglB*⁺, where RglB⁺ activity is weak. The Rgl restricting functions act independently and additively and show some quantitative and qualitative differences in strains K-12 and B. K-12 mutants with the RglA⁺ RglB⁻ phenotype restrict all Glu⁻ forms of T2, T4, and T6; the RglA⁻ RglB⁺ mutants restrict only Glu⁻ T2 and T4 and also may be responsible for the reduced plating efficiency of wild-type T2. Neither activity restricts wild-type T6 or T4 αgt^- βgt^- which, like wild-type T2, have about 25% unprotected HMC residues. In *E. coli* B the RglA⁺ function acts only on regions normally protected by glucose in all three phages, but restriction is more effective than in K-12 (compare T*6). The RglB⁺ activity of *E. coli* B acts on Glu⁻ T2 and T4 but is less effective than its counterpart in K-12. The virtually complete glucosylation of T4 αgt^- βgt^- does not spare this phage from restriction by the RglB⁺ activity. These observations suggest that restriction is not exerted on all Glu⁻ HMC residues but only at specific sites determined by the nucleotide sequences and the specificities of the separate Rgl functions. For example, the strain differences might be accounted for by differences in site specificity. The Rgl restricting functions appear to be specific for HMC DNA phage since the cytosine-containing phages grow normally.

Mutants exhibiting the various Rgl phenotypes shown in Table 1 were isolated in a serial fashion

(Geogopoulos and Revel, 1971; Revel, 1967). Mutagenesis with nitrosoguanidine and selection of nibbled colonies with T6 *gt* yielded the RglA⁻ RglB⁺ derivatives. Further mutagenesis of these single mutants and selection with T2 *gt* gave completely permissive RglA⁻ RglB⁻ strains. Strains with the phenotype RglA⁺ RglB⁻ were selected from the permissive host as revertants characterized by their inability to grow T6 *gt*. This phenotype could also be derived from the wild-type RglA⁺ RglB⁺ after mutagenesis and selection with T*4. Only a few mutations have been mapped (Revel, unpublished data). Mutations in strain K-12 giving the RglB phenotype map at about 98.5 min on the *E. coli* map, showing 25% cotransduction with the *serB* locus and 90% linkage with the *hsd* locus which controls host-specific restriction and modification in *E. coli* strains B and K-12 (Arber, 1974). There is no evidence of interaction between the *hsd* and *rgl* restriction systems since mutations in either one do not affect the other (Revel, 1967; Revel and Luria, 1970; Wood, 1966). Three-factor transductional analysis yields the probable order *rglB*, *hsd*, *serB* and confirms the independence of the *rglB* and *hsd* functions (Revel, unpublished data). An *rglB* gene of *E. coli* B maps in the same region and is transduced into a K-12 RglA⁻ RglB⁻ host with *serB* with a linkage of 10% along with, but separable from, the *hsd* locus. Mutations giving the RglA phenotype in strain K-12 are 70% cotransduced with *purB* and map at about 25.5 min towards *trp*. Mutations with the RglA⁻ phenotype in *E. coli* B, almost always associated with a requirement for thiamine, are unmapped.

It is important to emphasize that there has been no in vivo complementation analysis to determine whether mutants with similar phenotypes and map locations may be altered in separate genes. However, all of the work discussed below on the mechanism, localization, and control of Rgl restriction in strain K-12 in vivo and in vitro has used a single set of mutants derived from the original Lederberg *E. coli* K-12 strain (RglA⁺ RglB⁻). Although many wild-type K-12 strains exhibit a similar phenotype, some common laboratory strains such as C600, CR63, and Hayes Hfr are RglA⁻ RglB⁺. The strains K802 *hsdR* and K803 *hsdS* exhibit a RglA⁻ RglB⁻ phenotype, as does their parent K704 *hsdR*⁻ *S*⁻ (Wood, 1966). The derivation of strain K704 from a cross between Cavalli Hfr and C600 suggests that RglA⁻ RglB⁻ strains may also exist "naturally."

Mechanism of Rgl Restriction In Vivo

The specificity of the restriction reaction revealed by genetic analysis, as well as analogy to the in vitro analysis of restriction endonucleolytic activities directed against unmodified lambda DNA (Arber, 1974; Meselson et al., 1972), suggests that the initial Rgl restriction activity in vivo should be a limited site-specific cleavage which is followed by secondary exonucleolytic breakdown of large intermediate fragments to the acid-soluble material associated with the restriction phenotype. This has been found to be true. Early efforts to define an HMC-specific nuclease in wild-type hosts, or to reveal an association of restricting functions with a loss or change in abundance of known *E. coli* nucleases, were unsuccessful (Eigner and Block, 1968; B. Molholt, Ph.D. thesis, University of Indiana, Bloomington, 1967; Molholt and Fraser, 1968; Revel, 1967; Richardson, 1966). The opportunity to test the two-step pathway idea came with the observation that unmodified lambda phage is restricted but its DNA is not degraded in *E. coli* K-12 *recB recC* (ExoV⁻) hosts (Simmon and Lederberg, 1972). Dharmalingam and Goldberg (1976a) showed that nonglucosylated T4 behaved the same way: growth was inhibited in Rgl⁺ ExoV⁻ hosts, but no DNA breakdown occurred. Instead, large acid-insoluble fragments about one-fifth the size of the T4 genome accummulated. These fragments were not seen in infections of Rgl⁺ ExoV⁺ hosts, where exonucleolytic breakdown of Glu⁻ DNA was evident. Furthermore, Rgl cleavage is a prerequisite for exonuclease V breakdown. In Glu⁻ phage-infected Rgl⁻ ExoV⁺ cells, the unmodified uncleaved T4 DNA remained largely intact, presumably protected from exonuclease V attack at the ends of the genome by bound gp2 and gp64 (Dharmalingam and Goldberg, 1976a; Silverstein and Goldberg, 1976; Hamilton and Goldberg, unpublished data). The cleavage was shown to be Rgl⁺ specific, rather than due to a T4-induced single-strand nicking activity such as T4 endonuclease I (Altman and Meselson, 1970). Neutral sucrose gradient analysis demonstrated the Glu⁻ T4 fragments generated in a Rgl⁺ ExoV⁻ host in the presence of chloramphenicol to inhibit formation of the T4 nuclease (Dharmalingam and Goldberg, 1976b).

These experiments define a sequential pathway for the degradation of nonglucosylated T4 DNA in restricting hosts: a few double-strand breaks are followed by exonucleolytic breakdown of the unmodified DNA. Whether the initial endonucleolytic cleavage has type II specificity, where cleavage and recognition occur at the same site, or is a type I or type III mechanism, with double-strand cuts occurring at some distance from the recognition site (Arber, 1974; Yuan and Hamilton, 1982), is not known. The former would generate discrete fragments representing different parts of the genome, whereas the latter would result in sequence heterogeneity among the primary restriction fragments. DNA agarose gel analysis and terminal sequencing would be useful to distinguish these possibilities.

Mechanism of Rgl Restrictions In Vitro

In vivo experiments do not address the question of whether the *rgl* genes defined genetically identify the structural genes for the restriction endonucleases or the specificity factors that direct these enzymes to recognize HMC DNA. The existence of temperature-sensitive mutations affecting the RglA phenotype (Revel, 1967) does not resolve this issue, since the sensitive protein could be either the endonuclease, an associated specificity subunit, or some other cell component with which the Rgl system interacts.

Initial in vitro experiments (Fleischman and Richardson, 1971) used toluene permeabilization of cells to examine the ability of restricting (Rgl⁺) and permissive (Rgl⁻) bacterial strains to synthesize DNA in the presence of Mg^{2+}, ATP, dATP, dGTP, dTTP, and either dCTP or HMdCTP. Rgl⁺ cells stably incorporated radioactive dTTP into *E. coli* DNA in the presence of dCTP but were inhibited 90 to 100% when HMdCTP was the substrate. In contrast, Rgl⁻ cells synthesized DNA with either dCTP or HMdCTP. Both RglA and RglB functions had to be defective for successful DNA synthesis in this system. The incorporation of HMdCTP into newly replicating DNA was shown to be a prerequisite for expression of Rgl⁺ restricting activity: preincubation with HMdCTP under replicating conditions, but not in the absence of the remaining triphosphates, prevented subsequent incorporation of dCTP by Rgl⁺ cells. This suggests that newly synthesized HMC DNA is recognized and broken down in this system. Apparently HMC present in only one strand of DNA is sufficient to allow recognition by the Rgl⁺ functions. Interestingly, DNA repair synthesis occurred in toluenized cells of both Rgl⁻ and Rgl⁻ strains irrespective of whether HMdCTP or dCTP was the substrate present in the reaction mixture.

An RglB HMC DNA specific endonuclease activity requiring ATP and Mg^{2+} was demonstrated to be present in cell-free extracts of wild-type *E. coli* K-12 and an *rglA rglB*⁺ mutant, but absent in strains lacking the RglB⁺ function (*rglA*⁺ *rglB* or *rglA rglB*) (Fleischman et al., 1976). These preparations degraded bacteriophage fd double-stranded circular DNA (replicative form I [RFI]) when HMC substituted for cytosine in one strand, but fd RFI DNAs containing only cytosine, or in which the HMC residues were modified with glucose added by T2 α-glucosyltransferase, were not substrates for this activity. When RglB⁺ cell-free extracts were derived from *recB*⁺ *recC*⁻ (ExoV⁻) cells, HMC fd DNA was degraded to acid-soluble material. In RglB⁻ extracts derived from a *recB recC* host, the accumulation of large acid-insoluble fragments could be demonstrated by alkaline sucrose gradient centrifugation. Thus the RglB⁺-controlled reaction, while specific for HMC fd DNA, cleaves at only a small number of specific sites. The ultimate breakdown of HMC-containing DNA, in vitro as in vivo, occurs in two steps: an initial RglB⁺-controlled double-strand cleavage of HMC fd DNA, followed by exonucleolytic breakdown of the fragments to acid-soluble nucleotides. This system formed the basis of an assay to purify an RglB protein, measuring the ability to complement exonuclease V activity provided by an *rglA rglB recB*⁻ *recC*⁺ extract to yield acid-soluble breakdown products from HMC fd DNA. Alas, the 250-fold purified RglB protein had no endonucleolytic activity when it was incubated with HMC fd DNA in the absence of the *rglA rglB recB*⁺ *recC*⁺ extract and the products were examined on sucrose gradients. No further characterization of the protein was reported, except its inability to act as an exonuclease. A requirement for other non-dialyzable heat-sensitive components of the permissive cell extract suggests that a second protein is required for RglB activity. The *hsd* restriction endonucleases of *E. coli* strains B and K-12 have been shown by both biochemical and genetic analysis to be composed of three nonidentical subunits determining the restriction endonuclease (*hsdR*), the modification activity (*hsdM*), and the specific recognition element (*hsdS*) (Arber, 1974; Yuan and Hamilton, 1982). Possibly the RglB⁺ function is composed of two subunits, one an endonuclease and the other a specific recognition factor, active only in a complex. Mutations in both genes could give the same phenotype and be closely linked on the genetic map. Extensive in vivo and in vitro complementation analysis of many more RglB⁻ mutants and biochemical purification of the RglB enzyme will be needed to establish whether the RglB protein described here is the endonuclease or specificity subunit, and how the individual proteins function separately or together in RglB restriction.

Notable in these cell-free extract experiments was the absence of RglA activity previously detected as an inhibition of DNA replication in toluene-permeabilized RglA⁺ restricting cells (Fleischman and Richardson, 1971). The RglA activity may have been lost or inactivated in the preparation of the extracts because it is an integral part of the membrane. Alternatively, association with the membrane may be essential for RglA function. A third possibility is that the synthetic HMC fd DNA may have lacked the appropriate HMC sites; whereas the HMC substitution in one strand of the fd RFI duplex is sufficient for RglB recognition, RglA may require this base in both strands. Perhaps circular molecules of Glu⁻ T4 DNA created by religation of *Eco*RI restriction fragments would be a more appropriate substrate.

Localization and Control of Rgl Functions In Vivo

It will be recalled that the two genetically defined restricting functions have different specificities: the RglA function restricts Glu⁻ T2, T4, and T6; the RglB function restricts Glu⁻ T2 and T4. Where are these activities located in the cell, and how are they regulated and controlled? Paradoxically, all T-even wild-type phage (Glu⁺) replicate and produce a normal burst size of unmodified T* (Glu⁻) progeny in *galU* hosts that restrict infecting Glu⁻ phage growth (Fukasawa, 1964; Fukasawa and Saito, 1964). Fukasawa (1964) proposed a membrane location for the restricting functions, perhaps at the point of phage entry, to explain the protection of intracellular Glu⁻ progeny DNA. The apparently conflicting observation that DNA replication in vitro in toluenized K-12 cells is inhibited by both the RglA and RglB functions could be reconciled with this model because of the known association of DNA replication complexes with the cell membrane and the failure of the Rgl⁺ functions to inhibit DNA repair synthesis in vitro (Fleischman and Richardson, 1971). Support for a membrane location of RglA came from experiments showing that nonglucosylated T6 *gt* superinfecting phage DNA is protected from restriction in *E. coli* B by a primary infecting T4 *imm*⁻ phage. (T4 *imm*⁻ phage are deficient in both superinfection exclusion and superinfection breakdown activities [Vallée and de Lapeyrière, 1975].) Since protein synthesis by the primary phage was not required, either for inhibition of DNA breakdown or for successful passage of unmodified T6, Hewlett and Mathews (1975) suggested that simple attachment of the primary phage to the cell membrane was sufficient to inactivate the RglA function. A more extensive analysis by Goldberg and his colleagues (see below) confirmed these observations for RglA⁺ and showed, in addition, that the RglB⁺ function, which affects only nonglucosylated T2 and T4, is controlled in a different manner. They proposed that though the RglA activity may reside in the membrane, the RglB restricting activity is probably located in the cytoplasm, a suggestion supported by the in vitro analysis of

RglB⁺-directed endonucleolytic cleavage of HMC fd DNA (Fleischman et al., 1976).

Silverstein and Goldberg (1976) observed that the ability of superinfecting Glu⁻ T4 to escape restriction-induced DNA breakdown, and successfully to complement T4 amber mutations in *E. coli*, required protein synthesis by the primary T4 phage. To resolve this apparent conflict with the data of Hewlett and Mathews (1975), Dharmalingam and Goldberg (1976a; 1976b) examined the effects of primary T4 *imm*⁻ infection on the fate of superinfecting Glu⁻ T4 and Glu⁻ T6 in the presence and absence of chloramphenicol in various K-12 mutants (*rglA rglB*, *rglA*⁺ *rglB*, *rglA rglB*⁺, and *rglA*⁺ *rglB*⁺) by measuring the breakdown of superinfecting Glu⁻ DNA. Without primary phage infection, Glu⁻ T4 DNA is degraded in cells having either the RglA⁺ or RglB⁺ phenotype, or both restricting functions, but not in permissive RglA⁻ RglB⁻ cells; Glu⁻ T6 DNA is degraded only in RglA⁺ hosts. In the superinfection experiments, breakdown occurred only with Glu⁻ T4, a target of both RglA and RglB restricting functions, and then only in RglB⁺ cells, and only in the absence of primary phage expression. Glu⁻ T6, sensitive only to RglA, was not broken down in either RglA⁺ or RglB⁺ hosts in the presence or absence of chloramphenicol. These results show that the RglA and RglB functions are differentially localized and regulated but behave similarly in *E. coli* B and K-12. The lack of breakdown of either Glu⁻ T4 or Glu⁻ T6 in all hosts in the absence of chloramphenicol, where the primary phage is expressed, is expected since the phage induces an inhibitor of exonuclease V (Behme et al., 1976; Sakaki, 1974; Tanner and Oishi, 1971). On the other hand, breakdown of Glu⁻ T4 by RglB⁺ cells in the presence of chloramphenicol implies, according to the two-step degradation pathway, prior RglB⁺ cleavage. Neutral sucrose gradient analysis of the initial cleavage products of RglB⁺ activity on Glu⁻ T4 in an Rgl⁻ ExoV⁻ host confirmed this deduction directly. While in the absence of primary infection Glu⁻ T4 DNA is cut to large fragments, primary infection with T4 *imm*⁻ phage protects superinfecting Glu⁻ T4 from RglB⁺ cleavage. This protection is in turn abolished by inhibition of primary phage expression. The protection ability appears with the kinetics of phage early functions. Thus the primary phage makes an early protein that inhibits RglB⁺ cleavage activity.

Dharmalingam and Goldberg have called this inhibitor of RglB function Arn (antirestriction nuclease). They propose that the differential regulation and location of RglA and RglB activities provide an explanation for the ability of Glu⁻ progeny phage to replicate in *galU* restricting host mutants. Thus the RglA function would be inactivated by infecting Glu⁺ phage attachment to the membrane, and parental Glu⁺ phage expression of the *arn* gene would provide an inhibitor of RglB⁺ activity within the cytoplasm, affording protection for Glu⁻ replicating DNA. This protection hypothesis has been reinforced by the identification and cloning of the T4 *arn* gene (Dharmalingam et al., 1982). The gene was identified by screening T4 deletion mutants for the inability to give a burst on a restricting *galU* host. The absence of an *arn* gene in phage mutants lacking the 55.5- to 58.5-kilobase region of the T4 genome was verified by the failure of T4 *imm arn* phage to protect superinfecting Glu⁻ T4 from RglB⁺-controlled cleavage. A functional *arn* gene was cloned on plasmid pBR325, and the 0.8-kilobase DNA insert was shown to be homologous with T4 DNA absent in the *arn* deletion mutant phage.

Phage-encoded inhibitors of nuclease activities are not unusual and have been described for lambda, T3, T7, T5, and the *Bacillus* phage NR2 (see, for review, Krüger and Schroeder, 1981). The mechanism of T4 Arn inhibition of host RglB activity, as well as the regulation of *arn* gene expression, remains to be clarified. Does Arn act by direct binding to the RglB restriction endonuclease, as has been demonstrated for the action of the T7 0.3 protein inhibitor of the host-specific restriction endonuclease (Mark and Studier, 1981), and if so, to which of the hypothetical subunits does it bind? Or does it function indirectly? In terms of T4 survival, Arn appears to be useful only under a special circumstance, namely, for glucosylated phage infecting restricting hosts unable to provide UDPG for the glucosylation reaction. Dharmalingam and Goldberg (1979b) have shown that the *arn* gene is a direct target for RglB restriction (see below). Thus entering Glu⁻ phage would be restricted, either at the point of entry by the RglA membrane function or by immediate RglB inactivation of the *arn* gene whose function is essential to protect replicating Glu⁻ progeny DNA. Phenotypically Glu⁻ progeny (T*) of a Glu⁻ parent growing in this strain will at least have the opportunity to escape and try to find a more hospitable permissive *galU*⁻ host.

Many strains appear to exist naturally that represent the variety of mutant phenotypes that have been isolated from *E. coli* K-12. Hewlett and Mathews (1975) were careful to state that their discussion and conclusions concerning RglA activity were limited to *E. coli* B. They observed that T4 *imm*⁻ primary infection of a particular K-12 strain, W3110, in contrast to the results reported for the K-12 Lederberg strain by Dharmalingam and Goldberg (1976a), did not protect a Glu⁻ T6 phage carrying two amber mutations in the αgt gene (T6 *αgt am*16 *am*10) from restriction-associated DNA breakdown. This may correlate with the stronger inhibition of *E. coli* DNA synthesis in the presence of HMdCTP seen in toluenized W3110 cells compared with other toluenized K-12 cells (Fleischman and Richardson, 1971). A possible explanation is that strain W3110 may have another Rgl activity directed against Glu⁻ T6 (and possibly Glu⁻ T2 and Glu⁻ T4), revealed only under conditions that inactivate the RglA function. This postulated activity must be modulated in the infected cells since all T-even phages give a full burst of Glu⁻ T* phage in the *galU* derivative of strain W3110 (=W4597). Possibly, like P1 restriction of non-glucosylated T-even phage DNA, this proposed Rgl activity might recognize HMC sites only in the total absence of glucosylation, as occurs in the T6 *αgt* double amber mutant but not in T* phage DNA (Revel and Georgopoulos, 1969; Revel and Luria, 1970). In fact, the restricting activity might be the remnant of a P1 prophage. Alternatively, the newly revealed restricting activity in strain W3110 could be the RglA restriction complex associated with membrane components in such a way that simple phage attachment does not suffice to inactivate it in the wild-type strain but may do so in the *galU* derivative. Recent work on T4 receptors in the cell membrane of K-12 strains illustrates the complexity of membrane component interactions affecting the process of phage infection (Yu and Mizushima, 1982).

A variety of conditions can affect Rgl restriction. UV irradiation, old age (Luria and Human, 1952), or brief heating at 50°C (Revel, unpublished data) renders restricting hosts permissive for Glu$^-$ T-even phage. The UV inactivation phenomenon has been reinvestigated recently in the context of UV-induced SOS functions (Dharmalingam and Goldberg, 1980). UV irradiation is known to induce the *recA* protease and an inhibitor of exonuclease V activity as well as various DNA repair functions, presumably mediated by single-stranded DNA (Witkin, 1976). UV-induced alleviation of the Rgl system could be due either to prevention of the Rgl cleavage reaction by RecA protease activity or to repair of the Rgl-generated double-strand breaks. It was found that *tif sfi* cells, induced at 42°C for SOS functions, overproduce RecA protein but fail to make the inhibitor of exonuclease V and are still restricting for Glu$^-$ T4. Thus alleviation probably cannot be attributed to RecA destruction of the Rgl proteins. It is proposed that the UV-induced inhibitor of exonuclease V made in Rgl$^+$ (*tif*$^+$) cells blocks the Rgl DNA breakdown pathway, and the large fragments of Rgl cleavage action are repaired by UV-induced DNA repair mechanisms to restore a normal T4 genome. On the other side of the coin, Rgl-restricted Glu$^-$ T4 DNA can apparently induce SOS functions, as indicated by filamentation and increased mutagenesis, presumably by the generation of exonuclease V degradaion products, perhaps single-strand gaps, from the Rgl cleavage fragments.

Other Restriction Endonucleases Act on T4 DNA

Both modified Glu$^+$ and unmodified Glu$^-$ T-even phage DNAs are immune to most of the known restriction endonucleases. Kaplan and Nierlich (1975) first showed that *Eco*RI restriction endonuclease (which recognizes GAATTC) cleaves Glu$^-$ T4 αgt^- βgt^- DNA, but not the wild-type T4 αgt^+ βgt^+ DNA, in vitro. McNicol and Revel (unpublished data, 1977) observed that *Eco*RI also fails to cut single *gt* mutant T4 phage DNAs, T4 αgt^- βgt^- and T4 αgt^- βgt^+. The *Eco*RI* form of this restriction enzyme exhibits reduced specificity and recognizes AATT. *Eco*RI* also fails to cleave wild-type T4 DNA, but cuts the totally unmodified αgt^- βgt^- more extensively and, in addition, cuts the partially modified DNA derived from both T4 single *gt* mutants, yielding about 30 fragments greater than 1 kilobase, 14 of which migrated similarly on agarose gels. This is consistent with the observation that though T4 αgt^- βgt^+ appears to be 100% glucosylated by chemical analysis, some sites are still available for RglB$^+$ restriction (Table 1). These sites would normally be protected from *Eco*RI* by α-glucosylation in wild-type T4 DNA. The non-glucosylated *gt* derivatives of T2, T4, and T6 can be restricted with characteristic DNA breakdown by P1 r^+ m^+ but not by P1 r^- m^+ in vivo and are protected by chemically undetectable amounts of glucose. It has been suggested that this protection may reflect the mode of action of the P1 restriction enzyme on this DNA (Revel and Georgopoulos, 1969; Revel and Luria, 1970). Ishaq and Kaji (1980) have reported a plasmid-controlled restriction enzyme, RTS-1, that is specific for T4 Glu$^+$ DNA and does not act on T4 Glu$^-$ or T7 DNA in vivo or in vitro. The in vitro products of limited endonucleolytic cleavage are heterogeneous. *Taq*I (T$^\vee$CGA), *Eco*RV (GATAT$^\vee$C), and *Aha*III (AAA$^\vee$TTT) (Roberts, 1982) act on both Glu$^-$ and Glu$^+$ T4 DNA.

EXPRESSION OF THE Glu$^-$ GENOME IN PERMISSIVE AND RESTRICTING HOSTS

Early Gene Expression

How do the lack of glucosylation and the restriction degradation of DNA affect T4 gene expression? Sodium dodecyl sulfate (SDS)-polyacrylamide gels and functional tests have been used to explore this question. At early times the presence or absence of glucose does not significantly alter the pattern of T4 gene expression by host RNA polymerase in permissive hosts. Differences induced in the restrictive host can be attributed to the sequential Rgl cleavage and exonuclease V breakdown of the unmodified genome. Dharmalingam and Goldberg (1979a; 1979b) have defined three classes of genes differentially affected by the restriction pathway (Table 2). Thus cleavage alone, by destroying the structural continuity of the genome, suffices to block chromosome replication and the expression of some genes. In an Rgl$^+$ ExoV$^-$ host, DNA synthesis is inhibited and both the *r*IIB and the *am* genes are specifically but differentially inactivated. The double-strand break could affect gene expression either directly, if it occurs within the structural gene or its regulatory promoter region, or indirectly by interruption of operon function. Inactivation of the *r*IIB gene function may be an example of the latter. Normally *r*IIB is under dual transcriptional control: it is transcribed early as a delayed early gene by transcriptional readthrough from an early promoter for the *r*IIA gene: later *r*IIB transcription is initiated from its own middle promoter while transcription from the early promoter is turned off. SDS-polyacrylamide gel analysis of early proteins in Glu$^-$-infected Rgl$^+$ ExoV$^-$ hosts revealed an abnormality in *r*IIB synthesis. Whereas the *r*IIA protein is made at early times, *r*IIB is not, but the *r*IIB gene is expressed later when its own promoter can be recognized. Dharmalingam and Goldberg propose that an Rgl cleavage site upstream from the middle promoter interrupts early operon continuity and function, preventing expression of *r*IIB from the early but not from the middle promoter. The *am* structural gene or its regulatory region may be a direct target site for Rgl cleavage in Rgl$^+$ ExoV$^-$ hosts. This must be cleavage specifically directed by the RglB host function since, as discussed earlier, superinfecting T4 Glu$^-$ DNA is sensitized to ExoV$^+$ exonucleolytic breakdown only in RglB$^+$ hosts when primary phage expression does not occur. It is not known which restricting activity is responsible for inactivation of the *r*IIB gene function.

dCMP hydroxymethylase and dTMP kinase, studied by earlier investigators using Rgl$^+$ ExoV$^+$ hosts (Fukasawa, 1964; Hattman, 1964; Hattman et al., 1966), and T6 immunity and the inhibitor of exonuclease V (Dharmalingam and Goldberg, 1979b), are examples of a second class of genes located close to the Rgl cleavage site and inactivated by further loss of genome integrity by exonuclease V breakdown. It was

TABLE 2. Early function of nonglucosylated T-even phages in Rgl^+ $ExoV^-$ and Rgl^- $ExoV^-$ hosts

Class	Gene, function	Host[a]	
		Rgl^+ $ExoV^-$	Rgl^- $ExoV^-$
1	DNA synthesis	−	−
	*r*IIB membrane protein:		
	Early promoter	−	
	Middle promoter	+	
	arn, RglB inhibitor	−	−
2	T6 *imm*	+	−
	1; deoxyribonucleotide kinase		±
	42; hydroxymethylase	+	±
	44; DNA accessory protein		−
	T2 *αgt*; α-glucosyltransferase		±
	arx; inhibitor of exonuclease V	+	−
	Host killing	+	±
	Turn off host proteins	+	±
	Regulation of early enzyme translation	+	
	Turn off UV-extended early synthesis		−
	T4 exclusion of T2		−
	Glu^- T4 multiplicity activation		−
	Glu^- T6 multiplicity activation		−
3	Glu^- T2 multiplicity activation		+
	T2 *imm*	+	+
	T4 *imm*	+	+

[a] +, Function detected; −, function not detected; ±, function variable (see text).

shown in the earlier work that expression of the dCMP hydroxymethylase and kinase genes is specifically influenced by the multiplicity of infection of Glu^- phage in restricting hosts: at high multiplicity the enzyme activities achieved normal wild-type levels for Glu^- derivatives from all three phages. This might be the result expected if exonuclease V were limiting and were titrated by an increased number of ends exposed by cleavage, sparing some fragments from breakdown. Alternatively, prevention of initial cleavage of some genomes, leaving them intact, or recombination of fragments to reestablish chromosome integrity might be responsible for this effect. Multiplicity reactivation of restricted genomes has been observed for T*2 but not for T*4 or T*6 (Hattman, 1964), the differences being attributed to different specificities of the Rgl degrading mechanism (Revel and Luria, 1970). Perhaps the inhibitor of exonuclease V is not destroyed in Glu^- T2 because there is no cleavage site nearby. The presence of the exonuclease V inhibitor would cause a more severe limitation of exonuclease V in cells infected with T*2 than with T*4 or T*6. Phage genes that cause host killing and the turnoff of host protein synthesis are also not direct targets of initial cleavage but function less effectively in the Rgl^- $ExoV^-$ restriction hosts.

A third class of genes represented by T4 *imm* and T2 *imm*, expressed even when the cleaved fragments are broken down, must be located further away from the specific recognition or cleavage sites. Lack of function in the restricting host where the entire degradation pathway is intact is not sufficient to assign a gene to class 1 or 2. *regA* may be a class 2 gene if the failure of turnoff of host translation by UV irradiation reflects inactivation of the *regA* gene function.

Late Gene Expression

Late T4 gene expression is significantly altered by lack of glucose modification in both permissive and restricting hosts (Dharmalingam and Goldberg, 1979a). Normally the late genes are transcribed by a highly T4-modified host RNA polymerase from a "competent" DNA structure, and transcription is tightly coupled to DNA replication (Rabussay and Geiduschek, 1977a). In permissive hosts Glu^- T4 late gene expression, in contrast to that of Glu^+ T4, is not only uncoupled from chromosome replication, allowing synthesis of proteins not found in Glu^+ T4 DNA-negative (DO) mutant infections, but also unusual since gp32 is overproduced and new proteins unique to the Glu^- T4 template are made as well. The two effects may well be correlated: gaps in the template Glu^- DNA (Dharmalingam and Goldberg, 1976b) could titrate the autoregulatory protein gp32 (Gold et al., 1976) and create sites for initiation of new late transcripts (Cox and Conway, 1973; Hall et al., 1970; Sauerbier and Brautigam, 1970). Some new proteins may result from defective processing. In restricting hosts, restriction cleavage adds further complexity: some proteins are missing, while others specific to the cleaved Glu^- nonreplicating genome are made. In brief, the patterns of late T4 proteins on SDS gels are complex but suggest that deficient glucosylation, impairment of replication, and restriction cleavage may affect late gene expression independently.

It has long been known that T-even *gt* mutants grow poorly, giving only about 20% the wild-type burst in permissive hosts (Revel, 1967). The presence or absence of glucose on parental DNA is irrelevant, but the state of glucosylation of progeny DNA is crucial (Georgopoulos and Revel, 1971). The profound alterations of late protein synthesis exhibited by Glu^- phage in permissive hosts could change the relative concentrations of various structural proteins, affecting the delicate balance required for proper virion assembly. The question remains: why do Glu^- wild phage give a full burst of Glu^- progeny on a *galU* host? That any Glu^- progeny are produced in a restricting *galU* host has been clarified by the discovery that, as discussed above, T4 induces an inhibitor of RglB activity, Arn. Arn antagonizes RglB restricting functions located in the cytoplasm, providing a convenient explanation for the protection of intracellular Glu^- DNA from RglB-controlled cleavage. It does not, however, address the question of why the growth of Glu^- T* is so much more efficient than the growth of *gt* phages. Certainly they should exhibit the same bizarre pattern of protein synthesis and should behave similarly with respect to morphogenesis. Although the *galU* host W4597 is notoriously leaky (Hattman and Fukasawa, 1963) (see Table 1), the DNA of T* progeny of $B/4_0$ is nearly as glucose deficient as *gt* DNA; differences can be detected only by the ability of the former to escape restriction by P1 (Revel and Geogopoulos, 1969). The wild-type Glu^- T* progeny do have the necessary glucosyltransferases to carry out some modification. This must be very specific and appears to be sufficient to allow normal progeny production. It would be of interest to examine late protein synthesis of wild-type phage in a permissive *galU* host to clarify this issue. The localization of a few glucose groups on T*4 DNA might be detected in heteroduplexes of T* Glu^- T4

DNA and Glu$^-$ T2 DNA derived from a double *gt* mutant, by antibodies specific for α- or β-glucose (McNicol and Goldberg, 1973).

Negative Effect of β-Glucosylation

T4 uses the host RNA polymerase throughout infection, modifying it during development to accomodate its own transcriptional and developmental program. Snyder and Montgomery (1974; Montgomery and Snyder, 1973) have reported a mutation of RNA polymerase which delays T4 growth, causing a cold-sensitive delay in DNA replication and affecting late gene expression as well. The most frequent class of phage mutants (*gor*) that suppress this effect are deficient in glucosylation (αgt^+ βgt^-). Since an αgt^- βgt^- double mutant grew no better than the αgt^+ βgt^- phage when the mutation was moved to an Rgl$^-$ host, it appears that only glycosylation of adjacent HMC residues not susceptible to α-glucosylation (de Waard et al., 1967) interferes with growth on the mutant host. Glucosylation of parental DNA, provided by growth before infection or de novo early in infection, was sufficient to delay phage growth. One of the models suggested that HMC-rich regions normally glucosylated may contain recognition signals for RNA polymerase. Unmodified by glucose, they may closely resemble host recognition sites, eliminating the need for a particular T4 modification of the polymerase. In effect, the absence of β-glucosylation bypasses the need for a specific RNA polymerase modifying factor. The very earliest phage promoters, those read immediately after infection by host polymerase, may be thymidine rich and not subject to β-glucosylation and would not require a modified polymerase. Other T4 *gor* mutants that suppress the *rifB2* mutation in the host RNA polymerase have mutations that map in gene 45 and perhaps in gene 55.

GLUCOSE-STIMULATED RECOMBINATION

Recombination of various genetic markers in crosses between nonglucosylated T4 *gt* mutants in Rgl$^-$ hosts occurs at about the same frequency as with the corresponding gt^+ phage (Georgopoulos, 1968; Georgopoulos, thesis). However, a recent report documents a region-specific special glucosyl-stimulated recombination system in T4 (Levy and Goldberg, 1980a, 1980b, 1980c). Early mapping experiments indicated that T4 contains several regions exhibiting higher recombination frequencies than expected from the physical map size (Mosig, 1966). The most prominent distortion is in the T4 genome region corresponding to the promoter-distal end of gene 34, adjacent to gene 35 (Beckendorf and Wilson, 1972). Comparing the recombination frequency of two amber mutations in the gene 34–35 region to that of two amber mutations in the *r*II region, Levy and Goldberg (1980a) found that only the former was stimulated by glucose. Maximum stimulation occurred when both parental and progeny DNA were glucosylated. This region-specific recombination could be distinguished from normal recombination mechanisms by its sensitivity to heat. A specific nicking enzyme might cut only on Glu$^+$ DNA in this region, creating ends and gaps that are recombinogenic. Alternatively, a nicking enzyme might generate region-specific breaks in both Glu$^+$ and Glu$^-$ DNA. If the gaps or single-stranded DNA ends are more readily repaired in Glu$^-$ DNA or, conversely, rendered more sensitive to endonucleolytic cleavage and degradation, potential recombinants might be generated only by the nicked Glu$^+$ DNA. The results of three-factor crosses, examining the frequency of an unselected exchange of markers among selected recombinants, support the notion that region-specific nicking is nondiscriminant, occurring in both Glu$^+$ and Glu$^-$ DNA. In Glu$^-$ crosses, longer heteroduplex regions are formed than in Glu$^+$ crosses; in addition there appears to be efficient mismatch repair. Why such repair should favor Glu$^-$ DNA over Glu$^+$ DNA is unclear. Perhaps the *E. coli* repair machinery works more effectively with Glu$^-$ DNA; in particular, the host 5′-3′ editing function, not supplied by T4 polymerase, may prefer a Glu$^-$ substrate. It was suggested that the delayed maturation of Glu$^-$ phage would expose Glu$^-$ recombination intermediates to repair machinery for a longer time, a hypothesis which fails to account for the stimulation seen at early times. A third possibility is that heteroduplex regions in Glu$^+$ DNA are shorter than in Glu$^-$ DNA. Possibly a combination of these factors is involved.

This system bears a superficial resemblance to *chi*-stimulated recombination in lambda (Lam et al., 1974), but the enzymes involved are different. Whereas *chi*-mediated recombination requires the host *recB-recC* recombination system, the T4 special recombination system does not. Stimulation of the system by mutations in *E. coli recL*, *recK*, and *uvrE*, genes whose products normally function in UV repair by sealing gaps, suggests that the persistence of unrepaired nicks might account for high recombinogenicity. Of interest in this connection is work by Ramig quoted in Russell (1974) to the effect that substitution of the T4 34-35 join region by the homologous region of T2 eliminates the junctional recombination gradient described by Beckendorf and Wilson (1972).

SUMMARY AND CONCLUSIONS

Modification and restriction of T-even phages was first described 30 years ago, the first description of any restriction-modification system. Though an understanding of the mechanism of modification rapidly followed the discoveries of HMC, glucosylation of HMC, and phage-induced glucosyltransferases, the mechanism of host-controlled restriction and the impact of modification on phage development have not yet been determined. Progress has been made, however.

Restriction and modification in bacteria, as currently understood, serves principally as a mechanism for evolving the species. A methylase modifies two adenines in a specific DNA site (nucleotide sequence) which would itself be cleaved by a restriction endonuclease, or lead to cleavage at a distance, if unmodified. Central to the bacterial restriction-modification systems is coordinated control: DNA modification and restriction work in concert to achieve the common goal of protection of endogenous DNA and exclusion of exogenous DNA. This may be achieved by separate methylases and restriction endonucleases exhibiting the same site specificity, or alternatively, the modifying and restricting functions may exist as a multifunctional complex composed of R, M, and S subunits coordinately regulated by the methylation status of the recognized DNA site (Yuan and Hamilton, 1982).

In contrast, it appears that in the T-even phage-host relationship restriction and modification have evolved as mutually antagonistic modes of protection under separate but modulated controls. Modification is phage specific but only somewhat site specific. It consists of glucosylation of most of the HMC residues that substitute for cytosine in T-even DNA, yielding a unique glucosylation pattern for each phage DNA. Modification is primarily under phage control. Glucosylation is catalyzed by phage-induced glucosyltransferases but is subject to modulation by the host which must supply the prerequisite glucosyl donor, UDPG, for the reaction.

Restriction (Rgl) is under host control, presumably developed by bacteria to exclude those phages with HMC DNA that escaped the conventional restriction-modification surveillance. Rgl restriction acts on unmodified HMC DNA either during injection or soon after entry into the cell. Early genetic analysis revealed that Rgl functions are under multigenic control. Mutants defined at least two Rgl systems that act independently and exhibit different and limited specificities: the RglA function restricts all Glu^- T-even phages; the RglB activity recognizes only Glu^- T2 and Glu^- T4; not all free HMC residues are recognized. The mechanism of restriction has now been demonstrated directly in vivo and in vitro to be a specific and limited endonucleolytic cleavage of HMC DNA to large fragments which are subsequently broken down by exonuclease V to give the acid-soluble degradation products associated with the Rgl phenotypes. Additional experiments, while answering some questions, have revealed a greater genetic complexity than was previously imagined.

The RglB function, active in cell-free extracts, is probably located in the cytoplasm, and its function is modulated by a phage-induced inhibitor, Arn. The RglB restriction endonuclease has not been isolated, but a protein component defined by the *rglB1* allele has been purified by assaying its ability to complement an *rglA rglB1* extract. The requirement for additional proteins to achieve cleavage of HMC fd DNA in vitro implies that at least two genes control RglB restriction activity. Nothing is known about the site specificity. The reaction could occur by a simple type II mechanism where recognition and cleavage are coincident, or double-strand breaks could result from either a type I (*hsd*) or type III (P1) mechanism where cleavage occurs some 30 to 3,000 bases from the point of recognition. Multiple subunits and the involvement of ATP would favor the latter and predict that the primary cleavage products would be heterogeneous.

RglA restriction, inactive in cell-free extracts and inactivated in vivo by primary phage infection in the absence of protein synthesis, probably resides in the membrane. It is not known whether the genetically defined RglA gene encodes the restriction endonuclease, its specificity subunit, or one or more membrane proteins with which restriction enzyme must interact. There has been no biochemical analysis of this activity beyond the initial observation that in toluenized cells it is capable of inhibiting *E. coli* DNA synthesis when HMdCTP is the substrate in place of dCTP. The question of localization within the membrane is intriguing: is the RglA activity associated with the inner or the outer membrane, or does it contribute to the pore structure through which phage DNA penetrates into the cell? The almost invariable association of the $RglA^-$ phenotype of *E. coli* B with a requirement for thiamine and failure to find thiamine-positive revertants suggests that both defects may result from a deletion. This strain would be of particular interest to examine for abnormalities in membrane proteins.

The possibility of additional Rgl restricting systems is suggested both by the variety of Rgl phenotypes that exist naturally and by the failure of a primary infecting phage to prevent cleavage of Glu^- T6 *gt* in strain W3110. This newly revealed activity could be remnant of a P1 prophage. Alternatively, it could be due to the RglA system, associated in such a way with other membrane components that primary phage attachment is not detrimental.

Differential localization and control of Rgl restriction systems provides a satisfying explanation for the ability of Glu^+ phage to replicate and produce unmodified progeny in a *galU* restricting host. Although neither the modulation of restriction by phages nor the modulation of glucosylation by the host cells is of direct benefit to a particular Glu^- phage or to a particular host cell, these mechanisms do have survival benefit for the species.

It appears that the evolution of DNA glucosylation may have been accompanied by subtle changes in the T4 transcriptional apparatus. In *gt* mutant infections the normal tight regulation of late protein synthesis is disrupted. In certain RNA polymerase mutants the presence of β-glucosylation has a detrimental effect on both DNA synthesis and late transcription. Thus it is perhaps not surprising that *gt* mutants have a low burst size. The reason why equally nonglucosylated T^* phages give a full yield remains an open question.

REGULATION OF GENE EXPRESSION

Transcription

Phage-Evoked Changes in RNA Polymerase

DIETMAR RABUSSAY

Bethesda Research Laboratories, Gaithersburg, Maryland 20877

Much of the regulation of gene expression in T4-infected cells occurs at the transcriptional level. In *Escherichia coli*, transcription (i.e., the DNA-programmed synthesis of stable RNAs and mRNAs) is catalyzed by a single enzyme, the DNA-dependent RNA polymerase (RNAP). Different coliphages vary in their dependence on the host RNAP for the transcription of their own DNA. Some, such as phage λ, depend on the host enzyme in its original form throughout the infective cycle. Phage N4 supplies its own preformed RNAP from the very start of the infection (Falco et al., 1980; Rothman-Denes and Schito, 1974). Phages T3 and T7 use the host RNAP early in infection for the transcription of a subclass of viral genes, one of which codes for a phage-specific RNAP that selectively transcribes the rest of the genome (reviewed by Chamberlin, 1982). Bacteriophage T5 and the T-even phages use yet another strategy: the basic host RNAP is employed throughout infection, but this enzyme is considerably modified, in several steps, during phage development.

In the case of bacteriophage T4, none of the changes in *E. coli* RNAP that occur during the prereplicative period, i.e., before the onset of DNA replication, has been shown to be essential for this period of T4 development. However, at the end of the prereplicative period, RNAP starts to undergo some changes that are essential for late gene expression.

I want to emphasize at the outset that essentially at any time after T4 infection, different RNAP complexes of different subunit composition are believed to coexist in the infected cell. The reasons for this assumption will be discussed below. The simultaneous presence of different RNAP complexes is probably necessary because different classes of genes served by different types of promoters are transcribed (or blocked) simultaneously.

In this chapter, I shall summarize information about the changes in *E. coli* RNAP that occur after T4 infection and discuss their potential function in gene regulation. Reviews covering the subject addressed in this chapter have been published previously (Khesin et al., 1980; Rabussay, 1982a; Rabussay, 1982c; Rabussay and Geiduschek, 1977a).

CHANGES IN RNA POLYMERASE STRUCTURE

Essentially all studies of T4-evoked changes in the host RNAP have been carried out in *E. coli*. A comparison with the fate of the transcriptases of other enteric bacteria which support growth of T4 has not been carried out but could be particularly informative about the function of the changes observed in the *E. coli* enzyme.

E. coli RNAP consists of a core of relatively tightly associated proteins forming a complex with the subunit composition $\beta'\beta\alpha_2$. The molecular sizes of these subunits are 165, 151, and 36.5 kilodaltons (K), respectively (reviewed by Chamberlin, 1982). This enzyme is capable of initiating RNA synthesis at nicks or ends of double-stranded DNA and can elongate RNA chains until stopped by a termination signal. The core enzyme interacts with, or binds, a number of other cellular proteins, the most prominent of which is the σ protein (molecular size, 70K) (Burton et al., 1981). The complex $\beta'\beta\alpha_2\sigma$, known as RNAP holoenzyme, specifically and selectively initiates transcription at promoters that share structural features (Rosenberg and Court, 1979). Initiation of transcription by the holoenzyme involves local unwinding of about 10 base pairs of the double helix at the promoter. The σ protein is specifically required for this step. After the nascent RNA chain has reached a length of about 10 nucleotides, σ is released, and the core RNAP continues until a termination signal is reached (for reviews, see Losick and Chamberlin, 1976). The existence of more than one σ factor which might confer different initiation specificities to the core (as in *Bacillus subtilis*; for reviews, see Geiduschek and Ito, 1982; Losick and Pero, 1981) has not been demonstrated in uninfected *E. coli*. However, other RNAP-binding components (such as the *nusA* protein, termination factor rho, and ppGpp) and their effects on transcription specificity and efficiency have been characterized to various extents. In addition, a considerable number of other *E. coli* proteins displaying some affinity for RNAP have been identified, but their functions are largely unclear (Greenblatt and Li, 1981; Ratner, 1974a; Ratner, 1976; for a review see Lathe, 1978). The point I want to make is that *E. coli* RNAP displays a high degree of plasticity (and a corresponding degree of functional flexibility) in forming various transcription complexes for different transcriptional demands. This plasticity will also be hinted at as we look at the structural changes of RNAP after T4 infection and their possible functional implications.

The conservation of all core host RNAP subunits after T4 infection has been demonstrated by radioactive isotope labeling experiments (Goff and Weber, 1970; Schachner and Zillig, 1971). The continuing role of subunit β in transcription is also implied by the unchanged rifampin response phenotype after T4 infection (DiMauro et al., 1969; Haselkorn et al., 1969; Zillig et al., 1970). However, chemical modifications of the *E. coli* subunits and the addition of T4-specific subunits occur in a defined time sequence after infection. Table 1 summarizes the better-characterized changes in chronological order, and Fig. 1 presents an overview of the presumed RNAP complexes present during different times in the T4 vegetative period.

The first change, termed alteration, is complete within 30 s after infection at 30°C (Seifert et al., 1969). It results in the covalent attachment of an adenosine-5′-diphospho-5″-β-D-ribosyl (ADP-ribosyl) residue to one of the two α subunits of RNAP (Rohrer et al., 1975). Free RNAP as well as enzyme engaged in transcription must be subject to this alteration. This process is catalyzed by a protein which is a component of the phage head and which must be transferred to the infected cell at the time of infection. The 70K altering enzyme has been purified to near homogeneity from T4 phage (Rohrer et al., 1975). The alteration reaction involves the transfer of the ADP-ribosyl moiety from the aromatic nitrogen of oxidized NAD to a guanido nitrogen of the arginyl residue 265 of one of the α subunits of a RNAP protomer (Ovchinnikov et al., 1977; Rohrer et al., 1975). One or two other arginyl residues of the same α subunit also became altered with low frequency. The second α subunit apparently is inaccessible to the altering enzyme, whose specificity is not high: many other host proteins, including a fraction of the σ protein, also become ADP-ribosylated (Rabussay et al., 1972; Rohrer et al., 1975). This lack of specificity makes it especially difficult to establish the significance of the alteration process for transcriptional regulation. The altering enzyme is coded for by the T4 gene *alt*, whose product is synthesized from around 3 min after infection to the end of the lytic cycle (Goff, 1979; Horvitz, 1974a; Horvitz, 1974b). The primary gene *alt* product has no ADP-ribosyl transferase activity and is probably activated by proteolytic modification during phage assembly. This can be concluded from the kinetics of appearance and disappearance of the altering activity in the infected cell: maximal altering activity is observed around 2 min after infection, yet 8 min later, altering activity originating from the infecting phage is essentially no longer detectable, although gp*alt* is continuously synthesized (Horvitz, 1974a; Horvitz, 1974b; Rohrer et al., 1975). Moreover, as seen in experiments involving T4 *mod*$^-$ infections, alteration of the RNAP α subunit is reversed about 10 min after infection, presumably by a T4-induced protein (W. Zillig, personal communication).

The purified altered enzyme has a markedly lower σ content than unmodified holoenzyme. Since the only known difference between altered and unmodified RNAP is the ADP-ribosylation of one α subunit, the apparently weaker binding between σ and the altered core must be attributed to ADP-ribosylation.

A second ADP-ribosylation process, which results in the covalent modification of both α subunits of RNAP, begins 1 to 2 min after infection. This process, termed

TABLE 1. T4-evoked changes of *E. coli* RNAP

Change	Time first detectable in RNAP (min after infection at 30°C)	Effect or function
ADP-ribosylation of one of the two α subunits (alteration)	<0.5	Lowers affinity for σ; participation in host shutoff?
Phosphorylation, adenylylation, or ADP-ribosylation of a fraction of σ	<0.5	Unknown
ADP-ribosylation of both α subunits (modification)	1.5–2.0	Lowers affinity for σ; selective shutoff of some early genes around 4 min; participation in host shutoff?
Binding of 10K protein	<5	σ antagonist
Binding of 15K protein	5	T4 gene 60 codes for a subunit of T4 DNA topoisomerase and may also code for the 15K protein
Binding of gp33 (12K)	5–10	Positive control of late transcription
Binding of gp55 (22K)	5–10	Positive control of late transcription
Interaction with gp45 (27K)	(Does not copurify with RNAP)	Gp45 is a component of the core of the T4 replisome and is also directly involved in late transcription

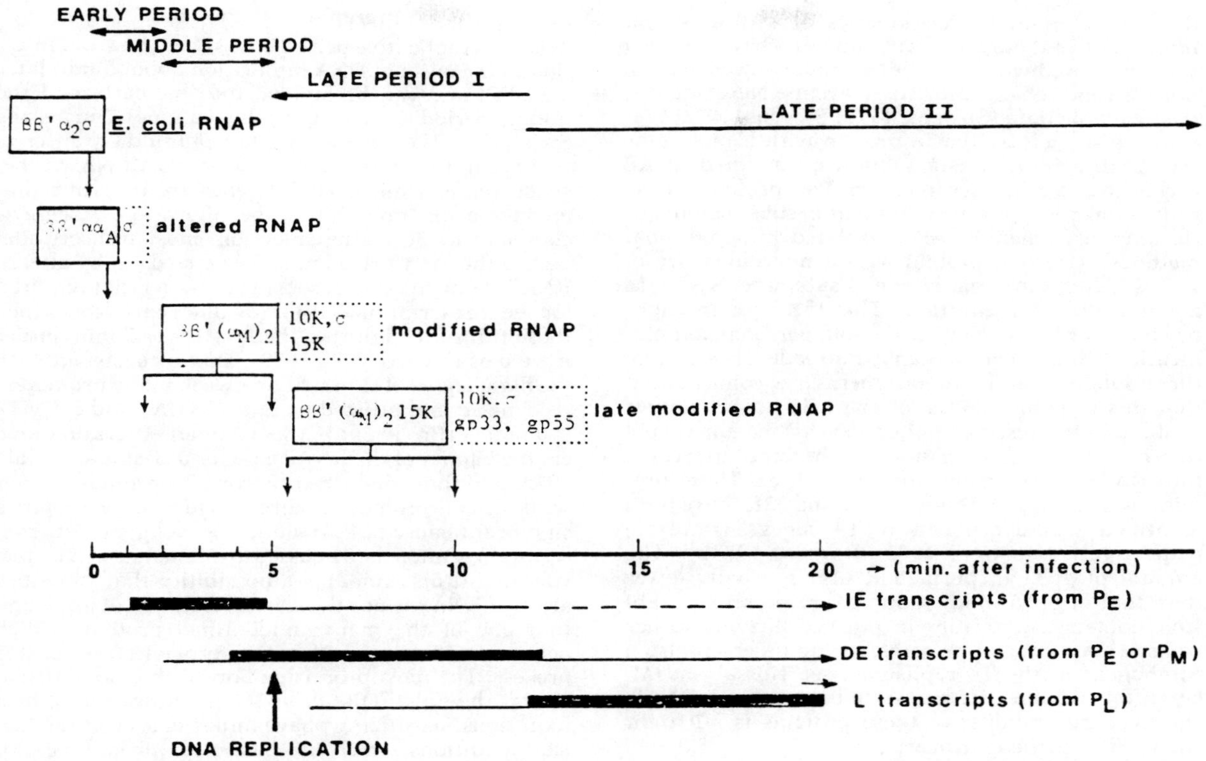

FIG. 1. Changes in RNA polymerase during different periods of T4 development (from Rabussay, 1982c). The times given refer to infections at 30°C. The stoichiometric subunit compositions of unmodified, altered, modified, and late-modified core RNA polymerases are given in solid-line boxes. Subunits within dotted-line boxes are present in substoichiometric amounts in RNA polymerases purified by standard procedures. Solid arrows protruding from boxes point to the time of completion of the changes in question; dashed arrows mark the onset of the corresponding changes. For example, changes leading to modified RNAP start at about 1.5 min after infection and are complete by 5 min. The lower part of the figure shows (greatly simplified) the timing of major changes in the transcriptional pattern of T4 development. Thin lines indicate low, and thick lines indicate high, rates of transcription of the corresponding gene class. DNA replication starts at about 5 min after infection. Abbreviations: α_A, altered α; α_M, modified α; P_E, P_M, and P_L, early, middle, and late promoters, respectively; IE, DE, and L, immediate early, delayed early, and late transcripts, respectively. For other symbols and abbreviations, see the text.

modification, is catalyzed by the T4 gene *mod* protein (27K) which has been purified about 100-fold from T4-infected cells (Goff, 1974; Horvitz, 1974a; Horvitz, 1974b; Skorko et al., 1977). The highly specific action of the modifying enzyme is strictly limited to both α subunits of RNAP and results in the irreversible ADP-ribosylation of the arginyl residue in position 265 (Goff, 1974; Rohrer et al., 1975). The modification reaction does not depend on previous alteration. Modification of the α subunits reaches a maximum at about 5 min after infection (at 30°C), but the achieved level of completion may vary. Measurements of modification kinetics based on the analysis of purified RNA polymerases show essentially complete modification of all α subunits by 4 min after infection (Seifert et al., 1969); on the other hand, analysis of immune precipitates, obtained with antibodies against RNAP from crude lysates of T4-infected cells, indicates incomplete modification even later in infection (Horvitz, 1974a; Horvitz, 1974b). It is not clear to what extent this discrepancy is due to differences in the genetic backgrounds of the host and phage strains used, the physiological conditions before and after infection, or the procedures employed for RNAP preparation and analysis.

Changes in the antigenicity and tryptic peptide patterns of subunits β and β′ after infection were reported some years ago (Schachner and Zillig, 1971). It was subsequently realized that these changes might have been caused by one or several of the T4-specific RNAP subunits which were present as contaminants in the original β and β′ subunit preparations. It was also observed that a host mutant with a temperature-sensitive β′ subunit became temperature insensitive during the late period of T4 infection (Khesin, 1970). However, this has never been shown to be a direct effect of a change in the β′ subunit itself.

T4-evoked changes in RNAP also include changed affinity for the σ subunit and the integration of de novo synthesized T4-specific subunits into modified host RNAP (Table 1, Fig. 1). We shall discuss the addition of T4-specific subunits next and the changed binding of σ later. The kinetics of synthesis and integration of these T4-specific subunits have not (yet) been measured with sufficient accuracy to construct a precise time schedule for these events. Nevertheless, the available data (Stevens, 1970; Stevens, 1972; Stevens, 1974) suggest that a protein with a molecular size of 10K appears as a subunit of modified RNAP earlier than 5 min after infection. This protein, whose gene is unknown, is always present in substoichiometric amounts when RNAP is isolated by traditional methods. A second protein with a molecular size of 15K is found somewhat later as a subunit of RNAP (at about 5 min after infection). This 15K protein might be coded for by gene 60 (C. G. Goff, personal communication). Since gene 60 is known to code for one of the three subunits of T4 topoisomerase, a confirmation that this gene also codes for the 15K RNAP subunit could have interesting implications for the function of this protein. Two more subunits become integrated into RNAP, presumably soon thereafter. These proteins, with molecular sizes of 12K and 22K, have been identified as the products of T4 genes 33 and 55, respectively (Horvitz, 1973; Ratner, 1974b). The amount of gp33 in purified RNAP is usually lower than that of gp55; both subunits are present in substoichiometric quantities in purified enzyme. Other transcription control proteins, including gp45 (also a component of the T4 replisome: see Nossal and Alberts, this volume) are known to interact with RNAP. However, the binding of these proteins is not tight enough to effect copurification.

POTENTIAL INVOLVEMENT OF RNA POLYMERASE CHANGES IN GENE REGULATION

Figure 1 gives a schematic and much-simplified overview of the principal facts about transcriptional regulation. (i) Three major classes of T4 transcription units (early, middle, and late) have been distinguished; each of these starts to be expressed at a different time after infection, is served by a distinct class of promoters, and requires a different transcription apparatus. (An additional class of late promoters has recently been found; Geiduschek et al., this volume.) (ii) Starting about 1.5 min after infection, different major gene classes are transcribed concurrently. (iii) Different RNAP complexes with different subunit compositions probably exist side by side at almost any time after infection. (The sole exception may be the complete conversion to altered RNAP, $\beta'\beta\alpha\alpha_{ADPR}\sigma$, between 0.5 and 1.5 min after infection). The different core polymerases (shown within the solid-line boxes in Fig. 1) may be associated with one or several of the subunits, present in substoichiometric amounts in purified RNAP (these subunits are shown in the dotted-line boxes in Fig. 1). For example, modified RNAP probably exists in different forms, two in which the modified core is associated with σ (±15K) and two in which the modified core is associated with the 10K protein (±15K).

In the following sections, I shall discuss the proven or implied roles of the different changes of RNAP in prereplicative and postreplicative transcription regulation.

Prereplicative Period

The prereplicative period spans the time from infection to the onset of DNA replication about 5 min later (at 30°C) and can be divided into the early and the middle period (Brody et al., this volume). During the early period, i.e., before the start of middle transcription around 1.7 min after infection, host RNA synthesis decreases while early T4 genes are transcribed at an increasing rate. Since the alteration process is complete by 30 s after infection, most transcription during the early period must be carried out by altered RNAP. As mentioned earlier, alteration is not required for the transcription of T4 immediate early genes (i.e., those transcribed during the first 1.5 or 2 min) either in vivo or in vitro (Goff, 1979). This is consistent with the high degree of similarity between many promoters recognized and utilized by *E. coli* RNAP and early T4 promoters (Brody et al., this volume). Alteration lowers the affinity of the RNAP core for σ (Rabussay et al., 1972) and thus might favor strong T4 promoters over weak *E. coli* promoters; direct evidence for this as a significant cause of host shutoff is lacking (host transcription shutoff is discussed by Snustad et al., this volume). It also remains a possibility that the alteration of RNAP is not the critical function of gp*alt* and that one of the many unidentified proteins which becomes ADP-ribosylated is the crucial target of this process. The most promising approach to an elucidation of the function of *alt* is the identification of host mutations, sensitizing phage mutations, or physiological conditions which produce clear phenotypes for mutations in *alt*.

The start of the middle period is signaled by the appearance of delayed early transcripts which arise either from the readthrough of RNAP from immediate early into delayed early genes across potential rho-sensitive termination sites, or from initiation at middle promoters (Brody et al., this volume). By this time (about 1.7 min after infection), a considerable amount of RNAP is present in the modified form. The first regulatory event, readthrough across immediate early-delayed early barriers, seems to be prohibited, rather than facilitated, by modification of RNAP: at physiological ionic strength (ca. 0.17 M), T4-modified RNAP is terminated more efficiently by termination factor rho than is unmodified RNAP. The same is also true of altered RNAP (Schäfer and Zillig, 1973; Rabussay, unpublished). Although such an effect seems, at first glance, to run counter to the desired regulatory result, it is conceivable that this phenomenon serves to prevent an overcommitment of T4 to other than immediate early gene expression: if translation does not occur at a high level, transcription can stop at the end of immediate early genes and not proceed into adjacent delayed early genes.

Modification of RNAP also seems to be involved in other negative regulation: host transcription continues to decline during the middle period, and a group of immediate early genes is turned off at about 4 min after infection. Mailhammer et al. (1975) measured the initiation efficiency of in vitro reconstituted unmodified and modified RNA polymerases on several *E. coli* promoters and on total T4 DNA. The results indicate that modification of the α subunits is directly responsible for a lower rate of initiation at the tested *E. coli* promoters, whereas overall transcription efficiency on T4 DNA is not measurably affected. This suggests an involvement of α modification in host shutoff, although presently available in vivo results do not support this notion (Goff and Setzer, 1980). A related series of experiments has been carried out more recently by Goldfarb (1981b) and Goldfarb and Palm (1981). These authors studied the effect of α modification on transcription initiation efficiency at a number of well-mapped T4 early promoters in vitro. Surprisingly, they found a subset of early promoters (all of which were recognized by unmodified holoenzyme) that was not utilized by T4-modified RNAP. This finding could explain the shutoff of a group of immediate early genes around 4 min after infection. However, it also raises the question what the differences between the two subgroups of T4 early promoters are and what their exact structural and functional relationships with *E. coli* promoters might be.

The functional properties of modified RNAP displayed in vitro are compatible with a role in negative regulation. Khesin and co-workers have shown that T2-modified RNAP, as compared with unmodified *E. coli* RNAP, is characterized by a requirement for relatively high temperatures for promoter opening, a slower opening and closing of promoters, and a lower stability of its complexes with DNA (Khesin et al., 1976; reviewed by Khesin et al., 1980). The observed differences are clearly visible at or above physiological ionic strength but are less apparent at ionic strengths below 0.1 M. All of these properties of modified RNAP are consistent with a weaker binding between the modified core and subunit σ. There is suggestive evidence that the affinity of σ for the core (and thus, presumably, the efficiency of interaction during RNA chain initiation) is basically set by the ADP-ribosylation of the α subunits and in addition is flexibly regulated by the 10K protein. The negative effect of ADP-ribosylation on σ binding was discussed above. The 10K protein has been purified and has been characterized as an anti-σ activity (Stevens, 1976; Stevens, 1977; Stevens and Rhoton, 1975). The 10K protein inhibits the formation of open promoter complexes and thus the initiation of RNA synthesis (Stevens, 1977). This effect has a relatively sharp ionic strength optimum (0.2 M) and is abolished by the nonionic detergent Triton X-405. Recent in vitro experiments have shown that the inhibitory activity of the 10K protein can be reversed by excess amounts of σ. Conspicuously, T4-modified RNAP shows a higher relative and absolute susceptibility to the anti-σ activity of the 10K protein than unmodified RNAP does. In addition, preliminary experiments have shown that the fluorescence quench observed upon interaction of the core and σ is partially reversed by the addition of the 10K protein (H. Heumann and D. Rabussay, unpublished results). These results directly support the notion of the 10K protein being a σ antagonist. As indicated by their extensive copurification, σ and then 10K protein certainly interact with each other. Anti-σ probably also interacts with core RNAP, as indicated by the results of Khesin et al. (1972) and by its presence in purified modified and late modified RNAP in the absence of corresponding amounts of subunit σ.

In conclusion, T4 preserves and uses not only the host core RNAP but also the host σ subunit. However, the efficiency of σ action seems to be controlled in a subtle way. There is no convincing evidence for the function of T4-specific σ subunits in prereplicative transcription control.

Neither alteration nor modification of the α subunits is essential for T4 growth on a number of bacterial hosts tested (Horvitz, 1974a; Horvitz, 1974b). Even a double mutant, *alt*⁻ *mod*⁻, does not show a significant effect on T4 multiplication under normal laboratory conditions (Goff and Setzer, 1980). One can imagine that under unfavorable conditions the presumed functions of alteration and modification (i.e., the reduction of host transcription and the shutdown of phage genes whose products are no longer needed) do confer a significant advantage upon the phage. Moreover, a crucial experiment regarding the function and essentiality of *alt* and *mod* is presently not possible because of the unavailability of suitable multiple mutants (e.g., mutants in *alt* or *mod* and the gene coding for the 10K protein). It is possible that the 10K protein can compensate for a lack of alteration or modification of α, especially if the amount of 10K protein is regulated. Thus, the function of *alt* and *mod* should be assessed in a 10K-deficient genetic background.

Postreplicative Period

The postreplicative period starts at about 5 min after infection (30°C) when T4 DNA replication is first initiated. Transcription of true late genes starts 1 to 2 min later (Young et al., 1980). By about 9 min after T4 infection, host transcription is no longer detectable (Kennell, 1968; Kennell, 1970), and several minutes thereafter, a major shift in T4 transcription occurs in which true late transcripts become the dominant species and a number of prereplicative genes are shut off.

The transcription of T4 late genes depends on modifications in the RNAP and a special competent (processed) DNA template which is usually created in the course of T4 DNA replication. In addition, late transcription is coupled to DNA replication via at least one protein, gp45. As mentioned above, this protein is a component of the core complex of the T4 replisome and is also directly involved in T4 late transcription (Alberts et al., 1975; Wu, Geiduschek, and Cascino, 1975). The exact spatial relationship between the T4 replisome and late transcription complexes is not known. It is therefore difficult to assess whether the same gp45 molecule is part of both complexes. Although gp45 does not copurify with RNAP, there is sufficient genetic and biochemical evidence for its interaction with modified and late-modified RNAP: T4 mutations (*com*, *gor*), which allow phage growth on late transcription-defective hosts (*tabD*, certain Rif^r), map in genes 45 and 55 (Coppo et al., 1975a; Coppo et al., 1975b; Snyder and Montgomery, 1974). In addition, gp45 specifically binds to immobilized modified RNAP but not to unmodified core or holoenzyme (Ratner, 1974a). This result suggests that ADP-ribosylation of the α subunits is crucial for substantial gp45 binding since all other T4-specific RNA-binding proteins seem to bind well to matrix-bound unmodified and modified RNAP. If this interpretation is correct, a very weak interaction between unmodified core and gp45 must suffice for late transcription since *mod* is not essential for T4 growth.

The changes in RNAP reaching completion within the first minutes of the late period (see Table 1 and Fig. 1) are of different importance for the transcription of T4 late genes. ADP-ribosylation of α is clearly not essential for late transcription in vivo (Goff and Setzer, 1980; Horvitz, 1974a; Horvitz, 1974b) or in vitro (D. Rabussay, unpublished results). The requirement for the 10K and 15K proteins for late transcription, or T4 development in general, has not been tested since no mutants are available. However, gp45, gp33, and gp55 have been shown to be necessary for, and directly involved in, this process: gp45 and gp55 are continuously required (Bolle et al., 1968b; Coppo et al., 1975a; Coppo et al., 1975b; Pulitzer, 1970; Rabussay and Geiduschek, 1977b; Wu, Geiduschek, and Cascino, 1975). The requirement for gp33 seems less stringent than that for gp55: even double amber mutants in gene 33 show a leaky phenotype (Horvitz, 1973; Rabussay and Geiduschek, 1977b; Wu and Geiduschek, 1975). Besides regulating late transcription positively, genes 33 and 55 seem to regulate certain early transcription units negatively (Bolund, 1973; Sköld, 1970). A particularly intriguing aspect of negative transcriptional regulation is the effect of mutations in genes 33 and 55 on the initiation of T4 DNA replication. Lesions in these genes have been known to revert the arrest of DNA replication caused by lesions in recombination functions (Shah and Berger, 1971; Shalitin and Naot, 1971; Warner and Hobbs, 1967). It has recently been found that this is probably due to the continued synthesis of an RNA primer for DNA replication in 33^- or 55^- infections. Binding of gp33 and gp55 to RNAP thus seems to control a shift from an initially active mechanism of T4 DNA replication (depending on the synthesis of RNA primers by RNAP at preferred origins) to a different mechanism which takes over a few minutes later (using recombinational intermediates for initiation at a multitude of secondary origins) (Luder and Mosig, 1982; Mosig, this volume).

Next, two aspects of postreplicative transcription will be discussed: possible functions of the RNA polymerase subunits in the mechanism of late transcription, and a model for the concurrent transcription of different classes of transcription units.

Very little is known as yet about the mechanism of initiation for late transcription. However, during the past few years several late promoters have been mapped and sequenced. A consensus in late promoters serving late structural genes has been identified consisting of a relatively long, extremely AT-rich stretch of DNA closely upstream from the transcription initiation site. No consensus sequence has been found in the −35 region of these promoters (see Christensen and Young, this volume). Thus, these late promoters, which will be referred to as type I late promotors, are considerably different from other known procaryotic promoters, implying the existence of special initiation factors for their recognition and utilization. Most recently, a different type of late promoter (type II) governing the initiation of certain anti-late transcripts has been detected. These promoters display their own characteristic consensus sequence (Geiduschek et al., this volume). At present, we cannot even guess which of the RNAP-binding proteins are involved in the recognition and opening of these different late promoters. For reasons discussed earlier in this paper, it seems very unlikely that the 10K protein functions as a σ-like protein, and recent in vitro experiments (Geiduschek et al., this volume) suggest that gp45 is also not a σ analog. Moreover, the probability that host σ participates in late promoter utilization appears remote, for the following reasons. (i) As already discussed, σ binds weakly to modified core, probably competes with T4-specific proteins for binding, and is present in much smaller quantities in purified late-modified RNAP than any of the other core-binding proteins. This has to be viewed in light of the fact that late transcripts represent the predominantly synthesized species during most of the postreplicative period (Bolle et al., 1968a; Bolle et al., 1968b). (ii) In vitro experiments with a cellophane-supported crude lysate system have shown that excess amounts of purified σ boost early RNA synthesis but reduce the level of late transcription (D. Rabussay, unpublished results). (iii) Initiation of transcription at defined late T4 promoters in vitro has been demonstrated to depend on late-modified RNAP which contained, at most, traces of σ. Unmodified holoenzyme or core enzyme or late-modified enzyme from T4 *am*292 (gene 55^--infected cells did not initiate RNA synthesis at these promoters (G. A. Kassavetis, T. Elliott, D. Rabussay, and E. P. Geiduschek, Cell, in press). Recent in vivo experiments with a temperature-sensitive mutant carrying a defect in *rpoD* (the gene coding for σ) show that upon shifting to the nonpermissive temperature, transcription of both early and late genes decreases (Zograff, 1981). However, a direct and specific effect of σ inactivation on late transcription has not been demonstrated. Transcription shutoff in this system after raising the temperature is slow, which makes it particularly difficult to distinguish between effects on early and late transcription.

Exclusion of the subunits just mentioned leaves gp33, gp55, and the 15K protein as candidates for σ-like initiation factors. Among these, gp33 resembles host σ most closely because of its relatively loose binding to the core (see below). It would be interesting to know whether gp33, like σ, is released after initiation and to what extent other analogies can be drawn between host σ and some of the T4-specific RNAP-binding proteins. With improved detection methods and a relatively simple in vitro system for late transcription now available (Kassavetis et al., in press), the functional analysis of these subunits has become experimentally accessible.

An interesting and important problem of postreplicative transcription is posed by the concurrent expression of different classes of transcription units served by different types of promoters. The promoters which are active during the first part of the postreplicative period (late period I in Fig. 1) include early, middle, and two types of late promoters. Early promoters may not be active during late period II. Initiation of transcription at early and middle promoters definitely requires a host σ subunit; in addition, middle promoter utilization is greatly facilitated by gp*mot*. ADP-ribosylation or T4-specific RNAP subunits are probably not required for the activation of these promoters (see Brody et al., this volume). However, judging from the presence of one tightly bound molecule of 15K protein per RNAP core protomer, it is reasonable to assume that during the late period even early and middle transcripts are synthesized by an RNAP complex containing the 15K subunit. As mentioned above, the utilization of late promoters requires gp33, gp55, and gp45. However, we do not know which of these three proteins are necessary for initiation at type I and type II late promoters.

How is the ratio of early, middle, and late gene transcription controlled during the postreplicative period? A key determinant must be the efficiency of interaction between the core and σ. Subunit σ can be purified from cells even late after T4 infection in active form and in quantities comparable to those found in uninfected cells (Rabussay et al., 1972; Stevens, 1972). Assuming, in a first approximation, that binding of σ to the core will lead to initiation at early or middle promoters, we conclude that access of σ to the core must be controlled to prevent an overabundance of early and middle transcription at the expense of late RNA synthesis (compare the effect of increasing amounts of σ on the ratio of pre- and postreplicative transcripts in vitro, as discussed above). The working model we prefer is one in which σ and different T4-specific subunits, or sets of subunits, interact flexibility with the modified core. This implies the existence of a heterogeneous population of RNAP complexes within the cell late after T4 infection. (In support of this view, the substoichiometric presence of some of the T4-specific subunits in purified RNAP complexes is striking, but not convincing, evidence.) Each type of complex would be specifically suited for the recognition and utilization of a corresponding type of promoter. The presumed particular roles of the T4-specific subunits in early, middle, and late gene transcription have already been discussed above.

Returning to the central importance of controlling σ-core interactions, we want to point out that, in fact, several mechanisms seem to participate in this effort: ADP-ribosylation of the α subunits lowers the affinity between the core and σ, the 10K protein displaces or binds σ (see above), and gp33 seems to compete with σ for the same binding site at the core RNAP (Ratner, 1974a). Like σ and the 10K protein, gp33 can be removed from the T4-modified enzyme by phosphocellulose chromatography (Horvitz, 1973). This still leaves us with the question whether, or to what extent, the action of σ is actually regulated during the late period. It is conceivable that σ action is regulated indirectly via the synthesis of gp33, the 10K protein, or both. It might also be significant that the amount of gp33 in the RNAP complex is decreased by defects in gp55 (Ratner, 1974a). Thus, binding of gp33 could depend on the amount of gp55 already present in the RNAP complex. Finally, it should be interesting to find out whether σ, as an exception among the host proteins, can be synthesized after T4 infection under special conditions.

CONCLUSION

Changes in *E. coli* RNAP occurring at defined times after T4 infection have been documented and analyzed extensively. The changes include two different steps of ADP-ribosylation of the α subunits, diminished interaction between the RNAP core and σ, and the binding of four T4-specific subunits, with different affinities, to the modified core complex. The requirement for two of these new subunits, gp33 and gp55, in late gene expression was established some time ago. The functions of the other changes are finally coming into clearer focus. An important emerging insight is the continued use of the host σ subunit in the transcription of early and middle genes throughout the infection cycle and the delicate and subtle protein interactions involved in controlling σ action. Many questions regarding the function of the different changes remain to be answered. The function of the nonessential ADP-ribosylations are particularly difficult to unveil. Genetic studies involving related phage and host functions, in vivo studies under nonoptimal growth conditions, and more specific in vitro assay systems will be required. The genes for two of the T4-specific RNAP subunits have not yet been identified; it is unknown whether these proteins are essential for T4 development.

It is assumed that RNAP complexes of different subunit composition exist side by side in the infected cell and that each distinct complex is responsible for the recognition of, and initiation at, a corresponding promoter structure. With recent advances in the performance and analysis of middle and late gene transcription in vitro, an analysis of the mechanism of action of the different RNAP complexes seems now feasible. The expected results should not only yield insights into novel transcription mechanisms but should also facilitate an understanding of the regulatory processes that govern gene expression during T4 development.

I thank Peter Geiduschek and the editors for discussions and comments.

Regulation of Transcription of Prereplicative Genes

EDWARD BRODY,[1] DIETMAR RABUSSAY,[2] AND DWIGHT H. HALL[3]

Institut de Biologie Physico-Chimique, F-75005 Paris, France[1]; *Bethesda Research Laboratories, Gaithersburg, Maryland 20877*[2]; *and School of Biology, Georgia Institute of Technology, Atlanta, Georgia 30332*[3]

Prereplicative RNA synthesis in T4-infected cells is controlled in a complex and interesting fashion. Unlike bacteriophages T7 and λ, T4 has so many and such unstable prereplicative transcripts that isolation and sequencing of individual RNA molecules has been a difficult task. Now that T4 DNA fragments can be cloned to provide probes of individual transcription units, the precision of T4 transcription analysis should soon equal that attained with the smaller bacteriophages.

Despite this complexity, patterns of transcription and classes of RNA molecules have been clearly delineated in T4-infected cells. This has been possible because a few T4-coded molecules change the transcription specificity of a large number of promoters and terminators in a highly ordered manner.

Before we concern ourselves with the regulatory mechanisms involved in prereplicative transcription, we want to summarize the temporal pattern of gene expression and clarify the nomenclature to be used in this chapter. Immediately after T4 infection, a class of genes referred to as immediate early (IE) genes starts to be transcribed. RNA synthesis in T4-infected cells proceeds at an average rate of 15 to 20 nucleotides per s, and translation is usually closely coupled to transcription. One group of IE genes gets shut off after about 4 min, and another group is turned off around 12 min after infection (all times are given for infections at 30°C). Transcription of a second class of genes, delayed early (DE) genes, commences at about 2 min after infection. A number of DE genes also get switched off around 12 min (the time when true late transcription becomes dominant), and the rest are switched off at different times later in infection or continue to be expressed throughout the lytic cycle. Transcription of true late genes starts shortly after the onset of T4 DNA replication (which occurs around 5 min after infection) and continues for the rest of the vegetative period.

Transcription of IE genes is initiated at early promoters by unmodified host RNA polymerase or its altered form (see Rabussay, this volume). Protein synthesis is not required for the transcription of this class of genes, i.e., IE genes are also transcribed in the presence of drugs (or under conditions) which inhibit translation. On the other hand, transcription of DE genes does require T4 protein synthesis. A substantial amount of evidence is now available suggesting that two different mechanisms regulate DE transcription; one involves anti-termination, i.e., readthrough across potential transcription termination sites located at the junctions between IE and DE genes, whereas the other results in the activation of a new class of promoters, the middle promoters. Although the former mechanism may not obligatorily depend on a positive regulatory gene product, there is evidence that the latter does depend on T4 gp*mot*. In addition, middle promoter activation may depend on a change in template structure.

Owing to historical reasons and the complex organization of prereplicative transcription units, the terminology employed can be confusing, and we want to clarify it at the outset. There are two types of prereplicative transcription units on the T4 genome: early and middle. Early transcription units are served by early promoters and encompass IE and DE genes.

IE genes are proximal, and DE genes are distal, to early promoters. IE and DE genes are separated by potential transcription termination sites. Middle transcription units are served by middle promoters and may encompass middle genes (which are exclusively accessible from middle promoters) and IE and DE genes (which are also accessible from early promoters). Early and middle transcription units overlap to a great extent, perhaps entirely (leaving aside their relative utilization). Thus, many prereplicative genes can be expressed in two modes, an early mode (when transcription is initiated at early promoters) and a middle mode (when transcription starts at middle promoters). The organization of prereplicative transcription units will be discussed in more detail in the next section.

Prereplicative transcription units are found mainly in two large blocks on the T4 chromosome, between 158 and 75 kilobases (kb) and between 123 and 147 kb on the standard T4 map. Together, they make up more than half of the T4 genome. In addition, there exist prereplicative RNA molecules that are complementary to RNA transcribed from late genes. These are called anti-late RNA species. The function of anti-late RNA is not known, but this class of molecules is subject to many or all of the regulatory steps which control prereplicative mRNA species, and they will be discussed in that context.

The extensive overlap of early and middle transcription units has made the study of the latter difficult. It is easy to eliminate middle-mode transcription from T4-infected cells. This allows a study of early transcription units. The opposite situation, elimination of early transcription to study pure middle-mode transcription, is more difficult but has recently become possible both in vivo and in vitro. Since detailed reviews of T4 gene regulation have been published (Rabussay, 1982a; Rabussay and Geiduschek, 1977a), we have tended to emphasize very recent findings bearing on the control of prereplicative mRNA synthesis.

EARLY TRANSCRIPTION UNITS

Polycistronic Nature

Early T4 transcription units are somewhat like constitutive bacterial operons. They contain promoters which are recognized by *Escherichia coli* RNA polymerase holoenzyme and polycistronic coding units of various lengths, and they end with termination sequences most of which are thought to be rho-independent, i.e., termination factor rho is not required for stopping transcription and releasing the RNA product. Early and middle transcription units are oriented in the same direction on the T4 genome so that virtually all RNA synthesis during the prereplicative period is *l* strand-specific, or counterclockwise on the standard T4 genetic map (O'Farrell et al., 1980; Wood and Revel, 1976). A small amount of *r* strand-specific RNA is made during the prereplicative period (Geiduschek and Grau, 1970; Guha et al., 1971; Notani, 1973). Its appearance is blocked when infection is carried out in the presence of chloramphenicol (CAM). Its significance and regulation are not yet known. An example of an early transcription unit is shown in Fig. 1. It starts near the beginning of gene 39, continues through genes 60, *r*IIA, and *r*IIB, and terminates distal to the *r*IIB gene (in the region known as D1). This transcription unit covers a distance of about 6.3 kb. Since RNA chains grow at a rate of 18 nucleotides per s at 30°C, it takes almost 6 min for RNA polymerase to traverse this transcription unit. Another example is the early transcription unit containing the two clusters of tRNA genes. It is served by two early promoters about 500 base pairs (bp) apart, followed by an early transcription unit of about 3.3 kb (Goldfarb, 1981b; Goldfarb and Daniel, 1980; Goldfarb and Daniel, 1981a).

As RNA polymerase molecules traverse early transcription units, they generate polycistronic RNA molecules. Promoter-proximal genes are transcribed before promoter-distal genes in the prereplicative period (Black and Gold, 1971; Brody and Geiduschek, 1970; Brody, Sederoff, Bolle, and Epstein, 1970; Grasso and Buchanan, 1969; Milanesi et al., 1969; Milanesi et al., 1970; Salser et al., 1970; Sederoff, Bolle, Goodman, and Epstein, 1971), and there is a general correspondence between in vivo and in vitro topography of prereplicative transcription (see Kutter and Rüger, this volume). There are exceptions to this general rule, and many of them are now known to be due to the overlapping of middle and early transcription units. (Middle promoters are not recognized in most in vitro systems; see below). Elimination of the middle mode in T4 infection reinforces the correspondence between topography and kinetics of prereplicative genes (Daegelen et al., 1982a; Daegelen et al., 1982b; C. Thermes and E. Brody, submitted for publication). A recent experiment from Gold's laboratory is instructive. The *r*IIB protein is made both from a middle-mode mRNA whose 5′ end is close to the ATG initiation codon of the *r*IIB gene and from a polycistronic early mRNA starting farther upstream of *r*IIB (and perhaps also from a third type of RNA initiated near the beginning of the *r*IIA gene; see Fig. 1). Amber and ochre mutations in *r*IIA (the gene immediately upstream from *r*IIB) do not normally cause a polarity effect for *r*IIB synthesis. Elimination of middle-mode RNA synthesis makes infection with these same amber and ochre *r*IIA mutants exhibit polarity for *r*IIB expression (R. Sweeney and L. Gold, personal communication). Elimination of middle-mode expression reveals the polycistronic nature of many early transcription units.

Initiation at Early Promoters

Early promoters are recognized on bare T4 DNA by *E. coli* RNA polymerase holoenzyme (Brody, Diggelmann, and Geiduschek, 1970; Brody and Geiduschek, 1970; O'Farrell and Gold, 1973b; Travers, 1970a). The σ subunit is necessary for this recognition, for the utilization of these promoters, or for both (Bautz and Bautz, 1970; Bautz et al., 1969; Travers, 1969; Travers, 1970a; Travers, 1970b). A few early promoters have been sequenced (Table 1) and seem to show

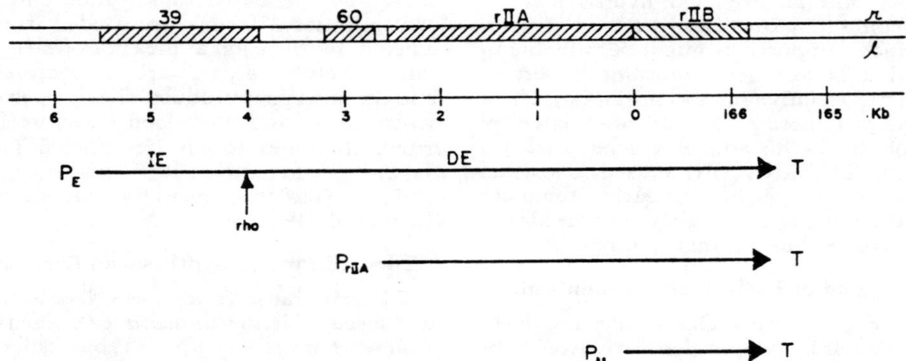

FIG. 1. Structure and transcription of the *r*II region of T4 DNA. The T4 genes are drawn to scale and placed on the appropriate segment of the T4 genetic map. All transcription is *l* strand specific. The early promoter P_E is located close to the beginning of gene 39. The rho site which marks the IE-DE division on this early transcription unit may be in the coding sequence of the gene 39 protein (Thermes, unpublished observation) about 1.6 kb downstream from the early promoter. The *r*IIA promoter (see text) must be very close to the beginning of the *r*IIA gene. It is presumed to code for a dicistronic message encoding both *r*IIA and *r*IIB sequences, but there are in fact no data demonstrating this. The middle promoter, P_M, which needs the gp*mot* for its utilization, lies in a region of the *r*IIA gene which has been sequenced by Pribnow et al. (1981). This middle-mode RNA codes only for the *r*IIB protein. It is assumed that these three RNA species all terminate at the same point beyond the *r*IIB gene but before the first early promoter (not shown) in the D2 region of the chromosome (Goldfarb and Burger, 1981; Goldfarb and Palm, 1981; Sederoff, Bolle, and Epstein, 1971; Sederoff, Bolle, Goodman, and Epstein, 1971).

TABLE 1. Sequences of some presumed early and middle T4 promoters[a]

Gene	Sequence in region: −35[b]	−10
Early promoters		
30	T T T G A C T G A G C T	T A T A A T
ORF2 (next to gene 30)[c]	T A T T A A G C C C G G	T A T A A T
*ip*III	T A C T T G A A T A G A	T A A A A T
frd	T T G T G A A A A A G T C T G	T A T T A T
Middle promoters		
1	A G A A G T T T A A T G C T T C	T A T A A T
*r*IIB	A T C A A A T A A T G C T T C A	T A A A A T
45	T T T A A C G T T A T T G C T T	T A T A A T
32	C T C A T A T T G C T T A	T A T T A T
43	T A A G C A A G G C T T C G G C	T A T A A T
E. coli promoters (consensus)[d]	T T G A C A	T A T A A T

[a] Sequences of the nontranscribed strand of early and middle T4 promoters or presumed promoters are shown.
[b] A simple bar above the sequence represents a possible homology to the *E. coli* consensus sequence. The conserved sequence seen in the four middle promoters is doubly underlined. The arrow represents a highly conserved A at the −37 position in these middle promoters. References and discussion of the possible significance of these sequences are given in the text.
[c] ORF, Open reading frame.
[d] See Rosenberg and Court, 1979.

strong homology with the −10 and −35 regions of *E. coli* and other early bacteriophage promoters (Rosenberg and Court, 1979). No extensive attempt has yet been made to compare early promoters with and without glucosyl groups on their hydroxymethylcytosine residues or to compare substitution of cytosine for hydroxymethylcytosine in such promoters; only two promoters have been identified whose recognition in vitro differs significantly between deoxycytosine-containing DNA and glucosylated hydroxymethyl-deoxycytosine-containing DNA (see Kutter and Rüger, this volume). Such comparisons might be valuable in determining what factors are important in setting promoter strengths (relative rates of initiation). The α subunits of RNA polymerase are ADP-ribosylated by the products of the T4 *alt* and *mod* genes early in infection. The modified RNA polymerase has lost the in vitro capacity to recognize some early promoters (Goldfarb, 1981b), but this is probably not true for the altered enzyme (see Rabussay, this volume).

Termination at End of Early Transcription Units

Termination of early RNA chains has not been thoroughly studied. It is thought that early transcription units are discrete and that efficient termination takes place at the end of each polycistronic transcription unit. A number of stem-loop structures, followed by runs of thymines, characteristic of rho-independent transcription termination signals, have been found at the ends of T4 genes (but not necessarily early transcription units) (Krisch and Allet, 1982; Oliver and Crowther, 1981; Spicer et al., 1982). If similar sites exist at the end of each early transcription unit, they do not function with 100% efficiency in vitro. Rho-independent termination is seen in vitro at 0.2 M NaCl, but the RNA molecules generated are up to 7 to 12 kb long, longer than the average in vivo early transcription unit (Milanesi et al., 1970; Millette et al., 1970; Richardson, 1970), and some are much longer than the early mRNA isolated from T4-infected cells. Several termini have recently been mapped (see Kutter and Rüger, this volume). The largest of these in vitro synthesized RNA molecules must represent more than one early transcription unit. It is possible that additional factors are needed in vivo for 100% efficient termination at the ends of early transcription units; likely candidates are the bacterial rho protein and the *nusA* gene product. The *trp* operon terminator, which also has a stem-loop structure followed by a run of thymines, is only 25% efficient in vitro but can be brought to close to 100% efficiency by addition of rho and NusA proteins to the reaction mixture (Farnham et al., 1982).

Effect of Protein Synthesis on Early Transcription

T4 early transcription units, like bacterial operons, are susceptible to rho-mediated polarity of transcription when protein synthesis is blocked. Unlike bacterial operons, but like bacteriophages λ (Ward and Gottesman, 1982) and P4 (Sauer et al., 1981), T4 has developed a mechanism (or two) for overcoming this kind of polarity.

When T4 infects *E. coli* in the presence of CAM, early transcription units produce truncated RNA molecules which correspond to their promoter-proximal regions (Brody, Sederoff, Bolle, and Epstein, 1970; Daegelen and Brody, 1976; Daegelen et al., 1982a; Daegelen et al., 1982b; Milanesi et al., 1969; Milanesi et al., 1970;

Salser et al., 1970; Sederoff, Bolle, and Epstein, 1971; Sederoff, Bolle, Goodman, and Epstein, 1971; Young, 1970; Young, 1975; Young et al., 1980). For example, in the *r*II early transcription unit (Fig. 1), gene 39-specific RNA is synthesized under these conditions (Young et al., 1980), but almost no *r*IIA or *r*IIB RNA can be detected (Daegelen and Brody, 1976; Daegelen et al., 1982a; Daegelen et al., 1982b; Sederoff, Bolle, and Epstein, 1971; Sederoff, Bolle, Goodman, and Epstein, 1971; Young et al., 1980). The collection of CAM-insensitive, promoter-proximal transcripts is called IE RNA. Promoter-distal, CAM-sensitive RNA is called DE RNA. IE RNA corresponds to RNA made on early transcription units during the first 90 s or so of a normal infection at 30°C (Brody, Sederoff, Bolle, and Epstein, 1970; Salser et al., 1970); IE RNA is also transcribed from T4 DNA in vitro before DE RNA (Milanesi et al., 1969; Milanesi et al., 1970; O'Farrell and Gold, 1973b). Assuming an RNA chain elongation rate of 18 nucleotides per s, IE RNA should be about 1.6 kb long, and this is the maximum length found for T4 IE RNA (Brody, Sederoff, Bolle, and Epstein, 1970; Sederoff, Bolle, Goodman, and Epstein, 1971). IE RNA from the D2 region of the chromosome (between genes *r*IIB and 52) sediments as a discrete peak in sucrose gradients, smaller than the average IE molecule (Sederoff, Bolle, Goodman, and Epstein, 1971). This was the first indication that IE RNA molecules terminate at precise sites in early transcription units.

Inhibition of protein synthesis by puromycin or K^+ starvation also limits early transcription to IE species (Grasso and Buchanan, 1969; Peterson et al., 1972). Studies with amino acid starvation of an auxotrophic strain gave conflicting results. Witmer et al. (1975) reported that T4 infection of a *leu*$^-$ strain in the absence of the required amino acid allowed synthesis of DE RNA. However, the rate of amino acid incorporation into protein in their experiments was 10 to 12% of normal, which probably accounts for the fact that DE RNA (perhaps even from middle-mode synthesis) was found. Previously, Legault-Demare et al. (1969) had shown that starvation of a *leu*$^-$ strain for leucine (apparently under stricter conditions) before T4 infection led to the inhibition of DE RNA synthesis. This raises the question (not yet answered) as to how much protein synthesis is needed to allow RNA polymerase to transcribe past IE-DE junctions. We shall return to this problem in the next section.

The requirement of protein synthesis for DE RNA synthesis is twofold. On the one hand, it is known that elongation of chains from IE to DE regions takes place in vivo (Black and Gold, 1971; Brody, Sederoff, Bolle, and Epstein, 1970; Schmidt et al., 1970; Sederoff, Bolle, Goodman, and Epstein, 1971), which implies that protein synthesis is necessary for anti-termination to occur at IE-DE junctions. On the other, a T4 gene product, gp*mot*, is necessary for initiation of RNA synthesis to take place in DE regions of early transcription units. This delayed initiation constitutes the middle mode of T4 RNA synthesis, and we shall come back to it further on.

Anti-Termination in Early Transcription Units

The CAM effect on early transcription is due to rho-mediated polarity, analogous to that seen in bacterial operons (Imamoto, 1973; Morse, 1970) and in bacteriophage λ (Kourilsky et al., 1971). Lack of translation of IE RNA allows rho factor to induce termination of RNA chains at rho-sensitive sites. Translation of the IE RNA allows RNA polymerase to transcribe past these potential *rho* sites into DE regions. Formally, this is analogous to transcription on the two early transcription units of bacteriophage λ, in which *rho* terminates RNA chains prematurely in the absence of protein synthesis; synthesis of the λ IE gpN allows RNA polymerase to transcribe past these sites. It is still not known whether T4 codes for an N-like anti-terminator protein.

The first suggestion that the *rho* gene is responsible for CAM-induced polarity in T4-infected cells came from in vitro experiments (Goldberg, 1970; Jayaraman, 1972; Richardson, 1970) which showed that rho factor could induce termination of T4 RNA synthesis at sites which corresponded more or less to IE-DE junctions. The expectation that host mutants defective in the rho function would suppress the CAM-induced polarity was not immediately fulfilled (Linder and Sköld, 1977; Witmer et al., 1975; Young, 1975), presumably because the *rho* mutants used contained too much residual rho activity. This residual activity was eliminated in conditionally lethal *rho* mutants (Das et al., 1976; Inoko et al., 1977) such as the *rho ts*15 allele. In such cells, T4 infection in the presence of CAM allowed synthesis of almost normal amounts of DE RNA (Daegelen et al., 1982a; Daegelen et al., 1982b). For example, RNA extracted after infection of these rho-defective hosts in the presence of CAM contained the same levels of *r*IIA and *r*IIB RNA as were found in the early mode after a normal *rho*$^+$ infection.

A different experimental approach has led to the same conclusion. Host mutants have been isolated which block T4 growth at specific developmental steps (Stitt et al., 1980; Takahashi et al., 1975). One type of such mutations, called *tabC* by the Naples group and HDF by the Pasadena group, maps in the *E. coli rho* gene. The mutant rho in these hosts seems to be insensitive to the normal T4 mechanism inducing anti-termination (Caruso et al., 1979; Pulitzer et al., 1979; Stitt et al., 1980). The most striking proof of this is that infection of *tabC* strains, without inhibitors of protein synthesis, gives only IE RNA as long as middle-mode RNA synthesis has been removed by a *mot* mutation in the infecting phage.

If wild-type rho terminates transcription when protein synthesis is reduced, what is the mechanism that normally overcomes this potential polarity in T4-infected cells? Where is the T4 equivalent of the λ N gene? It is possible to isolate T4 mutants, called *com*Cα (Caruso et al., 1979; Pulitzer et al., 1979; Takahashi and Yoshikawa, 1979) or *go*F (Stitt et al., 1980) mutants, which are capable of giving almost wild-type burst sizes on *tabC* hosts. The mutant product of the *com*Cα allele seems to act as an anti-terminator in infections of *tabC* bacteria. But the wild-type product of the *com*Cα gene does not seem to be equivalent to the λ N protein. The *com*Cα locus has been mapped, and at least part of this gene is eliminated by two large deletions mapping between genes 39 and 56 (Stitt et al., 1980; Takahashi and Yoshikawa, 1979). Infection of *E. coli* wild-type hosts with T4 carrying either of these deletions does not limit transcription to IE RNA. One might argue that there is still *com*Cα activity in these deletions since the *com*Cα locus is close to one end point of the deletions. Alternatively, it is possible that middle-mode RNA masks the N-like defect in these deletion mutants. Further experiments on *com*Cα will be needed to understand what function this gene codes for in wild-type T4.

Transcription beyond potential *rho* sites may not require an N-like gene but merely translation of IE RNA, irrespective of the protein products. The λ N protein is an anti-terminator, not just a rho antagonist, since it permits RNA polymerase to traverse rho-independent, as well as rho-dependent, termination sites (Ward and Gottesman, 1982). It is possible that all that is needed to keep rho from acting at T4 IE-DE junctions is the movement of ribosomes along nascent IE RNA; that is, coupled transcription-translation. This would make T4 early transcription units analogous to polycistronic *E. coli* operons, in which coupled transcription-translation seems to be sufficient to overcome potential polarity (Adhya and Gottesman, 1978). There is suggestive experimental evidence for this point of view. L. Gold (manuscript in preparation) used six amino acid analogs which are incorporated into proteins with high efficiency but yield inactive proteins. He verified that a number of early enzymes had undetectable activity after infection of *E. coli* by T4 in the presence of these amino acid analogs. Protein products could be analyzed by sodium dodecyl sulfate-polyacrylamide electrophoresis. Although band widths were slightly enlarged and blurred (as would be expected), the pattern allowed identification of T4 proteins. DE as well as IE proteins were seen, suggesting that transcription beyond potential rho sites did not depend on the synthesis of functional T4 IE proteins. Similar results were found when T4 infection was carried out in the presence of the miscoding antibiotic neamine (Thermes and Brody, unpublished results).

To resolve the problem posed by these apparently contradictory results, we want to emphasize two points. First, there may be a minority of transcription units which do depend on the action of T4-specific anti-termination protein(s). A good candidate for such a transcription unit is the T4 tRNA cluster (Goldfarb and Daniel, 1980; for discussion, see Rabussay, 1982a). Second, for the majority of potential termination sites, translation per se is probably sufficient to overcome rho action. However, the rate of translation must be crucial. We now come back to the conflicting reports (Legault-Demare et al., 1969; Witmer et al., 1975) on whether or not leucine starvation of *leu*$^-$ bacteria leads to polarity of T4 early transcription. As mentioned, there must be some protein synthesis rate at which ribosomal movement protects potential rho sites from being acted upon, but this rate has not yet been determined. Even if it is translation per se that overcomes rho action, it is likely that the IE-DE junction does represent a barrier in T4 development and that the observed block is not just a pharmacological effect induced by the addition of an antibiotic. One has only to imagine that in natural (as opposed to laboratory) environments, protein synthesis limits T4 development. At low protein synthesis rates, *rho* action directs all protein synthesis capacity toward IE species, one of which, *mot*, leads to the middle mode of RNA synthesis (Rabussay, 1982a).

Experiments of M. Uzan and E. Brody (manuscript in preparation) reinforce this idea. In *E. coli* K-12 (but not in *E. coli* B), excess valine in the growth medium leads to a block in isoleucine biosynthesis (Leavitt and Umbarger, 1962). Excess serine, methionine, and glycine blocks protein synthesis, and this block is also relieved by isoleucine (Uzan and Danchin, 1978). When the *E. coli* K-12 strain CP78 was infected under either of the conditions leading to a block in isoleucine biosynthesis, phage production was drastically reduced and DNA synthesis was greatly delayed, but the incorporation of labeled phenylalanine into proteins was still 20 to 30% the normal value (or the value obtained when isoleucine was added along with the inhibitory amino acids). However, sodium dodecyl sulfate-polyacrylamide electrophoresis of the proteins made under inhibiting conditions showed a distribution and relative kinetics like those seen after infection of *tabC* strains by T4 (Caruso et al., 1979). In the presence of isoleucine, the normal early pattern was restored. Thus, under these conditions, it seems that even 20 to 30% of the normal protein synthesis rate does not suffice to overcome rho action completely.

Whether T4-coded anti-terminators exist or not, it is clear that T4 loses its susceptibility to CAM-induced polarity during the prereplicative period of infection (Brody, Sederoff, Bolle, and Epstein, 1970; Sakiyama and Buchanan, 1972; Salser et al., 1970; Young, 1970). CAM inhibits protein synthesis as well when it is added during or immediately after infection as it does when it is added before infection. Nonetheless, the ability of CAM to induce polarity is lost between 1.5 and 4 min after infection at 30°C. This is, at least in part, a direct effect and not entirely due to masking by middle-mode RNA synthesis, which depends on the IE protein, gp*mot*.

Resistance to CAM-induced polarity does develop more slowly after *mot*$^-$ infections than after wild-type infections, but it develops to the same level in both (Thermes and Brody, submitted for publication). Thus, termination sites change during the course of T4 infection in such a way that rho cannot act at these sites, even when protein synthesis is eliminated. This T4-induced resistance to termination is remarkable in two respects. First, the resistance to CAM-induced polarity acts with different kinetics on different rho sites on the T4 genome. Second, the change in a given rho site from CAM sensitive to CAM insensitive is stable during many rounds of transcription and completely irreversible. These two properties are unlike those of N-induced anti-termination in bacteriophage λ infections.

These properties put important constraints on the possible explanations of T4-induced anti-termination. If coupled transcription-translation changes rho sites from CAM sensitive to CAM insensitive, the variable kinetics and irreversibility of anti-termination suggest the following working model. In each transcription unit, rho sites remain sensitive to CAM-induced polarity until the RNA polymerase-nascent RNA-ribosome complex has moved beyond them. When the transcription-translation complex moves past a rho site, the site adopts a CAM-insensitive state. One possibility is that a host or phage protein binds to a potential transcriptional pause site in the T4 DNA, but that the DNA has to be slightly unwound by the passing RNA polymerase for this binding to occur. This protein, irreversibly bound, would decrease pausing and thus eliminate polarity. One might alternatively imagine a host RNA polymerase-binding protein playing the same role.

Other Effects of Rho on T4 Development

T4 grows poorly on *rho* conditional lethal mutants (Daegelen et al., 1982a; Inoko et al., 1977; Ito and Sekiguchi, 1978; Ito et al., 1978; Zograff and Gintsburg, 1980). Zograff and Gintsburg (1980) have shown that T4 and T2 DNA replication is inhibited in *rho* mutants. They suggest that this is a direct effect of rho, since they can show that the early proteins synthesized in the *rho*-defective host are capable of catalyzing DNA synthesis when a functional *rho* allele is provided in *trans*. This is a striking result which needs to be investigated in more detail. The fact that a class of *com*C mutants maps in gene 45 (Takahashi and Yoshikawa, 1979) may reflect the same involvement of *rho* in replication, possibly by affecting RNA primer synthesis by RNA polymerase (gene 45 protein is one of the DNA polymerase accessory proteins and interacts also with T4-modified RNA polymerase).

T4 protein synthesis is somewhat abnormal in *rho* conditional lethal mutants. Premature synthesis of late proteins during the early period has been seen (Daegelen et al., 1982a; Ito and Sekiguchi, 1978; Ito et al., 1978; Zograff and Gintsburg, 1980). The premature synthesis of gp7 and gp*alt* have been shown to be under *mot* control in these hosts (Daegelen et al., 1982a). There is a small but significant increase in *r* strand transcription during the early period when *rho* hosts are infected with T4 (P. Daegelen and E. Brody, unpublished results). These observations could reflect transcriptional readthrough from early into late regions of the T4 genome in the absence of the *rho* function.

MIDDLE-MODE TRANSCRIPTION

Definition and Analysis of Middle-Mode RNA Synthesis

Initiation of RNA synthesis in the early mode seems to be derived from interactions between unprocessed T4 DNA and host RNA polymerase (see Rabussay, this volume). Initiation at a different type of promoter was first demonstrated by Schmidt et al. (1970), who showed that *r*IIB RNA synthesis started before the *r*IIB gene could have been reached by chain elongation of an RNA starting at an IE promoter. Travers found an activity in T4-infected cells that directed *E. coli* core RNA polymerase to initiate at regions that were not IE promoters (Travers, 1969; Travers, 1970a). O'Farrell and Gold (1973a) showed that a number of T4 proteins were under the control of promoters that were not recognized until 1 to 2 min after infection. The synthesis of these same proteins was subsequently shown to be under the control of the T4 gp*mot* (Mattson et al., 1974; Mattson et al., 1978), suggesting that the T4 gp*mot* controlled initiation of RNA synthesis at a group of promoters on T4 DNA not ordinarily recognized or utilized by *E. coli* RNA polymerase.

Irradiation of DNA with UV light destroys its ability to be transcribed, presumably because pyrimidine dimers act as transcription termination signals. The UV sensitivity of a gene has been shown to be proportional to the distance from its promoter (Bräutigam and Sauerbier, 1974). By the use of this rationale, it has been found that promoter proximities change for genes 43 and 45 within the first minutes after T4 wild-type infection, i.e., transcription of these genes switches from more distant (early) promoters to nearby (middle) promoters. Promoters farther upstream also continue to be used after *mot*$^-$ infections (Hercules and Sauerbier, 1973; Hercules and Sauerbier, 1974; Sauerbier et al., 1976). With cloned T4 DNA sequences as hybridization probes, it has been possible to show that mot acts at the level of transcription. Gene 43 RNA is under *mot* control in the prereplicative period, as is the RNA from an early region close to the late gene 24 (Daegelen et al., 1982b; T. Mattson, personal communication). Gene 1 RNA is also, to a large extent, under *mot* control (Thermes and Brody, submitted for publication). Protein analysis and measures of enzymatic activity had shown that gp1 and gp43 were under middle-mode control (Linder and Sköld, 1980; Mattson et al., 1978; Natale and Buchanan, 1974; Sauerbier et al., 1976). Not all T4 proteins whose appearance is *mot* dependent show a corresponding *mot* dependence in their mRNA levels. Gene product *r*IIB is a case in point.

Although the appearance of this protein is under *mot* control, *r*IIB RNA levels found in early pulse-labeled preparations decrease only by one-third when *mot*$^-$ phage are used instead of wild-type phage (Daegelen et al., 1982b; Mattson, personal communication; Sweeney and Gold, personal communication). The early mode *r*IIB RNA (see Fig. 1) which persists in *mot*$^-$ infections must be a poor template for *r*IIB protein synthesis. Translational discrimination against other early-mode RNA molecules has been suggested (Daegelen et al., 1982b); early lysozyme mRNA, which is not translated at all, may be the most extreme example (Jayaraman and Goldberg, 1970; Kasai and Bautz, 1969). Such discrimination may occur by base pairing between sequences necessary for translation (e.g., Shine-Dalgarno sequences) and their complements existing upstream in polycistronic mRNA (Gold et al., 1981; Merril, Gottesman, and Adhya, 1981).

Because of the fact that early and middle transcription units overlap, it is difficult to ascertain which T4 proteins are under *mot* control. Weak middle promoters would have their effects masked by strong early promoters in the overlapping transcription units. Comparing the effects of *mot*$^+$ and *mot*$^-$ infections in *tabC* hosts (blocked in early-mode DE expression) suggests that almost all DE genes have a middle (*mot*-dependent) mode of expression (Pulitzer et al., 1979).

Characteristics of *mot* Mutants

The standard *mot* mutation, *ts*G1, was selected for its ability to increase T4 yield in λ-lysogenic cells infected at a restrictive temperature with a T4 double mutant carrying temperature-sensitive mutations in genes 42 and *r*IIA (Mattson et al., 1974). A nonsense mutation, *am*G1, has been isolated by using a marker escue procedure with *ts*G1, but most studies have ›een done with *ts*G1 because it displays a stronger ›henotype than *am*G1 (Mattson et al., 1978). Another et of *mot* mutants, including *far*P14 and *far*P85, was solated on the basis of resistance to folate analogs Johnson and Hall, 1973). Homyk et al. (1976) de-.cribed a third set of *mot* mutants, called *sip*, which ›artially suppress the inability of *r*II mutants to grow n λ lysogens and which are similar to suppressors of II mutations described by Freedman and Brenner 1972). Both *ts*G1 and *sip* mutations reduce the lethali-y of *r*II mutations by increasing the efficiency of ›roductive infection, but they do not increase the mall burst size of productively infected cells.

All these mutations (*mot*, *far*, and *sip*) map near ach other between genes 52 and *t* (Chace and Hall, 975; Hall and Snyder, 1981; Mattson et al., 1974). The results of complementation experiments, i.e., nixed infections with two different *mot* mutants, ndicate that all these mutations are in a single *mot* ene (Hall and Snyder, 1981; Sauerbier et al., 1976). The *mot* gene function is not essential for phage rowth on most host strains. The fact that *am*G1 rows better than *ts*G1 suggests that it is better not to ave a *mot* gene product than to have a defective gene roduct. However, the normal *mot* gene function is equired for efficient growth on *E. coli* CTr5X and *abG* strains. The efficiency of plating of *mot* mutants relative to T4 wild-type) is reduced up to 40-fold on *E. oli* CTr5X (Hall and Snyder, 1981) and up to 10^5-fold n *tabG* hosts (Pulitzer et al., 1979).

The *mot* mutations alter the expression of many enes. Mutations representing each of the three types *ts*G1, *far*P14, and *sip*2) have been shown to have xtended synthesis of the *r*IIA gene product and re-luced synthesis of the products of genes 32, 43, and 45 Chace and Hall, 1975; Hall and Snyder, 1981; Matt-on et al., 1978). The effect on expression of many arly genes is similar but has not been tested with all ypes of *mot* mutants. Both *far* and *sip* mutations ause overproduction of dihydrofolate reductase (*ts*G1 and *am*G1 have not been tested), which sufficiently xplains how they confer resistance to folate analogs hat inhibit this enzyme. The only major known differ-nce in gene expression among *mot* mutants is that sG1 causes delayed synthesis of the *r*IIB gene prod-ict, and *far*P14 causes overproduction of it (Chace and Hall, 1975; Mattson et al., 1978). The basis for this lifference is not known, but it could indicate a second-ary function for the *mot* gene product which is differ-ntially affected in *ts*G1 and *far*P14.

The apparent lack of complementation is not due to he dominance of one of the mutations because they ave all been shown to be recessive. For example, in a nixed infection with amber mutants in gene 43 and *ip*2 (*mot*), the synthesis of gene 43 product is not lelayed (Hall and Snyder, 1981). In such an infection, he gene 43 product must be made from the *sip* nutant DNA. This shows that *sip*2 is recessive and hat the *sip*⁻ product can act in *trans* to turn on the xpression of gene 43.

It is not clear why *mot* mutations reduce the lethali-y of *r*II mutations in λ lysogens. Homyk et al. (1976) letected partial suppression of *r*II deletions and sug-gested that *sip* mutations might somehow compensate or the complete absence of *r*II products. Since the *r*II products are membrane components (Huang, 1975), other membrane proteins might partially substitute for them. In fact, Chace and Hall (1975) observed that the products of genes 39 and 52, which are also membrane proteins (Huang, 1975), are overproduced by some *mot* mutants. The products of genes 39 and 52 are also needed for normal synthesis of DNA (Epstein et al., 1963) and are subunits of a T4 DNA topoisomer-ase (Liu, Liu, and Alberts, 1979; Stetler et al., 1979).

It is not known why *mot* mutants grow poorly on *E. coli* CTr5X. This strain is a derivative of a clinical isolate carrying an amber mutation in an unidentified gene and is nonpermissive for certain T4 mutants in genes *pse*T (5'-polynucleotide kinase 3'-phosphatase) and 63 (RNA ligase and tail fiber attachment activity) (see Snyder, this volume). Considerable evidence ex-ists that gp*pse*T and RNA ligase are involved in the creation of a particular DNA structure (or structures) which seem(s) to be important for proper T4 DNA replication, late gene transcription, and DNA packag-ing (Depew and Cozzarelli, 1974; Sirotkin et al., 1978; reviewed in Rabussay, 1982a). Interestingly, these defects disappear at elevated growth temperatures of about 42°C (Champness and Snyder, 1982). It is as-sumed that most *E. coli* strains contain gene prod-uct(s) which can substitute for gp*pse*T and RNA ligase and that *E. coli* CTr5X and a group of *E. coli* mutants called *lit* (late inhibitor of T4) are deficient in these host functions (Champness and Snyder, 1982; Snyder, this volume). A detailed comparison of the properties of T4 mutants in *mot*, *pse*T, and RNA ligase on CTr5X and *lit* hosts may yield interesting insights into the functions coded by these genes and their possible relations to each other.

Pulitzer et al. (1979) isolated *tabG* mutants of *E. coli* by selecting for bacterial mutants that would signifi-cantly reduce the plating efficiency of *ts*G1 (*mot*). The *tabG* mutants (probably defective in the β subunit of RNA polymerase) also fail to support growth of mu-tants in genes 45 and 55. The latter are required for late transcription (see Rabussay, this volume). The enhanced lethality of *mot* in *tabG* cells may be an indirect effect because prereplicative gene expression of *mot* mutants in *tabG* seems to be similar to that observed in *tab*⁺ cells, although this needs to be studied in more detail. The restriction of growth of *mot* mutants on *tabG* cells may be due to additive delays in late gene expression or to an altered expres-sion of some key early protein not easily seen in polyacrylamide gels.

Homyk et al. (1976) described T4 L mutations that reverse the effects of *sip* mutations. Hall and Snyder (1981) have shown that the mutation L_2 effectively suppresses all the phenotypes of *sip*2 by changing the expression of early genes back to normal. Since *mot* mutations seem to affect transcription by locally al-tering the structure of the T4 DNA (see below), a suppressor mutation like L_2 might cause a compensa-tory change in the DNA structure, thereby restoring normal transcription. Preliminary results indicate that L_2 maps near the *r*II genes. Many genes in this region have products that are involved in membrane functions (Wood and Revel, 1976) but a function in T4 development for the gene affected by the L_2 mutation has not been identified. The L_2 single mutant is difficult to distinguish from wild-type T4. The only phenotype observed is partial resistance to a folate analog.

Middle Promoters and Mechanism of Action of gp*mot*

A number of T4 early and middle promoters have been mapped and sequenced in recent years (Krisch and Allet, 1982; Owen et al., 1983; Krayev, Zimin, Tanyashin, and Skryabin, personal communication; Pribnow et al., 1981; S. Purohit and C. K. Mathews, personal communication; Spicer and Konigsberg, this volume; Spicer et al., 1982; J. Velten, Ph.D. thesis, University of California, San Diego, 1981). Partial sequences are shown in Table 1. We begin with the caveat that the evidence indicating that these sequences are in fact promoters varies widely; in some cases, the promoters are only presumed to exist because of their proximity to Shine-Dalgarno sequences and initiator ATGs for the corresponding genes. The −10 regions (Pribnow boxes) of all nine prereplicative promoters show strong homology to the consensus −10 region of *E. coli* promoters (Rosenberg and Court, 1979). The five middle promoters have hostlike −10 regions; this is to be contrasted to *Bacillus subtilis* bacteriophage SPO1 middle-mode promoters, directed by a new RNA polymerase subunit, SPO1 gp28, whose −10 regions have no homology with SPO1 early −10 regions (see Geiduschek and Ito, 1982, for a review). The −35 regions of these T4 promoters are more difficult to evaluate. The −35 region of the gene 30 early promoter is very close to the *E. coli* consensus sequence. From the sequence data, one would imagine that the promoter for gene 30 (a known IE gene; see Young et al., 1980) would be recognized by *E. coli* RNA polymerase. The same seems to be true for the −35 region of a presumed *ip*III promoter (Owen et al., 1983); *ip*III is also known to be an IE gene (Black and Gold, 1971; Brody et al., 1971). There is an open reading frame just before the gene 30 sequence (Krayev, Zimin, Tanyashin, and Skryabin, personal communication). This region is unmapped genetically, but the open reading frame has a good Shine-Dalgarno sequence before the initiating ATG triplet. About 30 bp upstream, there is a −10 region identical to that found in front of gene 30. The corresponding −35 region is TTAAGC, which has some homology with the consensus sequence recognized by the *E. coli* RNA polymerase (Rosenberg and Court, 1979). The gene and control region for dihydrofolate reductase (*frd*) and part of the gene for thymidylate synthase (*td*) have been sequenced by Purohit and Mathews (personal communication). Upstream from the *frd* coding region, a potential early promoter sequence has been identified. Gene *frd* is known to be expressed with IE characteristics (Trimble et al., 1972). The coding regions of gene *frd* and the adjacent gene downstream, *td*, overlap by 4 bp. Well upstream from the translation initiation site for *td*, within the coding region of *frd*, are two potential promoter sequences (not shown in Table 1). Both contain a clear −10 consensus sequence and display a sequence slightly related to the −35 consensus sequence of *E. coli* in the correct position. In addition, one of them contains a sequence similar to the A^{T}_{A}TGCTT sequence found in middle promoters (see Table 1 and below). Gene *td* is expressed with DE characteristics (Trimble et al., 1972).

Potential promoters of two other prereplicative genes (genes 44 and 62) have been sequenced (see Spicer and Konigsberg, this volume). Both show obvious −10 consensus boxes and reasonable −35 *E. coli* consensus sequences. Gene 44 also shows a sequence reminiscent of the middle consensus sequence but in the −20 rather than in the −35 region.

The −35 regions of the presumed middle promoters show various degrees of homology with those of *E. coli* promoters. The −35 regions of genes 1, 45, and possibly 43 show some homology with the *E. coli* consensus sequence, whereas those of genes *r*IIB and 32 show little, if any, homology. There is, however, a new consensus sequence in these −35 regions. The sequence A^{T}_{A}TGCTT (gene 43 has a G in position 3) starts at position −35 (genes 32 and 43), −33 (*r*IIB), −32 (gene 1), and −30 (gene 45). If these sequences indeed represent middle promoters, this sequence might comprise part of the *mot* recognition site. To verify that these sequences are in fact middle promoters, it will be necessary to map the 5′ ends of RNA molecules that are under *mot* control. Sweeney and Gold (personal communication) have done this for *r*IIB RNA (see Singer et al., this volume). We shall discuss below the possibility that the gene 1 promoter might contain both an early and a middle signal.

It has been difficult to assay middle-mode RNA because of the overlap of early and middle transcription units. In vitro RNA synthesis or coupled transcription-translation with purified T4 DNA usually yields only early-mode transcripts (Brody and Geiduschek, 1970; Jayaraman, 1972; Milanesi et al., 1969; Milanesi et al., 1970; O'Farrell and Gold, 1973a; O'Farrell and Gold, 1973b). In contrast, Trimble and Maley (1975) reported that *E. coli* RNA polymerase could synthesize middle RNA in vitro, but only under low-ionic-strength conditions. Travers (1970a) found an activity in T4-infected cells which stimulated *E. coli* core RNA polymerase to transcribe DE regions of T4 DNA. Recently, an in vitro system has been developed that allows *mot*-dependent synthesis of DE RNA (Thermes et al., 1976). Early-mode DE RNA synthesis is blocked by rifampin-inactivated *E. coli* RNA polymerase, and *mot*-dependent DE RNA synthesis occurs when one adds rifampin-resistant RNA polymerase to this system (de Franciscis and Brody, 1982). *mot* seems to interact with the DNA template in such a way that middle promoters can be recognized by either *E. coli* RNA polymerase or by T4-modified RNA polymerase (de Franciscis et al., 1982). In addition to *mot*, the *E. coli* σ subunit seems to be necessary for in vitro recognition of middle promoters. (It has also been suggested that the σ subunit is necessary for T4 RNA synthesis throughout the T4 growth cycle; Zograff, 1981). De Franciscis et al. (1982) isolated a DNA protein complex from infected cells which showed *mot*-dependent DE RNA synthesis when transcribed with *E. coli* RNA polymerase. The *mot* dependence of middle promoter recognition seems itself to be dependent on σ-core interactions of the RNA polymerase used for synthesis. When σ-core interactions were strong (*E. coli* holoenzyme), *mot* dependence was bypassed at 0.05 M NaCl, but *mot* was necessary at higher ionic strengths. When σ-core interactions were weak (T4 enzyme), *mot* was necessary even at low ionic strength. When σ-core interactions were abolished (T4 enzyme at 0.2 M NaCl), *mot* could not act to direct middle-mode RNA synthesis.

Recently, Uzan and Brody (manuscript in preparation) have identified the *mot* protein and have shown that it is a component of this same DNA-protein complex which is capable of middle transcription. In *ts mot* mutant infections, the *mot* protein does not bind to this DNA-protein complex at nonpermissive temperatures. Gp*mot* is an IE protein with a molecular weight of 24,500 whose synthesis is rapidly inhibited about 5 min after infection at 30°C. The protein is stable and seems to bind to a DNA-agarose column. It is not one of the T4-coded proteins that binds strongly to the host RNA polymerase (see Rabussay, this volume).

We propose the following mechanism of action for *mot*-controlled initiation of middle-mode RNA synthesis. Gp*mot* is one of the earliest proteins made after T4 infection. Gp*mot* interacts with T4 DNA, perhaps in part at A$^{T}_{A}$TGCTT sites; this interaction allows RNA polymerase to initiate at what otherwise would not be initiation sites in vivo for RNA polymerase. The *mot* dependence would be enhanced by the modification of α subunits of RNA polymerase which occurs early in phage development and which weakens σ-core interactions (Khesin et al., 1976; Rabussay, this volume; Schachner et al., 1971). This model suggests that *mot* may act in a way analogous to the λ *c*II protein (Shimatake and Rosenberg, 1981).

There are, however, elements of information that are difficult to reconcile with this simple notion. (i) Sweeney and Gold have shown that *r*IIB RNA synthesis can be affected by mutation in a region that is about 70 bp upstream from the *mot*-dependent RNA start site (see Singer et al., this volume). (ii) J. Velten and J. Abelson (personal communication) have shown that *E. coli* RNA polymerase binds to and initiates from the gene 1 promoter in vitro; this reaction takes place without the *mot* protein. A high proportion of gene 1 RNA made in vitro is terminated 210 bp downstream at what appears to be a typical rho-independent terminator. Nonetheless, gene 1 protein does not seem to be made in vivo unless mot is present. (iii) Daegelen et al. (1982b) have suggested that *mot* plays a role in early-mode anti-termination as well as in initiation of middle-mode RNA synthesis. For example, in *rho*$^{+}$ strains, no gene 43-specific RNA could be found early after infection by *mot*$^{-}$ bacteriophage. We emphasize that this result depends on the time at which pulse-labeling is carried out. A later pulse showed early-mode gene 43 RNA. When the same *mot*$^{-}$ phage were used to infect *rho*$^{-}$ bacteria, about 25% of the wild-type level of gene 43 RNA was detected in a similar early pulse-labeling experiment. Interestingly, the *rho*$^{+}$ or *rho*$^{-}$ genotype did not appreciably affect the *mot* dependence of gene 43 protein synthesis, supporting the notion that early-mode gene 43 RNA is relatively poorly translated (see above). The 25% of wild-type gene 43 mRNA was about the same amount that was found by early-mode readthrough of *rho* sites as determined by experiments in which *rho*$^{-}$ bacteria were infected in the presence of CAM. This result could mean that *mot*, in addition to being needed for initiation of 75% of the gene 43-specific RNA (middle mode) is also needed to help overcome rho action for the 25% of gene 43 RNA that is synthesized in the rho-sensitive mode. Alternatively, one could argue that rho removal simply allows readthrough from early transcription units farther upstream. This, however, would imply that rho is necessary for efficient termination between early transcription units.

ANTI-LATE RNA

Anti-late RNA is transcribed, like the overwhelming majority of early and middle RNA species, from the *l* strand of T4 DNA. Its distinguishing characteristic is that this RNA class is capable of hybridizing with late RNA, giving rise to RNA-RNA duplexes (Geiduschek and Grau, 1970). Although it accounts for only 2 to 3% of total prereplicative RNA, it comes from widely dispersed regions of the T4 chromosome (Notani, 1973; Young et al., 1980). At least part of this RNA seems to be the result of initiation of transcription from the *l* strand in late regions (T. Elliott and E. P. Geiduschek, personal communication).

When rifampin is added to cells 1 min after T4 infection, middle-mode proteins, but not early-mode proteins, are poorly synthesized during the prereplicative period (O'Farrell and Gold, 1973a). Rifampin addition at 1 min does not inhibit the appearance of anti-late RNA (Frederick and Snyder, 1977), which suggests that early promoters can be used to generate these RNA molecules. When infection is carried out in the presence of CAM, no anti-late RNA is detected (Geiduschek and Grau, 1970; Notani, 1973). This is due, at least in part, to rho-induced termination, since infection of *rho*$^{-}$ strains in the presence of CAM leads to the synthesis of normal amounts of anti-late RNA (Daegelen et al., 1982b). These results suggest that anti-late RNA can be generated via initiation at IE promoter(s) and readthrough past a potential rho site which must be located just after the promoter(s) (compare Kutter and Rüger, this volume). (Alternatively, the IE part of such a transcript might not be capable of complexing with *r*-strand transcripts.) However, anti-late regulation has more levels of complexity. Infection of bacteria with *mot*$^{-}$ bacteriophage leads to a long delay in anti-late RNA production, characteristic of middle-mode RNA synthesis (Daegelen et al., 1982b; Frederick and Snyder, 1977). The simplest interpretation is that anti-late RNA is expressed in an early and a middle mode, like many normal early genes. Another possibility is that mot acts to abolish rho action. Further progress on individual anti-late transcripts is needed to decide which interpretation is correct. In addition, some anti-late RNA has recently been shown to depend on the products of T4 genes 33, 45, and 55 and thus share control features of late gene expression (Geiduschek et al., this volume).

The fact that *rho* intervenes in early-mode anti-late synthesis seems paradoxical because it implies that translation of this anti-mRNA ordinarily occurs in the cell (see discussion of early-mode anti-termination above). In fact, anti-late RNA has the same half-life as early messenger RNA (Notani, 1973). It is also found in polysomes (Helland, 1979). Recently, Cascino et al. (1981) have found open reading frames in a number of eucaryotic anti-message molecules; they have suggested that T4 anti-late RNA may, in fact, code for protein (personal communication).

OTHER CONTROLS ON PREREPLICATIVE TRANSCRIPTION

Early and middle transcription, as outlined here, do not constitute all prereplicative transcription. There are a few transcription events that seem to be under still another type of control. For example, Selzer et al. (1978) cloned a 2-kb fragment which covered the beginning of the *r*IIA gene and which could not have included any known early or middle promoter. This fragment expressed *r*IIA RNA in uninfected cells from a promoter which must have been very close to the beginning of the *r*IIA gene (Selzer et al., 1981). Furthermore, this *r*IIA RNA was translated to give high levels of the corresponding fragment of the *r*IIA protein. Is this *r*IIA promoter actually used in T4-infected cells? There are results that suggest that it is. Although the bulk of *r*IIA RNA is expressed as an early-mode DE gene, there is always a small amount of *r*IIA RNA expressed when T4 infection takes place in the presence of CAM. More important, Daegelen and Brody (1976) found that the usual DE kinetics of *r*IIA RNA synthesis were transformed into IE kinetics if *r*II genes were introduced by superinfection into *E. coli* previously infected by T4 carrying *r*II deletions. This suggested that after infection T4 develops the capacity to initiate at a promoter close to the beginning of the *r*IIA gene. Recent experiments (Daegelen and Brody, manuscript in preparation) show that when *r*II genes are introduced into cells by superinfection, *r*IIB expression depends on *mot* activity, but *r*IIA expression does not. Thus, *r*IIA seems to have a promoter that is not used immediately after infection but which can be activated by T4 development. Unlike middle-mode promoters, the *r*IIA promoter does not need *mot* for its activation. It is possible that some of the RNA molecules having the *r*IIA and *r*IIB message covalently linked (Sederoff, Bolle, Goodman, and Epstein, 1971; Young and Menard, 1981; see Christensen and Young, this volume) start at this promoter.

What might control utilization of the *r*IIA promoter? J. F. Pulitzer, A. Coppo, and co-workers (personal communication) have shown that some *mot*-independent gene expression is under control of a region that maps between genes 39 and 56. In particular, *r*IIA RNA synthesis seems to be very sensitive to the elimination of genes in this region. Both *mod* (controlling ADP-ribosylation of the α subunits of RNA polymerase) and *com*Cα (see above) map in this region. Their data suggest that more than one gene is involved in this control. It is not certain whether these three different types of experiments are all measuring the activity of the same *r*IIA control region. However, taken together, they imply a new kind of control element in prereplicative T4 transcription.

Gene 43 protein synthesis is autoregulated. Mutants defective in gene 43 activity overproduce the inactive protein (Russel, 1973). This autoregulation is transcriptional (Krisch et al., 1977). The overproduction of gene 43 RNA in a gene 43 mutant infection depends on *mot* activity (Mattson et al., 1978).

Gene 1 RNA synthesis has a number of unusual features. It is transcribed primarily as a middle-mode RNA (Cohen et al., 1974; Linder and Sköld, 1980; Mattson et al., 1978; O'Farrell and Gold, 1973a; O'Farrell and Gold, 1973b; Thermes and Brody, submitted for publication). Sakiyama and Buchanan (1973) found two size classes of gene 1 mRNA in T4-infected cells. One of these (12S) was concluded to be monocistronic, and the other (15S) could have been di- or polycistronic. Natale et al. (1975) found only the larger class in infected cells. Gene 1 mRNA is not made in the presence of CAM (Linder and Sköld, 1977; Sakiyama and Buchanan, 1972). Moreover, in vivo, the early-mode kinetics of gene 1 suggest it takes a long time for the polymerase to reach a rho-sensitive site, and that gene 1 RNA is DE in its early mode (Thermes and Brody, submitted for publication). Despite its *mot* dependence in vivo, gene 1 RNA can be transcribed from an early promoter in vitro (Natale and Buchanan, 1972; Natale and Buchanan, 1977; Velten, Ph.D. thesis). This early promoter seems to be very close to the gene 1 coding sequence (Natale et. al., 1975; Trimble and Maley, 1975; Trimble et al., 1972). Actually, as suggested by the sequence of the gene 1 promoter (Table 1), the early and middle promoters might overlap. It originally appeared that if the early promoter should be far upstream of gene 1, it would have to be in an anti-late region since the late gene 3 has been reported to be only about 200 bp upstream from gene 1 (Velten, Ph.D. thesis). However, recent evidence of Goldberg (this volume) indicates that genes 3 and 2/64 are in fact transcribed in the early direction, like lysozyme and gene 1; the kinetics have not, however, been determined, and no conclusive statements can be made about possible early transcripts from this region. How can all these results be reconciled? One interesting possibility presents itself. Velten (Ph.D. thesis) had found that a strong terminator exists 210 bp downstream from the gene 1 promoter. He found evidence of pausing at this terminator in vitro. If *mot* facilitates both initiation at the promoter and anti-termination past this terminator, the various results might be reconciled. This would imply that pausing of RNA polymerase (cf. Kingston and Chamberlin, 1981) may be an important factor in gene 1 transcription (Velten, Ph.D. thesis). Careful mapping of in vivo and in vitro transcripts in *mot*$^+$ and *mot*$^-$ infections will be needed to resolve this problem.

CONCLUSION

T4 prereplicative transcription takes place in a cell that starts out using its RNA polymerase holoenzyme, but this enzyme is chemically altered immediately after infection by gp*alt*. The transcription taking place during the first 1.5 min seems to have the same initiation specificity as is observed with bare T4 DNA when it is transcribed by purified *E. coli* RNA polymerase holoenzyme. Both the RNA polymerase and the DNA undergo changes starting about 2 min after infection. *mot* interaction with T4 DNA seems to impart new initiation specificity to this template. It may also change termination specificity. Independently, coupled transcription-translation, but not necessarily a particular T4 protein, renders transcription insensitive to *rho*-induced polarity. In addition, other T4 products seem to change initiation specificity, but it is not known whether these depend on changes in the RNA polymerase or in the template properties of the DNA. The prereplicative period of T4 growth remains a rich system for the study of developmental changes in transcription specificity.

Characterization of T4 Transcripts

A. C. CHRISTENSEN† AND E. T. YOUNG

Department of Biochemistry, University of Washington, Seattle, Washington 98195

The complex developmental program of bacteriophage T4 is regulated at the level of transcription. Although this basic premise was substantiated long ago, and various classes of T4 genes were recognized, a detailed understanding of the mechanisms underlying these changes has been elusive. The study of T4 transcription received a much-needed stimulus when the first cloned fragments of T4 DNA were isolated (Mattson et al., 1978; Selzer et al., 1978). Direct and technically simple methods became available to analyze the transcripts from specific genes and transcription units. The timing of transcription, the sizes of transcripts, and the locations of the 5' and 3' ends of transcripts have been determined for a number of genes.

TIMING OF TRANSCRIPTION

Early work based on analyses of global transcription or of individual mRNAs assayed indirectly revealed several regulatory classes of T4 genes (see Brody et al., this volume; Rabussay, this volume). This early work was confirmed by analyzing labeled T4 RNA by hybridization to cloned fragments of known T4 genes (Young et al., 1980). This catalog revealed the same four classes of transcripts that had been deduced by less direct techniques and provided additional data on the relative rates of synthesis and levels of different mRNAs. The transcription patterns of representative immediate-early (gene 30), delayed-early (gene 43), quasi-late ($tRNA^{Arg}$), and late (gene 23) genes are shown in Fig. 1. In addition to gene 30, genes 39, 52, 40 or 41, 42 or β*gt* and tRNA (*arg* cluster) were classified as immediate early. Genes *r*IIA, *r*IIB, and 43 were classified as delayed early, and the tRNA (*arg* cluster) genes were classified as quasi-late.

This work also revealed the location of anti-late transcripts in several late regions and suggested the location of early genes in the *w*-25 region of the T4 genome. Within the class of late genes there appeared to be a difference of several minutes in the time at which transcription could first be detected, suggesting that some late genes are farther from their promoters than other late genes.

SIZES OF TRANSCRIPTS

Transcriptional regulation of the T4 early genes is obviously very complex (see Brody et al., this volume). However, several aspects of current models of regulation of early genes can be tested by determining the sizes of the transcripts produced by these genes. In the absence of processing of the primary transcript, the size of an RNA, coupled with determination of the gene(s) it represents, would indicate the approximate initiation and termination sites on the T4 genome. In practice, the rapid turnover of procaryotic mRNA complicates this analysis. Moreover, it is not known whether T4 mRNAs are processed, as is known to occur for bacteriophage T7 mRNAs.

In an attempt to identify and map early and late transcription units, T4 RNA was fractionated by preparative gel electrophoresis (Young and Menard, 1981; Young et al., 1981). The fractionated RNA was used for in vitro translation, and the labeled proteins synthesized in vitro were identified by sodium dodecyl sulfate-gel electrophoresis. Transcripts were also analyzed by hybridization to cloned T4 genes. The size of a large number of identified and unidentified transcripts was determined. Transcripts from genes 32, *r*IIA, *r*IIB, 52, *ip*III, 21, 22, 23, and several unidentified genes occurred as multiple species, some of which were presumably polycistronic messages. Except for the cases of *r*IIA, *r*IIB, and 21, 22, and 23, the genetic composition of the polycistronic transcripts was not determined. As predicted by early work (Kasai and Bautz, 1969), *r*IIB activity was present on a polycistronic mRNA with *r*IIA. Unexpectedly, *r*IIB activity was also present on not one, but two, other smaller messages. At least the larger of these two *r*IIB transcripts probably represents cotranscription of *r*IIB and the D1 region downstream from *r*IIB.

The two gene 32 messages are present in both early and late RNA, consistent with the known stability of gene 32 RNA. These gene 32 mRNAs are probably initiated at two upstream promoter sequences and terminated at a common site at the end of gene 32 (Krisch and Allet, 1982; H. Krisch, personal communication).

The gene *ip*III appeared to be encoded in three or four different mRNAs, and these RNAs were also present at both early and late times after infection.

Why T4 goes to the trouble of producing multiple mRNA species is unclear. It could be to amplify the level of proteins needed in large quantities. Another possibility is that this mechanism serves some regulatory function—for example, to produce two gene products in close proximity to one another either physically or temporally. In some cases, the different promoters serving the same gene are distinguished by the RNA polymerase present in the cell (e.g., *r*IIB monocistronic versus polycistronic mRNA), but in other cases they are not (e.g., the two promoters in the D region [Goldfarb and Burger, 1981] are recognized by the unmodified *Escherichia coli* RNA polymerase).

These conclusions are in general agreement with other studies which indicated that early T4 transcripts are regulated by changes in promoter utilization (mediated by the *mot* gene product [Mattson et al., 1978] and possibly by polymerase changes) and by changes in termination mediated by *E. coli* rho factor

†Present address: Department of Biology, University of North Carolina, Chapel Hill, NC 27514.

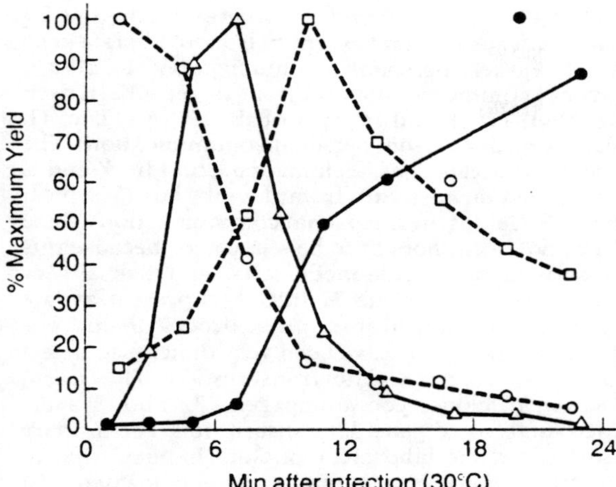

FIG. 1. Kinetics of T4 transcription. Representatives of each of the classes of prereplicative genes, immediate early (gene 30) (○), delayed early (gene 43) (△) and quasi-late (tRNA^{Arg}) (□), as well as a late transcript (gene 23) (●) are displayed. The data are taken from Young et al. (1980) and represent the hybridization of pulse-labeled RNA to DNA from plasmids containing T4 genes.

(Daegelen et al., 1982a; Daegelen et al., 1982b). However, the regulation of the size of specific transcripts by either of these mechanisms has not been directly studied by analysis of RNA isolated from the appropriate mutant-infected cells. This is a fruitful field for future studies. Even more useful will be the direct determination of the 5′ and 3′ ends of early T4 mRNAs. Since a large number of cloned T4 early genes are available, this research should be fairly straightforward. Analysis of the early genes *r*IIB and 32 is described in other chapters of this book. Analysis of late genes is described below. In addition, localization of strong binding sites or transcription start sites for unmodified and T4-modified *E. coli* RNA polymerase should allow the identification of different classes of T4 promoters. Transcription start sites for unmodified polymerase are reported in the chapter by Kutter and Rüger in this volume.

LOCATING 5′ AND 3′ ENDS OF TRANSCRIPTS

The locations of a small number of T4 transcripts have been determined by S1 mapping. The method was originally devised by Berk and Sharp (1977) and was modified for single-stranded DNA probes by Nasmyth et al. (1980). Briefly, it involves protecting a 5′ end-labeled DNA probe from S1 nuclease digestion by hybridization to RNA. The procedure creates labeled fragments of DNA whose 5′ ends are known from the procedure used to construct and label the probe and whose 3′ ends correspond to the locations of the 5′ ends of the mRNA to which it was hybridized. The length of the DNA fragment reveals the distance between the labeling site and the 5′ end of the mRNA. If the same labeled probe is subjected to the sequencing reactions of Maxam and Gilbert and the samples are resolved on the same gel, the exact position of the S1 nuclease cleavage site can be determined. This method has the advantage of requiring no RNA purification since the specificity is provided by the DNA probe used. In addition, much of the late region of the T4 genome has been cloned (Mattson et al., 1977; Selzer et al., 1978), so the construction of probes is easy and rapid.

As discussed in the previous section, previous work had indicated that genes 21, 22, and 23 are parts of a complex transcription unit. Young et al. (1981) found that gene 23 is transcribed as a monocistronic message and as the distal gene of one or two polycistronic messages which also encode gene 22 and perhaps gene 21. Kassavetis and Geiduschek (1982) found mRNA molecules that start just upstream from genes 22 and 23 as well as mRNAs that start farther upstream and probably include gene 21. It has also been suggested, on the basis of polarity of amber mutations on downstream genes, that genes 36, 37, and 38 are cotranscribed (King and Laemmli, 1971). All this implies that there are promoters in front of genes 21, 22, 23, and 36. S1 mapping of these genes and the lysozyme (*e*) gene was carried out. The 5′ ends of the messages were located, and the sequences upstream were compared. The 3′ end of the gene 23 transcripts was also located by a complementary technique which involves labeling the DNA probe at the 3′ end. Some of these results have been described by Christensen and Young (1982b).

Locations of 5′ Ends

The 5′ ends of five T4 late transcripts have been located with respect to the primary DNA sequence. Of these five, two (genes 22 and 23) have been shown to have 5′-triphosphate termini by in vitro capping of the mRNAs (Kassavetis and Geiduschek, 1982). These authors have also found that there is a cappable transcript extending from upstream of gene 22 to at least the middle of gene 23. This transcript is probably the mRNA whose 5′ end we have located upstream from gene 21. Just upstream from gene 36 is a sequence resembling a transcription terminator (see below) which probably serves to terminate a gene 34-gene 35 transcript and suggests that transcription initiation occurs between this terminator and the beginning of gene 36. Therefore, the 5′ ends of the gene 22 and 23 transcripts must arise through transcription initiation, and the 5′ ends of the gene 21 and 36 transcripts probably do too. Although evidence of RNA polymerase binding to these sites does not exist (and, given the difficulty of achieving in vitro late transcription, may not exist for some time), we will be bold and refer to these regions as late promoters.

What is a late promoter? In other words, what sequences are necessary and sufficient for late transcription? As a first step toward answering these questions, we have compared the DNA sequences of the five late promoters that we have studied (Fig. 2). The only common feature is the sequence TATAAATA just upstream from the initiation sites. There is also strong homology to the sequence CTATT just to the right of the TATAAATA sequence. We have called the TATAAATACTATT sequence the "juke box," and homologies to it are underlined in Fig. 2. In either direction from the juke box, including the −35 region,

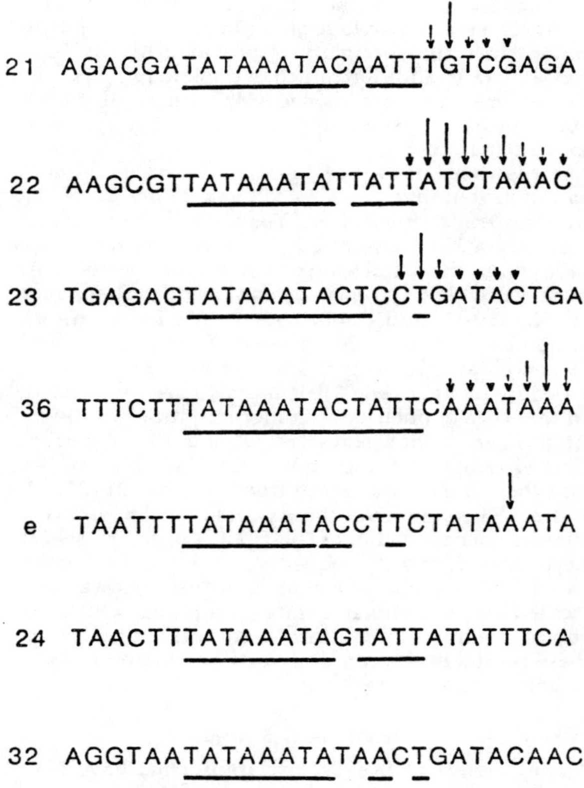

Juke Box TATAAATACTATT

FIG. 2. DNA sequences near the 5′ ends of T4 late mRNAs. Only the mRNA identical strand is shown. The sequences are aligned to emphasize the homology. The lengths of the arrows indicate the relative proportions of the mRNA molecules ending at each position for genes 21, 22, 23, and 36 (Christensen and Young, 1982b). The approximate 5′ end of the gene *e* late mRNA is shown (A. Christensen, T. Young, G. Stormo, and L. Gold, submitted for publication; the sequence is from Owen et al., 1983). The homologous sequences upstream from genes 24 and 32 are also shown. The juke box is given at the bottom of the figure, and homologies to it are underlined in the sequences.

the five sequences are quite dissimilar. This is at least in part because the five promoters are in different environments. For example, the gene 23 juke box is in the coding region of gene 22, and surrounding sequences are constrained by the need to encode a functional gp22. The gene 36 juke box is in intercistronic DNA, between coding regions for genes 35 and 36, but immediately upstream from the juke box is what appears to be a transcription terminator (see above). Sequences downstream from the juke boxes constitute the 5′ leader sequences of mRNAs, so these sequences are selected for their particular roles in mRNA stability and translational efficiency.

The 100% conserved portion of the juke box (TATAAATA) is not found anywhere else in the T4 DNA sequenced to date except in front of the late gene 24 (M. Parker, personal communication; T. Elliott, personal communication), in front of gene 32 (Krisch and Allet, 1982), and in front of the tRNA cluster (J. Broida and J. Abelson, personal communication). The gene 24 transcript has been mapped, and its 5′ end is located just downstream from the juke box (T. Elliott and E. P. Geiduschek, personal communication). Gene 32 has not been thought to be a late gene because it is transcribed as middle-mode RNA and has a very stable message, so gene 32 mRNA is present at late times whether late transcription occurs or not. For this reason, it would have been very difficult to detect gene 32-specific true late transcription. The recent isolation of clones containing gene 32 allows a detailed analysis of gene 32 transcription. Preliminary results from the laboratory of Gold indicate that at late times there is a cappable 5′ end located just downstream from the gene 32 juke box that is not observed in a 33^- 55^- infection, in which late transcription does not occur (R. Sweeney and N. Guild, personal communication). This transcript is found in addition to the middle-mode transcript, which appears to be initiated farther upstream (Krisch and Allet, 1982; H. Krisch, personal communication). The juke box in front of the tRNAs is of unknown function, but it is known that the tRNAs are transcribed at late times as well as early. Dual promoters, one middle and one late, may be a general feature of the so-called quasi-late genes.

In summary, there appears to be a highly conserved sequence which occurs just upstream from promoter-proximal late and quasi-late genes and nowhere else in the T4 DNA sequenced to date. It is reasonable to conclude that this sequence is involved in the specific recognition of late promoters by RNA polymerase and may be essential for late transcription.

Locations of 3′ Ends

Pribnow (1979) and Rosenberg and Court (1979) have compiled sequences associated with transcription termination. They have found that all these sequences have a region of hyphenated dyad symmetry. mRNAs containing these sequences would be capable of forming stable stem-and-loop structures. In most cases, the stem-and-loop sequence is followed by a run of thymines in the mRNA identical strand. The 3′ end of the RNA is found to be in the run of uracil residues that result from the transcription of this sequence. Two terminators which are absolutely dependent on *rho* differ by not having a run of thymines. The hypothesis is that the stem-and-loop structure causes the RNA polymerase to pause at that site, and either a run of thymines or factor *rho* destabilizes the DNA-RNA polymerase complex and termination occurs. It is clear that *rho* plays a role in transcription termination after T4 infection and that there are also *rho*-independent terminators in the T4 genome (Daegelen et al., 1982a; Daegelen et al., 1982b; Rabussay, 1982b). Structures similar to *E. coli* terminators have been found in T4 at several locations: near gene 1 (J. Velten, Ph.D. thesis, University of California, San Diego, 1981), downstream from gene 32 (Krisch and Allet, 1982), downstream from gene 35 (Oliver and

Crowther, 1981), and downstream from gene 23 (M. Parker, A. Christensen, A. Doermann and T. Young, manuscript in preparation). All of these sequences are found in regions in which termination is thought to occur in vivo. The gene 1 terminator is thought to be *rho* dependent because inhibition of translation increases transcription termination at this site. It appears to act as an attenuator for gene 1 transcription in vivo (Velten, thesis). This sequence does not have a run of thymines. The stem-and-loop structures which may form in these four cases are shown in Fig. 3.

We find that in vivo mRNA has 3′ ends located within the run of thymines just after the gene 23 stem-and-loop structure (Parker et al., manuscript in preparation). This, along with in vitro evidence for termination at the gene 1 stem-and-loop sequence (Velten, thesis), suggests that transcription termination signals in T4 are very similar to the termination signals used by *E. coli*.

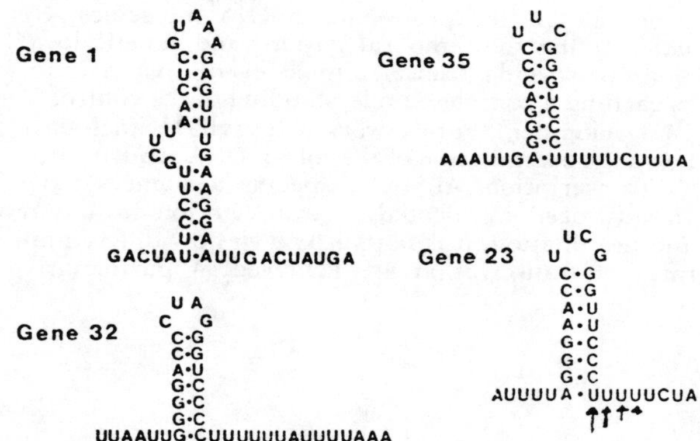

FIG. 3. Terminators in T4: genes 1, 32, 35, and 23. The stem-and-loop structures expected from the sequences of the putative terminators found near the 5′ end of gene 1 and at the 3′ end of genes 32 and 35 and at the actual site of gene 23 termination are shown. The sequences are from Velten (thesis) for gene 1: Krisch and Allet (1982) for gene 32; Oliver and Crowther (1981) for gene 35; and A. C. Christensen (Ph.D. thesis, University of Washington, Seattle, 1982) for gene 23. The arrows indicate the 3′ ends actually found for gene 23.

EXPRESSION OF CLONED LATE GENES

Because of the difficulties involved in achieving in vitro transcription of T4 late genes, several groups have attempted to analyze late transcription by analyzing the expression of late genes cloned into plasmid of phage vectors (Jacobs and Geiduschek, 1981; Jacobs et al., 1981; Oliver et al., 1981; Vorozheikina et al., 1980). Expression of late genes has primarily been assayed by complementation of an infecting amber mutant phage or analysis of protein synthesis on sodium dodecyl sulfate-polyacrylamide gels. These studies have shown that some cloned late genes are expressed in uninfected cells to some extent. Expression seems to be increased by infection with a helper T4 phage. To distinguish expression of the plasmid-borne gene from the phage gene, the helper phage carries an amber mutation in the gene being analyzed. The most detailed study of this problem is reported in the two papers by Jacobs and co-workers (Jacobs and Geiduschek, 1981; Jacobs et al., 1981). These workers found that complementation of amber mutant phage by a wild-type cloned gene was relatively inefficient unless the helper phage was defective for endonucleases II and IV (*denA* and *denB*, respectively). Complementation was further increased by mutations in the *alc* gene (the *alc* gene restricts transcription of cytosine-containing DNA; see Snustad et al., this volume). Jacobs and co-workers also found that complementation was dependent on genes 55, 43, 32, and 46. T4 late gene expression is normally dependent on genes 55 (RNA polymerase accessory protein), 43 (DNA polymerase), and 32 (helix-destabilizing protein), but not on gene 46 (DNA exonuclease).

The interpretation of the results obtained in these studies is complicated by two factors. First, no one has shown that initiation of transcription of plasmid-borne T4 genes occurs at the same site that is used when the gene is present in the phage T4 chromosome. It seems highly unlikely that T4 late promoters would be recognized in uninfected cells, considering both the sequences of late promoters (see above) and the fact that late genes are not transcribed in vivo or in vitro by the unmodified *E. coli* RNA polymerase. Second, recombination between plasmid and phage may be occurring in infected cells. Jacobs et al. (1981) attempt to show that recombination is not responsible for complementation by showing that the burst contains very few wild-type phage. This only shows that the extent of marker rescue is low and does not address the possible occurrence of single crossover events between the mutant phage chromosome and the wild-type gene in the plasmid. Such an event would result in integration of the plasmid sequences into the phage genome. The gene would then be present in two copies, one mutant and one wild type, flanking the vector sequences. A T4 DNA molecule with a large plasmid insertion could not be packaged into a phage particle without the concomitant deletion of several kilobases of T4 DNA. It has recently been shown that T4 can transduce bacterial or plasmid genes (Wilson et al., 1979) and that the transducing phage particles contain a large number of tandemly repeated copies of plasmid sequences integrated into the genome (Takahashi and Saito, 1982c). These recombinant phage can only be detected by transducing ability since they contain only a fraction (if any) of the phage genome and thus cannot productively infect cells. It would be instructive to analyze a burst of phage from one of these complementation experiments for the presence of plasmid pBR322 sequences in the phage DNA or for the ability to transduce the pBR322 ampicillin resistance gene. That gene 46 is required for complementation (Jacobs and Geiduschek, 1981) suggests that recombination is required for the expression of cloned late genes, or at least that expression of late genes from plasmids is fundamentally different from expression of late genes from the phage genome.

CONCLUSIONS

Several experimental approaches have been used in recent years to study the control mechanisms of T4 transcription. Most of these have relied on cloned T4 genes as specific probes for mRNA molecules. By determining the temporal pattern and genetic locations of specific transcriptional events, we are approaching a complete understanding of the control of T4 development. Future work will probably include a thorough examination of the role of DNA sequences in T4 transcription. Although specific sequences have already been implicated in transcription initiation and termination, it is not yet known how differential rates of transcription are achieved. A particularly exciting finding is that some genes have promoters of more than one temporal class. This strategy undoubtedly allows fine control over transcription rates at various stages of infection and is probably responsible for the highly complex pattern of gene expression exhibited by T4.

We are grateful to Michael Parker, Gus Doermann, Gary Stormo Rosemary Sweeney, Nancy Guild, Larry Gold, Joel Broida, John Abelson, Tom Elliott, Peter Geiduschek, Gerry Smith, and Henry Krisch for communicating results before publication.

Regulation of Late Gene Expression

E. PETER GEIDUSCHEK, THOMAS ELLIOTT, AND GEORGE A. KASSAVETIS

Department of Biology, University of California, San Diego, La Jolla, California 92093

In this chapter, we summarize what is known about the components of T4 late gene expression. Those facts that have been known for some time are presented without much discussion, but some recent findings are described more fully. Some subtopics are also discussed in other chapters in this volume.

REGULATORY PROTEINS INTERACTING WITH RNA POLYMERASE AND THE DNA TEMPLATE FOR LATE TRANSCRIPTION

Three T4-coded proteins, the gene 33, 45, and 55 products, are known to be normally required for late gene expression. All three interact with RNA polymerase, as described by Rabussay (this volume). Gp33 and gp55 copurify with this enzyme, along with two other (presumably) phage-coded proteins, as already mentioned. Gp45 does not, but genetic and biochemical evidence points to its being an RNA polymerase-binding protein which interacts with gp55 and with the RNA polymerase from phage-infected cells (Coppo et al., 1975a; Ratner, 1974a).

The DNA template of T4 late gene transcription must normally contain hydroxymethylcytosine (HMC) in place of cytosine. That restriction is enforced by a protein that is encoded by the T4 *alc/unf* gene (Snyder et al., 1976; abbreviated below as *alc*) whose mechanism of action is not known. The *alc* gene product, which must be a very early T4-coded protein because it can block essentially all transcription of cytosine-containing DNA (Kutter et al., 1981), is discussed in some detail by Snustad et al. (this volume). In transcriptionally normal hosts, glucosylation of HMC-containing T4 DNA and late transcription are compatible. However, an *Escherichia coli* RNA polymerase mutation which confers a delicately poised state of nonpermissiveness for T4 has been described. The failure of T4 late transcription conferred by this mutation can be relieved by blocking DNA β-glucosylation (Montgomery and Snyder, 1973; Snyder, 1972). It has also been suggested that at least some late gene transcription from unglucosylated DNA might be independent of the function of gp33 (Dharmalingam and Goldberg, 1979a).

Perhaps the longest-known fact about T4 late transcription is that there is some connection with DNA replication (Luria, 1962; Wiberg et al., 1962). If replication is never allowed to start, the block to late gene transcription normally is almost complete. Blocking ongoing replication also suppresses late gene expression (Lembach et al., 1969; Riva et al., 1970a), although the residual rate of late transcription is not simply proportional to the residual rate of DNA replication or to the size of the DNA pool. In one quantitative analysis, an approximately 10-fold decrease in the rate of late RNA synthesis was found when DNA synthesis was blocked by shifting a temperature-sensitive T4 DNA polymerase mutant to the nonpermissive temperature (Riva et al., 1970a). Blocking ongoing phage T4 DNA synthesis with metabolic inhibitors generates less extreme effects on late transcription in vivo than does the inactivation of the T4 DNA polymerase (Rabussay and Geiduschek, 1979b).

That serves to emphasize how important it is to be able to distinguish intrinsic effects of concurrent DNA replication on late transcription from effects of the individual DNA replication proteins. For example, since the gene 45 protein, which is required for late transcription, is also a central component of the replisome, interacting with most of the other proteins of that assembly (see Nossal and Alberts, this volume), T4 replication proteins might affect late transcription directly, through their interaction with gp45. One set of in vitro experiments bearing (albeit indirectly) on this subject was done with a cellophane disk system for in vitro T4 late transcription, in which de novo initiation of late transcription was independent of concurrent DNA replication in vitro but depended on the presence of phage DNA which had previously undergone replication in vivo. In this system, temperature-sensitive T4 DNA polymerase still conferred temperature sensitivity on the initiation of late transcription in vitro, but a temperature-sensitive T4 gene 42 protein (dCMP hydroxymethylase) had little effect (Rabussay and Geiduschek, 1979b).

That there is, in addition, a structural constraint on the T4 DNA template for late transcription was first suggested by the finding that late transcription could be uncoupled from concurrent replication in certain T4 multiple mutants (altered in DNA polymerase, DNA ligase, and gene 46 exonuclease function) which accompany the blocking of replication with the introduction of breaks into the unreplicated DNA (Cascino et al., 1970; Riva et al., 1970b; Wu et al., 1975). Such uncoupled late transcription is initiated at the normal gene 23 late promoter (Christensen and Young, 1982b) and is absolutely dependent on gp55 and gp45. It is not absolutely dependent on any other replication protein, although mutations in certain replication protein genes do generate quantitative effects on this replication-independent late gene expression (Wu et al., 1975). The rather general model that has been proposed is that T4 late genes must be made competent for transcription, that competence is normally conferred by concurrent replication and is probably transient but may be conferred in other ways, and that some replication proteins might play a structural role in defining competent DNA. For example, one concrete model of DNA template competence proposed that late promoters cannot be recognized in intact, double-stranded DNA and that they are normally only utilized when made accessible at replication forks, but that promoter melting can also be facilitated by nicks or gaps in the DNA (Rabussay and Geiduschek, 1977a). Recent experiments that are described below show that this model must be at least partly incorrect.

In summary, the replication coupling of late gene expression normally has two manifestations: an essentially complete absence of late transcription if replication cannot start, and a large quantitative effect if replication is allowed to start and is subsequently blocked. In the latter case, it is important to distinguish the separate contributions to replication coupling that (some) replication proteins and DNA structure probably make.

We would like to discuss an intriguing recent observation on T4 late transcription in this context of template competence. *E. coli lit* mutants are nonpermissive for T4 wild-type phage. The *lit* defect results in blocked T4 late gene expression. T4 *gol* mutants, which compensate for the defect (and thus grow on Lit^- bacteria) have been found. The intriguing property of *gol* mutants is that they exert their effect on gene expression only in *cis* (in mixed infections) and that the *cis* effect extends to late genes that are so far away from the *gol* site as to be genetically (i.e., recombinationally) unlinked. Thus, the *gol* site in some ways resembles a eucaryotic enhancer sequence (Banerji et al., 1981; Champness and Snyder, 1982). The extensive domain of the *cis* effect is difficult to explain but might suggest that, at least in this genetic background, the DNA template for late transcription is chosen quite early in infection. All of six *gol* mutants that have been examined thus far map within 40 nucleotides near the gene 22 (N)-proximal end of gene 23 (W. Champness and L. Snyder, submitted for publication).

T4 LATE PROMOTERS

As also discussed in the chapter by Christensen and Young (this volume), several T4 late promoters have now been mapped. The comparison of DNA sequences with the positions of sites of transcriptional initiation identifies an extraordinarily AT-rich upstream consensus sequence, TATAAATActatt, which spans the position of the *E. coli* promoter −10 consensus sequence. (The lower-case letters designate those nucleotides that are not shared by every one of the seven consensus sequences that have been identified so far.) The T4 late promoters apparently do not share a −35 consensus sequence. It is curious that the absolutely conserved sequence does not include at least one C. We think that this means that the *alc*-dictated HMC requirement involves something other than late promoter recognition.

These new findings appear to introduce an element of simplicity and conventionality into the subject of late gene regulation: among the plethora of T4-specified RNA polymerase-binding proteins, one or more is the long-postulated σ analog (Travers, 1970b). Given the differences between the consensus sequences of T4 late promoters and bacterial promoters, it is possible that the T4-specified analog and σ confer considerably different enzyme-DNA interactions on the RNA polymerase core. Competence might refer to promoter activation, although it seems remarkable that so AT-rich a binding site should require activation. The analysis of T4 late gene regulation will now tend to focus on these promoters and on the others that will soon show up in further sequencing.

Some recent observations balance the apparent simplicity of this result with complicating evidence for a hitherto unrecognized class of T4 late transcripts with interesting properties. The existence of RNA complementary to T4 late transcripts (anti-late RNA) has been known for some time (and is discussed by Brody et al., this volume). It has been thought that anti-late RNA primarily is middle RNA. Recently, anti-late transcripts that are coordinately regulated with late RNA have been found. Three such transcripts have been mapped to the vicinity of genes 22, 23, and 24. The transcripts are under gp55, gp45, and gp33 control, are replication dependent, and, in the one case studied thus far, first appear just slightly before, or at the same time as, the corresponding conventional late transcripts. The positions of the 5′ ends of the three mapped transcripts are remarkable. Each is positioned to generate convergent transcription across a late promoter. These transcripts are always much less abundant than gene 22, 23, or 24 mRNA. They do not show the late promoter consensus sequence but do share a shorter and looser consensus sequence of their own. The question of whether the analyzed 5′ ends of these newly found late transcripts are due to nucleolytic processing or transcriptional initiation is not yet completely settled, but the evidence that transcriptional initiation does occur at one of these sites is relatively strong. The possibility that late transcriptional regulation involves more than one initiation specificity is therefore suggested by these new findings (T. Elliott, G. A. Kassavetis, and E. P. Geiduschek, manuscripts in preparation). The dual-strand transcription pattern, reminiscent of transcription in the regions of the ColE1-like replicative origins (Selzer et al., 1983; Tomizawa and Itoh, 1982), is also intriguing.

EXPRESSION OF CLONED T4 LATE GENES

When it was found that the cloned T4 late genes (Mattson et al., 1977; Selzer et al., 1978; Velten et al., 1976; Wilson et al., 1977) could complement mutations in infecting phage, there was hope that analysis of that complementation would provide a radical clarification or simplification of notions regarding late gene regulation. That hope has not yet been realized; indeed, some new puzzles have been contributed. The principal questions are simple to state. (i) Are late promoters active on plasmids? (ii) What, if anything, is required to activate them? In the following paragraphs, we summarize the current status of these questions.

Several cloned T4 late proteins can be made in uninfected *E. coli* (Jacobs et al., 1981; Oliver et al., 1981). It has been suggested that late promoters might be utilized in uninfected bacteria (Oliver et al., 1981). Our own guess is that expression in uninfected cells is from *E. coli*-type promoters, either in the cloned fragments or in the vector. This is almost certainly the case for gene 23 in the experiments of Jacobs et al. Uninfected *E. coli* are competent to translate T4 late RNA (Goldman and Lodish, 1972; Wilhelm and Haselkorn, 1971). Expression of these genes before infection certainly complicates the direct interpretation of complementation experiments in terms of the regulated expression of T4 late genes. The worst case appears to have been encountered by Oliver et al. (1981), who found essentially no additional expression of the cloned genes 10 and 11 after T4 infection. However, the T4 genetic background of these experiments (*den*$^+$ *alc*$^+$; dCTPase$^-$) permits degradation of host DNA after infection and expresses the *alc* function and therefore represents the most severe possible bias against postinfection expression of the T4 late genes.

On the other hand, expression of cloned gene 23 after phage infection has been documented, and its regulation has been studied. Even these findings have received divergent interpretations (see Snustad et al. and Christensen and Young, this volume). For the sake of clarity, we shall summarize the findings with T4 genetic backgrounds in which host DNA is not degraded (*den*$^-$), in which T4 DNA replication leads to incorporation of dCMP into DNA (56$^-$), and in which late gene expression in cytosine-containing T4 DNA is allowed (*alc*$^-$). Activity of the plasmid-borne genes was first analyzed in terms of complementation, then in terms of protein synthesis, and more recently in terms of RNA synthesis (Elliott et al., manuscripts in preparation; Jacobs and Geiduschek, 1981, referring to work by Mattson and co-workers; Jacobs et al., 1981). Synthesis of gp23 required gp55 and was partly independent of replication (and of the function of certain replication proteins) but also partly dependent on the function of gp46. The partial replication independence suggested that it should be possible to selectively analyze the activity of cloned late promoters by S1 mapping analysis of RNA (Berk and Sharp, 1978) in cells infected with replication-negative phage (i.e., *denA*$^-$, *denB*$^-$, *alc*$^-$, 56$^-$, 43$^-$ genetic backgrounds). Under these circumstances, the utilization of the gene 23 promoter contributed by the plasmid depended on the function of gp55 and was at least partly independent on gp46 function. Contention regarding the interpretation of the earlier findings had focused on gp46, the involvement of which had first been found by Mattson and co-workers (as referred to in Brody et al., this volume).

The T4 gene 46 is required for recombination (as discussed by Mosig, this volume, and Warner and Snustad, this volume) and is associated with an exonuclease function (Mickelson and Wiberg, 1981). Gp46 has never been recognized as essential for expression of T4 chromosome-carried late genes. The interpretation offered for the gp46 dependence of late genes on plasmids was that late promoter activity might rigorously require the integration of a plasmid-carried late gene into the T4 chromosome. The ability of T4 phage to transduce bacterial and plasmid genes (Wilson et al., 1979) and the ability of plasmids to recombine and probably to replicate after T4 infection, forming repetitive, linear structures (Takahashi and Saito, 1982c), adds general plausibility to that suggestion without, however, proving it. Stripped of the abundant technical detail, this interpretation is tantamount to insisting that the structural requirements of template competence for late gene expression are so special that they can only be achieved on the replicating T4 chromosome. In light of the gp43 and gp46 independence of the plasmid-borne gene 23 promoter, we favor the alternative view that the gene 23 promoter can be activated on the plasmids which have been used. With regard to DNA competence, we guess that the activation of T4 late genes does have a DNA structural component but that activation can be accomplished via multiple, alternative pathways. In our view, the partial gp46 involvement with the expression of the plasmid-carried gene 23 promoter in vivo merely suggests the existence of a novel activation mechanism.

Can the vector influence late promoter activation and are any late promoters inactive in at least some plasmids after infection by T4 *den*$^-$ *alc*$^-$ phage? Knowing the answers to these questions would certainly help us to integrate the *gol* phenomenon with other studies of the DNA template for late transcription.

LATE TRANSCRIPTION IN VITRO AND CONCLUDING COMMENTS

There is a long history of attempts to demonstrate T4 late transcription in vitro. The first experiments with crude extracts quickly followed the discovery of early and late RNA (Khesin et al., 1962; Khesin et al., 1963), and numerous, only partly published, experiments with more or less highly purified RNA polymerase followed (Hall et al., 1970; Khesin, 1970; Stevens, 1974). All failed to reproduce the predominantly *r* strand-specific, highly asymmetric transcription characteristic of the T4-infected cell. Some of these experiments showed transcriptional properties that differed from those of *E. coli* RNA polymerase holoenzyme and included production of a low but distinct proportion of *r* strand-specific T4 RNA. However, when tested, this *r* strand RNA production was found to be concomitant with symmetric transcription (Hall et al., 1970; reviewed by Khesin et al., 1980) eventually attributed to the anti-σ activity of the 10-kilodalton protein (Khesin et al., 1980; Stevens, 1974; see Rabussay, this volume). Thus, it was not possible to associate novel properties of these enzymes conclusively with late gene-specific transcription. The exceptions were a crude system that gave extremely low incorporation and consequently was of limited use (Snyder and Geiduschek, 1968) and an uncorroborated observation of *r* strand transcription unaccompanied by assays of symmetry (Travers, 1970b). Eventually, an in vitro system based on concentrated, gently prepared lysates (Schaller et al., 1972) of T4-infected cells supported on cellophane disks gave the sought-for highly selective late transcription under suitably adjusted conditions (Rabussay and Geiduschek, 1977b; Rabussay and Geiduschek, 1979a). As already stated, late transcription in this system was dependent on prior in vivo replication of T4 DNA. The competence for late transcription was also rather sensitive to mechanical disruption. These properties of the in vitro system were consistent with the general notions about DNA competence that have already been described. The cellophane disk system was used to show that the T4 gene 55 product was required for late transcription in vitro (Rabussay and Geiduschek, 1977b).

The recent mapping of several late promoters (Christensen and Young, 1982b; Christensen and

Young, this volume; Kassavetis and Geiduschek, 1982; G. A. Kassavetis, T. Elliott, D. P. Rabussay, and E. P. Geiduschek, Cell, in press) has made it possible to apply much more sensitive and incisive methods to the analysis of late T4 transcripts. The question of selective transcription in vitro has, accordingly, been taken up again, with remarkably straightforward and encouraging results. It has been reported that T4-modified late RNA polymerase recognizes the gene 23 and 24 promoters in underwound plasmid DNA and even in mature (linear) glucosylated HMC-containing phage DNA. Recognition of these two promoters is specific to RNA polymerase containing gp55. Heparin-resistant binary RNA polymerase complexes can form at the gene 23 promoter in underwound plasmid DNA. Relaxed DNA is a much poorer template than underwound DNA (Kassavetis et al., in press). The barely detectable signals for gene 23 promoter utilization in linear, double-stranded DNA explain why it would not have been possible to demonstrate specific late transcription by the older analytical methods.

These new findings have general implications for the regulation of T4 late genes, as follows.

(i) Evidently, late promoter recognition by T4-modified RNA polymerase can operate in double-stranded DNA. That is consistent with the supposition that the DNA template competence or activation requirement for late gene expression probably is not absolute. In a preceding section, we argued that an approximately 10-fold effect of ongoing replication on late transcription must probably be divided between effects of individual replication proteins (particularly the T4 DNA polymerase) on late transcription and effects of DNA structure on the same process.

(ii) This late promoter utilization in vitro apparently does not require gp45; yet the gp45 requirement for late gene expression in vivo is general and absolute. Apparently, the purified RNA polymerase-based late transcription system is, in that crucial respect, incomplete. It is conceivable that gp45 and other replication proteins, such as gp43, which can interact with RNA polymerase through gp45, will turn out to affect kinetic properties of transcription initiation at late promoters or to improve discrimination between late promoters and other sites. Such speculations as these are now likely to prove amenable to direct experimental tests.

We thank D. P. Rabussay for helpful comments on the text.

Our research on this subject has been supported by a grant from the National Institute of General Medical Sciences. G.A.K. gratefully acknowledges a postdoctoral fellowship of the National Institute of General Medical Sciences, and T.E was supported by a predoctoral traineeship from the U.S. Public Health Services.

Processing and Translation

Translational Regulation in T4 Phage Development†

JOHN S. WIBERG[1] AND JIM D. KARAM[2]

Department of Radiation Biology and Biophysics, University of Rochester School of Medicine and Dentistry, Rochester, New York 14642,[1] and Department of Biochemistry, Medical University of South Carolina, Charleston, South Carolina 29425[2]

Transcriptional control of T4 phage gene expression has been the subject of intensive study over the last 15 years. Although many important questions about the selectivity of transcription in the T4-infected cell remain unanswered, the outlines of this process, particularly the temporal sequence of appearance of T4 gene products, are reasonably well defined and have provided a sensible context within which other levels of T4 gene control have been recognized and are being investigated. In addition to the transcriptional mechanisms that control phage development, some T4 genetic clusters are subject to regulation at the levels of RNA processing/modification, translation, and assembly of the protein complexes that ultimately determine levels of biological activity for individual phage-induced enzymes and structural proteins. That is, in essence, T4 utilizes much of the same types of regulatory processes that service the more complex genomes of its *Escherichia coli* host and other procaryotes.

The realization that T4, a DNA virus, utilizes nontranscriptional regulatory mechanisms to a significant degree came about only relatively recently, and the available evidence indicates that this phage offers a model system that is uniquely suited for studies on translational regulation and control of gene product activity via protein and protein-nucleic acid complex assemblies. The chapters in this volume on morphogenesis and on genes and gene products cite many examples of gene control at the post-translational level, including protein cleavage (processing) and protein-protein and protein-nucleic acid interactions in DNA transactions and phage morphogenesis. The focus of this chapter is control of T4 mRNA utilization in protein biosynthesis. Experience with a variety of genetic systems has shown that the selectivity of translation is influenced by a number of factors, including mRNA concentration, mRNA conformation, and diffusible substances (e.g., protein repressors) that interact with the translational machinery. Such factors are known to play important roles in the control of translation in T4-infected cells, and we shall discuss specific examples. We also attempt to point out similarities between the T4 system and other systems in which translational regulation is being studied. We should mention at the onset, however, that most of what is known about the control of translation in T4 infections pertains to the early phase (prereplicative and replicative periods) of phage development. This constricts our view of the extent to which T4 relies on translational regulatory mechanisms in relation to other phage-mediated control processes.

A recent review by Gold et al. (1981) and the book edited by Grunberg-Manago and Safer (1982) are excellent sources of information about translational regulatory mechanisms in procaryotes and eucaryotes.

T4-INDUCED MODIFICATION OF THE HOST TRANSLATIONAL APPARATUS

Effects on Unknown Targets

The earliest indication of post-transcriptional control by T-even phage was apparently observed by Hall et al. (1964); they showed by DNA-RNA hybridization that "all species of early RNA are present at the late time" (19 min), yet this "early T2 RNA does not function as messenger after 15 min." They concluded that "one must therefore envision a regulatory control of the function of T2 RNA as well as its synthesis." Bautz et al. (1966) came to the same conclusion from related observations on T4 lysozyme (*e* gene product). Workers in other laboratories extended these seminal observations by noting various discrepancies between in vivo levels of a given T4 mRNA and the rate of synthesis of the corresponding protein; in these cases, however, the arguments are more convincing because functional (i.e., translatable in vitro) mRNA was measured. Those studying T4 lysozyme were Salser et al. (1969), Wilhelm and Haselkorn (1969), and Zetter and Cohen (1974). Those studying T4 deoxyribonucleotide kinase, an early enzyme, were Cohen (1972), Cohen et al. (1972), Sakiyama and Buchanan (1971), and Zetter and Cohen (1974). Cohen (1972) studied T4 dCMP hydroxymethylase. Witmer, in his review (1976) of the regulation of T4 gene expression, cited the work of many laboratories that showed that there is an early, untranslated burst of *e* mRNA synthesis in T4$^+$ infections, followed by a later, translated burst. He concluded that the blockade of early *e* mRNA translation is a function of the mRNA and not of the translational machinery; the basis of the early blockade remains unexplained. The T4 *regA* protein (discussed elsewhere in this chapter) probably acts as a translational

† Report no. UR-3490-2227 of the University of Rochester Department of Radiation Biology and Biophysics.

repressor (Karam et al., 1981), and, under appropriate conditions, *regA* mutants overproduce all three of these enzymes (Wiberg et al., 1973). However, it is not known whether the *regA* gene is involved in the abovementioned discrepancies between enzyme synthesis and mRNA level. On the other hand, Trimble and Maley (1976) showed, by comparing *regA*$^+$ and *regA*$^-$ infections, that the *regA* gene is responsible for such discrepancies in DNA$^-$ infections (where the *regA* effects are most pronounced) for kinase and hydroxymethylase: in the *regA*$^+$ state, functional mRNA persists after the synthesis of enzyme stops.

Foreign mRNAs. T4 blocks translation of the mRNA of lambda bacteriophage (Pearson and Snyder, 1980). In a study that asked why the RNA phage M12 cannot replicate in cells superinfected with T4, Hattman and Hofschneider (1968) concluded that T4 blocks translation of the RNA phage while leaving intact the preexisting polysomes containing the parental M12 RNA. In simultaneous, mixed infection by T4 and M12, the parental M12 RNA formed polysomes almost as efficiently as in control infections, despite no production of M12 RNA polymerase. In contrast, Goldman and Lodish (1971) found that the RNA of the RNA phage f2 (very closely related to M12) was released from polysomes shortly after superinfection by T4, whereas f2 RNA became extensively fragmented. The basis of the discrepancy in results between these two laboratories is unclear; we return to this subject below.

Kennell (1970) showed that although significant amounts of mRNA were transcribed from the *E. coli lac* operon during the first few minutes after infection by T4, most of this mRNA was not associated with ribosomes. In contrast, virtually all *lac* mRNA in uninfected bacteria was associated with ribosomes. Furthermore, T4 infection prevented further growth of preinduced *lac* polysomes. It was concluded that T4 infection interferes, within seconds, with the reassociation of ribosomes to host mRNA. Fabricant and Kennell (1970) found that T2 ghosts (DNA-less particles resulting from osmotic shock of whole phage) kill most cells to which they attach, apparently by unrepairable leakage of cell components, but that in the surviving 11% of the cells, host protein synthesis was blocked for 30 min and then resumed. During this 30-min period, mRNA from the lactose operon could be induced but was not translated. They suggested that progenitors of modern T-even phage used only this phage attachment effect to shut off protein synthesis, but evolved mechanisms of the type implied by the work of Hsu and Weiss (1969) (see below) to inhibit specifically host protein synthesis before the host could recover.

Several T4 mutations affecting the structure of T4 tRNAs have been described that result in suppression of T4 nonsense mutations (Schmidt and Apirion, this volume). However, a novel, multinonsense suppressor in T4 has been reported by Ribolini and Baylor (1975). This mutation, *psu*$^+$*SB*, was mapped tentatively between genes 31 and 32, far from either the T4 tRNA gene cluster or the *mb* mutation, which is thought to function in a tRNA-processing step (Wilson and Abelson, 1972). Ribolini and Baylor suggested two models. In model A, the nonmutated form of the *psu*$^+$*SB* gene codes for a tRNA-modifying enzyme. In model B, the *psu*$^+$*SB* gene codes for a ribosomal protein and is analogous to the *ram* locus in *E. coli* (Rosset and Gorini, 1969); i.e., the *psu*$^+$*SB* mutation increases ribosomal ambiguity. It is tempting to speculate that such an effect might be related to the observation of Artman and Werthamer (1974) (see below) that T4 infection causes rapid development of resistance to streptomycin, an antibiotic that binds to ribosomal protein S12 of *E. coli* and can induce ribosomal ambiguity (Garvin et al., 1973).

Effects on Ribosomes and Associated Factors

A number of laboratories have reported changes in the properties of *E. coli* ribosomes or their associated factors after infection by T4. Much of this work describes changes in translational specificity among various mRNAs, especially those of *E. coli*, T4, and RNA phages. As we describe below, there is disagreement about the existence and interpretation of such effects. Among the RNA phages, f2, R17, MS2, and M12 are very closely related (Horiuchi, 1975), so results with one are generally assumed to apply to the others.

Hsu and Weiss (1969) discovered that ribosomes from T4-infected cells were much less efficient for in vitro translation of MS2 RNA and *E. coli* mRNA than were ribosomes from uninfected *E. coli*; nonetheless, translation of T4 mRNA or poly(U) was similar for both kinds of ribosomes. Chloramphenicol prevented the development of this early change in ribosomes, which other evidence argued was in a factor extractable by high concentrations of salt. Schedl et al. (1970) made a complementary observation with the related RNA phage f2: a factor removed from *E. coli* ribosomes by a high salt concentration stimulated the translation of f2 RNA by ribosomes from T4-infected cells; they suggested that a T4 enzyme modifies a host translation factor. Their results also implied that, at least in vitro, the host factor can compete with the T4-modified factor.

Dube and Rudland (1970) found that ribosomes from T4-infected cells bind poorly to R17, f2, and *E. coli* mRNA but bind as well as uninfected-cell ribosomes to T4 mRNA; the effect was localized to the high-salt-wash (initiation factor) fraction. This work was amplified by Steitz et al. (1970), who showed that, after infection by T4 phage, ribosomes selectively lose the ability to bind efficiently to the R17 RNA initiation sites of the coat protein and RNA synthetase, but not the A protein. This change resides in the initiation factor fraction. Pollack et al. (1970) concluded that it is the IF3 component in the factor fraction that confers on T4 ribosomes their specificity for T4 late mRNA relative to MS2 RNA and T4 early mRNA. This same group (Revel et al., 1970) separated IF3 into two subfractions that differed in their activities with T4 late mRNA relative to MS2 RNA.

Groner, Pollack, Berissi, and Revel (1972a) reported the purification to homogeneity from *E. coli* ribosome-associated factors of an interference factor (factor *i*) that inhibits the translation of MS2 RNA without quantitatively affecting that of T4 mRNA. Factor *i* binds to IF3 and inhibits initiation of translation of the MS2 coat protein gene, but stimulates initiation from the MS2 synthetase gene; factor *i* stimulated translation of certain T4 genes at the expense of others

(Groner, Pollack, Berissi, and Revel, 1972b). Groner et al. suggested that previous reports of heterogeneity in IF3 activity might result, at least in part, from interaction with such *i* factors. However, Jay and Kaempfer (1974, 1975) argued strongly that factor *i* represses translation of R17 RNA not by binding to IF3, but by binding to the RNA, specifically, to the coat protein gene. They found (1974) that IF3, even when present in 10-fold molar excess over factor *i*, failed to relieve the inhibition of initiation on R17 RNA. In contrast, Revel, Pollack, Groner, et al. (1973) found that the inhibitory effect of factor *i* on the translation of MS2 coat protein was overcome by excess IF3.

Perhaps factor *i* (i.e., *i*α [see below]) can act on both mRNA and IF3, with the relative bias depending on other conditions or factors. It should be noted that factor *i* is one of the three host-coded subunits of Qβ phage RNA replicase (Groner, Scheps, Kamen, et al., 1972) and is identical to ribosomal protein S1 (Wahba et al., 1974). Lee-Huang and Ochoa (1973) separated *E. coli* IF3 into two fractions; IF3α is specific for MS2 phage, *E. coli*, and early T4 mRNA and has low activity with T4 late mRNA, whereas the reverse is true of IF3β. Lee-Huang and Ochoa (1972) also found two *i* factors in *E. coli*; *i*α inhibited functions of IF3α, whereas *i*β inhibited functions of IF3β. Factor *i*α is the *i* factor described by Groner, Pollack, Berissi, and Revel (1972b). The ratio of IF3β to IF3α was reported to increase considerably after T4 infection (Lee-Huang and Ochoa, 1971). Revel, Groner, Pollack, et al. (1973) and Revel, Pollack, Groner, et al. (1973) isolated three *i* factors from *E. coli* and characterized some of their cistron specificities on MS2 RNA, T4, and T7 early and late mRNAs. They argued that these factors can explain the heterogeneity of IF3 toward various mRNAs. They also reported that two of the *i* factors are chemically modified after T4 infection (M. Revel, Y. Groner, Y. Pollack, H. Zeller, D. Cnaani, and U. Nudel, Abstr. 9th Int. Congr. Biochem., Stockholm, 1973, p. 133). Wahba (personal communication cited in Rabussay and Geiduschek, 1977a) purified an altered S1 protein (i.e., an altered *i*α) from T4-infected cells and found that it retained the translation-stimulating activity of S1 with S1-depleted ribosomes but lacked the translation-inhibiting property of excess S1.

Singer and Conway (1973), noting the already-large literature implying that T4 regulates translation by changes in the levels of different IF3 initiation factors, and noting the contradictory conclusions of Goldman and Lodish (1972), paid particular attention to possible effects of a number of experimental variables. Their results suggested that either dialysis of ribosomes after preincubation or reconstitution of high-salt-washed ribosomes with their high-salt-wash fraction resulted in an alteration of the ribosome's translational activity. Observing appropriate precautions, they confirmed both the negligible translation of f2 RNA and the normal translation of late T4 mRNA by ribosomes from T4-infected cells when isolated with their initiation factors still attached. With such ribosomes, formylmethionyl tRNA did not bind in response to low concentrations of f2 RNA but did bind in response to high concentrations of f2 RNA; yet translation did not occur at the high concentrations.

Disagreement on T4 effects. Acceptance of the validity of some of these claims for T4-induced changes in the translational apparatus has not been universal. As implied above, Goldman and Lodish (1972) saw no change in the specificity of initiation factors after T4 infection and concluded that T4 infection causes a nonspecific decrease in the efficiency of initiation factors to translate natural message. These authors (1975) offered evidence that simple competition between f2 RNA and T4 mRNA, in the absence of changes in the translational apparatus, may be adequate to explain how T4 superinfection blocks translation of f2 RNA and possibly host mRNA; Lodish (1976) extensively discussed mRNA competition in a general review article. Spremulli et al. (1974) also saw no alteration in messenger specificity of IF3 upon T4 infection of several strains of *E. coli*. It is certainly easier (if not fairer) to explain away a failure to confirm a positive observation as being due to experimental problems than to discount the positive observation itself, particularly where the positive observation was made in a number of independent laboratories. For example, Haselkorn and Rothman-Denes (1973) suggested that both the low efficiency and the absence of specificity of initiation by T4 ribosomes seen by Goldman and Lodish (1972) may be due to losses of IF3 during preparation of the systems for protein synthesis. The complexity of ribosomes, especially their sensitivity to variations in handling, is well known (van Holde and Hill, 1974); e.g., simple dilution (or excess washing) of ribosomes can cause loss of initiation factor IF2 (Chu and Mazumder, 1974). Thus, it seems best to reserve judgment on this controversy until new approaches, especially genetic ones, are used. For instance, if T4 mutants can be found that block the development of a specific effect on translation (e.g., the purported change in specificity of mRNA recognition) and simultaneously result in the loss of a specific T4 protein (as from the ribosome), the credibility of the purported effect will be enhanced (see below). Furthermore, physiological studies of such mutants should reveal the in vivo significance of the protein.

Artman and Werthamer (1974) demonstrated that T4 infection rapidly confers resistance to streptomycin in *E. coli*. They showed that, relative to ribosomes from uninfected cells, ribosomes isolated from T4-infected cells exhibit a marked decrease in ability to bind streptomycin and are virtually resistant to the ability of the drug to block the T4 mRNA-directed binding of formylmethionyl tRNA. They concluded that T4 infection alters the ribosome. The T4 *regA* gene is not responsible for this development of resistance to streptomycin (Wiberg and Cardillo, unpublished data).

Gausing (1972) cited many reports that the rate of protein synthesis decreases by about 60% within minutes after T4 infection and showed that the effect is on the rate of peptide chain growth. The proportion of ribosomes in polysomes did not change after infection. Since the rate of RNA chain growth also decreases by about 50% (Bremer and Yuan, 1968), it was not clear whether the effect on peptide chain growth was direct or indirect, via coupling of transcription with translation. Pennica and Cohen (1978) observed a second, abrupt decrease in the rate of protein synthesis, at 10 min after T4 infection. This decrease required the expression of T4 late genes and was attributed to a change in the initiation-factor-free ribosomes. Since the sedimentation profile of the rRNAs on sucrose gradients did not change, it was judged probable that the alteration in ribosome function is due to a change in the protein portion of the ribosome.

T4 proteins on ribosomes. The demonstration by several laboratories that T4-specific proteins are found on ribosomes has provided further presumptive evidence for T4 regulation of translation. Smith and Haselkorn (1969) found significant amounts of T4-specific, ^{35}S-labeled protein tightly bound to both 30 and 50S ribosomes displayed in sucrose gradients. Dube and Rudland (1970) used disc gel electrophoresis in sodium docedyl sulfate (SDS) to show that, after T4 infection, two new, ^{35}S-labeled proteins (about 37,000 and 28,000 daltons) are found in the ribosomal factor fraction, and a major one (about 48,000 daltons) is associated with salt-washed ribosomes. Rahmsdorf et al. (1973) observed three or four T4 proteins on factor-free ribosomes isolated from ^{14}C-amino acid-labeled, T4-infected cells; the proteins were separated on urea-containing polyacrylamide gels without SDS, so their sizes were not determined. Wiberg and Cardillo (unpublished data) have observed two very small T4 proteins (<10,000 daltons) tightly bound to sucrose gradient-fractionated ribosomes isolated from ^{14}C-amino acid-labeled, T4-infected cells and displayed by SDS-polyacrylamide gel electrophoresis. They are early proteins, as they are produced in the absence of late-gene expression. Snustad (personal communication) has confirmed this observation and, in addition, has observed three larger T4 proteins on ribosomes, at least two of which appear to be smaller than those of Dube and Rudland (1970). Snustad's five proteins are all on the 30S ribosomal subunit. Thus, at least seven T4 proteins have been found tightly bound to ribosomes.

Wiberg and Laiken (unpublished data) have found that the following T4 deletion mutations do not cause loss of either of the two smallest T4 proteins from ribosomes: (32–63)1, (32–63)9, and (39–56)12 (Homyk and Weil, 1974). Snustad (personal communication) has found that the following T4 deletion mutations do not cause loss of any of his five T4 proteins from ribosomes: PT8 (Bruner et al., 1972), *far*P13 (Chace and Hall, 1975), (39–56)12 (Homyk and Weil, 1974), and *ip*^{0}, a triple amber mutant defective in *ip*I, *ip*II, and *ip*III. Once the genes controlling these ribosome-bound T4 proteins are identified, the way will be open to learning their functions.

Chace and Hall (1975) isolated T4 *regB* mutants on the basis of their resistance to folate analogs. The mutants overproduce dihydrofolate reductase by producing it for longer than normal, and do so even in the presence of rifampin, indicating that the overproduction is a post-transcriptional event. Champney (1980) has isolated and characterized temperature-sensitive mutants of *E. coli* that have alterations in ribosomal protein S10, S15, or L22. D. Hall and C. E. Smith (Abstr. 11th Int. Congr. Biochem., Toronto, 1979, p. 86) found that T4$^+$ can grow at high temperature in some of these *E. coli* mutants if the early part of the infection is at low temperature; this suggested that T4 protein made early in the infection alters the ribosome. Three different *regB* mutants (all deletions) grow very poorly at high temperature on two of Champney's mutants of *E. coli* that contain different alterations in the L22 protein. After infection of one of these ribosome mutants at high temperature, there is reduced expression of delayed-early genes by both T4$^+$ and a *regB* mutant, but there is reduced expression of an immediate-early gene only by the *regB* mutant (Hall, personal communication). These results support the previous suggestion that the *regB* product may affect ribosomes.

CONSTITUTIVE AND REGULATED TRANSLATION OF T4 mRNA

It is known for practically all of the bacteria and phages that have been so studied that not all mRNA species of an organism are translated at the same rates either in vivo or in vitro. At least two elements of mRNA structure contribute to this selectivity of translation: intrinsic strengths of different ribosomal binding regions on the mRNA, and the accessibility of these regions to ribosomes. In T4-infected cells, mRNA availability for translation is also regulated by control of the temporal sequence of appearance of mRNA, i.e., transcriptional control (Brody et al., this volume; Rabussay, this volume), and by differential mRNA degradation. In a later section we describe T4 mutants that exhibit reduced levels of mRNA decay and altered competition in the mRNA pool for translational sites of the infected host. Here, we comment on the intrinsic features of mRNA structure that can contribute to differential translational rates.

Ribosomal binding regions for mRNA species from a variety of procaryotic sources have now been sequenced (see Gold et al., 1981, Steitz, 1979, for recent reviews). The efficiencies of these regions in initiating translation by ribosomes have been evaluated within the context of the positions of two nucleotide sequence domains: the initiation codon (AUG, sometimes GUG, and occasionally UUG) and a three- to nine-base-long sequence that is located 5′ to the initiator codon and is complementary to part of the 3′-hydroxyl terminus of 16S rRNA, i.e., the Shine-Dalgarno sequence (Shine and Dalgarno, 1974; Shine and Dalgarno, 1975 [see Table 2]). The two sequences and the spacing ("window") between them are very important determinants for initiation, though not the only ones. Mutations in both sequences can severely reduce or eliminate translation of the message (Belin et al., 1979; Dunn et al., 1978; Hoess et al., 1980; Nelson et al., 1981). The space between the two sequences seems to vary little among mRNAs (five to nine nucleotides), and mutations that alter this distance also reduce translation (Gold et al., 1981). In an analysis of the ribosome-binding region of T4 *r*IIB mRNA, L. Gold and his coworkers (Nelson et al., 1981; Pribnow et al., 1981; Singer et al., 1981) noted that mutations which narrowed the window between the Shine-Dalgarno sequence and the AUG initiator (from the normal five bases down to four or three bases) severely reduced translation. Mutations that increased the spacing by one or two bases appeared to have only small effects on *r*IIB translation; the effects of further expansion of this window could not be tested because the appropriate mutants were not available. Such mutants, however, are being generated in Gold's laboratory by the use of in vitro mutagenesis techniques with recombinant DNA clones of specific segments of the T4 *rIIB* translation initiation region (Singer et al., this volume). In addition, the Gold group used a novel statistical approach to analyze the ribosome-binding sequences from a large number of sources, asking whether other determinants besides the Shine-Dalgarno sequence, initiator, and window exist. Their analyses suggest that such determinants do occur 5' and 3' to the Shine-Dalgarno sequence and initiator AUG, respectively (Stormo et al., 1982). Such analyses should prove helpful in planning the genetic studies that will be necessary to test mRNA sequence domains for roles in the initiation process. The T4 *rII* region is one of the best currently available genetic systems for planning such studies.

In addition to primary structure determinants, secondary (and probably higher-order) structures can make an initiator more or less accessible to ribosomes and thus influence the efficiency of mRNA utilization. The effects vary from one mRNA species to another, depending on size and interactions between sequences occurring at long distances from each other on the mRNA. The genomes of the RNA phages are well-studied examples. Intramolecular interactions that determine the shape of the MS2 RNA molecule restrict initiation of translation to the middle (coat protein) cistron, which is apparently the only one of the phage cistrons left accessible to ribosomes (Fiers et al., 1975; Fiers et al., 1976; Lodish, 1975). Translation of the other cistrons is initiated as middle-cistron translation unfolds the molecule. So, intramolecular interactions can play a regulatory role in determining the temporal expression of genetic information on polycistronic mRNA. Such interactions might also prevent potential false starts in internal coding sequences that resemble normal initiator regions of cistrons (Lodish, 1970). Documented cases of regulation of T4 mRNA translation by nucleic acid folding and unfolding are not abundant, although we suspect that such controls do have roles in T4 phage development. For example, the T4 *rIIB* gene is transcribed in two modes that yield polycistronic and monocistric mRNA species (Brody et al., this volume). The monocistronic transcript is the more active of the two in *rIIB* protein synthesis. Similarly, overlapping modes of transcription for different genetic clusters on T4 DNA, which become activated at different times after phage infection, may be a means of regulating, sequentially, the accessibility of initiator regions of specific mRNAs to ribosomes.

In summary, the intrinsic structural features of an mRNA species are important determinants of its functional potential, but the extent to which this potential is realized also depends on interactions involving the structural features of other mRNA species in the same translational pool. We expect translation of all mRNAs to be subject to this form of regulation.

SPECIFIC TRANSLATIONAL REPRESSORS AND THE COMPONENTS OF REPRESSION

For many procaryotic mRNAs the level of translation may be determined largely by the intrinsic strength of the ribosome-binding site, without intervention by specific regulatory molecules. Some mRNAs, however, are known to be specifically regulated. The list of such mRNAs is small, but the realization that protein repressors of specific translation even exist is rather recent, and it may turn out that such regulators of gene expression are ubiquitous. Two systems of specific translational repression are being studied in T4: autogenous translational repression by the T4 helix-destabilizing protein produced by gene 32 (Alberts and Frey, 1970; Gold et al., 1976; Krisch et al., 1974; Lemaire et al., 1978; Russel et al., 1976) and repression of translation of many T4-induced early mRNAs mediated by the product of phage gene *regA* (Karam and Bowles, 1974; Karam et al., 1977; Karam et al., 1981; Sauerbier and Hercules, 1973; Trimble and Maley, 1976; Wiberg et al., 1973). The gene *regA* protein is also subject to autogenous regulation (Cardillo et al., 1979), but whether this is at the level of transcription or translation is not known. Regulation of the T4 gene 32 protein and its biological implications are discussed in detail elsewhere (von Hippel et al., this volume), so we will only comment on it briefly here. We also find it useful to discuss translational regulation in the gene 32 and *regA* systems in comparison to other cases of translational repression that have been identified in *E. coli* and the RNA phages.

In *E. coli*, groups of ribosomal proteins are encoded by different operons, some of which also specify nonribosomal components (Nomura et al., 1977). Some of the ribosomal proteins repress, at the translational level, the operons (or portions of the operons) that encode them (see Nomura et al., 1982, for a recent review and references). This "feedback" regulation coordinates the rate of ribosome assembly with the rates of synthesis of the ribosomal proteins and maintains the proper molar ratios of these proteins relative to one another. The 16 and 23S rRNAs serve as primary ligands for key regulatory ribosomal proteins. When the available sites on these ligands are filled (ribosome assembly), free regulatory proteins interact with target sites on their secondary ligands (their own mRNAs) and repress further accumulation of unassembled ribosomal components.

Many insights about the nature of interactions between translational regulatory molecules and RNA have been derived from studies with RNA phages. In the case of MS2 (R17) phage, for example, coat protein (middle-cistron product) accumulates after infection and represses replicase synthesis (distal-cistron product), thus providing a balance between the amounts of RNA and coat to be assembled into phage (Lodish, 1975). It has been shown that coat protein can bind to a region 5' to the initiation codon of the replicase gene and protects this region from RNase T1 attack (Bernardi and Spahr, 1972). Physicochemical measurements on the isolated coat-protected region indicated that it can exist in a structure consisting of two hairpins, one of which holds the replicase initiation domain (Shine-Dalgarno sequence and AUG) in a sequestered state (Gralla et al., 1974). Binding of coat protein to this loop may stabilize it so that the replicase initiation region remains inaccessible to ribosomes. The other hairpin loop (5' to the replicase AUG–Shine-Dalgarno loop) seems to play a role in facilitating ribosome entry into the replicase initiation region (Borisova et al., 1979). So, the long-range and short-range intramolecular interactions that determine the shape of mRNA molecules and the accessibility of ribosome-binding sites can be influenced via the intervention of repressor proteins.

Although secondary and tertiary structural features constitute an essential element in the repression mechanisms used by RNA phages, it is also possible to use the lack of secondary structure as the feature recognized by a specific repressor. The one known probable example is the gene 32 system of T4 phage. Synthesis of the T4 gene 32 protein is autogenously regulated at the translational level (Gold et al., 1976; Krisch et al., 1974; Lemaire et al., 1978; Russel et al., 1976). This protein is known to bind preferentially and cooperatively to single-stranded nucleic acid (Alberts and Frey, 1970) and to play important roles in phage DNA replication, recombination, and repair (see Lemaire et al., 1978, and Williams and Konigsberg, this volume, for references). The available evidence on gene 32 autoregulation and the recently derived sequence of the ribosome-binding domain of this gene (Krisch et al., 1980) are consistent with the following model that was proposed by L. Gold and coworkers (Gold et al., 1981; Lemaire et al., 1978). In the infected cell, the gene 32 protein binds primarily to single-stranded DNA. When all such DNA is covered, the protein begins to find single-stranded targets on mRNA, with the transcript of gene 32 constituting an especially good target because the nucleotide sequence surrounding the translation initiation region contains a long, unstructured domain that can allow translational repression to occur via cooperative binding of gene 32 protein to it (Krisch and Allet, 1982; Krisch et al., 1980). These interactions of the helix-destabilizing protein allow for control of the involvement of this protein in phage DNA replication, genetic recombination, and DNA repair. The region surrounding the gene 32 mRNA initiator also contains several repeats of the sequence UUAAA (Krisch and Allet, 1982; Krisch et al., 1980), which may mean that there is sequence recognition by the gene 32 protein and that the "unstructured" nature of the region makes this sequence accessible to the repressor.

HIGHER ORDERS OF SPECIFIC TRANSLATIONAL REPRESSION: THE T4 *regA* GENE

In the systems of translational repression so far mentioned, each protein repressor was observed to be specific to only one molecular species of mRNA. For example, in the most complex case, synthesis and assembly of ribosomal proteins, no polycistronic mRNAs from two different operons shared the same repressor (Nomura et al., 1982). Are there translational repressors that recognize different molecular species of mRNA? The protein product of T4 gene *regA* may be just that kind of repressor.

The T4 *regA* gene product, a peptide of about 12,000 daltons, is an early phage function that plays a role in the utilization of a subpopulation of T4 early transcripts and also displays autogenous regulation (Cardillo et al., 1979; Karam and Bowles, 1974; Wiberg et al., 1973). Loss of this phage function results in functional stabilization of many of these early mRNAs and in hyperproduction of their translational products. There is also underproduction of some T4-induced early proteins in *regA*$^-$ infections (Karam and Bowles, 1974; Wiberg et al., 1973). Table 1 lists the T4-induced early proteins whose synthesis is known to be affected and those whose synthesis is known to be unaffected in *regA*$^-$ infections. T4 *regA* mutations affect the expression of many unlinked T4 early genes (Table 1; also refer to T4 map, Kutter and Rüger, this volume) without affecting the transcription of these genes (Karam and Bowles, 1974; Sauerbier and Hercules, 1973; Trimble and Maley, 1976). It should be emphasized, however, that much of what we know about the effects of *regA*$^-$ mutations is based on results from physiological experiments that, in some cases, did not distinguish too clearly between specific and indirect effects of these mutations on the synthesis of individual phage-induced proteins (see footnotes to Table 1).

A possible role for the *regA* protein is that of a translational repressor that recognizes a structural feature common to all the mRNAs subject to *regA*-mediated control. Recent studies by Karam et al. (1981) strongly suggest that this structural feature is a short nucleotide sequence within the ribosome-binding domain of a *regA*-controlled mRNA and that no higher-order structural features of mRNA are required in the recognition. By using a genetic approach, these investigators identified the sequence 5'-(AUG)UACAAU-3' in the initiation region of T4 *r*IIB mRNA as being part of the "target" for *regA*-mediated control of *r*IIB protein synthesis (the initiation codon is shown in parentheses). DNA sequencing determinations revealed a similar sequence, 5'-AUUAC(AUG)-3', in the initiation region of the *regA*-regulated T4 gene 45, but no genetic evidence is available in this case to implicate the sequence in translational control (Spicer et al., 1982). The DNA sequence information derived by Spicer and Konigsberg (this volume) also suggests that gene 44 harbors a variation of this sequence, but that gene 43, which is not sensitive to *regA*-mediated control, does not. The T4 gene 32 initiator region, which is regulated by gene 32 protein and not by *regA* protein, also does not bear a *regA*-like target as compared with the gene 45, gene 44, and *r*IIB sequences. Table 2 compares sequences of ribosome-binding regions for some T4 early mRNAs in relation to the response to *regA*$^+$ gene function. Clearly, adequate identification of *regA* target sites on T4 mRNAs will require genetic localization (precisely mapped target lesions) as well as nucleotide sequence determination. So far, this has been accomplished only in one case, that of T4 *r*IIB mRNA (Karam et al., 1981). The *r*IIB target is short (7 to 10 nucleotides long) and is AU rich. Since T4 DNA is rich in AT base pairs, it is likely that nucleotide sequences similar to the *r*IIB *regA* target sequence will occur frequently in phage-induced mRNAs, and there must be some means of preventing unwanted interactions between *regA* protein and such sequences. Possibly, unwanted target sequences are sequestered by secondary and higher-order mRNA structures, or diffusible factors may work in conjunction with *regA* protein to enhance its selectivity for true targets.

The genetic and nucleotide sequencing data so far accumulated suggest that although *regA*-sensitive targets on mRNAs may have similar sequences, it is almost certain that these targets will exhibit function-

TABLE 1. Effect of *regA* mutations on production of various T4 proteins

T4 gene[a]	Protein	Effect[b]	References[c]
1	Deoxyribonucleotide kinase	+	14, 15
30	DNA ligase	−	8, 13
32	Helix destabilizing	0	4, 7, 8
39	DNA topoisomerase	+[d]	7, 11
41	DNA helicase/primase	−	8
42	dCMP hydroxymethylase	+	4, 7, 14, 15
43	DNA polymerase	0 to −	5, 7, 8, 10, 15
44	DNA polymerase accessory	+	1, 7
45	DNA polymerase accessory	+	1, 7, 8, 15
46	DNA exonuclease	+[d]	7, 11, 14, 15
52	DNA topoisomerase	+[d]	7
56	dCTPase	+	4, 14, 15
62	DNA polymerase accessory	+	1, 2
63	RNA ligase and tail fiber attachment enzyme	+, −	6[e]
cd	dCMP deaminase	+	14
dam	DNA-adenine methylase	+	16
e	Lysozyme[f]	+	15
α-gt	α-Glucosyltransferase	+	16
β-gt	β-Glucosyltransferase	+	16
*ip*III	Internal protein III	− to weak +	4, 7, 8[g]
nrdA	Ribonucleotide reductase	0	3
nrdB	Ribonucleotide reductase	0	3
nrdC	Thioredoxin	+	3, 12
pk (*pseT*)	5′-Polynucleotide kinase and 3′ phosphatase	0	9
*r*IIA		+	7, 8, 14, 15
*r*IIB		+	7, 8, 14
td	dTMP synthetase	−	15

[a] All except gene *e* have been tested in the absence of late gene expression (i.e., usually DNA⁻).

[b] +, Overproduced; −, underproduced; 0, no effect.

[c] (1) B. Alberts, personal communication cited in Wiberg et al., 1973; (2) Barry et al., 1973; (3) O. Berglund, personal communication; (4) Cardillo et al., 1979; (5) K. Hercules, personal communication; (6) Higgins et al., 1977; (7) Karam and Bowles, 1974; (8) Karam et al., 1977; (9) P. McIntosh, personal communication; (10) N. Nossal, personal communication; (11) Sauerbier and Hercules, 1973; (12) Sjöberg and Söderberg, 1976; (13) Swift and Wiberg, unpublished data cited in Wiberg et al., 1973; (14) Trimble and Maley, 1976; (15) Wiberg et al., 1973; (16) H. Witmer, personal communication.

[d] In the studies by Karam and Bowles (1974) and Karam et al. (1977), *regA*⁻ mutations caused a general improvement in early-protein synthesis, but the effects on synthesis of the gene 44, gene 45, gene 62, *r*IIA, *r*IIB, and certain other phage-induced proteins were conspicuously strong and were clearly accompanied by functional stabilization of the transcripts for these genes (see also Karam et al., 1979). Karam interprets the effects on synthesis of the gene 39, gene 52, and gene 46 proteins to be due to indirect consequences of *regA*⁻ infections, i.e., healthier protein-synthesizing machinery and a competitive advantage for transcripts of these genes over some of the other mRNAs in the pool.

[e] R. Gumport (personal communication) and Wiberg and Laiken (unpublished) observe an underproduction of T4-induced RNA ligase in *regA*⁻ infections, whereas Higgins et al. (1977) and Karam et al. (1977) reported that this enzyme is overproduced in such infections. The reasons for the disagreement between the results from the different laboratories are not clear, but the SDS gel assays that were used by at least one of these groups (Karam et al., 1977) may not have resolved the gene 63 protein from neighboring hyperproduced proteins well enough to unambiguously show the effect on synthesis of this protein.

[f] Strong effect seen with late gene expression.

[g] In Fig. 4 of Karam et al., 1977, two of three *regA* mutants appeared to underproduce internal protein III at 42°C.

al differences. That this is probably the case is indicated by results from physiological studies with different T4 *regA* mutants. An examination of a variety of such mutants revealed that some differ qualitatively in the degree to which they alter the patterns of hyperproduced phage proteins in *regA*⁻ infections (Cardillo et al., 1979; Karam et al., 1977; Karam et al., 1982). An example is depicted in Fig. 1. Note that the T4 *regA md*73 lesion exhibited stronger effects on *r*IIB and weaker effects on gene 45 protein synthesis than did the *regA md*67 phage mutation. Such differences between different *regA* mutations probably reflect differential responses of the different mRNA *regA* targets; in this case, *r*IIB versus gene 45.

Biological significance of the *regA* gene. The likelihood that translational control by the T4 *regA* protein involves recognition of a consensus target sequence on several mRNAs makes this phage function a particularly interesting one, as it may be part of a coordinated mechanism for control of all early development in T4 phage infections. The biological significance of the *regA* gene, however, remains largely a mystery. In attempting to find explanations for the "global" nature of this regulatory mechanism in phage development, several properties of the *regA* gene and its protein product will have to be related to one another and to events that, at first glance, appear to be far removed from translational regulation.

(i) The T4 *regA* gene maps within a cluster of essential genes that encode DNA replication enzymes (Karam and Bowles, 1974; Wiberg et al., 1976 [see T4 map, Kutter and Rüger, this volume]). Defects in this gene, however, cause no adverse effects on phage DNA replication, although molar ratios of some replication enzymes become altered in *regA*⁻ infections. Does the *regA* protein participate in construction of the DNA

TABLE 2. Nucleotide sequences of known and possible *regA* target sites on some T4 mRNAs

T4 gene	Translation initiation region[a]	Response to *regA* control[b]
*r*IIB	CUAA UAA GGA AAA UU (AUG) UAC AAU AUU AAAU	+
45	AUUG AAG GAA AUU AC (AUG) AAA CUG UCU AAAG	+
44	UUGA AUG AGG AAA UU (AUG) AUU ACU GUA AAUG	+
43	AACU AAG GAA UAU CU (AUG) AAA GAA UUU UAUC	0
e	UACU UAG GAG GUA UU (AUG) AAU AUA UUU GAAA	Unusual
*ip*III	AUUU AAA GGA AAC AU (AUG) AAA ACA UAU CAAG	0
32	UUA AAA AGG AAA UAA AA (AUG) UUU AAA CGU AAAU	0

[a] The nucleotide sequences shown were compiled from several sources: gene *r*IIB, Belin et al., 1979, and Pribnow et al., 1981; genes 45 and 44, Spicer and Konigsberg, this volume; gene 43, Spicer and Konigsberg, this volume, and Trojanowska and Karam, unpublished data; genes *e* and *ip*III, G. R. Smith, personal communication; gene 32, Krisch et al., 1980. The initiation codon is in parentheses, the Shine-Dalgarno sequence is overscored, and the presumed *regA* target sequence is underscored.
[b] +, Overproduced; 0, no effect.

replication complex in vivo? At growth temperatures between 43 and 45°C (but not at 30 to 37°C), *regA*⁻ mutants exhibit DNA synthetic rates of about 80% of normal rates, but DNA packaging and viable-phage production are reduced severalfold; furthermore, certain of the proteins in each of the three morphogenic pathways are underproduced, and proteolytic cleavage of head-related proteins is greatly reduced (Mowrey-McKee and Wiberg, unpublished data; Wiberg et al., 1973). Could the *regA* protein be a component of the multienzyme complexes that couple phage DNA replication to late transcription and DNA packaging? Alternatively, are there temperature-dependent differential effects on competition between T4-induced early and late mRNAs?

(ii) T4 *regA* mutants are partially defective in the ability to carry out host DNA breakdown; *regA* mutants are typically hydroxyurea sensitive (Wiberg et al., 1973). Is the *regA* protein directly involved in host nucleoid destruction, or do *regA*⁻ mutations simply cause underproduction of T4 unfoldase or DNase?

(iii) The T4 *regA* protein appears to be associated with an RNA-cell membrane component in *regA*⁺ infections (Karam et al., 1982). What is the RNA component? Chromatographic and electrophoretic analyses of radiolabeled *regA* protein suggest that it is a moderately basic protein (pI ~ 8.2) (Alford and Karam, unpublished data). Does the protein bind directly to an RNA target?

(iv) The recently derived nucleotide sequence of the T4 *regA* gene (Spicer and Konigsberg, personal communications; Trojanowska and Karam, unpublished data) reveals no recognizable *r*IIB-like *regA* target sequence in the presumed ribosome initiation region of the *regA* transcript. What, then, is the target for autogenous control of T4 *regA* protein biosynthesis, and is it indeed on the *regA* transcript?

(v) Derepression of translation of *regA*-regulated mRNAs in *regA*⁻ infections is accompanied by message stabilization (Campbell and Gold, personal communications; Karam et al., 1977). Does *regA* protein initiate mRNA breakdown directly at the target sequence, or is mRNA decay in *regA*⁺ infections an indirect consequence of the inhibition of specific translation? The latter possibility is currently favored because at least in one case (the T4 gene 1 deoxyribonucleoside monophosphate kinase transcript), the mRNA was found to be intrinsically stable as well as sensitive to inhibition by *regA*⁺ function (Trimble and Maley, 1976).

(vi) Since some T4-induced enzymes are underproduced in *regA*⁻ infections, it is conceivable that *regA*

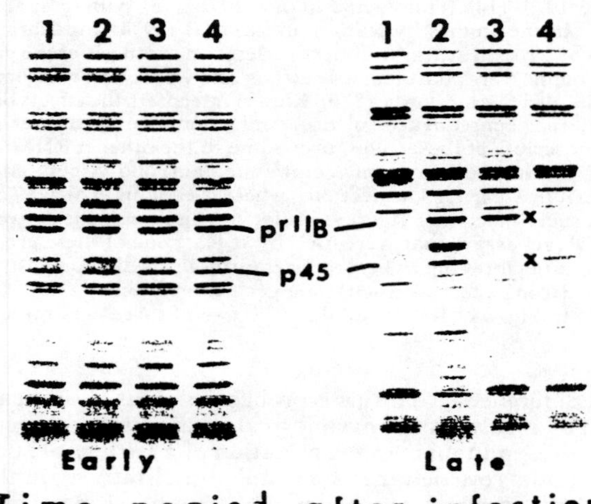

FIG. 1. Differential responses of the T4 gene 45 and *r*IIB *regA* targets to different *regA*⁻ mutations. T4 *regA*⁺ and *regA*⁻ mutants bearing *am* mutations in genes 33, 42, 44, and 55 were used to infect *E. coli* B^E (Sup^0) cultures. Growth was at 30°C in M9 medium. At 5 to 10 min (early period) and 25 to 30 min (late period) after infection, samples of the infected cultures were labeled with ^{14}C-amino acids and then analyzed by SDS gel electrophoresis and autoradiography as described by Karam et al. (1977). The T4 *regA* alleles examined were *regA*⁺ (lanes 1), *regAR*9 (lanes 2), *regA md*73 (lanes 3), and *regA md*67 (lanes 4). ×, Differences in rates of synthesis of the *r*IIB and 45 proteins with the *regA md*73 and *regA md*67 infections.

protein binding to certain mRNAs results in stimulation rather than repression of translation (Wiberg et al., 1973). Is this the case, or can the underproduction be accounted for by mRNA competition for the limited number of ribosomes in the infected cell?

In one model for the role of *regA* protein in phage development, Campbell and Gold (1982) proposed that this protein binds to a site on the RNA primer for DNA replication and assists in the formation of the T4 "replisome." The primer is envisaged to be the primary ligand for *regA* protein; it sequesters the protein temporarily such that synthesis of some phage-induced enzymes, particularly the nucleotide-synthesizing enzymes, is derepressed. Removal of the primer and release of *regA* protein after the necessary enzymes are made available and DNA replication is underway establishes repression of unwanted mRNAs (the secondary ligands).

Another model, suggested by Trojanowska et al. (Abstr. Cold Spring Harbor Bacteriophage Meet., Cold Spring Harbor, N.Y., 1982, p. 35), envisages a different type of role for the *regA* protein and is based on the premise that rapid transcription of the T4 genome after its entry into the host cell may necessitate the use of translational fine-tuning mechanisms for optimal utilization of genetic information (Karam et al., 1982). The major role of the *regA* protein may be in regulating the relative concentrations of the various proteins that form the multienzyme complexes needed in DNA replication, biosynthesis and utilization of DNA precursors, and other cell membrane-associated transactions. By controlling one or more components in each of these processes, the *regA* protein could regulate much of early development. In this model, the primary ligand for *regA* protein is an RNA component of the host membrane-nucleoid complex, which is proposed to sequester *regA* protein shortly after the protein is synthesized. This would allow synthesis of certain essential T4-induced enzymes at a derepressed level. Eventual dissolution of the host nucleoid, a process that may involve direct participation by *regA* protein in addition to several other phage-induced enzymes (Snustad et al., this volume), may release the protein to repress and initiate decay of the mRNAs that are no longer needed for providing certain early enzymes.

FUTURE DIRECTIONS

Genes 32 and *regA* constitute two important translational regulatory functions in the life cycle of T4 phage, and analysis of the mechanisms that underlie the physiological roles of these two genes is likely to enhance our general understanding of procaryotic translational control considerably. So far, progress in elucidating the role of T4 gene 32 protein has been more rapid than in the case of the *regA* protein, partly because the helix-destabilizing protein can be overproduced and isolated in highly purified form much more easily than *regA* protein can. Once the *regA* protein purification hurdle has been cleared, it should be possible to test in vitro the activities of this regulatory substance on the physical and biological properties of specific mRNAs. It will also be important to characterize a large number of *regA* target sites on different mRNAs to gain insights about the differential effects that the *regA* protein exhibits on the translation of individual species within the T4 mRNA population. It is interesting to note that most of the translational repressors that have so far been identified serve as regulators of their own synthesis. This may be indicative of an evolutionary advantage for such repressors in procaryotes.

It is almost certain that post-transcriptional control functions other than gene 32 and *regA* proteins operate in T4-infected cells. Some of these functions may, like gene 32 and *regA* proteins, be directed at specific mRNA species; others may be directed at the host translational machinery where they exert differential effects on mRNA selection. Much of our insight about the existence of these translational regulatory factors comes from in vitro studies on isolated ribosomal and soluble fractions from T4-infected and uninfected *E. coli* as well as from studies on biological activities of T4-induced mRNAs isolated at different stages in the phage growth cycle. Considering the complexity of in vitro work with ribosomes (i.e., their sensitivity to small changes in pH, ionic conditions, etc.), clear answers about the effects of T4 infection on the host translational machinery may be impossible to attain without focused genetic studies. At present, there exists a large collection of interesting observations in need of explanations. Greater understanding of the intricacies of ribosome function in the T4-infected cell will probably derive from ongoing studies aimed at identifying all ribosome-associated, T4-induced proteins and the genes that encode them, followed by detailed physiological and biochemical analyses of mutations in these genes.

J.S.W. was supported by Public Health Service grant GM-21999 from the National Institute of General Medical Sciences, and under contract with the U.S. Department of Energy at the University of Rochester Department of Radiation Biology and Biophysics. J.D.K. was supported by Public Health Service grant GM-18842 from the National Institute of General Medical Sciences and by National Science Foundation grant PCM 82-01869.

Autoregulation of Expression of T4 Gene 32: a Quantitative Analysis

PETER H. VON HIPPEL,[1] STEPHEN C. KOWALCZYKOWSKI,[1]† NILS LONBERG,[1]‡ JOHN W. NEWPORT,[1]§ LELAND S. PAUL,[1] GARY D. STORMO,[2] AND LARRY GOLD[2]

Institute of Molecular Biology and Department of Chemistry, University of Oregon, Eugene, Oregon 97431,[1] and Department of Molecular, Cellular and Developmental Biology, University of Colorado, Boulder, Colorado 80309[2]

Gene expression, manifested as the orderly production of specific proteins of appropriate types and amounts in defined progressions, is regulated at virtually every step of mRNA synthesis and its translation into protein.

Exploration of the Jacob-Monod (1961) operon model, and its progressive modification and expansion as more complex patterns of control have been revealed experimentally, has demonstrated that differential regulation of the synthesis of particular mRNA molecules occurs at initiation, during elongation, and at termination of transcription. Control always involves some form of feedback, most simply through the metabolite-level-dependent binding of constitutively synthesized repressor or activator proteins to regulatory sites on the DNA (e.g., see Savageau, 1979). However, control may also be autogenous in nature in that such repressor or activator proteins may directly modulate the expression of their own structural genes, at either the transcriptional or the translational level (Goldberger, 1974; Savageau, 1979).

Steitz (1979) and Gold et al. (1981) have recently reviewed a large body of evidence that demonstrates that gene regulation at the initiation step of translation is quite general. In these systems the binding of control proteins or the presence of elements of mRNA secondary structure can modulate the access of ribosomes to ribosomal initiation sites and thus regulate the relative levels of expression of mRNA sequences coding for various proteins. In a particularly simple (in concept) version of such a system, the protein at issue binds specifically and reversibly to its own ribosomal initiation site ("translational operator") and thus controls its own synthesis. The autoregulation, by this mechanism, of the production of gene 32 protein in T4 phage infection represents the best-understood example of the operation of such a system and is described in this chapter.

In principle such control is particularly appropriate for the regulation of the free intracellular concentrations of proteins required in considerable quantity as structural elements of multiprotein "organelles" such as DNA replication complexes, ribosomes, etc. The structural protein is produced as needed and is incorporated into the organelle, typically by a self-assembly process based on coupled equilibria. When the organelle has been saturated with the protein at issue, further synthesis leads to an increase in the free intracellular concentration of this species. Ultimately a critical concentration is passed, and the protein binds to a regulatory site on its own mRNA, leading to reversible shutoff of synthesis. Economy of protein design suggests that this latter binding should involve generally the same interactions, and the same binding site(s), as are used in the functional binding of the protein as a structural component.

Clearly, while simple in concept, this scenario can lead to difficulties if the binding of the protein is relatively nonspecific, since (for example) this can result in uncontrolled binding to, and perhaps premature shutoff of, initiation sites for unrelated proteins as well. The necessary discrimination involves finely tuned systems of coupled binding equilibria. The gene 32 protein system is particularly well suited for demonstrating the possibilities and problems inherent in such control mechanisms. A similar treatment will be presented elsewhere, in which the principles of the gene 32 protein autoregulation system are also extended to show how these same approaches might be used to explain the autoregulation of synthesis of coordinately regulated *systems* of proteins, such as those involved in the structure and assembly of the ribosome (Fairfield and von Hippel, manuscript in preparation).

AUTOREGULATION OF T4-CODED GENE 32 PROTEIN SYNTHESIS

Gene 32 protein is an essential component of the T4 DNA replication, recombination, and repair systems (for recent reviews, see Doherty et al., *in* J. F. Kane, ed., *Multifunctional Proteins*, in press; Williams and Konigsberg, this volume). It plays a "structural" (as opposed to a catalytic) role, binding in saturating amounts to the single-stranded DNA (ssDNA) that is transiently produced in the essential intermediate stages of these processes. Genetic and biochemical studies have shown that the total amount of gene 32 protein produced in a phage infection depends directly on the amount of intracellular ssDNA present (Gold et al., 1976; Krisch et al., 1974). It has also been shown that the synthesis of gene 32 protein is regulated at the translational level (Lemaire et al., 1978; Russel et al., 1976).

In effect, intracellular control of the free concentration of gene 32 protein involves an orderly progression of binding events. All ssDNA sequences are saturated as the level of free protein increases initially. Only after this process is complete does the free intracellular protein concentration rise to a threshold level high enough to permit binding to the gene 32 mRNA "translational operator" site (Russel et al., 1976), resulting in the specific cessation ("repression") of gene 32 protein synthesis. In vitro experiments have shown that this level of free protein concentration is

† Present address: Department of Molecular Biology, Northwestern School of Medicine, Chicago, IL 60611.

‡ Present address: Department of Biochemistry and Molecular Biology, Harvard University, Cambridge, MA 02138.

§ Present address: Department of Biochemistry and Biophysics, University of California, San Francisco, CA 94143.

not sufficient to permit binding to translational initiation sites of other T4 mRNAs (Lemaire et al., 1978), to permit binding to the very large reservoir of double-stranded DNA present in the cell (Jensen et al., 1976; Newport et al., 1981), or to prevent the reannealing of double-stranded DNA after the replication process is complete (von Hippel et al., 1982).

A combination of biochemical (Lemaire et al., 1978) and physical chemical (Jensen et al., 1976; Kelly et al., 1976; Kowalczykowski, Lonberg, Newport, and von Hippel, 1981; Newport et al., 1981) experiments has provided the necessary data for a quantitative molecular description of the autoregulatory cycle responsible for the establishment and maintenance of physiological levels of gene 32 protein in T4 infection of *Escherichia coli*. These studies are summarized briefly below and are described in full detail by von Hippel et al. (1982).

Binding Parameters for Gene 32 Protein

The binding of a protein to a nucleic acid lattice can be described by three thermodynamic constants (McGhee and von Hippel, 1974): the binding-site size (n; in units of nucleotide residues covered per protein monomer bound), the intrinsic association constant (K; in units of M^{-1}), and the cooperativity parameter (ω; unitless). These parameters have been measured for the binding of gene 32 protein to a variety of single-stranded deoxyribose- and ribose-containing homo- and heteropolynucleotides as a function of salt concentration and temperature (Kowalczykowski et al., 1981; Newport et al., 1981). The results show that n is constant at 7 (± 1) nucleotide residues, that ω is constant at $\sim 2 \times 10^3$, and that K varies with nucleotide composition of the lattice, salt concentration, and temperature. These measurements have permitted us to calculate values of the effective affinity constant of gene 32 protein binding in the cooperative polynucleotide binding mode ($K\omega$) to ssDNA and RNA sequences either of known sequence or of average T4 DNA composition, under physiological conditions. We define these conditions as a temperature of 37°C and a salt concentration of 0.23 M NaCl; this salt concentration has been shown to be approximately equivalent, in terms of the strength of protein-nucleic acid binding interactions, to the actual intracellular ionic environment (Kao-Huang et al., 1977).

In Vitro Repression Experiments

Lemaire et al. (1978) have conducted experiments that demonstrate the translational repression of gene 32 protein in vitro, using a cell-free translation system containing a crude RNA preparation from T4-infected *E. coli* cells, and ribosomes, tRNA, and supernatant proteins derived from uninfected *E. coli*. The results of these experiments may be summarized as follows. (i) Gene 32 protein binds preferentially to a specific component of the RNA derived from T4-infected cells. Since shutoff is specific for the synthesis of gene 32 protein, this component must be a portion of the gene 32 mRNA. (ii) The abruptness with which shutoff occurs as a function of added gene 32 protein suggests that this repression (and the binding of the protein to the gene 32 mRNA that is assumed to be responsible for it) is cooperative in gene 32 protein concentration. (iii) ssDNA effectively binds gene 32 protein more tightly than does either ssRNA in general or the gene 32 mRNA translational operator site. (iv) The binding affinity of gene 32 protein for the gene 32 mRNA operator is larger than that for most other RNA constituents in the system and is comparable to that of (unstructured) poly(rU). (v) Double-stranded DNA, and also the other components of the cell-free translation system, bind gene 32 protein less strongly than does the gene 32 mRNA operator. (vi) The addition of gene 32 protein to levels that are three- to fourfold greater than required to halt gene 32 protein synthesis does shut off the synthesis of other T4 proteins in the cell-free translation system, suggesting that the gene 32 mRNA operator site differs only quantitatively (in terms of gene 32 protein binding) from translational control sites on other T4 mRNAs. These and other data can also be used to estimate that the free intracellular gene 32 protein concentration maintained in vivo (during T4 infection) is ~ 3 μM (von Hippel et al., 1982).

Calculation of In Vivo Gene 32 Protein Binding (Titration) Curves for Various Structured and Unstructured Nucleic Acid Targets

By using the known binding parameters for gene 32 protein to various nucleic acid sequences, titration curves for the binding of gene 32 protein to various potential nucleic acid targets under physiological conditions have been calculated (von Hippel et al., 1982). The results are fully and quantitatively compatible with the experimental facts outlined above and, together with the sequencing data of Krisch and coworkers (Krisch and Allet, 1982; Krisch et al., 1980), have permitted the definition of the gene 32 mRNA translational operator site.

A two-state calculation was used initially to determine the expected levels of binding of gene 32 protein to unstructured ssDNA and RNA lattices. The results showed that long ssDNA lattices of average T4 composition, unencumbered by secondary structure, would be expected to saturate at ~ 0.01 μM free gene 32 protein, whereas comparable RNA lattices would saturate at ~ 0.3 μM protein. Both types of lattice should thus be fully saturated at physiological gene 32 protein concentrations.

However, most nucleic acid sequences in the cell are partially or completely involved in secondary structure. As a consequence, the favorable (to binding) free-energy change (ΔG_{bind}) involved in the interaction of gene 32 protein with single-stranded lattices will be opposed by the conformational free energy ($\Delta G°_{conf}$) favoring the maintenance of partially double-stranded structures. This conformational free energy can be estimated by using the approach and parameters developed by Tinoco et al. (1973). As a consequence, higher free gene 32 protein concentrations are needed to saturate such initially structured nucleic acid lattices. Such calculations reveal (data not shown) that, because of its tighter binding to ssDNA lattices, physiological concentrations of gene 32 protein will saturate DNA lattices containing stem-loop structures with as much as 70% of the sequences involved in base pairing. Thus, virtually all secondary structure that might be expected to develop adventitiously in single-stranded regions during DNA replication should be removable by gene 32 protein at the controlled free in vivo concentration.

The situation for mRNA should be quite different. A variety of lines of evidence (see Gold et al. [1981] for a summary) suggests that mRNA secondary structure is crucial for biological activity and thus should not be "melted" by gene 32 protein. The calculated results (Fig. 1) are fully compatible with this expectation, showing that because of the lesser (relative to ssDNA) affinity of gene 32 protein for ssRNA, only very weak elements of mRNA secondary structure should be melted at physiological gene 32 protein concentrations.

Finite Nucleic Acid Lattice Effects

To this point, the calculations described above were carried out by using a two-state "infinite lattice" model. In this model it is assumed that the stem-loop regions (for example) for which binding curves are being calculated are already flanked by gene 32 protein-complexed sites. This means that every protein monomer bound will contribute a full "unit" of both intrinsic binding affinity (K) and binding cooperativity (ω) to the interaction. Thus

$$\theta = \frac{(K_{conf})(K_{bind})[P]^m}{1 + (K_{conf})(K_{bind})[P]^m} \tag{1}$$

where θ = the fraction of the lattice sites under consideration that have been saturated at free protein concentration [P], m = the length of the lattice sequence under consideration in protein monomer units ($m = N/n$; where n = the protein site size and N = the lattice segment length in nucleotide residues), K_{conf} = $[NA_{ss}]/[NA_{ds}]$ ($[NA_{ss}]$ and $[NA_{ds}]$ represent, respectively, the molar concentrations of open [single-stranded] and duplex [base-paired] nucleic acid lattice, in units of nucleotide residues), and

$$K_{bind} = (K\omega)_1(K\omega)_2 \ldots (K\omega)_m = \prod_{i=1}^{i=m} (K\omega)_i \tag{2}$$

We note that $K_{bind} = (K\omega)^m$ for infinite lattices of constant composition.

This model is quite appropriate for considering the titration by gene 32 protein of an mRNA segment containing a weak stem-loop structure (hairpin) or for "filling in" a single-stranded lattice segment comprising the transient "single-stranded window" in a moving DNA replication fork, but it is less valid for estimating the degree of saturation (within an mRNA molecule) of single-stranded regions that are flanked by elements of secondary structure too stable to be melted at the physiological gene 32 protein concentration. For such regions a finite lattice calculation needs to be made, where

$$K_{bind} = K_1(K\omega)_2(K\omega)_3 \ldots (K\omega)_m = K_1 \prod_{i=2}^{i=m} (K\omega)_i \tag{3}$$

We note that the finite lattice binding definition of K_{bind} (equation 3) differs from that for infinite lattice binding (equation 2) only by the loss of one "unit of ω", but for short sequences this loss can make an enormous difference in the resulting titration curve (Fig. 2). Figure 2 thus shows that, due to this finite-lattice effect, even totally unstructured mRNA sequences (of average T4 composition) do not bind gene

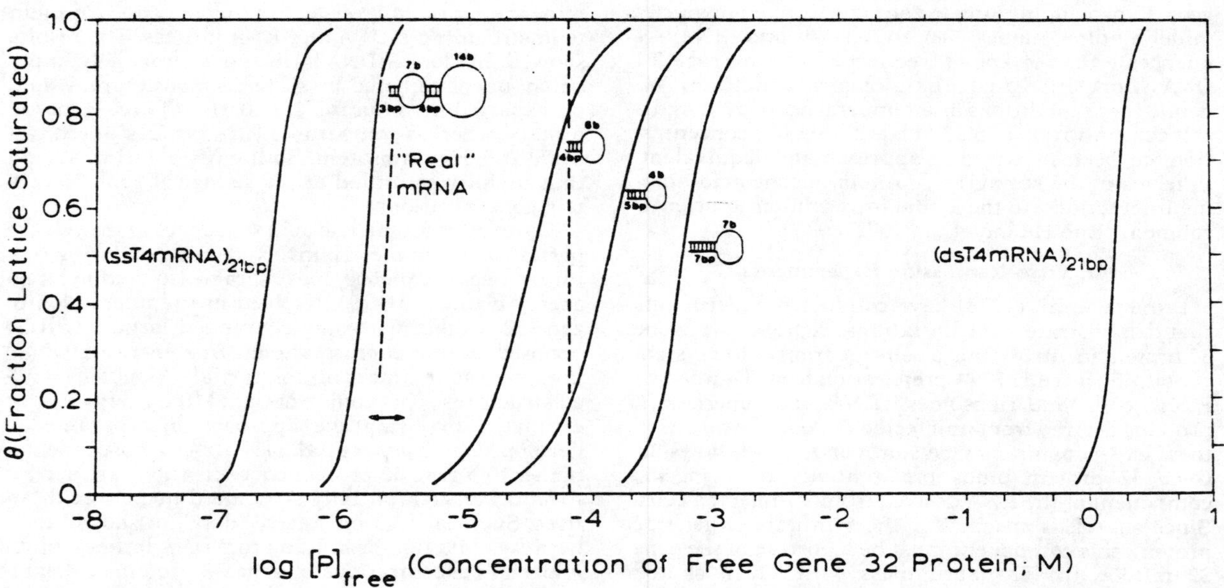

FIG. 1. Binding curves for the melting and complexation by gene 32 protein of various hypothetical initially looped and bulged T4 mRNA structures, plotted as a function of free gene 32 protein. The titration curves correspond, respectively, to the indicated stem-loop (and/or bulge) structures. The sloped dashed line labeled "real mRNA" is the approximate binding isotherm for the gene 32 mRNA control site, as estimated from the experiments of Lemaire et al. (1978). (Figure from von Hippel et al., 1982.)

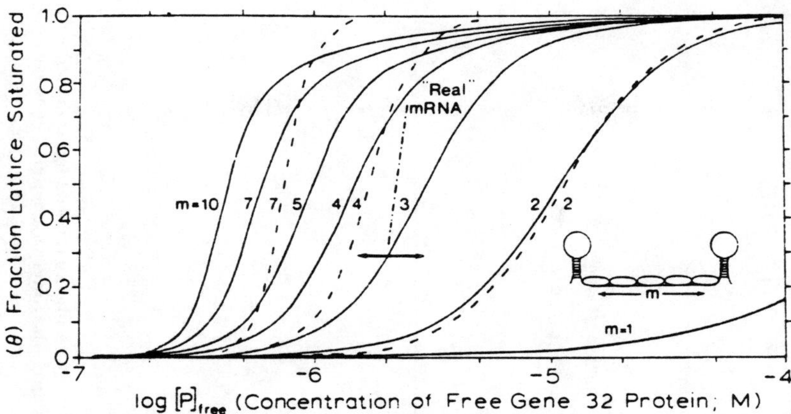

FIG. 2. Binding curves for the finite mRNA lattices of varying length. The dashed curves represent the two-state approximation, calculated as outlined in the text. The solid curves were calculated by the "exact" method of Epstein (1978); for further details see Newport et al. (1981). The lengths of the lattices are defined in units (m) of protein monomer binding sites. The site size of gene 32 protein binding cooperatively in the polynucleotide binding mode is seven nucleotide residues. Thus, the lengths of the respective finite lattices, in units of nucleotide residues, are $7m$. (Figure from von Hippel et al., 1982.)

32 protein under physiological conditions and protein concentrations if they have a lattice length (m) of less than 4 (~28 nucleotide residues). Furthermore, also due to this effect, even longer regions containing elements of weak secondary structure will remain uncomplexed. We expect that under physiological conditions the average mRNA molecule will be highly structured; thus, sequences that are sufficiently unstructured to bind gene 32 protein under intracellular conditions may be relatively rare.

The Gene 32 mRNA Translational Operator Site

The calculations above suggest that, in principle, the simplest way to define the gene 32 protein translational operator site, and to ensure that it saturates at lower free gene 32 protein concentrations than do control sequences on other T4 mRNAs, is to have the gene 32 mRNA operator consist of a uniquely unstructured segment, as originally proposed by Russel et al. (1976). The combination of the availability of $K\omega$ values for all of the relevant nucleic acid lattices, the recent sequencing of gene 32 mRNA by Krisch et al. (1980) and Krisch and Allet (1982), and the availability of a large T4 DNA sequence library (Gold et al., 1981; Schneider et al., 1982; Stormo and Schneider, unpublished data) now makes it possible to test this suggestion quantitatively.

The sequence surrounding the initiation codon of the gene 32 message is shown in Fig. 3. In mRNA sequences this region contains the information for translational initiation, and thus this general sequence clearly is the most logical candidate for the gene 32 mRNA translational operator site. This view is based on the simplest translational repression model, in which gene 32 protein (as repressor) competes with the ribosome for this operator-initiator site.

The sequence of gene 32 mRNA in the vicinity of the initiation codon is remarkable, even for a phage containing 66% adenine-plus-thymine residues. As Fig. 3 shows, the ribosome-binding site region contains a stretch of 40 nucleotides (residues 33 to 72, inclusive) in which the only bases other than A or U are the three nearly essential G residues of the Shine-Dalgarno sequence and the initiation codon (Gold et al., 1981). Values of $\Delta G°_{conf}$ have been computed for a variety of arbitrary segments within the gene 32 initiation sequence to determine whether an unstructured domain of sufficient length to serve as an operator site could exist in this region within the quantitative constraints outlined above. Some of the results are shown in Fig. 3. In essence, it was found that the longest (unstructured and partially structured) potential operator sequence that can be saturated under intracellular conditions and at the regulated gene 32 protein concentration is represented by line D in Fig. 3. This sequence is shown in the bound conformation (complete with stable flanking hairpins) at the bottom of the figure; it binds nine gene 32 protein monomers!

Is the Gene 32 mRNA Operator Sequence Unique?

It was also, of course, necessary to determine whether the proposed gene 32 mRNA operator sequence defined in Fig. 3 is unique. To this end calculations were carried out using the entire catalog of T4 nucleic acid sequences. The results showed that the proposed gene 32 mRNA operator has much less secondary structure than virtually any other sequences within the T4 sequence catalog (~5% of the total T4 genome). A comparison with more than 10 other T4 ribosome-binding sites showed none to be as unstructured as the proposed gene 32 mRNA operator (von Hippel et al., 1982).

The T4 Gene 32 mRNA Autogenous Regulatory System

The conclusions outlined above are summarized in Fig. 4 for the actual T4 system. Figure 4 shows, as required, that the actual ssDNA sequences of the T4 DNA replication complex (and presumably also those for T4 DNA recombination and repair systems) are met phe lys arg lys ser thr ala glu leu ala ala gln met ala lys leu asn gly (etc.)

GCTCATGAGGTAAAGTGTCATAGCACCAACTGTTAATTAAATTAAATTAAAAAGGAAATAAAAATGTTTAAACGTAAATCTACTGCTGAACTCGCTGCACAAATGGCTAAACTGAATGGCAATAAAGGTTTTTCTTCTGAAGATAAAGGCGAGT

aaa bbbb bbbbaaaccd ee ffffff ffffff eed cc ggggggghhh ii iihhh jjj jjj gggggg

$\Delta G^\circ_{conf(a-b)}$ = -5.2 $\Delta G^\circ_{conf(c-f)}$ = -3.6 $\Delta G^\circ_{conf(g-j)}$ = -14.6

line	nucleotide residues(N)	protein monomers(m)	ΔG°_{conf}
B	18	2	0
C	39	5	-2.4
D	65	9	-3.6
E	89	12	-8.8
F	130	18	-18.2

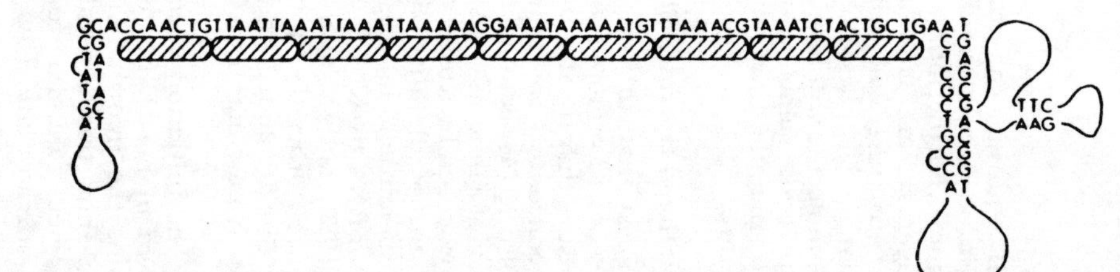

FIG. 3. Sequence and conformational stability of the putative gene 32 mRNA operator site and vicinity. At the top is the DNA sequence (noncoding strand only; the sequence as written corresponds to mRNA when T is replaced by U), with the beginning of the gene 32 protein sequence written above. The lower-case letters below the DNA sequence correspond to possible base-pairing interactions; i.e., the bases marked *aaa* can pair with the subsequent *aaa* sequence to form the stem of a hairpin structure, etc. The lines (labeled B through F) correspond to the segments tested as potential operator sites (see text). The structure at the bottom is the preferred operator sequence, drawn in a gene 32 protein-saturated conformation showing the proposed flanking hairpin termini. (Figure from von Hippel et al., 1982.)

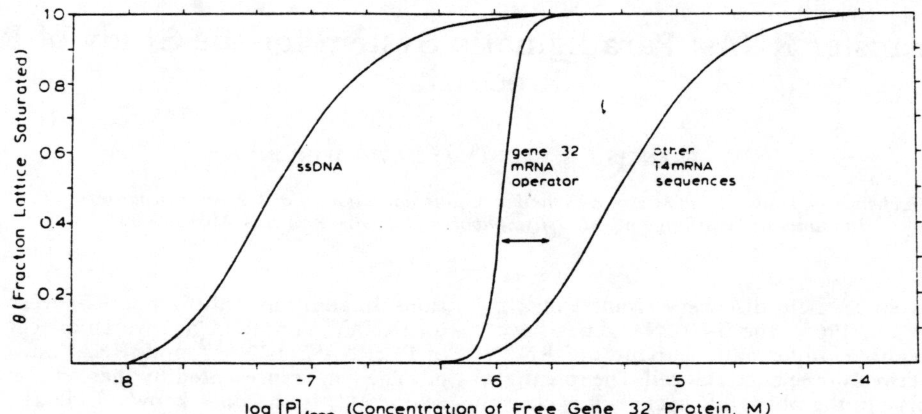

FIG. 4. Binding curves summarizing the gene 32 protein autoregulatory system. The ssDNA curve is calculated by using the real T4 DNA sequences with 50-residue lattice length (N) replication window and the infinite-lattice calculation mode. The gene 32 mRNA operator curve is calculated for the putative operator structure (Fig. 3, line D) shown at the bottom of Fig. 3. The "other mRNA" curve is calculated by using real T4 sequences with $N = 50$ and the finite-lattice approach. (Figure from von Hippel et al., 1982.)

saturated with gene 32 protein at concentrations well below the autoregulated value. The proposed gene 32 mRNA translational operator site then saturates quite sharply (cooperatively) at free protein concentrations just below the autoregulated level. As also required, other T4 mRNA initiation (ribosomal binding) sequences are not appreciably complexed at the maintained intracellular free gene 32 protein concentration.

SUMMARY AND OUTLOOK

The results presented here show that the regulation of expression of gene 32 of phage T4 can be modeled, using physical chemical binding data, to provide a quantitative and functionally economical picture of this system that is fully consistent with available biochemical and genetic information. The same approach, suitably modified to base control on "heteroprotein" (rather than on "homoprotein") cooperativity, appears to provide a useful way of thinking about the assembly and the autoregulation of synthesis of the components of the *E. coli* ribosome (Fairfield and von Hippel, manuscript in preparation; von Hippel and Fairfield, 1982) and, perhaps, the T4 replisome (Campbell and Gold, 1982). As further quantitative information is obtained about other T4 proteins, it may turn out that related approaches will help to explain their regulation as well (see, for example, the *regA* system [Karam et al., 1981; Karam and Wiberg, this volume]). However, it is already clear that the gene 32 system, per se, represents the simplest possible prototype; the control of expression of most of the other T4 genes will probably be much more interrelated and thus much more complex in quantitative detail.

This work was supported in part by Public Health Service research grants GM-15792 and GM-29158 (to P.H.v.H.) and GM-19963 (to L.G.) from the National Institutes of Health. J.W.N. and L.S.P. were predoctoral trainees on Public Health Service training grants GM-07759 and GM-70015.

We are very grateful to Henry Krisch and his colleagues for making the DNA sequence of the gene 32 region of phage T4 available to us before publication.

T4 Transfer RNAs: Paradigmatic System for the Study of RNA Processing

FRANCIS J. SCHMIDT[1] AND DAVID APIRION[2]

Department of Biochemistry, University of Missouri-Columbia, Columbia, Missouri 65212,[1] and Department of Microbiology and Immunology, Washington University School of Medicine, St. Louis, Missouri 63110[2]

In the years since their discovery (Daniel et al., 1968; Weiss et al., 1968), the T4 tRNA genes have provided a wealth of information about the RNA-processing functions of the bacterial cell. The role that these tRNAs play in the phage life cycle is less clear. Although they serve essential functions under some conditions, they are not required for growth in common laboratory strains of *Escherichia coli*. Nevertheless, since all of the T-even and several T-odd phages code for tRNAs, it is almost an article of biological faith that these RNAs have persevered for a reason. Recent studies have pointed out that there are reasons for the existence of phage tRNA species, but the detailed processes through which they fulfill functions are not completely defined. This chapter discusses some of the "natural history" of phage tRNAs but concentrates on the processing pathways by which these tRNAs are synthesized. We emphasize their application in understanding more widespread cellular processes.

T4 tRNA SPECIES

Identity and Structure

When bacteriophage T4 infects *E. coli*, eight new T4-coded species of tRNA are synthesized (McClain et al., 1972) (Fig. 1 and Table 1). The phage-coded tRNAs accept the amino acids glutamine, leucine, glycine, proline, serine, theonine, isoleucine, and arginine. They are arranged in this order from the 5′ to the 3′ end in the genome. The tRNA genes are followed by and coordinately expressed with genes for species 2(D) and species 1(C), two small RNAs of unknown function.

Suppressors

Genetic analysis of T4 tRNAs depends largely on the use of mutant tRNAs capable of suppressing nonsense mutations that cause premature chain termination of T4 protein genes. In T4, four independent informational suppressors have been isolated by Wilson, McClain, and co-workers. Numerous suppressor-negative point or deletion mutants in the tRNA genes are available as well (Comer et al., 1974; Foss et al., 1979; Kao and McClain, 1977; McClain, 1970; Wilson and Kells, 1972). Each of these four suppressors alters a different T4 tRNA species. Coupled with a mutant in $tRNA^{Ile}$ (see below), this means that five of eight tRNAs can be manipulated genetically. As might be expected, suppressor-negative point mutations in tRNA structures can interfere with correct processing of the appropriate precursors. These altered forms can then be analyzed to give some insight into the structural requirements of the tRNA biosynthetic reactions. In addition, unlinked antisuppressor mutations (of $tRNA^{Ser}_{UAG}$ or $tRNA^{Ser}_{UAA}$) which affect the synthesis of T4 $tRNA^{Pro}$, $tRNA^{Ser}$, and $tRNA^{Ile}$ have been isolated. The gene represented by these mutations, termed *mb* or *M1*, is the only known T4-coded function required for tRNA biosynthesis; all other su^- mutations appear to map within the tRNA transcription unit (McClain et al., 1973; Wilson et al., 1972).

Interaction Between Host and T4 tRNA Metabolism

Why does bacteriophage T4 have tRNAs in the first place? One clue is provided by consideration of the genetic code: since T4 DNA has a lower G+C content than *E. coli*, specialized tRNAs may be adapted to complement the available coding capacity of the host. As the first approximation, at least, this is correct.

When phage T2 infects *E. coli*, a single isoacceptor of host tRNA ($tRNA^{Leu}_1$) is cleaved by a phage-encoded endonuclease. This tRNA recognizes CUG, a rare codon in T2 DNA (Kano-Sueoka and Sueoka, 1968). In contrast, the T2 and T4 $tRNA^{Leu}$ species recognize the codon UUG, a less-abundant tRNA isoacceptor in the host. The model, which still seems valid, is that phage-coded tRNA species and the phage-coded alterations of host tRNA species serve to rapidly change over a cell's protein synthetic machinery, adapting it to serve viral purposes (Scherberg and Weiss, 1972; Sueoka and Kano-Sueoka, 1970).

Further analyses of these phenomena have been greatly aided by the availability of the CT (California Institute of Technology) collection of 600 natural *E. coli* isolates. Wilson (1973) found that of 36 isolates able to support the growth of wild-type T4, 4 would not grow T4 whose tRNA genes were deleted. This natural host range system has been exploited in several ways to study tRNA function in vivo.

Guthrie and McClain (1973) isolated mutants of T4 which did not grow on the isolate CT439. Some of these mutations mapped in the T4 $tRNA^{Ile}$ sequence; the mutation prevented synthesis of mature tRNA by interfering with its biosynthesis. Although at first sight the addition of a phage-coded $tRNA^{Ile}$ able to read the codon AUA (Guthrie and McClain, 1979; Scherberg and Weiss, 1972) rather than the major host isoleucine codons AUU and AUC would fit into this model, the situation is probably more complex. Isolate CT439 contains as much of the isoacceptor $tRNA^{Ile}_{AUA}$ as the laboratory B strain of *E. coli*, yet only the former restricts T4 phage mutated in the $tRNA^{Ile}$ gene (S. Nishimura, personal communication cited in Guthrie and McClain, 1979). Therefore, the functional requirement for this tRNA may be due to some other biochemical process. The nature of this process is

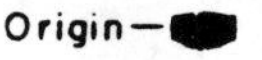

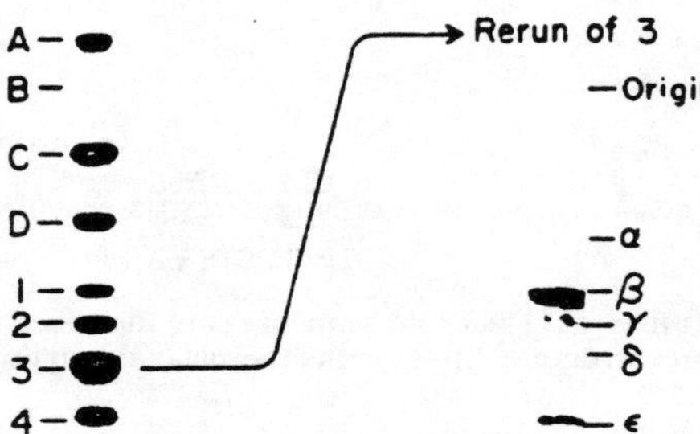

FIG. 1. Eight tRNAs synthesized by T4. Bands A and B are precursors to $tRNA^{Pro}$ and $tRNA^{Ser}$ and to $tRNA^{Thr}$ and $tRNA^{Ile}$, respectively. They and $tRNA^{Ser}$ (band 1), $tRNA^{Leu}$ (band 2), and $tRNA^{Gly}$ (band 4), along with stable RNA species C and D (1 and 2 in the nomenclature of Abelson et al., 1975), are separated by electrophoresis in a 10% polyacrylamide gel. Further electrophoresis of band 3 in a 20% gel separates it into five other tRNA species (see Table 1). From Guthrie et al., 1975.

unclear, but results discussed below point out that T4 $tRNA^{Ile}$ is unique in several respects.

The use of natural CT isolates which restrict certain T4 mutants has also been fruitful in defining some roles for other RNA metabolic functions. *E. coli* CTr5x (a B hybrid strain) and its parent strain CT196 restrict the growth of T4 mutants deficient in RNA ligase (*rli*) and polynucleotide kinase (*pnk*) (Depew and Cozzarelli, 1974; Sirotkin et al., 1978). The growth restriction seems to be due to the activity of a T4-coded endonuclease which cleaves the anticodon loop of host $tRNA^{Ile}_{AU(C,U)}$ (David et al., 1982). The tRNA of nonrestricting hosts is unaffected. In the restricting host, $tRNA^{Ile}_{AU(C,U)}$ does not contain a modified base in the third position of the anticodon; thus, its susceptibility to the T4-coded endonuclease may be a function of its lack of modification. (Note that T4 $tRNA^{Ile}$ is always modified [see below].) In this view, RNA ligase and polynucleotide kinase might serve as RNA repair enzymes (for further discussion, see Snyder, this volume).

A final aspect of the influence of T4 on host RNA metabolism is the alteration of the host valyl-tRNA synthetase by a T4-coded peptide. This peptide, the product of the *vs* gene, associates with and confers increased heat stability on the enzyme (Marchin et al., 1972; McClain et al., 1975). No restrictive hosts have been found for phage which lack this function, and its metabolic role is unclear (Comer et al., 1975). However, a recently described *E. coli* strain is nonpermissive for wild-type T4 but permissive for phage mutants impaired in the alteration process (Marchin, 1980).

SEQUENCE ORGANIZATION OF THE T4 tRNA GENE CLUSTER AND ITS IMPLICATIONS FOR tRNA BIOSYNTHESIS

Although many details of tRNA biosynthesis were known previously from analysis of individual RNAs before DNA sequence analysis began, the recently determined DNA sequence permits a review of the tRNA biosynthetic pathways in a more logical fashion. Sequence analysis of DNA from the tRNA gene cluster has been done by Fukada and Abelson (1980) and by Mazzara et al. (1981); the two sets of data obtained by the two groups using two methodologies are in excellent agreement. Before the DNA sequencing effort, most of the stable T4 RNAs had been sequenced by fingerprinting or in vitro transcription techniques (Table 1).

Cloning Studies

The T4 tRNAs are clustered in a single region on the chromosome, whose transcription is counterclockwise on the standard T4 map. In the map in Fig. 2, the tRNA genes are flanked by gene 57 on the left and gene *e* on the right (Wilson et al., 1972). Transcription of this region proceeds from a single complex promoter about 900 base pairs upstream from the 5′-proximal tRNA. The DNA between the promoter and the first tRNA gene codes for an internal protein (*ipI*). The tRNA genes are arranged in two groups: the first one contains tRNAs Gln, Leu, Gly, Pro, Ser, Thr, and Ile, and the second, about 500 nucleotides downstream, contains $tRNA^{Arg}$, species 2, and species 1 small

TABLE 1. T4-coded low-molecular-weight stable RNA species

RNA	Allele	References	
		Sequence	Genetics
Species 1(C)		Paddock and Abelson, 1973	
Species 2(D)		Fukada and Abelson, 1980; Mazzara et al., 1981	
$tRNA^{Ser}$	psu_a^+, psu, psu_1^+	Barrell et al., 1973	McClain, 1970; McClain et al., 1973; Wilson and Kells, 1972
$tRNA^{Leu}$	psu_3^+	Pinkerton et al., 1973	Foss et al., 1979
$tRNA^{Thr}$ (α)		Guthrie et al., 1979	
$tRNA^{Pro}$ (β)		Barrell et al., 1973	
$tRNA^{Gln}$ (γ)	psu_2^+	Seidman et al., 1974	Comer et al., 1974
$tRNA^{Ile}$ (δ)	*HA1*	Guthrie and McClain, 1979	Guthrie and McClain, 1973
$tRNA^{Arg}$ (ε)	psu_4^+	Mazzara et al., 1977	Kao and McClain, 1977
$tRNA^{Gly}$		Barrell et al., 1973; Stahl et al., 1974	

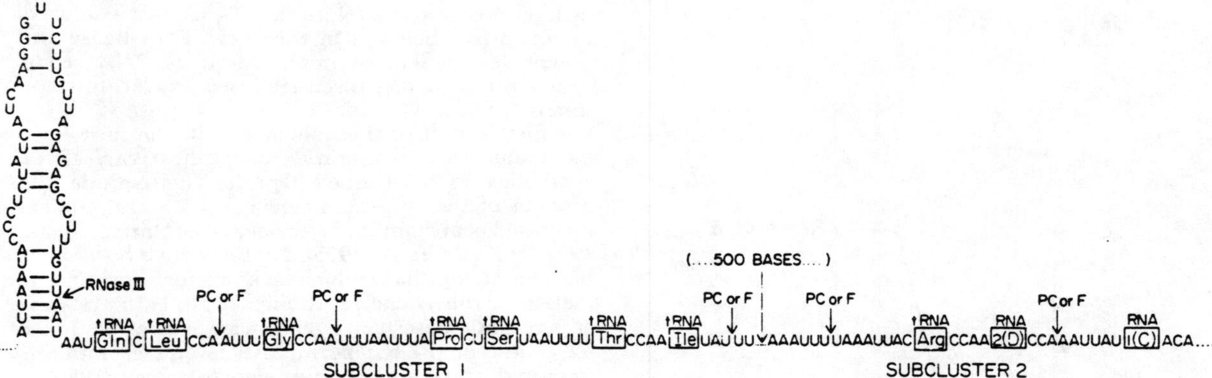

FIG. 2. Organization of the T4 tRNA genes. Transcription is from left to right. The tRNA sequences are identified by boxes, and the interstitial sequences are shown, arranged in a probable secondary structure. Sites of nuclease cleavage on the primary transcript are indicated.

RNAs. The RNA is terminated about 80 nucleotides from the 3′ end of species 1 RNA.

Cloning of the entire operon in plasmid vectors has not been achieved, arguing that excess expression of DNA from this region of the map is lethal to *E. coli*. When clones of the 5′ end of the gene cluster were isolated in λgt as minute plaques, spontaneous, faster-growing deletions and insertions arose at high frequency (Fukada and Abelson, 1980). These events removed the RNA polymerase-binding site of the tRNA gene cluster and about 500 base pairs of DNA downstream from this promoter. It has been suggested that lethality is due to overproduction of the T4 internal protein I, a DNA-binding protein found in the phage head. The gene for internal protein I apparently lies downstream from the promoter and would be removed by the deletions in the λgt clones. The T4 tRNA gene cluster thus furnishes a useful system for study of the processing of composite transcripts leading to both proteins and stable RNA. Although the upstream portion of the T4 tRNA cluster is difficult to clone, the downstream cluster of $tRNA^{Arg}$ and stable RNA species 1(C) and 2(D) has been isolated several times (Fukada, Gossens, and Abelson, 1980; Mazzara et al., 1981).

Transcription

Transcription of the T4 gene cluster can be done with purified RNA polymerase holoenzyme, nucleoside triphosphates, and phage DNA; no accessory factors or phage-coded products are necessary for this reaction. Several transcripts have been isolated by Daniel and co-workers (Goldfarb and Daniel, 1980; Goldfarb and Daniel, 1981a); these could then be processed by an S-100 supernatant from uninfected *E. coli* cells.

In vitro processing of the largest product yielded 9 of the 10 stable RNA species (all except $tRNA^{Ile}$ [see below]), showing that this transcript continued through both sets of tRNA genes. In contrast to this work, other workers reported that a coupled in vitro transcription-processing extract from uninfected cells was unable to synthesize $tRNA^{Pro}$ or $tRNA^{Ser}$ in addition to $tRNA^{Ile}$ (Kaplan and Nierlich, 1974). In vivo, synthesis of these three tRNAs requires the T4 *mb* (*M1*) gene product, and it was proposed that failure to synthesize these tRNAs reflected an in vitro requirement as well. An alternative explanation, that these tRNAs are unstable in the processing extract, seems to be favored by the results obtained with isolated precursors (Goldfarb and Daniel, 1980). Two other, shorter, transcripts of this region were also isolated; these yielded neither $tRNA^{Ile}$ nor RNAs from subcluster II. These latter RNA species, therefore, probably represent the products of termination after (or within) subcluster I (Goldfarb and Daniel, 1981a).

The nucleotide sequence (Fukada and Abelson, 1980; Mazzara et al., 1981) of the DNA corresponding to the stable RNAs (Fig. 2) shows the following. (i) The individual RNA species are separated by only a few (1 to 10) base pairs in the DNA. (ii) The sequences of individual RNA species reflect their modes of biosynthesis (see below for a description of the individual pathways). (iii) There is a strong termination point 77 base pairs downstream from the species 1(C) gene (Pragai and Apirion, 1982); this terminator is preceded by a sequence capable of forming a stable ($\Delta G^{0\prime}$ = −20.7 kcal [ca. −88.6 kJ] mol^{-1}) stem and loop. This terminator could explain why in vitro transcription of T4 DNA does not proceed into gene *e* (lysozyme). Since there is evidence that gene *e* is synthesized in phage-infected cells as a 6-kilobase transcript (Brody et al., 1971), i.e., encompassing the entire tRNA transcription unit, it is possible that this termination is not 100% efficient in vivo. Indeed, there is evidence for a certain amount of read-through (Pragai and Apirion, 1982). Figure 3a shows this terminator. (iv) A second potential terminator occurs within the distal half of the $tRNA^{Ile}$ gene, where a weaker ($\Delta G^{0\prime}$ = −3.3 kcal [ca. −14 kJ] mol^{-1}) stem-and-loop structure can precede a run of four to five T residues. It is too early to conclude that this terminator is responsible for the *rho*-dependent termination of transcription that Goldfarb and Daniel (1981a) observed in vitro between tRNA subclusters I and II, but its presence may account for some unusual features of $tRNA^{Ile}$ synthesis (see below). Figure 3 shows this stem and loop and the $tRNA^{Ile}$ structure.

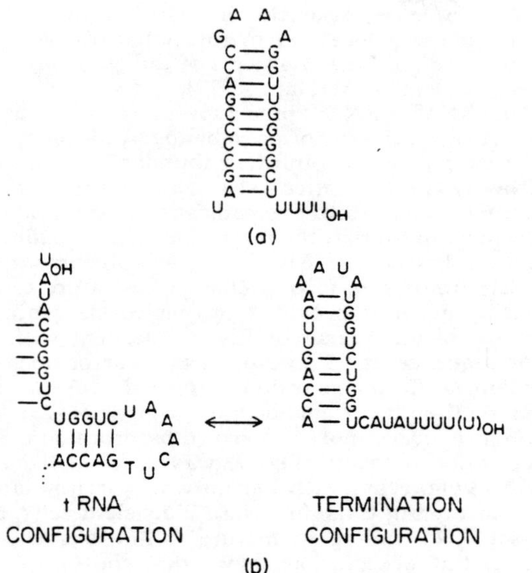

FIG. 3. Potential transcription terminators in the tRNA transcriptional unit. (a) A strong, *rho*-independent terminator distal to the gene for species 1(C) RNA. (b) A potential terminator within the T4 $tRNA^{Ile}$ sequence. The 3′ end of the $tRNA^{Ile}$ gene is followed by a U-rich sequence that is folded into a different conformation from that of the tRNA.

PROCESSING OF T4 tRNAS

Enzymes Required for tRNA Processing

Once (or while) the primary transcript is synthesized, host enzymes are required to mature the transcription product into tRNA and stable RNA species. These enzymes catalyze three types of reactions: cutting and trimming of the RNA, chain extension, and base modification (not necessarily in this order). Genetic and sequence experiments show several features of this processing. (i) T4 $tRNA^{Gln}$, $tRNA^{Pro}$, $tRNA^{Ser}$, and $tRNA^{Ile}$ genes do not contain the CCA_{OH} terminus common to all tRNAs, so this sequence must be added by processing. Since *E. coli* mutants deficient in tRNA nucleotidyltransferase (the *cca* locus) do not permit expression of $tRNA^{Ser}$ or $tRNA^{Gln}$ suppressors, this host enzyme is involved in CCA synthesis in vivo (Deutscher et al., 1974). (ii) Expression of $tRNA^{Ser}$ (but not $tRNA^{Gln}$, $tRNA^{Leu}$, or $tRNA^{Arg}$) suppressors is deficient in strain BN, which has been identified as deficient in a cellular exonuclease (Maisurian and Buyanovskaya, 1973; Schmidt and McClain, 1978b; Seidman, Schmidt, Foss, and McClain, 1975). This is consistent with the DNA and RNA sequence data, which show extra nucleotides at the 3′ end of this tRNA. (iii) Expression of the suppressor form of $tRNA^{Gln}$ is abolished in hosts deficient in *E. coli* RNase III (McClain, 1979); therefore, this enzyme must act near the proximal part of the tRNA gene cluster. The requirement for other enzymatic activities is indicated by in vivo accumulation of the appropriate tRNA precursors in the absence of the enzymatic activity. Temperature-sensitive RNase P mutants (Abelson et al., 1975; Guthrie et al., 1975) and RNase E mutants interfere with tRNA synthesis at high temperature, and so these enzymes must also be directly or indirectly required for tRNA synthesis. Finally, an enzyme has been partially purified by Goldfarb and Daniel (1981b) and termed RNase PC. This enzyme, when allowed to digest primary transcript produced in vitro, produced molecules which could be the $tRNA^{Pro}$-$tRNA^{Ser}$ and $tRNA^{Gln}$-$tRNA^{Leu}$ dimeric precursors. In addition, $tRNA^{Arg}$, species 1(C), and species 2(D) precursors were formed, although the cleavages were not 100% faithful. Another enzyme, RNase F, has been isolated which could introduce a cleavage between species 1 and the termination stem-and-loop structure (Watson and Apirion, 1981). These last two activities, RNases PC and F, could be identical.

From Transcript to Precursor

Analysis of the processing of the T4 tRNA species, like that of many other large RNAs, is complicated by the involvement of multiple enzymes acting on a multidomain structure. As a result, processing blocks (especially in vitro) are not always absolute, even in the complete absence of the relevant activity. Nevertheless, some conclusions can be drawn.

RNase III. Synthesis of the immature 5′ end of the $tRNA^{Gln}$-$tRNA^{Leu}$ precursor RNA is accomplished by RNase III. A long hairpin double-stranded region of the transcript can be formed. Sequence analysis of precursor forms synthesized in an *rnc* host shows that, in common with many other RNase III sites, cleavage occurs within this hairpin distal to a pyrimidine-rich sequence (Fig. 2) (Pragai and Apirion, 1981). In the presence of RNase III, about 50 nucleotides are removed in vitro or in vivo to give a precursor whose further maturation is independent of RNase III. The RNA on which RNase III acts is itself derived by an undetermined enzymatic step that separates the mRNA portion of the transcript from the nonmessenger portion of the transcript. At present we believe that this is accomplished by nonspecific degradative enzymes that are prevented from further digestion of the transcript by the special secondary-tertiary structural features of the nonmessenger part of the RNA. In the absence of RNase III, the 50-odd nucleotides that are usually removed by a specific RNase III cleavage are probably digested by a nonspecific nuclease(s). This happens since this RNA cannot be processed by the enzymes that normally do so after the cleavage by RNase III. Whereas the RNase III cleavage is six nucleotides from the 5′ end of $tRNA^{Gln}$ (the first tRNA in the cluster), the nonspecific nuclease(s) leads to a cleavage which is only two nucleotides from the 5′ end of $tRNA^{Gln}$. Subsequently another enzyme, probably RNase F or RNase PC, cleaves the transcript, leading to a $tRNA^{Gln\text{-}Leu}$ dimer. In the rnc^+ (RNase III^+) cells this dimer contains six extra nucleotides at its 5′ end and has two sites for RNase P at the 5′ end of each of the tRNAs. The site at the 5′ end of $tRNA^{Leu}$ is cleaved preferentially. In the shorter precursor that accumulates in the absence of RNase III, both of these sites are still accessible to RNase P but with reduced efficiency. The efficiency of the RNase P cleavage at the 5′ end of $tRNA^{Gln}$ is, however, sufficiently low that the degradative enzymes usually manage to degrade this $tRNA^{Gln}$ precursor before it can be processed by

RNase P (Gurevitz and Apirion, unpublished data). These findings explain why it is only the accumulation of the first tRNA in the cluster, $tRNA^{Gln}$, that is reduced in the RNase III^- host (McClain, 1979; Pragai and Apirion, 1981).

The processing to yield the 3′ end of the $tRNA^{Gln}$-$tRNA^{Leu}$ precursor is accomplished by another enzyme. This enzyme could be the PC nuclease described by Goldfarb and Daniel (1981b), which may be related to RNase F found by Watson and Apirion (1981).

RNase E. When an *rne* mutant deficient in RNase E is infected with T4, a large composite tRNA precursor accumulates whose 5′ end is identical to that produced by RNase III (Pragai and Apirion, 1982). In itself this might be taken to indicate that RNase E cleaves the RNase III cleavage product, but cell extract prepared from an *rne* mutant was still capable of correct processing to give the $tRNA^{Gln}$-$tRNA^{Leu}$ dimer. Two interpretations of this result are possible: (i) RNase E is normally required for correct processing of the RNase III cleavage product, but another enzyme(s) can do this in vitro; or (ii) RNase E is an accessory enzyme in this pathway. The possibilities cannot be distinguished with certainty, but the absence of RNase E appears to affect RNase P activity (Jain, Gurewiz, and Apirion, 1982). In vitro data, alone, cannot be used to predict processing pathways, however, and the question is still unresolved (see below).

The generation of other tRNA precursors, i.e., $tRNA^{Pro}$-$tRNA^{Ser}$, pre-$tRNA^{Arg}$, precursors to stable RNAs 1(C) and 2(D), and probably pre-$tRNA^{Gly}$, can be accomplished by nuclease PC in vitro, although these product RNAs have not been characterized by sequence analysis. The in vitro synthesis of $tRNA^{Thr}$-$tRNA^{Ile}$ dimeric precursor by a purified enzyme has not been reported (Goldfarb and Daniel, 1981b).

From Precursor to tRNA

When wild-type *E. coli* cells are infected with T4, several tRNA precursors in addition to the mature RNA species can be detected. Historically these precursors served as the first substrates for the elucidation of tRNA biosynthetic reactions, and many of the concepts which arose have guided further analysis of tRNA biosynthetic reactions.

Structurally the T4 tRNA precursors can be envisioned as a series of tRNA-like domains joined by linker regions. Evidence for this comes from the inability of base-change mutations to interfere with base modification of more than one tRNA in a dimer (McClain and Seidman, 1975) and from in vitro studies using S1 nuclease (Manale et al., 1979) as a structural probe. Furthermore, processing enzymes fail to recognize oligonucleotides spanning the cleavage site within a precursor (Schmidt et al., 1976).

Each of the similar T4 tRNA domains, however, possesses a unique maturation pathway, determined by combinations of requirements for removal of 3′ nucleotides either distal to or in place of the CCA_{OH} sequence, for synthesis of the CCA_{OH} sequence in whole or in part, and for RNase P to act on a CCA-containing precursor molecule. As the gene sequences are different, so too are the maturation pathways.

The $tRNA^{Pro}$-$tRNA^{Ser}$ precursor RNA. The $tRNA^{Pro}$-$tRNA^{Ser}$ precursor was the first tRNA dimer whose nucleotide sequence and maturation pathway were known in detail; many aspects of this pathway have been reviewed by McClain (1977).

The $tRNA^{Pro}$-$tRNA^{Ser}$ precursor is a relatively abundant RNA species in normal labeling of T4-infected *E. coli*, being about one-third as abundant as a mature tRNA by 15 min after infection. Other T4 tRNA precursors are present in significantly lower amounts. This must mean that the metabolic step immediately after synthesis of the $tRNA^{Pro}$-$tRNA^{Ser}$ precursor must be rate limiting in vivo. A clue to the nature of this limiting event is provided by nucleotide sequence analysis of this precursor RNA: consistent with the gene sequence, the 3′ end of this precursor does not contain a CCA_{OH} terminus. Instead, UAA_{OH} and shorter 3′ ends are found (Barrell et al., 1973). The CCA_{OH} sequence must be derived by processing. The maturation pathway (Fig. 4) was deduced from an analysis of precursor RNA forms which accumulate in various mutant cells. In RNase P-deficient cells, precursor RNA ending in mature CCA_{OH} sequence is found; this observation shows that the CCA_{OH} sequence can be added before RNase P cleavage (Seidman and McClain, 1975). In cells deficient in tRNA nucleotidyltransferase, a shortened 3′ sequence ending in G_{OH}, i.e., with neither UAA_{OH} or CCA_{OH}, is found, and further precursor maturation is blocked (Schmidt, 1975; Seidman and McClain, 1975). In *E. coli* BN, the UAA_{OH} form of precursor RNA predominates and maturation is blocked. Strain BN was shown to be lacking a specific nuclease capable of removing extra 3′ nucleotides from immature tRNA species (Schmidt and McClain, 1978a; Seidman, Schmidt, Foss, and McClain, 1975).

This scheme (Fig. 4) makes several predictions about the enzymology of tRNA processing which were confirmed in vitro and in vivo. First, tRNA nucleotidyltransferase is capable of accurate synthesis of CCA_{OH} on the precursor RNA (Schmidt, 1975). Second, since the $GUAA_{OH}$ and G_{OH}, but not the CAA_{OH}, forms of precursor RNA can be seen in labeled wild-type cells, the latter precursor should be a better substrate for RNase P than the former two species; this was shown by kinetic experiments to be the case (Schmidt et al., 1976). Finally, a precursor whose CCA sequence is encoded in the genome should not require tRNA nucleotidyltransferase or BN RNase for maturation. Fortuitously, such a species exists: phage T2 codes for an identical $tRNA^{Pro}$-$tRNA^{Ser}$ whose CCA sequence is genomically encoded: its cleavage in vivo does not depend on the presence of cellular BN RNase or tRNA nucleotidyltransferase (Seidman, Barrell, and McClain, 1975; Seidman, Schmidt, Foss, and McClain, 1975).

Once the 3′ end of $tRNA^{Ser}$ is matured to CCA_{OH}, RNase P acts to cleave the precursor RNA. In vitro, cleavage at the interstitial site is about 2.5 times faster than at the 5′ site. This difference is magnified in vivo, so an extensive search for $tRNA^{Pro}$-$tRNA^{Ser}$ dimers with cleaved 5′ ends was not successful (Schmidt and McClain, 1978b). On the other hand, RNase P cleavage of the monomeric tRNA precursor to generate the mature 5′ end of $tRNA^{Pro}$ occurs without the necessity for 3′ maturation, as evidenced by the presence of T2 $tRNA^{Pro}$ with a mature 5′ but not 3′ end in *cca* or BN hosts (T2 $tRNA^{Ser}$ is synthesized accurately in these mutant hosts [Seidman, Barrell, and McClain, 1975]). This facile cleavage of a 3′-immature monomeric precursor contrasts sharply with the restricted cleavage of a 3′-immature dimer; the structural features recognized in these two reactions are unknown.

To summarize: maturation of T4 $tRNA^{Pro}$ and $tRNA^{Ser}$ from their common precursor requires the participation of three host enzymes. Two of these enzymes, BN RNase and tRNA nucleotidyltransferase, are required for 3′-CCA_{OH} synthesis, whereas RNase P is solely responsible for 5′-terminal maturation. A final step, methylation at the 2′-hydroxyl group of G_{18} in $tRNA^{Ser}$, apparently occurs on the mature tRNA chain but has not been characterized.

The $tRNA^{Gln}$-$tRNA^{Leu}$ precursor. The $tRNA^{GLn}$-$tRNA^{Leu}$ precursor RNA is present only in transient amounts; unlike the $tRNA^{Pro}$-$tRNA^{Ser}$ precursor, it does not accumulate in hosts lacking either RNase BN or tRNA nucleotidyltransferase.

Both of these tRNAs are available in suppressor form, which permits in vivo estimation of the enzymes required for tRNA synthesis. As noted above, $tRNA^{Gln}_{Su+}$ but not $tRNA^{Leu}_{Su+}$ requires RNase III activity for its expression; this enzyme is required to generate the 5′ end of the dimeric precursor from the primary transcript. Further, $tRNA^{Gln}_{Su+}$ expression is restricted by *cca* hosts (Deutscher et al., 1974) but not by those lacking RNase BN, whereas $tRNA^{Leu}_{Su+}$ requires neither of these activities for its expression (Foss et al., 1979). These observations permit the following conclusions about these tRNA species: (i) neither tRNA has extra nucleotides in place of the CCA sequence; and (ii) the CCA sequence of $tRNA^{Leu}$ is genomically encoded, but that of $tRNA^{Gln}$ is synthesized by tRNA nucleotidyltransferase. These conclusions were borne out by the DNA and RNA sequence analysis (Fig. 2).

The $tRNA^{Gln}$-$tRNA^{Leu}$ precursor RNA is present only in very small amounts in wild-type *E. coli*, consistent with the notion that synthesis of the CCA_{OH} sequence is the rate-limiting step in processing the $tRNA^{Pro}$-$tRNA^{Ser}$ precursor RNA. (Since the 3′-CCA sequence of $tRNA^{Leu}$ is genomically encoded, cleavage of its precursor is faster than that of the other tRNA dimers.) RNase P is required for the cleavage of this precursor RNA, however, a fact which permitted Guthrie (1975) to determine its sequence and in vitro processing pathway. Purified RNase P cleaved the dimer accurately to generate mature 5′ termini of both $tRNA^{Leu}$ and $tRNA^{Gln}$. Further, the interstitial $tRNA^{Leu}$ cleavage occurred faster than the $tRNA^{Gln}$ cleavage; whether this is due to the state of the respective 3′-terminal sequences (CCA_{OH} versus C_{OH}) has not been determined. This feature of RNase P cleavage has also been noted in the in vitro processing of the $tRNA^{Pro}$-$tRNA^{Ser}$ precursor RNA (see above).

T4 $tRNA^{Gln}$ requires tRNA nucleotidyltransferase to synthesize its 3′ end. Like the 3′-terminal maturation of $tRNA^{Pro}$, this can occur after RNase P cleavage, since McClain et al. (1978) found that $tRNA^{Gln}$ with a

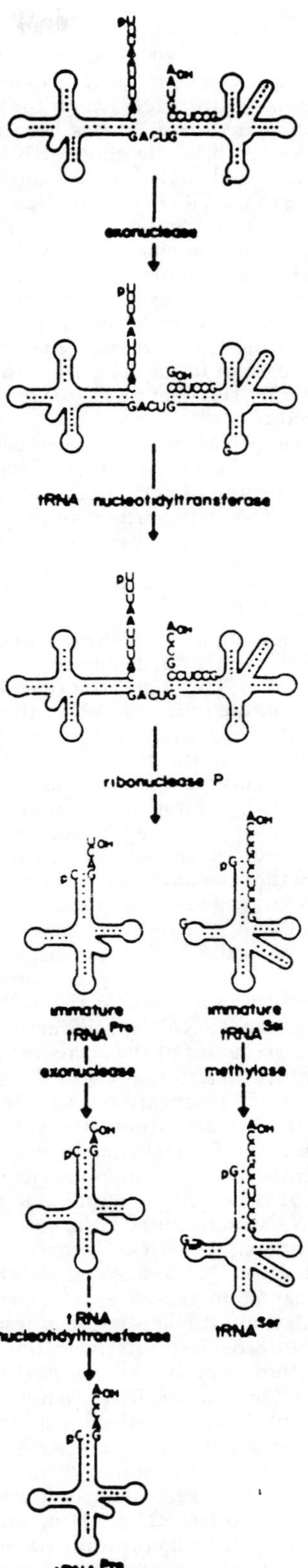

FIG. 4. Processing pathway for the synthesis of T4 $tRNA^{Pro}$ and $tRNA^{Ser}$ from their common precursor (species A in Fig. 1). See the text for details. Reprinted from McClain (1977) with permission of the American Chemical Society.

mature 5′ end and an immature 3′ end was synthesized in *cca* hosts.

The tRNA^{Thr}-tRNA^{Ile} precursor. T4 tRNA^{Ile} is unusual in several aspects. Its coding specificity is such that it recognizes AUA exclusively. Since T4 DNA is more AT-rich than *E. coli*, it has been hypothesized that this tRNA replaces the major tRNA$^{Ile}_{AU(C,U)}$ of the host (Scherberg and Weiss, 1972). Surprisingly, the gene sequence of the T4 tRNA gene cluster shows that the ability to read AUA arises from modification of the C in the anticodon sequence CAU which would normally read the methionine codon AUG (Fukada and Abelson, 1980). In contrast to other modified bases, this nucleotide is always found in full molar yield, presumably ensuring that missense reading of methionine codons cannot occur. The T4-encoded nuclease found recently by David et al. (1982) may edit out the undermodified tRNA^{Ile}.

Furthermore, tRNA^{Ile} is normally underproduced in phage-infected cells, even compared with tRNA^{Thr}, which is in the same dimeric precursor (Guthrie and Scholla, 1980). Two explanations of this phenomenon have been proposed which invoke either processing or transcriptional events to account for the underproduction of tRNA^{Ile}. Both explanations have features which commend them.

An explanation involving aberrant processing of the tRNA^{Thr}-tRNA^{Ile} dimer has been proposed by Guthrie and Scholla (1980). They observed that this precursor does not accumulate in *E. coli* A49, a thermosensitive RNase P mutant, at high temperature. This is so despite the fact that the tRNA^{Thr}-tRNA^{Ile} precursor accumulates in wild-type, BN, and *cca* hosts (Guthrie and Scholla, 1980; Schmidt et al., unpublished data; Seidman, Schmidt, Foss, and McClain, 1975). The immature 3′ ends of tRNA^{Ile} precursors in mutant hosts indicate that the pathway for synthesizing these tRNAs can be arranged in the same form as that for tRNA^{Pro} and tRNA^{Ser}. However, these blockages are not complete: in both BN and *cca* hosts, 3′-immature, 5′-mature monomeric tRNA^{Ile} species are synthesized, albeit inefficiently, and mature tRNA^{Thr} is made as well. The immature tRNA^{Ile} species arises by cleavage of the dimeric precursor at the correct interstitial site before 3′ maturation. Since tRNA^{Thr} requires only RNase P for its 5′ maturation and RNase D or its homolog for its 3′ maturation, it is synthesized normally after RNase P cleavage. In *rnp* cells at the restrictive temperature, a monomeric precursor to tRNA^{Thr} can be observed; it contains a 5′-immature, 3′-mature tRNA which can be accurately processed in vitro. The synthesis of tRNA^{Thr}, therefore, seems to require RNase P for 5′ cleavage and either RNase D or some other enzyme to generate its 3′ terminus.

This explanation still leaves the question: how is tRNA^{Ile} synthesized? Extracts from uninfected cells deficient in RNase P activity were used to cleave the tRNA^{Thr}-tRNA^{Ile} precursor RNA. While some of the dimer was cleaved to tRNA-sized product, this product was deficient in tRNA^{Ile}; i.e., tRNA^{Thr} sequences appeared in about a 2:1 molar ratio over tRNA^{Ile}. Guthrie and Scholla (1980) concluded that another RNase activity besides RNase P could cleave the precursor with the concomitant degradation of tRNA^{Ile}; note that this cannot be the T4-coded tRNA^{Ile} nuclease of David et al. (1982). The relationship among the cellular and phage-coded nucleases is unclear.

A second explanation, not necessarily incompatible with the one described above, invokes a transcriptional component to account for the relative underproduction of tRNA^{Ile}. Goldfarb and Daniel (1982) observed that transcription of subcluster I of the tRNA genes terminated in a *rho*-dependent reaction after (or in) tRNA^{Ile}. They also noted that tRNA^{Ile} was not synthesized from this terminated transcript. It is possible that this underproduction of tRNA^{Ile} is due to preferential degradation of the unmodified tRNA^{Ile}, but it is also possible that the in vitro terminated transcript did not contain a complete tRNA^{Ile} sequence at all. In support of this possibility, Mazzara et al. (1981) found, in a computerized but not exhaustive search of the gene sequence from the tRNA cluster, that a potential terminator could be formed within tRNA^{Ile}. This terminator structure is shown in Fig. 3b, and it is apparent on examination that it is mutually incompatible with the normal cloverleaf folding of tRNA^{Ile}. Attenuation (mechanisms of gene regulation involving mutually incompatible RNA secondary structures, one of which is a transcription terminator) has been proposed to account for the regulation of other operons besides the classic *trp* case (Bertrand et al., 1975). The present mechanism, if true, would be an example of an attenuation mechanism in a tRNA operon. In other attenuation mechanisms, translation apparently discriminates between the two structures. Since this region of the tRNA gene cluster is apparently not translated, the extent of tRNA base modification can perhaps discriminate between two alternate foldings of the nascent RNA chain.

Cluster II: tRNA^{Arg} and stable RNAs 1(C) and 2(D). Precursors to the adjacent cluster II RNAs can be generated by RNase PC cleavage of an in vitro transcript that can be subsequently matured to the final species by crude cellular S-100 extracts (Goldfarb and Daniel, 1981b). Examination of the DNA sequence indicates that the 3′-CCA ends of tRNA^{Arg} and species 2(D) are genomically encoded, but that of species 1(C) is derived by processing (Fig. 2). The maturation of species 1(C) probably follows a pathway similar to that of a tRNA whose CCA_{OH} sequence is not genomically encoded. After RNase PC or RNase F cleaves the primary transcript, the 3′ CA_{OH} is synthesized by the sequential actions of RNase BN and tRNA nucleotidyltransferase. The 3′-terminal CCA sequences of tRNA^{Arg} and species 2(D) are genomically encoded; maturation of these RNAs does not, therefore, require tRNA nucleotidyltransferase or RNase BN activities. The 5′ ends of stable RNA species 1(C) and 2(D) are probably generated by RNase P (see Fig. 7). Precursor RNA to tRNA^{Arg} does not accumulate in RNase P-deficient hosts. The maturation pathway has not been elucidated but may be similar in some respects to the metabolism of the pre-tRNA^{Thr}-tRNA^{Ile} dimer.

Monomeric precursor to tRNA^{Gly}. tRNA^{Gly} is unique among those of cluster I in that its cleavage from the primary transcript is as a monomer. An inspection of the gene sequence shows that this tRNA is flanked at both the 5′ and 3′ ends by AT-rich sequences. Other tRNAs are either preceded or followed by these sequences (but not both). Presumably the enzyme(s) required for cleavage of the transcript

TABLE 2. Host processing functions involved in T4 tRNA synthesis

Enzyme	RNAs affected	Function
RNase P	All	5′ Cleavage to generate mature tRNAs
RNase BN	$tRNA^{Pro}$, $tRNA^{Ser}$, $tRNA^{Ile}$, species 1(C)	Removal of extra nucleotides from 3′ end of RNAs without encoded CCA
RNase PC (or F?)	All	Cleavage of primary transcript
RNase III	$tRNA^{Gln}$	Cleavage of primary transcript at 5′ end
RNase D or homolog	$tRNA^{Leu}$, $tRNA^{Gly}$, $tRNA^{Thr}$, $tRNA^{Arg}$, species 2(D), T2 $tRNA^{Ser}$	Removal of 3′ nucleotides from 3′ end of RNAs of encoded CCA sequence
RNase E	Species 1(C)	Accessory cleavage factor in processing large precursor molecule
tRNA nucleotidyltransferase	$tRNA^{Pro}$, $tRNA^{Ser}$, $tRNA^{Gln}$, $tRNA^{Ile}$	Synthesis of 3′-terminal CCA sequence

must recognize these sequences (by base composition or structure). Since $tRNA^{Gly}$ has a CCA sequence encoded, precursor maturation requires only the removal of an extra nucleotide before or after RNase P cleavage.

Enzymatic Features of T4 tRNA Biosynthesis

A summary of the host enzymes involved in T4 tRNA synthesis is shown in Table 2. All of the characterized enzymes are host encoded, even though one, the BN RNase, is apparently dispensable in normal laboratory growth. Two phage functions are known which have an effect on processing. One gene near *rII*, called *mb* or *M1* (McClain et al., 1973; Wilson and Abelson, 1972), is required for the synthesis of species 2(D), $tRNA^{Pro}$, $tRNA^{Ser}$, and $tRNA^{Ile}$. These RNA species also require host RNase BN to remove extra nucleotides from their 3′ ends. If the *mb* (*M1*) gene product were required for the efficient action of RNase BN, infection of wild-type cells with *mb* (*M1*) mutant phage should show the same labeling pattern as infection of host BN by wild-type phage. This, however, is not the case: all low-molecular-weight RNAs are reduced in amount, and pre-$tRNA^{Pro}$-$tRNA^{Ser}$ does not accumulate. It is known, however, that the *mb* (*M1*) gene product is *trans*-active in mixed infections and requires phage protein synthesis for activity (Abelson et al., 1975; Schmidt and McClain, unpublished data). Fukada and Abelson (1980) have hypothesized that the *mb* (*M1*) protein is a nuclease inhibitor.

A second phage-coded function which may affect tRNA processing is the T4-induced modification of RNase D which was described by Deutscher and coworkers (Cudny et al., 1980). The modification results in a larger-molecular-weight RNase D, but detailed kinetic studies of the modified enzyme have not been reported. The *M1* (*mb*) gene does not seem to be involved in the modification of RNase D.

Other Phages Encoding tRNA

The synthesis of tRNAs by naturally occurring phages is a phenomenon so far limited to the T phages and their close relatives. Transfer and low-molecular-weight RNAs are encoded by T-even phages T2, T4, T6, and RB69. Although the tRNA-coding capacities of these species vary widely, it is of interest that the stable RNA species 2(D) and species 1(C) are conserved; the functional significance of this is unknown (Likover-Moen et al., 1978). Some inferences about the evolutionary plasticity of this region of the genome as it affects processing can be drawn from comparative patterns of RNA synthesis. First, the CCA sequence may be evolved independently of the rest of the tRNA genes. Both T2 and T4 code for $tRNA^{Ser}$ species; in T4 the CCA must be derived by processing, whereas in T2 it is encoded (Seidman, Barrell, and McClain, 1975; Seidman, Schmidt, Foss, and McClain, 1975). Second,

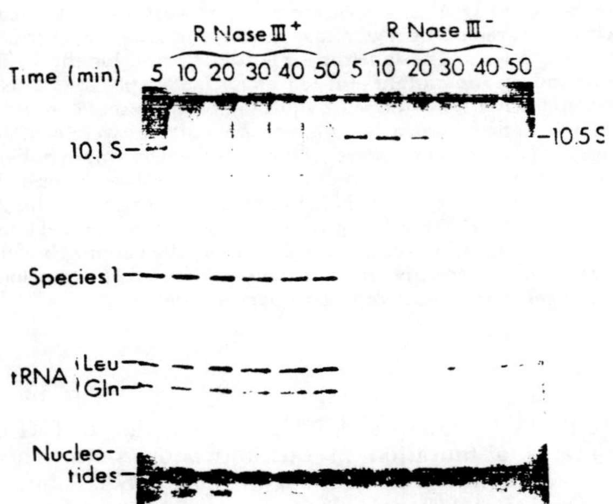

FIG. 5. Small RNA molecules in *rnc*⁺ (N2076, RNase III⁺) and *rnc* (N2077, RNase III⁻) strains after infection with T4Δ27. The cells were grown in Tris-based medium containing 0.6% (wt/vol) peptone at 37°C. $^{32}P_i$ (0.2 mCi/ml) was added to the cultures 5 min after infection. Samples were withdrawn at the indicated times after labeling. An autoradiogram of the 10% portion of the gel is shown. The top one-fifth of the gel, which consists of 5% (wt/vol) polyacrylamide, was removed after the gel was dried. The gel contained 7 M urea, and each slot was loaded with about 250,000 cpm. Below the $tRNA^{Gln}$ band there are two smaller RNAs (X1 and X2); the one which migrates slower is referred to as X1.

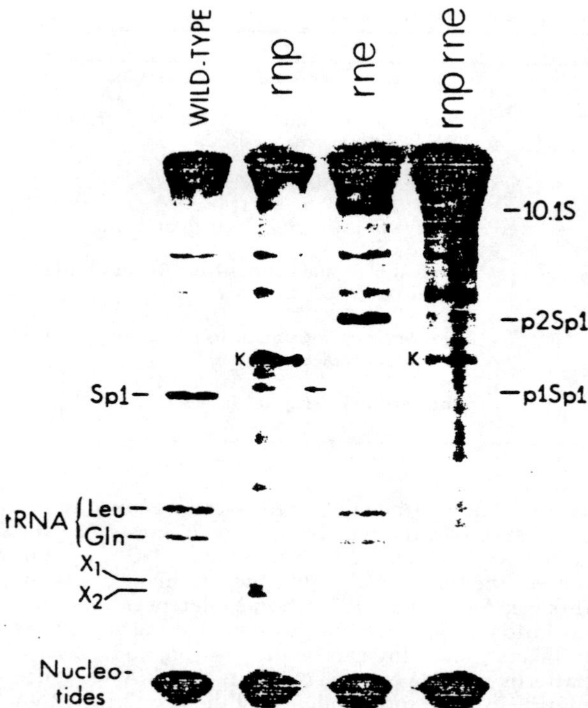

FIG. 6. Synthesis of RNA after infection of various host mutants with phage T4Δ27. Cultures of the wild type (N3433) and of *rnp* (N2021), *rne* (N3431), and *rne rnp* (N3522) mutants were grown at 30°C and at an absorbance at 560 nm of about 0 to 3 U (≅3 × 10^8 cells/ml); the cultures were then shifted to 43°C. After 30 min the cells were infected with T4Δ27 (15 phage per cell). After 4 min the cultures were superinfected with the same number of phage, and 1 min later $^{32}P_i$ (1 mCi/ml) was added. The labeling time was 15 min for the wild type and the *rnp* mutant and 30 min for the other two strains. Incorporation was stopped by adding 80% ethanol containing 1% diethyl pyrocarbonate, and the cultures were centrifuged. The resultant pellets were treated with sample buffer containing 0 to 2% sodium dodecyl sulfate and were applied directly onto the gel after being heated in boiling water for 2 min. The 5%–10% tandem polyacrylamide gel contained 0 to 2% sodium dodecyl sulfate and 7 M urea. The running buffer contained Tris and glycine. The figure shows only the portion of the gel that contains the 10% acrylamide.

the tRNA^{Gln} sequence of T2(H), but not that of T2(L), contains a mutation in the anticodon stem. This altered sequence results in an unstable tRNA^{Gln}, but a dimeric tRNA^{Gln}-tRNA^{Leu} precursor RNA can be observed in T2(H)-infected *rnpA*49 cells (Likover-Moen et al., 1978). This means that the degradative mechanism that edits defective tRNA^{Gln} must act primarily on monomer-sized RNAs. This may be relevant to the explanation of the unusual metabolism of tRNA^{Ile} from T6; as in the case of T2 tRNA^{Ser}, the 3′-CCA sequence of this tRNA is derived transcriptionally (McClain et al., 1979).

T5 and its close relative BF23 synthesize a large enough variety of tRNA species to account for nearly the entire genetic code. Whether this capacity is related to phage virulence is unknown. DNA sequence analyses of these tRNA genes now under way may define the enzymatic requirements for biosynthesis of these tRNAs more precisely (Hunt et al., 1980; Ozeki et al., 1975).

NEW CONCEPTS IN RNA PROCESSING DERIVED FROM STUDIES ON T4 tRNA

T4Δ27 System

While the overall T4 tRNA cluster is a very useful tool for the study of RNA processing, use of the Δ27 deletion (Wilson et al., 1972) can further simplify the system. This is an internal deletion that is missing 7 of the 10 genes in the tRNA cluster (Abelson et al., 1975). The first two genes (coding for tRNA^{Gln} and tRNA^{Leu}) and the last [coding for species 1(C) RNA] remain intact. Studies of processing of RNA from T4Δ27 led to a number of new concepts in RNA processing and RNA maturation (see above and below).

Interactions Between tRNA-Processing Enzymes

Synthesis of tRNA and species 1(C) RNA in T4Δ27-infected cells begins from a primary transcript encompassing all of these RNA species (Goldfarb and Daniel, 1981b). In a wild-type host, mature RNAs are made; when T4Δ27 infects cells deficient in RNase P (*rnp*), RNase III (*rnc*), or RNase E (*rne*), various precursors accumulate (Fig. 5 and 6). The identities of these RNAs suggest a specific model for processing (Fig. 7) (Pragai and Apirion, 1981; Pragai and Apirion, 1982). In a key experiment in the development of this model, 10.1S RNA, the product of RNase III cleavage at the 5′ end and transcription termination at the 3′ end, accumulated in an RNase E-deficient host. This might be taken to imply that cleavage of 10.1S RNA to generate pre-1(C) RNA is carried out by RNase E. However, cell extracts lacking RNase E activity were still capable of cleaving the 10.1S RNA. Further, RNase E itself could not carry out this cleavage (Roy and Apirion, unpublished data). The results are most easily explained by invoking an effect of RNase E on another processing enzyme, perhaps RNase F. Consistent with this is the accumulation of another RNA,

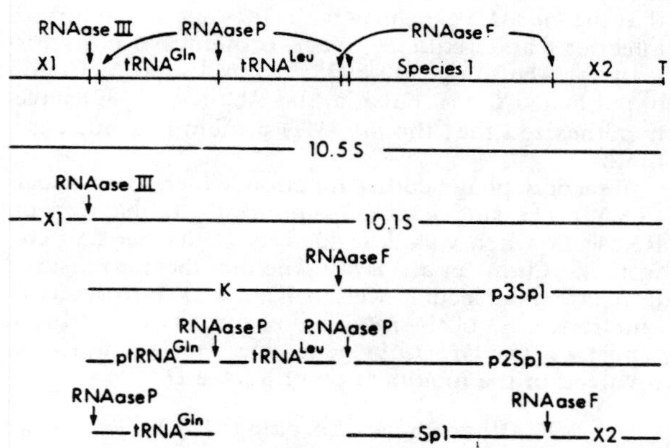

FIG. 7. Schematic processing pathway of RNA from the tRNA gene cluster in T4Δ27. The precursors, 10.5S, 10.1S, K, and p2Sp1 RNA, or the final products, species 1, tRNA^{Leu}, tRNA^{Gln}, X1, and X2 RNA, can be observed in Fig. 5 and 6.

p2Sp1 RNA, in an *rne* mutant. This precursor to stable RNA species 1(C) does not contain a site for RNase E cleavage; therefore, the *rne* mutation must exert a secondary effect on another processing enzyme, again perhaps RNase F.

RNase P and the Possibility of a Processing Complex

Extracts from an *rne* or an *rnc* mutant cannot express RNase P activity in vitro, under certain conditions, when the substrate is the dimeric precursor of $tRNA^{Gln}$-$tRNA^{Leu}$ (Jain, Gurewiz, and Apirion, 1982; Pragai and Apirion, 1982). Can one explain these results in terms of a macromolecular interaction between the components of the cellular RNA-processing system? No definitive proof is yet available. However, under some conditions, RNase III, RNase P, and RNase E activities can cosediment as a unit which is larger than each of them separately (Jain, Gurewiz, and Apirion, 1982). Work from our laboratories using a cloned gene which synthesized the RNA component of *E. coli* RNase P has shown unexpected anomalous results in that this cloned gene could not complement the *rnpB* mutation which results in a greatly reduced level of this RNA (Jain, Pragai, and Apirion, 1982; Motamedi et al., 1982). These observations can be explained if the *rnpB* gene codes for a protein in the RNA-processing complex and a mutation in this protein makes the RNA moiety of the RNase P more accessible to cellular nucleases. Finally, mutants that affect the processing at the 3′ end of T4 species 1(C) RNA have been found; surprisingly, these new mutants affect RNase P activity. We suggest from all of these observations that the processing enzymes of the bacterial cell may be arranged in a loose complex, perhaps held together by the RNA component of RNase P. The use of T4 tRNAs to diagnose multiple RNA-processing events could be very instrumental in characterizing the organization of such a complex.

Work in our laboratories is sponsored by Public Health Service grants GM26756 (to F.J.S.) and GM19821 (to D.A.) from the National Institutes of Health.

MORPHOGENESIS

Morphogenesis of the T4 Head

LINDSAY W. BLACK[1] AND MICHAEL K. SHOWE[2]

Department of Biological Chemistry, University of Maryland School of Medicine, Baltimore, Maryland 21201,[1] and Waksman Institute of Microbiology, Rutgers University, Piscataway, New Jersey 08854[2]

Bacteriophage T4 head formation occurs by one of the most complex assembly processes so far analyzed at the molecular level. The capsid, which encloses a single linear DNA duplex, is constructed basically from a single major shell protein according to the icosahedral design found in simple spherical viruses (see Eiserling, this volume). However, unlike the small icosahedral and helical viruses, in which the nucleic acid and the capsid subunits can self-assemble (Bancroft, 1970; Hohn and Hohn, 1970), the T4 head is produced in discrete stages under the control of numerous genes. These produce both enzymes and transient structural proteins not found in the finished head.

Studies of phage T4 head morphogenesis have resulted in the discovery of a number of novel assembly mechanisms: (i) the prohead core or scaffold, an obligatory precursor structure eliminated from the final structure, which controls its assembly, size, and shape; (ii) virus-specific proteolysis which controls the assembly pathway; (iii) packaging of DNA within a preformed prohead to condense and terminate the concatemeric, replicating DNA; and (iv) an assembly pathway controlled in time by specific sequential protein interactions which elaborate enzymatic activities. While the descriptive outline and, to a lesser degree, some of the basic mechanisms of the T4 assembly system are now generally agreed upon, there remain a number of challenging, unsolved phenomena which are attractive for investigations of basic biochemical mechanisms. These include shape and size determination, DNA condensation and higher-order structure, and genetic control of flexible tertiary and quaternary protein structure.

Before describing the relatively recent studies which are the basis for our current view of the mechanisms by which the T4 head is assembled and matured, we recapitulate some older work which, in retrospect, served to divide head morphogenesis into a series of distinct problems.

EARLY OBSERVATIONS AND DEFINITION OF THE MORPHOGENETIC PROBLEMS

Vegetative Phage: Multiplication Requires Assembly

The birth of phage assembly studies followed the general acceptance of the idea that vegetative phage are different from, and a precursor to, infective phage. This idea, that intracellular phage differed from mature phage, developed from genetic experiments (Hershey and Rotman, 1949; Luria, 1947), changes in radiation sensitivity during infection (Luria and Latarjet, 1947), and the discovery of infectivity eclipse (Doermann, 1948). Anderson (1949) showed by electron micrographs that when T2 phage adsorbed on the outside of *Escherichia coli*, their heads emptied, and he suggested that the contents entered the cell. Herriott (1951) showed that the head contents included the phage DNA. By 1951, Doermann (1951a) interpreted the results of his studies on intracellular phage growth and recombination in the following terms:

> The major conclusion appears to be that the infecting particles are not recoverable as such. . . . one must therefore conclude that the phages of the T-even series do not multiply by binary fission and further, that the first completed phage particles are not themselves the parents of those viruses which appear later within the same cell.

Finally, the experiments of Hershey and Chase (1952) showed that only the injection of the phage DNA was necessary for the production of progeny. The "separation of the phage into genetic and non-genetic parts," implied that at some stage in intracellular growth progeny phage would have to be assembled de novo.

The idea that the DNA and the protein of the phage might be synthesized independently and then combined received support from several laboratories at about this time, but the conclusion was drawn with some reluctance. Levinthal and Fisher (1952) found that "doughnut"-shaped membranes appeared in premature lysates of a T2-infected culture at a time which suggested that they were precursors to mature phage. The doughnuts were shown by Anderson (1953) to be empty phage heads. In 1948, Foster had reported that proflavin reversibly inhibited a late step in phage production, and DeMars et al. (1953) found that its effect was to cause infected cells to accumulate DNA-free empty heads and tail-like structures instead of phage at the time when morphologically normal phage appear in untreated cultures. They observed:

> Both our experiments and those of Maaloe and coworkers lead us to the conclusion that the various premature phage elements that can be isolated from infected bacteria contain sulfur (=protein) but little or no phosphorus (=nucleic

acid) until the very last moments before complete maturation. This is surprising. . . . Either these parts are formed separately from the nucleic-acid containing parts and combine with them at the end of maturation, or the two are so loosely combined in the immature forms that release from bacteria causes their separation.

For several years the relationship between DNA replication and condensation and head formation was obscured by the ambiguous status of the empty head membranes found in lysates of either normal or proflavin-treated cultures. Kellenberger and Sechaud (1957) pointed out that the available evidence was consistent with their being not head precursors, but either abortive products or the breakdown products of full but fragile immature heads. The latter interpretation received support when Koch and Hershey (1959) concluded that much of the protein label isolated from cultures producing bacteriophage T2 was derived from fragile precursors not stable to the chloroform lysis isolation procedures they used.

At this time, Kellenberger et al. (1959) described the formation of highly electron dense phage head-shaped structures from thin sections of infected cells fixed to give the best possible preservation of the intracellular DNA. They suggested that these were DNA condensates which became coated with the head membrane and finally stabilized as mature phage by addition of the tail. However, improved sectioning techniques allowed the later demonstration (Kellenberger et al., 1968) that the condensates were not naked, but already covered by a head membrane. Subsequent studies have shown that association of the head with DNA is one of the last steps in head morphogenesis.

Form Determination and the Conditional-Lethal Mutants

Starting from the suggestion by Crick and Watson (1956) that simple viruses are built from a single type of protein subunit, Caspar and Klug (1962) proposed that the so-called spherical viruses would all be found to have icosahedral symmetry. They suggested that a small icosahedral virus shell should be capable of self-assembly. This has now been demonstrated to be the case for a number of RNA viruses (Bancroft, 1970; Hohn and Hohn, 1970).

It was immediately recognized that the assembly of the bacteriophage T4 head was a more complicated problem (Caspar and Klug, 1962; Kellenberger, 1961). Since the T4 head is not isometric, it was not clear that it would have icosahedral symmetry, and it was argued that additional information, possibly contained in a "morphopoietic" (shape-determining) core would be required to specify the shape of the head membrane.

The discovery of the conditional-lethal mutants in bacteriophage T4 (Epstein et al., 1963) allowed a change in approach to the study of head assembly. Seven genes were shown to be required to produce a morphologically normal head; the question of whether a single type of subunit could assemble to form the T4 head shell was replaced by an effort to identify the role of each of these gene products in head formation. Sarabhai et al. (1964) showed that gene 23 coded for the major shell protein. In a series of publications (Favre et al., 1965; Kellenberger et al., 1968; Laemmli, Béguin, and Gujer-Kellenberger, 1970; Laemmli, Molbert, Showe, and Kellenberger, 1970), Kellenberger and his students and collaborators described in detail the aberrant structures produced by mutants in other genes (Fig. 1). The structures themselves could be ordered with respect to gene products required to produce them. Amorphous lumps, open-ended tubes (polyheads), and membrane-bound aberrant heads lacking DNA (tau particles), required a sequence of increasing numbers of gene products for their production (Laemmli, Molbert, Showe, and Kellenberger, 1970). The order of gene product action suggested by the sequence is in most cases consistent with current, more detailed knowledge of gene product function.

The head-related assemblies provided by the conditional-lethal mutants were also analyzed to give information on the structure of the normal head and on the composition of intermediates on the path of head formation. Polyheads were shown to have their subunits arranged in the form of a hexagonal lattice (Finch et al., 1964; Kellenberger and Boy de la Tour, 1965). This information was used by Moody (1965) in support of a prolate icosahedron for the structure of the normal T4 head. The scaffolding function of gp22 was first inferred from its presence in tau particles (Showe and Black, 1973), since it is not a component of the mature head.

Complementation In Vitro: Y Genes and X Genes, Form Versus Function

In 1966 Edgar and Wood showed that for many pairs of conditional-lethal mutants, defective lysates could be mixed to produce active phage in vitro. Their report extended the number of essential T4 genes to 66 and classified many of the mutants for their ability to produce either heads or tails which were competent for assembly in vitro. Further studies (Edgar and Lielausis, 1968; King, 1968) identified a total of 18 genes as required for formation of a competent head. They could be classified into broad groups by several characteristics. The Y genes were those previously shown to be required to make a morphologically normal head (genes 20, 21, 22, 23, 24, 31, and 40). The X gene mutants produced normal-appearing heads of which variable fractions contained DNA when isolated. Of them, mutants in genes 2, 4, 50, 64, and 65 produced substantial numbers of full heads, whereas gene 16, gene 17, and gene 49 mutants produced apparently empty heads. While none of these X gene mutants could serve as a head donor in vitro, mutants in the two "head completion" genes (genes 13 and 14) produced full but fragile heads which could.

These studies showed a striking difference between the head assembly reactions and those of the tail and tail fiber. Although complementation in vitro was immediately demonstrated among many of the genes involved in the latter, no head assembly reactions, only activation of an assembled head, could be shown. While mutations in most tail genes usually result in the accumulation of normal precursors which can be complemented in vitro, head gene mutants usually produce aberrantly assembled structures which cannot. These differences affected the approaches taken in studying head and tail assembly. Pathways of gene product action were rapidly established for the tail and tail fibers by complementation of crude lysates in

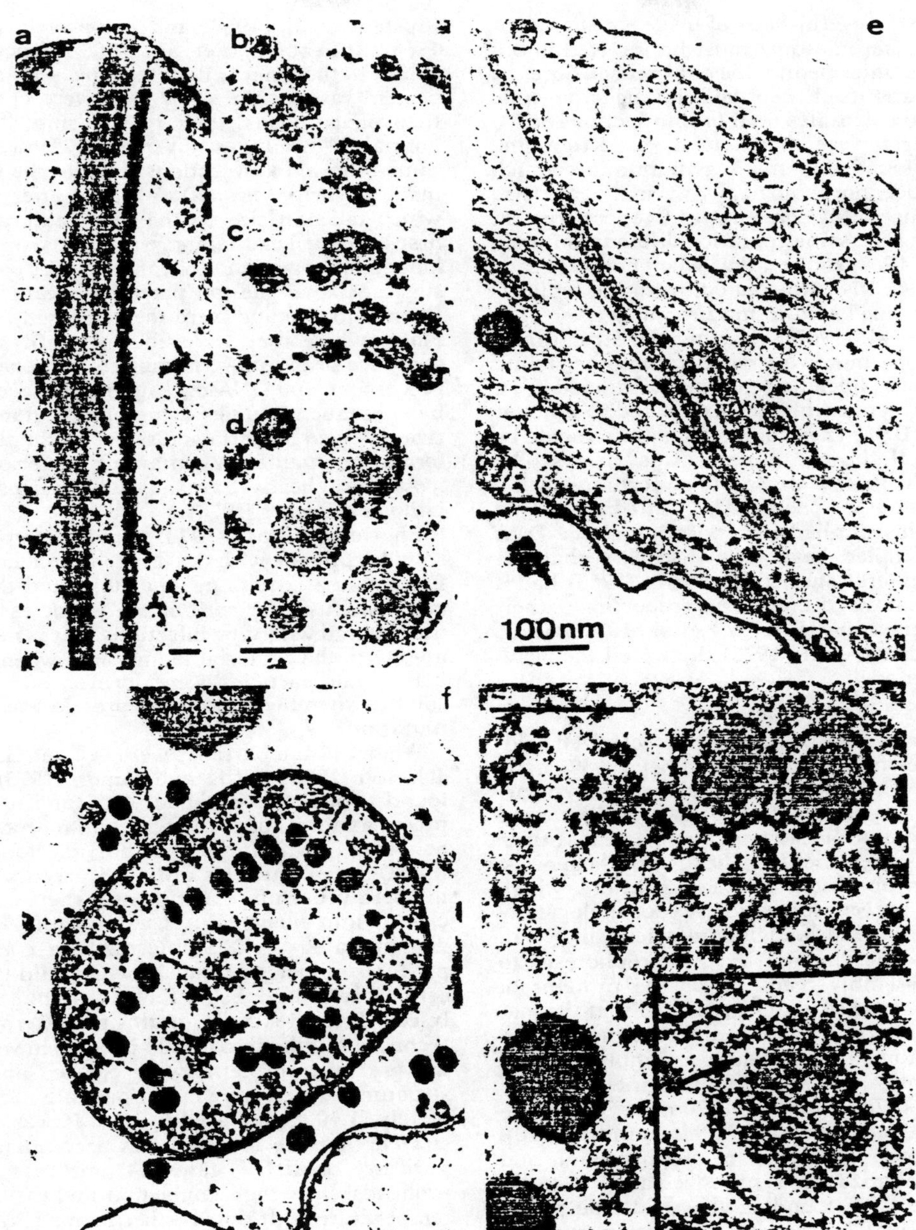

FIG. 1. Structures produced in mutant- and wild-type-infected bacteria as seen in thin sections. (a) Polyheads and multilayered polyheads from a gene 20 mutant. (b–d) Polyheads seen in cross section. (e) Tau particles lining the inner bacterial membrane in a gene 21 mutant infection. A few polyheads with well-defined cores are also seen in the center of the cell. (f) Wild-type T2 infection showing a number of phages with their tails oriented towards the inner membrane. A single tau particle attached to the membrane is also visible. (g) Wild-type T4 infection showing a completed phage and two tau particles on the membrane. Inset is a tau particle at higher magnification with the arrow pointing out the region at which the core and shell are in close contact. Panels a through e are from Kellenberger et al. (1968), panel f is from Simon (1969), and panel g is from Simon (1972). Bars, 100 nm.

vitro, using plaque formation as an assay. Studies of head formation have depended on the isolation and characterization of intermediates and structures produced by mutants. Even now, the sequence of action of head gene products is known only for a few proteins. In fact, the strictly ordered sequence of protein additions characteristic of tail assembly may not apply to head assembly (Fig. 2).

DNA Packaging

Although it had initially been proposed that the head shell assembled around a DNA condensate, (Kellenberger, 1961), several lines of evidence suggested that the phage chromosome must be inserted into a preformed head intermediate, several years before the intermediate was itself identified. At the same time that the DNA condensates seen in thin sections of T4-infected bacteria were shown to have surrounding shells, it was found that the Y gene mutants which failed to produce morphologically normal heads also failed to produce DNA condensates (Kellenberger et al., 1968). Using the same conditional-lethal mutants, Frankel (1966, 1968) showed that when normal heads were not formed, DNA was not cut from the replicating pool to the size found in mature phage.

Streisinger et al. (1967), studying the structure of the circularly permuted phage chromosome, found that the genomes of deletion mutants contained extra copies of other genes which compensated in amount for the deleted DNA. They suggested that this could be accounted for if the chromosome was always of a constant length and that this could be accomplished if it was packaged as a "one-headful" unit from a long concatemer. This implied incorporation of the DNA into a preassembled head. The discovery that the "petite" phages first studied by Mosig (1963) had a head volume that was decreased in proportion to their shortened chromosome (Eiserling et al., 1970; Mosig, Carnighan, Bibring, et al., 1972) supported the headful packaging model.

Protein Cleavage

In 1966, Champe and Eddleman characterized acid-soluble peptides first found by Hershey (1957) in T-even phage heads. They further showed by pulse-chase labeling experiments that the peptides were derived from acid-insoluble precursors and that their production was dependent on head morphogenesis, because it was prevented by the mutations which were known to prevent formation of morphologically normal heads. In 1970, four laboratories which had been using polyacrylamide gel electrophoresis to study phage assembly reactions or the proteins of purified head-related structures reported that the major capsid protein, gp23, had a different mobility in mature heads than it had before assembly or after aberrant assembly in Y gene mutant-infected cells (Dickson et al., 1970; Hosoda and Cone, 1970; Kellenberger and Kellenberger, 1970; Laemmli, 1970). Pulse-chase isotope labeling of wild-type-infected cells showed that gp23 was synthesized as a precursor larger than the form found in the wild-type phage capsid. Laemmli (1970) showed, in addition, that the product of gene 24 and an internal protein were similarly shortened and that the product of gene 22 disappeared, as though it was cleaved to small fragments. All cleavage was blocked by a mutation in any one of the Y genes. The cleaved form of gp23 was found only in cold sodium dodecyl sulfate (SDS)-resistant form, i.e., most likely as mature capsids. These results suggested that cleavage might take place in a large structure requiring the participation of so many known head gene products that it might well be the head itself.

The Prohead

Headlike structures, called tau particles, were observed in T4 21^- mutant-infected bacteria in 1963 (Epstein et al., 1963) and extensively described in 1968 (Kellenberger et al., 1968). Since they were bound to the inner cell membrane, seemed to contain protein rather than DNA, and were clearly smaller than phage heads, their relationship, if any, to the normal T4 head was not obvious. However, in 1972 Simon showed thin sections of wild-type-infected cells made at early times (10 to 14 min after infection), in which tau particles on the cell membrane were the first ordered head-related structures seen. From the kinetics of appearance of various types of particles, he suggested that the tau particles became emptied on the membrane and were then released into the cytoplasm where they were filled with DNA. This basic sequence of events—assembly of a prohead, prohead processing, and DNA packaging—has been confirmed by subsequent studies using both mutant- and wild-type-infected bacteria and seems to characterize head formation in all DNA-containing bacteriophages with icosahedral heads (Earnshaw and Casjens, 1980; Murialdo and Becker, 1978).

Showe and Black (1973) found that gp22 formed a soluble complex with the three internal head proteins. From this association, they suggested that it was a component of the core of tau particles in both 21^- mutant and wild-type infections. It was mainly associated with the bacterial membranes when membrane-bound tau particles were formed but was mainly soluble when the major shell protein was removed by mutation.

When 21^- tau particles were isolated (Laemmli and Johnson, 1973; Luftig and Lundh, 1973), they were found to sediment at about 400S and to contain the precursor forms of gp23 and internal protein III (IPIII) as well as gp22. It was not possible to demonstrate that 21^- tau particles are head precursors, since in temperature shift experiments, they could not be chased into phage. But Laemmli and Favre (1973) demonstrated a 400S intermediate containing the products of genes 22 and 23 when wild-type-infected cells were pulse-labeled for only 1 min before being lysed. The radioactivity in this peak (termed prohead I) chased rapidly into a 350S particle in which most of the precursor proteins were cleaved and then into structures which were either partially (550S) or completely (1,100S) filled with DNA. It was also demonstrated that the tau particles accumulated by temperature-sensitive gene 24 mutants are active proheads, since they can be matured to filled heads in temperature shift-down experiments (Bijlenga et al., 1974).

The prohead has been a focal point for many studies of T4 head morphogenesis since it is the end product of the shape-determining assembly reactions and the starting point for the cleavage and DNA-packaging reactions which lead to the mature head.

PROHEAD ASSEMBLY

The T4 prohead is composed of about 3,000 molecules, including at least 11 different proteins whose numbers in the prohead have been determined. It is constructed in two or possibly three layers, each of which appears to originate in a common initiation complex where the prohead is anchored to the inner bacterial membrane. Each of the six essential gene

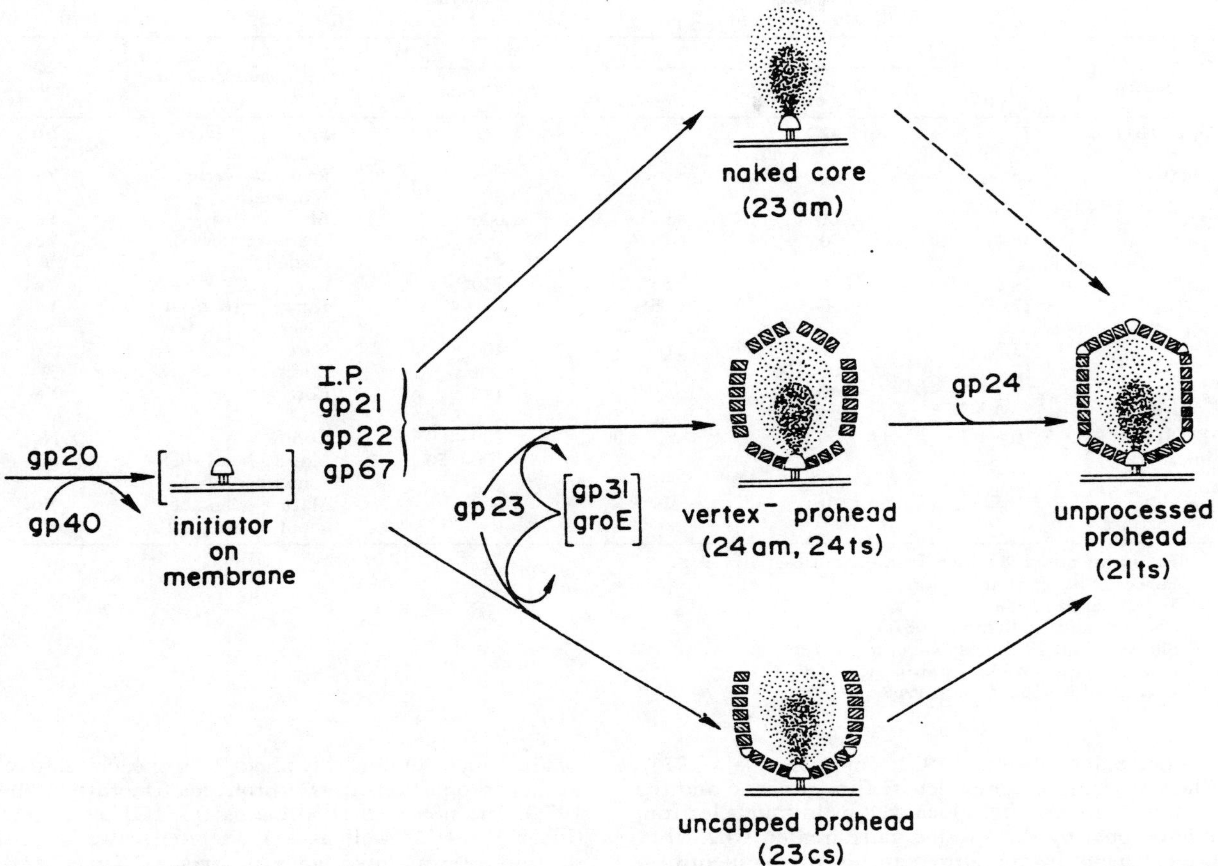

FIG. 2. Prohead assembly. A series of possible precursors to the unprocessed prohead is shown. Brackets indicate that the intermediate is only inferred from genetic data. Parentheses enclose the T4 gene number and type of mutant which accumulates the structure shown above the designation. Only the vertex-lacking prohead has been rigorously shown to be maturable to infectious phage, but it has not been demonstrated in wild-type-infected bacteria. Naked cores and uncapped proheads accumulate in wild-type-infected bacteria grown at 19°C, as well as in the mutants shown.

products which are prohead components has been localized in one of the layers or in the initiator, and structures have been proposed for the outermost layer (the shell) and the middle layer (the core). The innermost layer (the kernel) probably consists of a single gene product (gp21). The model we present below is based mainly on inferences from structures produced by mutants and the reassociation properties of the prohead proteins studied in vitro. No intermediate earlier than the unprocessed prohead (Fig. 2), has been demonstrated to be a precursor to the normal head.

Quantitation and Location of the Known Prohead Components

Wild-type proheads have not been isolated in sufficient quantity for a detailed determination of their protein composition. This is only known for the aberrant or immature proheads which accumulate in various mutants. The best composition data are from 21^- particles, since these are the most stable. Because they are mutated in the gene which codes for the processing protease (Showe et al., 1976a; Showe et al., 1976b), it seems likely that their composition is the same as that of the unprocessed wild-type prohead. The compositions of proheads and mature heads are summarized in Table 1.

Of the 11 protein species demonstrated in the prohead, 6 are genetically essential head gene products, 4 are nonessential, and 1 is not yet genetically identified. The shell (Fig. 2), is composed of hexamers of the major capsid protein, gp23, except at the 12 icosahedral vertices. According to the structure determined for the T-even head, there should be about 1,000 molecules of gp23 per prohead. (For T4, Aebi et al. [1974] would calculate 960 subunits in the form of hexamers, based on a measured length of 110 nm for the mature head. For phage T2, Branton and Klug [1975] propose 840 subunits in the form of hexamers, based on counts of morphological subunits on the surface of phage prepared by freeze-etching.)

At the membrane attachment (proximal) vertex of the prohead shell is a ring of gp20 (Driedonks et al., 1981; Müller-Salamin et al., 1977; Onorato et al., 1978; R. A. Driedonks and J. Caldenty, J. Mol. Biol., in press). At the other 11 vertices are pentamers of gp24,

TABLE 1. Major protein components of the T4 prohead and mature head[a]

Gene product	Prohead Mol wt (×10³)	Prohead No. of copies	Mature head Mol wt (×10³)	Mature head No. of copies	Location in prohead	Essential
gp*alt* (B1)	75	42	67	43	Proximal vertex, core	No
gp20	65	30	65	15	Proximal vertex of shell	Yes
gp23	56[b]	1,000	48.7[b]	1,000	Shell	Yes
gp24	48.4[c]	70	46	65	Distal vertices of shell	Yes
gp22	29.8[b]	600	2.5[d]	120[d]	Core	Yes
gp21	27.5[e]	50–100	18.5[e]	1–4[e]	Kernel, proximal vertex	Yes
IPIII	23.5	370	21	400	Core	No
17K	17	200–300		0	Core	?
gp67 (PIP)	9.1[f]	330–380	3.9[d]	140[d]	Core	Yes
IPI, IPII	10, 11.7[e]	750	8.9, 10[c]	200–400[g]	Core	No
hoc		0	39.1[c]	100–150	Mature head shell not in prohead	No
soc		0	9.7[c]	1,000	Mature head shell not in prohead	No

[a] Except as noted, data are from Onorato et al. (1978).
[b] Parker et al., in preparation.
[c] Tsugita et al., 1980.
[d] Champe and Eddleman, 1967.
[e] Showe et al., 1976a; Showe et al., 1976b.
[f] Volker, Gafner, Bickle, and Showe, 1982.
[g] Black and Ahmad-Zadeh, 1969.

(Müller-Salamin et al., 1977; Onorato et al., 1978). These two minor components of the prohead and the mature head have been localized by immunoelectron-microscopy, by the specific gaps created when they are extracted from mature capsids, by the identification of distal-vertex gaps in 24⁻ proheads, and by the direct visualization of gp20 rings on 21⁻ proheads (Fig. 3 and 4).

The remaining proteins are inside the prohead shell. In the absence of other criteria, we have assigned them to the core or to the membrane attachment vertex where gp20 is localized based on the following. Proteins found in the cores of open-ended polyheads (which lack vertices) are considered to be distributed throughout the core, as are the proteins which are easily lost from the fragile misshaped proheads isolated from temperature-sensitive gene 23 mutants. Conversely, proteins not found in core-containing polyheads (van Driel and Couture, 1978a), and which tend to remain associated with the temperature-sensitive proheads even when they have lost their visible cores (Onorato et al., 1978), are likely to be associated with the gp20, or the proximal vertex. These two criteria are mutually consistent. Not surprisingly, proteins present in relatively large numbers (200 or more molecules per prohead) are found to be core proteins, whereas proteins with fewer than 100 copies appear to be proximal-vertex proteins.

Two essential gene products, gp22 and gp67, are probably distributed throughout the core since they are both major constituents of both prohead and polyhead cores (Laemmli, Paulson, and Hitchins, 1974; Laemmli and Quittner, 1974; Showe and Black, 1973; van Driel and Couture, 1978a). They are both apparently lost during the isolation of *ts*21 *ts*23 proheads, which contain only about 15% of each relative to their proportions in *ts*21 proheads (Onorato et al., 1978). The nonessential proteins IPI, IPII, and IPIII (Black, 1974) as well as a 17,000-molecular-weight protein of unknown genetic origin (Kurtz and Champe, 1977) behave like gp22 and are therefore probably distributed throughout the core. Gp21, the remaining essential protein, has been tentatively localized in the center of the core, where it forms a dense kernel visible in *ts*21 proheads but absent in *am*21 proheads (van Driel et al., 1980). It is absent from polyheads and tightly bound to the *ts*21 proheads, suggesting that it is also associated with the proximal vertex of the prohead. The nonessential protein gp*alt* is distributed like gp21 and so it probably also is associated with the proximal vertex, consistent with its low copy number in the prohead.

Activation of the Major Prohead Protein, gp23

Mutants in gene 31 fail to produce any ordered assemblies of gp23, which instead accumulates as lumps on the inner bacterial membrane (Laemmli, Béguin, and Gujer-Kellenberger, 1970). This is one of two Y head genes whose product is not a prohead component. The assembly of head-related structures not containing gp23 is unaffected by these mutations (Traub et al., 1981). Gp31 has been purified and shown to be a soluble protein with a molecular weight of 16,000 (Castillo and Black, 1978).

A phenotype similar to that of 31⁻ mutants is displayed by certain bacterial mutants with mutations in two or three genes near *purA*. For an extensive discussion of these host functions and their interactions with T4 functions, see Simon and Binkowski,

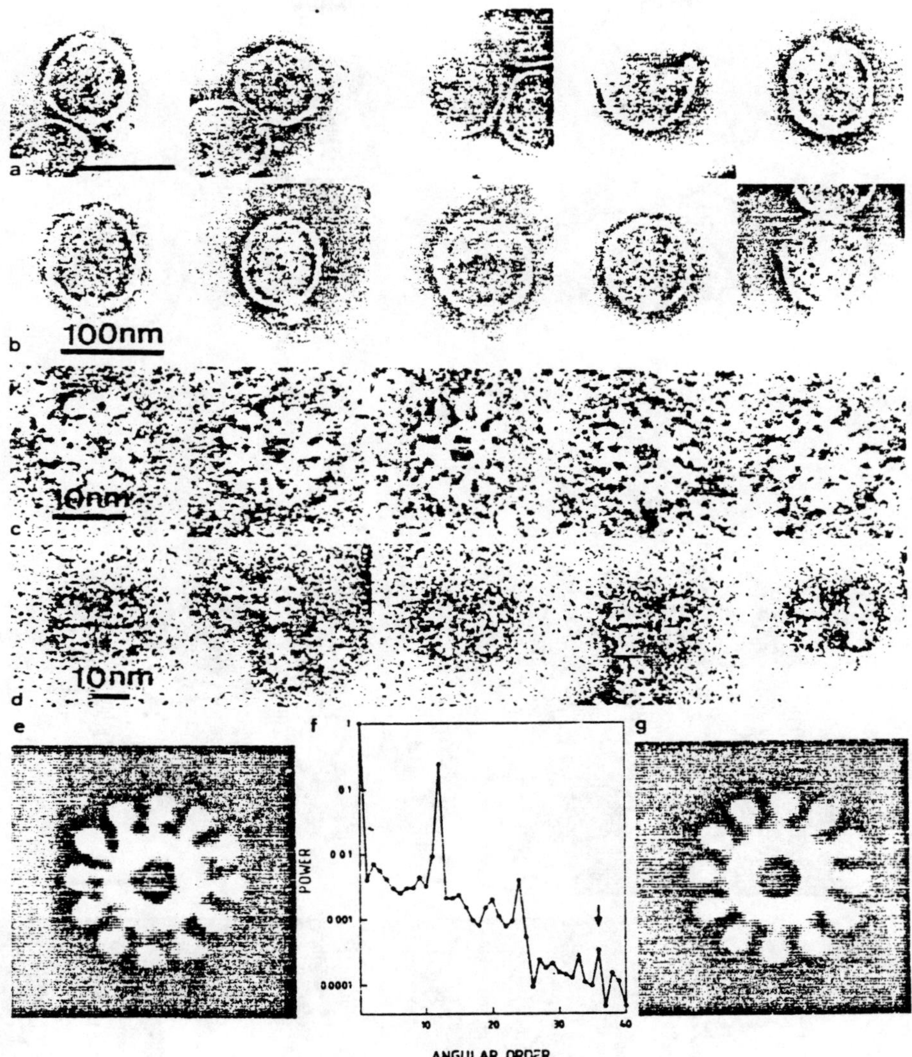

FIG. 3. Location and structure of gp20. (a) Proheads selected to show the substructure of the plug at the proximal vertex. Note that the plug appears to connect the core to the shell. (b) Proheads decorated with ferritin-labeled anti-gp20 antibodies. (c) gp20 rings at high magnification. (d) Side views of gp20 double rings. Long chains of rings in this orientation are also observed. (e) An average image of 10 particles viewed as in panel c. (g) A filtered image of panel e. (f) A power spectrum of panel g showing the rings to be 12-fold symmetric. Panels a and b are from Driedonks and Caldentey (in press), and panels c through g are from Driedonks et al. (1981).

this volume. There is some genetic and physical evidence for a physical interaction between gp23 and gp31. Castillo and Black (1978) demonstrated a decrease of soluble gp31 antigen when gp23 lumps are formed. And recently, a gene 23 mutation has been isolated which bypasses the need for both *groE* and gene 31 functions (L. D. Simon and B. Randolph, submitted for publication).

Mutants altered in the essential core protein gp22 form polyheads, whereas mutations altering the nonessential or semiessential core protein IPIII give rise to a half-normal burst of phage and produce polyheads from the remainder of the gp23. Double mutants which lack both gp22 and IPIII have a phenotype like that of *groE* and gene 31 mutants; i.e., they produce only lumps of precipitated gp23 (Black and Brown, 1976; Showe and Black, 1973).

None of the above-described host or phage proteins is required for the formation of gp23 assemblies in vitro. Pure gp23 obtained by the low-salt depolymerization of polyheads will reform polyheads when the ionic strength is gradually raised (van Driel, 1977; van Driel and Couture, 1978a). This property suggests that gp31 and gp*groE* might act at a step before gp23 forms an ordered structure, i.e., during the folding of the polypeptide chain as, or soon after, it leaves the ribosome. Both gp31 and IPIII are early proteins, assuring that a supply of both will be present in the

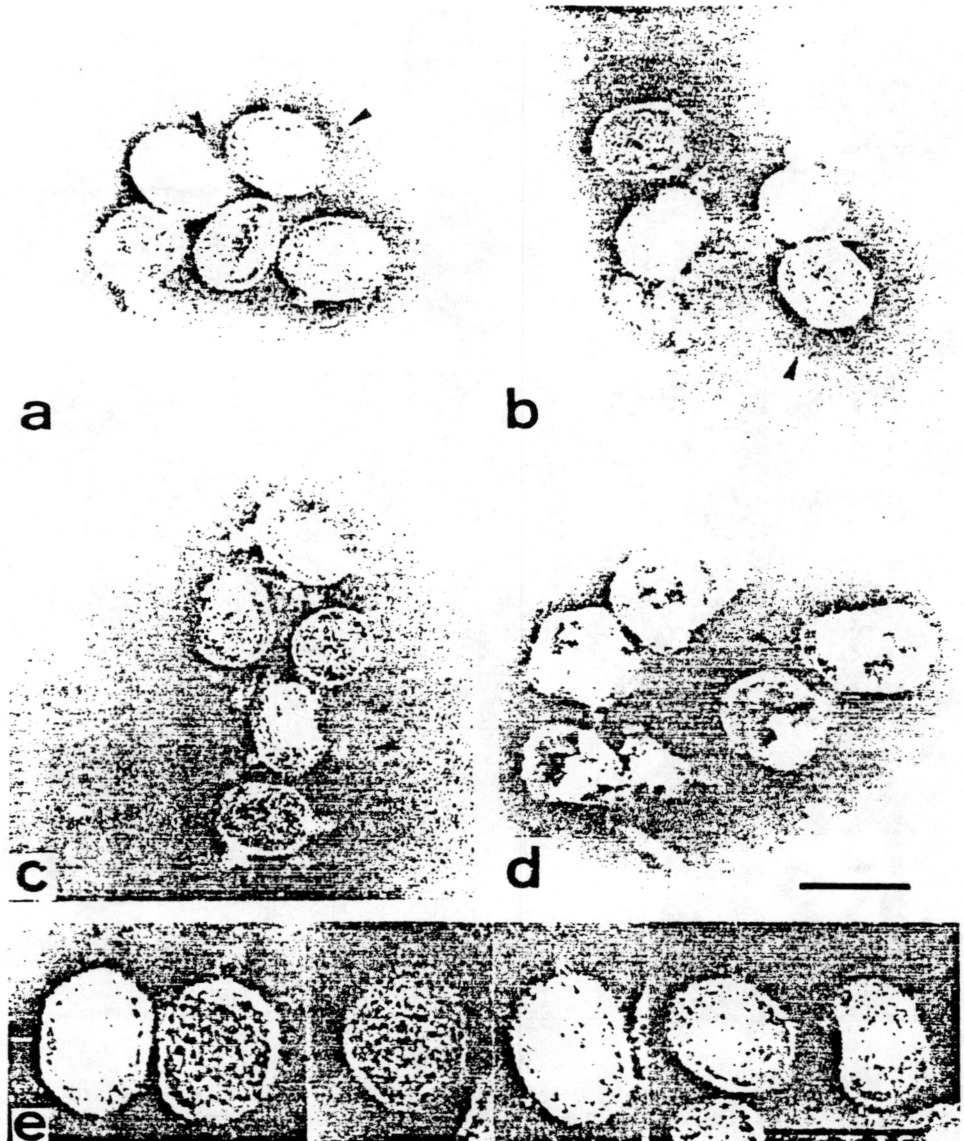

FIG. 4. Location of gp24 on mutant proheads. The proheads were incubated with rabbit antibody to purified gp24 and, in the case of panel c, post-labeled with ferritin-conjugated goat anti-rabbit immunoglobulin. (a) Gene 21 (tsN8) proheads. (b) Gene 21 (amE322) proheads. (c) Gene 21 (tsN8) proheads. (d) Misshapen "crummy" proheads from the gene 23 mutant tsA78. (e) Preparation shown in panel a at higher magnification. The arrows point out some of the antibody-decorated vertices. Bars, 100 nm; the bar in panel d applies to panels a through d. From Müller-Salamin et al. (1977).

cell when the synthesis of gp23 begins. These results suggest that a gp31-gp*groE* complex might act as a carrier for gp23 from the ribosome to either the prohead core or a core subunit. Laemmli, Béguin, and Gujer-Kellenberger (1970) suggested a carrier function for gp31 alone before the discovery of the *groE* gene and its genetic interactions with gene 31.

A requirement for carrier functions is also suggested by the solubility properties of gp23 vis à vis the intracellular salt concentration. In vitro, gp23 begins to form a precipitate and polymerizes to form hexamers and polyheads at a protein concentration of 1 mg/ml when the salt concentration is as low as 0.01 M Mg^{2+} and 0.01 M K^+ at pH 6, or 0.1 M K^+ at pH 7 (van Driel, 1980). In *E. coli*, the intracellular salt concentration is probably closer to 0.3 M K^+ (Christian and Waltho, 1962; Epstein and Schultz, 1965; Kellenberger, 1980), but neither precipitation nor polyhead formation occurs in gene 20 mutant-infected bacteria until the gp23 concentration is about fivefold higher

(Laemmli and Eiserling, 1968), at which point polyhead formation begins. Therefore, free gp23 is probably not an intermediate in vivo.

The form in which gp23 is added to the prohead is not known. In vitro, core proteins and gp23 obtained by low-salt dissociation of polyheads sediment independently under salt conditions which promote the reassembly of polyheads (van Driel, 1980). It has been reported that the gp23 found soluble in freshly isolated crude lysates sediments as a 4.3S monomer (Yanagida, 1972), but it is possible that the conditions of centrifugation or a low protein concentration might have promoted dissociation of oligomers. Gp*groEL* is found in the cell as a double-ringed quadridecamer, with a diameter of 12.9 nm (Hohn et al., 1979). Its dimensions and multimeric structure seem more appropriate to interaction with hexamers of gp23 (diameter, about 11.0 nm), although this function is only speculative.

Prohead Initiation

Initiation is a discrete assembly step. Several types of observation support the hypothesis that there are rate-controlling initiation steps which regulate the synthesis of proheads and other head-related structures such as polyheads. (i) The formation of open-ended polyheads begins in 20^- mutant infections only after a lag compared with prohead formation in wild-type or 21^- mutant infections (Laemmli and Eiserling, 1968; Laemmli, Molbert, Showe, and Kellenberger, 1970). The lag entails a buildup of gp23 subunits, and it is extended if this accumulation of subunits is slowed. This suggests that gp20 (or a structure which includes it) can initiate prohead shell formation at lower precursor concentrations than are required for the formation of polyhead shells. (ii) The formation of polyheads is also regulated by an initiation step. Polyheads shorter than the full length of the cell are rarely observed, suggesting that their formation consists of a slow initial process followed by a rapid polymerization of core and shell subunits until the tube is as long as the cell (Laemmli, Molbert, Showe, and Kellenberger, 1970). It has been suggested (Katsura, 1978) that the initiator for polyhead assembly might be a band of hexamers which closes to form a ring. Such a ring would have to contain about 20 hexamers before it could close and more might normally have to be incorporated before the ends find each other. This should be a much slower process than that suggested for prohead initiation (see below), even if it takes place on a core. (iii) Gene 24 mutants form large numbers of proheads and also form polyheads without the lag characteristic of 20^- mutants (Laemmli, Molbert, Showe, and Kellenberger, 1970). However, it has been shown that most of the 24^- "polyheads" are either greatly extended proheads or have an aberrant prohead or a cap at one end (Fig. 5), suggesting that they are initiated as proheads rather than as open-ended polyheads (Onorato et al., 1978). The kinetics of 24^- polyhead formation therefore confirm the hypothesis that it is the initiation step which limits the rate of polyhead formation. When their initiation takes place at a membrane-bound cap like that of a prohead, polyheads are formed simultaneously with the formation of proheads.

In sum, initiation rates apparently determine the gp23 assembly made. It appears that efficiently initiated structures use up the precursor proteins before the less rapidly initiated assemblies can be started. In wild-type-infected cells, prohead formation channels the flow of precursors into phage heads and inhibits the nonproductive formation of polyheads.

Gene products required for initiation. Three gene products, gp22, gp20, and gp40, are implicated in prohead initiation, because mutations in their genes produce polyheads from gp23 instead of proheads (Laemmli, Molbert, Showe, and Kellenberger, 1970). Their known locations in the prohead or genetic interrelationships confirm them in this function. Gp22 is the major core protein of proheads and polyheads. When gp23 is not present, gp22 is isolated as a soluble complex with the phage internal protein precursors (Showe and Black, 1973) (see above). Gp20, the proximal-vertex protein of the prohead, is not soluble but is found bound either to proheads or to bacterial membranes (Brown and Eiserling, 1979a; Brown and Eiserling, 1979b; Hsiao and Black, 1978b). Gp40 is not a prohead component but is found exclusively bound to bacterial membrane. It is essential for phage production only at high temperature (Hsiao and Black, 1978a, Brown and Eiserling, 1979a), and seems to assist the gp20 function since gene 40 amber mutants can be bypassed by a missense mutation in gene 20 (Hsiao and Black, 1978b).

Gp20 is in contact with the prohead core and the shell at the membrane-bound vertex and appears to anchor the prohead to the bacterial membrane. All polyheads, even core-containing ones, lack both gp20 and the fivefold axis of symmetry which distinguish the icosahedral proheads from polyheads. (Capped polyheads, considered ipso facto to be giant proheads, are not found in 20^- mutants.) It was suggested that the rate-limiting prohead initiation step is the establishment of the first pentameric vertex of the prohead shell by gp20 on the membrane (Onorato et al., 1978; Showe and Onorato, 1978). Core and shell initiation can now be experimentally distinguished. Paradoxically, some of the observations discussed below suggest that a 12-fold-symmetric gp20 initiator serves to initiate the 5-fold-symmetric shell rather than the 6-fold-symmetric core.

Separation of core and shell initiation. van Driel and Couture (1978a, 1978b) have shown that prohead-shaped shell-less cores can be formed in vitro from a mixture of core proteins in the presence of "initiator." This initiator is a Sepharose 6B void-volume fraction derived from dissociated proheads; it is highly enriched in gp20. This fraction also allows the formation of prohead-like structures when gp23 is included with the core proteins, but cannot induce the formation of closed shells from gp23 alone, requiring core proteins as well. Incomplete shells frequently surround the core, and these are always bound to the presumptive gp20-containing initiator. Accordingly, it was suggested that gp20 first initiates core formation, and shell initiation follows. This role for gp20 in core initiation would be consistent with the finding that the cores of giant proheads have sixfold axial symmetry (Paulson and Laemmli, 1977), which matches the symmetry of the gp20 dodecamer ring (Driedonks et al., 1981).

Using new fixation techniques, Traub et al. (1981)

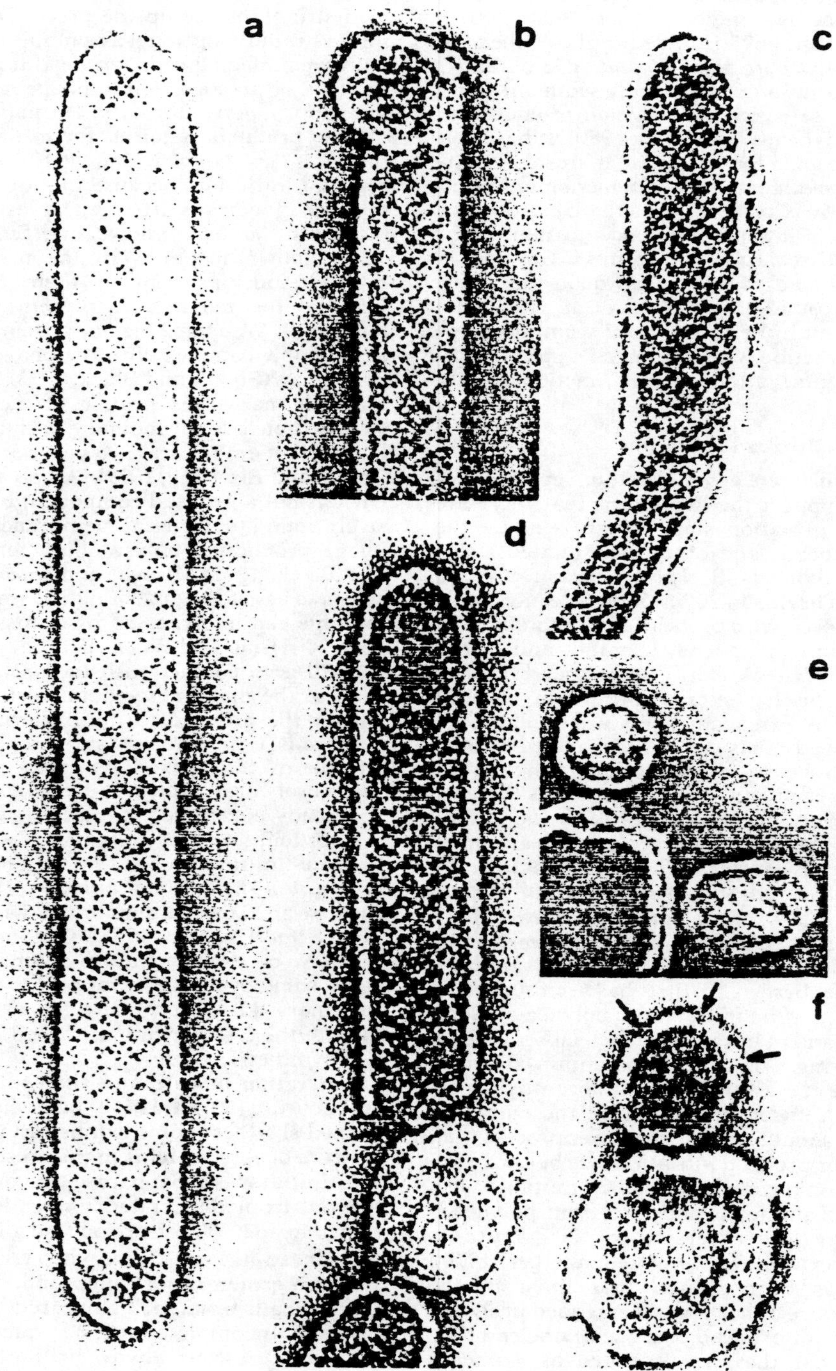

FIG. 5. Proheads and polyheads produced by a gene 24 amber mutant. (a) Giant prohead. (b) Polyhead growing from distal vertices of an uncapped prohead. (c) A pitch change has resulted in the growth of a polyhead (narrow tube) which was initiated as a giant prohead. (d) Giant proheads attached to a membrane vesicle. The arrow points to the proximal vertex plug (gp20) attaching the core to the vesicle even when the shell is broken away. (e) A normal and an isometric prohead after gp24 addition in vitro to fill in the vertices. (f) A prohead on a vesicle before gp24 addition. Arrows point to gaps visible at some of the vertex positions. From Showe and Onorato (1978) and Onorato et al. (1978).

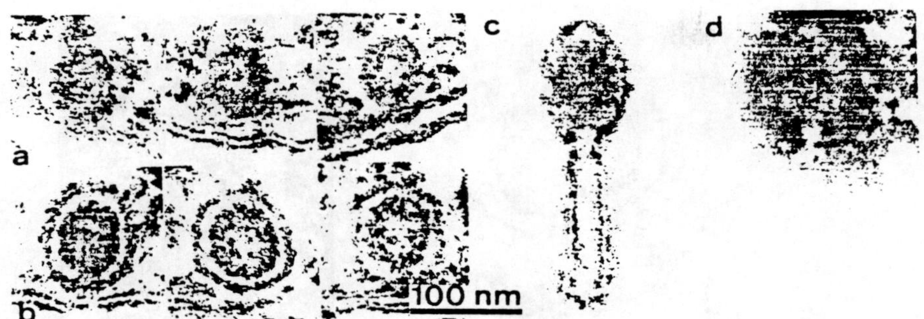

FIG. 6. Naked prohead cores assembled in vivo. (a) Naked cores from a mutant in the shell protein gene 23. (b) Gene 24 mutant proheads for comparison with panel a. Panels a and b were prepared as a combined imbedding before thin sectioning. (c and d) Negatively stained naked cores produced by gene 20 mutant E481. In this mutant the cores detach from the membrane and may attach a tail. From Traub et al. (1981).

have observed shell-less, membrane-bound cores in thin sections of 23^- amber mutant-infected cells (Fig. 6). These are absent, as expected, in *am*23 *am*22 infections; surprisingly, they are observed in *am*20 *am*23 infections, except for the most amino-terminal gene 20 amber mutant, in which about one third of the cores become detached from the membrane. Traub et al. conclude that gp20 is not required to initiate core formation, but only shell formation. They suggest that some part of the neck, or head-tail connector, identified by Coombs and Eiserling (1977), has this core initiation function. This structure also contains proteins in multiples of six, matching the symmetry of the core.

It should be pointed out that gp20 is large enough so that an amber fragment might lose the binding site for gp23 but still be able to bind to the membrane and initiate core formation. A second-site mutation, possibly in gene 40, allows phage to be produced by using the gp20 amber fragment B8, which is about 8,000 molecular-weight units shorter than wild-type gp20 (Volker and Showe, unpublished data).

If the core is initiated on a neck which need not include gp20, the latter could be added to the neck only to serve as an anchor for the shell. But gp20 clearly is also required to anchor the core to the membrane, since 20^- cores were found free in the bacterial cytoplasm when the most amino-terminal gene 20 amber mutant was used. Driedonks et al. (1981) has suggested that naked cores are aberrant structures analogous to polyheads; i.e., they form late in infection, using a non-physiological initiator. A decision on this question should be possible by using purified gp20 to try to initiate core or prohead formation in vitro or by order-of-function studies in vivo.

Mechanism of Prohead Elongation and Form Determination

Determination of the mechanism of assembly of the T4 head has been seen as a challenge, in part because it was argued that regulation of the shape of the prolate T4 capsid must be more complex than form determination of phage or viruses with heads which are regular icosahedra. However, there is still no direct evidence for a unique sequence of gene product addition during normal prohead formation. The *ts*24 prohead, which lacks all of the vertices except the one bound to the membrane, can be matured to viable phage (Bijlenga et al., 1973), but this particle has not been demonstrated to be an intermediate in wild-type infections. Yanagida et al. (M. Yanagida, Y. Suzuki, and T. Toda, Adv. Biophys., in press) have observed both naked cores and partially completed proheads, or "cups" (Fig. 7), in wild-type-infected cells grown at 19°C. These structures also accumulate in certain cold-sensitive mutants, but have not been shown to be converted to viable phage. A satisfactory explanation for form determination of the T4 head will probably depend on a more detailed model for the mechanism of head assembly than we now have.

Variation in prohead form. The prohead shell and core proteins can assemble to form proheads with various lengths, all of which can be matured to infective phage. Wild-type infections produce small numbers of phage with isometric and intermediate-length heads (petites), as well as normally elongated ones (Mosig, 1963; Mosig, Carnighan, Bibring, et al., 1972). Mutations in certain head genes (Table 2) can greatly increase the percentage of phage with these shorter heads (to up to 70% [Eiserling et al., 1970]) and can also lead to the production of extra-long heads (Doermann et al., 1973a; Doermann et al., 1973b) termed giants, or lollipops. (The short-headed phage contain partial chromosomes and so are only infective when two or more infect simultaneously.) Heads which are longer or shorter than normal are also produced by mutants which probably indirectly affect head gene proteins (Chao et al., 1974; Wever et al., 1981) and by growth of wild-type T4 in the presence of amino acid analogs (Cummings et al., 1973; Cummings and Bolin, 1976) or at high temperature (Wever et al., 1981).

Variation in head width is a much rarer occurrence than length variation. Phage heads which are isometric, but as wide as the normal prolate head is long, have been described (Doermann, personal communication; Paulson et al., 1976), and they are always associated with a mutation in gene 22. Gene 67 mutants produce aberrant proheads which appear to vary in both length and width, but these structures have not yet been extensively characterized (Volker, Gafner, Bickle, and Showe, 1982; Volker, Kuhn, Showe, and Bickle, 1982).

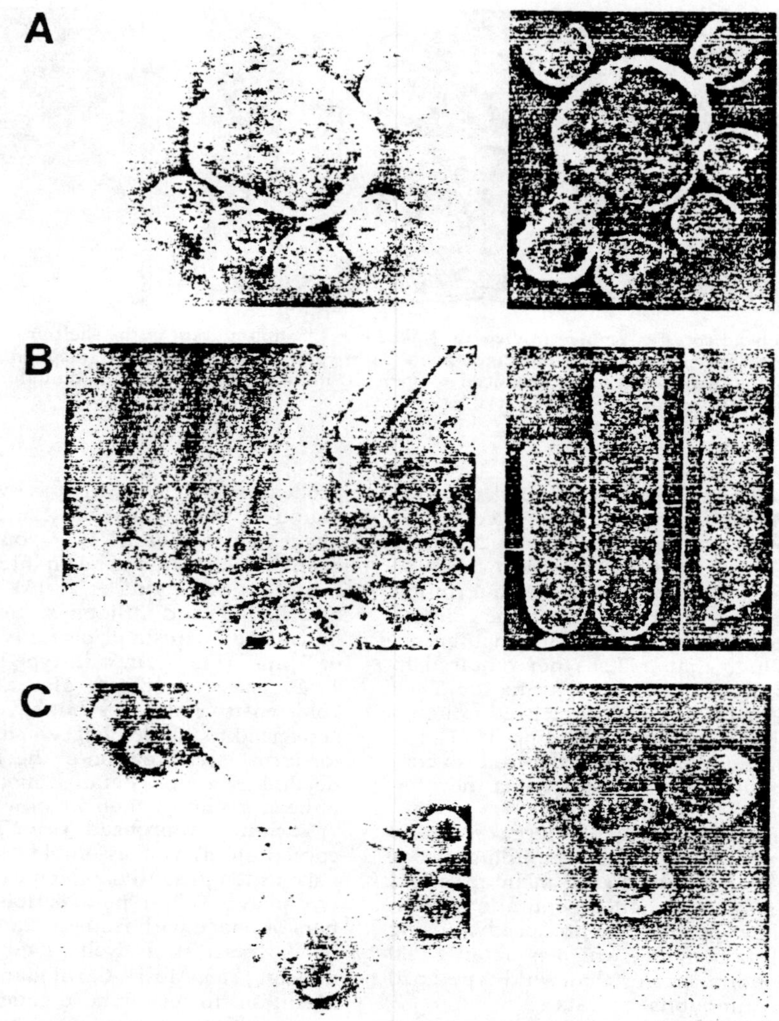

FIG. 7. Head-related structures produced by bacteria infected with cold-sensitive gene 23 mutants. (A) Uncapped proheads from mutant 23(cs51). (B) Proheads and uncapped giant proheads from mutant 23(cs53). (C) Proheads from mutant 23(cs780) which have aberrant distal caps. This mutant is a revertant of tsA78. These proheads should be compared to those of Fig. 4d which are known to have a shell of uncleaved gp23 which is in the expanded or "transformed" state characteristic of mature heads. From Yanagida et al. (in press).

In addition to regular icosahedra and prolate icosahedra, T4 heads with less regular structure occur occasionally in stocks. These variants may have biprolate heads with one, two, or three tails, or they may have normally shaped heads with extra tails or a tail at one of the nonapical vertices (Boy de la Tour and Kellenberger, 1965). The numbers of these particles may be increased by growth with putrescine or cadaverine, which also increase the number of petite phage produced (Cummings and Bolin, 1976). Similarly, temperature-sensitive gene 22 mutants grown at permissive temperature have been observed to produce multitailed, wide, and biprolate phage as well as large numbers of petites. (Paulson et al., 1976).

Mechanisms proposed for form determination during prohead assembly have usually attempted to account both for the shape of the normal T4 head and for the form variants found in both wild-type and mutant strains (Cummings and Bolin, 1976; Paulson et al., 1976; Showe and Onorato, 1978). However, it has also been suggested that petite and giant phage are assembled by a different mechanism from that which gives rise to the normal prolate head (van Driel and Couture, 1978b).

Evidence for assembly mechanisms requiring specific interactions between shell proteins and core proteins. Since mutations in both shell and core protein genes can lead to the production of prohead size variants (Table 2), it has been assumed that both the shell and the core contribute size-determining information during assembly. Doherty (1982a, 1982b) has provided evidence for specific, length-determin-

TABLE 2. Mutations affecting the form of the T4 head[a]

Gene	Gene product function	Type of mutation	Phenotype	Reference
20	Shell and/or core initiator	*ts*	Polyheads and petite heads	Volker, 1981
67	Major core component	Frameshift	Proheads of various size with no cores or unstable cores	Volker et al.[b]
22	Major core component	*am* (suppressed)	Petite phage	F. A. Eiserling, personal communication M. K. Showe, unpublished data
		ts (suppressed)	Petite and multitailed phage	Paulson et al., 1976
		ts	Wide-headed phage	Doermann, personal communication
23	Major shell protein	Missense (*ts*)	Petite phage	Eiserling et al., 1970
		Missense (*ts*)	Petite phage	McNicol et al., 1977
		Missense (*ptg*)	Petite and giant phage	Doermann et al., 1973
		ts	Misshapen, expanded proheads	Aebi et al., 1974
		cs	Uncapped proheads and giant proheads, proheads with apparently expanded distal caps	Yanagida, personal communication
24	Shell vertex protein	*ts* (intermediate temperature)	Giant phage	Bijlenga et al., 1976
		am	Giant proheads	Onorato et al., 1978
*ip*III	Semiessential core protein	*am*	Petite phage	Black and Brown, 1976

[a] Only mutations affecting head form are listed. Other phenotypes, e.g., those which prevent prohead formation altogether or interfere with prohead maturation or DNA packaging, are described in the text.
[b] Volker, Gafner, Bickle, and Showe, 1982; Volker, Kuhn, Showe, and Bickle, 1982.

ing interactions among these gene products by analyzing second-site revertants of the mutant in gene 23, *ptg*19-80. This strain, which produces intermediate-length petites and a few giants was found to carry two mutations: a gene 23 mutation, 19-80*c*, which alone produces more petite particles than the original isolate, and *sup-or*, a partial suppressor probably located in the gene for IPIII. Additional mutations which cause more complete suppression of the *ptg* phenotype were obtained as pairs, one in gene 22 and one in gene 24, both required for suppression. The suppressor pairs alone have no phenotype, producing a normal burst as judged by plaque morphology. The two pairs tested were found to be allele specific. They suppress only *ptg*19-80 or other mutations in isolates which cannot be distinguished from the *ptg*19-80 strain by recombination and fail to suppress *ptg* mutations located at seven other sites in gene 23. This result shows that at least one of the two protein-protein interactions (gp22-gp23 or gp24-gp23) is both specific and required to determine the length of the normal head. Whether two or more specific interactions are involved is not clear, since the suppressor mutations were only tested pairwise as they were isolated, against other *ptg* mutations.

Evidence against specific core-shell interactions in length determination. As mentioned above, "naked" cores are assembled both in vivo and in vitro when gp23 is absent. These have the size and shape of the cores found in 21⁻ or 24⁻ mutant proheads (Traub et al., 1981; van Driel and Couture, 1978b), suggesting that the shell proteins are not required to determine the form of the normal head despite the fact that mutations in the shell protein genes lead to heads of altered length.

One way in which a scaffolding protein might function to establish the curvature of the shell would be to form a core protein-shell protein complex which would then polymerize to form the prohead. However, core and shell proteins from dissociated proheads do not reassociate in this fashion in vitro. On the contrary, several of the core proteins form a 6S complex (Showe and Black, 1973; van Driel, 1980), the shell proteins gp23 and gp24 form hexamers and pentamers, respectively, and these oligomers sediment independently in the ultracentrifuge under conditions in which proheads are formed (van Driel, 1980).

Determination of the width of the prohead. Mutations leading to width variation have only been obtained in genes 22 and 67, both essential core proteins (Table 2). This suggests that the core is the primary determinant of the width, or "T number," of the normal prohead (Caspar and Klug, 1962). Polyheads lacking one of the core components gp22 or IPIII are narrower than those produced by mutants with normal core proteins (Steven, Aebi, and Showe, 1976), leading to the suggestion that the core places a lower limit on the prohead diameter (Showe and Onorato, 1978). van Driel and Couture (1978b) suggested that shell proteins assemble around a preformed core, which they showed could assemble in the absence of shell proteins. All of these observations are consistent with the normal width being determined exclusively by the structure of the core.

Models suggested to explain determination of prohead length. Kellenberger (1969, 1972b) suggested several mechanisms which might function in determining the size of large icosahedral viral capsids or the length of helically symmetric structures: cumulated strain, vernier, and form-determining core. The cumulated strain model proposed that each subunit successively added to the growing structure is increasingly deformed, until finally growth stops because subunits will no longer bind. There is no evidence yet for or against this type of mechanism. Elongation limited by a vernier mechanism was supported by Cummings and Bolin, (1976), and Paulson et al. (1976) proposed specifically that a vernier might operate between the core and the shell of the prohead. However, a direct test of the mechanism, based on the measured periodicity of the polyhead shell made from mutant gp23, proved negative (Paulson and Laemmli, 1977). The mutant gp23 which gives rise to petite proheads makes polyheads with the normal lattice constant. The suggestion of a form-determining role for the core appears now to be correct. However, the way the form of the core is itself determined is still unknown.

The realization (Steven, Aebi, and Showe, 1976) that gp23 lattices have an intrinsic curvature greater than that of the cylindrical part of the prohead led to the proposal of a kinetic model for head length determination (Showe and Onorato, 1978). It was proposed that the shell tends to close into an isometric prohead, but is prevented by the growth of the core. Reduced amounts of gp22 should result in production of petite proheads because the core grows more slowly, whereas reduction of gp23 should decrease the number of petites. Gene dosage experiments support this model, but no more than 10% petite particles can be produced by multiply infecting wild-type and gene 22 amber mutants to reduce the levels of gp22 (Showe, unpublished data). More convincing results are obtained with the petite mutant E920g. Its burst is normally 60 to 70% petite phage, and this can be raised to 90% by coinfection with E920g-*am*22 (Showe, unpublished data) or reduced to 0% when the concentration of the mutant gp23 is reduced by 85% by coinfecting with a gene 23 amber mutant (Eiserling et al., 1970; Showe and Onorato, 1978).

Although the concentrations of core and shell proteins can be manipulated to change the ratios between normal and petite phage, it is not possible to produce slightly longer phage heads by this means as the kinetic model predicts. This is understandable if the wild-type core has a structure with a defined length, as suggested by the experiments of van Driel and Couture, (1978b). Proheads longer than normal are produced in substantial numbers by mutations which eliminate the vertex protein gp24 (Bijlenga et al., 1976, Cummings et al., 1977; Onorato et al., 1978). Since these are about 10 times the normal length rather than slightly longer than the normal head, this finding is probably better interpreted as a defect in the assembly process which leads to uncontrolled elongation, rather than as an alteration in a mechanism which regulates head length. In contrast, elongated phage or proheads produced as a result of a mutation in gene 23 appear to show distributions of lengths which are allele specific and frequently center around five normal head lengths (Doermann et al., 1973a; Doermann et al., 1973b; Yanagida, personal communication). The length distribution of the giant phage heads produced by growth of T4 in the presence of canavanine can also be shown to be dependent on gene dosage. The most striking effect is a reduction in average length from over 10 times normal to less than 5 times normal when gp22 is limited by coinfection with an *am*22 mutant (Cummings et al., 1977).

Reconciliation of genetic and biochemical data for length determination. Length determination of the normal head dictated by the structure of the normal prolate core can be reconciled with size variation induced by mutations in shell protein genes if the core and the shell grow concurrently. In this case gene dosage effects must be seen as modifying a predetermined structure, rather than as being the primary determinant of the structure.

Giant proheads were suggested to arise through failure of distal-cap formation (Showe and Onorato, 1978). This has been shown to occur frequently in gene 24 mutants which lack the vertex protein. The distal cap must be inherently less stable than the one bound to the membrane if gp24 is not present, since the distal cap has a gap at the apex which is occupied by gp20 on the membrane-bound cap. Yanagida (personal communication) has found cold-sensitive gene 23 mutants which produce both giant and normal-length proheads without distal caps. This mechanism for giant formation requires the assumption that the core structure can be extended when the shell is open, since giant proheads have cores throughout their length.

Gene 24 giants are limited in length only by the size of the cell (Bijlenga, et al., 1976). The giants produced by the *ptg* mutations in gene 23 are, in general, shorter than 24^- giants (Doermann et al. 1973a; Doermann et al., 1973b). They can be explained if the gp23 of which they are composed has decreased affinity for gp24 (allowing core and shell extension), but has also a higher intrinsic curvature, resulting in the production of giant proheads longer than wild type, but shorter than the 24^- giants. The finding that second-site revertants of the *ptg* mutants consist of paired mutations in genes 22 and 24 is at least consistent with this interpretation. The mutation in gene 22 would reverse the *pt* phenotype, and the mutation in gene 24 would allow the vertex protein to bind with normal affinity to the mutated gp23, allowing the normal-length proheads to mature and preventing the formation of giant phage.

Gene 23 mutants which produce petite proheads are likely to arise from an increased intrinsic curvature of their gp23, as previously suggested to account for both the *pt* phenotype and the narrower polyheads produced from this protein (Paulson and Laemmli, 1977; Showe and Onorato, 1978). *pt* mutant E920g, which does not produce any giants (as do the *ptg* mutants), is presumed to make an altered gp23 of higher curvature which retains the normal affinity for gp24. The increased tendency to curve of the altered protein is confirmed by the demonstration that each of two independently isolated *pt* mutants is able to substitute gp23 for gp24 at the icosahedral vertices (McNicol et al., 1977; Showe and Onorato, 1978).

The production of petite phage by core gene mutations has been explained as a decrease in the rate of core formation relative to shell formation (Showe and

Onorato, 1978). This explanation is consistent with the available gene dosage experiments (Cummings and Bolin, 1976; Cummings et al., 1977).

Engel et al. (1982) have recently proposed a model for the surface structure of the normal prolate core. Although their model deals only with one core component, gp22, it suggests that the length of the core is limited by the way the chains of gp22 are packed. Further studies on form determination of the T4 head clearly require a better understanding of the core structure. In particular, it seems important to establish whether mutants in core proteins (gp22, gp67, IPIII) which produce proheads of altered size can produce naked cores of altered size as well, or whether petite prohead production requires concurrent assembly of the core and shell.

PROHEAD MATURATION

When assembly of the T4 prohead is complete, all of its proteins except gp20 are cleaved. In the case of gp23, IPI, IPII, and IPIII, most of the protein remains associated with the prohead, and analysis has shown that the cleavage removes an amino-terminal peptide (Isobe et al., 1976b). For two other head proteins which also lose peptides, gp24 and gp*alt*, the locations of the cleaved bond(s) are not known. The remaining components, gp67, gp21, gp22, and 16K, are extensively digested. All of the peptide fragments are lost from the prohead except for the two acidic internal peptides found in the phage head, i.e., peptide II and peptide VII, derived from gp67 and gp22, respectively. This cleavage must be efficiently controlled, since newly synthesized precursors are not cleaved before they are assembled.

The cleaved prohead undergoes a series of changes which transforms the rather unstable particle into the chemically resistant capsid (Fig. 8). The cleaved prohead is released from the membrane, and the gp23 undergoes one or more conformational changes which alter the chemical, physical, and antigenic properties of the shell. In wild-type-infected cells, this transformation is probably accompanied by DNA packaging, but transformation of cleaved mutant proheads or of polyheads is a spontaneous process and need not be accompanied by packaging either in vivo or in vitro.

Cleavage: Genetic Control and Enzymology of T4 Prehead Proteinase

Identification of the proteinase responsible for prohead cleavage. Mutations in the Y group of head genes which prevent proper head assembly also prevent nearly all of the maturation cleavage reactions from taking place in vivo (Champe and Eddleman, 1967; Laemmli, 1970). However, a phage-specific proteinase activity not found in extracts of uninfected *E. coli* can be detected in lysates of most of these mutants. This activity was demonstrated by two types of assay: the cleavage of the phage internal peptides from a mixture of radioactive phage proteins (Giri et al., 1976; Goldstein and Champe, 1974) and the preferential cleavage of purified core protein gp22 (Onorato and Showe, 1975). These two assays give equivalent results: although most Y mutant lysates are highly active, gene 21 mutant lysates are inactive. Lysates

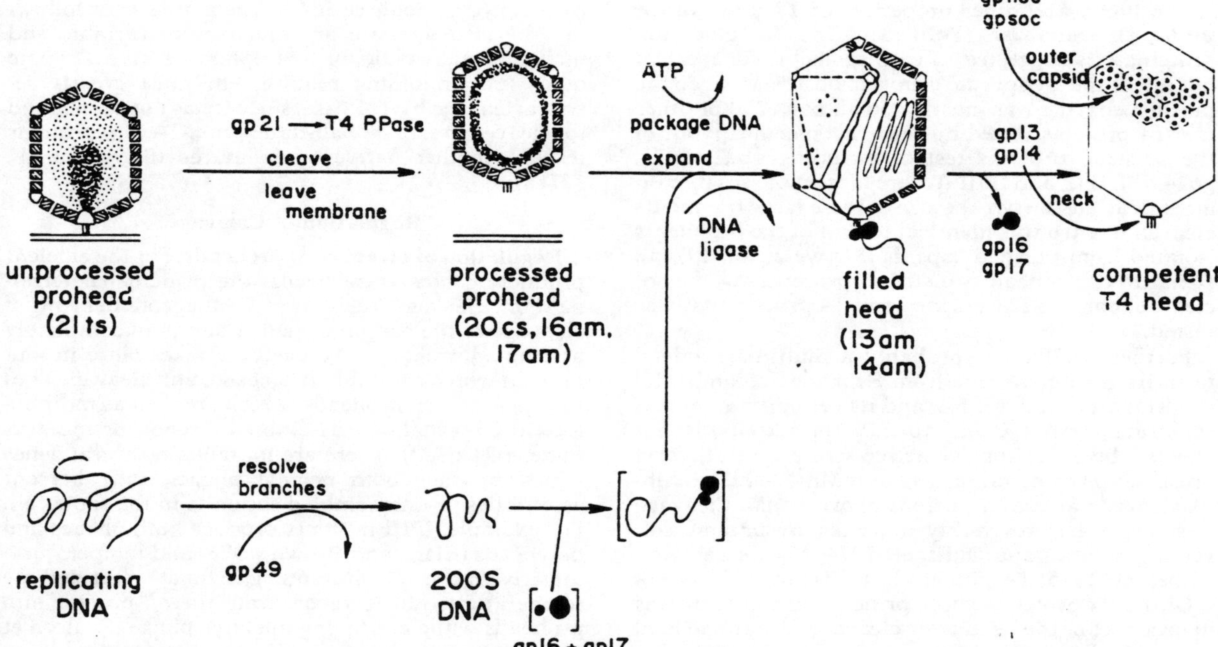

FIG. 8. Prohead maturation. Parallel pathways are shown for maturation of phage DNA from the replicating pool and prohead processing. These join in wild-type infections with the attachment of the prohead to the vegetative DNA. As in Fig. 1, brackets indicate that the intermediate is inferred from genetic experiments. Parentheses enclose the genes in which mutants accumulate the illustrated intermediate.

from strains which mature normal heads have little activity, as do those from strains carrying amino-terminal amber mutations in gene 22. The activity was also shown to be associated with purified mutant *ts*23 proheads and was able to cleave in vitro their gp23 to the shortened form found in the mature capsid, (Onorato and Showe, 1975). Laemmli and Quittner (1974) showed that the gp23 of partially purified polyheads would also cleave in vitro if it was prepared from a strain which is wild type for gene 21. Cleavage of IPIII in crude lysates was also shown (Bachrach and Benchetrit, 1974). All of these results suggested that proheads contain a phage-coded proteinase either coded for or under the control of gene 21.

This interpretation was confirmed by purification of the gp22-cleaving activity from a lysate of a gene 24 amber mutant-infected culture (Showe et al., 1976a; Showe et al., 1976b). The purified enzyme has a molecular weight of 18,500 on SDS-polyacrylamide gels. Antibodies prepared against it cross-react with a 27,500-molecular weight polypeptide shown in vitro (Showe et al., 1976b) and in vivo (Showe, 1979) to be a precursor to the 18,500-molecular-weight antigen. Immune replicas of polyacrylamide gels were used to show that the 27,500-molecular-weight polypeptide is the product of gene 21: 21^- amber mutants lack it and produce instead shorter cross-reacting species whose length is a function of their genetic map position. The identification of gp21 in crude lysates as a protein of about 26,000 molecular weight was made independently by Castillo et al. (1977); Laemmli and Johnson (1973), pointed out a band of this size as one of a number of differences between *ts*21 and *am*21 proheads.

Specificity and other properties of T4 prehead (or prohead) proteinase (T4PPase). The 18,500-dalton proteinase is suggested to be responsible for most, if not all, of the T4-specific cleavage observed, since the purified enzyme was inactive on all bacterial or other non-T4 proteins tested but cleaved specifically all of the prohead proteins tested, including gp22, gp23, gp24, IPI, IPII, and IPIII. Where large moieties remain after cleavage in vivo, the size of these large fragments cleaved in vitro was identical to that of the fragments isolated from purified capsids (Showe et al., 1976a). Denatured prohead substrate proteins were not cleaved, nor was any purified non-T4 protein that was tested.

Purified T4PPase is probably a multimer, judged from its elution volume from Sephadex columns. Its isoelectric point is pH 5.3, and its pH optimum on all substrates tested is 7.5 to 8.0. Its activity is not affected by inhibitors of active-serine or sulfhydryl proteases or by metal ions, except Mn^{2+}, which inhibits cleavage at concentrations above 1 mM. Cleavage is strongly but reversibly inhibited by organic solvents (Laemmli and Quittner, 1974; Showe and Kellenberger, 1975; Tsugita et al., 1975).

Of the six proteins whose principal moiety remains in or a part of the head after cleavage, the amino acid sequence through the cleavage site of the precursor has been determined for four. The cleaved forms of IPI, IPII, IPIII, and gp23 isolated from mature capsids all have alanine amino termini, and the sequences of their respective precursors all show a glutamic acid to be the carboxyl side of the cleaved bond (Isobe et al., 1976a, Isobe et al., 1976b; Isobe et al., 1977; Isobe et al., 1978; Tsugita et al., 1975). For gp23 and IPIII, T4PPase has been shown to cleave the same peptide bond in vitro (Showe et al., 1976a; Showe, 1979).

Proteins and peptides which are cleaved to small fragments during maturation show the primary specificity of T4PPase to be wider than the above-described result suggests. The 67-residue peptide removed from gp23 by a Glu-Ala cleavage is further cleaved in vitro at a Glu-Gly bond and at a Glu-Asn bond (Showe, 1979). Peptide II is cleaved from gp67 in vitro (Showe, unpublished data) and in vivo (Kurtz and Champe, 1977; Volker, Gafner, Bickler, and Showe, 1982) at a Glu-Gly bond. The sequences determined for gp22 (M. Parker and A. Christensen, personal communication) and peptide VII (van Eerd et al., 1977) show that the latter is cleaved from gp22 at Glu-Lys and Glu-Gln dipeptide bonds. Although the latter two specificities have not been demonstrated with purified T4PPase in vitro, the fact that glutamic acid is the carboxyl donor of every bond cleaved during prohead maturation strongly suggests that T4PPase is the only enzyme functioning inside the prohead. Other, probably host, proteases may function to further degrade some peptides after they leave the head. The report that denatured gp22 can be cleaved to yield peptide VII by crude lysates (Kurtz and Champe, 1977) is consistent with an initial cleavage under these conditions by a host enzyme. Denatured gp22 is not a substrate for purified T4PPase (Showe et al., 1976a).

Table 3 summarizes the primary sequence around known cleavage sites for T4PPase. A common requirement for cleavage is that the carboxyl side of the cleaved bond be glutamic acid, immediately preceded by two hydrophobic residues. The amino acids following the cleavage site are much more variable and include basic, acidic, or hydrophobic residues. Some other protein-folding requirement must greatly restrict cleavage by T4PPase, since it has not been found to cleave any of a half-dozen non-T-even proteins tested, whether native or denatured (Showe et al., 1976a).

Regulation of Cleavage

Regulation of cleavage of proheads. For the efficient production of mature heads, the principal determinant of cleavage regulation is the confinement of cleavage of the core and shell subunits to completely assembled proheads. No cleavage takes place in mutants in which assembly is blocked, and cleavage does take place when proheads, which are seen as morphologically normal, are assembled (except for mutants in gene 21 itself). There are mutants in several genes which produce both normal phages and aberrant assemblies or unassembled subunits in the same cell. For example, IPIII mutants produce both phages and polyheads (Black and Brown, 1976) and temperature-sensitive gene 22 mutants grown at intermediate temperatures fail to incorporate the *alt* protein into proheads while producing infective phage (Paulson et al., 1976). In both cases, only the precursors assembled into proheads are cleaved, while the unassembled or improperly assembled proteins remain uncleaved (Paulson et al., 1976; Steven, Couture, Aebi, and Showe, 1976).

TABLE 3. Specificity of the proteolytic cleavage which accompanies T4 prohead maturation

Gene product	Cleavage site (*)	References[a]
gp23	Pro Leu Leu Glu* Ala Glu Ile Gly	Isobe et al., 1976
IPI	Thr Ile Thr Glu* Ala Thr Leu Thr	Isobe et al., 1977
IPII	Phe Ile Ala Glu* Ala Arg Val Gly	Isobe et al., 1976a
IPIII	Phe Ile Ala Glu* Ala Thr Val Val	Isobe et al., 1976
gp23	Pro Leu Leu Glu* Gly Glu Gly Leu	Isobe et al., 1976; Showe, 1979
gp23	Lys Ile Phe Glu* Asn Glu Gln Lys	Isobe et al., 1976; Showe, 1979
gp22	Lys Ile Ala Glu* Lys Ala Glu Glu	Parker et al., in preparation
gp22	Lys Ile Ala Glu* Gln Ala Ser Lys	Parker et al., in preparation
gp67	Phe Leu Ile Glu* Gly Glu Glu Pro	Volker et al., 1982

[a] The references given describe the amino acid sequence in the neighborhood of the cleavage site, from either protein or DNA sequencing. Other references, describing the identification of precursor-product relationships, are given in the text.

The explanation proposed for this regulation (Showe et al., 1976b; Showe, 1979) is that gp21 assembles with the prohead as an inactive zymogen which is only activated inside the structure on completion of the shell. Proteolysis of the core and shell is accompanied by protease autodigestion. This explanation is consistent with the following observations. Gp21 itself cannot be demonstrated in vitro to have any proteolytic activity. In Y gene mutant infections, it is slowly cleaved through a series of inactive intermediates to T4PPase, but in wild-type infections, it is cleaved directly and rapidly to the active 18,500-molecular-weight polypeptide (Showe, 1979). In vitro, T4PPase can be shown to cleave gp21 to the 18,500-molecular-weight species. The purified enzyme self-digests in vitro. Finally, *ts*21 proheads contain about 50 molecules of gp21, but mature heads contain only 2 to 4 molecules of the 18,500-molecular-weight antigen and no gp21.

How the gp21-to-T4PPase conversion is started is not known. Although the first molecule of T4PPase should be able to activate the remaining gp21 in the prohead, it is not clear how completion of the prohead can accomplish the activation of the first molecule of protease unless, like some other proteinases, gp21 can self-activate (Al-Janabi et al., 1972).

Regulation of cleavage in mutant-infected *E. coli*. Gp21 is slowly cleaved to T4PPase in Y gene-infected cells (Showe et al., 1976b; Showe, 1979), but no cleavage of any other proteins takes place unless the cells are lysed. This is particularly striking in the case of 24^- mutants. These accumulate unprocessed proheads in which, at late times after infection, nearly all of the gp21 has been converted to T4PPase. These are stable to isolation only at pH 6 (Onorato et al., 1978). The proteinase in them can be activated by the addition of gp24. This completes the vertices, and cleavage and maturation of the proheads to capsids follow. If, instead of adding gp24, the pH is raised to 7.6, the proheads disintegrate and the dissociated subunits are cleaved. A possible explanation for these observations is that although gp21 is cleaved to T4PPase in Y gene mutant lysates, it remains inactive until the cleaved peptides are removed. In vivo, this would require completion of the shell; in vitro, raising the pH or dilution would have this effect.

Gp*lip*. A 35,000-molecular-weight protein inhibitor of T4PPase (late inhibitor of proteinase) is present in Y gene mutant lysates (Showe et al., 1976a). It is coded for by a late T4 gene (Showe, unpublished data). Amber mutants have now been obtained to locate the gene and to characterize its function (U. Küri, thesis, Universität Basel, 1979). The gene is located between the head gene 24 and the tail gene 25, in the map order 24-*hoc*-*lip*-*uvs*W-25. The *lip* product is nonessential for T4 growth, since an amber mutant which produces no detectable cross-reacting *lip* antigen reduces the burst size by only one-third. Gp*lip* is probably a minor prohead component, since (i) it is cleaved like other prohead proteins if and only if proheads mature, (ii) a failure to obtain recombinants between *lip* mutants and gene 22 mutants suggests that gp*lip* and gp22 interact directly, and (iii) the inhibition of T4PPase activity by gp*lip* in vitro suggests a specific interaction between it and gp21.

Enzyme Activation by Maturation Cleavage

Limited proteolysis is a widespread mechanism for enzyme activation, so it is reasonable to ask whether any of the prohead proteins has an enzymatic activity which is activated by its cleavage.

The activation of gp21 to T4PPase is discussed above. Its inactivation by autodigestion might also be important as a control mechanism, since otherwise active T4PPase would be injected into the cell with the DNA.

Two cleaved internal proteins have been shown to have functions after their injection with the phage DNA. IPI, nonessential in most laboratory strains of *E. coli* (Black, 1974) prevents restriction of the infecting DNA in *E. coli* CT596 (Black and Abremski, 1974). The injected cleaved *alt* protein ADP-ribosylates the subunit of the host RNA polymerase in the first few minutes after T4 infection (Horvitz, 1974a, Horvitz, 1974b; Seifert et al., 1969; Seifert et al., 1971). The activity, as measured in vitro, is maximal at 1 min after infection (Skorko et al., 1977). It is not known, however, whether cleavage is required for the activity of either of these two proteins or incidental to it. Since IPI function is required in CT596 before expression of the infecting DNA, newly synthesized IPI would not be effective even if the uncleaved protein is active in vivo, and there is no known assay for its function in vitro.

It is not known whether gp*alt* is synthesized early or late, and its map position offers no clue (Goff, 1979), so the rapid decrease in *alt* activity after infection is not evidence against activity of newly synthesized but uncleaved protein. *alt* activity has not been assayed in vitro with lysates of Y gene mutants which do not cleave the *alt* protein.

Black and his co-workers (Black et al., 1981; Manne et al., 1982) have recently purified a late protein which has ATPase and endonuclease activities (see below). Both the antigen and the activities are missing from lysates of mutants in which the maturation cleavages do not occur. The enzyme appears to be a minor cleavage product derived from the major capsid precursor gp23 (Rao and Black, manuscript in preparation).

DNA PACKAGING

In outline, DNA packaging is the same among all the complex, double-stranded DNA-containing bacteriophages: DNA fills a preformed DNA-free prohead. In the case of T4, this can be understood to take place by the interaction of three general classes of components: (i) the mature, processed prohead, activated for packaging according to the previously outlined assembly and processing scheme; (ii) concatemeric DNA, which is terminated in headful packaging to yield circularly permuted and terminally redundant DNA; and (iii) soluble, packaging-linkage proteins which link the DNA to the prohead at the DNA entrance vertex or connector and act in packaging and termination.

Packaging and Replication

Early genetic observations of phage T4 led to the concept of headful packaging, i.e., that the length of the chromosome is determined by the volume of the preformed head shell into which it is inserted (Streisinger et al., 1967). Evidence that DNA packaging was coupled directly to the cutting of concatemeric DNA to the mature length originated in the studies of Frankel (1966, 1968) and Fujisawa and Minagawa (1971). In addition, the first biochemical evidence in any phage system for headful packaging came from a demonstration that T4 DNA packaging could resume into partially packaged, *ts*49 heads reversibly arrested during filling (Luftig et al., 1971). However, T4 DNA packaging initially proved more difficult to uncouple from replication than that of other phages, and a number of studies suggested a dependence of packaging or head formation upon concurrent DNA replication (Chao et al., 1974; Kuhl and Hofschneider, 1969; Luftig and Lundh, 1973). More recent evidence (Black, 1981b; Hsiao and Black, 1977; Wagner and Laemmli, 1976) now suggests that these observations should probably be interpreted as resulting from the indirect, secondary effects of replication on late transcription. In addition, it is possible that repair synthesis of DNA is needed to produce a DNA substrate efficient in DNA packaging in certain late mutants (e.g., 49^- mutants [Laemmli, Teaff, and D'Ambrosia, 1974]). Older observations may also reflect regulatory interactions between DNA metabolism and packaging in vivo. In any case, it is clear that T4 DNA polymerase (gp43) and, therefore, DNA replication are not required for DNA packaging (Hsiao and Black, 1977; Hsiao and Black, 1978c).

Prohead Structure and DNA Packaging: Packaging Intermediates and Prohead Expansion

Laemmli and Favre (1973) have demonstrated radioisotopically labeled precursors to the T4 head in wild-type-infected cells. The first, an unprocessed particle termed prohead I, is probably essentially identical in structure to the tau particles which accumulate in temperature-sensitive gene 24 mutants. These were also shown (Bijlenga et al., 1973) to be precursors to full heads in temperature shift experiments. Two other precursor species, prohead II and prohead III, could be detected in wild-type infections. Prohead II was found not to be attached to the replicative DNA, whereas prohead III was attached and contained about 50% of the full DNA complement. Prohead III was in this respect similar to the partially filled heads which accumulate in 49^- infections (Luftig et al., 1971). Proheads II and III were reported to contain the cleaved form of the major capsid protein, gp23, but uncleaved major core proteins gp22 and IPIII. It was concluded that cleavage of the core proteins follows cleavage of shell proteins and accompanies packaging of DNA, and the cleavage of the core proteins to the internal peptides was suggested to drive the packaging (Laemmli, 1975; Laemmli and Favre, 1973; Wagner and Laemmli, 1976). However, the protein compositions determined for proheads II and III were based on short pulse-chase labeling experiments and have not been substantiated by subsequent characterization of these intermediates or of related mutant proheads. Therefore, their compositions and the implied sequence of shell cleavage, core cleavage, and DNA packaging cannot be considered definitive. In particular, DNA packaging into mutant proheads does not require concomitant cleavage of core proteins, since efficient filling of a fully processed prohead has been demonstrated in vivo (Hsiao and Black, 1977; Hsiao and Black, 1978c). Rather, it appears that packaging is initiated into the cleaved prohead which detaches from the bacterial membrane, and which contains only cleaved forms of all the precursor proteins (Hsiao and Black, 1977). Such particles, which accumulate in *cs*20 infection, can be efficiently matured to active heads in vivo in the absence of T4 prohead proteinase activity. Morphologically identical proheads which accumulate in 16^- and 17^- mutant infections can also be filled with DNA in vitro (Black, 1981) (*ts*16 and *ts*17 particles cannot be matured in vivo). Therefore, it appears likely that there are only two discrete, functional prohead precursors: (i) the core-containing unprocessed prohead (i.e., prohead I and tau particles), and (ii) the processed prohead, activated for packaging, which is probably not yet expanded (Fig. 8).

Gene 16^-, 17^-, and *cs*20 mutant infections accumulate two morphologically distinct particles (Fig. 9 and 10) of essentially identical protein composition, small and large processed proheads (Carrascosa and Kellenberger, 1978; Hsiao and Black, 1977; Wunderli et al., 1977). They are found in the cytoplasm of the infected cells. DNA is packaged mainly, if not exclusively, into the majority small particles in *cs*20 infections, and packaging is accompanied by an expansion to the large particles or to full heads (Zachary and Black, unpublished data). This coupling of packaging to expansion has previously been demonstrated for other bacteriophages in which cleavage either does not occur or is much less extensive than it is for T4 (Kellenberger, 1980).

The expansion of the prohead involves extensive conformational rearrangement of the shell protein

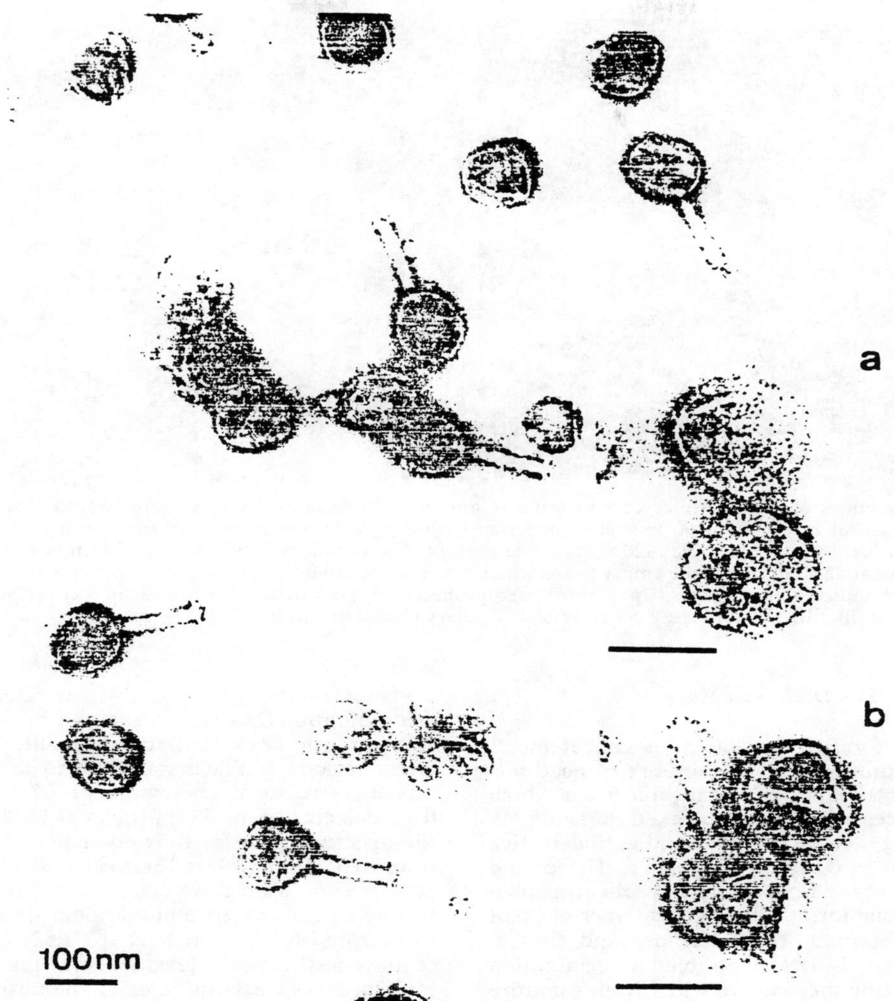

FIG. 9. Lysate from *am*17 mutant-infected bacteria. (a) Higher magnification of the rough-appearing cleaved, unexpanded proheads. (b) Higher magnification of the cleaved, expanded proheads, which frequently have attached tails. The unexpanded heads appear larger than the expanded ones because they flatten on the electron microscope grid. Bars, 100 nm. From Carrascosa and Kellenberger (1978).

gp23, discussed below. Although expansion apparently normally accompanies DNA packaging, it can occur in vivo or in vitro in the absence of packaging (Carrascosa, 1978; Hsiao and Black, 1977; Steven, Couture, Aebir, and Showe, 1976). Moreover, a significant amount of packaging can also occur into *ts*49 heads which appear to be already expanded (Luftig et al., 1971; Wunderli et al., 1977). Since expansion and packaging are to this extent independent, the mechanistic relationship between them is not yet clear.

The existence of functional giant and petite phages is of consequence for possible packaging mechanisms as well as for assembly mechanisms. The petite phages contain permuted, incomplete chromosomes of about two-thirds the wild-type length (Eiserling et al., 1970; Mosig, 1963), whereas giants contain multiple genome equivalents packaged as a single DNA molecule (Cummings et al., 1973; Doermann et al., 1973b). The existence of these particles suggests that any shell-DNA contact is relatively nonspecific and that the packaged DNA can assume, at a fixed DNA density, a variety of curvatures within the head. Study of the giant phage head shows the condensed DNA to be oriented parallel to the long axis of the head in a spool-like arrangement (Earnshaw, King, Harrison, and Eiserling, 1978).

An end in the concatemeric DNA, rather than a loop, is apparently packaged into the 49^- partially filled head (Kemper and Brown, 1976; Wagner and Laemmli, 1976). Consistent with these observations, T4 DNA has been found by glycosylation labeling experiments to enter the *ts*49 head in a linear fashion from one end (Black and Silverman, 1978). Both ends appear to be situated in proximity to the entrance vertex, since the first packaged end also appears to be the first ejected (Black and Silverman, 1978). This apparent order is opposite to that reported for most phages (Murialdo and Becker, 1978).

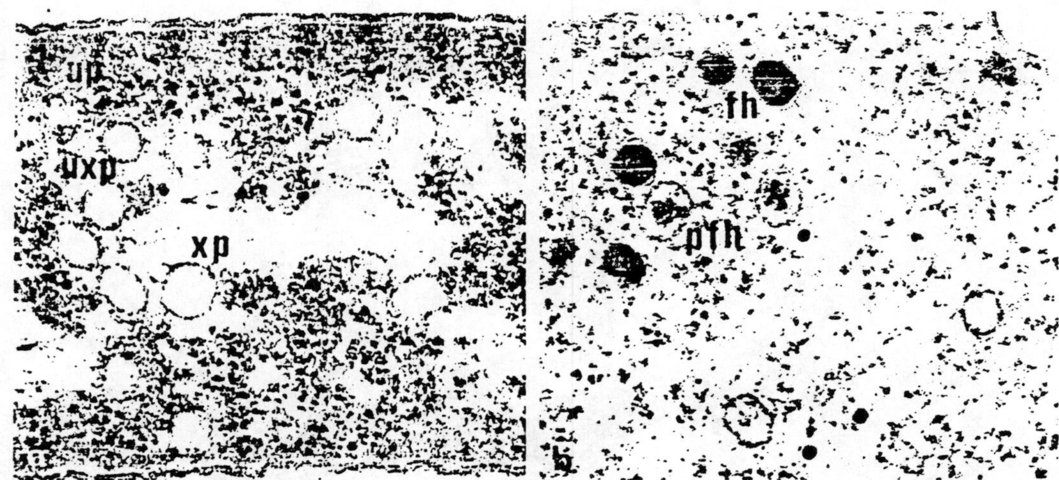

FIG. 10. Thin sections of bacteria infected with gene 16 and gene 20 mutants, showing both unexpanded and expanded processed proheads. (a) *E. coli* P301 90 min after infection with a gene 16 amber mutant. (b) *E. coli* N1252 (DNA ligase temperature sensitive) infected with the *cs*20 *ts*30 double mutant. The sample was taken after 85 min postinfection at 20°C followed by 35 min at 42°C. Filling of the empty heads which accumulate at low temperature is inhibited by the mutations in the host and bacteriophage ligase genes. Up, Unprocessed prohead; uxp, unexpanded prohead; xp, expanded prohead; pfh, partially filled head; fh, filled head. Figure courtesy of A. Zachary (Zachary and Black, 1981).

The DNA Substrate

The substrate for DNA packaging is concatemeric DNA which accumulates in the absence of head formation (Frankel, 1966) or in mutant infections which accumulate processed proheads blocked immediately before packaging, i.e., in 16⁻, 17⁻, and *cs*20 defective heads (Fujisawa and Minagawa, 1971; Hsiao and Black, 1977). All of the DNA in the T4 replicating pool may be part of one intracellular concatemer of great complexity (Huberman, 1968; Kemper and Brown, 1976). Packaging of T4 DNA is coupled to termination by a headful cutting mechanism which yields mature DNA of 166 ± 2 kilobases, which is circularly permuted and terminally redundant; there is apparently about 1% variability in DNA content (Kim and Davidson, 1974). Density variants of T4, as a consequence of deletions or additions, are not found. Rather, additions lead to compensating deletions, consistent with a strict headful packaging mechanism (Homyk and Weil, 1974; Streisinger et al., 1967). Similarly, petite or giant phage heads package DNA from the intracellular concatemer to DNA densities similar to that of the standard head size (Doermann et al., 1973b; Mosig, 1968). Although the ends of the mature DNA are widely distributed over the T4 genome (MacHattie et al., 1967; Mosig, 1968; Mosig et al., 1971), it is uncertain whether these ends are truly random or whether T4 initiates headful packaging from one or more preferential sites, as, for example, sequential P22 headful packaging is initiated from a *pac* site on the concatemer (Tye et al., 1974). The high T4 recombination rate and the packaging of some petite heads in each burst, as well as variability in headful packaging, might obscure the regularity of a processive, *pac*-site-type mechanism for T4. However, from the finding that certain mutations in gene 32 are compensated by mutations in gene 17, it has been argued that packaging sites in the T4 DNA are generated at sites of recombination (Mosig, Ghosal, and Bock, 1981).

Lesions in DNA structure can interfere with or arrest packaging. The first of these to be characterized was the structural aberration in T4 DNA resulting from defects in gene 49 (Luftig et al., 1971). Gene 49 is the structural gene for T4 endonuclease VII, which has been well characterized (Frankel et al., 1971; Kemper et al., 1981b; Minagawa and Ryo, 1978). This enzyme is able to remove recombinational (Holliday) structures from DNA (Mizuuchi et al., 1982). In the absence of functional gene 49 product, very fast sedimenting branched DNA accumulates in the absence of heads (Kemper and Brown, 1976), whereas in the presence of heads, such DNA is partially packaged, presumably until a branch structure interrupts packaging; restoration of gene 49 function allows the arrested packaging to be completed as the branches are removed. For a time it was supposed that the gene 49 nuclease might function as a packaging terminase, but recent studies seem to rule this out: (i) a number of extragenic suppressors in genes affecting recombination eliminate the need for gene 49 function (Dewey and Frankel, 1975; Wakem and Ebisuzaki, 1981); and (ii) order-of-function studies demonstrate that gene 49 function can be eliminated before DNA packaging is initiated (Hsiao and Black, 1978c; A. P. Schmidt, thesis, Villanova University, Villanova, Pa., 1977). Therefore one can conclude that gene 49 nuclease is needed to remove recombinational branches which interfere with packaging, but cannot be responsible for headful termination of the DNA.

Order-of-function studies demonstrate a requirement for DNA ligase (gene 30) in packaging, whereas DNA polymerase (gene 43) is not required (Hsiao and Black, 1977; Hsiao and Black, 1978c). A number of *ts* mutations in genes 32, 41, 42, 45, and 49 also have been shown not to significantly affect packaging, although only the *ts*49 was demonstrated to have the properties necessary to allow a conclusion about the non-essentiality of this gene product in the packaging step (Hsiao and Black, 1977; Schmidt, 1977; Black, unpublished data). Both T4 and *E. coli* DNA ligase can act in packaging, and in the absence of ligase activity, packaging is almost completely defective (Fig. 10). Partially packaged heads accumulate, and these can be completed upon restoration of ligase activity. Ligase-deficient, partially packaged heads appear to be less uniformly packaged than 49^- heads (Zachary and Black, 1981). UV damage to T4 DNA appears to exert a kinetic effect on head filling. Thymine dimers can be packaged, but at a reduced rate in vivo (Zachary and Black, manuscript in preparation).

It is obvious why unresolved recombinational branch structures in the DNA would arrest packaging. However, since such branches in the DNA should be resolved by gene 49 nuclease in the absence of ligase (Mizuuchi et al., 1982), the explanation of the ligase requirement for packaging in vivo is uncertain. Nicks in the DNA might not be tolerated by the packaging apparatus, ligase might actually participate directly in packaging (e.g., in an essential nicking-resealing reaction for DNA translocation [Black and Silverman, 1978]), or a protein might bind strongly to nicks and thereby block packaging. Further information about structural requirements in duplex DNA for packaging should shed light on the mechanism and also reveal the interaction of packaging with other intracellular DNA processes.

DNA Packaging-Linkage Proteins

Two late, nonstructural gene products, gp16 and gp17, are directly implicated in both packaging and in linking DNA to the prohead. Processed proheads which accumulate in infections by these mutants are not linked to the concatemeric DNA (Laemmli and Favre, 1973; Luftig and Ganz, 1972). Packaging of DNA in vitro into 16^- or 17^- mutant proheads requires the addition of these gene products (Black, 1981b; Black et al., 1981). They appear to be soluble DNA-binding proteins which are not tightly associated with the prohead (Rao and Black, unpublished data); gp17 is a protein of 67,000 molecular weight (Vanderslice and Yegian, 1974; Black, unpublished data). Some mutations in gene 17 suppress acridine inhibition of DNA packaging (Piechowski and Susman, 1967). In the presence of acridine, processed proheads and partially DNA-filled heads accumulate in vivo (Schärli and Kellenberger, 1980; Wagner and Laemmli, 1979). Gp17 is absolutely required for packaging, whereas gp16 is partially dispensable at very late times after infection (Granboulan et al., 1971; Wunderli et al., 1977). Gp17 appears to function continuously throughout filling, since some temperature-sensitive mutants in this gene accumulate partially filled heads, whereas others fail to initiate packaging (Wagner and Laemmli, 1979; Wunderli et al., 1977).

Genetic evidence strongly suggests a physical interaction of gp17 with the gp20 connector structure located at the DNA entrance vertex of the prohead. Specific, second-site mutations in gene 17 can overcome the block to initiation of packaging into the processed prohead containing the abnormal *cs*20 connector (Hsiao and Black, 1977). Specific incompatibility between the *cs*20 mutation and various substitutions at amber sites in genes 16 and 17 also supports the notion of specific gene product interaction; genes 16 and 17 also appear to act at the same step by order-of-function analyses (Black, unpublished data). Therefore, gp16 and gp17 are implicated both in DNA packaging and in linking concatemeric DNA to the prohead at the DNA entrance-connector vertex. By analogy to phage P22, these gene products could act as a terminase in making an initiating, packaging cut in the concatemer (see below).

A new T4 enzyme has recently been discovered and proposed to function in DNA packaging (Black et al., 1981; Manne et al., 1982). This enzyme is both a highly active, DNA-dependent ATPase and a relatively sequence-nonspecific endonuclease, which makes double-strand breaks in DNA. It has been purified to apparent homogeneity from infected bacteria as a 40,000-molecular-weight, soluble protein. The enzyme is apparently derived from the major T4 capsid protein precursor (gp23) as a minor cleavage product. This conclusion is drawn from the following lines of evidence. Enzyme activity is not detected in infections which fail to synthesize late proteins or fail to synthesize or cleave gp23. Immunological evidence suggests a close structural relationship between gp23 and the enzyme, since antibody against the enzyme reacts with cleaved gp23, and antibody against polyheads made of cleaved gp23 reacts with the enzyme. In addition, similar one- and two-dimensional peptide maps are obtained from the enzyme, gp23, and cleaved gp23 (Rao and Black, in preparation). The genetic and structural relationships of the enzyme to the major T4 capsid protein in conjunction with its enzymatic properties strongly suggest a function in DNA packaging (see below).

Packaging In Vitro

Processed T4 proheads can be converted to viable phage in vitro under conditions which permit the demonstration of packaging of exogenous DNA (Black, 1981b). The packaging reaction requires, in addition to mature or concatemeric DNA, an extract containing a processed prohead, gp16, gp17, and ATP. Isolated proheads are also active for packaging in vitro (Rao and Black, in preparation). Although the genetically demonstrable efficiency of packaging is low (about 10^{-5} with respect to DNA or proheads), many additional biologically inactive heads are filled which may be defective at a step other than the packaging reaction. The high level of recombination observed and the dependence of packaging on some T4 early functions suggest that exogenous mature DNA may be packaged after recombination into endogenous concatemer (Black, 1981b and unpublished data). These results suggest that the requirements for packaging T4 DNA in vitro are similar to those for the corresponding reaction for other phages, e.g., lambda, T7, and P22; they are also consistent with the known requirements for filling the T4 head in vivo.

Proposed DNA Packaging Mechanisms and Termination of Packaging

Major reasons for studying DNA packaging include the possibility of a mechanistic relationship to chromosome condensation and the possibility that a unitary mechanism will apply to all the large duplex DNA-containing bacteriophages. Thus, a number of general models have been proposed for the basic energetic mechanism accounting for DNA packaging (Earnshaw and Casjens, 1980). In the light of present information, some of these appear improbable, if not eliminated entirely, for phage T4. The proposed models fall into a number of categories.

In the first category are models which propose that prohead core components drive packaging. (i) It was originally proposed that basic, DNA-binding internal proteins of T4 acted as a condensing principle for DNA packaging (Kellenberger, 1961). However, the basic DNA-binding internal proteins of phage T4, which in fact have histone-like primary sequences (Isobe et al., 1976a; Isobe et al., 1976b; Isobe et al., 1977), are nonessential for packaging (Black, 1974; Showe and Black, 1973). Internal-protein mutants are affected in early stages of head assembly or in infectivity, not in DNA packaging. (ii) It was proposed (Laemmli, 1975; Laemmli and Favre, 1973; Wagner and Laemmli, 1976) that the driving force for packaging was cleavage of the essential core protein, gp22, to the acid-soluble peptides, a process which was claimed to accompany DNA packaging in the head maturation pathway (see above). It was suggested that this could produce a psi-like DNA condensate (Lerman, 1971) within the head. A related model coupled packaging to exit of the scaffolding protein from the prohead of phage P22 (King and Casjens, 1974). However, this core cleavage model, as stated (Wagner and Laemmli, 1976), appears to be eliminated for T4 by order-of-function studies (Hsiao and Black, 1977) and characterization of the prohead structure which is active for packaging (Black, 1981b). These experiments demonstrate that cleavage of the core can be completed before the initiation of DNA packaging.

A second class of models has proposed that the expansion of the prohead which accompanies packaging is energetically coupled to packaging. Pressure resulting from the expansion of the capsid has been proposed to suck in the DNA (Hohn et al., 1974; Serwer, 1975). There is no evidence for or against this idea for phage T4.

In the third category of models, it is proposed that DNA packaging is driven by ATP hydrolysis. ATP is now known to be required to package DNA in vitro in all phage systems. However, there is no evidence that ATP hydrolysis is energetically coupled to packaging, since cleavage of the concatemeric DNA by packaging terminases requires ATP (Murialdo and Becker, 1978). The most specific models for ATP-driven packaging were proposed by Hendrix (1978) and Black and Silverman (1978). A rotating, ATP-driven connector at the site of the sixfold-fivefold symmetry mismatch at the DNA entrance vertex of the head has been proposed to drive DNA into the head (Hendrix, 1978). Probably the only evidence for any phage that the connector structure is actually involved in packaging comes from the T4 *cs*20 mutation in the connector protein (Hsiao and Black, 1977). However, only *cs*20 mutations blocking initiation of packaging have been found; once packaging is initiated, it continues to completion in the mutant even at the nonpermissive temperature. The *cs*20 defect can also be overcome by compensating mutations in the packaging-linkage proteins (Black and Silverman, 1978; Hsiao and Black, 1977). These properties suggest that the connector functions as an anchoring site for the packaging-linkage proteins rather than as a dynamic machine for moving DNA into the head.

An ATP-driven DNA translocase acting at the DNA entrance vertex of the prohead has been proposed to drive DNA packaging (Black and Silverman, 1978). This model proposed that DNA translocation would be coupled to introduction of superhelical turns into the DNA at this vertex, which, if there was continuous duplex DNA, would lead to a translational torque along the duplex axis. This model thus attempted to explain the inhibition of packaging by DNA ligase deficiency and possibly by nicks in the DNA (Black, 1981a; Black and Silverman, 1978).

The finding that there is a minor cleavage product of the major capsid protein gp23 which has DNA-related enzymatic activities (see above) suggests that it is responsible for DNA packaging by functioning directly and enzymatically as a DNA translocase. It is reasonable also to speculate that the anchored gp23 in the processed prohead lattice also expresses transiently the enzymatic functions which can be detected in the soluble form of the processed protein (ATP- and DNA-binding activities). These functions of the structural gp23 in the head would most likely be involved in binding DNA and organizing it within the head. Later, in the structural organization of cleaved gp23 which results from head expansion, these DNA binding sites could be withdrawn from the DNA so that it could be ejected. In this view, the function of head expansion for DNA packaging is primarily to eliminate the ATP- and DNA-binding enzymatic activities of the cleaved form of the major capsid protein which are required for packaging.

A question closely related to that of the packaging mechanism is how headful-measuring termination of concatemeric DNA is achieved. In phages with unique DNA end sequences (e.g., phage lambda), cutting is carried out at specific *cos* sites by a sequence-specific terminase. Whether cutting actually occurs at a *cos* site also depends upon a DNA structure which is determined by the degree of head filling (Murialdo and Becker, 1978). In phages such as T4 and P22, which lack sequence-specified DNA ends, cutting is determined by sequence-independent headful-measuring termination. The mechanism of this termination is unknown, but presumably is related to the way DNA is packaged and its structure within the head. Since, as discussed above, the gene 49 endonuclease has been ruled out as the T4 terminase, there is no clear candidate for this activity. The most likely possibilities appear to be either that gp16 and gp17 act both as the initiating and the measuring terminase, or that gp16 and gp17 act as the initiating terminase to produce a free end from the concatemer while the gp23-derived endonuclease activity acts in headful termination upon completion of head filling (Black et al., 1981; Manne et al., 1982). A specific model for headful termination based on characterization of this enzyme activity proposed that cleavage is activated by cessation of ATP hydrolysis and by a structural change in the DNA, such as supercoiling, which persists when the DNA translocation is halted

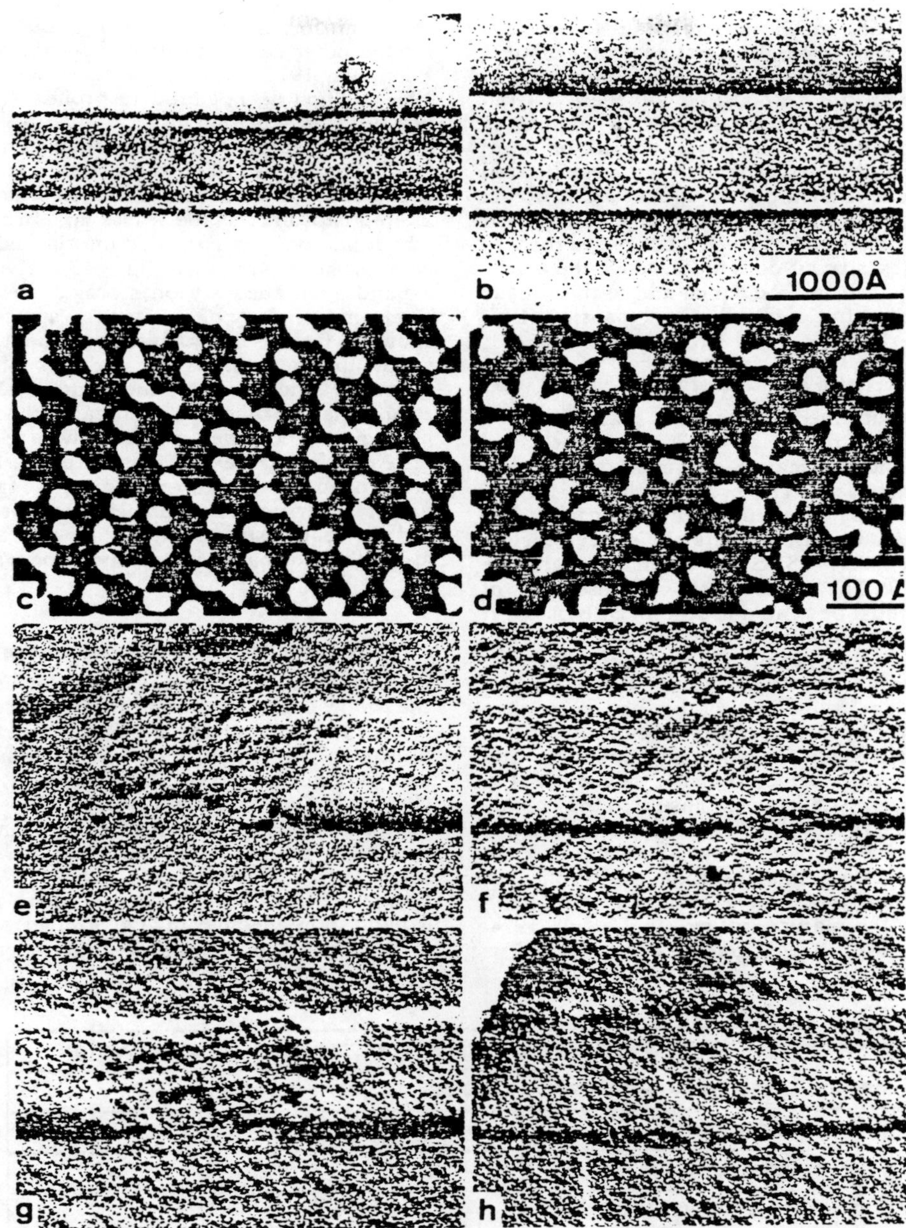

FIG. 11. Micrographs and optically filtered images of negatively stained polyheads and micrographs of freeze-dried and shadowed polyheads. (a) Uncleaved, unexpanded polyhead, negatively stained. (b) Cleaved, expanded polyhead, negatively stained. (c) Optically filtered image of uncleaved, unexpanded polyhead. (d) Optically filtered image of cleaved, expanded polyhead. (e and g) Uncleaved, unexpanded polyheads prepared by freeze-drying and shadowing to show the surface topography. (f and h) Cleaved, expanded polyheads, freeze-dried and shadowed. Broken polyhead fragments have been chosen so that the inner and outer surfaces can be compared. Note the rough interior and smooth exterior in panels e and g and the reverse in panels f and h. From Kistler et al. (1978). Panels c and d courtesy of A. Steven.

at the completion of filling. Studies using purified components are required to decide among these and other possible mechanisms.

SHELL TRANSFORMATION AND CAPSOMER MATURATION

For many phages, including T7, P22, and lambda as well as T4, DNA packaging is associated with an increase in size of the prohead (Murialdo and Becker, 1978). In vivo, this transformation of the T4 prohead shell to the shell of the mature head follows cleavage of the prohead proteins and entails striking changes in the properties of the shell. These can be appreciated by comparing T4 proheads with capsids. Besides the difference in protein composition of the two shells (Table 1), the mature capsid is 15% larger (Kellenberger et al., 1968), much more resistant to dissociation (Kellenberger, 1968), and antigenically distinct (Favre et al., 1965; Yanagida, 1972). The differences in the shells which are responsible for these altered properties have been mainly studied by comparing polyheads with either cleaved polyheads or giant phage, since the surface morphology of these extended structures can be determined by optical diffraction of electron micrographs (Finch et al., 1964; Yanagida et al., 1970). T2 is also a more convenient phage than T4 for some studies, since it does not produce the nonessential proteins hoc and soc which add to the outside of the T4 capsid after cleavage and expansion (Ishii and Yanagida, 1975).

Shell Characteristics and Transformation In Vitro

Although an intermediate state has been identified and studied (Laemmli et al., 1976), the uncleaved prohead shell and the cleaved, expanded mature shell are best characterized. The prohead shell is fragile, being dissociated by distilled water (van Driel, 1977). When stabilized by 0.1 M salts, it is nevertheless dissociated by 0.1% SDS at room temperature (Laemmli, 1970). The mature shell, composed only of cleaved gp23, must be heated to 53°C in SDS before it dissociates (Steven, Couture, Aebi, and Showe, 1976). Other changes (Fig. 11 and 12) which define the transformed state in proheads or expanded polyheads include a decrease in thickness of the shell from 4.3 to 2.6 nm, a change in surface morphology from protrusions inside the shell to protrusions outside, and changes in antigenicity, including the movement of at least one antigenic site from the inside to the outside of the shell (Kistler et al., 1978). The increase in the size of the capsid is seen to be a reflection of an increase in the lattice constant of the shell from 11.0 nm for the prohead-type lattice (Yanagida et al., 1970) to 13.0 nm for the mature lattice (Aebi et al., 1974; DeRosier and Klug, 1972; Yanagida et al., 1972). The shell transformation must also produce binding sites for the outer capsid proteins hoc and soc (Ishii and Yanagida, 1975; Steven, Couture, Aebi, and Showe, 1976), which do not bind to proheads or uncleaved polyheads (see below).

An intermediate state in the shell transformation of polyheads cleaved in vitro was described by Laemmli et al. (1976). They observed a class of polyheads with altered capsomer shape whose gp23 was cleaved but which had not yet expanded and lacked the stability of a prohead-type shell (Fig. 13). These polyheads expand spontaneously on storage. The altered capsomer morphology of this type of structure has been shown to be caused by a conformational change of the gp23, and not simply by the loss of the cleaved fragment, since cleaved, unexpanded polyheads have a capsomer morphology identical to that of uncleaved polyheads when they are fixed with glutaraldehyde instead of formaldehyde (Kistler et al., 1982). A similar dependence of capsomer morphology on fixation for certain trypsinized polyheads and polyheads composed of *ts*gp23 suggests that their altered appearance also reflects a change in conformation which is reversible by fixation.

Mutations Which Affect Transformation

The intermediate, "cleaved but anchored" state of the prohead shell first identified in cleaved polyheads is characteristic of the proheads isolated from mutants in genes 16 and 17 and the *cs*20 mutant (Steven and Carrascosa, 1979). These mutants produce fully cleaved but unstable and unexpanded particles (Fig. 10), which do not incorporate DNA and are not attached to the pool of replicating DNA (Hsiao and Black, 1978c; Laemmli and Favre, 1973; Luftig and Ganz, 1972). Although some of these proheads can expand spontaneously in vivo or in vitro, the coupling of the defect in packaging with the production of

SPECIMEN →	CLASS I POLYHEADS		CLASS III POLYHEADS		SHEETS
SURFACE LABELLED →	INSIDE	OUTSIDE	INSIDE	OUTSIDE	
FAB – FRAGMENTS PREABSORBED WITH: NO PREABSORPTION	+	–	–	+	+
CLASS I POLYHEADS	–	–	–	+	+
CLASS III POLYHEADS	–	–	–	–	+
SHEETS	–	–	–	–	–
SUMMARY: SURFACE LOCATION OF DETECTED ANTIGENIC DETERMINANTS	 A		A,B		A,B,C

FIG. 12. Antigenic sites on class I (uncleaved, unexpanded) and class III (cleaved, expanded) polyheads and on sheets composed of a proteolytic fragment of gp23. Fab fragments were prepared from antibodies raised against the sheets and used in absorption and labeling experiments as shown. From Kistler et al. (1978).

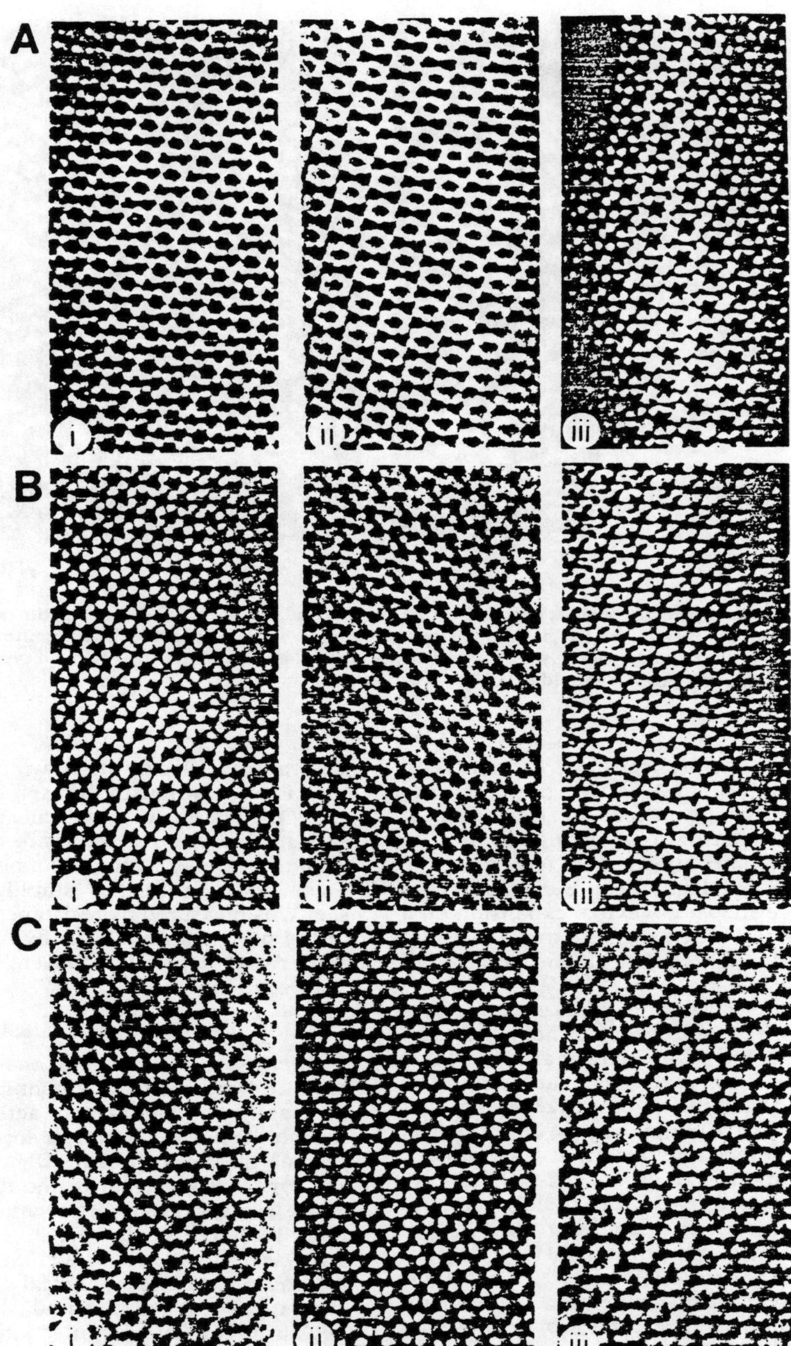

FIG. 13. Filtered images of T4 polyheads. (A) Uncleaved, unexpanded polyheads. (B) Cleaved, unexpanded (i.e., cleaved but anchored) polyheads. (C) Cleaved, expanded polyheads. For details of sample preparation and staining see Laemmli et al. (1976).

mainly unexpanded proheads suggests that it is DNA packaging which, in vivo, triggers expansion.

The unprocessed proheads produced by mutants with defects in gene 21 or 24 are unexpanded. Although processing is normally a requirement for expansion, 21^- proheads can be expanded by treatment with a nonionic detergent (Onorato et al., 1978). This appears to be similar to the expansion of P22 proheads and shells by treatment with SDS (Earnshaw et al., 1976; Earnshaw and King, 1978) and to the expansion

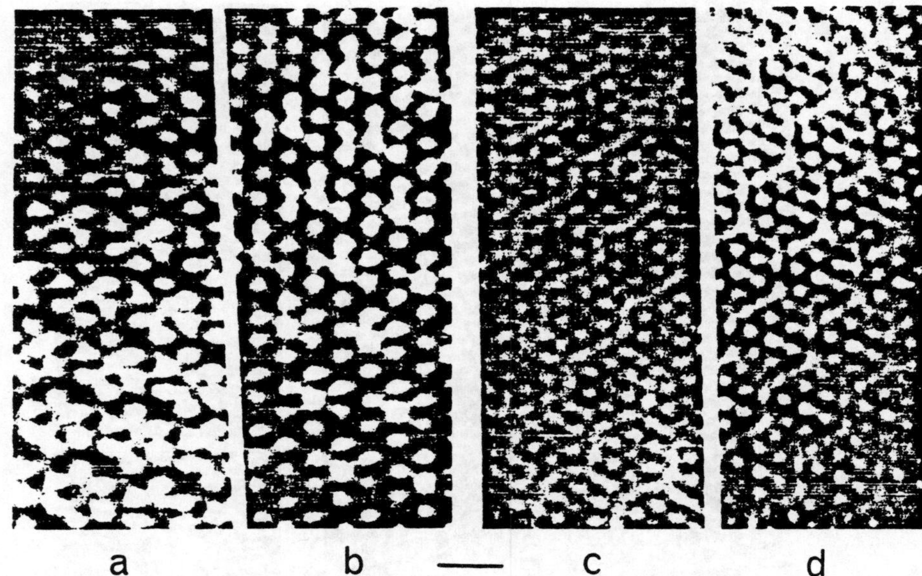

FIG. 14. Optically filtered images of giant heads of wild-type T4D and mutants in the nonessential outer capsid proteins hoc and soc. (a) hoc soc double mutant which has a shell composed only of cleaved gp23. (b) soc mutant whose shell contains cleaved gp23 and hoc. (c) hoc mutant whose shell contains only cleaved gp23 and soc. (d) T4D wild type with a shell composed of cleaved gp23, hoc, and soc. From Yanagida (1977).

of lambda polyheads and proheads by treatment with 4 M urea (Kunzler and Hohn, 1978; Wurtz et al., 1976). How these treatments trigger expansion is not clear. It is possible that a conformational change in most of the shell protein subunits allows them to relax to the more stable, transformed state. But Steven and Carrascosa (1979) have shown a wave of expansion on a giant cleaved particle, which suggests that expansion of the T4 prohead might begin at one point from which it is propagated.

Three temperature-sensitive gene 23 mutants produce aberrant misshapen proheads at nonpermissive temperature (Aebi et al., 1974; Favre et al., 1965). One of these, A78, has been extensively analyzed. The proheads produced from the mutant gp23 are expanded even though they have not undergone any cleavage. Their lumpy appearance may result, at least in part, from the incorporation of an excess gp24 (Onorato et al., 1978), possibly resulting in a particle with more than 12 vertices. Polyheads made from the A78 protein have shells in the transformed state (i.e., they are SDS resistant and have a 12.9-nm lattice constant) when they are produced at nonpermissive temperature. At intermediate temperatures, core-containing polyheads made from this protein are found to be unexpanded, whereas 22^- polyheads without cores are expanded, suggesting that the core actively anchors the shell in the unexpanded state in this mutant (Kistler et al., 1982).

Addition of Outer Capsid Proteins

The discovery of the structural role of the nonessential surface proteins hoc and soc (Fig. 14) was crucial to solving the structure of the T4 head surface lattice (Ishii and Yanagida, 1975; Ishii and Yanagida, 1977). The capsomer maturation, which is begun when the shell is transformed to the expanded state, is completed by the binding of a molecule of hoc at the center of each hexamer and six molecules of soc at positions of local threefold or twofold symmetry (Aebi et al., 1977). Their addition further stabilizes the transformed shells, as measured by resistance to high temperature in the presence of SDS or by alkali resistance (Ishii and Yanagida, 1977; Steven, Couture, Aebi, and Showe, 1976).

HEAD MATURATION

Mutants in a number of the X genes (see above) produce heads which are nonfunctional as judged by their ability to complement active tails in vitro, even though the heads are morphologically normal and the proteolytic cleavage and DNA-packaging reactions are apparently normal in the mutant strains (Edgar and Lielausis, 1968; Granboulan et al., 1971; King, 1968; Wunderli et al., 1977).

It was originally reported that mutations in genes 2, 64, and 50 lead to abnormal cleavage of gp23 and other head proteins (Laemmli, 1970; Vanderslice and Yegian, 1974). Later detailed kinetic measurements of gp23 cleavage in these X gene mutant infections did not support these qualitative observations, although reduced cleavage of gp23 was detected among some of the double mutants, probably due to the accumulation of some abnormal head structures in which no cleavage occurs (Wunderli et al., 1977). The exact role of these X genes (genes 2, 4, 13, 14, 50, 64, and 65) in head maturation is ill defined, as is the stage of assembly when the products act or are added to the head. It is known that gp13 and gp14 can be added to completed heads from 13^- and 14^- infections to activate them for tail joining in in vitro complementation mixtures (Edgar and Lielausis, 1968; Hamilton and Luftig, 1972). Gp13 and gp14 are also components of the necked tail, which can be disconnected from the mature phage (Coombs and Eiserling, 1977). Since there are other, unidentified minor proteins in such necked tails (Coombs and Eiserling, 1977), it is likely that some of these head completion gene products are also part of the head-tail junction. In fact, some of these head completion genes affect the stability of the head-tail junction in vivo (Granboulan et al., 1971).

Such proteins could be added to the neck early (as part of the gp20 initiator structure on the membrane) or late (after DNA packaging). Normal protein cleavage in these mutants would argue for late addition, but the question remains open. In the absence of gene 2 function, morphologically normal, noninfectious particles are produced (King, 1968), which, however, can successfully infect an exonuclease V-deficient host (Silverstein and Goldberg, 1976). These experiments led to the hypothesis that gp2 is injected along with the DNA to protect it from nuclease attack, although gp2 has not been definitely located within the particle. Similarly, IPI incorporation into the head is required to allow successful infection of some natural *E. coli* isolates; in its absence, infection is followed by DNA breakdown (Abremski and Black, 1979; Black and Abremski, 1974).

We thank Alasdair Steven, Arthur Zachary, and Louise Showe for comments on the manuscript, and Edouard Kellenberger and Michel Wurtz for providing many of the electron micrographs from the archives at the Biocentrum in Basel.

Research reported from our laboratories was supported by grants from the National Institute of General Medical Sciences and the National Institute of Allergy and Infectious Diseases.

T4 Tail Morphogenesis

PETER B. BERGET[1] AND JONATHAN KING[2]

Department of Biochemistry and Molecular Biology, The University of Texas Medical School, and Graduate School of Biomedical Sciences, Houston, Texas 77225,[1] and Department of Biology, Massachusetts Institute of Technology, Cambridge, Massachusetts 02139[2]

The most distinctive feature of the familiar T4 image is the complex contractile tail apparatus, responsible for delivering the phage chromosome into the host cell. Composed of more than 20 different protein species, the tail has been instructive in understanding the genetic control of organelle assembly, the regulation of protein-protein interactions during morphogenesis, and the initiation and termination of repeated structures (Fig. 1). Important features of the assembly process include (i) regulation of the assembly process at the level of the interactions between proteins in the pathway, rather than at the level of genome expression; (ii) critical regulatory interactions in the assembly process between the growing structure and soluble subunits, rather than interactions between soluble subunits themselves, and (iii) regulation by switching subunits from a nonreactive precursor conformation into a reactive conformation upon incorporation into the growing structure (Caspar, 1980; King, 1980).

TAIL GENES

Table 1 lists the genes and gene products which are required for tail assembly. Nearly half of the genes known to be required for T4 particle formation are involved in the assembly of the tail structure. Twenty-two genes code for proteins which are incorporated into the tail structure, and two code for catalytic factors required for complete assembly. Nine of these genes map in one cluster, genes 53, 5, 6, 7, 8, 9, 10, 11, and 12. Eight of the genes map in a second cluster, 25-26-51-27-28-29-48-54, separated from the first cluster by the major head gene cluster. Genes 18 and 19, which code for the major tail sheath and tail tube subunits (King and Laemmli, 1973), map as a separate pair next to the head cluster, and two more genes, 3 and 15, are flanked by head genes. The *frd* and *td* genes map in one of the early regions. We discuss the possible significance of these gene clusters later on.

Table 1 also lists the functions or structural locations of these gene products, where known, and the "small" molecules known to be components of the T4 tail. In addition, the stoichiometries in terms of copies per phage particle and polypeptide molecular weights of these proteins are given. The variation in these stoichiometries points out that the assembly of this structure involves the interaction of diverse oligomers of gene products and differs from the assembly of ribosomes, where most structural components are found in single copies per structural unit. Tail polypeptide chains cover a large molecular weight range from 15,000 to 140,000 g/mol (King and Mykolajewycz, 1973).

STRUCTURE OF THE TAIL

The complete tail is defined as the structure accumulating in cells blocked in head morphogenesis. An electron micrograph of sucrose gradient-purified T4 tails is shown in Fig. 2. These tails do not have tail fibers associated with them; they have the extended sheath covering the tail tube, and a short extension of the tail tube projects from the end of the sheath (King, 1968). A few baseplate-tail tube structures are also seen in Fig. 2. At late times in mutant-infected cells, additional proteins add to the proximal end of the complete tail, forming a neck structure normally found only in the mature phage (Coombs and Eiserling, 1977; see Eiserling, this volume).

Contractile Sheath

The sheath is composed of 24 annuli spaced 40 Å (4 nm) apart with six subunits per annulus. The spatial disposition of these subunits is known from three-dimensional reconstructions from electron micrographs (DeRosier and Klug, 1968). These subunits are bonded to each other radially within the same annulus and also between annuli. There is probably some weak interaction between sheath subunits and tube subunits since during assembly the sheath subunits do not polymerize onto the baseplate unless the tail tube has initiated. The sheath subunits are much less tightly bound to each other in the extended sheath than in the contracted sheath (To et al., 1969). The polymerized sheath subunits on partially assembled tails are in equilibrium with unassembled subunits, with the partially assembled structure being somewhat cold sensitive (King, 1968). The stabilization of the mature sheath depends on the stabilization of the last annulus when the tail is completed (King, 1971).

Sheath contraction. Considerable effort has been put into defining the mechanism of sheath contraction and the nature of the transition from the extended to the contracted state (Moody, 1973). Two points are important for purposes of understanding the assembly process: the sheath appears to be built in a high-energy state such that once triggered, contraction is spontaneous; and almost all of the sheath annuli must slide past the tube annuli during contraction (see Fig. 7 of Eiserling, this volume). However, for mechanical force to be generated the proximal sheath annuli must stay bound to the proximal end of the tube, and the distal sheath annuli must remain bound to the baseplate. The contracted sheath is very tightly bonded; it remains intact when the entire rest of the phage, including the tube and capsid, has been dissociated under denaturing conditions (To et al., 1969). The propagation of the contraction proceeds from the

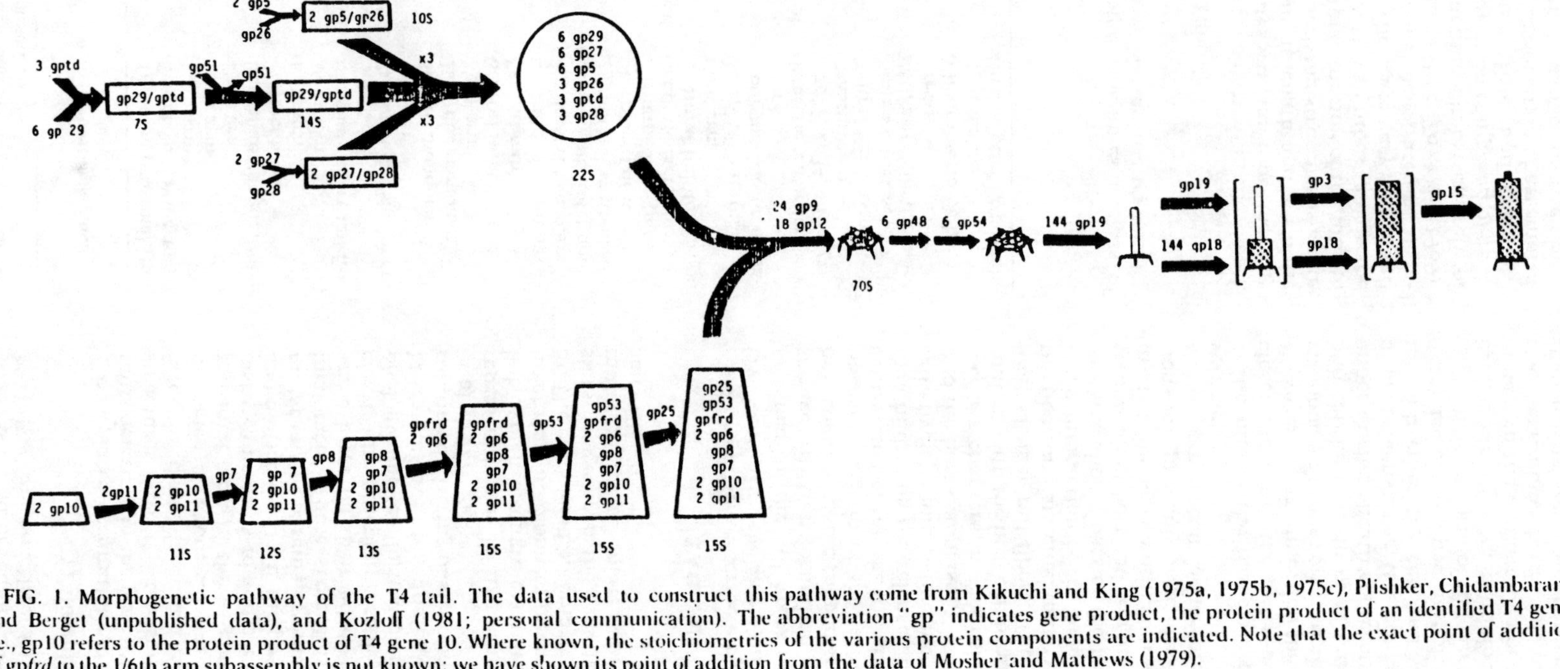

FIG. 1. Morphogenetic pathway of the T4 tail. The data used to construct this pathway come from Kikuchi and King (1975a, 1975b, 1975c), Plishker, Chidambaram, and Berget (unpublished data), and Kozloff (1981; personal communication). The abbreviation "gp" indicates gene product, the protein product of an identified T4 gene, i.e., gp10 refers to the protein product of T4 gene 10. Where known, the stoichiometries of the various protein components are indicated. Note that the exact point of addition of gp*frd* to the 1/6th arm subassembly is not known; we have shown its point of addition from the data of Mosher and Mathews (1979).

TABLE 1. Genes, gene products and other factors required for tail assembly

Gene	Gene product mol wt (× 10^3)	Copies per phage	Function/location
3	29	?	Tail tube proximal tip
5	44	6	Center hub, lysozyme
6	85	12	1/6th arm
7	140	6	1/6th arm
8	46	6	1/6th arm
9	34	24	Baseplate, long tail fiber attachment
10	88	12	1/6th arm
11	24	12	1/6th arm
12	55	18	Baseplate short tail fibers
15	35	?	Proximal tail sheath stabilizer
18	80	144	Sheath monomer
19	21	144	Tail tube monomer
25	15	6	1/6th arm
26	41	3	Central hub
27	49	6	Central hub
28	24	3	Central hub, gamma-glutamyl hydrolase
29	77	6	Central hub, folyl polyglutamyl synthetase
48	44	6	Baseplate tail tube length "determiner"?
51	?	Nonstructural	Catalytic factor for central hub formation
53	23	6	1/6th arm
54	36	6	Baseplate, tail tube polymerization initiator?
57	6	Nonstructural	Catalytic factor for gp12, gp34, and gp37 maturation
td	29	3	Central hub, thymidylate synthetase
frd	20	6	1/6th arm, dihydrofolate reductase
—	1.1	6	Dihydropteridine hexaglutamate
—	0.065	6	Zinc, in gene 12 protein

baseplate up the sheath and resembles a class of crystal dislocations known as Martensitic transitions (Olson and Hyman, 1983).

Tail Tube

Cells infected with mutants defective in the sheath protein accumulate baseplates with tail tubes (see Fig. 2). The tail tubes are 95 nm in length and are quite smooth. However, their underlying structure is annular, and the ratio of the annular spacing of the tube to the sheath is 1:1 (Moody, 1971). Tail tubes can also be isolated as free structures. These are very stable and often have their central channel penetrated by negative stain (Duda and Eiserling, 1982; To et al., 1969). Isolated tail tubes are composed almost entirely of gp19 and lack the small projection composed of gp15 present on the sheathed tail (King and Laemmli, 1973). Subunits isolated by dissociation of assembled tubes behave differently from precursor unassembled tube subunits; the former polymerize spontaneously, whereas the latter require a baseplate template (King, 1971; Poglazov and Nicolskaya, 1969; To et al., 1969).

Baseplate

Baseplates accumulate in cells infected with mutants defective in the gene 19 tail tube subunit. These structures have the tail spikes projecting from one face, the distal face, and a small platform with a diameter of about 20 nm on the other face, the proximal face. In the assembly of the baseplate the last two proteins to add are the products of gene 48 and 54 (see Fig. 1). If the gp48 protein is removed by mutation, the baseplates dimerize between their proximal faces. These structures sit on electron microscope grids on their edges (Fig. 3), revealing more clearly the baseplate profile (King, 1971). The morphology and conformational rearrangement of the baseplate have been studied in considerable detail by rotational filtering of electron micrographs (Crowther et al., 1977; Fig. 4). The baseplate has an outer rim representing the joining of the six vertices, a central hub which strongly excludes stain, and a complex inner ring interconnecting the hub and outer rim (Fig. 5). The completed baseplates only rarely transform to stars. However, if the gene 12 protein is removed by mutation, the baseplates transform spontaneously to stars (Fig. 6).

TAIL FUNCTION

As originally described by Simon and Anderson (1967a; 1967b), the injection process involves (i) attachment of the tail fibers and baseplate to the cell surface, (ii) transformation of the baseplate from a hexagonal to a star-shaped structure, simultaneously releasing the tip of the tail tube and initiating the contraction of the sheath, and (iii) contraction of the sheath, driving the released tail tube tip through the cell envelope and delivering the phage chromosome into the host cytoplasm.

Other chapters in this volume describe the initial contact with the host cell through the distal tips of the long, slender tail fibers. The proximal ends of these tail fibers are attached to the vertices of the baseplate. Upon interaction of the long fibers with their receptors, the short fibers of the baseplate interact with a second set of host cell receptors. Since particles lacking tail fibers are not infectious, both sets of interactions are probably needed to activate the baseplate and trigger the contraction process (see Goldberg, this volume).

The hub of the baseplate presumably covers and closes the tip of the tail tube in the mature phage. However, during injection the tube penetrates the cell envelope and must open to allow DNA exit. Thus the hub can be thought of as a diaphragm which opens either during sheath contraction or after penetration of the cell envelope. It is not known which proteins of the hub enter the cell and which remain with the expanded star-shaped baseplate.

The mature form of the tail which has to be assembled is clearly a metastable structure, which must remain in the metastable uncontracted form through sewer and stomach and still maintain the ability to be efficiently triggered upon interaction with the bacterial receptors. Some of the proteins of the baseplate are

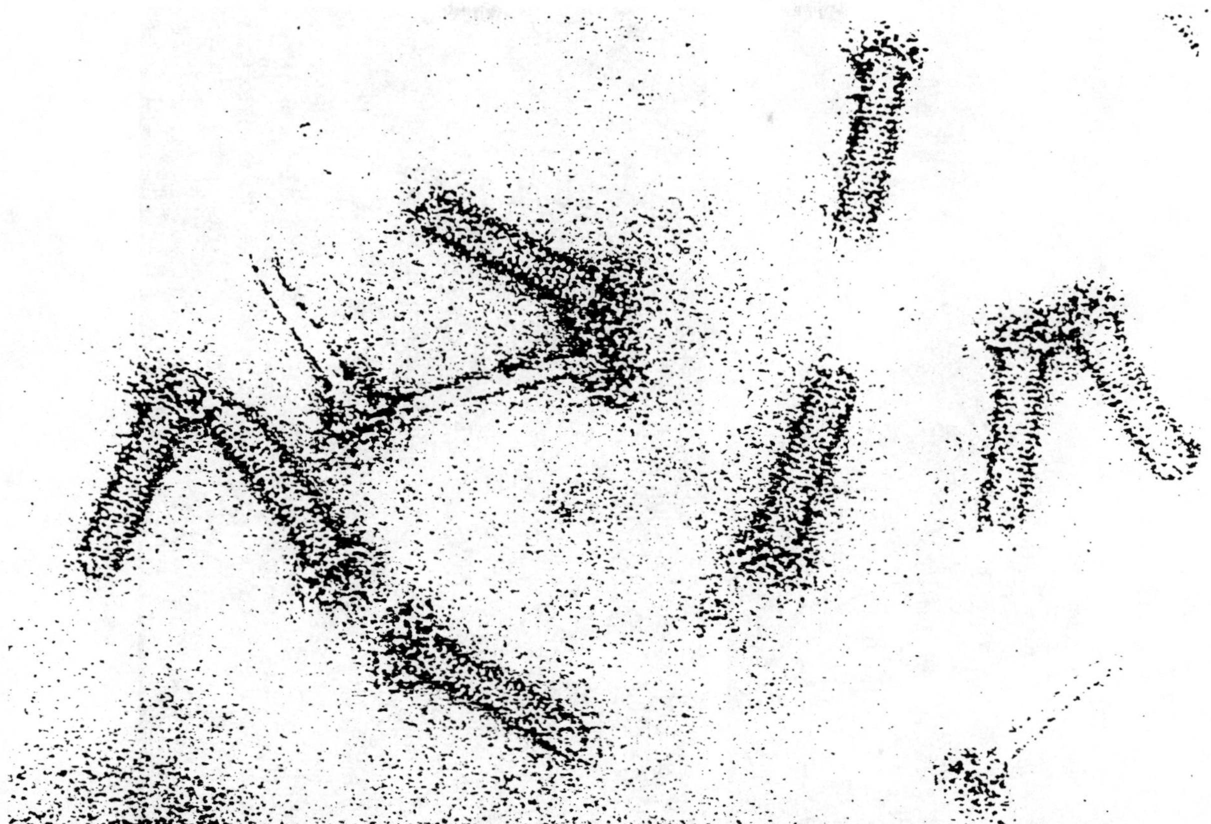

FIG. 2. Precursor T4 tails. Tails were isolated from su^- cells infected with phage carrying an amber mutation in gene 23, defective in capsid assembly. The tails were isolated by sucrose gradient centrifugation and negatively stained with uranyl acetate for electron microscopy. Some tails have lost their sheath, revealing the tail tube.

probably involved in this balance between stability and contractility (Crowther, 1980).

It is not clear whether the end of the DNA that first gets injected resides in the head or is already threaded down the tail of the mature particle. If the end of the DNA is still completely in the head in the mature phage, the injection process also requires the opening of a valve or channel between the tail and the head. Alternatively, this channel may open upon the joining of the head to the tail during assembly so that the DNA is prethreaded through the tail tube (see Goldberg, this volume).

PROTEIN LOCATION AND FUNCTION

Though we focus below on the assembly pathway for the tail, some of the tail gene products have been identified with functional activities during the infection process. Others have been shown to have unusual or unexpected properties which deserve special mention. We briefly summarize this information.

The gene 9 protein is located at the baseplate vertices and is required for attachment of the tail fiber (Crowther et al., 1977; King, 1968). It plays a role on the triggering of baseplate expansion (Crowther, 1980).

The gene 11 protein forms the tips of the six spikes that project from the baseplate and plays a role in the attachment and injection process (Crowther et al., 1977; King, 1968).

The gene 12 protein is located on the distal face of the baseplate facing the cell and may be the morphological short fibers (Crowther et al., 1977; Kells and Haselkorn, 1974). Complete phage particles assemble in its absence. These particles can absorb to cells but do not kill them (King, 1968). The gene 12 protein is a Zn^{2+}-containing metalloprotein (Zorzopulos and Kozloff, 1978). The maturation of the gene 12 protein requires the catalytic activity of the gene 57 protein for its maturation into its trimeric structure (Kells and Haselkorn, 1974; King and Laemmli, 1973). The requirement for gene 57 activity is shared with two of the tail fiber proteins, gp34 and gp37 (King and Laemmli, 1971).

The gene 5 protein has recently been shown to possess a "lysozyme" activity (Kao and McClain, 1980b). This protein is a structural component of the central hub of the baseplate, and it may function in the cell wall penetration of the tail tube during DNA injection.

The gene 28 protein has been shown recently to be a structural component of the T4 baseplate (Kozloff and Zorzopulos, 1981) and to possess a gamma-glutamyl carboxypeptidase enzymatic activity (Kozloff and

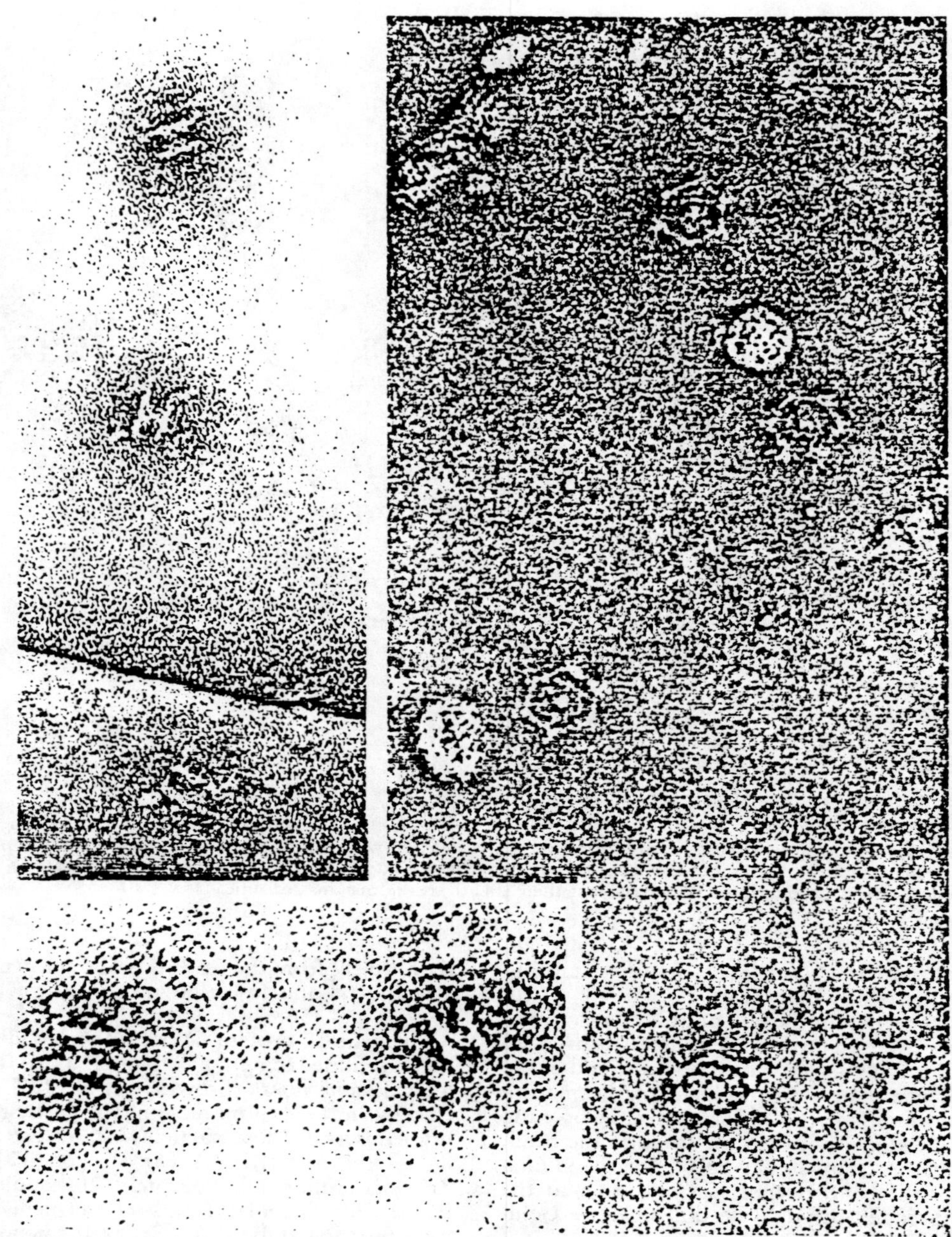

FIG. 3. Baseplates from 48^--infected cells. Baseplates were isolated by sucrose gradient centrifugation from cells infected with phage carrying an amber mutation in gene 48. The dimers sediment faster than the monomers, at about 105S (King, 1971).

Lute, 1981). This enzymatic activity seems to play a role in the formation of the unusual folate molecules found in T4-infected cells (Kozloff and Lute, 1965; Kozloff, Verses, Luter, and Crosby, 1970; Nakamura and Kozloff, 1978; Kozloff, this volume). Furthermore, it has been suggested that this protein, in cooperation with other folate binding proteins found in the T4 baseplate, may provide the binding force required to hold the baseplate subassemblies together (Kozloff, 1981).

The gene 29 protein has recently been reported to be a folyl polyglutamyl synthetase (Kozloff, this volume) which is responsible for part of the alteration in folate metabolism after infection by T4. It is also one of the major polypeptides of the T4 baseplate central hub (Kikuchi and King, 1975b).

The products of genes *td* and *frd* are the T4-induced enzymes thymidylate synthetase and dihydrofolate reductase, respectively. Both of these gene products are found as structural components of the T4 base-

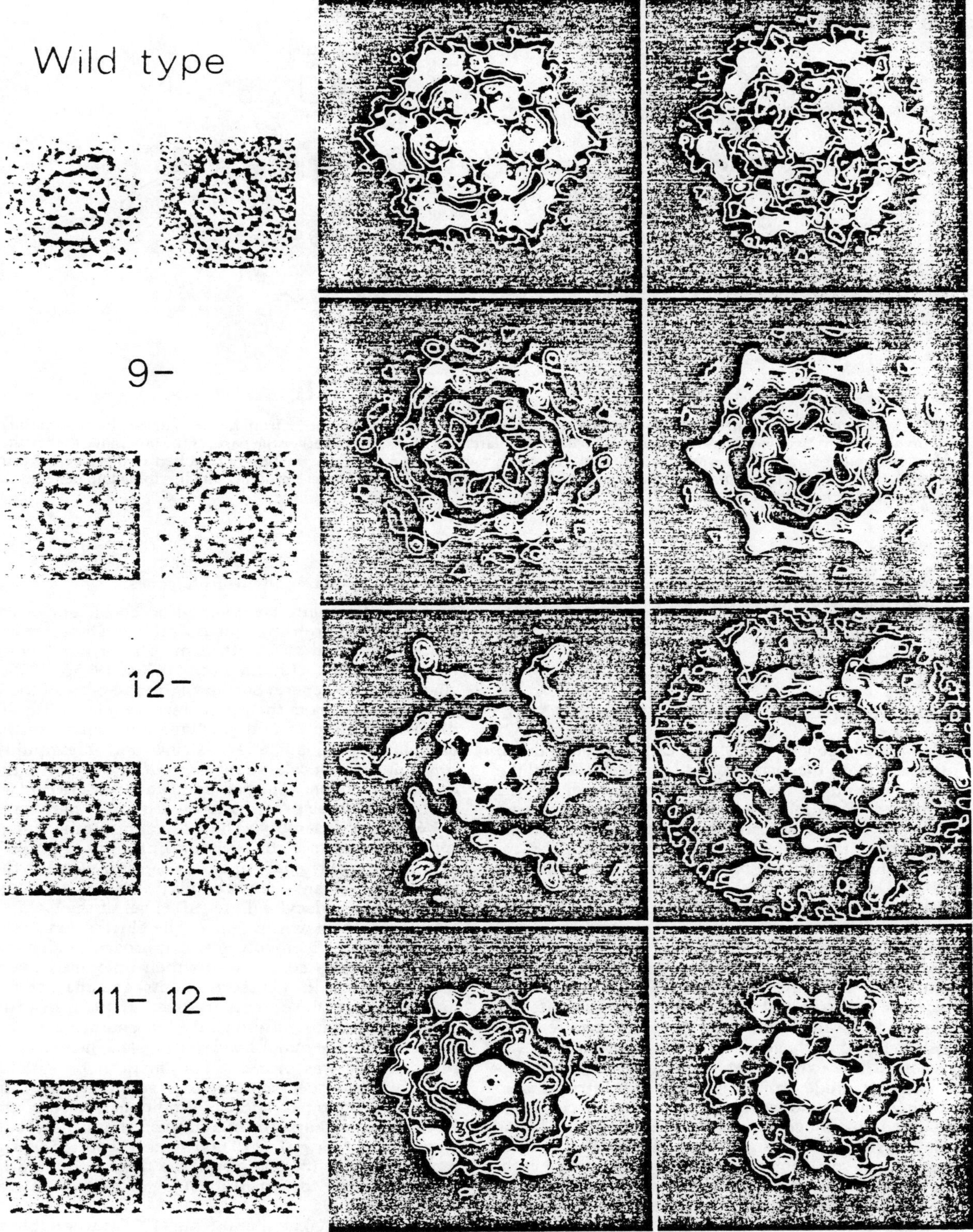

FIG. 4. Rotationally filtered baseplate images. Negatively stained electron micrograph images of various classes of baseplates were rotationally filtered by the procedures of Crowther and Amos (1971). The proteins missing from the baseplates are shown above the micrographs. The images of the 12^- baseplates suggest structures in transition from the hexagon to the star form. From Crowther et al. (1977).

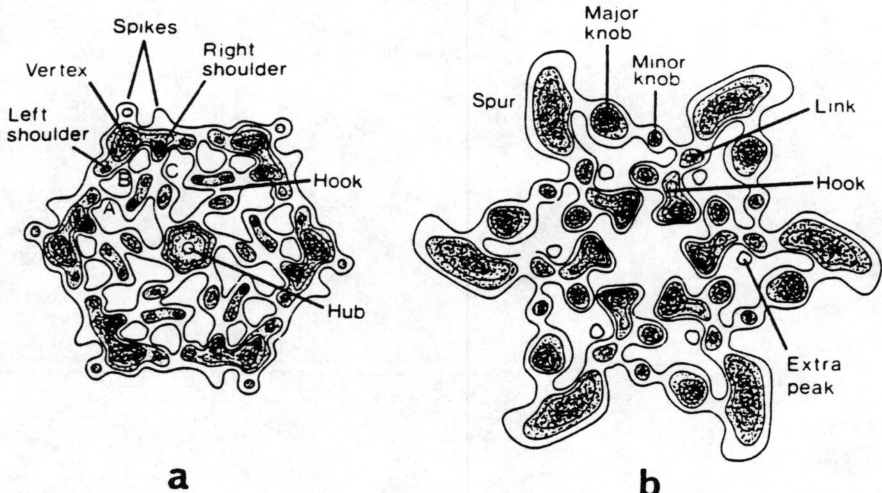

FIG. 5. Schematic diagram of baseplate hexagon and star forms. The transformation from hexagon to star form is complex and involves rearrangement of most of the mass of the baseplate. It is not known whether the parts of the hub move to the outer radius or whether some proteins are actually lost. The star drawing does not include the thin fibrils often seen projecting from the vertices of stars derived from complete baseplates. These fibrils may be the gene 12 protein, which is folded up beneath the baseplate in the hexagonal state.

plate. Recently substantial evidence has been presented that these proteins reside in the central hub and 1/6th arm subassemblies of the baseplate, respectively (Kozloff, 1981; Kozloff and Zorzopulos, 1981; Mosher and Mathews, 1979).

MORPHOGENETIC PATHWAY OF THE T4 TAIL

The morphogenesis of the tail can be broken down into two stages, the formation of the baseplate and the polymerization of the tail tube and sheath on the completed baseplate (Fig. 1). Of the 22 proteins found in the T4 tail, 18 are structural components of the baseplate (King and Mykolajewycz, 1973). The remaining four structural proteins are found in the tail tube and sheath.

Much of the work on T4 tail assembly involved in vitro complementation analysis (Edgar and Lielausis, 1968; Edgar and Wood, 1966). The complementation reactions in the tail pathway proceed relatively efficiently, but certain features must be kept in mind. Since 144 tube subunits and 144 sheath subunits are required to assemble one tail, the reactions are quite sensitive to the phage protein concentration in the extracts. These experiments require very concentrated extracts made from mutant phage strains also carrying a gene *t* amber mutation, delaying cell lysis, and extending phage maturation past the normal lysis time (Josslin, 1970). In addition, the sheath subunits aggregate at very high concentration into polysheath, in which form they are not available for sheath assembly. The tube subunits are also somewhat unstable in the crude extract during the incubation (King, unpublished data). Thus very concentrated infected-cell extracts are required, and these have to be kept cold and used quickly after thawing.

Baseplate Morphogenesis

The T4 baseplate (sedimentation coefficient of ca. 70S) is formed from two subassemblies. These are the central hub and the 1/6th arm. The original work reported by Kikuchi and King (1975a, 1975b, 1975c) elucidated the general outline of the assembly of these two structures from their precursor proteins. With the discovery that gp26 and gp28 are components of the baseplate (Kozloff, 1981; Kozloff and Zorzopulos, 1981) and the localization of gp*frd* and gp*td* in the assembly scheme (Mosher and Mathews, 1979), we have added these proteins to the assembly pathway.

1/6th arm assembly. The baseplate 1/6th arm is constructed from eight structural proteins (gp6, gp7, gp8, gp10, gp11, gp25, gp53, and gp*frd*) without the requirement of any known catalytic factors. A tracing of a sodium dodecyl sulfate (SDS) gel of the isolated 1/6th arms is shown in Fig. 7. The elucidation of the assembly pathway involved two approaches: identifying subassembly complexes by their unique sedimentation behavior in sucrose gradients, and characterizing these complexes with respect to their in vitro complementation activities. By incorporating amber mutations in the major head proteins, the head assembly pathway was blocked. This made it possible to identify phage-specific complexes sedimenting in the 10 to 20S region of sucrose gradients. Many of the 1/6th arm proteins are of relatively high molecular weight, such as gp7 (140,000 g/mol), gp6 (85,000 g/mol), and gp10 (88,000 g/mol), so that they could be unambiguously identified in SDS-gel analysis of gradient fractions.

In addition, Kikuchi and King (1975a) were able to obtain in vitro complementation between all pairs of extracts infected with mutants defective in the 1/6th arm genes. The positive complementation results es-

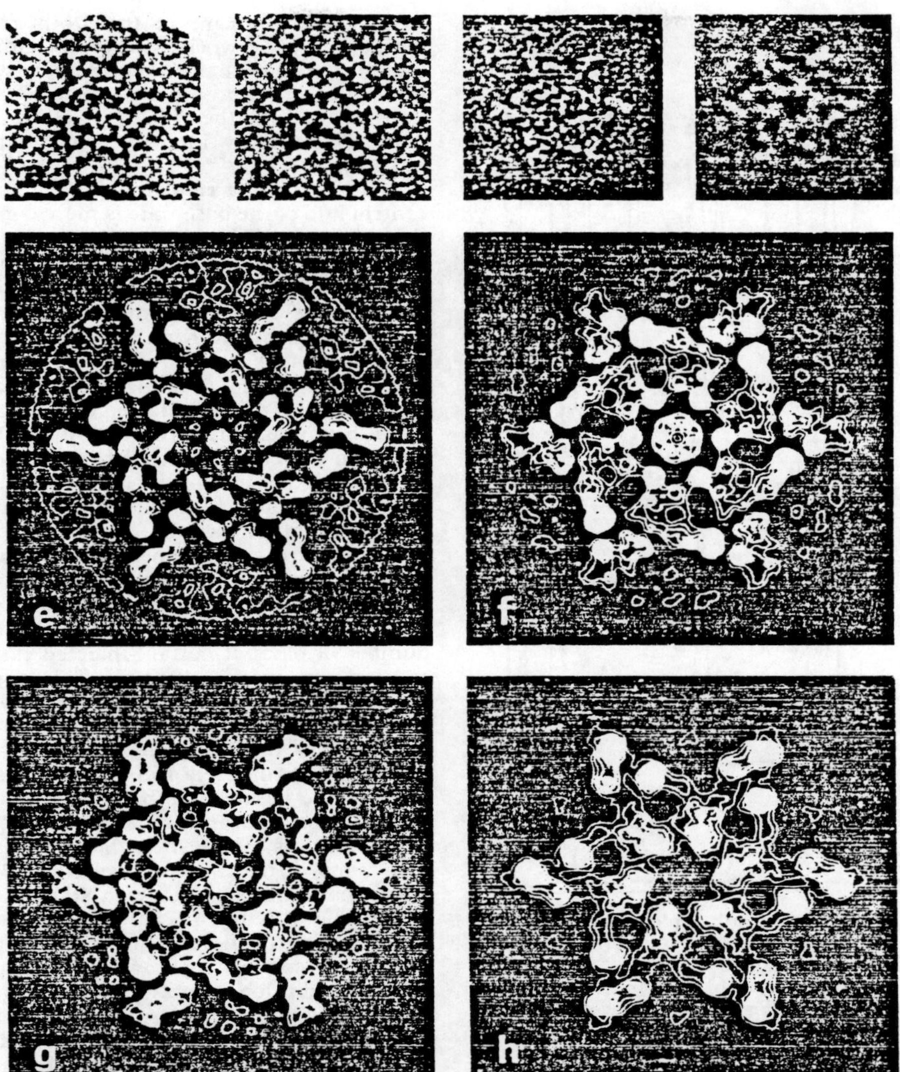

FIG. 6. Baseplate stars from 12$^-$-infected cells. The electron micrographs were taken of the peak fraction of baseplates isolated from 12$^-$-infected cells. These sedimented at 63S and were probably hexagons. The transformation to star shape presumably takes place during preparation of the electron microscope grid. These baseplates lack the morphological fibrils seen on stars from complete baseplates, which probably represent the gene 12 protein (Kells and Haselkorn, 1974). a–d, Electron microscope images; e–h, the respective rotationally filtered images. From Crowther et al. (1977).

tablished that the baseplate proteins accumulated in precursor form in the extracts. Kikuchi and King (1975a) were able to recover these activities from sucrose gradients and to characterize directly the precursor complexes accumulating in the mutant-infected cells.

Baseplate 1/6th arm structures accumulate as a 15S complex in cells infected with mutants defective in hub assembly (e.g., 5$^-$, 27$^-$, or 51$^-$ mutants). These structures slowly aggregate into aberrant sixfold symmetrical assemblies closely resembling complete baseplates, but lacking the hub proteins and unable to serve as a substrate for baseplate completion or tail tube polymerization (Fig. 8). These structures are further discussed below. The next to last step in the formation of this arm complex is the addition of gp53. If this protein is removed by mutation, a 15S complex accumulates, containing all of the 1/6th arm proteins except gp53 and gp25 (Fig. 7). These do not polymerize, indicating that gp53 plays a critical role in the radial interactions.

The 1/6th arm proteins assemble in a strictly sequential fashion, as shown in Fig. 1. The assembly of each structural protein, except gp11, into the growing structure is a prerequisite for the next protein in the pathway. Thus each protein seems to modify the growing 1/6th arm structure so that it is then a substrate for the next protein addition reaction. In this pathway the first two proteins to interact are the products of genes 10 and 11. The gene 11 protein

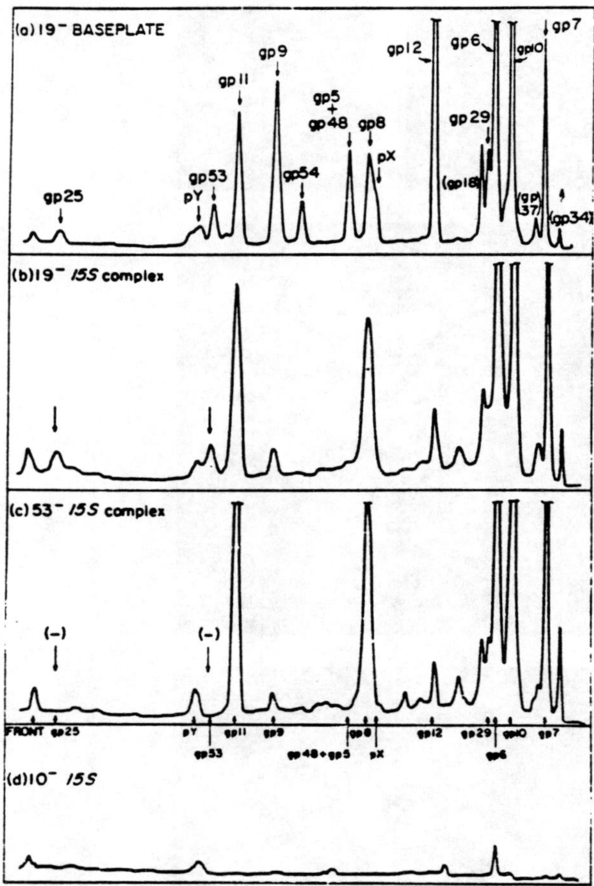

FIG. 7. Protein composition of baseplates and baseplate precursors. These tracings of SDS-gel patterns show the protein compositions of complete baseplates (a), the 15S precursor from the same lysate (b), and the 15S complex from 53^--infected cells (c). The infected cells were labeled with ^{14}C-mixed amino acids, and lysates were fractionated by sucrose gradient centrifugation. The peak fractions of the relevant distributions were analyzed. The bottom panel (d) shows a control from a 10^- lysate, unable to form the 15S complex. pX is now known to be gp27 (Berget and Warner, 1975), and pY is probably gp*frd* (Mosher and Mathews, 1979).

forms the tips of the morphological spikes on the distal face of the baseplate (Crowther et al., 1977), indicating that the arm assembly pathway proceeds from the outer vertex towards the hub.

Although the addition of gp11 to gp10 is not an obligatory step in the assembly of the 1/6th arms (Edgar and Lielausis, 1968), in the absence of gp7 all of the gp10 and gp11 is found in a 10/11 complex (Berget and King, 1978). The product of gene 10 can be thought of as the keystone protein in the assembly of the 1/6th arms. In the absence of gp10, no 1/6th arm-related assemblies are formed (Kikuchi and King, 1975a). The gene 7 protein adds to the 10/11 complex to make the 7/10/11 complex. Then, in sequential order, the products of genes 8, 6, 53, and 25 add to the growing 1/6th arm structure.

The product of T4 gene *frd* (dihydrofolate reductase) is also a component of the 1/6th arm structures. Mathews and co-workers have determined that it is present in the 1/6th arm intermediate which accumulates in cells infected with a gene 53 amber mutant (Mosher and Mathews, 1979). However, the exact point of assembly into the 1/6th arm structure is unknown.

Assembly of the central hub. The assembly of the central hub of the baseplate is more complicated than that of the 1/6th arm. In addition, the much smaller mass of the hub, the smaller size of the hub proteins, and the presence of aberrant side reactions have rendered its assembly somewhat more difficult to dissect (Kikuchi and King, 1975b). The central hub is comprised of six structural proteins (gp5, gp26, gp27, gp28, gp29, and gp*td*). These structures accumulate as a 22S complex in cells infected with mutants defective in 1/6th arm assembly (i.e., 10^- mutants). They have not been directly visualized by electron microscopy. In addition, because of the much smaller mass of the hub, it has only been identified through in vitro complementation assays.

The in vitro complementation between different pairs of mutants in the hub complex was much less efficient than the complementation reactions in the 1/6th arm pathway (Kikuchi and King, 1975c). A number of pairs of mutant-infected extracts complemented very weakly, if at all. This might be due to polarity effects (Stahl et al., 1970) or to the nature of the pathway itself. A further feature limiting the efficiency of the complementation reactions is the ability of the complete 15S arm complexes to polymerize in the absence of the arm into aberrant baseplate-like structures such as those shown in Fig. 8. These are less stable than the complete baseplates, and many of them are recovered as 40S structures which have transformed to a starlike configuration (Kikuchi and King, 1975b). The 40S structures are missing some of the proteins found in the hexagonal 70S forms, namely, gp9 and gp12 (Fig. 9). It seems that the addition of gp9 and gp12 to the aberrant hexagons is inefficient, resulting in baseplates lacking these proteins which transform to stars.

The 22S complex appears to form through the interaction of two smaller complexes, a 7S gene 29 complex and a 10S complex composed of gp5 and gp27. It is not clear whether the formation of the 22S complex is a radial polymerization process or the interaction of two already cyclical complexes.

Recent results from Kozloff's laboratory have extended the original work of Kikuchi and King (1975c) by showing that the products of genes 26, 28, and *td* are structural components of the baseplate central hub and by determining where these proteins are involved in the assembly pathway of the central hub (Kozloff, 1981; Kozloff, personal communication). To locate these structural proteins they have used an in vitro "chase" technique which allows the detection of proteins which occur in a few copies per phage particle. Through the use of this technique, protein complexes which were not detected in the original study by Kikuchi and King were identified. The central hub assembly pathway currently suggested by Kozloff is incorporated in Fig. 1 (Kozloff, personal communication). In this pathway gp29 and gp*td* interact to form a 29/*td* complex which sediments at 7S. Gp51 catalyzes

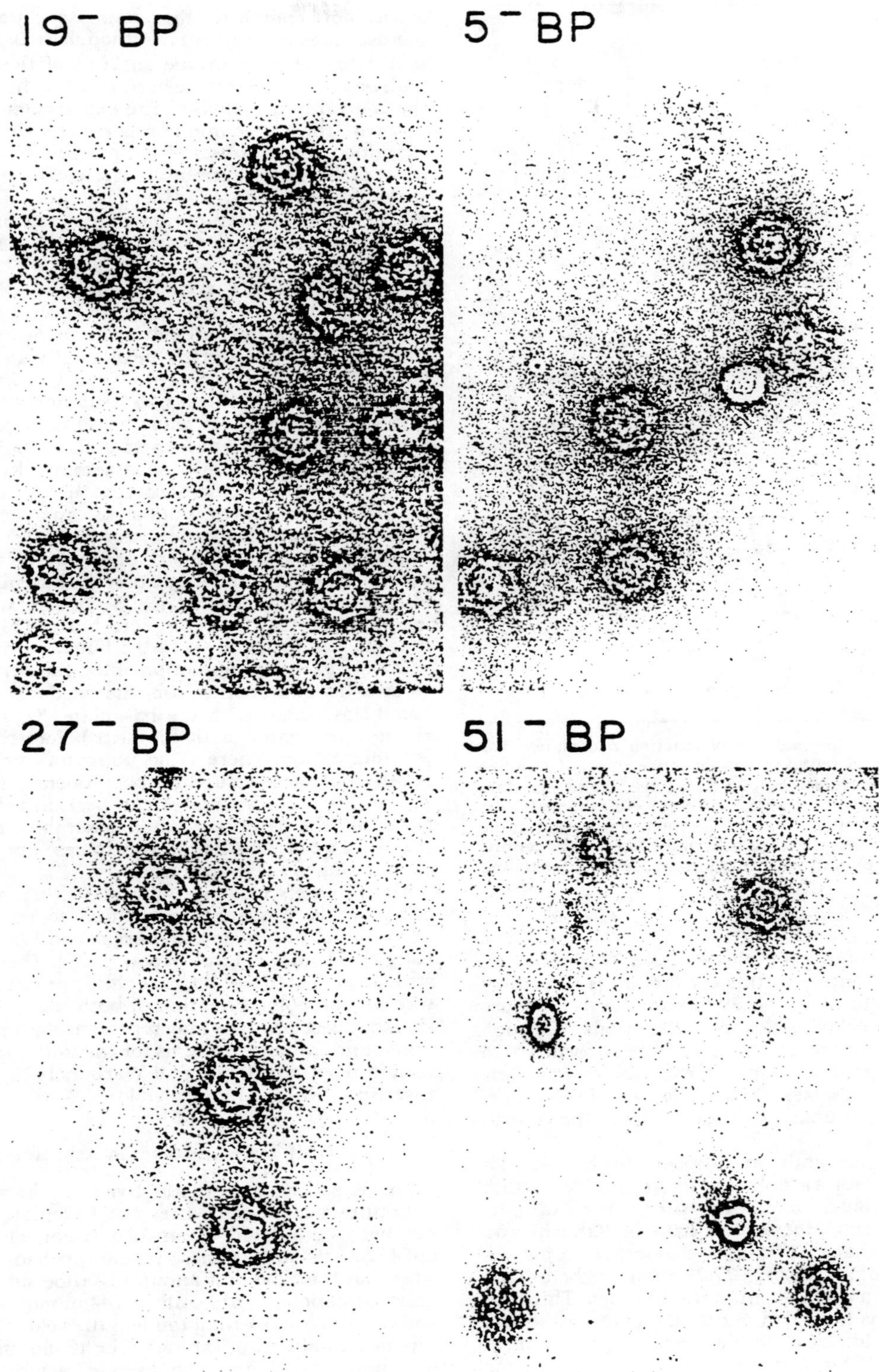

FIG. 8. Inactive baseplates. The top left panel shows complete baseplates from 19⁻-infected cells. Note the exclusion of negative stain from the very center of the structures. The other three panels show structures accumulating in cells defective in various steps in hub assembly. The centers of these structures are penetrated by the stain. These baseplates are inactive as phage precursors when tested by in vitro complementation (Kikuchi and King, 1975b).

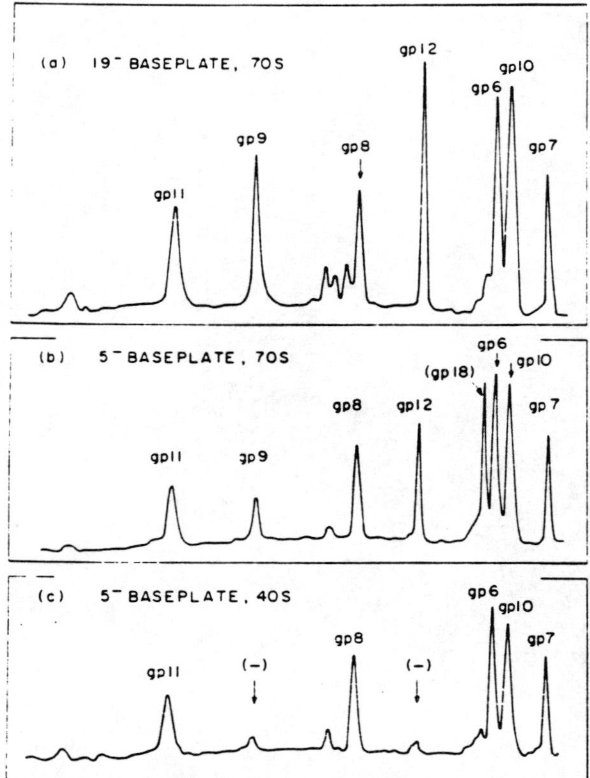

FIG. 9. Protein composition of inactive baseplates. The inactive baseplates found in hub-defective infected cells are isolated as two populations, 70S hexagons (b) and 40S stars (c). The 70S structures from 5^--infected cells (shown in Fig. 8) do not incorporate gp48 or gp54 (which are present in the lysates) and are therefore not substrates for tube assembly. The stars lack gp9 and gp12 in addition. From Kikuchi and King (1975b).

the conversion of this complex to a 14S 29/*td* complex. The molecular nature of this conversion is unknown; however, this form of the 29/*td* complex is the substrate for the ensuing assembly reactions. Through separate assembly reactions gp5 and gp26 form a 5/26 complex and gp27 and gp28 form a 27/28 complex. These three complexes, 29/*td*, 5/26, and 27/28, then coassemble in an unknown order to form the central hub structure.

The central hub pathway proposed by Kozloff and co-workers is not completely in accord with that proposed by Kikuchi and King (1975c). For example, the in vitro complementation data of Kikuchi and King suggest that gp27 and gp5 interact to form a complex (presumably a 5/26/27/28 complex) before the interaction with a gp29-containing complex. This type of interaction is not ruled out by the pathway shown here, but no additional data are available to confirm this proposal. Clearly, many proteins are interacting in complex ways in this pathway. Some of these proteins are being utilized in metabolic pathways as enzymes at the same time as they are being withdrawn as structural components of virus particle. The experiments of Kikuchi and King relied on the ability of assembly complexes to remain associated through sucrose density gradients. Although it is clear that such was the case for the analysis of the 1/6th arm pathway, it may not have been true for the analysis of the central hub pathway. The experiments using the in vitro "chase" technique reported by Kozloff (1981) examined only a subset of the possible pairwise mixtures of detective extracts to derive the information summarized in Fig. 1. A serious effort needs to be made, using a combination of the techniques now available, to further characterize the central hub pathway.

Assembly of the sixfold symmetrical baseplate structure. Once the 1/6th arm and central hub subassemblies are completed, they spontaneously assemble to form the characteristic hexagonally shaped baseplate. This intriguing assembly reaction involves the interaction of one central hub structure with six 1/6th arm structures in what must be a radial polymerization reaction: six 1/6th arm assemblies polymerizing around one central hub subassembly. Kikuchi and King (1975b) demonstrated that in the absence of central hub structures, 1/6th arms can assemble slowly into baseplate-like structures which are missing the central density characteristic of baseplates containing central hubs (Fig. 8). These aberrant baseplate assemblies can also be formed in 25^- lysates which have incomplete arms but complete hubs. This suggests that gp25 links the arms to the hubs.

During or upon completion of the coassembly of the 1/6th arms and central hub, the products of genes 9 and 12 associate with the baseplate. These proteins are not associated with the arm before their radial polymerization. There is no obligatory order to the addition of these structural components, as both $9^+/12^-$ and $9^-/12^+$ baseplates can be isolated (Fig. 10). Note that the baseplate can be assembled from arms that lack gp11, the tips of the spikes, but that gp12 cannot add to these structures.

The products of genes 48 and 54 then add to the baseplate in a sequential fashion to complete the assembly of the baseplate (Berget and Warner, 1975; King, 1971; Meezan and Wood, 1971). These last two structural proteins of the baseplate prepare this structure for the polymerization of both the tail tube and sheath. These proteins may represent the length determiner and initiator proteins for the tail tube polymerization reaction (Berget and Warner, 1975; Duda and Eiserling, 1982).

Tail Tube and Sheath Assembly

The assembly of the distinctive tail tube and sheath structures occurs only after the baseplate structure has been completed via incorporation of gp48 and gp54. One or both of these species probably form the sites that bind the first annulus of tube subunits. The polymerization of gp19 utilizes 144 monomers of gp19 and terminates at a total tail length of exactly 100 nm. The proximal end of the tail tube is modified by the incorporation of gp3, which may act as a "glue" protein between the tail tube and the sheath, connecting these two tubular structures only at this end of the tail (King, 1971). During tail tube polymerization, sheath monomers, gp18, start polymerizing around the tail tube and continue to polymerize until the

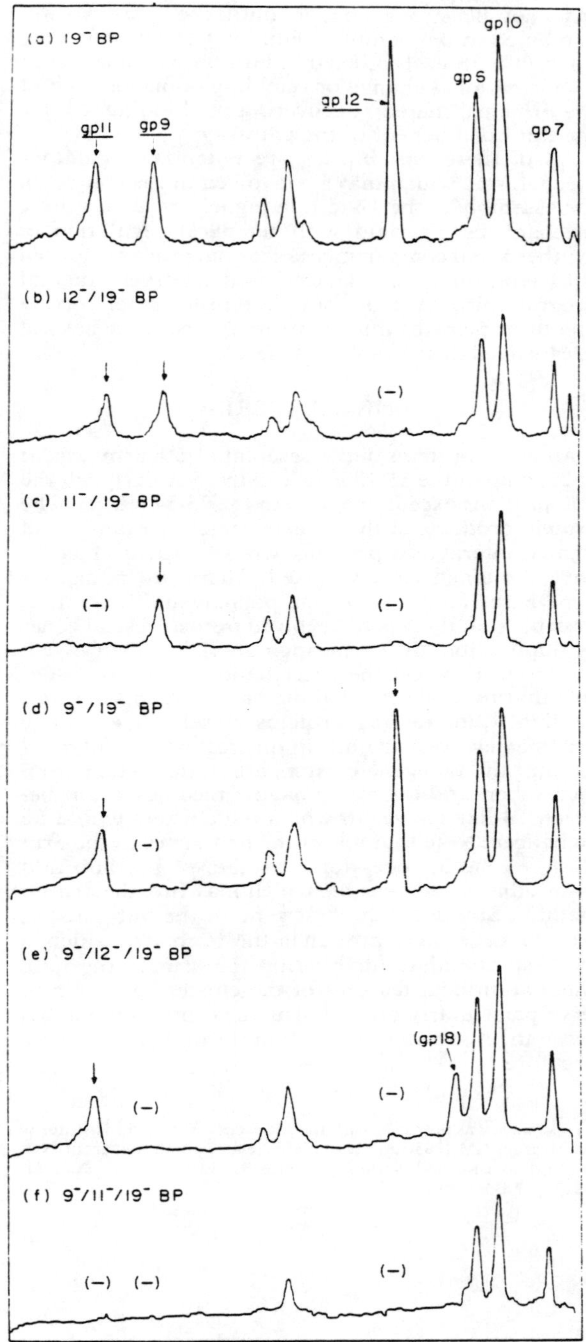

FIG. 10. Independence of gp9 addition from gp11/g12 addition. The tracings are of SDS-gel patterns of radioactively labeled baseplates isolated from mutant-infected cells.

sheath reaches the proximal end of the tail tube. At this point the tail structure is completed by the addition of gp15, which stabilizes the sheath structure (King and Mykolajewycz, 1973) and produces the "connector" structure required for T4 head attachment.

The polymerization of the tail tube is very tightly regulated at the assembly level. No case has been reported of the polymerization of precursor tail tube subunits into tubes in the absence of the baseplate, either in vivo (King, 1971) or in vitro with purified tube subunits and baseplates (Wagenknecht and Bloomfield, 1977). The first set of tube monomers is clearly bound by the baseplate. However, each newly formed annulus must constitute the active site for binding of additional monomers from solution. This is analogous to the growth of *Salmonella* flagella by polar addition of monomeric flagellin subunits to the distal tip of the growing flagella (Asakura et al., 1968). In this case the subunits undergo a substantial conformational change upon binding (Uratani et al., 1972). We assume that the tail tube utilizes the same mechanism.

Tail sheath polymerization follows the same pattern but is less tightly controlled. In the absence of a proper substrate for tail sheath polymerization, an aberrant form of polymerized sheath monomers called polysheath forms. Moody (1973) proposed that these structures are analogous to the packing of the sheath subunits in the contracted sheath, but polysheath is relatively easily dissociated into subunits, unlike the contracted sheath.

Tail Length Determination

The assembly experiments reveal that the length of the sheath is determined by the length of the tail tube. If tail tube length is limited due to a shortage in supply of tube subunits [e.g., *ts*19⁻ mutants at intermediate temperatures], the sheath subunits polymerize only as far as the partially formed tube (King, 1971). Two mechanisms have been proposed for the regulation of the length of the T4 tail. Kellenberger (1972b) suggests that the tail tube monomers polymerize because the total free energy of a monomer is lowered through intersubunit bond formation. However, to achieve the most favorable bonding, each subunit may have to become slightly deformed. As more subunits are added, the deformation strain per subunit increases until finally the positive free energy accumulated equals the negative free energy associated with intersubunit bonding. Thus it then becomes unfavorable to add any more subunits, and the polymerization reaction stops.

The feasibility of such an "accumulated strain" hypothesis has been investigated by Wagenknecht and Bloomfield (1975). Their theoretical analysis shows that such a model is thermodynamically feasible for producing very narrow length distributions. In a further study Wagenknecht and Bloomfield (1977) have shown that gp19 monomers can be polymerized onto baseplates in a partially purified in vitro system yielding tail tubes with a length distribution narrowly centered around 100 nm, the same value found in vivo. Thus, depending on the purity of their system, few intracellular factors other than baseplates and gp19 are required to produce tail tubes of the correct length.

A second hypothesis is that there may be "tape measure" molecules along which the gp19 monomers polymerize (King, 1968; King, 1971). The length of this molecule then determines the length of the tail tube. The natural candidate would be either gp48 or gp54. Recently it has been demonstrated that the product of gene 48 can be localized to the tail tubes which have been "broken off" tail tube-baseplate complexes (Duda and Eiserling, 1982). By electron microscopy, such tail tubes have been shown to have their central cavity filled with material that excludes stain. Occasionally tail tubes can be seen which have material protruding from one end and stain penetrating into the other. This suggests that these tail tubes may be filled with gp48, and it seems likely, given its position in baseplate assembly, gp48 may indeed be a length-determining molecule for the T4 tail (see Eiserling, this volume). The sensitivity of baseplates to proteases (Kikuchi and King, unpublished data; Bloomfield, personal communication) also suggests that a protein may be involved in tube initiation and length determination. Work on phage lambda suggests that gpH, a protein product found in the mature tail, acts as a tape measure for tail length determination (P. Youderian, Ph.D. thesis, Massachusetts Institute of Technology, Cambridge, 1978; Youderian and King, manuscript in preparation; R. Hendrix, personal communication).

PROTEIN-DETERMINED ORDERED ASSEMBLY REACTIONS

The assembly of the 1/6th arm of the baseplate is a straightforward example of a pathway controlled by protein-protein interactions. Four of the six assembly reactions in this pathway are ordered by the previous interaction of a structural protein with the growing complex (the additions of gp8, gp6, gp53, and gp25). Nothing is known about the molecular mechanisms governing the assembly of this structure; however, several formal possibilities exist for the protein-determined ordering of this pathway.

Modifications in the primary structure of a component in this pathway by proteolysis or some other covalent modification could change the nature of an incorporated protein so as to provide an otherwise obscured binding site for the next protein in the pathway. Proteolytic cleavage of the proteins involved in T4 tail formation has not been observed in pulse-chase experiments, but subtle changes cannot be ruled out. Second, two different structural components of the 1/6th arm may each contribute one half of a binding site for the next protein in the pathway. The interaction of each of these proteins with the third may not be strong enough until the first two are combined into a complex. Finally, two proteins may interact with each other in a fashion which causes a conformational change or refolding of one or both of the proteins, thereby uncovering the binding site for the third component in the pathway.

All of these possibilities are potential regulatory mechanisms which may be involved in the control of the assembly of the T4 tail. In higher organisms these mechanisms, overlaid with temporal regulation of synthesis and compartmentalization, probably form a subset of regulatory devices used in developmental programming. Yet the simplest problem of sorting any three proteins into an assembly sequence has not been solved at the molecular level.

GENE CLUSTERING

All of the proteins of the baseplate 1/6th arm, except gp25, map in the 53-12 gene cluster. Similarly, all the hub proteins except gp5 map in the 25-54 cluster. The protein products of these genes are clear examples of tightly interacting proteins whose structural genes have remained closely linked. There are numerous hypotheses for such linkage phenomena. Since it is possible that these genes evolved from ancestral genes by duplication, the clustering may reflect their evolutionary history. On the other hand, it seems reasonable that natural selection may act to keep the genes for tightly interacting proteins closely linked since they may evolve as a unit. In this regard it is interesting that the two gene clusters are quite distant from each other, and one can imagine recombination between T4-like phages in which the clusters would be exchanged. Note that one of the first genes in the arm cluster, gene 5, codes for a protein of the hub and maps adjacent to gene 53, which may link the arm to the hub. Similarly, the first gene in the hub cluster, gene 25, codes for a protein of the 1/6th arm which is the best candidate for binding the arm to the hub. These additional features of the clustering may preserve particularly critical structural interactions between the hub and 1/6th arms in the overall baseplate structure.

This paper was prepared with the support of National Institutes of Health grant GM-17980 and National Science Foundation grant PCM 80-11661 to J.K., and National Science Foundation grant PCM 81-04523 to P.B.B.

Long Tail Fibers: Genes, Proteins, Assembly, and Structure

W. B. WOOD[1] AND R. A. CROWTHER[2]

Department of Molecular, Cellular & Developmental Biology, University of Colorado, Boulder, Colorado 80309[1] *and MRC Laboratory of Molecular Biology, Cambridge CB2 2QH, England*[2]

The primary adsorption organelles of T4 are the six long tail fibers attached to the baseplate (see Goldberg, this volume). Kellenberger et al. (1965) first showed that the tail fibers are essential for infection and reviewed earlier work on their properties. As expected, the tail fibers carry the primary host-range determinants at their distal ends. In structure they appear to be slender, quite rigid rods about 50 Å (5 nm) in diameter, with a proximal knob at the end that attaches to the baseplate and a distal tip that tapers slightly and sometimes appears bent in electron micrographs (Fig. 1 and cover photograph of this volume). The proximal and distal halves of the fiber are each about 80 nm long, and in the infectious phage they are joined at an angle of about 160°, which becomes acute as the tail fibers bend during the process of irreversible adsorption (Goldberg, this volume).

The tail fibers can take on at least two configurations: extended, when they are attached only to the baseplate by the flexible joints at their proximal ends, or retracted, when they are bound up along the tail sheath and the head of the phage particle. Retraction involves interaction with the whiskers (Conley and Wood, 1975) that extend outward from the collar region (Coombs and Eiserling, 1977; Yanagida and Ahmad-Zadeh, 1970) and probably also with the sheath and possibly the head (for discussion, see Conley and Wood, 1975).

Transitions between these two configurations strongly affect both physical and biological properties of the phage (Kellenberger et al., 1965). Extension markedly changes its hydrodynamic properties; for example, the sedimentation coefficient decreases from about 1,000S in the retracted state to 750S in the extended state (Conley and Wood, 1975). Because the distal half fiber carries substantial net positive charge, the dipole moment of the particle also increases with extension (Baran and Bloomfield, 1978; Bloomfield, this volume). Biologically, the degree of extension controls the infectivity of the virus, since only fibers in the extended configuration can cooperatively attach to host cells. Factors that promote retraction and therefore lower infectivity are low pH, low temperature, low ionic strength, and in some strains absence of tryptophan or other cofactors (Conley and Wood, 1975; Kellenberger et al., 1965). These effects probably are biologically significant in that they inhibit infection under environmental conditions that are unfavorable for phage multiplication.

The tail fibers are potent immunogens. When a rabbit is hyperimmunized with purified phage particles, most of the phage killing activity of the resulting serum is directed against the tail fibers (Edgar and Lielausis, 1965). Blocking of this activity provides a convenient serological assay for tail fiber antigens.

The tail fibers are of interest for a variety of reasons. They are composed of fibrous proteins of unusual structure; their assembly involves the participation of novel accessory proteins that appear to catalyze noncovalent association of the structural proteins; and their function represents an experimentally accessible example of specific virus-host interaction as well as the initiation mechanism for the elaborate triggering and injection process of T4 infection. Taking advantage of genetic, serological, functional, and structural analysis, much has been learned about the various tail fiber components, although many details remain to be explained. This chapter briefly reviews the genetic control of tail fiber structure, the steps in tail fiber assembly and attachment, and recent fine-structure analysis of assembled fibers and nucleotide sequences of tail fiber genes. For reviews on the extensive early work on tail fiber functions in adsorption, see Stent (1963) and Kellenberger et al. (1965). Other recent reviews of tail fiber assembly in the context of phage morphogenesis may be found in Wood and King (1979) and Wood (1980).

GENETIC CONTROL OF TAIL FIBER STRUCTURE, ASSEMBLY, AND ATTACHMENT

Tail Fiber Antigens

Tail fiber structure and assembly are controlled by the five contiguous genes 34, 35, 36, 37, and 38 and the unlinked gene 57 (Epstein et al., 1963). Attachment of fibers to the phage particle is controlled by two additional genes, 63 and *wac*. The positions of these genes are shown on the T4 map of Kutter and Rüger (this volume).

Early serological analysis of defective lysates by serum blocking defined three classes of tail fiber antigens: A, missing from gene 34-defective lysates; B, missing from 36-defective and 37-defective lysates and low (~10% of normal levels) in 38-defective lysates; and C, missing in 37-defective lysates and low in 38-defective lysates. All blocking antigens are low in 57-defective lysates, and all are present at normal levels in 35-defective lysates (Edgar and Lielausis, 1965). Tail fiber precursors isolated from 34-defective, 35-defective, and 36-defective, but not 37-defective lysates were shown to block killing of phage by isolated *Escherichia coli* lipopolysaccharide, suggesting that C antigen, under the control of gene 37, interacted with host cell wall components and therefore was distally located (Wilson et al., 1970). Antibody labeling of whole phage with adsorbed sera showed clearly that the A antigens are confined to the proxi-

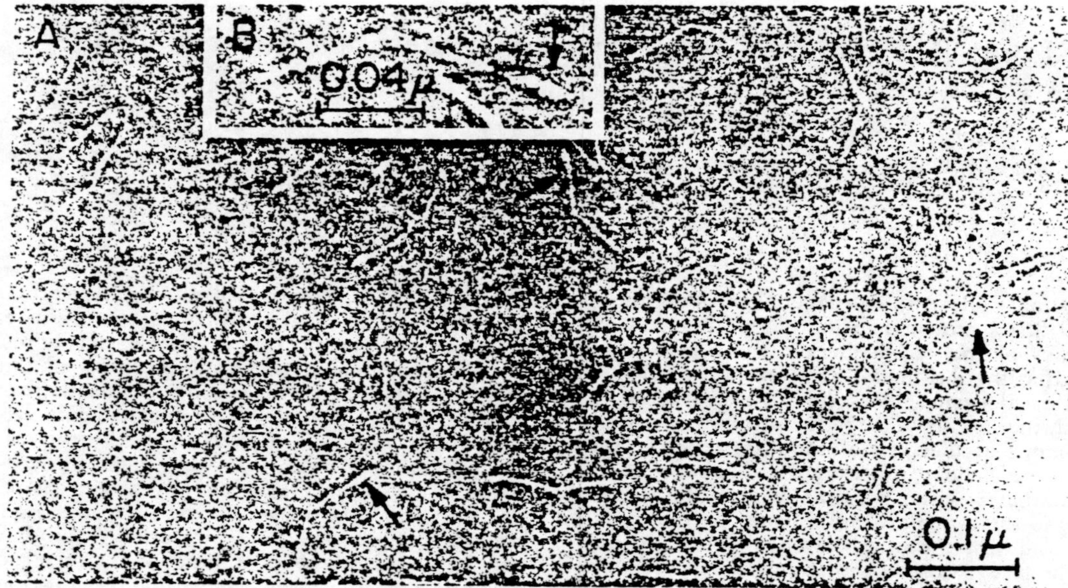

FIG. 1. (A) Electron micrograph of purified whole tail fibers and proximal half fibers showing knobbed ends (arrows). (B) Enlargement of a single whole fiber (From Ward et al., 1970.)

mal half fiber and the C antigens to the distal half (Yanagida and Ahmad-Zadeh, 1970). Together these findings indicated that gp34 is the major component of the proximal half fiber and that gp36, gp37, and gp38 are involved in distal half fiber formation. A structural role had been suspected for gp37 on the basis of known gene 37 mutations affecting host range (reviewed in Kellenberger et al., 1965).

Structural Proteins

By electron microscopy, gene 34, 35, 36, 37, 38, and 57 mutants were shown to produce tail-fiberless phage particles (Epstein et al., 1963) and, except for gene 57 mutants, short rods that appeared to be half fibers (Eiserling et al., 1967; Ward et al., 1970). With identification of the corresponding gene products by denaturing polyacrylamide gel electrophoresis (Laemmli, 1970) and development of techniques for separating and analyzing the protein composition of tail fiber-related components in mutant lysates (King and Wood, 1969; Ward et al., 1970), further experiments (King and Laemmli, 1971; Ward and Dickson, 1971) showed that the proximal half fiber is composed entirely of two copies of gp34 (M_r = 150,000), whereas the distal half fiber is composed of two copies of gp37 (M_r = 110,000) as the major component and in addition (Dickson, 1973) contains two copies of gp36 (M_r = 40,000) and one copy of gp35 (M_r = 24,000). (Some of these molecular weight values have been refined recently through DNA sequence analysis; see Nucleotide Sequence Analysis, below.)

The orientation of gp34 has been inferred from the properties of gene 34 *ts* mutants. The *ts*N1 mutation, which maps at the extreme promoter-proximal end of gene 34, does not affect A antigen production (Edgar and Lielausis, 1965) or formation of whole tail fibers, but does prevent fiber attachment to the baseplate (King and Wood, 1969). Several *ts* mutations near the promoter-distal end of the gene cause substantial reductions in A antigen production (Edgar and Lielausis, 1965), and one of them, *ts*B45, causes phage formed at permissive temperature to lose their distal half fibers rapidly when incubated with 1% sodium dodecyl sulfate (SDS) under conditions that do not rapidly inactivate wild-type phage (B. S. Seed, Ph.D. thesis, California Institute of Technology, Pasadena, 1980; B. S. Seed, G. J. Del Zoppo, and W. B. Wood, manuscript in preparation). These findings suggest that the two amino termini of gp34 lie at the proximal end of the half fiber near the baseplate attachment site, and that the two carboxyl termini lie at the distal end near the site of attachment of the distal half fiber. Other experiments by Beckendorf (1973), using antibody specific to different regions of folded gp37 to decorate whole phage in electron micrographs, showed that in the distal half fiber as well, the amino-terminal ends of gp37 are proximal and the carboxy-terminal ends are distal, and that the intervening polypeptide is roughly colinear with the half fiber structure. These experiments are described in more detail in Fine Structure, below.

Accessory Proteins in Assembly

The products of genes 38 and 57 are absent from the phage particle, the completed tail fiber, and all of the assembly precursors found in the various tail fiber mutants (Dickson, 1973). These gene products therefore appear to be accessory proteins, required for tail fiber assembly but not included in the structure. Function of gp38 is required for formation of the distal half fiber only, whereas function of gp57 is required for formation of both halves of the fiber.

The gp38 protein has an M_r of 26,000 (Dickson, 1973). For the gp57 protein, several different M_r values ranging from 5,000 to 18,000 have been reported. These discrepancies were resolved by the finding that *am* mutations in gene 57 cause the disappearance from infected-cell lysates of two polypeptides, one of M_r about 6,000 and the other of M_r about 18,000. Their relationship remains unclear, but the smaller of the two is capable of supplying gp57 function (Herrmann and Wood, 1981).

Tail Fiber Attachment

Gene 63 mutants and *wac* mutants make tail fiber antigens and assembled tail fibers in normal amounts, but attach fibers to the baseplate at subnormal rates (Wood and Conley, 1979; Wood and Henninger, 1969). At temperatures below 25°C, gene 63 mutants do not form plaques, but at 42°C they make plaques only slightly smaller than normal. The gp63 polypeptide (M_r = 42,000), which probably functions as a dimer (Wood and Henninger, 1969), has been partially purified (Wood et al., 1978) for use in studies of tail fiber attachment in vitro (see Attachment, below). This protein is not found in complete phage, tail-fiberless particles, or tail fibers. Therefore, like gp38 and gp57, it is an accessory protein for tail fiber attachment.

Mutants defective in the *wac* gene were first characterized as phages that showed impaired tail fiber attachment (Dewey et al., 1974) and remained infectious in the presence of polyethylene glycol, which causes normal phage to undergo whisker-mediated retraction of the tail fibers, thereby preventing infection (Follansbee et al., 1974). The *wac* mutants lack whiskers, but nevertheless attach tail fibers at a sufficient rate in vivo to form small plaques. Polyethylene glycol-resistant mutants with phenotypes similar to *wac* also result from missense mutations in gene 36, suggesting that the whiskers interact with the gp36 component of the distal half tail fiber in attachment and retraction (Conley and Wood, 1975; Follansbee et al., 1974; Wood and Conley, 1979). The gp*wac* protein (M_r = 54,000) is present in about 36 copies per phage particle and probably is responsible for not only the whiskers but also the structure referred to as the collar in electron micrographs of T4 (Coombs and Eiserling, 1977). In a sense, gp*wac* is also an accessory protein for tail fiber attachment in that it is not a structural component of either the tail fiber or the baseplate. Thus efficient tail fiber acquisition during T4 morphogenesis requires the function of eight genes, four of which (34 through 37) code for structural proteins of the tail fiber, and the other four of which (38, 57, 63, and *wac*) code for accessory proteins involved in tail fiber assembly (38, 57) and attachment (63, *wac*).

ASSEMBLY OF THE TAIL FIBER

Proximal Half Fiber

Characterization of mutant phenotypes by serology, electron microscopy, and analysis of precursors for protein composition, combined with results of in vitro complementation of mutant-infected-cell extracts (Edgar and Wood, 1966), led to formulation of the sequential order of steps in tail fiber assembly as diagrammed in Fig. 2 (King and Wood, 1969).

Least is known about the role of gp57, which is required not only for dimerization of gp34 and gp37 to form the precursors of the proximal and distal half fibers, respectively, but also for formation of the short tail fibers from gp12 monomers (Kells and Haselkorn, 1974). The dimerization of gp34 probably is initiated at the carboxy-terminal ends of the polypeptides, based on the observations that amino-terminal missense in gp34 does not affect A antigen formation, whereas carboxy-terminal missense decreases it markedly and nonsense mutations, even at the extreme carboxyl terminus, eliminate it entirely. Dimerization of gp34 appears to complete proximal half fiber assembly.

Distal Half Fiber

In distal half fiber assembly, the gp57-mediated step is placed first because it is the only one in the process that has never been demonstrated in vitro; that is, a 57-defective extract cannot complement a 37-defective extract, whereas a 38-defective extract can do so (Bishop and Wood, 1976). In gene 57 mutant infections, gp37 and gp34 (as well as gp12) remain unassembled, based on ease of monomer extraction by SDS and absence of serum blocking antigens, and appear to be associated with the cell debris fraction in extracts (King and Laemmli, 1971; Ward and Dickson, 1971). The gp57 protein may act to prevent deleterious interactions of the nascent fiber polypeptides with host cell components or otherwise to affect the internal environment of the host cell in some way that promotes dimerization. In support of this view, Revel et al. (1976) were able to isolate mutant *E. coli* host strains in which T4 gene 57 function is no longer required for plaque formation.

In gene 38 mutants, proximal half fiber assembly is normal, but distal half fiber assembly appears blocked as in gene 57 mutants, with gp37 remaining unassembled and apparently associated with the cell debris fraction in extracts (King and Laemmli, 1971; Ward and Dickson, 1971). In vitro complementation of 37-defective and 38-defective extracts can be demonstrated, but the reaction proceeds only to the extent of about 1% of gp37 conversion to the antigenic dimer form (Wood and Bishop, 1973), and the instability of 38 activity in extracts has hampered its purification and further study. However, cross-strain complementation experiments showed that the gene 38 functions of T2 and T4 are specific for their respective gene 37 products (Russell, 1974), and extension of these analyses to T2-T4 hybrids demonstrated that this specificity is confined to the carboxy-terminal portion of the gp37 polypeptide (Beckendorf, 1973; Beckendorf et al., 1973). A missense mutation (*ts*3813) that partially suppresses gene 38 defects, allowing low levels of distal half fiber and active phage production in the absence of gene 38 function, maps near the promoter-distal end of gene 37 (Bishop and Wood, 1976) in the *d* segment as defined by Beckendorf et al. (1973; see Fig. 6). More precise characterization and mapping of mutations in this region has shown that *ts*3813 maps in the promoter-proximal (*d*1) region of this segment, while a host range mutation (*h*G3) maps in the more promoter-distal region of this segment (*d*2). At restrictive temperature, *ts* mutations in the *d*1 region prevent C antigen formation, whereas mutations in the *d*2

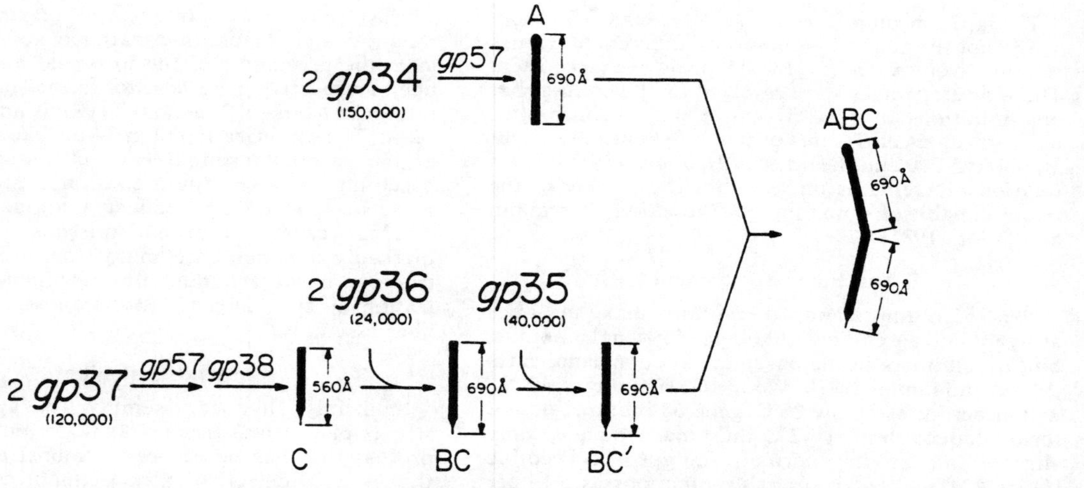

FIG. 2. Pathway of tail fiber assembly. See text for explanation. The completed distal half fiber is designated BC′ to indicate that it is structurally different but serologically indistinguishable from its precursor BC. (From Wood, 1979.) (10 Å = 1 nm.)

region do not (M. Snyder, Ph.D. thesis, University of Colorado, Boulder, 1982; M. Snyder and W. B. Wood, manuscript in preparation). These observations support the view that the accessory protein gp38 somehow promotes gp37 dimerization by specific interaction with a region near the carboxy-terminal ends of two gp37 polypeptides. A speculative suggestion for the mechanism of this interaction is shown in Fig. 3.

In the subsequent step of distal half fiber assembly, two molecules of gp37 are added to the structure. As a result the half fiber becomes longer by about 20%, and one or more new serum blocking antigens (B antigen) are formed (Ward et al., 1970). This addition is thought to occur at the proximal end of the structure because neither distal tip morphology (Ward et al., 1970) nor function (Wilson et al., 1970) appears to change, and because decoration experiments show that B antigen is located near the central kink of the tail fiber on completed phage (Yanagida and Ahmad-Zadeh, 1970). In vitro complementation between fresh 36-defective and 37-defective extracts proceeds efficiently; however, the C-half fibers in 36-defective extracts become inactivated with time due to an apparently specific proteolytic cleavage near one end of the gp37 polypeptide (R. J. Bishop, S. Aley, and W. B. Wood, unpublished data).

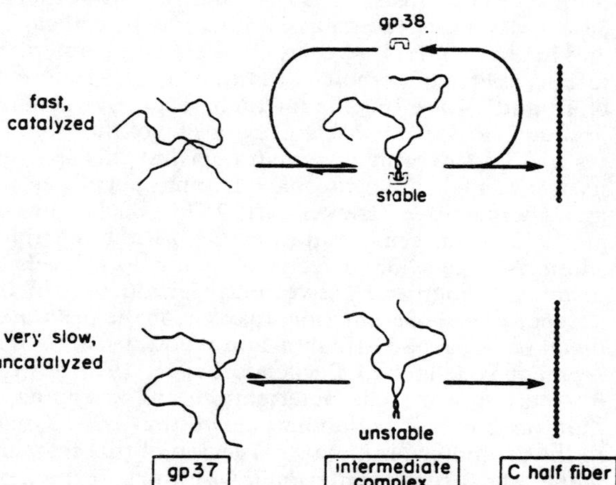

FIG. 3. Possible mechanism for gp38 action in gp37 dimerization. (From Wood, 1979.)

The addition of gene 35 product does not appear to change either the morphology or the serum blocking properties of the distal half fiber, but its function is required for joining to the proximal half fiber. The position of gp35 action in the pathway was originally assigned from in vitro complementation results showing that 35-complementing activity was stable in a 34-defective extract but not in 36-defective or 37-defective extracts, suggesting that gp35 was stabilized by interaction with a completed BC half fiber (King and Wood, 1969). Subsequent analysis of half-fiber precursors for protein composition showed that one molecule of gp35 is added subsequent to gp36 addition (Dickson, 1973). Presumably, gp35 resides at the proximal end of the half fiber since it is implicated in joining to the proximal half fiber, but there is little direct evidence on this point (Earnshaw et al., 1979).

Joining of the Proximal and Distal Half Fibers

Because no additional mutations have been identified that prevent joining of the distal and proximal half fibers, this reaction is presumed to occur spontaneously although it has not been studied in detail. Based on analysis of second-site revertants of *ts*B45, a promoter-distal gene 34 mutation that affects stability of the proximal-distal joint, Seed (thesis) and coworkers (Seed et al., in preparation) have obtained indirect evidence for interaction of the gp34 carboxy-terminal and gp37 amino-terminal regions.

Several puzzling questions remain unanswered regarding the joining reaction. In lysates made with phage particle-defective mutants as well as with wild-type phage, only about a third of the free tail fiber antigen is present as whole fibers, and the rest is present as half fibers (King and Wood, 1969; Ward et al., 1970). This situation does not represent an equilibrium between assembled and unassembled forms, because whole fibers isolated from the mixture do not dissociate under physiological conditions. The free half fibers probably are not inactive precursors, because at least in some extracts nearly 100% of the antigenic species can be converted by in vitro complementation to whole fibers attached to infectious phage. Moreover, in crude 37-defective extracts the A antigen, although present in expected amounts, cannot be seen in electron micrographs as a half fiber until aging or manipulation of the extract during half-fiber purification (Ward et al., 1970). A possible explanation for these observations may be that gp34 dimers originally accumulate in an antigenic but loosely folded form, whose conversion to a half fiber and union with the distal half fiber can be promoted by interaction with the phage particle. This notion is supported by Terzaghi's evidence (1971) that joining of the two halves to each other and of the proximal half to the baseplate may occur after at least transient association of the distal half fiber with the phage particle (Terzaghi et al., 1979; Wood and Conley, 1979). Quite possibly, this pathway is an alternative route to tail fiber completion and attachment. It cannot be the only pathway, because both active proximal half fibers visible by electron microscopy and active whole fibers can form in the absence of particles, and isolated whole fibers can attach to particles to produce active phage (Ward et al., 1970; Wood and King, 1969).

ATTACHMENT OF TAIL FIBERS TO THE PHAGE BASEPLATE

Properties of the Reaction

Attachment of tail fibers to the baseplate is probably the terminal step in phage assembly (Fig. 4). Efficient attachment requires the participation of gp63 and the whiskers (gp*wac*), although neither of these proteins is essential. The failure of tail fibers to attach efficiently to free tails is explained by the requirement for whiskers, which are not added until after head-tail joining (Coombs and Eiserling, 1977). This feature of the pathway helps to insure that all completed particles acquire tail fibers and become infectious without requiring production of tail fibers in large excess over tails, which are made in about fivefold excess over heads (S. Ward, unpublished data).

The tail fiber attachment reaction proceeds efficiently in vitro (Edgar and Wood, 1966; Wood and Henninger, 1969) and has been extensively studied. Rates equal to those observed in crude extracts are obtained with tail-fiberless particles, tail fibers, and gp63, all purified to near homogeneity, suggesting that no soluble cofactors or other proteins are necessary (Wood et al., 1978). The only other requirements for the reaction are ionic; the highest rates are obtained in the presence of about 1 M NH_4^+ or other nonmetallic cations, but in their absence a somewhat lower rate can be obtained by using Mg^{2+} at an optimum concentration of about 0.02 M (Wood and Henninger, 1969).

Kinetic analysis of the reaction provided evidence that tail fibers attach not cooperatively, but rather

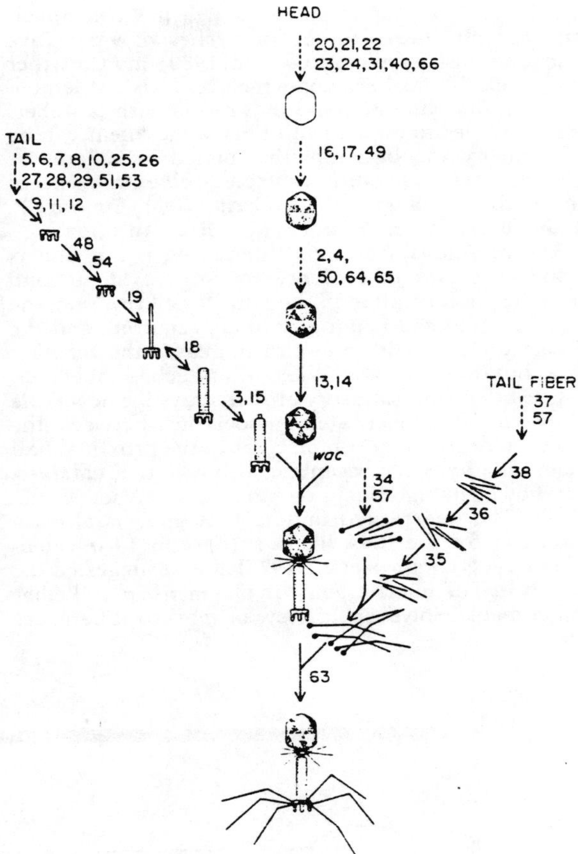

FIG. 4. Pathway of terminal steps in phage assembly. (Adapted from Wood, 1979.)

randomly and one at a time to baseplate sites in the particle population (Wood and Henninger, 1969), in agreement with earlier electron micrographs of phages produced in vitro during tail fiber attachment (Edgar and Wood, 1966) and in vivo by "leaky" tail fiber mutants (Eiserling et al., 1967). More recent physical measurements on tail fiber distributions support this mechanism (Baran and Bloomfield, 1978; Bloomfield, this volume).

Role of gp63

In the absence of gp63, tail fibers attach slowly in vitro to tail-fiberless but otherwise complete particles; addition of gp63 increases the rate of reaction up to 100-fold (Wood and Henninger, 1969; Wood et al., 1978). The accessory protein also stimulates the probably nonphysiological attachment of proximal half fibers to the baseplate, but not the subsequent addition of distal half fibers to these particles, indicating that gp63 probably acts only to promote the baseplate attachment reaction but not the interaction of the distal half fiber with the proximal half fiber or the whiskers (Wood et al., 1978).

The site of tail fiber attachment to the baseplate is almost certainly gp9, which occupies a peripheral position in the baseplate outer wedges (Crowther et al., 1977). The postulated identity of the attachment site as gp9, based on the properties of 9-defective particles (discussed by Crowther [1980] and Crowther et al. [1977]) has been confirmed recently by demonstrating that an antiserum that reacts with tail-fiberless particles to prevent tail fiber attachment can be blocked by incubation with isolated baseplates or with any defective mutant-infected-cell extract except those that lack gp9 (M. A. Urig, S. M. Brown, P. Tedesco, and W. B. Wood, J. Mol. Biol., in press).

The mechanism of gp63 interaction is not understood. The lack of evidence for any covalent bond breakage or formation in the attachment reaction, the finding that gp63 is not absolutely required, and the observation that its presence increases the reaction rate but not the final yield of attached fibers all suggest that this accessory protein plays the novel role of promoting noncovalent association between the amino termini of gp34 molecules in the proximal half fiber and gp9 in the baseplate, although true catalysis has not been rigorously demonstrated (Wood et al., 1978). The unexpected finding that gp63 is also responsible for the RNA ligase activity in T4-infected-cell extracts (Snopek et al., 1977) at first suggested the possibility of nucleic acid involvement in tail fiber attachment. However, in view of marked differences in the requirements for the two reactions, as well as the isolation of gene 63 mutants that affect RNA ligase activity but not tail fiber attachment activity (Runnels et al., 1982), a more likely explanation is that gp63 is a bifunctional protein that promotes two physiologically unrelated reactions. A likely hypothesis for its role in tail fiber attachment could be that it stabilizes an energetically unfavorable gp9 or gp34 configuration which is an intermediate in the attachment reaction, and in so doing decreases the conformational activation energy and thereby increases the rate of the reaction.

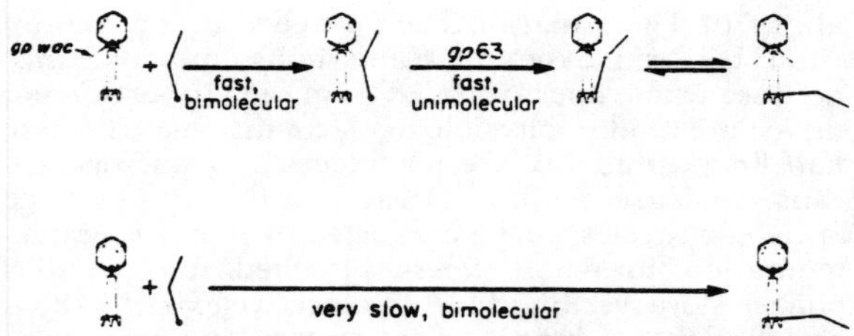

gp wac
fast, bimolecular
gp63
fast, unimolecular
very slow, bimolecular

FIG. 5. Proposed roles for whiskers (gp*wac*) and gp63 in tail fiber attachment. (From Wood, 1979.)

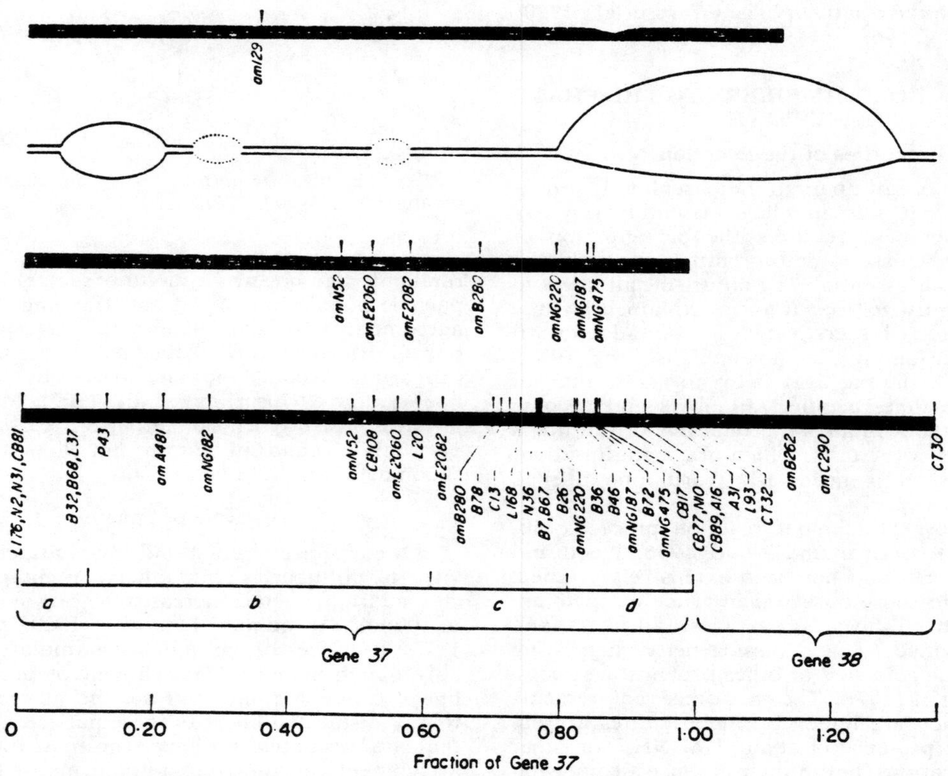

FIG. 6. Alignment of recombinational, translational, and heteroduplex maps of gene 37. Maps from top to bottom are: translational map of T2 gene 37; diagram of T2-T4 heteroduplex of DNA from the gene 37-38 region (dotted lines indicate regions of partial homology seen sometimes as duplexes); translational map of T4 gene 37; recombinational map of T4 genes 37 and 38 *am* and *ts* markers; boundaries and functional segments of T4 gene 37; linear scale showing fractions of T4 gene 37. (From Beckendorf et al., 1973.)

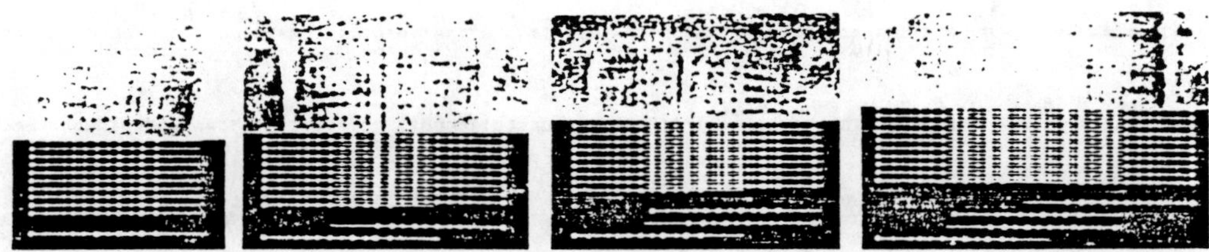

FIG. 7. Electron micrographs and computer simulations of the different types of banded aggregates (singlet, doublet, triplet, and quadruplet) of distal half tail fibers (Earnshaw et al., 1979). The upper picture of each pair shows a micrograph of a negatively stained aggregate, and the lower shows a computer simulation of the pattern using a beaded domain model for the fiber with appropriate overlaps.

Role of the Whiskers

Phages without whiskers attach tail fibers abnormally slowly both in vivo and in vitro if gp*wac* is absent from the extract supplying tail fibers (Terzaghi et al., 1979; Wood and Conley, 1979). The rate-enhancing effect of whiskers depends on their interaction with gp36 in the distal half fiber in a manner probably identical to that of the interaction controlling tail fiber retraction in complete phage (Conley and Wood, 1975).

The mechanisms of action of gp63 and whiskers suggested by the above observations are summarized in Fig. 5. We propose that the bimolecular reaction of tail fiber attachment in the absence of the accessory proteins may be so slow because of the improbability of a precisely required collision angle and a conformational activation energy barrier in the gp9-gp34 reaction. The cloud of whiskers around a particle could provide a less sterically demanding interaction and a larger target size, leading to a rapid bimolecular reaction. Attachment to the baseplate then would require only an internal (unimolecular) reaction, increased in rate by the catalytic action of gp63. For further discussion of this reaction, see Bloomfield and Prager (1979) and Bloomfield, this volume.

FINE STRUCTURE OF ASSEMBLED DISTAL HALF FIBERS

Genetic and Antibody-Decoration Experiments

The first analysis of the distal half-fiber fine structure was made by heteroduplex mapping and antibody decoration of T2-T4 hybrids (Beckendorf, 1973; Beckendorf et al., 1973). Crosses between T2 and T4 phage mutant in genes 37 and 38 divide gene 37 into four segments (labeled *a*, *b*, *c*, and *d* in Fig. 6), which show different frequencies of T2-T4 recombination. These crosses show that the two functional specializations of gp37, namely, interaction with gp38 during assembly and attachment to the bacterial surface during infection, are determined by a single segment (*d*) at the carboxyl end of P37. As mentioned above (Assembly of the Tail Fiber), more recent results subdivide this segment into a more amino-terminal region of gp38 interaction and a more carboxy-terminal region for host interaction (Snyder, thesis). In this segment of gene 37 and in all of gene 38 there is no recombination between T2 and T4. The rest of gene 37 contains a segment (*b*) giving little T2-T4 recombination, flanked by two small segments (*a* and *c*) with relatively high T2-T4 recombination.

From electron micrographs of T2/T4 heteroduplex DNA, a heteroduplex map can be constructed. Such a map (Fig. 6) shows four heterologous loops in the region of genes 37 and 38. When genes 37 and 38 are aligned with this heteroduplex pattern, based on heteroduplex patterns of T2-T4 hybrids, regions of low T2-T4 recombination correspond to regions of T2-T4 heterology. Regions with relatively high recombination are homologous.

The M_r of T2 gp37 determined on SDS-polyacrylamide gels is about 13,000-fold greater than that of T4 gp37. Analysis of the hybrid gene products from T2-T4 hybrid phage shows that this molecular weight difference, like the functional differences, is determined by the carboxy-terminal segment. The M_r values for gp37 amber fragments provide a translation map with which the various amber mutations can be physically positioned within the gene and the genetic and physical maps can be aligned (Fig. 6). In this way the crossover points in various different hybrid T2-T4 gene 37 species may be localized within particular segments of the polypeptide chain.

A T4-specific anti-BC serum was used by Beckendorf (1973) to analyze these T2-T4 hybrids both by serum blocking experiments and by electron microscopy of antibody-decorated phage. The experiments showed that C antigen could be divided into four or five subclasses which have different specificities and correspond to different parts of gp37. More importantly, they showed that the polypeptide chain is grossly colinear with the assembled half fiber and has its carboxyl terminus near the distal thin tip and its

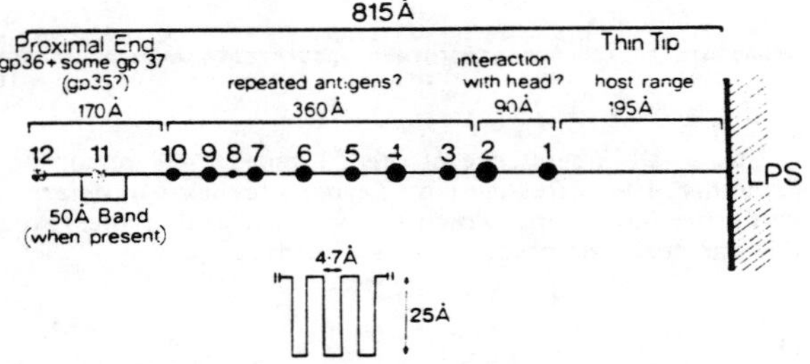

FIG. 8. Model for the structure of a distal half fiber based on the work of Beckendorf (1973) and Earnshaw et al. (1979). See text for discussion. The inset shows the way in which a polypeptide chain in the cross-β configuration would be oriented in the fiber. LPS, Lipopolysaccharide. (From Earnshaw et al., 1979.) (10 Å = 1 nm.)

```
 S  F  S  E  V  S  R  N  G  G  I  S  K  P  A  E  F  G  V  N  G  I  R  V  N  Y  I  C  E  S  A  S  P  P  D  I  M  V  L  P
AAGCTTTTCTGAAGTATCAAGAAATGGCGGCATTTCGAAACCTGCTGAATTTGGCGTCAATGGTATTCGTGTTAATTATATCTGCGAATCCGCTTCACCTCCGGATATAATGGTACTTCC
Hin     10        20        30        40        50        60        70        80        90        100       110       120
                                                RI*

 T  Q  A  S  S  K  T  G  K  V  F  G  Q  E  F  R  E  V  *
TACGCAAGCATCGTCTAAAACTGGTAAAGTGTTTGGGCAAGAATTTAGAGAAGTTTAAATTGAGGGAGCCTTCGGGTTCCCTTTTTCTTTATAAATACTATTAAAATAAAGGGGCATACA
      130       140       150       160       170       180       190       200       210       220       230       240
                                        RI*A'

M  A  D  L  K  V  G  S  T  T  G  G  S  V  I  W  H  Q  G  N  F  P  L  N  P  A  G  D  D  V  L  Y  K  S  F  K  I  Y  S  E
ATGGCTGATTTAAAAGTAGGTTCAACAACTGGAGGCTCTGTCATTTGGCATCAAGGAAATTTTCCATTGAATCCAGCCGGTGACGATGTACTCTATAAATCATTTAAAATATATTCAGAA
|     250       260       270       280       290       300       310       320       330       340       350       360
Gene 36 --->                                  amE1

Y  N  K  P  Q  A  A  D  N  D  F  V  S  K  A  N  G  G  T  Y  A  S  K  V  T  F  N  A  G  I  Q  V  P  Y  A  P  N  I  M  S
TATAACAAACCACAAGCTGCTGATAACGATTTCGTTTCTAAAGCTAATGGTGGTACTTATGCATCAAAGGTAACATTTAACGCTGGCATTCAAGTCCCATATGCTCCAAACATCATGAGC
      370       380       390       400       410       420       430       440       450       460       470       480
           Alu                  Alu

P  C  G  I  Y  G  G  N  G  D  G  A  T  F  D  K  A  N  I  D  I  V  S  W  Y  G  V  G  F  K  S  S  F  G  S  T  G  R  T  V
CCATGCGGGATTTATGGGGGTAACGGTGATGGTGCTACTTTTGATAAAGCAAATATCGATATTGTTTCATGGTATGGCGTAGGATTTAAATCGTCATTTGGTTCAACAGGCCGAACTGTT
      490       500       510       520       530       540       550       560       570       580       590       600
                                                                      amE 302                               Hae

V  I  N  T  R  N  G  D  I  N  T  K  G  V  V  S  A  A  G  Q  V  R  S  G  A  A  A  P  I  A  A  N  D  L  T  R  K  D  Y  V
GTAATTAATACACGCAATGGTGATATTAACACAAAAGGTGTTGTGTCGGCAGCTGGTCAAGTAAGAAGTGGTGCGGCTGCTCCTATAGCAGCGAATGACCTTACTAGAAAGGACTATGTT
      610       620       630       640       650       660       670       680       690       700       710       720
                                                 Alu

D  G  A  I  N  T  V  T  A  N  A  N  S  R  V  L  R  S  G  D  T  M  T  G  N  L  T  A  P  N  F  F  S  Q  N  P  A  S  Q  P
GATGGAGCAATAAATACTGTTACTGCAAATGCAAACTCTAGGGTGCTACGGTCTGGTGACACCATGACAGGTAATTTAACAGCGCCAAACTTTTTCTCGCAGAATCCTGCATCTCAACCC
      730       740       750       760       770       780       790       800       810       820       830       840

S  H  V  P  R  F  D  Q  I  V  I  K  D  S  V  Q  D  F  G  Y  Y  *
                                                                          M  A  T  L  K  Q  I  Q  F  K  R  S  K  I  A  G
TCACACGTTCCACGATTTGACCAAATCGTAATTAAGGATTCTGTTCAAGATTTCGGCTATTATTAAGAGGACTTATGGCTACTTTAAAACAAATACAATTTAAAAGAAGCAAAATCGCAG
      850       860       870       880       890       900       910       920       930       940       950       960
                                                                          Gene 37 --->

 T  R  P  A  A  S  V  L  A  E  G  E  L  A  I  N  L  K  D  R  T  I  F  T  K  D  D  S  G  N  I  I  D  L  G  F  A  K  G  G
GAACACGTCCTGCTGCTTCAGTATTAGCCGAAGGTGAATTGGCTATAAACTTAAAAGATAGAACAATTTTTACTAAAGATGATTCAGGAAATATCATCGATCTAGGTTTTGCTAAAGGCG
      970       980       990       1000      1010      1020      1030      1040      1050      1060      1070      1080
                                                                                                SauA1/A2

Q  V  D  G  N  V  T  I  N  G  L  L  R  L  N  G  D  Y  V  Q  T  G  G  M  T  V  N  G  P  I  G  S  T  D  G  V  T  G  K  I
GGCAAGTTGATGGCAACGTTACTATTAACGGACTTTTGAGATTAAATGGCGATTATGTACAAACAGGTGGAATGACTGTAAACGGACCCATTGGTTCTACTGATGGCGTCACTGGAAAAA
      1090      1100      1110      1120      1130      1140      1150      1160      1170      1180      1190      1200

  F  R  S  T  Q  G  S  F  Y  A  R  A  T  N  D  T  S  N  A  H  L  W  F  E  N  A  D  G  T  E  R  G  V  I  Y  A  R  P  Q  T
TTTTCAGATCTACACAGGGTTCATTTTATGCAAGAGCAACAAACGATACTTCAAATGCCCATTTATGGTTTGAAAATGCCGATGGCACTGAACGTGGCGTTATATATGCTCGCCCTCAAA
      1210      1220      1230      1240      1250      1260      1270      1280      1290      1300      1310      1320
  SauA2/B  amC281                                      amA481
    Bgl

 T  T  D  G  E  I  R  L  R  V  R  Q  G  T  G  S  T  A  N  S  E  F  Y  F  R  S  I  N  G  G  E  F  Q  A  N  R  I  L  A  S
CTACAACTGACGGTGAAATACGCCTTAGGGTTAGACAAGGAACAGGAAGCACTGCCAACAGTGAATTCTATTTCCGCTCTATAAATGGAGGCGAATTTCAGGCTAACCGTATTTTAGCAT
      1330      1340      1350      1360      1370      1380      1390      1400      1410      1420      1430      1440
                                                                RI*A/B
                                                                  RI

  D  S  L  V  T  K  R  I  A  V  D  T  V  I  H  D  A  K  A  F  G  Q  Y  D  S  H  S  L  V  N  Y  V  Y  P  G  T  G  E  T  N
CAGATTCGTTAGTAACAAAACGCATTGCGGTTGATACCGTTATTCATGATGCCAAAGCATTTGGACAATATGATTCTCACTCTTTGGTTAATTATGTTTATCCTGGAACCGGTGAAACAA
      1450      1460      1470      1480      1490      1500      1510      1520      1530      1540      1550      1560

 G  V  N  Y  L  R  K  V  R  A  K  S  G  G  T  I  Y  H  E  I  V  T  A  Q  T  G  L  A  D  E  V  S  W  W  S  G  D  T  P  V
ATGGTGTAAACTATCTTCGTAAAGTTCGCGCTAAGTCCGGTGGTACAATTTATCATGAAATTGTTACTGCACAAACAGGCCTGGCTGATGAAGTTTCTTGGTGGTCTGGTGATACACCAG
      1570      1580      1590      1600      1610      1620      1630      1640      1650      1660      1670      1680

 F  K  L  Y  G  I  R  D  D  G  R  M  I  I  R  N  S  L  A  L  G  T  F  T  T  N  F  P  S  S  D  Y  G  N  V  G  V  M  G  D
TATTTAAACTATACGGTATTCGTGACGATGGCAGAATGATTATCCGTAATAGCCTTGCATTAGGTACATTCACTACAAATTTCCCGTCTAGTGATTATGGCAACGTCGGTGTAATGGGCG
      1690      1700      1710      1720      1730      1740      1750      1760      1770      1780      1790      1800

 K  Y  L  V  L  G  D  T  V  T  G  L  S  Y  K  K  T  G  V  F  D  L  V  G  G  G  Y  S  V  A  S  I  T  P  D  S  F  R  S  T
ATAAGTATCTTGTTCTCGGCGACACTGTAACTGGCTTGTCATACAAAAAAACTGGTGTATTTGATCTAGTTGGCGGTGGATATTCTGTTGCTTCTATTACTCCTGACAGTTTCCGTAGTA
      1810      1820      1830      1840      1850      1860      1870      1880      1890      1900      1910      1920
                                                            SauB/C
```

FIG. 9. DNA sequence of the T4 gene 35–38 region and the amino acid sequence of the proteins encoded (Oliver and Crowther, 1981). Restriction enzyme sites used in determining the sequence are underlined. The tryptophan (W) and CAG-glutamine (Q) codons (which are potential amber sites) are also underlined, and the name of the amber mutant is given where this has been unambiguously established.

amino terminus near the proximal location of gp36. This finding raised the question of how the polypeptide chain is folded within the half fiber since an α-helix, which might have been expected from the morphology, would produce a structure several times too long.

Analysis of Distal Half Fiber Aggregates

A partial answer to this question was provided by Earnshaw et al. (1979), who studied the regular aggregates of distal half fibers which formed when concentrated 34-defective lysates were precipitated with am-

```
 R  K  G  I  F  G  R  S  E  D  Q  G  A  T  W  I  M  P  G  T  N  A  A  L  L  S  V  Q  T  G  A  D  N  N  N  A  G  D  G  G
CTCGTAAAGGTATATTTGGTCGTTCTGAGGACCAAGGCGCAACTTGGATAATGCCTGGTACAAATGCTGCTCTCTTGTCTGTTCAAACACAAGCTGATAATAACAATGCTGGAGACGGAC
     1930      1940      1950      1960      1970      1980      1990      2000      2010      2020      2030      2040

 T  H  I  G  Y  N  A  G  G  K  M  N  H  Y  F  R  G  T  G  Q  M  N  I  N  T  Q  Q  G  M  E  I  N  P  G  I  L  K  L  V  T
AAACCCATATCGGGTACAATGCTGGCGGTAAAATGAACCACTATTTCCGTGGTACAGGTCAGATGAATATCAATACCCAACAAGGTATGGAAATTAACCCGGGTATTTTGAAATTGGTAA
     2050      2060      2070      2080      2090      2100      2110      2120      2130      2140      2150      2160

 G  S  N  N  V  Q  F  Y  A  D  G  T  I  S  S  I  Q  P  I  K  L  D  N  E  I  F  L  T  K  S  N  N  T  A  G  L  K  F  G  A
CTGGCTCTAATAATGTACAATTTTACGCTGACGGAACTATTTCTTCCATTCAACCTATTAAATTAGATAACGAGATATTTTTAACTAAATCTAATAATACTGCGGGTCTTAAATTTGGAG
     2170      2180      2190      2200      2210      2220      2230      2240      2250      2260      2270      2280

 P  S  Q  V  D  G  T  R  T  I  Q  W  N  G  G  T  R  E  G  Q  N  K  N  Y  V  I  I  K  A  W  G  N  S  F  N  A  T  G  D  R
CTCCTAGCCAAGTTGATGGCACAAGGACTATCCAATGGAACGGTGGTACTCGCGAAGGACAGAATAAAAACTATGTGATTATTAAAGCATGGGGTAACTCATTTAATGCCACTGGTGATA
     2290      2300      2310      2320      2330      2340      2350      2360      2370      2380      2390      2400

 S  R  E  T  V  F  Q  V  S  D  S  Q  G  Y  Y  F  Y  A  H  R  K  A  P  T  G  D  E  T  I  G  R  I  E  A  Q  F  A  G  D  V
GATCTCGCGAAACGGTTTTCCAAGTATCAGATAGTCAAGGATATTATTTTTATGCTCATCGTAAAGCTCCAACCGGCGACGAAACTATTGGACGTATTGAAGCTCAATTTGCTGGGGATG
     2410      2420      2430      2440      2450      2460      2470      2480      2490      2500      2510      2520
SauC/D

 Y  A  K  G  I  I  A  N  G  N  F  R  V  V  G  S  S  A  L  A  G  N  V  T  M  S  N  G  L  F  V  Q  G  G  S  S  I  T  G  Q
TTTATGCTAAAGGTATTATTGCCAACGGAAATTTTAGAGTTGTTGGGTCAAGCGCTTTAGCCGGCAATGTTACTATGTCTAACGGTTTGTTTGTCCAAGGTGGTTCTTCTATTACTGGAC
     2530      2540      2550      2560      2570      2580      2590      2600      2610      2620      2630      2640

 V  K  I  G  G  T  A  N  A  L  R  I  W  N  A  E  Y  G  A  I  F  R  R  S  E  S  N  F  Y  I  I  P  T  N  Q  N  E  G  E  S
AAGTTAAAATTGGCGGAACAGCAAACGCACTGAGAATTTGGAACGCTGAATATGGTGCTATTTTCCGTCGTTCGGAAAGTAACTTTTATATTATTCCAACCAATCAAAATGAAGGAGAAA
     2650      2660      2670      2680      2690      2700      2710      2720      2730      2740      2750      2760
                           RI*C/D amE2082

 G  D  I  H  S  S  L  R  P  V  R  I  G  L  N  D  G  M  V  G  L  G  R  D  S  F  I  V  D  Q  N  N  A  L  T  T  I  N  S  N
GTGGAGACATTCACAGCTCTTTGAGACCTGTGAGAATAGGATTAAACGATGGCATGGTTGGGTTAGGAAGAGATTCTTTTATAGTAGATCAAAATAATGCTTTAACTACGATAAACAGTA
     2770      2780      2790      2800      2810      2820      2830      2840      2850      2860      2870      2880
                                                                                      SauD/E

 S  R  I  N  A  N  F  R  M  Q  L  G  Q  S  A  Y  I  D  A  E  C  T  D  A  V  R  P  A  G  A  G  S  F  A  S  Q  N  N  E  D
ACTCTCGCATTAATGCCAACTTTAGAATGCAATTGGGGCAGTCGGCATACATTGATGCAGAATGTACTGATGCTGTTCGCCCGGCGGGTGCAGGTTCATTTGCTTCCCAGAATAATGAAG
     2890      2900      2910      2920      2930      2940      2950      2960      2970      2980      2990      3000

 V  R  A  P  F  Y  M  N  I  D  R  T  D  A  S  A  Y  V  P  I  L  K  Q  R  Y  V  Q  G  N  G  C  Y  S  L  G  T  L  I  N  N
ACGTCCGTGCGCCGTTCTATATGAATATTGATAGAACTGATGCTAGTGCATATGTTCCTATTTTGAAACAACGTTATGTTCAAGGCAATGGCTGCTATTCATTAGGGACTTTAATTAATA
     3010      3020      3030      3040      3050      3060      3070      3080      3090      3100      3110      3120

 G  N  F  R  V  H  Y  H  G  G  G  D  N  G  S  T  G  P  Q  T  A  D  F  G  W  E  F  I  K  N  G  D  F  I  S  P  R  D  L  I
ATGGTAATTTCCGAGTTCATTACCATGGCGGCGGAGATAACGGTTCTACAGGTCCACAGACTGCTGATTTTGGATGGGAATTTATTAAAAACGGTGATTTTATTTCACCTCGCGATTTAA
     3130      3140      3150      3160      3170      3180      3190      3200      3210      3220      3230      3240
                                                                                RI*D/E

 A  G  K  V  R  F  D  R  T  G  N  I  T  G  G  S  G  N  F  A  N  L  N  S  T  I  E  S  L  K  T  D  I  M  S  S  Y  P  I  G
TAGCAGGCAAAGTCAGATTTGATAGAACTGGTAATATCACTGGTGGTTCTGGTAATTTTGCTAACTTAAACAGTACAATTGAATCACTTAAAACTGATATCATGTCGAGTTACCCAATTG
     3250      3260      3270      3280      3290      3300      3310      3320      3330      3340      3350      3360

 A  P  I  P  W  P  S  D  S  V  P  A  G  F  A  L  M  E  G  Q  T  F  D  K  S  A  Y  P  K  L  A  V  A  Y  P  S  G  V  I  P
GTGCTCCGATTCCTTGGCCGAGTGATTCAGTTCCTGCTGGATTTGCTTTGATGGAAGGTCAGACCTTTGATAAGTCCGCATATCCAAAGTTAGCTGTTGCATATCCTAGCGGTGTTATTC
     3370      3380      3390      3400      3410      3420      3430      3440      3450      3460      3470      3480
   amNG187                                     amNG475

 D  M  R  G  Q  T  I  K  G  K  P  S  G  R  A  V  L  S  A  E  A  D  G  V  K  A  H  S  H  S  A  S  A  S  S  T  D  L  G  T
CAGATATGCGCGGGCAAACTATCAAGGGTAAACCAAGTGGTCGTGCTGTTTTGAGCGCTGAGGCAGATGGTGTTAAGGCTCATAGCCATAGTGCATCGGCTTCAAGTACTGACTTAGGTA
     3490      3500      3510      3520      3530      3540      3550      3560      3570      3580      3590      3600

 K  T  T  S  S  F  D  Y  G  T  K  G  T  N  S  T  G  G  H  T  H  S  G  S  G  S  T  S  T  N  G  E  H  S  H  Y  I  E  A  W
CTAAAACCACATCAAGCTTTGACTATGGTACGAAGGGAACTAACAGTACGGGTGGACACACTCACTCTGGTAGTGGTTCTACTAGCACAAATGGTGAGCACAGCCACTACATCGAGGCAT
     3610      3620      3630      3640      3650      3660      3670      3680      3690      3700      3710      3720
        Hin

 N  G  T  G  V  G  G  N  K  M  S  S  Y  A  I  S  Y  R  A  G  G  S  N  T  N  A  A  G  N  H  S  H  T  F  S  F  G  T  S  S
GGAATGGTACTGGTGTAGGTGGTAATAAGATGTCATCATATGCCATATCATACAGGGCGGGTGGGAGTAACACTAATGCAGCAGGGAACCACAGTCACACTTTCTCTTTTGGGACTAGCA
     3730      3740      3750      3760      3770      3780      3790      3800      3810      3820      3830      3840

 A  G  D  H  S  H  S  V  G  I  G  A  H  T  H  T  V  A  I  G  S  H  G  H  T  I  T  V  N  S  T  G  N  T  E  N  T  V  K  N
GTGCTGGCGACCATTCCCACTCTGTAGGTATTGGTGCTCATACCCACACGGTAGCAATTGGATCACATGGTCATACTATCACTGTAAATAGTACAGGTAATACAGAAAACACGGTTAAAA
     3850      3860      3870      3880      3890      3900      3910      3920      3930      3940      3950      3960
                                                   SauE/F

                                                         M  K  I  Y  H  Y  Y  F  D  T  K  E  F  Y  K  E  E  N  Y  K
 I  A  F  N  Y  I  V  R  L  A  *
ACATTGCTTTTAACTATATCGTTCGTTTAGCATAAGGAGAGGGGCTTCGGCCCTTCTAAATATGAAAATATATCATTATTATTTGACACTAAAGAATTTTACAAAGAAGAAAATTACAA
     3970      3980      3990      4000      4010      4020      4030      4040      4050      4060      4070      4080
                                                         |
                                                      Gene 38 --->

 P  V  K  G  L  G  L  P  A  H  S  T  I  K  K  P  L  E  P  K  E  G  Y  A  V  V  F  D  E  R  T  G  D  W  I  Y  E  E  D  H
ACCGGTTAAAGGCCTCGGTCTTCCTGCTCATTCAACAATTAAAAAACCTTTAGAACCTAAAGAAGGATACGCGGTTGTATTTGATGAACGTACTCAGGATTGGATTTATGAAGAAGACCA
     4090      4100      4110      4120      4130      4140      4150      4160      4170      4180      4190      4200
                                                                                           amB262 amHL612

 R  G  K  R  A  W  T  F  N  K  E  E  I  F  I  S  D  I  G  S  P  V  G  I  T  F  D  E  P  G  E  F  D  I  W  T  D  D  G  W
TCGCGGAAAACGCGCATGGACTTTTAATAAAGAARAAATTTTTATAAGTGACATTGGAAGCCCGGTTGGTATAACTTTCGATGAGCCCGGCGAATTTGATATATGGACTGATGACGGTTG
     4210      4220      4230      4240      4250      4260      4270      4280      4290      4300      4310      4320
          amC290                                                                                  amN62         amH41

 K  E  D  E  T  Y  K  R  V  L  I  R  N  R  K  I  E  E  L  Y  K  E  F  Q  V  L  N  N  M  I  E  A
GAAAGAAGACGAAACATATAAGCGAGTTTTAATTCGTAATAGAAAAATTGAAGAATTATATAAAGAGTTCCAAGTTTTAAATAATATGATTGAAGCTT
     4330      4340      4350      4360      4370      4380      4390      4400      4410      2         12        22
                                                                                      Hin
```

FIG. 9. *Continued*

monium sulfate. Various classes of aggregate-containing, parallel or antiparallel overlapping fibers, termed singlet, doublet, triplet, and quadruplet types (Fig. 7), each exhibited a characteristic transverse banding pattern when viewed by negative staining in an electron microscope. The singlets contained a single parallel set of half fibers with their thin tips generally ending in diffuse material, apparently containing lipopolysaccharide. The doublets contained two overlapping antiparallel sets of half fibers with the thin tips of each set buried in the aggregate. Triplets contained an additional set of fibers with their thin tips sticking out at one end, and quadruplets were like two overlapping doublets with all the thin tips buried. Analysis of the aggregates was carried out by computer simulation using as a model for the half fiber a series of beaded domains of different size and spacing. A single model for the fiber, together with some simple rules for the overlap, reproduced all the observed patterns to a good approximation (Fig. 7). The beaded appearance with regularly spaced domains of varying size is sometimes seen in individual fibers, though less clearly than in the banded aggregates.

X-ray diffraction patterns from wet but unoriented pellets of aggregates showed a ring corresponding to a spacing of 0.47 nm in the specimen. Electron diffraction patterns from glucose-embedded, partially oriented clumps of aggregates had a sharp ring at the same spacing, which in this case showed orientation effects, being stronger in the direction corresponding to the fiber axis. Such an oriented ring at 0.47-nm spacing is characteristic of a cross-β fold in which the polypeptide chain zig-zags from side to side roughly perpendicular to the fiber axis, with successive antiparallel β-strands separated by 0.47 nm.

The overall picture that has emerged from these studies is summarized in Fig. 8. The model shows four rather different regions along the half fiber. At the left is a region of about 17 nm where little structural detail is seen except for two ill-defined stain-excluding blobs, which must consist at least in part of gp36. This region also binds extra material (possibly gp*wac*), seen as a strong band at the ends of doublet and quadruplet aggregates and at one end of singlets and triplets. The next region extends for about 36 nm and consists of eight beaded domains about 5 nm apart which probably contain the repeated antigenic sites found by Beckendorf (1973). The next region is about 9 nm long and contains two large domains, which may interact with the head of the phage when the fibers are retracted. Finally, there is the thin tip 19.5 nm long, carrying the fiber determinants responsible for binding to host lipopolysaccharide. The total length of the distal half fiber is 81.5 nm. Within this domain-like structure the polypeptide chain is envisaged as zig-zagging from side to side in a cross-β fold roughly normal to the fiber axis (inset to Fig. 8). This type of folding would certainly accommodate the known length of polypeptide chain within the observed dimensions of the fiber, in the colinear manner required by Beckendorf's observations (1973).

NUCLEOTIDE SEQUENCE ANALYSIS OF GENES 36 AND 37

The nucleotide sequence of an *Eco*RI fragment of T4 DNA covering genes 36 and 37 and parts of genes 35 and 38 was determined by Oliver and Crowther (1981), using M13 subcloning and rapid dideoxy sequencing (Fig. 9). The presence of many genetic markers in genes 36 and 37 allowed positive identification and ordering of the DNA subfragments with respect to the genetic map. The sequence contains two long open reading frames coding for polypeptides of 221 and 1,026 amino residues, corresponding to genes 36 and 37, respectively. The sites of a number of amber mutations can be unambiguously identified, and the sequence can be aligned with the genetic map. In all cases there is a potential amber codon close to the position of known mutations.

Surprisingly, the gp37 amino acid sequence derived from the nucleotide sequence does not show any strong indications of repetitive features or duplications. An extended structure like the tail fiber might well have evolved by duplication of segments, but if this were the case, subsequent evolution has largely obscured the events. Application of the empirical rules for secondary structure prediction (Chou and Fasman, 1978a; 1978b) indicates that a large proportion of the polypeptide chain is in the form of β-strand. There are several stretches of sequence that exhibit alternating regions of high probability for β-strand and β-turn, as expected for an antiparallel β-structure. However, the pattern is very irregular, with predicted β-strand lengths varying between 7 and 15 residues. It thus seems likely that each globular domain of the fiber is formed by opposed regions from the two gp37 polypeptides, each folded in a rather irregular antiparallel β-structure but with the strands running approximately perpendicular to the axis of the fiber.

CONCLUSION

The tail fiber is operationally a rather complex organelle, yet simple enough that its assembly, attachment, and function should eventually be fully understandable. At present, however, many questions remain to be answered. Several of these have been raised in the preceding paragraphs; a few more are mentioned below.

Although a considerable amount is now understood about the structure of the assembled distal half fiber, it is still not clear how a cross-β folded domain-like structure confers stiffness on the fiber. Current models for the control of triggering of tail contraction (Crowther, 1980; Goldberg, this volume) require that the tail fibers, when bound to the bacterial surface, activate the baseplate. This means that the fibers must be able to transmit a mechanical force, which probably changes the configuration of gp9 in the baseplate. To do so, the fibers must be mechanically stiff, as also appears to be the case from micrographs of isolated half fibers, which almost never appear bent.

Little is known about the structure or function of the proximal half fiber and its interaction with the distal half fiber and baseplate. In particular, does the proximal half fiber function simply as an inert spacer, or more actively as a transducer of conformational changes triggered by interaction of the distal half fiber tip with the host cell surface?

The most unusual aspect of tail fiber assembly and attachment is the novel, probably catalytic involvement of four accessory proteins, each apparently acting by a different mechanism. Of general biochemical interest is the question of how these proteins promote the noncovalent association of structural components. Of general biological interest is the extent to which such accessory proteins will be found to play roles in other systems of intracellular or extracellular macromolecular assembly.

Research from the laboratory of W.B.W. was supported by Public Health Service grants AI-09238 and AI-14994 from the National Institute of Allergy and Infectious Diseases.

Physical Studies of Morphogenetic Reactions

VICTOR A. BLOOMFIELD

Department of Biochemistry, University of Minnesota, St. Paul, Minnesota 55108

INTRODUCTION

Goals of the Physical Approach to Morphogenesis

T4 assembly reactions present a striking array of interactions between proteins, nucleic acids, and small ions. The major challenge to physical biochemists working in this area is to understand these interactions at the same level of molecular detail as has been achieved for simpler protein and nucleoprotein complexes, and ultimately for small organic and inorganic molecules. That is, one wishes to define the structures of reactants, products, and intermediate assembly states at atomic resolution; to know the relative energies of these states; and to understand the way structure and energetics determine the dynamics of assembly reactions. Another factor, which does not generally arise in the study of simpler molecules, is the need to understand the regulation of interactions either by extrinsic molecules or by the intrinsic properties of the biopolymer species and their complexes.

Although the great majority of information on T4 assembly pathways has been obtained by in vitro complementation with crude lysates, work with purified species is necessary for mechanistic studies of individual assembly steps. Most of the standard methods for structure determination (with the notable exception of electron microscopy) require pure compounds for unambiguous interpretation. Measurement of thermodynamic properties requires establishment of equilibrium, whereas infectivity assays needed to determine complementation are generally irreversible. Likewise, while complementation can be used to assay assembly dynamics, attempts to determine the effect of pH, temperature, or ions on the kinetics of a particular reaction may affect other, coupled, reactions as well.

In this chapter I review two examples of an approach to quantitative characterization of assembly dynamics using purified components. These are the last two steps in T4 assembly: joining of heads with tails and attachment of tail fibers to fiberless particles. I pay particular attention to the use of dynamic laser light scattering to follow the kinetics of assembly of these large particles, and to the construction of detailed kinetic theories to interpret the rate measurements. These two topics are emphasized, and treated below in a general context, because they should be of broad utility in many studies of phage assembly.

Dynamic Laser Light Scattering

Among the many techniques that have been used to study phage assembly, light scattering holds some special advantages. It allows direct measurement of size changes during assembly. It is rapid, so fairly fast reactions can be studied. It is most sensitive to the largest particles in solution, so the presence of monomeric proteins does not complicate the study of larger complexes. It requires rather small amounts of material and is nondestructive. And it can be combined with other solution techniques, such as sedimentation or electrophoresis, to provide a more complete picture of particle structure.

In the most familiar manifestation of the technique, the total intensity I of light scattered from a solution of concentration c is used to determine the molecular weight M, the radius of gyration R_g, and the second virial coefficient B of the macromolecule (Tanford, 1961)

$$I = KcM(1 - q^2R_g^2/3 + \ldots)/(1 + 2BMc + \ldots) \quad (1)$$

K includes the incident intensity, the refractive index increment, and various optical and geometrical constants. q is the magnitude of the scattering vector; $q = (4\pi n/\lambda)\sin(\theta/2)$, where n is the solution refractive index, λ is the laser wavelength in vacuo, and θ is the scattering angle. M and R_g give information on particle mass and size and their changes during assembly, whereas B reflects interaction between particles. By combining total-intensity light scattering with stopped-flow kinetics techniques, it is possible to follow assembly reactions with half-lives as short as 10 ms.

More recently, dynamic laser light scattering, or quasielastic light scattering (QLS), has become a powerful technique to measure the translational diffusion coefficient D and other dynamical properties of biological macromolecules (Berne and Pecora, 1976; Bloomfield and Lim, 1978). In contrast to total-intensity light scattering, QLS measures the fluctuations of the scattered light, correlating I values that are separated by characteristic times on the order of microseconds to milliseconds.

The extent of correlation, described by the homodyne correlation function $g(t)$, measures the rapidity of Brownian motion, and, hence, the diffusion coefficient. If only translational diffusion of a single species is important, then

$$g(t) = \langle I(0)I(t)\rangle = 1 + \exp(-2q^2Dt) \quad (2)$$

From this it can be seen that the correlation decays more rapidly with larger D. Since D is inversely proportional to the hydrodynamic radius R_h according to the Stokes-Einstein equation $D = kT/6\pi\eta R_h$, QLS presents a convenient way to measure size changes during assembly. (For particles such as phage and their assembly intermediates, whose shapes are more complicated than spheres, frictional properties may be accurately calculated by modern hydrodynamic theory [Garcia de la Torre and Bloomfield, 1981].)

If the solution contains a mixture of scattering

particles, then equation 2 is generalized to

$$g(t) - 1 = [\Sigma A_i \exp(-q^2 D_i t)]^2 \quad (3)$$

where the A's are the fractional scattering powers, proportional to Mc, of each component in solution. Since QLS measurements can be made with T4 phage solutions in 10 to 15 s, it is possible to use equation 3 to measure A's and c's as functions of reaction time for rapid assembly reactions such as head-tail joining (Aksiyote-Benbasat and Bloomfield, 1975; Aksiyote-Benbasat and Bloomfield, 1981).

Yet another form of light scattering that has applicability to phage assembly is electrophoretic light scattering. The original, and most elegant, version of this technique involves measurement of the Doppler shift in the scattered light produced when a charged macromolecule acquires a non-zero drift velocity in the presence of an electric field (Ware, 1974). Under favorable circumstances, both D and the electrophoretic mobility can be measured rapidly and simultaneously. However, for most species smaller than whole cells, diffusional broadening of the light-scattering spectrum requires that measurements be made at very low scattering angles, introducing a degree of experimental difficulty that has discouraged most QLS laboratories from taking up this technique.

Therefore, my co-workers and I (Lim et al., 1977) devised another type of electrophoretic light scattering, which combines QLS with conventional band electrophoresis in a sucrose density gradient. After electrophoresis has occurred for a time sufficient for the band to move an appreciable distance, and for different species within the band to separate, the column is scanned with a laser beam. At a given height, or distance travelled from the origin, scattering above solvent background indicates the presence of a macromolecular species, whose mobility μ is determined from the height divided by the elapsed time and the field strength. The total I indicates the c of species with that μ, and QLS gives D. The effective charge Z, in units of proton charge e, is then obtained from

$$\mu = (Ze/f)\phi(\kappa R_h) \quad (4)$$

where f is the frictional coefficient kT/D, and ϕ is a function of ionic strength through the Debye-Huckel inverse length κ. As described below, we have used this techique to characterize the attachment of T4 tail fibers to fiberless particles (Baran and Bloomfield, 1978).

Diffusion-Controlled Reactions

One may distinguish two major classes of chemical reactions: unimolecular rearrangements and dissociations, and bimolecular associations. Within the bimolecular class, a mechanistic division can be made, depending on whether the rate-limiting step is the activation energy-limited reorganization of the initial encounter complex or the diffusion together of the reactants in the proper orientation. Although bimolecular phage assembly steps obviously involve the formation of strong noncovalent bonds, several of them—head-tail joining and perhaps tail fiber attachment—appear to be diffusion controlled. Since these reactions are relatively amenable to both experimental study and theoretical understanding, we have concentrated on them in our research.

The simplest sort of diffusion-controlled reaction is one between two spheres A and B, each of which is uniformly reactive over its surface. The bimolecular rate constant for this model, devised by von Smoluchowski (1917) to explain the kinetics of colloidal flocculation, is

$$k^* = (4\pi N/1{,}000)(D_A + D_B)(R_A + R_B) \text{ liter/mol-s} \quad (5)$$

where R and D are the radii and diffusion coefficients of the two species, and N is Avogadro's number. If A and B are of equal size, and if D is calculated from R by the Stokes-Einstein equation, we find that $k^* = 6.5 \times 10^9$ in water at 20°C. Since the half-life of a bimolecular reaction in which both reactants are present at the same molar concentration [A] is $t_{1/2} = 1/k[A]$, we may estimate that in a typical *Escherichia coli* cell infected with T4 to give a burst size of 100, the half-life for head-tail joining is about 0.2 ms.

This is a lower limit, since electrostatic and especially orientational factors will tend to lower the rate constant significantly. To take these into account, we may approximate the effective rate constant k as

$$k = f\delta F k^* \quad (6)$$

where f denotes the electrostatic enhancement/retardation factor, F is the fraction of surface area available for reaction, and δ is the enhancement due to rotational diffusion. These are difficult factors to evaluate, especially for structurally complex, highly charged reactants like phage parts, and they are not strictly independent as implied by the factorization in equation 6. However, some estimates may be made.

The simplest realistic treatment of electrostatic effects is that for two spheres interacting through a screened Coulomb potential

$$V = (Z_A Z_B e^2/\varepsilon) \exp(-\kappa r)/r \quad (7)$$

at center-to-center distance r. The factor f is (Debye, 1942)

$$f = \left[a \int_{\infty}^{\infty} \exp(V/kT) r^2 dr \right]^{-1} \quad (8)$$

where a is the distance of closest approach. For reactants of the same charge, $V > 0$ and therefore $f < 1$, so the reaction is retarded by coulombic repulsion. For reactants of opposite charge, the rate will be enhanced.

If only a fraction of the reactant surfaces is reactive, then k^* will be reduced by that fraction F. However, rotational diffusion will generally speed up such reactions, since the proper relative orientation need not be achieved by translational diffusion alone. Depending on the relative magnitudes of translational and rotational diffusion coefficients, and the angle subtended by the reactive patch on each surface, the rotational enhancement δ can be as much as 2 to 10 or more. A good recent treatment of this complicated topic, which compares and unifies previous work, is that of Shoup et al. (1981).

KINETICS OF HEAD-TAIL JOINING

The experimental and theoretical strategies outlined above are illustrated by our studies of the head-tail joining reaction

$$H + T \rightarrow HT$$

in phage T4D (Aksiyote-Benbasat and Bloomfield, 1975; Aksiyote-Benbasat and Bloomfield, 1981). Since the diffusion coefficient of the heads, $D_H = 3.6 \times 10^{-8}$ cm^2/s, is greater than that of the head-tail complex (HT), $D_{HT} = 3.14 \times 10^{-8}$ cm^2/s, the average $\langle D \rangle$ of the reacting mixture decreases measurably as attachment proceeds.

We have used QLS to determine the bimolecular rate constant for head-tail joining over a wide range of temperature, pH, and ionic conditions. To summarize the results: concentration variation showed that the reaction is indeed second order. The activation energy is consistent with a diffusion-controlled reaction. The reaction involves an ionizable group with a pK of 6.8 and brings together head and tail regions of opposite charge. Magnesium ions seem to be required only for stability of the reactants, but do not otherwise affect the rate.

Kinetic Analysis

The strong scattering power of phage heads is crucial to this approach. Experiments were performed at H and T concentrations of around 5×10^{10} particles per ml (8×10^{-11} M) so that the reaction half-life was lengthened to about 500 s. (Compare this with the microsecond to millisecond time scale usually associated with diffusion-controlled reactions.) Thus, $\langle D \rangle$ changed very little in the time required to make a QLS measurement.

Purified heads (osmotic shock resistant) and tails were mixed under conditions where tails are slightly in excess of heads, to ensure full conversion to HT particles. The standard integrated second-order rate law is then

$$kt = 1/(T^* - H^*) \times \ln\{H^*(T^* - [HT])/T^*(H^* - [HT])\} \quad (9)$$

where the *'s denote initial concentrations. The only unknown in the equation, [HT], was determined from the definition of the z-average $\langle D \rangle$, which after some rearrangement gives

$$[HT] = H^*/\{1 + (M_{HT}/M_H)^2[(D_{HT} - \langle D \rangle)/(\langle D \rangle - D_H]\} \quad (10)$$

The logarithmic term in equation 9, plotted versus t, gave a straight line through the origin whose slope was $k(T^* - H^*)$. Values of k were typically on the order of 10^7 liter/mol-s, a large value suggestive of a diffusion-controlled reaction. The variation of k with H^* was as expected for a second-order reaction.

Temperature Dependence

Between 10 and 37°C the rate constant at pH 6.9 increased from 8×10^6 to 1.5×10^7 liter/mol-s. An Arrhenius plot of k versus $1/T$ was linear, giving an activation energy $\Delta E^\ddagger$ of 4.1 ± 0.1 kcal/mol (ca. 17 ± 0.4 kJ/mol). This value is close to that expected for viscous flow in water, further supporting the idea that head-tail joining is diffusion controlled. The activation entropy $\Delta S^\ddagger$ is −12.6 cal/mol-degree (ca. −52.7 J/mol-degree), indicating that the activated complex is substantially restricted relative to the reactants. This likely reflects the severe geometric constraints on the relative orientation of head and tail required for attachment. We may use another concept to express the same idea. The steric factor p, defined empirically as the ratio of the observed k to that expected for complete reactivity in the encounter complex and defined mathematically as $p = \exp(\Delta S^\ddagger/RT)$, is 1/567. As we shall see below, the orientational constraints are even more severe than implied by this value, since electrostatic and rotational diffusion effects accelerate the reaction.

Salt Dependence

The rate constant decreases monotonically with increasing ionic strength between 0.05 and 0.23 M. This result is quite surprising, since both heads and tails have net negative charge, so that increasing salt would have been expected to screen coulombic repulsions. It implies that either the reactive vertex of the head or the end of the tail distal to the baseplate is positively charged.

There is also a decrease in k at ionic strengths below 0.05 M, which is explained by the instability of tails in low salt concentrations. The rate constant is independent of magnesium ion concentration above 5 mM. Below that concentration, the heads are unstable and slowly release their DNA, even when magnesium is replaced by putrescine.

pH Dependence

In buffer with 0.1 M ionic strength at 30°C, the rate constant decreases from 2×10^7 to 0.5×10^7 liter/mol-s over the pH range 5.8 to 7.5 and levels off between pH 7.5 and 8.3. The pH-rate profile in this region is fit well by a simple titration curve of a group with pK 6.8, such as a histidine. There is no evidence of cooperativity in the pH profile, as might have been expected given the close interaction of six tail protein subunits and five head vertex subunits in the joining complex.

The reaction rate decreases below pH 5.8, apparently due to loss of activity of the heads, which were shown by QLS and total-intensity light scattering to leak DNA at pH 5.4.

Interpretation of Kinetic Results

The head-tail joining reaction appears to be diffusion controlled, judged by its temperature dependence and high rate. Yet the rate constant is about 675-fold lower than would have been estimated from equation 5, using $D_H = 3.6 \times 10^{-8}$ and $D_T = 6.0 \times 10^{-8}$ cm^2/s (Aksiyote-Benbasat and Bloomfield, 1982) and setting the center-to-center distance $R_H + R_T$ equal to the head radius plus one-half of the tail length, 950 Å (95 nm). This estimate is in good agreement with the steric factor p estimated from the activation entropy. We attribute this reduction in rate constant to steric constraints. Only 1 of the 12 head vertices and 1 of the 2 tail ends are reactive. The long axes of head and tail must be nearly colinear for reaction to occur, and matching the fivefold symmetry of the head with the sixfold symmetry of the tail may require additional alignment. These factors could easily cause a reduction of the observed magnitude, or even greater.

In fact, both the electrostatic and the rotational factors discussed in the introduction to this chapter accelerate the reaction, so the steric reduction is considerably greater than estimated above. Rotational diffusion contributes roughly a factor $\delta = 2$. To estimate the electrostatic contribution, we used a simple screened Coulomb potential as indicated by equations 6 and 7. This is clearly oversimplified, given the complicated geometry of the reactants, but, lacking a detailed picture of the charge distribution on the head and tail, a more sophisticated treatment would be unjustified. The rate-ionic strength profile indicates that the product of charges in the reacting regions of head and tail, $Z_H Z_T$, is in the range of -30 to -120. This is consistent with one or two charges (of unknown sign) on each of the five head vertex subunits and the six tail subunits. Choosing a charge product of -60, we calculated an electrostatic acceleration at 0.08 M ionic strength and 293°K of $f = 9.1$. With the observed $k/k^* = 1/675$, this gives a steric retardation of $F = 1/12{,}300$.

To check the consistency of this approach, we used F to estimate the reactive area on heads and tails. That is, F is the product of solid angles, divided by 4π, subtended by the reactive patch on each reactant. This yields a reactive patch area of 2.45×10^4 Å^2, corresponding to a radius of 88 Å. This is gratifyingly close to the radius of the tail.

TAIL FIBER ATTACHMENT

The last step in T4 assembly is the attachment of tail fibers to the baseplate of the head-tail particle. This reaction is considerably more complicated than head-tail joining. Although the attachment is noncovalent, it is catalyzed by a nonstructural protein, gp63 (Wood and Henninger, 1969), and whiskers, the *wac* protein, are necessary for efficient attachment (Conley and Wood, 1975). Fiber attachment poses a host of interesting questions, of which we address three: how to measure the elementary event, i.e., the attachment of a single fiber; whether fiber attachment is cooperative; and how the whiskers may be acting to facilitate fiber attachment.

Electrophoretic Light Scattering Detection of Partially Fibered Phage

To understand the chemical mechanism of the reaction

$$F + HT\text{-}F_{i-1} \rightarrow HT\text{-}F_i \ (i = 1, \ldots, 6)$$

one must be able to isolate, or at least detect and measure, phage containing specific numbers of fibers (F). Electron micrographs of in vitro complementation mixtures show substantial quantities of particles with fewer than six fibers, indicating that the reaction is not strongly cooperative (Edgar and Wood, 1966). However, because of the small hydrodynamic differences between fiberless particles and complete T4, and thus even smaller differences between the intermediate species with one to five fibers, techniques based on differences in hydrodynamic radius, such as sedimentation and column chromatography, cannot be used to separate these species.

To resolve intermediate-fibered forms, we were able to use an apparatus (Lim et al., 1977) that simultaneously measures electrophoretic mobility μ, diffusion coefficient, and relative concentration. It combines band electrophoresis in a vertical, sucrose gradient-stabilized column, with QLS determination of D of the species within the band. The entire electrophoresis cell is scanned through the laser beam of the QLS apparatus by a vertical translation stage. Total-intensity light scattering at each point gives the concentration of the virus species that has moved to that position under the influence of the electric field; the distance travelled divided by the time and the field gives the mobility.

In vitro assembly reactions using a tail fiber extract and purified HT particles were conducted with 8×10^{11} HT per ml and various dilutions of fibers. Components were incubated for 3 h at 30°C, and then the reaction was quenched by immersion of the reaction tube in an ice bath. The infectious titer increased as more fibers were added and attached, but generally did not exceed 50% of the HT present. The discrepancy between this result and earlier claims of 100% maximal complementation appears to arise from the use of extinction coefficients based on PFU per milliliter (Edgar and Wood, 1966; Wood and Henninger, 1969) and on total particles per milliliter (our work). Since these differ by a factor of two, the results are in fact consistent.

A typical electrophoretic light scattering result is shown in Fig. 1 (Baran and Bloomfield, 1978). In this

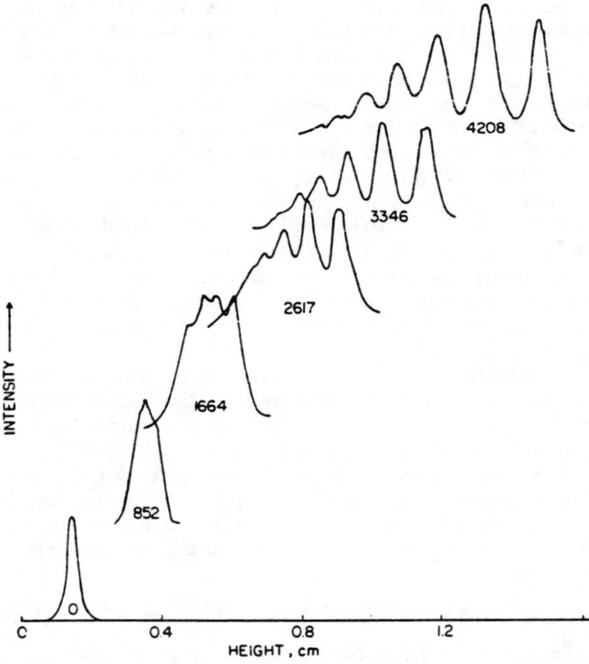

FIG. 1. The total intensity of scattered light as a function of scanning height in electrophoresis cell for partially complemented phage (8% of total particles infective). The number under each curve gives the cumulative time, in seconds, of application of the field (7.5 V/cm). Intensity, in arbitrary units, is proportional to concentration. From right to left, zero to six fibers are attached. From Baran and Bloomfield (1978); reprinted with permission.

sample, in which 8% of the HT particles became infective, there are clearly five electrophoretically distinct species present, with a hint of two more at the trailing edge. Since seven species are possible—fiberless particles and particles with one to six fibers—a maximum of seven peaks is expected, and no more were ever observed. The leading and trailing peaks correspond to fiberless particles and to complete, six-fibered phage, respectively. It is therefore reasonable to assign the second-fastest peak to HT-F_1, the third-fastest peak to HT-F_2, etc.

The good separation in these experiments arises because the addition of fibers causes both an increase in frictional drag (decrease in D) and a decrease in net negative charge, hence a relatively large decrease in μ. The addition of the first fiber causes a 1.5% decrease in D, but a 10% decrease in μ. A complete table of D and μ values for partially fibered species, along with the Stokes radii and charges estimated from these values, is given in the original paper (Baran and Bloomfield, 1978).

Noncooperativity of the Attachment Reaction

With this ability to measure the relative amounts of partially fibered phage, we were able to study the relationship between infectivity and number of fibers. Wood and Henninger (1969) showed that particles with three or fewer fibers infect bacteria less efficiently than particles with four to six fibers. Fitting the kinetics of fiber complementation to a quantitative model, they deduced that particles with zero or one fiber were uninfectious, whereas the two-fibered species was 5% and the three-fibered species was 50% as infectious as intact phage. Electrophoretic light scattering confirms that a low-titer sample is composed primarily of particles with zero to two fibers, whereas in high-titer samples, there are significant quantities of particles with four to six fibers attached.

It is interesting to note that Kellenberger et al. (1965), examining shadowed preparations of normal wild-type phage, found an average of only about three fibers per phage, indicating that our inability to drive the attachment reaction to completion even with a large excess of fibers may not be an artefact of the purified system.

From a qualitative inspection of a distribution such as that in Fig. 1, it is apparent that the reaction is not highly cooperative. If it were, one would observe particles with zero and six fibers, with few or no intermediate species. It is possible to make a more quantitative test for random attachment. In that case, the probability of finding K fibers attached to N baseplate binding sites is given by the binomial distribution

$$P_N(K) = [N!/K!(N - K)!]p^K(1 - p)^{(N-K)} \qquad (11)$$

where p is the probability of attaching one fiber to a baseplate site. From here on, we use $N = 6$. To determine p, one measures the average number $\langle K\rangle$ of fibers per particle: $p = 6\langle K\rangle$. Figure 2 shows comparisons between observed and calculated distributions for four samples covering a range of fiber inputs. Agreement with the random, noncooperative model is generally excellent.

If the second-order rate constant for attachment of a

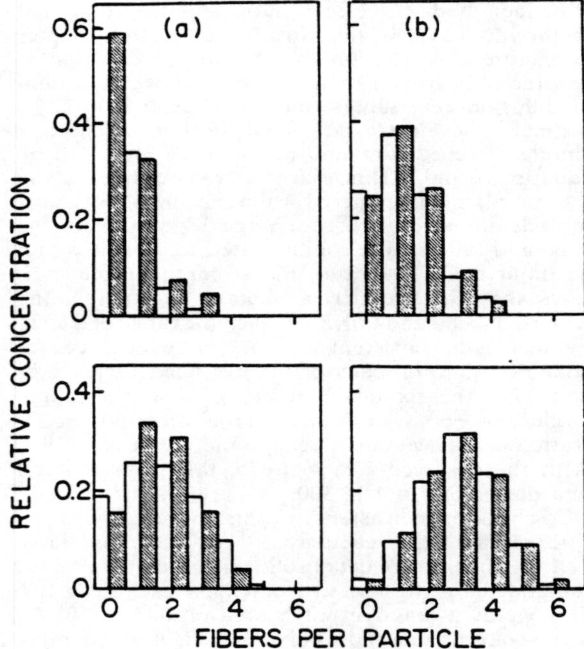

FIG. 2. Distribution of fibers per T4 particle as a function of the number added per baseplate site: (a) 0.3; (b) 0.8; (c) 1.5; (d) 3. The hatched bars are calculated from the binomial distribution. From Baran and Bloomfield (1978), reprinted with permission.

fiber to a baseplate site is k, then the probability of binding at time t is

$$p = 1 - \exp(-kt) \qquad (12)$$

This leads to an expression for the time evolution of the distribution of fibered species

$$p(K,t) = [6!/K!(6 - K)!][1 - \exp(-kt)]^K[\exp(-kt)]^{6-K} \qquad (13)$$

which for short times ($kt \ll 1$) becomes

$$p(K,t) = [6!/K!(6 - K)!](kt)^K[\exp(-kt)]^{6-K} \qquad (14)$$

This is similar to the equation obtained by Wood and Henninger (1969), with two exceptions. The first difference, a trivial notational one, is that they have $6k$ where we have k; this comes from their definition of the elementary rate constant as that for the reaction of a fiber with the empty baseplate containing six sites. The second difference is the statistical factor $6!/(6 - K)!$, which arises from the observation that if K sites are already filled, there are $(6 - K)$ ways to fill the next site. The neglect of this factor leads to rather large discrepancies for large K's and short times. At longer times, agreement is better.

Kinetic-Model Interpretation of the Role of Whiskers

Given a way to measure fiber attachment directly, attention turns to the mechanism of the reaction. An important observation is that attachment is considerably accelerated by whiskers attached to the tail at the head-tail junction (Bishop et al., 1974; Conley and Wood, 1975). It has been proposed that these long, fibrous proteins serve as jigs to facilitate proper alignment of fibers with the phage body (Dewey et al., 1974; Terzaghi, 1971). We have examined theoretically some aspects of this idea (Bloomfield and Prager, 1979). We found that whiskers might accelerate fiber attachment by converting the reaction pathway from an orientationally very restrictive translational diffusional approach between fiber and baseplate to formation of a looser complex between fiber and whisker followed by a rapid rotational diffusion of the fiber end to the baseplate site.

In the absence of whiskers, the simplest reaction model is the irreversible attachment of a fiber (F) to a baseplate site (B) to form a B-F complex

$$B + F \rightarrow B\text{-}F \tag{15}$$

With whiskers (W) attached, the fiberless phage is denoted BW. We assume that a whisker binds reversibly to a fiber, probably near the kink. The tethered fiber then undergoes rotational diffusion until it encounters and reacts with the baseplate site

$$BW + F \rightleftharpoons BW\text{-}F \rightarrow WB\text{-}F \tag{16}$$

The rate of formation of B-F in mechanism 15 is

$$d[B\text{-}F]/dt = k_1[B][F] \tag{17}$$

whereas steady-state analysis of mechanism 16 yields

$$d[WB\text{-}F]/dt = [k_2k_3/(k_{-2} + k_3)][BW][F] \tag{18}$$

With [BW] = [B], the acceleration in the presence of whiskers is $(k_2/k_1)[k_3/(k_{-2} + k_3)]$. Because the factor in square brackets is <1, the acceleration must result from $k_2/k_1 \gg 1$.

This quantity is the ratio of second-order rate constants for interaction of fibers with whiskers and with baseplates. One plausible difference between these interactions is their steric specificity. The insertion of the fiber end into the baseplate site is likely to require precise alignment in a tight cone of solid angle θ_{BF}, whereas binding of the fiber knee to the whisker tip might take place over a larger range of alignment angles θ_{WF}. For small angles, the steric and rotational dependences $F\delta$ in equation 6 are extremely sensitive to angle, varying approximately as θ^3 (Schurr and Schmitz, 1976). Thus a 3- to 4-fold ratio of θ's may lead to a 27- to 64-fold ratio of k's. This appears to be in the range observed (Bishop et al., 1974). If, in addition, reaction 15 has a significant activation energy, while BW-F is a looser complex, then k_2/k_1 may be even greater.

Even if $k_2/k_1 \gg 1$, whiskers cannot accelerate fiber attachment if $k_3/(k_{-2} + k_3) \ll 1$; that is, if $k_{-2} \gg k_3$. Noting that $k_2/k_{-2} = K$, the equilibrium constant for fiber-whisker attachment, we may make a rough estimate of k_{-2}. Given the likely orientational constraints on the attachment reaction, even if it is diffusion controlled, k_2 will probably be less than 10^8 liter/mol-s. If the whisker-fiber bond is relatively weak, say with a free energy $\Delta G°$ of -5 kcal/mol (ca. -21 kJ/mol), then K will be around 4×10^3 liter/mol, so k_{-2} might well be in the range of 2×10^4 s^{-1}. Thus, k_3 should not be much smaller than this. In fact, one can show (Bloomfield and Prager, 1979) that k_3 is in the range of the rotational diffusion coefficient of the fibers, approximately 5,500 s^{-1}, so the condition is met.

OUTLOOK FOR PHYSICAL STUDIES

Light Scattering and Other Physical Techniques

Light scattering in its various forms has proved to be a powerful method for studying phage assembly It is rapid enough to measure kinetics, sensitive enough to detect low concentrations of virus particles, and soundly enough based on theory to allow confident deductions from experiment. For example, Kunzler and Hohn (1978) have used it to study stages in the morphogenesis of lambda phage heads. Like all physical techniques, however, it is most reliable and enlightening when used along with other methods.

These may be divided into two main classes. Solution techniques are most powerful for studying assembly equilibria and dynamics, and conformational changes under native conditions. Microscopy and diffraction techniques, on the other hand, are more static but define more precisely the complexity of biomolecular structures. Ideally, the two complement each other. A good example is our recent hydrodynamic study of the T4 tail and baseplate (Aksiyote-Benbasat and Bloomfield, 1982), in which diffusion and sedimentation measurements were used to refine the hydrated dimensions of these structures in solution, given a knowledge of the structural components and their arrangement from electron microscopy.

In the past, solution techniques have included mainly various types of sedimentation, light scattering, and electrophoresis. In the future, we may anticipate increasing use of methods such as transient electric birefringence, triplet anisotropy decay, and saturation transfer electron paramagnetic resonance, which measure rotational motion on the microsecond to millisecond time scale and are more sensitive to size changes than QLS, which measures mainly translation.

Negative-staining electron microscopy has certainly been the most powerful source of information to date about the structures of assembly intermediates. New and refined sample preparation methods should assure the continued importance of single image electron microscopy. However, image reconstruction techniques are also being applied increasingly to ordered arrays of phage subunits (Crowther and Klug, 1975), which should ultimately provide substantially higher-resolution structural information. In the same class we might include low-angle X-ray scattering, which has provided information on DNA packaging inside heads (Earnshaw and Harrison, 1977). It is not impossible that phage as complicated as T4 might be crystallized (although the odds are better with more isometric particles [however, see Speyer and Khairallah, 1973]) and that atomic-resolution structures might be obtained by X-ray crystallography, but a more hopeful route to at least some types of atomic-level structural and dynamic information is through solid state nuclear magnetic resonance techniques (Opella et al., 1980).

The Crucial Role of Biochemical Preparations

While physical chemical studies of phage assembly processes hold the promise of obtaining much more detailed and quantitative information than is available by in vitro complementation studies with crude lysates, they are also much more difficult. This is not so much because of the complexities of the instrumentation or the underlying theory, but rather because of the requirements for high-quality biochemical preparations. The components must be biologically active, pure, and available in suitable amounts (typically micrograms) for physical studies. Only the first of these is generally required in studies using crude extracts. Since most T4 work has been done with this (admittedly highly productive) approach, relatively few procedures have been worked out for large-scale purification of phage components. The problem is particularly severe for minor phage components, which may play important regulatory roles while comprising only a few tenths of 1% of the phage mass. Affinity chromatography and gene cloning techniques will undoubtedly play an increasingly important role here. For now, it will suffice to conclude that physical chemists interested in the assembly of T4 and other phages will have to interact closely with phage biochemists and geneticists, not only to define the nature of the systems and phenomena to be understood, but also to obtain the materials with which to do their work.

The preparation of this review, and much of the work described, was supported by Public Health Service research grant GM 17855 from the National Institutes of Health.

STRUCTURE, ORGANIZATION, AND MANIPULATION OF THE GENOME

Map of the T4 Genome and Its Transcription Control Sites

ELIZABETH KUTTER[1] AND WOLFGANG RÜGER[2]

The Evergreen State College, Olympia, Washington 98505[1] and Arbeitsgruppe Molekulare Genetik, Ruhr-Universität Bochum, D-4630 Bochum 1, Federal Republic of Germany[2]

Recent analysis of T4 transcription patterns, gene sequences, origins of replication, and related data have taken advantage of the extensive restriction mapping of T4 and its correlation with the genetic map. Here we summarize the current data, including the locations of those genes and specific transcripts that have now been mapped to within a few hundred base pairs (bp). This is a continuing project; therefore, any additional sequence data, clone analyses, and promoter mapping data will be greatly appreciated by the authors. Sequence data should also be submitted to the T4 sequence bank being maintained by Larry Gold, University of Colorado, Boulder, Colo. 80309.

T4 RESTRICTION MAP: STRUCTURE AND LIMITATIONS

Construction of the Map

A number of laboratories have worked on restriction maps of T4 over the last few years (Carlson and Nicolaisen, 1979; Kiko et al., 1979; Kutter et al., 1980; Marsh and Hepburn, 1981; Niggeman et al., 1981; O'Farrell et al., 1980; Rüger et al., 1979; Takahashi et al., 1979; E. Kutter, W. Rüger, R. Marsh, P. O'Farrell, and S. Redmon-Payne, *N.I.H. Genetic Map Book*, in press); this work is drawn together here with the kind cooperation of all concerned. Table 1 shows the restriction enzymes whose cleavage patterns have been examined in detail. The sites cleaved by *Bam*HI, *Bgl*II, *Kpn*I, *Sal*I, *Sma*I, and *Xho*I have all been mapped relative to each other by several groups, as indicated, with good general agreement. We have used all the available mapping and clone analysis data to position these sites as precisely as possible relative to one another (Table 2) and then to position the sites for other enzymes relative to them. The sites of *Xba*I and *Eco*RI relative to the above should also be quite accurate (within about 0.2 kilobases [kb]); each has been mapped by at least two groups, and most discrepancies have been resolved by reanalysis or by rechecking data. The relative locations of the *Bgl*I, *Hae*II, *Cla*I, *Hin*dIII, *Pst*I, *Pvu*I, *Sac*II, and *Xma*III sites are potentially less precise (except where they have been analyzed on cloned fragments) since each has been mapped in only one laboratory (Table 1) and only relative to the sites cut by the better-studied enzymes, not relative to each other generally. Thus, occasional errors and inversions of closely spaced sites for two enzymes are to be expected. Furthermore, for any of the enzymes, two sites separated by less than 200 bp may have been mistaken for a single site. For instance, acrylamide gel analysis by Rüger (unpublished data) has recently indicated the existence of an additional 90-bp *Bgl*II fragment (band 12) which is still unmapped and a 90-bp *Xba* fragment (band 22) which appears to map adjacent to band 16. *Eco*RI also gives several additional bands of 95 to 130 bp. Similar analysis needs to be extended to the other enzymes.

When available, analysis of clones has been taken into consideration in assigning map positions. However, in the few cases in which this produced conflicts in assignments, the cloning and sequencing data were given lower priority, since many T4 workers have reported observing rearrangements such as deletions in conjunction with cloning. Also, a single sequencing error could apparently produce or eliminate a restriction site; in particular, such problems might occur if only one of the two strands has been sequenced.

Assignment of the Map Origin

Although T4 DNA is linear, its genetic map is circular owing to the circular permutation of endpoints in any given population of phage; the actual molecules in phage particles are 170 kb in length, with a 2.3% terminal redundancy. The *r*IIA-*r*IIB junction has traditionally been used as the origin of the standard genetic map since it was well characterized physically and genetically; distances are given in kb clockwise from this junction (cf. Wood and Revel, 1976; Mosig, appendix, this volume). Here (Fig. 1 and Table 2), the physical position of the *r*IIA-*r*IIB junction has been established very precisely by using the following information: (i) the clonal analysis of this region by Selzer et al. (1978); (ii) the comparative restriction analysis of three mutants carrying different deletions in the region (Table 2), namely SaΔ9, *r*II NB5060, and the mutant of Bruner "*r*IIH23" (Carlson, 1980; Niggemann et al., 1981; O'Farrell et al., 1980); (iii) a fine-structure analysis of the appropriate *Eco*RI band isolated from SaΔ9 DNA by using *Xho*I, *Hae*II, and *Hin*dIII (Niggemann et al., 1981); and (iv) the

TABLE 1. Restriction enzyme sites mapped

Enzyme	Recognition sequence	No. of sites	References[a]
*Bam*HI	G ↓ GATCC	1	5, 7, 9, 10
*Sac*II	CCGC ↓ GG	1	11
*Xma*III	C ↓ GGCCG	2	11
*Pvu*I	CGAT ↓ CG	4	5
*Sma*I	CCC ↓ GGG	5	3, 7, 9
*Xma*I	C ↓ CCGGG		
*Bgl*I	GCCN_5GGC	7	2
*Kpn*I	GGTAC ↓ C	7	1, 5, 7–9
*Sal*I	G ↓ TCGAC	8	1, 5, 7–9
*Bgl*II	A ↓ GATCT	14	5, 7–9
*Xho*I	C ↓ TCGAG	17	7–9
*Xba*I	T ↓ CTAGA	24	5, 6
*Eco*RV	GATAT ↓ C	28	4, 12
*Pst*I	CTGCA ↓ G	32	7
*Cla*I	AT ↓ CGAT	34	4
*Hae*II	PuGCGC ↓ Py	40	6
*Eco*RI	G ↓ AATTC	70	6, 7
*Hin*dIII	A ↓ AGCTT	78	7

[a] References: (1) Carlson and Nicolaisen, 1979; (2) Carlson, 1980; (3) Kiko et al., 1979; (4) Liebig and Rüger, unpublished data; (5) Marsh and Hepburn, 1981; (6) Niggemann et al., 1981; (7) O'Farrell et al., 1980; (8) Rüger et al., 1979; (9) M. Shimizu, H. Takahashi, and H. Saito, personal communication; (10) Wilson et al., 1980; (11) G. Wilson, personal communication; (12) N. P. Kuzmin, V. M. Kryukov, V. I. Tanyashin, and A. A. Bayer, Dokl. Biol. Sci. (Engl. Trans. Dokl. Akad. Nauk SSSR), in press.

sequence of the 870-bp *Hin*dIII fragment spanning the junction (Pribnow et al., 1981). The precise dividing line is set just before the first *r*IIB codon.

Strain Considerations and Band Labeling

Some confusion has arisen from the differences in nomenclature and in band-number assignments resulting from variations in strains being mapped and in degrees of resolution. The strain differences arise primarily from the fact that most restriction enzymes, including all of those mapped here, cleave the normal glucosylated hydroxymethyldeoxycytosine (HMdC)-containing DNA very poorly, if at all. Thus, all of the mapping work to date has used one or another of the special deoxycytosine-containing T4 strains (T4dC strains) strains discussed in the chapter by Snustad et al. (this volume) and its appendix which are capable of making cytosine-containing T4 DNA (T4 dC-DNA). All such strains are missing endonuclease IV (endoIV) and dCTPase, and most also lack dCMP hydroxymethylase (HMase), endoII, and/or gp*alc*. Various mapping efforts have used strains carrying different deletions in the *r*II-*ac* region to remove endoIV, the lack of which causes no direct phenotypic variations. We have now renumbered the bands (Table 2, column 3) to the order expected for undeleted T4 DNA. To do so, we have used the data of Carlson (1980), Takahashi et al. (1979), Rüger (unpublished data), and Wilson (personal communication) comparing deleted phage with those carrying an endoIV point mutation to determine restriction sites for various enzymes within the region normally deleted. (Note that no information is available for *Hin*dIII between 162.7 and 165.1 kb or between 47.2 and 57.9 kb, and no direct correlation has been attempted between *Hin*dIII gel band patterns and the map given here.)

Strain differences also raise the possibility of differences in cleavage sites, particularly since most of the strains are hybrids between T4D (source of the amber mutations) and T4B (for the *r*II deletions). The few apparent observed variations might be attributable to such strain differences. For example, the *Hae*II site given at 15.8 kb, as determined by Niggemann et al. (1981) and Rüger (unpublished data) by using the strain {dCTPase, HMase, *denB*-SaΔ9}, is not present in the sequence of the region as determined by Macdonald and Mosig (see Mosig, this volume) by using the T4dC strain [*alc*GT7, dCTPase, HMase, *denB*-*r*IINB5060]. Also, a 6.8-kb *Hae*II fragment appears in the strain {*unf*39, dCTPase, HMase, endoIV (pt)} DNA which is not observed in other strains; its map position has not yet been determined (Rüger, unpublished results). The *Hae*II sites observed could potentially be affected by the host used, since McClelland et al. (1981) have shown that some *Hae*II sites are sensitive to different host methylation systems (see also below).

The possible effects of using different T4 strains with different deletions should be particularly kept in mind in using Table 3, which presents the map data in order of expected fragment size with undeleted DNA to facilitate applications to cloning and hybridization to blots. Furthermore, interpretable band patterns are obtainable only from DNA in which cytosine virtually completely replaces hydroxymethylcytosine, i.e., strains carrying a mutation in gene 42 (HMase), gene 1 (dHMP kinase) or somewhat less satisfactorily, endoII (see Kutter and Snyder, this volume), although other strains are useful for many applications. Most of the mapping and cloning work to date, including that summarized here, has been done with T4dC strains {endoIV-*ac*SaΔ9, dCTPase, HMase}, {*alc*GT7, endoIV-*r*IIΔNB5060, dCTPase, HMase} or {*alc*/*unf*39, endoIV (pt), dCTPase, HMase}; strains carrying dHMP kinase mutations, which give better phage production, may be used increasingly in the future.

CLEAVAGE OF HMdC-CONTAINING T4 DNA

As mentioned above, most restriction enzymes cleave normal HMdC-containing T4 DNA very poorly, if at all. Strains producing DNA in which cytosine replaces only part of the hydroxymethylcytosine (T4dC strains) have proven useful for many cloning purposes, allowing the production of a truly random set of partial cleavage products. *Eco*RI and *Xba*I will cleave unglucosylated HMdC-DNA, but they do so incompletely.

Three enzymes have recently been reported to cleave normal T4 DNA: *Taq*I (T ↓ CGA) and *Aha*III (TTT ↓ AAA), both of which produce mainly very small fragments (less than 2 kb) with the largest 5 to 6 kb (D. Coit and B. Alberts, personal communication; K. Kreuzer, B. Alberts, and N. Brown, personal communication), and *Eco*RV (GATAT ↓ C), which produces about 24 bands (V. Tanyashin, personal communication). These are proving useful for a variety of applications. Unfortunately, however, the restriction patterns are somewhat different when one compares T4 dC-DNA and HMdC-DNA, particularly for low-molecular-weight fragments (Fig. 2). Kreuzer (personal communication) has found evidence that this is due, in part at

TABLE 2. Map of restriction sites, band numbers and sizes, and well-characterized genetic loci[a]

SITE	ENZYME	BAND	SIZE/KB.
27.45	ECO1	30.0	1.50
26.68	HIN3	0.0	2.22
26.10	ECO1	32.3	1.35
25.80	HAE2	9.0	5.25
25.50	HIN3	0.0	1.18
24.80	SAL1	7.0	9.20
24.30	ECO1	26.1	1.80
23.40	PST1	0.0	6.71
23.05	HIN3	0.0	2.45
23.00	HIN3	0.0	0.05
22.85	CLA1	8.2	5.56
22.45	HIN3	0.0	0.55
22.40	PST1	0.0	1.00
22.37	XBA1	4.0	12.33
22.01	BGL2	3.1	17.39
22.00	CLA1	21.0	0.85
21.64	HIN3	0.0	0.81
21.15	ECO1	17.1	3.15
20.75	SAL1	9.0	4.05
20.63	PST1	0.0	1.77
20.60	HIN3	0.0	1.04
20.30	XHO1	8.0	8.90
20.06	HIN3	0.0	0.54
18.68	HAE2	7.0	7.12
17.98	HIN3	0.0	2.06
17.90	XBA1	11.0	4.47
16.24	ECO1	6.1	4.91
16.19	HIN3	0.0	1.79
15.80+	HAE2	17.1	2.88
15.77	XBA1	17.0	2.13
15.59	ECO1	38.0	0.65
15.05+	HAE2	0.0	0.75
14.75	ECO1	0.0	0.84
14.10	HAE2	0.0	0.95
13.30	HIN3	0.0	2.89
13.10	XBA1	15.0	2.67
12.35	HIN3	0.0	0.95
11.85	ECO1	18.0	2.90
10.90	ECO5	1.1	25.55
10.80	HAE2	15.0	3.30
10.70	CLA1	3.0	11.30
10.60	ECO1	32.2	1.25
10.40	KPN1	3.0	36.80
10.30	HIN3	0.0	2.05
9.00	ECO1	28.0	1.60
8.80	HIN3	0.0	1.50
8.80	KPN1	7.0	1.60
7.80	BGL2	4.0	14.21
7.75	ECO1	32.1	1.25
7.30	ECO5	13.0	3.60
7.05	ECO1	0.0	0.70
5.90	ECO5	19.0	1.40
5.80	HIN3	0.0	3.00
5.55	ECO1	29.1	1.50
5.00	HAE2	8.0	5.80
5.00	CLA1	8.1	5.70
4.90	HIN3	0.0	0.90
3.90	ECO5	17.0	2.00
3.30	HIN3	0.0	1.60
2.85	ECO1	19.1	2.70
2.80	XMA3	1.0	120.25
2.50	ECO1	0.0	0.35
1.40	CLA1	10.3	3.60
1.10	XHO1	1.0	19.20
0.85	HIN3	0.0	2.45
0.80	HAE2	12.0	4.20
0.43	HIN3	0.0	0.42
0.02	BGL1	2.0	40.88

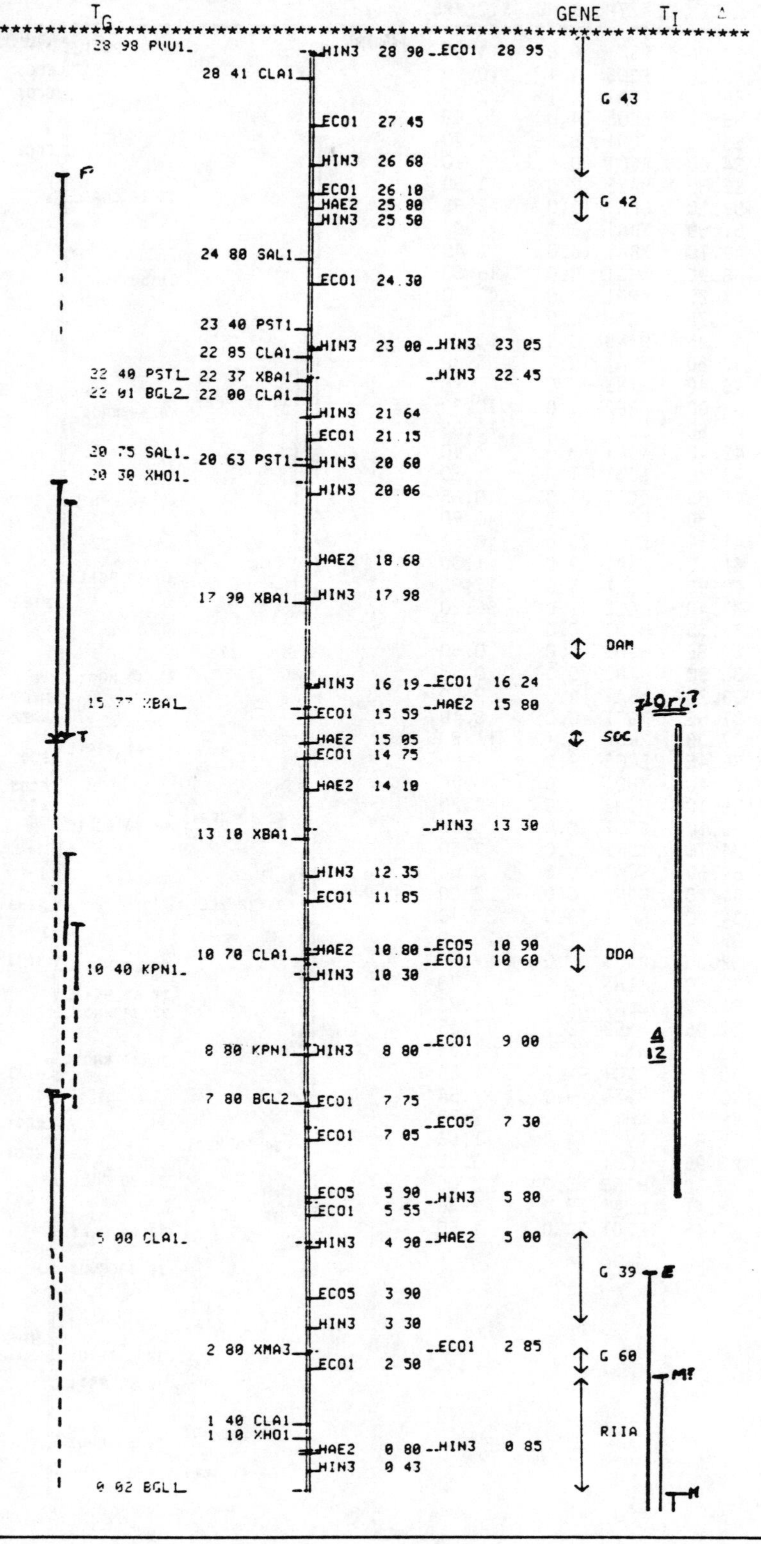

TABLE 2—*Continued*

SITE	ENZYME/BAND		SIZE/KB
56.00	PST1	0.0	11.90
56.00	ECO5	4.1	10.90
55.75	CLA1	12.1	2.95
55.55	ECO5	24.0	0.45
55.10	ECO1	8.0	4.70
54.00	ECO1	0.0	1.10
53.30	HAE2	3.0	11.80
53.10	CLA1	13.0	2.65
51.50	XBA1	2.1	14.90
49.10	XBA1	16.0	2.40
48.90	XHO1	3.0	15.50
47.20	KPN1	2.0	40.00
45.75	PST1	0.0	10.25
45.25*	HIN3	0.0	13.25
43.80	XBA1	10.0	5.30
43.40	HIN3	0.0	1.85
43.00	HAE2	4.0	10.30
42.50	ECO5	2.0	13.05
42.40	XBA1	18.0	1.40
42.20	ECO1	1.0	11.80
41.65	ECO5	21.0	0.85
41.30	ECO1	0.0	0.90
41.14	ECO5	23.0	0.51
41.10	XBA1	19.0	1.30
40.90	BGL1	5.0	17.80
39.40	BGL2	1.0	56.50
39.35	HIN3	0.0	4.05
38.85	HIN3	0.0	0.50
38.30	HIN3	0.0	0.55
38.20	XBA1	14.0	2.90
37.65	PST1	0.0	8.10
37.30	XHO1	5.1	11.60
36.45	ECO5	9.0	4.69
36.40	XHO1	13.0	0.90
36.10	HIN3	0.0	2.20
35.10	ECO1	2.0	6.20
34.70	XBA1	13.0	3.50
34.50	ECO1	0.0	0.60
34.10	HIN3	0.0	2.00
34.00	SAL1	3.0	23.90
33.00	CLA1	1.0	20.10
32.30	PVU1	2.0	67.30
32.30	HIN3	0.0	1.80
31.27	CLA1	16.0	1.73
31.05	HAE2	2.1	11.95
30.95	HIN3	0.0	1.35
30.64	PVU1	3.2	1.66
30.11	PST1	0.0	7.54
29.20	XHO1	9.1	7.20
28.98	PVU1	4.0	1.66
28.95	ECO1	4.0	5.55
28.90	HIN3	0.0	2.05
28.41	CLA1	12.2	2.86
27.45	ECO1	30.0	1.50

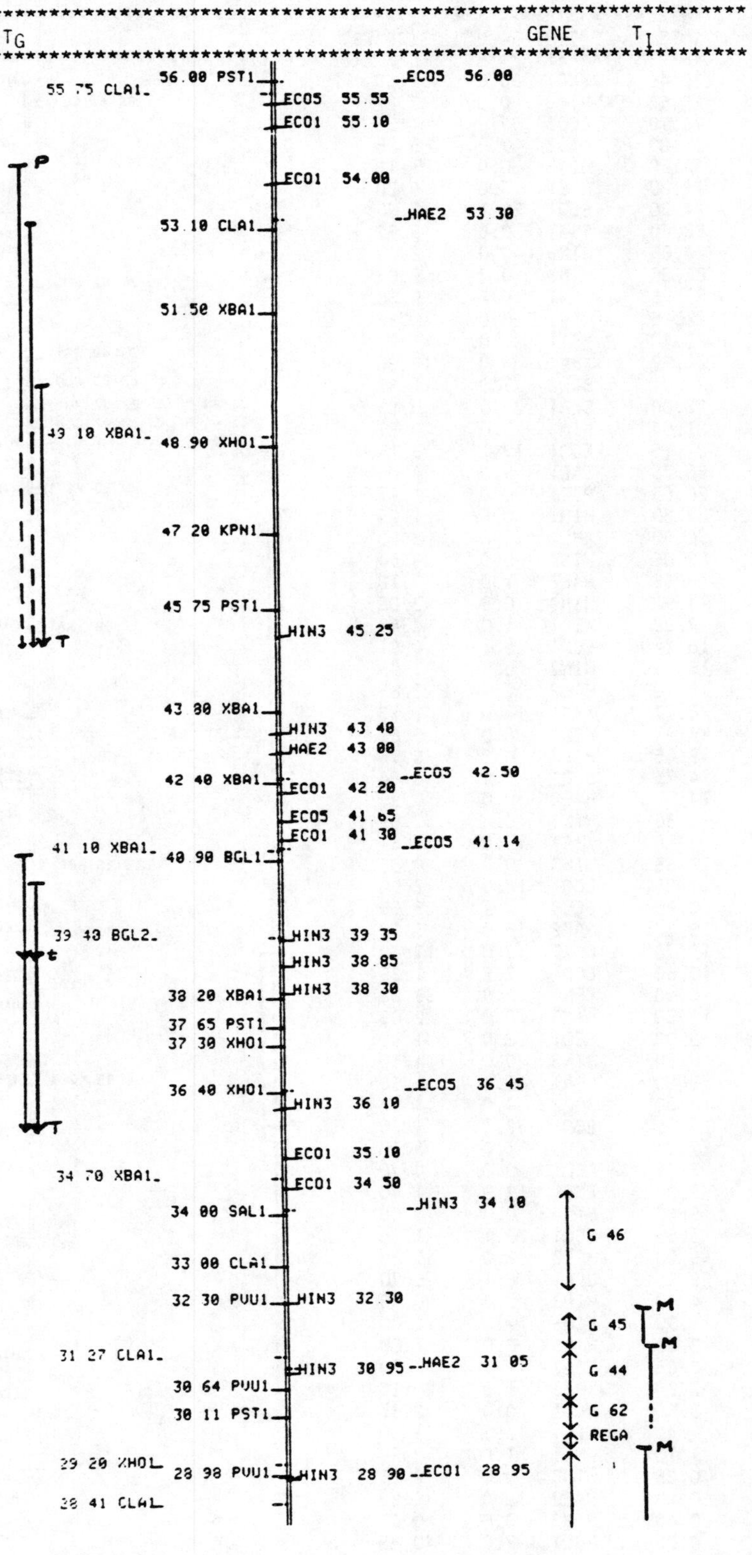

T_G GENE T_I

56.00 PST1 — ECO5 56.00
55.75 CLA1
ECO5 55.55
ECO1 55.10
P
ECO1 54.00
HAE2 53.30
53.10 CLA1
51.50 XBA1
49.10 XBA1
48.90 XHO1
47.20 KPN1
45.75 PST1
HIN3 45.25
T
43.80 XBA1
HIN3 43.40
HAE2 43.00
42.40 XBA1 — ECO1 42.20 — ECO5 42.50
ECO5 41.65
ECO1 41.30 — ECO5 41.14
41.10 XBA1
40.90 BGL1
39.40 BGL2 — HIN3 39.35
t
HIN3 38.85
38.20 XBA1 — HIN3 38.30
37.65 PST1
37.30 XHO1
36.40 XHO1 — ECO5 36.45
HIN3 36.10
T
ECO1 35.10
34.70 XBA1 — ECO1 34.50
34.00 SAL1 — HIN3 34.10
G 46
33.00 CLA1
32.30 PVU1 — HIN3 32.30
M
G 45
M
31.27 CLA1 — HIN3 30.95 — HAE2 31.05
G 44
30.64 PVU1
30.11 PST1
G 62
REGA
M
29.20 XHO1
28.98 PVU1 — HIN3 28.90 — ECO1 28.95
28.41 CLA1

TABLE 2—*Continued*

SITE	ENZYME	BAND	SIZE/KB
83.60	HAE2	19.2	2.00
82.70	ECO1	11.0	3.85
82.70	HAE2	0.0	0.90
82.60	ECO5	14.0	2.60
82.50	PST1	0.0	4.50
82.00	CLA1	2.0	16.30
80.80	ECO1	25.0	1.90
80.30	HIN3	0.0	11.00
79.70+	HAE2	16.1	3.00
79.70	BGL1	3.0	23.08
79.50	XBA1	5.0	11.00
78.80	HAE2	0.0	0.90
78.30	CLA1	10.1	3.70
78.00	XHO1	2.0	17.00
77.40	ECO5	8.0	5.20
77.20	PST1	0.0	5.30
76.90	HAE2	20.1	1.90
75.01	ECO1	7.0	4.79
74.33	PST1	0.0	2.87
73.47	PST1	0.0	0.86
73.36	HAE2	14.0	3.54
73.17	SAL1	1.0	38.68
72.05	ECO5	7.2	5.35
71.39	ECO5	22.0	0.66
71.29	HAE2	19.1	2.07
70.40	ECO1	5.1	5.61
70.10	CLA1	5.0	8.20
69.85	ECO1	0.0	0.55
69.50	HIN3	0.0	10.80
69.40	PST1	0.0	4.07
69.00	HIN3	0.0	0.50
68.50	XHO1	7.0	9.50
68.10	HAE2	13.0	3.19
67.90	PST1	0.0	1.50
66.90	ECO5	10.0	4.49
66.40	XBA1	3.0	13.10
65.90	ECO1	10.0	3.95
65.52	HIN3	0.0	3.48
65.30	HIN3	0.0	0.22
65.10	HAE2	16.2	3.00
64.70	HIN3	0.0	0.60
64.40	XHO1	12.0	4.10
64.20	ECO1	27.1	1.70
63.90	HIN3	0.0	0.80
63.10	HIN3	0.0	0.80
62.20	ECO1	24.0	2.00
62.00	HIN3	0.0	1.10
61.60	ECO1	0.0	0.60
60.10	HIN3	0.0	1.90
59.80	ECO1	26.2	1.80
59.70	CLA1	4.1	10.40
58.70	BGL1	4.0	21.00
58.70	CLA1	19.0	1.00
58.50	HIN3	0.0	1.60
57.90	SAL1	5.0	15.27
56.00	PST1	0.0	11.90

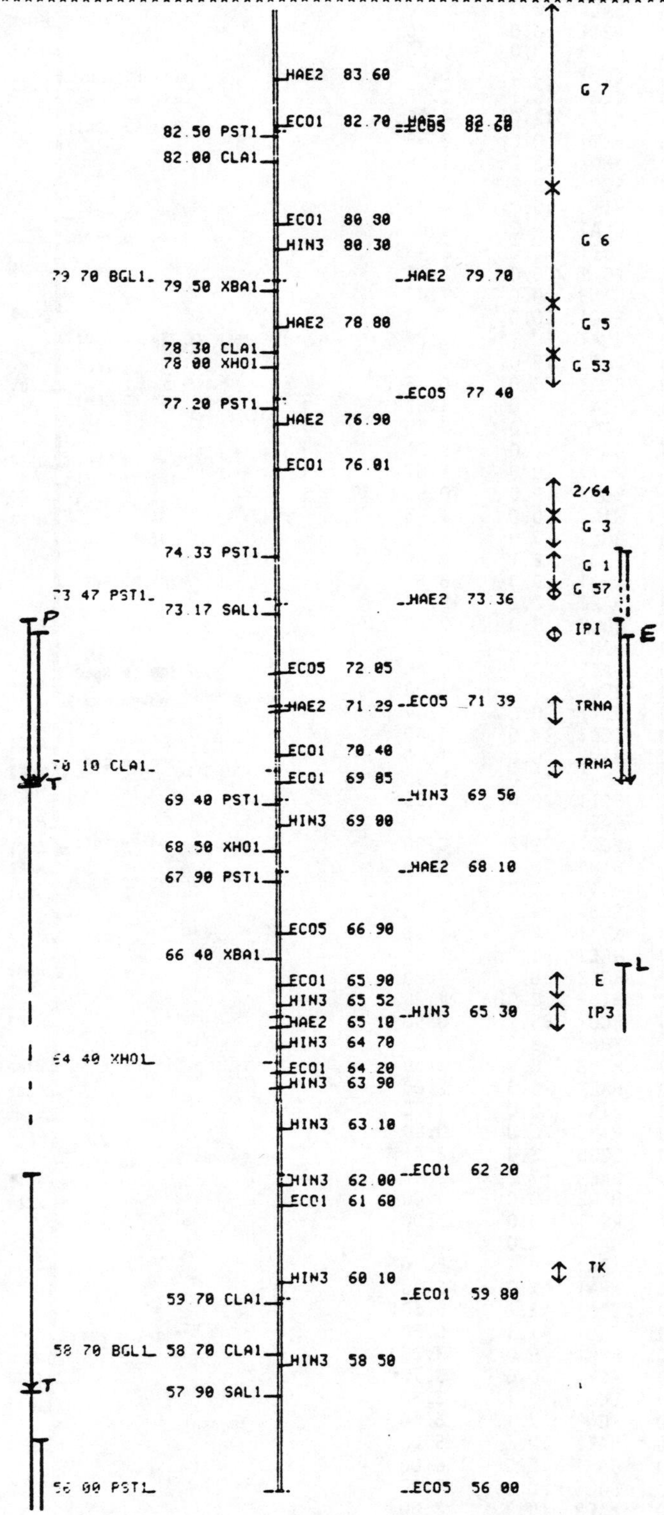

T_G GENE T_I

HAE2 83.60
82.50 PST1 ECO1 82.70 HAE2 82.70 ECO5 82.60
82.00 CLA1
G 7
ECO1 80.80
HIN3 80.30
G 6
79.70 BGL1 79.50 XBA1 HAE2 79.70
HAE2 78.80
G 5
78.30 CLA1
78.00 XHO1
G 53
ECO5 77.40
77.20 PST1
HAE2 76.90
ECO1 76.01
2/64
G 3
74.33 PST1
G 1
G 57
73.47 PST1 73.17 SAL1 HAE2 73.36
P
IPI
E
ECO5 72.05
HAE2 71.29 ECO5 71.39
TRNA
ECO1 70.40
70.10 CLA1 ECO1 69.85
TRNA
T
69.40 PST1 HIN3 69.50
HIN3 69.00
68.50 XHO1
67.90 PST1 HAE2 68.10
ECO5 66.90
66.40 XBA1
L
ECO1 65.90
E
HIN3 65.52
HAE2 65.10 HIN3 65.30
IP3
HIN3 64.70
64.40 XHO1 ECO1 64.20
HIN3 63.90
HIN3 63.10
HIN3 62.00 ECO1 62.20
ECO1 61.60
HIN3 60.10
TK
59.70 CLA1 ECO1 59.80
58.70 BGL1 58.70 CLA1 HIN3 58.50
T
57.90 SAL1
56.00 PST1 ECO5 56.00

TABLE 2—*Continued*

SITE	ENZYME/BAND		SIZE/KB
111.85	SAL1	2.0	31.70
111.70	PST1	0.0	2.40
111.65	HIN3	0.0	0.75
111.60	ECO1	3.0	6.05
111.40	CLA1	15.1	1.90
110.85	XBA1	12.0	4.20
109.15	PST1	0.0	2.55
108.70	XHO1	5.2	11.40
108.10	ECO1	12.2	3.50
107.18	HIN3	0.0	4.47
106.75	CLA1	9.2	4.65
106.72	PST1	0.0	2.43
106.58	PST1	0.0	0.14
105.78	HIN3	0.0	1.40
105.76	ECO1	20.0	2.34
105.48	CLA1	17.1	1.27
104.44	CLA1	18.0	1.04
104.28	CLA1	23.0	0.16
104.26	HIN3	0.0	1.52
104.21	ECO5	5.0	9.19
104.04	BGL1	1.0	49.31
103.91	PST1	0.0	2.67
103.40	BGL2	5.0	10.60
103.11	HIN3	0.0	1.15
102.78	BGL1	7.0	1.26
102.10	ECO1	12.1	3.66
101.90	XHO1	9.3	6.80
101.00	HAE2	2.2	12.00
100.10	BGL2	9.0	3.30
99.90	HIN3	0.0	3.21
99.80	HAE2	22.1	1.20
99.60	PVU1	1.0	95.38
99.50	PST1	0.0	4.41
98.70	ECO1	13.0	3.40
98.30	HIN3	0.0	1.60
98.30	CLA1	7.0	5.98
98.15	PST1	0.0	1.35
98.10	HAE2	21.1	1.70
98.10	ECO1	39.2	0.60
97.10	HIN3	0.0	1.20
96.70	PST1	0.0	1.45
96.60	ECO1	29.1	1.50
96.60	XBA1	2.2	14.25
96.20	HAE2	20.2	1.90
95.90	BGL2	8.1	4.20
95.80	HIN3	0.0	1.30
95.70	ECO1	35.1	0.90
95.00	XHO1	9.2	6.90
94.90	HIN3	0.0	0.90
94.20	HAE2	19.3	2.00
93.20	HIN3	0.0	1.70
92.30	HIN3	0.0	0.90
91.80	ECO5	3.0	12.41
91.40+	HAE2	17.2	2.80
91.30	HIN3	0.0	1.00
90.70	PST1	0.0	6.00
90.60	HAE2	0.0	0.80
90.50	XBA1	9.1	6.10
90.40	KPN1	1.0	46.10
90.10	ECO1	5.2	5.60
89.60	ECO1	41.1	0.50
87.20	KPN1	6.0	3.20
87.00	PST1	0.0	3.70
86.55	ECO1	16.1	3.50
86.20	ECO5	7.1	5.60
85.60	HAE2	10.1	5.00
85.30	BAM1	0.0	166.00
85.20	ECO5	20.0	1.00
83.60	HAE2	19.2	2.00

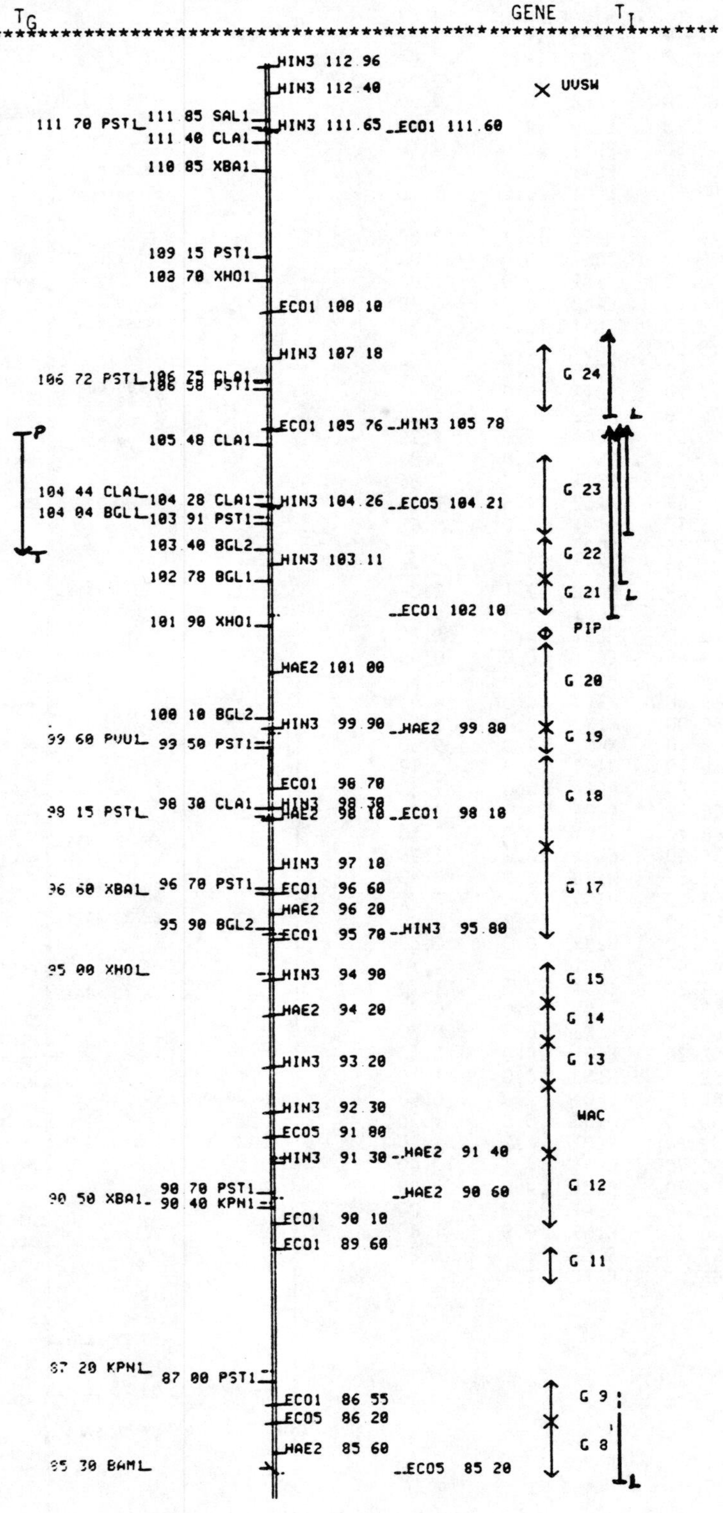

T_G GENE T_I

HIN3 112.96
HIN3 112.40 × UVSW
111 70 PST1 111 85 SAL1 HIN3 111.65 ECO1 111.60
111 40 CLA1
110 85 XBA1
109 15 PST1
108 70 XHO1
ECO1 108 10
HIN3 107 18
106 72 PST1 106 75 CLA1 106 58 PST1 G 24
ECO1 105 76 HIN3 105 78
P
105 48 CLA1
104 44 CLA1 104 28 CLA1 HIN3 104.26 ECO5 104.21 G 23
104 04 BGL1 103 91 PST1
103 40 BGL2 G 22
T
HIN3 103.11
102 78 BGL1 G 21 L
ECO1 102 10
101 90 XHO1 PIP
HAE2 101 00
G 20
100 10 BGL2 HIN3 99 90 HAE2 99 80
99 60 PVU1 99 50 PST1 G 19
ECO1 98 70
98 15 PST1 98 30 CLA1 HIN3 98 30 HAE2 98 10 ECO1 98 10 G 18
HIN3 97 10
96 60 XBA1 96 70 PST1 ECO1 96 60 G 17
HAE2 96 20
95 90 BGL2 ECO1 95 70 HIN3 95 80
95 00 XHO1 HIN3 94 90 G 15
HAE2 94 20 G 14
HIN3 93 20 G 13
HIN3 92 30
ECO5 91 80 WAC
HIN3 91 30 HAE2 91 40
90 50 XBA1 90 70 PST1 90 40 KPN1 HAE2 90 60 G 12
ECO1 90 10
ECO1 89 60
G 11
87 20 KPN1 87 00 PST1
ECO1 86 55 G 9
ECO5 86 20
HAE2 85 60 G 8
85 30 BAM1 ECO5 85 20 L

TABLE 2—*Continued*

SITE	ENZYME/BAND		SIZE/KB
140.00	PST1	0.0	10.10
139.85	ECO1	0.0	0.70
139.00	ECO5	16.0	2.10
138.90	HAE2	1.0	15.30
138.80	XHO1	4.1	12.85
137.70	HIN3	0.0	5.18
137.20	BGL2	3.2	17.40
136.80	ECO1	17.2	3.05
136.50	KPN1	4.0	23.80
135.25	CLA1	4.2	10.60
135.15	ECO1	27.2	1.65
134.90	CLA1	22.0	0.35
133.85	ECO1	31.0	1.30
133.50	XBA1	1.0	17.45
133.20	SMA1	2.0	15.50
131.65	ECO1	21.0	2.20
131.50	ECO1	46.0	0.15
131.00	HAE2	6.0	7.90
130.30	ECO1	33.0	1.20
129.10	BGL2	6.0	8.10
128.35	PST1	0.0	11.65
127.50	CLA1	6.0	7.40
127.20	ECO1	15.0	3.10
126.00	HIN3	0.0	11.70
126.00	XHO1	4.2	12.80
125.70	ECO1	29.3	1.50
125.25	ECO1	42.0	0.45
124.90	BGL2	8.2	4.20
124.75	ECO1	41.0	0.50
124.40	CLA1	11.1	3.10
124.30	XBA1	6.0	9.20
124.10	HIN3	0.0	1.90
123.05	XMA3	0.0	45.75
123.00	SBA1	20.0	1.30
122.90	HIN3	0.0	1.20
122.45	ECO1	22.0	2.30
122.10	HIN3	0.0	0.80
122.00	HAE2	5.0	9.00
121.20	HIN3	0.0	0.90
121.00	HAE2	23.0	1.00
120.85	PST1	0.0	7.50
120.10	XHO1	10.0	5.90
119.40	BGL2	7.2	5.50
119.30	BGL2	11.0	0.10
118.80	HAE2	18.1	2.20
118.30	PST1	0.0	2.55
118.00	HAE2	0.0	0.80
117.65	ECO1	6.2	4.80
116.05	PST1	0.0	2.25
115.80	ECO5	1.2	23.20
115.05	XBA1	7.0	7.95
114.50	HIN3	0.0	6.70
114.15	CLA1	4.3	10.25
114.10	PST1	0.0	1.95
114.00	BGL2	7.1	5.30
113.40	ECO5	15.0	2.40
113.30	CLA1	20.0	0.85
113.10	HIN3	0.0	1.40
113.00	HAE2	10.2	5.00
113.00	HIN3	0.0	0.10
112.96	HIN3	0.0	0.04
112.40	HIN3	0.0	0.56
111.85	SAL1	2.0	31.70

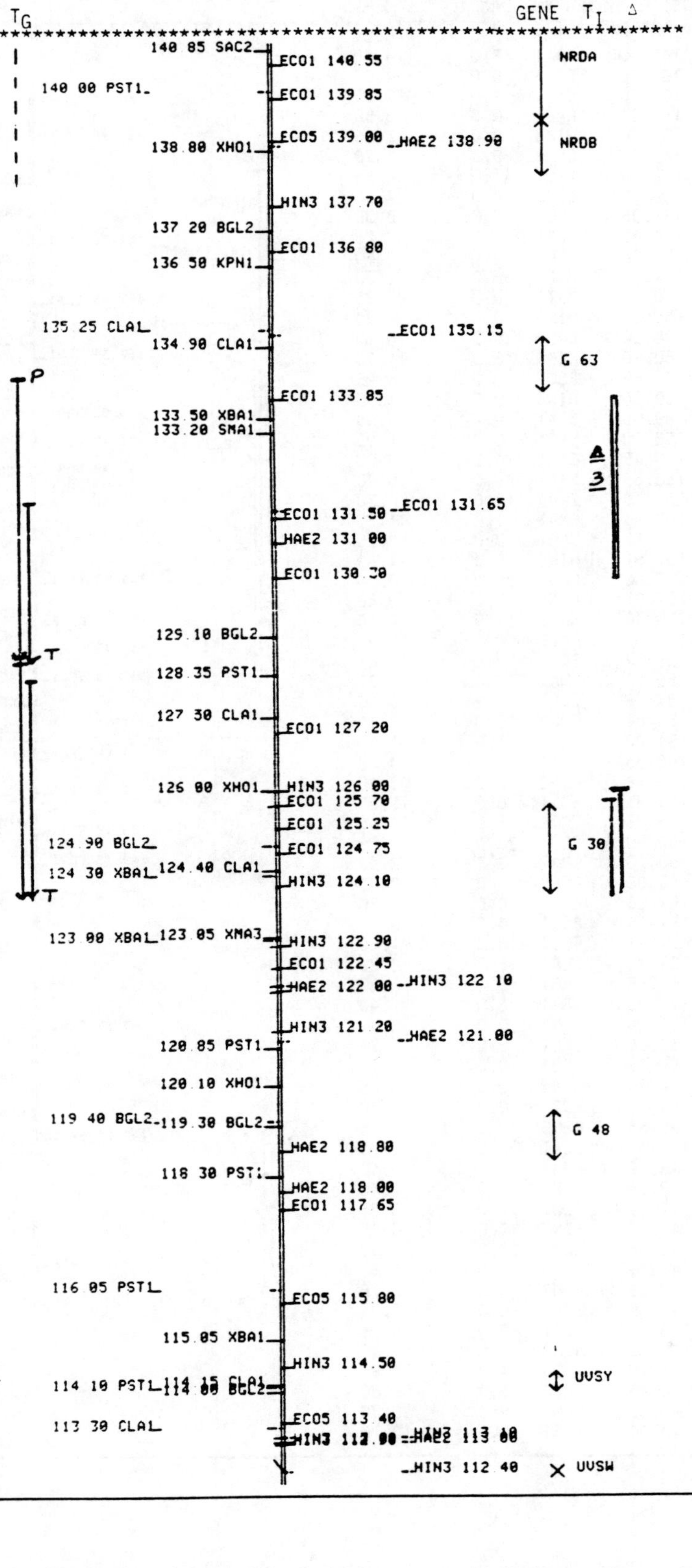

T_G GENE T_I △

NRDA
NRDB
G 63
G 30
G 48
UVSY
UVSW

TABLE 2—*Continued*

SITE	ENZYME	BAND	SIZE/KB
165.56	HIN3	0.0	0.87
165.00	XBA1	2.3	14.10
164.00	ECO1	0.0	4.50
163.50	SAL1	4.0	23.25
163.30	HAE2	0.0	3.50
162.70	HIN3	0.0	2.86
162.70*	CLA1	9.3	4.70
162.45	XHO1	11.0	4.65
162.30	ECO5	6.0	7.60
161.95	ECO1	0.0	2.05
161.45	CLA1	17.2	1.25
161.45	ECO1	0.0	0.50
161.10	HAE2	18.2	2.20
160.90	PST1	0.0	25.73
160.80	HIN3	0.0	1.90
160.60+	ECO1	0.0	0.85
160.30	KPN1	5.0	14.50
159.65	ECO1	0.0	0.95
159.30	HIN3	0.0	1.50
159.05	XBA1	9.2	5.95
158.60	ECO1	0.0	1.05
158.45	CLA1	11.2	3.00
158.35	XBA1	21.0	0.70
158.10	ECO5	12.0	4.20
157.90	SMA1	1.0	141.30
157.85	ECO1	0.0	0.75
157.81	HIN3	0.0	1.49
157.80	ECO5	25.0	0.30
157.01	HIN3	0.0	0.80
156.70+	HAE2	11.0	4.40
155.80	HAE2	0.0	0.90
155.79	BGL2	2.0	18.01
155.53	SMA1	4.0	2.37
154.78	ECO1	16.2	3.07
154.60	BGL2	10.0	1.19
154.20	HAE2	21.2	1.60
153.93	CLA1	9.1	4.52
153.39	HIN3	0.0	3.62
153.35	BGL1	6.0	12.67
153.35	ECO5	11.0	4.45
152.80	ECO1	23.0	1.98
152.10	SMA1	3.2	3.43
152.05	ECO1	0.0	0.75
151.85	ECO1	0.0	0.20
151.80	PST1	0.0	9.10
151.75	ECO5	18.0	1.60
151.70	CLA1	15.2	2.23
151.65	XHO1	6.0	10.80
150.95	XBA1	8.0	7.40
150.85	HIN3	0.0	2.54
150.76	SAL1	6.0	12.74
150.75	ECO1	0.0	1.10
150.30	HIN3	0.0	0.55
150.10	PST1	0.0	1.70
149.65	HIN3	0.0	0.65
149.45	CLA1	14.0	2.25
148.70	SMA1	3.1	3.40
148.25	HIN3	0.0	1.40
147.50	ECO1	14.0	3.25
147.30	HIN3	0.0	0.95
145.85	CLA1	10.2	3.60
144.75	HIN3	0.0	2.55
143.99	HIN3	0.0	0.76
143.55	SAL1	8.0	7.21
143.25	ECO1	9.0	4.25
142.88	HIN3	0.0	1.11
141.10	ECO5	4.2	10.65
140.85	SAC2	0.0	166.00
140.55	ECO1	19.2	2.70
140.00	PST1	0.0	10.10

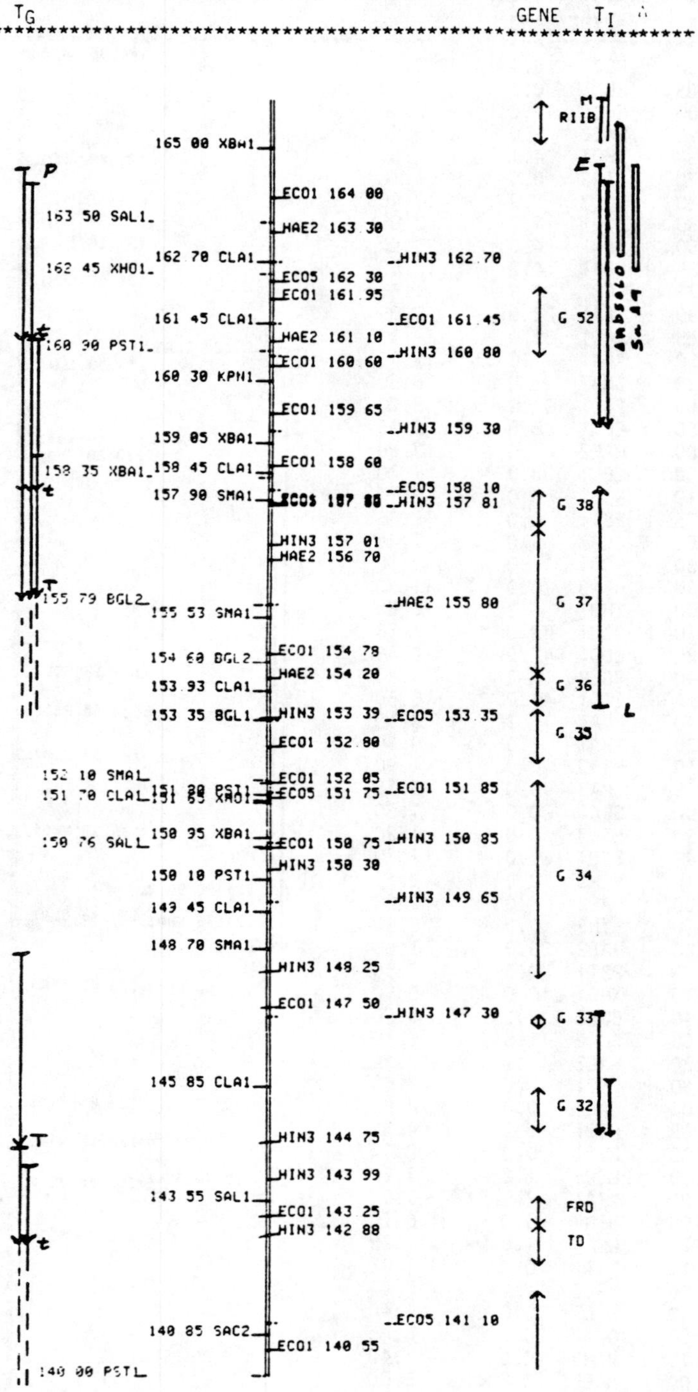

T_G

163 50 SAL1
162 45 XHO1
160 90 PST1
158 35 XBA1
155 79 BGL2
152 10 SMA1
151 70 CLA1
150 76 SAL1
140 00 PST1

165 00 XBA1
162 70 CLA1
161 45 CLA1
160 30 KPN1
159 05 XBA1
158 45 CLA1
157 90 SMA1
155 53 SMA1
154 60 BGL2
153 93 CLA1
153 35 BGL1
151 80 PST1
151 65 XHO1
150 95 XBA1
150 10 PST1
149 45 CLA1
148 70 SMA1
145 85 CLA1
143 55 SAL1
140 85 SAC2

ECO1 164 00
HAE2 163 30
ECO5 162 30
ECO1 161 95
HAE2 161 10
ECO1 160 60
ECO1 159 65
ECO1 158 60
ECO1 157 85
HIN3 157 01
HAE2 156 70
ECO1 154 78
HAE2 154 20
HIN3 153 39
ECO1 152 80
ECO1 152 05
ECO5 151 75
ECO1 150 75
HIN3 150 30
HIN3 148 25
ECO1 147 50
HIN3 144 75
HIN3 143 99
ECO1 143 25
HIN3 142 88
ECO1 140 55

HIN3 162 70
ECO1 161 45
HIN3 160 80
HIN3 159 30
ECO5 158 10
HIN3 157 81
HAE2 155 80
ECO5 153 35
ECO1 151 85
HIN3 150 85
HIN3 149 65
HIN3 147 30
ECO5 141 10

GENE T_I

RIIB
G 52
G 38
G 37
G 36
G 35
G 34
G 33
G 32
FRD
TD

least, to methylation of certain sites on dC-DNA but not HMdC-DNA; for example, the first *Taq*I site in the gene 37 sequence is apparently thus protected from cutting in T4 dC-DNA and in glucosylated HMdC-DNA but is fully cut when the DNA contains unglucosylated hydroxymethylcytosine. This site overlaps a *dam* methylation site (GATC). However, most *Taq* sites which he has examined are the same on the three DNAs. (His evidence also appears to indicate a normally protected *Xba* site at about 158.1 kb). In addition, homologous dC-, HMdC-, and glucosylated HMdC-containing fragments may migrate slightly differently on gels, with the extent of the difference affected by the percentage of G + C in the fragment. This probably is responsible for much of the difference seen in Fig. 2.

Mapping of T4 restriction targets has not been completed for *Taq*I or *Aha*III enzymes on either dC- or HMdC-DNA. However, A. Hack and W. Rüger have recently used Birnstiel analysis to chart tentatively most of the *Taq*I cuts in T4 dC-DNA (a total of almost 200); these data are available from either of the authors of this chapter. It should be noted that *Sal*I, *Xho*I, and *Cla*I sites represent subsets of *Taq*I sites. A virtually complete draft restriction map for *Eco*RV has been included in Tables 2 and 3.

COMPARISON WITH THE GENETIC MAP

Fig. 1 presents, in the traditional circular format, the genetic map of T4 correlated with the restriction map; for detailed references for each gene, see Wood and Revel (1976), Mosig (appendix, this volume) and appropriate chapters of this book. Thick bars indicate those genes which have been localized within a few hundred bp on the basis of cloning data; the appropriate references are listed in Table 4. These well-localized genes are also indicated on the right side of Table 2, along with bars indicating those regions which have been or are being sequenced (Seq?) and those transcripts which have been mapped (see below), all given relative to the restriction map.

As discussed by Doermann (this volume), a number of factors complicate the construction and interpretation of purely genetic maps of T4. T4 recombines very extensively; the work of Stahl et al. (1964) suggests that a mapping function giving perfect additivity would yield a total map of about 2,500 bp, equivalent to about 25 crossovers per progeny phage or 1% recombination per 70 bp, but there are clearly marked local variations in recombination frequency. Wood and Revel (1976) refined the relationship between physical and genetic distances by adjusting the map distances in a number of regions based on physical measurements of the sizes and locations of various deletions, as determined by heteroduplex mapping, and the known molecular weights of protein products. Mosig (1968; Mosig, appendix, this volume) developed a map of physical distances based on marker rescue with incomplete genomes, a technique indicating the approximate midpoints of transcription units; a further refinement by Childs (1971) allowed the relative positioning of markers within given genes. The correlation between these three adjusted maps and the one determined here by restriction mapping is generally good, seldom differing by more than 2 kb; the areas of local deviation are worth examining more closely for recombinational abnormalities.

TRANSCRIPTION PATTERNS

Promoters and Terminators Recognized by Unmodified *Escherichia coli* RNA Polymerase In Vitro

As discussed in detail by Brody et al. (this volume), prereplicative RNA synthesis is controlled in a complex and interesting fashion in T4-infected cells. The early promoters, recognized very soon after infection, are also recognized in vitro by the RNA polymerase from uninfected cells.

The availability of a detailed restriction map greatly facilitates the in vitro study of transcription of the T4 genome. For example, the RNA polymerase from uninfected *E. coli* was used by H. Gram, H. D. Liebig, A. Hack, and W. Rüger (manuscript in preparation) to transcribe both normal T4 DNA and T4 dC-DNA, and the ^{14}C-labeled transcripts were separated electrophoretically on polyacrylamide-agarose gels (Rüger, 1978; Summers et al., 1973). The main transcripts were eluted from the gel and hybridized to the restriction fragments generated with the endonucleases *Xba*I, *Hae*II, and *Eco*RI. Analysis of the resultant autoradiograms permitted determination of the approximate location of all major T4 in vitro transcripts larger than about 0.5 kb.

To localize these transcripts more precisely, sets of overlapping restriction fragments were isolated (Gram et al., manuscript in preparation) and transcribed, and the transcripts were analyzed on gels as above. Each template was then shortened in a defined manner with restriction enzymes, and the transcription and analysis processes were repeated. The chain lengths of the original transcript and the shorter runoff transcripts obtained in these experiments allowed precise localization of each promoter relative to a number of restriction sites and determination of the

[a] At the left, each cut site is indicated in kb from the *r*IIA-*r*IIB junction, along with the band number, where known, and band size of the fragment which begins at that cut site. An asterisk (*) indicates that the region immediately following has not been mapped for that enzyme; thus, it is not known whether there are additional cut sites between 47.20 and 57.90 kb or between 162.7 and 165.1 kb for *Hin*dIII. A plus sign (+) indicates some uncertainty in the site. In particular, the *Hae*II fragments shown on both sides of 79.7, 91.4, and 155.8 kb may be reversed, and there are some conflicting data in the *Hae*II map between 14.9 and 15.8 kb. Sites are indicated to the nearest 0.05 kb, except where sequencing or very detailed clone analyses permit them to be given to the nearest 0.01 kb. The right half of the map indicates cut sites proportional to map distances, along with a variety of genetic information: T_G, promoters and strong (T) and weak (t) terminators as observed by Gram et al. (in preparation) in vitro by the use of unmodified *E. coli* polymerase (see text). Gene, approximate extents and gene numbers for those genes whose map position is known from cloning work to within a few tenths of a kb (see Table 4 for references), along with the reported extents of in vivo transcripts (T_I). A few well-characterized deletions are indicated to the right under Δ by long, open boxes.

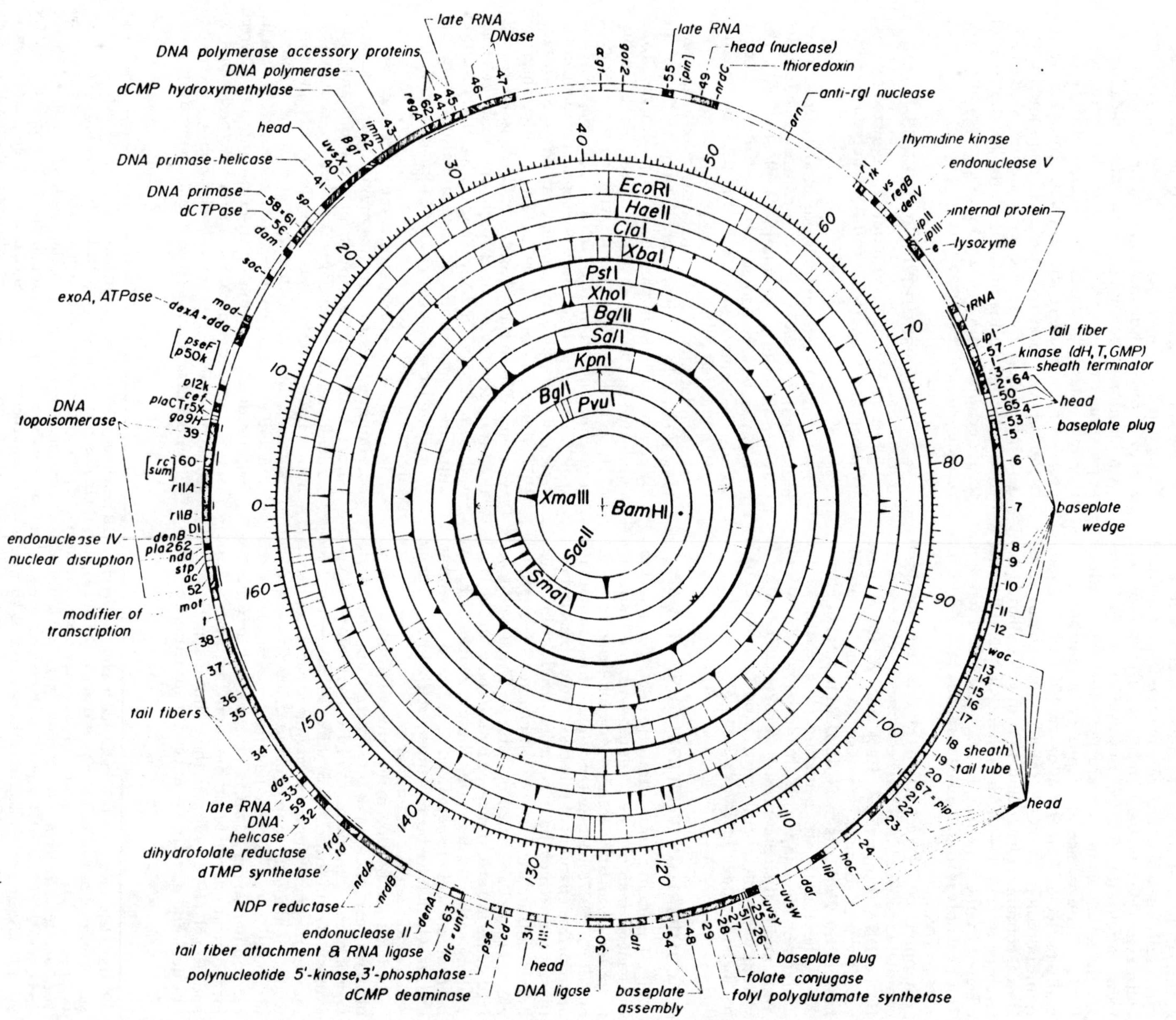

late RNA
DNase
DNA polymerase accessory proteins
DNA polymerase
dCMP hydroxymethylase
head
DNA primase-helicase
DNA primase
dCTPase
exoA, ATPase
DNA
topoisomerase
endonuclease IV
nuclear disruption
modifier of
transcription
tail fibers
late RNA
DNA
helicase
dihydrofolate reductase
dTMP synthetase
NDP reductase
endonuclease II
tail fiber attachment & RNA ligase
polynucleotide 5'-kinase,3'-phosphatase
dCMP deaminase
head
DNA ligase
baseplate
assembly
baseplate plug
folate conjugase
folyl polyglutamate synthetase
head
sheath
tail tube
baseplate
wedge
baseplate plug
head
tail fiber
kinase (dH,T,GMP)
sheath terminator
internal protein
lysozyme
endonuclease V
thymidine kinase
anti-rgl nuclease
late RNA
head (nuclease)
thioredoxin
EcoRI
HaeII
ClaI
XbaI
PstI
XhoI
BglII
SalI
KpnI
BglI
PvuI
XmaIII
BamHI
SacII
SmaI

◆ FIG. 1. Genetic map of bacteriophage T4, correlated with the restriction map. In those regions indicated by a heavy arc within the circle, restriction mapping data have been used to help locate the genes (see also Table 1). Throughout, more conventional genetics information and reported sizes of protein products have been used, where available (see Mosig, appendix, this volume). The thin lines just within the gene circle indicate those regions which have been or are being sequenced, as follows: 165.5 to 1.0 kb, Pribnow et al., 1981; A. Sugino, personal communication; 3 to 5 kb, W. Huang, personal communication; 15 to 17 kb, P. M. Macdonald and G. Mosig, manuscript in preparation; 18 to 23 kb, M. Nakanishi and B. Alberts, personal communication; 27 to 32 kb, Spicer and Konigsberg, this volume; 65 to 66 kb, J. E. Owen, D. W. Schultz, A. Taylor, and G. R. Smith, submitted for publication; 70 to 75 kb, J. Broida and J. Velten, personal communication; 74 to 76 kb, E. Goldberg, personal communication; 102 kb, Volker, Gafner, Bickle, and Showe, 1982; 103 to 107 kb, M. Parker and G. Doermann, personal communication; A. Christensen and E. T. Young, personal communication; 124 to 126 kb, V. Tanyashin, personal communication; 134 to 135 kb, S. Brown and W. B. Wood, personal communication; 143 to 144 kb, S. Purohit and C. Mathews, personal communication; 145 to 146 kb, Krisch and Allet, 1982; 154 to 158 kb, Oliver and Crowther, 1981; Wood and Crowther, this volume.

TABLE 3. Restriction fragments of bacteriophage T4, ordered by fragment size for each enzyme[a]

SITE	ENZYME/BAND		SIZE/KB
104.04	BGL1	1.0	49.31
0.02	BGL1	2.0	40.88
79.70	BGL1	3.0	23.08
58.70	BGL1	4.0	21.00
40.90	BGL1	5.0	17.80
153.35	BGL1	6.0	12.67
102.78	BGL1	7.0	1.26

SITE	ENZYME/BAND		SIZE/KB
39.40	BGL2	1.0	56.50
155.79	BGL2	2.0	18.01
137.20	BGL2	3.2	17.40
22.01	BGL2	3.1	17.39
7.80	BGL2	4.0	14.21
103.40	BGL2	5.0	10.60
129.10	BGL2	6.0	8.10
119.40	BGL2	7.2	5.50
114.00	BGL2	7.1	5.30
95.90	BGL2	8.1	4.20
124.90	BGL2	8.2	4.20
100.10	BGL2	9.0	3.30
154.60	BGL2	10.0	1.19
119.30	BGL2	11.0	0.10

SITE	ENZYME/BAND		SIZE/KB
10.90	ECO5	1.1	25.55
115.80	ECO5	1.2	23.20
42.50	ECO5	2.0	13.05
91.80	ECO5	3.0	12.41
56.00	ECO5	4.1	10.90
141.10	ECO5	4.2	10.65
104.21	ECO5	5.0	9.19
162.30	ECO5	6.0	7.60
86.20	ECO5	7.1	5.60
72.05	ECO5	7.2	5.35
77.40	ECO5	8.0	5.20
36.45	ECO5	9.0	4.69
66.90	ECO5	10.0	4.49
153.35	ECO5	11.0	4.45
158.10	ECO5	12.0	4.20
7.30	ECO5	13.0	3.60
82.60	ECO5	14.0	2.60
113.40	ECO5	15.0	2.40
139.00	ECO5	16.0	2.10
3.90	ECO5	17.0	2.00
151.75	ECO5	18.0	1.60
5.90	ECO5	19.0	1.40
85.20	ECO5	20.0	1.00
41.65	ECO5	21.0	0.85
71.39	ECO5	22.0	0.66
41.14	ECO5	23.0	0.51
55.55	ECO5	24.0	0.45
157.80	ECO5	25.0	0.30

SITE	ENZYME/BAND		SIZE/KB
33.00	CLA1	1.0	20.10
82.00	CLA1	2.0	16.30
10.70	CLA1	3.0	11.30
135.25	CLA1	4.2	10.60
59.70	CLA1	4.1	10.40
114.15	CLA1	4.3	10.25
70.10	CLA1	5.0	8.20
127.50	CLA1	6.0	7.40
98.30	CLA1	7.0	5.98
5.00	CLA1	8.1	5.70
22.85	CLA1	8.2	5.56
162.70	CLA1	9.3	4.70
106.75	CLA1	9.2	4.65
153.93	CLA1	9.1	4.52
78.30	CLA1	10.1	3.70
145.85	CLA1	10.2	3.60
1.40	CLA1	10.3	3.60
124.40	CLA1	11.1	3.10
158.45	CLA1	11.2	3.00
55.75	CLA1	12.1	2.95
28.41	CLA1	12.2	2.86
53.10	CLA1	13.0	2.65
149.45	CLA1	14.0	2.25
151.70	CLA1	15.2	2.23
111.40	CLA1	15.1	1.90
31.27	CLA1	16.0	1.73
105.48	CLA1	17.1	1.27
161.45	CLA1	17.2	1.25
104.44	CLA1	18.0	1.04
58.70	CLA1	19.0	1.00
22.00	CLA1	21.0	0.85
113.30	CLA1	20.0	0.85
134.90	CLA1	22.0	0.35
104.28	CLA1	23.0	0.16

SITE	ENZYME/BAND		SIZE/KB.
90.40	KPN1	1.0	46.10
47.20	KPN1	2.0	40.00
10.40	KPN1	3.0	36.80
136.50	KPN1	4.0	23.80
160.30	KPN1	5.0	14.50
87.20	KPN1	6.0	3.20
8.80	KPN1	7.0	1.60

SITE	ENZYME/BAND		SIZE/KB.
73.17	SAL1	1.0	38.68
111.85	SAL1	2.0	31.70
34.00	SAL1	3.0	23.90
163.50	SAL1	4.0	23.25
57.90	SAL1	5.0	15.27
150.76	SAL1	6.0	12.74
24.80	SAL1	7.0	9.20
143.55	SAL1	8.0	7.21
20.75	SAL1	9.0	4.05

SITE	ENZYME/BAND		SIZE/KB
157.90	SMA1	1.0	141.30
133.20	SMA1	2.0	15.50
152.10	SMA1	3.2	3.43
148.70	SMA1	3.1	3.40
155.53	SMA1	4.0	2.37

SITE	ENZYME/BAND		SIZE/KB
133.50	XBA1	1.0	17.45
51.50	XBA1	2.1	14.90
96.60	XBA1	2.2	14.25
165.00	XBA1	2.3	14.10
66.40	XBA1	3.0	13.10
22.37	XBA1	4.0	12.33
79.50	XBA1	5.0	11.00
124.30	XBA1	6.0	9.20
115.05	XBA1	7.0	7.95
150.95	XBA1	8.0	7.40
90.50	XBA1	9.1	6.10
159.05	XBA1	9.2	5.95
43.80	XBA1	10.0	5.30
17.90	XBA1	11.0	4.47
110.85	XBA1	12.0	4.20
34.70	XBA1	13.0	3.50
38.20	XBA1	14.0	2.90
13.10	XBA1	15.0	2.67
49.10	XBA1	16.0	2.40
15.77	XBA1	17.0	2.13
42.40	XBA1	18.0	1.40
41.10	XBA1	19.0	1.30
123.00	XBA1	20.0	1.30
158.35	XBA1	21.0	0.70

SITE	ENZYME/BAND		SIZE/KB.
1.10	XHO1	1.0	19.20
78.00	XHO1	2.0	17.00
48.90	XHO1	3.0	15.50
138.80	XHO1	4.1	12.85
126.00	XHO1	4.2	12.80
37.30	XHO1	5.1	11.60
108.70	XHO1	5.2	11.40
151.65	XHO1	6.0	10.80
68.50	XHO1	7.0	9.50
20.30	XHO1	8.0	8.90
29.20	XHO1	9.1	7.20
95.00	XHO1	9.2	6.90
101.90	XHO1	9.3	6.80
120.10	XHO1	10.0	5.90
162.45	XHO1	11.0	4.65
64.40	XHO1	12.0	4.10
36.40	XHO1	13.0	0.90

TABLE 3—*Continued*

SITE	ENZYME/BAND		SIZE/KB
45.25*	HIN3	0.0	13.25
126.00	HIN3	0.0	11.70
80.30	HIN3	0.0	11.00
69.50	HIN3	0.0	10.80
114.50	HIN3	0.0	6.70
137.70	HIN3	0.0	5.18
107.18	HIN3	0.0	4.47
39.35	HIN3	0.0	4.05
153.39	HIN3	0.0	3.62
65.52	HIN3	0.0	3.48
99.90	HIN3	0.0	3.21
5.80	HIN3	0.0	3.00
13.30	HIN3	0.0	2.89
162.70	HIN3	0.0	2.86
144.75	HIN3	0.0	2.55
150.85	HIN3	0.0	2.54
0.85	HIN3	0.0	2.45
23.05	HIN3	0.0	2.45
26.68	HIN3	0.0	2.22
36.10	HIN3	0.0	2.20
17.98	HIN3	0.0	2.08
28.90	HIN3	0.0	2.05
10.30	HIN3	0.0	2.05
34.10	HIN3	0.0	2.00
160.80	HIN3	0.0	1.90
60.10	HIN3	0.0	1.90
124.10	HIN3	0.0	1.90
43.40	HIN3	0.0	1.85
32.30	HIN3	0.0	1.80
16.19	HIN3	0.0	1.79
93.20	HIN3	0.0	1.70
58.50	HIN3	0.0	1.60
98.30	HIN3	0.0	1.60
3.30	HIN3	0.0	1.60
104.26	HIN3	0.0	1.52
8.80	HIN3	0.0	1.50
159.30	HIN3	0.0	1.50
157.81	HIN3	0.0	1.49
105.78	HIN3	0.0	1.40
148.25	HIN3	0.0	1.40
113.10	HIN3	0.0	1.40
30.95	HIN3	0.0	1.35
95.80	HIN3	0.0	1.30
97.10	HIN3	0.0	1.20
122.90	HIN3	0.0	1.20
25.50	HIN3	0.0	1.18
103.11	HIN3	0.0	1.15
142.88	HIN3	0.0	1.11
62.00	HIN3	0.0	1.10
20.60	HIN3	0.0	1.04
91.30	HIN3	0.0	1.00
12.35	HIN3	0.0	0.95
147.30	HIN3	0.0	0.95
92.30	HIN3	0.0	0.90
4.90	HIN3	0.0	0.90
94.90	HIN3	0.0	0.90
121.20	HIN3	0.0	0.90
165.56	HIN3	0.0	0.87
21.64	HIN3	0.0	0.81
63.10	HIN3	0.0	0.80
63.90	HIN3	0.0	0.80
122.10	HIN3	0.0	0.80
157.01	HIN3	0.0	0.80
143.99	HIN3	0.0	0.76
111.65	HIN3	0.0	0.75
149.65	HIN3	0.0	0.65
64.70	HIN3	0.0	0.60
112.40	HIN3	0.0	0.56
38.30	HIN3	0.0	0.55
150.30	HIN3	0.0	0.55
22.45	HIN3	0.0	0.55
20.06	HIN3	0.0	0.54
38.85	HIN3	0.0	0.50
69.00	HIN3	0.0	0.50
0.43	HIN3	0.0	0.42
65.30	HIN3	0.0	0.22
113.00	HIN3	0.0	0.10
23.00	HIN3	0.0	0.05
112.96	HIN3	0.0	0.04

SITE	ENZYME/BAND		SIZE/KB
42.20	ECO1	1.0	11.80
35.10	ECO1	2.0	6.20
111.60	ECO1	3.0	6.05
70.40	ECO1	5.1	5.61
90.10	ECO1	5.2	5.60
28.95	ECO1	4.0	5.55
16.24	ECO1	6.1	4.91
117.65	ECO1	6.2	4.80
76.01	ECO1	7.0	4.79
55.10	ECO1	8.0	4.70
164.00	ECO1	0.0	4.50
143.25	ECO1	9.0	4.25
65.90	ECO1	10.0	3.95
82.70	ECO1	11.0	3.85
102.10	ECO1	12.1	3.66
108.10	ECO1	12.2	3.50
98.70	ECO1	13.0	3.40
147.50	ECO1	14.0	3.25
21.15	ECO1	17.1	3.15
127.20	ECO1	15.0	3.10
154.78	ECO1	16.2	3.07
86.55	ECO1	16.1	3.05
136.80	ECO1	17.2	3.05
11.85	ECO1	18.0	2.90
2.85	ECO1	19.1	2.70
140.55	ECO1	19.2	2.70
105.76	ECO1	20.0	2.34
122.45	ECO1	22.0	2.30
131.65	ECO1	21.0	2.20
161.95	ECO1	0.0	2.05
62.20	ECO1	24.0	2.00
152.80	ECO1	23.0	1.98
80.80	ECO1	25.0	1.90
24.30	ECO1	26.1	1.80
59.80	ECO1	26.2	1.80
64.20	ECO1	27.1	1.70
135.15	ECO1	27.2	1.65
9.00	ECO1	28.0	1.60
5.55	ECO1	29.1	1.50
27.45	ECO1	30.0	1.50
96.60	ECO1	29.2	1.50
125.70	ECO1	29.3	1.50
26.10	ECO1	32.3	1.35
133.85	ECO1	31.0	1.30
7.75	ECO1	32.1	1.25
10.60	ECO1	32.2	1.25
130.30	ECO1	33.0	1.20
54.00	ECO1	0.0	1.10
150.75	ECO1	0.0	1.10
158.60	ECO1	0.0	1.05
159.65	ECO1	0.0	0.95
41.30	ECO1	0.0	0.90
95.70	ECO1	35.1	0.90
160.60	ECO1	0.0	0.85
14.75	ECO1	0.0	0.84
152.05	ECO1	0.0	0.75
157.85	ECO1	0.0	0.75
7.05	ECO1	0.0	0.70
139.85	ECO1	0.0	0.70
15.59	ECO1	38.0	0.65
34.50	ECO1	0.0	0.60
61.60	ECO1	0.0	0.60
98.10	ECO1	39.2	0.60
69.85	ECO1	0.0	0.55
89.60	ECO1	41.1	0.50
124.75	ECO1	41.2	0.50
161.45	ECO1	0.0	0.50
125.25	ECO1	42.0	0.45
2.50	ECO1	0.0	0.35
151.85	ECO1	0.0	0.20
131.50	ECO1	46.0	0.15

SITE	ENZYME/BAND		SIZE/KB
138.90	HAE2	1.0	15.30
101.00	HAE2	2.2	12.00
31.05	HAE2	2.1	11.95
53.30	HAE2	3.0	11.80
43.00	HAE2	4.0	10.30
122.00	HAE2	5.0	9.00
131.00	HAE2	6.0	7.90
18.68	HAE2	7.0	7.12
5.00	HAE2	8.0	5.80
25.80	HAE2	9.0	5.25
85.60	HAE2	10.1	5.00
113.00	HAE2	10.2	5.00
156.70	HAE2	11.0	4.40
0.80	HAE2	12.0	4.20
73.36	HAE2	14.1	3.54
163.30	HAE2	14.2	3.50
10.80	HAE2	15.0	3.30
68.10	HAE2	13.0	3.19
65.10	HAE2	16.2	3.00
79.70	HAE2	16.1	3.00
15.80	HAE2	17.1	2.88
91.40	HAE2	17.2	2.80
118.80	HAE2	18.1	2.20
161.10	HAE2	18.2	2.20
71.29	HAE2	19.1	2.07
83.60	HAE2	19.2	2.00
94.20	HAE2	19.3	2.00
76.90	HAE2	20.1	1.90
96.20	HAE2	20.2	1.90
98.10	HAE2	21.1	1.70
154.20	HAE2	21.2	1.60
99.80	HAE2	22.1	1.20
121.00	HAE2	23.0	1.00
14.10	HAE2	0.0	0.95
78.80	HAE2	0.0	0.90
155.80	HAE2	0.0	0.90
82.70	HAE2	0.0	0.90
90.60	HAE2	0.0	0.80
118.00	HAE2	0.0	0.80
15.05	HAE2	0.0	0.75

SITE	ENZYME/BAND		SIZE/KB
160.90	PST1	0.0	25.73
56.00	PST1	0.0	11.90
128.35	PST1	0.0	11.65
45.75	PST1	0.0	10.25
140.00	PST1	0.0	10.10
151.80	PST1	0.0	9.10
37.65	PST1	0.0	8.10
30.11	PST1	0.0	7.54
120.85	PST1	0.0	7.50
23.40	PST1	0.0	6.71
90.70	PST1	0.0	6.00
77.20	PST1	0.0	5.30
82.50	PST1	0.0	4.50
99.50	PST1	0.0	4.41
69.40	PST1	0.0	4.07
87.00	PST1	0.0	3.70
74.33	PST1	0.0	2.87
103.91	PST1	0.0	2.67
109.15	PST1	0.0	2.55
118.30	PST1	0.0	2.55
106.72	PST1	0.0	2.43
111.70	PST1	0.0	2.40
116.05	PST1	0.0	2.25
114.10	PST1	0.0	1.95
20.63	PST1	0.0	1.77
150.10	PST1	0.0	1.70
67.90	PST1	0.0	1.50
96.70	PST1	0.0	1.45
98.15	PST1	0.0	1.35
22.40	PST1	0.0	1.00
73.47	PST1	0.0	0.86
106.58	PST1	0.0	0.14

[a] In a few cases, the assigned band numbers are slightly out of order; it is not yet clear whether this is due to misassignment of band numbers or to small errors in cut-site positions. No band numbers are included for *Pst*I or *Hin*dIII because the sites were mapped by a modified Birnstiel technique rather than by direct correlation with band patterns.

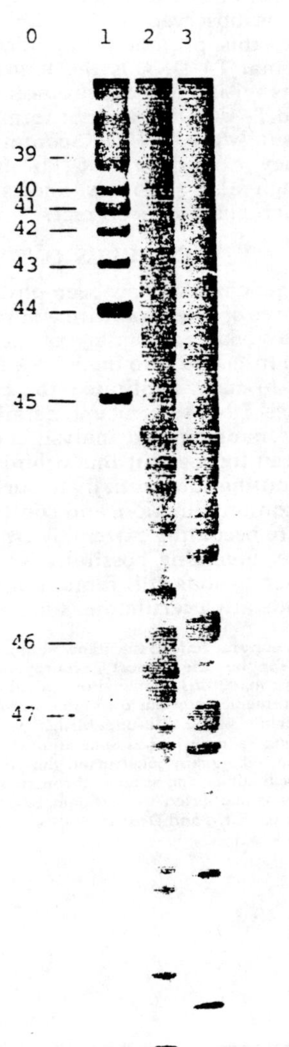

FIG. 2. Comparison of *Taq*I restriction patterns for small fragments from T4dC-DNA and normal glucosylated HMdC-DNA. The digestion was carried out for 48 h at 65°C, and the digest was run on a 6% polyacrylamide gel. Lane 0, number of *Eco*RI fragment; lane 1, T4dC-DNA {*alc*10, dCTPase, HMase, endoIV point mutant, endoII}, digested with *Eco*RI; lane 2, same DNA, digested with *Taq*I; lane 3, T4D⁻ DNA, digested with *Taq*I.

direction of transcription, as well as the positions of most of the rho-independent terminators (Table 2). A total of 31 strong promoter sites and 14 terminator sites have thus been mapped. Those in the tRNA and D regions seem to agree well with earlier reports (see Goldfarb, 1981a; Goldfarb and Burger, 1981), although most of their D-region transcripts appeared to read through the incomplete terminator indicated here (at about 161.1 kb) to one which they estimated at about 159.4 kb, which is not observed here; in contrast, additional terminators are observed here at 158.0 and 156.0 kb. The reasons for the differences are not clear; they may reflect, at least in part, the fact that apparent RNA size is being measured in one case and physical endpoints on the DNA in the other.

In addition to the strong transcripts mapped above, weak transcripts were observed when DNAs from the regions 43.8 to 49.3 kb (*Xba*I-10), 20.5 to 24.6 kb (*Sal*I-9), 120.1 to 126.0 kb (*Xho*I-10), and 165.0 to 7.8 kb (*Xba*I-*Bgl*II) were transcribed. They could not be mapped more precisely since the hybridization experiments were unsatisfactory with the small amount of labeled material available. The early regions at 26.4 to 35.5 kb (genes 43 to 47), 41.0 to 45.0 kb (probably near gene 49), and 134.4 to 136.8 kb (genes 63 and *denA*) are transcribed in vitro only weakly, if at all, by unmodified *E. coli* RNA polymerase.

No transcripts were found by these methods in the late regions at 73.0 to 103.4 kb or at 106.2 to 123.8 kb. Some readthrough is observed from the early region into the tail-fiber region. The only transcript initiated in the late regions under these experimental conditions is a 2.8-kb runoff transcript on *Bgl*II-5, hybridizing between 103.4 and 106.2 kb (genes 23 to 24). Transcription in this region is counterclockwise, as in all early regions. This transcript has also been observed in vivo (see Rabussay and Geiduschek, this volume).

In Vivo Promoters

A number of T4 early, middle, and late promoters have now been identified on the basis of S1 nuclease mapping and sequencing data. These are indicated, with appropriate references, on the map (Table 2) and are discussed in detail in the chapters in this book by Brody et al., Christensen and Young, Rabussay, Singer et al., Spicer and Konigsberg, and Wood and Crowther.

Discussion

Several observed features of the transcription pattern bear comment. (i) Throughout infection, overlapping transcripts are observed. (ii) Tandem promoters, first reported in the tRNA region by Goldfarb (1981a), seem to be a common feature of the T4 early regions. Six pairs of promoters observed here were only separated by 200 to 500 bp, and several others were fairly closely spaced. There are several possible explanations for this pattern. It may improve the efficiency with which T4 early genes are transcribed relative to host genes or those of coinfecting phage; this appears to be the function of the set of three tandem promoters that regulate T7 early transcription (Dunn and Studier, 1973; Minkley and Pribnow, 1973). Or it may be used to ensure that certain genes are transcribed in appropriate amounts, despite changes in the properties of the RNA polymerase. Goldfarb (1981b) suggests this possibility for the tRNA genes on the basis of the observation that one of the pair of promoters but not the other continues to be read by T4-modified *E. coli* polymerase. (iii) In most regions, promoters and terminators seem to be the same whether dC-containing or glucosylated HMdC-containing T4 DNA is used as the template. However, two of the regions are transcribed quite differently from dC- and HMdC-DNA.

TABLE 4. Well-characterized genetic loci[a]

Locus Start	End	Gene	References
*0.00	2.30	*rII*A	12, 19, 21
2.40	2.90	60	8
3.40	5.20	39	8, 27
10.45	11.00	*dda*	6
*15.03	15.27	*suc*	37
16.80	17.20	*dam*	36
25.60	26.20	42	32
*26.50	29.40	43	10, 22, 25
*29.45	29.75	*reg*A	10, 22
*29.81	30.37	62	10, 22
*30.37	31.33	44	10, 22
*31.38	32.06	45	10, 22, 25
32.50	34.40	46	10, 25
60.10	60.50	*tk*	14
*65.00	65.54	*ip*III	17, 23
*65.64	66.13	e	17
*69.95	70.30	tRNA	5, 26, 34
*70.99	71.56	tRNA	5, 26, 34
*72.63	72.91	*ip*I	34
*73.44	73.68	57	5, 26, 34
*73.68	74.41	1	26, 34
*74.50	75.10	3	26, 35
75.13	75.83	2764	35
77.60	78.25	53	32
78.25	79.25	5	32
79.25	81.50	6	32
*81.50	85.10	7	30, 32, 39
*85.10	86.20	8	30, 32, 29
86.20	87.00	9	32, 39
*88.90	89.60	11	32, 33
90.00	91.45	12	32
91.45	92.80	*wac*	32
92.80	93.65	13	32
93.65	94.40	14	32
94.40	95.20	15	32
95.70	97.50	17	32
97.50	99.35	18	32
99.35	99.90	19	32
99.90	101.55	20	32
*101.61	101.86	*pip*	28
102.10	102.80	21	2, 9
*102.80	103.63	22	2, 9, 18
*103.68	105.24	23	2, 9, 18
*106.10	107.40	24	2, 4, 9, 18
112.40	112.40	*uvsW*	3, 24
114.00	114.40	*uvsY*	3, 24
118.60	119.60	48	3, 24
*123.90	125.70	30	27, 31
134.00	135.10	63	1, 13
138.30	139.40	*nrd*B	7, 13
139.40	141.80	*nrd*A	7, 13
*142.30	143.08	*td*	7, 13, 38
*143.08	143.66	*frd*	7, 13, 38
*144.95	145.85	32	11, 13
147.05	147.35	33	11, 20
*148.10	152.10	34	20, 25
*152.45	153.56	35	16, 20
*153.63	154.30	36	16, 20
*154.31	157.38	37	16, 20
*157.41	158.16	38	16, 20
160.80	162.20	52	8, 12, 21
*165.10	166.00	*rII*B	12, 19, 21

[a] These genes have been mapped to within a few tenths of a kb by analysis of cloned fragments or by restriction analysis of strains carrying appropriate deletions. Those known most precisely are designated with an asterisk (*). References: (1) S. Brown and W. Wood, personal communication; (2) Christensen and Young, this volume; (3) J. DeVries and S. Wallace, submitted for publication; (4) Elliott and Geiduschek, personal communication; (5) Fukada et al., 1980; (6) Gauss et al., 1983; P. Gauss, personal communication; (7) Hänggi and Zachau, 1980; (8) W. M. Huang, personal communication; (9) Jacobs et al., 1981; (10) J. Karam, personal communication; (11) Krisch and Selzer, 1981; (12) Mattson et al., 1977; (13) Mileham et al., 1980; (14) A. Mileham, personal communication; (15) N. Murray, personal communication; (16) Oliver and Crowther, 1981; Wood and Crowther, this volume; (17) Owen et al., submitted for publication; (18) M. Parker and A. H. Doermann, personal communication; (19) Pribnow et al., 1981; (20) Revel, 1981; (21) Selzer et al., 1978; (22) Spicer and Konigsberg, this volume; (23) Stormo et al., 1982; (24) Takahashi and Saito, 1982b; (25) V. Tanyashin, personal communication; (26) Velten, 1980; (27) Velten and Abelson, 1980; (28) Volker et al., 1982; (29) Vorozheikina et al., 1980; (30) Wilson et al., 1980; (31) Wilson and Murray, 1979; (32) Wilson et al., 1977; (33) P. Berget, personal communication; (34) J. Broida, personal communication; (35) E. Goldberg, personal communication; (36) Hattman, this volume; Hattman, personal communication; (37) Ishii and Yanagida, 1977; P. MacDonald, E. Kutter, and G. Mosig, manuscript in preparation; (38) S. Purohuit and C. Mathews, personal communication; (39) Kuzmin, Tanyashin, and Baer, 1982.

First, when T4 dC-DNA is used as a template, a strong promoter is observed at 57.9 kb, making a 3.5-kb transcript; this promoter functions poorly in vitro when normal T4 DNA is the template. Second, the transcripts initiated at the promoter sites at 41.0 and 40.4 kb on T4 dC-DNA are not terminated at 38.9 kb, as they usually are on HMdC-containing T4 DNA, but rather they extend to 35.6 kb. Interestingly, this readthrough also probably affects gene *αgt*, if its genetic map position is correct.

CONCLUSIONS

The T4 genome has now been physically mapped to a high degree of precision with a number of restriction enzymes by taking advantage of special T4dC strains which can insert dC into their DNA in place of HMdC. This has, in turn, facilitated the clonal analysis of many genes, DNA sequencing, detailed promoter and messenger mapping, and analysis of origins of replication. As seen throughout this volume, these advances are contributing substantially to our understanding of T4 organization, function, and control. Some areas of the map are becoming extremely crowded (see Fig. 1), raising the intriguing possibility of overlap between genes; other regions still remain only poorly characterized, indicating fertile areas for future research.

We express our gratitude to the many members of the T4 community who have supplied the support, encouragement, and unpublished data that have made this compendium possible. Special thanks are due to certain members of our own laboratories: H. Gram, A. Hack, and H. D. Liebig, whose still-unpublished work is presented here; Steve Redmond-Payne, who has done all of the computer work; and Burt Guttman, who again constructed the circular map as well as providing much advice and support throughout.

The work was supported by National Science Foundation grant PCM-7905626 (to E.K.) and Deutsche Forschungsgemeinschaft grant RU123-13 (to W.R.).

Organization and Structure of Four T4 Genes Coding for DNA Replication Proteins

ELEANOR K. SPICER AND WILLIAM H. KONIGSBERG

Yale School of Medicine, New Haven, Connecticut 06510

Considerable progress has been made in the past few years towards understanding T4 DNA replication. This progress has been due to a large extent to the development of an in vitro T4 DNA replication system reconstituted from purified protein components (for review, see Nossal and Alberts, this volume). Using a reconstituted replication system, Alberts and coworkers have demonstrated that DNA synthesis with suitable templates can be carried out by the coordinated activity of seven proteins (Sinha et al., 1980). Five proteins are thought to be involved in leading-strand synthesis, namely the products of gene 43 (T4 DNA polymerase), gene 32 (helix-destabilizing protein), and genes 44, 45, and 62 (DNA polymerase accessory proteins), whereas lagging-strand synthesis requires, in addition, the products of genes 41 and 61 (RNA priming proteins) (Alberts et al., 1977; Nossal and Peterlin, 1979).

Apart from the protein product of gene 32 (gp32), very little is known about these replication proteins in terms of their chemistry and their mode of interaction with DNA and with each other in the replication complex. The primary structure of gp32 has been determined (Williams et al., 1980), and studies of its structure and function have progressed sufficiently for its functional domains to be described in some detail (Williams and Konigsberg, 1981; Williams and Konigsberg, this volume). To achieve this level of understanding with the other T4 replication proteins requires a corresponding amount of information about each. The products of genes 43, 44, 45, and 62 have been purified, facilitating efforts to define the functional activity of each in the replication complex and permitting protein chemistry studies of each (Barry et al., 1973; Nossal, 1979). The objective of current studies has been to determine the primary structure of the four proteins required in addition to gp32 for leading-strand synthesis (gp43, gp44, gp45, and gp62) by sequencing their respective genes. Knowledge of the primary structures of these proteins will form the necessary foundation for interpreting future chemical modification and intermolecular cross-linking studies aimed at locating protein-protein and protein-DNA contacts within the replication complex. Ultimately, one would like to know the three-dimensional structure of each protein, their topological arrangement at an advancing replication fork, and the functional contribution of each protein to DNA replication.

The basis of our current understanding of T4 DNA replication stems initially from genetic studies of the T4 replication genes. Mutations in genes coding for each of the five proteins required for leading-strand synthesis in vitro leads to inhibition of DNA synthesis in vivo, demonstrating the essential role that each protein plays in replication. Knowledge of the primary structures of these proteins will permit identification of the amino acid substitutions which abolish or alter the polypeptide functions. Correlation of these changes with information gained from chemical modification studies of the proteins should help in defining the functionally important regions of each protein. This approach will be especially important in structure and function studies of the multifunctional T4 DNA polymerase. Mutations have been isolated throughout gene 43 which in some cases increase, and in other cases decrease, the fidelity of DNA synthesis by gp43. By determining the amino acid substitutions of a number of such altered polymerases, one can begin to delineate the important functional domains of this complex polypeptide.

An additional objective of these studies of the DNA replication genes is to learn more about the mechanisms that control their expression. Genetic and biochemical studies have demonstrated that expression of the early genes is controlled in some cases at the transcriptional level and in other instances at the level of translation (for review, see Rabussay, 1982a). The structural studies of the four replication genes which are presented in this chapter should help to define some of the characteristics of these control mechanisms.

ORGANIZATION OF REPLICATION GENES

The genes coding for the replication proteins gp43, gp44, gp45, and gp62 map within a 6.5-kilobase (kb) region of the T4 genome (Fig. 1). Included in this region is the *regA* gene, sandwiched between genes 43 and 62. Genetic analysis has revealed the order of these five genes (for review, see Wood and Revel, 1976), and restriction analysis and cloning studies have further defined their location on the physical map of the T4 genome (see Kutter and Rüger, this volume). Also shown in Fig. 1 are the sizes of the protein products of these genes. The clustering of the four replication protein genes (g43, g44, g45, and g62 in Fig. 1) may facilitate a coordinate assembly of their gene products into a functional replication complex. Alternatively, the organization of these genes may be related to the mechanisms controlling their temporal expression. Whatever the reason, the close proximity of these genes has greatly facilitated their isolation as well as their subsequent cloning and sequence analysis.

To date, approximately 60% of the 6.5-kb region shown in Fig. 1 has been sequenced. The sequencing method employed involved purification of DNA restriction enzyme fragments, cloning of the fragments

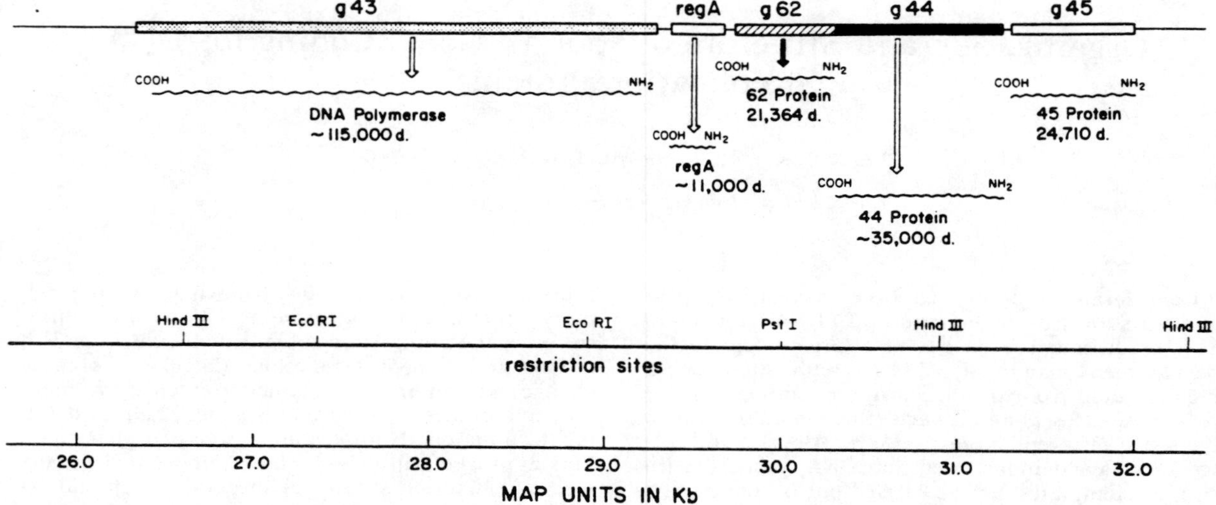

FIG. 1. DNA replication region of the T4 genome. The map units refer to the distance from the *r*IIA-*r*IIB junction in kb (Kutter and Rüger, this volume). The sizes of the protein products are from the following sources: gp44, gp45, and gp62 from the nucleotide sequence data presented here; *regA* protein from Cardillo et al. (1979); DNA polymerase from Goulian et al. (1968).

into M13 vectors (Messing et al., 1981) and subsequent sequence analysis by the method of Sanger, Nicklen, and Coulson (1977). In the case of genes 45, 44, and 62 and the 5′ end of gene 43, the T4 DNA fragments were obtained from a λ-T4 hybrid phage, λ806-17, constructed by Wilson et al. (1977). The remaining 2 kb, spanning almost all of gene 43, was isolated from chimeric plasmids of T4 DNA and pBH20 (plasmids pMU333 and pMU222) constructed by J. Karam (personal communication). Approximately 300 nucleotides which lie at the 3′ end of gene 43 have not as yet been cloned into plasmid or phage vectors. The strategy used in sequencing each of the four genes is briefly described below, along with a description of what has been learned about each of the protein products from these structural studies.

GENE 45

The product of gene 45 has two essential functions for the life cycle of bacteriophage T4. It is required for DNA replication and for expression of the late T4 genes (Epstein et al., 1963; Warner and Hobbs, 1967; Wu and Geiduschek, 1975). The precise role of gp45 in DNA replication is not entirely clear. The gp44, gp45, and gp 62 have been termed DNA polymerase accessory proteins (Piperno and Alberts, 1978). In vitro, a complex formed between gp44 and gp62 exhibits a DNA-dependent ATPase activity which is stimulated by gp45 (Piperno et al., 1978). This ATPase activity is required for the increased rate and processivity of polymerization observed upon the addition of the accessory proteins to an in vitro DNA replication assay (Huang et al., 1981; Piperno and Alberts, 1978). Although no association of gp45 with DNA alone or with T4 DNA polymerase has been demonstrated, gp45 does interact with single-stranded DNA-cellulose in the presence of both gp32 and the gp44-gp62 protein complex (D. Mace and B. M. Alberts, J. Mol. Biol., in press). Thus, the action of gp45 may be potentiated via protein-protein interactions rather than directly by gp45-DNA or by gp45-deoxynucleotide triphosphate interactions.

The exact role that gp45 plays in late mRNA synthesis is also not yet clear. There is evidence that in vitro gp45 can bind to T4-modified *Escherichia coli* RNA polymerase (Ratner, 1974a), but the effect of this association on promoter selection by RNA polymerase (RNAP) has not been well defined. Perhaps gp45 acts as an RNAP-modifying factor, switching the promoter sequences recognized by RNAP in a fashion analogous to that which operates in SPO1-infected cells (Losick and Pero, 1981). The observation that late transcription is dependent upon DNA replication and the fact that gp45 is found in both the DNA replication complex and in an RNAP complex suggests that gp45 may play a role in coordinating these two processes (Rabussay, 1982a).

The nucleotide sequence of gene 45 has been determined in its entirety (Spicer et al., 1982). Interpretation of the nucleotide sequence data was facilitated by the availability of purified gp45 (provided by N. Nossal, National Institutes of Health). The location of the structural gene for gp45 and determination of the correct reading frame of the gene was accomplished by correlating the DNA sequence information with extensive amino acid sequence data for gp45. Figure 2 presents the complete nucleotide sequence of gene 45 and the derived amino acid sequence of gp45. The sequence of the NH_2-terminal 24 amino acids of gp45 was determined by automated sequencing of the intact protein. In all, a total of 92% of the amino acid residues predicted from the DNA sequence was confirmed by either direct protein sequencing of tryptic peptides (indicated by solid underlining in Fig. 2) or by matching the predicted amino acid compositions with those determined for isolated tryptic peptides

```
ATG AAA CTG TCT AAA GAT ACT ACT GCT CTG CTT AAA AAT TTC GCT ACT ATT AAC TCT GGT ATT ATG CTT AAA TCC
MET-LYS-LEU-SER-LYS-ASP-THR-THR-ALA-LEU-LEU-LYS-ASN-PHE-ALA-THR-ILE-ASN-SER-GLY-ILE-MET-LEU-LYS-SER-

GGT CAA TTT ATT ATG ACT CGC GCA GTT AAT GGT ACA ACT TAT GCG GAA GCA AAT ATT TCT GAC GTT ATT GAT TTT
GLY-GLN-PHE-ILE-MET-THR-ARG-ALA-VAL-ASN-GLY-THR-THR-TYR-ALA-GLU-ALA-ASN-ILE-SER-ASP-VAL-ILE-ASP-PHE-

GAT GTA GCA ATT TAC GAT TTG AAC GGT TTT CTC GGT ATT CTG TCT TTA GTT AAT GAT GCA GAA ATT TCC CAG TCA
ASP-VAL-ALA-ILE-TYR-ASP-LEU-ASN-GLY-PHE-LEU-GLY-ILE-LEU-SER-LEU-VAL-ASN-ASP-ALA-GLU-ILE-SER-GLN-SER-

GAA GAT GGA AAT ATT AAA ATT GCT GAT GCC CGC TCA ACA ATT TTT TGG CGA GCA GCC GAT CCG AGT ACA GAT GTT
GLU-ASP-GLY-ASN-ILE-LYS-ILE-ALA-ASP-ALA-ARG-SER-THR-ILE-PHE-TRP-PRO-ALA-ALA-ASP-PRO-SER-THR-VAL-VAL-

GCT CCT AAT AAA CCA ATT CCA TTC CCG GTA GCA TCT GCT GTT ACT GAA ATT AAA GCT GAA GAC CTT CAA CAG CTG
ALA-PRO-ASN-LYS-PRO-ASN-PRO-PHE-PRO-VAL-ALA-SER-ALA-VAL-THR-GLU-ILE-LYS-ALA-GLU-ASP-LEU-GLN-GLN-LEU-

TTG CGT CTA TCT CGT GGT CTG CAA ATT GAT ACA ATT GCT ATC ACG GTA AAA GAA GGT AAA ATC GTA ATT AAC GGT
LEU-ARG-VAL-SER-ARG-GLY-LEU-GLN-ILE-ASP-THR-ILE-ALA-ILE-THR-VAL-LYS-GLU-GLY-LYS-ILE-VAL-ILE-ASN-GLY-

TTT AAT AAA GTA GAA GAT TCT GCT CTG ACC CGT GTT AAA TAT TCT TTG ACT CTT GGT GAT TAT GAT GGT GAA AAT
PHE-ASN-LYS-VAL-GLU-ASP-SER-ALA-LEU-THR-ARG-VAL-LYS-TYR-SER-LEU-THR-LEU-GLY-ASP-TYR-ASP-GLY-GLU-ASN-

ACA TTT AAT TTC ATT ATC AAT ATG GCA AAT ATG AAA ATG CAA CCA GGA AAT TAT AAA CTT CTG CTT TGG GCA AAA
THR-PHE-ASN-PHE-ILE-ILE-ASN-MET-ALA-ASN-MET-LYS-MET-GLN-PRO-GLY-ASN-TYR-LYS-LEU-LEU-LEU-TRP-ALA-LYS-

GGT AAA CAA GGT GCT GCT AAA TTT GAA GGT GAA CAC GCG AAT TAT GTG GTA GCT CTT GAA GCT GAT TCT ACC CAC
GLY-LYS-GLN-GLY-ALA-ALA-LYS-PHE-GLU-GLY-GLU-HIS-ALA-ASN-TYR-VAL-VAL-ALA-LEU-GLU-ALA-ASP-SER-THR-HIS-

GAT TTT TAA TAG
ASP-PHE-END-END-
```

FIG. 2. Complete nucleotide sequence of gene 45 and predicted sequence of gp45 (Spicer et al., 1982). Solid underlining indicates sequence confirmed by protein sequencing; broken underlining indicates tryptic peptides whose compositions match those predicted from the nucleotide sequence.

(indicated by broken underlining in Fig. 2). Gp45 is composed of 227 amino acids, with a calculated molecular weight of 24,710, in excellent agreement with the values of 24,500 and 27,000 which we and Barry et al. (1973), respectively, estimated from sodium dodecyl sulfate-polyacrylamide gel electrophoretic analysis.

The secondary structure of gp45, predicted by application of the rules of Chou and Fasman (1978b), is shown in Fig. 3. According to this analysis, gp45 should consist of 30% α helix, 25% β sheet, and 45% random coil conformations. The α-helical and β-sheet conformations are distributed throughout the structure, as are the 28 positively charged and 23 negatively charged amino acid residues.

GENES 44 AND 62

Like the gp45, the products of genes 44 and 62 play essential roles in the replication of T4 DNA in vivo (Epstein et al., 1963). The purification of gp44 and gp62 using complementation assays has permitted the study of the specific roles of these proteins in DNA synthesis in vitro. As noted earlier, the gp44-gp62 complex has a DNA-dependent nucleotidase activity which is specific for ATP or dATP and which is stimulated by gp45 (Alberts et al., 1977; Nossal, 1979. Piperno and Alberts, 1978). Addition of the accessory proteins to an in vitro synthesis reaction with primed DNA templates increases the rate and average length of the product made by DNA polymerase (Alberts et

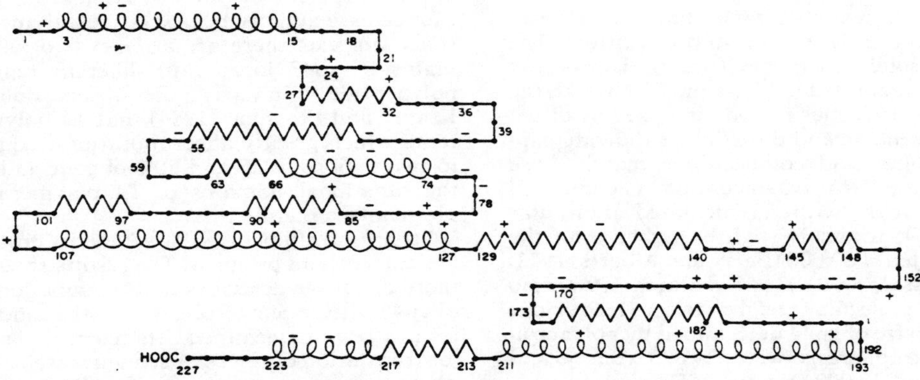

FIG. 3. Predicted secondary structure of gp45. Symbols: + and −, positively and negatively charged residues; ℓℓℓ, α helix; ∿∿, β pleated sheets; ●–●, random coil conformations.

al., 1977; Piperno and Alberts, 1978). In addition, the accessory proteins are required for strand displacement synthesis, which originates at single-stranded breaks in duplex DNA. Based on these observations, it has been suggested that the accessory proteins utilize ATP hydrolysis to form a specific protein complex with gp43 which facilitates the movement of polymerase along the template and decreases the rate of dissociation of polymerase from the template (Alberts et al., 1977). Furthermore, the accessory proteins may play a role in unwinding the helix or may enhance the action of gp32 in strand separation (Nossal and Peterlin, 1979). Finally, the formation of the accessory protein-polymerase complex may increase the fidelity of DNA replication since the addition of accessory proteins to T4 DNA polymerase increases its 3'→5' exonuclease degradation ratio fivefold (Alberts et al., 1980).

Gp44 and gp62 copurify from T4-infected cells as a tight complex with many subunits. The molar ratio of gp44 to gp62 in this complex appears to be variable; Barry and Alberts (1972) found the ratio to be 4:2, but Nossal (1979) purified a complex (used in these studies) with a ratio of 3.6:1. To obtain protein chemical data on the individual polypeptides, reduced carboxymethylated gp44-gp62 complex was applied to a Sephacryl S-300 column equilibrated with 6 M guanidine-hydrochloride. Under these conditions, the two proteins are fully dissociated and thus separate on the basis of their molecular weights. Both gp44 and gp62 have now been purified in sufficient quantity to permit tryptic and chymotryptic digestion to be carried out on each protein and to allow the corresponding peptides to be separated by high-pressure liquid chromatography. The amino acid compositions of many peptides from both gp44 and gp62 have been determined, and in some cases, the peptides have been subjected to direct protein sequencing by the solid-phase method.

The complete nucleotide sequence has been determined for genes 44 and 62. The combined coding region of genes 44 and 62 is 1,510 base pairs (bp). The protein chemistry data have been used to corroborate the nucleotide sequence data of the respective genes. Figures 4 and 5 present the complete nucleotide sequence of genes 44 and 62 and the amino acid sequence of gp44 (E. Spicer, K. R. Williams, N. Nossal, and W. H. Konigsberg, manuscript in preparation) and of gp62 (M. Quinones, K. R. Williams, N. Nossal, G. N. Godson, and E. K. Spicer, in preparation). The results of the protein chemistry studies of gp44 and gp62 are also summarized in Fig. 4 and 5, in which the solid underlining indicates regions analyzed by direct amino acid sequencing and dotted lines indicate peptides whose amino acid compositions match those predicted from the DNA sequence. Gp62 is composed of 187 amino acids with a calculated molecular weight of 21,364, which is slightly higher than the reported estimates of 20,000 (Barry and Alberts, 1972) and 17,000 (Nossal, 1979). Gp44 contains 319 amino acids and has a molecular weight of 35,731. Previous estimates of the size of gp44 determined by polyacrylamide gel electrophoresis were 34,000 (Barry and Alberts, 1972) and 32,000 (Nossal, 1979).

The ATPase activity of the DNA polymerase accessory proteins has been shown to be carried out by the gp44-gp62 complex; however, the question which of the two proteins binds ATP remains unanswered. Recently, Walker et al. (1982) reported amino acid homologies in six ATP-binding enzymes which they suggested might be related to the presence of a common adenine nucleotide binding fold in those proteins. Comparison of those conserved sequences with the sequence of gp44 reveals some similarities (K. Williams, personal communication). The sequence Gly-Thr-Gly-Lys-Thr (residues 53 to 57 in gp44) is found in the *E. coli* ATP synthase α subunit and is highly conserved in the five other proteins examined by Walker et al. In addition, there is a sequence common to ATP-binding proteins, Gly-X-X-X-hydrophobic-hydrophobic-hydrophobic-hydrophobic-Asp, which is present in gp44 (residues 99 through 107). The fact that these sequence homologies are present in gp44 but not in gp62 suggests that the ATPase activity is carried out by gp44.

GENE 43

The product of gene 43 is a DNA polymerase which is required for both initiation and maintenance of viral DNA replication (Epstein et al., 1963; Warner and Hobbs, 1967). T4 DNA polymerase resembles *E. coli* DNA polymerase I in its size (110,000 to 115,000 daltons) and its dual activities of 5'→3' polymerization and 3'→5' exonuclease digestion (Goulian et al., 1968). However, the two polymerases differ in two important respects. (i) T4 polymerase lacks a 5'→3' exonuclease activity, and (ii) *E. coli* polymerase I seems to function as a repair enzyme (Kelly et al., 1969), whereas T4 polymerase most likely is the sole viral replication enzyme.

The exonuclease activity of T4 polymerase appears to play a proofreading function during DNA synthesis by removing improperly base-paired nucleotides at the 3' end of a growing chain (Brutlag and Kornberg, 1972; Huang and Lehman, 1972). The isolation of gp43 mutants which possess greatly increased or decreased mutation frequencies led to the hypothesis that the fidelity of polymerization is, at least in part, a function of the relative levels of polymerase and proofreading activities (Bessman et al., 1974). The finding by Nossal and Hershfield (1971) that a gene 43-coded amber fragment of polymerase (*am*B22) retains its exonuclease activity but lacks the polymerization activity suggests there are at least two functional domains of gp43 located in different regions of the polypeptide. Similarly, the observation by Reha-Krantz and Bessman (1981) that T4 polymerase mutants which greatly affect mutation frequencies map in two regions at 25 and 80% of gene 43 implies that the functional domains of T4 polymerase may be physically quite distinct.

Structural studies of gene 43 and its protein product are currently in progress. The results should permit a more definitive description of the functional domains of gp43. At the 5' end of gene 43, the sequence of 450 bp has been determined. The correct reading frame for translation of the DNA sequence was evident from direct protein sequencing of the NH_2-terminal end of purified gp43 (K. Williams, personal communication). The DNA sequence of the 5' end of gene 43 and the

```
ATG ATT ACT GTA AAT GAA AAA GAA CAC ATT CTT GAA CAG AAA TAT CGT CCA TCT ACT ATC
Met-Ile-Thr-Val-Asn-Glu-Lys-Glu-His-Ile-Leu-Glu-Gln-Lys-Tyr-Arg-Pro-Ser-Thr-Ile-
                                                                             20
GAT GAA TGT ATT CTT CCC GCT TTT GAT AAA GAA ACC TTT AAA TCT ATT ACA AGT AAA GGT
Asp-Glu-Cys-Ile-Leu-Pro-Ala-Phe-Asp-Lys-Glu-Thr-Phe-Lys-Ser-Ile-Thr-Ser-Lys-Gly-
                                                                             40
AAG ATT CCA CAT ATT ATT CTT CAT TCT CCT TCT CCA GGA ACA GGT AAA ACA ACT GTA GCA
Lys-Ile-Pro-His-Ile-Ile-Leu-His-Ser-Pro-Ser-Pro-Gly-Thr-Gly-Lys-Thr-Thr-Val-Ala-
                                                                             60
AAA GCA TTA TGT CAT GAT GTA AAT GCT GAT ATG ATG TTT GTG AAT GGG TCA GAT TGT AAA
Lys-Ala-Leu-Cys-His-Asp-Val-Asn-Ala-Asp-Met-Met-Phe-Val-Asn-Gly-Ser-Asp-Cys-Lys-
                                                                             80
ATT GAT TTC GTT CGT GGT CCT TTG ACT AAT TTT GCC AGC GCC GCT TCA TTT GAT GGT CGT
Ile-Asp-Phe-Val-Arg-Gly-Pro-Leu-Thr-Asn-Phe-Ala-Ser-Ala-Ala-Ser-Phe-Asp-Gly-Arg-
                                                                            100
CAA AAA GTA ATC GTT ATT GAT GAA TTT GAC CGT TCA GGG TTA GCA GAG TCT CAG CGA CAT
Gln-Lys-Val-Ile-Val-Ile-Asp-Glu-Phe-Asp-Arg-Ser-Gly-Leu-Ala-Glu-Ser-Gln-Arg-His-
                                                                            120
CTT CGT TCC TTT ATG GAA GCT TAT AGT TCA AAC TCT AGT ATT ATT ATT ACT GCT AAT AAT
Leu-Arg-Ser-Phe-Met-Glu-Ala-Tyr-Ser-Ser-Asn-Cys-Ser-Ile-Ile-Ile-Thr-Ala-Asn-Asn-
                                                                            140
ATT GAT GGT ATT ATT AAA CCG CTT CAG TCA CGC TGC CGA GTT ATT ACA TTC GGT CAA CCA
Ile-Asp-Gly-Ile-Ile-Lys-Pro-Leu-Gln-Ser-Arg-Cys-Arg-Val-Ile-Thr-Phe-Gly-Gln-Pro-
                                                                            160
ACT GAT GAA GAT AAA ATT GAA ATG ATG AAG CAG ATG ATT CGT GGA TTG ACT GAA ATC TGC
Thr-Asp-Glu-Asp-Lys-Ile-Glu-Met-Met-Lys-Gln-Met-Ile-Arg-Arg-Leu-Thr-Glu-Ile-Cys-
                                                                            180
AAG CAT GAA GGA ATT GCT ATA GCT GAT ATG AAA GTT GTA GCA GCT TTG GTT AAA AAG AAT
Lys-His-Glu-Gly-Ile-Ala-Ile-Ala-Asp-Met-Lys-Val-Val-Ala-Ala-Leu-Val-Lys-Lys-Asn-
                                                                            200
TTT CCT GAT TTT CGT AAA ACT ATT GGC GAG CTC GAT ACT TAT TCG TCT AAA GGT GTT TTG
Phe-Pro-Asp-Phe-Arg-Lys-Thr-Ile-Gly-Glu-Leu-Asp-Ser-Tyr-Ser-Ser-Lys-Gly-Val-Leu-
                                                                            220
GAT GCT GGT ATT TTA TCA CTG GTT ACT AAC GAT CGT GGT GCT ATT GAT GAT GTT CTT GAG
Asp-Ala-Gly-Ile-Leu-Ser-Leu-Val-Thr-Asn-Asp-Arg-Gly-Ala-Ile-Asp-Asp-Val-Leu-Glu-
                                                                            240
TCT CTC AAA AAT AAA GAT GTT AAA CAA CTC AGA GCT TTA GCA CCA AAA TAT GCG GCT GAT
Ser-Leu-Lys-Asn-Lys-Asp-Val-Lys-Gln-Leu-Arg-Ala-Leu-Ala-Pro-Lys-Tyr-Ala-Ala-Asp-
                                                                            260
TAT TCG TGG TTC GTG GGT AAA CTT GCC GAA GAA ATC TAT TCA GCT GTA ACT CCA CAA AGT
Tyr-Ser-Trp-Phe-Val-Gly-Lys-Leu-Ala-Glu-Glu-Ile-Tyr-Ser-Arg-Val-Thr-Pro-Gln-Ser-
                                                                            280
ATT ATT CGT ATG TAC GAA ATT GTC GGC GAA AAT AAT CAG TAT CAT GGT ATT GCA GCT AAT
Ile-Ile-Arg-Met-Tyr-Glu-Ile-Val-Gly-Glu-Asn-Asn-Gln-Tyr-His-Gly-Ile-Ala-Ala-Asn-
                                                                            300
ACT GAA TTG CAT TTA GCT TAT CTT TTC ATT CAA TTA GCA TGC GAA ATG CAG TGG AAG TGA
Thr-Glu-Leu-His-Leu-Ala-Tyr-Leu-Phe-Ile-Gln-Leu-Ala-Cys-Glu-Met-Gln-Trp-Lys-COOH
                                                                         319
```

FIG. 4. Nucleotide sequence of gene 44 and predicted amino acid sequence of gp44. The sequences of numerous amino acid residues were confirmed by protein sequencing (solid underlining) or by matching tryptic peptide compositions (broken underlining).

predicted sequence of the NH_2-terminal 150 residues of gp43 are shown in Fig. 6, in which the underlined residues indicate amino acids confirmed by protein sequencing. An interesting feature of this sequence is the high content of aromatic residues (a total of 19 Phe + Tyr are predicted in the 150 residues), which is consistent with the relatively high concentration of these residues in the amino acid composition of the intact protein (unpublished observation).

The Klenow or large proteolytic fragment of *E. coli* polymerase I resembles T4 DNA polymerase in that it lacks the 5′→3′ exonuclease activity but possesses 5′→3′ synthetic and 3′→5′ exonucleolytic activities. However, a comparison between the 150 NH_2-terminal residues of gp43 and the amino acid sequence of Klenow polymerase I (Joyce et al., 1982) shows no amino acid homology.

CODON USAGE

Examination of the codon usage of a number of *E. coli* genes (see, for example, Post and Nomura, 1980), and ϕX174 genes (Sanger, Nicklen, and Coulson, 1977) demonstrates a pattern of nonrandom codon usage in these genes (for review, see Grantham et al., 1981). The nucleotide sequences of the four T4 replication genes also demonstrate a marked nonrandom usage of codons. Table 1 summarizes the codons used in genes 44, 45, and 62 and the coding region that has been sequenced in gene 43. For comparison, the codon

```
ATG AGC TTA TTT AAA GAT GAT ATT CAA TTA AAC GAG CAT CAA GTT GCT TGG TAT TCA AAA
Met-Ser-Leu-Phe-Lys-Asp-Asp-Ile-Gln-Leu-Asn-Glu-His-Gln-Val-Ala-Trp-Tyr-Ser-Lys-
                                                                             20
GAT TGG ACA GCT GTC CAA TCC GCT GCT GAT TCT TTT AAA GAA AAA GCA GAA AAT GAA TTT
Asp-Trp-Thr-Ala-Val-Gln-Ser-Ala-Ala-Asp-Ser-Phe-Lys-Glu-Lys-Ala-Glu-Asn-Glu-Phe-
                                                                             40
TTT GAA ATA ATT GGA GCT ATT AAT AAT AAA ACT AAA TGC TCT ATT GCT CAA AAA GAT TAT
Phe-Glu-Ile-Ile-Gly-Ala-Ile-Asn-Asn-Lys-Thr-Lys-Cys-Ser-Ile-Ala-Gln-Lys-Asp-Tyr-
                                                                             60
TCA AAA TTC ATG GTT GAA AAT GCA TTA TCA GAA TTT CCA GAG TGT ATG CCA GCT GTA TAT
Ser-Lys-Phe-Met-Val-Glu-Asn-Ala-Leu-Ser-Gln-Phe-Pro-Glu-Cys-Met-Pro-Ala-Val-Tyr-
                                                                             80
GCT ATG AAT TTA ATT GGA TCA GGC TTA AGT GAT GAA GCT CAT TTT AAT TAT CTA ATG GCT
Ala-Met-Asn-Leu-Ile-Gly-Ser-Gly-Leu-Ser-Asp-Glu-Ala-His-Phe-Asn-Tyr-Leu-Met-Ala-
                                                                            100
GCA GTT CCT CGT GGT AAA AGA TAT GGT AAA TGG GCA AAA CTG GTT GAA GAT TCC ACC GAA
Ala-Val-Pro-Arg-Gly-Lys-Arg-Tyr-Gly-Lys-Trp-Ala-Lys-Leu-Val-Glu-Asp-Ser-Thr-Glu-
                                                                            120
GTA TTG ATT ATT AAG TTA CTT GCT AAG CGG TAT CAA GTT AAT ACA AAT GAT GCA ATT AAC
Val-Leu-Ile-Ile-Lys-Leu-Leu-Ala-Lys-Arg-Tyr-Gln-Val-Asn-Thr-Asn-Asp-Ala-Ile-Asn-
                                                                            140
TAT AAA TCA ATT CTT ACT AAA AAT GGA AAA CTT CCT TTA GTA TTA AAA GAA CTA AAA GGT
Tyr-Lys-Ser-Ile-Leu-Thr-Lys-Asn-Gly-Lys-Leu-Pro-Leu-Val-Leu-Lys-Glu-Leu-Lys-Gly-
                                                                            160
TTA GTC ACG GAT GAT TTT TTG AAA GAA GTG ACT AAG AAC GTA AAA GAA CAG AAA CAA CTC
Leu-Val-Thr-Asp-Asp-Phe-Leu-Lys-Glu-Val-Thr-Lys-Asn-Val-Lys-Glu-Gln-Lys-Gln-Leu-
                                                                            180
AAA AAA CTA GCA TTG GAA TGG TAA
Lys-Lys-Leu-Ala-Leu-Glu-Trp COOH
                        187
```

FIG. 5. Nucleotide sequence of gene 62 and predicted amino acid sequence of gp62. Solid underlining shows residues confirmed by protein sequencing; broken underlining shows residues matched by tryptic peptide compositions.

```
ATG AAA GAA TTT TAT ATC TCT ATT GAA ACA GTC GGA AAT AAC ATT GTT GAA CGT TAT ATT
Met-Lys-Glu-Phe-Tyr-Ile-Ser-Ile-Glu-Thr-Val-Gly-Asn-Asn-Ile-Val-Glu-Arg-Tyr-Ile-
                                                                             20
GAT GAA AAT GGA AAG GAA CGT ACC CGT GAA GTA GAA TAT CTT CCA ACT ATG TTT AGG CAT
Asp-Glu-Asn-Gly-Lys-Glu-Arg-Thr-Arg-Glu-Val-Glu-Tyr-Leu-Pro-Thr-Met-Phe-Arg-His-
                                                                             40
TGT AAG GAA GAG TCA AAA TAC AAA GAC ATC TAT GGT AAA AAC TGC GCT CCT CAA AAA TTT
Cys-Lys-Glu-Glu-Ser-Lys-Tyr-Lys-Asp-Ile-Tyr-Gly-Lys-Asn-Cys-Ala-Pro-Gln-Lys-Phe-
                                                                             60
CCA TCA ATG AAA GAT GCT CGA GAT TGG ATG AAG CGA ATG GAA GAC ATC GGT CTC GAA GCT
Pro-Ser-Met-Lys-Asp-Ala-Arg-Asp-Trp-Met-Lys-Arg-Met-Glu-Asp-Ile-Gly-Leu-Glu-Ala-
                                                                             80
CTC GGT ATG AAC GAT TTT AAA CTC GTC TAT ATA AGT GAT ACA TAT GGT TCA GAA ATT GTT
Leu-Gly-Met-Asn-Asp-Phe-Lys-Leu-Ala-Tyr-Ile-Ser-Asp-Thr-Tyr-Gly-Ser-Glu-Ile-Val-
                                                                            100
TAT GAC CGA AAA TTT GTT CGT GTA GCT AAC TGT GAC ATT GAG GTT ACT GGT GAT AAA TTT
Tyr-Asp-Arg-Lys-Phe-Val-Arg-Val-Ala-Asn-Cys-Asp-Ile-Glu-Val-Thr-Gly-Asp-Lys-Phe-
                                                                            120
CCT GAC CCA ATG AAA GCA GAA TAT GAA ATT GAT GCT ATC ACT CAT TAC GAT TCA ATT GAC
Pro-Asp-Pro-Met-Lys-Ala-Glu-Tyr-Glu-Ile-Asp-Ala-Ile-Thr-His-Tyr-Asp-Ser-Ile-Asp-
                                                                            140
GAT CGT TTT TAT GTT TTC GAC CTT TTG AAT
Asp-Arg-Phe-Tyr-Val-Phe-Asp-Leu-Leu-Asn
                                    150
```

FIG. 6. NH_2-terminal sequence of DNA polymerase predicted from the nucleotide sequence of gene 43. The sequence of the first 22 amino acids was determined by direct protein sequencing of gp43.

TABLE 1. Use of codons in T4 replication genes

Codon	No. of codons in:						Codon	No. of codons in:					
	Gene 45	Gene 44	Gene 62	Gene 43[a]	Sum[b]	R-proteins[c]		Gene 45	Gene 44	Gene 62	Gene 43	Sum	R-proteins
Phe TTT	8	9	7	7	31	9	Tyr TAT	5	9	7	9	30	4
Phe TTC	3	4	1	1	9	17	Tyr TAC	1	1	0	2	4	13
*Leu TTA[d]	1	6	9	0	16	4	TAA	1	0	1	0	2	7
Leu TTG	3	5	3	1	12	3	TAG	1	0	0	0	1	1
Leu CTT	7	8	3	2	20	4	His CAT	0	7	2	2	11	4
Leu CTC	1	3	1	3	8	3	His CAC	2	1	0	0	3	9
Leu CTA	1	0	3	0	4	0	*Gln CAA	5	5	7	1	18	8
Leu CTG	7	2	1	0	10	67	*Gln CAG	2	6	1	0	9	26
Ile ATT	17	27	9	7	60	16	Asn AAT	13	11	9	3	36	4
Ile ATC	3	4	0	4	11	37	Asn AAC	3	2	3	4	12	32
*Ile ATA	0	1	1	1	3	1	Lys AAA	16	20	21	10	67	77
Met ATG	6	10	5	7	28	24	Lys AAG	0	5	3	3	11	28
Val GTT	6	9	5	5	25	47	Asp GAT	15	20	10	9	54	14
Val GTC	0	1	2	1	4	8	Asp GAC	2	1	0	7	10	31
Val GTA	6	6	4	2	18	42	Glu GAA	11	17	12	13	53	58
Val GTG	1	2	1	0	4	17	Glu GAG	0	3	2	2	7	15
Ser TCT	9	7	2	1	19	24	Cys TGT	0	4	1	2	7	1
Ser TCC	2	1	2	0	5	21	Cys TGC	0	3	1	1	5	6
Ser TCA	2	6	5	4	17	1	TGA	0	1	0	0	1	1
Ser TCG	0	2	0	0	2	0	Trp TGG	2	2	4	1	9	4
Pro CCT	1	3	2	2	8	4	Arg CGT	3	9	1	5	18	46
Pro CCC	0	1	0	0	1	1	Arg CGC	2	1	0	0	3	24
Pro CCA	3	6	2	3	14	6	Arg CGA	1	3	0	3	7	0
Pro CCG	2	1	0	0	3	32	Arg CGG	0	0	1	0	1	1
Thr ACT	7	11	3	3	24	31	Ser AGT	1	5	1	1	8	4
Thr ACC	2	1	1	1	5	21	Ser AGC	0	1	1	0	2	7
*Thr ACA	5	4	2	2	13	3	*Arg AGA	0	1	1	0	2	1
*Thr ACG	1	0	1	0	2	2	*Arg AGG	0	0	0	1	1	0
Ala GCT	12	14	11	6	43	75	Gly GGT	13	11	3	5	32	44
Ala GCC	2	3	0	0	5	13	Gly GGC	0	2	1	0	3	35
Ala GCA	8	8	6	1	23	42	Gly GGA	2	2	3	2	9	1
Ala GCG	2	1	0	0	3	27	Gly GGG	0	2	0	0	2	0

[a] First 450 nucleotides of gene 43.
[b] Sum of codons used in the four replication genes, including the partial sequence of gene 43.
[c] Post and Nomura, 1980.
[d] *, Codon recognized by T4 tRNA.

usage of a composite of ribosomal proteins (R-proteins) reported by Post and Nomura (1980) is given. In general, for the R-proteins, the codons which are used preferentially are those recognized by the most abundant isoaccepting species of tRNA. In many cases, the codons used by the four T4 replication genes appear to have the same preference. The use of ATT and ATC codons (recognized by abundant tRNAs) instead of ATA (recognized by a minor tRNA species) for Ile is a good example. Superimposed upon this bias, there appears to be a natural preference for codons using A and T residues, most likely owing to the AT-rich composition of T4 DNA. Both R-proteins and ϕX174 genes show a preference for T in the third codon. The T4 genes shown here exhibit an even higher bias for T: 48.3% of the codons terminate in T, compared with 42.9% termination in T for ϕX174 genes, 30% for R-proteins, and 39% for procaryotic phage genes in general (Kohli and Grosjean, 1981). In all, 78.3% of the T4 codons terminate in A or T. A particularly good example of this AT constraint is seen in the use of codons for Ser. The R-proteins show a bias for the abundant-tRNA codons TCT and TCC, whereas the T4 genes show a bias for TCT and TCA. The result is that translation of these T4 genes requires participation of the less abundant tRNA that recognizes TCA. This may be one of the controls used by T4 to regulate the amount of gene product formed.

An important factor in the consideration of translation efficiency and codon usage is the fact that T4 synthesizes eight of its own tRNAs (see Schmidt and Apirion, this volume). All of the T4 tRNAs have been sequenced (Barrell et al., 1974; Guthrie et al., 1978; Mazzara et al., 1977; Paddock and Abelson, 1975; Pinkerton et al., 1973; Seidman et al., 1974; Stahl et al., 1974), and it has been found that the T4 tRNAs recognize the codons least frequently used by *E. coli*. The codons known to be recognized by T4 tRNAs are indicated by an asterisk (*) in Table 1. Inspection of the codon usage of the replication genes indicates that the T4 tRNAs may not be used to a significant extent in the translation of these early genes. For example, the CGT and CGA codons recognized by *E. coli* $tRNA^{Arg}$ are used most often by the T4 replication genes, whereas the AGA and AGG codons recognized by T4 $tRNA^{Arg}$ (Mazzara et al., 1977) are used sparingly. Similarly, the codon AUA, thought to be recognized by T4 $tRNA^{Ile}$ (Fukada and Abelson, 1980), is used less frequently than the codons ATT and ATC, which are recognized by abundant *E. coli* tRNAs. Presumably the replication genes are transcribed at about the same time that the T4 tRNAs are being transcribed and processed (Fukada and Abelson, 1980), so that translation of the replication gene mRNAs must involve predominantly the host tRNAs.

REGULATION OF EXPRESSION OF REPLICATION GENES

The DNA replication genes belong to the early or prereplicative class of genes, which are expressed soon after phage infection. The initiation and duration of expression of these four protein products appears to be regulated by a variety of mechanisms which include control of transcription initiation and termination as well as translation initiation and termination (see Brody et al., this volume, and Wiberg and Karam, this volume).

Transcriptional Control

The mRNA hybridization studies of Guha et al. (1971) and Notani (1973) indicate that most T4 early mRNA is transcribed from the *l* strand of DNA, in a counterclockwise direction. The DNA sequence analysis of the replication genes confirm that all four genes are transcribed in that direction, as indicated in Fig. 1. Early studies on the locations of promoters suggested that genes 45, 44, and 62 were cotranscribed from a single promoter lying proximal to gene 45 (Hercules and Sauerbier, 1973; Stahl et al., 1970). However, the isolation by Bowles and Karam (1979) of a *cis*-acting mutation mapping between genes 44 and 45 which results in hyperproduction of gp44 and gp62 but does not affect gp45 production suggests there is a regulatory region in front of gene 44 (see also Karam et al., 1979). In addition, there is considerable evidence that gene 43 is transcribed from its own promoter into a monocistronic mRNA (Hercules and Sauerbier, 1973; Ito et al., 1978; Young and Menard, 1981).

Although these four prereplicative genes are transcribed by the host *E. coli* RNAP, it is uncertain whether any nucleotide sequences associated with promoter selection by RNAP should be expected to be present in the T4 promoters. These genes are transcribed by RNAP which has been modified by ADP-ribosylation of one or both α subunits (see Rabussay and Geiduschek, this volume). Modification of both α subunits has been shown to affect the promoter selection and the transcription efficiency of RNAP (Goldfarb and Palm, 1981); however, the effect of this modification on replication genes transcription is undefined. In addition, the product of the T4 *mot* gene influences the expression of these genes, controlling in a positive way expression of genes 45 and 43 and in a negative way expression of gene 44 (Brody, Rabussay, and Hall, this volume; Chace and Hall, 1975; Mattson et al., 1974; Mattson et al., 1978). Possible mechanisms of action of the *mot* protein include a direct effect on RNAP in its promoter recognition, a modulation of attenuation of transcription, or an effect on mRNA processing (Mattson et al., 1978). The studies of Linder and Sköld (1980) lend support to the hypothesis that the *mot* protein influences promoter selection. Additionally, a study of gene 43 transcription and translation in *mot*⁻ and *rho*⁻ mutants led Daegelen et al. (1982b) to conclude that *mot* has two activities. One influences promoter recognition by RNAP, and the other acts as a rho antagonist, allowing transcription of gene 43 to proceed through an attenuator or termination site. These hypotheses concerning promoter location and regulation are summarized in Fig. 7.

The level of T4 DNA polymerase production, in addition to being dependent upon the *mot* protein, is self-regulated, as demonstrated by the fact that T4 phage carrying mutations in gene 43 greatly overproduce a defective gp43 (Russell, 1973). This regulation occurs at the level of transcription since the overproduction of DNA polymerase in temperature-sensitive mutants of gene 43, which is observed upon a temperature shift up, is inhibited by rifampin (Krisch et al., 1977). The amount of gp43 produced is also influenced by mutations in genes 44 and 45, leading to the suggestion that autoregulation of gene 43 expression requires gp43 to be associated in a complex with at least gp44 and gp45 (Miller et al., 1981).

Since the DNA regions preceding genes 43, 44, and 45 have been sequenced, we can examine the potential regulatory regions for sequences which might be homologous to each other and which might have sequences in common with those found in *E. coli* promoters. The nucleotide sequences upstream from the four replication genes are shown in Figure 8, along with the nucleotide sequences associated with *E. coli* RNAP recognition (−35) and binding (−10) sites (Pribnow, 1979; Rosenberg and Court, 1979). Each of the four T4 genes contains sequences which are homologous to *E. coli* promoter sequences in the −10 region. The most striking homology is found in front of gene 45, where 10 out of 13 nucleotides are identical to a consensus sequence derived from the known sequence in the −10 region of 46 promoters (Rosenberg and Court, 1979). The regions 5′ proximal to genes 44 and 43 also contain homologous sequences, including perfect Pribnow box sequences, 5′-TATAAT-3′ (Pribnow, 1975). Gene 62 contains sequences homologous to the *E. coli* −10 region but is not generally thought to possess its own promoter (Hercules and Sauerbier, 1973; Karam et al., 1979). Thus, either the homology alone is not sufficient for RNAP binding, or gene 62 is transcribed from its own promoter (early in infection?) as well as being transcribed from an upstream promoter (later in infection), as suggested by Chace and Hall (1975). In addition to the homology to *E. coli* −10 regions, there is a sequence 5′-GGCTT-3′ (at various distances before the consensus sequence) which is present in genes 44 and 43, is partially repeated in gene 45, and is absent from gene 62.

In the −35 regions, there is a weak homology in gene 45 to conserved *E. coli* sequences, but no homology is found in genes 43 and 44. Thus, it may be that the −35 region does not play a crucial role in transcrip-

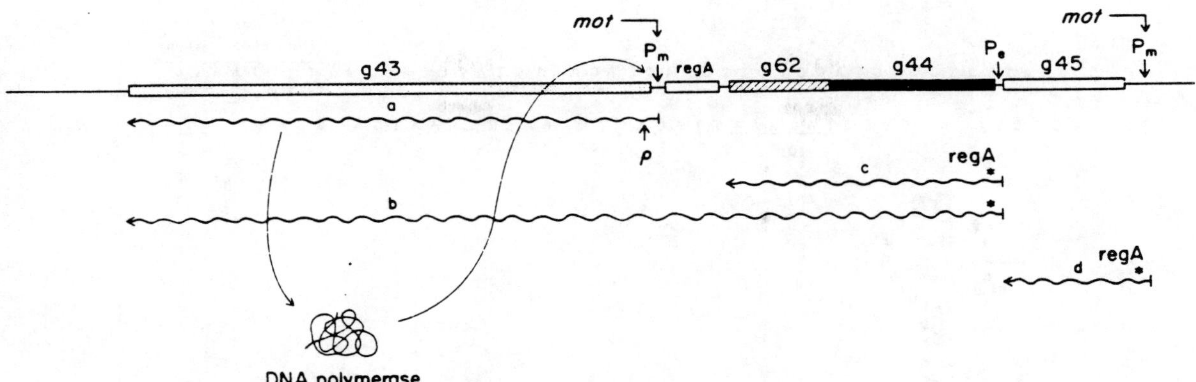

FIG. 7. Schematic summary of proposed mechanisms controlling replication genes expression. P_e indicates an immediate early promoter recognized by *E. coli* RNAP, which is independent of *mot*. P_m indicates a middle early promoter which is dependent upon *mot* for transcription. The wavy lines indicate length and direction of mRNA synthesis. Daegelen et al. (1982b) proposed that transcript *a* accounts for 75% of the cellular level of gene 43 mRNA and the *b* transcript contributes 25%. Transcripts *c* and *d* are suggested by the genetic studies of Bowles and Karam (1979). Translation of transcripts *c* and *d* may be controlled by the product of the *regA* gene (Karam and Bowles, 1974; Trimble and Maley, 1976; Wiberg et al., 1973). ρ indicates a potential site for rho-mediated termination which the *mot* protein may modulate (Daegelen et al., 1982b).

tion of these T4 early genes and that other sequences serve as recognition signals for RNAP for these middle-mode promoters (see also Brody et al., this volume). It is interesting to note that the sequence 5′-TATAAATACT-3′, which was found by Christensen and Young (1982a) to be present in the transcription initiation regions of two T4 late genes (genes 23 and 36), is not present in the regions preceding these four genes. Studies are in progress to learn whether or not these putative promoter regions of genes 43, 44, and 45 do in fact serve as transcription initiation sites. Using RNAP purified from uninfected *E. coli*, we have demonstrated (unpublished observations) that a restriction fragment carrying the putative gene 45 and gene 44 promoters is selectively retained on nitrocellulose filters by RNAP (Reznikoff, 1976). However, no strong promoters have been found in vitro in this region (see Kutter and Rüger, this volume).

The signals required for termination of transcription are not as thoroughly defined as those which function in initiation of transcription. However, in general, rho-independent terminator sequences have the following common features (Adhya and Gottesman, 1978; Platt, 1981; Rosenberg and Court, 1979). (i) The nucleotide sequence before the termination site is higher in G + C content than normal; (ii) an inverted repeat sequence occurs upstream from the termination site, allowing a stem-and-loop structure to form in the mRNA; and (iii) the terminated RNA transcript ends in a run of U residues. Based on these observations of known terminator sites, the sequence at the 3′ end of gene 45 could serve as a moderately

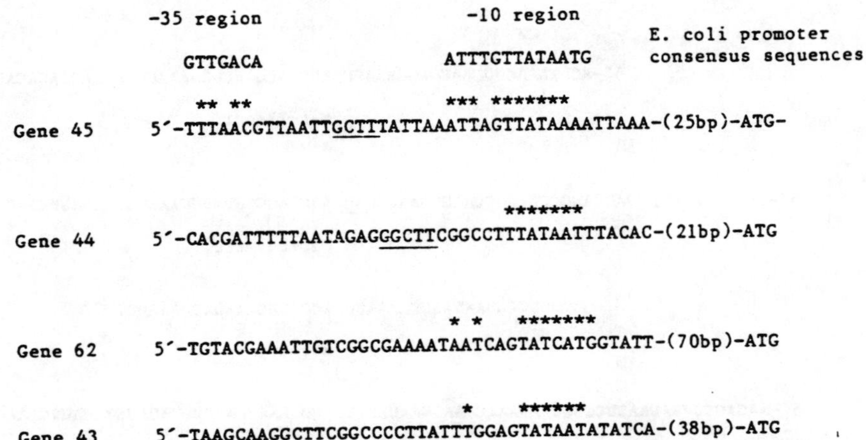

FIG. 8. Nucleotide sequences located 5′ proximal to the structural genes for the four replication proteins. The sequences shown are variable distances from the start of the structural genes (indicated in parentheses) and are aligned in the potential −10 regions. The asterisks (*) indicate homology with the *E. coli* promoters consensus sequences (Rosenberg and Court, 1979), given on the first line.

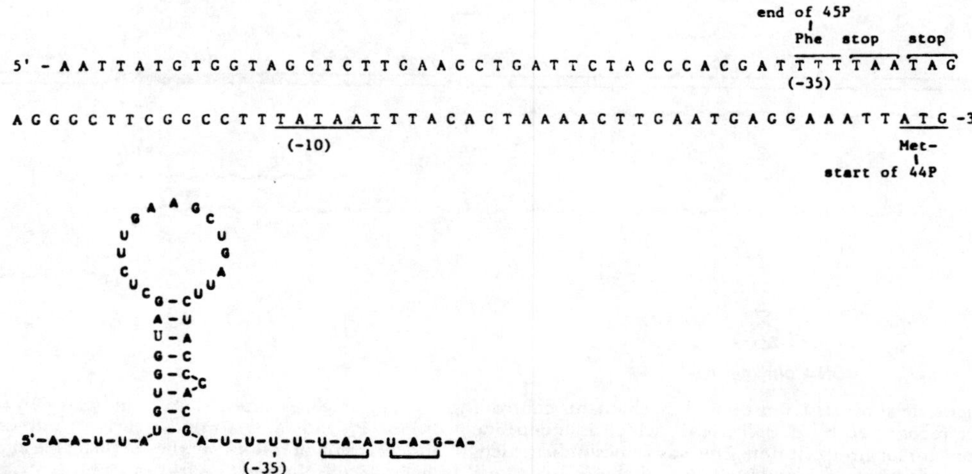

FIG. 9. Nucleotide sequence at the junction of genes 45 and 44. The first two lines give the sequence of the noncoding strand of gene 45 DNA, continuing past the end of the structural gene for gp45 into the proposed promoter for gene 44. The third line shows the potential secondary structure of the 3′ end of gene 45 mRNA. The tandem stop codons at the presumed end of the transcript are underlined. The COOH-terminal residue of gp45 is Phe, coded for by the terminal codon UUU.

effective transcription termination site (Spicer et al., 1982). As shown in Fig. 9 the mRNA from this region of gene 45 is rich in G + C residues, and it contains a short region of dyad symmetry which could enable a stem-and-loop structure to form. Following this putative structure is a run of 10 A + U residues, including 5 tandem U's. The potential Pribnow box (−10) sequence preceding gene 44 is located immediately adjacent to the potential gene 45 transcription termination region, as shown in Fig. 9. Thus, it is possible that the distal region of gene 45 contains overlapping promoter and terminator sequences. The nucleotide sequence 3′ distal to gene 62 does not exhibit any of the characteristics of a rho-independent transcription termination region. Thus, transcription in this case may proceed through gene *regA* and possibly gene 43, or transcription termination may be mediated by rho factor at the 3′ end of gene 62.

Translational Control

Although the expression of gene 43 appears to be controlled predominantly by factors which affect its transcription, the levels of expression of genes 44, 45, and 62 are subject to translational as well as transcriptional control. Mutants in the T4 *regA* gene fail to

gene 45	5′-AUCUCAUUUGAAUUGAAGGAAAUUAC AUG AAA CUGUCUAAAGAUACUACUGCUCUGCUUAAA-3′	Met-Lys-etc.
gene 44	5′-ACUAAAACUUGAAUGAGGAAAUU AUG AUU ACUGUAAAUGAAAAAGAACACAUUCUUGA-3′	Met-Ile-etc
gene 62	5′-UUUUCAUUCAAUUAGCAUGCGAAAUGCAGUGGAAGUGAU AUG AGC UUAUUUAAAGAUGAUAUUCAAUUAAACGAG	Met-Ser-etc.
rIIB	5'-GCGGCCUAAUAAGGAAAAUU AUG UAC AAUAUUAAAUGC-3'	Met-Tyr
gene 43	5′-AAGAGCCUAAUAACUCGGGCUAUAAACUAAGGAAUAUCU AUG AAA GAAUUUUAUAUCUCUAUUGAAACAGUCGGA	Met-Lys-etc.

FIG. 10. Nucleotide sequence of predicted translation initiation regions of T4 replication genes. The potential ribosome binding site (Shine and Dalgarno, 1975) on each mRNA is underlined. For comparison, the *r*IIB mRNA nucleotide sequence spanning the initiation codon is given (Karam et al., 1981).

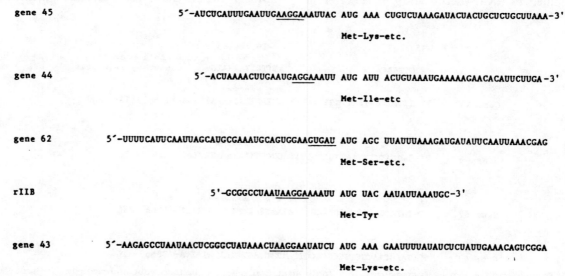

shut off the synthesis of a class of early T4 proteins (Karam and Bowles, 1973; Trimble and Maley, 1976. Wiberg et al., 1973), which leads to overproduction of gp44, gp62, and gp45 as well as of the products of genes 42, 56, *r*IIA, *r*IIB, and others (see Wiberg and Karam, this volume). Since the overexpression of these genes is observed in the presence of rifampin (Trimble and Maley, 1976), it seems likely that the product of *regA* exerts its effect by influencing the translational efficiency of the mRNA of these genes. The autoregulation of gene 32 is similar in that gp32 binds to gene 32 mRNA, blocking translation initiation under conditions of excess cellular concentrations of gp32 (Gold et al., 1976; Krisch et al., 1974; Krisch et al., 1977). In this case, gp32 has a specific affinity for its own mRNA, which may be related to the absence of secondary structure in its mRNA in the region of the translation start site (Lemaire et al., 1978) or to a preferential affinity for repeated AT-rich sequences flanking the ribosome binding site (Krisch and Allet, 1982).

The *regA* protein differs from gp32, however, in that it has a pleiotropic effect on a set of genes. Presumably, the mRNAs of these genes possess some common features which serve as a target for the *regA* protein. Karam and co-workers (Gold et al., 1981; Karam et al. 1981; Wiberg and Karam, this volume) have examined mutations in the *r*IIB gene which render it insensitive to *regA* control, and they have found that such mutations map within a short region overlapping the *r*IIB initiation codon. It was concluded from these studies that the *regA* protein affects the initiation of translation, perhaps by blocking ribosome binding or inhibiting ribosome movement. A possible mechanism would be for *regA* protein to bind to a specific nucleotide sequence, in the region of the ribosome binding site, on each of its target mRNAs. However, comparison of the nucleotide sequences of the translation initiation regions of genes 44, 45, 62, and *r*IIB, (Fig. 10) does not reveal a conserved nucleotide sequence. In addition, there is no predictable common secondary structure for these mRNAs. Thus, it appears that *regA* action does not involve recognition of a single specific nucleotide sequence. Clearly, direct experiments involving *regA* protein-mRNA interactions are needed to elucidate further the mechanism of action of this interesting translational regulator.

Mapping of Mutations and Construction of Multi-Mutant Genomes

A. H. DOERMANN

Department of Genetics, University of Washington, Seattle, Washington 98195

This chapter's on mutant construction
And mapping by UV destruction.
Just blast them to bits
With enough lethal hits,
And figure it out by deduction.

—Alan Christensen

With its relatively large genome (166,000 nucleotide pairs), T4 is less dependent on host genes for multiplication than simpler phages are (see also Binkowski and Simon, this volume; Drake and Ripley, this volume). By 1976, mutations had served to identify more than 135 genes in the genetic complement of T4 (Wood and Revel, 1976), and that number is still increasing slowly (for example, see Mosig, 1982; Völker et al., 1982a). Of the genes recognized in 1976, 65 are known to be indispensable for completion of the T4 growth cycle. This large component of identified essential genes provides a powerful advantage for biological and biochemical studies. It is, however, not always easy to exploit that advantage because of the high frequency of genetic recombination that characterizes T4 multiplication. It is sometimes difficult, for example, to locate the map position of a new mutation, particularly when no clue to its whereabouts is available at the outset. And in the face of high crossover frequencies, it frequently proves difficult to introduce an additional mutation into a genome already carrying mutations in several genes.

It is important to realize that the particles emerging from a bacterium infected with two T4 strains have not resulted from a single mating of two parental genomes. Rather, they are the offspring from perhaps 10 parental genomes multiplying in one cell and producing several hundred progeny phage particles. Within this mini-population, each genome that materializes in the progeny will have had multiple opportunities for genetic exchange (Visconti and Delbrück, 1953). To complicate the picture even further, the crossovers that occur in the construction of a progeny genome are not randomly distributed around the genetic map but are found in tight clusters. This clustering is usually called high negative interference (Chase and Doermann, 1958), being the opposite of what is experienced with eucaryotes, in which one crossover generally interferes with a second in the immediate vicinity. The highly nonrandom distribution of crossovers implies that a genetic map which reflects distances between mutations cannot be built up simply by progressively adding up adjacent recombination values regardless of their sizes. If that were done, the number of crossovers that occur in long intervals would be underestimated because the multiple crossovers resulting from clustering would be scored as having one crossover if the number is odd and none if it is even. Stahl et al. (1964) attempted to deal with the problem of determining the total map length. They constructed and compared eight mapping functions designed to convert recombination frequencies to map values. Although each of their functions still gave significant deviation from perfect additivity, their results suggest that a mapping function that gives perfect additivity would yield a total map of about 2,500 map units. A progeny T4 particle from a biparental cross would, on the average, have experienced about 25 crossovers throughout its genome based on those calculations. In any event, the number of crossovers per genome is not small.

Difficulties that arise in both mapping and construction of multimutant strains are, of course, to a large extent a function of the types of mutation under study. Those mutations for which highly restrictive plating conditions are known (suppressor sensitive, *r*II, some temperature sensitive, etc.) are generally easy to manipulate. Other mutations in essential genes, not so easily identified, are, however, coming more and more into use because qualitative changes in functional properties of proteins promise to elucidate more intimate details of the biological system in which these gene products play a role. In addition, much work today involves alterations in nonessential genes. Mutations in these categories may require special procedures for handling them genetically.

This chapter is designed to discuss techniques for dealing with problems of the types mentioned in the preceding paragraphs. As such, it does not concern itself with precise theoretical considerations but rather addresses two practical objectives. The first is to provide a set of directions with enough methodological detail to permit locating a particular mutation in the T4 genome with a reasonable degree of precision. Such methodology for producing adequate mapping information is a prerequisite for manipulating even the simplest multimarked genomes. Its availability makes it possible to achieve the second aim, which is to describe procedures for inserting an additional mutation into a genome while minimizing the likelihood of losing any of the desired markers already present. The UV rescue technique plays an important role in achieving both goals and will therefore be described in particular detail.

MAPPING A MUTATION FOR WHICH NO CLUE TO ITS LOCATION IS AVAILABLE

The procedure adopted for finding the general location of a new mutation will, of course, vary, depending on the phenotype that distinguishes it from its wild-type allele. When a conditional-lethal mutation is under consideration, the procedure simply involves crossing it to a variety of well-distributed amber or temperature-sensitive mutations and scoring the frequency of wild-type recombinants among the total progeny. One such procedure has been detailed by Edgar and Lielausis (1964), and logical variations of that protocol can adapt it to the requirements of a particular situation.

An alternative procedure that may, in some circumstances, be preferable would rely on a replica-plating technique, such as the method described by Doermann and Boehner (1970), which makes it feasible to analyze crosses that include multiple amber mutations. For that method, a separate test plate is made for each amber included in the cross. The basic ingredients of such a plate test are amber-restrictive plating bacteria, for example *Escherichia coli* B (*su*), and a tester phage strain that carries only the one amber mutation to be scored on that plate. When phage from a cross-progeny plaque are transferred (via sterile toothpick or a replicator device) to the test plate, two results are possible. The transfer spot will, upon incubation, become turbid if the transfer phage carries the same amber mutation as the tester phage. Alternatively, the spot will remain clear if it carries the am^- allele. (The tester phage provides the wild-type alleles of all other ambers in the cross.) The rest of the genotype of the phage that gave rise to that progeny plaque can be determined by transfers from the same plaques to other test plates. Although this approach may not always save time and work, it does have merit in specialized situations. Replica-plating methods can be used, for instance, when it is desirable to construct a genome that carries a number of amber mutations in addition to the unmapped mutation. It seems worthwhile to illustrate with an example how multimutant genome construction and mapping can both be served by the replication procedure.

The mutation to be mapped is a bypass-24 mutation. The original bypass-24 (*byp*24-1) mutation was discovered and characterized by McNicol et al. (1977), who showed that it can be identified by its property of rendering the normally essential T4 protein gp24 dispensable to the multiplication of the phage. In consequence of that fact, it can be distinguished from its wild-type allele in cross progeny as long as both parental genomes carry the same gene 24 amber mutation, in which case only the plaques of progeny carrying *byp*24 in their genomes will make clear spots in the gene 24 replica-plating test. A new *byp*24 mutation, *byp*24-14, was isolated in this laboratory and is used here to illustrate the utility of the replication procedure for mapping and strain construction.

T4 *am*E355(24) *byp*24-14 was crossed with T4 *am*B-82(*e*) *am*N132(8) *am*E355(24) *am*N54(31) *am*A453(32) *am*H17(52) *am*E51(56) *am*N130(46). The latter parent was chosen because the eight ambers in it are distributed around the genome about as uniformly as possible. (The fact that *am*E355 is present in both parents and simply serves as a recognition marker for *byp*24 leaves a doubly long interval between the gene 8 and gene 31 amber mutations, but that was nevertheless the most suitable stock available containing a gene 24 amber.) The cross procedure used followed the method of Chase and Doermann (1958), except that the amber-permissive strain *E. coli* CR63 (su^+) served as the host. Platings were also made against that bacterium. A total of 189 progeny plaques were tested for genotype by the procedure of Doermann and Boehner (1970). The segregants found are summarized according to genotype in Table 1.

The position of *byp*24-14 can be assigned to a segment bounded by two adjacent amber mutations from the data given in Table 1. A first step is simply to count the number of recombinations that occurred between the *byp*24-14 locus and each of the other marked locations. Table 2 shows that *byp*24-14 is closer to genes 8 (*am*N132) and 31 (*am*N54) than to any of the others. Because it shows 22% recombination with *am*N132(8) and 25% with *am*N54(31), it appears to be located about midway between those two markers.

The location can be verified by considering recombinations between *byp*24-14 and every pair of adjacent ambers used in the cross. The genetic map of T4 is circular (Streisinger et al., 1964; Wood and Revel, 1976), which implies that all recombinant genomes have resulted from an even number of crossovers. The lowest frequency of recombination of one marker with two others will then define the class that includes the nearest marker on the right of the target marker and the nearest marker on the left. That pair places the greatest restriction on the location of the two crossovers. The frequency of recombination with each pair of adjacent ambers has been collected from the data in Table 1 and is summarized in Table 3. It is seen that the lowest frequency of exchange of *byp*24-14 with two adjacent amber sites was observed with *am*N132(8) and *am*N54(31), confirming that the bypass mutation is located between those two markers.

To narrow the location of a mutation to one or several genes, a second cross was made, but with the second parent marked with several ambers distributed in the centrally located third of the segment bounded by genes 8 and 31, in addition to *am*E355(24). The data (not shown here) indicate that the mutation is in or near gene 23. A fair number of multiple-amber strains that could be useful in first- and second-stage mapping are available in the collection of the author and may be obtained on request.

The foregoing procedure for mapping can, if appropriately planned at the outset, be advantageous for the construction of multimarked strains. If, for instance, it is desirable to combine *byp*24-14 with amber mutations in various locations, the multimarked parent for the original cross can be chosen with that in mind. In the cross that was actually done, 50 of the possible segregant genotypes that include *byp*24-14 were encountered. By selecting two of those for a second cross, virtually any missing combination of markers could be readily constructed. For example, the most improbable recombinant that includes *byp*24-14 would be the one marked with the four ambers *am*N132(8), *am*N54(31), *am*H17(52), and *am*N130(46). That combination, which would have required a crossover in each of the eight marked intervals and which may be symbolized by + − + − + − + −, is not present among the first 189 segregants listed in Table 1. It would, however, be rather easily constructed by crossing two recombinants that did appear in that progeny, namely, (+ − + − + + + +) × (+ − − − + − + −). Here several combinations of only two crossovers are needed to produce the genotype desired, and replication tests of only three markers (*byp*24-14, *am*H17, and *am*N130) are necessary to identify it.

Methods described to this point can be used to locate the approximate map positions of most muta-

TABLE 1. Genotypes of 189 segregants from an eight-factor mapping cross to identify the T4 map region in which *byp*24-14 is located

	Gene[a]									Gene								
Class	*am*8-82(*e*)	*am*N132(8)	*byp*24-14	*am*N54(31)	*am*A453(32)	*am*H17(52)	*am*E51(56)	*am*N130(46)	No.	*am*8-82(*e*)	*am*N132(8)	*byp*24-14	*am*N54(31)	*am*A453(32)	*am*H17(52)	*am*E51(56)	*am*N130(46)	No.
Parental	+	+	+	+	+	+	+	+	17	−	−	−	−	−	−	−	−	25
Two-crossover	+	+	+	+	+	+	+	−	3	−	−	−	−	−	−	−	+	2
	+	+	+	+	+	+	−	+	4	−	−	−	−	−	−	+	−	2
	+	+	+	+	+	+	−	−	2	−	−	−	−	−	−	+	+	7
	+	+	+	+	+	−	+	+	4	−	−	−	−	−	+	−	−	1
	+	+	+	+	+	−	−	+	2	−	−	−	−	−	+	+	+	2
	+	+	+	+	+	−	−	−	7	−	−	−	−	+	−	−	−	4
	+	+	+	+	−	+	+	+	3	−	−	−	−	+	+	−	−	1
	+	+	+	+	−	−	−	+	3	−	−	−	−	+	+	+	+	1
	+	+	+	+	−	−	−	−	1	−	−	−	+	−	−	−	−	1
	+	+	+	−	−	−	+	+	1	−	−	−	+	+	−	−	−	2
	+	+	+	−	−	−	−	+	1	−	−	−	+	+	+	+	+	3
	+	+	+	−	−	−	−	−	2	−	−	+	−	−	−	−	−	2
	+	+	−	+	+	+	+	+	4	−	−	+	+	+	−	−	−	1
	+	+	−	−	+	+	+	+	1	−	−	+	+	+	+	−	−	2
	+	+	−	−	−	+	+	+	3	−	−	+	+	+	+	+	+	1
	+	+	−	−	−	−	+	+	1	−	+	−	−	−	−	−	−	1
	+	−	+	+	+	+	+	+	2	−	+	+	+	+	−	−	−	2
	+	−	−	+	+	+	+	+	1	−	+	+	+	+	+	−	−	1
										−	+	+	+	+	+	+	−	5
										−	+	+	+	+	+	+	+	3
Four-crossover	+	+	+	+	−	+	+	−	2	−	−	−	−	+	−	+	−	1
	+	+	+	+	−	+	−	+	1	−	−	−	+	−	−	−	+	1
	+	+	+	+	−	+	−	−	1	−	−	−	+	−	−	+	+	2
	+	+	+	−	+	−	+	+	2	−	−	−	+	+	−	+	−	1
	+	+	+	−	+	−	−	−	2	−	−	−	+	+	−	+	+	1
	+	+	−	+	+	+	−	−	1	−	−	+	−	−	−	−	+	1
	+	+	−	+	+	+	−	−	1	−	−	+	+	−	−	−	+	1
	+	+	−	−	+	−	+	+	1	−	−	+	+	−	+	−	−	1
	+	−	+	+	+	+	−	+	1	−	−	+	+	+	−	−	+	1
	+	−	+	+	+	+	−	−	1	−	−	+	+	+	−	+	−	1
	+	−	+	+	+	−	−	+	1	−	−	+	+	+	+	−	+	1
	+	−	+	−	+	+	+	+	1	−	+	−	−	−	−	+	−	1
	+	−	+	−	−	−	+	+	1	−	+	−	−	−	+	−	−	1
	+	−	−	+	+	+	−	+	1	−	+	−	−	−	+	+	−	1
	+	−	−	+	+	−	−	+	2	−	+	−	+	+	+	+	+	1
	+	−	−	+	−	−	−	−	1	−	+	+	−	−	−	+	−	2
	+	−	−	−	+	+	+	−	1	−	+	+	−	−	−	+	+	1
	+	−	−	−	+	+	−	−	1	−	+	+	−	−	+	−	−	1
	+	−	−	−	+	−	−	−	1	−	+	+	−	−	+	+	−	1
	+	−	−	−	−	+	+	−	1	−	+	+	+	−	−	−	+	1
										−	+	+	+	−	−	+	+	2
										−	+	+	+	+	−	+	+	1
										−	+	+	+	+	+	−	+	1
Six-crossover	+	+	−	+	−	+	−	+	2	−	−	+	−	−	+	−	+	1
	+	+	−	+	−	+	−	−	1	−	−	+	−	+	−	−	+	1
	+	−	−	−	+	−	+	−	1	−	+	+	−	+	−	+	+	1
										−	+	+	+	−	+	−	+	1

[a] Symbols: + denotes that the replication test spot was clear, indicating the presence of the am^{+} allele in the segregant phage being tested; − denotes that the test spot was turbid, indicating the amber allele; in the case of the *byp*24 test, the clear spot (+) indicates *byp*24 and the turbid spot (−) indicates $byp24^{+}$. Numbers in parentheses after the mutation symbols identify the gene in which the mutation is located.

tions that are readily identified. They are, however, not always adaptable to precision mapping, and mutations less easily recognized also present complications. The UV rescue procedure, which is discussed in the following section, frequently provides a solution to such problems.

UV RESCUE PROCEDURE

Cross-reactivation (marker rescue) has been defined as "rescue of genetic markers from inviable phage particles in mixed infection with viable particles" (Adams, 1959). It was first carried out by Luria (1947)

TABLE 2. Two-factor recombination frequencies used to locate the map position of *byp*24-14

Marker pair	Recombinants[a]	
	No.	%
byp and 8-82	62	33
byp and N132	42	22
byp and N54	47	25
byp and A453	61	32
byp and H17	74	39
byp and E51	86	46
byp and N130	76	40
8-82 and N132	45	24
N132 and N54	53	28
N54 and A453	45	24

[a] Data taken from Table 1.

and Luria and Dulbecco (1949) in experiments that complemented the original work on multiplicity reactivation of UV-inactivated phage. More extensive studies of marker rescue from UV-irradiated phage were made by Doermann et al. (1955) and Doermann (1961). A brief yet comprehensive review of both multiplicity reactivation and cross-reactivation may be found in the monograph of Harm on *Biological Effects of Ultraviolet Radiation* (1980). For present purposes, it seems necessary to discuss only the UV rescue experiment appropriate to mapping and multimutant construction, and it seems most useful to describe a practical procedure at the outset.

The UV rescue technique is based on the following principle. A readily selectable marker, for example an am^{+}, is used to permit recovering from irradiated genomes the segment that includes the marker. The recovered section can then be analyzed genetically for mutations corescued with the am^{+}. The size of the rescued segment can, of course, be influenced by altering the UV dose. Increasing the dose shortens the average length of the rescued segment; decreasing the dose makes it longer. UV rescue in genetic manipulations with T4 owes its practicality mainly to two features of that phage: (i) the high recombination frequency that characterizes T4 crosses, particularly the high negative interference (tendency for crossover clustering) that is observed over short genetic distances (Chase and Doermann, 1958); and (ii) the wide distribution of known conditional-lethal mutations in the T4 genome (Wood and Revel, 1976). This combination of characteristics allows high doses of radiation to be employed and still permits selective rescue of short genetic segments from the UV-irradiated genome. The overall plan is to make a mixed infection with several particles of each of two parental phage types. One, the helper phage, is not irradiated, but its genome carries a conditional-lethal mutation. The other, the donor phage, is irradiated. Its genome contains the wild-type site allele of the conditional-lethal mutation of the helper and is the selectable marker mentioned above. The infection is made in a host permissive for the helper, and the culture is incubated until lysis liberates the progeny. The lysate is plated on a host that is restrictive for the conditional-lethal genomes so that plaque formation is limited to those particles that have acquired from the donor genome the wild-type allele of the conditional-lethal mutation. When the irradiated phage has a recognizable mutation that is not restricted by the plating conditions, the frequency of its appearance among the plaque-former progeny will be directly related to its proximity to the location of the conditional-lethal mutation that has been chosen to mark the helper parent. A detailed protocol for a UV rescue experiment is given in the legend to Fig. 1, but many modifications can be introduced to fit the procedure to a particular objective.

Those who read the background literature will note that the earlier rescue experiments differ from the present experiments in two important respects. In the earlier experiments, the mixedly infected bacteria were infected with a *single* donor particle and plated *before* lysis because the objective was only to estimate the probability of rescue of a particular site or segment from the irradiated genome. In the present context, one aim is to obtain genetically homogeneous phage strains from rescue plaques. This is achieved by *postlysis* platings. The higher multiplicity of infection of the donor phage increases the likelihood of recovering a rescue phage at any particular UV dose, thus increasing the fraction, among the total phage plated, of phage that can form plaques on *E. coli* B (su^{-}). This allows higher doses of UV radiation to be employed.

Several factors are significant for predicting the most effective UV dose range for any particular rescue experiment. When, for example, corescue of two mutations, each identifiable by an efficient selective system, is called for, the genetic map distance separating them plays an important role. As the distance increases, the dose must be correspondingly reduced. That assumes even more importance if one donor mutation must be recognized by plaque morphology, which limits the number of rescue progeny that can be examined. The dose is even more restricted if the plaque of the rescued mutant cannot be distinguished from a plaque made by its wild-type counterpart and additional, possibly cumbersome, genetic tests are required to identify it. Thus, the ideal dose range cannot be prescribed a priori for every situation. Nevertheless, some guidelines can be obtained from the data given in Fig. 1. That figure represents the simplest kind of rescue experiment, namely an experiment with irradiated wild-type phage as the donor parent. Rescue crosses were carried out by using five different helpers at each dose. The graph shows the frequency of wild-type phage, normalized to zero UV dose, among all progeny phage produced in a particular rescue. The values are plotted against the UV dose.

TABLE 3. Recombination frequencies of *byp*24-14 with coupled pairs of adjacent ambers

Amber pair	No. of recombinants[a]
8-82 and N132	29
N132 and N54	17
N54 and A453	32
A453 and H17	35
H17 and E51	49
E51 and N130	54
N130 and 8-82	35

[a] Data taken from Table 1.

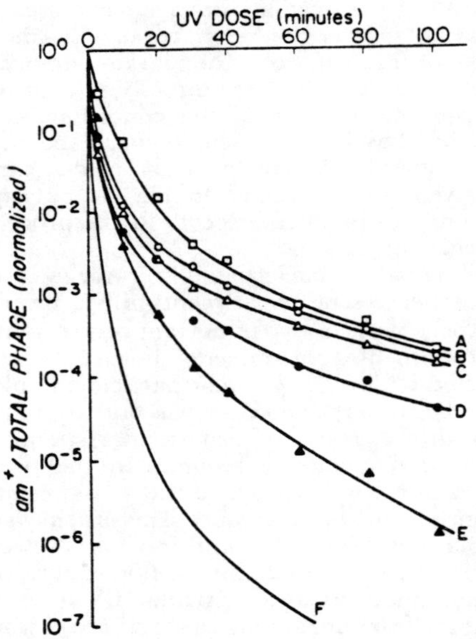

FIG. 1. Frequency of UV rescue as a function of dose. The experiment on which this figure is based is summarized in the following protocol, which may, with proper quantitative modifications, be used for any UV rescue crosses in which ambers are the conditional-lethal mutations in the helper phage. For this experiment, a suspension of wild-type T4 at a concentration of 2.4×10^9 particles per ml was irradiated in a UV-transparent medium (M9 medium lacking glucose [Adams, 1959]) by means of a low-pressure mercury vapor lamp. It should be pointed out that a single wild-type suspension was exposed to UV radiation, and samples were withdrawn at intervals. For any UV dose, a sample of the irradiated input phage was used with each helper, thus making all UV-irradiated input identical at a particular dose. Whenever the irradiated phage was mixed with host bacteria, yellow light was used to avoid photoreactivation. The number of PLH was calculated from the slope of the low-dose survival curve (survival greater than 5×10^{-4}, which avoids multiplicity reactivation on the assay plates) by target theory (Harm, 1980, chapter 4). The dose rate found was 3.4 PLH/min. For the rescue cross, exponentially dividing *E. coli* CR63 (su^+) in H-broth (Chase and Doermann, 1958) at 10^8 cells per ml were aerated at 30°C with a mixture of 5 helper phage and 2.4 donor phage per bacterium. After an adsorption period of 10 min, a 25-fold dilution was made, and aeration was continued for 70 min more, at which time chloroform was added to each culture to assure lysis of any remaining infectious centers. The titer of each lysate was determined by plating on strain CR63 for total phage and on strain B for am^+ phage. The fractions, B plaques/CR63 plaques were normalized by dividing them by the corresponding zero-dose ratios. The values obtained in that way have been plotted against UV dose in minutes. The helper phages used carried the following amber mutations: A, *am*A455(34); B, *am*B22(43); C, *am*B17(23); D, *am*B17(23) *am*B272(23); E, *am*B17(23) *am*E1236(23). F is the curve predicted for the simultaneous rescue of two am^+ sites that are widely separated genetically. It is assumed that they are rescued independently of each other and that each is rescued with the same dose dependence as *am*B17⁺.

Details of the experiment are given in the figure legend.

In analyzing the data, attention is first directed to Fig. 1, curves A, B, and C, in which the helper carried a single amber mutation in genes 34 (A), 43 (B), and 23 (C). These three ambers were selected for the differences they might be expected to display in rescue crosses. The gene 34 amber, *am*A455, is located in the region of gene 34 that exhibits an unusually high frequency of recombination for its physical length (Beckendorf and Wilson, 1972). Thus, rescue of that recombinogenic segment might be enhanced. Womack (1965) carried out a UV rescue experiment with 43 widely distributed ambers and two *r*II mutants using a single UV dose to the wild-type donor (51 phage-lethal hits [PLH]). She found that rescues at that dose varied among the markers over about a 10-fold range and that *am*A455⁺ was among the group more readily rescued. Her results showed that *am*B22(43)⁺ was about average, whereas *am*B17(23)⁺ was among the least frequently rescued loci. The results shown in Fig. 1 agree in general with the data of Womack, but in the present context, the differences become negligibly small at high UV doses. When doses lower than 100 PLH are contemplated, the region of the chromosome that is targeted can be taken into consideration by estimating the relative sensitivity from the paper of Womack.

The reason for including experiments with double-amber helpers in Fig. 1 is to provide at least some information about the effect of the distance separating two markers on the frequency of their corescue. When two markers are genetically far apart, they are rescued more or less independently of each other, and the simultaneous rescue frequency is approximately equal to the product of their individual rescue frequencies (Doermann et al., 1955). Curve F in Fig. 1 represents the result that is to be expected if the helper phage carries two amber mutations that are widely separated in the T4 genome. Curve D shows the frequency of corescue of two am^+ sites that are separated by only 201 nucleotide pairs (Fig. 2; see also M. L. Parker, A. Christensen, E. T. Young, and A. H. Doermann, manuscript in preparation). It is seen that at the 60-min UV dose, for example, the rescue frequency of *am*B17⁺ alone (curve C) was about 3.6×10^{-4}, whereas corescue of *am*B17⁺ and *am*B272⁺ was about 1.2×10^{-4}. Among the genomes that rescued *am*B17⁺, about one-third also rescued a site located 201 nucleotides away. The *am*E1236⁺ marker is separated from *am*B17⁺ by 1,221 nucleotide pairs and was corescued only 4% of the time after the 60-min UV dose, but still about 100 times more frequently than an independently rescued marker would be. Similar reasoning can be applied to mutations which display selectable or nonselectable phenotypes, thereby predicting a dose at which they would be corescued with an efficiency suitable for the objective of a particular experiment.

FINE-STRUCTURE MAPPING

Undoubtedly, the analysis of the *r*IIA and *r*IIB genes of T4 made by Benzer (1959 and 1961) is, and will long remain, the most thorough fine-structure genetic study. Within those two genes, Benzer identified 80 segments that could be distinguished by deletions, and in the 80 segments he mapped 311 nonallelic point mutations of T4B (Benzer, 1962). A combination of several favorable characteristics of *r*II mutants

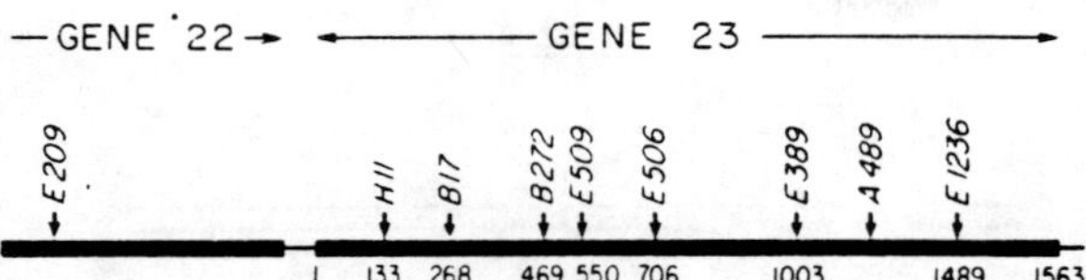

FIG. 2. Map of the region from gene 22 to gene 23. The amber mutations are located in the order determined by Celis et al. (1973) and Sarabhai et al. (1964). The *am*E1236 location has been added by us. The order has been completely verified with three-point tests by Sarabhai et al. (1964) and in my laboratory. The mutation *am*E509 has been shown to be at the same site as *am*H32 in the Celis map and *am*E389 at the same site as *am*H36. The numbers under the map indicate the nucleotide position in gene 23 of the first nucleotide of the corresponding amber codon (Parker et al., manuscript in preparation). The nucleotide location of *am*A489 has been determined since this manuscript was submitted. That amber codon begins at nucleotide 1192.

underlies the Benzer achievement: the mutant plaques are clearly distinguishable from wild-type plaques on the permissive host, *E. coli* B; *r*II mutants are unable to multiply on lambda lysogens on which the wild-type and other categories of *r* mutant do multiply efficiently; and the *r*II gene products are completely dispensable in the permissive host. These properties are shared by some other T4 genes, such as the lysozyme gene *e* (Streisinger et al., 1961; Wood and Revel, 1976), in which deletions, ambers, and plaque morphology mutations have also been described. Nevertheless, no other gene compares with the *r*II genes in the efficiency with which the system can be genetically analyzed. In fact, the vast majority of T4 genes in which conditional-lethal mutations are known are unconditionally lethal if deleted, thus eliminating deletion mapping for studying that class of genes.

To do intragenic mapping when deletion mapping is not available, the simplest procedure, when it can be used, is to make two-factor crosses and to establish from the recombination values a map based on the principle of linearity of marker distribution. It should, however, be noted that the additivity of the short genetic intervals involved in intragenic mapping will be poor owing to the high negative interference that is observed over such genetic segments (Chase and Doermann, 1958). Such a map of the amber mutations in gene 23 was made by Sarabhai et al. (1964) and extended by Celis et al. (1973). They assigned positions to 13 amber mutations on the basis of two-factor crosses in which they could score the wild-type recombinant by platings on the amber-restrictive host *E. coli* B (su^-). In addition, they verified most of the order with more definitive three-factor crosses.

It is not, however, always possible to use methods that selectively identify a recombinant genotype. The *bvp*24-14 mutation can again serve as an example. As discussed above, it could readily be mapped to gene 23 or its vicinity by the replica-plating technique. Because of the phenotypic nature of that mutation, it could not be located more precisely by selective plating procedures, and replica-plating methods would have required testing unrealistic numbers of progeny plaques. The solution was to employ the UV rescue procedure after irradiating *bvp*24-14 *am*E355(24) *am*E303(24). For mapping this mutation, paired rescue crosses employing two categories of helper are needed. The first helper has two (the experiment could be done with only one) ambers in the general region in which *bvp*24-14 is expected to map. Crossing it with the irradiated donor estimates the simultaneous rescue frequency of the am^- sites at the selected UV dose, regardless of whether the bypass mutation is corescued. In a second cross, the helper carries the same ambers plus one or more additional ambers in gene 24 (preferably the same amber[s] present in the irradiated *bvp*24-14 stock). For progeny phage to make plaques on *E. coli* B (su^-) when the latter helper is used, it is necessary for the bypass mutation to be corescued from the irradiated donor together with the am^- alleles. The quotient of the frequency in the second cross divided by the frequency in the first gives the corescue frequency for any particular UV dose. If the bypass mutation is near the segment rescued by the helper in the first cross, the quotient will be high, and it will drop off as the distance between the two increases.

That type of experiment (Fig. 3) was carried out with *bvp*24-14 and another mutant, *bvp*24-4. A set of seven helper pairs was used. A map of the ambers used in this mapping experiment is given in Fig. 2. Each pair had two ambers in the gene 22 to 23 segment of the T4 genome, with one member of the pair also carrying two ambers in gene 24. The rescue crosses are referred to as the two-amber cross (no ambers in gene 24) and the four-amber cross (with two ambers in gene 24). Table 4 summarizes the results, and additional technical details are given in footnote *a* to that table. It is clear that *bvp*24-4 maps very close to *am*E389 in gene 23 since it is almost always corescued with *am*E389$^-$ (the four-amber/two-amber quotient is 1.01 for the *am*E506-*am*E389 double and 0.98 for the *am*E389-*am*A489 double). Corescue with markers to either side is significantly less frequent.

The mapping data (Table 4) for *bvp*24-14 indicate a type of complicating factor that may be encountered in UV rescue analysis. The map position of that mutation is not as clearly defined, in part because maximum four-amber/two-amber quotients fail to reach unity. The reason for this is unclear. It appears to be related to the fact that the bypass-24 phenotype of this mutation is cold sensitive. The efficiency of plating of *bvp*24-14 *am*E355(24) *am*E303(24) is less than 0.01 when plates are incubated at 30°C. In contrast, the bypass-24 phenotype of *bvp*24-4 is not cold sensitive, and four-amber/two-amber quotients equal to 1 are found with that mutation. The relationship with cold sensitivity is supported by the observation that platings made at 37°C gave rescue efficiencies lower by at least a factor of 2 than those given in Table 4, which, for bypass 24-14, were incubated at 41°C. It is, however, not simply a question of relative plating efficiencies at the two temperatures, because the efficiency of plating of the unirradiated stock of *bvp*24-14 *am*E355(24) *am*E303(24) is equal at the two temperatures. The following hypothesis, which has not yet been tested experimentally, is suggested. Presumably, a substantial fraction of UV rescue genomes after high-dose exposures are internal heterozygotes for the rescued segment. In the present case, one DNA

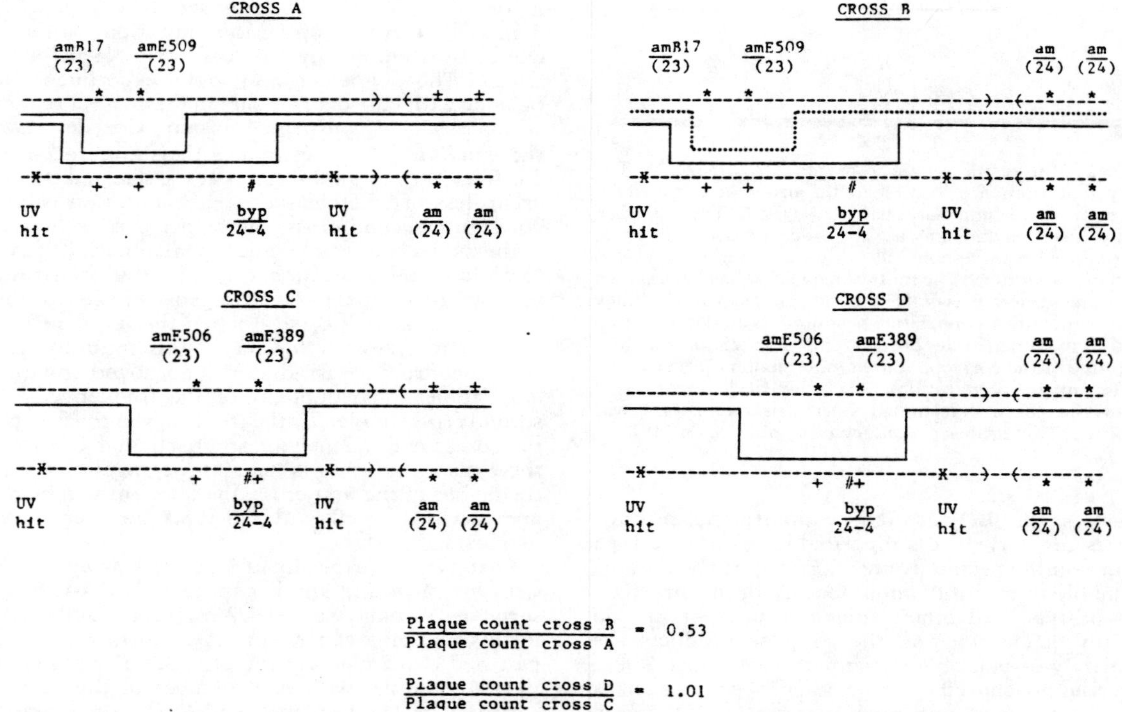

FIG. 3. Mapping by UV rescue. The crosses diagrammed represent a single *byp*24-4 *am*E355(24) *am*E303(24) genome in two pairs of rescue situations. The helper parents in both crosses of a pair carry the same two gene 23 amber mutations, and the two pairs differ from each other in the distance that separates the marked gene 23 segment from the bypass mutation. In the first pair (crosses A and B), that distance is significant, whereas in the second pair, the bypass mutation is shown to lie within the segment. In the second cross in each pair, the helper phage has two amber mutations in gene 24 as well as in gene 23. All progeny platings are made versus the amber-restrictive host, *E. coli* B (*su*⁻). Solid lines in the diagrams represent those segregants from each cross that can make plaques. Neither parent (broken lines) can do so because the helper carries amber mutations and the donor has been heavily irradiated (247 PLH). In crosses A and C, plaque formation requires only that the two *am*(23) sites of the helper be replaced by the *am*⁺ sites from the irradiated donor. In crosses B and D, however, it is necessary in addition to corescue the *byp* mutation to circumvent the *am*(24) mutations, here present in both parents. In cross D, corescue will occur in virtually all cases because the bypass mutation lies between the ambers. The ratio of the plaque count in cross D to the plaque count in cross C will be approximately 1. In cross B, however, the bypass mutation will, in many cases, not have been corescued with the two *am*⁺ sites (for example, the segregant indicated by the dotted line in cross B). In such a situation, the ratio of the plaque count in cross B to the plaque count in cross A would be less than one, and it would approach zero as the distance between the gene 23 ambers and the bypass mutation becomes large. The data in Table 4 give a more complete series of paired crosses, of which two are diagrammed here.

strand would carry *byp*, and the other would carry *byp*⁺. Infection of the amber-restrictive host by such a heterozygote would result in production of both *byp* and *byp*⁺ gene products. The admixture of *byp*⁺ protein might well interfere with the *byp* protein in effectively replacing the gene 24 function. The idea that mixing mutant and nonmutant protein might hinder effective functioning was, to our knowledge, first suggested by Kellenberger (personal communication), who called it the poisoning effect. In spite of the difficulty noted, it seems quite certain that *byp*24-14 is located in the gene 22-proximal portion of gene 23, most likely between *am*H11 and *am*B17, although it could be in the interval from *am*E209 to *am*H11.

UV rescue mapping has proved useful in several other applications, of which two will be mentioned. Snyder (Ph.D. thesis, University of Colorado, Boulder, 1982) used UV rescue with double-amber helpers to confirm her hypothesis that the carboxy-terminal 20% of T4 gp37 is subdivided into two functional domains. She predicted the segments into which two mutations would fall based on their phenotypes. Her prediction was confirmed by comparison of UV rescue frequencies with several double-amber helpers that defined different critical fractions of the terminal segment of gene 37.

In another application, Doherty (1982a) faced the problem of locating the positions of mutations that suppress the phenotype of the *ptg*19-80c missense mutation. Mutations of the *ptg* class are located in gene 23, which encodes the major structural protein of the T4 capsid. They cause loss of the normal fidelity in control of capsid length (Doermann et al., 1973b), which results in the production of some extremely long capsids (giants) and a majority of capsids that are shorter than normal (petites). The problem was difficult for at least two reasons: (i) the suppressor mutations themselves could only be identified in the presence of the suppressible *ptg* mutation; (ii) two suppressors, which he called *x* and *y*, in different

TABLE 4. UV rescue mapping of *byp24-4* and *byp24-14*[a]

Helper genotype	Four-amber/two-amber quotient	
	byp24-4 *am*E355 *am*E303[b]	*byp24-14* *am*E355 *am*E303[c]
*am*E209(22) *am*H11(23)		0.35
*am*H11(23) *am*B17(23)	0.38	0.46, 0.51
*am*B17(23) *am*E509(23)	0.53	0.32
*am*E509(23) *am*E506(23)	0.64	0.13
*am*E506(23) *am*E389(23)	1.01	0.07
*am*E389(23) *am*A489(23)	0.98	0.03
*am*A489(23) *am*E1236(23)	0.38	0.0004

[a] This experiment follows the general protocol described in the legend to Fig. 1 except for the genetic constitutions of donor and helper phages. The donor phages, *byp24-4 am*E355(24) *am*E303(24) and *byp24-14 am*E355(24) *am*E303(24), received UV doses of 247 and 266 PLH, respectively. The helper phage genotypes are given in the table, which lists only the two-amber genotypes. The approximate locations of these ambers may be seen in Fig. 2. The four-amber helpers were the same, except that they also carried *am*E355 and *am*E303 in gene 24. Platings from each cross were made against the amber-restrictive host *E. coli* B (su^-) after chloroform treatment. The titers found in the four-amber crosses were divided by those of the corresponding two-amber crosses, giving the four-amber/two-amber quotients listed in the table. (I thank David S. McPheeters for providing the data given in this table.)

[b] Plates incubated at 37°C.

[c] Plates incubated at 41°C.

locations were both required to elicit detectable suppression. By constructing a suitable set of five multiple-amber strains and using them as helpers in UV rescue crosses, he was able to show that x is located in or near gene 22 and that y is in or near gene 24.

Finally, we have found that UV-rescue mapping provides a method for rather precise intragenic mapping of mutations whose reversion frequencies are high. With such mutations the problem is encountered that stocks used for crosses contain many phage which cannot be distinguished from wild type. The progeny of those parents in the cross make low recombination estimates unreliable or impossible to obtain. With the UV-rescue procedure, even a fairly high frequency of revertants in a stock (for example, 10% or even more) would not be expected to interfere with locating a mutation with respect to the amber mutations represented in the helper phage.

MAPPING THE ULTRAFINE GENE STRUCTURE

Benzer's analysis of the *r*II genes of T4 (Benzer, 1959; Benzer, 1961) provided very precise information about the locations of the various *r*II mutations. Nevertheless, Tessman (1965), by devising a more sensitive variation of the Benzer system of scoring recombinants, was able to separate mutations which Benzer had originally assigned to a single genetic site. He achieved this by plating the unlysed infectious centers rather than the progenies of his crosses. That technique enabled him to place more progeny phage in total on a plate without obliterating the bacterial lawn. In addition, it localized in single plaques the large clones of wild-type phage that arise from revertants in the parental stocks. When very closely linked markers are crossed, the clone size of recombinants averages barely more than 1 in the single bursts in which they occur (Stahl, 1956). The minimum nonzero recombination frequencies observed by Benzer clustered around 0.02%. With his more sensitive technique, Tessman established recombination values more than 2 orders of magnitude lower. He called the map that he constructed with these values the ultrafine-structure map. The values obtained by Benzer, when compared with the total *r*II-gene map and then to the total T4 map and its 166,000-nucleotide genome, already suggested that, at most, several nucleotides separated mutations showing 0.02% recombination (see also Drake and Ripley, this volume). From his analysis, Tessman concluded that extreme variation occurs in the frequencies of crossovers that occur for different pairs of adjacent nucleotides. If the frequencies were uniform, a minimum crossover frequency could be established from the lowest values observed experimentally, and that frequency, multiplied by the number of nucleotides in the T4 genome, would equal the total genetic map of that phage. Tessman, however, found a recombination value of $7 \times 10^{-6} \pm 1 \times 10^{-6}$% for the cross *r*NB419 × *r*NB433. The total map of T4 should then be 7×10^{-6}% times 1.66×10^5 adjacent nucleotide intervals, which equals 1.2% recombination or 1.2 map units. As pointed out earlier in this chapter, estimates of the genetic map of T4 are more than 1,000 times greater. In the interval between *r*NB419 and *r*NB33, recombination must be at least 1,000 times less likely than between many of the other pairs of adjacent nucleotides. Thus, the analysis of Tessman implies wide deviation from additivity of map values in the ultrafine structure range and predicts significant complications in relating mutations to DNA sequence by the map saturation approach.

Because the technology for determining the nucleotide sequence of DNA has, in the past decade, developed to such a high degree of effectiveness (Maxam and Gilbert, 1980; Sanger et al., 1977), the procedure of choice today for identifying a mutation with a particular nucleotide change is to compare the nucleotide sequences of mutant and wild-type DNAs. Those techniques are beyond the scope of this chapter, but Kutter and Rüger (this volume) deal with restriction mapping of T4 DNA, and their discussion should be consulted if T4 cloning and sequencing is contemplated. DNA sequencing alone, however, is not an adequate mapping tool for two reasons. First, the number of nucleotides that must be sequenced is impractically large, even if the mutation in question is already known to be in a particular gene. Sequencing should, therefore, be supplemented by genetically localizing the mutation to a shorter segment. Second, and perhaps more important, the nucleotide alteration found by sequencing must be shown to be responsible for the phenotypic effect being scrutinized. It should be noted that most mutations in T4 have been induced by a more or less vigorous mutagenic treatment designed to cause nucleotide changes. Not all nucleotide changes that result in amino acid substitution produce a phenotypically recognized effect. It is not possible to ascribe a phenotypic effect to a nucleotide change from sequencing alone unless the whole mutated gene sequence is shown to have only one nucleotide difference from the wild-type sequence.

Identification of a nucleotide substitution with a particular mutation is facilitated when the sequence locations of rather closely spaced amber mutations in the gene are known in advance. Because the amber triplet is known and is unique, misidentifications are virtually eliminated. Once the amber locations are known, another mutation can be mapped to a short polynucleotide span by UV rescue mapping against the ambers as described above, and that stretch can then be sequenced in the mutant. Similarly, marker rescue from short cloned DNA fragments rather than from UV-irradiated phage can also prove useful for localizing particular mutations. Furthermore, in cases in which phenotypic expression depends on multiple mutations, rescue from cloned fragments may help to clarify complex situations.

USE OF UV RESCUE IN STRAIN CONSTRUCTION

Preparation of Isogenic Stocks

For correct interpretation of both genetic and biochemical T4 experimental results, it is always helpful, and sometimes crucial, to use stocks that carry no unsuspected mutations. When two or more strains are involved in an experiment, the ideal situation is to have stocks that are isogenic except for the mutation(s) under study. If the problem calls for the isolation of new mutations, complications arising from genetic heterogeneity can be largely avoided if the appropriate genetic segment has been or can be cloned. The fragment can then be mutagenized apart from the rest of the T4 genome, as was done by Völker and Showe (1980). The great majority of mutant strains now available, however, have been obtained after exposure of the whole T4 genome to highly mutagenic regimes, and a simple method for cleaning them up is needed. The classical method of successive backcrosses to an original genotype was used in modified form by Streisinger (1956) in transferring the UV sensitivity gene *u* (now called v^+) from T4 into T2. He made a series of eight successive backcrosses of the hybrid T2 *u* to a standard T2 genome. A simpler alternative now available is to employ the UV rescue technique, with which a single cross can replace the whole series of eight, provided a high enough UV dose can be utilized. The mutant would be irradiated and rescued by a helper genotype selected for the purpose. Some idea of the effectiveness of UV doses over a wide range can be obtained by referring to Fig. 1. Even when no selective system is available, recovery of the target mutation would not be difficult. After a 10-min dose of UV radiation (ca. 35 PLH), the frequency of rescue of a site such as *am*B17$^+$ was about 10^{-2} (Fig. 1, curve C), whereas another unwanted site at the distance of *am*E1236$^+$ (1,221 nucleotides away) was corescued in less than 50% of the cases (curve E). The corescue of any marker that maps far away would occur only about 1% of the time (Fig. 1, curve F). When a selective system can be employed (a combination of amber and temperature-sensitive mutations generally makes that possible), much higher doses can be employed, substantially enhancing the effectiveness of the method. At a dose of 200 PLH (ca. 60 min), for example, *am*E1236$^+$ would be corescued with *am*B17$^+$ in only about 5% of the cases; and even *am*B272$^+$, which is only 201 nucleotides away from *am*B17$^+$, is corescued with it only one time out of three. By starting with a set of isogenic helpers that could be obtained by rescuing the selector mutations into a common wild-type genome, it would not be difficult to introduce a set of mutations into an isogenic genetic background.

Preparation of Stocks with Markers that Are Difficult to Recognize

When mutations are unambiguously recognizable by a simple test (for example, identification of a specific *r*II mutation by replication on a lambda lysogen or a specific amber on an su^- bacterial host [detailed procedures described by Doermann and Boehner, 1970]), construction of multiple-mutant combinations causes little difficulty. Strains of T4 carrying as many as 24 conditional-lethal mutations have been built up in the laboratory of the author. That procedure is, however, insufficient in situations in which one or more of the critical mutations cannot be recognized by replica-plating tests alone. The UV rescue technique can often be adapted to the solution of such problems. Two cases are discussed.

The first case deals again with the *x* and *y* mutations, which, as mentioned in an earlier section, suppress the phenotypic expression of *ptg*19-80c mutations in T4 gene 23, and which Doherty (1982a) was able to map to the vicinities of genes 22 and 24, respectively. By another application of the technique, Doherty (1982b) was able to answer a second and equally difficult question. The known *ptg* mutations occur at eight different sites in gene 23. Doherty wished to test whether *x* and *y* suppress *ptg* mutations in general or whether they display specificity, suppressing the phenotype of a smaller set, for example *ptg*19-80 only, or perhaps all the mutations located at that site. To answer that question, it was necessary to produce recombinants that include *x*, *y*, and a *ptg* mutation representing each of the other sites. This was no simple task for two reasons. One is that the *ptg* mutations are all located between *x* and *y*, and therefore, the genotype *x ptg y* would be formed only by a double crossover in a rather short map interval. Second, it could not be predicted whether suppression of the *ptg* phenotype would be observed, and therefore, the appropriate recombinant might well be overlooked because the *x* and *y* mutations have no independently recognizable phenotypes. It proved possible, however, to obtain an unambiguous decision by using the UV rescue procedure. From the mapping experiments mentioned in a previous section, Doherty already had available a segregant of the genotype *x am*E506(23) *y*, and UV-irradiated *ptg* strains representing each of the eight sites were crossed to that strain. Plaques could be formed only by those progeny phage in which the *am*E506 of the helper was replaced by its am^+ allele from the UV-irradiated *ptg* strains. It could be shown that the plaque-former phage included a substantial fraction of segregants containing the *x ptg y* genotype. Suppression of the *ptg* phenotype could then be recognized because it results in plaque enlargement to a size intermediate between *ptg* and wild-type phage. Using this technique, Doherty (1982b) was able to show that the *x* and *y* mutations originally isolated in combination with *ptg*19-80c were highly specific: they suppressed only the phenotypes of those *ptg* mutations that map at the site occupied by *ptg*19-80c.

A major potential use of this technique is to permit the introduction of mutations in almost any gene into a genetic background that is appropriate for cloning. This necessity arises from the fact that in T4 DNA, glucosylated hydroxymethylcytosine (HMC) normally replaces all cytosine residues, rendering it resistant to the action of nearly all restriction enzymes. However, a genome with mutations in at least four T4 genes can multiply in the appropriate host, producing phage in whose DNA 95 to 100% of the HMC residues are replaced by cytosine (dC-DNA) (Snustad et al., this volume; Snyder et al., 1976). It is important that the host have no restriction-modification system (r^- m^-), and because two of the required mutations are ambers, it must also be su^- to prevent them from functioning. The genes whose functions must be inactivated are *denB*, *alc*, 56, and 42 (or 1) or *denA*. Because these genes are distributed over much of the total T4 map, one or more of these mutations is replaced by its wild-type allele in most of the segregants of a cross. Introduction of a new mutation into this genetic background (for example, in preparation for cloning or mapping to a restriction fragment) can, therefore, be difficult. The UV rescue technique can be effectively applied to reduce that difficulty. In principle, it is necessary only to irradiate the strain that has the mutation to be inserted and then to rescue it by using the unirradiated four-mutant dC-DNA phage. When the new mutation can be identified readily by plaque morphology, temperature sensitivity, or other means, it can be introduced in a single step. If, alternatively, its identification poses difficulty, a more complicated procedure may be required. In preparation, it would be advisable to map the mutation and then select a nearby pair of ambers that bracket it. A double-amber strain carrying that pair would first be irradiated and crossed with the four-mutant parent in an r^- m^- su^+ host such as *E. coli* K803, plating the segregants on similar cells. The suppressor enables phage carrying the amber pair to make plaques, but it also suppresses the gene 42 and gene 56 ambers, so that the progeny contain HMC. The plaques can be tested by replication (Doermann and Boehner, 1970) to identify the minority that carry the added ambers. Even at a dose as low as 40 PLH, less than 10% of the progeny of the cross will have lost any of the dC-DNA markers (compare Fig. 1). By replication tests, the gene 42 and gene 56 ambers could, of course, be readily verified, but the other three would prove more difficult to check.

Once the two ambers have been added to the dC-DNA set of mutations, the hard-to-identify mutant strain can be irradiated and crossed with the multimutant strain, again by using an su^- host. This progeny is plated on r^- m^- su^- bacteria. The vast majority of phage that can form plaques will have retained the dC-DNA set, and all will have replaced the added pair of ambers with their am^+ alleles. A substantial fraction of those will have brought the new mutation along with the flanking am^+ loci. That fraction can be maximized by using flanking ambers that are as close as possible to the new mutation.

Unfortunately, conditional-lethal amber mutations generally do not permit phage production under conditions necessary for making deoxycytosine-containing phage. However, as long as the ambers do not seriously impair DNA synthesis, the intracellular T4 dC-DNA can be extracted and used for such purposes as cloning; under such conditions, the *alc* mutation is also unnecessary.

CONCLUSION

Genetic mapping of T4 and multimutant construction depend on reliable analyses of cross progeny genotypes and are mutually interdependent. The solution to achieving the desired objective depends on devising an effective combination of replication procedures and UV rescue technique with the more classical methods involving selective plating and plaque morphology distinctions.

Lawrence Sandler and Michael Parker read this manuscript, and I am grateful for their helpful suggestions. Betty Kutter gave this manuscript far more attention than editorial supervision requires, and that too is deeply appreciated.

The experimental work reported here was supported by Public Health Service grant GM13280 from the National Institute of General Medical Sciences.

The Analysis of Mutation in Bacteriophage T4: Delights, Dilemmas, and Disasters

JOHN W. DRAKE AND LYNN S. RIPLEY

Laboratory of Genetics, National Institute of Environmental Health Sciences, Research Triangle Park, North Carolina 27709

A considerable portion of the theory of mutation is founded upon studies conducted with bacteriophage T4. Our present purpose is to provide a guide to the several genetic systems of T4 appropriate to the analysis of mutation, indicating briefly their advantages and their pitfalls. In many cases we will draw upon unpublished personal experiences, particularly in matters pertaining to the art of artefact avoidance, and thus will be unable to provide useful literature references. For brevity in other matters, we will mostly cite recent references, review articles, or chapters in this book where those provide adequate entries to the literature. Two general sources of information about mutagenesis in T4 are an out-of-print and somewhat out-of-date book (Drake, 1970) and a review article (Drake and Baltz, 1976). However, the pitfalls of mutagenesis research are often remarkably similar from organism to organism, and the reader is therefore also directed to Auerbach's (1976) extensive comments upon a number of cellular systems.

WHY USE T4?

Physical and Chemical Properties

Bacteriophage T4 is among the most readily manipulated of laboratory organisms, and its extensive use has provided a rich variety of well-developed procedures. A striking general advantage of T4 for studies of mutation is the ease with which truly huge numbers of phage particles can be obtained and subjected to selective procedures to ferret out rare mutated genomes. For instance, a 50-ml cell suspension seeded with the progeny of a single plaque will yield as many as 10^{13} particles after only a few hours of incubation, or 5 mg of T4 DNA by the next day.

T4 stocks are very stable. They can be stored at 4°C with only small losses of titer over several years. Such stocks may often be used directly for qualitative or exploratory experiments, but should usually be regrown if more than a few months old before they are used for more critical purposes. Note that base substitution mutations slowly accumulate in stored stocks, especially at low pH values (see Bingham et al., 1976, and references therein). Note also that both the initial titers and the stabilities of T4 stocks may be very significantly reduced as a result of mutagenic treatments or of mutations affecting somatic proteins. In our experience, stocks treated with UV light or gamma rays are stable both in total titer and in frequency of induced mutations, a situation highly advantageous for mutation studies; but particles treated with chemical mutagens, such as nitrous acid, hydroxylamine, and alkylating agents, are labile and must be used immediately, since even overnight storage at 4°C results in dramatic losses of titer and changes of mutant frequency. Thus, mutagenized phage particles should be assayed immediately unless prior tests have shown that changes occurred in neither total titer nor mutant frequency after storage.

The DNA of T4 differs notably from that of cellular organisms in its replacement of cytosine by glycosylated 5-hydroxymethylcytosine (5HMC), and this difference is frequently important in mutation studies. For instance, 5HMC effectively isolates T4 DNA from the action of many host enzymes during the course of intracellular reproduction, strongly influences the reactivity of T4 DNA with many chemical and physical mutagens, and renders T4 DNA resistant to most restriction enzymes. Although 5HMC may present either an advantage or a disadvantage in particular experiments, it should be noted that T4 mutants are available that make DNA containing conventional cytosine (see Snustad et al., this volume). For brevity in describing mutational pathways, we will generally symbolize 5HMC as C in this chapter, except where the difference is of immediate significance.

Biological Properties

Although most bacteriophages depend extensively upon host enzymes for their replication, repair, and recombination, T4 is largely independent of such enzymes for its DNA metabolism; it encodes its own. This renders T4 unsuitable as a probe of the cellular systems, but sometimes makes it particularly well suited for studies of mutagens that evoke the action of complex and overlapping host repair systems.

Recombination in T4 is independent of the host *recA* system and proceeds at the greatest rate reported in any organism. A rough rule is that 1 centimorgan (1% recombinants) corresponds to a marker separation of 100 to 200 base pairs; however, marker and position effects often produce twofold or greater variations. This rapid rate provides both advantages and disadvantages to the experimentalist. The main advantage is the ease with which recombinants can be recovered from crosses, even crosses using physically closely linked markers for which there is no good selective system for the desired rcombinant. Thus, induced mutations may easily be backcrossed away from contaminating or irrelevant coincidental mutations as close as about 10 base pairs. (Although recombination frequencies are approximately proportional to physical distances down to about 10 base pairs, at shorter distances they then fall off very sharply and unpredictably.)

One disadvantage of the high rate of recombination in T4 is the extensive dispersion of parental-strand fragments among progeny genomes. This can confound experiments that would use isotopic labeling, or DNA lesions, to trace strands of DNA during intracellular growth. Such recombination cannot, unfortunately, be substantially blocked by mutations, since the greatest effects of all known antirecombinogenic mutations are less than an order of magnitude, and are usually accompanied by substantially reduced burst sizes (see Mosig, this volume).

This high recombination frequency also leads to high frequencies of heteroduplex and terminal redundancy heterozygotes. Thus, considerable care must be taken in circumstances where it is important to distinguish between heterozygotes of mutational origin and those of recombinational origin (see, for example, Lindstrom and Drake, 1970).

Although the synthesis, recombination, and much of the repair of T4 DNA is autonomous, its transcription and translation depend heavily upon host systems (see Binkowski and Simon, this volume). An important advantage of this dichotomy is the ease with which chain termination codons can be used to study mutation. Virtually the entire range of genes whose functions are required during viral multiplication can be marked with such codons, whose reversion and interconversion are often convenient measures of mutation. Suppressing and nonsuppressing (but otherwise isogenic) hosts provide simple differential plating systems. However, T4 encodes a few tRNA species, some of which can mutate to produce informational suppressors (see Schmidt and Apirion, this volume). Fortunately, these suppressors are effective only upon transcripts from late genes. Thus, they are not recovered among "revertants" of nonsense codons in early genes, although they may appear among the corresponding "revertants" of late genes.

Bacteriophage systems in general offer opportunities to compare mutagenic treatments of DNA in the intracellular state versus the extracellular state. This has been particularly useful for studying such mutagens as *N*-methyl-*N*′-nitro-*N*-nitrosoguanidine and nitrosomethylurea, which produce strikingly different results depending upon this very parameter (Ripley, 1981b).

Bacteriophage T4 offers additional options, namely, alternative ways to recover extracellularly treated genomes. For instance, when a mutagen kills extensively due to its effects upon phage somatic proteins, often the genome can still be assayed at useful efficiencies by converting the phages to π particles able to infect spheroplasts; π particles are produced by exposing T4 to 6 M urea (see, for instance, Baltz, 1976). Indeed, the method can be directly applied to estimate the fraction of lethal hits attributable to nongenomic damage, a value that can be further compared with measures based upon cross-reactivation or multiplicity reactivation.

T4 DNA also lends itself to both transfection assays (Baltz, 1971; Benzinger et al., 1975) and transformation assays (see Baltz and Drake, 1972 and Baltz, 1976). Although transfection is inefficient, particularly if the DNA is damaged during the course of mutagenic treatment, efficient transformation can be achieved by exposing spheroplasts to both π particles and T4 DNA. The subpopulation of transformed genotypes can be readily selected when the π particles carry mutations in one or more essential genes closely linked to the region of interest, and the transforming DNA carries the corresponding wild-type allele(s): selective plating of the progeny carrying the wild-type allele(s) then specifically enriches for transformants of the desired region. Either single-stranded or double-stranded DNA fragments as short as 100 to 1,000 bases can be used in the transformation assay. Larger pieces are recovered at a higher efficiency, however, making analysis of heterogeneous DNAs more difficult. The potential of this assay has not frequently been exploited in mutation studies, partly because of uncertainties about the metabolic steps between the entry of the DNA into the spheroplast and the ultimately emerging progeny and partly because the product of transformation and mutation frequencies (particularly reversion frequencies) is sometimes too low.

MAJOR SYSTEMS

General Considerations

Two of the most important characteristics of a mutation system are the kinds of DNA alterations that produce the mutant phenotype and the specific properties of that mutant phenotype. Genetic systems that permit the identification of both forward and reverse mutations can have a substantial advantage, since, as described below, the applications of these two methods of mutation measurement have different strengths and weaknesses.

A substantial advantage of measuring forward mutation frequencies often is its simplicity. However, the sensitivity of such measurements (i.e., the factor of increase over the spontaneous background) is frequently less than in reversion systems. In compensation, a single forward mutation measurement detects most or all types of mutations simultaneously. Thus, the simple detection of a mutant frequency increase reveals a mutagenic effect, but does not identify the DNA changes involved. Therefore, forward mutation is the measurement of choice in the absence of information about the type(s) of mutation to be induced or where no appropriate reversion system is available, as in the case of deletion mutations.

When forward mutation frequencies are increased, analysis of the responsible DNA changes can proceed in two ways. Classically, the approach has involved determining the characteristics of the induced mutants and then deducing the forward molecular pathways of their induction. Transition and frameshift mutations are specifically recognized by the application of certain well-characterized mutagens to determine which provoke reversion of the mutant. The forward production of A·T → G·C transitions produces mutants reverted by base analogs and by hydroxylamine (a mutagen that induces primarily G·C → A·T transitions). G·C → A·T transitions produce mutants reverted by base analogs but not by hydroxylamine. Frameshift mutations are identified by their reversion after proflavin, but not base analog, treatments. Mutants that fail to revert (even spontaneously) are classified as deletions. The major difficulties with this mode of analysis are its occasionally incorrect classification of mutants and its inability to resolve certain classes of mutations (such as transversions). Incorrect classifications may result from two different limitations of the system: the failure of certain mutants to respond in the expected way to particular mutagens and, more seriously, the phenotypic reversion of certain mutants by mutational pathways that do not restore the wild genotype. Transversion mutations provide a particularly serious problem. Because no mutagen has been identified that produces primarily transversion mutations, transversions cannot be directly recognized; thus, they tend either to be incorrectly classified as transition mutations for reasons described above or to fall into a motley class of "other" mutants that revert spontaneously, but do not clearly respond to any of the diagnostic mutagens. The limitations of this general mode of analysis are likely to be largely overcome with the rapid evolution of DNA sequencing technology, which should soon allow a definitive identification of mutant sequences with an investment of time comparable to that now required for classical reversion analysis.

Reverse mutation systems can offer substantial advantages in measuring frequencies and specificities of mutation. They are usually more rapid than forward mutation tests as probes of mutagen specificity and are therefore appropriate when the specificities of mutagens are to be examined or compared. Their major advantage is their high specificity. This specificity often results in much greater increases over background mutation frequencies than are seen in forward mutation; this is particularly true when the mutagen produces a narrow spectrum of mutation types that are not major components of the spontaneous spectrum. A general limitation of the reversion approach is the limited number of mutation sites investigated. This often does not permit identification of neighboring-sequence determinants of mutation and sometimes does not include mutants responding along all possible pathways. Reversion analysis is rarely suitable for the analysis of deletions, although some can be detected in certain tests for frameshift mutations. Reversion analysis is also not well suited for identifying unexpected mutation types or rates that may occur in only a limited subset of DNA sequences that are unlikely to be represented in a small set of tester strains. Reversion might, on the other hand, be very useful in further studies of such mutations once they have been identified.

When a comprehensive picture of mutagen specificity is desired, the above compilation should strongly suggest to the reader that both forward and reverse mutation systems be exploited and that DNA sequencing be included in the attack.

The phenotypic traits that have been most valuable in mutation studies have been plaque morphology, conditional lethality, resistance to inactivation, and various combinations of these.

Plaque morphology mutants are screened visually. They often require much individual effort in order to obtain sufficient mutants to ensure accuracy, and the scoring may be sensitive to experimental conditions. For instance, plaque morphology is often sensitive to background genetic markers that render plaques larger or smaller or to mutagenic treatments that result in irregular plaque sizes and shapes. It is generally desirable to reduce the background level of plaque irregularity by preadsorbing particles to host cells before plating, thus removing the variation among plated particles in the time of their first contact with a host cell.

The scoring of mutants or revertants by selective plating is often easier than the visual scoring of morphology variants. The range of differential plating systems is wide and includes classical host range systems, resistance to acridine in the plating medium, suppression of chain-terminating codons, heat- or cold-sensitive systems, and special systems, such as the *r*II and *e* genes. The last two are particularly useful because they provide a rich supply of mapped and otherwise well-characterized mutants, DNA sequence information is becoming available, and a large fraction of all possible experimental errors have already been committed. However, all differential plating systems are prone to serious plating-density artefacts, a ubiquitous bête noire of the profession, to be discussed later in this chapter.

e as in Lysozyme

The *e* gene is located at about 1:30 on most T4 maps (see Kutter and Rüger, this volume). It is the structural gene for the lysozyme that is required for phage release from infected cells, it has been sequenced, and considerable information is available concerning mutational perturbations of its function (see Grütter et al., this volume).

Because of its relatively low molecular weight, lysozyme diffuses from plaques well ahead of phage particles. When plates are exposed to chloroform vapors, rendering uninfected cells susceptible to the enzyme, wild-type plaques become surrounded by a large halo, whereas leaky *e* mutants and some e^+ pseudorevertants produce smaller halos. Thus, alleles producing less than the wild-type level of enzyme activity can be treated as plaque morphology variants.

This approach does not, however, identify null mutants, since they produce neither plaques nor halos: plaque development aborts just at the time when the first round of phage would normally be released. Instead, null mutants are harvested by their inability to lyse cells: infected cells are allowed to lyse normally in the presence of anti-T4 serum, after which the minority of nonlysed but infected cells are collected, washed, and artificially lysed. Null mutants can form plaques on plates supplemented with chicken egg white lysozyme. The assay is somewhat tedious, however, and plating conditions must be critically adjusted for different cell types (see, for instance, Drake, 1967).

An advantage of the lysozyme system is the availability of methods to reduce spontaneous mutational backgrounds to the level of mutants arising in the previous round of growth. To obtain an e^+ stock depleted of *e* mutants, for instance, one need only singly infect (multiplicity of infection of 0.1 or less) and harvest only the spontaneously released progeny (omitting any chloroform step and restricting growth to the first round of infection). Since this procedure necessarily results in stocks of low titer, extensive concentration and purification may be required. To obtain *e* stocks depleted of e^+ revertants (see, for instance, the footnote to the footnote of Table 1 of Okada et al., 1972), cells are multiply infected and incubated well beyond the minimum latent period; the nonbursting cells (containing only *e* genotypes) can then be gently concentrated and artificially lysed.

This is a particularly important procedure, since *e* mutants grow at a selective disadvantage compared with e^+ revertants, even in lysozyme-supplemented media.

A complication of the *e* system is caused by the extracistronic suppression of lysozyme mutants by mutations inactivating the *sp* (spackle) gene located at about 10:30 on the genetic map (Emrich, 1968). However, the combination of *sp* and *e* produces only partially wild phenotypes, and the complication can often be circumvented by lowering plating temperatures to 30°C, a temperature that is nonpermissive for the suppressed mutant.

A major historical advantage of the *e* system was the ease with which wild-type and revertant or pseudorevertant enzymes could be purified from lysates and subjected to amino acid sequence analyses. This advantage has been, or soon will be, overtaken by the delights of DNA sequencing. Its continuing attraction as a system for mutation analysis probably now rests mainly on the opportunities it provides the protein chemist (see Grütter et al., this volume).

3,600 Characters in Search of a Function

Considerable information on the physiology, DNA sequence, and mutational pliability of the *r*II system is provided by Singer et al. (this volume) and references therein. Located at about 8:30 or 9:00 on the T4 map, it is composed of two cistrons that are easily the most mutationally saturated genes of any organism. Its mutations produce dramatic effects upon plaque morphology and are highly pleiotropic with respect to numerous other aspects of phage reproduction. Its two proteins are readily extracted from the membrane of the infected cell and displayed on gels. Despite this wealth of information, there is no real understanding of the biochemical basis of *r*II function(s).

The R plaque morphology produced by *r*II mutants (and by *r* mutations at other loci as well) is large and sharp-edged, in contrast to the smaller, fuzzy-edged wild-type plaque. The mutant plaque is explicable by the loss of lysis inhibition, in which multiple infection delays cell lysis and thereby greatly increases the burst size. The adaptive value of lysis inhibition is manifest in both liquid and semisolid media, where final titers of r^+ phages are at least an order of magnitude higher despite their smaller plaque diameters.

When R plaques are picked from *Escherichia coli* B lawns, they can be further subdivided into three main sets. Those that produce R^+ plaques on BB or nonlysogenic K strains and are unable to grow on λ lysogens are *r*II mutants. Those that produce R plaques on BB or K strains and can grow on λ lysogens are *r*I mutants. Those producing R^+ plaques on BB cells and small or infrequent plaques on λ lysogens are mostly leaky or rapidly reverting *r*II mutants. A small fraction of R plaques detected on B lawns are mutated at other loci, but have been little studied. The three main groups of mutants are similarly frequent, within factors of a few-fold that depend upon the mutagenic conditions: spontaneous and proflavin-induced mutants tend to be enriched in *r*II mutants compared with the other two classes, whereas the converse is true for mutants induced by agents as diverse as base analogs and gamma irradiation. The reason(s) for these differences are not understood, and almost nothing is known about the properties of the *r*I cistron except its location at about 2:00 to 2:30 on standard maps.

From the standpoint of the student of mutation, the crucial properties of *r*II mutants are their R plaque morphology on B cells, which facilitates screening for forward mutants, and their inability to grow on λ lysogens, which facilitates both reversion and mapping measurements. The λ block, caused in unknown ways by the rex^+ prophage genes, permits the selective detection of r^+ particles at frequencies of 10^{-8} or lower, since about 10^8 *r*II particles may be plated on a single dish without decreasing the plating efficiency of the r^+ genotype.

A number of additional properties of the *r*II system are also of major importance for mutation studies. First, the rII^+ function must be expressed before DNA replication in a λ lysogen if the infection is to succeed, a property that can be used to distinguish between mutations arising in the transcribed and complementary strands of DNA (see below). Second, over 100 base pairs at the beginning of the *r*IIB cistron are dispensable for growth in a λ lysogen: the plastic nature of this region permits unique studies of frameshift, deletion, and duplication mutagenesis not feasible in regions where amino acid sequence requirements are more rigid. Third, there are strong indications that missense mutations (those producing amino acid substitutions) are well tolerated throughout much of the *r*II polypeptides, producing no mutant phenotype. The proportion of chain termination mutations among transition mutations is very high, much higher than the 5% expected if missense were as deleterious as nonsense; perhaps as many as 90% of missense mutations go undetected. Fourth, the *r*II proteins do not appear to be intimately involved in the fidelity of DNA synthesis. This is a common problem with early T4 proteins, and its absence is a fortunate property of a gene used to monitor mutation frequencies. However, *r*II mutations do interact with several other T4 and host functions, as may be seen sporadically in several chapters in this volume. The most important are the interactions with DNA-binding proteins and DNA ligases, so care must be taken to avoid selection artefacts when coupling mutations in these genes with *r*II mutations. Fifth, the high resolving power of recombination in T4, when coupled with DNA sequencing and dense mapping of mutants in certain regions, sometimes permits the assignment of nonsense mutations to specific base pairs by recombination tests alone (Pribnow et al., 1981).

A set of *r*II mutants exists with phenotypes intermediate between those of the wild type and the full mutant. They often appear as second-site mutations, or as same-site but different-base-pair mutations, that incompletely restore wild-type function during the course of reversion. Conversely, they can also be recovered as second-site mutations that combine with a leaky *r*II mutation to produce a much more mutant phenotype, but that are only faintly mutant themselves when separated from the first mutation (Koch and Drake, 1970). Mutants of both types display some degree of R plaque morphology, but usually grow well on λ lysogens. It is usually the case that more r^+

function is needed to achieve lysis inhibition than to overcome the λ block. This low level of *r*II protein required for growth on λ lysogens is further revealed by the efficient plating of *r*II ochre mutants on weak ochre suppressors (where chain elongation probabilities may be 5% or less). Finally, although extracistronic suppressors of *r*II mutations have not been found when plating for revertants on λ lysogens, mutations in non-*r*II genes sometimes weakly promote plaque formation by leaky *r*II mutations, enhance transmission coefficients, or suppress the R phenotype on B cells (see Hall and Snyder, 1981, and references therein).

Reversion of Amber Mutations

Amber mutations are available in most T4 genes required for viral reproduction. (Ochre mutations are rarely recovered in genes whose proteins are required in large amounts because of the inefficiency of ochre suppressors. Opal mutations are rare because they have not often been sought and are often leaky.) The reversion of amber mutations may be used to assess mutation rates. Amber mutations in non-*r*II genes should be useful if functional interactions are suspected between the *r*II locus and the mutagenic conditions or if late rather than early genes are desired in order to avoid problems of mutation expression as described below. However, only certain base pair substitution pathways can be monitored. Furthermore, the T4 tRNA genes can generate informational suppressors that may act upon amber mutations in late, but not early, genes (see Schmidt and Apirion, this volume). It should also be noted that amber mutations in early genes, even when suppressed sufficiently for good growth, may affect the fidelity of DNA synthesis (Reha-Krantz and Bessman, 1977; Watanabe and Goodman, 1978).

SOME SPECIAL MEASUREMENTS

We describe here certain procedures that can generate very detailed information about the mutation process. For the most part they have been applied exclusively by using the *r*II system. However, one important procedure, the depletion of preexisting mutants from a stock, as described above for the lysozyme system, does not work in the *r*II system.

Analyses of Specific Mutational Pathways

Base substitution pathways can be conveniently measured by the reversion and interconversion of nonsense codons. In describing these systems, we will use the conventional language of mRNA (amber = UAG, ochre = UAA, opal = UGA), leaving the reader to carry out reverse transcription where appropriate.

The conversion of an ochre mutation to an amber or an opal mutation is an unambiguous test for A·T → G·C transitions. It is also a simple test, requiring only that the stock be plated on the amber or opal suppressor and that sufficient plaques be picked and spot tested to determine the frequency of convertants against the background of revertants. (Such spot testing is generally required for this measurement because differential plating is usually an insufficiently sensitive measure of convertant frequencies.) The specific test for G·C → A·T transitions, however, is somewhat more difficult, for two reasons. First, the amber → ochre conversion cannot be monitored, because all known bacterial ochre suppressors also suppress amber mutations. The other G·C → A·T conversion, namely, opal → ochre, is rendered difficult by the general leakiness of opal mutations. Thus, a subset of opal mutations that are of below-average leakiness must be sought out (see, for instance, Ronen and Rahat, 1976).

Transversion mutations can be measured by using a special set of *r*II nonsense mutations whose transition and transversion revertants can be phenotypically distinguished (Ripley, 1975). These tests measure transversions at a number of A·T base pairs, and, in those instances where such transversions do not lead to viable revertants, transversions at G·C sites can be specifically detected.

Frameshift mutations are most conveniently studied in the first hundred or so base pairs of the *r*IIB cistron (see Pribnow et al., 1981; Ripley and Shoemaker, 1982; and Singer et al., this volume). In this region, amino acid composition of the *r*IIB protein is largely irrelevant to growth on λ lysogens. Thus, frameshift suppressors of frameshift mutations of opposite sign can be collected over rather extensive distances, limited only by barriers consisting of nonsense codons in the altered reading frame. The suppressing frameshift mutations are selected with the same ease that is characteristic of reversion systems, but are in an important sense really a set of forward mutations arising in a region large enough to offer the richness of DNA sequence complexity that is the main advantage of forward mutation studies. This method of identifying frameshift mutations also circumvents reliance on proflavin revertibility as a sole criterion. Furthermore, the target is sufficiently small to facilitate the application of DNA sequencing methods to large numbers of mutations.

Almost all deletion mutations of more than a few base pairs are unable to revert at detectable frequencies, and this trait can easily be adapted to their detection among a set of *r*II mutations. (The frameshift system described above, however, can detect moderately large additions and deletions within the target region if they have the appropriate reading frame characteristics.) The extents of deletions can then be estimated by recombination tests against an array of well-mapped point mutations. Large deletions extending outside of the *r*II region are recoverable if they exit the *r*IIB cistron, since it is followed by several genes not required under ordinary growth conditions; but those exiting the *r*IIA cistron are generally not recoverable because of the proximity of essential genetic information (see Pribnow et al., 1981, and references therein). Very large deletions can be recovered in the intervals between genes 39 and 56 and between 41 and *e*; but their identification and sizing is difficult and requires physical methods.

The induction of large duplications has rarely been studied in T4, but a selective scheme is available in the *r*II locus (Weil et al., 1965). Since duplications are highly unstable under most circumstances, it should be possible to identify them among sets of *r*II mutants simply by screening for those with very large reversion rates (e.g., revertant frequencies on the order of 0.1). To date, however, all such mutants of spontaneous origin that have been further characterized as to size with the "5-fluorodeoxyuridine test" described below have behaved like point mutations.

Correlating DNA Lesions with Mutations

The modeling of many mutagenic processes depends upon establishing the relationship between DNA lesions produced by a mutagen and the specificity of the ensuing mutations. It is generally important to know which base of the base pair is affected. When the modified base is immediately transcribed as another base, the *r*II system can distinguish which base of a base pair constitutes the mutagenic target. First, mutants are chosen that revert by the same pathway, for instance, G·C → A·T. The orientation of their reverting base pairs with respect to the transcribed strand versus the complementary (nontranscribed) strand may then be established. If the mutating codon (for instance, UGA → UAA) is known, the orientation is established directly. In other cases, calibration with a previously well-characterized mutagen (for instance, hydroxylamine, which specifically modifies the C residue of the G·C base pair) permits an identification of the orientation by a comparison of induced revertant (or convertant) frequencies after direct plating of the mutagenized population on the nonpermissive host, with the induced frequency after the population has first grown for a single cycle on a permissive host. Mutants having the mutagenized base on the transcribed strand will exhibit a slightly decreased frequency from that level (by about twofold) after being passaged due to resolution of the mutational heteroduplex heterozygote. Mutants with the target base on the nontranscribed strand should exhibit an increased revertant frequency only after being passaged. In practice, the choice of tester mutants is critical to the test. Most mutants produce some "immediate" response. In those instances where a strong delayed response follows, the weak immediate response may be attributed to leakiness of the *r*II mutation. However, when the delayed response is not dramatically stronger, it may reflect production of a suppressor mutation at a nearby base pair having the opposite orientation, rather than production of the mutation at the nontranscribed base. Consistent results with several mutations in both orientations should therefore be obtained before drawing strong conclusions about altered bases on the basis of this test. Note that a DNA lesion that depends upon postreplication repair for conversion to a mutation (as expected, for example, with mutagens acting via the T4 *WXY* system) will behave as though the lesion occupied the nontranscribed strand, regardless of its true location. Although no examples have surfaced yet, should an altered base have a different specificity for transcription than for replication, aberrant assignment of the target base could then occur.

Dual-Host Procedures

A mutant's ability to grow on one host and not on another can often be converted into a characteristic plaque morphology, allowing visual identification. For example, amber mutations can be selected as cloudy plaques (due to incomplete lysis) on plates seeded with two hosts, one an amber suppressor and the other not. In practice, such dual-host procedures work best when the plaques are large; when applied to *r*II mutations, for instance, a B strain is used as the permissive host. Plaque sizes can also be enlarged by other methods (e.g., see Schnitzlein et al., 1974) that could be helpful in other systems. There usually exist somewhat narrow optima for both the number and the ratio of the two host cells, as well as for the time of incubation. Thus, calibration with a standard mutant of the type desired should be done, if possible. This technique can be sufficiently powerful to detect even rare mutants, for instance, at a frequency of 10^{-5} or less (Conkling et al., 1980). It may also be useful in constructing double mutants by recombination. For instance, amber-frameshift double *r*II mutants can be distinguished from the parental single mutants by plating them on a mixture of permissive B and amber-suppressing K(λ) lysogens: cloudy R plaques contain either the frameshift parent or the desired double mutants and can be further distinguished by spot testing for their content of revertants, which will be detectable only among the single mutants. (In a similar vein, although in a single-host context, it is possible to distinguish multiple-amber mutants from their single-amber progenitors because of the smaller plaque sizes of the former on amber suppressors.)

Somewhat different plaque morphologies are usually produced in dual-host procedures when the two cell types are mixed in the soft top agar, compared with being added in separate layers. In the latter procedure, the time of addition of the second layer is critical. In either case it is often desirable to enhance the adsorption of the virus particles to the permissive host at the first round of infection by preadsorption.

Heterozygosity and Analysis of Mutation

Most mutations arise through mechanisms that generate a heterozygote as an early step of the process. Since the properties of this heterozygote can provide important clues to the mutagenic process, the efficiency with which heterozygotes are scored is an important consideration in any measurement of mutation frequencies. In the *r*II system, heterozygous plaques can often be detected phenotypically on B cells by their irregular shape; although visual scoring is always less sensitive than picking and replating plaques, it is far easier and, if carefully conducted, usually sufficiently accurate. (Care should be taken to use fresh plates, optimal phage and cell densities, and preadsorption. The plates should be scored after about 10 to 12 h of incubation.) Heterozygotes are most readily detected when the plaque contains approximately equal numbers of *r* and r^+ particles, but even plaques containing small fractions of *r* particles can be recognized by using special plating techniques (Drake, 1966a). These techniques are important when examining mutagens that operate, or may operate, through the *WXY* system (see also Drake, 1982).

The more detailed analysis of mutational heterozygosity may be helpful to probe the nature of the lesion responsible for a mutation. For example, a single-burst or pseudo-single-burst analysis can provide a profile of the mutant burst size distribution, which may in turn provide information about the mispairing probability of the lesion or the intervention of repair processes. Mutagenized particles are adsorbed to permissive host cells at low multiplicities, and the complexes are diluted and distributed in tubes at concentrations empirically determined to produce on the order of one mutant clone per 10 tubes. (Usually, each tube will also contain many nonmutant clones.) After lysis, each tube is plated in its entirety on the selective host, so that the size of the mutant clone is determined. Parallel measurements are made of the total number of progeny. An unrepaired DNA lesion that produces mutant progeny at every round of replication will generate clones of mean size equal to half the total burst size, whereas a DNA lesion that rarely produces mutant progeny will generate much smaller clones, whose mean size is half the reciprocal of the number of rounds of DNA replication (usually around 6%). When combined with measurements of mutant frequencies per initial treated particle, such data can provide minimum estimates of the number of DNA lesions produced by the mutagenic treatment.

Heterozygosity can also be used, in a much different way, to estimate the size of a mutation. When a cross is performed between r and r^+ particles, a small percentage of the progeny form irregular plaques containing progeny of both phenotypes. Some (but by no means all) of these plaques arise from recombinational heteroduplex heterozygotes, wherein the hybrid DNA that mediates genetic recombination overlaps the marker site. When the thymidylate synthetase inhibitor 5-fluorodeoxyuridine is used in such a cross to inhibit DNA synthesis, such recombinants accumulate to high levels and are not resolved to homozygotes by DNA replication until they are later plated in the absence of 5-fluorodeoxyuridine. The frequency of heterozygotes depends strongly upon the nature of the mutation: only point mutations (base pair substitutions and frameshift mutations) generate high frequencies, whereas deletions of about 50 base pairs or less exhibit no increase over the background in the absence of 5-fluorodeoxyuridine; mutants of intermediate size produce intermediate frequencies (see Drake, 1966b).

DNA Sequence Analysis

The ultimate description of a mutant is its DNA sequence, which, when compared with the sequence at the original site, may provide substantial insights into the mutation process. In such studies one usually seeks the sequence not of a single mutant, but of many. Thus, the application of DNA sequencing to mutational analysis involves the repeated analysis of the same gene (or a portion thereof). Such repetition offers opportunities for efficiency not offered in other DNA sequencing applications. Indeed, this efficiency is essential: characterizing mutational specificity requires the sequences of many mutants.

Strategies that are helpful to maximize efficiency aim to minimize the time required to prepare the DNA from each individual mutant and to enrich the DNA sequence of interest. The analysis of mutant T4 DNA sequences was delayed by the presence of glucosylated 5HMC in the DNA. This modified base inhibits digestion by most restriction enzymes, but several exceptional enzymes have now been identified. The *Taq*I enzyme has been used most extensively. Although limit digests are not easily obtained, gel fractionation followed by Southern blotting permits the direct identification of deletions or additions in the *r*II region. Furthermore, gel fractionation of the *Taq*I digest provides a substantial enrichment for *r*II sequences, thus permitting efficient cloning into M13 or other suitable vectors. The M13 and dideoxy sequencing approaches avoid the requirement for preparative strand separations of each mutant DNA sequenced, and only one or a few primers must be prepared for the sequencing of many or all of the mutants.

The problem of sequencing large numbers of mutants clearly highlights the importance of a strategy that does not require using cytosine-containing T4 DNA in the initial step, since moving mutants into such a genetic background is excessively slow. However, the alternative strategy, inducing the mutations in a cystosine-containing DNA T4 genome, is dangerous. Pyrimidine metabolism is profoundly altered in these strains, and such perturbations are themselves often mutagenic (Bernstein et al., 1972; Smith et al., 1973; Williams and Drake, 1977). Furthermore, both the frequency and the specificity of mutations might be greatly influenced by host enzymes that can act on cytosine-containing T4 DNA, but not on 5HMC-containing DNA.

ARTEFACT ALLEY

Although an artefact occasionally provides serendipitous insights into unimagined solutions, it more often ends in fiasco. The complexity of the taxonomy of mutation is compounded by the vast number of ways that a genome can go wrong, a number far greater than the ways it can go right. Thus, the taxonomy of mutation artefact is the potentially boundless list of all of the ways that the study of wrongness can go wrong. However, there is a rather small set of artefacts that recur so regularly that each generation must be specifically warned.

Differential Survival and Reproduction

The frequency of mutants in a population is a function of the mutation rate, the total size of the population, and the relative survivals or reproductive efficiencies of the two or several mutationally related alleles. Differential survivals of the preexisting genotypes or different growth rates of the mutants and wild type produce errors in calculating either absolute or relative mutation rates unless appropriate corrections are made. Control experiments should always be included, preferably strictly in parallel, until a system is sufficiently characterized to render them no longer necessary. The conventional and usually most appropriate safeguard is the reconstruction selection control: organisms of the genotype whose appearance is being monitored are added to the starting population at frequencies well above those expected to occur through mutation, and the mixture is run through the same protocol as is the experimental population. Then, even when differential survival or reproduction does occur, it becomes possible to factor it away from the true mutational response (or nonresponse).

Differential survivals or growth constants determined in other labortories, or at very different times in a single laboratory, are dangerous guides; reasonably current values should be used. Even in the unlikely case that medium and growth conditions do not change, organisms do.

The major limitation of this control experiment is that it does not address the behavior of mutational heterozygotes. For instance, *WXY* mutagenesis at a nontemplating site might produce an intermediate whose further replication is delayed until released from some block to DNA synthesis, or the heterozygote might be especially favored or disfavored for encapsulation. It is not clear to us what to do about the question of mutational heterozygote selection.

Plating Density Artefacts

Most microbial systems for the analysis of mutation are subject to what are usually called plating density artefacts. The population is plated at high density on selective plates to score mutants and at low density on nonselective plates to score total organisms. The plating efficiencies under the two conditions are likely to be unequal, and variably unequal between the treated and control populations. The net effect may be either an increase or a decrease in apparent mutant frequencies.

There are at least two types of such artefacts in the commonly used T4 mutation screens: reactivation effects, acting to enhance apparent mutant frequencies, and overcrowding effects, acting to reduce apparent mutant frequencies.

Each type of effect is readily seen in both forward and reverse mutation screens. When plating even untreated, fully viable populations of mutants for revertants, there usually exist combinations of numbers of plating cells and phage that, when exceeded, results in fewer plaques. If huge numbers of cells are used to seed plates, then, because T4 does not replicate well in stationary-phase cells, plaques become very small, or invisible. If excessive numbers of phage particles are plated, cell killing occurs; this is perfectly obvious when the lawn is obliterated, but may be inobvious when it merely results in severalfold fewer plaques. When one is screening for plaque morphology variants, plaque crowding will quickly interfere with scoring, but this should be immediately obvious.

Many modes of inactivation act specifically upon the T4 genome, rather than upon its somatic proteins. Very often, such inactivated particles can inject their DNAs and initiate many aspects of the infective cycle. When multiple infection occurs, functional complementation may produce a successful cycle of growth, and recombination may produce both live and dead progeny particles. These interactions are commonly detected as multiplicity reactivation, in which two or more "dead" particles (unable to propagate in single infections) produce a burst containing viable progeny, and as cross-reactivation, in which a live particle and a dead particle coinfect a cell, and the progeny contain markers originating from the dead particle.

Reactivation artefacts are possible whenever the phage population contains particles that are inactive because of DNA damage; they become near certainties when the surviving fraction falls below about 1%. Exceptions to this rule occur, however, when the lethal damage resides mostly in the protein of the phage (for instance, after heat or antibody inactivation) or when genomic killing prevents injection (for instance, after ionizing radiation).

A reasonably reliable preliminary indicator of reactivation artefacts is an inactivation curve that bends upwards in its lower reaches. Unless specific care is taken, for instance, UV inactivation curves tend to bend upward at survivals below about 10^{-3} to 10^{-4}, whereas X-ray inactivation curves tend to remain straight for at least several more decades. Another indicator of trouble is highly variable mutant or revertant frequencies at survivals below about 10^{-2}.

Attempts are often made to obviate reactivation artefacts by holding multiplicities of infection below 0.05 or less, although this can become somewhat difficult when survivals are low, since the relevant multiplicity is that of total, not surviving, particles, and the product of mutation frequency, surviving fraction and total organisms can rather quickly fall below one per plate, making scoring tedious. However, even multiplicities of 0.05 or less may permit enough multiple infections to lead to significant levels of reactivation. And neither calculations nor reconstruction controls are apt to be satisfactory, since they both require extrapolations to regions well outside of those where the measurements are performed.

The correct way to defend against reactivation artefacts is to repeat the mutation frequency determinations at a number of decreasing multiplicities of infection until a clearly direct dependence is observed between numbers of mutants scored and numbers plated (with, of course, unit slope). The difficulty with this test is its demands upon the investigator. Suppose, for instance, that a putative induced mutation rate of 10^{-7} is being measured, up from the background rate of 10^{-8}. Suppose further that the mutagenic treatment has inactivated the phages to a survival of 10^{-3} and that no more than 5×10^{8} cells can be plated without serious loss of efficiency. These are all reasonable numbers. If one begins the dilution experiments at a multiplicity of infection of 0.01, therefore, the number of plaques per plate will be five: not too bad; we will pour 20 plates and have a mutant number accurate to perhaps 10% (but see below). However, let us now decrease by 100-fold the number of particles plated, a factor probably necessary to eliminate the chance that reactivation will occur to a significant extent. Then the number of plaques per plate will be 0.05 (or fewer, if reactivation has been occurring), and we will need to pour about 2,000 plates per experimental point to obtain modest accuracy. This degree of care is infrequently exercised.

Reversion systems should therefore only be used when inactivation does not exceed about 50% or when the question merits carrying out the appropriate controls. However, the induction of an absolute increase in the concentration of mutants may be taken as a strong indication that the increase is real, regardless of the amount of killing.

Plaque Size Artefacts

The survivors of inactivating treatments usually produce smaller plaques than do untreated T4 particles. The unusual plating conditions associated with scoring mutants sometimes themselves also reduce plaque sizes. There is, from time to time, a danger that a significant fraction of plaques will become too small to be seen. The chief indicator of this danger is a gradation of plaque sizes down to those difficult to see. The control is simply to enlarge plaque sizes generally (see Drake, 1966a, and Schnitzlein et al., 1974). This can be achieved by enhancing phage diffusion, for instance, by the use of minimal concentrations of high-quality agar or slowing the growth of the host, either by adopting an intrinsically slowly growing strain or by nutritional control. For instance, plating host cells defective in thymidine kinase on a low-thymine medium will produce a very thin lawn; but since T4 encodes its own thymidine kinase, it happily multiplies to form large plaques ringed by a region of cell growth due to thymidine that diffuses out of the plaque.

Stochastic Artefacts

Most students of mutation eventually encounter an experiment yielding a very significantly increased mutant frequency that never repeats itself. The frequency of such events is at least equal to the probability level adopted as a criterion of significance and is actually much greater because of sources of variation other than the stochastic.

There is little one can do to avoid occasionally missing a mutational response. However, all positive results should be replicated, both to avoid stochastic artefacts and to avoid being misled by experimental errors. Negative results that are to be taken seriously should likewise be confirmed.

The standard measure of accuracy in most microbial experiments is based on the reasonable assumption that replica counts will follow a Poisson distribution. Thus, the variance will be the square root of the mean. It is a common observation, in our laboratory as well as others, that the actual variability between parallel measurements, including the treatment and pipetting steps, is about twice this value. Perhaps those who take more care to count more plaques take similarly greater care in their other operations.

Use of Two-Dimensional Polyacrylamide Gels to Identify T4 Prereplicative Proteins

RAE LYN BURKE,[1]† TIM FORMOSA,[1] KATHLEEN S. COOK,[2] AUDREY F. SEASHOLTZ,[2] JUNKO HOSODA,[3] AND HERB MOISE[3]

Department of Biochemistry and Biophysics, University of California at San Francisco, San Francisco, California, 94143[1]*; Department of Biological Chemistry, The University of Michigan, Ann Arbor, Michigan, 48109*[2]*; and Lawrence Berkeley Laboratory, Biology and Medicine Division, University of California at Berkeley, Berkeley, California 94720*[3]

Polyacrylamide gel electrophoresis is an extremely useful analytical tool for the analysis of complex mixtures of proteins. Components can be separated, identified, and quantitated with a single gel analysis. This technique has been used to study many aspects of T4 development, including phage morphogenesis (Castillo et al., 1977; Laemmli, 1970; Vanderslice and Yegian, 1974), T4 gene regulation (Hosoda and Levinthal, 1968; Kutter et al., 1975; O'Farrell and Gold, 1973a), and biochemical properties of specific T4 proteins (Huang, 1975; Huang and Buchanan, 1974; Manoil et al., 1977).

Using a one-dimensional discontinuous sodium dodecyl sulfate (SDS) system (Laemmli, 1970), which separates denatured proteins based on their size, O'Farrell et al., (1973) resolved approximately 33 bands representing bacteriophage T4 prereplicative proteins. Of these, 25 were characterized as missing in specific nonsense or deletion mutant T4-infected cells, and 19 were identified as the products of known T4 genes. This one-dimensional resolution has been extended by employing a gradient acrylamide gel with a superimposed sucrose gradient and with an increased gel length so that a total of 58 to 64 separate prereplicative proteins are visible (Morton et al., 1978).

How many genes of this transcriptional class are likely to be encoded by the T4 phage DNA? About 88 early and middle genes have already been identified genetically, and the regions of the T4 DNA expressed as early or middle functions encompass about 111 kilobases (Kutter and Rüger, this volume; Mosig, appendix, this volume; Wood and Revel, 1976). Assuming that an average gene covers 1 kilobase (sufficient to encode a protein of about 40,000 daltons) and that T4 genes do not overlap, we can estimate that there are about 111 T4 early genes (or even more if the average molecular weight is smaller than 40,000; see below). Therefore, it is evident that not all T4 prereplicative proteins can be resolved by one-dimensional polyacrylamide gel electrophoresis.

The two-dimensional polyacrylamide gel electrophoresis technique developed by O'Farrell (1975) combined isoelectric focusing in the first dimension with SDS gel electrophoresis in the second to generate a theoretical resolution capacity of 5,000 proteins. However, only proteins with isoelectric points in the pH range 4 to 7 are resolved; proteins with basic isoelectric points are excluded from the first dimension. This two-dimensional electrophoretic fractionation has been employed to identify T4 late proteins, particularly gp23 (Castillo et al., 1977), and recently to resolve 45 separate T4 phage coat proteins (Gersten et al., 1981), although none of these were identified as specific gene products.

A more recent modification of this gel system employing a nonequilibrium pH gradient electrophoresis (NEPHGE) gel in the first dimension displays proteins with isoelectric points ranging from pH 3 to pH 10. For *Escherichia coli*, approximately 15 to 30% more proteins are thus resolved. We have used these NEPHGE gels to identify particular T4 prereplicative proteins (Alberts et al., 1980; Alberts et al., 1983; Cook and Seasholtz, 1982). In each laboratory, independently, these experiments relied on the initial establishment of a standard gel pattern and the identification of the positions or coordinates of protein spots corresponding to specific prereplicative gene products. Here we wish to add the identification of more gene products, present and compare standard gel patterns obtained in three different laboratories, and discuss the effects of various methods of sample preparation on the components solubilized and resolved.

TECHNIQUE

Procedural details appear to affect significantly the solubilization and resolution of a number of proteins. One method is therefore presented here in some detail, and some comparisons among the methods used in the three laboratories are summarized below.

Sample Preparation

Materials were obtained from the sources noted by Cook and Seasholtz (1982). *E. coli* B was grown at 37°C in M9 minimal medium (Miller, 1972) supplemented with 20 μg of tryptophan per ml to a cell density of 5 × 10^8 cells per ml. A 1-ml portion of the culture was transferred to a sterile 30-ml Corex tube at 37°C and was infected with the appropriate phage stock at a multiplicity of infection of 10. Mixed ^{14}C-labeled amino acids (10 μCi) were split into four aliquots and were added at 3.5, 4.5, 5.5, and 6.5 min postinfection (p.i.). A total of 20 to 50% of the radioisotope was incorporated under these conditions. At 7.5 min p.i., Casamino Acids were added to a final concentration of 1%, and the incubation was continued for 1 min longer. The culture was chilled briefly on ice and then was centrifuged for 10 min at 12,000 × *g*. The supernatant was discarded, and 100 μl of sonication buffer (100 mM Tris-hydrochloride [pH 7.4], 5 mM $MgCl_2$, 50 μg of pancreatic RNase per ml) was added to the pellet. Phenylmethylsulfonyl fluoride was added to

† Present address: Chiron Corp., Emeryville, CA 94608.

inhibit proteases (1 μl of a 100 mM stock). The solution was transferred to a 1.5-ml Eppendorf tube and was sonicated for 1.5 min with three 30-s blasts interspersed with 1 min of chilling on ice. Sonication was carried out at 150 W from the large probe of a Biosonic (or equivalent) sonicator. For this sonication, the capped tube was held directly against the probe suspended in a beaker of ice water. DNase I and micrococcal nuclease were each added to a final concentration of 50 μg/ml, and $CaCl_2$ was added to 10 mM. The sample was held at 4°C for 10 min, and 120 mg of solid urea was added along with 0.1 ml of lysis buffer (9.5 M urea, 5% Nonidet P-40 [NP-40], 4% ampholines at pH 5 to 7, 1% ampholines at pH 3.5 to 10, 12.5% mercaptoethanol). The sample was quick-frozen in a dry ice-ethanol bath and was stored at −70°C.

Electrophoretic Procedure

For the first dimension, nonequilibrium pH gradient gels were poured to a height of 12 cm as described (O'Farrell et al., 1977). These gels contained a final concentration of 1.6% ampholines at pH 5 to 7 and 0.4% ampholines at pH 3.5 to 10. A sample prepared as above and containing approximately 5×10^5 acid-precipitable cpm and 1 μg of total protein in a volume of 10 to 50 μl was loaded onto each gel. The gels were then electrophoresed toward the cathode (with the basic proteins leading the separation) at 400 V (constant voltage) for 3 h for a total of 1,200 V · h. Electrophoresis for longer times improves the resolution of the more acidic species, but some of the very basic proteins (such as gp*ip*III and gp61) may be lost. The gels were carefully extruded from the glass tubes, equilibrated in SDS running buffer as described (O'Farrell, 1975), and stored frozen at −70°C. For the second dimension, an SDS polyacrylamide gel 0.75 mm thick with a 2.5-cm 4% stacking gel and a 12-cm 12.5% separating gel (O'Farrell, 1975) was used. The gels were electrophoresed for approximately 4 h at 20 mA of constant current until a marker of bromophenol blue dye traveled 10 cm into the separating gel. The gels were fixed in methanol-acetic acid-water (5:1:5) and were stained in the same solution containing 0.2% Coomassie blue. Gels were destained by treatment for 30 min in methanol-acetic acid-water (5:1:5) and then overnight in 7% acetic acid. When necessary to enhance the detection sensitivity to aid in the identifica-

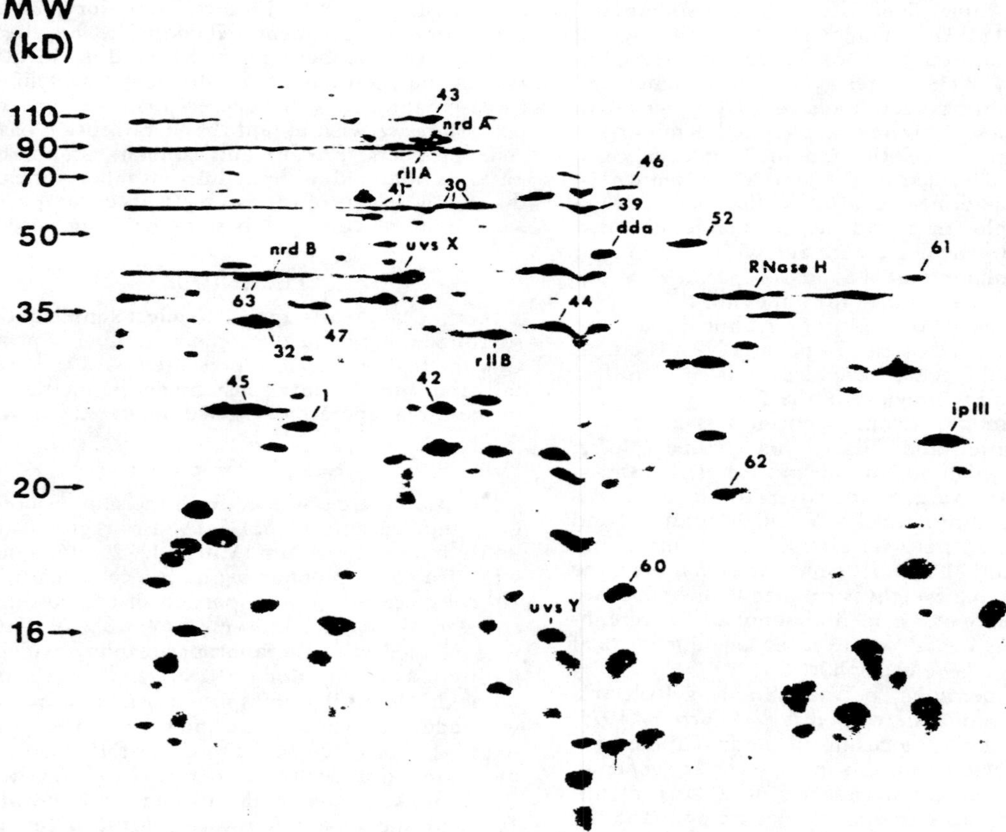

FIG. 1. Autoradiogram of a two-dimensional gel profile of ^{14}C-labeled T4 prereplicative proteins. The molecular weight scale was established with the following proteins as standards: gp43, 110,000; gp30, 60,000; gp*dda*, 47,000; X' protein, 40,000; gp32, 34,500; gp45, 25,000; gp62, 20,000; gp60, 18,000; gp*uvsY*, 16,000. These sizes were determined on a one-dimensional gel with molecular weight standards phosphorylase A, 96,000; bovine serum albumin, 68,000; ovalbumin, 44,000; gp32, 34,500; gp45; 24,700; myoglobin, 16,700; and lysozyme, 14,300. The pH gradient increases from left to right, with acidic proteins on the left.

tion of host proteins, gels were silver stained (Merril, Goldman, Sedman, and Ebert, 1981). For autoradiography, the gels were dried on Whatman 3MM paper and exposed to Kodak X-Omat No Screen film. To increase sensitivity, the gels were occasionally fluorographed (Bonner and Laskey, 1974). We have noticed that the prior use of silver staining reduces the efficiency of fluorography. This electrophoresis procedure was used to generate the gel pattern displayed in Fig. 1.

An alternative procedure, used to prepare samples for the gel pattern shown in Fig. 2, has been published (Cook and Seasholtz, 1982). This method differs from the one presented here in the following details: (i) infected cells were labeled with $^{35}SO_4$ at 30°C from 2 to 8 min p.i.; (ii) cell pellets from unlabeled cells were added to the labeled portion at a ratio of 6.6:1 before sonication; (iii) sonicated cell suspensions were centrifuged at 5,000 × *g* for 5 min to remove unbroken cells and debris; and (iv) these cell extracts were not treated with nucleases or the protease inhibitor phenylmethylsulfonyl fluoride. The electrophoresis conditions were also slightly different, as follows: (i) ampholines at pH 3.5 to 10 were added to a final concentration of 2% for the first-dimension NEPHGE gel, and no ampholines at pH 5 to 7 were included; (ii) NEPHGE gels were electrophoresed for 4 h at 500 V for a total of 2,000 V · h; (iii) for the second dimension, a 10% polyacrylamide gel was used; and (iv) the gel sample contained about 10^5 cpm and 10 to 100 μg of protein rather than 5 × 10^5 cpm and 1 μg of total protein.

A method to extend the analysis to lower-molecular-weight proteins has been developed by Hosoda and Moise. A sample gel pattern is shown in Fig. 3. For this gel, infected cells were labeled at 30°C from 8 to 13 min p.i. The sample preparation differs from the two presented above and is a modification of that used by Ames and Nikaido (1976). A cell pellet equivalent to 5 × 10^8 cells was lysed by direct boiling for 5 to 10 min in 50 μl of lysis buffer (63 mM Tris-hydrochloride [pH 6.8], 2% SDS, 5% 2-mercaptoethanol). Urea (50 mg) was added, along with 4 μl of NP-40 and 100 μl of NEPHGE buffer (9.5 M urea, 5% [vol/vol] 2-mercaptoethanol, 8% [wt/vol] NP-40, 2% [wt/vol] ampholines at pH 3.5 to 10). In this protocol, the final ratio of NP-40 to SDS was 8; the excess NP-40 was added to displace SDS bound to proteins before isoelectric focusing. The NEPHGE gels were then run for 4 h at 500 V. Electrophoresis in the second dimension was the same as that described above, except that the separating gel was 18 cm long and consisted of a 10 to 17.5% acrylamide gradient with a superimposed 5 to 20% sucrose gradient (Morton et al., 1978).

IDENTIFICATION OF SPECIFIC PROTEINS

Shown in Fig. 1 is the two-dimensional NEPHGE pattern of ^{14}C-labeled proteins from T4D-infected cells at early times after infection. Approximately 96 T4 prereplicative proteins are resolved in this particular pattern. A total of 25 have been identified as the products of specific genes, as indicated, by at least one of three methods: (i) detection of a spot which is absent in the profile obtained from bacteria infected with an appropriate T4 (nonsense or deletion) mutant(s) and which reappears when the T4 amber mutant is grown in a host cell carrying a suppressor tRNA; (ii) demonstration of coelectrophoresis of a particular protein from the T4-infected cell with a bonafide purified protein; and (iii) comparison with the protein profile generated in the two-dimensional analysis shown in Fig. 2. The corresponding molecular weights and the exact basis for the identification of each spot are given in Table 1. A number of additional proteins are identified in Snustad et al. (this volume), Fig. 9.

FIG. 2. Two-dimensional gel profile of ^{35}S-labeled early T4 proteins synthesized after infection by *am*A453 (gene 32). The pH of the gradient increases from left to right. The numbered spots correspond to the products of the following genes: (1) 43, (2) *nrdA*, (3) *r*IIA, (4) 46, (5) 30, (6) 41, (7) 39, (8) 52, (9) *nrdB*, (10) 32, (11) 44, (12) *r*IIB, (13) *td* (thymidylate synthetase), (14) 42, (15) 1, (16) *ip*III, and (17) 60. The protein spots are numbered on the upper right side except when arrows have been used. Of these 17 proteins, 14 were identified by two-dimensional gel analysis of lysates from amber mutant-infected cells; *r*IIA, *r*IIB, and *td* were identified from deletion mutant-infected cells (Cook and Seasholtz, 1982).

Some proteins generate multiple spots in the first dimension, including gp63 (RNA ligase), gp41 (replication component), and gp30 (DNA ligase), probably owing to the presence of modified species or covalent enzyme-adenylate complexes in the case of the ligases. Other proteins, for example gp*dda*, run reproducibly as double spots in the second dimension. The source of this heterogeneity is not clear, but it could arise during electrophoresis from incomplete protein denaturation or from protein modification. It has been noted that the inclusion of a free radical scavenger such as 0.1 mM sodium thioglycolate to the cathode buffer reservoir blocks the alteration of tryptophan, histidine, and methionine side chains during electrophoresis (M. W. Hunkapiller, E. Lujan, F. Ostrander, and L. E. Hood, Methods Enzymol., in press). Such an addition reduces gp61 to a single spot in the second dimension but has no similar effect on dda.

The gel shown in Fig. 1 has been autoradiographed long enough to reveal gp61. The abundance of this protein in vivo has been estimated to be 0.06% of the total T4 protein by weight or 0.0006% of the total cellular protein, corresponding to 160 molecules per cell (R. Burke and B. Alberts, unpublished observa-

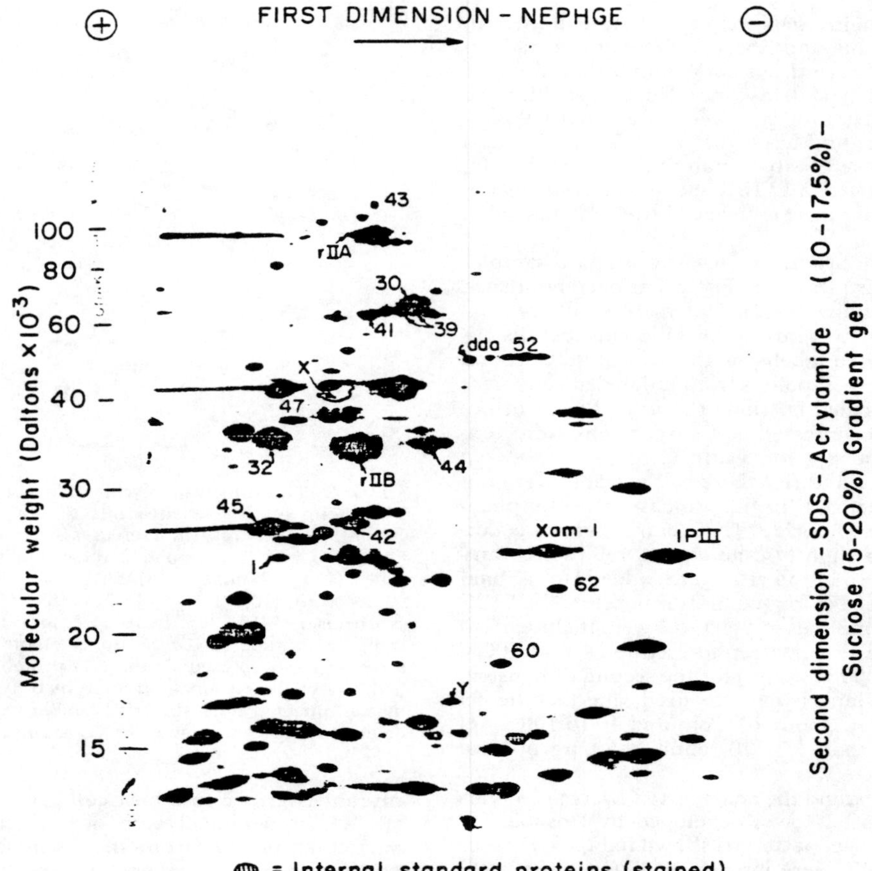

FIG. 3. Fluorogram of a two-dimensional gel pattern of a ^{14}C-labeled lysate from T4 *uvsX* (*am*1)-infected *E. coli*. A total of 19 spots are identified, of which 7 were determined by direct analysis of the protein patterns from amber mutant-infected cells. These proteins were gp30 (*am*H39X), gp32 (*am*H18), gp39 (*am*N116), gp45 (*am*E10), gp47 (*am*H3x5), *uvsX* (X*am*1 from Mark Conkling), and *uvsY* (Y*am*1 from Mark Conkling). Other identifications were done indirectly by comparing this gel profile with that shown in Fig. 1. The crosshatched circles mark the positions of unlabeled stained proteins added to the cell lysate as internal standards by using 2 μg of protein per spot of 32* III protein, a tryptic fragment of gp32 (molecular weight, 26,000) soybean trypsin inhibitor (21,500), and bovine hemoglobin α chain (15,800) and β chain (15,100). (Actually, this sample of hemoglobin resolved into three spots as marked). The open circle indicates where gp*uvsX*, if present, would run.

tions). Some minor spots of lower intensity than gp61 may represent major *E. coli* proteins which are labeled in the uninfected cells. We calculate that even if 99.95% of the bacterial cells were infected by phage, an *E. coli* protein which represents 1% of the total cellular protein would contain only fivefold fewer cpm than gp61 owing to the labeling of uninfected cells. To detect labeled host proteins, the autoradiogram may be aligned with the Coomassie blue or silver stained pattern which represents major *E. coli* proteins.

The profile in Fig. 1 is similar enough to that shown in Fig. 2, obtained by Cook and Seasholtz (1982), to permit the cross-identification of a given protein between one gel and the other. On this very clear pattern, over 65 spots are resolved, and 17 are identified and designated by number with the corresponding genes given in the figure legend. One protein which we were not able to cross-identify in Fig. 1 was thymidylate synthetase. In addition, gp61 is observed in Fig. 1 but is missing here. This absence may have resulted from the increased electrophoresis time in the first dimension or from a shorter autoradiographic exposure. However, a closer comparison of the two profiles shows many differences, such as missing spots or major alterations in intensity. These variations most likely arise from the different labeling regimes, 3.5 to 7.5 min p.i. at 37°C for Fig. 1 compared with 2 to 8 min p.i. at 30°C for Fig. 2, since the transcriptional pattern changes so rapidly during this phase of the infection (O'Farrell and Gold, 1973a).

ANALYSIS OF LOW-MOLECULAR-WEIGHT PROTEINS

Previously, E. Kutter had observed over 50 proteins with molecular weights below 14,000 on two-dimensional gels (see Snustad et al., this volume). To improve the resolution of such proteins, two of us (J.H.

TABLE 1. Basis of identification of T4 proteins for the two-dimensional gel of Fig. 1

Gene	Protein activity	How identified	Source
43	DNA polymerase	Mutant *am*B22	Lab stock
nrdA	Ribonucleoside diphosphate reductase, subunit	Comparison with Fig. 2	
*r*IIA	Membrane protein	Mutant *r*1589	Peter Gauss[a] (*r*IIA-*r*IIB fusion)
46	Exonuclease	Mutant *am*B14	Junko Hosoda
41	RNA priming protein	Mutant *am*N81	Lab stock
30	DNA ligase	Purified protein	
39	DNA topoisomerase subunit	Purified topoisomerase	
52	DNA topoisomerase subunit	Purified topoisomerase	
dda	DNA helicase	(i) Mutant Δ(39–56)12 *am*L148; (ii) purified protein	Peter Gauss
uvsX	Recombination	*X am* mutant	Mark Conkling[b]
61	RNA priming protein	(i) Mutant *am*E219; (ii) purified protein	Jim Karam[c]
nrd B	Ribonucleoside diphosphate reductase, subunit	Comparison with Fig. 2	
63	RNA ligase	Purified protein	Olke Ulenbeck[d]
Unknown	RNase H[e]	Purified protein	
47	Exonuclease	Mutant *am*NG163	Junko Hosoda
32	Helix-destabilizing protein	Mutant *am*A453	Lab stock
44	Polymerase accessory protein	Purified protein	
*r*IIB	Membrane protein	Mutant *r*1589	Peter Gauss
45	Polymerase accessory protein	Purified protein	
1	Deoxynucleotide monophosphate kinase	Purified protein	
*ip*III	Internal capsid protein	Purified protein	Lindsay Black[f]
62	DNA polymerase accessory protein	Purified protein	
60	DNA topoisomerase subunit	Purified topoisomerase	
uvsY	Recombination protein	*Y am* mutant	Susan Wallace[g]

[a] University of Colorado, Boulder.
[b] Massachusetts General Hospital, Boston.
[c] Medical University of South Carolina, Charleston.
[d] University of Illinois, Urbana; provided purified protein.
[e] Unpublished observations of Vicki Chandler and Bruce Alberts.
[f] University of Maryland, Baltimore; provided purified protein.
[g] Valhalla University, Valhalla, N.Y.

and H.M.) have employed a double-gradient acrylamide-sucrose gel in the second dimension. An example of the enhanced resolution obtained with this method is shown in Fig. 3. The lower-molecular-weight proteins generate much sharper spots, and the size range of the fractionation is extended to molecular weights well below 10,000. The number of these small gene products is surprising; of the 194 proteins resolved in this gel pattern, 45% have molecular weights less than 20,000, and 50 have molecular weights less than 15,000. Many of the smallest appear to be extremely abundant. This profile, with many basic proteins and a predominant abundance of low-molecular-weight proteins, differs dramatically from those generated by *E. coli* and by eucaryotic cells (O'Farrell, 1975). Only 16 T4 prereplicative proteins with molecular weights smaller than 15,000 are listed by Wood and Revel (1976). In addition, gp*uvsY* is documented here. The *reg*A protein has been reported to have a molecular weight of 12,000 (Cardillo et al., 1979; J. Karam, personal communication), and two small ribosome-associated proteins with molecular weights smaller than 10,000 have been observed (J. S. Wiberg, personal communication), as well as an exonuclease V-inactivating protein with a molecular weight of 12,000 (Behme et al., 1976) and an anti-restriction nuclease (*arn*) protein (Dharmalingam et al., 1982; Revel, this volume) for a total of 22 identified very small T4-encoded proteins. We speculate that owing to their small size, many of the genes coding for small proteins may not have been targets for mutagenesis. Also, a number of them are apparently missing in specific deletion mutants and may represent uncharacterized nonessential genes (E. Kutter, personal communication). However, we realize that some of these minor proteins may not represent primary gene products but rather may arise from intracellular processing or as artifacts of sample preparation or electrophoresis. A single charge change (owing, for example, to deamidation) or a partial denaturation could result in a major change in the electrophoretic migration of such a small protein.

DISCUSSION

We have all observed small variations in the patterns as a result of sample solubilization conditions. For example, an increase in the number of very basic proteins solubilized is observed when protamines and high salt are added to displace DNA-bound proteins instead of using a DNase digestion (Sanders et al., 1980). The extraction of proteins by boiling in an SDS buffer (Fig. 3) releases substantially more of some membrane proteins (for example, gp*r*IIA, gp*r*IIB, gp46, gp47, and an unidentified spot to the acidic side of gp32) than the NP-40 sonication method does. However, other proteins, notably acidic species, disappear from the pattern (Cook and Seasholtz, 1982). It should also be noted that the apparent charge of some spots, for instance that of *uvsY*, is altered by SDS treatment. However, for any one given method, the patterns are quite reproducible.

The final method of choice is dictated by ease of sample handling and the specific proteins of interest. Encouragingly, the patterns obtained by these different methods are generally similar enough to permit the cross-identification of this core group of proteins which can serve as a constellation to aid further mapping of prereplicative proteins and facilitate the application of two-dimensional gel analysis to various physiological studies.

We are grateful to Bruce Alberts and to G. Robert Greenberg for providing support, advice, and encouragement. We also thank Betty Kutter for her help with the initial gel profiles.

The work was supported in part by Public Health Service grant GM24020 from the National Institutes of Health in the laboratory of Bruce Alberts; by Public Health Service grants GM25973 and GM29025 from the National Institutes of Health and by National Science Foundation grant PCM77-20291 in the laboratory of G. Robert Greenberg; and by Public Health Service grant GM23563 from the National Institutes of Health in the laboratory of Junko Hosoda.

SOME COMPLEXITIES OF T4 GENES, GENE PRODUCTS, AND GENE PRODUCT INTERACTIONS

The *r*II Genes: a History and a Prospectus

BRITTA SWEBILIUS SINGER, SIDNEY T. SHINEDLING, AND LARRY GOLD

Department of Molecular, Cellular and Developmental Biology, University of Colorado, Boulder, Colorado 80309

WHAT DO THE *r*II GENE PRODUCTS DO?

We have known for 25 years that two genes comprise the *r*II locus (Benzer, 1955), and we have "seen" the *r*IIA and *r*IIB proteins on sodium dodecyl sulfate gels for 10 years (O'Farrell et al., 1973). Nevertheless, we have no satisfactory idea what the *r*II proteins do. No purified *r*IIA or *r*IIB protein has been used successfully in any in vitro assay. However, a large amount of literature once was concerned with the role of the *r*II proteins during phage infection; we review that literature here.

Role of the *r*II Proteins in the Membrane

At first the *r*II proteins were thought to play an important role in membrane function. Mutants altered in either *r*IIA or *r*IIB yield large, sharp-edged plaques on lawns of *Escherichia coli* B (Hershey, 1946a). In liquid media, cell lysis (using *E. coli* B) occurs much more rapidly after infection with *r*II mutants than after infection with wild-type phage. When cells infected by wild-type phage are opened, both *r*II proteins are found in the inner, cytoplasmic membrane of the infected cells (Ennis and Kievitt, 1973; Takacs and Rosenbusch, 1975; Weintraub and Frankel, 1972). Few other early T4 proteins partition as completely into the membrane, or remain there after attempted extractions with a variety of chemicals (Takacs and Rosenbusch, 1975).

The hypothesis that *r*II gene products function in the membrane was strengthened by the finding that *r*II mutants suppress lysis-defective T4 mutants in the *t* gene (Josslin, 1971). Gene *t* is thought to encode a lysis function. The *r*II proteins might further slow lysis of t^- phage by stabilizing the membrane (Josslin, 1971). We isolated a bacterial strain (called TabR [Nelson et al., 1982]) which appears to have a defective membrane and which shows a strong requirement for the *r*II proteins for phage production. When the *r*II proteins are missing, the infected cells lyse very early. Again, one might argue that loss of a membrane component (either of the *r*II proteins) destabilizes the weakened cell envelope of the infected TabR mutant (Nelson et al., 1982).

Role of the *r*II Proteins in Overcoming *rex* Exclusion

We have not yet mentioned the restrictive host used for all classic genetic studies of *r*II mutants. *E. coli* lysogenic for phage lambda restrict *r*II mutants (Benzer, 1955). Restriction is caused by the *rex* genes (there are two, A and B, adjacent to *c*I in the immunity region, and both are required [Matz et al., 1982]). Rex^- lysogens do not restrict *r*II mutants. No functions of lambda from outside the immunity region are needed for restriction; plasmids carrying only the lambda *c*I and *rex* genes convert bacteria to hosts restrictive for *r*II mutants (L. Gold and A. Peskin, unpublished data). The lambda *rex* genes are defined primarily by their exclusion of T4 *r*II mutants (*rex* = *r*II exclusion); thus, *rex* activities shed no light on *r*II functions. However, some investigators have suggested that at least one of the *rex* gene products is a membrane protein (Campbell and Rolfe, 1975). During restrictive infections (of lambda lysogens by *r*II mutants), a generalized metabolic arrest occurs after about 10 min (Garen, 1961).

Some remarkable experiments were carried out by Sekiguchi (1966). He found that restricted infections were made permissive when monovalent cations were eliminated from the medium. That is, a lambda lysogen is restrictive for an *r*II mutant (even an *r*IIA-*r*IIB deletion) only if the monovalent cation concentration is above several millimolar. At pH 7.5, in the absence of monovalent cations, rII^- and rII^- infections of lambda lysogens yield nearly equivalent bursts. We have repeated all of Sekiguchi's experiments, using an *E. coli* B strain made lysogenic for lambda, and all of the results were reobtained (L. Gold and M. Cobianchi, unpublished data). Furthermore, a salt-refractory period begins several minutes after the *r*II gene products would appear in a wild-type infection (L. Gold and M. Cobianchi, unpublished data). Historically, these results have led everyone to believe that rII^- infections of a lambda lysogen have an impaired membrane.

Thorough documentation of this view has been provided by Duckworth et al. (1981), who at the same time emphasize Mg^{2+} suppression of the restricted infection. Our reading of the same literature has brought us to a different conclusion: *rex*-mediated restriction is a function of high monovalent cation concentration, not the presence of Mg^{2+}. In fact, it was shown that sucrose or Tris was as efficacious as Mg^{2+} in allowing *r*II$^-$ progeny development on a lysogenic host, as long as the monovalent cation concentration was kept low (Sekiguchi, 1966). In addition, both Mg^{2+} and polyamines allow some *r*II$^-$ phage growth on a *rex*$^+$ host, even in the presence of a high salt concentration (Ferroluzzi-Ames and Ames, 1965; Garen, 1961).

The Sekiguchi data allow two interpretations, each a restatement of his observations. The usual interpretation is that *r*II$^-$ infections of *rex*$^+$ lysogens are poisoned at high monovalent cation levels by rapid intrusion of medium components into the cell. This intrusion is considered to be related to leakage caused by *r*II$^-$ infections (Ferroluzzi-Ames and Ames, 1965). A second interpretation is that lambda lysogens are phenotypically Rex$^-$ in low monovalent cation-containing media. That is, were we to assay *rex* activity in a manner independent of the *r*II genotype of infecting T4, *rex* activity would be monovalent cation dependent. In this view the cationic constitution of the medium affects *rex*, not *r*II. Lambda lysogens (if *rex*$^+$) also exclude other, nonimmune lambdoid phages (Toothman and Herskowitz, 1980a). These restricted infections also become permissive if Mg^{2+} replaces monovalent cations in the medium (Toothman and Herskowitz, 1980c). Furthermore, other conditions restrictive for *r*II$^-$ infections (a TabR host [Nelson et al., 1982]) are not reversed by low salt concentrations (Nelson and Gold, unpublished data).

We are not the first authors to wonder whether protein activity might be dependent on monovalent cations. Kohno and Roth (1979) have published a remarkable paper showing that all *ts* missense mutations in the *Salmonella typhimurium his* operon can be phenotypically suppressed by modest concentrations of NaCl. These *ts* mutations were selected only as temperature-sensitive histidine auxotrophs. Kohno and Roth note that conventional wisdom about salt suppression would center on membrane defects; however, the *his* proteins in question are soluble proteins. Kohno and Roth have argued that proteins may be stabilized by salt in the medium, as they are in vitro by salts added to solutions containing the enzymes. Is it not possible, therefore, that one or both of the *rex* gene products are inactivated by the absence of salt in the medium? If so, the activity of *rex* (and, by inference, of *r*II) need not be confined to the membrane. The restrictive phenotypes (see above) would all be explained by indirect metabolic interactions. In this view, it is noteworthy that at least one lambdoid phage gene that overcomes *rex*-mediated exclusion is involved in phage DNA replication (Toothman and Herskowitz, 1980b).

Role of the *r*II Proteins in Replication

What, then, do we know about the *r*II gene products and DNA replication? Mutations of T4 DNA ligase are suppressed by *r*II mutations, as long as functional host DNA ligase is available (Berger and Kozinski, 1969; Gellert and Bullock, 1970; Karam, 1969; Krisch et al., 1971). Deletions of *r*II revert on our new restrictive host (TabR) by elimination of the nonessential T4 gene for thymidine kinase (Nelson et al., 1982); we do not understand this result, intriguing though it is. Deletions of *r*II never revert when the plating host is a lambda lysogen (Benzer, 1955; Nelson et al., 1982). When the Alberts laboratory purified "replication complexes" from T4-infected cells (including any proteins bound to newly replicated or parental DNA), the *r*IIA and *r*IIB proteins were enriched in the complex (Manoil et al., 1977). The complex was reported to be free of membranes. Both the *r*IIA and *r*IIB proteins have been shown to bind to DNA-cellulose, even when other important replication proteins are missing from the extracts (Huang and Buchanan, 1974).

Most importantly, the *r*IIB gene is activated, transcriptionally, by the T4 *mot* gene product (Mattson et al., 1974; Mattson et al., 1978) and is repressed, translationally, by the T4 *regA* gene product (Karam and Bowles, 1974). The genes whose transcription is stimulated by *mot* are a crucial set of genes involved in DNA replication (genes 43, 45, 32, and 1), as are the larger set of genes repressed by *regA* (genes 1, *cd*, 56, 42, *r*IIA, *r*IIB, 44, 62, 45, and *regA*) (Campbell and Gold, 1982; Karam and Bowles, 1974; Wiberg et al., 1973). We have presented a guess about *r*II function, derived mostly from thinking about regulatory patterns (Campbell and Gold, 1982). This argument would be overwhelming were it not for the obvious tautology (E. Brody, personal communication); one must search the early regions of the T4 genome with great care to find functions not involved in DNA or deoxyribonucleotide synthesis!

It is clear, finally, that neither the phenomenology of *r*II$^-$ infections nor the behavior of the *r*II gene products during various subcellular fractionation attempts is going to define *r*II function. Isolation of the proteins and enzymatic characterization based on guesses and existing clues will be necessary. Let us go back, then, to the *r*II cistrons as genetic elements. Let us trace their history and the uses to which they were put in the golden age of molecular biology. We will then conclude with a section on the uses of *r*IIA and *r*IIB in studies of gene expression during T4 development.

DEVELOPMENT OF *r*II AS A GENETIC SYSTEM

Hershey's discovery of *r* plaques in bacteriophage T2 opened up the field of phage genetics. He showed that plaque morphology was a heritable trait and that both forward and backward mutations could be detected (1946a, 1946b). Forty years later it seems incredible that perhaps his most significant contribution in those early experiments and the experiments of Delbrück and Bailey (1946) was the demonstration that it is possible to do a genetic cross between differently marked phage. The first experiments were interpreted to be akin to bacterial transformations; that the reassortment of phenotypes might be attributed to "transfers, or even exchanges, of genetic materials" was considered as an alternate explanation. Hershey and Rotman (1949) demonstrated that there are three "classes" of *r* mutants in T2 and measured their linkage relations to each other and to a host range mutant. The title of this paper was "Genetic Recombination between Host-Range and Plaque-Type Mutants of Bacteriophage in Single Bacterial Cells." Phage genetics had arrived.

The discovery that made *r*II a genetic system of choice was Benzer's happy finding that *r*II mutants are unable to grow on lambda lysogens. Because nonlysogens are permissive hosts, all *r*II mutants, even deletions, are conditional lethals. Intrinsic conditional lethality is a boon to the geneticist, but it is a feature shared by many other systems, e.g., various auxotrophies of bacteria. The advantage to finding such a condition in phage is that T4 has no mating types. Newly discovered mutations can be crossed to each other without the necessity of first putting them into an appropriate genetic background. And, of course, the fact that on *E. coli* B, *r* mutants have a distinctive visible phenotype made it easy to find new mutants.

Structure of *r*II

Structure-function relationship. In 1955, Benzer showed that the *r* mutations of one cluster in T2 and the *r* mutations that Doermann and Hill (1953) mapped to linkage group II were all restricted by K(λ). In fact, two-thirds of all the *r* mutants isolated were restricted by K(λ), and the mutations mapped in this region. Restriction gave Benzer a handle both for functional analysis and for fine-structure mapping of the *r* mutations of that region (1955). He performed complementation tests and found that the mutations fell into two groups, *r*IIA and *r*IIB. These he termed cistrons (Benzer, 1957) to distinguish functional units defined by complementation from other items of genetic currency—those defined by recombination or mutation. Fine-structure mapping revealed that all of the mutations could be described as a linear array and that all of the mutations of the *r*IIA group mapped to one side of the mutations of the *r*IIB group (see below). He further noted that different mutations reverted at different rates, and some reverted not at all. These observations became the basis for Benzer's elucidation of the topology and topography of the gene. Among mutations chosen for extreme stability to reversion, some gave no recombinants with a number of other mutations which were, however, capable of recombining with each other. Benzer (1955) wrote, "These results can be understood if it is assumed that each mutation extends over a certain length of the chromosome, and production of wild type requires recombination *between* those lengths." Phage genetics had come a long way in a decade. The fact that a plausible and elegant structure of DNA had been elucidated (Watson and Crick, 1953a; Watson and Crick, 1953b) must have made Benzer's work easily comprehensible.

Topology. Benzer's next task was to prove that the gene is a linear structure. Since DNA was known to be the genetic material and since DNA has a linear structure, it seemed reasonable that the gene should have a linear structure also. Because of the approximate additivity of recombination frequencies, quantitative arguments (referred to above) could be made that the gene is a linear array of mutable sites. Benzer wanted a qualitative proof. He wrote, "A critical examination of the question should be made from the point of view of *topology*, since it is a matter of how the parts of the structure are *connected* to each other, rather than of the distances between them" (Benzer, 1959). He used nonreverting mutations so that any wild-type phage that emerged from a cross could be attributed only to recombination. As Benzer (1957) pointed out, nonreverting mutations that fail to recombine with several point mutations are deletions, although that had not yet been independently demonstrated. In his classic paper, "On the Topology of the Genetic Fine Structure," Benzer (1959) showed that the set of partially overlapping *r*II deletions he used can only be described as a linear structure ($P \ll 10^{-17}$). However, in regions where the deletions did not partially overlap, special branched structures could not be ruled out. Furthermore, the orientation of such regions cannot be known by deletion mapping. By 1961, Benzer had the whole *r*II region sorted out, with the exception of the distal end of *r*IIB, a region with no partially overlapping deletion that orients it with respect to the rest of *r*II. Parma et al. (1979) showed by duplication mapping (analogous to deletion mapping but harder to do) that the distal end of *r*IIB should be turned around relative to the rest of the published map.

Topography. Deletion mapping made possible Benzer's next tour de force, "On the Topography of the Genetic Fine Structure" (1961). Over 2,400 induced and spontaneous mutations were mapped to 308 sites within *r*IIA and *r*IIB. Some sites had only one isolate; other sites were considerably hotter (one site had more than 500 isolates). Not all of the mutations mapping at a site were necessarily identical to each other. Benzer (1957) showed that mutations at one of the hottest sites fall into two classes on the basis of reversion frequency. Pribnow et al. (1981) extended this analysis by suggesting that the three hot spots in question are all located at sites where there are six adjacent A · T base pairs in the wild-type sequence, inferring that high-reverting mutations have gained an A · T, whereas low-reverting mutations have lost an A · T. We have confirmed this by sequencing the DNA (unpublished data). At other sites, variants could be identified on the basis of ambivalence, i.e., the ability to plate on one but not another lambda lysogen (Benzer and Champe, 1961). Tessman (1965), using ultrahigh-sensitivity crosses, showed that one of the hot spots is actually composed of three different mutations. These three mutations also have different ambivalence from one another and, as we now know (Nelson et al., 1981), are amber, UGA, and ochre derived from three adjacent base pairs (the wild-type codons are UGGCAA). Ambivalence therefore reflects the use of lambda lysogens carrying different suppressor tRNAs (see below).

The topography of *r*II has been a source of interest for decades. As early as 1958, Benzer and Freese noted that the spectrum of mutations induced by 5-bromouracil (BU) is different from the spontaneous spectrum. They pointed out:

> Were it not for the high resolution of our genetic techniques, it might have been erroneously concluded that 5-bromouracil acts aspecifically, since, among all the induced mutants of the *r* phenotype, the proportion of rII mutants and even the ratio of A cistron to B cistron mutants are quite comparable to the spontaneous values. It is only at finer resolution that the different effects are revealed, and it is seen that mutations are, for the most part, produced at specific places in the genetic structure.

The specificity of dozens of chemical mutagens has

since been studied by using the *r*II system (for reviews, see Drake, 1970, and Drake and Ripley, this volume). Mutators may be thought of as another sort of mutagen. The *r*II system has been used to study fidelity of replication in vivo since Speyer's insight that missense in the DNA polymerase might affect mutation rates (Speyer, 1965). Replicative fidelity has been reviewed by Drake (1970, 1973) and by Sinha and Goodman (this volume).

The phenomenon of hot spotting has also been a source of fascination, but, with the exception of the proposal of Streisinger and his colleagues that frameshifts are most likely to occur in sequences of iterated base pairs (Okada et al., 1972; Streisinger et al., 1966; for *r*II, see Pribnow et al., 1981), most *r*II hot spots are not understood mechanistically. Koch (1971) observed that one amber codon is somewhat more mutable than the corresponding ochre codon; a subtle effect of a temperature-sensitive mutation on reversion of a nearby amber codon has been measured (Conkling et al., 1980). We have shown that the sequence UGGCAA (see above) is a hot spot when placed in a different context (Singer, Mol. Gen. Genet., in press). In addition, we have noted that deletions that are repeatedly isolated probably occur because directly repeated sequences allow semilegitimate recombination (Pribnow et al., 1981, Singer et al., 1982). We suspect that individual hot spots will be derived from one of a set of conditions.

Size of the *r*II cistrons. Finally we come to the question of how many sites there are within the *r*II region, or, to ask a related and somewhat more contemporary question, how many base pairs are there within *r*IIA and *r*IIB? Benzer (1961) assumed that sites represented by one or two mutations were equally mutable. Fitting his distribution to a Poisson distribution, he calculated that there were at least 120 sites not yet found. Adding this number to the 308 sites that had been found gave 428 as a minimum estimate of the number of sites in the *r*II cistrons. The size of *r*II in base pairs is considerably larger. Drake (1970) applied the mapping function of Stahl et al. (1964) to the data of Edgar et al. (1962) and concluded there are 1,750 to 2,000 base pairs in the cistrons. Bujard et al., (1970) examined heteroduplexes between *r*II deletions and wild-type DNA in the electron microscope. They calculated that *r*IIA is about 1,800 base pairs and that *r*IIB is about 850 base pairs. We have estimated the size of deletions by sequence analysis and restriction digests (Pribnow et al., 1981). We noted that the ratio of base pairs per site is relatively constant for different regions of *r*II, the average being 9.7 base pairs per site. This works out to 2,350 base pairs for *r*IIA and 1,250 for rIIB (there are 370 sites in the complete Benzer collection [Benzer, 1961; S. Champe, personal communication]). These estimates are roughly in accord with the size of the *r*IIA and *r*IIB proteins (O'Farrell et al., 1973; Pribnow et al., 1981). The almost 10-fold discrepancy between Benzer's estimate and ours suggests that even the sites represented by only one or two isolates are hotter than average.

The Code

The unravelling of the genetic code for proteins through the efforts of molecular geneticists is a familiar but still-engrossing tale and one in which *r*II played a large role. The unravelling of the code went hand in hand with an elucidation of the basic mechanisms of mutagenesis. Crick thought that the latter was the more significant contribution because the biochemists would have figured out the code regardless (Judson, 1979). No matter—it makes a good story either way.

Early analysts of the nature of the genetic code were informed by cryptography as well as by a knowledge of the structure of DNA. It was clear, long before it was demonstrated experimentally, that there must be a colinear relation between the gene and the polypeptide chain. How then was information encoded and information transfer achieved? One early conceptual difficulty was the poor correspondence between the numbers of codons and amino acids. If the codon has two nucleotides there are only 16 codons available for 20 amino acids (too few), whereas if the codon has three nucleotides there are 64 codons for 20 amino acids (far too many). Another difficulty was the size consideration. The spacing between amino acids in peptides is similar to the spacing between base pairs in DNA. How could amino acids interact with nucleotides in DNA in a ratio of one to three and be polymerized into polypeptides? Crick et al. (1957) proposed criteria for a code in which only 20 triplets were sense (encoded an amino acid) and the balance were nonsense (encoded no amino acid)—"the commaless code." They showed that there are at least eight substantially different codes that meet the criteria proposed. The codes were elegant but wrong. To describe how such a code might be used, the authors suggested

> that a single chain of RNA, held in a regular configuration, is the template. Let the intermediates in protein synthesis be 20 distinct molecules, each consisting of a trinucleotide chemically attached to one amino acid. The bases of each trinucleotide are chosen according to the code given above. Let these intermediate molecules combine, by hydrogen bonding between bases, with the RNA template and there await polymerization.

While work on the code was proceeding by ratiocination, data were accumulating on the action of mutagens. Benzer and Freese (1958) showed that the spectrum of mutations induced by BU differs from the spontaneous spectrum. Freese (1959a) extended this observation to include 2-aminopurine, showing partial overlap with the BU and 5-bromodeoxyuridine spectra but no congruence with the spontaneous spectrum. Brenner et al. (1958) compared mutations isolated in response to proflavin with spontaneous and BU-induced mutations. They found no correspondence between proflavin- and BU-induced mutations and only partial coincidence between proflavin-induced and spontaneous mutations. All of the mutagen-induced mutations were revertible, but proflavin-induced mutations were not revertible by base analogs (and vice versa). These results led Freese (1959b) to the ingenious proposal that base analogs induce transitions. He suggested that 2-aminopurine is more likely to induce one transition and BU the other one, although at that point it could not be known which was which. The consequence of this kind of mutation for coding is that any transition could be corrected by the reverse transition at the same site.

Thus, revertants would be true revertants. He attributed the balance of mutations to transversion, but Brenner et al. (1961) had a better idea.

Brenner et al. (1961) noted that while either transition or transversion would cause substitution of one amino acid for another, insertion or deletion of a base pair would lead to a much more drastic alteration, such as a considerable alteration in amino acid sequence. The hypothesis of colinearity of gene and polypeptide enjoyed great currency at the time but had not been proved. Hopes of solving the problem by using the *r*II system were dashed when the *r*II protein was nowhere to be found. Consequently, Brenner and Barnett turned to genes *h* and *o* (now designated 37 and 24, respectively [W. B. Wood, personal communication]), both known to encode structural proteins. They found that spontaneous mutations in these genes were revertible by base analogs but not by proflavin (in contrast to the majority of *r*II mutations [Freese, 1959b]). Moreover, it was virtually impossible to induce mutations in these genes by the action of proflavin. The difference is between genes for essential proteins (*h* and *o*) and genes for nonessential proteins (*r*IIA and *r*IIB). Since *h* and *o* encode proteins present in the phage particle, only mutants with mutations that result in a slightly altered protein can survive. Phage lacking the protein altogether, or having a highly garbled analog of the protein, are not viable. These data, already in hand, lent support to the idea that proflavin-induced mutations would lead to a drastically altered protein by virtue of an addition or deletion of a base pair. This notion then led to the suggestion that reversion by closely linked suppressors (the idea had heretofore been dismissed because of lack of evidence) might be possible for proflavin-induced mutations. That suggestion led to the elucidation of the general nature of the genetic code for proteins (Crick et al., 1961).

Crick et al. (1961) backcrossed revertants of three different proflavin-induced mutations to wild type to segregate intragenic suppressors of the original *r*. The suppressors of P13 (renamed FC0) were most easily segregated, and so work concentrated on those suppressors and on suppressors of those suppressors. FC0 is located within the dispensable region of *r*IIB (see below); since greater lengths of garbled protein could be tolerated than in essential regions of the gene, the suppressors could be located farther away and thus were easier to separate from the original mutation. FC0 was labeled (+) as if it inserted a base. Suppressors of FC0 were designated (−), and suppressors of those suppressors were shown to behave like (+). In general, (+)(−) and (−)(+) combinations are pseudowild whereas (−)(−) and (+)(+) are mutant. Furthermore, many combinations of (+)(+)(+), and (−)(−)(−) are pseudowild. Thus the coding ratio is three or (less likely) some multiple of three. Some (+)(−) combinations are *r* rather than pseudowild because the shift in the reading frame produces nonsense. In the code proposed by Crick et al. (1957), only one-third of the triplets encoded amino acids, and the rest were nonsense. The finding here that nonsense is relatively rare leads to the conclusion that the code is degenerate, i.e., that at least some amino acids are encoded by more than one triplet. A further conclusion from this study is that protein synthesis begins at a fixed place and continues in-frame without punctuation. This code is not as elegant as the one proposed earlier, but it works. These studies were extended considerably, and some anomalous results were explored and largely explained (Barnett et al., 1967) (see below).

In parallel with this work, deletion 1589 and the phenomenon of ambivalence were leading Champe and Benzer to a clear formulation of the nature of nonsense. Benzer and Champe (1961) noted that some mutants were more active on one lysogen than on another and that many of these same mutants are phenotypically suppressed when grown in the presence of 5-fluorouracil. They suggested that the difference between the two strains

> may reside in slight differences in their mechanisms for translating genetic information. A prediction of the model would be that suppressor mutations, while specific in action on only certain mutations within a cistron, might nevertheless be capable of affecting some mutations in various other unrelated cistrons.

The deletion 1589 spans the intercistronic divide, deleting about 770 base pairs of *r*IIA and 230 base pairs of *r*IIB. Although it has no *r*IIA activity, it complements *r*IIB mutants as if its *r*IIB function were intact. Champe and Benzer (1962a) postulated that deletion 1589 results in a fused polypeptide that has *r*IIB but not *r*IIA activity. The deletion 1589 thus defines a region of *r*IIB that is dispensable The paradox confronting Champe and Benzer was that there were at least 21 point mutation sites located within the dispensable region. They argued that mutations mapping in this region must prevent the expression of the remainder of the cistron. Because 8 of the 21 mutations in the dispensable region were revertible by base analogs, they suggested that

> the genetic code is such that base pair substitutions are capable of converting sense to nonsense. In other words, the code is not completely degenerate; not *every* triplet necessarily corresponds to an amino acid.

Benzer and Champe (1962) made double mutants between deletion 1589 and several different *r*IIA$^-$ point mutations. Some of the *r*IIA$^-$ point mutations had no effect on the *r*IIB activity of deletion 1589. These they identified as missense. Other mutations eliminated the *r*IIB activity of deletion 1589. These point mutants belonged to ambivalent subset I; i.e., they are able to grow on the lysogen KB-3 but not on KB. Moreover, growth on KB-3 restored *r*IIB activity to the double mutants. Thus, KB-3 contained a suppressor of certain *r*IIA$^-$ alleles, such that nonsense became sense. Benzer and Champe postulated, correctly, that "the action of the suppressor mutation in KB-3 could be explained by the appearance of a new or modified sRNA adaptor that fits a coding unit absent from the KB dictionary."

Mutants of ambivalent subset I came to be known as ambers, and the suppressor in KB-3 was called an amber suppressor. The next nonsense mutants found were designated ochres. From there it was possible to bootstrap from nonsense to suppressor, to identify the nonsense codons as UAA, UAG, and UGA, and to show that nonsense codons exert their effect at the level of translation, not transcription (Brenner and Beckwith, 1965; Brenner and Stretton, 1965; Brenner et al.,

1965; Sambrook et al., 1967). Finally, Crick and Brenner (1967) deduced the absolute frame of frameshift mutations in their early study. They were right (Pribnow et al., 1981). This series of experiments and deductions has been described as "the exuberant last exfoliation of molecular genetics at its rococo extreme" (Judson, 1979).

RECENT USES OF THE *r*II CISTRONS: GENE EXPRESSION

Transcription

The *r*IIB gene is a classic middle gene (Brody et al., this volume). Rifampicin added 1 min postinfection eliminates some specific *r*IIB transcripts from appearing, as does the genetic elimination of the T4 *mot* gene product (O'Farrell and Gold, 1973a; Mattson et al., 1974; Mattson et al., 1978). We have located the 5′ end of the middle *r*IIB transcript precisely and have shown that transcript to be *mot* dependent (R. Sweeney and L. Gold, manuscript in preparation; R. Sweeney, Ph.D. thesis, University of Colorado, Boulder, 1982). The 5′ end falls within the coding region of the adjacent *r*IIA gene; hence, *r*IIA mutations have been extremely useful in our studies of the *r*IIB promoter. Parenthetically, *r*IIB promoter mutations yield enough *r*IIB protein for complementation; had this not been the case, Benzer would have faced a peculiar set of point mutations (Pribnow et al., 1981). We have tentatively identified a *mot* activation site ca. 70 base pairs from the initiating base pair; the great distance suggests that *mot* activation will be mechanistically amusing. Furthermore, this site (defined by two mutations) may be moved 12 base pairs closer to the initiating base pair, and the middle promoter functions normally. A full description of this work is in preparation (Sweeney, Ph.D. thesis; Sweeney et al., manuscript in preparation).

Translation

Ribosome-binding-site alterations. Some years ago we set out to isolate quantity mutations that diminished *r*IIB translation. The wild-type *r*IIB sequence (as mRNA) includes an obvious structure (Fig. 1) (Pribnow et al., 1981; Singer et al., 1981). The message made using the *mot*-dependent middle promoter starts at −121. The UAAUAA (starting at −13) (Fig. 1) is the *r*IIA translational stop signal. Nelson et al.

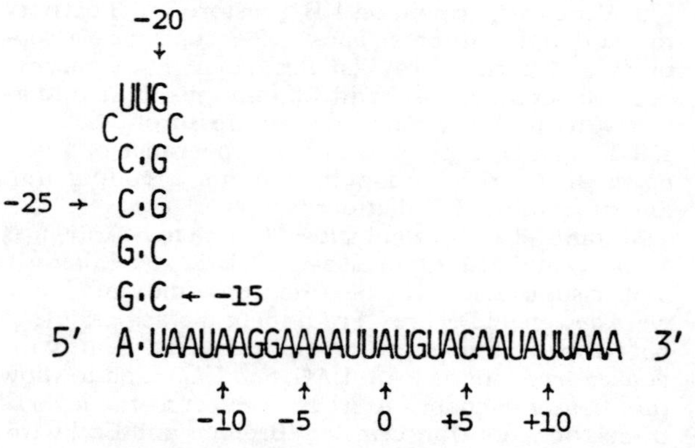

```
          -20
           ↓
          UUG
         C   C
          C·G
  -25 →   C·G
          G·C
          G·C ← -15
5′  A·UAAUAAGGAAAAUUAUGUACAAUAUUAAA  3′
        ↑    ↑    ↑    ↑    ↑
      -10   -5    0   +5  +10
```

FIG. 1. Portion of wild-type *r*IIB mRNA sequence.

TABLE 1. Mutants defective in translation

Mutant	Phenotype	Sequence alteration[a]	Rationalization
HD263	*ts* for translation (Belin et al., 1979; Singer and Gold, 1976)	G(+2) → A	Alters initiation codon
HD263rev10	Pseudowild *ts*⁺ revertant of HD263 (Singer and Gold, 1976)	G(+2) → A* U(−1) → G	Creates new initiation codon
zAP10	Translational down	A(0) → G*	Changes initiation codon
zEM72	Translational down	G(−7) → A*	Ruins Shine-Dalgarno region
zHA87	Translational down	C(−16) → U* C(−25) → U	Destroys hairpin
zHA104	Translational down	G(−17) → A* G(−26) → A	Destroys hairpin

[a] Asterisks indicate mutant sequences that will be reported elsewhere (Pribnow et al., submitted for publication).

(1981) used a powerful selection for translational defectives. These mutants were collected, the mutations were mapped, and the levels of protein synthesized from early and early-plus-middle messages were determined (Singer et al., 1981). The more interesting messages have now been sequenced (in preparation), and some of these are shown in Table 1. These data strengthen the conclusion that initiation of protein synthesis utilizes an AUG and a Shine-Dalgarno region (Gold et al., 1981). In addition, the two double mutants (zHA87 and zHA104) suggest that the hairpin facilitates *r*IIB translation (Singer et al., 1981).

Translational reinitiation. Sarabhai and Brenner (1967) studied translational reinitiation in the early portion of the *r*IIB mRNA. Barnett et al. (1967) noted several other domains within *r*IIB at which translational reinitiation could occur. We showed that these interpretations of elegant genetics were correct (Napoli et al., 1981; Singer and Gold, 1976). We have now sequenced (manuscript in preparation) three mutations that allow reinitiation if a ribosome terminates nearby. The three different mutations fall within the sequence shown in Fig. 2. The alteration within codon 10 (Fig. 2) creates an initiation codon (that functions for reinitiation only [Napoli et al., 1981; Sarabhai and Brenner, 1967]; no Shine-Dalgarno region exists 5′ to the new AUG); the alterations within codon 11 create acceptable Shine-Dalgarno regions (GGA or GAG) that apparently drive reinitiation (only) at a CUG or UUG just 3′ to the mutations.

Regulation by the *regA* gene product. The expression of *r*IIB is activated (transcriptionally) by the *mot* gene product and is repressed (translationally) by the *regA* gene product (Karam and Bowles, 1974; Trimble and Maley, 1976; Wiberg et al., 1973). We characterized *r*IIB mutations that rendered *r*IIB expression

rIIB codon =		10	11	12	13	14	15	16	
	5'	AAC	GAA	CAA	GCU	GAA	AUU	GUU	3'
reinitiation mutants		U	GG						

FIG. 2. Mutations creating translational reinitiation sequences in *r*IIB mRNA. A complete report of these mutations will be published elsewhere (Pribnow et al., submitted for publication).

insensitive to *regA*-mediated repression (Campbell and Gold, 1982; Karam et al., 1981). These mutations all cluster around the initiating AUG of *r*IIB. Based on these observations we have deduced a "*regA* recognition site" of

5′ AUGUACAAU 3′
0 +5

regA-insensitive mutations include A(0) to G, G(2) to A, UAC(3–5) to AAA, and AA(6,7) to AAA. However, the mutation U(3) to UU remains *regA* sensitive, as though recognition of the 5′ and 3′ parts of the "operator" is a bit flexible with respect to spacing. Complex alternatives to the simple idea of competition between the *regA* protein and ribosomes can be imagined; only in vitro repression and binding studies with purified *regA* protein will be conclusive.

PROSPECTUS

We have been systematically sequencing *r*IIA and *r*IIB mutations that either amuse us or are likely to be useful to others. Since *r*II genetics remain very powerful, we believe that many questions about T4 (mutagenesis, replication fidelity, recombination, gene expression, etc.) will be asked and answered through exploitation of both classical and new (i.e., recombinant DNA-mediated) *r*II technology. Our list of sequenced mutations will be published soon (Pribnow et al., manuscript in preparation). Up-to-date lists will continuously be available to everyone, as are the strains of Benzer and Barnett et al. in our collection. We hope people will ask for these materials, as well as for mutants we have selected in the course of our work.

ADDENDUM IN PROOF

We thank Sewell Champe for reminding us of the still-unexplained fact that lysis inhibition during infection by rII^- phage requires superinfection.

Interactions of T4-Induced Enzymes In Vivo: Inferences from Intergenic Suppression

JIM D. KARAM, MARIA TROJANOWSKA, AND MELANIE BAUCOM

Department of Biochemistry, Medical University of South Carolina, Charleston, South Carolina 29425

Biological systems utilize two types of processes to regulate the frequencies of collision between proteins, nucleic acids, and substrates in biochemical reactions. First, transcriptional and translational mechanisms determine the intracellular concentrations of enzymes and structural proteins and RNA. Second, compartmentation of these macromolecules can increase or decrease the effective concentrations of the participants in a biochemical transition. In eucaryotic cells compartmentation is often achieved via the physical barriers that separate intracellular organelles. In addition, a compartmentation of sorts can be achieved via the assembly of multienzyme complexes and the specific immobilization of proteins on cell membranes near the sites of action of these complexes. Although bacteria lack intracellular organelles, there is considerable evidence that such essential processes as DNA replication, transcription, and translation are carried out by multiprotein complexes that are capable of recycling their components in coordination with the different steps of the processes they catalyze. It also appears that these complexes and their components are not always freely diffusible in the cell. Studies with T4 phage are contributing extensively to our understanding about the assembly and regulation of such complexes. T4 is readily amenable to biochemical, physiological, and genetic studies because of its dramatic inhibitory effect on host macromolecular biosynthetic processes and its genetically well-characterized chromosome.

Many major metabolic events in T4-infected cells are closely interrelated: the synthesis of DNA precursors and DNA replication appear to be connected together functionally and structurally via shared components of multienzyme complexes (Mathews and Allen, this volume); DNA replication and recombination are almost inseparable (Mosig, this volume); DNA replication proteins play roles in transcription during the late stages of phage growth (Rabussay, this volume); DNA replication, recombination, and phage morphogenesis are coupled (Black and Showe, this volume; Mosig, this volume); and so on. Because of these interrelationships we expect, and usually find, that alterations in the in vivo level of an essential enzyme, e.g., a replication protein, can have profound effects not only on the process serviced by this protein (e.g., DNA replication), but also on other essential phage functions (e.g., "late" transcription and phage morphogenesis). Usually, extensive analyses by both genetic and biochemical tools are needed to characterize the functional relatedness between phage-induced proteins; however, balanced approaches are not always possible. In discussing intergenic suppression for the study of protein interactions in T4-infected cells, we try to emphasize genetics as a sensitive tool by which interactions can be first detected or recognized. We also cite examples in which the distinction between physical interactions (formation of multiprotein complexes) and functional interactions (cooperation between gene products) could be derived either primarily from genetic evidence or from biochemical confirmation of genetic models.

INFORMATIONAL SUPPRESSION AS A MEANS OF MANIPULATING THE QUALITY AND INTRACELLULAR CONCENTRATIONS OF T4-INDUCED PROTEINS

In *Escherichia coli* and T4-infected *E. coli* nonsense mutations are suppressible at the translational level by certain mutant tRNA species that readily recognize specific protein chain-terminating codons in the mRNA (Gorini, 1970). The known tRNA suppressors of the UAA (ochre), UAG (amber), and UGA (opal) nonsense codons are listed in Table 1. Ochre, amber (*am*), and opal mutations are also suppressible by translational ambiguity, i.e., the misreading of codons by normal tRNA species (Gorini, 1974; Kurland, 1979; Steege and Soll, 1979). The principles that underly the two types of suppression are virtually the same in that the frequency of insertion of an amino acid in response to a nonsense codon depends on several factors, including the strength of the codon-anticodon interaction, the location of the codon in the mRNA sequence (the reading context), the physiological environment of the ribosome (ionic conditions, temperature), and conditions that affect the fidelity of amino acid activation or the interaction between an amino acyl-tRNA and the ribosome. Translational ambiguity, however, does not signify random mistranslation of a codon by the tRNA pool in the cell; rather, it involves weak interactions with a specific number of tRNAs that bear a relationship to the codon being misread.

Missense, e.g., temperature-sensitive (*ts*), mutations are also suppressible by tRNA suppressors and by ribosomal ambiguity (Hill, 1975); however, the usefulness of such mutants in genetic assays can be limited in that the altered codons (amino acid substitutions) of missense mutants are usually not easy to characterize. Conceivably, many of the *ts* T4 mutants that have been isolated after treatments with mutagenic agents harbor more than one lesion, and these lesions together contribute to the temperature lability of the gene product under study; the presence of such multiple lesions cannot be verified by simple backcrosses, as the mutations may not exhibit measurable phenotypes when separated from one another. In contrast, nonsense mutations can be defined precisely via their characteristic behavior in response to tRNA suppressors (Table 1) and can be easily "purified" away from

TABLE 1. tRNA nonsense suppressors of T4 and *E. coli*

Nonsense triplet decoded	Suppressor designation and efficiency (%)	Suppressor gene	
		E. coli[a]	T4
UAG (amber)	SU$^+$1, Ser (60)[b]	*supD*	tRNA^{Ser} (45–70)[c]
	SU$^+$2, Gln (30)[b]	*supE*	
	SU$^+$3, Tyr (50)[b]	*supF*, *tyrT*	
	SU$^+$6, Leu (55)[b]	*supP*	
	SU$^+$7, Gln (76)[b]; low level of Trp[a]	*supU*, *trpT*	
UAA (ochre) and UAG (amber)	SU$^+$4, Tyr (12[UAA]; 16[UAG])[b]	*supF*, *tyrT*	
	SU$^+$5, Lys (6[UAA])[b]	*supG*	
	SU$^+$8, Gln and Trp (5[UAG])[b]	*supU*, *trpT*	tRNA^{Gln} (<17)[d]
	SU$^+$B, Gln	*supB*, *glnV*	
	SU$^+$β, Lys (8[UAA])[e]	*supL*	
UGA (opal)	SU$^+$9 Trp (30)[f]	*supU*, *trpT*	

[a] See Ozeki et al. (1980) for references and Backman and Low (1980) for nomenclature and map positions.
[b] See Smith (1972) for references.
[c] McClain et al. (1973); Wilson and Kells (1972).
[d] Comer et al., 1974.
[e] Kaplan, 1971.
[f] Model et al., 1969.

secondary lesions by genetic backcrosses. These considerations have been discussed by Karam and O'Donnell (1973) and are important when assessing the effects of specific amino acid substitutions in the gene product and when second-site mutations are sought as suppressors of specific defects in the phage genome. Given the variety of amino acids that can be inserted by suppression of the UAG codon (Table 1) and the large number of T4 *am* mutants that have been isolated, it is possible to generate variants of many T4 gene products that are specifically altered at known locations in the polypeptide chain. These can include fully active, partially active, temperature-labile, and inactive forms of the T4 gene product under investigation. In addition, because different suppressors exhibit different decoding efficiencies, it is sometimes possible to manipulate gene product dosage (enzyme concentrations) in the cell and to derive insights about biological role and activity from the effects of suppression. Such manipulations are relatively easy to carry out with essential T4 gene functions because conditional-lethal mutations in such functions can be assayed by simple plating techniques.

CIRCUMVENTING THE EFFECTS OF PRIMARY LESIONS IN ESSENTIAL T4-INDUCED ENZYMES

Informational suppression (decoding) of T4 nonsense and missense mutations can be mediated by both host- and phage-derived mutant tRNAs (Schmidt and Apirion, this volume) and, possibly, via phage-induced modification of the host translational machinery (Wiberg and Karam, this volume). All such mechanisms overcome a genetic defect by "correcting" the primary structure of the mutant polypeptide. Bacteria and their viruses can utilize other forms of suppression that do not act directly on the primary lesion (Gorini, 1970; Hartman and Roth, 1973). Table 2 lists some types of indirect suppressors that have been detected in the T4 system by the use of relatively simple plating techniques. Usually, a rapid screening of the plating characteristics of several *ts* and nonsense mutants of a particular T4 gene under permissive, semipermissive, and nonpermissive conditions can give clues about the nature of the requirement for the gene function in phage development and whether this requirement can be circumvented. For example, the T4-induced DNA ligase (gene 30) is normally required for phage DNA replication and recombination, but is dispensable in T4 *r*II$^-$ mutant infections (Berger and Kozinski, 1969; Ebisuzaki and Campbell, 1969; Karam, 1969). Single and double *am*, as well as *ts*, mutants of T4 gene 30 all plate with efficiencies near 10^{-4} under nonpermissive conditions, with most plaques being formed by gene 30 *r*II double defectives (Heere and Karam, 1975; Karam, 1969). The frequency of *r*II forward mutations in T4 stocks is 0.01 to 0.1% (Benzer, 1955). So, T4 *r*II$^-$ mutations can be isolated as suppressors of gene 30 defects (Table 2). Another example is the isolation of T4 *regA* mutations (Table 2; see also Wiberg and Karam, this volume) as suppressors of "leaky" mutants in T4 DNA replication genes 62 and 44 (Karam and Bowles, 1974; Karam et al., 1977). In this case, T4 *regA*$^-$ *ts* gene 44 and *regA*$^-$ *am* gene 62 double mutants were isolated as plaque formers under plating conditions that were semipermissive for the *am* and *ts* mutants used. These double mutants still failed to plate under the nonpermissive conditions for the single *am* and *ts* mutants. That is, the T4 gene 62 and 44 functions are not circumvented by *regA* gene defects; rather, *regA*$^-$ mutations lead to hyperproduction of undersynthesized (poorly suppressed *am*) or incompletely active (partially *ts*) enzymes to levels that can support near normal phage DNA replication. Wiberg and Karam (this volume) provide a more detailed treatment on the T4 *regA* gene function.

As may be inferred from the examples listed in Table 2, plating assays rarely provide enough information to explain the mechanisms of suppression, but can provide the initial insights that justify subsequent biochemical characterization of these mechanisms. In the remainder of this chapter we discuss intergenic suppression within the context of the types of functional and physical interactions that underlie the biological roles of phage-induced proteins.

GENETIC INTERACTIONS IN T4 DNA TRANSCRIPTION

Table 3 lists some of the T4-related interactions of *E. coli* RNA polymerase. Much of what we know about T4-induced modifications of the *E. coli* polymerase comes from careful biochemical studies that were carried out on purified modified enzyme (see Rabussay, 1982c, and this volume). Phage mutants, however, have been very useful in the identification of several of the T4-induced proteins that become associated with the host enzyme and alter its specificity for promoters or that bring about chemical modification

TABLE 2. Pathways of indirect suppression of genetic deficiencies in T4

Genetic location of primary lesion (T4 gene)	Genetic location of suppressor (T4 or *E. coli* gene) (references[a])	Probable pathway or mechanism of suppression
*r*II	*r*II (1, 8, 36, 40)	Intragenic suppression: reconstitution of biological activity via second-site mutations in the same gene, e.g., certain double-missense mutants, reinitiation mutants, and mutants in which one frameshift mutation corrects the frame of reading created by another
e	*e* (43)	
23	22 and 24 (11, 32)	Allele-specific, positive and negative, intergenic suppression: reconstitution or inhibition of biological activity via interaction of one defective gene function with another, e.g., interaction of two mutant polypeptides leading to normal function
24	23 (11, 32)	
31	23 (38); *E. coli groE* (44)	
32	52, 39, 60, and 58–61 (34); 30 and *r*II (33); 17 (35); *dda(sud)* (14, 28); *E. coli lig* (24)	
38	37 (3)	
42	43 (6)	
43	*E. coli optA* (14)	
45	44 (23); *E. coli tabD* (*rnpC*?) (7)	
55	*E. coli tabD* (*rnpC*?) (7)	
63	*t*, 5, or 8 (17)	
frd	41 or 61 (16)	
30	*E. coli lig* (15)	Providing a functional substitute for the defective or missing phage gene product
39, 52, or 60	*E. coli gyrA* (31)	
dda	*E. coli optA* (14)	
30	*r*II (2, 19, 20); *denA* (45)	Removal of a detrimental process or metabolite(s)
56, 56/1, or 56/42	*denB* and *alc* together (42)	
Accumulation of single-stranded DNA	32[b]	
62 or 44	*regA* (21, 22)	Hyperproduction of an undersynthesized or partially active essential protein
44	*hp6* site (4)	
*r*IIA/42	*mot* (29, 30)	
17	[*psu*⁻*SB*] (39)	Largely uncharacterized mechanisms
30	*sum* (5); *su*30 (24, 25)	
46 and 47	*das* (*suα*) (18, 26)	
49	*fdsA* (*uvsX*?); *fdsB* (*uvsY*?) (10)	
59	*dar* (46)	
e	*r*IV (12, 37); *st*III (27)	
gor	*E. coli rnpB* (41)	
pseT	*stp* (9)	
*r*II	*sip* (13)	
t (*st*II)	*st*III (27)	
"*pla*"	Host genes[c]	

[a] (1) Barnette et al., 1967; (2) Berger and Koziniski, 1969; (3) Bishop and Wood, 1976; (4) Bowles and Karam, 1979; (5) Chan and Ebisuzaki, 1973; (6) Chao et al., 1977; (7) Coppo et al., 1975a, 1975b; (8) Crick et al., 1961; (9) Depew et al., 1975; (10) Dewey and Frankel, 1975; (11) Doherty, 1982b; (12) Emrick, 1968; (13) Freedman and Brenner, 1972; (14) Gauss et al., 1982; (15) Gellert and Bullock, 1970; (16) Hall and McDonald, 1981 and personal communications; (17) Hall et al., 1980; (18) Hercules and Wiberg, 1971; (19) Karam, 1969; (20) Karam and Barker, 1971; (21) Karam and Bowles, 1974; (22) Karam et al., 1977; (23) Karam, Bowles, and Leach, 1979; (24) Karam, Leach, and Heere, 1979; (25) Krylov, 1972; (26) Krylov and Plotnikova, 1971; (27) Krylov and Yankovsky, 1975; (28) Little, 1973; (29) Mattson et al., 1974; (30) Mattson et al., 1978; (31) McCarthy, 1979; (32) McNicol et al., 1977; (33) Mosig and Breschkin, 1975; (34) Mosig, Luder, Garcia, et al., 1979; (35) Mosig, Ghosal, and Bock, 1981; (36) Nelson et al., 1981; (37) Okamoto and Yutsudo, 1974; (38) Revel et al., 1980; (39) Ribolini and Baylor, 1975; (40) Sarabhai and Brenner, 1974; (41) Snyder and Montgomery, 1974; (42) Snyder et al., 1976; (43) Terzaghi et al., 1976; (44) Tilly et al., 1981; (45) Warner, 1971; (46) Wu, Wu, and Yeh, 1975.

[b] Many physiological conditions resulting from specific genetic deficiencies result in abnormal accumulation of single-stranded DNA, which is susceptible to nuclease attack. Such conditions result in overproduction of gene 32 protein, which covers single-stranded DNA and presumably protects this DNA from nucleases (see Lemaire et al., 1978, for references).

[c] Various types of T4 lesions cause an inability to form plaques (*pla*) on specific *E. coli* hosts (see Wood and Revel, 1976, for references).

TABLE 3. T4-mediated interactions and alterations of *E. coli* RNA polymerase (RNAP)

T4-mediated process	T4 gene product involved	Effect and RNAP subunit involved	References[a]
Protein-protein interactions	33	Required for late promoter recognition; co-isolates with core RNAP	(6, 12, 13, 18)
	55	Required for late promoter recognition; co-isolates with core RNAP; interacts with β′ subunit	(1, 2, 12, 13, 18)
	45	Required for late transcription; interacts with β′ subunit; retained by T4-modified RNAP affinity column	(1, 2, 12, 13, 19)
	10K (gene Unknown)	Antagonizes σ protein	(10)
	15K (gene 60?)	Role of RNAP in T4 DNA replication?	(5)
Modification of RNAP subunits	*alt*	Reversible ADP-ribosylation of one α subunit; lowered affinity of RNAP to σ protein	(4, 7, 14, 15)
	mod	Irreversible ADP-ribosylation of both α subunits; lowered affinity of RNAP to σ protein	(4, 7, 14, 15)
Functional interactions	*alc*	Transcription of cytosine-containing templates	(17)
	DO (DNA replication proteins)	Transcription of late phage genes	(11)
	gor	Involvement of RNAP in T4 DNA replication; (interaction with β subunit?)	(16)
	mot	Utilization of T4 middle promoters; interaction with rho protein	(3, 8, 9)

[a] (1) Coppo et al., 1975b; (2) Coppo et al., 1975a; (3) Daegelen et al., 1982b; (4) Goff, 1974; (5) Goff, cited in Rabussay, 1982c; (6) Horvitz, 1973; (7) Horvitz, 1974a; (8) Mattson et al., 1974; (9) Mattson et al., 1978; (10) Rabussay, 1982c; (11) Rabussay and Geiduschek, 1979b; (12) Ratner, 1974a; (13) Ratner, 1974b; (14) Rohrer et al., 1975; (15) Skorko et al., 1977; (16) Snyder and Montgomery, 1974; (17) Snyder et al., 1976; (18) Stevens, 1972; (19) Wu, Geiduschek, and Cascino, 1975.

of its subunits. Genetic studies also contributed to our initial insights that, in addition to the coupling between DNA replication and late-gene transcription, certain DNA replication proteins participate directly in transcription of phage late-gene functions (Rabussay and Geiduschek, 1979b; Wu, Geiduschek, and Cascino, 1975). Recently, J. Pulitzer and his co-workers isolated *E. coli* RNA polymerase mutants (named *tabD*) that specifically restricted the growth of T4 *ts* mutants in genes 55 and 45 at the permissive temperature for these mutants in wild-type *E. coli* hosts (Coppo et al., 1975a; Coppo et al., 1975b). The T4 gene 55 protein has been shown to copurify with RNA polymerase from T4-infected cells and to be essential for late-gene expression in infected cells (Pulitzer and Geiduschek, 1970; Stevens, 1972). In addition, Ratner (1974a, 1974b) showed that an *E. coli* RNA polymerase affinity column binds the T4 gene 55 protein and that T4-modified RNA polymerase binds the gene 45 protein. So, the gene 45 protein, a DNA replication enzyme, is implicated in transcription. Interestingly, *E. coli tabD* mutants selected for a *ts* mutant in gene 45 restrict late phage transcription in infections with the mutant without restricting its replication. This result implies that the gene 45 protein is a bifunctional enzyme that plays two separate roles in phage development.

Other interactions of T4-induced enzymes in late-gene expression involve the requirement for a hydroxymethylcytosine (HMC)-substituted phage DNA template. The presence of HMC and glucose residues in T4 DNA protects this DNA against host restriction systems (Revel, 1967) and against phage-induced nucleases, particularly the *denB* product, endonuclease IV (Bruner et al., 1972; Kutter et al., 1975). Table 4 summarizes some properties of T4 mutants defective in their abilities to introduce HMC residues into phage DNA. This topic is discussed further by Snustad (this volume). In hosts lacking the restriction systems against T4, phage DNA containing cytosine and HMC (about 50% each) can be synthesized in large quantity, provided that the phage is defective in genes 56 (dCTPase) and *denB* (endonuclease IV). T4 56^- $denB^-$ infections fail to produce viable phage, however, because of their inability to utilize cytosine-containing phage DNA for late transcription. This block can be overcome by mutations in the T4 *alc* gene (Kutter et al., 1981; Snyder et al., 1976; Tables 2 and 4), an incompletely characterized phage function that appears to play roles in the inhibition of transcription of cytosine-containing DNA templates (including the host genome) and in the phage-induced unfolding of the host chromosome (Pearson and Snyder, 1980; Sirotkin et al., 1977). T4 56^- $denB^-$ alc^- infections yield reduced bursts of viable phage as compared with infections with wild-type phage, and the phage DNA made under these conditions still contains about 40% HMC residues, probably because the gene 42 hydroxymethylase in such mutants uses the dCMP generated from host DNA breakdown ($denA^+$). Some HMdCMP is also made in T4 56^- $denA^-$ $denB^-$ alc^- infections because of the leakiness of 56^- and $denA^-$ lesions. T4 56^- 42^- $denB^-$ alc^- mutants have been isolated as pseudorevertants of 56^- 42^- $denB^-$ phage; these exhibit reduced DNA synthesis and sharply reduced phage yields, but synthesize HMC-free phage DNA (Morton et al., 1978; Wilson et al., 1977). So, the requirement for an HMC-substituted DNA template in T4 late-gene expression can be circumvented; however, suppression via mutations in

TABLE 4. Growth characteristics of T4 mutants defective in synthesis of HMC-containing DNA[a]

Defective T4 genes	DNA synthesis	Phage production
56[b]	0	0
56 42[b]	0	0
56 *denB*[c]	+	0
56 42 *denB*[c]	+	0
56 *denB alc*[d]	+	+
56 42 *denB alc*[d]	+	+

[a] Growth properties are for phage mutants grown in *E. coli* hosts lacking the *r*(B,K) *m*(B,K) and *rgl* restriction systems.

[b] In T4 56⁻ (dCTPase defective) infections, phage-induced host DNA breakdown provides some dCMP for the gene 42 dCMP hydroxymethylase. Phage DNA containing both cytosine and HMC (about 50% each) is made under these conditions, but it is degraded by phage-induced nucleases because of its cytosine content. Loss of dCMP hydroxymethylase (42⁻) further restricts entry of HMC residues into phage DNA, but the HMC-free, cytosine-containing DNA made in T4 56⁻ 42⁻ infections is still degraded. Formation of HMdCMP in T4 56⁻ infections can also be restricted by the use of T4 *denA*⁻ mutations, which block host DNA breakdown (see Snyder et al., 1976, for references).

[c] T4 endonuclease IV (*denB*) plays a major role in degradation of cytosine-containing phage DNA. Although T4 56⁻ *denB*⁻ and 56⁻ 42⁻ *denB*⁻ infections yield substantial amounts of phage DNA, this DNA is not a competent template for transcription, and no phage are produced by these infections (Snyder et al., 1976; Kutter et al., 1981).

[d] T4 *alc*⁻ mutations allow transcription of late phage functions from cytosine-containing phage DNA templates (Snyder et al., 1976). T4 56⁻ *denB*⁻ *alc*⁻ infections produce more phage than do 56⁻ 42⁻ *denB*⁻ *alc*⁻ infections, but phage from the 42⁻ infections are completely HMC free.

the *alc* gene does not occur without some cost to phage yield. Additional phage-phage and phage-host gene interactions affecting transcription are discussed by Brody et al. (this voume).

REPLACEMENT OF DEFECTIVE PHAGE FUNCTIONS BY HOST GENE PRODUCTS

The definition of an essential T4 gene function incorporates in it the failure of the *E. coli* host to support phage growth when the gene function is eliminated. Some T4 genes are essential for growth in one *E. coli* host, but not in another (several examples are cited in the review by Wood and Revel, 1976). One of several possible reasons for such host range effects in T4 infections is the ability of certain hosts, or the same host under certain conditions, to provide a substitute for a missing phage function. T4 and its *E. coli* host carry out very similar biochemical transactions, especially with regard to DNA replication, repair, recombination, transcription, and other metabolic events involving nucleic acid precursors. For example several of the *E. coli* enzymes for synthesis of DNA and RNA precursors are easily utilized by T4 in vivo, although the phage usually requires (and synthesizes) higher levels of these activities than can be provided from host sources (Mathews and Allen, this volume). On the other hand, many of the T4-induced enzymes involved in phage DNA replication and recombination appear to be specifically suited for propagation of their own genetic apparatus. For example, Mosig et al. (Mosig, Bowden, and Bock, 1972; Mosig, Luder, Garcia, et al., 1979) have examined the ability of various *E. coli* DNA replication enzymes to substitute for T4 DNA replication enzymes in vivo and found that such substitutions can occur in several cases, but either inefficiently or under very special conditions, e.g., the simultaneous elimination of several phage functions. There are, however, examples of very efficient utilization of host "replication" enzymes as replacements for defective phage functions. One such example is substitution of the *E. coli* DNA ligase (*lig* gene) for the T4 gene 30 DNA ligase (Gellert and Bullock, 1970). *E. coli* mutants that overproduce DNA ligase can support the growth of T4 defective in gene 30 ligase production. In addition, T4 gene 30 mutants that are also defective in gene *r*IIA or *r*IIB can grow at near normal levels in *E. coli* K-12 cells harboring normal levels of the host DNA ligase (Berger and Kozinski, 1969; Ebisuzaki and Campbell, 1969; Karam, 1969). The mechanism of the suppression by T4 *r*II gene mutations of gene 30 defects is not completely understood, but is related to reduced breakdown of phage DNA in *r*II⁻ infections (Karam and Barker, 1971). It appears that, normally, T4 requires high levels of DNA ligase to counteract the effects of endonucleolytic cleavage of phage DNA, especially during the early stages of establishing the DNA replication pool (Kozinski, 1968). Conditions can be created (*r*II⁻ infections) whereby either nucleolytic attack on phage DNA is kept to a minimum or host repair enzymes are channeled for use on the phage DNA undergoing the nucleolytic attack. Interestingly, multiple mutants of T4 completely defective in genes 30 and *r*II, but also carrying lesions at certain other phage loci, including gene *su*30 (Krylov, 1972) and gene 32 (helix-destabilizing protein) are capable of replicating and producing significant phage yields when grown under conditions of sharply reduced host ligase (Karam, Leach, and Heere, 1979). So, it appears that the environment of the T4-infected cell can be adjusted to make substitution of phage by host DNA ligase very efficient. Other examples of host-phage enzyme substitutions in T4 DNA replication and recombination are discussed below.

The ability of *E. coli* enzymes to substitute for defective T4 enzymes may or may not necessitate specific physical interaction between host and phage-derived proteins. For example, the use of some host enzymes in the biosynthesis of T4 DNA precursors probably does not involve the formation of complexes between these enzymes and phage gene products (Mathews et al., 1979). The substitution of *E. coli* DNA ligase for the T4 DNA ligase, on the other hand, probably does involve specific physical interactions between the host enzyme and protein components of the T4 DNA replication/recombination apparatus (Karam, Leach, and Heere, 1979). In this case, a specific lesion in T4 gene 32 (helix-destabilizing protein) was required to allow substitution from the specific host ligase deficiency (the *lig*2 allele). Such an allele-specific suppression is probably indicative of the central role that the T4 gene 32 protein plays in facilitating interactions among components of the phage replication/recombination complex. It should be emphasized, however, that complex formation between *E. coli* DNA ligase and T4-derived components may be possible only when certain phage gene products (the T4 ligase and *r*II proteins) are defective and do not place the host enzyme at a competitive disadvantage for access into the T4 DNA replication/recombination complex. A similar situation has been described by Mosig, Bowden, and Bock (1972) for substitution for certain *E. coli* DNA replication enzymes for defective T4 enzymes. They observed that utilization of the *E. coli dnaC* and *dnaG* proteins for T4 DNA replication required an absence of the DNA-binding activity of gene 32 protein. Possibly, a completely functional gene 32 protein blocks access of these host proteins to the T4 DNA and the DNA replication complex.

GENETIC INTERACTIONS IN T4 DNA REPLICATION AND RECOMBINATION

There is ample genetic and biochemical evidence that the T4-induced DNA replication proteins function in multienzyme complex form (see chapters in this volume by Mathews and Allen, Nossal and Alberts, and Mosig for examples and references). It is not clear, however, what the overall organization of this complex is and whether all of its components have yet been identified, either genetically or biochemically. There are strong indications that recombinational events constitute an integral part of normal T4 DNA replication, especially at high multiplicities of infection and during the intermediate and late stages of the phage growth cycle (Luder and Mosig, 1982). One attractive notion envisages T4 DNA replication, recombination, repair, certain types of phage transcription, DNA precursor biosynthesis, and DNA packaging during phage morphogenesis to be all interconnected via several multienzyme complexes that share components and recycle these components in coordination with the needs of the phage growth cycle. Biochemical confirmation of this notion has been slow in coming because, as pointed out by Mathews et al. (1979) and Mosig, Luder, Garcia, et al. (1979), transient and weak protein-protein interactions that probably underlie the recycling of enzymes in the complexes of T4 DNA metabolism make direct isolation and biochemical characterization of these complexes difficult; however, many insights about the changing structures of such complexes are being derived from genetic studies. The T4 gene 32 protein has been receiving special attention in these studies because of its activity as a single-stranded DNA-binding enzyme (SSB) and as a helix-destabilizing enzyme (Alberts and Frey, 1970) and its ability to bind a number of T4-induced enzymes known to be involved in replication or recombinational events (Alberts et al., 1983; Huberman et al., 1971). The gene 32 protein may play a central role in coordinating the assembly of DNA replication/recombination multienzyme complexes from phage and host components. The known interactions of this protein are summarized in Table 5. Mosig, Luder, Garcia, et al. (1979) have been assaying in vivo enzyme interactions in T4 DNA replication and recombination by searching for and characterizing three types of mutations in T4 gene 32; (i) mutations that inactivate some functions of the gene, but leave others intact; (ii) second-site mutations (suppressors) that alleviate the effects of lesions in this gene; and (iii) second-site mutations that worsen the effects of specific gene 32 lesions (negative suppression). They detected all three classes of phenotypes among *ts* and *am* mutants of T4 gene 32 and proposed a model in which different domains along the gene 32 protein are used to interact

TABLE 5. *E. coli* and T4-induced gene products that interact with the T4 gene 32 (SSB) protein

T4-induced protein	*E. coli* proteins
DNA polymerase (gene 43)[a,b]	32K protein (gene unknown)[a]
SSB protein (gene 32)[a]	DNA ligase (*lig*)[c]
DNA polymerase accessory protein 45 (gene 45)[a]	Minor amounts of several unidentified proteins[a]
Recombination exonuclease 46/47 (genes 46 and 47)[a]	
uvsX protein (*uvsX*)[a]	
uvsY protein (*uvsY*)[a]	
DNA helicase (*dda*)[a,c]	
RNase H (gene unknown)[a]	
Minor amounts of several unidentified proteins[a]	
DNA topoisomerase (genes 39, 52, and 60)[c]	
RNA priming protein 61 (gene 61)[c]	
DNA packaging protein 17 (gene 17)[c]	
*r*II proteins (*r*IIA and *r*IIB genes)[c]	

[a] Interactions inferred from results of chromatography of extracts from T4-infected cells on gene 32 protein affinity columns (T. Minagawa quoted by Alberts et al., 1983).

[b] Burke et al., 1980; Huberman et al., 1971.

[c] Interactions inferred from genetic data (see Table 2 for references).

with different components of the phage DNA replication and recombination systems (DNA, membrane proteins, nucleases, and replication and repair enzymes). Recently, they also detected suppression of certain gene 32 mutations by specific *ts* gene 17 mutations (Mosig, Ghosal, and Bock, 1981). T4 gene 17 is a late phage gene function that encodes a protein involved in DNA packaging. So, it appears that gene 32 protein and gene 17 protein interact in vivo and that the gene 32 protein plays an important role in the coupling of T4 DNA replication to phage maturation. The models for interaction of the gene 32 protein in DNA replication, recombination, and packaging have been discussed by Mosig, Ghosal, and Bock (1981) and are diagrammed in Fig. 2 of the chapter by Mosig in this volume.

Regarding the coupling of the T4 DNA replication/recombination complex to DNA precursor biosynthesis and transcription of late phage genes, it should be emphasized that T4 DNA polymerase (gene 43) a component of this complex, is known to interact with the nucleotide biosynthetic apparatus (Mathews and Allen, this volume) and that some gene 43 mutations behave as suppressors of dCMP hydroxymethylase deficiencies (Chao et al., 1977). There are also several indications that the hydroxymethylase is directly involved in control of DNA replication (Tomich et al., 1974; Wovcha et al., 1973). In addition, as discussed above, *E. coli* RNA polymerase, T4 gene 45 protein, and active phage DNA replication are all essential components of the transcriptional' process for late phage gene functions. It has been shown that the T4 gene 45 protein can be retained by affinity columns charged with either gene 32 protein (Alberts et al., 1983) or T4-modified *E. coli* RNA polymerase (Ratner, 1974a, 1974b). Possibly, such interactions provide the means for coupling the replication complex to the "late transcription complex," and it should be worthwhile to search for defects in the transcriptional steps mediated by gene 45 protein that are suppressible by specific gene 32 alterations.

Alberts et al. (1983) observed that gene 32 protein affinity columns show specific binding to the phage gene 43 DNA polymerase and to the gene *dda* protein (DNA-dependent ATPase, DNA helicase). A recent study by Gauss et al. (1983) provides a new insight about the biological significance of these interactions. It was observed that *E. coli optA* mutants fail to support the growth of a *dda* mutant and of several T4 gene 43 mutants. Also, a pseudorevertant of a T4 gene 32 mutant was shown to carry a mutation in the T4 *sud* region (Little, 1973) where gene *dda* maps. This suggests that the phage DNA helicase interacts with the gene 32 protein. It appears that the *E. coli optA* gene product is a functional counterpart of the T4 *dda* protein and that T4 proteins *dda*, gp43, and gp32 interact to effect DNA unwinding during replication. Some T4 gene 43 mutants may be unable to translocate along the DNA template and need the assistance provided by the DNA-dependent ATPase activities of *dda* and *optA* proteins.

GENETIC INTERACTIONS IN T4 MORPHOGENESIS

Many of the interactions among components of the T4 DNA replication/recombination complex appear to be necessarily weak or transient. By contrast, the nucleoprotein complex that constitutes a T4 phage particle is held together by strong, stable interactions. Both types of complexes harbor major and minor components that are amenable to genetic analysis, although there may be an important difference between these complexes in terms of the severity of genetic lesions that they can tolerate within their components before structural or functional integrity is detectably destabilized. Probably, the stability of interactions in the assembly of a T4 phage particle offers resilience to some changes in amino acid sequence of its component structural proteins. Nevertheless, genetic assays are proving to be very valuable in analysis of the protein-protein interactions of T4 morphogenesis.

In T4 head assembly several proteins, including the products of phage genes 22, 23, 24, *ip*I, *ip*II, and *ip*III, are cleaved to specific fragments that constitute components of the mature head particle (see Black and Showe, this volume, for details). The gene 23 and gene 24 proteins each lose a portion of their N-terminal sequences in the processing reactions, although processed and unprocessed gene 24 protein is found in phage heads. McNicol et al. (1977) isolated a T4 gene 23 mutation that affected the N-terminal portion of the gene 23 protein and that made assembly of viable phage particles possible in the absence of gene 24 protein. The processed gene 23 protein is normally present in about 1,000 copies per head as compared with about 30 copies of processed gene 24 protein. Since the two processed peptides have similar molecular weights, it is possible that the gene 23 mutation that bypasses the gene 24 protein requirement allows substitution of processed gene 23 protein for a missing processed gene 24 protein. In addition, McNicol et al. observed that gene 24 defects can abolish the processing of the other head proteins, including gene 23 protein. The gene 23 mutation that bypasses the requirement for gene 24 protein leads to recovery of the processing activity. So, it appears that proper orientation of head components is necessary for recognition by processing enzymes and that certain alterations in gene 23 protein (the major structural head component) can lead to a recovery of proper conformations for the interacting proteins of the phage head. Other evidence for an interaction between the T4 gene 23 and gene 24 functions comes from work by Doherty (1982b) who isolated pseudorevertants of a gene 23 mutant that contained, in addition to the gene 23 lesion, two mutations, one in gene 22 and the other in gene 24. Both mutations were required for suppression, and neither exhibited effects on phage growth and morphogenesis in the absence of the gene 23 mutation. Also, suppression was specific to the gene 23 mutant allele used in the isolation of pseudorevertants. The results from these genetic studies are suggestive of physical interactions between the protein products of T4 genes 22, 23, and 24.

Genetic evidence also implicates *E. coli* host gene products in control of T4 morphogenesis (Coppo et al., 1973; Georgopoulos et al., 1972; Pulitzer and Yanagida, 1971; Revel et al., 1980; Simon et al., 1979; Takano and Kakefuda, 1972: Tilly et al., 1981; discussed further by Binkowski and Simon, this volume). Much of the evidence consists of the isolation of host mutants that specifically inhibit T4 morphogenesis; in some cases, phage mutants are isolated that are able to bypass the block created in the mutant host. Revel et al. (1980), for example, isolated a large number of *E. coli* mutants that adsorbed and were killed by wild-type T4 but nevertheless failed to support viable phage production. These mutants were grouped into seven classes, four of which specifically affected phage head assembly at the step controlled by T4 gene 31. At least two host genes seem to be involved, and phage mutations that overcome (suppress) the block in the mutant hosts mapped either in gene 31 or in gene 23. The phage suppressor mutations exhibited some degree of allele specificity, suggesting that a complex of gene 31 and gene 23 proteins with host proteins is formed during the early stages of phage head morphogenesis. One of the host proteins involved is probably the product of *E. coli* gene *groE* (Tilly et al., 1981). In a very recent study, Simon and Randolph (personal communication) isolated T4 mutants (named *byp*31) that mapped in gene 23 and that made the gene 31 protein dispensable in head assembly. These mutants also overcome the need for the host *groE* function. It is possible that the T4 gene 31 and *E. coli groE* proteins provide a scaffold for associations between gene 23 protein molecules and that certain alterations in the conformation of gene 23 protein allow it to assemble without the need for this scaffold. The T4 gene 31 protein may also play a role in control of transcription of phage genes (Simon et al., 1979; Stitt et al., 1980).

Another example of interactions among the T4 morphogenesis functions involves the product of gene 63, a bifunctional protein that bears RNA ligase activity and that promotes attachment of tail fibers to baseplates during phage assembly. Suppressors that lead to a bypass of the gene 63 function map in lysis gene *t*, baseplate gene 5, and tail sheath gene 18 (Hall et al.,

1980). Probably, the basis for the suppression is an improvement in the interaction between the components involved in tail fiber attachment such that catalysis by gene 63 protein becomes unnecessary. A similar type of mechanism may underly the suppression of gene 38 mutants by gene 37 (tail fiber) mutations (Bishop and Wood, 1976); in this case, it appears that the gene 38 protein promotes dimerization of gene 37 protein subunits and that certain alterations in the gene 37 protein allow it to dimerize without assistance.

CONTROL OF PROTEIN INTERACTIONS BY GENETIC LINKAGE

A conspicuous feature of the T4 genetic map is the clustering of genes with related functions (Kutter and Rüger, this volume). To some degree this clustering is related to control of phage genome transcription: most genes that are transcribed counterclockwise (early genes) cluster on one side of the circular T4 map, and most genes that are transcribed clockwise (late genes) map on the other side. However, there are indications that many subclusters of T4 early and late genes are not internally regulated as single units. For example, the cluster consisting of T4 genes 45, 44, 62, *regA*, 43, and 42 may harbor as many as five independently recognized promoters (gene 45, genes 44 and 62, *regA*, gene 43, and gene 42) as well as separate transcriptional and translational control mechanisms: the transcripts for genes 45, 44 and 62, 42, and possibly *regA* are subject to translational repression by *regA* protein (Wiberg and Karam, this volume), whereas gene 43 is under autogenous control (Russel, 1973), probably at the transcriptional level (Krisch et al., 1976; Miller et al., 1981). We know that these genes encode proteins that interact functionally in vivo either as components of the T4 DNA replication complex or in the coordination of nucleotide precursor biosynthesis with replication (see above). Conceivably, close linkage between genes, whether or not it provides shared regulatory signals (promoters, operators, or ribosome initiation sites), determines localization of transcription and translation of a genetic segment and, as a consequence, facilitates interaction of the protein products from this segment. One possible example in support of this notion is the interaction between the T4 gene 45 protein and the gene 44/62 ATPase-dATPase, which serves to enhance the processivity of T4 DNA polymerase (gene 43) on DNA templates in replication (see Alberts et al., 1983, for a recent discussion). It has been observed that a nonsense mutation in T4 gene 45 exerts a *cis*-dominant inhibitory effect on the biological activity of a partially defective gene 44 mutant (Stahl et al., 1970), without affecting the synthesis of gene 44 protein (Karam et al., 1979). It appears that the protein fragment encoded by the gene 45 mutant "poisons" the mutant gene 44 protein, but only when both are encoded by the same genome. This may mean that protein assembly into multienzyme complexes is, at least in part, determined by genetic linkage.

CONCLUDING REMARKS

The use of genetic assays for the study of in vivo interactions by T4-induced enzymes and structural proteins almost always entails the examination of an abnormal physiological condition under which the interactions are detected. So, one must also always consider whether such interactions can occur in normal infections. The strength of the genetic approach is not so much in defining interactions precisely (which it can sometimes do), but rather in revealing the various possible means by which gene products can interact. Biochemical characterization of interactions inferred from genetics is always necessary, although the complexity or fragility of enzyme and nucleic acid associations in T4-infected cells has hindered progress on the in vitro reconstitution and manipulation of the major processes that control phage development. In addition, genetic assays frequently fail at detecting important interactions when these are controlled by gene functions that are duplicated from host or phage genome sources. Some participants in complex processes (e.g., DNA replication) may primarily play optimizing roles that cannot be recognized via the isolation of single and double mutants. In such cases, biochemical assays can be more powerful. For example, Alberts et al. (1983) used protein affinity columns to search for T4-induced and host-derived proteins that can interact with the phage gene 32 and gene 45 proteins. In addition to the interactions predicted from genetic and other biochemical evidence, they detected specific binding with a number of as-yet-unidentified phage and *E. coli* proteins.

To derive the most from genetic assays, it is desirable to have available a large number of well-characterized mutations in the genes of interest. This is especially important when trying to distinguish between allele-specific and gene-specific effects. Despite the major efforts that have been mounted over the last 20 years to isolate T4 mutants, many interesting T4 genes remain represented by no more than one or two mutations. It appears, however, that with the advent of recombinant DNA technology and with improvements in methods for manipulating interactions between T4 and its *E. coli* host, it should become easier to selectively obtain defined lesions in desired locations of the T4 DNA. In particular, the method for the isolation of *E. coli* "*tab*" mutants (Takahashi, et al., 1975) is being used to develop hosts that are specifically restrictive to mutations of T4 genes of interest, e.g., the *r*II gene and gene 32 (Doherty et al., 1982; Nelson et al., 1982; Nelson and Gold, 1982). These hosts can be used for large-scale screening of T4 mutants. Also, in principle, bacteria harboring recombinant plasmids in which essential T4 genes are expressed can be used as permissive hosts in the isolation of a variety of "lethal" mutations in these genes. The exploitation of these methods will probably provide many tools for future studies on the control of gene expression at the level of multiprotein complex assembly.

We thank Peter Gauss and Gisela Mosig for many interesting discussions and helpful suggestions for this manuscript.

Our work is supported by grants from the National Institute of General Medical Sciences (grant no. GM18842) and the National Science Foundation (grant no. PCM 82-01869). M.B. is a predoctoral student of the Molecular and Cellular Biology and Pathobiology Program and is recipient of a predoctoral stipend from the College of Graduate Studies, Medical University of South Carolina.

Host Functions That Affect T4 Reproduction

GLORIA BINKOWSKI AND LEE D. SIMON

The Waksman Institute of Microbiology, Rutgers, The State University of New Jersey, New Brunswick, New Jersey 08903

Initially, it may seem surprising that a virus with 150 to 200 genes is highly dependent on various host cell functions for its survival. This surprise may be even greater when one considers that many T4 bacteriophage genes are not required for phage replication in normal laboratory strains of *Escherichia coli*. Particular genes of this nonessential group are, however, required for replication in one or another unusual *E. coli* host strains which have been isolated, often from hospitals. The point here is that even though T4 carries genes that enable it to replicate in many different types of *E. coli* strains, it nevertheless requires a number of host cell functions. The host functions upon which T4 depends are probably universal among *E. coli* strains and perhaps also among other gram-negative bacteria, and thus, they do not impose a limitation upon the range of host cells in which T4 can replicate. Many different phages appear to be dependent upon one or more of the same host functions as T4. This fact suggests that dependencies upon these functions may have evolved long ago in one or more common ancestors of these diverse phages.

A classical method of dissecting and elucidating complex biological processes involves the isolation and characterization of mutants that affect these processes. This approach has been used by a number of investigators to probe the various types of complex interactions between T4 phage and its host bacterium, *E. coli*. Because these interactions generally involve many phage and bacterial components interacting with each other, the precise molecular bases of these phenomena are still somewhat obscure. Much, however, has been learned about these interactions, and in the following pages we shall present and discuss representative examples of interactions between T4 and its host, *E. coli*, during phage production.

HOST DEFECTIVE *E. COLI rho* MUTANTS

General Background

T4 phage infection of a type of host defective mutant first described by Simon et al. (1974) is characterized by a low rate of T4 DNA synthesis, the production of empty T4 heads, the absence of phage tails or tail fibers, and the absence of certain T4 proteins. This host defective *E. coli* mutant adsorbs and is killed by T4 particles, but it fails to produce viable T4 phage (Simon et al., 1974). After the characterization of this *E. coli* mutant, referred to as *hd*590, other phenotypically similar host defective mutants (designated *tabC* and *HDF*) were also reported (Caruso, et al., 1979; Stitt, et al., 1980; Takahashi, 1978; B. Stitt, Ph.D. thesis, California Institute of Technology, Pasadena, 1978).

Some properties of these bacterial mutants seemed similar to properties of *E. coli nitA* mutants in the *rho* gene. First, *nitA* mutants may block T4 production. Second, specific T4 proteins are either present in reduced amounts or are essentially absent from T4-infected $nitA^-$ cells. Third, the proportion of filamentous cells is much higher in cultures of *nitA* mutants than in $nitA^+$ control cultures. Fourth, *nitA* mutants have difficulty in adapting to growth in minimal media (Inoko et al., 1977).

More recently, it has been shown that the mutations responsible for the *hd*590, *tabC*, and *HDF* host defects also appear to be in the *E. coli rho* gene (Caruso et al., 1979; Simon et al., 1979; Stitt et al., 1980), closely linked by P1 transduction to *ilv*. Host defective *E. coli* strains have been characterized as *rho* mutants by the map positions of the relevant mutations and by the ability of the strain to propagate N^- λ phage mutants (Reyes et al., 1976; Simon et al., 1979).

The rho protein is involved in transcription termination in *E. coli* (Adhya and Gottesman, 1978; Roberts, 1969; Roberts, 1976), and many *rho* mutants defective in this activity have been described (Adhya and Gottesman, 1978; Das et al., 1976; Das et al., 1979; Korn and Yanofsky, 1976; Ward and Gottesman, 1982). Mutations in the *rho* gene can cause a variety of effects in *E. coli* such as poor growth in minimal media, filamentation, hyperdegradation of abnormal proteins, inability to propagate certain bacteriophages, and defects in oxidative phosphorylation, in recombination, and in repair of DNA damaged by UV light (Das et al., 1976; Das et al., 1979; Inoko et al., 1977; Simon et al., 1979).

E. coli rho strains defective for T4 production have been characterized in detail (Caruso et al., 1979; Simon et al., 1974; Simon et al., 1979; Stitt et al., 1980; Takahashi, 1978). Many of these *rho* host defective mutants are temperature sensitive or cold sensitive both for bacterial growth and for phage development. At nonpermissive temperatures, the efficiencies of plating (EOP) for T4 range from 2 to 7 orders of magnitude lower on the particular *rho* mutants than on isogenic rho^+ host cells. A few *rho* host defective mutants, such as *tabC803*, are neither temperature sensitive for bacterial growth nor permissive for T4 production at any temperature tested (Caruso et al., 1979; Takahashi, 1978). A cold-sensitive mutant, *E. coli tabC cs*110, showed an increasing EOP for T4 as temperature was increased, but even the highest EOP on this strain was 100-fold lower than that on a rho^+ host (Caruso et al., 1979). Two other host defective *rho* mutants, *HDF*3.41 and *hd*590, are temperature sensitive for bacterial growth, but T4 production in these strains is blocked at both low and high temperatures, although the block is more pronounced at high temperatures (Simon et al., 1974; Stitt et al., 1980). The variations in phenotype are probably the result of the different *rho* alleles that cause the host defective characteristics of these strains. However, it should be noted that many of the host defective *rho* strains were isolated after mutagenesis with nitrosoguanidine, which can cause clustered mutations (Cerda-Olmedo et al., 1968; Guerola et al., 1971). Fine-structure mapping of these host defective mutations has not been done, and it is conceivable, although unlikely, that mutations near but not in the *rho* gene are responsible for the host defective phenotypes of these strains. Since the best available evidence suggests that *rho* mutations are, in fact, responsible for the host defective character of these strains, we shall continue to refer to them as *rho* mutants.

When examined by electron microscopy, lysates of T4$^+$-infected *rho* host mutants are found to contain empty heads, no tails, and no baseplate structures (Simon et al., 1974; Stitt et al., 1980; Takahashi, 1978). Electrophoretic patterns of T4 proteins synthesized at various times after infection of the host defective *rho* mutants show that early proteins such as gp43, gp*r*IIA, and several other unidentified early and quasilate proteins are present in reduced levels (Caruso et al., 1979; Stitt et al., 1980). Other early proteins, such as gp*r*IIB, gp32, and *ip*III, are overproduced in these strains (Caruso et al., 1979; Stitt et al., 1980). This overproduction may be the result of a delay in shutoff of early protein synthesis similar to that observed in the *E. coli tabC803* mutant (Caruso et al., 1979). Many late proteins, such as the major capsid protein gp23 and the tail sheath protein gp18, are present in relatively normal amounts; however, at least four late proteins, including gp34 (tail fiber), gp7 (baseplate), gp37 (tail fiber), and gp*wac* (whisker), are either entirely absent or are present at reduced levels (Caruso et al., 1979; Simon et al., 1974; Stitt et al., 1980). Hybridization competition experiments indicate that the RNA in a T4-infected *tabC* strain contains an essentially normal ratio of early and late T4 transcripts (Takahashi, 1978).

T4 DNA synthesis occurs at a reduced level and is delayed in most host defective *rho* mutants. At the nonpermissive temperatures for phage development, the amounts of T4 DNA synthesized in these mutants varies from approximately 20 to 60% of that synthesized in wild-type hosts (Caruso et al., 1979; Simon et al., 1974; Stitt et al., 1980; Takahashi, 1978). Thus, the reduced amount of T4 DNA synthesized in these *rho* host defective mutants cannot, in itself, account for the drastically lowered EOP of T4$^+$ particles. In fact, *E. coli tabC803* permits synthesis of substantial amounts of T4$^+$ DNA, although this mutant has one of the lowest EOPs for T4$^+$ phage (Caruso et al., 1979).

Another characteristic of these host defective *E. coli* mutants is that lysis after T4$^+$ infection is delayed, although production of T4 lysozyme appears to be essentially normal (Stitt et al., 1980). Stitt and colleagues suggested that the first step in lysis, namely the breakdown of the inner cell membrane of a *rho* mutant, may not occur as it should, and thus, lysis could be delayed. Involvement of the *E. coli* inner membrane in the host defective phenotype was first suggested for *hd*590 after it was demonstrated that low concentrations of EDTA, which react primarily with components of the cell envelope, increase phage production up to 100-fold in *hd*590 but have relatively little effect on phage production in the wild-type host cell (Simon et al., 1974)

Involvement of Rho Protein in Bacteriophage Development

Transcriptional control of phage development. Rho protein plays an important role in the control of transcription of λ phage genes. A very brief summary of that role may suggest, by analogy, possible effects of rho protein on T4 production. In λ phage infections of *rho*$^+$ *E. coli* strains, transcription does not proceed past rho-dependent transcription termination sites in the λ genome unless sufficient λ N protein, which is transcribed immediately after infection, is synthesized (Herskowitz and Hagen, 1980). Protein N of λ suppresses transcription termination and positively controls phage development (Ward and Gottesman, 1982). The presence of protein N enables transcription to continue past the rho-dependent termination sites into the delayed early genes that are required for phage development (Herskowitz and Hagen, 1980).

N protein is capable of preventing transcription termination (anti-terminating) at both rho-dependent and rho-independent termination sites. Thus, N protein does not function simply to inactivate the rho protein (Gottesman et al., 1980). Anti-termination of λ phage transcription by protein N is a complex process that involves interactions among many factors, such as ribosomal proteins, a *cis*-acting site in the DNA template or mRNA, RNA polymerase, and other host transcription factors (Ward and Gottesman, 1982). Protein N may stabilize the RNA polymerase complex on the DNA template and thereby suppress transcription termination (Ward and Gottesman, 1982).

The control of phage development by regulation of transcription termination is not as well understood for T4 as for λ phage. The RNA transcribed from T4 promoters can be classified as early, quasilate, or late, depending upon the time of its appearance after infection (Salser et al., 1970). Certain early genes, referred to as immediate early genes, are transcribed from early promoters within 90 s after infection (Brody, Sederoff, Bolle, and Epstein, 1970; see also Brody et al., this volume). However, the distinctions among early, delayed early, and quasilate genes are somewhat blurred in that various early genes may be transcribed both from early promoters and also from middle promoters. For instance, in vivo, the *r*IIB transcript is found at the distal end of an mRNA molecule initiated at an early promoter or at the proximal end of an mRNA molecule initiated at a middle promoter (Schmidt et al., 1970; Sederoff, Bolle, Goodman, and Epstein, 1971). A T4 protein encoded by the *mot* gene seems to be responsible for activating or positively controlling such middle promoters (Mattson et al., 1974; Mattson et al., 1978; Pulitzer, et al., 1979). Recent studies suggest that *mot* encodes a protein which may have topoisomerase activity (Brody et al., this volume; Kreuzer and Huang, this volume). It is known that transcription of particular genes may be affected by the extent of supercoiling of the relevant DNA (Smith et al., 1978). It seems likely that *mot* affects transcription by changing the degree of supercoiling of T4 DNA.

The *rho* host defective mutants have been useful in illuminating the dependence of T4 phage prereplicative transcription both on the host *rho* function and on the T4 *mot* function. Results of infection of *rho tabC* bacteria by the T4 *mot ts*G1 mutant suggest that T4 prereplicative transcription depends upon the dual functioning of the *rho* and *mot* genes (Pulitzer et al., 1979). *rho* function negatively controls transcription, whereas the *mot* function appears to control transcription positively. The transcription of most delayed early genes seems to be controlled either by a promoter that is *mot* independent and *rho* sensitive or by a promoter that is *mot* dependent and *rho* insensitive (Pulitzer et al., 1979).

In rho^+ *E. coli*, when chloramphenicol is added to inhibit protein synthesis at the time of T4 infection, only that T4 RNA defined as immediate early RNA is synthesized (Brody and Geiduschek, 1970; Grasso and Buchanan, 1969; Salser et al., 1970). However, in T4 infection of *E. coli rho ts*15, a *rho* mutant that fails to terminate transcription, synthesis of T4 mRNA in the presence of chloramphenicol is not confined to the immediate early class of mRNA molecules (Brody, 1978). For example, transcripts encoding the delayed early genes deoxycytidine triphosphatase and UDPglucose–DNA-β-glucosyltransferase are synthesized in an *E. coli rho ts*15 strain in the presence of chloramphenicol, whereas these same T4 genes are not transcribed in T4-infected *E. coli* rho^- strains in the presence of chloramphenicol (Linder and Skold, 1980). In vitro studies have also demonstrated that transcription from immediate early promoters can be initiated in a system that contains purified *E. coli* RNA polymerase and T4 DNA (Richardson, 1970). In this system, the RNA that is synthesized in the presence of rho protein is confined to immediate early RNA (Richardson, 1970); it corresponds to the RNA synthesized in rho^+ cells in vivo in the presence of chloramphenicol (O'Farrell and Gold, 1973a; Schmidt et al., 1970; Sederoff, Bolle, Goodman, and Epstein, 1971). In vitro, in the absence of the rho protein, both immediate early and delayed early RNA are synthesized; the messages for these delayed early genes are located at the promoter-distal ends of the transcripts (O'Farrell and Gold, 1973a; Schmidt et al., 1970; Sederoff, Bolle, Goodman, and Epstein, 1971). These results support a model in which rho protein can terminate transcription early in T4 phage development. One or more anti-terminator phage proteins are postulated to be necessary for the transcription of delayed early genes which are under the exclusive control of immediate early T4 promoters (Pulitzer et al., 1979; Schmidt et al., 1970; Stitt et al., 1980).

Not all *rho* mutants are defective in the propagation of T4. Originally, *E. coli rho* mutants were isolated as suppressors of polarity (Beckwith, 1963); the *rho ts*15 mutation is capable of suppressing polarity with high efficiency, and it is very deficient in transcription termination. T4 phage have a nearly normal EOP on *E. coli rho ts*15 (Zograff and Gintsburg, 1980; K. Tomczak, M.S. thesis, Rutgers University, New Brunswick, N.J., 1979).

In contrast, *rho* host defective mutants such as *E. coli HD*F026, on which $T4^+$ phage have a very low EOP, are only weakly defective in their ability to terminate transcription (Simon et al., 1979). Thus, it is not the absence of *rho*-mediated transcription termination that blocks T4 production in the *rho* host defective strains. If, as has been suggested by Caruso et al. (1979), Pulitzer et al. (1979), and Stitt et al. (1980), a T4 protein(s) must interact with the rho protein to modulate its transcription termination activity, then in a *rho* host defective strain, the mutant rho protein may be insensitive to the T4 modulation system.

In the case of λ transcription, studies of various *E. coli rho* mutants define two distinct effects of different *rho* mutations on the requirement for N protein activity in λ phage production (Das et al., 1976; Ward and Gottesman, 1982). A cell carrying the *rho ts*15 mutation propagates λ phage efficiently, even in the absence of protein N (Das et al., 1976), whereas the *rho* mutant *HD*F026 propagates λ phage very inefficiently, even in the presence of protein N (Simon et al., 1979; Ward and Gottesman, 1982). Thus, λ protein N is not essential in *rho* mutants that are very defective in transcription termination. On the other hand, in *rho* mutants such as *HD*F026, protein N is unable to modulate rho-mediated transcription termination, and consequently λ phage production is blocked. These observations are consistent with the suggestion (Caruso et al., 1979; Pulitzer et al., 1979; Stitt et al., 1980) that *rho* mutants insensitive to phage antitermination systems may prevent normal T4 transcription and T4 phage production.

Effects of hyperproteolytic activity on phage production. Specific *rho* mutants, however, may affect T4 phage production in ways other than by altered transcription termination activity. Several *rho* mutants, including some that are host defective for T4, display abnormalities with regard to protein degradation. These *rho* mutants, including *E. coli* strains *HD*F026 (Stitt et al., 1980), *hd*590 (Simon et al., 1974), and *rho ts*15 (Das et al., 1979), exhibit elevated levels of degradation of abnormal proteins and have been characterized as hyperproteolytic (Simon et al., 1979). The average levels of degradation of normal protein in such *rho* mutants are similar to those in *rho* *E. coli* cells (Simon et al., 1979).

The elevated levels of degradation of abnormal proteins in these *rho* mutants may, in fact, be related to the inability of T4 to replicate in particular strains. Studies of *E. coli* strains carrying both *lon* and *rho* mutations have focused on this point. The *lon* gene encodes an ATP-dependent protease (Charette et al., 1981) which is responsible, in part, for the degradation of abnormal proteins in *E. coli* (Gottesman and Zipser, 1978; Shineberg and Zipser, 1973). The EOP of $T4^+$ on the *rho* (*HD*F026) host defective mutant is about 10^{-7}, whereas the EOP of $T4^+$ on the *rho lon* double mutant is 0.7 (Simon et al., 1979). Thus, the *lon* mutation substantially suppresses the host defective phenotype of the *HD*F026 *rho* mutation. Furthermore, the EOP of $T4^+$ on the *tabC803 rho* strain is increased approximately 50-fold by the addition of a *lon* mutation to the bacterial strain (Caruso et al., 1979). At least one *rho* host defective mutation, however, does not appear to show any suppression by *lon* mutations (Caruso et al., 1979). Since the only known function of the *lon* protease in *E. coli* involves the breakdown of proteins, the ability of a *lon* mutation to suppress a host defective *rho* mutation suggests that proteolytic activity may, at least in some instances, be involved in the *rho* host defect. The hyperproteolytic activity of the *rho* cells is reduced by the *lon* mutation (Simon et al., 1979). This reduction in proteolytic activity apparently enables T4 replication to proceed in some host defective *rho* strains. However, the reduced proteolytic activity does not completely compensate for the effects of the *rho* host defective mutation since T4$^+$ burst sizes, EOP, or both remained reduced to some extent in these *lon rho* bacteria. Altered transcription termination in these *E. coli* double mutants and proteolytic activities other than those mitigated by a *lon* mutation also may be significant in the ability of T4$^+$ phage to replicate in *E. coli rho* host defective strains.

The overall level of protein degradation probably does not account for the inability of T4 to replicate in certain *rho* mutants. T4$^+$ phage can replicate in *E. coli rho ts*15 (Zograff and Gintsburg, 1980; K. Tomczak, M.S. thesis), which has a higher level of degradation of abnormal proteins than do some other *rho* mutants (e.g., *HDF*3.41) in which T4$^+$ phage do not replicate (Simon et al., 1979; L. D. Simon, unpublished observations). This observation and the fact that T4$^+$ phage do inhibit degradation of abnormal proteins (see below) in *E. coli* host defective *rho* mutants (Simon et al., 1979) lead to the conclusion that factors other than the overall level of protein degradation are of critical importance in the generation of the *rho* host defective phenotype.

Compensation for *rho* Host Defective Mutations by T4 Mutations

There exist T4 phage mutants that form plaques with much higher efficiencies than do T4$^+$ particles on the *rho* host defective strains (Caruso et al., 1979; Simon et al., 1974; Stitt et al.. 1980; Takahashi, 1978). These phage mutants also form plaques with normal efficiencies on *rho*$^-$ strains (Caruso et al., 1979; Simon et al., 1974; Stitt et al., 1980; Takahashi, 1978). Such T4 mutants, originally isolated on the *rho* mutants *hd*590, *HDF*, or *tabC*, have been designated as *go*590, *go*F, or *comC*, respectively (Caruso et al.. 1979; Simon et al., 1974; Stitt et al., 1980; Takahashi, 1978). These T4 phage carry mutations in one of several genetic loci. The predominant class of T4 mutants that can replicate in the host defective *rho* strains maps near gene 39 (Caruso et al., 1979; Simon et al., 1974; Stitt et al., 1980; Takahashi, 1978) in a nonessential region of the T4 chromosome which is defined by *del*(39-56)12 (Homyk and Weil, 1974); another class of these T4 mutants maps in gene 31 (Simon et al., 1974); a third class appears to map between genes 31 and 33 (Stitt et al., 1980); a fourth class maps near, but not in, gene *e* (Stitt et al., 1980); and a fifth class maps in gene 45 (Coppo et al., 1975b; Takahashi, 1978).

In infections of *rho* host defective bacteria by appropriate T4 mutants, the initiation of T4 DNA synthesis usually is normal or only very slightly delayed, and the rate of DNA synthesis is higher than in analogous T4$^+$ infections. As expected, the synthesis of T4 proteins is also more nearly normal (Caruso et al., 1979; Stitt et al., 1980). For example, in T4 *comC*α infections of *E. coli tabC* bacteria, gp43 and gp*r*IIA are synthesized at the same time and to the same extent as in T4$^+$ infections of *E. coli tabC*$^+$ cells (Caruso et al., 1979).

Of the T4 mutants that can replicate in *rho* host defective strains, not all of the phage can replicate in all of the *rho* host defective strains. Some of the T4 mutants (e.g., *go*590-1 and *go*F3.03-2) can replicate well only in specific *rho* host defective strains (i.e., *E. coli hd*590 and *E. coli HDF*3.03, respectively) in addition to *rho*$^+$ strains (Stitt et al., 1980). Although some other T4 mutants, particularly those with mutations near gene 39, can replicate in all *rho* host defective strains, allele specificity seems to exist between some *rho* alleles and certain T4 genes.

With regard to the function of the rho protein in T4 infection, the sites of some of the T4 mutations that suppress the *rho* host defective phenotype are rather intriguing and puzzling. The T4 *go*590-1 phage mutation maps in gene 31 (Simon et al., 1974). The only established function of gp31 is related to T4 capsid assembly, which occurs on the bacterial inner membrane (Simon, 1972). It may not be coincidental, therefore, that the inner membrane of *hd*590 is implicated in the *rho* host defective phenotype (Simon et al., 1974). T4 capsids consist of late T4 proteins, but gp31, in contrast, is synthesized early after infection (Castillo et al., 1977). T4 *byp*31 (bypass 31) mutations eliminate the requirement for gp31 in T4 phage production (L. D. Simon, manuscript in preparation). These *byp*31 mutations are in gene 23, which encodes the major T4 capsid protein. *byp*31 mutations, however, do not permit T4 to replicate in *rho* host defective strains (L. D. Simon, manuscript in preparation). We suggest that gp31 may be involved both in early transcriptional events and in T4 capsid morphogenesis. Gene 31 is not closely linked to other capsid genes; rather, gene 31 is near a small cluster of genes, some of which appear to have two distinct functions. Specifically, gene 63 encodes a protein with RNA ligase activity (Silber et al., 1972; Snopek, et al., 1977) and with an activity that promotes tail fiber addition to fiberless phage particles (Runnels et al., 1982; Wood and Bishop, 1973; Wood and Henninger, 1969); also, gene *pse*T encodes a protein with 5′-polynucleotide kinase (Novogrodsky and Hurwitz, 1966; Richardson, 1970; Sirotkin et al., 1978) and 3′-phosphatase activities (Becker and Hurwitz, 1967; Cameron and Uhlenbeck, 1977; Depew and Cozzarelli, 1974). If gp31 actually does function both in transcriptional control and in capsid morphogenesis, then the *byp*31 experiments mentioned above suggest that the transcriptional role of gp31 is not absolutely required in T4 infections of *rho*$^+$ hosts.

The *go*F1 and *com*Cα mutations do not map at precisely the same site, but the mutations apparently are in the same previously undefined and unidentified gene covered by *del*(39-56)12 (Stitt et al., 1980; Takahashi, 1978; Takahashi and Yoshikawa, 1979). The gene is located in a region of the T4 genome that is nonessential for replication in *E. coli rho*$^-$ strains. If, as proposed (Stitt et al., 1980), the *go*F1 gene product is a modulator of rho protein transcription termination activity, then the modulation function of the *go*F1 gene product is not required under normal conditions. The goF1 gene function is also nonessential for replication in bacteria carrying the *rho ts*15 mutation (L. D. Simon, unpublished observations).

In mixed infections of T4$^+$ with *go*F1 particles or T4 *del*(39-56)12 with *go*F1 particles in an *E. coli rho* host defective strain, the *go*F1 allele seems to act in *cis* so that T4 particles carrying the *goF1* allele predominate among the T4 progeny (Stitt et al., 1980). The explanation of this phenomenon, although presently unknown, will undoubtedly relate to the molecular basis of the host defective phenotypes of the above-mentioned *rho* mutants.

Another class of T4 mutants able to suppress *rho* host defects also has mutations in gene(s) that may be nonessential. These mutations map near gene *e* in a region of the T4 chromosome known to contain many nonessential genes (Kim and Davidson, 1974).

A mutation in T4 gene 45 can also suppress the *rho* host defective phenotype of *E. coli tabC* mutants (Coppo et al., 1975b; Takahashi, 1978). Gp45 is required for late transcription of the T4 genome (Wu and Geiduschek, 1975), and gp45 is able to bind to the T4-modified *E. coli* RNA polymerase transcription complex (Ratner, 1974). In *rho* host defective cells, the altered gp45 may allow transcription to proceed beyond *rho*-sensitive termination sites in the T4 genome.

Although phages T2 and T4 are very closely related to each other (Kim and Davidson, 1974; Russell, 1974), their abilities to replicate in *rho* strains are quite different. The efficiency of plating of T2$^+$ phage on *E. coli rho ts*15 is several orders of magnitude lower than that of T4$^+$ phage, which form plaques with an EOP of nearly 1.0 on *rho ts*15 (Zograff and Gintsburg, 1980; L. D. Simon, unpublished observations). Furthermore, the efficiency of plating of T2 particles is far lower (by at least 3 orders of magnitude) than that of T4 particles on *rho* host defective *E. coli* mutants (Das et al., 1979; L. D. Simon, unpublished observations). Although homologous with each other to a great extent, the genomes of T2 and T4 have regions of nonhomology, particularly in the early genes and in nonessential regions of the phage genomes (Kim and Davidson, 1974). It is notable that one specific region of nonhomology exists near the site of the T4 *goF1* and T4 *comCα* mutations.

At the molecular level, no specific mechanisms have been demonstrated for modulation of rho protein activity by T4 early proteins. The location of phage mutations in several genes that can suppress the *rho* host defective phenotype may be an indication of the complexities both of the transcription termination process and of the regulation of gene expression through transcription in T4 phage development.

HOST CELL FUNCTIONS IN T4 PHAGE MORPHOGENESIS

Role of the Cell Membrane in T4 Phage Assembly

The involvement of the cell membrane in T4 phage morphogenesis was first inferred from results of studies of T4 phages with mutations in gene 31 (Kellenberger, 1966; Simon, 1972). In the mature phage particle, gp23, the major capsid protein (Sarabhai et al., 1964) is present in a cleaved form and is referred to as gp32* (Laemmli, 1970; Laemmli and Favre, 1973). Gp31 is required for the assembly of the capsid; however, it is not incorporated into the mature phage particle (Castillo et al., 1977). Gp31, therefore, appears to act catalytically (Castillo et al., 1977; Laemmli et al., 1970; Snustad, 1968) in T4 capsid morphogenesis. No head-related structures are formed in the absence of functional gp31 (Coppo et al., 1973; Epstein et al., 1963; Laemmli et al., 1970); only lumps or aggregates of gp23 associated with the bacterial inner membrane are observed in thin sections through infected cells (Kellenberger, 1966; Simon, 1972). In T4 31$^-$ infections, a large percentage of the gp23 synthesized sediments with the cell envelope fraction after lysis (Laemmli et al., 1970). Other specific phage proteins such as gp20 and gp40, which are involved in capsid morphogenesis, also seem to associate with the bacterial cell membrane (Brown and Eiserling, 1979b; Driedonks et al., 1981; Hsiao and Black, 1978a, 1978b). The membrane-associated functions of gp20 and gp40 are described elsewhere (Black and Showe, this volume). Results of temperature shift experiments on bacteria infected with the T4 gene 31 mutant *ts*A70 suggest that the gp23 lump material that accumulates on the inner cell membrane at the restrictive temperature can be utilized in the production of heads after the infected cells have been shifted to permissive temperature. In these experiments, phage particles are produced even when chloramphenicol is added at the time of the temperature shift to prevent further protein synthesis (Coppo et al., 1973; Laemmli et al., 1970). Evidence from electron microscope studies of thin sections through wild-type *E. coli* cells infected with T4 *ts*A70 phage particles indicates that after the shift to permissive temperature, many preheads, associated with the cell membrane, arise from the lump material; furthermore, the amount of lump material present in the cells simultaneously decreases (Simon, 1972).

Data from in vivo studies of normal T4$^+$ infections also indicate that the assembly of T4 capsids occurs on the inner membrane of the host cell. Examination of thin sections through T4$^+$-infected *E. coli* B bacteria at various times after infection suggests that T4 capsids assemble according to the following sequence of events: (i) lumps form on the cell membrane; (ii) preheads form on the cell membrane from the lump material; (iii) preheads give rise to empty heads; (iv) empty heads leave the cell membrane and then fill with DNA in the central region of the cell (Simon, 1972).

In addition to serving as a site for T4 capsid assembly, the inner *E. coli* membrane also appears to be involved in the assembly of long and short tail fibers (Edgar and Lielausis, 1965; Kells and Haselkorn, 1974; Mason and Haselkorn, 1972) and of the baseplate (Edgar and Lielausis, 1965; Kells and Haselkorn, 1974; Mason and Haselkorn, 1972; Simon, 1969). A rationale for the use of the cell membrane as a surface on which T4 morphogenetic reactions occur has been proposed (Simon, 1969). In this proposal, the point is made that the assembly of complex structures which depends upon the ordered interactions of many different proteins is likely to be very highly dependent on the concentrations of the reactants. The cell membrane would function by localizing the various proteins involved in these assembly reactions on its surface; the membrane would thus facilitate in vivo assembly by reducing the high concentration dependence that otherwise may exist for such complex reactions.

Host Mutants that Block T4 Capsid Morphogenesis

General background. Infection of particular *E. coli* host mutants by T4$^+$ particles can produce a phenotype similar to that of T4 31$^-$ infections of wild-type

E. coli bacteria. T4[+] infection of these host defective mutants results in the production of normal tails and tail fibers and in normal lysis. All T4 proteins are synthesized, but gp23 is not processed to its mature form, and heads and head-related structures are not assembled (Coppo et al., 1973; Georgopoulos et al., 1972; Revel et al., 1980; Simon et al., 1975; Takano and Kakefuda, 1972; B. Stitt, Ph.D. thesis). Amorphous aggregates of capsid proteins form on the inner surface of the bacterial cell membrane (Takano and Kakefuda, 1972).

E. coli mutants that block T4 capsid development at the level of gene 31 function carry mutations that usually map near *purA*. Numerous such *E. coli* mutants, designated *groE* (Georgopoulos et al., 1972; Georgopoulos and Eisen, 1974), *mop* (Takano and Kakefuda, 1972), *tabB* (Coppo et al., 1973), and *hdh* (further classified as *HD*A, *HD*B, *HDD*, and *HD*AD) (Revel et al., 1980) with this host defective phenotype have been isolated. Some of these mutants are temperature sensitive or cold sensitive for T4 production (Coppo et al., 1973; Georgopoulos et al., 1972; Revel et al., 1980). Growth characteristics and other aspects of the physiology of the uninfected host defective cells are quite diverse. At high temperatures, one of the mutants, *E. coli groE*$_L$*44* (formerly referred to as *groE*A44), filaments and is unable to divide (Georgopoulos and Eisen, 1974; Georgopoulos et al., 1972). Another type of physiological change has been observed in the host defective *E. coli mop* mutant. The *mop* allele appears to alter the permeability of the *E. coli* inner membrane to certain tripeptides; furthermore, an unidentified *E. coli* protein present in *mop*$^+$ cell membranes is absent in *mop*$^-$ cell membranes. Therefore, Takano et al. have suggested that the *mop* gene may have a membrane-related function (Takano and Kakefuda, 1972). *E. coli* strains *groE*$_L$44, *mop*, *tabB*, and *hdh* probably all carry mutant alleles of the *groE* gene. This conclusion is derived from the observations that the host defective phenotypes of these strains are essentially similar to each other and that the host defective mutations are closely linked to each other on the *E. coli* chromosome. A mutation in another *E. coli* gene designated *fatA*, which is unlinked to the *groE* gene on the *E. coli* chromosome, causes alterations in the lipid composition of the cell membrane and also blocks T4 capsid development at the same step as do *E. coli groE* mutants (Simon et al., 1975).

The *groE*$_L$ and *groE*$_S$ proteins. Recently, the *groE* locus has been found to consist of two closely linked genes encoding the proteins designated gp*groE*$_L$ and gp*groE*$_S$. The protein gp*groE*$_S$ has been identified on sodium dodecyl sulfate gels as a polypeptide with a molecular weight of 15,000 (Tilly et al., 1981). *E. coli groE*$_S$ mutants have phenotypes similar to those of *groE*$_L$ mutants with respect to temperature-sensitive bacterial growth and to defects in λ phage capsid assembly (Tilly et al., 1981). However, the *groE*$_S$ mutants which have been examined are permissive for T4[+] production; thus, gp*groE*$_S$ does not appear to be involved in T4 head morphogenesis (C. Georgopoulos, personal communication). The *groE*$_L$ protein is a soluble protein with a molecular weight of 65,000 (Georgopoulos and Hohn, 1978; Hendrix, 1979; Hendrix and Tsui, 1978; Hohn et al., 1979). Although a soluble protein, gp*groE*$_L$ is rich in the hydrophobic amino acids leucine and valine (Hendrix, 1979). The protein forms an oligomer that is cylindrical in shape with sevenfold symmetry and is comprised of 14 gp*groE*$_L$ monomeric subunits (Hendrix, 1979; Hohn et al., 1979). It seems that oligomers with sevenfold symmetry are uncommon among proteins (Hendrix, 1979); two other proteins which exhibit this symmetry are pyridine nucleotide transhydrogenase in *Pseudomonas* spp. (Wermuth and Kaplan, 1976) and a gp*groE*$_L$-like protein which has recently been isolated from *Bacillus subtilis* (Carrascosa et al., 1982).

Gp*groE*$_S$ and gp*groE*$_L$ are heat shock proteins in uninfected *E. coli* bacteria. In *E. coli* cells growing at a rate of 0.4 generations per h at 37°C with acetate as an energy source, gp*groE*$_L$ accounts for 0.8% of total cell protein. At the same temperature, in cells growing at 2 generations per h with glucose as an energy source, gp*groE*$_L$ comprises 1.6% of total cell protein (Peterson et al., 1978). Under steady-state conditions with glucose as an energy source at 46°C, gp*groE*$_L$ comprises 12% of total cell protein (Herendeen et al., 1979).

Changes in the temperature at which the cells are growing quickly and dramatically affect the rates of synthesis of the *groE*$_L$ and *groE*$_S$ proteins (Lemaux et al., 1978); upon a rapid temperature shift from 30 to 42°C, the rate of gp*groE*$_L$ synthesis increases severalfold (Yamamori and Yura, 1982). The synthesis of heat shock proteins, including the *groE*$_L$ and probably the *groE*$_S$ proteins, is regulated by the *E. coli hin* gene (Yamamori and Yura, 1982). Heat shock proteins appear to confer upon bacteria the ability to survive at elevated temperatures. Depending upon growth conditions, gp*groE*$_L$ may be the most abundant protein in *E. coli* (Herendeen et al., 1979).

In monomeric form, gp*groE*$_L$ appears to be associated with ribosomes (Subramanian et al., 1976; Subramanian et al., 1979). Also, a protein identical to gp*groE*$_L$ copurifies with RNA polymerase (Hendrix, 1979; Ishihama et al., 1976; Neidhardt, et al., 1981). Since the level of gp*groE*$_L$ in *E. coli* bacteria is quite high, especially at elevated temperatures, it would not be surprising if the *groE*$_L$ protein were associated with cellular components other than ribosomes and RNA polymerase complexes.

Both the *E. coli* inner cell membrane and the *groE*$_L$ protein participate in an early step of capsid morphogenesis. Although the *groE*$_L$ protein has been isolated in the soluble fraction of the cell, *groE*$_L$ protein is rich in hydrophobic amino acids. Evidence from studies of *mop* and *fatA* host defective *E. coli* mutants suggests that the alterations of the cell membrane may affect T4 capsid assembly at the level of gp31 and gp*groE*$_L$ function. These observations lead to the speculation that the *groE*$_L$ protein may functionally associate with the *E. coli* inner membrane. Electron micrographs show that gp23 may bind to the *E. coli* inner membrane before the assembly of T4 capsids or preheads begins. Thus, gp*groE*$_L$, gp31, and gp23 may interact with each other on the cell membrane to promote T4 capsid assembly.

The morphogenesis of T-even phages, λ, R17, ϕ80, and T5 phages involves the *groE* proteins (Coppo et al., 1973; Georgopoulos et al., 1972; Revel et al., 1980; Sternberg, 1973a; Takano and Kakefuda, 1972). Phage T4 requires a *groE* protein function for cleavage of a tail protein (Zweig and Cummings, 1973). Both T-even and λ phages require *groE* proteins for early steps in head assembly, even though T-even capsid assembly occurs on the cell membrane (Simon, 1972), and λ phage head assembly occurs in the cell cytoplasm (Zachary et al., 1976). In λ-infected *E. coli* cells, the level of *groE*$_L$ protein doubles between 10 and 20 min after infection (Drahos and Hendrix, 1982). Although synthesis of most host proteins ceases soon after T4 infection, it would be of interest to examine specifically the effect of T4 infection on the synthesis of *groE*$_L$ protein in host bacteria.

Gp*groE*$_L$-like proteins may function in the morphogenesis of virions other than coliphages. A protein that physically resembles and serologically cross-reacts with gp*groE*$_L$ has been isolated from *B. subtilis* (Carrascosa et al., 1982). In ϕ29 phage-infected *B. subtilis* cells, this gp*groE*$_L$-like protein has been shown to cosediment with phage protein gp10, a protein involved in the production of the ϕ29 capsid (Carrascosa et al., 1982).

Interactions among proteins gp23, gp31, and gp*groE*$_L$. T4 phage mutants able to replicate in *E. coli* *groE*$_L^-$ strains have been isolated. Different *groE*$_L$ mutants are not identically suppressed by various classes of T4 phage mutants. The host defective phenotypes of some *groE*$_L$ host mutants are suppressed both by certain T4 mutations in T4 gene 23 and by other mutations in T4 gene 31. Other *E. coli groE*$_L$ strains are suppressed by specific mutations in T4 gene 31 and not by mutations in gene 23. Still others are suppressed only by particular mutations in gene 23 (Revel et al., 1980; Takahashi et al., 1975). Thus, T4 particles with mutations in gene 31 or in gene 23 which were selected for their ability to propagate on a particular *groE*$_L$ strain may not be able to form plaques on various other *groE*$_L$ mutants (Revel et al., 1980; Takahashi et al., 1975). Negative complementation also seems to exist between particular T4 gene 23 and gene 31 mutants and specific *E. coli groE*$_L$ strains; that is, certain T4 phage mutants that can replicate in specific *E. coli groE*$_L$ host defective strains form plaques with lower efficiencies than do T4$^+$ particles in certain other *groE*$_L$ strains (Coppo et al., 1973; Revel et al., 1980; Takahashi et al., 1975). Also, T4 phage particles with either the *ts*A70 or *ts*A56 mutation in gene 31 are unable to propagate in a particular *groE*$_L$ mutant, *E. coli tabB212*, at a temperature which is permissive for T4$^+$ infection of *tabB212* and for *ts*A70 or *ts*A56 infection of *groE*$_L^+$ strains (Coppo et al., 1973). Other experiments have shown that in the cases studied, the level of gp31 activity is not involved in the suppression of the *groE*$_L$ phenotype by T4 gene 31 mutants (Coppo et al., 1973; Takahashi et al., 1975).

The allele specificity between T4 mutants in genes 23 or 31 and the *E. coli groE*$_L$ strains that they suppress implies that direct physical interactions among the products of these genes are required in the course of T4 head morphogenesis (Georgopoulos et al., 1972; Revel et al., 1980; Takahashi et al., 1975; Takano and Kukefuda, 1972). Neither *ts*A56 with *tabB212* nor *ts*A70 with *tabB212* infection is permissive for phage production at any temperature (Coppo et al., 1973; Takahashi et al., 1975), apparently because the abnormal gp31 and gp*groE*$_L$ proteins cannot interact with each other to promote the association of gp23 molecules that is required to form the lattice of the phage capsid.

Certain T4 gene 23 mutants are able to suppress particular *groE*$_L$ host defective mutations in the presence of normal gp31 (Revel et al., 1980; Takahashi et al., 1975; Takano and Kukefuda, 1972). In the case of λ phage, certain mutations in gene *E*, which encodes the major capsid protein, can also suppress particular *groE* mutations that block λ phage capsid morphogenesis (Georgopoulos et al., 1973; Sternberg, 1973b). Some of these λ gene *E* mutations result in a decrease in the amount of gp*E* synthesized (Sternberg, 1973b), and it appears, therefore, that a change in the ratio of gp*E* to another protein, perhaps to gp*groE*, may be responsible for the suppression of the *groE* mutation (Georgopoulos et al., 1972). T4 gene 23 mutations probably do not suppress *groE* mutations by changing the ratio of gp23 to another protein; the amount of gp23 produced in infections by gene 23 mutants that suppress the *groE*$_L$ phenotype is similar to that produced in T4$^+$ infections (Revel et al., 1980).

Another type of mutation, *byp*31, in T4 gene 23 bypasses the requirements both for gp31 and for normal *groE*$_L$ protein in capsid assembly (L. D. Simon, manuscript in preparation). Regardless of the presence or absence of gp31, T4 *byp*31 mutants form plaques on all *groE*$_L$ hosts tested including *E. coli groE*$_L$*44* and *E. coli tabB212* as well as on the *fatA* mutant *E. coli hd*B3-1 (L. D. Simon, manuscript in preparation). There is no apparent allele specificity between the *byp*31 allele and *groE*$_L$ alleles. Rather, the type of change in gp23 caused by the *byp*31 mutation seems to eliminate the requirement for *groE*$_L$ and for gene 31 functions in T4 capsid morphogenesis. These data support the suggestion that gp23, gp31, and gp*groE*$_L$ normally interact during T4 capsid morphogenesis.

Tail Fiber Assembly

Indirect evidence suggests that T4 tail fibers may be assembled on the bacterial inner membrane. In T4 57$^-$ infections, gp37 and gp34 of the long tail fiber and gp12 of the short tail fiber of the baseplate bind in monomeric form to the cell membrane (Edgar and Lielausis, 1963; Kells and Haselkorn, 1974; Mason and Haselkorn, 1972). In T4$^+$ infections of wild-type cells, an oligomer of gp12 is also bound to the cell membrane (Edgar and Lielausis, 1963; Kells and Haselkorn, 1974; Mason and Haselkorn, 1972). These observations are consistent with electron microscopic evidence which suggests that tail fibers, baseplates, or both are assembled on the cell membrane. In thin sections through infected *E. coli* cells, many intracellular T4 phage are oriented radially with heads extending toward the center of the bacterial cell and with their baseplates or tail fibers touching the cell membrane (Simon, 1969).

Two steps of tail fiber assembly are affected by *E. coli* functions. In T4$^+$ infection of an *E. coli* mutant, designated *tabA*, the major components of the tail fiber are made, but functional tail fibers are not produced at 42°C, and the phage particles formed are not infective (Pulitzer and Yanagida, 1971). Synthesis of an early T4 protein at nonpermissive temperature causes the block to be irreversible in shift-down experiments to a permissive temperature.

Certain *E. coli* mutants bypass the requirement for T4 gp57 function to convert monomers of long tail fiber proteins gp34 and gp37 into dimers and to convert gp12 monomers into trimers (Revel et al., 1976). The bypass mutation confers temperature sensitivity upon cells that carry the *byp*57 allele. Also the bypass mutation is specific for gene 57. The *E. coli byp*57 gene and various alleles of gene 57 examined do not interact with each other in an allele-specific manner. Therefore, the gene product of the host *byp*57 allele probably does not physically interact with T4 gp57 (Revel et al., 1976).

Since the tail fibers of intracellular phage appear to be associated with the *E. coli* inner membrane, the mutations responsible for the *tabA* and *byp*57 host phenotypes might affect a membrane protein or a membrane function. However, the mutations responsible for the two host phenotypes have not been mapped, and the functions of these genes for *E. coli* bacteria have not been identified.

EFFECTS OF T4 PHAGE INFECTION ON *E. COLI* PROTEIN DEGRADATION

In *E. coli* cells, normal proteins generally have much longer half-lives than either proteins with abnormal conformations or protein fragments (Goldberg and St. John, 1976; Mount, 1980). Nonsense fragments of *E. coli* proteins usually are so unstable that they are not detectable in the cell (Goldschmidt, 1970; Lin and Zabin, 1972). Nonsense fragments of T4 phage proteins in infected *E. coli* cells, however, are found in amounts similar to those of normal T4 proteins (Celis et al., 1973). Protein degradation studies have shown that in T4-infected cells, the turnover of normal *E. coli* proteins is unaffected, whereas the turnover of abnormal proteins is substantially, if not completely, inhibited (Simon et al., 1978).

The T4-induced inhibition of abnormal protein degradation in the cell is not simply a result of phage adsorption to the bacterial surface and injection of phage DNA (Simon et al., 1978). The inhibition of degradation of abnormal proteins is dependent upon the synthesis of an early T4 protein. Chloramphenicol, when added to block protein synthesis at the time of infection, can completely eliminate this inhibition; nevertheless, the inhibition occurs in T4 33^- or T4 55^- infections in which essentially no late proteins are made (Simon et al., 1978). The T4 gene responsible for this function is not located in deleted regions of the chromosome of various T4 deletion mutants (Chace and Hall, 1975; Depew et al., 1975; Homyk and Weil, 1974; Johnson and Hall, 1973; Wilson and Abelson, 1972; Wilson et al., 1972). These deletion mutants inhibit degradation of abnormal proteins to the same extent as do $T4^+$ particles (L. D. Simon, unpublished observations). Further, all early T4 amber mutants in genes 31, 39, 41, 52, 58, 59, and 60 that we have tested under nonpermissive conditions behave similarly to wild-type phage in reducing the turnover of abnormal proteins. Recently, we have cloned the T4 *pin* gene (*p*roteolysis *in*hibition), which functions to block the turnover of abnormal proteins in *E. coli* cells. When carried on expression plasmids, the T4 *pin* gene functions to inhibit breakdown of abnormal proteins in *E. coli* transformants (Simon et al., 1983).

The ability to shut off the host system that degrades abnormal proteins seems to be common among the T-even bacteriophages. The significance of this ability is not clearly established at present. This type of phage function, however, is not universal among coliphages. λ phage and ϕX174 phage infections of *E. coli* have little, if any, effect on the degradation of abnormal proteins. Phage P22, which replicates in *Salmonella typhimurium*, phages ϕ-1, SP3, SPO1, SPP1, and PBS1, which replicate in *B. subtilis*, and phage nt-1, which replicates in *Vibrio natreigens*, have very little or no inhibitory effect on host degradation of abnormal proteins (K. Tomczak, M.S. thesis; L. D. Simon, unpublished observations).

SOME OTHER EXAMPLES OF *E. COLI*-T4 INTERACTIONS

A Host Strain that Blocks the Replication of T4 IPI Mutants

T4 internal protein I (*ip*I), which is nonessential for propagation of T4 in standard *E. coli* strains (Abremski and Black, 1979; Black, 1974; Black and Abremski, 1974), is required for propagation of T4 in *E. coli* CT596 (Abremski and Black, 1979). $T4^+$ can propagate normally in CT596; however, in T4 *ip*I^- infections, the phage DNA is degraded after injection into the host cell (Abremski and Black, 1979), and no T4 protein or T4 mRNA is synthesized (Abremski and Black, 1979; Black and Abremski, 1974). Thus, *ip*I can function directly or indirectly to protect T4 DNA from nuclease activity in CT596. CT596 is also defective in the propagation of T4 with mutations in genes *r*IIA and *r*IIB, and in the propagation of λ, P1, T2, T5, and T6 phages (Black and Abremski, 1974). A defective prophage which is probably carried in CT596 may account for the block in T4 IPI^- phage production (Black and Abremski, 1974); however, the molecular aspects of this block are not known.

T4 Mutants that Form Plaques More Efficiently on *E. coli* K-12 Strains than on *E. coli* B strains

The plaque-forming ability of certain T4 mutants on *E. coli* K-12 strains when compared with plaque-forming ability on *E. coli* B strains suggests that certain steps in phage production involve host factors that may differ between K-12 and B strains. Certain mutants in either genes 8 or 53, which encode components of the outer wedge of the baseplate (Kikuchi and King, 1975a), are able to produce many more viable phage in K-12 strains than in B strains (Georgopoulos et al., 1977). Differences in translation, transcription, cell membrane-associated factors involved in T4 assembly, and protein degradation between K-12 and B strains could account for the block in the production of these T4 mutants in the B strains.

In infections of *E. coli* by T4 mutants, referred to as DNA-delay mutants in genes 39, 52, or 60, synthesis of T4 DNA is delayed, and burst size is reduced, especially at low temperatures. This delay appears to be exacerbated in *E. coli* B hosts (Mufti and Bernstein, 1974; Yegian et al., 1971). A host membrane function has been implicated in this block in T4 DNA synthesis; early in infection T4 DNA replication may be associated with the membrane and in infections by T4 phage defective in gene 39, the phage DNA may dissociate from the membrane prematurely (Frankel et al., 1968). The proteins encoded by genes 39, 52, and 60 are subunits of a T4 topoisomerase (Liu, Liu, and Alberts, 1979; Liu et al., 1980), the activity of which can, to some extent, be replaced by that of *E. coli* gyrase (McCarthy, 1979). Gyrase from *E. coli* B may be less able than gyrase from *E. coli* K-12 to replace the T4 topoisomerase, and therefore, this proposed difference may account for the DNA delay phenotype seen in *E. coli* B hosts.

OUTLOOK

In the preceding discussion, several unresolved questions pertaining to the complex interactions between bacteriophage T4 and its host bacterium, *E. coli*, are brought into focus. With regard to *rho* function and T4 production, it is evident that although normal *rho* activity is not essential for T4 plaque formation, in the absence of proper *rho* function T4 burst size is reduced. Rho protein thus appears to play an important role in T4 infections. Certain *E. coli rho* mutants severely block T4 production. How these *rho* mutants prevent T4 production is not clear at present. Some evidence suggests that the block is at the level of transcriptional control; other evidence suggests that the block results from degradation of proteins. Both suggestions may, in fact, be correct. An understanding of the molecular basis of the host defective *E. coli rho* mutants and of the role of *rho* protein in T4 production still must be obtained.

Various other as yet unresolved questions involve the role of the *E. coli* $groE_L$ protein in T4 capsid morphogenesis. T4 capsids assemble on the *E. coli* inner membrane; nevertheless, it remains unknown where in the cell gp$groE_L$ and gp31 interact with gp23 during the process of head morphogenesis. Furthermore, the fundamental question of what, at the molecular level, gp31 and gp$groE_L$ actually do to promote T4 head assembly is completely unanswered at the present time.

The molecular biology of the *E. coli* cell membrane interactions with the T4 proteins involved in baseplate, tail fiber, and capsid assembly also is obscure and must be studied further.

These are a few of the questions and areas of investigation that hold promise for significant developments toward an understanding of virus-host cell interactions during virus production.

T4 Polynucleotide Kinase and RNA Ligase

LARRY SNYDER

Department of Microbiology and Public Health, Michigan State University, East Lansing, Michigan 48824

Many T4-encoded enzymes have been assigned physiological functions, but two of the most interesting and useful, the polynucleotide kinase and the RNA ligase, remain enigmatic. There is no apparent need for an RNA ligase activity in procaryotes since there is no evidence for intervening sequences in RNA requiring cleavage and subsequent religation for their removal. Hence, it is hoped that studies of the role of these enzymes will lead to the uncovering of heretofore unknown functions in bacteria. This short review is an attempt to summarize some of the observations concerning the polynucleotide kinase and RNA ligase. For a more comprehensive treatment, the reader is referred to recent reviews (Gumport and Uhlenbeck, 1981; Higgins and Cozzarelli, 1979; Kleppe and Lillehaug, 1979).

DISCOVERY AND SPECIFICITY OF THE ENZYMES

Polynucleotide kinase was discovered independently by Richardson (1965) and Novogrodsky and Hurwitz (1966) as an enzyme which could catalyze the transfer of the γ-phosphate of ATP to a 5′-hydroxyl terminus of either DNA or RNA. Since then, the details and specificity of the reaction have been studied extensively, and I can attempt only a brief summary.

The kinase activity exhibits a remarkable lack of specificity, both in the donor of the γ-phosphate and in the polynucleotide recipient. All of the deoxy or ribonucleoside triphosphates can probably serve equally well as donors of the γ-phosphate (Sano, 1976). DNA and RNA can serve equally well as the recipients, but there is a certain minimal length. Nucleosides are not substrates, but a 3′-phosphate mononucleotide will be phosphorylated to yield the 3′,5′-diphosphate (Lillehaug and Kleppe, 1975). Originally, it was reported that some 3′-monophosphate mononucleotides were better substrates than others (Lillehaug and Kleppe, 1975), but with the finding that the polynucleotide kinase is also a 3′-phosphatase (see below), this should be reinvestigated.

The polynucleotide kinase prefers single-stranded 5′-hydroxyl termini, so blunt-ended or recessed 5′ termini with 3′ overhangs are poor substrates. So are single-strand nicks in DNA, at least at lower ATP concentrations (Lillehaug et al., 1976).

The polynucleotide kinase protein also has an indigenous 3′-phosphatase activity (Cameron and Uhlenbeck, 1977; Sirotkin et al., 1978). This activity was first discovered by Becker and Hurwitz (1967) but was not known to be part of polynucleotide kinase until much later. Unlike the 5′-kinase activity, the 3′-phosphatase strongly prefers DNA to RNA and will use a 3′-phosphate mononucleotide as substrate to yield the nucleoside (Becker and Hurwitz, 1967; Cameron and Uhlenbeck, 1977). This activity is also a cyclic 2′,3′-phosphatase (O. C. Uhlenbeck, personal communication), and it would be ironic, in the light of recent events (see below), if the preference for DNA exists because 3′-phosphate-terminated DNA more closely resembles, fortuitously, the real substrate, and not because DNA is the real substrate in vivo.

The 5′-kinase and 3′-phosphatase activities of the polynucleotide kinase have different active centers, so they are not manifestations of the same reaction (Cameron et al., 1978; Sirotkin et al., 1978). However, as I shall discuss below, the two activities probably participate in the same sequence of reactions in vivo.

The RNA ligase was also discovered in the laboratory of Hurwitz (Silber et al., 1972) as an enzyme which can cyclize 5′-phosphate- and 3′-hydroxy-terminated poly(A). As expected, it will also join two polynucleotides, one terminated in a 5′ phosphate (the donor) and one terminated in a 3′ hydroxyl (the acceptor). The reaction uses ATP, but only to activate the donor by the addition of AMP in a phosphodiester linkage (Kaufmann and Littauer, 1974). Once the donor has been activated by the formation of a donor-AMP complex, the ATP can be dispensed with (England et al., 1977; Ohtsuka et al., 1976; Sugino, Snopek, and Cozzarelli, 1977). The presence of a 3′-hydroxyl-terminated acceptor is required for the donor activation step but need not be the same molecule which participates in the subsequent ligation reaction (Sugino, Snopek, and Cozzarelli, 1977).

The RNA ligase strongly prefers RNA for the cyclization reaction but will also cyclize DNA substrates such as poly(dT) (Snopek et al., 1976). Most of the specificity for RNA resides in the 3′-hydroxyl acceptor molecule, and many substrates, including DNA, which can be activated by the addition of AMP, will serve equally well as donor molecules (England et al., 1977; Snopek et al., 1976).

The RNA ligase protein has also been shown to stimulate blunt-end joining by the T4 DNA ligase (Sugino, Goodman, Heynecker, et al., 1977) and is often used for this purpose. This stimulation is not due to any known enzymatic activity of the RNA ligase and is quite likely due to the presence of the protein itself.

The composite activities of polynucleotide kinase and RNA ligase suggest that the two enzymes may be involved in the same reaction in vivo. The 5′-kinase and 3′-phosphatase activities of the polynucleotide kinase can convert a 3′-phosphate- (or 2′3′-cyclic phosphate-) and 5′-hydroxyl-terminated polynucleotide into a substrate for RNA ligase. It is worth noting the remarkable similarity between the composite activities of the T4-induced enzymes and the eucaryotic RNA splicing enzymes. The eucaryotic systems cleave

RNA next to an intron to yield the cyclic 2',3'-phosphate. A phosphatase cleaves the cyclic phosphate to yield the 2' phosphate, and the 5' hydroxyl is phosphorylated with ATP to yield a 5'-phosphate terminus (Knapp et al., 1977). The two termini are then ligated, yielding a polynucleotide with a 2' phosphate branch (Konarska et al., 1982; J. N. Abelson, personal communication). The phage enzymes have all of these activities except the intron-specific endonuclease. Another difference is that the T4-encoded phosphatase cleaves both the 2' and 3' linkages so that the ligated RNA does not have a 2' phosphate branch (Uhlenbeck, personal communication). But the similarities between the phage-induced enzymes and the eucaryotic RNA splicing systems are such as to suggest that procaryotes may have genes with introns, after all. I shall return to this point later.

STRUCTURAL GENES

Polynucleotide Kinase Is Encoded by the *pseT* Gene

The *pseT* gene of T4 was found by Depew and Cozzarelli (1974) as the gene which encodes the T4-induced 3'-phosphatase. To find the original 3'-phosphatase-deficient mutant, they screened 150 isolates of heavily mutagenized T4 for the failure to induce the enzyme. They named the original mutant *pseT*1, for *p*ho*s*phatas*e* *t*hree-prime.

The *pseT*1 mutant multiplied normally on most strains of *Escherichia coli*, suggesting that the 3'-phosphatase activity is not required for T4 development in these strains. However, by screening the so-called Cal Tech clinical strains of *E. coli*, they found a nonpermissive host, CT196. This host also exhibits a reduced plating efficiency for the parent, *pseT*$^+$ T4, so they crossed it with various Hfr strains of *E. coli* K-12 and found a hybrid strain, *E. coli* CTr5x, which restricts the *pseT*1 mutant but not *pseT*$^+$ T4. It was then a relatively easy matter to map the *pseT*1 mutation and isolate another mutation in the *pseT* gene, *pseT*2.

Essentially the same strategy had been used earlier by Chan and Ebisuzaki (1970) to isolate T4 mutants deficient in polynucleotide kinase. However, they could find no phenotypes associated with polynucleotide kinase deficiency and did not map the mutations.

The discovery that the purified polynucleotide 5'-kinase is also a 3'-phosphatase (Cameron and Uhlenbeck, 1977) pointed to the *pseT* gene as the structural gene for polynucleotide kinase. The evidence is almost indisputable (Sirotkin et al., 1978). Most *pseT* mutations, isolated because they prevent T4 multiplication in *E. coli* CTr5x and placed in the *pseT* gene by complementation, inactivate both polynucleotide kinase and 3'-phosphatase. One of the original Chan and Ebisuzaki mutations, 7-1, inactivates 3'-phosphatase and has a *pseT* mutation. The other, 7A, may not inactivate the kinase in vivo but only in vitro (Sirotkin et al., 1978).

In retrospect, Depew and Cozzarelli missed discovering that *pseT* is the gene for polynucleotide kinase because of an ironic coincidence. They assayed polynucleotide kinase after infection by the *pseT*1 mutant and found it to be normal (R. Depew, personal communication). To date, *pseT*1 is the only known *pseT* mutation which inactivates the 3'-phosphatase activity, but not the 5'-kinase activity, of the product of the *pseT* gene.

RNA Ligase Is Encoded by Gene 63

The gene for RNA ligase was identified by Snopek and Cozzarelli. Using the same strategy which yielded the original *pseT* mutant, they isolated five T4 mutants which failed to induce wild-type levels of RNA ligase. They used the biochemical assay for RNA ligase to map one of the mutants to the gene 63 region. In a collaboration with the laboratory of Wood, they showed that gene 63 is the structural gene for the RNA ligase (Snopek et al., 1977). The assignment of RNA ligase to gene 63 presented a dilemma. The product of gene 63 already had a known function; it was required, at lower temperatures, for the attachment of tail fibers to baseplates during morphogenesis (see Wood and Crowther, this volume). Hence, it ranked a number, 63, and was identified in the original conditional-lethal mutant collection (Epstein et al., 1963).

The tail fiber attachment and RNA ligase activities of the gene 63 product are apparently independent activities of the same protein. The two activities have very different requirements in vitro (Wood et al., 1978). Furthermore, many mutations which inactivate RNA ligase leave tail fiber attachment intact (Runnels et al., 1982; Snopek et al., 1977). Finally, tail fiber attachment and RNA ligase deficiencies are suppressed by different extracistronic mutations (Hall et al., 1980; Runnels et al., 1982) (see below).

GENETIC AND PHYSICAL MAP OF THE GENE 63-*pseT* REGION

The *pseT* gene and gene 63 are closely linked, but three-factor crosses place the *unf/alc* gene between them. However, the genetic crosses do not establish the distance between the genes, and there could be other genes besides *unf/alc* between *pseT* and gene 63. In fact, the physical map supports this prediction (Mileham et al., 1980; Snustad et al., this volume). The entire region is contained on an 11.5-kilobase (kb) *Hin*dIII fragment, and the *pseT*-gene 63 region is probably encompassed by four *Eco*RI fragments of 1.1, 2.2, 0.12, and 1.3 kb. The 1.1-kb and 2.2-kb fragments have been cloned, but not the 1.3-kb fragment. The *pseT* gene maps at the junction of the 2.2-, 0.12-, 1.3-kb fragments (N. Murray, personal communication). These three fragments are eliminated in a *pseT* deletion mutant, *pseT*Δ1 (Mileham et al., 1980), which appears to terminate a short distance into the *alc* gene (see Snustad et al., this volume) and does not extend into gene 63. By default, the 1.3-kb fragment is thought to contain at least part of gene 63 because neither of these two cloned fragments nor the fragment beyond the 1.3-kb fragment rescues an amber mutation in gene 63 (Mileham et al., 1980). This would leave 2.2 kb of DNA to encode part of all of the *unf/alc* gene. It seems unlikely that there are extensive noncoding sequences on this fragment or that the *unf/alc* product is greater than 100 kilodaltons. Presumably, there is at least one other gene on this fragment.

The presence of other gene(s) between *pseT* and 63 (or just beyond *pseT*) is also suggested by the behavior of *pseT*Δ1 and other deletions like it, which exhibit phenotypes not exhibited by *alc* mutants or *pseT* point mutants. The deletion mutant is restricted on host *lit* mutants (see below) at higher temperatures than either *pseT* or gene 63 point mutants (see below).

Furthermore, when the T4 *pseT*Δ1 mutant mutiplies, it accumulates second-site mutations in *gol*, *stp*, gene 57, and another gene adjacent to gene 57 (K. Sirotkin, W. C. Champness, and L. R. Snyder, unpublished observations). Point mutations in *pseT*, gene 63, or *unf/alc* do not accumulate such mutations. Both the *gol* site and the product of the *stp* gene interact somehow with the polynucleotide kinase and RNA ligase (see below), but the significance of the gene 57 and other mutations is totally obscure. The responsible gene or genes deleted in *pseT*Δ1 appear to be interesting regarding the function of the RNA ligase and polynucleotide kinase.

STUDIES OF THE PHYSIOLOGICAL ROLE OF THE ENZYMES

As suggested by their biochemical activities, the polynucleotide kinase and RNA ligase are almost certainly involved in the same reaction in vivo. The T4 *pseT* gene and gene 63 are closely linked, as are many T4 genes of related function. Furthermore, *E. coli* CTr5x, the restrictive host for *pseT* mutants, is also a restrictive host for *rli* mutants (Runnels et al., 1982), and the defects during phage development in *E. coli* CTr5x are very similar. Finally, *rli* and *pseT* mutations share a common suppressor. Depew and Cozzarelli (1974) had shown that mutations which inactivate a T4 gene, *stp*, close to *r*II, suppress the defects of *pseT*$^-$ T4 in *E. coli* CTr5x, and *stp* mutations also suppress *rli* mutations (Runnels et al., 1982).

The defects exhibited by *pseT*$^-$ and *rli*$^-$ T4 in the nonpermissive host, *E. coli* CTr5x, offer clues to the function of the polynucleotide kinase and RNA ligase. Depew and Cozzarelli (1974) found that *pseT* mutations reduced the rate of T4 DNA replication by about one-half in *E. coli* CTr5x, and the DNA made was somewhat shorter than normal. They also postulated a role for the 3′-phosphatase in T4 DNA packaging into virions. Subsequent studies (Runnels et al., 1982; Sirotkin et al., 1978) have confirmed the defect in T4 DNA replication but have also revealed a defect in T4 late gene expression for both *pseT* and *rli* mutants. At least some of the defect in late gene expression is at the level of transcription, as determined by the hybridization of labeled RNA to cloned late genes of T4 (L. Snyder, unpublished observations). Because the defect in late gene expression is probably sufficient to explain the lack of phage production, there is no reason to postulate defects in DNA packaging, etc.

The defects exhibited by *rli*$^-$ and *pseT*$^-$ T4 in T4 DNA replication and late expression are most severe at lower temperatures. At 30°C, T4 DNA replication is reduced to about 10% of normal, and there is almost no synthesis of late proteins. At 42°C, both T4 DNA replication and late gene expression are normal.

Interestingly, the defect in T4 DNA replication occurs before any apparent effect on gene expression. This suggests a direct role for the polynucleotide kinase and RNA ligase (or the product of the reaction they promote) in T4 DNA replication as opposed to an indirect role vis-à-vis processing of a tRNA or mRNA involved in translation. However, the defects could be due to failure to synthesize a protein which cannot be easily detected on one-dimensional polyacrylamide gels.

It is interesting that *pseT* and *rli* mutations prevent the expression of all genes late in infection (Runnels et al., 1982) and not just the so-called true late genes, those whose synthesis is coupled to replication. This is in contrast to the situation when T4 DNA replication is prevented by other means, e.g., mutations in the replication genes. Then the early gene products are synthesized much later than normal. The effect of *pseT* and *rli* mutations in *E. coli* CTr5x is reminiscent of the effect on late gene expression when T4 DNA replicates to contain cytosine and the *unf/alc* gene is active (Snustad et al., this volume; Snyder et al., 1976) or when the host *lit* gene is mutated (Cooley et al., 1979) (see below). The T4 *unf/alc* gene product and host *lit* mutations also prevent late gene expression only at lower temperatures. These coincidences, plus the genetic evidence linking *pseT*, *rli*, and *unf/alc* with the host gene *lit* (see below), suggest that these gene products all affect the same requirement for T4 gene expression.

Another phenotype associated with *rli*$^-$ mutants also indicates a role in DNA metabolism. The phenotype suggests an interaction between the T4 RNA ligase and T4 DNA ligase in vivo. S. Brown and W. B. Wood (personal communication) have found that some *rli* mutations make a temperature-sensitive mutation in gene 30 (DNA ligase) more restrictive. Surprisingly, the restriction is not due to a defect in DNA replication, gene expression, or DNA packaging, but to a defect in T4 head morphogenesis. The mutant infections resemble gene 24 mutant infections in that preheads accumulate. One wonders if these phenotypes are related to the above-mentioned in vitro stimulation by RNA ligase of blunt-end joining by T4 DNA ligase.

The group of Kaufmann has made some intriguing observations related to the functions of RNA ligase and polynucleotide kinase. In earlier experiments, they permeabilized T4-infected *E. coli* and labeled with [γ^{32}P]ATP. They found extensive labeling of RNA, presumably by the T4-induced polynucleotide kinase. They treated the RNA with alkaline phosphatase and found that some of the label remained with RNA, presumably because of ligation by the T4-induced RNA ligase. The label was in small tRNA-sized RNAs, presumably of host origin (David et al., 1979).

More recently, Kaufmann and his collaborators have found what are, apparently, cleavage products of a tRNA-like RNA after T4 infection. These cleavage products only appear after infection of *E. coli* CTr5x and its derivatives and not after infection of other *E. coli* B and K-12 strains (David et al., 1982a). These cleavage products can be assembled to form two RNAs of tRNA-like structure. One of these has been sequenced and has an anticodon for isoleucine (David et al., 1982a). It is intriguing that the cleavage occurs next to the anticodon in the same position as the cleavage during the removal of the intervening sequences in some species of yeast tRNA (Goodman et al., 1977).

A link between the cleavage products and the function of RNA ligase and polynucleotide kinase is indicated because the fragments only appear in *E. coli* CTr5x or in strains into which the restricting locus of strain CTr5x has been moved by transduction (see below). The fragments persist later into infection by *pseT*$^-$ and *rli*$^-$ mutants than after infection by wild-type T4 (David et al., 1982b). Finally, and perhaps most conclusively, the cleavage products do not appear after infection by *stp*⁻ mutants (Kaufmann, personal communication). Remember that *stp*⁻ mutations are the extracistronic suppressor of *pseT*⁻ and *rli*⁻ mutations in *E. coli* CTr5x.

The work of Kaufmann and his collaborators suggests a simple explanation for the restriction of *pseT*⁻ and *rli*⁻ mutants of T4 in *E. coli* CTr5x. According to this explanation, *E. coli* CTr5x has a different major isoleucine tRNA than do most other *E. coli* strains. After T4 infection, this tRNA is cleaved by the product of the *stp* gene. The cleavage will block translation, and therefore gene expression, unless the tRNA is religated by the polynucleotide kinase and RNA ligase. However, this explanation does not account for a number of observations. One is the temperature dependence of the restriction. It is hard to believe that the tRNA is needed for translation at 30°C but not at 42°C. But even if the cleavage of a tRNA does not easily account for all of the phenotypes, the observation is very important because it is the first clear demonstration of an in vivo activity promoted by polynucleotide kinase and RNA ligase.

GENETIC ANALYSIS OF RESTRICTIVE HOSTS

Important insights could be gained from an understanding of why only certain hosts are restrictive for *pseT*⁻ and *rli*⁻ mutants of T4. Is this because the host codes for enzymes analogous to the phage polynucleotide kinase and RNA ligase? It may be relevant that Abelson and his collaborators have recently obtained evidence for an RNA ligase activity in uninfected bacteria. The substrate they used was the cleaved tyrosine tRNA of yeast with the intervening sequence removed. The enzyme so far has not worked with other substrates such as poly(A) (J. Abelson, personal communication). This would explain how the enzyme has escaped detection this long. It is possible that *E. coli* CTr5x may have suffered a mutation in the gene for the host RNA ligase. Alternatively, the responsible *E. coli* CTr5x gene may code for the substrate of the RNA ligase, such as the tRNA-like RNA of Kaufmann and his collaborators.

To help answer questions such as these, we have mapped the locus in *E. coli* CTr5x which causes the restriction of *pseT*⁻ and *rli*⁻ T4 (M. Abdul Jabbar and L. Snyder, manuscript in preparation). The locus, which we have named *prr* for *p*olynucleotide kinase *R*NA ligase *r*estrictive, maps to 29 min on the *E. coli* map and can be transduced into other strains of *E. coli*, in which it exhibits similar, if somewhat less restrictive, phenotypes. Interestingly, both λ bacteriophage and T4 with cytosine substituted for hydroxymethylcytosine in its DNA also fail to multiply on strains with the *prr* locus. The frequency with which λ plaques appear on *E. coli* with the *prr* locus varies greatly from one strain of λ to another, but the plating efficiency is usually between about 10^{-1} and 10^{-3}. The λ in these plaques can now multiply normally in *E. coli* which are either *prr*⁺ or *prr* mutant. However, if they have been propagated on *prr*⁺ *E. coli*, they are again restricted in *E. coli* with the *prr* locus. Thus, they are not true mutants but are undergoing a reversible DNA modification of some sort. It is not yet clear whether the modification occurs in the *prr*⁺ or *prr* mutant *E. coli* (Abdul Jabbar and Snyder, to be published).

The restriction of T4 with cytosine-containing DNA is also subject to a DNA modification since isogenic strains with hydroxymethylcytosine are not restricted. One is reminded that the product of the *unf/alc* gene, which maps between *pseT* and gene 63 and blocks transcription from T4 cytosine-containing DNA but not T4 hydroxymethylcytosine-containing DNA (Snustad et al., this volume; Snyder et al., 1976). This may extend the analogy between the *pseT* and *rli* region of T4 and the *prr* locus of *E. coli*.

The restriction of λ and T4 cytosine-containing DNA by the host *prr* locus behaves superficially like a classical restriction-modification system, but the consequences are very different. The DNA of the restricted phage enters the cell but is not degraded, and no phage early gene products are synthesized. Hence, the host cells survive (Abdul Jabbar and Snyder, to be published). It is still possible that the *prr* locus and the gene responsible for the restriction of λ and T4 cytosine-containing DNA are different genes but very closely linked. Even if so, the phenomenon is very intriguing because there is no precedent in procaryotic molecular biology in which DNA is in the cell but ignored by the transcription and translation apparatus.

In addition to mapping the restricting locus in *E. coli* CTr5x, we attempted to isolate similar restricting mutants in *E. coli* K-12 (Cooley et al., 1979). These mutants were unstable and accumulated suppressors in a new *E. coli* gene, *lit*, at 25 min on the genetic map. We think, but have not proven, that the *lit* gene product interacts somehow with the *prr* locus we mapped in *E. coli* CTr5x. We have preliminary evidence that the restricting locus in the K-12 mutants also map between *his* and *trp* (W. Cooley Champness and L. Snyder, unpublished observations). Perhaps *E. coli* CTr5x is a partial *lit* mutant, although this has not been demonstrated.

The *E. coli lit* gene is of interest quite apart from its possible relationship to the T4-induced polynucleotide kinase and RNA ligase enzymes. At lower temperatures, even wild-type T4 fail to multiply on a host *lit* mutant because of a severe defect in late gene expression. The defect is caused by a *cis*-acting site on T4 DNA, the *gol* site (for grow on *lit*) which lies within the coding sequence for gene 23. The wild-type form of the *gol* site interferes with late gene expression of a *lit* host, and mutations in the site (*gol* mutations) overcome this effect. It is not yet clear whether *gol* mutations inactivate the site or alter it so it can function in a *lit* host. However, it is probably relevant that *gol*⁺ T4 interfere with gene expression by *gol* mutant T4 mixed in infections (Cooley Champness and Snyder, 1982), as though the mutant *gol* site is still required for gene expression. The *gol* site is quite large. A number of *gol* mutations have been located by DNA sequencing, and they are due to base pair transitions as much as 40 base pairs apart.

The *gol* site even exerts an effect on plasmids in which it has been cloned. This is apparent from transformation experiments with plasmids containing T4 DNA inserts from the *gol* region. If *lit*⁺ *E. coli* are transformed by plasmids containing a T4 DNA insert including the *gol*⁺ site and a selectable antibiotic resistance marker, normal frequencies of antibiotic-resistant transformants arise. However, if the recipients are *lit*, there are very few transformants, and the few are mostly due to lit^+ revertants or deletions of the *gol* site on the plasmid. There is little doubt that it is the *gol* site itself that interferes with plasmid transformation of *lit* cells. If the T4 DNA insert is from a *gol* mutant rather than gol^+ T4, then transformation is normal. Conversely, about one-fourth of the mutations induced in a plasmid with the gol^+ site, which permit plasmid transformation of *lit* recipients, also confer the *gol* phenotype when crossed into the bacteriophage (Cooley Champness and Snyder, manuscript in preparation). The other three-fourths may inactivate gp23 encoded by the same stretch of DNA and are thus lethal when crossed into the bacteriophage. The observation that the *gol* site is active in plasmids in the absence of the rest of the T4 genome should help answer the question of how one site can prevent the expression of the entire genome.

Other *E. coli* mutants have been isolated which restrict *rli* mutants of T4 (Nelson et al., 1982). These mutants, called *tabR*, were isolated as restricting *r*II mutants, but they restrict *rli* mutants even more completely. A random search revealed that mutations in a number of T4 genes can cause restriction, but *r*II and gene 63 mutations were most prominent.

The effects of *tabR* mutations on T4 development are very different from the effects of the *prr* locus. The cells lyse very early after infection by an rli^- mutant. Therefore, it becomes problematic whether they would also exhibit defects in T4 DNA replication and late gene expression.

There are a number of interesting features of *tabR* mutants which may bear on the question at hand. They are methionine auxotrophs. Furthermore, the *tabR* mutations probably cause membrane defects which could explain the premature lysis. It is intriguing that the *r*II genes and *stp*, the suppressor of rli^- and $pseT^-$ mutants, are both in the so-called membrane region of the T4 genome, which encodes many proteins which become part of, or alter, the host membrane. Unfortunately, *tabR* mutations have not yet been mapped, so their relationship to the *prr* locus is unknown.

CONCLUSION

The evidence concerning the function of the RNA ligase and polynucleotide kinase is causing a schism between those who believe the conclusions of biochemical experiments and those who believe genetic experiments. On the one hand, the biochemical evidence points more and more to a role in RNA processing for the two enzymes. The striking similarities between the phage-induced enzymes and the eucaryotic RNA splicing enzymes, the cleavage of a tRNA-like RNA after infection of *E. coli* CTr5x, and the discovery of an RNA ligase activity in uninfected bacteria which uses, as substrate, no less than a yeast tRNA with the intervening sequence removed, make the discovery of RNA splicing in procaryotes seem imminent. On the other hand, the genetic and physiological evidence suggests a direct involvement with DNA through *cis*-acting sites which can prevent gene expression and the possible involvement of the cellular membrane. How does one bridge this schism? One possibility is that it is not the enzymes themselves which interact with DNA, but rather an RNA which they process. The RNA could interact with certain sites on DNA, of which the *gol* site is the prototype, and in this way exert its effects on T4 DNA replication and late gene expression. This need not be the only RNA which is processed by the enzymes, but it may be the only one required for T4 development. Of course, the genetic and physiological experiments could be deceptive, and all of the observed phenotypes could be secondary consequences of the failure to process an mRNA or tRNA properly. Even if so, the studies of the biological role of the T4-induced RNA ligase and polynucleotide kinase seem to be leading to the discovery of heretofore unknown reactions in procaryotes.

Portions of this work performed in our laboratory were supported by Public Health Service grant 1 ROI GM28001 from the National Institutes of Health.

Structure, Function, and Evolution of the Lysozyme from Bacteriophage T4

M. G. GRÜTTER,† L. H. WEAVER, T. M. GRAY, AND B. W. MATTHEWS

Institute of Molecular Biology and Department of Physics, University of Oregon, Eugene, Oregon 97403

Bacteriophage T4 lysozyme is, for several reasons, an attractive candidate for structural study. It has been well characterized biochemically, and mutant enzymes can be obtained which differ from the wild-type enzyme in stability or in intrinsic catalytic activity (see Tsugita, 1971, for a review). Also, there is interest in comparing the properties of the phage enzyme with those of lysozymes from other sources. The function of the enzyme is to hydrolyze the glycosidic linkages in the bacterial cell wall, leading to cell lysis. Phage lysozyme cleaves the same glycosidic bond as does hen egg-white lysozyme, but is 250-fold more active toward *Escherichia coli* cell walls (e.g., see Tsugita, 1971).

Some properties of hen egg, goose, and phage lysozymes are summarized in Table 1. It is now known that the hen egg and goose enzymes are representatives of two distinct types of avian lysozymes (e.g., see Canfield et al., 1971; Simpson et al., 1980). The amino acid sequence of each lysozyme is nonhomologous with those of the others.

STRUCTURAL STUDIES

The structure of phage lysozyme was initially determined at 2.4-Å (0.24-nm) resolution (Matthews and Remington, 1974; Remington et al., 1978) and has recently been confirmed by high-resolution refinement (Weaver and Matthews, unpublished data). Structural studies of the goose lysozyme are in progress (Grütter et al., 1979), and the structure of hen egg-white lysozyme was determined a number of years ago by Phillips and co-workers (Blake et al., 1965).

Phage lysozyme has a bilobal structure, with the active site located at the junction of the two domains (Fig. 1). A detailed description of the structure has been given by Remington et al. (1978).

Although not at first obvious, it has now become apparent that there are similarities between the three-dimensional backbone structures of hen egg-white lysozyme and phage lysozyme (Remington and Matthews, 1978; Rossmann and Argos, 1976). The extent of this correspondence is shown in two different ways in Fig. 2 and 3.

Figure 2 shows the backbone conformation of phage lysozyme and the position of a bound oligosaccharide, as inferred from crystallographic experiments (Anderson et al., 1981). The solid black connections indicate the part of the phage lysozyme molecule which can be approximately superimposed in three dimensions onto corresponding parts of the backbone of hen lysozyme (Matthews, Grütter, Anderson, and Remington, 1981; Rossmann and Argos, 1976). In Fig. 3 the comparison between the two lysozymes is displayed diagrammatically, with the connected solid bars showing the respective parts of the two molecules which can be approximately superimposed in three dimensions. Altogether, 78 alpha-carbon atoms of the two structures can be regarded as "equivalent," and the distances between these atoms are displayed in Fig. 3.

Strikingly, it is found that the same transformation which relates the backbones of the two lysozymes also superimposes their respective active sites. In fact, it is possible to superimpose corresponding elements of the two active sites quite precisely (Fig. 4). The series of parallels which we see between the two lysozymes has led us to hypothesize that these two enzymes have arisen by divergent evolution from the same precursor (Matthews, Remington, Grütter, and Anderson, 1981). It is interesting to note that the part of hen egg-white lysozyme which is most similar to phage lysozyme is coded by two complete exons (Fig. 3) (Jung et al., 1980), suggesting that these two exons may have been involved in the evolution of the respective lysozymes (Artymiuk et al., 1981; Matthews, Remington, Grütter, and Anderson, 1981). The sequence of the phage lysozyme gene has been determined (Owen et al., 1983) and shows no apparent homology with that of the hen egg-white lysozyme gene (A. Taylor, personal communication).

MUTANT LYSOZYMES

As a by-product of early studies by Streisinger and colleagues, directed toward in vitro demonstration of the nature of the genetic code, many mutations in the lysozyme gene have been identified (e.g., see Streisinger et al., 1966). In a number of cases, the resultant changes in the amino acid sequence of the protein have been determined. These are summarized in Fig. 5. The behavior of these mutant lysozymes in terms of the three-dimensional structure of the lysozyme molecule has been discussed by Remington et al. (1978).

TABLE 1. Chicken, T4 phage, and goose lysozymes[a]

Lysozyme	Mol wt	No. of amino acid residues	No. of Cys residues
Hen egg white	14,500	130	8 (4 S-S)
T4	18,700	164	2 (2 S-H)
Goose egg white	20,400	185	4 (2 S-S)

[a] All cleave the same bond between *N*-acetylmuramic acid and *N*-acetylglucosamine. All three amino acid sequences are nonhomologous.

† Present address: Abteilung Biophysikalische Chemie, Biozentrum der Universitat Basel, CH-4056 Basel, Switzerland.

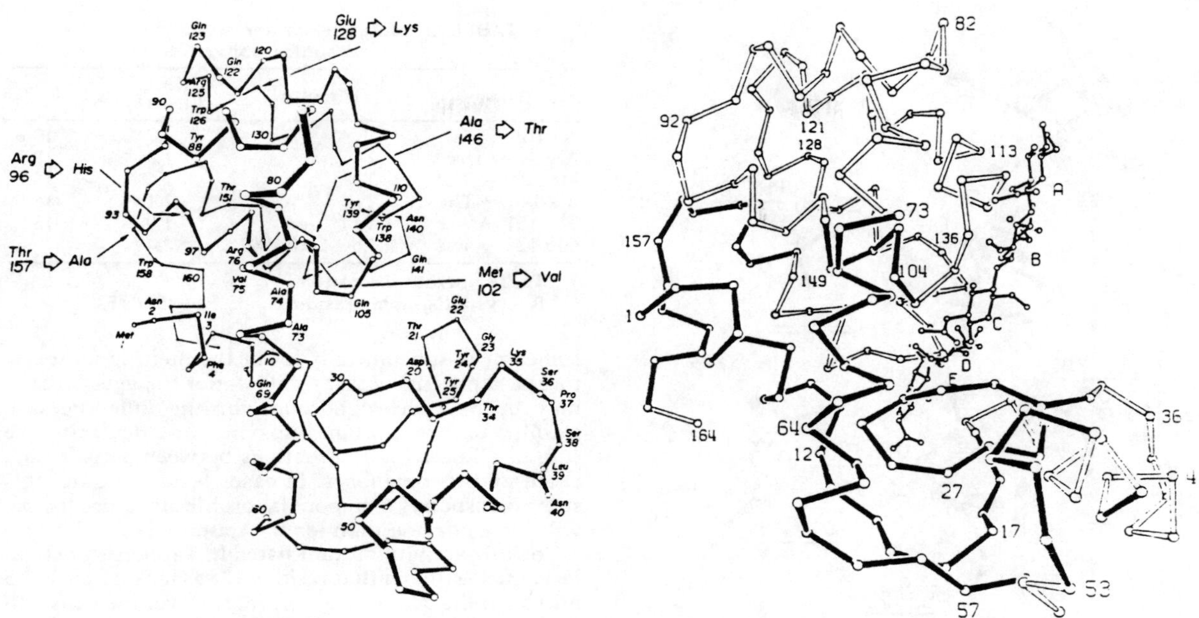

FIG. 1. Backbone of phage lysozyme, showing the amino acid substitutions in the respective mutant lysozymes that we have studied crystallographically. The other named residues indicate the locations of known mutations in the lysozyme gene (see Fig. 5). The active site is located within the pronounced cleft in the middle of the molecule.

FIG. 2. Backbone of T4 phage lysozyme showing the active-site cleft and the position occupied by a bound oligosaccharide, as inferred from crystallographic and model-building experiments. The part of the structure drawn solid is equivalent to part of the hen egg-white lysozyme backbone, whereas the backbone drawn with open connections has no counterpart in hen egg-white lysozyme (see Fig. 3).

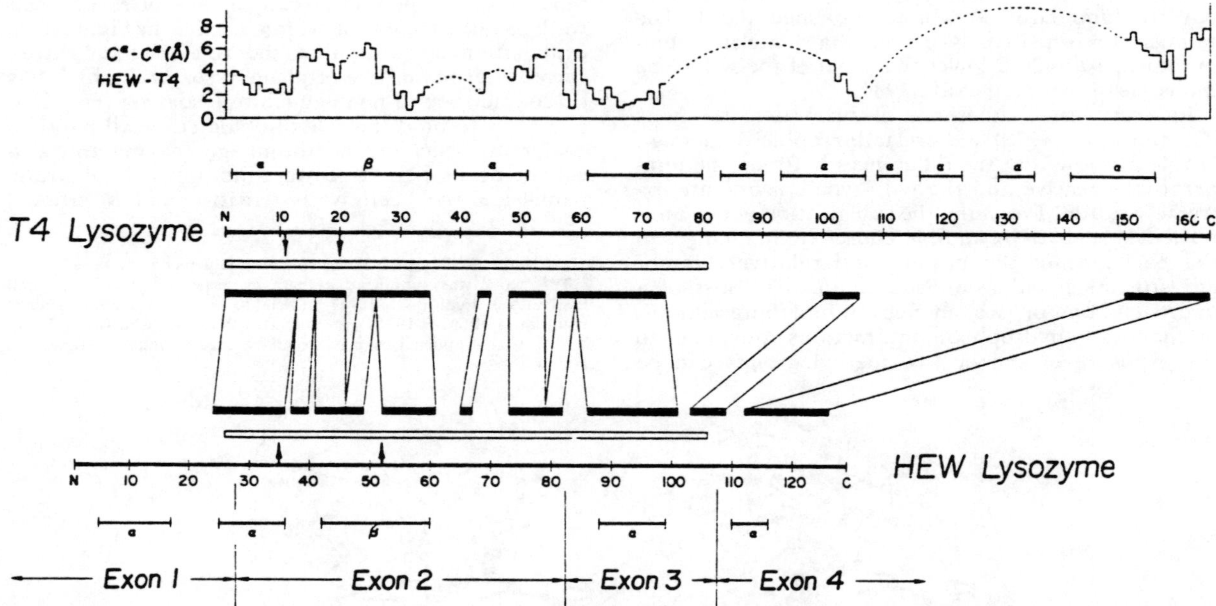

FIG. 3. Schematic drawing showing the parts of the backbone of phage lysozyme which correspond to parts of the backbone of hen egg-white (HEW) lysozyme. The equivalent alpha-carbon atoms in the two molecules are indicated by the connected solid bars (after Rossmann and Argos, 1976). The distances between the individual pairs of equivalent alpha-carbon atoms are displayed at the top of the figure. The open bars show the locations of the 80-residue segments of phage and hen egg-white lysozyme which agree best (after Remington and Matthews, 1978). Also shown are the exon boundaries for the hen egg-white lysozyme gene.

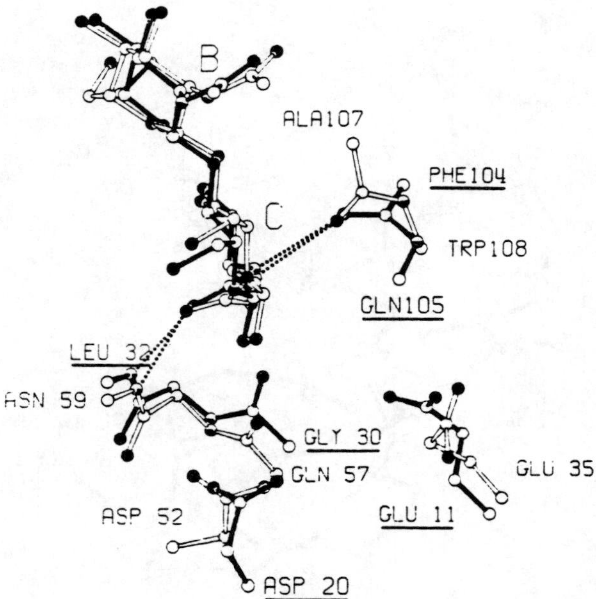

FIG. 4. Part of the active site of phage lysozyme (solid bonds), including a disaccharide bound in the B and C subsites, superimposed on the corresponding elements of hen egg-white lysozyme (open bonds). The 54 corresponding atoms shown in the figure superimpose with a root-mean-square agreement of 1.35 Å (0.135 nm). (From Matthews, Remington, Grütter, and Anderson, 1981.)

More recently, we have studied five point-mutant lysozymes in detail, four that are temperature sensitive and one with low activity (Fig. 1 and Table 2) (Grütter and Gray et al., unpublished data). The temperature-sensitive lysozymes have melting temperatures 10 to 20°C lower than that of the wild-type molecule (Schellman et al., 1981).

In every case, difference electron density maps (Grütter et al., 1979) and crystallographic refinement (Table 2) show that the differences between the temperature-sensitive and the wild-type enzyme are extremely subtle. Typically, the substitution of one amino acid side chain by another causes slight changes in the backbone of the protein and relatively minor adjustments in adjacent side chains. The loss of a hydrogen bond or two, changes in ionic interactions, or changes in hydrophobic interactions equivalent to the exposure of one or two methyl groups can be

TABLE 2. Mutant lysozymes studied by X-ray crystallography

Mutant	Phenotype[a]	Resolution (Å)	R[b] (%)
Native		1.7	19.6
Arg 96 → His	Ts	1.9	19.5
Met 102 → Val	Ts	2.1	18.1
Ala 146 → Thr	Ts	2.1	18.0
Thr 157 → Ala	Ts	1.9	18.9
Glu 128 → Lys	Low activity	2.4	

[a] Ts, Temperature sensitive.
[b] R (crystallographic residual) = $\Sigma|F_0 - F_c|/\Sigma|F_0|$.

sufficient to substantially lower the melting temperature of a protein. Energy calculations suggest that it may be possible to account for the differences in stability of these mutant lysozymes by calculating the sum of all possible interactions between a given side chain and its neighbors. In cases tested to date, this sum corresponds to a more favorable structure for the wild-type enzyme than for the mutants.

The low-activity mutant listed in Table 2 is particularly interesting in that residue 128 is located 25 Å (2.5 nm) from the active-site cleft (Fig. 1), yet the substitution of a glutamate by a lysine at this position reduces the enzymatic activity to 4% that of the native enzyme. We have shown crystallographically that this loss of activity is not due to some large change in the enzyme structure; in fact, the only structural change which does occur is limited to the immediate environment of the amino acid substitution (Grütter and Matthews, 1982).

To rationalize this unusual behavior, we suggest that Glu 128 in the native enzyme participates in the binding of the peptidoglycan linkage of *E. coli* cell walls, as illustrated diagrammatically in Fig. 6. This explanation suggests that the role of the "extra" carboxy-terminal lobe of phage lysozyme, which has no counterpart in hen egg-white lysozyme (Fig. 2, 3, and 6), is to bind the cross-linking cell wall peptide, conferring specificity to the phage enzyme and also enhancing its activity toward its optimal substrate, namely, *E. coli* cell walls (Grütter and Matthews, 1982).

We thank G. Elizabeth Pluhar for excellent technical assistance.
This work was supported in part by grants from the National Institutes of Health (GM21967 and GM20066), the National Science Foundation (PCM-8014311), and the M. J. Murdock Charitable Trust. T.M.G. was supported by a National Science Foundation Graduate Fellowship.

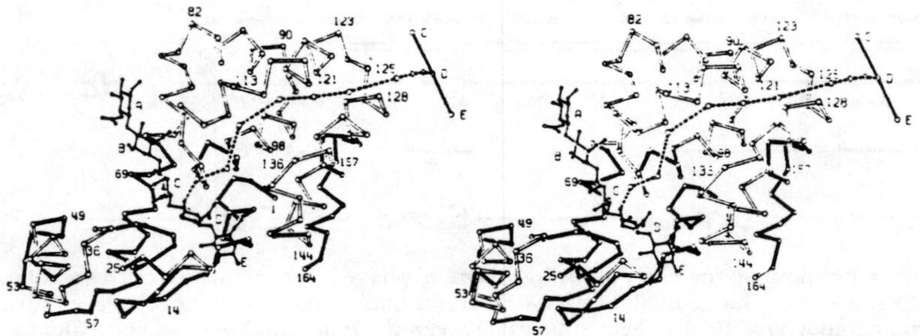

FIG. 6. Stereo drawing showing the proposed location of the cross-linking peptidoglycan linkage (broken line) and its interaction with the C-terminal lobe of phage lysozyme. The solid connections show the parts of phage lysozyme which correspond to parts of hen egg-white lysozyme (see Fig. 2).

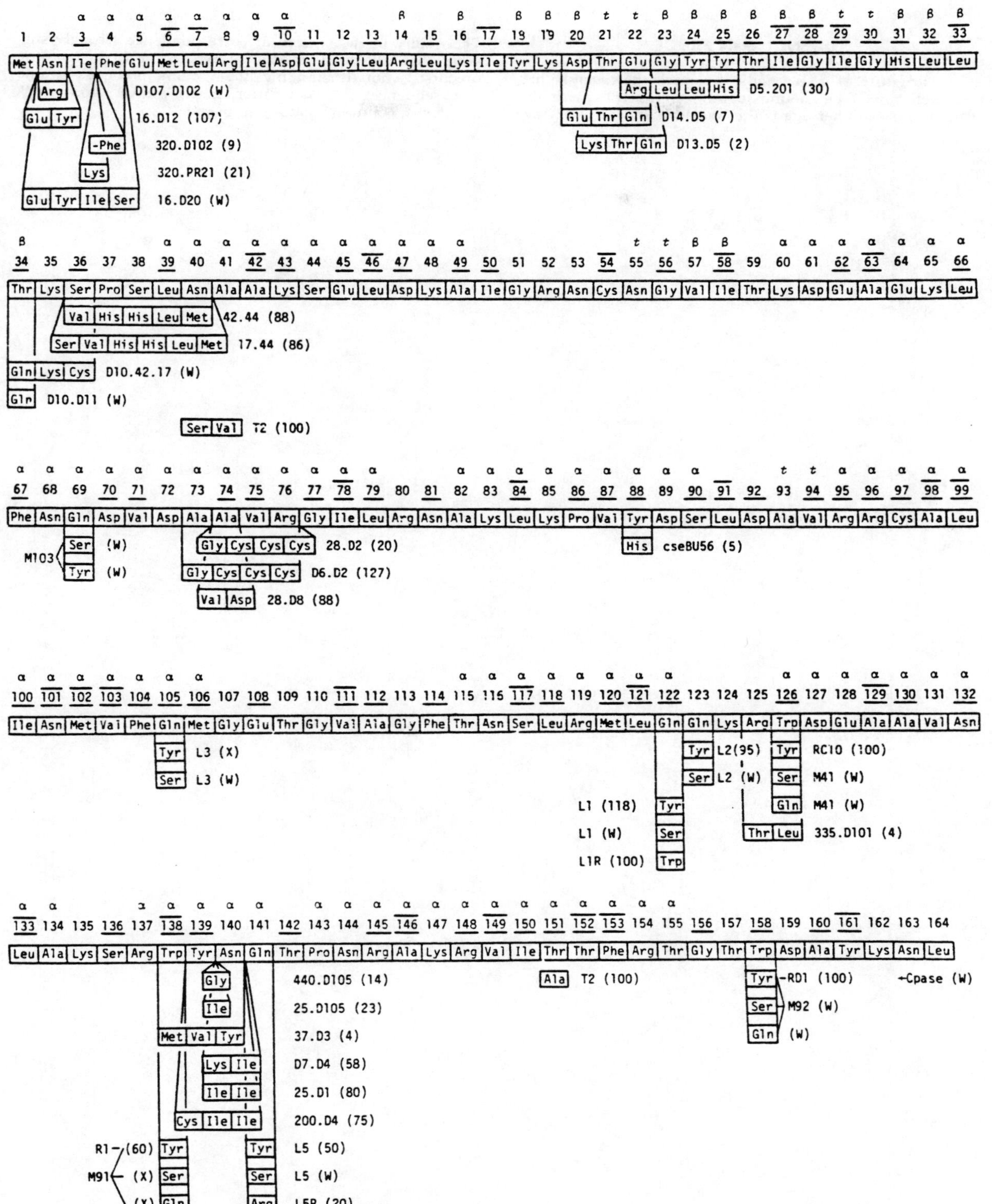

FIG. 5. Mutant lysozymes. The location and identity of the respective mutations are shown; L indicates an ochre mutant, and M indicates an amber mutant. Specific activities relative to wild-type lysozyme are shown within parentheses (see text); W indicates activity similar to wild type, and X indicates very low activity. The residues located in helices, β-sheet, or turns are labeled, respectively, α, β, or *t*. Where the amino acid side chain is partly exposed to solvent, the residue number is underlined, whereas residues which are essentially internal are indicated with a line above and below the number. Residues with unmarked numbers are essentially fully exposed to the solvent. (From Remington et al., 1978.)

ADDENDUM IN PROOF

The structure of goose egg-white lysozyme recently has been determined and found to have similarities with the structures of both hen egg-white lysozyme and T4 lysozyme. The nature of the structural correspondence strongly suggests that all three lysozymes evolved from a common precursor, notwithstanding their nonhomologous amino acid sequences (M. G. Grütter, L. H. Weaver, and B. W. Matthews, Nature [London], in press).

Postscript

CHRISTOPHER K. MATHEWS

Department of Biochemistry and Biophysics, Oregon State University, Corvallis, Oregon 97331

That remarkable beast, T4 phage
Has intrigued many scientists sage
Since the years just post-war
Even now, friend T4
In some circles is still all the rage

The need for a multiauthored book on T4 has been apparent at least since 1971, when a comparable volume, *The Bacteriophage Lambda*, appeared for T4's distant relative. That year also witnessed the publication of my own monograph, "Bacteriophage Biochemistry," an experience which convinced me that it was virtually impossible, even then, for one person to review the entire phage literature. In the late 1970s several events occurred to reinvigorate and unite the community of T4 phage researchers. These included scientific developments which presented abundant new research opportunities, establishment of a Division of Bacteriophage Biology in the American Society for Microbiology, and establishment of a biennial phage meeting, on the West Coast, devoted largely to T4. At the 1981 meeting Betty Kutter suggested, and all agreed, that it was time to stop talking about a book and start writing it. Hence, the present volume.

Our purposes in writing this book were to celebrate T4 as the organism centrally involved during the seminal developments of molecular biology, to provide a detailed overview of the field for our own students and other scientists either new to or departed from T4, and to identify the most interesting questions and research opportunities approachable with T4. To be sure, the early roles of T4 in biological research have been chronicled before, notably in Stent's 1963 book, *The Molecular Biology of Bacterial Viruses*; in the 1966 festschrift, *Phage and the Origins of Molecular Biology*; and in Cohen's (1968) *Virus-Induced Enzymes*. However, the story bears retelling, and this is done particularly elegantly in Doermann's prefatory chapter and in the chapter by Singer et al. There we see how much T4 helped shape our ideas about gene and chromosome structure, mRNA, genetic regulation, host-induced modification and restriction, gene and gene product interactions, the genetic code, morphogenetic principles, and preemption of host functions by virus infection.

We see in other chapters how T4 has become retooled and updated for contemporary investigations. In particular, as discussed by Kutter and Rüger, the availability of T4 with cytosine-substituted DNA has made accessible the power of restriction mapping, gene cloning, and DNA sequencing. We see the mutant collections which allowed identification of proteins on one-dimensional gels now applied to the beautiful and much more informative two-dimensional gels depicted in the article by Burke et al. We see in several chapters the power of image enhancement techniques, which allow ever-more-penetrating analyses of viral structure and assembly. We see the venerable *r*II system updated by DNA sequencing and trotted forth again for molecular analyses of mutagenesis, recombination, and regulation, as described by Singer et al. and by Drake and Ripley. In Doermann's chapter we see continued refinement of that hoariest of phage techniques, the genetic cross, for construction and analysis of complex genomes.

Nearly four decades after Seymour Cohen described the first biochemical experiments on phage reproduction, T4 still provides the system of choice for exploring many fundamental questions. Examples include the explorations of replication mechanisms for duplex DNA, described by Nossal and Alberts; studies on replication fidelity, as discussed by Sinha and Goodman; biological functions of topoisomerases, as described by Kreuzer and Huang; translational autoregulation of gene expression, as put forth by von Hippel et al.; protein structure-function relationships, as described by Grütter et al.; and mechanism and rationale for steps in RNA processing, as discussed by Schmidt and Apirion. In the latter regard, one must acknowledge that to date T4 has contributed little to our understanding of gene splicing. However, according to Snyder's chapter, T4 must have RNA-splicing reactions waiting to be discovered! Of course, T4 also serves the entire molecular biology growth industry with a cornucopia of enzymes of known and unknown function: DNA polymerase, polynucleotide kinase, DNA ligase (why does only the T4 enzyme blunt-end ligate?), protease inhibitor, and RNA ligase, to name a few.

While T4 is lighting the way to answering questions we all acknowledge as important, it continues to throw maddening puzzles in the faces of its friends and explorers. They are maddening because they derive from bizarre sets of observations and because they provide absolutely no assurance that the investigator who pursues them will discover general biological truths. Why, for example, does T4 have three evidently independent mechanisms for inactivation and degradation of the host cell chromosome, as discussed by Snustad et al.? Why does T4 use two perfectly respectable enzymes of DNA precursor synthesis also as structural elements of the baseplate, as discussed by Kozloff? What is the mechanism of lysis inhibition, how do the *r*II proteins function to control lysis (or anything else), and what is this mysterious power of the lambda *rex* gene to restrict growth of *r*II mutants, as chronicled by Singer et al.? Why does T4 specify so many low-molecular-weight proteins (see the gel patterns in the chapter by Burke et al.)? What are the roles of phage proteins that are isolated in association with host cell membrane, or ribosomes, or RNA polymerase? How do phage ghosts affect bacterial metabolism? How can a site (*gol*), acting in *cis* within an apparently unrelated gene, affect expression of many other T4 genes, as related by Snyder? What are the functions of enzymes such as polynucleotide kinase-phosphatase, and what sense does it make for RNA ligase and tail fiber attachment activity to reside within the same protein, also as discussed by Snyder? Does *alc/unf* have something important to tell us, aside from how to clone T4 genes (such as the state of the host nucleoid; see Snustad et al.)? What is the physical state of intracellular T4 DNA?

T4 continues as a seductive research assistant, both because it presents chances to cleanly ask significant questions and because it keeps throwing up questions which may not be easily answered and whose answers may or may not have general significance. T4 is seductive also in the sense that once one gets into T4 it is hard to get out. For all who have left T4 to study tumor virology, immunogenetics, or neurobiology, there are nearly as many who, like me, keep at least one foot firmly planted in the T4 camp.

Part of the appeal of phage work lies in the ability to interact with the community of T4 researchers. These constitute a worldwide network of extraordinarily stimulating, helpful, and enthusiastic scientists. My chief regret, as an editor of this book, is that we could not have engaged all of them in writing the book, especially a number of outstanding scientists from outside the United States. A few of these have been brought in as coauthors, and we hope that the work of the others is appropriately represented through discussion and citations. As editors we also take pride in the large number of chapters (16 of 35) which represent collaborations among two, or even three, leading laboratories in a subfield. While we were not always successful in inducing actual or potential competitors to collaborate, those successes that were achieved underscore the vitality of contemporary T4 research. These joint efforts, we feel, should enhance the value of the book.

To the greatest extent possible we have made this book a production of the T4 community. We attempted to maximize the number of contributors and minimize editorial changes, such that each chapter would represent the perspectives of its authors. I think we can take satisfaction in the high quality of work described in these chapters and in the continuing vigor of research with an old and trusted friend, bacteriophage T4.

Appendix: T4 Genes and Gene Products

GISELA MOSIG

Department of Molecular Biology, Vanderbilt University, Nashville, Tennessee 37235

The total genetic information of T4 is stored in the sequence of 166,000 base pairs (Kim and Davidson, 1974; Wood and Revel, 1976). Mature DNA molecules are linear, but their genes are circularly permuted, and each chromosome carries a 3% terminal redundancy; thus, recombination frequencies yield a circular map (Fig. 1) (Stahl et al., 1964; Streisinger et al., 1964; Streisinger et al., 1967). A map of recombinational distances (Edgar and Wood, 1966) is reproduced in the inner circle. It is aligned with another map of now known T4 genes (outer circle) at the *r*IIA-*r*IIB junction. In the outer circle, distances between genes marked with a small *x* have been calculated from the frequencies with which ends of small, incomplete T4 chromosomes are cut during maturation (Childs, 1971; Mosig, 1968; Mosig, 1976; Mosig et al., 1971). (The original data were adjusted for the corrected distance between *r*II and *e* [Kim and Davidson, 1974] since *r*I was used as one reference marker in those studies.) Positions of other genes were determined from recombination frequencies (Edgar et al., 1964; Stahl et al., 1964; and references below). Genes whose relative positions are uncertain are shown inside the outer circle in the map and in parentheses in the table. The scale (middle circle) corresponds to the scale of the accompanying T4 restriction map (Kutter and Rüger, this volume) to facilitate comparisons.

Recombinational distances (inner circle) are obviously different, in certain areas, from physical distances (Mosig, 1968). Most pronounced are differences in genes 34 and 35 and around gene 43. On the other hand, distances based on cutting of ends are remarkably similar to distances on the restriction map. Minor discrepancies exist in the region of genes 46 and 47, 5 through 10, and between *r*II and gene 37. There is a weak preference for cutting ends in the latter region (Mosig et al., 1971).

Table 1 shows all reported sizes of gene products and their cleavage products, except when the molecular weight is known from a protein sequence. Cleavages are indicated by arrows. Sizes of transcription units, directions, and modes of control of transcription are summarized elsewhere (Champe, 1974; Hercules and Sauerbier, 1974; Mosig, 1968; Mosig, 1976; O'Farrell and Gold, 1973; Rabussay and Geiduschek, 1977a; Stahl et al., 1970; Wood and Revel, 1976; and Rabussay, Brody et al., Christensen and Young, and Geiduschek et al., all this volume).

Origins of replication (arrowheads) are discussed by Kozinski and by Mosig (both this volume) and by King and Huang, 1982.

Thanks are due to many colleagues who communicated unpublished data, particularly to Betty Kutter for her unflagging encouragement.

Work in our laboratory has been supported by Public Health Service grant GM 13221 from the National Institutes of Health.

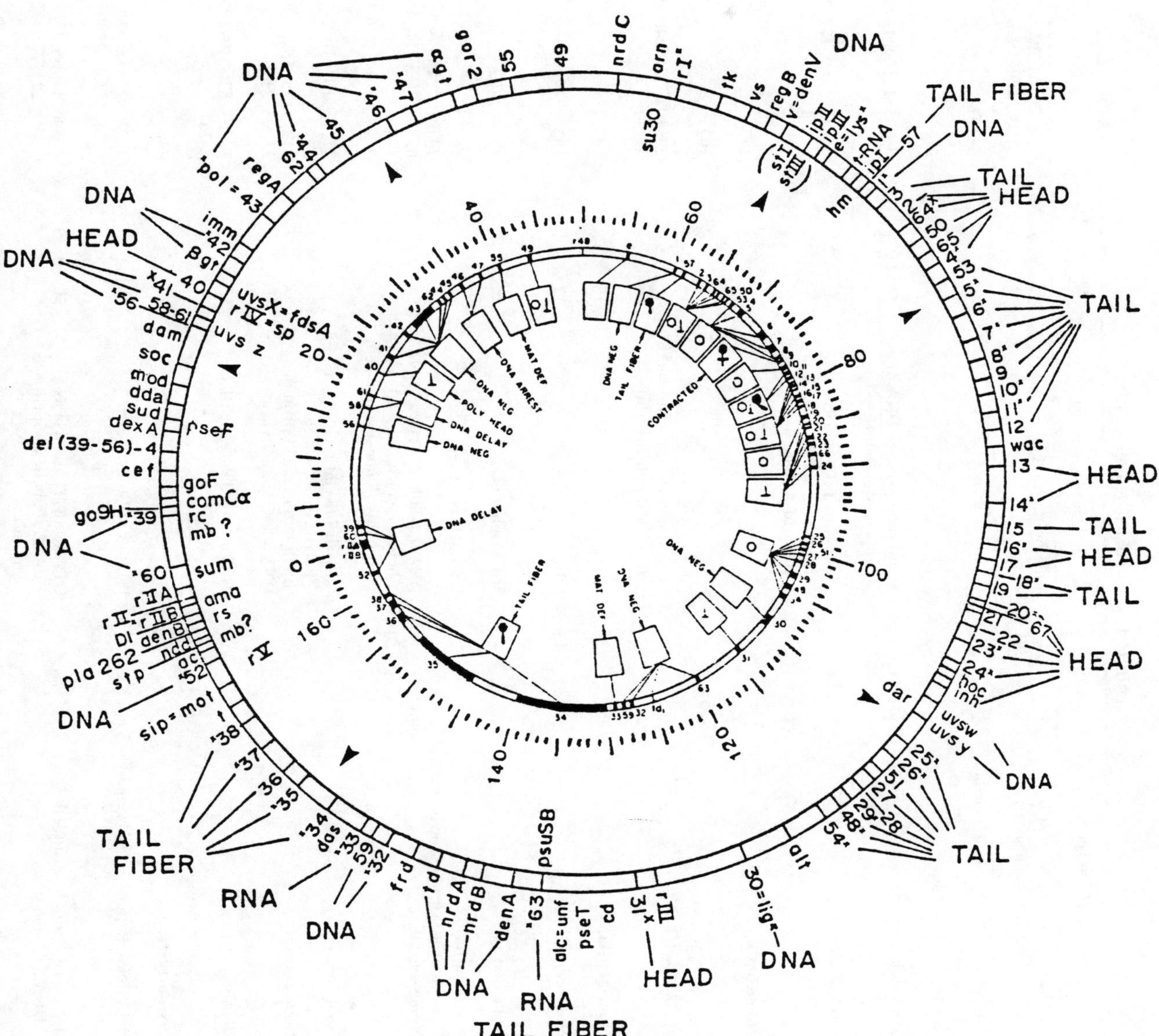

FIG. 1. Recombinational map of T4 genes.

TABLE 1. Genes of bacteriophage T4[a]

Gene	Mutant phenotype	Restrictive host	Function or product	Mol wt of product (× 10³)	References[b]
*r*IIA	Rapid lysis, suppressor of *lig* and some 32 , nonessential	*rex*⁺-λ lysogens, *tabR*	Membrane protein	72, 74, 83, 95	7, 8, 26, 34, 39c, 72, 85, 101, 108, 184, 190b, 244, 265
(*m* = *sum*)	Suppressor of *lig* . nonessential				34, 278
(*rc*)	Acriflavine resistance				205
60	DNA delay	S/6, 25°C	Membrane protein, DNA topoisomerase subunit	16, 18	39c, 111, 159, 187, 217a, 286
39	DNA delay	S/6, 25°C	Membrane protein, DNA topoisomerase subunit, DNA-dependent ATPase	63 → 22	66, 159, 169, 187, 194, 198, 238, 244, 286
(*go*9H)	Nonessential				115
pla CTr5x		CTr5x			102
*go*F *comC*α	Nonessential		Modulation of transcription termination factor rho		29, 102, 201, 240, 245a, 279, 290
cef	Nonessential	*roc*		18	102, 214
del (39–56) (segment 4)	Nonessential		(?)	12	102
(*pseF*)	Nonessential		5′-Phosphatase		45, 46, 102
dexA	Nonessential		Exonuclease A		102, 264
sud	Nonessential, suppressor of 32				74a, 102, 157
dda	Nonessential	*optA*	DNA-dependent ATPase, DNA helicase *dexA*, *sud*, and *dda* probably define a single gene	56	6, 74a, 102, 150, 151
mod	Nonessential		Adenylribosylation of RNA polymerase		77, 78, 102, 105
soc	Nonessential		Small outer capsid protein	9	113, 113a, 164a
dam			DNA adenine methylase	15	23, 95, 217a
56	DNA negative		dCTPase, dUTPase, dCDPase, dUDPase	15	152a, 152b, 194, 235, 262, 267
uvsZ	DNA arrest				44a
58–61	DNA delay hyper-rec, UV sensitive	S/6, 25°C	Primase subunit	40, 42	4, 22, 92, 158, 160, 163, 192, 223, 223a, 223b
41	DNA arrest, single-stranded DNA, UV sensitive		GTPase, dGTPase, ATPase, dATPase, helicase, primase subunit	58, 60, 63, 66	4, 39c, 44, 66, 158, 161, 192, 194, 223, 223b, 244, 255a

TABLE 1.—*Continued*

Gene	Mutant phenotype	Restrictive host	Function or product	Mol wt of product (× 10^3)	References[b]
40	Polyheads		Helper of head vertex assembly	14, 18	15, 16, 24, 25, 110, 155
(*sp* = *r*IV)	Suppressor of *e*⁻, rapid lysis				64, 149
(*uvsX* = *fdsA*)	UV sensitive, recombination deficient, suppressor of 49		*recA*-like protein	38	38, 44, 48, 55, 94, 179, 258
β*gt*	No β-glucosylation of HMC DNA		β-Glucosyltransferase	46	75, 210, 278
42	DNA negative		dCMP-hydroxymethylase	25	66, 194, 267, 280
imm	Immunity to superinfection exclusion			77, 45	36, 41, 252, 289
43 = *pol*	DNA negative		DNA polymerase	112	4, 47, 66, 83, 194, 261
regA			Translational regulation of (early) protein synthesis	12	28, 120, 121, 268
62 }	DNA negative		DNA polymerase accessory proteins, ssDNA-dependent rATPase, dATPase	21.4	4, 66, 137a, 158
44 }				35.9	194, 223b
45	DNA negative, no late mRNA		Accessory protein of DNA and RNA polymerases	24.7	4, 66, 158, 164, 194, 235a
46 }	DNA arrest, recombination deficient, reduced host degradation		Control of recombination nuclease	68, 71, 60	9, 22, 39c, 43, 66, 136, 258, 264a, 269
47 }				37	
α*gt*	No α-glucosylation of DNA		α-Glucosyltransferase		75, 106, 210
gor	Suppressor of RNA polymerase defect				234
55	No late mRNA		Peptide associated with RNA polymerase	17, 22	19, 194, 206, 239
49	Partially filled heads, highly branched DNA		Endonuclease VII, resolvase of Holliday structures	42, 50	48, 49, 60, 61, 123–126, 178, 191
nrdC	Nonessential		Thioredoxin	10	247
arn	DNA degradation under certain conditions		Anti-restriction nuclease (vs. *E. coli rglB*)		50

(Su30)	Nonessential, suppressor of *lig*				145
*r*I	Nonessential, rapid lysis				7, 53
tk	Nonessential		Thymidine kinase		31
vs	Nonessential		Modifier of valyl-tRNA synthetase		165, 171
regB	Nonessential, folate analog resistant		Regulation of gene expression		32
v = *denV*	UV sensitive		Endonuclease V, *N*-glycosidase	18	82, 94, 162, 172, 189, 204, 218, 255, 264a, 278
(*st*I)	Nonessential, star				144
(*st*III)	Nonessential, suppressor of *e* and *t*				148
*ip*II	Nonessential		Internal protein II	11.7 → 10	15, 109, 114, 219
*ip*III	Nonessential		Internal protein III	21 → 18	15, 114, 219, 253
e = *lys*	No lysis		Endolysin	19	242, 251
(*go*F3.03.2)	Grow on *HDF* hosts				240
tRNAs 1 2	Nonessential (suppressors of nonsense mutations)	CT439	tRNA precursors		1, 70, 74, 80, 87, 117, 168, 170, 200, 271
arg	psu_4 , *op*				
gly					
ser	psu_a^+, psu_b^+, psu_1^+				
pro					
ileu					
thr					
leu	psu_3^+				
gln	psu_2^+				
*ip*I		CT596	Internal protein I	10	2, 17, 18, 114, 219
57	Poor tail fiber assembly, bypassed in certain host mutants		Morphogenetic catalyst of long and short tail fibers	18, 6	51, 59, 99, 100, 122, 134, 211
(*hm*)	Nonessential, mutator				55
1	DNA negative		dHMP-kinase	22, 25	39c, 57, 66, 194
3	Unstable tails		Tail tube, sheath	29	66, 132

TABLE 1.—*Continued*

Gene	Mutant phenotype	Restrictive host	Function or product	Mol wt of product ($\times 10^3$)	References[b]
2, 64	Inactive filled heads, noninfectious particles; 2 and 64 mutations are probably in the same gene		Head completion, terminal DNA protecting protein	25	60, 61, 66, 78a, 131, 224, 225
50	Inactive filled heads, noninfectious particles		Head completion		60, 61, 131
65	Inactive filled heads, noninfectious particles		Head completion		60, 61, 131
4	Inactive or empty capsids		Head completion		60, 61, 66
53	Defective tails		Baseplate	23	60, 61, 95, 128
5	Defective tails		Baseplate, central plug, lysis of cell walls	37	60, 61. 66, 118, 253
6	Defective tails, permit fiberless plating		Baseplate outer wedge	78, 85	42, 60, 61, 66, 135, 253
7	Defective tails, permit fiberless plating		Baseplate outer wedge	127, 140	42, 60, 61, 66, 135, 253
8	Defective tails		Baseplate outer wedge	39, 46	60, 61, 66, 135, 253
9	Defective tails, fiberless particles		Baseplate completion	30, 34	61, 66, 131, 135, 253
10	Defective tails		Baseplate outer wedge	88, 90	60, 61, 66, 135, 253
11	Defective tails		Baseplate outer wedge	25, 26	60, 61, 66. 135, 253
12	Defective tails		Short tail fibers	57, 58	122, 131, 246, 253, 273
wac	Nonessential		Whisker antigen	52	49, 69, 253
13			Head completion	33	60, 61, 66, 253
14			Head completion	30	60, 61, 66, 253
15			Tail completion, collar connector	32, 35	60, 61, 131, 134, 253
16	DNA maturation defective, empty heads		Head filling		16, 60, 61, 66, 163a

17 = *q*	Quinacrine resistant, DNA maturation defective, suppressors of certain 32 or 20 mutants	Head filling	69	16, 60, 61, 66, 132, 186, 199
18	Defective tails	Tail sheath protein	70, 80	60, 61, 66, 131, 134, 250, 253
19	Defective tails	Tail core protein	20, 21	60, 61, 66, 134, 253
20	Polyheads	Head plug protein (connector to neck)	63, 65, 67	56, 61, 66, 68, 153, 155, 253
67 = *pip*	Head defect	Core protein, precursor to internal peptides	9.1	152, 197, 257
21	Faulty heads	Maturation protease, head assembly core	27.5, 21.5; small peptides	61, 66, 68, 155, 196, 197, 219, 249, 254, 274
22	Faulty heads	Head assembly core (later degraded)	30, 31, 32	61, 66, 68, 81, 114, 153, 155, 196, 219, 253
23	No or faulty heads	Major head subunit	55 → 46, 56 → 46, 55 → 47	52, 54, 61, 63, 68, 153, 253
gol (in 23)	Grow on *lit* hosts (CTr5x)			33a
24 = *os*	Faulty heads, osmotic shock resistance	Vertex head subunit	47 → 45	61, 66, 68, 155, 156, 173, 253
		One of the products of genes 20–24 is cleaved to a packaging-related, DNA-dependent ATPase endonuclease	43	164b
hoc-eph	Nonessential	Minor capsid protein	40	37, 284
inh		Inhibitor of gene 21 protease	35	221
(*dar*)	Suppressor of 59			107, 281, 282
uvsW	UV sensitive, recombination deficient			43, 92, 175, 245
uvsY = *fdsB*	UV sensitive, DNA synthesis reduced, recombination deficient, suppressor of 49		16	26a, 43, 48, 92, 174, 230, 245, 258
25	Tail defects	Baseplate, outer wedge	15	60, 61, 66, 128
26	Tail defects	Baseplate, central plug	41(?)	60, 61, 66, 123. 139–141
51	Tail defects	Baseplate plug		60, 61, 66, 123, 139

TABLE 1.—*Continued*

Gene	Mutant phenotype	Restrictive host	Function or product	Mol wt of product (× 10^3)	References[b]
27	Tail defects, permit fiberless plating		Baseplate, central plug	48	42, 60, 61, 66, 129
28	Tail defects		Baseplate, distal surface of central plug, cleavage enzyme for folylpolyglutamates	24, 25	60, 61, 66, 129, 140, 141
29	Tail defects		Baseplate, central plug	77	60, 61, 66, 129, 135, 253
48	Tail defects		Baseplate, surface	37	10, 60, 61, 66, 253
54	Tail defects		Baseplate protein	36	10, 60, 61, 66, 128
alt	Nonessential		Adenylribosylation of RNA polymerase (packaged and injected with DNA)	61, 79	77, 78, 105
30 = *lig*	DNA arrest, hyper-*rec*		DNA ligase	59, 68	22, 39c, 67, 107, 194, 273a
rIII	Nonessential, rapid lysis				7, 53
31	Capsid protein lumps, suppressor of mutations in host defective, *groE*, and two other host genes		Organizer of gp23	16	30, 61, 66, 68, 76, 154, 212, 227, 274, 279
cd	Nonessential		dCMP deaminase		39a, 88, 89, 247
pseT	Nonessential	CTr5x (*lit*)	Deoxyribonucleotide 3′-phosphatase–5′-polynucleotide kinase	30	40, 45, 215, 229
alc = *unf*	Allows late transcription of cytosine-containing DNA		Unfolding of host DNA	18	136, 215, 228, 233, 235
63	Poor tail fiber attachment		RNA ligase, helper of tail fiber attachment	42	84, 173a, 213, 215, 231, 253, 277, 278
(*psu*⁺ SB)	Suppressor of nonsense mutations				213
denA	Nonessential, defective in host DNA degradation		Endonuclease II		136, 260
nrdB	Nonessential		Ribonucleotide reductase B subunit	35, 40	11, 12, 39b, 102, 287, 288
nrdA	Nonessential		Ribonucleotide reductase A subunit	80, 85	

td	Nonessential	Thymidylate synthetase	Thymidylate synthetase, baseplate central plug component	29, 30	27, 89, 141, 238
frd			Dihydrofolate reductase, baseplate wedge component	19, 22, 23	27, 89, 93, 180, 202, 287
32	DNA arrest, recombination deficient, UV sensitive	*tab-32*	ssDNA-binding = helix-destabilizing protein	33.5	3, 4, 20, 66, 138, 143, 185, 190a, 216, 248, 270, 281
59	DNA arrest				283
33	No late RNA synthesis		RNA polymerase-associated peptide	10	19, 66, 206, 239
das = *su*	Suppressor of 46 and 47				98, 147, 176
34	Fiberless particles		Proximal tail fiber subunit (A antigen)	145, 150	51, 58, 59, 61, 66, 133, 207, 253, 259, 275
35	Fiberless particles		Hinge tail fiber subunit	39	51, 58, 59, 61, 66, 133, 207, 253, 259, 275
36	Fiberless particles		Small distal tail fiber subunit	24.3	51, 58, 59, 61, 66, 133, 195a, 207, 253. 259, 275
37	Fiberless particles, host range		Large distal tail fiber subunit	112.8	51, 58, 59, 61, 66, 133, 195a, 207, 253, 259, 275
38	Fiberless particles		Assembly catalyst of gp37	26, 27, 28	51, 58, 59, 61, 66, 133, 134, 207, 253, 259, 275
t = *st*II	Lysis defective, suppressor of gene 63 and *r*II mutations				90, 116, 147
mot = *sip*	Regulation of middle gene expression, suppressor of *r*II in K(λ)		Expression from middle promoters		73, 91, 102, 103, 167, 201
(*r*V)	Temperature dependent, rapid lysis				149
52	DNA delay	S/6, 25°C	Membrane protein, DNA topoisomerase subunit	51	39c, 112, 159, 163, 187, 194, 238, 286
(*mb*)	Nonessential (may be located clockwise from *r*II)		Modifier of suppressor tRNA		271
ac	Nonessential, acriflavine resistant				222
(*ama*) (*rs*)	Nonessential, acriflavine resistant				205

TABLE 1.—*Continued*

Gene	Mutant phenotype	Restrictive host	Function or product	Mol wt of product (× 10^3)	References[b]
stp	Nonessential, suppressor of *pseT* mutation				46
ndd	Nuclear disruption defective	CT447		15	136, 137, 232
*pla*262	Nonessential	CT262			46
denB	Nonessential		Endonuclease IV		136, 256
D1	Nonessential				5, 79
rIIB	Nonessential, rapid lysis, suppressor of *lig* and some 32	rex^+ λ lysogens, *tabR*	Membrane protein	33, 41	7, 8, 85, 101, 108, 109a, 112, 184, 194, 244, 265

[a] Updated April 1983 by Gisela Mosig. ssDNA, Single-stranded DNA.

[b] Reference list:

1. Abelson et al., 1975.
2. Abremski and Black, 1979.
3. Alberts and Frey, 1970.
4. Alberts et al., 1980.
5. Bautz and Bautz, 1967.
6. Behme and Ebisuzaki, 1975.
7. Benzer, 1957.
8. Berger and Kozinski, 1969.
9. Berger et al., 1969.
10. Berget and Warner, 1975.
11. Berglund, 1972.
12. Berglund and Sjoberg, 1970.
13. Bernstein, 1968.
14. Bijlenga et al., 1976.
15. Black, 1974.
16. Black et al., 1981.
17. Black and Abremski, 1974.
18. Black and Ahmad-Zadeh, 1971.
19. Bolle et al., 1968a.
20. Breschkin and Mosig, 1977a.
21. Brody and Geiduschek, 1970.
22. Broker, 1973.
23. Brooks and Hattman, 1973.
24. Brown and Eiserling, 1979a.
25. Brown and Eiserling, 1979b.
26. Bujard et al., 1970.
26a. Burke et al., this volume.
27. Capco and Mathews, 1973.
28. Cardillo et al., 1979.
29. Caruso et al., 1979.
30. Castillo et al., 1977.
31. Chace and Hall, 1973.
32. Chace and Hall, 1975.
33. Champe, 1974.
33a. Champness and Snyder, 1982.
34. Chan and Ebisuzaki, 1973.
35. Childs, 1971.
36. Childs, 1973.
37. Childs, 1980a.
38. Childs, 1980b.
39. Childs and Birnboim, 1975.
39a. Chiu et al., 1977.
39b. Chiu et al., 1980.
39c. Cook, and Seasholtz, 1982.
40. Cooley et al. 1979.
41. Cornett and Vallee, 1973
42. Crowther, 1980.
43. Cunningham and Berger, 1977.
44. Cupido et al., 1980.
44a. Cupido, et al., 1982.
45. Depew and Cozzarelli, 1974.
46. Depew et al., 1975.
47. DeWaard et al., 1965.
48. Dewey and Frankel, 1975.
49. Dewey et al., 1974.
50. Dharmalingam et al., 1982.
51. Dickson, 1973.
52. Dickson et al., 1970.
53. Doermann, 1953.
54. Doermann, et al., 1973.
55. Drake, 1973.
56. Dredonks, 1981.
57. Duckworth and Bessman, 1967.
58. Edgar and Lielausis, 1964.
59. Edgar and Lielausis, 1965.
60. Edgar and Lielausis, 1968.

61. Edgar and Wood, 1966.
62. Edgar et al., 1964.
63. Eiserling et al., 1970.
63a. Elliott et al., 1973.
64. Emrich, 1968.
65. Ennis and Kievitt, 1973.
66. Epstein et al., 1963.
67. Fareed and Richardson, 1967.
68. Favre et al., 1965.
69. Follansbee et al., 1974.
70. Foss et al., 1979.
71. Frankel, 1966.
72. Frankel et al., 1971.
73. Freedman and Brenner, 1972.
74. Fukada and Abelson, 1980.
74a. Gauss and Gold, 1983.
75. Georgopoulos, 1968.
76. Georgopoulos et al., 1972.
77. Goff, 1979.
78. Goff and Setzer, 1980.
78a. Goldberg, this volume.
79. Goldfarb and Burger, 1981.
80. Goldfarb and Daniel, 1981a.
81. Goldstein and Champe, 1974.
82. Gordon and Haseltine, 1980.
83. Goulian et al., 1968.
84. Gumport et al., 1980.
85. Gussin and Peterson, 1972.
86. Guthrie and McClain, 1979.
87. Guthrie et al., 1975.
88. Hall and Tessman, 1966.
89. Hall et al., 1967.
90. Hall et al., 1980.
91. Hall and Snyder, 1981.
92. Hamlett and Berger, 1975.
93. Hanggi and Zachau, 1980.
94. Harm, 1963.
95. Hattman, this volume.
96. Hercules and Sauerbier, 1974.
97. Hercules and Wiberg, 1971.
98. Hercules et al., 1971.
99. Herrmann and Wood, 1981.
100. Herrmann, 1982.
101. Hershey, 1953.
102. Homyk and Weil, 1974.
103. Homyk et al., 1976.
104. Horvitz, 1973.
105. Horvitz, 1974b.
106. Hosoda, 1967.
107. Hosoda and Mathews, 1971.
108. Howard, 1967.
109. Howard et al., 1972.
110. Hsiao and Black, 1978a.
111. Huang, 1979a.
112. Huang and Buchanan, 1974.
113. Ishii and Yanagida, 1975.
113a. Ishii and Yanagida, 1977.
114. Isobe et al., 1976a.
115. Jensen and Susman, 1980.
116. Josslin, 1970.
117. Kao and McClain, 1977.
118. Kao and McClain, 1980b.
119. Karam and Barker, 1971.
120. Karam and Bowles, 1974.
121. Karam et al., 1981.
122. Kells and Haselkorn, 1974.
123. Kemper and Brown, 1976.
124. Kemper and Garabett, 1981.
125. Kemper et al., 1981a.
126. Kemper et al., 1981b.
127. Kikuchi and King, 1975a.
128. Kikuchi and King, 1975b.
129. Kikuchi and King, 1975c.
130. Kim and Davidson, 1974.
131. King, 1968.
132. King, 1971.
133. King and Wood, 1969.
134. King and Laemmli, 1971.
135. King and Laemmli, 1973.
136. Koerner and Snustad, 1979.
137. Koerner et al., 1979.
137a. Konigsberg and Williams, this volume.
138. Kozinski and Felgenhauer, 1967.
139. Kozloff and Lute, 1973.
140. Kozloff and Lute, 1981.
141. Kozloff and Zorzopulos, 1981.
142. Kozloff et al., 1977.
143. Krisch et al., 1977.
144. Krylov, 1971.
145. Krylov, 1972.
146. Krylov and Plotnikova, 1972a.
147. Krylov and Plotnikova, 1972b.
148. Krylov and Yankovsky, 1975.
149. Krylov and Zapadnaya, 1965.
150. Kuhn et al., 1979a.
151. Kuhn et al., 1979b.
152. Kurtz and Champe, 1977.
152a. Kutter and Wiberg, 1968.
152b. Kutter and Wiberg, 1969.
152c. Kutter et al., 1981.
152d. Kuzmin et al., 1982.
153. Laemmli, 1970.
154. Laemmli et al., 1970a.
155. Laemmli et al., 1970b.
156. Leibo et al., 1979.
157. Little, 1973.

158. Liu et al., 1979a.
159. Liu et al., 1979b.
160. Liu and Alberts, 1981a.
161. Liu and Alberts, 1981b.
162. Lloyd and Hanawalt, 1981.
163. Luder, Ph.D. thesis, Vanderbilt University, Nashville, Tenn., 1981.
163a.Luftig and Ganz, 1972b.
163b.Luftig et al., 1971.
164. Macchiato et al., 1979.
164a.Macdonald, Kutter, and Mosig, manuscript in preparation.
164b.Manne et al., 1982.
165. Marchin et al., 1980.
166. Marsh et al., 1971.
167. Mattson et al., 1974.
168. Mazzara et al., 1981.
169. McCarthy, 1979.
170. MClain, 1970.
171. McClain et al., 1975.
172. McMillan et al., 1981.
173. McNicol et al., 1977.
173a.Mei et al., 1982.
174. Melamede and Wallace, 1977.
175. Melamede and Wallace, 1980a.
176. Mickelson and Wiberg, 1981.
177. Mileham et al., 1980.
178. Minagawa and Ryo, 1979.
179. Minagawa, personal communication.
180. Mosher and Mathews, 1979.
181. Mosig, 1968.
182. Mosig et al., 1971.
183. Mosig, 1976.
184. Mosig et al., 1977.
185. Mosig et al., 1979a.
186. Mosig et al., 1981.
187. Mufti and Bernstein, 1974.
188. Muller-Salamin et al., 1977.
189. Nakabeppu and Sekiguchi, 1981.
190. Nelson and Butler, 1974.
190a.Nelson and Gold, 1982.
190b.Nelson et al., 1982.
191. Nishimoto et al., 1979.
192. Nossal, 1980.
193. O'Farrell and Gold, 1973a.
194. O'Farrell et al., 1973.
195. Oishi, 1968.
195a.Oliver and Crowther, 1981.
195b.Olson, G. B., and H. Hartman. J. Phys. Paris Colloq. 43:C4-855
196. Onorato and Showe, 1975.
197. Onorato et al., 1978.
198. Panuska and Goldthwait, 1980.
199. Piechowski and Susman, 1967.
200. Plunkett et al., 1981.
201. Pulitzer et al., 1979.
202. Purohit et al., 1981.
203. Rabussay and Geiduschek, 1977a.
204. Radany and Friedberg, 1980.
205. Rappaport et al., 1974.
206. Ratner, 1974.b
207. Revel, 1981b.
208. Revel and Hattman, 1971.
209. Revel and Lielausis, 1978.
210. Revel and Luria, 1970.
211. Revel et al., 1976.
212. Revel et al., 1980.
213. Ribolini and Baylor, 1975.
214. Rodriguez, Ph. D. thesis, Vanderbilt University, Nashville, Tenn., 1976.
215. Runnels et al., 1982.
216. Russel et al., 1976.
217. Sarabhai et al., 1964.
217a.Seasholtz and Greenberg, 1983.
218. Seawell et al., 1980.
219. Showe and Black, 1973.
220. Showe et al., 1976.
221. Showe, personal communication.
222. Silver, 1965.
223. Silver and Nossal 1979.
223a.Silver and Nossal, 1982.
223b.Silver et al., 1980.
224. Silverstein and Goldberg, 1976a.
225. Silverstein and Goldberg, 1976b.
226. Simon and Tessman, 1963.
227. Simon et al., 1974.
228. Sirotkin et al., 1977.
229. Sirotkin et al., 1978.
230. Smith and Symonds, 1973.
231. Snopek et al., 1977.
232. Snustad and Conroy, 1974.
233. Snustad et al., 1976.
234. Snyder and Montgomery, 1974.
235. Snyder et al., 1976.
235a.Spicer et al., 1982.
236. Stahl et al., 1964.
237. Stahl et al., 1970.
238. Stetler et al., 1979.
239. Stevens, 1972.
240. Stitt et al.,1980.
241. Streisinger, Edgar, and Denhardt, 1964.
242. Streisinger, Mukai, et al., 1964.
243. Streisinger et al., 1967.
244. Takacs and Rosenbusch, 1975.
245. Takahashi and Saito, 1982.
245a.Takahashi and Yoshikawa, 1979.
246. Terzaghi et al., 1979.
247. Tessman and Greenberg, 1972.
248. Tomizawa et al., 1966.
249. Traub et al., 1981.

250. Tschopp et al., 1979.
251. Tsugita et al., 1968.
252. Vallee and Cornett, 1972.
253. Vanderslice and Yegian, 1974.
254. van Driel et al., 1980.
255. van Minderhout et al., 1979.
255a. Venkatesan et al., 1982.
256. Vetter and Sadowski, 1974.
257. Volker et al., 1982.
258. Wakem and Ebisuzaki, 1981.
259. Ward and Dickson, 1971.
260. Warner, 1971.
261. Warner and Barnes, 1966.
262. Warner and Duncan, 1978.
263. Warner et al., 1970.
264. Warner et al., 1972.
264a. Warner et al., 1981.
265. Weintraub and Frankel, 1972.
266. Wiberg, 1966.
267. Wiberg et al., 1962.
268. Wiberg et al., 1973.
269. Wiberg et al., 1981.
270. Williams et al., 1981.
271. Wilson and Abelson, 1972.
272. Wilson et al., 1972.
273. Wilson, 1973.
273a. Wilson and Murray, 1979.
274. Wood, 1979.
275. Wood and Bishop, 1973.
276. Wood and Conley, 1979.
277. Wood and Henninger, 1969.
278. Wood and Revel, 1976.
279. Wood et al., 1973.
280. Wovcha et al., 1973.
281. Wu and Yeh, 1973.
282. Wu and Yeh, 1978.
283. Wu et al., 1972.
284. Yamaguchi and Yanagida, 1980.
285. Yasuda and Sekiguchi, 1970b.
286. Yegian et al., 1971.
287. Yeh and Tessman, 1972.
288. Yeh et al., 1969.
289. Yutsudo, 1979.
290. Zograff and Gintsburg, 1980.

LITERATURE CITED

Abelson, J., K. Fukada, P. Johnson, H. Lamfrom, D. P. Nierlich, A. Otsuka, G. V. Paddock, T. C. Pinkerton, A. Sarabhai, S. Stahl, J. H. Wilson, and H. Yesian. 1975. Bacteriophage T4 tRNAs: structure, genetics, biosynthesis. Brookhaven Symp. Biol. **26**:77–87.

Abremski, K., and L. W. Black. 1979. The function of bacteriophage T4 internal protein I in a restrictive strain of *Escherichia coli*. Virology **97**:439–447.

Adams, M. H. 1959. Bacteriophages. Interscience Publishers, Inc., New York.

Adesnik, M., and C. Levinthal. 1970. RNA metabolism in T4-infected *Escherichia coli*. J. Mol. Biol. **48**:187–208.

Adhya, S., and M. Gottesman. 1978. Control of transcription termination. Annu. Rev. Biochem. **47**:967–996.

Admiraal, G., and J. E. Mellema. 1976. The structure of the contractile sheath of bacteriophage mu. J. Ultrastruct. Res. **56**:48–52.

Aebi, U., R. Bijlenga, B. ten Heggeler, J. Kistler, A. C. Steven, and P. R. Smith. 1976. Comparison of the structural and chemical composition of giant T-even phage heads. J. Supramol. Struct. **5**:475–495.

Aebi, U., R. Bijlenga, J. van der Broek, R. van der Broek, F. Eiserling, C. Kellenberger, E. Kellenberger, V. Mesyanzihinov, L. Muller, M. Showe, R. Smith, and A. Steven. 1974. The transformation of tau-particles into T4 heads. II. Transformations of the surface lattice and related observations on form determination. J. Supramol. Struct. **2**:253–275.

Aebi, U. R., B. ten Heggeler, L. Onorato, J. Kistler, and M. K. Showe. 1977. A new method for localizing proteins in periodic structures: Fab fragment labeling combined with image processing of electron micrographs. Proc. Natl. Acad. Sci. U.S.A. **74**:5514–5518.

Aebi, U. R., R. van Driel, R. K. L. Bijlenga, B. ten Heggeler, R. van der Broek, A. Steven, and P. R. Smith. 1977. Capsid fine structure of T-even bacteriophages. Binding and localization of two dispensable capsid proteins into P23 surface lattice. J. Mol. Biol. **110**:687–698.

Aksiyote-Benbasat, J., and V. A. Bloomfield. 1975. Joining of bacteriophage T4D heads and tails: a kinetic study by inelastic light scattering. J. Mol. Biol. **95**:335–357.

Aksiyote-Benbasat, J., and V. A. Bloomfield. 1981. Kinetics of head-tail joining in bacteriophage T4D studied by quasi-elastic light scattering: effects of temperature, pH, and ionic strength. Biochemistry **20**:5018–5025.

Aksiyote-Benbasat, J., and V. A. Bloomfield. 1982. Hydrodynamics, size, and shape of bacteriophage T4D tails and baseplates. Biopolymers **21**:797–804.

Alberts, B. M., F. J. Amodio, M. Jenkins, E. D. Guttman, and F. L. Ferris. 1969. Studies with DNA-cellulose chromatography. I. DNA-binding proteins from *Escherichia coli*. Cold Spring Harbor Symp. Quant. Biol. **33**:289–305.

Alberts, B. M., J. Barry, P. Bedinger, R. L. Burke, U. Hibner, C.-C. Liu, and R. Sheridan. 1980. Mechanistic studies of DNA replication and genetic recombination. ICN-UCLA Symp. Mol. Cell. Biol. **19**:449–471.

Alberts, B. M., J. Barry, P. Bedinger, T. Formosa, C. V. Jongeneel, and K. N. Kreuzer. 1983. Studies on DNA replication in the T4 bacteriophage in vitro system. Cold Spring Harbor Symp. Quant. Biol. **47**:655–668.

Alberts, B. M., J. Barry, M. Bittner, M. Davies, H. Hama-Inaba, C. Liu, D. Mace, L. Moran, C. Morris, J. Piperno, and N. Sinha. 1977. In vitro DNA replication catalyzed by six purified T4 bacteriophage proteins, p. 31–63. *In* H. J. Vogel (ed.), Nucleic acid-protein recognition. Academic Press, Inc., New York.

Alberts, B. M., and L. Frey. 1970. T4 bacteriophage gene 32: a structural protein in the replication and recombination of DNA. Nature (London) **227**:1313–1318.

Alberts, B. M., C. F. Morris, D. Mace, N. Sinha, M. Bittner, and L. Moran. 1975. Reconstruction of the T4 bacteriophage DNA replication apparatus from purified components, p. 241–269. *In* M. Goulian and C. F. Fox (ed.), DNA synthesis and its regulation. W. A. Benjamin, Inc., Menlo Park, Calif.

Alberts, B. M., and R. Sternglanz. 1977. Recent excitement in the DNA replication problem. Nature (London) **269**:655–661.

Al-Janabi, J., J. A. Hartsuck, and J. Tang. 1972. Kinetics and mechanism of pepsinogen activation. J. Biol. Chem. **247**:4628–4632.

Allen, J. R., J. W. Booth, D. A. Goldman, G. W. Lasser, and C. K. Mathews. 1983. T4 phage deoxyribonucleotide-synthesizing multienzyme complex. J. Biol. Chem. **258**:5746–5753.

Allen, J. R., G. P. V. Reddy, G. W. Lasser, and C. K. Mathews. 1980. T4 ribonucleotide reductase. Physical and kinetic linkage to other enzymes of deoxyribonucleotide biosynthesis. J. Biol. Chem. **225**:7583–7588.

Altman, S., and L. S. Lerman. 1970. Kinetics and intermediates in the intracellular synthesis of bacteriophage T4 deoxyribonucleic acid. J. Mol. Biol. **50**:235–261.

Altman, S., and M. Meselson. 1970. A T4-induced endonuclease which attacks T4 DNA. Proc. Natl. Acad. Sci. U.S.A. **66**:716–721.

Ames, G. F., and K. Nikaido. 1976. Two-dimensional gel electrophoresis of membrane proteins. Biochemistry **15**:616–623.

Amos, L. A., and A. Klug. 1975. Three dimensional image reconstructions of the contractile tail of the T4 bacteriophage. J. Mol. Biol. **99**:51–73.

Anderson, C. W., and J. Eigner. 1971. Breakdown and exclusion of superinfecting T-even bacteriophage in *Escherichia coli*. J. Virol. **8**:869–886.

Anderson, R., and J. Coleman. 1975. Physicochemical properties of DNA binding proteins: gene 32 of T4 and *Escherichia coli* unwinding protein. Biochemistry **14**:5485–5491.

Anderson, T. F. 1949. The reaction of bacterial viruses with their host cells. Bot. Rev. **15**:464–505.

Anderson, T. F. 1950. Destruction of bacterial viruses by osmotic shock. J. Appl. Phys. **21**:70–82.

Anderson, T. F. 1951. Techniques for the preservation of three dimensional structure in preparing specimens for the electron microscope. Trans. N.Y. Acad. Sci. Ser. II **13**:130–134.

Anderson, T. F. 1952. Stereoscopic studies of cells and viruses in the electron microscope. Am. Nat. **86**:91–100.

Anderson, T. F. 1953. The morphology and osmotic properties of bacteriophage systems. Cold Spring Harbor Symp. Quant. Biol. **18**:197–203.

Anderson, T. F., C. Rappaport, and N. A. Muscatine. 1953. On the structure and osmotic properties of phage particles. Ann. Inst. Pasteur Paris **84**:7–14.

Anderson, W. F., M. G. Grutter, S. J. Remington, and B. W. Matthews. 1981. Crystallographic determination of the mode of binding of oligosaccharides to T4 bacteriophage lysozyme. Implications for the mechanism of catalysis. J. Mol. Biol. **147**:523–543.

Anraku, N., and I. R. Lehman. 1969. Enzymic joining of polynucleotides. VII. Role of the T4-induced ligase in the formation of recombinant molecules. J. Mol. Biol. **46**:467–479.

Anraku, N., and J. Tomizawa. 1965. Molecular mechanisms of genetic recombination of bacteriophage. V. Two kinds of joining of parental DNA molecules. J. Mol. Biol. **12**:805–815.

Aposhian, H. V., and A. Kornberg. 1962. Enzymatic synthesis of deoxyribonucleic acid. IX. The polymerase formed after T2 bacteriophage infection of *Escherichia coli*: a new enzyme. J. Biol. Chem. **237**:519–525.

Arai, K., and A. Kornberg. 1981. Unique primed start of phage φX174 DNA replication and mobility of the primosome in a direction opposite chain synthesis. Proc. Natl. Acad. Sci. U.S.A. **78**:69–73.

Arai, K., R. Low, and A. Kornberg. 1981. Movement and site selection for priming by the primosome in phage φ174 DNA replication. Proc. Natl. Acad. Sci. U.S.A. **78**:707–711.

Araki, H., and H. Ogawa. 1981. A T7 mutant defective in DNA-binding protein. Mol. Gen. Genet. **183**:66–73.

Arber, W. 1974. DNA modification and restriction. Prog. Nucleic Acid Res. Mol. Biol. **14**:1–37.

Arscott, P. G., and E. B. Goldberg. 1976. Cooperative action of the T4 tail fibers and baseplate in triggering conformational changes and in determining host range. Virology **69**:15–22.

Artman, M., and S. Werthamer. 1974. Transition from streptomycin-sensitive to streptomycin-resistant protein synthesis during bacteriophage T4 development. Biochem. Biophys. Res. Commun. **59**:75–81.

Artymiuk, P. J., C. C. F. Blake, and A. E. Stippel. 1981. Genes pieced together—exons delineate homologous structures of diverged lysozymes. Nature News Views **290**:287–288.

Asakura, S. 1970. Polymerization of flagellin and polymorphism of flagella. Adv. Biophys. **1**:99–155.

Asakura, S., G. Eguchi, and T. Iino. 1968. Unidirectional growth of *Salmonella* flagella in vitro. J. Mol. Biol. **35**:225–236.

Auerbach, C. 1976. Mutation research. Problems, results and perspective. John Wiley & Sons, Inc., New York.

Avery, O. T., C. M. McLeod, and M. McCarty. 1944. Studies on the chemical nature of the substance inducing transformation of pneumococcal types. Induction of transformation by a deoxyribonucleic acid fraction isolated from pneumococcus type III. J. Exp. Med. **79**:137–158.

Bächi, C. B., J. Reiser, and V. Pirrotta. 1979. Methylation and cleavage sequences of the *Eco*P1 restriction-modification enzyme. J. Mol. Biol. **128:**143–163.

Bachman, B. J., and K. B. Low. 1980. Linkage map of *Escherichia coli* K-12, edition 6. Microbiol. Rev. **44:**1–56.

Bachrach, U., and L. Benchetrit. 1974. Studies on phage internal proteins. II. Cleavage of a precursor of internal protein during the morphogenesis of bacteriophage T4. Virology **59:**51–58.

Baldy, M. W. 1968. Repair and recombination in phage T4. II. Genes affecting UV sensitivity. Cold Spring Harbor Symp. Quant Biol. **33:**333–338.

Baldy, M. W. 1970. The UV sensitivity of some early-function temperature-sensitive mutants of phage T4. Virology **40:**272–287.

Baldy, M. W., B. Strom, and H. Bernstein. 1971. Repair of alkylated bacteriophage T4 deoxyribonucleic acid by a mechanism involving polynucleotide ligase. J. Virol. **7:**407–408.

Baltz, R. H. 1971. Infectious DNA of bacteriophage T4. J. Mol. Biol. **62:**425–437.

Baltz, R. H. 1976. Biological properties of an improved transformation assay for native and denatured T4 DNA. Virology **70:**52–64.

Baltz, R. H., and J. W. Drake. 1972. Bacteriophage T4 transformation: an assay for mutations in vitro. Virology **49:**462–474.

Bancroft, J. B. 1970. The self-assembly of spherical plant viruses. Adv. Virus Res. **16:**99–133.

Banerji, J., S. Rusconi, and W. Schaffner. 1981. Expression of a β-globin gene is enhanced by remote SV40 DNA sequences. Cell **27:**299–308.

Baran, G. J., and V. A. Bloomfield. 1978. Tail fiber attachment in bacteriophage T4D studied by quasi-elastic light scattering-band electrophoresis. Biopolymers **17:**2015–2028.

Baril, E., B. Baril, H. Elford, and R. B. Luftig. 1973. Aggregated mammalian enzymes in deoxyribonucleotide and DNA replication, p. 275–291. *In* A. R. Kolber and M. Kohiyama (ed.), Mechanism and regulation of DNA replication. Plenum Publishing Corp., New York.

Barnett, L., S. Brenner, F. H. C. Crick, R. G. Shulman, and R. J. Watts-Tobin. 1967. Phase-shift and other mutants in the first part of the rIIB cistron of bacteriophage T4. Philos. Trans. R. Soc. London Ser. B **252:**487–560.

Barrell, B. G., A. R. Coulson, and W. H. McClain. 1973. Nucleotide sequence of glycine transfer RNA coded by bacteriophage T4. FEBS Lett. **37:**64–69.

Barrell, B. G., J. G. Seidman, C. Guthrie, and W. H. McClain. 1974. Transfer RNA biosynthesis: the nucleotide sequence of a precursor to serine and proline transfer RNAs. Proc. Natl. Acad. Sci. U.S.A. **71:**413–416.

Barricelli, N. A. 1956. A "chromosomic" recombination theory for multiplicity reactivation in phages. Acta Biotheor. **11:**107–120.

Barricelli, N. A. 1960. An analytical approach to the problems of phage recombination and reproduction. I. Multiplicity reactivation and the nature of radiation damages. Virology **11:**99–135.

Barricelli, N. A., and A. H. Doermann. 1961. An analytical approach to the problems of phage recombination and reproduction. II. Cross reactivation. Virology **13:**460–476.

Barrington, L., and L. M. Kozloff. 1954. Action of T2r+ bacteriophage on host cell membranes. Science **120:**110–111.

Barry, J., and B. Alberts. 1972. Bacteriophage T4 DNA replication: purification of the complex specified by T4 genes 44 and 62. Proc. Natl. Acad. Sci. U.S.A. **69:**2717–2721.

Barry, J., H. Hama-Inaba, L. Moran, B. M. Alberts, and J. Wiberg. 1973. Proteins of the bacteriophage replication apparatus, p. 195–215. *In* R. D. Wells and R. B. Inman (ed.), DNA synthesis in vitro. University Park Press, Baltimore.

Baumann, L., W. C. Benz, A. Wright, and E. B. Goldberg. 1970. Inactivation of urea-treated phage T4 by phosphatidylglycerol. Virology **41:**356–364.

Bautz, E. K. F., and F. A. Bautz. 1970. Initiation of RNA synthesis: the function of sigma in binding of RNA polymerase to promoter sites. Nature (London) **226:**1219–1222.

Bautz, E. K. F., F. A. Bautz, and J. J. Dunn. 1969. *E. coli* sigma factor: a positive control element in phage T4 development. Nature (London) **223:**1022–1024.

Bautz, E. K. F., T. Kasai, E. Reilly, and F. A. Bautz. 1966. Gene-specific mRNA. II. Regulation of mRNA synthesis in *E. coli* after infection with bacteriophage T4. Proc. Natl. Acad. Sci. U.S.A. **55:**1081–1088.

Bautz, F. A., and E. K. F. Bautz. 1967. Mapping of deletions in a nonessential region of the phage T4 genome. J. Mol. Biol. **28:**345–355.

Bayer, M. E. 1968. Adsorption of bacteriophage to adhesions between cell wall and membrane of *Escherichia coli*. J. Virol. **2:**346–356.

Beadle, G. W. 1945. Genetics and metabolism in *Neurospora*. Physiol. Rev. **25:**643–663.

Beadle, G. W. 1948. Some recent developments in chemical genetics. Fortschr. Chem. Org. Naturst. **5:**300–330.

Beckendorf, S. K. 1973. Structure of the distal half of the bacteriophage T4 tail fiber. J. Mol. Biol. **73:**37–53.

Beckendorf, S. K., J. S. Kim, and I. Lielausis. 1973. Structure of bacteriophage T4 genes 37 and 38. J. Mol. Biol. **73:**17–35.

Beckendorf, S. K., and J. H. Wilson. 1972. A recombination gradient in bacteriophage T4 gene 34. Virology **50:**315–321.

Becker, A., and J. Hurwitz. 1967. The enzymatic cleavage of phosphate termini from polynucleotides. J. Biol. Chem. **242:**936–950.

Becker, E. F, Jr., B. K. Zimmerman, and E. P. Geiduschek. 1964. Structure and function of cross-linked DNA. I. Reversible denaturation and *Bacillus subtilis* transformation. J. Mol. Biol. **8:**377–391.

Beckey, A. D., J. L. Wulff, and C. F. Earhart. 1974. Early synthesis of membrane protein after bacteriophage T4 infection. J. Virol. **14:**886–894.

Beckwith, J. 1963. Restoration of operon activity by suppressors. Biochim. Biophys. Acta **76:**162–164.

Behme, M. T., and K. Ebisuzaki. 1975. Characterization of a bacteriophage T4 mutant lacking DNA-dependent ATPase. J. Virol. **15:**50–54.

Behme, M. T., G. D. Lilley, and K. Ebisuzaki. 1976. Postinfection control by bacteriophage T4 of *Escherichia coli recBC* nuclease activity. J. Virol. **18:**20–25.

Belfort, M., A. Moelleken, G. F. Maley, and F. Maley. 1983. Purification and properties of T4 phage thymidylate synthetase produced by the cloned gene in an amplification vector. J. Biol. Chem. **258:**2045–2051.

Belin, D., J. Hedgepeth, G. B. Selzer, and R. H. Epstein. 1979. Temperature-sensitive mutation in the initiation codon of the *r*IIB gene of bacteriophage T4. Proc. Natl. Acad. Sci. U.S.A. **76:**700–704.

Bello, L. J., and M. J. Bessman. 1963a. The enzymology of virus-infected bacteria. IV. Purification and properties of the deoxynucleotide kinase induced by bacteriophage T2. J. Biol. Chem. **238:**1777–1787.

Bello, L. J., and M. J. Bessman. 1963b. The enzymology of normal and virus-infected bacteria. V. Phosphorylation of hydroxymethyl-dCDP and dTDP in normal and bacteriophage-infected *E. coli*. Biochim. Biophys. Acta **72:**647–652.

Benbasat, J. A., and V. Bloomfield. 1975. Joining of bacteriophage T4D heads and tails: a kinetic study by inelastic light scattering. J. Mol. Biol. **95:**335–357.

Benz, W. C., and H. Berger. 1973. Selective allele loss in mixed infections with T4 bacteriophage. Genetics **73:**1–11.

Benz, W. C., and E. B. Goldberg. 1973. Interactions between modified phage T4 particles and spheroplasts. Virology **53:**225–235.

Benzer, S. 1955. Fine structure of a genetic region in bacteriophage. Proc. Natl. Acad. Sci. U.S.A. **41:**344–354.

Benzer, S. 1957. The elementary units of heredity, p. 70–93. *In* B. Glass (ed.), The chemical basis of heredity. The John Hopkins Press, Baltimore.

Benzer, S. 1959. On the topology of the genetic fine structure. Proc. Natl. Acad. Sci. U.S.A. **45:**1607–1620.

Benzer, S. 1961. On the topography of the genetic fine structure. Proc. Natl. Acad. Sci. U.S.A. **47:**403–415.

Benzer, S. 1962. The fine structure of the gene. Sci. Am. **206:**70–84.

Benzer, S., and S. P. Champe. 1961. Ambivalent *r*II mutants of phage T4. Proc. Natl. Acad. Sci. U.S.A. **47:**1025–1038.

Benzer, S., and S. P. Champe. 1962. A change from nonsense to sense in the genetic code. Proc. Natl. Acad. Sci. U.S.A. **48:**1114–1121.

Benzer, S., and E. Freese. 1958. Induction of specific mutations with 5-bromouracil. Proc. Natl. Acad. Sci. U.S.A. **44:**112–119.

Benzinger, R., L. W. Enquist, and A. Skalka. 1975. Transfection of *Escherichia coli* spheroplasts. V. Activity of the *rec*BC nuclease in rec^+ and rec^- spheroplasts measured with different forms of bacteriophage DNA. J. Virol. **15:**861–871.

Berger, H., and W. C. Benz. 1975. Repair of heteroduplex DNA in bacteriophage T4, p. 149–154. *In* P. C. Hanawalt and R. B. Setlow (ed.), Molecular mechanisms for repair of DNA. Plenum Publishing Corp., New York.

Berger, H., and A. W. Kozinski. 1969. Suppression of T4D ligase mutations by rIIA and rIIB mutations. Proc. Natl. Acad. Sci. U.S.A. **64:**897–904.

Berger, H., and D. Pardoll. 1976. Evidence that mismatched bases in heteroduplex T4 bacteriophage are recognized in vivo. J. Virol. **20:**441–445.

Berger, H., A. J. Warren, and K. E. Fry. 1969. Variations in genetic recombination due to amber mutants in T4D bacteriophage. J. Virol. **3:**171–175.

Berget, P. B., and J. King. 1978. The isolation and characterization of precursors in T4 baseplate assembly. The complex of gene 10 and gene 11 products. J. Mol. Biol. **124:**469–486.

Berget, P. B., and H. R. Warner. 1975. Identification of P48 and P54 as components of bacteriophage T4 baseplates. J. Virol. **16:**1669–1677.

Berglund, O. 1972. Ribonucleoside diphosphate reductase induced by bacteriophage T4. I. Purification and characterization. II. Allosteric regulation of substrate specificity and catalytic activity. J. Biol. Chem. **247:**7270–7281.

Berglund, O. 1975. Ribonucleoside diphosphate reductase induced by bacteriophage T4. III. Isolation and characterization of proteins B1 and B2. J. Biol. Chem. **250:**7450–7455.

Berglund, O., and A. Holmgren. 1975. Thioredoxin reductase-mediated hydrogen transfer from *E. coli* thioredoxin-(SH)2 to phage T4 thioredoxin-S2. J. Biol. Chem. **250:**2778–2782.

Berglund, O., O. Karlstrom, and P. Reichard. 1969. A new ribonucleotide reductase system after infection with phage T4. Proc. Natl. Acad. Sci. U.S.A. **62:**2735–2739.

Berglund, O., and B. M. Sjöberg. 1970. A thioredoxin induced by bacteriophage T4. II. Purification and characterization. J. Biol. Chem. **245:**6030–6035.

Berk, A. J., and P. A. Sharp. 1977. Sizing and mapping of early adenovirus mRNAs by gel electrophoresis of S1 endonuclease-digested hybrids. Cell **12:**721–732.

Berk, A. J., and P. A. Sharp. 1978. Spliced early mRNAs of simian virus 40. Proc. Natl. Acad. Sci. U.S.A. **75:**1274–1278.

Bernardi, A., and P. F. Spahr. 1972. Nucleotide sequence at the binding site for coat protein on RNA of bacteriophage R17. Proc. Natl. Acad. Sci. U.S.A. **69:**3033–3037.

Berne, B. J., and R. Pecora. 1976 Dynamic light scattering with applications to chemistry, biology, and physics. John Wiley & Sons, Inc., New York.

Berns, K., and C. Thomas, Jr. 1961. A study of single polynucleotide chains derived from T2 and T4 bacteriophage. J. Mol. Biol. **3:**289–300.

Bernstein, C. 1979. Why are babies young? Meiosis may prevent aging of the germ line. Perspect. Biol. Med. **22:**539–544.

Bernstein, C. 1981. Deoxyribonucleic acid repair in bacteriophage T4. Microbiol. Rev. **45:**72–98.

Bernstein, C., and H. Bernstein. 1974. Coiled rings of DNA released from cells infected with bacteriophage T7 or T4 or from uninfected *Escherichia coli*. J. Virol. **13:**1346–1355.

Bernstein, C., H. Bernstein, S. Mufti, and B. Strom. 1972. Stimulation of mutation on phage T4 by lesions in gene 32 and by thymidine imbalance. Mutat. Res. **16:**113–119.

Bernstein, C., D. Morgan, H. L. Gensler, S. Schneider, and G. E. Holmes. 1976. The dependence of HNO_2 mutagenesis in phage T4 on ligase and the lack of dependence of 2AP mutagenesis on repair functions. Mol. Gen. Genet. **148:**213–220.

Bernstein, H. 1968. Repair and recombination in phage T4. I. Genes affecting recombination. Cold Spring Harbor Symp. Quant. Biol. **33:**325–332.

Bernstein, H. 1971. Reversion of frameshift mutations stimulated by lesions in early function genes of bacteriophage T4. J. Virol. **7:**460–466.

Bernstein, H. 1977. Germ line recombination may be primarily a manifestation of DNA repair process. J. Theor. Biol. **69:**371–380.

Bernstein, H., G. S. Byers, and R. E. Michod. 1981. Evolution of sexual reproduction: importance of DNA repair, complementation and variation. Am. Nat. **117:**537–549.

Bertrand, K., I. Korn, F. Lee, T. Platt, C. L. Squires, C. Squires, and C. Yanofsky. 1975. New features of the regulation of the tryptophan operon. Science **189:**22–26.

Bessman, M. J. 1959. Deoxyribonucleotide kinases in normal and virus-infected *E. coli*. J. Biol. Chem. **234:**2735–2739.

Bessman, M. J., N. Muzyczka, M. F. Goodman, and R. L. Schnaar. 1974. Studies on the biochemical basis of spontaneous mutation. J. Mol. Biol. **88:**409–421.

Bijlenga, R. K. L., U. Aebi, and E. Kellenberger. 1976. Properties and structure of a gene 24-controlled T4 giant phage. J. Mol. Biol. **103:**469–498.

Bijlenga, R. K. L., D. Scraba, and E. Kellenberger. 1973. Studies on the morphogenesis of T-even phage. IX. Tau-particles: their morphology, kinetics of appearance and possible precursor functions. Virology **56:**250–267.

Bijlenga, R. K. L., R. van der Broek, and E. Kellenberger. 1974. The transformation of tau-particles into T4 heads. J. Supramol. Struct. **2:**45–59.

Bingham, P. M., R. H. Baltz, L. S. Ripley, and J. W. Drake. 1976. Heat mutagenesis in bacteriophage T4: the transversion pathway. Proc. Natl. Acad. Sci. U.S.A. **73:**4159–4163.

Bishop, R. J., M. P. Conley, and W. B. Wood. 1974. Assembly and attachment of bacteriophage T4 tail fibers. J. Supramol. Struct. **2:**196–201.

Bishop, R. J., and W. B. Wood. 1976. Genetic analysis of T4 tail fiber assembly. I. A gene 37 mutation that allows bypass of gene 38 function. Virology **72:**244–254.

Bittner, M., R. Burke, and B. Alberts. 1979. Purification of the T4 gene 32 protein free from detectable deoxyribonuclease activities. J. Biol. Chem. **254:**9565–9572.

Black, L. W. 1974. Bacteriophage T4 internal protein mutants: isolation and properties. Virology **60:**166–179.

Black, L. W. 1981a. The mechanism of bacteriophage DNA encapsidation, p. 97–110. *In* M. DuBow (ed.), bacteriophage assembly. Alan R. Liss, Inc., New York.

Black, L. W. 1981b. In vitro packaging of bacteriophage T4 DNA. Virology **113:**336–344.

Black, L. W., and K. Abremski. 1974. Restriction of phage T4 internal protein I mutants by a strain of *Escherichia coli*. Virology **60:**180–191.

Black, L. W., and Ahmad-Zadeh. 1971. Internal proteins of bacteriophage T4: their characterization and relation to head structure and assembly. J. Mol. Biol. **57:**71–92.

Black, L. W., and D. T. Brown. 1976. Head morphologies in bacteriophage T4 head and internal protein mutant infections. J. Virol. **17:**894–905.

Black, L. W., and L. M. Gold. 1971. Pre-replicative development of the bacteriophage T4: RNA and protein and synthesis in vivo and in vitro. J. Mol. Biol. **60:**365–388.

Black, L. W., and D. J. Silverman. 1978. Model for DNA packaging into bacteriophage T4 heads. J. Virol. **28:**643–655.

Black, L. W., A. L. Zachary, and V. Manne. 1981. Studies of the mechanism of bacteriophage T4 DNA encapsidation, p. 111–126. *In* M. DuBow (ed.), Bacteriophage assembly. Alan R. Liss, Inc., New York.

Blake, C. C. F., D. F. Koenig, G. A. Mair, A. C. T. North, D. C. Phillips, and V. R. Sarma. 1965. Structure of hen egg white lysozyme. Nature (London) **206:**757–761.

Bleichrodt, J. F., and W. S. D. Roos-Verheij. 1979. γ-Ray mutagenesis in bacteriophage T4 is not enhanced by oxygen. Mutat. Res. **61:**121–127.

Bloomfield, V. A., and T. K. Lim. 1978. Quasi-elastic laser light scattering. Methods Enzymol. **48:**415–494.

Bloomfield, V. A., and S. Prager. 1979. Diffusion-controlled reactions on spherical surfaces: application to bacteriophage tail fiber attachment. Biophys. J. **27:**447–453.

Bolle, A., R. H. Epstein, W. Salser, and E. P. Geiduschek. 1968a. Transcription during bacteriophage T4 development: synthesis and relative stability of early and late RNA. J. Mol. Biol. **31:**325–348.

Bolle, A., R. H. Epstein, W. Salser, and E. P. Geiduschek. 1968b. Transcription during bacteriophage T4 development: requirements for late messenger synthesis. J. Mol. Biol. **33:**339–362.

Bolund, C. 1973. Influence of gene 55 on the regulation of synthesis of some early enzymes in bacteriophage T4-infected *E. coli*. J. Virol. **12:**49–57.

Bonifas, V., and E. Kellenberger. 1955. Etude de l'action des membranes du bacteriophage T2 sur *Escherichia coli*. Biochim. Biophys. Acta **16:**330–338.

Bonner, W. M., and R. A. Laskey. 1974. A film detection method for tritium-labeled proteins and nucleic acids in polyacrylamide gels. Eur. J. Biochem. **46:**83–88.

Boon, T., and N. D. Zinder. 1971. Genotypes produced by individual recombination events involving bacteriophage f1. J. Mol. Biol. **58:**133–151.

Borisova, G. P., T. M. Volkova, V. Berzin, G. Rosenthal, and E. J. Gren. 1979. The regulatory region of MS2 phage RNA replicase cistron. IV. Functional activity of specific MS2 RNA fragments in formation of the 70S initiation complex of protein biosynthesis. Nucleic Acids Res. **6:**1761–1774.

Botstein, D., and M. J. Matz. 1970. A recombination function essential to the growth of phage P22. J. Mol. Biol. **54:**417–440.

Botstein, D., C. H. Waddell, and J. King. 1973. Mechanism of head assembly and DNA encapsulation in *Salmonella* phage P22. I. Genes, proteins, structures, and DNA maturation. J. Mol. Biol. **80:**669–695.

Bowen, G. H. 1953a. Studies on ultraviolet irradiation phenomena—an approach to the problems of bacteriophage reproduction. Cold Spring Harbor Symp. Quant Biol. **18:**245–253.

Bowen, G. H. 1953b. Kinetic studies on the mechanism of photoreactivation of bacteriophage T2 inactivated by ultraviolet light. Ann. Inst. Pasteur Paris **84:**218–221.

Bowles, M., and J. Karam. 1979. Expression of bacteriophage T4 gene 45, 44 and 62. II. A possible regulatory site between genes 45 and 44. Virology **94:**204–207.

Boy de la Tour, E., and E. Kellenberger. 1965. Aberrant forms of the T-even phage head. Virology **27:**222–225.

Boyle, J. M. 1969. Radiation-sensitive mutants of T4D. II. T4y:

genetic characterization. Mutat. Res. 8:441–449.

Boyle, J. M., and N. Symonds. 1969. Radiation-sensitive mutants of T4D. I. T4v: a new radiation-sensitive mutant; effect of the mutation on radiation survival, growth and recombination. Mutat. Res. 8:431–439.

Bradley, D. E. 1967. Ultrastructure of bacteriophages and bacteriocins. Bacteriol. Rev. 31:230–314.

Branton, D., and A. Klug. 1975. Capsid geometry of bacteriophage T2: freeze-etching study. J. Mol. Biol. 92:559–565.

Bräutigam, A. R., and W. Sauerbier. 1974. Transcription unit mapping in bacteriophage T7. II. Proportionality of number of gene copies, mRNA, and gene products. J. Virol. 13:1110–1117.

Bremer, H., and D. Yuan. 1968. RNA chain growth-rate in *Escherichia coli*. J. Mol. Biol. 38:163–180.

Brenner, S. 1957. On the impossibility of all overlapping triplet codes in information transfer from nucleic acid to proteins. Proc. Natl. Acad. Sci. U.S.A. 43:687–694.

Brenner, S., L. Barnett, F. H. C. Crick, and A. Orgel. 1961. The theory of mutagenesis. J. Mol. Biol. 3:121–124.

Brenner, S., and J. R. Beckwith. 1965. Ochre mutants, a new class of suppressible nonsense mutants. J. Mol. Biol. 13:629–637.

Brenner, S., S. Benzer, and L. Barnett. 1958. Distribution of proflavin-induced mutations in the genetic fine structure. Nature (London) 182:983–985.

Brenner, S., G. Streisinger, R. W. Horne, S. P. Champe, L. Barnett, S. Benzer, and M. W. Rees. 1959. Structural components of bacteriophage. J. Mol. Biol. 1:281–292.

Brenner, S., and A. O. W. Stretton. 1965. Phase shifting of amber and ochre mutants. J. Mol. Biol. 13:944–946.

Brenner, S., A. O. W. Stretton, and S. Kaplan. 1965. Genetic code: the "nonsense" triplets for chain termination and their suppression. Nature (London) 206:994–998.

Breschkin, A. M., and G. Mosig. 1977a. Multiple interactions of DNA-binding protein in vivo. I. Gene 32 mutations of phage T4 inactivate different steps in DNA replication and recombination. J. Mol. Biol. 112:279–294.

Breschkin, A. M., and G. Mosig. 1977b. Multiple interactions of a DNA-binding protein in vivo. II. Effects of host mutations on DNA replication of phage T4 gene 32 mutants. J. Mol. Biol. 112:295–308.

Brody, E. N. 1978. The role of termination factor rho in the development of bacteriophage T4. Arch. Int. Phys. Biochim. 86:897–898.

Brody, E. N., L. W. Black, and L. M. Gold. 1971. Transcription and translation of sheared bacteriophage T4 DNA in vitro. J. Mol. Biol. 60:389–393.

Brody, E. N., H. Diggelmann, and E. P. Geiduschek. 1970. Transcription of the bacteriophage T4 template. Obligate synthesis of T4 prereplicative RNA in vitro. Biochemistry 9:1289–1299.

Brody, E. N., and E. P. Geiduschek. 1970. Transcription of the bacteriophage T4 template. Detailed comparison of in vitro and in vivo transcripts. Biochemistry 9:1300–1309.

Brody, E., R. Sederoff, A. Bolle, and R. H. Epstein. 1970. Early transcription in T4-infected cells. Cold Spring Harbor Symp. Quant. Biol. 35:203–211.

Broker, T. R. 1973. An electron microscopic analysis of pathways for bacteriophage T4 DNA recombination. J. Mol. Biol. 81:1–16.

Broker, T. R., and A. H. Doermann. 1975. Molecular and genetic recombination of bacteriophage T4. Annu. Rev. Genet. 9:213–244.

Broker, T. R., and I. R. Lehman. 1971. Branched DNA molecules: intermediates in T4 recombination. J. Mol. Biol. 60:131–149.

Brooks, J. E., and S. Hattman. 1973. Location of the DNA-adenine methylase gene on the genetic map of phage T2. Virology 55:285–288.

Brooks, J. E., and S. Hattman. 1978. In vitro methylation of bacteriophage lambda DNA by wild-type (dam^+) and mutant (dam^h) forms of phage T2 DNA adenine methylase. J. Mol. Biol. 126:381–394.

Brooks, P., and P. D. Lawley. 1963. Effects of alkylating agents on T2 and T4 bacteriophages. Biochem. J. 89:138–144.

Brown, P. O., and N. R. Cozzarelli. 1981. Catenation and knotting of duplex DNA by type 1 topoisomerases: a mechanistic parallel with type 2 topoisomerases. Proc. Natl. Acad. Sci. U.S.A. 78:843–847.

Brown, S. M., and F. A. Eiserling. 1979a. T4 gene 40 mutants. I. Isolation of new mutants. Virology 97:68–76.

Brown, S. M., and F. A. Eiserling. 1979b. T4 gene 40 mutants. II. Phenotypic properties. Virology 97:77–89.

Bruner, R., D. Solomon, and T. Berger. 1973. Presumptive D2a point mutants of bacteriophage T4. J. Virol. 12:946–947.

Bruner, R., A. Souther, and S. Suggs. 1972. Stability of cytosine-containing deoxyribonucleic acid after infection by certain T4 *r*II-D deletion mutants. J. Virol. 10:88–92.

Brutlag, D., and A. Kornberg. 1972. Enzymatic synthesis of deoxyribonucleic acid. XXXVI. A proofreading function for the 3′, 5′ exonuclease activity in DNA polymerase. J. Biol. Chem. 247:241–248.

Buckley, P., L. Kosturko, and A. W. Kozinski. 1972. In vivo production of the RNA-DNA copolymer after infection of *E. coli* by bacteriophage T4. Proc. Natl. Acad. Sci. U.S.A. 69:3165–3169.

Bujard, H., A. J. Mazaitis, and E. K. F. Bautz. 1970. The size of the *r*II region of bacteriophage T4. Virology 42:717–723.

Burke, R., B. Alberts, and J. Hosoda. 1980. Proteolytic removal of the COOH terminus of the T4 gene 32 helix-destabilizing protein alters the T4 in vitro replication complex. J. Biol. Chem. 255:11484–11493.

Burton, Z., R. R. Burgess, J. Lin, D. Moore, S. Holder, and C. A. Gross. 1981. The nucleotide sequence of the cloned *rpoD* gene for the RNA polymerase sigma subunit from *E. coli* K12. Nucleic Acids Res. 9:2889–2903.

Cairns, J., G. S. Stent, and J. D. Watson (ed.). 1966. Phage and the origins of molecular biology. Cold Spring Harbor Laboratory, Cold Spring Harbor, N.Y.

Cameron, V., D. Soltis, and O. C. Uhlenbeck. 1978. Polynucleotide kinase from a mutant which lacks the 3′-phosphatase activity. Nucleic Acids Res. 5:825–833.

Cameron, V., and O. C. Uhlenbeck. 1977. 3′-Phosphatase activity in T4 polynucleotide kinase. Biochemistry 16:5120–5126.

Campbell, A. 1961. Sensitive mutants of bacteriophage lambda. Virology 14:22–32.

Campbell, J. H., and B. G. Rolfe. 1975. Evidence for a dual control of the initiation of host-cell lysis caused by phage lambda. Mol. Gen. Genet. 139:1–8.

Campbell, K., and L. Gold. 1982. Construction of the bacteriophage T4 replication machine: synthesis of component proteins, p. 69–83. *In* M. Grunberg-Manago and B. Safer (ed.), Interactions of translational and transcriptional controls in the regulation of gene expression. Elsevier/North-Holland Publishing Co., Amsterdam.

Canfield, R. E., S. Kammesman, J. H. Sobel, and F. J. Morgan. 1971. Primary structure of lysozymes from man and goose. Nature (London) New Biol. 232:16–17.

Capco, G. R., and C. K. Mathews. 1973. Bacteriophage-coded thymidylate synthetase: evidence that the T4 enzyme is a capsid protein. Arch. Biochem. Biophys. 158:736–743.

Cardillo, T. S., E. F. Landry, and J. S. Wiberg. 1979. *regA* protein of bacteriophage T4D: identification, schedule of synthesis, and autogenous regulation. J. Virol. 32:905–916.

Carlson, K. 1973. Multiple initiation of bacteriophage T4 DNA replication: delaying effect of bromodeoxyuridine. J. Virol. 12:349–359.

Carlson, K. 1980. Correlation between genetic map and map of cleavage sites for sequence-specific endonucleases *Sal*I, *Kpn*I, *Bgl*I, and *Bam*HI in bacteriophage T4 cytosine-containing DNA. J. Virol. 36:1–17.

Carlson, K., and B. Nicolaisen. 1979. Cleavage map of bacteriophage T4 cytosine-containing DNA by sequence-specific endonucleases *Sal*I and *Kpn*I. J. Virol. 31:112–123.

Carrascosa, J. L. 1978. Head maturation pathway of bacteriophage T4 and T2. IV. In vitro transformation of T4 head-related particles produced by mutants in gene 17 to capsid-like structures. J. Virol. 26:420–428.

Carrascosa, J. L., J. M. Carazo, and N. Garcia. 1983. Structural localization of the proteins of the head to tail connecting region of bacteriophage 29. Virology 124:133–143.

Carrascosa, J. L., J. A. Garcia, and M. Salas. 1982. A protein similar to *Escherichia coli groEL* is present in *Bacillus subtilis*. J. Mol. Biol. 158:731–737.

Carrascosa, J. L., and E. Kellenberger. 1978. Head maturation pathway of bacteriophage T4 and T2. III. Isolation and characterization of particles produced by mutants in gene 17. J. Virol. 25:831–844.

Carroll, R., K. Neet, and D. Goldthwait. 1975. Studies of the self-association of bacteriophage T4 gene 32 protein by equilibrium sedimentation. J. Mol. Biol. 91:275–291.

Caruso, M., A. Coppo, A. Manzi, and J. Pulitzer. 1979. Host-virus interactions in the control of T4 prereplicative transcription. I. *tabC* (*rho*) mutants. J. Mol. Biol. 135:950–977.

Cascino, A., M. Cipollaro, A. M. Guerrini, G. Mastrocinque, A. Spena, and V. Scarlato. 1981. Coding capacity of complementary DNA strands. Nucleic Acids Res. 9:1499–1518.

Cascino, A., S. Riva, and E. P. Geiduschek. 1970. DNA ligation and the coupling of late T4 transcription to replication. Cold Spring Harbor Symp. Quant. Biol. 35:213–220.

Casjens, S. 1974. Bacteriophage lambda FII gene protein: role in head assembly. J. Mol. Biol. 90:1–23.

Casna, M. J., and D. A. Shub. 1982. Bacteriophage T4 as a generalized DNA cloning vehicle. Gene 18:297–307.

Caspar, D. L. D. 1980. Movement and self-control in protein assemblies. Quasi-equivalence revisited. Biophys. J. 32:103–135.

Caspar, D. L. D., and A. Klug. 1962. Physical principles in the construction of regular viruses. Cold Spring Harbor Symp. Quant. Biol. 27:1–24.

Castillo, C. J., and L. W. Black. 1978. Purification and properties of the bacteriophage T4 gene 31 protein required for prehead assembly. J. Biol. Chem. **253**:2132–2139.

Castillo, C. J., C. L. Hsiao, P. Coon, and L. W. Black. 1977. Identification and properties of bacteriophage T4 capsid-formation gene products. J. Mol. Biol. **110**:585–601.

Celis, J. E., J. D. Smith, and S. Brenner. 1973. Correlation between genetic and translational maps of gene 23 in bacteriophage T4. Nature (London) New Biol. **241**:130–132.

Cerda-Olmedo, E., P. C. Hanawalt, and N. Guerola. 1968. Mutagenesis of the replication point by nitrosoguanidine: map and pattern of replication of the *Escherichia coli*. J. Mol. Biol. **33**:705–719.

Chace, K. V., and D. H. Hall. 1973. Isolation of mutants of bacteriophage T4 unable to induce thymidine kinase activity. J. Virol. **12**:343–348.

Chace, K. V., and D. H. Hall. 1975. Characterization of new regulatory mutants of bacteriophage T4. II. New class of mutants. J. Virol. **15**:929–945.

Challberg, M. D., and P. T. Englund. 1979. The effect of template secondary structure on vaccinia DNA polymerase. J. Biol. Chem. **254**:7820–7826.

Chamberlin, M. 1982. Bacterial DNA-dependent RNA polymerases, p. 68–86. *In* P. D. Boyer (ed.), The enzymes, vol. 15. Academic Press, Inc., New York.

Champe, S. P. 1974. Linkage map of bacteriophage T4, p. 644–647. *In* A. I. Laskin and H. A. Lechevalier (ed.), Handbook of microbiology, vol. 14. CRC Press, Cleveland.

Champe, S. P., and S. Benzer. 1962a. An active cistron fragment. J. Mol. Biol. **4**:288–292.

Champe, S. P., and S. Benzer. 1962b. Reversal of mutant phenotypes by 5-flourouracil: an approach to nucleotide sequences in messenger RNA. Proc. Natl. Acad. Sci. U.S.A. **48**:532–546.

Champe, S. P., and H. L. Eddleman. 1967. Polypeptides associated with morphogenetic defects in bacteriophage T4, p. 55–70. *In* J. S. Colter and W. Paranchych (ed.), The molecular biology of viruses. Academic Press, Inc., New York.

Champney, W. S. 1980. Protein synthesis defects in temperature sensitive mutants of *Escherichia coli* with altered ribosomal proteins. Biochim. Biophys. Acta **609**:464–474.

Chan, V. L., and K. Ebisuzaki. 1970. Polynucleotide kinase mutant of bacteriophage T4. Mol. Gen. Genet. **109**:162–168.

Chan, V. L., and K. Ebisuzaki. 1973. Intergenic suppression of amber polynucleotide ligase mutations in bacteriophage T4. Virology **53**:60–74.

Chao, J., L. Chao, and F. Speyer. 1974. Bacteriophage T4 head morphogenesis: host DNA enzymes affect the frequency of petite forms. J. Mol. Biol. **85**:41–50.

Chao, J., M. Leach, and J. Karam. 1977. In vivo functional interaction between DNA polymerase and dCMP-hydroxymethylase of bacteriophage T4. J. Virol. **24**:557–563

Charette, M., G. W. Henderson, and A. Markovitz. 1981. ATP hydrolysis-dependent protease activity of the *lon* (*capR*) protein of *Escherichia coli* K-12. Proc. Natl. Acad. Sci. U.S.A. **78**:4728–4732.

Chase, J. W., and C. C. Richardson. 1974. Exonuclease VII of *Escherichia coli*. Mechanism of action. J. Biol. Chem. **249**:4553–4561.

Chase, M. C., and A. H. Doermann. 1958. High negative interference over short segments of the genetic structure of bacteriophage T4. Genetics **43**:332–353.

Childs, J. D. 1971. A map of molecular distances between mutations of bacteriophage T4D. Dept. Genetics U.W. **67**:455–468.

Childs, J. D. 1973. Superinfection exclusion by incomplete genomes of bacteriophage T4. J. Virol. **11**:1–8.

Childs, J. D. 1980a. Effect of hoc protein on the electrophoretic mobility of intact bacteriophage T4 D particles in polyacrylamide gel electrophoresis. J. Mol. Biol. **141**:163–173.

Childs, J. D. 1980b. Isolation and genetic properties of a bacteriophage T4 *uvsX* mutant. Mutat. Res. **71**:1–14.

Childs, J. D., and H. C. Birnboim. 1975. Polyacrylamide gel electrophoresis of intact bacteriophage T4D particles. J. Virol. **16**:652–661.

Childs, J. D., M. C. Paterson, B. P. Smith, and N. E. Gentner. 1978. Evidence for a near UV-induced photoproduct of 5-hydroxymethylcytosine in bacteriophage T4 that can be recognized by endonuclease V. Mol. Gen. Genet. **167**:105–112.

Chiu, C.-S., K. S. Cook, and G. R. Greenberg. 1982. Characteristics of bacteriophage T4-induced complex synthesizing deoxyribonucleotides. J. Biol. Chem. **257**:15087–15097.

Chiu, C.-S., S. M. Cox, and G. R. Greenberg. 1980. Effect of bacteriophage T4 *nrd* mutants on deoxynucleotide synthesis in vivo. J. Biol. Chem. **255**:2747–2751.

Chiu, C.-S., and G. R. Greenberg. 1968. Evidence for a possible direct role of dCMP hydroxymethylase in T4 phage DNA synthesis. Cold Spring Harbor Symp. Quant. Biol. **33**:351–357.

Chiu, C.-S., and G. R. Greenberg. 1973. Mutagenic effects of temperature-sensitive mutants of gene 42 of bacteriophage T4. J. Virol. **12**:199–201.

Chiu, C.-S., T. Ruettinger, J. B. Flanegan, and G. R. Greenberg. 1977. Role of deoxycytidylate deaminase in deoxyribonucleotide synthesis in bacteriophage T4 DNA replication. J. Biol. Chem. **252**:8603–8608.

Chou, P. Y., and G. D. Fasman. 1978a. Empirical predictions of protein conformation. Annu. Rev. Biochem. **47**:251–276.

Chou, P. Y., and G. D. Fasman. 1978b. Prediction of the secondary structure of proteins from their amino acid sequence. Adv. Enzymol. **47**:45–148.

Christensen, A. C., and E. T. Young. 1982a. Bacteriophage T4 late gene transcription. Fed. Proc. **41**:1293.

Christensen, A. C., and E. T. Young. 1982b. T4 late transcripts are initiated near a conserved DNA sequence. Nature (London) **299**:369–371.

Christian, J. H. B., and J. A. Waltho. 1962. Solute concentration within cells of halophilic bacteria. Biochim. Biophys. Acta **65**:506–508.

Chu, J., and R. Mazumder. 1974. Release of IF2 from native ribosomes by dilution. FEBS Lett. **40**:334–338.

Ciarrocchi, G., and S. Linn. 1978. A cell free assay measuring repair DNA synthesis in human fibroblasts. Proc. Natl. Acad. Sci. U.S.A. **75**:1887–1891.

Clark, A. J., and A. D. Margulies. 1965. Isolation and characterization of recombinant-deficient mutants of *Escherichia coli* K-12. Proc. Natl Acad Sci. U.S.A. **53**:451–458.

Clayton, L. K., M. F. Goodman, E. W. Branscomb, and D. J. Galas. 1979. Error induction and correction by mutant and wild type T4 DNA polymerases. Kinetic error discrimination mechanisms. J. Biol. Chem. **254**:1902–1912.

Cohen, P. S. 1972. Translation regulation of deoxycytidylate hydroxymethylase and deoxynucleotide kinase synthesis in T4-infected *Escherichia coli*. Virology **47**:780–786.

Cohen, P. S., and H. L. Ennis. 1965. The requirement for potassium for bacteriophage T4 protein and deoxyribonucleic acid synthesis. Virology **27**:282–289.

Cohen, P. S., P. J. Natale, and J. M. Buchanan. 1974. Transcriptional regulation of T4 bacteriophage-specific enzymes synthesized in vitro. J. Virol. **14**:292–299.

Cohen, P. S., B. R. Zetter, and M. L. Walsh. 1972. Evidence that more deoxynucleotide kinase in mRNA is transcribed than translated during T4 infection of *Escherichia coli*. Virology **49**:808–810.

Cohen, S. S. 1947. The synthesis of bacterial viruses in infected cells. Cold Spring Harbor Symp. Quant. Biol. **12**:35–49.

Cohen, S. S. 1948. The synthesis of bacterial viruses. II. The origin of the phosphorus found in deoxyribonucleic acids of the T2 and T4 bacteriophages. J. Biol. Chem. **174**:295–303.

Cohen, S. S. 1968. Virus-induced enzymes. Columbia University Press, New York.

Cohen, S. S., and T. F. Anderson. 1946. Chemical studies on host-virus interactions. I. The effects of bacteriophage adsorption on multiplication of its host, *Escherichia coli*. Br. J. Exp. Med. **84**:511–523.

Coleman, J., and J. Oakley. 1980. Physical chemical studies of the structure and function of DNA binding (helix-destabilizing) proteins. Crit. Rev. Biochem. **7**:247–289.

Collinsworth, W. L., and C. K. Mathews. 1974. Biochemistry of DNA-defective amber mutants of bacteriophage T4. IV. DNA synthesis in plasmolyzed cells. J. Virol. **13**:908–915.

Comer, M. M., C. Guthrie, and W. H. McClain. 1974. An ochre suppressor of bacteriophage T4 that is associated with a transfer RNA. J. Mol. Biol. **90**:665–676.

Comer, M. M., and F. C. Neidhardt. 1975. Effect of T4 modification of host valyl-tRNA synthetase on enzyme action in vivo. Virology **67**:395–403.

Conkling, M. A., J. A. Grunau, and J. W. Drake. 1976. Gamma-ray mutagenesis in bacteriophage T4. Genetics **82**:565–575.

Conkling, M. A., R. E. Koch, and J. W. Drake. 1980. Determination of mutation rates in bacteriophage T4 by unneighborly base pairs: genetic analysis. J. Mol. Biol. **143**:303–315.

Conley, M. P., and W. B. Wood. 1975. Bacteriophage T4 whiskers: a rudimentary environment-sensing device. Proc. Natl. Acad. Sci. U.S.A. **72**:3701–3705.

Conrad, S. E., and J. L. Campbell. 1979. Role of plasmid-coded RNA and ribonuclease III in plasmid DNA replication. Cell **18**:61–71.

Cook, K. S., and A. F. Seasholtz. 1982. Identification of some bacteriophage T4 prereplicative proteins on two-dimensional gel patterns. J. Virol. **42**:767–772.

Cooley, W. C., K. Sirotkin, R. Green, and L. Snyder. 1979. A new gene of *Escherichia coli* K-12 whose product participates in T4 bacteriophage late gene expression: interaction of *lit* with the T4-induced polynucleotide 5′-kinase 3′-phosphatase. J. Bacteriol. **140**:83–91.

Cooley Champness, W., and L. Snyder. 1982. The *gol* site: a *cis*-acting bacteriophage T4 regulatory region that can affect expression of all the T4 late genes. J. Mol. Biol. **155**:395–407.

Coombs, D., and F. A. Eiserling. 1977. Studies on the structure, protein composition, and assembly of the neck of bacteriophage T4. J. Mol. Biol. **116**:375–405.

Coppo, A., A. Manzi, and J. F. Pulitzer. 1975b. Host mutant (*tabD*)-induced inhibition of bacteriophage T4 late transcription. II. Genetic characterization of mutants. J. Mol. Biol. **96**:601–624.

Coppo, A., A. Manzi, J. F. Pulitzer, and H. Takahashi. 1973. Abortive bacteriophage T4 head assembly in mutants of *Escherichia coli*. J. Mol. Biol. **76**:61–87.

Coppo, A., A. Manzi, J. F. Pulitzer, and H. Takahashi. 1975a. Host mutant (*tabD*)-induced inhibition of bacteriophage T4 late transcription. I. Isolation and phenotypic characterization of the mutants. J. Mol. Biol. **96**:579–600.

Cornett, J. B., and M. Vallée. 1973. The map position of the immunity (*imm*) gene of bacteriophage T4. Virology **51**:506–508.

Cowie, D. B., R. J. Avery, and S. P. Champe. 1971. DNA homology among the T-even bacteriophages. Virology **45**:30–37.

Cox, G. S., and T. W. Conway. 1973. Template properties of glucose deficient T-even bacteriophage DNA. J. Virol. **12**:1279–1287.

Cozzarelli, N. R. 1980. DNA gyrase and the supercoiling of DNA. Science **207**:953–960.

Cramer, W. A., and R. B. Uretz. 1966. Acridine orange-sensitive photoinactivation of T4 bacteriophage. II. Genetic studies with photoinactivated phage. Virology **29**:469–479.

Crawford, J. T., and E. B. Goldberg. 1977. The effect of baseplate mutations on the requirement for tail-fiber binding for irreversible adsorption of bacteriophage T4. J. Mol. Biol. **111**:305–313.

Crawford, J. T., and E. B. Goldberg. 1980. The function of tail fibers in triggering baseplate expansion of bacteriophage T4. J. Mol. Biol. **139**:679–690.

Crick, F. 1958. On protein synthesis. Symp. Soc. Exp. Biol. **12**:138–163.

Crick, F. H. C., L. Barnett, S. Brenner, and R. J. Watts-Tobin. 1961. General nature of the genetic code for proteins. Nature (London) **192**:1227–1232.

Crick, F. H. C., and S. Brenner. 1967. The absolute sign of certain phase-shift mutants in bacteriophage T4. J. Mol. Biol **26**:361–363.

Crick, F. H. C., J. S. Griffith, and L. E. Orgel. 1957. Codes without commas. Proc. Natl. Acad. Sci. U.S.A. **43**:416–421.

Crick, F. H. C., and J. D. Watson. 1957. The structure of small viruses. Nature (London) **177**:473–475.

Crick, F. H. C., and J. D. Watson. 1957. Virus structure: general principles. Ciba Found. Symp., p. 5–13.

Crowther, R. A. 1980. Mutants of bacteriophage T4 that produce infective fiberless particles. J. Mol. Biol. **137**:159–174.

Crowther, R. A., and L. A. Amos. 1971. Harmonic analysis of electron microscopic images with rotational symmetry. J. Mol. Biol. **60**:123–130.

Crowther, R. A., and A. Klug. 1975. Structural analysis of macromolecular assemblies by image reconstruction from electron micrographs. Annu. Rev. Biochem. **44**:161–182.

Crowther, R. A., E. V. Lenk, Y. Kikuchi, and J. King. 1977. Molecular reorganization in the hexagon to star transition of the baseplate of bacteriophage T4. J. Mol. Biol. **116**:489–523.

Cudny, M., P. Roy, and M. P. Deutscher. 1980. Alteration of *Escherichia coli* RNase D by infection with bacteriophage T4. Biochem. Biophys. Res. Commun. **98**:337–345.

Cummings, D. J., and R. W. Bolin. 1976. Head length control in T4 bacteriophage morphogenesis: effect of canavanine on assembly. Bacteriol. Rev. **40**:314–359.

Cummings, D. J., V. A. Chapman, and S. S. DeLong. 1977. Structural aberrations in T-even bacteriophage. IX. Effect of mixed infection on the production of giant bacteriophage. J. Virol. **22**:489–499.

Cummings, D. J., V. A. Chapman, S. S. Delong, and M. L. Couse. 1973. Structural aberrations in T-even bacteriophage. III. Induction of "lollipops" and their partial characterizations. Virology **54**:245–261.

Cunningham, R. P., and H. Berger. 1977. Mutations affecting recombination in bacteriophage T4D. I. Pathway analysis. Virology **80**:67–82.

Cunningham, R. P., and H. Berger. 1978. Mutations affecting genetic recombination in bacteriophage T4D. II. Genetic properties. Virology **88**:62–70.

Cupido, M., J. Grimbergen, and B. de Groot. 1980. Participation of bacteriophage T4 gene 41 in replication repair. Mutat. Res. **70**:131–138.

Cupido, M., O. Schreij-Visser, and P. van der Ree. 1982. A bacteriophage T4 defective in both DNA replication and replication repair. Mutat. Res. **93**:285–295.

Curtis, M., and B. Alberts. 1976. Studies on the structure of intracellular bacteriophage T4 DNA. J. Mol. Biol. **102**:793–816.

Daegelen, P., and E. Brody. 1976. Early bacteriophage T4 transcription. A diffusable product controls *r*IIA and *r*IIB RNA synthesis. J. Mol. Biol. **103**:127–142.

Daegelen, P., Y. D'Aubeuton-Carafa, and E. Brody. 1982a. The role of *rho* in bacteriophage T4 development. I. Control of growth and polarity. Virology **117**:105–120.

Daegelen, P., Y. D'Aubeuton-Carafa, and E. Brody. 1982b. The role of *rho* in bacteriophage T4 development. II. *mot*-dependent (Middle Mode) RNA synthesis. Virology **117**:121–134.

Daniel, V., S. Sarid, and U. Z. Littauer. 1968. Coding by T4 phage DNA of soluble RNA containing pseudouridylic acid. Proc. Natl. Acad. Sci. U.S.A. **60**:1403–1409.

Dannenberg, R., and G. Mosig. 1981. Semi-conservative DNA replication is initiated at a single site in recombination-deficient gene 32 mutants of bacteriophage T4. J. Virol. **40**:890–900.

Dannenberg, R., and G. Mosig. 1983. Early intermediates in bacteriophage T4 DNA replication and recombination. J. Virol. **45**:813–831.

Das, A., D. Court, and S. Adhya. 1976. Isolation and characterization of conditional lethal mutants of *Escherichia coli* defective in transcription termination factor rho. Proc. Natl. Acad. Sci. U.S.A. **73**:1959–1963.

Das, A., D. Court, and S. Adhya. 1979. Pleiotropic effect of *rho* mutation in *Escherichia coli*, p. 459–465. *In* M. Chakravorty (ed.), Molecular basis of host-virus interaction. Science Press, Princeton, N.J.

Das, S. K., and R. K. Fujimura. 1979. Processiveness of DNA polymerases, a comparative study using a simple procedure. J. Biol. Chem. **254**:1227–1232.

David, M., G. D. Borasio, and G. Kaufmann. 1982a. Bacteriophage T4-induced anticodon-loop nuclease detected in a host strain restrictive to RNA ligase mutants. Proc. Natl. Acad. Sci. U.S.A. **79**:7097–7101.

David, M., G. Borasio, and G. Kaufmann. 1982b. T4 bacteriophage-coded polynucleotide kinase and RNA ligase are involved in host tRNA alteration or repair. Virology **123**:480–483.

David, M., R. Vecksteín, and G. Kaufmann. 1979. RNA ligase reaction products in plasmolyzed *E. coli* cells infected by T4 bacteriophage. Proc. Natl. Acad. Sci. U.S.A. **76**:5430–5434.

Davis, K. J., and N. Symonds. 1974. The pathway of recombination in T4. A genetic study. Mol. Gen. Genet. **132**:173–180.

Dawes, J. 1975. Characterization of the bacteriophage T4 receptor sites. Nature (London) **256**:127–128.

Dawes, J., and E. B. Goldberg. 1973a. Function of baseplate components in bacteriophage T4 infection. I. Dihydrofolate reductase and dihydropteroylhexaglutamate. Virology **55**:380–390.

Dawes, J., and E. B. Goldberg. 1973b. Functions of baseplate components in bacteriophage T4 infection. II. Products of genes 5, 7, 8, and 10. Virology **55**:391–396.

Debye, P. 1942. Reaction rates in ionic solutions. Trans. Electrochem. Soc. **82**:265–271.

de Franciscis, V. and E. Brody. 1982. In vitro system for middle T4 RNA. I. Studies with *Escherichia coli* RNA polymerase. J. Biol. Chem. **257**:4087–4096.

de Franciscis, V., R. Favre, M. Uzan, J. Leautey, and E. Brody. 1982. In vitro system for T4 RNA. II. Studies with T4-modified RNA polymerase. J. Biol. Chem. **257**:4097–4101.

Delbrück, M. 1945. The burst-size distribution in the growth of bacterial viruses (bacteriophages). J. Bacteriol. **50**:131–135.

Delbrück, M. 1946. Bacterial viruses or bacteriophages. Biol. Rev. **21**:30–40.

Delbrück, M. 1948. Biochemical mutants of bacterial viruses. J. Bacteriol. **56**:1–16.

Delbrück, M., and W. T. Bailey. 1946. Induced mutations in bacterial viruses. Cold Spring Harbor Symp. Quant. Biol. **11**:33–37.

Delbrück, M., and S. E. Luria. 1942. Interference between bacterial viruses. I. Interference between two bacterial viruses acting upon the same host, and the mechanism of virus growth. Arch. Biochem. **1**:111–141.

Delihas, N. 1961. The ability of irradiated bacteriophage T2 to initiate the synthesis of deoxycytidylate hydroxymethylase in *Escherichia coli*. Virology **13**:242–248.

Delius, H., C. Howe, and A. W. Kozinski. 1971. Structure of the replicating DNA from bacteriophage T4. Proc. Natl. Acad. Sci. U.S.A. **68**:3049.

Delius, H., R. Mantell, and B. Alberts. 1972. Characterization by electron microscopy of the complex formed between T4 bacteriophage gene 32-protein and DNA. J. Mol. Biol. **67**:341–350.

DeMars, R. I., S. E. Luria, H. Fisher, and C. Levinthal. 1953. The production of incomplete bacteriophage particles by the action of proflavin and the properties of the particles. Ann. Inst. Pasteur

Paris **84:**113–128.

Demerec, M., and U. Fano. 1945. Bacteriophage-resistant mutants in *Escherichia coli*. Genetics **30:**119–136.

Demple, B., and S. Linn. 1980. DNA *N*-glycosylase and UV repair. Nature (London) **287:**203–208.

Depew, R. E., and N. R. Cozzarelli. 1974. Genetics and physiology of bacteriophage T4 3′-phosphatase: evidence for involvement of the enzyme in T4 DNA metabolism. J. Virol. **13:**888–897.

Depew, R. E., T. J. Snopek, and N. R. Cozzarelli. 1975. Characterization of a new class of deletions of the D region of the bacteriophage T4 genome. Virology **64:**144–152.

DeRosier, D. J., and A. Klug. 1968. Reconstruction of three-dimensional structures from electron micrographs. Nature (London) **217:**130–134.

DeRosier, D. J., and A. Klug. 1972. Structure of the tubular variants of the head of bacteriophage T4 (polyheads). I. Arrangement of subunits in some classes of polyheads. J. Mol. Biol. **65:**469–488.

Deutscher, M. P., J. Foulds, and W. H. McClain. 1974. Transfer ribonucleic acid nucleotidyltransferase plays an essential role in the normal growth of *Escherichia coli* and in the biosynthesis of some bacteriophage transfer ribonucleic acids. J. Biol. Chem. **249:**6696–6699.

de Waard, A., A. V. Paul, and I. R. Lehman. 1965. The structural gene for deoxyribonucleic acid polymerase in bacteriophage T4 and T5. Proc. Natl. Acad. Sci. U.S.A. **54:**1241–1248.

de Waard, A., T. E. C. M. Ubbink, and W. Beukman. 1967. On the specificity of bacteriophage-induced hydroxymethylcytosine glucosyltransferases. II. Specificities of hydroxymethylcytosine alpha- and beta-glucosyltransferases induced by bacteriophage T4. Eur. J. Biochem. **2:**303–308.

Dewey, M. J., and F. R. Frankel. 1975. Two suppressor loci for gene 49 mutations of bacteriophage T4. I. Genetic properties and DNA synthesis. Virology **68:**387–401.

Dewey, M. J., J. S. Wiberg, and F. R. Frankel. 1974. Genetic control of whisker antigen of bacteriophage T4D. J. Mol. Biol. **84:**625–634.

Dharmalingam, K., and E. B. Goldberg. 1976a. Mechanism localization and control of restriction cleavage of phage T4 and lambda chromosomes in vivo. Nature (London) **260:**406–410.

Dharmalingam, K., and E. B. Goldberg. 1976b. Phage-coded protein prevents restriction of unmodified progeny T4 DNA. Nature (London) **260:**454–456.

Dharmalingam, K., and E. B. Goldberg. 1979a. Restriction in vivo. III. General effects of glucosylation and restriction on phage T4 gene expression and replication. Virology **96:**393–403.

Dharmalingam, K., and E. B. Goldberg. 1979b. Restriction in vivo. IV. Effect of restriction of parental DNA on the expression of restriction alleviation systems in phage T4. Virology **96:**404–411.

Dharmalingam, K., and E. B. Goldberg. 1980. Restriction in vivo. V. Induction of SOS functions in *Escherichia coli* by restricted T4 phage DNA, and alleviation of restriction SOS functions. Mol. Gen. Genet. **178:**51–78.

Dharmalingam, K., H. R. Revel, and E. B. Goldberg. 1982. Physical mapping and cloning of bacteriophage T4 antirestriction endonuclease gene. J. Bacteriol. **149:**694–699.

Dickson, R. C. 1973. Assembly of bacteriophage T4 tail fibers. IV. Subunit composition of tail fibers and fiber precursors. J. Mol. Biol. **79:**633–647.

Dickson, R. C. 1974. Protein composition of the tail and contracted sheath of bacteriophage T4. Virology **59:**123–138.

Dickson, R. C., S. L. Barnes, and F. A. Eiserling. 1970. Structural protein of bacteriophage T4. J. Mol. Biol. **53:**461–473.

Dicou, L., and N. R. Cozzarelli. 1973. Bacteriophage T4-directed DNA synthesis in toluene-treated cells. J. Virol. **12:**1293–1302.

DiMauro, E., L. Snyder, D. Marino, A. Lamberti, A. Coppo, and G. P. Tocchini-Valentini. 1969. Rifampicin sensitivity of the components of DNA-dependent RNA polymerase. Nature (London) **222:**533–537.

Dirksen, M.-L., J. S. Wiberg, J. F. Koerner, and J. M. Buchanan. 1960. Effect of ultraviolet irradiation of bacteriophage T2 on enzyme synthesis in host cells. Proc. Natl. Acad. Sci. U.S.A. **46:**1425–1430.

Doermann, A. H. 1948. Intracellular growth of bacteriophage. Carnegie Inst. Washington Yearb. **47:**176–182.

Doermann, A. H. 1951a. The intracellular growth of bacteriophages. I. Liberation of intracellular phage T4 by premature lysis with another phage or with cyanide. J. Gen. Physiol. **35:**645–656.

Doermann, A. H. 1951b. Intracellular phage growth as measured by premature lysis. Fed. Proc. **10:**591–594.

Doermann, A. H. 1952. The intracellular growth of bacteriophages. I. Liberation of intracellular bacteriophage T4 by premature lysis with another phage or with cyanide. J. Gen. Physiol. **35:**645–656.

Doermann, A. H. 1953. The vegetative phase in the life cycle of bacteriophage: evidence for its occurrence and its genetic characterization. Cold Spring Harbor Symp. Quant. Biol. **183:**3–11.

Doermann, A. H. 1961. The analysis of ultraviolet lesions in bacteriophage T4 by cross reactivation. J. Cell. Comp. Physiol. **58**(Suppl. 1):79–94.

Doermann, A. H. 1973. T4 and the rolling circle model of replication. Annu. Rev. Genet. **7:**325–341.

Doermann, A. H., and L. Boehner. 1963. An experimental analysis of bacteriophage T4 heterozygotes. I. Mottled plaques from crosses involving six *r*II loci. Virology **21:**551–567.

Doermann, A. H., and L. Boehner. 1970. The identification of complex genotypes in bacteriophage T4. I. Methods. Genetics **66:**417–428.

Doermann, A. H., M. C. Chase, and F. W. Stahl. 1955. Genetic recombination and replication in bacteriophage. J. Cell. Comp. Physiol. **45:**51–74.

Doermann, A. H., and C. Dissosway. 1949. Intracellular growth and genetics of bacteriophage. Carnegie Inst. Washington Yearb. **48:**170–176.

Doermann, A. H., F. A. Eiserling, and L. Boehner. 1973a. Capsid length in bacteriophage T4 and its genetic control, p. 243–258. *In* C. F. Fox and W. S. Robinson (ed.), Virus research. Academic Press, Inc., New York.

Doermann, A. H., F. A. Eiserling, and L. Boehner. 1973b. Genetic control of capsid length in bacteriophage T4. I. Isolation and preliminary description of four new mutants. J. Virol. **12:**374–385.

Doermann, A. H., and M. B. Hill. 1953. Genetic structure of bacteriophage T4 as described by recombination studies of factors influencing plaque morphology. Genetics **38:**79–90.

Doermann, A. H., and D. H. Parma. 1967. Recombination in bacteriophage T4. J. Cell. Comp. Physiol **70**(Suppl. 1, no. 2):147–164.

Doherty, D. H. 1982a. Genetic studies on capsid-length determination in bacteriophage T4. I. Isolation and partial characterization of second-site revertants of a gene 32 mutation affecting capsid length. J. Virol. **43:**641–654.

Doherty, D. H. 1982b. Genetic studies on capsid-length determination in bacteriophage T4. II. Genetic evidence that specific protein-protein interactions are involved. J Virol. **43:**655–663.

Doherty, D. H., P. Gauss, and L. Gold. 1982. On the role of the single-stranded DNA binding protein of bacteriophage T4 in DNA metabolism. I. Isolation and genetic characterization of new mutations in gene 32 of bacteriophage T4. Mol. Gen. Genet. **188:**77–90.

Donelli, G., F. Guglielmi, and L. Pauletti. 1972. Structure and physiochemical properties of bacteriophage G. I. Arrangement of protein subunits and contraction process of tail sheath. J. Mol. Biol. **71:**113–125.

Dove, W. F. 1968. The extent of *r*II deletions in phage T4. Genet. Res. **11:**215–219.

Drahos, D. J., and R. W. Hendrix. 1982. Effect of bacteriophage lambda infection on synthesis of *groE* protein and other *Escherichia coli* proteins. J. Bacteriol. **149:**1050–1063.

Drake, J. W. 1966a. Ultraviolet mutagenesis in bacteriophage T4. I. Irradiation of extracellular phage particles. J. Bacteriol. **91:**1775–1780.

Drake, J. W. 1966b. Heteroduplex heterozygotes in bacteriophage T4 involving mutations of various dimensions. Proc. Natl. Acad. Sci. U.S.A. **55:**506–512.

Drake, J. W. 1967. The length of the homologous pairing region for genetic recombination in bacteriophage T4. Proc. Natl. Acad. Sci. U.S.A. **53:**962–966.

Drake, J. W. 1970. The molecular basis of mutation. Holden-Day, San Francisco.

Drake, J. W. 1973. The genetic control of spontaneous and induced mutation rates in bacteriophage T4. Genetics **73**(Suppl.):45–64.

Drake, J. W. 1982. Methyl methanesulfonate mutagenesis in bacteriophage T4. Genetics **102:**639–651.

Drake, J. W., and E. F. Allen. 1968. Antimutagenic DNA polymerases of bacteriophage T4. Cold Spring Harbor Symp. Quant. Biol. **33:**339–344.

Drake, J. W., E. F. Allen, S. A. Forsberg, R. Preparata, and E. Greening. 1969. Spontaneous mutation. Genetic control of mutation rates in bacteriophage T4. Nature (London) **221:**1128–1131

Drake, J. W., and R. H. Baltz. 1976. The biochemistry of mutagenesis. Annu. Rev. Biochem. **45:**11–37.

Drake, J. W., and E. O. Greening. 1970. Suppression of chemical mutagenesis in bacteriophage T4 by genetically modified DNA polymerases. Proc. Natl. Acad. Sci. U.S.A. **66:**823–834.

Driedonks, R. A. 1981. The quaternary structure of the T4 gene product 20 oligomer, p. 315–323. *In* M. S. DuBow (ed.), Bacteriophage assembly. Alan R. Liss, New York.

Driedonks, R. A., A. Engel, B. ten Heggeler, and R. van Driel. 1981. Gene 20 product of bacteriophage T4. Its purification and structure. J. Mol. Biol. **152:**641–662.

Drivdahl, R., M. Chamberlin, and B. Kutter. 1981. RNA polymerase activity in *E. coli* after infection with *alc* mutants of bacteriophage T4, p. 32. Program, Evergreen Bacteriophage T4 Meeting.

Drivdahl, R., and E. Kutter. 1983. Effect of the bacteriophage T4 *alc* gene on RNA polymerase activity. Fed. Proc. **42:**1881.

Dube, S. K., and P. S. Rudland. 1970. Control of translation by T4 phage: altered binding of disfavoured messengers. Nature (London) **226:**820–823.

Duckworth, D. H. 1970a. Biological activity of bacteriophage ghosts and "take-over" of host functions by bacteriophage. Bacteriol. Rev. **34:**344–363.

Duckworth, D. H. 1970b. The metabolism of T4 phage ghost-infected cells. I. Macromolecular synthesis and transport of nucleic acid and protein precursors. Virology **40:**673–684.

Duckworth, D. H. 1971. Inhibition of host deoxyribonucleic acid synthesis by T4 bacteriophage in the absence of protein synthesis. J. Virol. **8:**754–758.

Duckworth, D. H., and M. J. Bessman. 1965. Assay for the killing properties of T2 bacteriophage and their "ghosts." J. Bacteriol. **90:**724–728.

Duckworth, D. H., and M. J. Bessman. 1967. The enzymology of virus-infected bacteria. X. A biochemical-genetic study of the deoxynucleotide kinase induced by wild type and amber mutants of phage T4. J. Biol. Chem. **242:**2877–2885.

Duckworth, D. H., J. Glenn, and D. J. McCorquodale. 1981. Inhibition of bacteriophage replication by extrachromosomal genetic elements. Microbiol. Rev. **45:**52–71.

Duda, R., and F. A. Eiserling. 1982. Evidence for an internal component of the bacteriophage T4D tail core: a possible length-determining template. J. Virol. **43:**714–720.

Dulbecco, R. 1949. Reactivation of ultraviolet-inactivated bacteriophage by visible light. Nature (London) **163:**949–950.

Dulbecco, R. 1950. Experiments on photoreactivation of bacteriophages inactivated with ultraviolet radiation. J. Bacteriol. **59:**329–347.

Dulbecco, R. 1952. A critical test of the recombination theory of multiplicity reactivation. J. Bacteriol. **63:**199–207.

Duncan, B. K., P. A. Rockstroh, and H. R. Warner. 1978. *Escherichia coli* mutants deficient in uracil-DNA glycosylase. J. Bacteriol. **134:**1039–1045.

Dunn, D. B., and J. D. Smith. 1958. The occurrence of 6-methylaminopurine in deoxyribonucleic acids. Biochem. J. **68:**627–636.

Dunn, J. J., E. Buzash-Pallest, and F. W. Studier. 1978. Mutations of bacteriophage T7 that affect initiation of synthesis of the gene 0.3 protein. Proc. Natl. Acad. Sci. U.S.A. **75:**2741–2745.

Dunn, J. J., and F. W. Studier. 1973. T7 early mRNAs are generated by site-specific cleavages. Proc. Natl. Acad. Sci. U.S.A. **70:**1559.

Dunn, J. J., and F. Studier. 1981. Nucleotide sequence from the genetic left end of bacteriophage T7 DNA to the beginning of gene 4. J. Mol. Biol. **148:**303–330.

Dworsky, P., and M. Schaechter. 1973. Effect of rifampin on the structure and membrane attachment of the nucleoid of *Escherichia coli*. J. Bacteriol. **116:**1364–1374.

Earhart, C. F., C. Y. Tremblay, M. J. Daniels, and M. Schaechter. 1968. DNA replication studied by a new method for the isolation of cell membrane-DNA complexes. Cold Spring Harbor Symp. Quant. Biol. 33:707–710.

Earnshaw, W. C., and S. R. Casjens. 1980. DNA packaging by double-stranded DNA bacteriophages. Cell **21:**319–331.

Earnshaw, W. C., S. Casjens, and S. C. Harrison. 1976. Assembly of the head of bacteriophage P22: X-ray diffraction from heads, proheads, and related structures. J. Mol. Biol. **104:**387–410.

Earnshaw, W. C., E. B. Goldberg, and R. A. Crowther. 1979. The distal half of the tail fiber of bacteriophage T4. J. Mol. Biol. **132:**101–131.

Earnshaw, W. C., and S. Harrison. 1977. DNA arrangement in isometric phage heads. Nature (London) **268:**598–602.

Earnshaw, W. C., and J. King. 1978. Structure of P22 coat protein aggregates formed in the absence of the scaffolding protein. J. Mol. Biol. **126:**721–747.

Earnshaw, W. C., J. King, and F. A. Eiserling. 1978. The size of the bacteriophage T4 head in solution with comments about the dimension of virus particles as visualized by electron microscopy. J. Mol. Biol. **122:**247–253.

Earnshaw, W. C., J. King, S. C. Harrison, and F. A. Eiserling. 1978. The structural organization of DNA packaged within the heads of T4 wild-type, isometric, and giant bacteriophages. Cell **14:**559–568.

Ebisuzaki, K. 1966. Ultraviolet sensitivity and functional capacity in bacteriophage T4. J. Mol. Biol. **20:**545–558.

Ebisuzaki, K., and L. Campbell. 1969. On the role of ligase in genetic recombination in bacteriophage T4. Virology **38:**701–702.

Ebisuzaki, K., C. L. Dewey, and M. T. Behme. 1975. Pathways of DNA repair in phage T4. I. Methyl methanesulfonate sensitive mutant. Virology **64:**330–338.

Edgar, R. S. 1966. Conditional lethals, p. 166–172. *In* J. Cairns, G. S. Stent, and J. D. Watson (ed.), Phage and the origins of molecular biology. Cold Spring Harbor Laboratory, Cold Spring Harbor, N.Y.

Edgar, R. S., G. H. Denhardt, and R. T. Epstein. 1964. A comparative genetic study of conditional lethal mutations of bacteriophage T4D. Genetics **49:**635–648.

Edgar, R. S., R. P. Feynmann, S. Klein, I. Lielausis, and C. M. Steinberg. 1962. Mapping experiments with *r* mutants of bacteriophage T4D. Genetics **47:**179–186.

Edgar, R. S., and I. Lielausis. 1964. Temperature sensitive mutants of bacteriophage T4D: their isolation and genetic characterization. Genetics **49:**649–662.

Edgar, R. S., and I. Lielausis. 1965. Serological studies with mutants of phage T4D defective in genes determining tail fiber structure. Genetics **52:**1187–1200.

Edgar, R. S., and I. Lielausis. 1968. Some steps in the assembly of bacteriophage T4. J. Mol. Biol. **32:**263–276.

Edgar, R. S., and W. B. Wood. 1966. Morphogenesis of bacteriophage T4 in extracts of mutant-infected cells. Proc. Natl. Acad. Sci. U.S.A. **55:**498–505.

Eigner, J., and S. Block. 1968. Host-controlled modification of T-even bacteriophages: relation of four bacterial deoxyribonucleases to restriction. J. Virol. **2:**320–326.

Eiserling, F. A., A. Bolle, and R. H. Epstein. 1967. Electron microscopic study of mutants of bacteriophage T4D defective in tail fiber genes. Virology **33:**405–412.

Eiserling, F. A., E. P. Geiduschek, R. H. Epstein, and E. J. Metter. 1970. Capsid size and deoxyribonucleic acid length: the petite variant of bacteriophage T4. J. Virol. **6:**865–876.

Eliasson, R., and P. Reichard. 1979. Replication of polyoma DNA in isolated nuclei. VII. Initiator RNA synthesis during nucleotide depletion. J. Mol. Biol. **129:**393–409.

Elliott, J., C. Richter, A. Souther, and R. Bruner. 1973. Synthesis of bacteriophage and host DNA in toluene-treated cells prepared from T4-infected *Escherichia coli*: role of bacteriophage gene DZa. J. Virol. **12:**1253–1258.

Ellis, E. L., and M. Delbrück. 1939. The growth of bacteriophage. J. Gen. Physiol. **22:**365–384.

Emrich, J. 1968. Lysis of T4-infected bacteria in the absence of lysozyme. Virology **35:**158–165.

Emrich, J., and G. Streisinger. 1968. The role of phage lysozyme in the life cycle of phage T4. Virology **36:**387–391.

Engel, A., R. van Driel, and R. Driedonks. 1982. A proposed structure of the prolate phage T4 prehead core. J. Ultrastruct. Res. **80:**12–22.

England, T. E., R. I. Gumport, and O. C. Uhlenbeck. 1977. Dinucleoside pyrophosphates are substrates for T4 induced ligase. Proc. Natl. Acad. Sci. U.S.A. **74:**4839–4842.

Englund, P. T. 1971. The initial step of in vitro synthesis of deoxyribonucleic acid by T4 deoxyribonucleic acid polymerase. J. Biol. Chem. **246:**5684–5687.

Ennis, H. L., and K. D. Kievitt. 1973. Association of the *r*IIA protein with bacterial membrane. Proc. Natl. Acad. Sci. U.S.A. **70:**1468–1472.

Enquist, L. W., and A. Skalka. 1973. Replication of bacteriophage lambda dependent on the function of host and viral genes. I. Interaction of *red*, *gam* and *rec*. J. Mol. Biol. **75:**185–212.

Epstein, I. R. 1978. Cooperative and non-cooperative binding of large ligands to a finite one-dimensional lattice. A model for ligand-oligonucleotide interactions. Biophys. Chem. **8:**327–339.

Epstein, R. H. 1958. A study of multiplicity-reactivation in bacteriophage T4. I. Genetic and functional analysis of T4D-K12(lambda) complexes. Virology **6:**382–404.

Epstein, R. H., A. Bolle, C. Steinberg, E. Kellenberger, E. Boy de la Tour, R. Chevalley, R. Edgar, M. Susman, C. Denhardt and I. Lielausis. 1964. Physiological studies of conditional lethal mutants of bacteriophage T4D. Cold Spring Harbor Symp. Quant. Biol. **28:**375–392.

Epstein, W., and S. G. Schultz. 1965. Cation transport in *E. coli*. V. Regulation of cation content. J. Gen. Physiol. **49:**221–234.

Erikson, R. L., and W. Szybalski. 1964. The Cs_2SO_4 equilibrium gradient and its application for the study of T-even phage DNA: glucosylation and replication. Virology **22:**111–120.

Eriksson, S. A. 1975. Ribonucleotide reductase from *E. coli*: demonstration of a highly active form of the enzyme. Eur. J. Biochem. **56:**289–294.

Fabricant, R., and D. Kennell. 1970. Inhibition of host protein synthesis during infection of *Escherichia coli* by bacteriophage T4. III. Inhibition by ghosts. J. Virol. **6:**772–781.

Falco, S. C., W. Zehring, and L. B. Rothman-Denes. 1980. DNA-dependent RNA polymerase from bacteriophage N4 virions. Purification and characterization. J. Biol. Chem. **255:**4339–4347.

Fareed, G. C., and C. C. Richardson. 1967. Enzymatic breakage and joining of deoxyribonucleic acid. II. The structural gene for polynucleotide ligase in bacteriophage T4. Proc. Natl. Acad. Sci. U.S.A. **58:**665–672.

Farnham, P. J., J. Greenblatt, and T. Platt. 1982. Effects of *nus* A protein on transcription termination in the tryptophan operon of *Escherichia coli*. Cell **29:**945–951.

Favre, R., E. Boy de la Tour, N. Segre, and E. Kellenberger. 1965. Studies on the morphopoiesis of the head of phage T-even. I. Morphological, immunological, and genetic characterization of polyheads. J. Ultrastruct. Res. **13:**318–342.

Fersht, A. R. 1979. Fidelity of replication of phage φX174 DNA by DNA polymerase III holoenzyme. Spontaneous mutation by misincorporation. Proc. Natl. Acad. Sci. U.S.A. **76:**4946–4950.

Fersht, A. R., and J. W. Knill-Jones. 1981. DNA polymerase accuracy and spontaneous mutation rates: frequencies of purine-purine, purine-pyrimidine, and pyrimidine-pyrimidine mismatches during DNA replication. Proc. Natl. Acad. Sci. U.S.A. **78:**4251–4255.

Fersht, A. R., J. W. Knill-Jones, and W. C. Tsui. 1982. Kinetic basis for spontaneous mutation. Misinsertion frequencies, proofreading specificities and cost of proofreading by DNA polymerase of *E. coli*. J. Mol. Biol. **156:**37–52.

Ferroluzzi-Ames, G., and B. N. Ames. 1965. The multiplication of T4rII phage in *E. coli* K12(lambda) in the presence of polyamines. Biochem. Biophys. Res. Commun. **18:**639–647.

Ficht, T. A., and R. W. Moyer. 1980. Isolation and characterization of a putative bacteriophage T5 transcription-replication enzyme complex from infected *E. coli*. J. Biol. Chem. **255:**7040–7048.

Fiers, W., R. Contreras, F. Duerinck, G. Haegeman, D. Iserentant, J. Merregaert, W. Min Jou, F. Molemans, A. Raeymaekers, A. van den Berghe, G. Volkaert, and M. Ysebaert. 1976. Complete nucleotide sequence of bacteriophage MS2 RNA: primary and secondary structure of the replicase gene. Nature (London) **260:**500–507.

Fiers, W., R. Contreras, F. Duerinck, G. Haegeman, J. Merregaert, W. Min Jou, A. Raeymaekers, G. Volkaert, M. Ysebaert, J. van de Kerckhove, F. Nolf, and M. van Montagu. 1975. A protein gene of bacteriophage MS2. Nature (London) **256:**273–278.

Finch, J. T., A. Klug, and A. O. W. Stretton. 1964. The structure of the "polyheads" of T4 bacteriophage. J. Mol. Biol. **10:**570–575.

Fisher, R. A. 1930. The general theory of natural selection. Oxford University Press, Oxford.

Flaks, J. G., and S. S. Cohen. 1957. The enzymic synthesis of 5-hydroxymethyldeoxycytidylic acid. Biochim. Biophys. Acta **25:**667–669.

Flanegan, J. B., C.-S., Chiu, and G. R. Greenberg. 1977. Inhibitory effects of agents altering the structure of DNA on the synthesis of pyrimidine deoxyribonucleotides in bacteriophage T4 DNA replication. J. Biol. Chem. **252:**6031–6037.

Flanegan, J. B., and G. R. Greenberg. 1977. Regulation of deoxyribonucleotide biosynthesis during in vivo bacteriophage T4 replication. J. Biol. Chem. **252:**3019–3027.

Fleischman, R. A., J. L. Campbell, and C. C. Richardson. 1976. Modification and restriction of T-even bacteriophages. In vitro degradation of DNA containing 5-hydroxymethylcytosine. J. Biol. Chem. **251:**1561–1570.

Fleischman, R. A., and C. C. Richardson. 1971. Analysis of host range restriction in *Escherichia coli* treated with toluene. Proc. Natl. Acad. Sci. U.S.A. **68:**2527–2531.

Follansbee, S. E., R. W. Vanderslice, L. G. Chavez, and C. D. Yegian. 1974. A new set of adsorption mutants of bacteriophage T4D: identification of a new gene. Virology **58:**180–199.

Forrest, G. L., and D. J. Cummings. 1970. Head proteins from T-even bacteriophage. I. Molecular weight characterization. J. Virol. **5:**398–415.

Foss, K., S.-H. Kao, and W. H. McClain. 1979. Three suppressor forms of bacteriophage T4 leucine transfer RNA. J. Mol. Biol. **135:**1013–1021.

Foster, R. A. C. 1948. An analysis of proflavin on bacteriophage growth. J. Bacteriol. **56:**795–809.

Fox, M. S. 1966. On the mechanism of integration of transforming deoxyribonucleate. J. Gen. Physiol. **49:**183–196.

Francke, B. 1977. Cell-free synthesis of herpes simplex virus DNA: conditions for optimal synthesis. Biochemistry **16:**5655–5664.

Frankel, F. R. 1966. Studies on the nature of replicating DNA in T4-infected *Escherichia coli*. J. Mol. Biol. **18:**127–143.

Frankel, F. R. 1968. DNA replication after T4 infection. Cold Spring Harbor Symp. Quant. Biol. **33:**485–493.

Frankel, F. R., M. L. Batcheler, and C. K. Clark. 1971. The role of gene 49 in DNA replication and head morphogenesis in bacteriophage T4. J. Mol. Biol. **62:**439–463.

Frankel, F. R., C. Majumbar, S. Weintraub, and D. Frankel. 1968. DNA polymerase and the cell membrane after T4 infection. Cold Spring Harbor Symp. Quant. Biol. **33:**495–500.

Frederick, R. J., and L. Snyder. 1977. Regulation of anti-late RNA synthesis in bacteriophage T4: a delayed early control. J. Mol. Biol. **114:**461–476.

Freedman, R., and S. Brenner. 1972. Anomalously revertable *r*II mutants of phage T4. Genet. Res. **19:**165–171.

Freese, E. 1959a. The specific mutagenic effect of a base analogue on phage T4. J. Mol. Biol. **1:**87–105.

Freese, E. 1959b. The difference between spontaneous and base-analogue induced mutations of phage T4. Proc. Natl. Acad. Sci. U.S.A. **45:**622–633.

Freese, E. B., and E. F. Freese. 1967. On the specificity of DNA polymerase. Proc. Natl. Acad. Sci. U.S.A. **57:**650–657.

Freifelder, D. 1968. Physiochemical studies on X-ray inactivation of bacteriophages. Virology **36:**613–619.

French, R. C., and L. Siminovitch. 1955. The action of T2 bacteriophage ghosts on *Escherichia coli* B. Can. J. Microbiol. **1:**757–774.

Friedberg, E. C. 1972. Studies on the substrate specificity of the T4 excision repair endonuclease. Mutat. Res. **15:**113–123.

Friedberg, E. C. 1975. Dark repair in bacteriophage systems: overview, p. 125–133. *In* P. C. Hanawalt and R. B. Setlow (ed.), Molecular mechanisms for repair of DNA. Plenum Publishing Corp., New York.

Friedberg, E. C., C. T. M. Anderson, T. Bonura, R. Cone, E. H. Radany, and R. J. Renolds. 1981. Recent developments in the enzymology of excision repair. Prog. Nucleic Acid Res. Mol. Biol. **26:**197–215.

Friedberg, E. C., and D. A. Clayton. 1972. Electron microscopic studies on substrate specificity of T4 excision repair endonuclease. Nature (London) **237:**99–100.

Friedberg, E. C., and J. J. King. 1969. Endonucleolytic cleavage of UV-irradiated DNA controlled by the v^+ gene in phage T4. Biochem. Biophys. Res. Commun. **37:**646–651.

Friedberg, E. C., and J. J. King. 1971. Dark repair of ultraviolet-irradiated deoxyribonucleic acid by bacteriophage T4: purification and characterization of a dimer-specific phage-induced endonuclease. J. Bacteriol. **106:**500–507.

Friedberg, E. C., and I. R. Lehman. 1974. Excision of thymine dimers by proteolytic and amber fragments of *E. coli* DNA polymerase I. Biochem. Biophys. Res. Commun. **58:**132–139.

Friedberg, E. C., K. Minton, G. Pawl, and P. Verzola. 1974. Excision of thymine dimers in vitro by extracts of bacteriophage-infected *Escherichia coli*. J. Virol. **13:**953–959.

Fry, S. E. 1979. Stimulation of recombination in phage T4 by nitrous acid-induced lesions. J. Gen. Virol. **43:**719–722.

Fujimoto, D., P. R. Srinivasan, and E. Borek. 1965. On the nature of the deoxyribonucleic acid methylases. Biological evidence for the multiple nature of the enzymes. Biochemistry **4:**2849–2855.

Fujisawa, H., and T. Minagawa. 1971. Genetic control of the DNA maturation in the process of phage morphogenesis. Virology **45:**289–291.

Fukada, K., and J. Abelson. 1980. DNA sequence of a T4 transfer RNA gene cluster. J. Mol. Biol. **139:**377–391.

Fukada, K., L. Gossens, and J. Abelson. 1980. The cloning of a T4 tRNA gene cluster. J. Mol. Biol. **137:**213–234.

Fukada, K., A. Otsuka, and J. Abelson. 1980. A restriction map of the T4 transfer RNA gene cluster. J. Mol. Biol. **137:**191–211.

Fukasawa, T. 1964. The course of infection with abnormal bacteriophage T4 containing nonglucosylated DNA on *Escherichia coli* strains. J. Mol. Biol. **9:**525–536.

Fukasawa, T., and S. Saito. 1964. The courses of infection with T-even phages on mutants of *Escherichia coli* K12 defective in the synthesis of uridine diphosphoglucose. J. Mol. Biol. **8:**175–183.

Furth, M. E., C. McLeester, and W. F. Dove. 1978. Specificity determinants for bacteriophage lambda DNA replication. J. Mol. Biol. **126:**195–225.

Furth, M. E., J. L. Yates, and W. F. Dove. 1979. Positive and negative control of bacteriophage lambda DNA replication. Cold Spring Harbor Symp. Quant. Biol. **43:**147–153.

Gaertner, F. H. 1978. Unique catalytic properties of enzyme clusters. Trends Biochem. Res. March 1978, p. 63–65.

Galas, D. J., and E. W. Branscomb. 1978. Enzymatic determinants of DNA polymerase accuracy. Theory of coliphage T4 polymerase mechanisms. J. Mol. Biol. **124:**653–687.

Galas, D. J., and M. Chandler. 1981. On the molecular mechanisms of transposition. Proc. Natl. Acad. Sci. U.S.A. **78:**4858–4862.

Gamow, R. I., and L. M. Kozloff. 1968. Chemically induced cofactor requirement for bacteriophage T4D. J. Virol. **2:**480–487.

Ganesan, A. K. 1973. A method for detecting pyrimidine dimers in the DNA of bacteria irradiated with low doses of ultraviolet light. Proc. Natl. Acad. Sci. U.S.A. **70:**2753–2756.

Ganesan, A. K. 1974. Persistence of pyrimidine dimers during postreplication repair in ultraviolet light-irradiated *Escherichia coli* K12. J. Mol. Biol. **87**:103–119.

Garcia de la Torre, J., and V. A. Bloomfield. 1981. Hydrodynamic properties of complex rigid biological molecules: theory and applications. Q. Rev. Biophys. **14**:81–139.

Garen, A. 1961. Physiological effects of *r*II mutations in bacteriophage T4. Virology **14**:151–163.

Garen, A., and O. Siddiqi. 1962. Suppression of mutation in the alkaline phosphatase structural cistron of *E. coli*. Proc. Natl. Acad. Aci. U.S.A. **48**:1121–1127.

Garvin, R. T., R. Rosset, and L. Gorini. 1973. Ribosomal asembly influenced by growth in the presence of streptomycin. Proc. Natl. Acad. Sci. U.S.A. **70**:2762–2766.

Gates, F. T., III, and S. Linn. 1977a. Endonuclease V of *Escherichia coli*. J. Biol. Chem. **252**:1647–1653.

Gates, F. T., III, and S. Linn. 1977b. Endonuclease from *Escherichia coli* that acts specifically upon duplex DNA damaged by ultraviolet light, osmium tetroxide, acid, and X-rays. J. Biol. Chem. **252**:2802–2807.

Gausing, K. 1972. Efficiency of protein and messenger RNA synthesis in bacteriophage T4-infected cells of *Escherichia coli*. J. Mol. Biol. **71**:529–545.

Gauss, P., D. H. Doherty, and L. Gold. 1983. Bacterial and phage mutations that reveal helix-unwinding activities required for bacteriophage T4 DNA replication. Proc. Natl. Acad. Sci. U.S.A. **80**:1669–1673.

Gefter, M., R. Hausmann, M. Gold, and J. Hurwitz. 1966. The enzymatic methylation of ribonucleic and deoxyribonucleic acid. X. Bacteriophage T3 induced *S*-adenosyl-methionine cleavage. J. Biol. Chem. **241**:1995–2006.

Geider, K., I. Baumel, and T. F. Meyer. 1982. Intermediate stages in enzymatic replication of bacteriophage fd. J. Biol. Chem. **257**:6488–6493.

Geider, K., and H. Hoffman-Berling. 1981. Proteins controlling the helical structure of DNA. Annu. Rev. Biochem. **50**:233–260.

Geiduschek, E. P., and O. Grau. 1970. T4 anti-messenger, p. 190–203. *In* L. Silvestri (ed.), RNA polymerase and transcription. North-Holland Publishing Co., Amsterdam.

Geiduschek, E. P., and J. Ito. 1982. Regulatory mechanisms in the development of lytic bacteriophage, p. 203–245. *In* D. Dubnau (ed.), *Bacillus subtilis*: the molecular biology of the bacilli, vol. 1. Academic Press, Inc., New York.

Gellert, M. 1981. DNA topoisomerases. Annu. Rev. Biochem. **50**:879–910.

Gellert, M., and M. L. Bullock. 1970. DNA ligase mutants of *Escherichia coli*. Proc. Natl. Acad. Sci. U.S.A. **67**:1580–1587.

Gellert, M., K. Mizuuchi, M. H. O'Dea, and H. A. Nash. 1976. DNA gyrase: an enzyme that introduces superhelical turns into DNA. Proc. Natl. Acad. Sci. U.S.A. **73**:3872–3876.

Georgopoulos, C. P. 1967. Isolation and preliminary characterization of T4 mutants with non-glucosylated DNA. Biochem. Biophys. Res. Commun. **28**:179–184.

Georgopoulos, C. P. 1968. Location of glucosyl transferase genes on the genetic map of phage T4. Virology **34**:364–366.

Georgopoulos, C. P., and H. Eisen. 1974. Bacterial mutants which block phage assembly. J. Supramol. Struct. **2**:349–359.

Georgopoulos, C., M. Georgiou, G. Selzer, and H. Eisen. 1977. Bacteriophage T4 mutants which propagate on *E. coli* K-12 but not on *E. coli* B. Experientia **33**:1157–1158.

Georgopoulos, C. P., R. W. Hendrix, S. Casjens, and A. D. Kaiser. 1973. Host participation in bacteriophage lambda head assembly. J. Mol. Biol. **76**:45–60.

Georgopoulos, C. P., R. W. Hendrix, A. D. Kaiser, and W. B. Wood. 1972. Role of the host cell in bacteriophage morphogenesis: effects of a bacterial mutation on T4 head assembly. Nature (London) New Biol. **239**:38–41.

Georgopoulos, C. P., and B. Hohn. 1978. Identification of a host protein necessary for bacteriophage morphogenesis (the *groE* gene product). Proc. Natl. Acad. Sci. U.S.A. **75**:131–135.

Georgopoulos, C. P., and H. R. Revel. 1971. Studies with glucosyl transferase mutants of the T-even bacteriophages. Virology **44**:271–285.

Gersten, D. M., P. Kurian, G. Ledley, C. M. Park, and P. V. Suhocki. 1981. Analysis of bacteriophage T4 coat proteins by two-dimensional electrophoresis and computerized densitometry. Electrophoresis **2**:123–125.

Gilbert, W., and D. Dressler. 1968. DNA replication: the rolling circle model. Cold Spring Harbor Symp. Quant. Biol. **33**:473–484.

Gillin, F. D., and N. G. Nossal. 1976a. Control of mutation frequency by bacteriophage T4 DNA polymerase. I. The CB120 antimutator DNA polymerase is defective in strand displacement. J. Biol. Chem. **251**:5219–5224.

Gillin, F. D., and N. G. Nossal. 1976b. Control of mutation frequency by bacteriophage T4 DNA polymerase. II. Accuracy of nucleotide selection by the L88 mutator, CB120 antimutator, and wild type T4 DNA polymerase. J. Biol. Chem. **251**:5225–5232.

Giri, J. G., J. E. McCullough, and S. P. Champe. 1976. Identification of gene products essential for in vitro formation of the internal peptides of bacteriophage T4. J. Virol. **13**:419–427.

Glickman, B. W., and M. Radman. 1980. *Escherichia coli* mutator mutants deficient in methylation-instructed DNA mismatch correction. Proc. Natl. Acad. Sci. U.S.A. **77**:1063–1067.

Goff, C. G. 1974. Chemical structure of modification of the *Escherichia coli* ribonucleic acid polymerase alpha polypeptides induced by bacteriophage T4 infection. J. Biol. Chem. **249**:6181–6190.

Goff, C. G. 1979. Bacteriophage T4 *alt* gene maps between genes 30 and 54. J. Virol. **29**:1232–1234.

Goff, C. G., and J. Setzer. 1980. ADP ribosylation of *Escherichia coli* RNA polymerase is nonessential for bacteriophage T4 development. J. Virol. **33**:547–549.

Goff, C. G., and K. Weber. 1970. A T4-induced RNA polymerase α subunit modification. Cold Spring Harbor Symp. Quant. Biol. **35**:101–108.

Gold, L., P. O'Farrell, and M. Russell. 1976. Regulation of gene 32 expression during bacteriophage T4 infection of *Escherichia coli*. J. Biol. Chem. **251**:7251–7262.

Gold, L., D. Pribnow, T. Schneider, S. Shinedling, B. S. Singer, and G. Stormo. 1981. Translational initiation in prokaryotes. Annu. Rev. Microbiol. **35**:365–403.

Gold, M., R. Hausmann, U. Mitra, and J. Hurwitz. 1964. The enzymatic methylation of RNA and DNA. VIII. Effects of bacteriophage infection on the methylation enzymes. Proc. Natl. Acad. Sci. U.S.A. **52**:292–297.

Goldberg, A. L., and A. C. St. John. 1976. Intracellular protein degradation in mammalian and bacterial cells. Annu. Rev. Biochem. **45**:747–803.

Goldberg, A. R. 1970. Termination of in vitro RNA synthesis by rho factor. Cold Spring Harbor Symp. Quant. Biol. **35**:157–161.

Goldberg, E. B. 1980. Bacteriophage nucleic acid penetration, p. 115–141. *In* L. L. Randall and L. Phillipson (ed.), Receptors and recognition, series B, vol. 7: Virus receptors, part 1: Bacterial viruses. Chapman and Hall, London.

Goldberger, R. F. 1974. Autogenous regulation of gene expression. Science **183**:810–816.

Goldfarb, A. 1981a. In vitro transcription of bacteriophage T4 tRNA gene cluster from two different promoters. Nucleic Acids Res. **9**:519–527.

Goldfarb, A. 1981b. Changes in the promoter range of RNA polymerase resulting from bacteriophage T4-induced modification of core enzyme. Proc. Natl. Acad. Sci. U.S.A. **78**:3454–3458.

Goldfarb, A., and H. J. Burger. 1981. Mapping of in vitro transcription units and identification of primary transcripts of the D region of bacteriophage T4. Nucleic Acids Res. **9**:2791–2800.

Goldfarb, A., and V. Daniel. 1980. Transcriptional control of two gene subclusters in the tRNA operon of bacteriophage T4. Nature (London) **286**:418–420.

Goldfarb, A., and V. Daniel. 1981a. Mapping of transcription units in the bacteriophage T4 tRNA gene cluster. J. Mol. Biol. **146**:393–412.

Goldfarb, A., and V. Daniel. 1981b. An *Escherichia coli* endonuclease responsible for primary cleavage of in vitro transcripts of bacteriophage T4 tRNA gene cluster. Nucleic Acids Res. **8**:4501–4516.

Goldfarb, A., and P. Palm. 1981. Control of promoter utilization by bacteriophage T4-induced modification of RNA polymerase alpha-subunit. Nucleic Acids Res. **9**:4863–4878.

Goldman, E., and H. F. Lodish. 1971. Inhibition of replication of ribonucleic acid bacteriophage f2 by superinfection with bacteriophage T4. J. Virol. **8**:417–429.

Goldman, E., and H. F. Lodish. 1972. Specificity of protein synthesis by bacterial ribosomes and initiation factors: absence of change after T4 infection. J. Mol. Biol. **67**:35–47.

Goldman, E., and H. F. Lodish. 1973. T4 phage and T4 ghosts inhibit f2 phage replication by different mechanisms. J. Mol. Biol. **74**:151–161.

Goldman, E., and H. F. Lodish. 1975. Competition between bacteriophage f2 RNA and bacteriophage T4 messenger RNA. Biochem. Biophys. Res. Commun. **64**:663–672.

Goldschmidt, R. 1970. In vivo degradation of nonsense fragments in *E. coli*. Nature (London) **228**:1151–1154.

Goldstein, J., and S. P. Champe. 1974. T4-induced activity required for specific cleavage of bacteriophage protein' in vitro. J. Virol. **13**:419–427.

Goldstein, R., J. Lengyel, G. Pruss, K. Barrett, R. Calendar, and E. Six. 1974. Head size determination and morphogenesis of satellite phage P4. Curr. Top. Microbiol. Immunol. **68**:59–75.

Goodman, H. M., M. V. Olson, and B. D. Hall. 1977. Nucleotide sequence of a mutant eukaryotic gene: the yeast tyrosine ochre suppressor and *sup4-O*. Proc. Natl. Acad. Sci. U.S.A. **74**:5453–5457.

Goodman, M. F., W. C. Gore, N. Muzyczka, and M. J. Bessman. 1974. Studies on the biochemical basis of spontaneous mutation. III. Rate model for DNA polymerase-effected nucleotide misincorporation. J. Mol. Biol. **88**:423–435.

Goodman, M. F., R. Hopkins, and W. C. Gore. 1977. 2-Aminopurine-induced mutagenesis in T4 bacteriophage: a model relating mutation frequency to 2-aminopurine incorporation in DNA. Proc. Natl. Acad. Sci. U.S.A. **74**:4806–4810.

Goodman, M. F., R. Hopkins, S. M. Watanabe, L. K. Clayton, and S. Guidotti. 1980. On the molecular basis of mutagenesis: enzymology and genetic studies with bacteriophage T4 system. ICN-UCLA Symp. Mol. Cell. Biol. **19**:685–705.

Goodman, M. F., S. Keener, S. Guidotti, and E. W. Branscomb. 1983. On the enzymatic basis for mutagenesis by manganese. J. Biol. Chem. **258**:3469–3475.

Gordon, L. K., and W. A. Haseltine. 1980. Comparison of the cleavage of pyrimidine dimers by the bacteriophage T4 and *Micrococcus luteus* UV-specific endonucleases. J. Biol. Chem. **255**:12047–12050.

Gorini, L. 1970. Informational suppression. Annu. Rev. Genet. **4**:107–134.

Gorini, L. 1974. Streptomycin and misreading of the genetic code, p. 791–804. *In* M. Nomura, A. Tissieres, and P. Lengyel (ed.), Ribosomes. Cold Spring Harbor Laboratory, Cold Spring Harbor, N.Y.

Gottesman, M. E., S. Adhya, and A. Das. 1980. Transcription antitermination by bacteriophage lambda *N* gene product. J. Mol. Biol. **140**:57–75.

Gottesman, S., and D. Zipser. 1978. Deg phenotype of *Escherichia coli lon* mutants. J. Bacteriol. **133**:843–851.

Goulian, M., Z. Lucas, and A. Kornberg. 1968. Enzymatic synthesis of deoxyribonucleic acid. XXV. Purification and properties of DNA polymerase induced by infection with phage T4^{+}. J. Biol. Chem. **243**:627–638.

Gralla, J., J. A. Steitz, and D. M. Crothers. 1974. Direct physical evidence for secondary structure in an isolated fragment of R17 bacteriophage mRNA. Nature (London) **248**:204–208.

Granboulan, P., J. Séchaud, and E. Kellenberger. 1971. On the fragility of phage T4-related particles. Virology **46**:407–425.

Grantham, R., C. Gautier, M. Gouy, M. Jacobzone, and R. Mercier. 1981. Codon catalog usage is a genome strategy modulated for gene expressivity. Nucleic Acids Res. **9**:43–74.

Grasso, R. J., and J. Buchanan. 1969. Synthesis of early RNA in bacteriophage T4-infected *Escherichia coli* B. Nature (London) **224**:882–885.

Green, E. W., and M. Schaechter. 1972. The mode of segregation of the bacterial cell membrane. Proc. Natl. Acad. Sci. U.S.A. **69**:2312–2316.

Green, R. R., and J. W. Drake. 1974. Misrepair mutagenesis in bacteriophage T4. Genetics **78**:81–89.

Greenblatt, J., and J. Li. 1981. Interaction of the sigma factor and the *nusA* gene protein of *E. coli* with RNA polymerase in the initiation-termination cycle of transcription. Cell **24**:421–428.

Greenberg, G. R., R. L. Somerville, and S. de Wolf. 1962. Resolution of phage-initiated and normal host thymidylate synthetases of *Escherichia coli*. Proc. Natl. Acad. Sci. U.S.A. **48**:242–247.

Greve, J., M. Maestre, H. Moise, and J. Hosoda. 1978a. Circular dichroism studies of the interaction of a limited hydrolysate of T4 gene 32 protein with T4 DNA and poly[d(A-T)]. Biochemistry **17**:893–898.

Greve, J., M. Maestre, H. Moise, and J. Hosoda. 1978b. Circular dichroism study of the interaction between T4 gene 32 protein and polynucleotides. Biochemistry **17**:887–893.

Grindley, N. D. F., and D. J. Sherratt. 1979. Sequence analysis at IS*1* insertion sites: models for transposition. Cold Spring Harbor Symp. Quant. Biol. **43**:1257–1261.

Grinius, L. 1982. Promotive-force-dependent DNA transport across bacterial membranes, p. 129–132. *In* A. N. Martonosi (ed.), Membranes and transport, vol. 2. Plenum Publishing Corp., New York.

Gronenborn, A. M., and R. W. Davies. 1981. DNA binding by dihydrofolate reductase from *Lactobacillus casei*. J. Biol. Chem. **256**:12152–12155.

Groner, Y., Y. Pollack, H. Berissi, and M. Revel. 1972a. Characterization of cistron specific factors for the initiation of messenger RNA translation in *E. coli*. FEBS Lett. **21**:223–228.

Groner, Y., Y. Pollack, H. Berissi, and M. Revel. 1972b. Cistron specific translation control protein in *Escherichia coli*. Nature (London) New Biol. **239**:16–19.

Groner, Y., R. Scheps, E. Kamen, D. Kolakovsky, and M. Revel. 1972. Host subunit of Q-beta replicase is translation control factor I. Nature (London) New Biol. **239**:19–20.

Grossman, L., S. Riazzudin, W. A. Haseltine, and C. P. Lindan. 1979. Nucleotide excision repair of damaged DNA. Cold Spring Harbor Symp. Quant. Biol. **43**:947–955.

Grunberg-Manago, M., and B. Safer (ed.). 1982. Interaction of translational and transcriptional controls in the regulation of gene expression. Elsevier Biomedical, New York.

Grütter, M. G., R. B. Hawkes, and B. W. Matthews. 1979. Molecular basis of thermostability in the lysozyme from bacteriophage T4. Nature (London) **277**:667–669.

Grütter, M. G., and B. W. Matthews. 1982. Amino acid substitutions far from the active site of bacteriophage T4 lysozyme reduce catalytic activity and suggest that the C-terminal lobe of the enzyme participates in substrate binding. J. Mol. Biol. **154**:525–535.

Grütter, M. G., K. L. Rine, and B. W. Matthews. 1979. Crystallographic data for lysozyme from the egg white of the Embden goose. J. Mol. Biol. **135**:1029–1032.

Guerola, N., J. L. Ingraham, and E. Cerda-Olmedo. 1971. Induction of closely linked multiple mutations by nitrosoguanidine. Nature (London) New Biol. **230**:122–125.

Guha, A., W. Szybalski, W. Salser, A. Bolle, E. P. Geiduschek, and J. F. Pulitzer. 1971. Controls and polarity of transcription during bacteriophage T4 development. J. Mol. Biol. **59**:329–349.

Gumport, R. I., D. M. Hinton, V. S. Pyle, and R. W. Richardson. 1980. Nucleic Acids Symp. Ser. **7**:167–171.

Gumport, R. I., and O. C. Uhlenbeck. 1981. T4 RNA ligase as a nucleic acid synthesis and modification agent, p. 1–44. *In* J. G. Chirikjian and T. S. Papas (ed.), Gene amplification and analysis. North-Holland Publishing Co., Amsterdam.

Gupta, A., C. de Brosse, and S. J. Benkovic. 1982. Template-primer-dependent turnover of (S_p)-dATPαS by T4 DNA polymerase. The stereochemistry of the associated 3′-5′ exonuclease. J. Biol. Chem. **257**:7689–7692.

Gussin, G. N., and V. Peterson. 1972. Isolation and properties of *rex*$^-$ mutants of bacteriophage lambda. J. Virol. **10**:760–765.

Guthrie, C. 1975. The nucleotide sequence of the dimeric precursor to glutamine and leucine transfer RNAs coded by bacteriophage T4. J. Mol. Biol. **95**:529–547.

Guthrie, C., and W. H. McClain. 1973. Conditionally lethal mutants of bacteriophage T4 defective in production of a transfer RNA. J. Mol. Biol. **81**:135–137.

Guthrie, C., and W. H. McClain. 1979. Rare transfer ribonucleic acid essential for phage growth. Nucleotide sequence comparison of normal and mutant T4 isoleucine-accepting transfer ribonucleic acid. Biochemistry **18**:3786–3795.

Guthrie, C., and C. A. Scholla. 1980. Asymmetric maturation of a dimeric transfer RNA precursor. J. Mol. Biol. **139**:349–375.

Guthrie, C., C. A. Scholla, H. Yesian, and J. Abelson. 1978. The nucleotide sequence of threonine transfer RNA coded by bacteriophage T4. Nucleic Acids Res. **5**:1833–1844.

Guthrie, C., J. G. Seidman, M. M. Comer, R. M. Bock, F. J. Schmidt, B. G. Barrell, and W. H. McClain. 1975. The biology of bacteriophage T4 transfer RNAs. Brookhaven Symp. Biol. **26**:106–123.

Hadi, S.-M., and D. A. Goldthwait. 1971. Endonuclease II of *Escherichia coli*. Degradation of partially depurinated deoxyribonucleic acid. Biochemistry **10**:4986–4994.

Hagen, E. W., B. E. Reilly, M. E. Tosi, and D. L. Anderson. 1976. Analysis of gene function of bacteriophage ϕ29 of *Bacillus subtilis*: identification of cistrons essential for viral assembly. J. Virol. **19**:501–517.

Hall, B. D., K. Fields, and G. Hager. 1970. The influence of protein factors and DNA structure upon transcription specificity, p. 148–150. *In* L. G. Silvestri (ed.), RNA polymerase and transcription. North-Holland Publishing Co., Amsterdam.

Hall, B. D., A. P. Nygaard, and M. H. Green. 1964. Control of T2-specific RNA synthesis. J. Mol. Biol. **9**:143–153.

Hall, B. D., and S. Spiegelman. 1961. Sequence complementarity of T2-DNA and T2-specific RNA. Proc. Natl. Acad. Sci. U.S.A. **47**:137–146.

Hall, D. H. 1967. Mutants of bacteriophage T4 unable to induce dihydrofolate reductase activity. Proc. Natl. Acad. Sci. U.S.A. **58**:584–588.

Hall, D. H., and P. M. Macdonald. 1981. Evidence for interactions between proteins involved in initiation of bacteriophage T4 DNA replication. J. Supramol. Struct. Cell. Biochem. **5**(Suppl.):341.

Hall, D. H., R. G. Sargent, K. F. Trofatter, and D. L. Russell. 1980. Suppressors of mutations in the bacteriophage T4 gene coding for both RNA ligase and tail fiber attachment activities. J. Virol. **36**:103–108.

Hall, D. H., and R. D. Snyder. 1981. Suppressors of mutations in the *r*II gene of bacteriophage T4 affect promoter utilization. Genetics **97**:1–9.

Hall, D. H., and I. Tessman. 1966. T4 mutants unable to induce deoxycytidylate deaminase activity. Virology **29**:339–345.

Hall, D. H., I. Tessman, and O. Karlström. 1967. Linkage of T4 genes controlling a series of steps in pyrimidine biosynthesis. Virology **31**:442–448.

Hall, Z. W., and I. R. Lehman. 1968. An in vitro transversion by a mutationally altered T4-induced DNA polymerase. J. Mol. Biol. **36**:321–333.

Halpern, M. E., T. Mattson, and A. W. Kozinski. 1979. Origins of phage T4 DNA replication as revealed by hybridization to cloned genes. Proc. Natl. Acad. Sci. U.S.A. **76**:6137–6141.

Halpern, M., T. Mattson, and A. W. Kozinski. 1982. Late events in T4 bacteriophage replication. III. Specificity of DNA reinitiation as revealed by hybridization to cloned genetic fragments. J. Virol. **42**:422–431.

Hamilton, D. L., and R. B. Luftig. 1972. Bacteriophage T4 head morphogenesis. III. Some novel properties of gene 13-defective heads. Virology **9**:1047–1056.

Hamilton, S., and D. E. Pettijohn. 1976. Properties of condensed bacteriophage T4 DNA isolated from *Escherichia coli* infected with bacteriophage T4. J. Virol. **19**:1012–1027.

Hamlett, N. V., and H. Berger. 1975. Mutations altering genetic recombination and repair of DNA in bacteriophage T4. Virology **63**:539–567.

Hanawalt, P. C., P. K. Cooper, A. K. Ganesan, and C. A. Smith. 1979. DNA repair in bacteria and mammalian cells. Annu. Rev. Biochem. **48**:783–836.

Hänggi, U. J., and H. G. Zachau. 1980. Isolation and characterization of DNA fragments containing the dihydrofolate reductase gene of coliphage T4. Gene **9**:271–285.

Harm, W. 1956. On the mechanism of multiplicity reactivation in bacteriophage. Virology **2**:559–564.

Harm, W. 1958. Multiplicity reactivation, marker rescue, and genetic recombination in phage T4 following X-ray inactivation. Virology **5**:337–361.

Harm, W. 1961. Gene-controlled reactivation of ultraviolet-inactivated bacteriophage. J. Cell. Comp. Physiol. **58**:69–77.

Harm, W. 1963. Mutants of phage T4 with increased sensitivity to ultraviolet. Virology **19**:66–71.

Harm, W. 1964. On the control of UV-sensitivity of phage T4 by the gene *x*. Mutat. Res. **1**:344–354.

Harm, W. 1968. Recovery of UV-inactivated *E. coli* cells by the *v* gene of phage T4. Mutat. Res. **6**:175–179.

Harm, W. 1974. Recovery of phage T4 from nitrous acid damage. Mutat. Res. **24**:205–209.

Harm, W. 1980. Biological effects of ultraviolet radiation. Cambridge University Press, Cambridge.

Harrison, S. 1981. Molecular organization of virus particles: implications for assembly, p. 327–342. *In* M. S. DuBow (ed.), Bacteriophage assembly. Alan R. Liss Inc., New York.

Harshey, R. M., R. McKay, and A. I. Bukhari. 1982. DNA intermediates in transposition of phage Mu. Cell **29**:561–571.

Hartman, P., and J. Roth. 1973. Mechanisms of suppression. Adv. Genet. **17**:1–105.

Haselkorn, R., and L. B. Rothman-Denes. 1973. Protein synthesis. Annu. Rev. Biochem. **42**:397–438.

Haselkorn, R., M. Vogel, and R. D. Brown. 1969. Conservation of the rifamycin sensitivity of transcription during T4 development. Nature (London) **221**:836–838.

Haseltine, W. A., L. K. Gordon, C. P. Lindan, R. H. Grafstrom, N. S. Shaper, and L. Grossman. 1980. Cleavage of pyrimidine dimers in specific DNA sequences by a pyrimidine-dimer DNA-glycosylase of *M. luteus*. Nature (London) **285**:634–641.

Hattman, S. 1964. The functioning of T-even phages with unglucosylated DNA in restricting *Escherichia coli* host cells. Virology **24**:333–348.

Hattman, S. 1970. DNA methylation of T-even bacteriophages and their nonglucosylated mutants: its role in P1-directed restriction. Virology **42**:359–367.

Hattman, S. 1972. Methylation of adenine residues in bacteriophage T2 DNA containing cytosine in place of 5-hydroxymethylcytosine. Virology **49**:404–412.

Hattman, S. 1981. DNA methylation, p. 517–548. *In* P. D. Boyer (ed.), The enzymes, vol. 14. Academic Press, Inc., New York.

Hattman, S., J. E. Brooks, and M. Masurekar. 1978. Sequence specificity of the P1-modification methylase (M-Eco P1) and the DNA methylase (M-Eco dam) controlled by the *E. coli dam*-gene. J. Mol. Biol. **126**:367–380.

Hattman, S., and T. Fukasawa. 1963. Host-induced modification of the T-even phages due to defective glucosylation of their DNA. Proc. Natl. Acad. Sci. U.S.A. **50**:297–300.

Hattman, S., and P. H. Hofschneider. 1967. Interference of bacteriophage T4 in the reproduction of RNA-phage M12. J. Mol. Biol. **29**:173–190.

Hattman, S., and P. H. Hofschneider. 1968. Influence of T4 on the formation of RNA phage-specific polyribosomes and polymerase. J. Mol. Biol. **35**:513–522.

Hattman, S., H. R. Revel, and S. E. Luria. 1966. Enzyme synthesis directed by non-glucosylated T-even bacteriophages in restricting hosts. Virology **30**:427–438.

Hattman, S., H. van Ormondt, and A. de Waard. 1978. Sequence specificity of the wild-type (dam^+) and mutant (dam^h) forms of bacteriophage T2 DNA adenine methylase. J. Mol. Biol. **119**:361–376.

Hausmann, R., and M. Gold. 1966. The enzymatic methylation of ribonucleic acid and deoxyribonucleic acid. IX. Deoxyribonucleic acid methylase in bacteriophage-infected *Escherichia coli*. J. Biol. Chem. **241**:1985–1994.

Hayward, W. S., and M. H. Green. 1965. Inhibition of *Escherichia coli* and bacteriophage lambda messenger RNA synthesis by T4. Proc. Natl. Acad. Sci. U.S.A. **54**:1675–1678.

Heere, L. J., and J. D. Karam. 1975. Analysis of expressions of the *rII* gene function of bacteriophage T4. J. Mol. Biol. **16**:974–981.

Hehlmann, R., and S. Hattman. 1972. Mutants of bacteriophage T2gt with altered DNA methylase activity. J. Mol. Biol. **67**:351–360.

Helene, C. 1978. Workshop summary: mechanism and diversity of photoreactivations, p. 123–128. *In* P. C. Hanawalt, E. C. Friedberg, and C. F. Fox (ed.), DNA repair mechanisms. Academic Press, Inc., New York.

Helene, C., F. Toulme, M. Charlier, and M. Yaniv. 1976. Photosensitized splitting of thymine dimers in DNA by gene 32 protein from phage T4. Biochem. Biophys. Res. Commun. **71**:91–98.

Helland, D. E. 1979. Complementary transcribed T4 RNA. Association with the polysomes. Biochem. Biophys. Res. Commun. **88**:1185–1193.

Henderson, D., and J. Weil. 1975. Recombination-deficient deletions in bacteriophage and their interaction with *chi* mutations. Genetics **79**:143–174.

Henderson, L., T. Copeland, R. Sowder, G. Smythers, and S. Oroszlan. 1981. Primary structure of the low molecular weight nucleic acid-binding proteins of murine leukemia viruses. J. Biol. Chem. **256**:8400–8406.

Hendrix, R. W. 1978. Symmetry mismatch and DNA packaging in large bacteriophages. Proc. Natl. Acad. Sci. U.S.A. **75**:4779–4783.

Hendrix, R. W. 1979. Purification and properties of groE, a host protein involved in bacteriophage assembly. J. Mol. Biol. **129**:375–392.

Hendrix, R. W., and S. Casjens. 1974. Protein cleavage in bacteriophage λ tail assembly. Virology **61**:156–159.

Hendrix, R. W., and L. Tsui. 1978. Role of the host in virus assembly: cloning of the *Escherichia coli groE* gene and identification of its protein product. Proc. Natl. Acad. Sci. U.S.A. **75**:136–139.

Hercules, K., J. L. Munro, S. Mendelsohn, and J. S. Wiberg. 1971. Mutants in a nonessential gene of bacteriophage T4 which are defective in the degradation of *Escherichia coli* deoxyribonucleic acid. J. Virol. **7**:95–105.

Hercules, K., and W. Sauerbier. 1973. Transcription units in bacteriophage T4. J. Virol. **12**:872–881.

Hercules, K., and W. Sauerbier. 1974. Two modes of in vivo transcription for genes 43 and 45 of phage T4. J. Virol. **14**:341–348.

Hercules, K., and J. S. Wiberg. 1971. Specific suppression of missense mutations in bacteriophage T4. J. Virol. **8**:603–612.

Herendeen, S. L., R. A. van Bogelen, and F. C. Neidhardt. 1979. Levels of major proteins of *Escherichia coli* during growth at different temperatures. J. Bacteriol. **139**:185–194.

Herman, R. E., and D. P. Snustad. 1982. Plasmid pR386 renders *Escherichia coli* cells restrictive to the growth of bacteriophage T4 *unf* mutants. J. Virol. **41**:330–333.

Herriott, R. M. 1951. Nucleic-acid-free T2 virus "ghosts" with specific biological action. J. Bacteriol. **61**:752–754.

Herriott, R. M., and J. L. Barlow. 1952. Preparation, purification, and properties of *E. coli* virus T2. J. Gen. Physiol. **36**:17–28.

Herriott, R. M., and J. L. Barlow. 1957. The protein coats or "ghosts" of coli phage T2. II. The biological functions. J. Gen. Physiol. **41**:307–331.

Herrmann, R. 1982. Nucleotide sequence of the bacteriophage T4 gene 57 and a deduced amino acid sequence. Nucleic Acids Res. **10**:1105–1112.

Herrmann, R., and W. B. Wood. 1981. Assembly of bacteriophage T4 tail fibers: identification and characterization of the nonstructural protein gp57. Mol. Gen. Genet. **184**:125–132.

Hershey, A. D. 1946a. Mutation of bacteriophage with respect to type of plaque. Genetics **31**:620–640.

Hershey, A. D. 1946b. Spontaneous mutations in bacterial viruses.

Cold Spring Harbor Symp. Quant. Biol. **11**:67–77.

Hershey, A. D. 1953. Inheritance in bacteriophage. Adv. Genet. **5**:89–115.

Hershey, A. D. 1957. Some minor components of bacteriophage T2 paricles. Virology **4**:237–254.

Hershey, A. D. 1958. The production of recombinants in phage crosses. Cold Spring Harbor Symp. Quant. Biol. **23**:19–46.

Hershey, A. D., and M. Chase. 1952. Independent functions of viral protein and nucleic acid in growth of bacteriophage. J. Gen. Physiol. **36**:39–56.

Hershey, A. D., J. Dixon, and M. Chase. 1953. Nucleic acid economy in bacteria infected with bacteriophage T2. I. Purine and pyrimidine composition. J. Gen. Physiol. **36**:777–789.

Hershey, A. D., A. Garen, D. K. Fraser, and J. D. Hudis. 1954. Growth and inheritance in bacteriophage. Carnegie Inst. Washington Yearb. **53**:210–225.

Hershey, A. D., and R. Rotman. 1948. Linkage among genes controlling inhibition of lysis in a bacterial virus. Proc. Natl. Acad. Sci. U.S.A. **34**:89–96.

Hershey, A. D., and R. Rotman. 1949. Genetic recombination between host range and plaque-type mutants of bacteriophage in single bacterial cells. Genetics **34**:44–71.

Hershfield, M. S. 1973. On the role of deoxyribonucleic acid polymerase in determining mutation rates. J. Biol. Chem. **248**:1417–1423.

Hershfield, M. S., and N. G. Nossal. 1972. Hydrolyis of template and newly synthesized deoxyribonucleic acid by the 3'-5' exonuclease activities of the T4 deoxyribonucleic acid polymerase. J. Biol. Chem. **247**:3393–3404.

Herskowitz, I., and D. Hagen. 1980. The lysis-lysogeny decision of phage lambda: explicit programming and responsiveness. Annu. Rev. Genet. **14**:399–445.

Herskowitz, I., and E. Signer. 1970. Control from the r strand of bacteriophage lambda. Cold Spring Harbor Symp. Quant. Biol **35**:355–368.

Hewlett, M. J., and C. K. Mathews. 1975. Bacteriophage-host interaction and restriction of nonglucosylated T6. J. Virol. **15**:776–784.

Hibner, U., and B. M. Alberts. 1980. Fidelity of DNA replication catalyzed in vitro on a natural DNA template by the T4 bacteriophage multienzyme complex. Nature (London) **285**:300–305.

Higgins, N. P., and N. R. Cozzarelli. 1979. DNA joining enzymes. A review. Methods Enzymol. **68**:50–71.

Higgins, N. P., A. P. Gaballe, T. J. Snopek, A. Sugino, and N. Cozzarelli. 1977. Bacteriophage T4 RNA ligase: preparation of a physically homogeneous nuclease-free enzyme from hyperproducing cells. Nucleic Acids Res. **4**:3175–3186.

Hill, C. W. 1975. Informational suppression of missense mutations. Cell **6**:419–427.

Hirashima, A., G. Childs, and M. Inouye. 1973. Differential inhibitory effects of antibiotics on the biosynthesis of envelope proteins of *Escherichia coli*. J. Mol. Biol. **79**:373–389.

Hoess, R. H., C. Foeller, K. Bidwell, and A. Landy. 1980. Site-specific recombination of bacteriophage: DNA sequence of regulatory regions and overlapping structural genes for *int* and *xis*. Proc. Natl. Acad. Sci. U.S.A. **77**:2482–2486.

Hohn, B., M. Wurtz, B. Klein, A. Lustig, and T. Hohn. 1974. Phage lambda DNA-packaging in vitro. J. Supramol. Struct. **2**:302–317.

Hohn, T., and B. Hohn. 1970. Structure and assembly of simple RNA bacteriophages. Adv. Virus Res. **16**:43–93.

Hohn, T., B. Hohn, A. Engel, M. Wurtz, and P. R. Smith. 1979. Isolation and characterization of the host protein groE involved in bacteriophage lambda assembly. J. Mol. Biol. **129**:359–373.

Hohn, T., and I. Katsura. 1977. Structure and assembly of bacteriophage lambda. Curr. Top. Microbiol. Immunol. **78**:69–110.

Hohn, T., M. Wurtz, and E. Engel. 1978. Sevenfold rotational symmetry of a protein complex. J. Ultrastruct. Res. **65**:90–93.

Holliday, R. 1964. A mechanism for gene conversion in fungi. Genet. Res. **5**:282–304.

Holmes, G. E., S. Schneider, C. Bernstein, and H. Bernstein. 1980. Recombinational repair of mitomycin C lesions in phage T4. Virology **103**:299–310.

Holmgren, A. 1976. Hydrogen donor system for *E. coli* ribonucleoside diphosphate reductase dependent on glutathione. Proc. Natl. Acad. Sci. U.S.A. **73**:2275–2279.

Holmgren, A. 1978. Glutathione-dependent enzyme reactions of the phage T4 ribonucleotide reductase system. J. Biol. Chem. **253**:7424–7430.

Homyk, T., A. Rodriguez, and J. Weil. 1976. Characterization of T4 mutants that partially suppress the inability of T4 *r*II to grow in lambda lysogens. Genetics **83**:477–487.

Homyk, T., Jr., and J. Weil. 1974. Deletion analysis of two nonessential regions of the T4 genome. Virology **61**:505–523.

Hopkins, R., and M. F. Goodman. 1979. Asymmetry in forming 2-aminopurine-hydroxymethylcytosine heteroduplexes: a model giving misincorporation frequencies and rounds of DNA replication from base-pair populations in vitro. J. Mol. Biol. **135**:1–22.

Horiuchi, K. 1975. Genetic studies of RNA phages, p. 29–50. *In* N. D. Zinder (ed.), RNA phages. Cold Spring Harbor Laboratory, Cold Spring Harbor, N.Y.

Horvitz, H. R. 1973. Polypeptide bound to the host RNA polymerase is specified by T4 control gene 33. Nature (London) New Biol. **244**:137–140.

Horvitz, H. R. 1974a. Control by bacteriophage T4 of two sequential phosphorylations of the alpha subunit of *Escherichia coli* RNA polymerase. J. Mol. Biol. **90**:727–738.

Horvitz, H. R. 1974b. Bacteriophage T4 mutants deficient in alteration and modification of the *Escherichia coli* RNA polymerase. J. Mol. Biol. **90**:739–750.

Hosoda, J. 1967. A mutant of bacteriophage T4 defective in alpha-glucosyl transferase. Biochem. Biophys. Res. Commun. **27**:294–298.

Hosoda, J., L. Burke, H. Moise, I. Kubota, and A. Tsugita. 1980. The control of T4 gene 32 helix-destabilizing protein activity in a DNA replication complex. ICN-UCLA Symp. Mol. Cell. Biol. **19**:505–514.

Hosoda, J., and R. Cone. 1970. Analysis of T4 proteins. I. Conversion of precursor proteins into lower molecular weight peptides during normal capsid formation. Proc. Natl. Acad. Sci. U.S.A. **66**:1275–1281.

Hosoda, J., and C. Levinthal. 1968. Protein synthesis by *Escherichia coli* infected with bacteriophage T4D. Virology **34**:709–727.

Hosoda, J., and E. Mathews. 1971. DNA replication in vivo by polynucleotide ligase defective mutants of T4. II. Effect of chloramphenicol and mutations in other genes. J. Mol. Biol. **55**:155–179.

Hosoda, J., E. Mathews, and B. Jansen. 1971. Role of genes 46 and 47 in bacteriophage T4 replication. I. In vivo deoxyribonucleic acid replication. J. Virol. **8**:372–387.

Hosoda, J., and H. Moise. 1978. Purification and physicochemical properties of limited proteolysis products of T4 helix-destabilizing protein (gene 32 protein). J. Biol. Chem. **253**:7547–7555.

Hosoda, J., B. Takacs, and C. Brack. 1974. Denaturation of T4 DNA by an in vitro processed gene 32 protein. FEBS Lett. **47**:338–342.

Hotchkiss, R. D. 1952. The role of deoxyribonucleotide in bacterial transformations, p. 426–436. *In* W. D. McElroy and B. Glass (ed.), Phosphorus metabolism, vol. II. Johns Hopkins Press, Baltimore.

Howard, B. D. 1967. Phage lambda mutants deficient in rII exclusion. Science **158**:1588–1589.

Howard, G. W., M. L. Wolin, and S. P. Champe. 1972. Diversity of phage internal components among members of the T-even group. Trans. N.Y. Acad. Sci. **34**:36–51.

Howard-Flanders, P. 1975. Repair by genetic recombination in bacteria, overview, p. 265–274. *In* P. C. Hanawalt and R. B. Setlow (ed.), Molecular mechanisms for repair of DNA. Plenum Publishing Corp., New York.

Howard-Flanders, P. 1982. Inducible repair of DNA. Sci. Am. **245**:72–80.

Howe, C. C., P. J. Buckley, K. M. Carlson, and A. W. Kozinski. 1973. Multiple and specific initiation of T4 DNA replication. J. Virol. **12**:130–148.

Hsiao, C. L., and L. W. Black. 1977. DNA packaging and pathway of bacteriophage T4 head assembly. Proc. Natl. Acad. Sci. U.S.A. **74**:3652–3656.

Hsiao, C. L., and L. W. Black. 1978a. Head morphogenesis of bacteriophage T4. I. Isolation and characterization of gene 40 mutants. Virology **91**:1–14.

Hsiao, C. L., and L. W. Black. 1978b. Head morphogenesis of bacteriophage T4. II. The role of gene 40 in initiating prehead assembly. Virology **91**:15–25.

Hsiao, C. L., and L. W. Black. 1978c. Head morphogenesis of bacteriophage T4. III. The role of gene 20 in DNA packaging. Virology **91**:26–38.

Hsu, W.-T., and S. B. Weiss. 1969. Selective translation of T4 template RNA by ribosomes from T4-infected *Escherichia coli*. Proc. Natl. Acad. Sci. U.S.A. **64**:345–351.

Huang, C. C., J. E. Hearst, and B. M. Alberts. 1981. Two types of replication proteins increase the rate at which T4 DNA polymerase traverses the helical regions in a single-stranded DNA template. J. Biol. Chem. **256**:4087–4094.

Huang, W. M. 1975. Membrane-associated proteins of T4-infected *Escherichia coli*. Virology **66**:508–521.

Huang, W. M. 1979a. Positive regulation of T even-phage DNA replication by the DNA delay protein of gene 39. Cold Spring Harbor Symp. Quant. Biol. **48**:495–499.

Huang, W. M. 1979b. Inhibition of initiation of bacteriophage T4 DNA replication by perturbation of *Escherichia coli* host membrane composition. J. Virol. **32**:917–924.

Huang, W. M., and J. M. Buchanan. 1974. Synergistic interactions of

T4 early proteins concerned with their binding to DNA. Proc. Natl. Acad. Sci. U.S.A. **71**:2226–2230.

Huang, W. M., and I. R. Lehman. 1972. On the exonuclease activity of phage T4 deoxyribonucleic acid polymerase. J. Biol. Chem. **247**:3139–3146.

Huang, W. M., D. Ngo, B. Olivera, and D. Mikkelson. 1976. T4 DNA replication in vitro on cellophane discs. Fed. Proc. **35**:1418.

Huberman, J. A. 1969. Visualization of replicating mammalian and T4 bacteriophage DNA. Cold Spring Harbor Symp. Quant. Biol. **33**:509–524.

Huberman, J., A. Kornberg, and B. Alberts. 1971. Stimulation of T4 bacteriophage DNA polymerase by the protein product of T4 gene 32. J. Mol. Biol. **62**:39–52.

Hunt, C., S. M. Desai, J. Vaughan, and S. Weiss. 1980. Bacteriophage T5 transfer RNA: isolation and characterization of tRNA species and refinement of the tRNA gene maps. J. Biol. Chem. **255**:3164–3173.

Hutchinson, N., T. Kazic, S. J. Lee, C. Rayssiguier, B. S. Emanuel, and A. W. Kozinski. 1978. Late replication and recombination in the vegetative pool of T4. Cold Spring Harbor Symp. Quant. Biol. **43**:517–523.

Imae, Y., and R. Okazaki. 1976. Replication of bacteriophage T4 DNA in vitro. J. Virol. **19**:435–445.

Imamoto, F. 1973. Diversity of regulation of genetic transcription. I. Effect of antibiotics which inhibit the process of translation on RNA metabolism in *Escherichia coli*. J. Mol. Biol. **74**:113–136.

Inoko, H., K. Shigesada, and M. Imai. 1977. Isolation and characterization of conditional-lethal *rho* mutants of *Escherichia coli*. Proc. Natl. Acad. Sci. U.S.A. **74**:1162–1166.

Isaackson, P. J., and G. R. Reeck. 1982. Removal of degradation products from calf thymus high mobility group nonhistone chromatin proteins by chromatography on immobilized double-stranded DNA. Biochim. Biophys. Acta **647**:378–380.

Isaacs, S. T., C.-K. J. Shen, J. E. Hearst, and H. Rapaport. 1977. Synthesis and characterization of new psoralen derivatives with superior photoreactivity with DNA and RNA. Biochemistry **16**:1058–1064.

Ishaq, M., and A. Kaji. 1980. Mechanism of T4 phage restriction by plasmid Rts-1: cleavage of T4 phage DNA by Rts-1 specific enzyme. J. Biol. Chem. **255**:4040–4047.

Ishihama, A., T. Ikeuchi, and T. Yura. 1976. A novel adenosine triphosphatase isolated from RNA preparations of *Escherichia coli*. J. Biochem. (Tokyo) **79**:917–925.

Ishii, T., Y. Yamaguchi, and M. Yanagida. 1978. Binding of the structural protein soc to the head shell of bacteriophage T4. J. Mol. Biol. **120**:533–544.

Ishii, T., and M. Yanagida. 1975. Molecular organization of the shell of T-even bacteriophage head. J. Mol. Biol. **97**:655–660.

Ishii, T., and M. Yanagida. 1977. The two dispensable structural proteins (soc and hoc) of the T4 phage capsid: their properties, isolation and characterization of defective mutants, and their binding with the defective heads in vitro. J. Mol. Biol. **109**:487–514.

Isobe, T., L. W. Black, and A. Tsugita. 1976a. Primary structure of bacteriophage T4 internal protein II and characterization of the cleavage upon phage maturation. J. Mol. Biol. **102**:349–365.

Isobe, T., L. W. Black, and A. Tsugita. 1976b. Protein cleavage during virus assembly: a novel specificity of assembly dependent cleavage in bacteriophage T4. Proc. Natl. Acad. Sci. U.S.A. **73**:4205–4209.

Isobe, T., L. W. Black, and A. Tsugita. 1977. Complete amino acid sequence of bacteriophage T4 internal protein I and its cleavage site on virus maturation. J. Mol. Biol. **110**:168–177.

Isobe, T., M. Yanagida, A. Boosman, and A. Tsugita. 1978. Characterization of the morphogenesis-dependent cleavage region of the major capsid protein (P23) of bacteriophage T4: sequence of an amber fragment of P23. J. Mol. Biol. **125**:339–356.

Ito, M., K. Hori, and M. Sekiguchi. 1978. The expression of bacteriophage T4 early genes. I. Four classes of patterns of expression in bacteriophage T4 early genes. J. Biochem. (Tokyo) **84**:864–871.

Ito, M., and M. Sekiguchi. 1978. The expression of bacteriophage T4 early genes. II. Roles of the beta subunit of RNA polymerase and termination factor rho of the host in the expression of T4 early genes. J. Biochem. (Tokyo) **84**:873–879.

Itoh, T., and J.-L. Tomizawa. 1980. Formation of an RNA primer for initiation of replication of ColE1 DNA by ribonuclease H. Proc. Natl. Acad. Sci. U.S.A. **76**:2450–2454.

Iwatsuki, N. 1977. Purification and properties of deoxythymidine kinase induced by bacteriophage T4 infection. J. Biochem. (Tokyo) **82**:1347–1359.

Jacob, F., S. Brenner, and F. Cuzin. 1963. On the regulation of DNA replication in bacteria. Cold Spring Harbor Symp. Quant. Biol. **28**:329–347.

Jacob, F., and J. Monod. 1961. Genetic regulatory mechanisms in the synthesis of proteins. J. Mol. Biol. **3**:318–356.

Jacobs, K. A., L. M. Albright, D. K. Shibata, and E. P. Geiduschek. 1981. Genetic complementation by cloned bacteriophage T4 late genes. J. Virol. **39**:31–45.

Jacobs, K. A., and E. P. Geiduschek. 1981. Regulation of expression of cloned bacteriophage T4 late gene 23. J. Virol. **39**:46–59.

Jain, S. A., P. Gurewiz, and D. A. Apirion. 1982. A small RNA that complements mutants in the RNA processing enzyme ribonuclease P. J. Mol. Biol. **162**:515–533.

Jain, S. A., B. Pragai, and D. Apirion. 1982. A possible complex containing RNA processing enzymes. Biochem. Biophys. Res. Commun. **106**:768–778.

Jarvik, J., and D. Botstein. 1975. Conditional-lethal mutations that suppress genetic defects in morphogenesis by altering structural proteins. Proc. Natl. Acad. Sci. U.S.A. **72**:2738–2742.

Jay, G., and R. Kaempfer. 1974. Host interference with viral gene expression: mode of action of bacterial factor i. J. Mol. Biol. **82**:192–212.

Jay, G., and R. Kaempfer. 1975. Translational repression of viral RNA by a host protein. J. Biol. Chem. **250**:5749–5755.

Jayaraman, R. 1972. Transcription of bacteriophage T4 DNA by *Escherichia coli* RNA polymerase in vitro: identification of some immediate-early and delayed-early genes. J. Mol. Biol. **70**:253–263.

Jayaraman, R., and E. B. Goldberg. 1970. Transcription of bacteriophage T4 genome in vivo. Cold Spring Harbor Symp. Quant. Biol. **35**:197–201.

Jazwinski, S. M., R. Marco, and A. Kornberg. 1975. The gene H spike protein of bacteriophage ϕX174 and S13. II. Relation to synthesis of the parental replicative form. Virology **66**:294–305.

Jensen, D., R. Kelly, and P. von Hippel. 1976. DNA melting proteins. II. Effects of bacteriophage T4 gene 32-protein binding on the conformation and stability of nucleic acid structures. J. Biol. Chem. **254**:7215–7228.

Jensen, J. L., and M. Susman. 1980. A mutant of *E. coli* that restricts growth of bacteriophage T4 at elevated temperatures. Genetics **94**:301–325.

Jesaitis, M. 1956. Differences in the chemical composition of the phage nucleic acids. Nature (London) **178**:637–641.

Jesaitis, M. A., and W. F. Goebel. 1953. The interaction between T4 phage and the specific lipocarbohydrate of phase II *Shigella sonnei*. Cold Spring Harbor Symp. Quant. Biol. **18**:205–208.

Johns, V., C. Bernstein, and H. Bernstein. 1978. Recombinational repair of alkylation lesions in phage T4. II. Ethyl methanesulfonate. Mol. Gen. Genet. **167**:197–207.

Johnson, J. R., G. M. Collins, M. L. Rementer, and D. H. Hall. 1976. Novel mechanism of resistance to folate analogs: ribonucleoside diphosphate reductase deficiency in bacteriophage T4. Antimicrob. Agents Chemother. **9**:292–300.

Johnson, J. R., and D. H. Hall. 1973. Isolation and characterization of mutants of bacteriophage T4 resistant to folate analogs. Virology **53**:413–426.

Jorgensen, S. E., and J. F. Koerner. 1966. Separation and characterization of deoxyribonucleases of *Escherichia coli* B. I. Chromatographic separation and properties of two deoxyribo-oligonucleotidases. J. Biol. Chem. **241**:3090–3096.

Josse, J., and A. Kornberg. 1962. Glucosylation of deoxyribonucleic acid. III. Alpha-and beta-glucosyl transferases from T4 infected *Escherichia coli*. J. Biol. Chem. **237**:1968–1976.

Josslin, R. 1970. The lysis mechanism of phage T4: mutants affecting lysis. Virology **40**:719–726.

Josslin, R. 1971. Physiological studies on the *t* gene defect in T4-infected *Escherichia coli*. Virology **44**:101–107.

Joyce, C. M., W. S. Kelley, and N. D. F. Grindley. 1982. Nucleotide sequence of the *E. coli polA* gene and primary structure of DNA polymerase I. J. Biol. Chem. **257**:1958–1964.

Judson, H. F. 1979. The eighth day of creation. Simon and Schuster, New York.

Jung, A., A. E. Sippel, M. Grez, and G. Schutz. 1980. Exons encode functional and structural units of chicken lysozyme. Proc. Natl. Acad. Sci. U.S.A. **77**:5759–5763.

Junghans, R. P., L. R. Boone, and A. M. Skalka. 1982. Retroviral DNA H structures: displacement-assimilation model of recombination. Cell **30**:53–62.

Kalasauskaite, E., and L. Grinius. 1979. The role of energy-yielding ATPase and respiratory chain at early stages of bacteriophage T4 infection. FEBS Lett. **99**:287–291.

Kalasauskaite, E., L. Grinius, D. Kadisaite, and A. Jesaitis. 1980. Electrochemical H^+ gradient but not phosphate potential is required for *Escherichia coli* infection by phage T4. FEBS Lett. **117**:232–236.

Kalasauskaite, E. V., D. L. Kadisaite, R. J. Daugelauicius, L. L. Grinius, and A. A. Jausaitis. 1983. Studies on energy supply for genetic processes: requirement for membrane potential in *Escherichia coli* infection by phage T4. Eur. J. Biochem. **130**:123–130.

Kanne, D., K. Straub, H. Rapaport, and J. E. Hearst. 1982. Psoralen-deoxyribonucleic acid photoreaction. Characterization of the monoaddition products from 8-methoxypsoralen and 4,5′,8-trimethylpsoralen. Biochemistry **21**:861–871.

Kanner, L. C., and L. M. Kozloff. 1964. The reaction of indole and T2 bacteriophage. Biochemistry **3**:215–223.

Kano-Sueoka, T., and N. Sueoka. 1968. Characterization of a modified leucyl-tRNA of *Escherichia coli* after bacteriophage T4 infection. J. Mol. Biol. **37**:475–491.

Kao, S. H., and W. H. McClain. 1977. U-G-A suppressor of bacteriophage T4 associated with arg-tRNA. J. Biol. Chem. **252**:8254–8257.

Kao, S. H., and W. H. McClain. 1980a. Baseplate protein of bacteriophage T4 with both structural and lytic functions. J. Virol. **34**:95–103.

Kao, S. H., and W. H. McClain. 1980b. Roles of T4 gene 5 and gene S in cell lysis. J. Virol. **34**:104–107.

Kao-Huang, Y., A. Revzin, A. P. Butler, P. O'Connor, D. Noble, and P. H. von Hippel. 1977. Nonspecific DNA binding of genome-regulating proteins as a biological control mechanism: measurement of DNA-bound *Escherichia coli lac* repressor in vivo. Proc. Natl. Acad. Sci. U.S.A. **74**:4228–4232.

Kaplan, D. A., and D. P. Nierlich. 1974. Initiation and transcription of a set of transfer ribonucleic acid genes in vitro. J. Biol. Chem. **250**:934–938.

Kaplan, D. A., and D. P. Nierlich. 1975. Cleavage of nonglucosylated bacteriophage T4 deoxyribonucleic acid by restriction endonuclease *Eco*RI. J. Biol. Chem. **250**:2395–2397.

Kaplan, S. 1971. Lysine suppressor in *Escherichia coli*. J. Bacteriol. **105**:984–987.

Karam, J. D. 1969. DNA replication by phage T4 *r*II mutants without polynucleotide ligase (gene 30). Biochem. Biophys. Res. Commun. **37**:416–422.

Karam, J. D., and B. Barker. 1971. Properties of bacteriophage T4 mutants defective in gene 30 (deoxyribonucleic acid ligase) and the *r*II gene. J. Virol. **7**:260–266.

Karam, J. D., and M. G. Bowles. 1974. Mutation to overproduction of bacteriophage T4 gene products. J. Virol. **13**:428–438.

Karam, J. D., M. Bowles, and M. Leach. 1979. Expression of bacteriophage T4 genes 45, 44 and 62. I. Discoordinate synthesis of the T4 45- and 44-proteins. Virology **94**:192–203.

Karam, J. D., M. Dawson, W. Gerald, M. Trojanowska, and C. Alford. 1982. Control of translation by the *regA* gene of T4 bacteriophage, p. 83–89. *In* M. Grunberg-Manago and B. Safer (ed.), Interaction of translational and transcriptional controls in the regulation of gene expression. Elsevier Biomedical, New York.

Karam, J. D., L. Gold, B. S. Singer, and M. Dawson. 1981. Translational regulation: identification of the site on bacteriophage T4 *r*IIB and mRNA recognized by the *regA* gene function. Proc. Natl. Acad. Sci. U.S.A. **78**:4669–4673.

Karam, J. D., M. Leach, and L. J. Heere. 1979. Functional interactions between the DNA ligase of *Escherichia coli* and components of the DNA metabolic apparatus of T4 bacteriophage. Genetics **91**:177–189.

Karam, J. D., C. McCulley, and M. Leach. 1977. Genetic control of mRNA decay in T4 phage-infected *Escherichia coli*. Virology **76**:685–700.

Karam, J. D., and P. V. O'Donnell. 1973. Suppression of amber mutations of bacteriophage T4 gene 43 (DNA polymerase) by translational ambiguity. J. Virol. **11**:933–945.

Karam, J. D., and J. F. Speyer. 1970. Reversible inactivation of T4 *ts* DNA polymerase mutants in vivo. Virology **42**:196–203.

Kasai, T., and E. K. F. Bautz. 1969. Regulation of gene-specific RNA synthesis in bacteriophage T4. J. Mol. Biol. **41**:401–417.

Kassavetis, G. A., and E. P. Geiduschek. 1982. Bacteriophage T4 late promoters: mapping 5′ ends of T4 gene 23 mRNAs. EMBO J. **1**:107–114.

Katritzky, A. R., and A. J. Waring. 1962. Tautomeric azines. I. Of 1-methyluracil and 5-bromo-1-methyluracil. J. Chem. Soc., p. 1540–1544.

Katsura, I. 1976. Morphogenesis of bacteriophage lambda tail: polymorphism in the assembly of the major tail protein. J. Mol. Biol. **107**:307–326.

Katsura, I. 1978. Structure and inherent properties of the bacteriophage lambda head shell. I. Polyheads produced by defective mutants in the major head protein. J. Mol. Biol. **121**:71–93.

Katsura, I. 1981a. Structure and function of the major tail protein of bacteriophage lambda. Mutants having small major tail protein molecules in their virion. J. Mol. Biol. **146**:493–512.

Katsura, I. 1981b. Intramolecular genetic analysis of lambda phage structural proteins, p. 327–342. *In* M. S. DuBow (ed.), Bacteriophage assembly. Alan R. Liss Inc., New York.

Katsura, I., and P. W. Kuhl. 1974. A regulator protein for the length determination of bacteriophage lambda tail. J. Supramol. Struct. **2**:239–252.

Katsura, I., and P. W. Kuhl. 1975a. Morphogenesis of the tail of bacteriophage lambda. II. In vitro formation and properties of phage particles with extra long tails. Virology **63**:238–251.

Katsura, I., and P. Kuhl. 1975b. Morphogenesis of the tail of bacteriophage lambda. III. Morphogenesis pathway. J. Mol. Biol. **91**:257–273.

Kaufmann, G., and U. Z. Littauer. 1974. Covalent joining of phenylalanine transfer ribonucleic acid half-molecules by T4 RNA ligase. Proc. Natl. Acad. Sci. U.S.A. **71**:3741–3745.

Kavenoff, R. 1972. Characterization of the *Bacillus subtilis* W23 genome by sedimentation. J. Mol. Biol. **72**:801–806.

Kellenberger, E. 1961. Vegetative bacteriophage and the maturation of the virus particles. Adv. Virus Res. **8**:1–61.

Kellenberger, E. 1966. Control mechanisms in bacteriophage morphopoiesis, p. 192–228. *In* G. E. W. Wolstenholme and M. O'Connor (ed.), Principles of biomolecular organization. Little, Brown & Co., Boston.

Kellenberger, E. 1968. Studies on the morphopoiesis of the head of phage T-even. V. The components of the T4 capsid-related structures. Virology **34**:549–561.

Kellenberger, E. 1969. Polymorphic assemblies of the same virus protein subunit, p. 349–366. *In* A. Engstrom and B. Strandberg (ed.), Symmetry and function of biological systems at the macromolecular level. John Wiley & Sons, Inc., New York.

Kellenberger, E. 1972a. Assembly in biological systems, p. 62. *In* The generation of subcellular structure. Elsevier/North-Holland Publishing Co., Amsterdam.

Kellenberger, E. 1972b. Assembly in biological systems and mechanisms of length determination in protein assemblies. Ciba Found. Symp. **7**:189–206 and 295–299.

Kellenberger, E. 1980. Control mechanisms in the morphogenesis of bacteriophage heads. Biosystems **12**:201–223.

Kellenberger, E., and W. Arber. 1955. Die Struktur des Schwanzes der Phagen T2 und T4 und der Mechanismus der irreversiblen Adsorbtion. Z. Naturforsch. Teil B **10**:698–704.

Kellenberger, E., A. Bolle, E. Boy de la Tour, R. H. Epstein, N. C. Franklin, N. K. Jerne, A. Reale-Scafati, J. Sechaud, I. Bendet, D. Goldstein, and M. A. Lauffer. 1965. Functions and properties related to the tail fibers of bacteriophage T4. Virology **26**:419–440.

Kellenberger, E., and E. Boy de la Tour. 1964. On the fine structure of normal and "polymerized" tail sheath of phage T4. J. Ultrastruct. Res. **11**:545–563.

Kellenberger, E., and E. Boy de la Tour. 1965. Studies on the morphopoiesis of the phage T-even. II. Observations on the fine structure of polyheads. J. Ultrastruct. Res. **13**:343–358.

Kellenberger, E., F. A. Eiserling, and E. Boy de la Tour. 1968. Studies on the morphopoiesis of the head of phage T-even. III. The cores of head-related structures. J. Ultrastruct. Res. **21**:335–360.

Kellenberger, E., and C. Kellenberger. 1970. On a modification of the gene product P23 according to its use as a subunit of either normal capsids of phage T4 or polyheads. FEBS Lett. **8**:140–144.

Kellenberger, E., and J. Séchaud. 1957. Electron microscopical studies of phage multiplication. II. Production of phage-related structures during multiplication of phages T2 and T4. Virology **3**:256–274.

Kellenberger, E., J. Séchaud, and A. Ryter. 1959. Electron microscopical studies of phage replication. IV. The establishment of the DNA pool of vegetative phage and the maturation of phage particles. Virology **8**:478–498.

Kellenberger, E., and M. Wurtz. 1982. The wrapping phenomenon in air-dried and negatively stained preparations. Ultramicroscopy **9**:139–150.

Kells, S. S., and R. Haselkorn. 1974. Bacteriophage T4 short tail fibers are the product of gene 12. J. Mol. Biol. **83**:473–485.

Kelly, R. B., M. R. Atkinson, J. A. Huberman, and A. Kornberg. 1969. Excision of thymine dimers and other mismatched sequences by DNA polymerase of *E. coli*. Nature (London) **224**:495–501.

Kelly, R. B., D. Jensen, and P. von Hippel. 1976. DNA "melting" proteins. IV. Fluorescence measurements of binding parameters for bacteriophage T4 gene 32-protein to mono-, oligo-, and polynucleotides. J. Biol. Chem. **251**:7240–7250.

Kelner, A. 1949. Effects of visible light on the recovery of *Streptomyces griseus* conidia from ultraviolet irradiation injury. Proc. Natl. Acad. Sci. U.S.A. **15**:73–79.

Kemper, B., and D. T. Brown. 1976. Function of gene 49 of bacteriophage T4. II. Analysis of intracellular development and the structure of very fast-sedimentating DNA. J. Virol. **18**:1000–1015.

Kemper, B., and M. Garabath. 1981. Studies on T4 head maturation.

I. Purification and characterization of gene-49-controlled endonuclease. Eur. J. Biochem. **115**:123–131.

Kemper, B., M. Garabath, and V. Courage. 1981a. Studies on T4 head maturation. II. Substrate specificity of gene-49-controlled endonuclease. Eur. J. Biochem. **115**:133–141.

Kemper, B., M. Garabath, and V. Courage. 1981b. Studies on the function of gene 49 controlled endonuclease of phage T4 (endonuclease VII), p. 157–166. *In* M. DuBow (ed.), Bacteriophage assembly. Alan R. Liss Inc., New York.

Kemper, B., and J. Hurwitz. 1973. Studies of T4-induced nucleases. Isolation and characterization of manganese-activated T4-induced endonuclease. J. Biol. Chem. **248**:91–99.

Kennell, D. 1968. Inhibition of host protein synthesis during infection of *Escherichia coli* by bacteriophage T4. I. Continued synthesis of host ribonucleic acid. J. Virol. **2**:1262–1271.

Kennell, D. 1970. Inhibition of host protein synthesis during infection of *Escherichia coli* by bacteriophage T4. II. Induction of host messenger ribonucleic acid and its exclusion from polysomes. J. Virol. **6**:208–217.

Khesin, R. B. 1970. Studies on the RNA synthesis and RNA polymerase in normal and phage-infected *E. coli* cells, p. 167–189. *In* L. Silvestri (ed.), RNA polymerase and transcription. North-Holland Publishing Co., Amsterdam.

Khesin, R. B., E. S. Bogdanova, A. D. Goldfarb, and Y. N. Zograff. 1972. Competition for the DNA template between RNA polymerase molecules from normal and phage-infected *E. coli*. Mol. Gen. Genet. **119**:299–314.

Khesin, R. B., Zh. M. Gorlenko, M. F. Shemyakin, I. A. Bass, and A. A. Prozorov. 1963. Connection between protein synthesis and regulation of messenger RNA's formation in *E. coli* B cells upon development of T4 phage. Biokhimiya **28**:1070–1086.

Khesin, R. B., V. G. Nikiforov, Yu. N. Zograff, O. N. Danilevskaya, E. S. Kalayaeva, V. M. Lipkin, N. N. Modyanov, A. D. Dmitriev, V. V. Velkov, and A. L. Gintsburg. 1976. Influence of mutations and phage infection on E. coli RNA polymerase p. 629–643. *In* R. Losick and M. Chamberlin (ed.), RNA polymerase. Cold Spring Harbor Laboratory, Cold Spring Harbor, N.Y.

Khesin, R. B., V. G. Nikiforov, and Yu. N. Zograff. 1980. A study of *E. coli* RNA polymerase, p. 267–318. *In* V. P. Skulachev (ed.), Biology reviews, section D, vol. I. Soviet Scientific Reviews, Moscow.

Khesin, R. B., M. F. Shemyakin, J. M. Gorlenko, S. L. Bogdanova, and T. P. Afanasieva. 1962. RNA polymerase in *E. coli* cells infected with T2 phage. Biokhimiya **27**:1092–1105.

Khorana, H. G., H. Buchi, H. Ghosh, N. Gupta, T. M. Jacob, H. Kossel, R. Morgan, S. A. Narang, E. Ohtsuka, and R. D. Wells. 1966. Polynucleotide synthesis and the genetic code. Cold Spring Harbor Symp. Quant. Biol. **31**:39–49.

Kiko, H., E. Niggemann, and W. Rüger. 1979. Physical mapping of the restriction fragments obtained from bacteriophage T4 dC-DNA with restriction endonucleases *Sma*I, *Kpn*I, and *Bgl*II. Mol. Gen. Genet. **172**:303–312.

Kikuchi, Y., and J. King. 1975a. Genetic control of bacteriophage T4 baseplate morphogenesis. I. Sequential assembly of the major precursor, in vivo and in vitro. J. Mol. Biol. **99**:645–672.

Kikuchi, Y., and J. King. 1975b. Genetic control of bacteriophage T4 baseplate morphogenesis. II. Mutants unable to form the central part of the baseplate. J. Mol. Biol. **99**:673–694.

Kikuchi, Y., and J. King. 1975c. Genetic control of bacteriophage T4 baseplate morphogenesis. III. Formation of the central plug and overall assembly pathway. J. Mol. Biol. **99**:695–716.

Kim, J.-S., and N. Davidson. 1974. Electron microscope heteroduplex studies of sequence relations of T2, T4, and T6 bacteriophage DNAs. Virology **57**:93–111.

King, G. J., and W. M. Huang. 1982. Identification of the origins of T4 DNA replication. Proc. Natl. Acad. Sci. U.S.A. **79**:7248–7252.

King, J. 1968. Assembly of the tail of bacteriophage T4. J. Mol. Biol. **32**:231–262.

King, J. 1971. Bacteriophage T4 tail assembly: four steps in core formation. J. Mol. Biol. **58**:693–709.

King, J. 1980. Regulation of protein interactions as revealed in phage morphogenesis, p. 101–134. *In* R. Goldberg (ed.), Biological regulation and development, vol. 2. Plenum Publishing Corp., New York.

King, J., D. Botstein, S. Casjens, W. Earnshaw, S. Harrison, and E. Lenk. 1976. Structure and assembly of the capsid of bacteriophage P22. Philos. Trans. R. Soc. London Ser. B. **276**:37–49.

King, J., and S. Casjens. 1974. Catalytic head assembling protein in virus morphogenesis. Nature (London) **251**:112–119.

King, J., and U. K. Laemmli. 1971. Polypeptides of the tail fibers of bacteriophage T4. J. Mol. Biol. **62**:465–477.

King, J., and U. K. Laemmli. 1973. Bacteriophage T4 tail assembly: structural proteins and their genetic identification. J. Mol. Biol. **75**:315–337.

King, J., and N. Mykolajewycz. 1973. Bacteriophage T4 tail assembly: proteins of sheath, core, and baseplate. J. Mol. Biol. **75**:339–358.

King, J., and W. B. Wood. 1969. Assembly of bacteriophage T4 tail fibers: the sequence of gene product interaction. J. Mol. Biol. **39**:533–601.

Kingston, R. E., and M. J. Chamberlin. 1981. Pausing and attenuation of in vitro transcription in the *rrnB* operon of *E. coli*. Cell **27**:523–531.

Kisliuk, R. L., Y. Gaumont, C. M. Baugh, J. H. Galivan, G. F. Maley, and F. Maley. 1979. Inhibition of thymidylate synthetase by poly-γ-glutamyl derivatives of folate and methotrexate, p. 431–436. *In* R. L. Kisliuk and G. M. Brown (ed.), Chemistry and biology of pteridines. Elsevier/North-Holland Publishing Co., New York.

Kistler, J., U. Aebi, L. Onorato, B. ten Heggeler, and M. K. Showe. 1978. Structural changes during the transformation of T4 polyheads. I. Characterization of the initial and final states by Fab-fragment labelling of freeze-dried and shadowed preparations. J. Mol. Biol. **126**:571–589.

Klein, A. 1965. Mechanismen der wirtskontrollierten Modifikation des Phagen T1. Z. Vererbungsl. **96**:346–363.

Kleppe, K., and J. R. Lillehaug. 1979. Polynucleotide kinase. Adv. Enzymol. **38**:245–275.

Klett, R. P., A. Cerami, and E. Reich. 1968. Exonuclease VI, a new nuclease associated with *E. coli* DNA polymerase. Proc. Natl. Acad. Sci. U.S.A. **60**:943–950.

Knapp, G., R. C. Ogden, C. L. Peebles, and J. Abelson. 1977. Splicing of yeast tRNA precursors: structure of the reaction intermediates. Cell **18**:37–45.

Koch, A. L., F. W. Putnam, and E. A. Evans, Jr. 1952. Biochemical studies of virus reproduction. VIII. Purine metabolism. J. Biol. Chem. **197**:113–120.

Koch, G., and A. D. Hershey. 1959. Synthesis of phage-precursor proteins in bacteria infected with T2. J. Mol. Biol. **1**:260–276.

Koch, R. E. 1971. The influence of neighboring base pairs upon base-pair substitution mutation rates. Proc. Natl. Acad. Sci. U.S.A. **68**:773–776.

Koch, R. E., and J. W. Drake. 1970. Cryptic mutants of bacteriophage T4. Genetics **65**:379–390.

Koch, R. E., M. K. McGaw, and J. W. Drake. 1976. Mutator mutations in bacteriophage T4 gene *32* (DNA unwinding protein). J. Virol. **19**:490–494.

Koerner, J. F., and D. P. Snustad. 1979. Shutoff of host macromolecular synthesis after T-even bacteriophage infection. Microbiol. Rev. **43**:199–223.

Koerner, J. F., S. K. Thies, and D. P. Snustad. 1979. Protein induced by bacteriophage T4 which is absent in *Escherichia coli* infected with nuclear disruption-deficient phage mutants. J. Virol. **31**:506–513.

Kogoma, T., T. A. Torrey, N. L. Subia, and G. G. Pickett. 1981. An alternative DNA initiation pathway in *E. coli*, p. 361–374. *In* D. Ray (ed.), The initiation of DNA replication. Academic Press, Inc., New York.

Kohli, J., and H. Grosjean. 1981. Usage of the three termination codons: compilation and analysis of the known eukaryotic and prokaryotic translation termination sequences. Mol. Gen. Genet. **182**:430–439.

Kohno, T., and J. Roth. 1979. Electrolyte effects on the activity of mutant enzymes in vivo and in vitro. Biochemistry **18**:1386–1392.

Konarska, M., W. Fillipowicz, and H. J. Gross. 1982. RNA ligation via 2' phosphomonoester 3'-5' phosphodiester linkage. Requirement of 2'-3' cyclic phosphate termini and involvement of a 5' hydroxyl polynucleotide kinase. Proc. Natl. Acad. Sci. U.S.A. **79**:1474–1478.

Korn, L. J., and C. Yanofsky. 1976. Polarity suppressors defective in transcription termination at the attenuator of the tryptophan operon of *Escherichia coli* have altered rho factor. J. Mol. Biol. **106**:231–241.

Kornberg, A. 1974. DNA synthesis, p. 240–241. W. H. Freeman & Co., San Francisco.

Kornberg, A. 1980. DNA replication. W. H. Freeman & Co.. San Francisco.

Kornberg, A. 1982. Supplement to DNA replication. W. H. Freeman & Co., San Francisco.

Kornberg, A., S. B. Zimmerman, S. R. Kornberg, and J. Josse. 1959. Enzymatic synthesis of DNA. VI. Influence of bacteriophage T2 on the synthesis pathway in host cells. Proc. Natl. Acad. Sci. U.S.A. **45**:772–785.

Kornberg, S. R., S. B. Zimmerman, and A. Kornberg. 1961. Glucosylation of deoxyribonucleic acid by enzymes from bacteriophage infected *Escherichia coli*. J. Biol. Chem. **236**:1487–1493.

Kosturko, L. D., and A. W. Kozinski. 1976. Late events in T4 bacteriophage production. I. Late DNA replication is primarily exponential. J. Virol. 17:794–800.

Kourilsky, P., L. F. Bourgignon, and F. Gros. 1971. Kinetics of viral transcription after induction of prophage, p. 647–666. *In* A. D. Hershey (ed.), The bacteriophage lambda. Cold Spring Harbor Laboratory, Cold Spring Harbor, N.Y.

Kowalczykowski, S., D. Bear, and P. H. von Hippel. 1981. Single-stranded DNA binding proteins, p. 373–444. *In* P. Boyer (ed.), The enzymes, vol. 14a. Academic Press, Inc., New York.

Kowalczykowski, S., N. Lonberg, J. Newport, and P. von Hippel. 1981. Interactions of bacteriophage T4-coded gene 32 protein with nucleic acids. I. Characterization of the binding interactions. J. Mol. Biol. 145:75–104.

Kozinski, A. W. 1961. Fragmentary transfer of ^{32}P-labeled parental DNA to progeny phage. Virology 13:124–128.

Kozinski, A. W. 1968. Molecular recombination in the ligase-negative T4 amber mutant. Cold Spring Harbor Symp. Quant. Biol. 33:375–391.

Kozinski, A. W. 1969. Unbiased participation of T4 phage DNA strands in replication. Biochem. Biophys. Res. Commun. 35:294–299.

Kozinski, A. W., and A. H. Doermann. 1975. Repetitive DNA replication of the incomplete genomes of phage T4 petite particles. Proc. Natl. Acad. Sci. U.S.A. 72:1734–1738.

Kozinski, A. W., A. H. Doermann, and P. B. Kozinski. 1976. Absence of interparental recombination in multiplicity reconstitution from incomplete bacteriophage T4 genomes. J. Virol. 18:873–884.

Kozinski, A. W., and Z. Z. Felgenhauer. 1967. Molecular recombination in T4 bacteriophage deoxyribonucleic acid. J. Virol. 1:1193–1202.

Kozinski, A. W., and L. D. Kosturko. 1976. Late events in T4 bacteriophage production. II. Giant bacteriophage contain concatemers generated by recombination. J. Virol. 17:801–804.

Kozinski, A. W., and P. B. Kozinski. 1963. Fragmentary transfer of ^{32}P-labeled parental DNA to progeny phage. II. Average size of the transferred parental fragment. Repair of the polynucleotide chain after fragmentation. Virology 20:213.

Kozinski, A. W., and P. B. Kozinski. 1968. Autonomous replication of short DNA fragments in the ligase negative T4 *am* H39X. Biochem. Biophys. Res. Commun. 33:670–674.

Kozinski, A. W., and P. B. Kozinski. 1969. Covalent repair of molecular recombinants in the ligase-negative amber mutant of T4 bacteriophage. J. Virol 2:85–88.

Kozinski, A. W., P. B. Kozinski, and R. James. 1967. Molecular recombination in T4 bacteriophage deoxyribonucleic acid. I. Tertiary structure of early replicative and recombining deoxyribonucleic acid. J. Virol. 1:758–770.

Kozinski, A. W., P. B. Kozinski, and P. Shannon. 1963. Replicative fragmentation in T4 phage: inhibition by chloramphenicol. Proc. Natl. Acad. Sci. U.S.A. 50:746–753.

Kozinski, A. W., and T. H. Lin. 1965. Early intracellular events in the replication of T4 phage DNA. I. Complex formation of replicative DNA. Proc. Natl. Acad. Sci. U.S.A. 54:273–278.

Kozinski, A. W., and S.-K. Ling. 1982. Genetic specificity of DNA synthesized in the absence of T4 gene 44 protein. J. Virol. 44:256–261.

Kozinski, A. W., S.-K. Ling, N. Hutchinson, M. E. Halpern, and T. Mattson. 1980. Differential amplification of specific areas of phage T4 genome as revealed by hybridization to cloned genetic segments. Proc. Natl. Acad. Sci. U.S.A. 77:5064.

Kozloff, L. M. 1953. Origin and fate of bacteriophage material. Cold Spring Harbor Symp. Quant. Biol. 18:209–220.

Kozloff, L. M. 1978. Properties of T4D bacteriophage grown in synthetic media containing Zn^{2+}, Co^{2+}, or Ni^{2+}. J. Biol. Chem. 253:1059–1064.

Kozloff, L. M. 1980. Folyl polyglutamate and folate-requiring enzyme as bacteriophage T4D baseplate structural components. Biosystems 12:239–247.

Kozloff, L. M. 1981. Composition of the T4D bacteriophage baseplate and the binding of the tail plug, p. 327–342. *In* M. DuBow (ed.), Bacteriophage assembly. Alan R. Liss, Inc., New York.

Kozloff, L. M., L. K. Crosby, and C. M. Baugh. 1979. Structural role of the polyglutamate portion of the folate found in T4D bacteriophage baseplate. J. Virol. 32:497–506.

Kozloff, L. M., L. K. Crosby, and M. Lute. 1975. Bacteriophage baseplate components. III. Location and properties of the phage structural thymidylate synthetase. J. Virol. 16:1409–1419.

Kozloff, L. M., L. K. Crosby, M. Lute, and D. H. Hall. 1975. Bacteriophage baseplate components. II. Binding and location of phage-induced dihydrofolate reductase. J. Virol. 16:1401–1408.

Kozloff, L. M., and M. Lute. 1959. A contractile protein in the tail of bacteriophage T2. J. Biol. Chem. 234:534–546.

Kozloff, L. M., and M. Lute. 1965. Folic acid, a structural component of T4 bacteriophage. J. Mol. Biol. 12:780–792.

Kozloff, L. M., and M. Lute. 1973. Bacteriophage tail components. IV. Pteroyl polyglutamate synthesis in T4D-infected *Escherichia coli* B. J. Virol. 11:630–636.

Kozloff, L. M., and M. Lute. 1977. Zinc, an essential component of the baseplates of T-even bacteriophages. J. Biol. Chem. 252:7715–7724.

Kozloff, L. M., and M. Lute. 1981. Dual functions of bacteriophage T4D gene 28 product. II. Folate and polyglutamate cleavage activity of uninfected and infected *Escherichia coli* cells and bacteriophage particles. J. Virol. 40:645–656.

Kozloff, L. M., M. Lute, and C. Baugh. 1973. Bacteriophage tail components. V. Complementation of T4D gene 28-infected bacterial extracts with pteroyl hexaglutamate. J. Virol. 11:637–641.

Kozloff, L. M., M. Lute, and L. K. Crosby. 1970. Bacteriophage tail components. III. Use of synthetic pteroyl hexaglutamate for T4D tail plate assembly. J. Virol. 6:754–759.

Kozloff, L. M., M. Lute, and L. K. Crosby. 1975. Bacteriophage T4 baseplate components. I. Binding and location of the folic acid. J. Virol. 16:1391–1400.

Kozloff, L. M., M. Lute, and L. K. Crosby. 1977. Bacteriophage T4 virion baseplate thymidylate synthetase and dihydrofolate reductase. J. Virol. 23:637–634.

Kozloff, L. M., M. Lute, L. K. Crosby, N. Rao, V. A. Chapman, and S. S. DeLong. 1970. Bacteriophage tail components. I. Pteroyl polyglutamates in T-even bacteriophages. J. Virol. 5:726–739.

Kozloff, L. M., M. Lute, and K. Henderson. 1957. Viral invasion. I. Rupture of thioester bonds in the bacteriophage tail. J. Biol. Chem. 228:511–528.

Kozloff, L. M., and F. W. Putnam. 1950. Biochemical studies of virus reproduction. III. The origin of virus phosphorus in the *Escherichia coli* T6 bacteriophage system. J. Biol. Chem. 182:229–2242.

Kozloff, L. M., C. Verses, M. Lute, and L. K. Crosby. 1970. Bacteriophage tail components. II. Dihydrofolate reductase in T4D bacteriophage. J. Virol. 5:740–753.

Kozloff, L. M., and J. Zorzopulos. 1978. Zinc uptake and incorporation into proteins in T4D bacteriophage infected *Escherichia coli*. J. Biol. Chem. 253:5548–5550.

Kozloff, L. M., and J. Zorzopulos. 1981. Dual functions of T4D gene 28 product: structural components of the viral tail baseplate central plug and cleavage enzyme for folyl polyglutamates. I. Identification of T4D gene 28 product in the tail plug. J. Virol. 40:635–644.

Krasin, F., S. Person, R. D. Ley, and F. Hutchinson. 1976. DNA crosslinks, single-strand breaks and effects on bacteriophage T4 survival from tritium decay of [2-^{3}H]adenine, [8-^{3}H]adenine and [8-H3]guanine. J. Mol. Biol. 101:197–209.

Krasnow, M. A., and N. R. Cozzarelli. 1982. Catenation of DNA rings by topoisomerases. Mechanism of control by spermidine. J. Biol. Chem. 257:2687–2693.

Krauss, S. W., B. D. Stollar, and M. Friedkin. 1973. Genetic and immunological studies of bacteriophage T4 thymidylate synthetase. J. Virol. 11:783–791.

Krell, H., H. Durwald, and H. Hoffman-Berling. 1979. A DNA-unwinding enzyme induced in bacteriophage T4-infected *Escherichia coli* cells. Eur. J. Biochem. 93:387–395.

Kreuzer, K. N., and C. V. Jongeneel. 1983. *E. coli* phage topoisomerase. Methods Enzymol. 100:144–160.

Krimm, S., and T. F. Anderson. 1967. Structure of normal and contracted tail sheaths of T4 bacteriophage. J. Mol. Biol. 27:197–202.

Krisch, H. M., and B. Allet. 1982. Nucleotide sequences involved in bacteriophage T4 gene 32 translational self-regulation. Proc. Natl. Acad. Sci. U.S.A. 79:4937–4941.

Krisch, H. M., A. Bolle, and R. Epstein. 1974. Regulation of the synthesis of bacteriophage T4 gene 32 protein. J. Mol. Biol. 88:89–104.

Krisch, H. M., R. M. Duvoisin, B. Allet, and R. H. Epstein. 1980. A chimeric plasmid containing gene 32 of bacteriophage T4D. ICN-UCLA Symp. Mol. Cell. Biol. 19:517–526.

Krisch, H. M., and G. B. Selzer. 1981. Construction and properties of a recombinant plasmid containing gene 32 of bacteriophage T4D. J. Mol. Biol. 148:199–218.

Krisch, H. M., D. B. Shah, and H. Berger. 1971. Replication and recombination in ligase-deficient *r*II bacteriophage T4D. J. Virol. 7:491–498.

Krisch, H. M., and G. van Houwe. 1976. Stimulation of the synthesis of bacteriophage T4 gene 32 protein by ultraviolet light irradiation. J. Mol. Biol. 108:67–81.

Krisch, H. M., G. van Houwe, D. Belin, W. Gibbs, and R. H. Epstein. 1977. Regulation of the expression of bacteriophage T4 genes 32 and 43. Virology 78:87–98.

Kruger, D. H., and C. Schroeder. 1981. Bacteriophage T3 and bacteriophage T7 virus-host cell interactions. Microbiol. Rev. **45**:9–51.

Krylov, V. N. 1971. Star mutants of the bacteriophage T4B. Genetika **7**:112–119.

Krylov, V. N. 1972. A mutation of T4B phage, which enhances suppression of ligase mutants with *r*II mutations. Virology **50**:291–293.

Krylov, V. N., and T. G. Plotnikova. 1971. A suppressor in the genome of phage T4 inhibiting phenotypic expression of mutations in genes 46 and 47. Genetics **47**:319–326.

Krylov, V. N., and T. G. Plotnikova. 1972a. Effect of gene-specified suppressor *su*α on the frequency of genetic recombination of T4B phage. Genetika **8**:60–64.

Krylov, V. N., and T. G. Plotnikova. 1972b. Genetic and physiological study of amber mutants in gene ST2 of T4B phage. Genetika **8**:85–95.

Krylov, V. N., and N. K. Yankovsky. 1975. Mutations in the new gene *st*III of bacteriophage T4B suppressing the lysis affect of gene *st*II and gene *e* mutants. J. Virol. **15**:22–26.

Krylov, V. N., and A. Zapadnaya. 1965. Bacteriophage T4B r mutations sensitive to temperature (r^{ts}). Genetika **1**:7–11.

Kudrna, R. D., J. Smith, S. Linn, and E. E. Penhoet. 1979. Survival of apurinic SV40 DNA in the D-complementation group of xeroderma pigmentosum. Mutat. Res. **62**:173–181.

Kuhn, B., M. Abdel-Monem, and H. Hoffmann-Berling. 1979a. DNA helicases. Cold Spring Harbor Symp. Quant. Biol. **43**:63–67.

Kuhn, B., M. Abdel-Monem, H. Krell, and H. Hoffmann-Berling. 1979b. Evidence for two mechanisms for DNA unwinding catalyzed by DNA helicases. J. Biol. Chem. **254**:11343–11350.

Kunkel, T. A., F. Eckstein, A. S. Mildvan, R. M. Koplitz, and L. A. Leob. 1981. Deoxynucleoside (1-thio) triphosphate prevent proofreading during in vitro DNA synthesis. Proc. Natl Acad. Sci. U.S.A. **78**:6734–6738.

Kunkel, T. A., and L. A. Loeb. 1979. On the fidelity of DNA replication. Effect of divalent metal ion activators and deoxynucleoside triphosphate pools on *in vitro* mutagenesis. J. Biol. Chem. **254**:5718–5725.

Kunkel, T. A., and L. A. Loeb. 1980. Fidelity of DNA replication. Accuracy of *E. coli* DNA polymerase in copying natural DNA in vitro. J. Biol. Chem. **255**:9961–9966.

Kunkel, T. A., R. R. Meyer, and L. A. Loeb. 1979. Single-strand binding protein enhances fidelity of DNA synthesis in vitro. Proc. Natl. Acad. Sci. U.S.A. **76**:6331–6335.

Kunzler, P., and T. Hohn. 1978. Stages of bacteriophage lambda head morphogenesis: physical analysis of particles in solution. J. Mol. Biol. **122**:191–215.

Kurland, C. G. 1979. On the accuracy of elongation, p. 597–614. *In* G. Chambliss, G. R. Craven, J. Davies, K. Davis, L. Kahan and M. Nomura (ed.), Ribosomes. University Park Press, Baltimore.

Kurosawa, Y., and T. Okazaki. 1979. Structure of the RNA portion of the RNA-linked DNA pieces in bacteriophage T4-infected *Escherichia coli* cells. J. Mol. Biol. **135**:841–861.

Kurtz, M. B., and S. Champe. 1977. Precursors of the T4 internal peptides. J. Virol. **22**:412–419.

Kutter, E., et al. 1982. Bacteriophage T4: restriction map correlated with the genetic map, p. 25–33. *In* Stephen O'Brien (ed.), NIH genetic maps, vol. II. National Institutes of Health, Bethesda, Md.

Kutter, E., A. Beug, R. Sluss, L. Jensen, and D. Bradley. 1975. The production of undegraded cytosine-containing DNA by bacteriophage T4 in the absence of dCTPase and endonucleases II and IV, and its effects on T4-directed protein synthesis. J. Mol. Biol. **99**:591–607.

Kutter, E., P. O'Farrell, and B. Guttman. 1980. Bacteriophage T4 restriction map, p. 33–40. *In* Stephen J. O'Brien (ed.), NIH genetic maps, vol. I. National Institutes of Health, Bethesda, Md.

Kutter, E., and J. Wiberg. 1969. Biological effects of substituting cytosine for 5-hydroxymethylcytosine in the DNA of bacteriophage T4. J. Virol. **4**:439–453.

Kutter, E. M., D. Bradley, R. Schenck, B. S. Guttman, and R. Laiken. 1981. Bacteriophage T4 *alc* gene product: general inhibitor of transcription from cytosine-containing DNA. J. Virol. **40**:822–829.

Kutter, E. M., and J. S. Wiberg. 1968. Degradation of cytosine-containing bacterial and bacteriophage DNA after infection of *E. coli* B with bacteriophage T4D wild type and with mutants defective in genes 46, 47, and 56. J. Mol. Biol. **38**:395–406.

Kuzmin, N., V. Tanyashin, and A. Baev. 1982. EcoRV restriction cleavage of T-even phage and cloning of the resultant fragments in pBR322. Proc. U.S.S.R. Acad. Sci. **265**:737–739. (In Russian).

Labedan, B., and E. B. Goldberg. 1979. Requirement for membrane potential in injection of phage T4 DNA. Proc. Natl. Acad. Sci. U.S.A. **76**:4669–4673.

Labedan, B., and E. B. Goldberg. 1982. DNA transport across bacterial membranes, p. 133–138. *In* A. N. Martonosi (ed.), Membranes and transport, vol. 2. Plenum Publishing Corp., New York.

Labedan, B., K. B. Heller, A. A. Jasaitis, T. H. Wilson, and E. B. Goldberg. 1980. A membrane potential threshold for phage T4 DNA injection. Biochem. Biophys. Res. Commun. **93**:625–630.

Labedan, B., and J. Legault-Demare. 1974. Evidence for heterogeneity in populations of T5 bacteriophage. J. Virol. **13**:1093–1100.

Labedan, B., and L. Letellier. 1981. Membrane potential changes during the first steps of coliphage infection. Proc. Natl. Acad. Sci. U.S.A. **78**:215–219.

Lacks, S., and B. Greenberg. 1977. Complementary specificity of restriction endonucleases of *Diplococcus pneumoniae* with respect to DNA methylation. J. Mol. Biol. **114**:153–168.

Laemmli, U. K. 1970. Cleavage of structural proteins during the assembly of the head of bacteriophage T4. Nature (London) **227**:680–685.

Laemmli, U. K. 1975. Characterization of DNA condensates induced by poly (ethylene oxide) and polylysine. Proc. Natl. Acad. Sci. U.S.A. **72**:4288–4292.

Laemmli, U. K., L. A. Amos, and A. Klug. 1976. Correlation between structural transformation and cleavage of the major head protein of T4 bacteriophage. Cell **7**:191–203.

Laemmli, U. K., F. Béguin, and G. Gujer-Kellenberger. 1970. A factor preventing the major head protein of bacteriophage T4 from random aggregation. J. Mol. Biol. **47**:69–85.

Laemmli, U. K., and F. A. Eiserling. 1968. Studies on the morphopoiesis of the head of phage T-even. IV. The formation of polyheads. Mol. Gen. Genet. **101**:333–345.

Laemmli, U. K., and M. Favre. 1973. Maturation of the head of bacteriophage T4. I. DNA packaging events. J. Mol. Biol. **80**:575–599.

Laemmli, U. K., and R. A. Johnson. 1973. Maturation of the head of bacteriophage T4 II. Head-related, aberrant τ-particles. J. Mol. Biol. **80**:601–611.

Laemmli, U. K., E. Molbert, M. Showe, and E. Kellenberger. 1970. Form determining function of the genes required for the assembly of the head of bacteriophage T4. J. Mol. Biol. **49**:99–113.

Laemmli, U. K., J. R. Paulson, and V. Hitchins. 1974. Maturation of the head of bacteriophage T4. J. Supramol. Struct. **2**:276–301.

Laemmli, U. K., and S. F. Quittner. 1974. Maturation of the head of bacteriophage T4. IV. The proteins of the core of the tubular polyheads and *in vitro* cleavage of the head proteins. Virology **62**:483–499.

Laemmli, U. K., N. Teaff, and J. D'Ambrosia. 1974. Maturation of the head of bacteriophage T4. III. DNA packaging into preformed heads. J. Mol. Biol. **88**:749–765.

Lake, J., and K. R. Leonard. 1974. Structure and protein distribution for the capsid of *Caulobacter crescentus* bacteriophage φCbK. J. Mol. Biol. **86**:499–518.

Lam, S. T., M. M. Stahl, K. D. McMilan, and F. W. Stahl. 1974. *rec*-mediated recombination hot spot activity in bacteriophage lambda. II. A mutation which causes hot spot activity. Genetics **77**:425–433.

Landy, A., and S. Spiegelman. 1968. Exhaustive hybridization and its application to an analysis of the ribonucleic acid synthesized in T4-infected cells. Biochemistry **7**:585–591.

Lanni, F., and Y. T. Lanni. 1953. Antigenic structure of bacteriophage. Cold Spring Harbor Symp. Quant. Biol. **18**:159–168.

Lark, K. G., C. A. Lark, and E. A. Meenen. 1981. *rec*-dependent DNA replication in *E. coli*: interaction between *rec*-dependent and normal DNA replication genes, p. 337–360. *In* D. Ray (ed.), The initiation of DNA replication. Academic Press, Inc., New York.

Lathe, R. 1978. RNA polymerase of *Escherichia coli*. Curr. Top. Microbiol. Immunol. **83**:37–91.

Leavitt, P. I., and H. E. Umbarger. 1962. Isoleucine and valine metabolism in *Escherichia coli*. XI. Valine inhibition of the growth of *Escherichia coli* strain K-12. J. Bacteriol. **83**:604–630.

Leder, P., and M. Nirenberg. 1964. RNA code words and protein synthesis. II. Nucleotide sequence of a valine RNA codeword. Proc. Natl. Acad. Sci. U.S.A. **52**:420–427.

Lee, D. D., and P. D. Sadowski. 1982. Bacteriophage T7 defective in the gene 6 endonuclease promotes site-specific cleavages of T7 DNA in vivo and in vitro. J. Virol. **44**:235–240.

Lee, N., and M. Inouye. 1974. Outer membrane proteins of *Escherichia coli*: biosynthesis and assembly. FEBS Lett. **39**:167–170.

Lee-Huang, S., and S. Ochoa. 1971. Messenger discriminating species of initiation factor F3. Nature (London) New Biol. **234**:236–239.

Lee-Huang, S., and S. Ochoa. 1972. Specific inhibitors of MS2 and late T4 RNA translation in *E. coli*. Biochem. Biophys. Res. Commun. **49**:371–376.

Lee-Huang, S., and S. Ochoa. 1973. Purification and properties of two messenger-discriminating species of *E. coli* initiation factor 3. Arch.

Biochem. Biophys. **156:**84–96.

Legault-Demare, L., A. Malhie, and F. Gros. 1969. Synthèse des messagers précoces phagique chez *Escherichia coli* infecte par T4 durant une carence specifique en aminoacide. Eur. J. Biochem. **8:**428–488.

Lehman, I. R. 1960. The deoxyribonucleases of *Escherichia coli*. I. Purification and properties of a phosphodiesterase. J. Biol. Chem. **235:**1497–1487.

Lehman, I. R., and R. M. Herriott. 1958. The protein coats or "ghosts" of coliphage T2. III. Metabolic studies of *Escherichia coli* B infected with T2 bacteriophage "ghosts". J. Gen. Physiol. **41:**1067–1082.

Lehman, I. R., and E. A. Pratt. 1960. On the structure of the glucosylated hydroxymethylcytosine nucleotides of coliphages T2, T4 and T6. J. Biol. Chem. **235:**3254–3258.

Lehman, I. R., and C. C. Richardson. 1964. The deoxyribonucleases of *Escherichia coli*. IV. An exonuclease activity present in purified preparations of deoxyribonucleic acid polymerase. J. Biol. Chem. **239:**233–241.

Lehman, I. R., G. G. Roussos, and E. A. Pratt. 1962. The deoxyribonucleases of *Escherichia coli*. II. Purification and properties of ribonucleic acid-inhibitable endonuclease. J. Biol. Chem. **237:**819–828.

Leibo, S. P., E. Kellenberger, C. Kellenberger-van der Kamp, T. G. Frey, and C. M. Steinberg. 1979. Gene 29-controlled osmotic shock resistance in bacteriophage T4: probable multiple gene functions. J. Virol. **30:**327–338.

Lemaire, G., L. Gold, and M. Yarus. 1978. Autogenous translational repression of bacteriophage T4 gene 32 expression in vitro. J. Mol. Biol. **126:**73–90.

Lemaux, P. G., S. L. Herendeen, P. Bloch, and F. C. Neidhardt. 1978. Transient rates of synthesis of individual polypeptides in *E. coli* following temperature shifts. Cell **13:**427–434.

Lembach, K. J., A. Kuninaka, and J. M. Buchanan. 1969. The relationship of DNA replication to the control of protein synthesis in protoplasts of T4-infected *Escherichia coli* B. Proc. Natl. Acad. Sci. U.S.A. **62:**446–453.

Leonard, K. R., A. K. Kleinschmidt, N. Agabian-Keshishian, L. Shapiro, and J. V. Maizel, Jr. 1972. Structural studies of the capsid of *Caulobacter crescentus* bacteriophage φCbK. J. Mol. Biol. **71:**201–216.

Leung, D., M. T. Behme, and K. Ebisuzaki. 1975. Effect of DNA delay mutations of bacteriophage T4 on genetic recombination. J. Virol. **16:**203–205.

Levin, D., and F. Hutchinson. 1973. Neutral sucrose sedimentation of very large DNA from *Bacillus subtilis*. I. Effect of random double-strand breaks and centrifuge speed on sedimentation. J. Mol. Biol. **75:**455–478.

Levinthal, C., and H. Fisher. 1952. The structural development of a bacterial virus. Biochim. Biophys. Acta **9:**419–429.

Levinthal, C., and J. Hosoda. 1953. Growth and recombination in bacterial viruses. Genetics **38:**500–511.

Levinthal, C., J. Hosoda, and D. Shub. 1967. The control of protein synthesis after phage infection, p. 71–87. *In* J. S. Colter and W. Paranchych (ed.), The molecular biology of viruses. Academic Press, Inc., New York.

Levy, J. N. 1975. Effects of radiophosphorous decay in bacteriophage T4D. I. The mechanism of phage inactivation. Virology **68:**1–13.

Levy, J. N., and E. B. Goldberg. 1980a. Region-specific recombination in phage T4. I. A special glucosyl-dependent recombination system. Genetics **94:**519–530.

Levy, J. N., and E. B. Goldberg. 1980b. Region-specific recombination in phage T4. II. Structure of the recombinants. Genetics **94:**531–547.

Levy, J. N., and E. B. Goldberg. 1980c. Region-specific recombination in phage T4. III. The effect of host mutations on glucosylation-dependent recombination in bacteriophage T4D. Genetics **94:**549–553.

Ley, R. D., and R. E. Krisch. 1974. Lethality and DNA breakage from ^{32}P and ^{33}P decay in bacteriophage T4. Int. J. Radiat. Biol. **25:**531–537.

Lichtenstein, J., and S. S. Cohen. 1960. Nucleotides derived from enzymatic digests of nucleic acids of T2, T4 and T6 bacteriophages. J. Biol. Chem. **235:**1134–1141.

Likover-Moen, T., J. G. Seidman, and W. H. McClain. 1978. A catalogue of transfer RNA-like molecules synthesized following infection of *Escherichia coli* by T-even bacteriophage. J. Biol. Chem. **253:**7910–7917.

Lillehaug, J. R., and K. Kleppe. 1975. Kinetics and specificity of T4 polynucleotide kinase. Biochemistry **14:**1221–1229.

Lillehaug, J. R., R. K. Kleppe, and K. Kleppe. 1976. Phosphorylation of double-stranded DNAs by T4 polynucleotide kinase. Biochemistry **15:**1858–1864.

Lim, T. K., G. J. Baran, and V. A. Bloomfield. 1977. Measurement of diffusion coefficient and electrophoretic mobility with a quasielectric light scattering-band electrophoresis apparatus. Biopolymers **16:**1473–1488.

Lin, S., and I. Zabin. 1972. β-Galactosidase. Rates of synthesis and degradation of incomplete chains. J. Biol. Chem. **247:**2205–2211.

Linder, C. H., and O. Sköld. 1977. Evidence for a diffusible T4 bacteriophage protein governing the initiation of delayed early RNA synthesis. J. Virol. **21:**7–15.

Linder, C. H., and O. Skold. 1980. Control of early gene expression of bacteriophage T4: involvement of the host *rho* factor and the *mot* gene of the bacteriophage. J. Virol. **33:**724–732.

Lindstrom, D. M., and J. W. Drake. 1970. The mechanics of frameshift mutagenesis in bacteriophage T4: the role of chromosome tips. Proc. Natl. Acad. Sci. U.S.A. **65:**617–624.

Ling, S. K., H. M. Vogelbacker, L. Restifo, T. Mattson, and A. W. Kozinski. 1981. UV-irradiated T4 bacteriophage DNA results in amplification of specific genetic areas. J. Virol. **40:**403–410.

Lippke, J. A., L. K. Gordon, D. E. Brash, and W. A. Haseltine. 1981. Distribution of UV light-induced damage in a defined sequence of human DNA: detection of alkaline-sensitive lesions at pyrimidine nucleoside-cytidine sequences. Proc. Natl. Acad. Sci. U.S.A. **78:**3388–3392.

Little, J. W. 1973. Mutants of bacteriophage T4 which allow amber mutants of gene 32 to grow in ochre-suppressing hosts. Virology **53:**47–59.

Liu, C.-C., and B. M. Alberts. 1980. Pentaribonucleotides of mixed sequence are synthesized and efficiently prime de novo DNA chain starts in the T4 bacteriophage DNA replication system. Proc. Natl. Acad. Sci. U.S.A. **77:**5698–5702.

Liu, C.-C., and B. M. Alberts. 1981a. Characterization of the DNA-dependent GTPase activity of T4 gene 41, an essential component of the T4 bacteriophage DNA replication apparatus. J. Biol. Chem. **256:**2813–2820.

Liu, C.-C., and B. M. Alberts. 1981b. Characterization of RNA primer synthesis in the T4 bacteriophage in vitro DNA replication system. J. Biol. Chem. **256:**2821–2829.

Liu, C.-C., R. L. Burke, U. Hibner, J. Barry, and B. M. Alberts. 1979. Probing DNA replication mechanisms with the T4 bacteriophage in vitro system. Cold Spring Harbor Symp. Quant. Biol. **43:**469–487.

Liu, L. F., C.-C. Liu, and B. M. Alberts. 1979. T4 DNA topoisomerase: a new ATP dependent enzyme essential for initiation of T4 bacteriophage DNA replication. Nature (London) **281:**456–461.

Liu, L. F., C.-C. Liu, and B. M. Alberts. 1980. Type II DNA topoisomerases: enzymes that can unknot a topologically knotted DNA molecule via a reversible double-strand break. Cell **19:**697–707.

Livingston, D. M., and C. C. Richardson. 1975. Deoxyribonucleic acid polymerase III of *Escherichia coli*. Characterization of associated endonuclease activities. J. Biol. Chem. **250:**470–478.

Livneh, Z., and I. R. Lehman. 1982. Recombinational bypass of pyrimidine dimers promoted by the *recA* protein of *Escherichia coli*. Proc. Natl. Acad. Sci. U.S.A. **79:**3171–3175.

Ljungquist, S. 1977. A new endonuclease from *Escherichia coli* B, acting at apurinic sites in DNA. J. Biol. Chem. **252:**2808–2814.

Lloyd, R. S., and P. C. Hanawalt. 1981. Expression of the *denV* gene of bacteriophage T4 cloned in *Escherichia coli*. Proc. Natl. Acad. Sci. U.S.A. **78:**2796–2800.

Lloyd, R. S., P. C. Hanawalt, and M. L. Dodson. 1980. Processive action of T4 endonuclease V on ultraviolet-irradiated DNA. Nucleic Acids Res. **8:**5113–5127.

Lo, K. Y., and M. J. Bessman. 1976. An antimutator deoxyribonucleic acid polymerase. J. Biol. Chem. **251:**2475–2479.

Lodish, H. 1970. Secondary structure of bacteriophage f2 ribonucleic acid and the initiation of in vitro protein biosynthesis. J. Mol. Biol. **50:**689–702.

Lodish, H. 1975. Regulation of in vitro protein synthesis by bacteriophage RNA tertiary structure, p. 301–318. *In* N. D. Zinder (ed.), RNA phages. Cold Spring Harbor Laboratory, Cold Spring Harbor, N.Y.

Lodish, H. 1976. Translational control of protein synthesis. Annu. Rev. Biochem. **45:**39–72.

Loeb, L. M., and T. A. Kunkel. 1982. Fidelity of DNA synthesis. Annu. Rev. Biochem. **52:**429–457.

Loeb, M. R. 1974. Bacteriophage T4-mediated release of envelope components from *Escherichia coli*. J. Virol. **13:**631–641.

Lohman, T. 1980. Kinetics of the T4 gene 32 protein-single stranded nucleic acid interaction. Biophys. J. **32:**458–460.

Lonberg, N., S. Kowalczykowski, L. Paul, and P. von Hippel. 1981. Interactions of bacteriophage T4-coded gene 32 protein with nucleic acids. III. Binding properties of two specific proteolytic digestion products of the protein (G32P*I and G32P*III). J. Mol. Biol. **145:**123–138.

Losick, R., and M. Chamberlin (ed.). 1976. RNA polymerase. Cold

Spring Harbor Laboratory, Cold Spring Harbor, N.Y.

Losick, R., and J. Pero. 1981. Cascades of sigma factors. Cell **25**:582–584.

Loveless, A. 1966. Genetic and allied effects of alkylating agents. Pennsylvania State University Press, University Park.

Luder, A., and G. Mosig. 1982. Two alternative mechanisms for initiation of DNA replication forks in bacteriophage T4: priming by RNA polymerase and by recombination. Proc. Natl. Acad. Sci. U.S.A. **79**:1101–1105.

Luftig, R. B., and C. Ganz. 1972a. Bacteriophage T4 head morphogenesis. II. Studies on the maturation of gene 49-defective head intermediates. J. Virol. **9**:377–389.

Luftig, R. B., and C. Ganz. 1972. Bacteriophage T4 head morphogenesis. IV. Comparison of gene 16-, 17-, and 49-defective head structures. J. Virol. **10**:545–554.

Luftig, R. B., and N. Lundh. 1973. Bacteriophage T4 head morphogenesis. Isolation, partial characterization, and fate of gene 21 defective tau-particles. Proc. Natl. Acad. Sci. U.S.A. **70**:1636–1640.

Luftig, R. B., W. B. Wood, and R. Okinaka. 1971. Bacteriophage T4 head morphogenesis. On the nature of gene 49 defective heads and their role as intermediates. J. Mol. Biol. **57**:555–573.

Lunn, C. A., and V. Pigiet. 1979. Characterization of a high activity form of ribonucleoside diphosphate reductase from *E. coli*. J. Biol. Chem. **254**:5008–5014.

Lunt, M. R., and E. A. Newton. 1965. Glucosylated nucleotide sequences from T-even bacteriophage deoxyribonucleic acids. Biochem. J. **95**:717–723.

Luria, S. E. 1945. Mutation of bacterial viruses affecting their host range. Genetics **30**:84–99.

Luria, S. E. 1947. Reactivation of irradiated bacteriophage by transfer of self-reproducting units. Proc. Natl. Acad. Sci. U.S.A. **33**:253–264.

Luria, S. E. 1949. Type hybrid bacteriophages. Rec. Genet. Soc. **18**:102.

Luria, S. E. 1950. Bacteriophage, an essay on virus reproduction. Science **111**:507–509.

Luria, S. E. 1953. Host-induced modification of viruses. Cold Spring Harbor Symp. Quant. Biol. **18**:237–244.

Luria, S. E. 1962. Genetics of bacteriophage. Annu. Rev. Microbiol. **16**:205–240.

Luria, S. E. 1970. The recognition of DNA in bacteria. Sci. Am. **222**:88–98.

Luria, S. E., and T. F. Anderson. 1942. Identification and characterization of bacteriophages with the electron microscope. Proc. Natl. Acad. Sci. U.S.A. **28**:127–130.

Luria, S. E., and R. Dulbecco. 1949. Genetic recombination leading to production of active bacteriophage from ultraviolet inactivated bacteriophage particles. Genetics **43**:93–125.

Luria, S. E., and M. L. Human. 1950. Chromatin staining of bacteria during bacteriophage infection. J. Bacteriol. **59**:551–560.

Luria, S. E., and M. L. Human. 1952. Nonhereditary, host-induced variation of bacterial viruses. J. Bacteriol. **64**:557–569.

Luria, S. E., and R. Latarjet. 1947. Ultraviolet irradiation of bacteriophage during intracellular growth. J. Bacteriol. **53**:149–163.

Macchiato, M. F., G. F. Grossi, and A. Cascino. 1979. Roles of gene 45 product into T4 DNA replication and late gene expression of: temperature reversibility effect. FEBS Lett. **104**:187–192.

Macdonald, P. M., and D. H. Hall. 1981. Mutations in bacteriophage T4 genes 41 and 61 cause folate analog resistance. J. Supramol. Struct. Cell. Biochem. **5**(Suppl.):341.

Macdonald, P. M., R. M. Seaby, W. Brown, and G. Mosig. 1983. Initiator DNA from a primary origin and induction of a secondary origin of bacteriophage T4 DNA replication, p. 111–116. *In* D. Schlessinger (ed.), Microbiology—1983. American Society for Microbiology, Washington, D.C.

Mailhammer, R., H.-L. Yang, G. Reiness, and G. Zubay. 1975. Effects of bacteriophage T4-induced modification of *Escherichia coli* RNA polymerase on gene expression in vitro. Proc. Natl. Acad. Sci. U.S.A. **72**:4828–4932.

Maisurian, A. N., and E. A. Buyanovskaya. 1973. Isolation of an *Escherichia coli* strain restricting bacteriophage suppressor. Mol. Gen. Genet. **120**:227–229.

Male, C. J., and L. M. Kozloff. 1973. Function of T4D structural dihydrofolate reductase in bacteriophage infection. J. Virol. **11**:840–847.

Maley, G. F., D. U. Guarino, and F. Maley. 1967. End product regulation of bacteriophage T2r$^+$-induced deoxycytidylate deaminase. J. Biol. Chem. **242**:3517–3524.

Maley, G. F., R. MacColl, and F. Maley. 1972. T2r$^+$ bacteriophage-induced enzymes. II. The subunit structure of deoxycytidylate deaminase. J. Biol. Chem. **247**:940–945.

Maley, G. F., and F. Maley. 1982. Allosteric transitions associated with the binding substrate and effector ligands to T2-phage-induced deoxycytidylate deaminase. Biochemistry **21**:3780–3785.

Manale, A., C. Guthrie, and D. Colby. 1979. S1 nuclease as a probe for the conformation of a dimeric tRNA precursor. Biochemistry **18**:77–83.

Manne, V., V. B. Rao, and L. W. Black. 1982. A bacteriophage T4 DNA packaging related DNA dependent ATPase-endonuclease. J. Biol. Chem. **257**:13223–13232.

Manoil, C., N. Sinha, and B. Alberts. 1977. Intracellular DNA-protein complexes from bacteriophage T4-infected cells isolated by a rapid two step procedure. J. Biol. Chem. **252**:2734–2741.

Manwaring, J. D., and J. A. Fuchs. 1979. Relationship between deoxyribonucleoside triphosphate pools and DNA synthesis in an *nrdA* mutant of *E. coli*. J. Bacteriol. **138**:245–248.

Marchin, G. L. 1980. Mutations in a nonessential viral gene permit bacteriophage T4 to form plaques on *Escherichia coli val-ts relA*. Science **209**:294–295.

Marchin, G. L., M. M. Comer, and F. C. Neidhardt. 1972. Viral modification of the valyl transfer ribonucleic acid synthetase of *Escherichia coli*. J. Biol. Chem. **247**:5132–5145.

Marinus, M. G., and N. R. Morris. 1973. Isolation of DNA methylase mutants of Escherichia coli K-12. J. Bacteriol. **114**:1143–1150.

Mark, K., and F. W. Studier. 1981. Purification of the 0.3 protein of bacteriophage T7, an inhibitor of the DNA restriction system of *Escherichia coli*. J. Biol. Chem. **256**:2573–2578.

Marsh, R. C., A. M. Breschkin, and G. Mosig. 1971. Origin and direction of bacteriophage T4 DNA replication. II. A gradient of marker frequencies in partially replicated T4 DNA as assayed by transformation. J. Mol. Biol. **60**:213–233.

Marsh, R. C., and M. L. Hepburn. 1981. Map of restriction sites on bacteriophage T4 cytosine-containing DNA for endonucleases *Bam*HI, *Bgl*II, *Kpn*I, *Pvu*I, *Sal*I, and *Xba*I. J. Virol. **38**:104–114.

Martin, R. 1977. A possible genetic mechanism of aging, rejuvenation, and recombination in germinal cells. ICN-UCLA Symp. Mol. Cell. Biol. **7**:355–373.

Masamune, Y., R. A. Fleischman, and C. C. Richardson. 1971. Enzymatic removal and replacement of nucleotides at single strand breaks in deoxyribonucleic acid. J. Biol. Chem. **246**:2680–2691.

Masamune, Y., and C. C. Richardson. 1971. Strand displacement during deoxyribonucleic acid synthesis at single strand breaks. J. Biol. Chem. **246**:2692–2701.

Mason, W. S., and R. Haselkorn. 1972. Product of T4 gene 12. J. Mol. Biol. **66**:445–469.

Mathews, C. K. 1967. Evidence that bacteriophage induced dihydrofolate reductase is a viral gene product. J. Biol. Chem. **242**:4083–4086.

Mathews, C. K. 1968. Biochemistry of DNA-defective amber mutants of bacteriophage T4. I. RNA metabolism. J. Biol. Chem. **243**:5610–5615.

Mathews, C. K. 1971a. Bacteriophage biochemistry. Van Nostrand Reinhold Co., New York.

Mathews, C. K. 1971b. Identity of genes coding for soluble and structural dihydrofolate reductase in bacteriophage T4. J. Virol. **7**:531–533.

Mathews, C. K. 1972. Biochemistry of DNA-defective amber mutants of bacteriophage T4. III. Nucleotide pools. J. Biol. Chem. **247**:7430–7438.

Mathews, C. K. 1976. Biochemistry of DNA-defective mutants of bacteriophage T4. Thymine nucleotide pool dynamics. Arch. Biochem. Biophys. **172**:178–187.

Mathews, C. K. 1977. Reproduction of large virulent bacteriophages. Compr. Virol. **7**:179–294.

Mathews, C. K., L. K. Crosby, and L. M. Kozloff. 1973. Inactivation of T4D bacteriophage by antiserum against bacteriophage dihydrofolate reductase. J. Virol. **12**:74–78.

Mathews, C. K., T. W. North, and G. P. V. Reddy. 1979. Multienzyme complexes in DNA precursor biosynthesis. Adv. Enzyme Regul. **17**:133–156.

Mathews, C. K., and N. K. Sinha. 1982. Are DNA precursors concentrated at replication sites? Proc. Natl. Acad. Sci. U.S.A. **79**:302–306.

Matthews, B. W., M. G. Grütter, W. F. Anderson, and S. J. Remington. 1981. Common precursor of lysozymes of hen egg white and bacteriophage T4. Nature (London) **290**:334–335.

Matthews, B. W., and S. J. Remington. 1974. Three dimensional structure of the lysozyme from bacteriophage T4. Proc. Natl. Acad. Sci. U.S.A. **71**:4178–4182.

Matthews, B. W., S. J. Remington, M. G. Grütter, and W. F. Anderson. 1981. Relation between hen egg white lysozyme and bacteriophage T4 lysozyme: evolutionary implications. J. Mol. Biol. **147**:545–558.

Mattson, T., J. Richardson, and D. Goodin. 1974. Mutant of bacteriophage T4D affecting expression of many early genes. Nature (London) **250**:48–50.

Mattson, T., G. van Houwe, A. Bolle, G. Selzer, and R. Epstein. 1977. Genetic identification of cloned fragments of bacteriophage T4 DNA and complementation by some clones containing early T4 genes. Mol. Gen. Genet. **154**:319–326.

Mattson, T., G. van Houwe, and R. H. Epstein. 1978. Isolation and characterization of conditional lethal mutations in the *mot* gene of bacteriophage T4. J. Mol. Biol. **126**:551–570.

Matz, K., M. Schmandt, and G. N. Gussin. 1982. The *rex* gene of bacteriophage lambda is really two genes. Genetics **102**:319–327.

Maxam, A., and W. Gilbert. 1980. Sequencing end-labeled DNA with base-specific chemical cleavage. Methods Enzymol. **65**:499–560.

Maynard-Smith, J. 1978. The evolution of sex. Cambridge University Press, Cambridge.

Maynard-Smith, S., and N. Symonds. 1973. Involvement of bacteriophage T4 in radiation repair. J. Mol. Biol. **74**:33–44.

Maynard-Smith, S., N. Symonds, and P. White. 1970. The Kornberg polymerase and the repair of irradiated T4 bacteriophage. J. Mol. Biol. **54**:391–393.

Mazzara, G. P., G. Plunkett, and W. H. McClain. 1981. DNA sequence of the transfer RNA region of bacteriophage T4: implications for transfer RNA synthesis. Proc. Natl. Acad. Sci. U.S.A. **78**:889–892.

Mazzara, G. P., J. G. Seidman, W. H. McClain, H. Yesian, J. Abelson, and C. Guthrie. 1977. Nucleotide sequence of an arginine transfer RNA from bacteriophage T4. J. Biol. Chem. **252**:8245–8253.

McCarthy, D. 1979. Gyrase-dependent initiation of bacteriophage T4 DNA replication: interaction of *Escherichia coli* gyrase with novobiocin, coumermycin and phage DNA-delay gene products. J. Mol. Biol. **127**:265–283.

McCarthy, D., C. Minner, H. Bernstein, and C. Bernstein. 1976. DNA elongation rates and growing point distributions of wild type T4 and DNA-delay amber mutant. J. Mol. Biol. **106**:963–981.

McClain, W. H. 1970. UAG suppressor coded by bacteriophage T4. FEBS Lett. **6**:99–101.

McClain, W. H. 1977. Seven terminal steps in a biosynthetic pathway leading from DNA to transfer RNA. Acc. Chem. Res. **10**:418–422.

McClain, W. H. 1979. A role for ribonuclease III in synthesis of bacteriophage T4 transfer RNAs. Biochem. Biophys. Res. Commun. **86**:718–724.

McClain, W. H., C. Guthrie, and B. G. Barrell. 1972. Eight transfer RNAs induced by infection of *Escherichia coli* with bacteriophage T4. Proc. Natl. Acad. Sci. U.S.A. **69**:3703–3707.

McClain, W. H., C. Guthrie, and B. G. Barrell. 1973. The *psu1*$^+$ amber suppressor gene of bacteriophage T4: identification of its amino acid and transfer RNA. J. Mol. Biol. **81**:157–171.

McClain, W. H., G. L. Marchin, F. C. Neidhardt, K. V. Chace, M. L. Rementer, and D. H. Hall. 1975. A gene of bacteriophage T4 controlling the modification of host valyl-tRNA synthetase. Virology **67**:385–394.

McClain, W. H., and J. G. Seidman. 1975. Genetic perturbations that reveal tertiary conformation of tRNA precursor molecules. Nature (London) **257**:106–110.

McClain, W. H., J. G. Seidman, and F. J. Schmidt. 1978. Evolution of the biosynthesis of 3′ terminal C-C-A residues in T-even bacteriophage transfer RNAs. J. Mol. Biol. **119**:519–536.

McClelland, M. 1981. The effect of sequence-specific DNA methylation on restriction endonuclease cleavage. Nucleic Acids Res. **9**:5859–5866.

McEntee, K., G. M. Weinstock, and I. R. Lehman. 1979. Initiation of general recombination catalyzed in vitro by the *recA* protein of *Escherichia coli*. Proc. Natl. Acad. Sci. U.S.A. **76**:2615–2619.

McGhee, J. D., and P. H. von Hippel. 1974. Theoretical aspects of DNA-protein interactions: co-operative and non-co-operative binding of large ligands to a one-dimensional homogeneous lattice. J. Mol. Biol. **86**:469–489.

McHenry, C., and A. Kornberg. 1981. DNA polymerase III holoenzyme, p. 39–50. *In* P. D. Boyer (ed.), The enzymes, vol. 14. Academic Press, Inc., New York.

McKay, D., and K. R. Williams. 1982. Crystallization of a tryptic core of the single-stranded DNA binding protein of bacteriophage T4. J. Mol. Biol. **160**:659–661.

McMillan, S., H. J. Edenberg, E. H. Radany, R. C. Friedberg, and E. C. Friedberg. 1981. The *denV* gene of bacteriophage T4 codes for both pyrimidine dimer-DNA glycosylase and aprymidinic endonuclease activities. J. Virol. **40**:211–223.

McNicol, L. 1973. Transforming ability of a T4 RNA-DNA copolymer. J. Virol. **12**:367–373.

McNicol, L. A., and E. B. Goldberg. 1973. An immunological characterization of glucosylation in bacteriophage T4. J. Mol. Biol. **76**:285–301.

McNicol, L. A., L. D. Simon, and L. W. Black. 1977. A mutation which bypasses the requirement for p24 in bacteriophage T4 capsid morphogenesis. J. Mol. Biol. **116**:261–283.

Meezan, E., and W. B. Wood. 1971. The sequence of gene product interaction in bacteriophage T4 core assembly. J. Mol. Biol. **58**:685–692.

Mei-hao, H., W. Ai, and H. Hui-fen. 1982. Purification of T4 RNA ligase by dextran blue-Sepharose 4B affinity chromatography. Anal. Biochem. **125**:1–5.

Meistrich, M. L. 1972. Contribution of thymine dimers to the ultraviolet light inactivation of mutants of bacteriophage T4. J. Mol. Biol. **66**:97–106.

Meistrich, M. L., and J. W. Drake. 1972. Mutagenic effects of thymine dimers in bacteriophage T4. J. Mol. Biol. **66**:107–114.

Melamede, R. J., and S. S. Wallace. 1977. Properties of the nonlethal recombinational repair *x* and *y* mutants of bacteriophage T4. II. DNA synthesis. J. Virol. **24**:28–40.

Melamede, R. J., and S. S. Wallace. 1978. The effect of exogenous deoxyribonucleosides on thymidine incorporation in T4-infected cells. FEBS Lett. **87**:12–16.

Melamede, R. J., and S. S. Wallace. 1980a. Properties of the nonlethal recombinational repair deficient mutants of bacteriophage T4. III. DNA replicative intermediates and T4w. Mol. Gen. Genet. **177**:501–509.

Melamede, R. J., and S. S. Wallace. 1980b. Phenotypic differences among the alleles of the T4 recombination deficient mutants. Mol. Gen. Genet. **179**:327–330.

Merril, C. R., D. Goldman, S. A. Sedman, and M. H. Ebert. 1981. Ultrasensitive stain for proteins in polyacrylamide gels shows regional variation in cerebrospinal fluid proteins. Science **211**:1437–1438.

Merril, C., M. Gottesman, and S. Adhya. 1981. *Escherichia coli gal* operon proteins made after prophage lambda induction. J. Bacteriol. **147**:875–887.

Meselson, M., R. Yuan, and J. Heywood. 1972. Restriction and modification of DNA. Annu. Rev. Biochem. **41**:447–466.

Messing, J., R. Crea, and P. Seeburg. 1981. A system for shotgun DNA sequencing. Nucleic Acids Res. **9**:309–321.

Meyer, T., and K. Geider. 1980. Replication of phage fd DNA with purified proteins, p. 579–588. *In* B. Alberts (ed.), Mechanistic studies of DNA replication and genetic recombination. Academic Press, Inc., New York.

Meyer, T., and K. Geider. 1982. Enzymatic synthesis of bacteriophage fd viral DNA. Nature (London) **296**:828–832.

Mickelson, C., and J. S. Wiberg. 1981. Membrane-associated DNase activity controlled by genes 46 and 47 of bacteriophage T4D and elevated DNase activity associated with the T4 *das* mutation. J. Virol. **40**:65–77.

Milanesi, G., E. N. Brody, and E. P. Geiduschek. 1969. Sequence of the in vitro transcription of T4 DNA. Nature (London) **221**:1014–1016.

Milanesi, G., E. N. Brody, O. Grau, and E. P. Geiduschek. 1970. Transcription of the bacteriophage T4 template in vitro: separation of delayed-early from immediate-early transcription. Proc. Natl. Acad. Sci. U.S.A. **66**:181–188.

Mileham, A. J., H. R. Revel, and N. E. Murray. 1980. Molecular cloning of the T4 genome: organization and expression of the frd-DNA ligase region. Mol. Gen. Genet. **179**:227–239.

Miller, J. H. 1972. Experiments in molecular genetics. Cold Spring Harbor Laboratory, Cold Spring Harbor, N.Y.

Miller, J. H. 1980. The *lacI* gene: its role in lac operon control and its use as a genetic system, p. 31–88. *In* J. H. Miller and W. S. Reznikoff (ed.), The operon. Cold Spring Harbor Laboratory, Cold Spring Harbor, N.Y.

Miller, R. C., Jr. 1970. Double-stranded scissions in bacteriophage T4 deoxyribonucleic acid as a result of ^{32}P decay. J. Virol. **5**:536–539.

Miller, R. C., Jr. 1972. Association of replicative T4 deoxyribonucleic acid and bacterial membranes. J. Virol. **10**:920–924.

Miller, R. C., Jr. 1975a. Replication and molecular recombination of T-even phage. Annu. Rev. Microbiol. **29**:355–376.

Miller, R. C., Jr. 1975b. T4 DNA polymerase (gene 43) is required in vivo for repair of gaps in recombinants. J. Virol. **15**:316–321.

Miller, R. C., Jr., and A. W. Kozinski. 1970. Early intracellular events in the replication of bacteriophage T4 deoxyribonucleic acid. V. Further studies on the T4 protein-deoxyribonucleic acid complex. J. Virol. **5**:490–501.

Miller, R. C., Jr., A. W. Kozinski, and S. Litwin. 1970. Molecular recombination in T4 bacteriophage deoxyribonucleic acid. III. Formation of long single strands during recombination. J. Virol. **5**:368–380.

Miller, R. C., E. T. Young, R. H. Epstein, H. M. Krisch, T. Mattson, and T. A. Bolle. 1981. Regulation of the synthesis of the T4 DNA polymerase (gene 43). Virology **110**:98–112.

Millette, R. L., C. D. Trotter, P. Herrlich, and M. Schweiger. 1970. In vitro synthesis, termination, and release of active messenger RNA. Cold Spring Harbor Symp. Quant. Biol. **35**:135–142.

Minagawa, T. 1977. Endonuclease of T4 ghosts. Virology **76**:234–245.

Minagawa, T., and T. Ryo. 1978. Substrate specificity of gene 49-controlled deoxyribonuclease of bacteriophage T4: special reference to DNA packaging. Virology **91**:222–233.

Minagawa, T., and Y. Ryo. 1979. Genetic control of formation of very fast sedimenting DNA of bacteriophage T4. Mol. Gen. Genet. **170**:113–115.

Minkley, E. G., and D. Pribnow. 1973. Transcription of the early region of bacteriophage T7: selective initiation with dinucleotides. J. Mol. Biol. **77**:255.

Miskimins, R., S. Schneider, V. Johns, and H. Bernstein. 1982. Topoisomerase involvement in multiplicity reactivation of phage T4. Genetics **101**:157–177.

Mizuuchi, K., B. Kemper, J. Hays, and R. A. Weisberg. 1982. T4 endonuclease VII cleaves Holliday structures. Cell **29**:357–365.

Model, P., R. E. Webster, and N. D. Zinder. 1969. The UGA codon in vitro: chain termination and suppression. J. Mol. Biol. **43**:117–190.

Moise, H., and J. Hosoda. 1976. T4 gene 32 protein model for control of activity at a replication fork. Nature (London) **259**:455–458.

Molholt, B., and B. de Groot. 1969. Double conditional lethality: temperature sensitive and amber mutations in the glucosyl transferase gene of bacteriophage T2. Eur. J. Biochem. **9**:222–228.

Molholt, B., and D. Fraser. 1968. Host-controlled restriction of T-even bacteriophages: relation of endonuclease I and T-even-induced nucleases to restriction. J. Virol. **2**:313–319.

Monod, J., and E. L. Wollman. 1947. L'inhibition de la croissance et de l'adaption enzymatique chez les bacteries infectées par le bacteriophage. Ann. Inst. Pasteur Paris **73**:937–956.

Montgomery, D. L., and L. R. Snyder. 1973. A negative effect of β-glucosylation on T4 growth in certain RNA polymerase mutants of *Escherichia coli*: genetic evidence implicating pyrimidine-rich sequences of DNA in transcription. Virology **53**:349–358.

Moody, M. F. 1965. The shape of the T-even bacteriophage head. Virology **26**:567–576.

Moody, M. F. 1971. Application of optical diffraction to helical structures in the bacteriophage tail. Philos. Trans. R. Soc. London Ser. B **261**:181–195.

Moody, M. F. 1973. Sheath of bacteriophage T4. III. Contraction mechanism deduced from partially contracted sheaths. J. Mol. Biol. **60**:613–635.

Moody, M. F., and L. Makowski. 1981. X-ray diffraction study of tail tubes from bacteriophage T2L. J. Mol. Biol. **150**:217–244.

Morris, C. F., H. Hama-Inaba, D. Mace, N. K. Sinha, and B. M. Alberts. 1979. Purification of the gene 43, 44, 45, and 62 proteins of bacteriophage T4 DNA replication apparatus. J. Biol. Chem. **254**:6787–6796.

Morris, C. F., L. A. Moran, and B. M. Alberts. 1979. Purification of the gene 41 protein of bacteriophage T4. J. Biol. Chem. **254**:6797–6802.

Morrison, A., and N. R. Cozzarelli. 1979. Site-specific cleavage of DNA by *E. coli* DNA gyrase. Cell **17**:175–184.

Morse, D. E. 1970. "Delayed-early" mRNA for the tryptophan operon? An effect of chloramphenicol. Cold Spring Harbor Symp. Quant. Biol. **35**:495–496.

Mortelmans, K., and E. C. Friedberg. 1972. Deoxyribonucleic acid repair in bacteriophage T4: observations on the role of the *x* and *y* genes and of host factors. J. Virol. **10**:730–736.

Morton, D., E. M. Kutter, and B. S. Guttman. 1978. Synthesis of T4 DNA and bacteriophage in the absence of dCMP hydroxymethylase. J. Virol. **28**:262–269.

Mosbaugh, D. W., and S. Linn. 1982. Characterization of the action of *Escherichia coli* DNA polymerase I at incisions produced by repair endodeoxyribonucleases. J. Biol. Chem. **257**:575–583.

Mosher, R. A., A. B. DiRenzo, and C. K. Mathews. 1977. Bacteriophage T4 virion dihydrofolate reductase: approaches to quantitation and assessment of function. J. Virol. **23**:645–658.

Mosher, R. A., and C. K. Mathews. 1979. Bacteriophage T4-coded dihydrofolate reductase: synthesis, turnover, and location of the virion protein. J. Virol. **31**:94–103.

Mosig, G. 1963. Genetic recombination in bacteriophage T4 during replication of DNA fragments. Cold Spring Harbor Symp. Quant. Biol. **28**:35–42.

Mosig, G. 1966. Distances separating genetic markers in T4 DNA. Proc. Natl. Acad. Sci. U.S.A. **56**:1177–1183.

Mosig, G. 1968. A map of distances along the DNA molecule of phage T4. Genetics **59**:137–151.

Mosig, G. 1970a. Preferred origin and direction of bacteriophage T4 DNA replication. I. A gradient of allele frequencies in crosses between normal and small T4 particles. J. Mol. Biol. **53**:503–514.

Mosig, G. 1970b. Recombination in bacteriophage T4. Adv. Genet. **15**:1–54.

Mosig, G. 1974. On the role of *Escherichia coli* DNA polymerase I and of T4 gene 32 protein in recombination of phage T4, p. 29–30. *In* R. F. Grell (ed.), Mechanisms in recombination. Plenum Publishing Corp., New York.

Mosig, G. 1976. Linkage map and genes of bacteriophage T4, p. 664–676. *In* G. B. Fassman (ed.), Handbook of biochemistry and molecular biology, 3rd ed. CRC Press, Cleveland.

Mosig, G. 1982. Genetic maps of bacteriophage T4. Genet. Maps **2**:15–24.

Mosig, G., S. Benedict, D. Ghosal, A. Luder, R. Dannenberg, and S. Bock. 1980. Genetic analysis of DNA replication in bacteriophage T4. ICN-UCLA Symp. Mol. Cell. Biol. **19**:527.

Mosig, G., W. Bergquist, and S. Bock. 1977. Multiple interactions of a DNA-binding protein in vivo. III. Phage T4 gene-32 mutations differentially affect insertion-type recombination and membrane properties. Genetics **86**:5–23.

Mosig, G., and S. Bock. 1976. Gene 32 of bacteriophage T4 moderates the activities of the T4 gene 46/47-controlled nuclease and of the *Escherichia coli* RecBC nuclease in vivo. J. Virol. **17**:756–761.

Mosig, G., D. W. Bowden, and S. Bock. 1972. *E. coli* DNA polymerase I and other host functions participate in T4 DNA replication and recombination. Nature (London) New Biol. **240**:12–16.

Mosig, G., and A. M. Breschkin. 1975. Genetic evidence for an additional function of phage T4 gene 32 protein: interaction with ligase. Proc. Natl. Acad. Sci. **72**:1226–1230.

Mosig, G., J. R. Carnigan, J. B. Bibring, R. Cole, H.-G. O. Bock, and S. Bock. 1972. Coordinate variation in lengths of DNA molecules and head lengths in morphological variants of bacteriophage T4. J. Virol. **9**:857–871.

Mosig, G., R. Dannenberg, D. Ghosal, A. Luder, S. Benedict, and S. Bock. 1979. General genetic recombination in bacteriophage T4. Stadler Genet. Symp. **11**:31–55.

Mosig, G., R. Ehring, W. Schliewen, and S. Bock. 1971. The patterns of recombination and segregation in terminal regions of T4 DNA molecules. Mol. Gen. Genet. **113**:51–91.

Mosig, G., D. Ghosal, and S. Bock. 1981. Interactions between the maturation protein gp17 and the single-stranded DNA binding protein gp32 initiate DNA packaging and compete with initiation of secondary DNA replication forks in phage T4, p. 139–150. *In* M. DuBow (ed.), Bacteriophage assembly. Alan R. Liss, Inc., New York.

Mosig, G., A. Luder, G. Garcia, R. Dannenberg, and S. Bock. 1979. In vivo interactions of genes and proteins in DNA replication and recombination of phage T4. Cold Spring Harbor Symp. Quant. Biol. **43**:501–515.

Mosig, G., A. Luder, L. Rowen, P. Macdonald, and S. Bock. 1981. On the role of recombination and topoisomerase in primary and secondary initiation of T4 DNA replication, p. 277–295. *In* D. Ray (ed.), The initiation of replication. Academic Press, Inc., New York.

Mosig, G., and R. Werner. 1969. On the replication of incomplete chromosomes of phage T4. Proc. Natl. Acad. Sci. U.S.A. **64**:747–754.

Motamedi, H., K. Lee, L. Nichols, and F. J. Schmidt. 1982. An RNA species involved in *Escherichia coli* ribonuclease P activity. Gene cloning and effect on transfer RNA synthesis in vivo. J. Mol. Biol. **162**:535–554.

Mount, D. W. 1980. The genetics of protein degradation in bacteria. Annu. Rev. Genet. **14**:279–319.

Mufti, S. 1979. Mutator effects of alleles of phage T4 genes 32, 41, 44, and 45 in the presence of antimutator polymerase. Virology **94**:1–9.

Mufti, S. 1980. The effect of ultraviolet mutagenesis of genes 32, 42, 44, and 45 alleles of phage T4 in the presence of wild-type or antimutator DNA polymerase. Virology **105**:345–356.

Mufti, S., and H. Bernstein. 1974. The DNA-delay mutants of bacteriophage T4. J. Virol. **14**:860–871.

Mukai, G. Streisinger, and N. Miller. 1967. The mechanisms of lysis in phage T4-infected cells. Virology **33**:398–404.

Muller, H. J. 1932. Some genetic aspects of sex. Am. Nat. **66**:118–138.

Müller-Salamin, L., L. Onorato, and M. K. Showe. 1977. Localization of minor protein components of the head of bacteriophage T4. J. Virol. **24**:121–134.

Murialdo, H., and A. Becker. 1978. Head morphogenesis of complex double-stranded deoxyribonucleic acid bacteriophages. Microbiol. Rev. **47**:529–576.

Murray, R. E., and C. K. Mathews. 1969. Biochemistry of DNA-defective amber mutants of bacteriophage T4. II. Intracellular DNA forms in infection by gene 44 mutants. J. Mol. Biol. **44**:249–262.

Murray, R. G. E., D. H. Gillen, and F. C. Heagy. 1950. Cytological changes in *Escherichia coli* produced by infection with phage T2. J. Bacteriol. **59**:603–615.

Muskavitch, K. T., and S. Linn. 1981. recBC-like enzymes: exonuclease V deoxyribonucleases, p. 233–250. *In* P. Boyer (ed.), The enzymes, vol. 14A. Academic Press, Inc., New York.

Muzyczka, N., R. L. Poland, and M. J. Bessman. 1972. Studies on the biochemical basis of spontaneous mutation. I. A comparison of the deoxyribonucleic acid polymerases of mutator, antimutator, and wild type strains of bacteriophage T4. J. Biol. Chem. **247**:7116–7122.

Nakabeppu, Y., and M. Sekiguchi. 1981. Physical association of pyrimidine dimer DNA glucosylase and apurinic/apyrimidinic DNA endonuclease essential for repair of ultraviolet-damaged DNA. Proc. Natl. Acad. Sci. U.S.A. **78**:2742–2746.

Nakabeppu, Y., K. Yamashita, and M. Sekiguchi. 1982. Purification and characterization of normal and mutant forms of T4 endonuclease V. J. Biol. Chem. **257**:2556–2562.

Nakamura, K., and L. M. Kozloff. 1978. Folate polyglutamates in T4D bacteriophage and T4D-infected *Escherichia coli*. Biochim. Biophys. Acta **540**:313–319.

Napoli, C., L. Gold, and B. S. Singer. 1981. Translational reinitiation in the *r*IIB cistron of bacteriophage T4. J. Mol. Biol. **149**:433–449.

Nasmyth, K. A., K. Tatchell, B. D. Hall, C. Astell, and M. Smith. 1980. Physical analysis of mating-type loci in *Saccharomyces cerevisiae*. Cold Spring Harbor Symp. Quant. Biol. **45**:961–981.

Natale, P. J., and J. M. Buchanan. 1972. DNA-directed synthesis in vitro of T4 phage-specific enzymes. Proc. Natl. Acad. Sci. U.S.A. **69**:2513–2517.

Natale, P. J., and J. M. Buchanan. 1974. Initiation characteristics for the synthesis of five T4 phage-specific messenger RNAs in vitro. Proc. Natl. Acad. Sci. U.S.A. **71**:422–426.

Natale, P. J., and J. M. Buchanan. 1977. Initiation of synthesis of messenger RNA of deoxynucleotide kinase by oligoribonucleotides. J. Biol. Chem. **252**:2304–2310.

Natale, P. J., C. Ireland, and J. M. Buchanan. 1975. Sizing of two bacteriophage T4-specific messenger ribonucleic acids formed in vitro. Biochem. Biophys. Res. Commun. **66**:1287–1293.

Neidhardt, F. C., T. A. Phillips, R. van Bogelen, M. W. Smith, Y. Georgalis, and A. R. Subramanian. 1981. Identity of the B56.5 protein, the A-protein, and the *groE* gene product of *Escherichia coli*. J. Bacteriol. **145**:513–520.

Nelson, M. A., M. Ericson, L. Gold, and J. F. Pulitzer. 1982. The isolation and characterization of TabR bacteria: hosts that restrict bacteriophage T4 *r*II mutants. Mol. Gen. Genet. **188**:60–68.

Nelson, M. A., and L. Gold. 1982. The isolation and characterization of bacterial strains (*tab* 32) that restrict bacteriophage T4 gene 32 mutants. Mol. Gen. Genet. **188**:69–76.

Nelson, M. A., B. S. Singer, L. Gold, and D. Pribnow. 1981. Mutations that detoxify an aberrant T4 membrane protein. J. Mol. Biol. **149**:377–403.

Nevers, P., and H.-C. Spatz. 1975. *Escherichia coli* mutants *uvrD* and *uvrE* deficient in gene conversions of lambda-heteroduplexes. Mol. Gen. Genet. **139**:233–243.

Newport, J. W., S. C. Kowalczykowski, N. Lonberg, L. S. Paul, and P. H. von Hippel. 1980. Molecular aspects of the interactions of T4 coded gene 32 protein and DNA polymerase (gene 43 protein) with nucleic acids, p. 485–505. *In* B. M. Alberts (ed.), Mechanistic studies of DNA replication and genetic recombination. Academic Press, Inc., New York.

Newport, J. W., N. Lonberg, S. C. Kowalczykowski, L. S. Paul, and P. von Hippel. 1981. Interactions of bacteriophage T4-coded gene 32 protein with nucleic acids. II. Specificity of binding to DNA and RNA. J. Mol. Biol. **145**:105–121.

Niggemann, F., I. Green, H.-P. Meyer, and W. Ruger. 1981. Physical mapping of bacteriophage T4. Mol. Gen. Genet. **184**:289–299.

Nirenberg, M., and J. Matthaei. 1961. The dependence of cell-free protein synthesis in *E. coli* upon naturally occurring or synthetic polyribonucleotides. Proc. Natl. Acad. Sci. U.S.A. **47**:1588–1602.

Nishida, Y., S. Yasuda, and M. Sekiguchi. 1976. Repair of DNA damaged by methyl methanesulfonate in bacteriophage T4. Biochim. Biophys. Acta **422**:208–215.

Nishimoto, H., M. Takayama, and T. Minagawa. 1979. Purification and some properties of deoxyribonuclease whose synthesis is controlled by gene 49 of bacteriophage T4. Eur. J. Biochem. **100**:433–440.

Nomura, M., and S. Benzer. 1961. The nature of "deletion" mutants in the rII region of phage T4. J. Mol. Biol. **3**:684–692.

Nomura, M., B. D. Hall, and S. Spiegelman. 1960. Characterization of RNA synthesized in *Escherichia coli* after bacteriophage T2 infection. J. Mol. Biol. **2**:306–326.

Nomura, M., S. Jinks-Robertson, and A. Miura. 1982. Regulation of ribosomal biosynthesis in *Escherichia coli*, p. 91–104. *In* M. Grunberg-Manago and B. Safer (ed.), Interaction of translational and transcriptional controls in the regulation of gene expression. Elsevier Biomedical, New York.

Nomura, M., K. Matsubara, K. Okamoto, and R. Fujimura. 1962. Inhibition of host nucleic acid and protein synthesis by bacteriophage T4: its relation to the physical and functional integrity of host chromosome. J. Mol. Biol. **5**:535–549.

Nomura, M., E. A. Morgan, and S. R. Jaskunas. 1977. Genetics of bacterial ribosomes. Annu. Rev. Genet. **11**:297–347.

Nomura, M., K. Okamoto, and K. Asano. 1962. RNA metabolism in *Escherichia coli* infected with bacteriophage T4: inhibition of host ribosomal and soluble RNA synthesis by phage and effect of chloromycetin. J. Mol. Biol. **4**:376–387.

Nomura, M., C. Witten, N. Mantei, and H. Echols. 1966. Inhibition of host nucleic acid synthesis by bacteriophage T4: effect of chloramphenicol at various multiplicities of infection. J. Mol. Biol. **17**:273–278.

Nonn, E., and C. Bernstein. 1977. Multiplicity reactivation and repair of nitrous acid-induced lesions in bacteriophage T4. J. Mol. Biol. **116**:31–47.

North, T. W., and C. K. Mathews. 1977. T4 phage-coded deoxycytidylate hydroxymethylase: purification and studies on intermolecular interactions. Biochem. Biophys. Res. Commun. **77**:898–904.

North, T. W., M. E. Stafford, and C. K. Mathews. 1976. Biochemistry of DNA-defective mutants of bacteriophage T4. VI. Biological functions of gene 42. J. Virol. **17**:973–982.

Nossal, N. G. 1974. DNA synthesis on a double-stranded DNA template by the T4 bacteriophage DNA polymerase and the T4 gene 32 DNA unwinding protein. J. Biol. Chem. **249**:5668–5676.

Nossal, N. G. 1979. DNA replication with bacteriophage T4 proteins. Purification of the proteins encoded by T4 genes 41, 45, 44 and 62 using a complementation assay. J. Biol. Chem. **254**:6026–6031.

Nossal, N. G. 1980. RNA priming of DNA replication by bacteriophage T4 proteins. J. Biol. Chem. **255**:2176–2182.

Nossal, N. G., and M. S. Hershfield. 1971. Nuclease activity in a fragment of bacteriophage T4 deoxyribonucleic acid polymerase induced by the amber mutant *am* B22. J. Biol. Chem. **246**:5414–4526.

Nossal, N. G., and M. S. Hershfield. 1973. Exonuclease activity of wild type and mutant T4 DNA polymerases: hydrolysis during DNA synthesis in vitro, p. 47–62. *In* R. D. Wells and R. B. Inman (ed.), DNA synthesis in vitro. University Park Press, Baltimore.

Nossal, N. G., and B. M. Peterlin. 1979. DNA replication by bacteriophage T4 proteins. The T4 43, 32, 44-62 and 45 proteins are required for strand displacement synthesis at nicks in duplex DNA. J. Biol. Chem. **254**:6032–6037.

Notani, G. W. 1973. Regulation of bacteriophage T4 gene expression. J. Mol. Biol. **73**:231–249.

Novogrodsky, A., and J. Hurwitz. 1966. The enzymatic phosphorylation of ribonucleic acid and deoxyribonucleic acid. I. Phosphorylation at 5' hydroxyl termini. J. Biol. Chem. **241**:2923–2932.

Nusslein-Crystalla, V., I. Niedenhof, and R. Rein. 1982. *dnaC*-dependent reconstitution of replication forks in *Escherichia coli* lysate. J. Bacteriol. **150**:286–292.

O'Donovan, G. A., G. A. Edlin, J. A. Fuchs, J. Neuhard, and E. Thomassen. 1971. Deoxycytidine triphosphate deaminase: characterization of an *Escherichia coli* mutant lacking the enzyme. J. Bacteriol. **105**:666–672.

O'Farrell, P. H. 1975. High resolution two-dimensional electrophoresis of proteins. J. Biol. Chem. **250**:4007–4021.

O'Farrell, P. H., E. Kutter, and M. Nakanish. 1980. A restriction map of the bacteriophage T4 genome. Mol. Gen. Genet. **179**:421–435.

O'Farrell, P. H., and P. Z. O'Farrell. 1977. Two-dimensional polyacrylamide gel electrophoretic fractionation. Methods Cell Biol. **16**:407–420.

O'Farrell, P. Z., and L. M. Gold. 1973a. Bacteriophage T4 gene expression: evidence for two classes of prereplicative cistrons. J. Biol. Chem. **248**:5502–5511.

O'Farrell, P. Z., and L. M. Gold. 1973b. Transcription and translation of pre-replicative bacteriophage T4 genes in vitro. J. Biol. Chem. **248**:5512–5519.

O'Farrell, P. Z., L. M. Gold, and W. M. Huang. 1973. The identification of prereplicative bacteriophage T4 proteins. J. Biol. Chem. **248**:5499–5505.

O'Farrell, P. Z., H. M. Goodman, and P. H. O'Farrell. 1977. High resolution two-dimensional electrophoresis of basic as well as acidic proteins. Cell **12**:1133–1142.

Ohshima, S., and M. Sekiguchi. 1972. Induction of a new enzyme activity to excise pyrimidine dimers in *Escherichia coli* infected with bacteriophage T4. Biochem. Biophys. Res. Commun. **47**:1126–1132.

Ohtsuka, E., S. Nishikawa, M. Sugiura, and M. Ikehara. 1976. Joining of ribonucleotides with T4 RNA ligase and identification of the oligonucleotide-adenylate intermediate. Nucleic Acids Res. **3**:1613–1623.

Oishi, M. 1968. Studies of DNA replication in vivo. III. Accumulation of single-stranded isolation product of DNA replication by conditional mutant strains of T4. Proc. Natl. Acad. Sci. U.S.A. **60**:1000–1006.

Okada, Y., G. Streisinger, J. E. Owen, J. Newton, A. Tsugita, and M. Inouye. 1972. Molecular basis of a mutational hot spot in the lysozyme gene of bacteriophage T4. Nature (London) **236**:338–341.

Okamoto, K., and M. Yutsudo. 1974. Participation of the *s* gene product of phage T4 in the establishment of resistance of T4 ghosts. Virology **58**:369–376.

Oleson, A. E., and J. F. Koerner. 1964. A deoxyribonuclease induced by infection with bacteriophage T2. J. Biol. Chem. **239**:2935–2943.

Oliver, D. B., and R. A. Crowther. 1981. DNA sequence of the tail fiber genes 36 and 37 of bacteriophage T4. J. Mol. Biol. **153**:545–568.

Oliver, D. B., and E. B. Goldberg. 1977. Protection of parental T4 DNA from a restriction exonuclease by the product of gene 2. J. Mol. Biol. **116**:877–881.

Oliver, D. B., M. H. Malamy, and E. B. Goldberg. 1981. Cloned genes for bacteriophage T4 late functions are expressed in *Escherichia coli*. J. Mol. Biol. **152**:267–283.

Olson, G. B., and H. Hartman. 1982. Martensite and life: displacive transformations as biological processes. J. Phys. Paris Colloq. **43**:C4-855.

Onorato, L., and M. K. Showe. 1975. Gene 21 protein-dependent proteolysis *in vitro* of purified gene 22 product of bacteriophage T4. J. Mol. Biol. **92**:395–412.

Onorato, L., B. Stirmer, and M. K. Showe. 1978. Isolation and characterization of bacteriophage T4 mutant preheads. J. Virol. **27**:409–426.

Opella, S. J., T. A. Cross, J. A. DiVerdi, and C. F. Sturm. 1980. Nuclear magnetic resonance of the filamentous bacteriophage fd. Biophys. J. **32**:531–548.

Ovchinnikov, Y. A., V. M. Lipkin, N. N. Modyanov, O. Y. Chertov, and Y. V. Smirnov. 1977. Primary structure of alpha-subunit of DNA-dependent RNA polymerase for Escherichia coli. FEBS Lett. **76**:108–111.

Owen, J. E., D. W. Schultz, A. Taylor, and G. R. Smith. 1983. Nucleotide sequence of the lysozyme gene of bacteriophage T4. Analysis of mutations involving repeated sequences. J. Mol. Biol. **165**:229–248.

Ozeki, H., H. Inokuchi, F. Yamao, M. Kodaira, H. Sakano, T. Ikemura, and Y. Shimura. 1980. Genetics of nonsense suppressor tRNAs in *Escherichia coli*, p. 341–362. *In* D. Soll, J. N. Abelson, and P. R. Schimmel (ed.), Transfer RNA: biological aspects. Cold Spring Harbor Laboratory, Cold Spring Harbor, N.Y.

Ozeki, H., H. Sakano, S. Yamada, T. Ikemura, and Y. Shimura. 1975. Temperature sensitive mutants of *Escherichia coli* defective in tRNA biosynthesis. Brookhaven Symp. Biol. **26**:89–105.

Paddock, G., and J. Abelson. 1973. Sequence of T4, T2 and T6 bacteriophage species I RNA and specific cleavage by an *E. coli* endonuclease. Nature (London) New Biol. **246**:2–6.

Paddock, G., and J. Abelson. 1975. Nucleotide sequence determination of bacteriophage T4 species I ribonucleic acid. J. Biol. Chem. **250**:4185–4206.

Paetkau, V., and G. Coy. 1972. On the purification of DNA-dependent RNA polymerase from *E. coli*: removal of an ATPase. Can. J. Biochem. **50**:142–150.

Panuska, J. R., and D. A. Goldthwait. 1980. A DNA-dependent ATPase from T4-infected *Escherichia coli*. J. Biol. Chem. **255**:5208–5214.

Parker, M. L., and F. A. Eiserling. 1983. Bacteriophage SPO1 structure and morphogenesis. I. Tail structure and length regulation. J. Virol. **46**:239–249.

Parma, D. H., M. Dill, and M. K. Slocum. 1979. Realignment of the genetic map of the terminus of the *r*IIB cistron of bacteriophage T4. Genetics **92**:711–720.

Parson, K. A., and D. P. Snustad. 1975. Host DNA degradation after infection of *Escherichia coli* with bacteriophage T4: dependence of the alternate pathway of degradation which occurs in the absence of both T4 endonuclease II and nuclear disruption on T4 endonuclease IV. J. Virol. **15**:221–224.

Patrick, M. H., and Rahn, R. O. 1976. Photochemistry of DNA and polynucleotides: photoproducts, p. 33–95. *In* S. Y. Wand (ed.), Photochemistry and photobiology of nucleic acids, vol. 2. Academic Press, Inc., New York.

Paulson, J. R., and U. K. Laemmli. 1977. Morphogenetic core of the bacteriophage T4 head. Structure of the core in polyheads. J. Mol. Biol. **111**:459–485.

Paulson, J. R., S. Lazaroff, and U. K. Laemmli. 1976. Head length determination in bacteriophage T4: the role of the core protein P22. J. Mol. Biol. **103**:155–174.

Pawl, G., R. Taylor, K. Minton, and E. C. Friedberg. 1976. Enzymes involved in thymine dimer excision in bacteriophage T4 infected *Escherichia coli*. J. Mol. Biol. **108**:99–109.

Pearson, R. E., and L. Snyder. 1980. Shutoff of lambda gene expression by bacteriophage T4: role of the T4 *alc* gene. J. Virol. **35**:194–202.

Pennica, D., and P. S. Cohen. 1978. Regulation of ribosome function following bacteriophage T4 infection. J. Mol. Biol. **122**:137–144.

Peterson, R. F., P. S. Cohen, and H. L. Ennis. 1972. Properties of phage T4 messenger RNA synthesized in the absence of protein synthesis. Virology **48**:201–206.

Peterson, S., P. L. Bloch, S. Reeh, and F. C. Neidhardt. 1978. Patterns of the amount of 140 individual proteins at different growth rates. Cell **14**:179–190.

Pettijohn, D. E., and R. Hecht. 1973. RNA molecules bound to the folded bacterial genome stabilize DNA folds and segregate domains of supercoiling. Cold Spring Harbor Symp. Quant. Biol. **38**:31–41.

Piechowski, M. M., and M. Susman. 1967. Acridine resistance in phage T4D. Genetics **56**:133–148.

Pinkerton, T. C., G. Paddock, and J. Abelson. 1973. Nucleotide sequence determination of bacteriophage T4 leucine transfer ribonucleic acid. J. Biol. Chem. **248**:6348–6365.

Piperno, J. R., and B. M. Alberts. 1978. An ATP stimulation of T4 DNA polymerase mediated via T4 gene 44/62 and 45 proteins. J. Biol. Chem. **253**:5174–5179.

Piperno, J. R., R. B. Kallen, and B. M. Alberts. 1978. Analysis of T4 DNA replication protein complex. J. Biol. Chem. **253**:5180–5185.

Pirie, N. W. 1946. The state of viruses in the infected cell. Cold Spring Harbor Symp. Quant. Biol. **11**:184–192.

Platt, T. 1981. Termination of transcription and its regulation in the tryptophan operon of *E. coli*. Cell **24**:10–23.

Plunkett, G., G. P. Mazzara, and W. H. McClain. 1981. Characterization of T4 band D RNA, a low molecular weight RNA of unknown function. Arch. Biochem. Biophys. **210**:298.

Poglazev, B. F., and T. I. Nicolskaya. 1969. Self-assembly of the protein of bacteriophage T2 tail cores. J. Mol. Biol. **43**:231–243.

Pollack, Y., Y. Groner, H. Aviv (Greenshpan), and M. Revel. 1970. Role of initiation factor B (F3) in the preferential translation of T4 late messenger RNA in T4 infected *E. coli*. FEBS Lett. **9**:218–221.

Pollock, P. N., and D. H. Duckworth. 1973. Outer-membrane proteins induced by T4 bacteriophage. Biochim. Biophys. Acta **322**:321–328.

Post, L. E., and M. Nomura. 1980. DNA sequences from the *str* operon of *Escherichia coli*. J. Biol. Chem. **255**:4660–4666.

Powling, A., and R. Knippers. 1976. Recombination of bacteriophage T4 in vivo. Mol. Gen. Genet. **149**:63–71.

Pragai, B., and D. Apirion. 1981. Processing of bacteriophage T4 tRNAs: the role of RNase III. J. Mol. Biol. **153**:619–630.

Pragai, B., and D. Apirion. 1982. Processing of bacteriophage T4 tRNAs: structural analysis and in vitro processing of precursors which accumulate in RNase E-strains. J. Mol. Biol. **154**:465–484.

Prashad, N., and J. Hosoda. 1972. Role of genes 46 and 47 in bacteriophage T4 reproduction. II. Formation of gaps on parental DNA of polynucleotide ligase defective mutants. J. Mol. Biol. **70**:617–635.

Prehm, P., B. Jann, K. Jann, G. Schmidt, and S. Stirm. 1975. On a bacteriophage T3 and T4 receptor region within the cell wall lipopolysaccharide of *Escherichia coli* B. J. Mol. Biol. **101**:277–281.

Pribnow, D. 1975. Bacteriophage T7 early promoters: nucleotide sequences of two RNA polymerase binding sites. J. Mol. Biol. **99**:419–443.

Pribnow, D. 1979. Genetic control systems, p. 219–277. *In* R. F. Goldberger (ed.), DNA in biological regulation and development. Plenum Publishing Corp., New York.

Pribnow, D., D. C. Sigurdson, L. Gold, B. S. Singer, C. Napoli, J. Brosius, T. J. Dull, and H. F. Noller. 1981. *r*II cistrons of bacteriophage T4. DNA sequence around the intercistronic divide and positions of genetic landmarks. J. Mol. Biol. **149**:337–376.

Priemer, M. M., and V. L. Chan. 1978. The effects of virus and host genes on recombination among ultraviolet-irradiatd bacteriophage T4. Virology **88**:338–347.

Pritchard, R. H. 1978. Control of DNA replication in bacteria, p. 1–26. *In* I. Molineux and M. Kohiyama (ed.), DNA synthesis present and future. Plenum Publishing Corp., New York.

Pulitzer, J. F. 1970. Function of T4 gene 55. I. Characterization of temperature sensitive mutations in the "maturation" gene 55. J. Mol. Biol. **49**:473–488.

Pulitzer, J. F., A. Coppo, and M. Caruso. 1979. Host-virus interactions in the control of T4 prereplicative transcription. II. Interaction between *tabC(rho)* mutants and T4 *mot* mutants. J. Mol. Biol. **135**:979–997.

Pulitzer, J. F., and E. P. Geiduschek. 1970. Function of T4 gene 55. II. RNA synthesis by temperature-sensitive gene 55 mutants. J. Mol. Biol. **49**:489–507.

Pulitzer, J. F., and M. Yanagida. 1971. Inactive T4 progeny virus formation in a temperature-sensitive mutant of *Escherichia coli* K-12. Virology **45**:539–554.

Purkey, R. M., and K. Ebisuzaki. 1977. Purification and properties of a DNA dependent ATPase induced by bacteriophage T4. Eur. J. Biochem. **75**:303–310.

Purohit, S., R. K. Bestwick, G. W. Lasser, C. M. Rogers, and C. K. Mathews. 1981. T4 phage-coded dihydrofolate reductase. Subunit composition and cloning of its structural gene. J. Biol. Chem. **256**:9121–9125.

Rabussay, D. 1982a. Bacteriophage T4 infection mechanisms, p. 219–331. *In* P. Cohen and S. van Heynigen (ed.), Molecular aspects of cellular regulation, vol. 2. Elsevier/North-Holland Biomedical Press, New York.

Rabussay, D. 1982c. Changes in *Escherichia coli* RNA polymerase after bacteriophage T4 infection. ASM News **48**:398–403.

Rabussay, D., and E. P. Geiduschek. 1977a. Regulation of gene action in the development of lytic bacteriophages. Compr. Virol. **8**:1–196.

Rabussay, D., and E. P. Geiduschek. 1977b. Phage T4-modified RNA polymerase transcribes T4 late genes in vitro. Proc. Natl. Acad. Sci. U.S.A. **74**:5305–5309.

Rabussay, D., and E. P. Geiduschek. 1979a. Construction and properties of a cell-free system for bacteriophage T4 late RNA synthesis. J. Biol. Chem. **254**:339–349.

Rabussay, D., and E. P. Geiduschek. 1979b. Relation between bacteriophage T4 DNA replication and late transcription in vitro and in vivo. Virology **99**:286–301.

Rabussay, D., R. Mailhammer, and W. Zillig. 1972. Regulation of transcription by T4 phage-induced chemical alteration and modification of transcriptase (EC 2.7.7.6), p. 213–227. *In* O. Wieland, E. Helmreich, and H. Holzer (ed.), Metabolic interconversion of enzymes. Springer-Verlag, Berlin, Heidelberg, New York.

Radany, E. H., and E. C. Friedberg. 1980. A pyrimidine dimer-DNA glycosylase activity associated with the *v* gene product of bacteriophage T4. Nature (London) **286**:182–185.

Radany, E. H., and E. C. Friedberg. 1982. Demonstrations of pyrimidine dimer-DNA glycosylase activity in vivo: bacteriophage T4-infected *Escherichia coli* as a model system. J. Virol. **41**:88–96.

Radding, C. M. 1978. Genetic recombination: strand transfer and mismatch repair. Annu. Rev. Biochem. **47**:847–880.

Radding, C. M. 1981. Recombination activities of *E. coli* recA protein. Cell **25**:3–4.

Rahmsdorf, H. J., P. Herrlich, S. H. Pai, M. Schweiger, and H. G. Wittmann. 1973. Ribosomes after injection with bacteriophage T4 and T7. Mol. Gen. Genet. **127**:259–271.

Rahn, R. O. 1979. Nondimer damage in deoxyribonucleic acid caused by ultraviolet irradiation. Photochem. Photobiol. Rev. **4**:267–330.

Rappaport, H., M. Russel, and M. Susman. 1974. Some acridine-resistant mutations of bacteriophage T4D. Genetics **78**:579–592.

Ratner, D. 1974a. The interactions of bacterial and phage proteins with immobilized *E. coli* RNA polymerase. J. Mol. Biol. **88**:373–383.

Ratner, D. 1974b. Bacteriophage T4 transcriptional control gene 55 codes for a protein bound to *Escherichia coli* RNA polymerase. J. Mol. Biol. **89**:803–807.

Ratner, D. 1976. The *rho* gene of E. coli maps at suA, p. 645–655. *In* R. Losick and M. Chamberlin (ed.), RNA polymerase. Cold Spring Harbor Laboratory, Cold Spring Harbor, N. Y.

Ray, P., N. K. Sinha, H. R. Warner, and D. P. Snustad. 1972. Genetic location of a mutant of bacteriophage T4 deficient in the ability to induce endonuclease II. J. Virol. **9**:184–186.

Ray, U., L. Bartenstein, and J. W. Drake. 1972. Inactivation of bacteriophage T4 by ethyl methane sulfonate: influence of host and viral genotypes. J. Virol. **9**:440–447.

Rayssiguier, C., A. W. Kozinski, and A. H. Doermann. 1980. Partial replicas of UV-irradiated bacteriophage T4 genomes and their role in multiplicity reactivation. J. Virol. **35**:451–465.

Rayssiguier, C., and P. R. R. Vigier. 1972. On the repair mechanism responsible for multiplicity reactivation in bacteriophage T4. Mol. Gen. Genet. **114**:140–145.

Rayssiguier, C., and P. R. R. Vigier. 1977. Genetic evidence for the existence of partial replicas of T4 genomes inactivated by irradiation under ultraviolet light. Virology **78**:442–452.

Reddy, G. P. V., and C. K. Mathews. 1978. Functional compartmentation of DNA precursors in T4 phage-infected bacteria. J. Biol. Chem. **353**:3461–3467.

Reddy, G. P. V., and A. B. Pardee. 1980. Multienzyme complex for metabolic channeling in mammalian DNA replication. Proc. Natl. Acad. Sci. U.S.A. **77**:3312–3316.

Reddy, G. P. V., A. Singh, M. E. Stafford, and C. K. Mathews. 1977. Enzyme associations in T4 phage DNA precursor synthesis. Proc. Natl. Acad. Sci. U.S.A. **74**:3152–3156.

Reha-Krantz, L. J., and M. J. Bessman. 1977. Studies on the biochemical basis of mutation. IV. Effect of amino acid substitution on the enzymatic and biological properties of bacteriophage T4 DNA polymerase. J. Mol. Biol. **116**:99–113.

Reha-Krantz, L. J., and M. J. Bessman. 1981. Studies on the biochemical basis of mutation. VI. Selection and characterization of a new bacteriophage T4 mutator DNA polymerase. J. Mol. Biol. **145**:677–695.

Reinberg, D., S. L. Zipursky, and J. Hurwitz. 1981. Separate requirements for leading and lagging strand DNA synthesis during OX A protein-dependent RF-RF DNA replication in vitro. J. Biol. Chem. **256**:13143–13151.

Remington, S. J., W. F. Anderson, J. Owen, L. F. ten Eyck, C. T. Grainger, and B. W. Matthews. 1978. Structure of the lysozyme from bacteriophage T4: an electron density map at 2.4A resolution. J. Mol. Biol. **118**:81–98.

Remington, S. J., and B. W. Matthews. 1978. A general method to assess similarity of protein structures, with applications to T4 bacteriophage lysozyme. Proc. Natl. Acad. Sci. U.S.A. **75**:2180–2184.

Revel, H. R. 1967. Restriction of nonglycosylated T-even bacteriophage: properties of permissive mutants of *E. coli* B and K12. Virology **31**:688–701.

Revel, H. R. 1981a. Molecular cloning of the T4 genome: organization and expression of the tail fiber gene cluster 34–38. Mol. Gen. Genet. **182**:445–455.

Revel, H. R. 1981b. Organization of the bacteriophage T4 tail fiber gene cluster 34–38, p. 353–364. *In* M. DuBow (ed.), Bacteriophage assembly. Alan R. Liss, New York.

Revel, H. R., and C. P. Georgopoulos. 1969. Restriction of non-glucosylated T-even bacteriophages by prophage P1. Virology **39**:1–17.

Revel, H. R., and S. Hattman. 1971. Mutants of T2gt with altered DNA methylase activity: relation to restriction by prophage P1. Virology **45**:484–495.

Revel, H. R., S. Hattman, and S. Luria. 1965. Mutants of bacteriophage T2 and T6 defective in alpha-glucosyl transferase. Biochem. Biophys. Res. Commun. **18**:545–550.

Revel, H. R., R. Herrmann, and R. J. Bishop. 1976. Genetic analysis of T4 tail fiber assembly. II. bacterial host mutants that allow bypass of T4 gene 57 function. Virology **72**:255–265.

Revel, H. R., and I. Lielausis. 1978. Revised location of the rIII gene on the genetic map of bacteriophage T4. J. Virol. **25**:439–441.

Revel, H. R., and S. Luria. 1970. DNA-glucosylation in T-even phage: genetic determination and role in phage-host interaction. Annu. Rev. Genet. **4**:177–192.

Revel, H. R., B. R. Stitt, I. Lielausis, and W. B. Wood. 1980. Role of the host cell in bacteriophage T4 development. I. Characterization of host mutants that block T4 head assembly. J. Virol. **33**:366–376.

Revel, M., H. Aviv, Y. Groner, and Y. Pollack. 1970. Fractionation of translation initiation factor IF3 into cistron-specific species. FEBS Lett. **9**:213–217.

Revel, M., Y. Groner, Y. Pollack, D. Cnaani, H. Zeller, and U. Nudel. 1973. Biochemical mechanism to control protein synthesis: mRNA specific initiation factors. Karolinska Symp. Res. Methods Reprod. Endocrinol. **6**:54–74.

Revel, M., Y. Pollack, Y. Groner, R. Scheps, H. Inouye, H. Berissi, and H. Zeller. 1973. IF3-interference factors: protein factors in *Escherichia coli* initiation of mRNA translation. Biochimie **55**:41–51.

Reyes, O., M. Gottesman, and S. Adhya. 1976. Suppression of polarity of insertion mutations in *gal* operon and *N* mutations in bacteriophage lambda. J. Bacteriol. **126**:1108–1112.

Reznikoff, W. S. 1976. Formation of the RNA polymerase-lac promoter open complex, p. 441–454. *In* R. Losick and M. Chamberlin (ed.), RNA polymerase. Cold Spring Harbor Laboratory, Cold Spring Harbor, N. Y.

Ribolini, A., and M. Baylor. 1975. Novel multinonsense suppressor in bacteriophage T4D. J. Mol. Biol. **98**:615–629.

Richardson, C. C. 1965. Phosphorylation of nucleic acid by an enzyme from T4 bacteriophage infected *E. coli*. Proc. Natl. Acad. Sci. U.S.A. **54**:158–165.

Richardson, C. C. 1966. Influence of glucosylation of deoxyribonucleic acid on hydrolysis by deoxyribonucleases of *Escherichia coli*. J. Biol. Chem. **241**:2084–2092.

Richardson, C. C., I. R. Lehman, and A. Kornberg. 1964. A deoxyribonucleic acid phosphatase-exonuclease from *Escherichia coli*. II. Characterization of the exonuclease activity. J. Biol. Chem. **239**:251–258.

Richardson, J. P. 1970. Rho factor function in T4 RNA transcription. Cold Spring Harbor Symp. Quant. Biol. **35**:127–133.

Ripley, L. S. 1975. Transversion mutagenesis in bacteriophage T4. Mol. Gen. Genet. **141**:23–40.

Ripley, L. S. 1981a. Influence of diverse gene 43 DNA polymerases on the incorporation and replication in vivo of 2-aminopurine at A-T base-pairs in bacteriophage T4. J. Mol. Biol. **150**:197–216.

Ripley, L. S. 1981b. Mutagenic specificity of nitrosomethylurea in bacteriophage T4. Mutat. Res. **83**:1–14.

Ripley, L. S., and N. B. Shoemaker. 1982. Polymerase infidelity and frameshift mutation, p. 161–178. *In* J. F. Lemonttand and W. M. Generoso (ed.), Molecular and cellular mechanisms of mutagenesis. Plenum Publishing Corp., New York.

Ritchie, D. A., A. T. Jamieson, and F. E. White. 1974. The induction of deoxythymidine kinase by bacteriophage T4. J. Gen. Virol. **24**:115–122.

Riva, S., A. Cascino, and E. P. Geiduschek. 1970a. Coupling of late transcription to viral replication in bacteriophage T4 development. J. Mol. Biol. **54**:85–102.

Riva, S., A. Cascino, and E. P. Geiduschek. 1970b. Uncoupling of late transcription from DNA replication in bacteriophage development. J. Mol. Biol. **54**:103–119.

Roberts, J. 1969. Termination factor for RNA synthesis. Nature (London) **244**:1168–1174.

Roberts, J. 1976. Transcription termination and its control in *E. coli*, p. 247–271. *In* R. Losick and M. Chamberlin (ed.), RNA polymerase. Cold Spring Harbor Laboratory, Cold Spring Harbor, New York.

Roberts, R. J. 1982. Restriction and modification enzymes and their recognition sequences. Nucleic Acids Res. **10**:117–144.

Rode, W., K. J. Scanlon, B. A. Moroson, and J. R. Bertino. 1980. Regulation of thymidylate synthetase in mouse leukemia cells (L1210). J. Biol. Chem. **255**:1305–1311.

Rohrer, H., W. Zillig, and R. Mailhammer. 1975. ADP-ribosylation of DNA-dependent RNA polymerase of *Escherichia coli* by an NAD^+: protein ADP-ribosyltransferase from bacteriophage T4. Eur. J. Biochem. **60**:227–238.

Roisin, M. P., and A. Kepes. 1978. Nucleoside diphosphokinase of *E. coli*, a periplasmic enzyme. Biochim. Biophys. Acta **526**:418–428.

Romaniuk, P. J., and F. Eckstein. 1982. A study of the mechanism of T4 DNA polymerase with diastereomeric phosphothiolate analogs of deoxyadenosine triphosphate. J. Biol. Chem. **257**:7684–7688.

Romano, L. J., and C. C. Richardson. 1979. Characterization of the ribonucleic acid primer and the deoxyribonucleic acid products synthesized by the DNA polymerase and gene 4 protein of bacteriophage T7. J. Biol. Chem. **254**:10483–10489.

Ronen, A. 1979. 2-Aminopurine. Mutat. Res. **69**:1–47.

Ronen, A., and A. Rahat. 1976. Mutagen specificity and position effects on mutation in T4 *rII* nonsense sites. Mutat. Res. **34**:21–34.

Rosenberg, M., and D. Court. 1979. Regulatory sequences involved in the promotion and termination of RNA transcription. Annu. Rev. Genet. **13**:319–353.

Rosenthal, D., and P. Reid. 1973. Rifampicin resistant DNA synthesis in phage T4 infected *Escherichia coli*. Biochem. Biophys. Res. Commun. **55**:993–1000.

Rosset, R., and L. Gorini. 1969. A ribosomal ambiguity mutation. J. Mol. Biol. **39**:94–112.

Rossmann, M. G., and P. Argos. 1976. Exploring structural homology of proteins. J. Mol. Biol. **105**:75–96.

Roth, A. C., N. G. Nossal, and P. T. Englund. 1982. Rapid hydrolysis of deoxynucleoside triphosphates accompanies DNA synthesis by T4 DNA polymerase and T4 accessory proteins 44/62 and 45. J. Biol. Chem. **257**:1267–1273.

Rothman-Denes, L. B., and G. C. Schito. 1974. Novel transcribing activities in N4-infected *E. coli*. Virology **60**:65–72.

Rubenstein, I., and S. B. Leighton. 1974. The influence of rotor speed on the sedimentation behavior in sucrose gradients of high molecular weight DNA's. Biophys. Chem. **1**:292–299.

Rüger, W. 1978. Transcription of bacteriophage T4 DNA in vitro: selective initiation with dinucleotides. Eur. J. Biochem. **88**:109–117.

Rüger, W., M. Neumann, U. Rohr, and E. Niggemann. 1979. The complete maps of *Bgl*I, *Sal*I, and *Xho*I restriction sites on T4 dC-DNA. Mol. Gen. Genet. **176**:417–425.

Runnels, J., and L. Snyder. 1978. Isolation of a bacterial host selective for bacteriophage T4 containing cytosine in its DNA. J. Virol. **27**:815–818.

Runnels, J. M., D. Soltis, T. Hey, and L. Snyder. 1982. Genetic and physiological studies of the role of RNA ligase of bacteriophage T4. J. Mol. Biol. **154**:273–286.

Ruska, H. 1940. Die Sichbarmachtung der Bakteriophagen lyse im Ubermikroskop. Naturwissenschaften **28**:45–52.

Russel, M. 1973. Control of bacteriophage T4 DNA synthesis. J. Mol. Biol. **79**:83–94.

Russel, M., L. Gold, H. Morrissett, and P. O'Farrel. 1976. Translational, autogenous regulation of gene 32 expression during bacteriophage T4 infection. J. Biol. Chem. **251**:7263–7270.

Russell, R. L. 1974. Comparative genetics of the T-even bacteriophages. Genetics **78**:967–988.

Sadowski, P. D., and I. Bakyta. 1972. T4 endonuclease IV. Improved purification procedure and resolution from T4 endonuclease III. J. Biol. Chem. **247**:405–412.

Sadowski, P. D., and J. Hurwitz. 1969a. Enzymatic breakage of deoxyribonucleic acid. I. Purification and properties of endonuclease II from T4 phage-infected *Escherichia coli*. J. Biol. Chem. **244**:6182–6191.

Sadowski, P. D., and J. Hurwitz. 1969b. Enzymatic breakage of deoxyribonucleic acid. II. Purification and properties of endonuclease IV from T4 phage-infected *Escherichia coli*. J. Biol Chem. **244**:6192–6198.

Sadowski, P. D., and D. Vetter. 1973. Control of T4 endonuclease IV by the D2a region of bacteriophage T4. Virology **54**:544–546.

Sadowski, P. D., H. R. Warner, K. Hercules, J. L. Munro, S. Mendelsohn, and J. S. Wiberg. 1971. Mutants of bacteriophage T4 defective in the induction of T4 endonuclease II. J. Biol. Chem. **246**:3431–3433.

Sahlin, M., A. Graslund, A. Ehrenberg, and B. M. Sjoberg. 1982. Structure of the tyrosyl radical in bacteriophage T4-induced ribonucleotide reductase. J. Biol. Chem. **257**:366–369.

Saigo, K. 1975. Tail-DNA connection and chromosome structure in bacteriophage T5. Virology **68**:154–165.

Saito, H., and C. C. Richardson. 1981. Genetic analysis of gene 1.2 of bacteriophage T7: isolation of a mutant of *Escherichia coli* unable to support the growth of T7 gene 1.2 mutants. J. Virol. **37**:343–351.

Saito, H., S. Tabor, F. Tamanoi, and C. C. Richardson. 1980. Nucleotide sequence of the primary origin of bacteriophage T7 DNA replication: relationship to adjacent genes and regulatory elements. Proc. Natl. Acad. Sci. U.S.A. **77**:3917–3921.

Sakaki, Y. 1974. Inactivation of the ATP-dependent DNase of *Escherichia coli* after infection with double-stranded DNA phages. J. Virol. **14**:1611–1612.

Sakiyama, S., and J. M. Buchanan. 1971. In vitro synthesis of deoxynucleotide kinase programmed by bacteriophage T4-RNA. Proc. Natl. Acad. Sci. U.S.A. **68**:1376–1380.

Sakiyama, S., and J. M. Buchanan. 1972. Control of the synthesis of T4 phage deoxynucleotide kinase messenger ribonucleic acid in vivo. J. Biol. Chem. **247**:7806–7814.

Sakiyama, S., and J. M. Buchanan. 1973. Relationship between molecular weight of T4 phage-induced deoxynucleotide kinase and the size of its messenger ribonucleic acid. J. Biol. Chem. **248**:3150–3154.

Salser, W., A. Bolle, and R. Epstein. 1970. Transcription during bacteriophage T4 development: a demonstration that distinct subclasses of the "early" RNA appear at different times and that some are "turned off" at late times. J. Mol. Biol. **49**:271–295.

Salser, W., R. F. Gesteland, and B. Ricard. 1969. Characterization of lysozyme synthesized in vitro. Cold Spring Harbor Symp. Quant. Biol. **34**:771–780.

Sambrook, J. F., D. P. Fan, and S. Brenner. 1967. A strong suppressor specific for UGA. Nature (London) **214**:452–453.

Sanders, M. M., V. E. Groppi, Jr., and E. T. Browning. 1980. Resolution of basic cellular protein including histone variants by two-dimensional electrophoresis: evaluation of lysine to arginine ratios and phosphorylation. Anal. Biochem. **103**:157–165.

Sanger, F., S. Nicklen, and A. R. Coulson. 1977. DNA sequencing with chain termination inhibitors. Proc. Natl. Acad. Sci. U.S.A. **74**:5463–5467.

Sanger, F., G. M. Air, B. G. Barrell, N. L. Brown, A. R. Coulson, J. C. Fiddes, C. A. Hutchison, P. M. Slocombe, and M. Smith. 1977. Nucleotide sequence of bacteriophage ϕX174 DNA. Nature (London) **265**:687–695.

Sano, H. 1976. Kinetic studies on the reaction catalyzed by polynucleotide kinase from phage T4 infected *E. coli*. Biochim. Biophys. Acta **422**:109–119.

Sarabhai, A., and S. Brenner. 1967. A mutant which reinitiates the polypeptide chain after chain termination. J. Mol. Biol. **27**:145–162.

Sarabhai, A. S., A. O. W. Stretton, and S. Brenner. 1964. Co-linearity of the gene with the polypeptide chain. Nature (London) **201**:13–17.

Sarkar, N., S. Sarkar, and L. M. Kozloff. 1964. Tail components of T2 bacteriophage. I. Properties of isolated contractile tail sheath. Biochemistry **3**:517–522.

Sato, K., and M. Sekiguchi. 1976. Studies on temperature-dependent ultraviolet light-sensitive mutants of bacteriophage T4: the structural gene for T4 endonuclease V. J. Mol. Biol. **102**:15–26.

Sato, S., C. A. Hutchison III, and J. I. Harris. 1977. A thermostable sequence-specific endonuclease from *Thermus aquaticus*. Proc. Natl. Acad. Sci. U.S.A. **74**:542–546.

Sauer, B., D. Ow., L. Ling, and R. Calendar. 1981. Mutants of satellite bacteriophage P4 that are defective in the suppression of transcriptional polarity. J. Mol. Biol. **145**:29–46.

Sauerbier, W., and A. R. Bräutigam. 1970. Control of gene function in bacteriophage T4. II. Synthesis of messenger ribonucleic acid and protein after interrupting deoxyribonucleic acid replication and glucosylation. J. Virol. **5**:179–187.

Sauerbier, W., and K. Hercules. 1973. Control of gene function in bacteriophage T4. IV. Post-transcriptional shutoff of early genes. J. Virol. **12**:538–547.

Sauerbier, W., K. Hercules, and D. H. Hall. 1967. Utilization of early promoters in mutant *far* P85 of bacteriophage T4. J. Virol. **19**:668–674.

Sauerbier, W., and M. Hirsch-Kauffmann. 1968. Transfer of ultraviolet light induced thymine dimer from parental to progeny DNA in bacteriophage T1 and T4. Biochem. Biophys. Res. Commun. **33**:32–37.

Savageau, M. A. 1979. Autogenous and classical regulation of gene expression: a general theory and experimental evidence, p. 57–108. *In* R. F. Goldberger (ed.), Biological regulation and development, vol. 1. Plenum Publishing Corp., New York.

Schaaper, R. M., and L. A. Loeb. 1981. Depurination causes mutations in SOS-induced cells. Proc. Natl. Acad. Sci. U.S.A. **78**:1773–1777.

Schachner, M., W. Seifert, and W. Zillig. 1971. A correlation of changes in host and T4 bacteriophage specific RNA synthesis with changes of DNA-dependent RNA polymerase in *E. coli* infected with bacteriophage T4. Eur. J. Biochem. **22**:520–528.

Schachner, M., and W. Zillig. 1971. Fingerprint maps of tryptic peptides from subunits of *Escherichia coli* and T4-modified DNA-dependent RNA polymerases. Eur. J. Biochem. **22**:513–519.

Schafer, R., and W. Zillig. 1973. The effects of ionic strength on termination of transcription of DNAs from bacteriophages T4, T5 and T7 by DNA-dependent RNA polymerase from *Escherichia coli* and the nature of termination by factor rho. Eur. J. Biochem. **33**:215–226.

Schaller, H., B. Otto, V. Nüsslein, J. Huf, R. Herrman, and F. Bonhoeffer. 1972. Deoxyribonucleic acid replication in vitro. J. Mol. Biol. **63**:183–200.

Schärli, C., and E. Kellenberger. 1980. Studies related to the head maturation pathway of bacteriophages T4 and T2. J. Virol. **33**:830–844.

Schedl, P. D., R. E. Singer, and T. W. Conway. 1970. A factor required for translation of bacteriophage f2 RNA in extracts of T4-infected cells. Biochem. Biophys. Res. Commun. **38**:631–637.

Schellman, J. A., M. Lindorfer, R. Hawkes, and M. Grütter. 1981. Mutations and protein stability. Biopolymers **20**:1989–1999.

Scherberg, N. M., and S. B. Weiss. 1972. T4 transfer RNAs: recognition and coding properties. Proc. Natl. Acad. Sci. U.S.A. **69**:1114–1118.

Scherzinger, E., E. Lanka, and G. Hillenbrand. 1977. Role of bacteriophage T7 DNA primase in the initiation of DNA strand synthesis. Nucleic Acids Res. **4**:4151–4163.

Schildkraut, C. L., K. L. Wierzchowski, J. Marmur, D. M. Green, and P. Doty. 1962. A study of the base sequence homology among the T series of bacteriophages. Virology **18**:43–55.

Schlagman, S., and S. Hattman. 1983. Molecular cloning of a functional *dam*$^+$ gene coding for T4 DNA-adenine methylase. Gene **22**:139–156.

Schmidt, D. A., A. J. Mazaitis, T. Kasai, and E. K. F. Bautz. 1970. Involvement of a phage T4 sigma factor and an antiterminator protein in the transcription of early T4 genes in vivo. Nature (London) **225**:1012–1016.

Schmidt, F. J. 1975. A novel function of *Escherichia coli* transfer RNA nucleotidyl transferase: biosynthesis of the C-C-A sequence in a phage T4 transfer RNA precursor. J. Mol. Biol. **250**:8399–8403.

Schmidt, F. J., and W. H. McClain. 1978a. An *Escherichia coli* ribonuclease which removes an extra nucleotide from a biosynthetic intermediate of bacteriophage T4 proline transfer RNA. Nucleic Acids Res. **5**:4129–4139.

Schmidt, F. J., and W. H. McClain. 1978b. Transfer RNA biosynthesis. Alternate orders of ribonuclease P cleavage occur in vitro but not in vivo. J. Biol. Chem. **253**:4730–4739.

Schmidt, F. J., J. G. Seidman, and R. M. Bock. 1976. Transfer ribonucleic acid biosynthesis: substrate specificity of ribonuclease P. J. Biol. Chem. **251**:2440–2445.

Schneider, S., C. Bernstein, and H. Bernstein. 1978. Recombinational repair of alkylation lesions in phage T4. I. *N*-methyl-*N*′-nitro-*N*-nitrosoguanidine. Mol. Gen. Genet. **167**:185–195.

Schneider, T. D., G. D. Stormo, J. S. Haemer, and L. Gold. 1982. A design for computer nucleic acid-sequence storage, retrieval, and manipulation. Nucleic Acids Res. **10**:3013–3024.

Schnitzlein, C. F., I. Albrecht, and J. W. Drake. 1974. Is bacteriophage T4 DNA polymerase involved in the repair of ultraviolet damage? Virology **59**:580–583.

Schurr, J. M., and K. S. Schmitz. 1976. Orientation constraints and rotational diffusion in bimolecular solution kinetics. A simplification. J. Phys. Chem. **80**:1934–1936.

Schwartz, M. 1980. Interaction of phages with their receptor proteins, p. 59–94. *In* L. Randall and L. Phillipson (ed.), Receptors and recognition, series B, vol. 7: Virus receptors, part 1: Bacterial viruses. Chapman and Hall, London.

Scocca, J. J., S. R. Panny, and M. J. Bessman. 1969. Studies of deoxycytidylate deaminase from T4-infected *E. coli*. J. Biol. Chem. **244**:3698–3706.

Scofield, M. S., W. L. Collinsworth, and C. K. Mathews. 1974. Continued synthesis of bacterial DNA after infection by bacteriophage T4. J. Virol. **13**:847–857.

Scott, F. W., and D. R. Forsdyke. 1980. Isotope dilution analysis of the effects of deoxyguanosine and deoxyadenosine on the incorporation of thymidine and deoxycytidine by hydroxyurea-treated thymus cells. Biochem. J. **190**:721–730.

Seasholtz, A. F., and G. R. Greenberg. 1983. Identification of bacteriophage T4 gene 60 product and a role for this protein in DNA topoisomerase. J. Biol. Chem. **258**:1221–1226.

Seawell, P. C., C. A. Smith, and A. K. Ganesan. 1980. *denV* gene of bacteriophage T4 determines a DNA glycosylase specific for pyrimidine dimers in DNA. J. Virol. **35**:790–797.

Sechaud, J., G. Streisinger, J. Emrich, J. Newton, H. Langford, H. Reinhold, and M. Stahl. 1965. Chromosome structure in phage T4. II. Terminal redundancy and heterozygosis. Proc. Natl. Acad. Sci. U.S.A. **54**:1333–1339.

Sederoff, R., A. Bolle, and R. H. Epstein. 1971. A method for the detection of specific T4 messenger RNAs by hybridization competition. Virology **45**:440–455.

Sederoff, R., A. Bolle, H. M. Goodman, and R. H. Epstein. 1971. Regulation of rII and region D transcription in T4 bacteriophage: a sucrose gradient analysis. Virology **46**:817–829.

Seeberg, E. 1978. Reconstitution of an *Escherichia coli* repair endonuclease activity from the separated *uvrA*$^-$ and *uvrB*$^+$/*uvrC*$^+$ gene products. Proc. Natl. Acad. Sci. U.S.A. **75**:2569–2573.

Seeberg, E. 1981. Multiprotein interactions in strand cleavage of DNA damaged by UV and chemicals. Prog. Nucleic Acid Res. Mol. Biol. **26**:217–226.

Seidman, J. G., B. G. Barrell, and W. H. McClain. 1975. Five steps in the conversion of a large precursor RNA into bacteriophage proline and serine transfer RNAs. J. Mol. Biol. **99**:733–760.

Seidman, J. G., M. M. Comer, and W. H. McClain. 1974. Nucleotide alterations in the bacteriophage T4 glutamine transfer RNA that affect ochre suppressor activity. J. Mol. Biol. **90**:677–689.

Seidman, J. G., and W. H. McClain. 1975. Three steps in conversion of large precursor RNA into serine and proline transfer RNAs. Proc. Natl. Acad. Sci. U.S.A. **72**:1491–1495.

Seidman, J. G., F. J. Schmidt, K. Foss, and W. H. McClain. 1975. A mutant of *Escherichia coli* defective in removing 3′ terminal nucleotides from some transfer RNA precursor molecules. Cell **5**:389–400.

Seifert, W., P. Qasba, G. Walter, P. Palm, M. Schachner, and W. Zillig. 1969. Kinetics of the alteration and modification of DNA-dependent RNA polymerase in T4-infected *E. coli* cells. Eur. J. Biochem. **9**:319–324.

Seifert, W., D. Rabussay, and W. Zillig. 1971. On the chemical nature of alteration and modification of DNA dependent RNA polymerase of *E. coli* after T4 infection. FEBS Lett. **16**:175–179.

Sekiguchi, M. 1966. Studies on the physiological defect in *r*II mutants of bacteriophage T4. J. Mol. Biol. **16**:503–522.

Sekiguchi, M., K. Shimizu, K. Sato, S. Yasuda, and S. Ohshima. 1975. Enzymic mechanism of excision-repair in T4-infected cells, p. 135–142. *In* P. C. Hanawalt and R. B. Setlow (ed.), Molecular mechanisms for repair of DNA. Plenum Publishing Corp., New York.

Sekiguchi, M., and Y. Takagi. 1960. Effect of mitomycin C on the synthesis of bacterial and viral deoxyribonucleic acid. Biochim. Biophys. Acta **41**:434–443.

Sekiguchi, M., S. Yasuda, S. Okubo, H. Nakayama, K. Shimada, and Y. Takagi. 1970. Mechanisms of repair of DNA in bacteriophage I. Excision of pyrimidine dimers from ultraviolet-irradiated DNA by an extract of T4-infected cells. J. Mol. Biol. **47**:321–242.

Sellin, H. G., P. R. Srinivasan, and E. Borek. 1966. Studies of phage-induced DNA methylase. J. Mol. Biol. **19**:219–222.

Selzer, G., D. Belin, A. Bolle, G. van Houwe, T. Mattson, and R. Epstein. 1981. In vivo expression of the *r*II region of bacteriophage T4 present in chimeric plasmids. Mol. Gen. Genet. **183**:505–513.

Selzer, G., A. Bolle, H. Krisch, and R. Epstein. 1978. Construction and properties of recombinant plasmids containing the *rII* genes of bacteriophage T4. Mol. Gen. Genet. **159**:301–309.

Selzer, G., T. Som, T. Itoh, and J. Tomizawa. 1983. The origin of replication of plasmid p15A and comparative studies on the nucleotide sequences around the origin of related plasmids. Cell **32:**119–129.

Serwer, P. 1975. Buoyant density sedimentation of macromolecules in sodium iothalamate density gradients. J. Mol. Biol. **92:**433–448.

Setlow, R. B., and W. L. Carrier. 1968. The excision of pyrimidine dimers in vivo and in vitro, p. 134–141. *In* W. J. Peacock and R. D. Brock (ed.), Replication and recombination of genetic material. Australian Academy of Science, Canberra.

Shah, D. B. 1976. Replication and recombination of gene 59 mutant of bacteriophage T4D. J. Virol. **17:**175–182.

Shah, D. B., and H. Berger. 1971. Replication of gene 46-47 amber mutants of bacteriophage T4D. J. Mol. Biol. **57:**17–34.

Shah, D. B., and L. DeLorenzo. 1977. Suppression of gene 49 mutations of bacteriophage T4 by a second mutation in gene X: structure of pseudo revertant DNA. J. Virol. **24:**794–804.

Shalitin, C., and Y. Naot. 1971. Role of gene 46 in bacteriophage T4 deoxyribonucleic acid synthesis. J. Virol. **8:**142–153.

Shane, B. 1980. Pteroylpoly (γ-glutamate) synthesis by *Corynebacterium* species. J. Biol. Chem. **255:**5655–5662.

Shapiro, D., and L. M. Kozloff. 1970. A critical C-terminal arginine residue necessary for bacteriophage T4D tail assembly. J. Mol. Biol. **51:**185–201.

Shapiro, J. A. 1979. Molecular model for the transposition and replication of bacteriophage Mu and other transposable elements. Proc. Natl. Acad. Sci. U.S.A. **76:**1933–1937.

Shcherbakov, V. P., L. A. Plugina, E. A. Kudryashova, O. I. Efremova, and S. T. Sizova. 1982. Marker-dependent recombination in T4 bacteriophage. I. Outline of the phenomenon and evidence suggesting a mismatch repair mechanism. Genetics **102:**615–625.

Shedlovsky, A., and S. Brenner. 1963. A chemical basis for the host-induced modification of T-even bacteriophages. Proc. Natl. Acad. Sci. U.S.A. **50:**300–305.

Shibata, T., C. DasGupta, R. P. Cunningham, and C. M. Radding. 1980. Homologous pairing in genetic recombination: formation of D loops by combined action of recA protein and a helix-destabilizing protein. Proc. Natl. Acad. Sci. U.S.A. **77:**2606–2610.

Shimatake, H., and M. Rosenberg. 1981. Purified lambda regulatory protein cII positively activates promoters for lysogenic development. Nature (London) **292:**128–132.

Shimizu, K., and M. Sekiguchi. 1976. 5′-3′-exonucleases of bacteriophage T4. J. Biol. Chem. **251:**2613–2619.

Shimizu, K., and M. Sekiguchi. 1979. Introduction of an active enzyme into permeable cells of *Escherichia coli*. Acquisition of ultraviolet light resistance by *uvr* mutants on introduction of T4 endonuclease V. Mol. Gen. Genet. **168:**37–47.

Shine, J., and L. Dalgarno. 1974. The 3′-terminal sequence of *Escherichia coli* 16S ribosomal RNA: complementary to nonsense triplets and ribosome binding sites. Proc. Natl. Acad. Sci. U.S.A. **71:**1342–1346.

Shine, J., and L. Dalgarno. 1975. Determinant of cistron specificity in bacterial ribosomes. Nature (London) **254:**34–38.

Shineberg, B., and D. Zipser. 1973. The *lon* gene and degradation of β-galactosidase nonsense fragments. J. Bacteriol. **116:**1469–1471.

Short, E. C., Jr., and J. F. Koerner. 1969. Separation and characterization of deoxyribonucleases of *Escherichia coli* B. II. Further purification and properties of an exonuclease induced by infection with bacteriophage T2. J. Biol. Chem. **244:**1487–1496.

Shoup, D., G. Lipari, and A. Szabo. 1981. Diffusion-controlled bimolecular reaction rates: the effect of rotational diffusion and orientation constraints. Biophys. J. **36:**697–714.

Showe, M. K. 1979. Limited proteolysis during the maturation of bacteriophage T4, p. 151–155. *In* Limited proteolysis in microorganisms: biological function, use in protein structural and functional studies. DHEW publication (NIH) 78-1591. U.S. Government Printing Office, Washington, D.C.

Showe, M. K., and L. W. Black. 1973. Assembly core of bacteriophage T4: an intermediate in head formation. Nature (London) **242:**70–72.

Showe, M. K., E. Isobe, and L. Onorato. 1976. Bacteriophage T4 prehead proteinase. I. Purification and properties of a bacteriophage enzyme which cleaves the capsid precursor proteins. J. Mol. Biol. **107:**35–54.

Showe, M. K., E. Isobe, and L. Onorato. 1976. Bacteriophage T4 prehead proteinase. II. Its cleavage from the product of gene 21 and regulation in phage-infected cells. J. Mol. Biol. **197:**55–69.

Showe, M. K., and E. Kellenberger. 1975. Control mechanisms in virus assembly, p. 407–438. *In* D. C. Burke and W. C. Russell (ed.), Control processes in virus multiplication. Cambridge University Press, Cambridge.

Showe, M. K., and L. Onorato. 1978. A kinetic model for form-determination of the head of bacteriophage T4. Proc. Natl. Acad. Sci. U.S.A. **75:**4165–4169.

Siebenlist, U., R. B. Simpson, and W. Gilbert. 1980. *E. coli* RNA polymerase interacts homologously with two different promoters. Cell **20:**269–281.

Siegel, P. J., and M. Schaechter. 1973. Bacteriophage T4 head maturation: release of progeny DNA from the host cell membrane. J. Virol. **11:**359–367.

Silber, R., V. G. Malathi, and J. Hurwitz. 1972. Purification and properties of bacteriophage T4 induced RNA ligase. Proc. Natl. Acad. Sci. U.S.A. **69:**3009–3013.

Silver, L. L., and N. G. Nossal. 1979. DNA replication by T4 bacteriophage proteins: role of the DNA-delay gene 61 in the chain-initiation reaction. Cold Spring Harbor Symp. Quant. Biol. **43:**489–494.

Silver, L. L., and N. G. Nossal. 1982. Purification of bacteriophage T4 gene 61 protein: a protein essential for synthesis of RNA primers in the T4 in vitro DNA replication system. J. Biol. Chem. **257:**11696–11705

Silver, L. L., M. Venkatesan, and N. G. Nossal. 1980. RNA priming and DNA synthesis by bacteriophage T4 proteins. ICN-UCLA Symp. Mol. Cell. Biol. **19:**000–000.

Silver, S. 1965. Acriflavin resistance: a bacteriophage mutation affecting the uptake of dye by the infected bacterial cells. Proc. Natl. Acad. Sci. U.S.A. **53:**24–30.

Silverstein, J. L., and E. B. Goldberg. 1976a. T4 DNA injection. I. Growth cycle of a gene 2 mutant. Virology **72:**195–211.

Silverstein, J. L., and E. B. Goldberg. 1976b. T4 DNA injection. II. Protection of entering DNA from host exonuclease V. Virology **72:**212–223.

Simmon, V. F., and S. Lederberg. 1972. Degradation of bacteriophage lambda deoxyribonucleic acid after restriction by *Escherichia coli* K-12. J. Bacteriol. **112:**161–169.

Simon, E. H., and I. Tessman. 1963. Thymidine-requiring mutants of phage T4. Proc. Natl. Acad. Sci. U.S.A. **50:**526–532.

Simon, L. D. 1969. The infection of *Escherichia coli* by T2 and T4 bacteriophages as seen in the electron microscope. III. Membrane-associated intracellular bacteriophages. Virology **38:**285–296.

Simon, L. D. 1972. Infection of *Escherichia coli* by T2 and T4 bacteriophages as seen in the electron microscope: T4 head morphogenesis. Proc. Natl. Acad. Sci. U.S.A. **69:**907–911.

Simon, L. D., and T. F. Anderson. 1967a. The infection of *Escherichia coli* by T2 and T4 bacteriophage as seen in the electron microscope. I. Attachment and penetration. Virology **32:**279–297.

Simon, L. D., and T. F. Anderson. 1967b. The infection of *Escherichia coli* by T2 and T4 bacteriophage as seen in the electron microscope. II. Structure and function of the baseplate. Virology **32:**298–305.

Simon, L. D., M. Gottesman, K. Tomczak, and S. Gottesman. 1979. Hyperdegradation of proteins in *Escherichia coli rho* mutants. Proc. Natl. Acad. Sci. U.S.A. **76:**1623–1627.

Simon, L. D., T. J. M. McLaughlin, D. Snover, J. Ou, C. Grisham, and M. Loeb. 1975. *E. coli* membrane lipid alteration affecting T4 capsid morphogenesis. Nature (London) **256:**379–383.

Simon, L. D., B. Randolph, N. Irwin, and G. Minkowski. 1983. Stabilization of proteins by a bacteriophage T4 gene cloned in *Escherichia coli*. Proc. Natl. Acad. Sci. U.S.A. **80:**2059–2067.

Simon, L. D., D. Snover, and A. H. Doermann. 1974. Bacterial mutation affecting T4 phage DNA synthesis and tail production. Nature (London) **252:**451–455.

Simon, L. D., J. G. Swan, and J. E. Flatgaard. 1970. Functional defects in T4 bacteriophages lacking the gene 11 and gene 12 products. Virology **41:**77–90.

Simon, L. D., K. Tomczak, and A. C. St. John. 1978. Bacteriophages inhibit degradation of abnormal proteins in *E. coli*. Nature (London) **275:**424–428.

Simpson, R. J., G. S. Begg, D. S. Dorow, and F. J. Morgan. Complete amino acid sequence of the goose-type lysozyme from the egg white of the black swan. Biochemistry **19:**1814–1819.

Singer, B. S., and L. Gold 1976. A mutation that confers temperature sensitivity on the translation of *r*IIB in bacteriophage T4. J. Mol. Biol. **103:**627–646.

Singer, B. S., L. Gold, P. Gauss, and D. H. Doherty. 1982. Determination of the amount of homology required for recombination in bacteriophage T4. Cell **31:**25–33.

Singer, B. S., L. Gold, S. T. Shinedling, M. Colkitt, L. R. Hunter, D. Pribnow, and M. A. Nelson, 1981. Analysis in vivo of translational mutants of the *r*IIB cistron of bacteriophage T4. J. Mol. Biol. **149:**405–432.

Singer, R. E., and T. W. Conway. 1973. Defective initiation of f2 RNA translation by ribosomes from bacteriophage T4-infected cells. Biochim. Biophys. Acta **331:**102–116.

Sinha, N. K., and M. D. Haimes. 1980. Probing the mechanism of transition and transversion mutagenesis using the bacteriophage T4 DNA replication apparatus in vitro. ICN-UCLA Symp. Mol. Cell. Biol. **19:**707–723.

Sinha, N. K., and M. D. Haimes. 1981. Molecular mechanisms of substitution mutagenesis. An experimental test of the Watson-Crick and Topal-Fresco models of base mispairing. J. Biol. Chem. **256**:10671–10683.

Sinha, N. K., C. F. Morris, and B. M. Alberts. 1980. Efficient in vitro replication of double stranded DNA templates by a purified T4 bacteriophage replication system. J. Biol. Chem. **255**:4290–4303.

Sinha, N. K., and D. P. Snustad. 1972. Mechanism of inhibition of deoxyribonucleic acid synthesis in *Escherichia coli* by hydroxyurea. J. Bacteriol. **112**:1321–1334.

Sinsheimar, R. L. 1954. Nucleotides from T2r$^+$ bacteriophage. Science **120**:551–553.

Sirotkin, K., W. Cooley, J. Runnels, and L. R. Snyder. 1978. A role in true-late gene expression for the T4 bacteriophage 5′ polynucleotide kinase 3′ phosphatase. J. Mol. Biol. **123**:221–233.

Sirotkin, K., J. Wei, and L. Snyder. 1977. T4 bacteriophage-coded RNA polymerase subunit blocks host transcription and unfolds the host chromosome. Nature (London) **265**:28–32.

Sirotnak, F. M., and R. W. McCuen. 1973. Hyperproduction of dihydrofolate reductase in *Diplococcus pnemnoniae* after mutation in the structural gene. Genetics **74**:543–556.

Sjoberg, B. M., and B. O. Soderberg. 1976. Thioredoxin induced by bacteriophage T4: crystallization and preliminary crysallographic data. J. Mol. Biol. **100**:415–419.

Skalka, A. 1974. A replicator's view of recombination (and repair), p. 421–432. *In* R. F. Grell (ed.), Mechanisms in recombination. Plenum Publishing Corp. New York.

Skold, O. 1970. Regulation of early RNA synthesis in bacteriophage T4-infected *Escherichia coli* cells. J. Mol. Biol. **53**:339–356.

Skoog, L., and G. Bjursell. 1974. Nuclear and cytoplasmic pools of deoxyribonucleoside triphosphates in Chinese hamster ovary cells. J. Biol. Chem. **249**:6434–6438.

Skorko, R., W. Zillig, H. Rohrer, H. Fujiki, and R. Mailhammer. 1977. Purification and properties of the NAD$^+$: protein ADP-ribosyltransferase responsible for the T4 phage-induced modification of the subunit of DNA-dependent RNA polymerase of *Escherichia coli*. Eur. J. Biochem. **79**:55–66.

Smit, J., Y. Kamio, and H. Nikaido. 1975. Outer membrane of *Salmonella typhimurium*: chemical analysis and freeze-fracture studies with lipopolysaccharide mutants. J. Bacteriol. **124**:942–958.

Smith, C. A., and P. C. Hanawalt. 1978. Phage T4 endonuclease V stimulates DNA repair replication in isolated nuclei from ultraviolet-irradiated human cells, including xeroderma pigmentosum fibroblasts. Proc. Natl. Acad. Sci. U.S.A. **75**:2598–2602.

Smith, C. L., M. Kubo, and F. Imamoto. 1978. Promoter-specific inhibition of transcription by antibiotics which act on DNA gyrase. Nature (London) **275**:420–423.

Smith, F. L., and R. Haselkorn. 1969. Proteins associated with ribosomes in T4-infected *E. coli*. Cold Spring Harbor Symp. Quant. Biol. **34**:91–94.

Smith, J. D. 1972. Genetics of transfer RNA. Annu. Rev. Genet. **6**:235–268.

Smith, K. C. 1978. Multiple pathways of DNA repair in bacteria and their roles in mutagenesis. Photochem. Photobiol. **28**:121–129.

Smith, K. C., and D. H. C. Meun. 1970. Repair of radiation-induced damage in *Escherichia coli*. I. Effect of *rec* mutations on post-replication repair of damage due to ultraviolet radiation. J. Mol. Biol **51**:459–472.

Smith, M. D., R. R. Green, L. S. Ripley, and J. W. Drake. 1973. Thymineless mutagenesis in bacteriophage T4. Genetics **74**:393–403.

Smith, P. R., U. Aebi, R. Josephs, and M. Kessel. 1976. Studies of the structure of the T4 bacteriophage tail sheath. J. Mol. Biol. **106**:243–275.

Smith, S. M., and N. Symonds. 1973. The unexpected location of a gene conferring abnormal radiation sensitivity on phage T4. Nature (London) **241**:395–396.

Snopek, T. J., A. Sugino, K. Agarwal, and N. R. Cozzarelli. 1976. Catalysis of DNA joining by bacteriophage T4 RNA ligase. Biochem. Biophys. Res. Commun. **68**:417–424.

Snopek, T. J., W. B. Wood, M. P. Conley, P. Chen, and N. R. Cozzarelli. 1977. Bacteriophage T4 RNA ligase is gene 63 product, the protein that promotes tail fiber attachment to the baseplate. Proc. Natl. Acad. Sci. U.S.A. **74**:3355–3359.

Snustad, D. P. 1968. Dominance interactions in *Escherichia coli* cells mixedly infected with T4D wild-type and amber mutants and their possible implications as to type of gene-product function: catalytic vs. stoichiometric. Virology **35**:550–563.

Snustad, D. P., and C. J. H. Bursch. 1977. Shutoff of host RNA synthesis in bacteriophage T4-infected *Escherichia coli* in the absence of host DNA degradation and nuclear disruption. J. Virol. **21**:1240–1242.

Snustad, D. P., C. J. H. Bursch, K. A. Parson, and S. H. Hefeneider. 1976. Mutants of bacteriophage T4 deficient in the ability to induce nuclear disruption: shutoff of host DNA and protein synthesis, gene dosage experiments, identification of a restrictive host, and possible biological significance. J. Virol. **18**:268–288.

Snustad, D. P., and L. M. Conroy. 1974. Mutants of bacteriophage T4 deficient in the ability to induce nuclear disruption. I. Isolation and genetic characterization. J. Mol. Biol. **89**:663–673.

Snustad, D. P., K. A. Parson, H. R. Warner, D. J. Tutas, J. M. Wehner, and J. F. Koerner. 1974. Mutants of bacteriophage T4 deficient in the ability to induce nuclear disruption. II. Physiological state of the host nucleoid in infected cells. J. Mol. Biol. **89**:675–687.

Snustad, D. P., M. A. Tigges, K. A. Parson, C. J. H. Bursch, F. M. Caron, J. F. Koerner, and D. J. Tutas. 1976. Identification and preliminary characterization of a mutant defective in the bacteriophage T4-induced unfolding of the *Escherichia coli* nucleoid. J. Virol. **17**:622–641.

Snustad, D. P., H. R. Warner, K. A. Parson, and D. L. Anderson. 1972. Nuclear disruption after infection of *Escherichia coli* with a bacteriophage T4 mutant unable to induce endonuclease II. J. Virol. **10**:124–133.

Snyder, L., and E. P. Geiduschek. 1968. *In vitro* synthesis of T4 late messenger RNA. Proc. Natl. Acad. Sci. U.S.A. **59**:459–466.

Snyder, L., L. Gold, and E. Kutter. 1976. A gene of bacteriophage T4 whose product prevents true late transcription on cytosine-containing T4 DNA. Proc. Natl. Acad. Sci. U.S.A. **73**:3098–3102.

Snyder, L. R. 1972. An RNA polymerase mutant of *Escherichia coli* defective in the T4 viral transcription program. Virology **50**:396–403.

Snyder, L. R., and D. L. Montgomery. 1974. Inhibition of T4 growth by an RNA polymerase mutation of *E. coli*: physiological and genetic analysis of the effects during phage development. Virology **62**:184–196.

Söderberg, B., B. Sjoberg, U. Sonnerstam, and C. Branden. 1978. Three-dimensional structure of thioredoxin induced by bacteriophage T4. Proc. Natl. Acad. Sci. U.S.A. **75**:5827–5830.

Souther, A., R. Bruner, and J. Elliott. 1972. Degradation of *Escherichia coli* chromosome after infection by bacteriophage T4: role of bacteriophage gene D2a. J. Virol. **10**:979–984.

Southern, E. M. 1975. Detection of specific sequences among DNA fragments separated by gel electrophoresis. J. Mol. Biol. **98**:503–517.

Speyer, J. F. 1965. Mutagenic DNA polymerase. Biochem. Biophys. Res. Commun. **21**:6–8.

Speyer, J. F., J. D. Karam, and A. B. Lenny. 1966. On the role of DNA polymerase in base selection. Cold Spring Harbor Symp. Quant. Biol. **31**:693–697.

Speyer, J. F., and L. H. Khairalla. 1973. Crystalline T4 bacteriophage. J. Mol. Biol. **76**:415–417.

Spicer, E. K., J. A. Noble, N. G. Nossal, W. H. Konigsberg, and K. R. Williams. 1982. Bacteriophage T4 gene 45: sequences of the structural gene and its protein product. J. Biol. Chem. **257**:8972–8979.

Spicer, E., K. R. Williams, and W. Konigsberg. 1979. T4 gene 32 protein trypsin-generated fragments: fluorescence measurements of DNA-binding parameters. J. Biol. Chem. **254**:6433–6436.

Spremulli, L. L., M. A. Haralson, and J. M. Ravel. 1974. Effect of T4 infection on initiation of protein synthesis and messenger specificity of initiation factor 3. Arch. Biochem. Biophys. **165**:581–587.

Stadler, L. J. 1954. The gene. Science **120**:811–819.

Stafford, M. E., G. P. V. Reddy, and C. K. Mathews. 1977. Further studies on bacteriophage T4 DNA synthesis in sucrose-plasmolysed cells. J. Virol. **23**:53–60.

Stahl, F. W. 1956. The effects of the decay of incorporated radioactive phosphorus on the genome of bacteriophage T4. Virology **2**:206–234.

Stahl, F. W. 1979a. Special sites in generalized recombination. Annu. Rev. Genet. **13**:7–24.

Stahl, F. W. 1979b. Genetic recombination. Thinking about it in phage and fungi. W. H. Freeman and Co., San Francisco.

Stahl, F. W., J. M. Crasemann, C. Yegian, M. M. Stahl, and A. Nakata. 1970. Co-transcribed cistrons in bacteriophage T4. Genetics **64**:157–170.

Stahl, F. W., R. S. Edgar, and J. Steinberg. 1964. The linkage map of bacteriophage T4. Genetics **50**:539–552.

Stahl, F. W., K. D. McMilin, M. M. Stahl, J. M. Crasemann, and S. Lam. 1973. The distribution of crossovers along unreplicated lambda bacteriophage chromosomes. Genetics **77**:395–408.

Stahl, F. W., K. D. McMilin, M. M. Stahl, and Y. Nozu. 1972. An enhancing role of DNA synthesis in formation of bacteriophage lambda recombinants. Proc. Natl. Acad. Sci. U.S.A. **69**:3598–3601.

Stahl, F. W., and M. M. Stahl. 1974. Red-mediated recombination in bacteriophage lambda, p. 407–419. *In* R. F. Grell (ed.), Mechanisms of recombination. Plenum Publishing Corp., New York.

Stahl, G., G. V. Paddock, and J. Abelson. 1974. Nucleotide sequence determination of bacteriophage T4 glycine transfer ribonucleic acid. Nucleic Acids Res. **1:**1287–1304.

Stalker, D. M., A. Shafferman, A. Tolun, R. Kolter., S. Yang, and D. R. Helinski. 1981. Direct repeats of nucleotide sequences are involved in plasmid replication and incompatibility, p. 113–124. *In* D. Ray (ed.), The initiation of DNA replication. Academic Press, Inc., New York.

Staudenbauer, W., E. Scherzinger, and E. Lanka. 1979. Replication of the colicin E1 plasmid in extracts of *Eschericnia coli*: uncoupling of leading strand from lagging strand synthesis. Mol. Gen. Genet. **177:**113–120.

Steege, D. A., and D. G. Soll. 1979. Suppression, p. 433–485. *In* R. F. Goldberger (ed.), Biological regulation and development, vol. 1. Plenum Publishing Corp. New York.

Steitz, J. A. 1979. Genetic signals and nucleotide sequences in messenger RNA, p. 349–399. *In* R. F. Goldberger (ed.), Biological regulation and development, vol. 1. Plenum Publishing Corp., New York.

Steitz, J. A., S. K. Dube, and P. S. Rudland. 1970. Control of translation by T4 phage: altered ribosome binding at R17 initiation sites. Nature (London) **226:**824–827.

Stent, G. S. 1963. Molecular biology of bacterial viruses. W. H. Freeman and Co., San Francisco.

Stent, G. S., and E. L. Wollman. 1952. On the two step nature of bacteriophage adsorption. Biochim. Biophys. Acta **8:**260–269.

Sternberg, N. 1973a. Properties of a mutant of *Escherichia coli* defective in bacteriophage lambda head formation (*groE*). I. Initial characterization. J. Mol. Biol. **76:**1–23.

Sternberg, N. 1973b. Properties of mutants of *Escherichia coli* defective in bacteriophage lambda head formation (*groE*). II. The propagation of phage lambda. J. Mol. Biol. **76:**25–44.

Stetler, G. L., G. J. King, and W. M. Huang. 1979. T4 DNA-delay proteins, required for specific DNA replication, form a complex that has ATP-dependent DNA topoisomerase activity. Proc. Natl. Acad. Sci. U.S.A. **76:**3737–3741.

Steven, A., and J. L. Carrascosa. 1979. Proteolytic cleavage and structural transformation: their relationship in bacteriophage T4 capsid maturation. J. Supramol. Struct. **10:**1–11.

Steven, A. C., U. Aebi, and M. K. Showe. 1976. Folding and capsomere morphology of the P23 surface shell of bacteriophage T4 polyheads from mutants in five different head genes. J. Mol. Biol. **102:**373–407.

Steven, A. C., E. Couture, U. Aebi, and M. K. Showe. 1976. Structure of T4 polyheads. II. A pathway of polyhead transformations as a model for T4 capsid maturation. J. Mol. Biol. **106:**187–221.

Steven, A. C., P. Serwer, M. E. Bisher, and B. L. Trus. 1983. Molecular architecture of bacteriophage T7 capsid. Virology **124:**109–120.

Stevens, A. 1970. An isotopic study of DNA-dependent RNA polymerase of *E. coli* following T4 infection. Biochem. Biophys. Res. Commun. **41:**367–373.

Stevens, A. 1972. New small polypeptides associated with DNA-dependent RNA polymerase of *Escherichia coli* after infection with bacteriophage T4. Proc. Natl. Acad. Sci. U.S.A. **69:**603–607.

Stevens, A. 1974. Deoxyribonucleic acid dependent ribonucleic acid polymerases from two T4 phage-infected systems. Biochemistry **13:**493–503.

Stevens, A. 1976. A salt-promoted inhibitor of RNA polymerase isolated from T4-infected *E. coli*, p. 617–627. *In* R. Losick and M. Chamberlin (ed.), RNA polymerase. Cold Spring Harbor Laboratories, Cold Spring Harbor, N. Y.

Stevens, A. 1977. Inhibition of DNA-enzyme binding by an RNA polymerase inhibitor from T4 phage-infected *Escherichia coli*. Biochim. Biophys. Acta **475:**193–196.

Stevens, A., and J. C. Rhoton. 1975. Characterization of an inhibitor causing potassium chloride sensitivity of an RNA polymerase from T4 phage-infected *Escherichia coli*. Biochemistry **14:**5074–5079.

Stitt, B., H. Revel, I. Lielausis, and W. B. Wood. 1980. Role of the host cell in bacteriophage T4 development. II. Characterization of host mutants that have pleiotropic effects on T4 growth. J. Virol. **35:**775–789.

Stone, K. W., and D. J. Cummings. 1972. Comparison of the internal proteins of the T-even bacteriophages. J. Mol. Biol. **64:**651–669.

Stonington, O. G., and D. E. Pettijohn. 1971. The folded genome of *Escherichia coli* isolated in a protein-DNA-RNA complex. Proc. Natl. Acad. Sci. U.S.A. **68:**6–9.

Stormo, G. D., T. D. Schneider, and L. M. Gold. 1982. Characterization of translational initiation sites in *E. coli*. Nucleic Acids Res. **10:**2971–2996.

Strauss, B., D. Scudiero, and E. Henderson. 1975. The nature of the alkylation lesion in mammalian cells, p. 13–24. *In* R. B. Setlow (ed.), Molecular mechanisms for repair of DNA. Plenum Publishing Corp., New York.

Streisinger, G. 1956. The genetic control of ultraviolet sensitivity levels in bacteriophages T2 and T4. Virology **2:**1–12.

Streisinger, G., and V. Bruce. 1960. Linkage of genetic markers in phages T2 and T4. Genetics **45:**1289–1296.

Streisinger, G., R. S. Edgar, and G. H. Denhardt. 1964. Chromosome structure in phage T4. I. Circularity of the linkage map. Proc. Natl. Acad. Sci. U.S.A. **51:**775–779.

Streisinger, G., J. Emrich, and M. M. Stahl. 1967. Chromosome structure in phage T4. III. Terminal redundancy and length determination. Proc. Natl. Acad. Sci. U.S.A. **57:**292–295.

Streisinger, G., F. Mukai, W. J. Dreyer, B. Miller, and S. Horiuchi. 1961. Mutations affecting the lysozyme of phage T4. Cold Spring Harbor Symp. Quant. Biol. **26:**25–30.

Streisinger, G., Y. Okada, J. Emrich, J. Newton, A. Tsugita, E. Terzaghi, and M. Inouye. 1966. Frameshift mutations and the genetic code. Cold Spring Harbor Symp. Quant. Biol. **31:**77–84.

Strike, P. 1978. Effects of phage infection on *Escherichia coli* excision repair measured in vitro, p. 271–275. *In* P. C. Hanawalt, E. C. Friedberg, and C. F. Fox (ed.), DNA repair mechanisms. Academic Press, Inc., New York.

Strike, P., H. O. Wilbraham, and E. Seeberg. 1981. Repair of psoralen near ultraviolet light damage in bacteriophage T3 and T4. Photochem. Photobiol. **33:**73–78.

Studier, F. W. 1975. Gene 0.3 of bacteriophage T7 acts to overcome the DNA restriction of the host. J. Mol. Biol. **94:**283–295.

Subramanian, A. R., C. Haase, and M. Giesen. 1976. Isolation and characterization of a growth-cycle reflecting high-molecular weight protein associated with *Escherichia coli* ribosomes. Eur. J. Biochem. **67:**591–601.

Subramanian, A. R., C. Haase, and M. Giesen. 1979. Comparison of ribosomal protein S1 and A-protein from Escherichia coli. FEBS Lett. **99:**357–360.

Sueoka, N., and T. Kano-Sueoka. 1979. Transfer RNA and cell differentiation. Prog. Nucleic Acids Res. Mol. Biol. **10:**23–53.

Sugino, A., H. M. Goodman, H. L. Heynecker, J. Shine, J. Boyer, and N. R. Cozzarelli. 1977. Interaction of bacteriophage T4 RNA and DNA ligases in joining of duplex DNA at base paired ends. J. Biol. Chem. **252:**3987–3994.

Sugino, A., N. P. Higgins, P. O. Brown, C. L. Peebles, and N. R. Cozzarelli. 1978. Energy coupling in DNA gyrase and the mechanism of action of novobiocin. Proc. Natl. Acad. Sci. U.S.A. **75:**4838–4842.

Sugino, A., T. J. Snopek, and N. R. Cozzarelli. 1977. Bacteriophage T4 ligase: reaction intermediate and interaction of substrates. J. Biol. Chem. **252:**1732–1738.

Summers, W. C., I. Brunovski, and R. W. Hyman. 1973. Process of infection with coliphage T7. VII. Characterization and mapping of the major in vivo transcription products of the early region. J. Mol. Biol. **74:**291–300.

Sutherland, B. M. 1978. Enzymatic photoreactivation of DNA, p. 113–122. *In* P. C. Hanawalt, E. C. Friedberg, and C. F. Fox (ed.), DNA repair mechanisms. Academic Press, Inc., New York.

Symonds, N. 1976. Recombination. MTP Int. Rev. Sci. **6:**165–189.

Symonds, N., H. Heindl, and P. White. 1973. Radiation sensitive mutants of phage T4: a comparative study. Mol. Gen. Genet. **120:**253–259.

Symonds, N., and D. A. Ritchie. 1961. Multiplicity reactivation after decay of incorporated radioactive phosphorous in phage T4. J. Mol Biol. **3:**61–70.

Symonds, N., K. A. Stacey, S. W. Glover, J. Schell, and S. Silver. 1963. The chemical basis for a case of host-induced modification in phage T2. Biochem. Biophys. Res. Commun. **12:**220–222.

Tabor, S., M. J. Engler, C. W. Fuller, R. L. Lechner, S. W. Mattson, L. J. Romano, H. Saito, F. Tamanoi, and C. C. Richardson. 1981. Initiation of bacteriophage T7 DNA replication, p. 387–408. *In* D. Ray (ed.), The initiation of DNA replication. Academic Press, Inc., New York.

Tabor, S., and C. C. Richardson. 1981. Template recognition sequence for RNA primer synthesis by gene 4 protein of bacteriophage T7. Proc. Natl. Acad. Sci. U.S.A. **78:**205–209.

Takacs, B. J., and J. P. Rosenbusch. 1975. Modification of *Escherichia coli* membranes in the prereplicative phase of T4 infection. Specificity of association and quantification of bound phage proteins. J. Biol. Chem. **250:**2339–2350.

Takagi, Y., M. Sekiguchi, S. Okubo, H. Nakayama, K. Shimada, S. Yasuda, T. Nishimoto, and H. Yoshihara. 1968. Nucleases specific for ultraviolet light-irradiated DNA and their possible role in dark repair. Cold Spring Harbor Symp. Quant. Biol. **33:**219–227.

Takahashi, H. 1978. Genetic and physiological characterization of *Escherichia coli* K-12 mutants (*tabC*) which induce the abortive infection of bacteriophage T4. Virology **87**:256–265.

Takahashi, H., A. Coppo, A. Manzi, G. Martire, and J. Pulitzer. 1975. Design of a system of conditional mutations (tab/k/com) affecting protein-protein interactions in bacteriophage infected *Escherichia coli*. J. Mol. Biol. **96**:563–578.

Takahashi, H., and H. Saito. 1982a. High-frequency transduction of pBR322 by cytosine-substituted T4 bacteriophage: evidence for encapsulation and transfer of head-to-tail plasmid concatemers. Plasmid **8**:29–35.

Takahashi, H., and H. Saito. 1982b. Cloning of *uvsW* and *uvsY* genes of bacteriophage T4. Virology **120**:122–129.

Takahashi, H., and H. Saito. 1982c. Mechanisms of pBR322 transduction mediated by cytosine-substituting T4 bacteriophage. Mol. Gen. Genet. **186**:497–500.

Takahashi, H., H. Saito, and Y. Ikeda. 1978. Viable T4 bacteriophage containing cytosine substituted DNA (T4dC phage). I. Behavior towards the restriction modification systems of *Escherichia coli* and derivation of a new T4 phage strain (T4dC) having the complete T4 genome. J. Gen. Appl. Microbiol. **24**:297–306.

Takahashi, H., M. Shimizu, H. Saito, and Y. Ikeda. 1979. Studies of viable bacteriophage containing cytosine-substituted DNA (T4dC phage). II. Cleavage of T4dC DNA by endonuclease *Sal*I and *Bam*HI. Mol. Gen. Genet. **168**:49–53.

Takahashi, H., and H. Yoshikawa. 1979. Genetic study of a new early gene, *comC*-α, of bacteriophage T4. Virology **95**:215–217.

Takano, T., and T. Kakefuda. 1972. Involvement of a bacterial factor in morphogenesis of bacteriophage capsid. Nature (London) New Biol. **239**:34–37.

Tanaka, K., H. Hayakawa, M. Sekiguchi, and Y. Okada. 1977. Specific action of T4 endonuclease V on damaged DNA in xeroderma pigmentosum cells in vivo. Proc. Natl. Acad. Sci. U.S.A. **74**:2958–2962.

Tanaka, J., and M. Sekiguchi. 1975. Action of exonuclease V (the recBC enzyme) on ultraviolet-irradiation DNA. Biochim. Biophys. Acta **383**:178–187.

Tanford, C. 1961. Physical chemistry of macromolecules, p. 275–316. John Wiley & Sons, Inc., New York.

Tanner, D., and M. Oishi. 1971. The effect of bacteriophage T4 infection on an ATP dependent DNase in *Escherichia coli*. Biochim. Biophys. Acta **28**:767–769.

Teifel, J., and H. Schmieger. 1981. The influence of host DNA replication on the formation of infectious and transducing Mu-particles. Mol. Gen. Genet. **184**:308–311.

Terzaghi, B. E., E. Terzaghi, and D. Coombs. 1979. Mutational alteration of the T4D tail fiber attachment process. J. Mol. Biol. **127**:1–14.

Terzaghi, E., 1971. Alternative pathways of tail fiber assembly in bacteriophage T4. J. Mol. Biol. **59**:319–327.

Terzaghi, E., Y. Okada, G. Streisinger, J. Emrich, M. Inouye, and A. Tsugita. 1966. Change of sequence of amino acids in phage T4 lysozyme by acridine-induced mutations. Proc. Natl. Acad. Sci. U.S.A. **56**:500–507.

Tessman, I. 1965. Genetic ultrafine structure in the T4rII region. Genetics **51**:63–75.

Tessman, I. 1968. Mutagenic treatment of double- and single-stranded phages T4 and S13 with hydroxylamine. Virology **35**:330–333.

Tessman, I., and D. B. Greenberg. 1972. Ribonucleotide reductase genes of phage T4: map location of the thioredoxin gene nrdC. Virology **49**:337–338.

Thelander, L., and P. Reichard. 1979. Reduction of ribonucleotides. Annu. Rev. Biochem. **48**:133–158.

Thermes, C., P. Daegelen, B. de Franciscis, and E. Brody. 1976. In vitro system for induction of delayed early RNA of bacteriophage T4. Proc. Natl. Acad. Sci. U.S.A. **73**:2569–2573.

Thomas, C. A., Jr. 1967. The rule of the ring. J. Cell. Physiol. **70**(Suppl. 1):13–34.

Tigges, M. A., C. J. H. Bursch, and D. P. Snustad. 1977. Slow switchover from host RNA synthesis to bacteriophage RNA synthesis after infection of *Escherichia coli* with a T4 mutant defective in the bacteriophage T4-induced unfolding of the host nucleoid. J. Virol. **24**:775–785.

Tikhonenko, A. S. 1970. Ultrastructure of bacterial viruses. Plenum Publishing Corp., New York.

Tilly, K. H., H. Murialdo, and C. Georgopoulos. 1981. Identification of a second *Escherichia coli groE* gene whose product is necessary for bacteriophage morphogenesis. Proc. Natl. Acad. Sci. U.S.A. **78**:1629–1633.

Tinoco, I., P. N. Borer, B. Dengler, M. D. Levine, O. C. Uhlenbeck, D. M. Crothers, and J. Gralla. 1973. Improved estimation of secondary structure in ribonucleic acids. Nature (London) New Biol. **246**:40–41.

To, C. M., E. Kellenberger, and A. Eisenstark. 1969. Disassembly of T-even bacteriophage into structural parts and subunits. J. Mol. Biol. **46**:493–511.

Tomich, P. K., C.-S. Chiu, M. G. Wovcha, and G. R. Greenburg. 1974. Evidence for a complex regulating the in vivo activities of early enzymes induced by bacteriophage T4. J. Biol. Chem. **249**:7613–7622.

Tomizawa, J., N. Anraku, and Y. Iwama. 1966. Molecular mechanisms of genetic recombination in bacteriophage. VI. A mutant defective in the joining of DNA molecules. J. Mol. Biol. **21**:247–253.

Tomizawa, J., and T. Itoh. 1982. The importance of RNA secondary structure in ColE1 primer formation. Cell **31**:575–583.

Toothman, P., and I. Herskowitz. 1980a. Rex-dependent exclusion of lambdoid phages. I. Prophage requirements for exclusion. Virology **102**:133–146.

Toothman, P., and I. Herskowitz. 1980b. Rex-dependent exclusion of lambdoid phages. II. Determinants of sensitivity to exclusion. Virology **102**:147–160.

Toothman, P., and I. Herskowitz. 1980c. Rex-dependent exclusion of lambdoid phages. III. Physiology of the abortive infection. Virology **102**:161–171.

Topal, M. D., S. DiGuiseppi, and N. K. Sinha. 1980. Molecular basis for substitution mutations. Effect of primer terminal and template residues on nucleotide selection by phage T4 DNA polymerase in vitro. J. Biol. Chem. **255**:11717–11724.

Topal, M. D., and J. R. Fresco. 1976. Complementary base pairing and the origin of substitution mutations. Nature (London) **263**:285–289.

Torrey, T. A., and T. Kogoma. 1982. Suppressor mutations (*rin*) that specifically suppress the *recA*$^+$ dependence of stable DNA replication in *Escherichia coli* K-12. Mol. Gen. Genet. **187**:225–230.

Tosi, M., B. Reilly, and D. Anderson. 1975. Morphogenesis of bacteriophage ϕ29 of *Bacillus subtilis*: cleavage and assembly of the neck appendage protein. J. Virol. **16**:1282–1295.

Traub, F., M. Maeder, and E. Kellenberger. 1981. Bacteriophage T4 head assembly. *In vivo* characterization of the morphopoietic core, p. 127–137. *In* M. DuBow (ed.), Bacteriophage assembly. Alan R. Liss, New York.

Travers, A. 1969. Bacteriophage sigma factor for RNA polymerase. Nature (London) **223**:1107–1110.

Travers, A. 1970a. Positive control of transcription by a bacteriophage sigma factor. Nature (London) **225**:1009–1012.

Travers, A. 1970b. RNA polymerase and T4 development. Cold Spring Harbor Symp. Quant. Biol. **35**:241–252.

Trimble, R. B., J. Galivan, and F. Maley. 1972. The temporal expression of T2r$^+$ bacteriophage genes in vivo and in vitro. Proc. Natl. Acad. Sci. U.S.A. **69**:1659–1663.

Trimble, R. B., and F. Maley. 1975. The influence of potassium and rifampicin on the expression of bacteriophage T2 pre-replicative genes in vitro and in vivo. Arch. Biochem. Biophys. **167**:377–387.

Trimble, R. B., and F. Maley. 1976. Level of specific prereplicative mRNA's during bacteriophage T4 *regA*$^-$, 43$^-$, and T4 43$^-$ infection of *Escherichia coli* B. J. Virol. **17**:538–549.

Tschopp, J., F. Arisaka, R. van Driel, and J. Engel. 1979. Purification, characterization and reassembly of the bacteriophage T4D tail sheath protein P18. J. Mol. Biol. **128**:247–258.

Tschopp, J., and P. R. Smith. 1978. Extra long T4 tails produced in in vitro conditions. J. Mol. Biol. **114**:281–286.

Tse, Y. C., and J. C. Wang. 1980. *E. coli* and *M. luteus* DNA topoisomerase I can catalyze catenation or decatenation of double stranded DNA rings. Cell **22**:269–276.

Tsugita, A. 1971. Phage lysozymes and other lytic enzymes, p. 343–411. *In* P. D. Boyer (ed.), The enzymes, vol. 5, 3rd ed. Academic Press, Inc., New York.

Tsugita, A., and J. Hosoda. 1978. DNA binding site of the helix-destabilizing protein gp32 from bacteriophage T4. J. Mol. Biol. **122**:255–258.

Tsugita, A., L. W. Black, and M. K. Showe. 1975. Protein cleavage during virus assembly: characterization of cleavage in T4 phage. J. Mol. Biol. **98**:271–275.

Tsugita, A., M. Inouye, E. Terzaghi, and G. Streisinger. 1968. Purification of bacteriophage T4 lysozyme. J. Biol. Chem. **243**:391–397.

Tsui, L., and R. W. Hendrix. 1980. Head-tail connector of bacteriophage lambda. J. Mol. Biol. **142**:419–438.

Tsurimoto, T., and K. Matsubara. 1981. Purified bacteriophage lambda O-protein binds to four repeating sequences at the lambda replication origin. Nucelic Acids Res. **9**:1789–1799.

Tutas, D. J., J. M. Wehner, and J. F. Koerner. 1974. Unfolding of the host genome after infection of *Escherichia coli* with bacteriophage T4. J. Virol **13**:548–550.

Tye, B.-K., J. A. Huberman, and D. Botstein. 1974. Non-random circular permutation of phage P22 DNA. J. Mol. Biol. **85**:501–532.

Uratani, Y., S. Asakura, and K. Imahori. 1972. A circular dichroism study of *Salmonella* flagellin: evidence for conformational change upon polymerization. J. Mol. Biol. **67**:85–98.

Uzan, M., and A. Danchin. 1978. Correlation between the serine sensitivity and the derepressibility of the *ilv* genes in *Escherichia coli relA*⁻ mutants. Mol. Gen. Genet. **165**:21–30.

Vallée, M., and J. B. Cornett. 1972. A new gene of bacteriophage T4 determining immunity against superinfecting ghosts and phage in T4 infected *Escherichia coli*. Virology **48**:777–784.

Vallée, M., and O. de Lapeyriere. 1975. The role of the genes *imm* and *s* in the development of immunity against T4 ghosts and exclusion of super infecting phage in *Escherichia coli* infected with T4. Virology **67**:219–233.

van Arkel, G. A., J. H. van de Pol, and J. A. Cohen. 1961. Genetic recombination and marker rescue of urea-disrupted bacteriophage T4 in spheroplasts of *E. coli*. Virology **13**:546–548.

van den Ende, P., and N. Symonds. 1972. The isolation and characterization of a T4 mutant partially defective in recombination. Mol. Gen. Genet. **115**:239–247.

Vanderslice, R. W., and C. D. Yegian. 1974. The identification of late bacteriophage T4 proteins on sodium dodecyl sulfate polyacrylamide gels. Virology **60**:265–275.

Van Driel, R. 1977. Assembly of bacteriophage T4 head-related structures. Assembly of polyheads *in vitro*. J. Mol. Biol. **114**:61–72.

Van Driel, R. 1980. Assembly of bacteriophage T4 head-related structures. IV. Isolation and association of T4 prehead proteins. J. Mol. Biol. **138**:27–42.

Van Driel, R., and E. Couture. 1978. Assembly of bacteriophage T4 head-related structures. II. *In vitro* assembly of prehead-like structures. J. Mol. Biol. **123**:115–128.

Van Driel, R., and E. Couture. 1978. Assembly of the scaffolding core of bacteriophage T4 preheads. J. Mol. Biol. **123**:713–719.

Van Driel, R., F. Traub, and M. K. Showe. 1980. Probable localization of the bacteriophage T4 prehead proteinase zymogen in the center of the prehead core. J. Virol. **36**:220–223.

Van Eerd, J. P., S. P. Champe, L. Yager, I. Kubota, and A. Tsugita. 1977. Primary structure of internal peptide VII of T-even bacteriophages. J. Mol. Biol. **117**:521–524.

van Holde, K. E., and W. E. Hill. 1974. General physical properties of ribosomes. *In* M. Nomura, A. Tissiers, and P. Lengyel (ed.), Ribosomes. Cold Spring Harbor Laboratory, Cold Spring Harbor, N.Y.

van Minderhout, L., and J. Grimbergen. 1975. A new type of UV-sensitive mutant of phage T4D. Mutat. Res. **29**:349–362.

van Minderhout, L., and J. Grimbergen. 1976. Evidence for a third type of UV repair in bacteriophage T4. Mutat. Res. **35**:161–166.

van Minderhout, L., J. Grimbergen, and B. de Groot. 1978. Non-essential UV-sensitive bacteriophage T4 mutants affecting early DNA synthesis: a third pathway of DNA repair. Mutat. Res. **52**:313–322.

van Minderhout, L., J. Grimbergen, and B. de Groot. 1979. Bacteriophage T4v mutants with a slow rate of thymine-dimer excision and a particular reactivation phenotype. Mutat. Res. **60**:253–262.

van Ormondt, H., J. Gorter, K. J. Havelaar, and A. de Waard. 1975. Specificity of a deoxyribonucleic acid transmethylase induced by bacteriophage T2. I. Nucleotide sequences isolated from *Micrococcus luteus* DNA methylated in vitro. Nucleic Acids Res. **2**:1391–1400.

Vanyushin, B. F., Ya. I. Buryanov, and A. N. Belozesky. 1971. Distribution of N6-methyladenine in DNA of T2 phage and its host *Escherichia coli* B. Nature (London) New Biol. **230**:25–27.

van Zeeland, A. A. 1978. Introduction of T4 endonuclease V into frozen and thawed mammalian cells for the determination of removal of UV induced photoproducts, p. 307–310. *In* P. C. Hanawalt, E. C. Friedberg, and C. F. Fox (ed.), DIVA repair mechanisms. Academic Press, Inc., New York.

Velten, J., and J. Abelson. 1980. The generation and analysis of clones containing bacteriophage T4 DNA fragments. J. Mol. Biol. **137**:235–248.

Velten, J., K. Fukada, and J. Abelson. 1976. *In vitro* construction of bacteriophage and plasmid DNA molecules containing DNA fragments from bacteriophage T4. Gene **1**:93–106.

Venkatesan, M., and N. C. Nossal. 1982. Bacteriophage T4 gene 44/62 and gene 45 polymerase accessory proteins stimulate hydrolysis of duplex DNA by T4 DNA polymerase. J. Biol. Chem. **257**:12435–12443.

Venkatesan, M., L. L. Silver, and N. C. Nossal. 1982. Bacteriophage T4 gene 41 protein, required for the synthesis of RNA primers, is also a DNA helicase. J. Biol. Chem. **257**:12426–12434.

Verly, W. G., and Y. Pacquette. 1972. Properties of the endonuclease for depurinated DNA from *Escherichia coli*. Can. J. Biochem. **50**:1199–1209.

Vetter, D., and P. D. Sadowski. 1974. Point mutants in the D2a region of bacteriophage T4 fail to induce T4 endonuclease IV. J. Virol. **14**:207–213.

Vielmetter, W., and E. Schuster. 1960. The base specificity of mutation induced by nitrous acid in phage T2. Biochem. Biophys. Res. Commun. **2**:324–328.

Viñuela, E., A. Camacho, F. Jimenez, J. L. Carrascosa, G. Ramirez, and M. Salas. 1976. Structure and assembly of phage ϕ29. Philos. Trans. R. Soc. London Ser. B. **276**:29–36.

Visconti, N., and M. Delbrück. 1953. The mechanism of genetic recombination in phage. Genetics **38**:5–33.

Volker, T. A., J. Gafner, T. A. Bickle, and M. K. Showe. 1982. Gene 67, a new essential bacteriophage T4 head gene codes for a pre-head core component, PIP. I. Genetic mapping and DNA sequence. J. Mol. Biol. **161**:479–489.

Volker, T. A., A. Kuhn, M. K. Showe, and T. A. Bickle. 1982. Gene 67, a new essential bacteriophage T4 head gene codes for a pre-head core component, PIP. II. The construction in vitro of unconditionally leathal mutants and their maintenance. J. Mol. Biol. **161**:491–504.

Volker, T. A., and M. K. Showe. 1980. Induction of mutants in specific genes of bacteriophage T4 using cloned restriction fragments and marker rescue. Mol. Gen. Genet. **177**:447–452.

Volkin, E. 1954. The linkage of glucose in coliphage nucleic acids. J. Am. Chem. Soc. **76**:5892–5893.

Volkin, E., and L. Astrachan. 1956. Phosphorus incorporation *Escherichia coli* ribonucleic acid after infection with bacteriophage T2. Virology **2**:149–161.

Volkin, E., L. Astrachan, and J. Countryman. 1958. Metabolism of RNA phosphorus in *Escherichia coli* infected with bacteriophage T7. Virology **6**:545–555.

von Hippel, P. H., S. C. Kowalczykowski, N. Lonenberg, J. N. Newport, L. S. Paul, G. D. Stormo, and L. Gold. 1982. Autoregulation of gene expression: quantitative evaluation of the expression and function of the bacteriophage T4 gene 32 (single stranded DNA binding) protein system. J. Mol. Biol. **162**:795–818.

von Smoluchowski, M. 1971. Mathematical theory of the kinetics of the coagulation of colloidal solutions. Z. Phys. Chem. **92**:129–168.

Vorozheikina, D., I. Glinskaite, L. Tikhomirova, and A. Bayev. 1980. Complementation of T4 phage am mutations by hybrid phages lambda-T4. Mol. Gen. Genet. **178**:655–661.

Wagenknecht, T., and V. A. Bloomfield. 1975. Equilibrium mechanisms of length regulation in linear protein aggregates. Biopolymers **14**:2297–2309.

Wagenknecht, T., and V. A. Bloomfield. 1977. In vitro polymerization of bacteriophage T4D core subunits. J. Mol. Biol. **116**:347–359.

Wagner, J., and U. K. Laemmli. 1976. Studies on the maturation of the head of bacteriophage T4. Philos. Trans. R. Soc. London Ser. B **276**:15–28.

Wagner, J. A., and U. K. Laemmli. 1979. Maturation of the head of bacteriophage T4. 9-Aminoacridine blocks a late step in DNA packaging. Virology **92**:219–229.

Wagner, R., Jr., and M. Meselson. 1976. Repair tracts in mismatched DNA heteroduplexes. Proc. Natl. Acad. Sci. U.S.A. **73**:4135–4139.

Wahba, A. J., M. J. Miller, A. Niveleu, T. A. Landers, G. G. Carmichael, K. Weber, D. A. Hawley, and L. I. Slobin. 1974. Subunit I of Q β replicase and 30S ribosomal protein S1 of *Escherichia coli*. Evidence for the identity of the two proteins. J. Biol. Chem. **249**:3314–3316.

Wais, A. C., and E. B. Goldberg. 1969. Growth and transformation of phage T4 in *E. coli* B/4. *Salmonella, Aerobacter, Proteus* and *Serratia*. Virology **39**:153–161.

Wakem, L. P., and K. Ebisuzaki. 1976. Pathways of DNA repair in phage T4. II. Sedimentation analysis of intracelluiar DNA in repair-defective mutants. Virology **73**:155–164.

Wakem, L. P., and K. Ebisuzaki. 1981. DNA repair-recombination functions in the DNA processing pathway of bacteriophage T4. Virology **112**:472–479.

Walker, J. E., M. Garaste, M. J. Brunswick, and M. J. Gay. 1982. EMBO J. **1**:945–951.

Walker, J. M., K. Gooderham, J. R. B. Hastings, E. Mayes, and E. Johns. 1980. The primary structures of non-histone chromosomal proteins HMG 1 and 2. FEBS Lett. **122**:264–270.

Wallace, S. S., and R. J. Melamede. 1972. Host- and phage-mediated repair of radiation damage in bacteriophage T4. J. Virol. **10**:1159–1169.

Ward, D. F., and M. E. Gottesman. 1982. Suppression of transcription termination by phage lambda. Science **216**:946–951.

Ward, J. F. 1975. Molecular mechanisms of radiation-induced damage to nucleic acids. Adv. Radiat. Biol. **5**:181–239.

Ward, S., and R. C. Dickson. 1971. Assembly of bacteriophage T4 tail fibers, III. Genetic control of the major tail fiber polypeptides. J. Mol. Biol. **62**:479–492.

Ward, S., R. B. Luftig, J. H. Wilson, H. Eddleman, H. Lyle, and W. B. Wood. 1970. Assembly of bacteriophage T4 tail fibers II. Isolation and characterization of tail fiber precursors. J. Mol. Biol. **54**:15–31.

Ware, B. R. 1974. Electrophoretic light scattering. Adv. Colloid Interface Sci. **4**:1–44.

Warner, H. R. 1971. Partial suppression of bacteriophage T4 ligase mutations by T4 endonuclease II deficiency: role of host ligase. J. Virol. **7**:534–536.

Warner, H. R. 1973. Properties of ribonucleoside diphosphate reductase in nucleotide-permeable cells. J. Bacteriol. **115**:18–22.

Warner, H. R., and J. E. Barnes. 1966. Deoxyribonucleic acid synthesis in *Escherichia coli* infected with some deoxyribonucleic acid polymerase-less mutants of bacteriophage T4. J. Virol. **28**:100–107.

Warner, H. R., L. M. Christensen, and M.-L. Persson. 1981. Evidence that the UV endonuclease activity induced by bacteriophage T4 contains both pyrmidine dimer-DNA glycosylase and apyrimidinic/apurinic endonuclease activities in the enzyme molecule. J. Virol. **40**:204–210.

Warner, H. R., B. F. Demple, W. A. Deutsch, C. M. Kane, and S. Linn. 1980. Apurinic/apyrimidinic endonucleases in repair of pyrimidine dimers and other lesions in DNA. Proc. Natl. Acad. Sci. U.S.A. **77**:4602–4606.

Warner, H. R., and B. K. Duncan. 1978. In vivo synthesis and properties of uracil-containing DNA. Nature (London) **272**:32–34.

Warner, H. R., and M. D. Hobbs. 1967. Incorporation of uracil-C^{14} into nucleic acids in *E. coli* infected with bacteriophage T4 and T4 amber mutants. Virology **33**:376–384.

Warner, H. R., and M. D. Hobbs. 1969. Effect of hydroxyurea on replication of bacteriophage T4 in *Escherichia coli*. J. Virol. **3**:331–336.

Warner, H. R., D. P. Snustad, S. E. Jorgensen, and J. F. Koerner. 1970. Isolation of bacteriophage T4 mutants defective in the ability to degrade host deoxyribonucleic acid. J. Virol. **5**:700–708.

Warner, H. R., D. P. Snustad, J. F. Koerner, and J. D. Childs. 1972. Identification and genetic characterization of mutants of bacteriophage T4 defective in the ability to induce exonuclease A. J. Virol. **9**:399–407.

Watanabe, S. M., and M. F. Goodman. 1978. Mutator and antimutator phenotypes of suppressed amber mutants in genes 32, 41, 44, 45, and 62 in bacteriophage T4. J. Virol. **25**:73–77.

Watanabe, S. M., and M. F. Goodman. 1981. On the molecular basis of transition mutations: the frequencies of forming and 2-aminopurine·cytosine adenine·cytosine base mispairs in vitro. Proc. Natl. Acad. Sci. U.S.A. **78**:2864–2868.

Watanabe, S. M., and M. F. Goodman. 1982. Kinetic measurement of 1-aminopurine·cytosine and 2-aminopurine·thymine base pairs as a test of DNA polymerase fidelity mechanisms. Proc. Natl. Acad. Sci. U.S.A. **79**:6429–6433.

Watson, J. D. 1972. Origin of concatemeric T7 DNA. Nature (London) New Biol. **239**:197–201.

Watson, J. D., and F. H. C. Crick. 1953a. A structure for deoxyribose nucleic acid. Nature (London) **171**:737–738.

Watson, J. D., and F. H. C. Crick. 1953b. Genetical implications of the structure of deoxyribonucleic acid. Nature (London) **171**:964–967.

Watson, J. D., and F. H. C. Crick. 1953c. The structure of DNA. Cold Spring Harbor Symp. Quant. Biol. **18**:123–131.

Watson, N., and D. Apirion. 1981. Ribonuclease F: a putative processing endonuclease from *Escherichia coli*. Biochem. Biophys. Res. Commun. **103**:543–551.

Weed, L. L., and S. S. Cohen. 1951. The utilization of host pyrimidines in the synthesis of bacterial viruses. J. Biol. Chem. **192**:693–700.

Weil, J., B. Terzaghi, and J. Crasemann. 1965. Partial diploidy in phage T4. Genetics **52**:683–693.

Weintraub, S. B., and F. R. Frankel. 1972. Identification of the T4rIIB gene product as a membrane protein. J. Mol. Biol. **70**:589–615.

Weiss, S. B., W.-T. Hsu, J. W. Foft, and N. M. Scherberg. 1968. Transfer RNA coded by the T4 bacteriophage genome. Proc. Natl. Acad. Sci. U.S.A. **61**:114–121.

Wermuth, B., and N. O. Kaplan. 1976. Pyridine nucleotide transhydrogenase from *Pseudomonas aeruginosa*: purification by affinity chromatography and physicochemical properties. Arch. Biochem. Biophys. **176**:136–143.

Werner, R. 1969. Initiation and propagation of growing points on the DNA of phage T4. Cold Spring Harbor Symp. Quant. Biol. **33**:501–507.

West, S. C., E. Cassuto, and P. Howard-Flanders. 1981. Heteroduplex formation by recA protein: polarity of strand exchange. Proc. Natl. Acad. Sci. U.S.A. **78**:6149–6153.

Wever, G. H., B. J. Thompson, R. M. Laiken, E. Ruby, and J. S. Wiberg. 1981. T4 head assembly and high temperature, p. 167–192. *In* M. DuBow (ed.), Bacteriophage assembly. Alan R. Liss, New York.

Weymouth, L. A., and L. A. Loeb. 1978. Mutagenesis during in vitro DNA synthesis. Proc. Natl. Acad. Sci. U.S.A. **75**:1924–1928.

Whitehouse, H. L. K. 1963. A theory of cross-over by means of hybrid deoxyribonucleic acid. Nature (London) **199**:1034–1040.

Wiberg, J. S. 1966. Mutants of bacteriophage T4 unable to cause breakdown of host DNA. Proc. Natl. Acad. Sci. U.S.A. **55**:614–621.

Wiberg, J. S. 1967. Amber mutants of bacteriophage T4 defective in dCTPase and dUTPase. J. Biol. Chem. **242**:5824–5829.

Wiberg, J. S., T. S. Cardillo, and C. Mickelson. 1981. Genetic and amber fragment maps of genes 46 and 47 of bacteriophage T4D. J. Virol. **40**:309–313.

Wiberg, J. S., M. L. Dirksen, R. H. Epstein, S. E. Luria, and J. M. Buchanan. 1962. Early enzyme synthesis and its control in *E. coli* infected with some amber mutants of bacteriophage T4. Proc. Natl. Acad. Sci. U.S.A. **48**:293–302.

Wiberg, J. S., S. Mendelsohn, V. Warner, K. Hercules, C. Aldrich, and L. Munro. 1973. SP62, a viable mutant of bacteriophage T4D defective in regulation of phage enzyme synthesis. J. Virol. **12**:775–1792.

Wickner, S. 1978. DNA replication proteins of *Escherichia coli*. Annu. Rev. Biochem. **47**:1163–1191.

Wickremasinghe, R. G., and A. V. Hoffbrand. 1980. Reduced rate of DNA replication fork movement in megaloblastic anemia. J. Clin. Invest. **65**:25–36.

Wilhelm, J. M., and R. Haselkorn. 1969. *In vitro* synthesis of T4 proteins: lysozyme and the products of genes 22 and 57. Cold Spring Harbor Symp. Quant. Biol. **34**:793–798.

Wilhelm, J. M., and R. Haselkorn. 1971. *In vitro* synthesis of T4 proteins: the products of genes 9, 18, 19, 23, 24, and 38. Virology **43**:198–208.

Williams, G. C. 1975. Sex and evolution. Princeton University Press, Princeton, N.J.

Williams, K. R., and W. Konigsberg. 1978. Structural changes in the T4 gene 32 protein induced by DNA and polynucleotides. J. Biol. Chem. **253**:2463–2470.

Williams, K. R., and W. Konigsberg. 1981. DNA helix-destabilizing proteins, p. 475–508. *In* J. G. Chirikjian and T. S. Papas (ed.), Gene amplification and analysis, vol. 2. Elsevier/North-Holland Publishing Co., Amsterdam.

Williams, K. R., J. J. L'Italian, R. A. Guggenheimer, L. Sillerud, E. Spicer, J. Chase, and W. Konigsberg. 1982. Comparative peptide mapping by HPLC: identification of single amino acid substitutions in temperature sensitive mutants, p. 499–507. *In* M. Elzinga (ed.), Methods in protein sequence analysis. Humana Press, Clifton, N.J.

Williams, K. R., M. B. LoPresti, and M. Setoguchi. 1981. Primary structure of the bacteriophage T4 DNA helix-destabilizing protein. J. Biol. Chem. **256**:1754–1762.

Williams, K. R., M. LoPresti, M. Setoguchi, and W. H. Konigsberg. 1980. Amino acid sequence of the DNA helix-destabilizer protein. Proc. Natl. Acad. Sci. U.S.A. **77**:4614–4617.

Williams, K. R., L. O. Sillerud, D. E. Schafer, and W. H. Konigsberg. 1979. DNA binding properties of the T4 DNA helix-destabilizing protein: a calorimetric study. J. Biol. Chem. **254**:6426–6432.

Williams, K. R., E. K. Spicer, M. B. LoPresti, R. Guggenheimer, and J. Chase. 1983. Limited proteolysis studies on the *Escherichia coli* single-stranded DNA binding protein: evidence for a functionally homologous domain in both the *Escherichia coli* and T4 DNA binding proteins. J. Biol. Chem. **258**:3346–3355.

Williams, R. C., and D. Fraser. 1953. Morphology of the seven T-bacteriophages. J. Bacteriol. **66**:458–464.

Williams, R. C., and D. Fraser. 1956. Structural and functional differentiation in T2 bacteriophage. Virology **2**:289–307.

Williams, R. C. and K. E. Richards. 1974. Capsid structure of bacteriophage lambda. J. Mol. Biol. **88**:547–550.

Williams, W. E., and J. W. Drake. 1977. Mutator mutations in bacteriophage T4 gene 42. Genetics **86**:501–511.

Wilson, G. G., and N. E. Murray. 1979. Molecular cloning of the DNA ligase gene from bacteriophage T4. I. Characterization of the recombinants. J. Mol. Biol. **132**:471–491.

Wilson, G. G., R. L. Neve, G. J. Edlin, and W. N. Konigsberg. 1980. BamH1 restriction site in the bacteriophage T4 chromosome is located in or near gene 8. Genetics **93**:285–296.

Wilson, G. G., V. I. Tanyashin, and N. E. Murray. 1977. Molecular cloning of fragments of bacteriophage T4 DNA. Mol. Gen. Genet. **156**:203–214.

Wilson, G. G., K. K. Y. Young, G. J. Edlin, and W. Konigsberg. 1979. High-frequency generalized transduction by bacteriophage T4. Nature (London) **280**:80–82.

Wilson, J. H. 1973. Function of the bacteriophage T4 transfer RNA's. J. Mol. Biol. **74**:753–766.

Wilson, J. H., and J. N. Abelson. 1972. Bacteriophage T4 transfer RNA. II. Mutants of T4 defective in the formation of functional suppressor transfer RNA. J. Mol. Biol. **69**:57–63.

Wilson, J. H., and S. Kells. 1972. Bacteriophage T4 transfer DNA. I.

Isolation and characterization of two phage-coded nonsense suppressors. J. Mol. Biol. **69**:39–56.

Wilson, J. H., J. Kim, and J. Abelson. 1972. Bacteriophage T4 transfer RNA. III. Clustering of the genes for the T4 transfer RNA's. J. Mol. Biol. **71**:547–556.

Wilson, J. H., R. B. Luftig, and W. Wood. 1970. Interaction of bacteriophage T4 tail fiber components with lipopolysaccharide fraction from *Escherichia coli*. J. Mol. Biol. **51**:423–434.

Winkler, U., H. E. Johns, and E. Kellenberger. 1962. Comparative study of some properties of bacteriophage T4D irradiated with monochromatic ultraviolet light. Virology **18**:343–358.

Wirak, D. D., and G. R. Greenberg. 1980. Role of bacteriophage T4 DNA-delay gene products in deoxyribonucleotide synthesis. J. Biol. Chem. **255**:1896–1904.

Witkin, E. M. 1976. Ultraviolet mutagenesis and inducible DNA repair in *Escherichia coli*. Bacteriol. Rev. **40**:869–907.

Witmer, H., A. Baros, D. Ende, and M. Dosmar. 1976. Control of synthesis of mRNA's for T4 bacteriophage-specific dihydrofolate reductase and deoxycytidylate hydroxymethylase. J. Virol. **19**:846–856.

Witmer, H., A. Baros, and J. Forbes. 1975. Effect of chloramphenicol and starvation for an essential amino acid on polypeptide and polyribonucleotide synthesis in *Escherichia coli* infected with bacteriophage T4. Arch. Biochem. Biophys. **169**:406–415.

Witmer, H. J. 1976. Regulation of bacteriophage T4 gene expression. Prog. Mol. Subcell. Biol. **4**:17–51.

Wolfenden, R. V. 1969. Tautomeric equilibria in inosine and adenosine. J. Mol. Biol. **40**:307–310.

Wollman, E., and E. Wollman. 1937. Les phases des bacteriophages (facteurs lysogenes). C.R. Seances Soc. Biol. Paris **124**:931–934.

Wollman, E., and E. Wollman. 1938. Recherches sur le phenomene de Twort d'Herelle (bacteriophagie on autolyse heredo-contagieuse). Ann. Inst. Pasteur Paris. **60**:13–58.

Womac, F. C. 1963. An analysis of single-burst progeny of bacteria singly infected with a bacteriophage heterozygote. Virology **21**:232–241.

Womack, F. C. 1965. Cross reactivation differences in bacteriophage T4D. Virology **26**:758–761.

Wood, W. B. 1966. Host specificity of DNA produced by *Escherichia coli*: bacterial mutations affecting the restriction and modification of DNA. J. Mol. Biol. **16**:118–133.

Wood, W. B. 1979. Bacteriophage T4 assembly and the morphogenesis of subcellular structure. Harvey Lect. **73**:203–223.

Wood, W. B. 1980. Bacteriophage T4 morphogenesis as a model for assembly of subcellular structure. Q. Rev. Biol. **55**:353–367.

Wood, W. B., and J. R. Bishop. 1973. Bacteriophage T4 tail fibers: structure and assembly of a viral organelle, p. 303–324. *In* C. F. Fox and W. F. Robinson (ed.), Virus research. Academic Press, Inc., New York.

Wood, W. B., and M. P. Conley. 1979. Attachment of tail fibers in bacteriophage T4 assembly: role of the phage whiskers. J. Mol. Biol. **127**:15–29.

Wood, W. B., M. P. Conley, H. L. Lyle, and R. C. Dickson. 1978. Attachment of tail fibers in bacteriophage T4 assembly. Purification, properties and site of action of the acccessory protein coded by gene 63. J. Biol. Chem. **253**:2437–2445.

Wood, W. B., R. C. Dickson, R. J. Bishop, and H. R. Revel. 1973. Self-assembly and non-self-assembly in bacteriophage T4 morphogenesis, p. 25–58. *In* R. Markham (ed.), Generation of subcellular structure, proceedings of the First John Innes Symposium. North-Holland Publishing Co., Amsterdam.

Wood, W. B., and M. Henninger. 1969. Attachment of tail fibers in bacteriophage T4 assembly: some properties of the reaction in vitro and its genetic control. J. Mol. Biol. **39**:608–618.

Wood, W. B., and J. King. 1979. Genetic control of complex bacteriophage assembly. Compr. Virol. **13**:581–633.

Wood, W. B., and H. R. Revel. 1976. The genome of bacteriophage T4. Bacteriol. Rev. **40**:847–868.

Worcel, A., and E. Burgo. 1972. On the structure of the folded chromosome of *Escherichia coli*. J. Mol. Biol. **271**:127–147.

Wovcha, M. G., C.-S. Chiu, P. K. Tomich, and G. R. Greenberg. 1976. Replicative bacteriophage DNA synthesis in plasmolyzed T4-infected cells: evidence for two independent pathways to DNA. J. Virol. **20**:142–156.

Wovcha, M. G., P. K. Tomich, C.-S. Chiu, and G. R. Greenberg. 1973. Direct participation of dCMP hydroxymethylase in synthesis of bacteriophage T4 DNA. Proc. Natl. Acad. Sci. U.S.A. **70**:2196–2200.

Wright, A., M. McConnel, and S. Kanegasaki. 1980. Lipopolysaccharide as a bacteriophage receptor, p. 27–57. *In* L. Randall and L. Phillipson (ed.), Receptors and recognition, series B, vol. 7: Virus receptors, part 1: Bacterial viruses. Chapman and Hall, London.

Wu, R, F. Ma, and Y.-C. Yeh. 1972. Suppression of DNA-arrested synthesis in mutants defective in gene 59 of bacteriophage T4. Virology **47**:147–156.

Wu, J., and Y. Yeh. 1973. Requirements of a functional gene 32 product of bacteriophage T4 in UV repair. J. Virol. **12**:758–765.

Wu, J., and Y. Yeh. 1975. New late gene, *dar*, involved in DNA replication of bacteriophage T4. I. Isolation, characterization, and genetic location. J. Virol. **15**:1096–1106.

Wu, J., and Y. Yeh. 1978. New late gene, *dar*, involved in the replication of bacteriophage T4 DNA. III. DNA replicative intermediates of T4 *dar* and a gene 59 mutant suppressed by *dar*. J. Virol. **27**:103–117.

Wu, J., and Y. Yeh. 1981. Requirement of bacteriophage T4 gene function for survival after X-ray irradiation. Bull. Inst. Zool. Acad. Sinica **20**:49–55.

Wu, R., and E. P. Geiduschek. 1975. The role of replication proteins in the regulation of bacteriophage T4 transcription. I. Gene 45 and HMC containing DNA. J. Mol. Biol. **96**:513–538.

Wu, R., E. P. Geiduschek, and A. Cascino. 1975. The role of replication proteins in the regulation of bacteriophage T4 transcription. II. Gene 45 and late transcription uncoupled from replication. J. Mol. Biol. **96**:539–562.

Wu, R., J. Wu, and Y. Yeh. 1975. Role of gene 59 of bacteriophage T4 in repair of UV-irradiated and alkylated DNA in vivo. J. Virol. **16**.5–16.

Wunderli, H., J. van den Broek, and E. Kellenberger. 1977. Studies related to the head-maturation pathway of bacteriophages T4 and T2. I. Morphology and kinetics of intracellular particles produced by mutants in the maturation genes. J. Supramol. Struct. **7**:135–161.

Wurtz, M., J. Kistler, and T. Hohn. 1976. Surface structure of in vitro assembled bacteriophage lambda polyheads. J. Mol. Biol. **101**:39–56.

Wyatt, G. R., and S. S. Cohen. 1952. A new pyrimidine base from bacteriophage nucleic acids. Nature (London) **170**:1072–1073.

Wyatt, G. R., and S. S. Cohen. 1953. The bases of the nucleic acids of some bacterial and animal viruses: the occurrence of 5-hydroxymethylcytosine. Biochem. J. **55**:774–782.

Yamaguchi, Y., and M. Yanagida. 1980. Head shell protein hoc alters the surface charge of bacteriophage T4. J. Mol. Biol. **141**:175–193.

Yamamori, T., and T. Yura. 1982. Genetic control of heat-shock protein synthesis and its bearing on growth and thermal resistance in *Escherichia coli* K-12. Proc. Natl. Acad. Sci. U.S.A. **79**:860–864.

Yamamoto, M., and H. Uchida. 1975. Organization and function of the tail of bacteriophage T4. II. Structural control of the tail contraction. J. Mol. Biol. **92**:207–223.

Yanagida, M. 1972. Identification of some antigenic precursors of bacteriophage T4. J. Mol. Biol. **65**:501–517.

Yanagida, M. 1977. Molecular organization of the shell of the T-even bacteriophage head. II. Arrangement of subunits in the head shell of giant phages. J. Mol. Biol. **109**:515–537.

Yanagida, M., and C. Ahmad-Zadeh. 1970. Determination of gene product positions in bacteriophage T4 by specific antibody association. J. Mol. Biol. **51**:411–421.

Yanagida, M., E. Boy de la Tour, C. Alff-Steinberger, and E. Kellenberger. 1970. Studies on the morphopoiesis of the head of bacteriophage T-even. VIII. Multilayered polyheads. J. Mol. Biol. **50**:35–58.

Yanofsky, C., B. C. Carlton, J. R. Guest, D. R. Helinski, and U. Henning. 1964. On the colinearity of gene structure and protein structure. Proc. Natl. Acad. Sci. U.S.A. **51**:266–272.

Yarosh, D. B. 1978. UV-induced mutation in bacteriophage T4. J. Virol. **26**:265–271.

Yarosh, D. B., V Johns., S. Mufti, C. Bernstein, and H. Bernstein. 1980. Inhibition of UV and psoralen-plus-light mutagenesis in phage T4 by gene 43 antimutator polymerase alleles. Photochem. Photobiol. **31**:341–350.

Yarosh, D. B., B. S. Rosenstein, and R. B. Setlow. 1981. Excision repair and patch size in UV-irradiated bacteriophage T4. J. Virol **40**:465–471.

Yasuda, S., and M. Sekiguchi. 1970a. Mechanisms of repair DNA in bacteriophage. II. Inability of ultraviolet-sensitive strains of bacteriophage in inducing an enzyme activity to excise pyrimidine dimers. J. Mol. Biol. **47**:243–255.

Yasuda, S., and M. Sekiguchi. 1970b. T4 endonuclease involved in repair of DNA. Proc. Natl. Acad. Sci. U.S.A. **67**:1839–1845.

Yee, J. K., and M. C. Marsh. 1981. Alignment of a restriction map with the genetic map of bacteriophage T4. J. Virol. **38**:115–124.

Yegian, C. D., M. Mueller, G. Selzer, V. Russo, and F. W. Stahl. 1971. Properties of DNA-delay mutants of bacteriophage T4. Virology **46**:900–919.

Yeh, Y. C., E. J. Dubovi, and I. Tessman. 1969. Control of pyrimidine biosynthesis by phage T4: mutants unable to catalyze the reduction of cytidine diphosphate. Virology **37**:615–623.

Yeh, Y. C., and I. Tessman. 1972. Control of pyrimidine biosynthesis by phage T4. II. In vitro complementation between ribonucleotide reductase mutants. Virology 47:767–772.

Young, E. T., 1970. Control of functional T4 messenger synthesis. Cold Spring Harbor Symp. Quant. Biol. **35**:189–195.

Young, E. T. 1975. Analysis of T4 chloramphenicol RNA by DNA:RNA hybridization and by cell-free protein synthesis and the effect of *E. coli* polarity suppressing alleles on its synthesis. J. Mol. Biol. **96**:393–424.

Young, E. T., T. Mattson, G. Selzer, G. van Houwe, A. Bolle, and R. Epstein. 1980. Bacteriophage T4 gene transcription studied by hybridization to cloned restriction fragments. J. Mol. Biol. **138**:423–445.

Young, E. T., and R. C. Menard. 1981. Sizes of bacteriophage T4 early mRNA's separated by preparative polyacrylamide gel electrophoresis and identified by in vitro translation and by hybridization to recombinant T4 plasmids. J. Virol. **40**:772–789.

Young, E. T., R. C. Menard, and J. Harada. 1981. Monocistronic and polycistronic bacteriophage T4 gene 23 messages. J. Virol. **40**:790–799.

Young, E. T., and G. van Houwe. 1970. Control of synthesis of glucosyl transferase and lysozyme messenger after T4 infection. J. Mol. Biol. **51**:605–619.

Yu, F., and S. Mizushima. 1982. Roles of lipopolysaccharide and outer membrane protein OmpC of *Escherichia coli* K-12 in the receptor function for bacteriophage T4. J. Bacteriol. **151**:718–722.

Yuan, R. 1981. Structure and mechanism of multifunctional restriction endonucleases. Annu. Rev. Biochem. **50**:285–315.

Yuan, R., and D. L. Hamilton. 1982. Restriction and modification of DNA by a complex protein. Am. Sci. **70**:61–69.

Yutsudo, M. 1979. Regulation of *imm* gene expression in bacteriophage T4-infected cells. J. Gen. Virol. **45**:351–359

Zachary, A., and L. W. Black. 1981. DNA ligase is required for encapsidation of bacteriophage T4 DNA. J. Mol. Biol. **149**:641–658.

Zachary, A., L. D. Simon, and S. Litwin. 1976. Lambda head morphogenesis as seen in the electron microscope. Virology **72**:429–442.

Zampieri, A., J. Greenberg, and G. Warren. 1968. Inactivating and mutagenic effects of 1-methyl-3-nitro-1-nitrosoguanidine on intracellular bacteriophage. J. Virol. **2**:901–904.

Zarybnicky, V. 1969. Mechanism of T-even DNA ejection. J. Theor. Biol. **22**:33–42.

Zerler, B. R., and S. S. Wallace. 1979. Repair of psoralen plus near ultraviolet light damage in bacteriophage T4. Photochem. Photobiol. **30**:413–416.

Zetter, B. R., and P. S. Cohen. 1974. Post-transcriptional regulation of T4 enzyme synthesis. Arch. Biochem. Biophys. **162**:560–567.

Zillig, W., E. Fuchs, P. Palm, D. Rabussay, and K. Zechel. 1970. On the different subunits of DNA-dependent RNA polymerase from *E. coli* and their role in the complex function of the enzyme, p. 167–189. *In* L. Silvestri (ed.), RNA polymerase and transcription. North-Holland Publishing Co., London, Amsterdam.

Zimmerman, S. B., S. R. Kornberg, and A. Kornberg. 1962. Glucosylation of deoxyribonucleic acid. II. Glucosyl transferases from T2 and T6 infected *Escherichia coli*. J. Biol. Chem. **237**:512–518.

Zograff, Y. N. 1981. On the role of the *Escherichia coli* RNA polymerase sigma factor in T4 phage development. Mol. Gen. Genet. **183**:557–558.

Zograff, Y. N. 1982. Influence of *Escherichia coli* mutation changing RNA polymerase sigma factor on growth of phage T4. (Translated from Russian.) Molecular Biology (Molekulyarnnaya Biologiya), vol. 16, no. 1, part 1. Consultants Bureau, New York.

Zograff, Y. N., and A. L. Gintsburg. 1980. Transcription termination factor rho and T-even phage development. Mol. Gen. Genet. **177**:699–705.

Zorzopulos, J., and L. M. Kozloff. 1978. Identification of T4D bacteriophage gene product 12 as the baseplate zinc metalloprotein. J. Biol. Chem. **253**:5593–5547.

Zweig, M., and D. J. Cummings. 1973. Cleavage of head and tail proteins during bacteriophage T5 assembly: selective host involvement in the cleavage of a tail protein. J. Mol. Biol. **80**:505–518.

Author Index